Materials Engineering and Sciences Division

Core Programming Topic at the 2011 AIChE Annual Meeting

Minneapolis, Minnesota, USA
16-21 October 2011

Volume 1 of 2

ISBN: 978-1-61839-729-4

Printed from e-media with permission by:

Curran Associates, Inc.
57 Morehouse Lane
Red Hook, NY 12571

Some format issues inherent in the e-media version may also appear in this print version.

Printed by Curran Associates, Inc. (2012)

For permission requests, please contact AIChE
at the address below.

AIChE
3 Park Avenue
New York, NY 10016-5991

Phone: (203) 702-7660
Fax: (203) 775-5177

www.aiche.org

Additional copies of this publication are available from:

Curran Associates, Inc.
57 Morehouse Lane
Red Hook, NY 12571 USA
Phone: 845-758-0400
Fax: 845-758-2634
Email: curran@proceedings.com
Web: www.proceedings.com

TABLE OF CONTENTS

VOLUME 1

(9a) Silica-Chitosan Nanocomposite for pH-Responsive, Stomach-Targeted Delivery of Therapeutics

Allan E. David[1], Arthur J. Yang[2] and Victor C. Yang[1], (1)Department of Pharmaceutical Sciences, The University of Michigan, Ann Arbor, MI, (2)Industrial Science & Technology Network, Inc, York, PA

It has already been confirmed that mucoadhesive drug delivery carriers, referred to as the Adhesive Micromatrix System, can adhere to the stomach wall in rats, thereby, remaining longer in the gastrointestinal (GI) tract. In recent years, chitosan and its derivatives have been widely assessed for the controlled release, or the delivery, of various drugs. Besides being biocompatible and biodegradable, chitosan offers advantages in drug delivery because of its permeation enhancement, mucoadhesiveness, and ability for sustained drug release. Illum et. al. reported that chitosan solutions, even at a low concentration (0.5%), are highly effective at increasing the adsorption of insulin across nasal mucosa in rats and sheep. It was suggested that the enhancement mechanism was a combination of bioadhesion and transient widening of the tight junction in a membrane. We intend to develop a chitosan drug delivery system that adheres to stomach mucosurface and delivers antibiotics closer to sites infected by *H Pylori*. However, due to its high solubility in acid, a composite form of chitosan is needed to prolong the residence and delivery time in stomach. The goal of this project is to develop a composite drug delivery system made by entrapping chitosan polymer with silica network. We demonstrated a process that an in-situ gelation of surface modified silica networks in the presence of chitosan created an interpenetrating network of silica and chitosan macromolecules. Silica gel is very stable in acid. The tight entanglement structure in a silica-chitosan composite could significantly retard chitosan's leaching under acidic environment. Therefore, in the gastric environment, this composite would control the drug release rate more effectively than bare chitosan.

In this study, amoxicillin, a small antibiotic drug, was selected to demonstrate the feasibility of using a silica-chitosan nanocomposite as a delivery device in the gastric cavity. The treatment efficacy could be substantially enhanced by the muco-adhesiveness of the silica-chitosan composite. Consequently, the aim of this study was to investigate the effect of the novel chitosan/silica nanocomposite compared with chitosan sponges on the mucosal adsorption of a model drug, amoxicillin, in vitro.

See more of this Session: Biocomposites - Bio-Derived and Bio-Inspired Composites
See more of this Group/Topical: Materials Engineering and Sciences Division

(9b) Structure-Function Relationships In Beetle Elytral Cuticle, a Biological Composite Material

Patricia A. Huber[1], Jessica Etsitty[2], Joseph Lomakin[1], Yasuyuki Arakane[3], Karl J. Kramer[4], Richard W. Beeman[4], Michael R. Kanost[3] and Stevin H. Gehrke[5], (1)Chemical & Petroleum Engineering, University of Kansas, Lawrence, KS, (2)Haskell Indian Nations Univeristy, Lawrence, KS, (3) Department of Biochemistry, Kansas State University, Manhattan, KS, (4)Grain Marketing and Production Research Center, Agricultural Research Service, US Department of Agriculture, Manhattan, KS, (5)Chemical & Petroleum Engineering and Pharmaceutical Chemistry, University of Kansas, Lawrence, KS

Insect cuticle is a widespread yet poorly-understood biological composite material with diverse and specialized mechanical properties. The main structural components, chitin and proteins, form a complex, fiber-reinforced network, while additional components such as catechols, enzymes and water modulate the interactions between the structural materials. We performed a comprehensive study of the mechanical properties of elytra from *Tribolium castaneum* and *Tenebrio molitor* as a function of maturation time. As the insects mature, the cuticle changes from a soft ($E=44\pm8$MPa) and ductile material to a stiff ($E=2400\pm1100$MPa) and brittle material. To understand the origin of such changes in mechanical properties, we have used dynamic mechanical analysis to examine the material interactions within this multicomponent biomaterial. Our results are consistent with the hypothesis that cuticle components become more cross-linked over time, given that the frequency dependence of the storage modulus decreases significantly. Dehydration also stiffens the cuticle and introduces additional physical interactions, but does not account for all the changes in mechanical properties seen upon tanning. Treating the fully mature cuticle with physical bond-breaking solvents does not completely reverse the properties to those of the immature state, suggesting that permanent, covalent interactions develop over time. RNA interference has also been used to support the hypothesized mechanism of enzymatic cross-linking leading to the development of mechanical properties.

In the design of bio-inspired composite materials, it is important to understand which features of the complex biological structures need to be replicated. Often, biomaterials have many layers of hierarchical structure from the macroscale to the molecular scale. For technological applications, the synthesis of biomimetic materials would be simplified if we knew exactly which features were essential for a particular function. For example, we have determined that molecular interactions such as protein cross-linking serve to increase the elastic response of cuticle over a broad range of frequencies. However, on a macroscale, the cuticle has rib structures that confer additional toughness by stopping the propagation of cracks. Knowledge of such structure-function relationships will allow for precise control over synthetic material properties. Our understanding of the structure-function properties of beetle elytral cuticle has inspired the design of high strength, biocompatible polymer networks made of agarose and polyethylene glycol (PEG) to be used as tissue engineering scaffolds. The structure of the interpenetrating hydrogel networks (IPNs) mimics the cuticle's combination of chitin fibers embedded in a protein matrix. The IPN has significantly greater toughness than either PEG or agarose alone. Structure-function characterization of other naturally efficient biological materials has the potential to improve the design of new bio-based composites by determining which aspects of the biological material are most essential to an analogous technological material.

See more of this Session: Biocomposites - Bio-Derived and Bio-Inspired Composites
See more of this Group/Topical: Materials Engineering and Sciences Division

Effect of Nanoclay Amount on the Properties of Biodegradable Polymer/Organoclay Nanocomposites Based on Cellulose Acetate and Poly(Butylene-Adipate-*co*-Terephthalate)

April Elizabeth Sloan, Department of Chemical & Biological Engineering, Missouri University of Science and Technology, Rolla, MO, Barbara Wheelden, Department of Chemical & Biomolecular Engineering, Ohio University, Athens, OH, Douglas Ludlow, Chemical & Biological Engineering, Missouri University of Science and Technology, Rolla, MO and Sunggyu Lee, Chemical and Biomolecular Engineering, Ohio University, Athens, OH

Abstract

With the ever increasing momentum of the 'Green Movement', a renewed interest in developing biodegradable polymers for a wide-range of industrial applications has been sparked. Although bio-based polymers offer a solution to industry's reliance on petroleum-based polymers, performance limitations, as opposed to their non-renewable counterparts, are often exhibited. The inclusion of a nanofiller, in contrary to conventional macro- or microscale particles, has been shown to counterbalance and often times enhance the properties of these newly formed nanocomposites.[1] In particular, well-dispersed nanosilicates (such as montmorillonite), in a polymer matrix have been shown to affect the properties in such ways as: increasing the storage modulus, increasing tensile and flexural properties, and decreasing permeability and flammability.[1,2]

Currently, cellulosic, starch, and soy-based plastics, along with polylactides (PLA) and hydroxyl alkanoates (PHAs) are some of the most widely known biopolymers found in industry and research. Cellulosic plastics, in particular the esters of cellulose, cellulose acetate (CA), cellulose acetate propionate (CAP), and cellulose acetate butyrate (CAB), are commercially utilized in a variety of fields ranging from coatings to films and membranes.[3,4] Within the last decade, CAB has evolved into one of the most sought after thermoplastic cellulose esters, and the effect of some of its properties, such as, high glass transition temperature (T_g) and stiffness on polymer blends, has been studied.[5]

Poly(Butylene-Adipate-*co*-Terephthalate) (PBAT), a fully biodegradable, aliphatic-aromatic copolyester, has recently been shown to enhance some properties (i.e. toughness, elongation, etc.) of some biodegradable polymer blends, including those with Poly(Lactic Acid).[6,7] Although PBAT is produced from fossil fuel sources, the blending with a bio-based polymer enhances the renewability of the final product.

Within this study, bio-nanocomposites (biopolymer matrix reinforced with nanoparticles) based on cellulose acetate butyrate (CAB), PBAT, and modified montmorillonite (MMT) are prepared using melt intercalation via twin-screw extrusion method. The effect of amount (0wt%-3wt%) of MMT on the morphology and both mechanical and rheological properties of varying CAB/PBAT blends (70/30; 50/50; and 30/70 (wt%/wt%)) bio-nanocomposites are investigated.

1. Petersson, L., Mathew, A.P., Oksman, K., "Dispersion and Properties of Cellulose Nanowhiskers and Layered Silicates in Cellulose Acetate Butyrate Nanocomposites". *J of Applied Polymer Science*, Vol. 112, 2001-2009, (2009)
2. Alexandre, M. and Dubois, P., "Polymer-Layered Silicate Nanocomposites: Preparation, Properties, and Uses of a New Class of Materials". *Mat Sci and Eng: R: Reports*, Vol. 28, 1-63, (2000)
3. Klemm, D., Heublein, B., Fink, H-P., Bohn, A., "Cellulose: Fascinating Biopolymer and Sustainable Raw Material". *Angewandte Chemie International Edition*, Vol. 44, 3358-3393, (2005)
4. Park, H-M., Misra, M., Drzal, L., Mohanty, A., ""Green" Nanocomposites from Cellulose Acetate Bioplastic and Clay: Effect of Eco-Friendly Triethyl Citrate Plasticizer". *Biomacromolecules*, Vol. 5, 2281-2288, (2004)
5. Kosaka, Pr., Kawano, Y., Fantini, M., Petri, D., "Structure and Properties of Maleated Linear Low-Density Polyethylene and Cellulose Acetate Butyrate Blends". *Macro Mat and Eng*, Vol. 291, 531-539, (2006)
6. Jiang, L., Wolcott, M., Zhang, J., "Study of Biodegradable Polylactide/Poly(Butylene adipate-co-terphthalate) Blends". *Biomacromolecules*, Vol. 7, 199-207, (2006)
7. Jiang, L., Zhang, M., "Properties of Poly(lactic acid)/Poly(butylenes adipate-co-terephthalate)/Nanoparticle Ternary Composites". *End Eng Chem Res*, Vol. 48, 7594-7602, (2009)

SOFT MATERIAL DESIGN USING PRINCIPAL COMPONENT ANALYSIS

Tao Wu[1], Sheldon Wang[2], and Barry Cohen[3]

[1,3]*Computer Science Department, New Jersey Institute of Technology, Newark, NJ*
[2]*McCoy School of Engineering, Midwestern State University, Wichita Falls, TX*
tw37@njit.edu, Sheldon.wang@mwsu.edu, barry.cohen@njit.edu

ABSTRACT

This paper presents a new computational framework for calculating the normal modes and interactions of proteins, macromolecular assemblies and surrounding solvents. The framework employs a combination of molecular dynamics simulation (MD) and principal component analysis (PCA). It enables the capture and visualization of the molecules' normal modes and interactions over time scales that are computationally challenging, providing a starting point for experimental and further computational studies of protein conformational changes. PCA reduces the dimensionality of MD atomic trajectory data and provides a concise way to visualize, analyze, and compare the motions observed over the course of a simulation. PCA involves diagonalization of the positional covariance matrix to identify an orthogonal set of eigenvectors or "modes" describing the direction of maximum variation in the observed conformational distribution. Slow conformational changes can be identified by projecting these dominant modes back to original trajectory data. In this work, the new multiscale methodology was applied to a relatively small mutant T4 phage lysozyme, establishing its equilibrium atomic thermal fluctuations and its inter-residue fluctuation correlations. These results were compared with published data obtained by NMA, by finite element methods, and by experiment. The eigenmodes captured are in quantitative agreement with previously published results.

1. Introduction

This paper presents a new computational framework for calculating the normal modes and interactions of proteins, macromolecular assemblies and surrounding solvents. The framework employs a combination of molecular dynamics simulation (MD) and principal component analysis (PCA). It enables the capture and visualization of the molecules' normal modes and interactions over time scales that are computationally challenging. It also provides a starting point for experimental and further computational studies of protein conformational changes.

Understanding a protein's mechanics is a prerequisite for insight into its biological function, since most proteins perform their functions through structural deformation, also called conformational change. Such conformational change has been modeled by atomic level simulation such as molecular dynamics (MD) simulation.

It is difficult to theoretically explain and predict the conformational changes in protein binding or interaction. Conformational change is not accessible to MD simulation available today due to its computational cost. It happens on millisecond scales (Elber 2005; Schlick *et al.* 1997) compared to MD simulation on nanosecond scale. Therefore other computational methods must be applied. One of these methods is normal mode analysis (NMA), which has proven useful for studying collective motion of biological macromolecules. In addition, NMA with coarse-grained modeling of protein structures of large proteins has been a computational alternative to atomic level simulation. Studies (Marques and Sanejouand 1995; Perahia and Mouawad 1995; Tama *et al.* 2000) show that some of the lowest-frequency normal modes of several proteins are strongly correlated with their large amplitude conformational changes.

Although NMA provides information about slow protein motion, its harmonic approximation limits its use to small amplitude motion in the potential energy surface associated with a local minimum (Barton *et al.* 2002). It is difficult, using NMA, to analyze such large scale effects as the hydrophobic effect (Ma 2005), which plays an important role in protein-protein interaction. A combination of MD simulation and principal component analysis (PCA) provides a systematic way to study protein slow motion. This method overcomes some of the difficulties of NMA.

[1] Researcher
[2,3] Professor

This paper presents a new computational framework for calculating the normal modes and interactions of proteins, macromolecular assemblies and surrounding solvents. The framework employs a combination of MD simulation and PCA. It enables the capture and visualization of the molecules' normal modes and interactions over time scales that are computationally challenging, providing a starting point for experimental and further computational studies of protein conformational changes. The new multiscale methodology was applied to a relatively small mutant T4 phage lysozyme. Results are compared with published data obtained by NMA and finite element methods. The eigenmodes captured are in quantitative agreement with previously published results.

2. Method and Theory

The T4 phage lysozyme molecular dynamics simulation utilizes the NAMD parallel molecular dynamics package version 2.6 with the CHARMM force field. The initial structure was determined by X-ray crystallography (Matsumura et al. 1989), retrieved from the Protein Data Bank (PDB ID 3LZM; residue numbers 1-164). First, the coordinates missing in the crystal structure are reconstructed. For the solvent system, the minimized structure is solvated in a periodic truncated cubic simulation box of 29,712 TIP3P water molecules, providing a minimum of 15Å of water between the protein surface and any box edge. The water box is rotated to minimize the system size. Sodium and chloride ions are added to neutralize the total system and achieve a salt concentration of 0.05 mol/L. Ion placement is random; minimum distances between ions and molecules as well as between any two ions are set to 5Å.

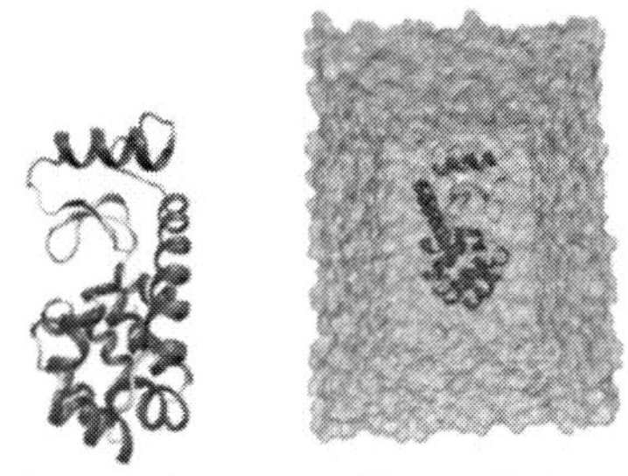

(a) Molecular structure in vacuum (b) Molecular structure in solvent

Figure 1: Schematic representation of the energy-minimized molecular structure

In the solvent simulation, energy minimization was performed in 20,000 steps. There were no energy changes during last 1,000 time-steps. This stage is shown in Fig. 1. Following minimization, the entire system was heated in 10 K increments up to 300 K with micro-canonical ensemble (NVE) equilibration per temperature step. The simulation was conducted at 300 K. Atomic trajectories were calculated in one femtosecond (fs) time steps using SHAKE constraints on hydrogen-heavy atom bonds. The total production simulation length is 15 nanosecond (ns). The first 9 ns were discarded as an "equilibration period." This extensive relaxation/equilibration period was necessary due to drift in the potential energy.

Principal component analysis involves diagonalization of the positional covariance matrix A to identify an orthogonal set of eigenvectors or "modes" describing the direction of maximum variation in the observed conformational distribution. Covariance matrix $\boldsymbol{A}$ is calculated with Eq. (1),

(1)

PCA diagonalization of the covariance matrix involves the following eigenvalue problem , where μ are the eigenvectors and λ are the eigenvalues of covariance matrix A.

Quasi-harmonic analysis (QHA) (Harris and Laughton 2007) and related principal component analysis techniques, based on the eigenvalue decomposition of an ensemble of protein conformation, have provided a useful method for identifying motion, particularly at long time scales (Ramanathan and Agarwal 2009). QHA captures the large-scale conformational fluctuation in a protein by diagonalizing the mass-weighted covariance matrix, known as the atomic fluctuation matrix ($F_{\alpha\beta}$). For a system with N atoms, $F_{\alpha\beta}$ is a $3N \times 3N$ symmetric matrix, defined as:

(2)

where α and β represent the 3N degrees of freedom in Cartesian space; m is the mass of the atom and the quantity within <> denotes an average over the ensemble of structures in MD simulation.

The inverse square roots of the eigenvalues determined by diagonalizing A or $F_{\alpha\beta}$ represent the frequencies associated with the protein eigenmode. The eigenvectors represent the displacement vectors of the individual atoms. The lowest frequencies correspond to large-scale cooperative motion in the protein; the higher frequencies represent localized motion. For a system with N atoms, there are $3N$-6 internal modes. However, due to computational costs, typically only a limited number of slow modes (lowest frequencies) are computed.

The overlap coefficients (OCs) of the a-carbon atoms are calculated to measure how well each PCA/QHA mode represents all motion (Bradley et al. 2008). PCA/QHA modes (v_i) are compared with a vector (Δx), that describes the displacement of a carbon atoms from the minimized starting simulation position. The structure is experimentally determined by X-ray crystallography. This vector is calculated following alignment of the minimized and crystal structures with the MD average structure to remove rotation and translation of the center of mass. The overlap coefficients are calculated by the angle between the two vectors:

(3)

Correlations among the fluctuations of residues, measured at alpha-carbon atoms, provide additional insight into protein function (Lange and Grubmüller 2006; Waight et al. 2010). An unweighted pair group method with arithmetic means (UPGMA) clustering algorithm is used to define groups of residues with similar correlation patterns under equilibrium fluctuation. The correlation is measured by constructing correlation matrices:

$$S_{ij} = \frac{\langle (\boldsymbol{r}_i - \langle \boldsymbol{r}_i \rangle)(\boldsymbol{r}_j - \langle \boldsymbol{r}_j \rangle) \rangle}{\sqrt{\langle (\boldsymbol{r}_i - \langle \boldsymbol{r}_i \rangle)(\boldsymbol{r}_i - \langle \boldsymbol{r}_i \rangle) \rangle \langle (\boldsymbol{r}_j - \langle \boldsymbol{r}_j \rangle)(\boldsymbol{r}_j - \langle \boldsymbol{r}_j \rangle) \rangle}},$$

The value of Sij ranges from -1.0 to 1.0, with positive and negative values indicating correlated and anti-correlated motion. ri and rj are the position of atoms i and j and <> denotes the trajectory average over snapshots. The matrix is calculated across all a carbon atoms of the protein.

A correlation analysis procedure similar to Bradley et al. (2008) is followed:

a. Construct a positional correlation matrix S_{ij} .
b. Compute the effective "distance" d_{ij}, based on the correlation measure.
c. Generate a "treelike" representation of correlated residues by the UPGMA method.
d. Do a depth-first search to choose the smallest clusters that contain at least one residue from each of the groups defined by the selection criteria. Experimental information plays an important role in the definition of these clusters.
e. Complete the cluster selection.

3. Result

Equilibrium thermal fluctuation of residues provides the first quantitative evaluation of the proposed MD-PCA framework and helps to understand protein function by local conformational flexibility. The root mean square fluctuation (RMSF) of a-carbon atoms is plotted as a function of residue number to show local conformational change. Atomic alpha-carbon fluctuations are compared to published results such as the All-Atom-Method(ATM), Rotational-Translational Block (RTB), Gaussian Network Model (GNM), and Finite Element Method (FEM) (shown in Fig. 2).

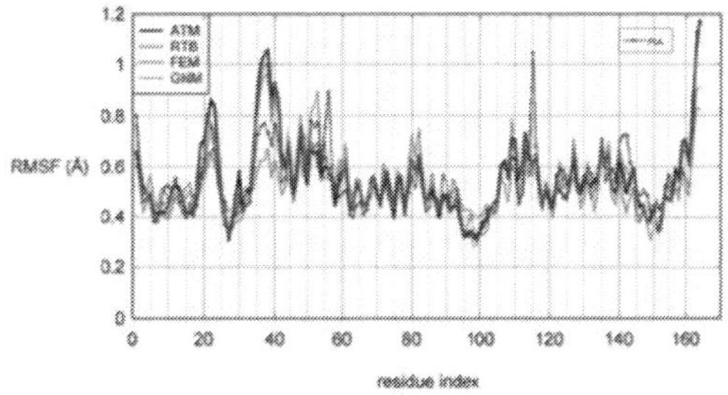

Figure 2: Methods used are: All-Atom-Method (ATM), Rotational-Translational Block (RTB), Gaussian Network Model (GNM), Finite Element Method (FEM), Data generated from simulation (PCA)

The T4 lysozyme shows flexibility at the solvent-exposed beta-hairpin turn between residues 20 and 24, also at the short loop between residues 35 and 38, and the solvent-exposed alpha-helix at residues 39-51 which can found in Fig. 2. These RMSF results are in quantitative agreement with other methods except at residue 115. This suggests that the alpha-helix containing residue 115 tends to bend at that spot. In the simulation without water, that alpha-helix does bend at residue 115 without solvent-exposed "protection".

With the calculated PCA/QHA modes placed into experimentally determined structural rearrangements, the overlap coefficients (OCs) were calculated. The results are shown in Fig. 3. The full basis set of 492 a-carbon PCA modes gives a very high overlap (*cos(θ)* = 1.000), as do all atom PCA/QHA, while carrying low error at 0:00001 for a-carbon PCA, 0:00006 for all atom PCA, and 0:0006 for all atom QHA. From this PCA/QHA basis set, a minimal set of modes was identified that represented the observed displacement. The first 10 modes were selected based on eigenvalue analysis.

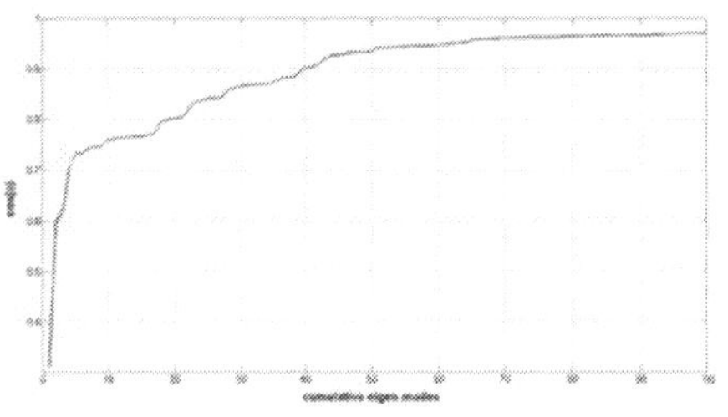

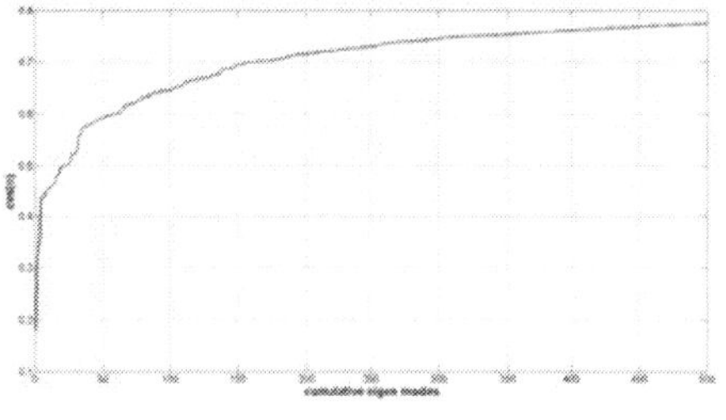

The first and largest all atom PCA mode accounts for 12.6% of the total motion, but has only a small overlap (*cos(q)* = 0.17) with the changes observed in the X-ray structure. This largest mode represents an asymmetric "twisting-like" mode (Fig. 4a, 4b). The second mode accounts for 7.3% of the total motion but has a greater overlap (*cos(q)* = -0.29). The lowest modes of proteins describe their most characteristic conformational changes, which are often directly related to their function. The lowest mode of the T4 lysozyme is a hinge bending mode similar to that of native hen egg lysozyme. Two lowest modes are presented in Figs. 4. The second mode (Figs. 4c, 4d) shows the hinge bending mode.

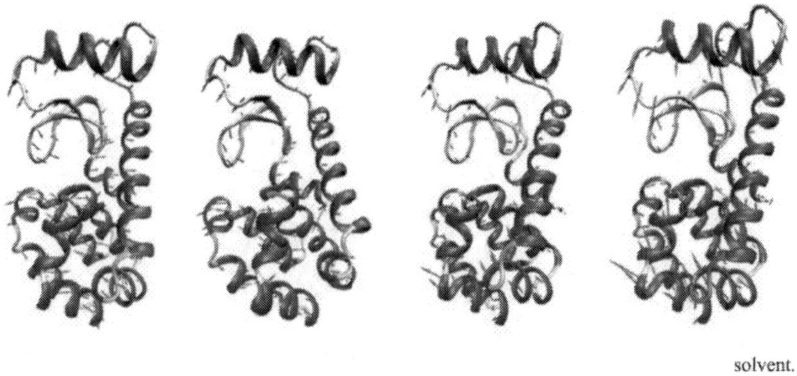

solvent.
d arrows
ent.
th QHA;

A cluster analysis of residues was produced from the correlation matrix. It depicts possible interactions among groups of residues. This method provides a starting point for experimental and computational studies designed to determine more fine-grained conformational changes in the protein.

In Fig. 5a, where results were calculated from the raw simulation data, the (anti-) correlation was weak. However, after projecting the first 10 eigenmodes to generate a new "cleaned" data set, the (anti-) correlation results shown in Fig 5b were in closer agreement with Bathe's (2008) result. However, more correlations among residue groups are found in this map, suggesting that more interaction have been identified with the PCA based method.

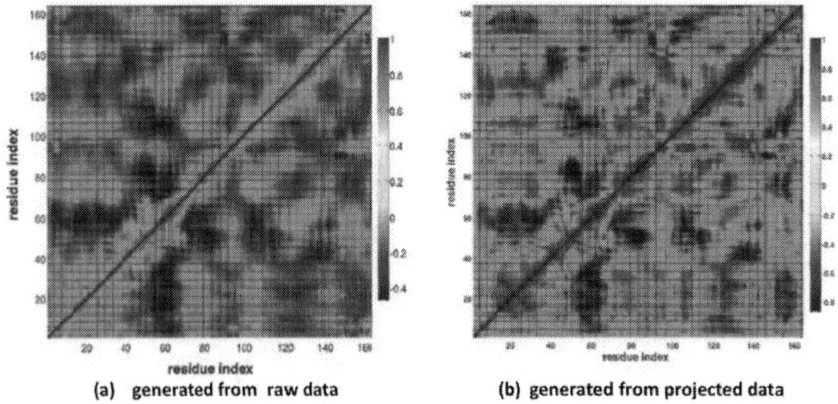

(a) generated from raw data

(b) generated from projected data

Figure 5: Correlated fluctuation of a-carbon atoms computed for T4 lysozyme

4. Conclusion and Future Work

An effective computational procedure based on MD-PCA framework is proposed to calculate the normal modes of proteins. QHA and PCA are compared, showing QHA to be better on slow motion analysis because it filters out the high frequency motion of hydrogen atoms. A method of clustering residue motions based on a correlation matrix is implemented. This provides a starting point for experimental and computational studies designed to further conformational changes in proteins. This work establishes the utility of the MD-PCA method to compute protein normal modes for a limited range of biomolecules. To evaluate its effectiveness fully, the present framework should be applied to a considerably broader set of proteins of variable structure and chemical composition.

REFERENCES

Barton, N. P.; Verma, C. S.; and Caves, L. S. D. (2002). Inherent flexibility of calmodulin domains: A normal-mode analysis study. *The Journal of Physical Chemistry B*, 106(42):11036–11040.

Bathe, Mark (2008). A finite element framework for computation of protein normal modes and mechanical response. *Proteins: Structure, Function, and Bioinformatics*, 70(4):1595–1609.

Bradley, Michael J.; Chivers, Peter T.; and Baker, Nathan A. (2008). Molecular dynamics simulation of the Escherichia coli NikR protein: equilibrium conformational fluctuations reveal interdomain allosteric communication pathways. *Journal of Molecular Biology*, 378(5):1155–1173.

Elber, Ron (2005). Long-timescale simulation methods. *Current Opinion in Structural Biology*, 15(2):151–156. Theory and simulation/Macromolecular assemblages.

Harris, Sarah A. and Laughton, Charles A. (2007). A simple physical description of DNA dynamics: quasi-harmonic analysis as a route to the configurational entropy. *Journal of Physics: Condensed Matter*.

Lange, Oliver F. and Grubmüller, Helmut (2006). Generalized correlation for biomolecular dynamics. *Proteins: Structure, Function, and Bioinformatics*, 62(4):1053–1061.

Ma, Jianpeng (2005). Usefulness and limitations of normal mode analysis in modeling dynamics of biomolecular complexes. *Structure*, 13(3):373–380.

Marques, Osni and Sanejouand, Yves-Henri (1995). Hinge-bending motion in citrate synthase arising from normal mode calculations. *Proteins: Structure, Function, and Genetics*, 23(4):557–560.

Matsumura, M.; Wozniak, J. A.; Sun, D. P.; and Matthews, B. W. (1989). Structural studies of mutants of T4 lysozyme that alter hydrophobic stabilization. *Journal of Biological Chemistry*, 264(27):16059–16066.

Perahia, David and Mouawad, Liliane (1995). Computation of low-frequency normal modes in macromolecules: Improvements to the method of diagonalization in a mixed basis and application to hemoglobin. Computers & Chemistry, 19(3):241–246. Third Conference on Computers in Chemistry.

Ryckaert, J.; Ciccotti, G.; and Berendsen, H. (1977). Numerical integration of the Cartesian equations of motion of a system with constraints: molecular dynamics of n-alkanes. Journal *of Computational Physics*, 23(3):327–341.

Schlick, Tamar; Barth, Eric; and Mandziuk, Margaret (1997). Biomolecular dynamics at long timesteps: Bridging the timescale gap between simulation and experimentation. *Annual Review of Biophysics and Biomolecular Structure*, 26(1):181–222.

Tama, Florence; Gadea, Florent Xavier; Marques, Osni; and Sanejouand, Yves-Henri (2000). Building-block approach for determining low-frequency normal modes of macromolecules. *Proteins: Structure, Function, and Genetics*, 41(1):1–7.

Waight, Andrew B.; Love, James; and Wang, Da-Neng (2010). Structure and mechanism of a pentameric formate channel. *Nat Struct Mol Biol*, 17(1):31–37.

Studies of Mechanical and Thermal Properties of Rice Husk Polypropylene (RHPP) Composites

1 ABSTRACT

In the present study, effect of additives such as talc, mica, $CaCO_3$, wollastonite, and zirconium oxide on mechanical and thermal properties of RHPP Composites was studied. Higher hardness and impact strength of these composites was noticed by use of these additives. Thermal properties of rice husk polypropylene composites (RHPP) composites modified with mica, talc, wollastonite, zirconium oxide, silica and $CaCO_3$ were studied by heat deflection temperature, thermo gravimetric analysis (TGA) and differential scanning calorimetry (DSC). Use of additives improves degradation temperature of composites without significant change in melting point. Considerable reduction in enthalpies of melting and degradation was noticed by addition of these additives in RHPP Composites. RHPP systems modified with maleated compatibilizer have better thermal stability compared with titanate modified composites. These properties could be utilized in various applications of building and automobile sectors.

2 LITERARURE REVIEW

Thermoplastics composites based on commodity plastics such as polyethylene, polypropylene, polystyrene etc. have gained lot of importance now days. These plastics can be reinforced with fillers such as $CaCO_3$, talc, mica, zirconia, wollastonite etc., to improve mechanical, thermal and chemical properties. Compared to thermoset composites, thermoplastic composites have important advantages of high production, less abrasion and reusability [1]. Use of fillers decreases the mechanical properties of polypropylene such as tensile strength and impact strength because of poor adhesion between filler and polymer matrix [2]. Compatiblizer (maleated polypropylene) and coupling agents are effective in improving the tensile strength, but no improvement in impact strength was noticed [3-5]. In the present study, additives/fillers such as $CaCO_3$, mica, talc, zirconium oxide, wollastonite are used with rice husk polypropylene composites and mechanical properties like hardness, impact strength, were analyzed to explore new applications of these composites [6-10].

Differential scanning calorimetry (DSC) is an important tool to study thermal behavior of polymers. Measurement of specific heat, glass transition temperature, melting, crystallization characteristics and thermal stability is possible by this instrument [11]. Thermo gravimetry and differential thermal analysis is very useful in understanding the thermal stability of poly vinyl chloride [12]. PVC/*cis*-PBR blends modified by PBR in 0–50 weight % were studied for thermal properties. Themo gravimetric analysis of blends indicates that all blends are more stable than PVC and stability increase with PBR content. Three parameters namely threshold degradation temperature (D_T), temperature at which rate of degradation was highest (D_M) and glass transition temperature (T_g) were measured to study the thermal stability [13]. Thermo gravimetric and differential scanning calorimetric analysis was carried out for flax and hemp fibers. Results prove that there is requirement of thermal stability for consolidation process of composite materials [14]. Effect of blend ratio, crosslink level and initiating system were studied on thermal degradation of natural rubber/polystyrene (NR/PS) interpenetrating polymer networks (IPNs). The thermo gravimetric analysis (TGA) shows better stability of IPNs compared with pure components. There is improvement in initial decomposition temperature and temperature for 50 % decomposition (T_{50}) with increase in concentration and cross linker level of plastic phase. Better mechanical strength of IPNs aged for 72 hr at 100°C was noticed due to cross linking on post curing [15]. Thermal characteristics of polypropylene/polyethylene (PP/PE) blend was studied by thermo gravimetric analysis (TGA) and differential scanning calorimetry (DSC). DSC is valuable for analyzing melting and crystallinity behavior of blend, whereas TGA for degradation study. An empirical relationship can estimate the overall crystalline percent, and for this information of blend ratios and densities of pure components are required. Calculation of crystalline by this method eliminates graphical integration process on determining areas under curves of thermo grams. Effect of temperature on induction time of PP/PE blend was studied, and it follows the trend of Arrhenius equation [16]. Synthesis and processing conditions affect structure of thermoplastic poly urethane (TPU) elastomers. Quality deviations observed from one production to another influence the structure of material in batch process. Synthesis and process conditions can control the properties of TPU. Melting behavior and residual crystalline after processing depend on stoichiometry and processing conditions as studied by DSC study [17]. Polypropylene was reinforced by layered silicate (montmorillonite) with MAPP as compatibilizer and thermal properties were studied. The composites show tactoid, intercalated and exfoliated structures side by side with different dominant states which depend on type of clay and processing conditions. System behaves as multi-component blends rather than binary blends because organic ions not only change the mixing behavior, but affect the material properties also [18]. Thermo gravimetric analysis (TGA) was used to study thermal behavior of liquefied wood (LW) reinforced low density polyethylene (LDPE), high density polyethylene (HDPE), and polypropylene (PP) composites. HDPE based composites have better thermal stability than PP. Melting temperature of polymers decrease with increase of LW loading. Enthalpy of virgin polymers also decreases with addition of LW [19].Microscopic and macroscopic properties of thermoplastic polyurethane (TPU) samples in hard and soft segments were analyzed. Higher glass transition temperature and lower angle of rebound result due to increase of hard segments content at the same molecular weight soft segments. Higher molecular weight of soft segments at same content of hard segments reduces glass transition temperature and angle of rebound [20]. Five samples of pure poly (vinyl chloride) (PVC) were obtained at different polymerization temperatures and influence of microstructure on thermal degradation behavior was studied. TGA study is used to detect two independent weight loss steps in dehydrochlorination process. Syndiotacticity tendency of PVC was seen due to relative contribution of these two steps [21]. Polypropylene was reinforced with fully exfoliated montmorillonite (MMT) and thermo oxidative /thermal stability was studied by DSC and TGA methods. Weight loss from TGA and enthalpy changes from DSC in a temperature range (from room temperature to 500 °C) was noticed. It was seen that, thermal degradation of polypropylene nano composites in oxygen-free atmosphere is strongly inhibited up to temperature of 420 C in contrast to pristine polypropylene which degrade at 300 C. Lower sensibility to thermal oxidation of nanocomposites was correlated with lower oxygen permeability which result from an increased diffusion path for oxygen as well as volatile decomposition products [22].

3. MATERIALS AND METHODS

3.1 Materials:

(i) Poly Propylene: PP was procured from M/S Reliance industries (Grade 110MA) and was analyzed for following properties.

- MFI : 11.50 gm/10 minutes
- Moisture Content : 0.039 %
- Tensile Strength at yield (D 638): 29.09 MPa
- Notched Izod impact strength (D 256): 20.42 J/m
- Hardness (ASTM 2240) : 63.75
- Abrasion resistance: (ASTM D 4060): 22.12 mg/1000 cycles

(ii) Rice Husk: It was obtained locally, grinded to fine powder (24 mesh). Following properties were measured:

- Bulk density (Kg/m^3): 104.4
- Moisture Content (%): 6.8
- Ash Content (%) : 14.4.

(iii) Compatibilizer: It was supplied by M/S Pluss Polymers P Ltd, Gurgaon. Compatibilizer grade was P 406 and percentage addition of compatibilizer was 3.0 % (of total weight of RHPP composite) in the composites as recommended by the supplier.

(iv) Additives: Mica, Talc, Wollastonite, Zirconium oxide, $CaCO_3$ were procured from M/S Power Inorganic, New Delhi India. Selection of these materials was based on their properties as mentioned below.

$CaCO_3$: Good dispersion characteristics, good impact strength, shrinkage reduction and better surface finish.

Talc: High compatibility, low wear resistance, good rigidity, and good creep resistance.

Mica: Dimensional stability, good rigidity and high heat deflection temperature.

Wollastonite: High aspect ratio, Low coefficient of expansion, low moisture absorption, good scratch resistance and excellent finish.

Zirconium oxide: Good wear resistance, hardness, better fracture toughness, and chemical inertness

3.2 Method:

Single screw extruder was used for mixing the above materials. Temperature setting in extruder was : 150 °C, 180°C, 215 °C, 210 °C. Granules of uniform composition and size were obtained, which were used for making the composite sheet by compression molding process . Mold used was rectangular in shape to produce sheet of 155 ×155 mm^2 size. Above mold was made keeping in view of requirement of different sizes of sample for doing different type of tests. Compression molding machine was made on and desired temperature was allowed to attain. Mold halves were cleaned and release agent (silicon) was applied on the mold surfaces. Plastic granules were kept in mold halves and mold was clamped in semi automatic compression molding machine at process conditions maintained below.

- Processing temperature: 200 °C
- Clamping pressure: 1.5 tons/in^2
- Heating Time: 15 minutes

After proper heating period, heating was switched off. Mold was cooled by circulating water for 20 minutes. Now mold was removed from the machine, two halves are opened and finally RHPP sheets were obtained. Various properties were checked by cutting the samples from this sheet as per the requirement.

Hardness was checked by Durometer hardness (SHORE D) tester as per the ASTM 2240): This test was conducted at SLIET, Longowal and sheet manufactured earlier (155 ×155 mm^2) was used as the specimen. Impact test was performed from CIPET, Amritsar as per the ASTM D256. Notched sample of 63.5 mm×12.7 mm ×3.2 mm (thickness) size was made for performing impact test. Heat deflection temperature was measured as per the ASTM 648 from CIPET Amritsar. Sample of 5x1/2x1/4 $inch^3$ size was taken and instrument used was ATS-FAAR (Italy made). Thermo gravimetric/ differential thermal analysis was performed as per ASTM E 1131 and differential scanning calorimetry as per the ASTM D 3418.

3.4 RESULTS AND DISCUSSION

Improvement in hardness and impact strength of RHPP composites was noticed by use of additives such as Mica, talc, wollastonite, zirconium oxide, $CaCO_3$. Effect of addition of additives on hardness of RHPP composites is given in table 1.

Table 1: Effect of Additives on Hardness of RHPP Composites

Varying Percentage of Additives →	0 [RHPP(70/30)]	1.25 %	2.50 %	3.75 %	5 %
Hardness (Talc)	71.80	73.94	75.65	77.84	79.78
Hardness ($CaCO_3$)	71.80	74.07	76.56	79.24	81.80
Hardness (Mica)	71.80	72.46	72.98	73.67	74.20
Hardness (Wollastonite)	71.80	74.24	76.78	79.26	81.46
Hardness (Zirconium Oxide)	71.80	74.88	77.60	81.34	83.94

Table 2 illustrates the effect of addition of additives on the impact strength of RHPP composites.

Table 2: Effect of Additives on Impact Strength of RHPP Composites

Varying percentage of Additives →	0 [RHPP(70/30)]	1.25 %	2.50 %	3.75 %	5 %
Impact Strength (J/m) Talc	13.70	17.10	18.24	19.36	20.53
Impact Strength (J/m) $CaCO_3$	13.70	14.45	15.24	15.96	16.67
Impact Strength (J/m) Mica	13.70	15.45	17.28	19.64	21.00
Impact Strength (J/m) Wollastonite	13.70	15.51	18.65	21.46	24.72

Impact Strength (J/m) Zirconium Oxide	13.70	14.29	16.57	19.84	21.70

In the figure 1, effect of percentage of various additives (mica, talc, wollastonite, zirconium oxide, and $CaCO_3$) on hardness of RHPP composites is shown in figure 1.

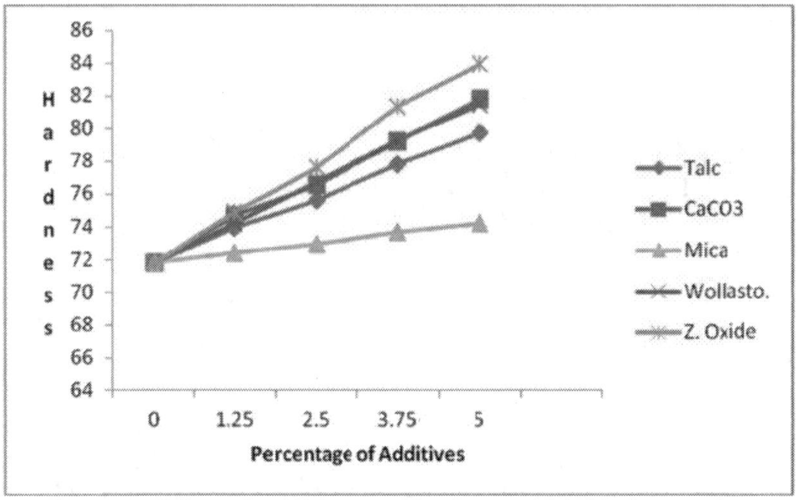

Figure 1: Effect of Additives on Hardness of RHPP Composites

It is seen from above figure that hardness increases with increase in percentage of all additives, maximum for zirconium oxide and minimum for mica. Improvement in hardness of RHPP composites modified $CaCO_3$ and wollastonite is almost same. Talc modified RHPP composite showed better hardness than mica modified. High density and brittle nature of additives compared with virgin polypropylene is responsible for getting better hardness of RHPP composites modified by these fillers. Beside this, there may be the formation of agglomerates, which have positive impact on hardness of RHPP composite.

Effect of additives such as mica, $CaCO_3$, wollastonite, talc and zirconium oxide on the impact strength of RHPP composites is illustrated in figure 2.

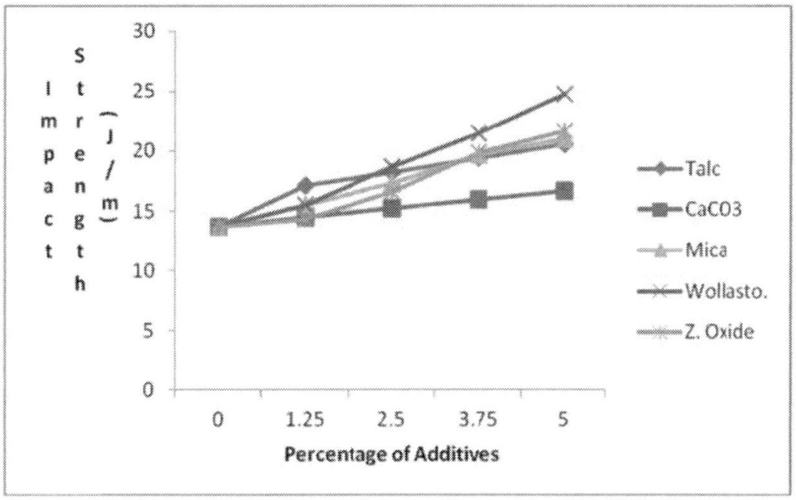

Figure 2: Effect of Additive on Impact Strength of RHPP Composites

It is concluded that impact strength increases by addition of these additives, maximum for wallastonite modified RHPP and lowest for $CaCO_3$ modified. Fine particle size of these fillers compared with rice husk is the main reason for getting higher impact strength in these composites. These additives fill or reduce empty space /void size present in the composites. There is decrease in stress concentration points which lead to higher impact strength in RHPP composites.

Effect of additives such as silica, mica and $CaCO_3$ on heat deflection temperature of RHPP composites was studied. Result for heat deflection temperature by addition of these additives are given in table 3.

Table 3: Effect of Additives on Heat Deflection Temperature of RHPP Composites

S. N.	Composition	Heat Deflection Temperature (°C)
1	RH/PP (30/70)	65.2
2	RH/PP/Silica (30/70/5)	67.7
3	RH/PP/$CaCO_3$ (30/70/15)	70.1
4	PP/RH/Mica (30/70/15)	72.4

It was observed that use of various additives improve heat deflection temperature of RHPP composites. Higher percentage of additives lead to formation of agglomerate and bring stiffening effect in composites. Further, nucleating action of filler in the RHPP composites at the initial stage might be responsible for higher value of HDT.

TG/DTA studied was conducted in the temperature range of 0-200 °C. Experiment was conducted in nitrogen atmosphere and results are reported in table 4.

Table 4: Effect of Additives on TG/DTA Studies of RHPP Composites

S. N.	Compositions	Weight Loss at 100° C (%)	Weight Loss at 199 ° C (%)	M P (° C)	Energy required for Melting (my/g)
1	PP/RH/CaCO3 70/30/15	99.7	99.5	166	51.1
2	PP/RH/Mica 70/30/15	99.6	99.3	166	39.4
3	PP/RH/Silica 70/30/5	99.1	98.4	166	45.7
4	PP/RH/Titanate 70/30/0.5	99.3	98.6	166	49.8
5	PP/RH/Titanate 70/30/0.7	99.7	99.4	167	61.7
6	PP/RH/Titanate 70/30/0.9	99.6	99.0	165	67.3

It is observed that no significant change in melting point (165-167°C) of RHPP composites takes place. It might be due to presence of traces of polypropylene in the final composite (which start melting in vicinity of 164 °C. There is decrease in the melting energies by addition of additives due to disordering in the system. Maximum energy required for melting is for $CaCO_3$ and minimum for mica. It indicates that bonding of $CaCO_3$ with RHPP composite is stronger as compared to mica based. It might be due to better dispersion and reactivity of $CaCO_3$ with RHPP composites. Energy required for melting of silica modified composite is between two, since silica is used in lower percentage as higher percentage of silica cause sticking problem in the composite .Weight loss is minimum for $CaCO_3$ modified RHPP composites as compared to mica and silica modified, indicating highest thermal stability of these composites with $CaCO_3$. It again confirms the stronger bonding of $CaCO_3$ with RHPP composites; therefore higher energy for melting was noticed earlier.

DTA and TGA studies of RHPP modified with varied percentage (0.5 to 0.9) of titanate coupling agent was conducted. Energy required for melting of RHPP composites modified with titanate (0.9 %) is highest indicating the good bond strength of RHPP composites. Though highest thermal stability was observed with titanate (0.7 %) modified RHPP composites, it might be due to some experimental error in the process during manufacturing of RHPP composites.

DSC test were performed at NIPER Chandigarh in the temperature range 0-500 °C with Mettler - Toledo AG instrument and results are reported in the table 5.

Table 5: Effect of Additives on DSC Studies of RHPP Composites

S. N.	Compositions	M P (°C)	Energy Required for Melting (mJ/g)	Degradation Temperature (°C)	Energy required for Degradation (mJ/g)
1	PP 100 %	164.38	101.21	390.9-473.23	612.58
2	PP/RH 70/30	163.88	67.91	421.58-479.98	505.54
3	PP/RH/Zirconia 70/30/15	163.03	53.16	418.02-481.65	321.18
4	PP/RH/Talc 70/30/15	164.46	9.22	422.85-481.68	319.52
5	PP/RH/Maleated PP Compatibilizer 70/30/3	169.98	61.43	419.88-481.56	444.43
6	PP/RH/Titanate Coupling Agent 70/30/0.9	162.71	41.70	412.55-484.03	275.24

DSC studies of PP and RHPP composite with various additives was conducted. It was found that PP modified with RH, zirconia and talc doesn't have much difference in melting point (162.71-169.98°C). Energy required for melting decrease since filler bring disordering in the system. Maximum decrease is for talc, it might be due to the lower decomposition temperature of talc. Same trend is observed with respect to degradation energy. Though degradation temperature increases compared with virgin PP, which might be due to combination of stiffness and nucleating action of fillers in initial stage in formation of RHPP composites.
Thermal properties of maleated compatibilizer and titanate coupling agent based RHPP was studied. Higher melting point (169.98°C) and degradation temperature was noticed with maleated PP modified RHPP composites showed better thermal stability compared with titanate modified. Maleated modified composites require higher energy for melting and degradation compared with titanate modified, confirming better bonding with RHPP composites.

5. CONCLUSION

Use of additives such as $CaCO_3$, mica, talc, wollastonite, and zirconium oxide with MAPP compatibilizer improve the hardness and impact strength of RHPP composites. Zirconium oxide based RHPP composite have maximum hardness. It is a very valuable property, since related with tensile strength and abrasive resistance of the materials. Previous studies indicate that use of compatibilizer/coupling agents did not had the positive effect on impact strength of composites. Additives such as $CaCO_3$, mica, wollastonite, talc and zirconium oxide, when used with rice husk proved very effective in enhancing the impact strength of these composites. Certain products where good impact can be advantageous are panel, door, shuttering, door frames, roofing, shelters, and slab.
In the present study, effect of additives such as mica, talc, wollastonite, zirconium oxide, silica and $CaCO_3$ on thermal properties of RHPP was studied by measurement of heat deflection

temperature measurement and thermal analysis techniques. Improvement in the HDT of RHPP composites was noticed by use of mica, silica and $CaCO_3$, maximum for mica. TG/DSCDTA study of additives modified RHPP composites results considerable reduction in enthalpy for melting and degradation as compared to unmodified RHPP composites, talc being the most effective. Use of these additives brings improvement in degradation temperature of composites. But no significant change in melting point of RHPP composite modified by various additives takes place. With increase in the percentage of coupling agent, increase in enthalpy of melting was studied. Thermal stability of RHPP systems modified with maleated compatabilizer is better compared with titanate modified composites.

5.6 REFERENCES

1. Dearmitt, D.C., and Breese K. D., Filled Polypropylene: A Cost-Performance Comparison of Common Fillers, *Plastics. Additives and Compounding*, Vol. 3 (9), 28-33 (2001)
2. Jilken, L. , Malhammar, G., and Selden R., (Effect of Mineral Fillers on the Impact and Tensile Properties of Polypropylene, *Polymer Testing* , Vol. 10 (5), 329-344 (1991).
3. Yang H. S. , Kim H. J., Son J. , Park,H.J., Lee B. J., and Hwang T. S., Rice-Husk Flour Filled Polypropylene Composites; Mechanical and Morphological Study, *Composite Structure*, Vol. 63 (3-4), 305-312 (2004).
4. Yang H. S., Kim H. J. , Park H.J. , Lee B. J. , and Hwang T.S., Effect of Compatibilizing Agents on Rice Husk Flour Reinforced Polypropylene Composites, *Composite Structures* , Vol. 7, 44-55 (2007) .
5. Fuad M. Y A. , Ismail Z. , Ishak Z. A. M., and Omar A. K. M., Application of Rice Husk Ash as Fillers in Polypropylene: Effect of Titanate, Zirconate and Silane Coupling Agents, *European Polymer Journal* , Vol. 31(9), 885-893 (1995).
6. Zuiderduin W.C. J. , Westzaan C. , Huetink J., and Gaymans. R. J., Toughening of Polypropylene with Calcium Carbonate Particles. *Polymer*, Vol. 44 (1), 261-275 (2003).
7. Premalal H. G. B. , Ismail H., and Baharin A., Comparison of Mechanical Properties of Rice Husk Powder Filled Polypropylene Composites with Talc Filled Polypropylene Composites, *Polymer Testing*, Vol. 21(7), 833-839 (2002).
8. Balkan O., and Demirer H., Mechanical Properties of Glass Bead- and Wollastonite-Filled Isotactic-Polypropylene Composites Modified with Thermoplastic Elastomers' *Polymer Composites* , Vol. 31 (7), 1285–1308 (2010).
9. Marine F. W. and Shepherd P. D., Mica Reinforced Plastics: A review Composites, *Composites*, Vol. 5 (5), 193-200 (1974).
10. Nourbakhsh A., Baghlani F. F., and Ashori A., Nano-SiO_2 filled rice husk/polypropylene composites: Physico-mechanical properties. *Industrial Crops and Products*, Vol. 33 (1) , 183-187 (2011).
11. Ghijsels A., and Waals F. M. T. A. M., Differential scanning calorimetry A powerful tool for the characterization of thermoplastics, *Polymer Testing*, Vol. (2), 149-160 (1980).
12. Manley T. R.,Guy A., and Clark L. J., Thermal analysis of the stability of polyvinylchloride, *ThermochimicaActa*, Vol. 93, 183-185 (1985).

13. Naqvi M. K., and Sen A. R., Thermal stability and thermal characteristics of PVC/cis-PBR blends, *Polymer Degradation and Stability*, Vol. 33 (3), 367-375 (1991).
14. Wielage B., Lampke T., Marx G., Nestler K, and Starke D., Thermo gravimetric and differential scanning calorimetric analysis of natural fibers and polypropylene, *ThermochimicaActa*, Vol. 337 (1-2), 169-177 (1999).
15. Mathew A. P., Packirisamy S., and Thomas S., Studies on the thermal stability of natural rubber/polystyrene interpenetrating polymer networks: thermogravimetric analysis, *Polymer Degradation and Stability*, Vol. 72 (3), 423-439 (2001).
16. Wong A. C. Y., and Lam F., Study of selected thermal characteristics of polypropylene/polyethylene binary blends using DSC and TGA, *Polymer Testing*, Vol. 21(6), 691-696 (2002).
17. Frick A., and Rochman A., Characterization of TPU-elastomers by thermal analysis (DSC), *Polymer Testing*, Vol. 23(4), 413-417 (2004).
18. Tidjani A., Wald O., Pohl M. M., Hentschel M. P., and Schartel B., Polypropylene–graft–maleic anhydride-nanocomposites: I—Characterization and thermal stability of nanocomposites produced under nitrogen and in air, *Polymer Degradation and Stability*, Vol. 82 (1), 133-140 (2003).
19. Doh G. H., Lee S.Y., Kang I. A., and Kong Y.T., Thermal behavior of liquefied wood polymer composites (LWPC), *Composite Structures*, Vol. 68 (1), 103-108 (2005).
20. Voda A., Beck K., Schauber T., Adler M., Dabisch T., Bescher M., Viol M., Demco D. E., and Blümich B., Investigation of soft segments of thermoplastic polyurethane by NMR, differential scanning calorimetry and rebound resilience, *Polymer Testing*, Vol.25 (2), 203-213(2006).
21. Gonzalez N., Mugica A., and Berridi M. J. F., Application of high resolution thermogravimetry to the study of thermal stability of poly(vinyl chloride) resins, *Polymer Degradation and Stability*, Vol. 91 (4), 629-633(2006).
22. Golebiewski J., and Galeski A., Thermal stability of nanoclay polypropylene composites by simultaneous DSC and TGA, *Composites Science and Technology*, Vol. 67 (15-16), 3442-3447 (2007).

(9f) Water Permeation In Polylactide and Polylactice/Montmorillonite Composites: Experimental Results Versus Theoretical Prediction

An Du[1], Genaro A. Gelves[2], Donghun Koo[1], Mary Ziegler[1], Uttandaraman Sundararaj[2] and Richard A. Cairncross[1], (1)Department of Chemical and Biological Engineering, Drexel University, Philadelphia, PA, (2)Department of Chemical and Petroleum Engineering, University of Calgary, Calgary, AB, Canada

Polylactide (PLA) is a well-known renewable and biodegradable polymer that has been widely used in medical and pharmaceutical applications. Uses of PLA in packaging are expanding but are limited by insufficient moisture barrier properties. PLA-clay nanocomposites produced by solvent-casting and melt-blending made a significant improvement in the moisture barrier properties of PLA. For example, a nanocomposite of PLA with 10 wt% motmorillonite reduced the water permeability in PLA by 35%. In this research, water permeation in PLA-montmorillonite composites at different concentrations of montmorillonite was investigated to compare experimental results with theoretical prediction. Specifically, a three-phase model that takes into account the interface region between montmorillonite and the polymer matrix was developed to determine the water permeability of the mixed matrix membrane. By fitting the model to the experimental results, the shape and orientation of montmorillonite particles were estimated and compared with results from wide angle X-ray diffraction (WAXD) and transmission electron microscopy (TEM) techniques.

See more of this Session: Biocomposites - Bio-Derived and Bio-Inspired Composites
See more of this Group/Topical: Materials Engineering and Sciences Division

(17a) Layer-by-Layer Assembled Macroscale Materials with Microscale 3D Topologies

Christine M. Andres and Nicholas A. Kotov, Chemical Engineering, University of Michigan, Ann Arbor, MI

Hybrid nanocomposites create functional materials with enhanced and diversified properties as compared to their bulk counterparts. Three-dimensional topology adds to these impressive properties by allowing for the addition of unique characteristics such as special deformation patterns, negative Poison's ratio, negative thermal coefficient, propagation of electromagnetic waves, biological, and mass transport properties. Typical methods for preparing materials with 3D microscale features are often restricted by limited material variability, lack of control over nanophase filling, challenging transitions to macroscale materials and feature sizes dependent on material delivery equipment. This work describes a new method to synthesize hierarchically ordered porous nanocomposites by using layer by layer assembly, the sequential adsorption of monoloyers attracted by chemical forces, for controlled introduction of nanocomposites into the void space of highly ordered arrays of uniform microspheres. Feature sizes of the inverted colloidal crystal structure obtained after dissolution of the array are controlled by the diameter of the microsphere, size and annealing time of the array, and number of LBL bilayers deposited on the surface. The method allows for 3D topologies consisting of a wide variety of materials to be created with excellent control over the interfacial connectivity between the inorganic building blocks and a polymer matrix, uniquely high contents of the inorganic phase, and simple transition to macroscale materials. Additionally, due to the morphological control of the technique, uniform composites or gradient structures with feature sizes below the typical dimensions obtained by free-form fabrication, direct write methods, and 3D lithography are possible. Establishing appropriate techniques to produce materials with hierarchical organization of materials involving nano-, micro-, and potentially millimeter scale features with fairly independent control at all levels allows for investigation of structural influences on macroscale material properties and for new practical applications due to the unusual combinations of properties which can be achieved by such controlled assembly of materials and 3D topologies.

See more of this Session: Composites I
See more of this Group/Topical: Materials Engineering and Sciences Division

(17b) Effect of Nanoparticle Alignment In PAM-MMT Nanocomposite Hydrogels On Protein Electrophoretic Separations

Jeffery W. Thompson[1], Holly A. Stretz[1], Pedro E. Arce[1], Harry J. Ploehn[2] and Hongsheng Gao[2], (1) Chemical Engineering, Tennessee Technological University, Cookeville, TN, (2)Chemical Engineering, University of South Carolina, Columbia, SC

Nanocomposite hydrogels are a class of materials containing a hydrogel matrix typically consisting of a polyamide and a nanophase which can encompass any number of organic, metallic, or ceramic particles. Furthermore, these nanophase materials may take any number of different morphologies and aspect ratios. While nanocomposite hydrogels have been previously selected primarily for their enhanced mechanical properties, other potential applications exist. As hydrogels are ubiquitous in electrophoretic separation, the ability to impact and potentially improve electrophoretic properties via the incorporation of nanoparticles could be another promising application of these nanocomposite hydrogels. In a recent publication, the impact of nanoparticle aspect ratio on electrophoretic transport in PAM-Au nanocomposite hydrogels has been shown using rod shaped particles which were randomly aligned (1). Another potential tunable parameter might be aligning anisotropic particles. Another recent publication has shown that montmorillonite nanoparticles may be aligned in the presence of a magnetic field exceeding 1.4 Tesla (2). Montmorillonite discs as both high in aspect ratio as well as having the ability to be aligned should be a great model nanoparticle for this system. This presentation will present information about the synthesis of these aligned nanomaterials using Oak Ridge National Laboratory's High- and Thermo-magnetic processing facility, characterization through electron microscopy, as well as information about the electrophoretic separation of two protein probes, ovalbumin and carbonic anhydrase over a range of compositions and two different orientations.

1) Thompson et.al. IECR 2010 *49* (23), pp 12104–12110.

2) Koerner et. al Chem Mater. 2005 17, pp 1990-1996.

See more of this Session: Composites I
See more of this Group/Topical: Materials Engineering and Sciences Division

(17c) Dynamics of Solid Microbubble Formation

Anaram Shahravan, Srinath Yelamarty and Themis Matsoukas, Department of Chemical Engineering, The Pennsylvania State University, University Park, PA

Hollow particles are usually synthesized by a templating approach in which a sacrificial core is coated with a suitable material and the core is then removed to produce a hollow particle. We report a novel technique that produces hollow particles that may be described as solid microbubbles, namely, hollow solid spheres, about 1 micron in size, surrounded by an ultra-thin solid layer. These microbubbles are formed under irradiation by an electron beam of precursor particles synthesized in a low-pressure plasma and their formation can be observed under TEM.

Toluene is introduced to a low-pressure plasma reactor that is capacitively coupled by a radio frequency generator (13.56 MHz). After formation of plasma particles they are taken to a transmission electron microscope (TEM) that works at 120 kV. Spherical plasma particles with diameter of 500±50 nm start to evolve, first grow and then shrink, over 8 to 12 seconds. Final size of the spherical particles are 1.6±0.6 microns in diameter and the average thickness of the hollow particle shells are 90 nm. We have introduced these particles to a field emission scanning electron microscope (FESEM) that works with lower accelerating voltage, 5 kV, and particles showed the same behavior and formed bubbles. These smart particles are even sensitive to other sources of energy such as argon laser that works at 120 mW.

See more of this Session: Composites I
See more of this Group/Topical: Materials Engineering and Sciences Division

(17d) Electrically Conductive Epoxy Nanocomposites

Ouassima Alloul[1], Jiahua Zhu[1], Suying Wei[2] and Zhanhu Guo[1], (1)Integrated Composites Laboratory (ICL), Dan F Smith Department of Chemical Engineering, Lamar University, Beaumont, TX, (2) Chemistry and Biochemistry, Lamar University, Beaumont, TX

Conductive polymer nanocomposites have attracted much interest recently owing to their versatile applications, such as battery cathodes, microelectronics and sensors, etc. In this work, we introduced different conductive nanofillers in epoxy matrix, such as carbon nanofibers (CNFs), Fe@FeO and Fe@C core@shell structured nanoparticles (1-3) to make conductive nanocomposites. The loading effect of the fillers on conductivity is studied and percolation threshold (P_c) differs from each other arising from the different filler nanostructures. Generally, CNFs obtain lower P_c due to the larger aspect ratio of its fiber structure than that of the nanoparticles. All these epoxy nanocomposites show enhanced mechanical strength (from tensile test and dynamic mechanical analysis) and thermal stability (thermalgravimetric analysis) than that of pure epoxy. To improve the CNFs dispersion and interfacial interaction with epoxy matrix, surface modification with amine terminated functional group is applied. Scanning electron microscopy (SEM) and transmission electron microscopy (TEM) are used to characterize the filler distribution and interfacial interaction.

References

(1) Rheological behaviors and electrical conductivity of epoxy resin nanocomposites suspended with in-situ stabilized carbon nanofibers, Polymer, doi:10.1016/j.polymer.2010.04.019 , 2010

(2) In situ stabilized carbon nanofiber (CNF) reinforced epoxy nanocomposites, Journal of Materials Chemistry, DOI: 10.1039/c0jm00063a ,2010

(3) Magnetic Epoxy Resin Nanocomposites Reinforced with Core-Shell Structured Fe@FeO Nanoparticles: Fabrication and Property Analysis, ACS Applied Materials & Interfaces, DOI: 10.1021/am100361h, 2010

See more of this Session: Composites I
See more of this Group/Topical: Materials Engineering and Sciences Division

(17e) PREPARATION of THERMO-RESPONSIVE FUNCTIONAL COMPOSITES by EMBEDDING TiO2/Fe3O4 NANOPARTICLES

Junichi Ida, Fumiko Matsushima, Masanori Ochi, Tatsushi Matsuyama and Hideo Yamamoto,
Department of Environmental Engineering for Symbiosis, Soka University, Hachioji, Tokyo, Japan

Thermo-responsive functional composites were prepared by embedding two kinds of functional inorganic particles within thermo-responsive microgel. In this study, poly(N-isopropylacrylamide) was used as thermo-responsive polymer, calcium alginate was used as structure support polymer and TiO_2 and Fe_3O_4 were used as functional inorganic nanoparticles respectively. In the experiment, thermo-responsive functional composites were prepared using single tube nozzle by modifying the simple process to prepare thermo-responsive capsule reported in our previous report[1]. In the process, two solutions were prepared such as a polymer solution and a solution for gelation reaction. In the polymer solution, NIPAM monomer, sodium alginate and a cross-linker agent were dissolved. TiO_2 and/or Fe_3O_4 were also dispersed in the polymer solution. The solution for gelation reaction was prepared by dissolving a gelation agent, a polymerization initiator and an polymerization accelerator into a buffer solution. When the droplets of the polymer solution containing functional particles were dripped into the solution for gelation reaction, gelation of alginate and polymerization of NIPAM and the cross-linker were occurred at the same time and gel beads with interpenetrating polymer network (IPN) structure were obtained instantaneously. The characterization was carried out by examining their thermo-responsive property, photocatalytic activity and localized heating property. The experimental results showed that TiO_2/Fe_3O_4 embedded microgels were obtained successfully. The obtained beads exhibited thermo-responsive volume change and photocatalytic activity. Localized heating of one thermo-responsive bead containing Fe_3O_4 was also achieved without raising a bulk solution temperature by focusing laser irradiation on the bead.

See more of this Session: Composites I
See more of this Group/Topical: Materials Engineering and Sciences Division

(17f) Preparation of Nano-Structured Hybrid Materials Via In Situ Synthesis

Lichen Xiang, Cody A. Gummelt, Matthew F. Milner, Jarett C. Martin, Cara S. Southworth and **Luyi Sun**, Department of Chemistry and Biochemistry, Texas State University-San Marcos, San Marcos, TX

Nanostructured polymer-inorganic composite materials, including polymer nanocomposites, layer-by-layer assembled thin films, have been extensively investigated over the past two decades and have found wide applications owing to their excellent performance. Either regular polymer nanocomposites or layer-by-layer assembled thin films are typically prepared using pre-synthesized nanofillers/nanoplatelets. Here, we report a new approach to prepare nanostructured composite materials via in situ synthesis of nanoplatelets within the polymer/monomer matrix. Alpha-zirconium phosphate (α-ZrP) was synthesized in a solution system containing a polymer (such as polyethylene glycol, PEG) or monomer. During the synthesis of α-ZrP, PEG chains were embedded into the α-ZrP interlayer space, leading to a larger interlayer distance, which is similar to the intercalated layered compound. Proper formulation ratio proved to be critical to avoid forming pristine α-ZrP, and avoid interfering the growth of the layered structure of α-ZrP. It has also been found that longer polymer chains are desirable for minimizing the formation of pristine α-ZrP, but would not affect the interlayer distance. All the PEG chains are perfectly parallel to the layer planes. Besides polymers, many other molecules have also been successfully embedded into the interlayer space to form an intercalated structure during in situ synthesis.

See more of this Session: Composites I
See more of this Group/Topical: Materials Engineering and Sciences Division

(17g) Predicting the Orientation of Concentrated Long Glass Fiber Suspensions In Simple Shear Flow: Application to Processing Flows

Kevin C. Ortman[1], D. G. Baird[1] and Peter Wapperom[2], (1)Chemical Engineering, Virginia Tech, Blacksburg, VA, (2)Mathematics Department, Virginia Polytechnic Institute, Blacksburg, VA

Great interest exists in adding long fibers into polymeric fluids due to the increase in properties associated with the composite, as compared to the neat resin. These properties, however, are dependent on the orientation of the fibers within the composite. In an effort to optimize industrial processing, optimize mold design, and maximize desired properties of the final part, it is highly desirable to be able to predict long fiber orientation as a function of processing conditions. For this reason, it is desirable to understand the link between the deformation response (rheology) of a long fiber reinforced thermoplastic and the effect it has on its underlying microstructure.

The purpose of this research is to understand the transient fiber orientation and the associated rheology of long glass fiber (> 1 mm) reinforced polypropylene, in a well-defined simple shear flow, to extend the knowledge gained from these fundamental experiments for the use of simulating (more complex) molding processes. Specifically, we are interested in associating the rheological behavior of long glass fiber reinforced polypropylene with the transient evolution of fiber orientation, in simple shear flow, to ultimately model fiber orientation in complex processing flows. A sliding plate rheometer was designed to measure stress growth in the startup and cessation of steady shear flow.

Two fiber orientation models were investigated to predict the transient orientation of the long glass fiber system. One model, the Folgar-Tucker model, has been particularly useful for predicting fiber orientation in short glass fiber systems and was used in this paper to assess its performance with long glass fibers. A second fiber orientation model, one that accounts for the flexibility of long fibers, was also investigated. The accuracy of both the Folgar-Tucker model and the semi-flexible orientation model, when used with the Lipscomb model for the stress tensor and a modified version of this tensor (one that tries to account for bending stresses), respectively, is evaluated by comparing orientation predictions against experimentally measured orientations.

See more of this Session: Composites I
See more of this Group/Topical: Materials Engineering and Sciences Division

(23a) Transport of Ions and Penetrants Through Structured Polymeric Matrices: Interplay of Structure and Dynamics of Polymers

Venkat Ganesan, Department of Chemical Engineering, The University of Texas at Austin, Austin, TX

In this talk, I will discuss our recent work in the context of the barrier properties and conductivities of structured polymeric matrices. While much earlier work has focused on these issues in the context of glassy homopolymeric membranes, there has been only a few isolated studies on the issues specific to structured polymeric matrices, such as those arising in block copolymers and nanocomposite membranes. In such materials, the polymeric matrix exhibits density (or volume fraction) inhomogeneities which also in turn leads to dynamical heterogeneities. Modeling and understanding the mechanistic features underlying the transport of ions and penetrants through such heterogeneous matrices is of interest to a wide range of applications, including, water purification, polymer batteries and fuel cell membranes. In our work, we have used a combination of coarse-grained simulations and nonequilibrium statistical mechanics models to shed light on the parameters influencing such transport properties. In this talk, I will present these results and discuss the ramifications for the design of batteries and polymer separation membranes.

See more of this Session: Emerging Areas In Polymer Science & Engineering Plenary
See more of this Group/Topical: Materials Engineering and Sciences Division

(23b) Structure Formation In Nanofibers: Modeling, Experiments and Applications

Yong Lak Joo, Chemical and Biomolecular Engineering, Cornell University, Ithaca, NY

Electrostatic fiber spinning or 'electrospinning' is a unique process for forming fibers with submicron scale diameters through the action of electrostatic force. The resulting nanofibers are collected as non-woven mats with extremely large surface to mass ratios which can be used in filtration, catalysis, and biomedical applications. We will present our recent work on the structure formation in nanofibers: i) modeling of electrically driven polymeric jets and structures in nanofibers, ii) experiments on the creation of hierarchical nanostructures in nanofibers, and iii) applications of nanostructured nanofibers. In the first part of the talk, we will present recent modeling efforts including non-isothermal modeling of polymer melts during electrospinning. Our recent *in-situ* infrared thermal measurements on polymer melt jets during electrospinning have revealed rapid quenching by ambient air, an order of magnitude faster than predicted by classical correlations. This drastic heat transfer enhancement which causes premature solidification of the melt jet has been described by electrohydrodynamic (EHD) effects. Analysis of EHD-driven air flow was performed and included into a comprehensive model for polymer melt electrospinning. The analysis was validated by excellent agreement of both predicted jet radius and temperature profiles with experimental results for electrospinning of Nylon-6, PP, and PLA melts. Based on this analysis, we present the concept of gas-assisted melt electrospinning which alleviates the undesirable quenching process and thus can increase the attenuation rate and resulting in production of sub-micron scale fibers. The second part of the talk deals with the formation of self-assembling structures of block copolymers such as PS-*b*-PI during electrospinning and the utilization of such confined assembly to control the distribution and location of nanoparticles in nanofibers. Transmission electron microscopy images of microtomed electrospun fibers reveal that magnetite nanoparticles were uniformly dispersed only in one of the domains in block copolymer nanofibers. We also present a coarse-grained Molecular Dynamics simulation study to elucidate the effect of nanoparticle inclusion and strong deformation on self assembly in block copolymer/nanoparticle composite nanofibers. The results confirm our electrospinning experiments that deformation can be used as a parameter to effectively control nanoparticle location and distribution. Finally, we have applied nanostructured nanofibers as an effective catalyst system for the hydrogen production via the alkaline hydrolysis of glucose. When coaxial nanofibers that consist of pure ceramic (silica, alumina, zirconia) nanofibers in the core and ceramic nanofibers containing metal (iron or nickel) nanocrystals in the sheath layer were compared with monoaxial inorganic nanofibers metal nanoparticles distributed throughout the nanofiber, a four-fold decrease in the catalyst for greater than 90% conversion was achieved by tuning the catalyst location to the surface of the nanofiber. Placing the catalyst selectively into the sheath layer also delayed the onset of side reactions such as thermal decomposition to form CH_4, CO and CO_2, and thus the maximum reaction temperature for pure hydrogen production could be elevated from 140°C to 200°C. The increased reaction temperature while maintaining the hydrogen selectivity in turn resulted in a substantial increase in the production rate of pure hydrogen.

(23c) Microstructured Block Copolymer Membranes for Lithium Batteries and Ethanol Separation

Nitash P. Balsara, Department of Chemical Engineering, University of California, Berkeley, Berkeley, CA

Nanostructured block copolymers enable the design of membranes with optimized transporting channels for a variety of applications. This presentation addresses the use of these membranse to transport lithium ions in an all solid lithium battery and to separate ethanol from a dilute aqueous stream. Previous studies of electrolytes showed an inverse correlation between ionic conductivity and shear modulus: increasing conductivity invariably resulted in a decrease in the shear modulus. Similarly, previous studies of alcohol-selective membranes showed an inverse correlation between alcohol separation factor (ratio of alcohol concentration in the exiting stream to that of the incoming stream) and total flux: increasing the separation factor invariably resulted in a decrease in total flux. Block copolymer membranes do not show these inverse correlations. Practical implications of our discoveries for enabling electric vehicles and bioethanol will be discussed. A large fraction of energy and materials that we use today are derived from industries that consume non-renewable resources. The two projects described here may enable a gradual transition toward to a more sustainable approach.

See more of this Session: Emerging Areas In Polymer Science & Engineering Plenary
See more of this Group/Topical: Materials Engineering and Sciences Division

(29a) Dynamics of Molecular Exchange In Solutions of Block Copolymer Micelles

Frank S. Bates[1], Soo-Hyung Choi[1] and Timothy P. Lodge[2], (1)Department of Chemical Engineering and Materials Science, University of Minnesota, Minneapolis, MN, (2)Department of Chemical Engineering and Materials Science and Department of Chemistry, University of Minnesota, Minneapolis, MN

Block copolymers offer certain advantages over traditional low molecular weight surfactants in a variety of engineering applications in both aqueous and organic media. Increasing the molecular weight of an amphiphilic molecule enhances interfacial activity but drastically reduces the critical micelle concentration, which leads to dramatically slower dynamics for molecular exchange resulting in non-equilibrium micelle configurations. Model poly(styrene-*b*-ethylene-*alt*-propylene) (PS-PEP) diblock copolymers have been prepared in normal and partially deuterated forms by anionic polymerization and catalytic hydrogenation and mixed in squalane (C_{30}), a model saturated oil, at concentrations between 0.5 and 15% by weight. Linear dynamic mechanical spectroscopy and small-angle x-ray scattering were employed to ascertain the state of order or disorder, size, size distribution and other structure features associated with the resulting spherical micelles, which contain a PS core surrounded by a solvated PEP corona. Time resolved small-angle neutron scattering measurements afforded access to the rate of block copolymer exchange at 1% and 15% loadings. These results demonstrate a remarkable and non-intuitive sensitivity of the molecular exchange process to relatively minor core block polydispersity and the state of micelle ordering.

See more of this Session: In Honor of Matthew Tirrell's 60th Birthday I
See more of this Group/Topical: Materials Engineering and Sciences Division

(29b) Simultaneous Electronic and Ionic Conduction In Block Copolymers

Anna E. Javier, Shrayesh N. Patel and **Nitash P. Balsara**, Department of Chemical Engineering, University of California, Berkeley, Berkeley, CA

Redox reactions occurring within battery electrodes require transport of both ions and electrons to the active centers. We report on the synthesis of a block copolymer, poly(3-hexyl thiophene)-*b*-poly (ethylene oxide) (P3HT-PEO), that exhibits simultaneous electronic and ionic conduction. Separate values of the electronic [(6.73±2.05) x 10^{-6} S/cm] and ionic [(1.09±0.469) x 10^{-4} S/cm] conductivity of a mixture of P3HT-PEO and a lithium salt were determined by a combination of AC impedance and DC techniques. The P3HT-PEO/salt mixture was used to make lithium battery cathodes, with the active material $LiFePO_4$ as the only other component. All-solid lithium batteries were assembled to demonstrate that the P3HT-PEO/salt mixture serves all of the transport and binding functions required in the cathode. The semiconducting nature of P3HT enables using cell potential to control charge transport within the electrode.

See more of this Session: In Honor of Matthew Tirrell's 60th Birthday I
See more of this Group/Topical: Materials Engineering and Sciences Division

(29c) Conformational Change of Bead-Spring Chain During Creep Process: Interplay of Eigenmodes

Hiroshi Watanabe, Institute for Chemical Research, Kyoto University, Kyoto, Japan

The bead-spring chain is a well-established model for non-entangled polymers and also serves as a fundamental model for entangled chains. The dynamic behavior of this model chain under strain-controlled conditions has been fully analyzed in the literature. However, the behavior of the chain during the stress-controlled creep process, in particular its conformational dynamics, was not well understood. In this talk, the conformational dynamics of the bead-spring chain is analyzed to demonstrate the importance of interplay of the motional eigenmodes during the creep process.

See more of this Session: In Honor of Matthew Tirrell's 60th Birthday I
See more of this Group/Topical: Materials Engineering and Sciences Division

(29d) Studies of Adhesion Between Vesicles with Weakly Attracting Bilayers

Johann Walter[1], Mansi Seth[1], John Frostad[1], Arun Ramachandran[2] and **L. Gary Leal**[1], (1)Department of Chemical Engineering, University of California, Santa Barbara, Santa Barbara, CA, (2)Department of Chemical Engineering and Applied Chemistry, University of Toronto, Toronto, ON, Canada

An important problem is to establish the conditions in which vesicles with weakly attractive bilayers adhere to each other in the presence of thermal agitation or flow. When they adhere, this may lead to the growth of multi-vesicle "particles" that are large enough to generate creaming or sedimentation instabilities, even in a suspension of submicron vesicles. In a recent study*, we considered the effect of vesicle deformability on the strength of adhesion for a pair of vesicles when the measured attractive force between supported bilayers (in an SFA, for example) is very weak, corresponding to an adhesion force between non-deforming spherical vesicles of O(1 kT). If we allow for deformation, adhesively interacting vesicles can form a flattened region in the contact zone, resulting in significantly enhanced adhesion compared to non-deforming vesicles. Furthermore, this deformation and thus the adhesive force will increase with time due to the osmotic expulsion of water driven by the tensed stretched membrane, and on an even longer time scale due to the equilibration of ions and solutes caused by the concentration differences created by the osmotic exchange of water. In the present work, we consider the dynamics of the adhesion between a pair of interacting vesicles, both from an experimental and theoretical point of view. The latter is based upon boundary integral simulations, while the former uses force measurements between a pair of giant unilamellar vesicles that are held at the tips of a pair of micropipettes by a weak aspiration pressure.

*"Adhesive Interactions Between Vesicles in the Strong Adhesion Limit", R. Ramachandran, T.H. Anderson, L.G. Leal and J.N. Israelachvili, Langmuir,**27,** 59-73 (2011).

See more of this Session: In Honor of Matthew Tirrell's 60th Birthday I
See more of this Group/Topical: Materials Engineering and Sciences Division

(41a) Keynote Speaker Abstract: Design of Hydrophobic Zeolites Containing Lewis Acid Active Sites for the Isomerization of Glucose In Aqueous Media

Mark E. Davis[1], Eranda Nikolla[1], Yuriy Roman[1], Manuel Moliner[2] and Sonjong Hwang[1], (1)Chemical Engineering, California Institute of Technology, Pasadena, CA, (2)Chemistry, Instituto de Tecnologia Quimica (UPV-CSIC), Valencia, Spain

We have reported that tin or titanium-containing zeolite beta catalysts (Sn-Beta or Ti-Beta) catalyze the isomerization of glucose to fructose in aqueous media at low pH in the presence of HCl and an inorganic salt (Proc. Natl. Acad. Sci., 107 (2010) 6164, ACS Catal, 1 (4), (2011) 408). Our current efforts are focused on obtaining a fundamental understanding of the nature of the active site in the Sn or Ti-Beta zeolite and the mechanism of the isomerization reaction of glucose to fructose. For example, we have utilized solid-state NMR experiments to probe the environment of tin in the zeolite. Sn-Beta zeolite is synthesized with labeled Sn-119 in order to provide a solid that can be effectively analyzed by solid-state NMR. We have also performed NMR experiments with deuterated and C-13 labeled glucose to study the interactions of the reactant with the tin active site in the zeolite. Detailed kinetic studies including isotope-labeling experiments have been conducted in order to obtain kinetic parameters such as activation energies and rate limiting steps for the isomerization of glucose to fructose. An overall picture of the state of the tin active center and the reaction pathway is emerging. In this presentation, we will discuss the important features of the catalytic materials that are necessary for obtaining reactivity. Control of the zeolite hydrophobicity and framework localization of either the tin or titanium are critical to achieving activity when the reaction medium is water.

See more of this Session: Novel Catalytic Materials I
See more of this Group/Topical: Catalysis and Reaction Engineering Division

(41b) Perovskites As Alternative Catalysts for Solid Oxide Fuel Cell Anodes: Effect of Dopants

Hyunkyu Choi, Chemical & Biomolecular Engineering, Ohio state university, Columbus, OH and Umit S. Ozkan, Chemical & biomolecular Engineering, Ohio state university, Columbus, OH

Perovskites as alternativecatalysts for solid oxide fuel cell anodes: Effect ofdopants

HyunkyuChoi, Umit S. Ozkan*

The OhioState University

HeterogeneousCatalysis Research Group

Departmentof Chemical and Biomolecular Engineering

Solid oxide fuel cells (SOFCs) show greatpotential for effectively generating clean power from a variety of fuels.Current state of the art anode catalysts based on Ni catalysts supported on yittria-stabilized zirconia (Ni/YSZ) deactivate in thepresence of coal-derived gas because they are highly susceptible to coking andsulfur poisoning. These processes decrease the anodic reaction rates, lower thepower densities, and lead to instability during long time operation. Therefore,the development of carbon and sulfur tolerant as well as highly activematerials suitable for anodes is essential to bring coal-gasfed SOFC systems closer to wide-spread application.

Ongoing research in our group examines developingperovskite-type catalysts by incorporating cerium tothe doped-lanthanum ferrites. The materials are prepared by sol gel techniques.The present work studies the kinetics of the hydrocarbon fuel oxidationreaction, the resistance to carbon coking and sulfur poisoning, the formationof oxygen vacancies, and the bulk and surface chemistry of cerium-doped perovskites. The phase transformation ofthese materials are characterized using in-situ X-ray diffraction (XRD).The activity as well as stability of these catalysts in the presence of $H_2S$is studied with time-on-stream reaction experiments for various anodereactions. Oxygen nonstoichiometry, an indicator ofoxygen vacancies, is vitally important for the diffusion of the oxide ion toreact in the three-phase boundary (TPB) and is determined by thermogravimetric analysis. Surface sites are probedusing methanol adsorption and diffuse reflectance infrared Fourier Transform(DRIFT) spectroscopy.

Session : Novel catalytic materials I

*Presenting author

331 KoffoltLaboratory•
140 West Nineteenth Avenue

Columbus, OH 43210

Phone: 614-292-4993

Fax: 614-292-3769

Email: choi.508@osu.edu

Umit S. Ozkan

333A KoffoltLaboratory•
140 West Nineteenth Avenue

Columbus, OH 43210

Phone: 614-292-6623

Fax: 614-292-3769

Email: ozkan@chbmeng.ohio-state.edu

See more of this Session: Novel Catalytic Materials I
See more of this Group/Topical: Catalysis and Reaction Engineering Division

(41c) Hybrid Organic-Inorganic Materials for Heterogeneous Catalysis: Development of Highly Structured Multifunctional Silicon Surfaces

Heidrun Gruber-Woelfler[1], Peter Feenstra[1], Georg J. Lichtenegger[1], Eleonora Polo[2] and Johannes G. Khinast[1], (1)Institute for Process and Particle Engineering, Graz University of Technology, Graz, Austria, (2)Dip. di Chimica, C.N.R.-ISOF Sez. Ferrara, Ferrara, Italy

Functionalization of single-crystal silicon surfaces is an area of intense research due to the potential use of hybrid organo-silicon materials as the basis for integrating various functionalities, such as molecular recognition devices and biomolecules, including proteins, DNA and carbohydrates. For instance, a wide variety of methods was developed to prepare organic monolayers of 1-alkenes and 1-alkynes on hydrogen-terminated silicon (H-Si).[1,2] Taking advantage of the strong Si-C bond, this kind of functionalization can also be used for the covalent immobilization of organometallic complexes onto silicon, thus leading to a powerful approach for the preparation of catalytically active materials, which can be repeatedly recycled.

We recently reported the covalent immobilization of a tethered ethylenebis(indenyl) (=EBI) ligand,[3] tethered EBI-zirconocenes[4] and of biomolecules[5] on H-terminated Si(111)-surfaces.

In this contribution we present the design of hybrid organic-inorganic silicon surfaces that include different kinds of organometallic catalysts in a highly structured arrangement.

The preparation of these multifunctional surfaces included in the first step the immobilization of an organic bis(oxazoline)-ligand on a H-terminated Si(111)-surface. The immobilization reaction was carried out via a UV-mediated hydrosilylation of the terminal double bond of the ligand and the Si-H group of the Si-wafer. In addition to being a clean and straightforward approach, this method lends itself to the sub-micron structuring of catalytic surfaces via photopatterning and photolithography. The immobilized ligand was then metalated with $Pd(OAc)_2$ to give a highly potential heterogeneous Pd-catalyst. On the same Si-wafer tethered titanocenes were attached to the surface applying again a hydrosilylation reaction. The resulting multifunctional surfaces were tested for the heterogeneous synthesis of active pharmaceutical ingredients, such as amines and substituted biphenyls.

(1) Buriak, J. M. *Chem.Rev.* **2002**, *102*, 1271-1308 and references therein.

(2) Scheres, L.; Giesbers, M.; Zuilhof, H. *Langmuir* **2010**, *26*, 4790-4795 and references therein

(3) Langner, A.; Panarello, A.; Rivillon, S.; Vassylyev, O.; Khinast, J.; Chabal, Y. J. *J.Am.Chem.Soc.*

2005, *127*, 12798-12799.

(4) Gruber-Woelfler, Rivillon, S.; Chabal, Y. J.; Schitter, G.; Polo, E.; Ringwald, M.; Khinast, J. G.. *Chem.Comm. (Hot Paper)* **2008**, 1329-1331

(5) Jeanquartier, C.; Schider, G.; Feichtenhofer, S.; Schwab, H.; Schennach, R.; Stettner, J.; Winkler, A.; Gruber-Woelfler, H.; Schitter, G.; Eder, R. J. P.; Khinast, J. G.. *Langmuir* **2008**, *24*, 13957-13961

See more of this Session: Novel Catalytic Materials I
See more of this Group/Topical: Catalysis and Reaction Engineering Division

(41d) Supports As Co-Catalysts for Activating Molecular Oxidation Catalysts

Nicholas Schoenfeldt[1], Andrew Korinda[1], Randall J. Meyer[2] and **Justin M. Notestein**[1], (1)Chemical and Biological Engineering, Northwestern University, Evanston, IL, (2)Department of Chemical Engineering, University of Illinois - Chicago, Chicago, IL

There is a wide class of non-heme oxidase enzymes thatconsist of histidine-coordinated first-row metal oxides, further activated bycarboxylates, and these structures have been extensively mimicked by solubleinorganic complexes. Many of these amine and pyridine-ligated complexes areactivated towards productive oxidation using H_2O_2 or $O_2$by coordination of carboxylates. As one example, dimeric1,4,7-triazacyclononane manganese oxides are known to be efficient and highlyactive homogeneous oxidation catalysts whose selectivity toward a number ofoxidation reactions (epoxidation / cis-dihydroxylation, alkane oxidation, alcoholoxidation) can be controlled by the addition of various acid co-catalysts.

Here we present a synthetic route to supported triazacyclononanemanganese oxides through an *in-situ* assembly method whereby the surfaceactive-site is synthesized under reaction conditions by simply combining apreformed Mn dimer and the carboxylate modified metal oxide co-catalytic supports.Self-assembly from MeCN/H_2O_2 solutions ontoorgano-functionalized oxide supports provides an inherent advantage bysupplying a co-catalyst which also acts as the support tether, thus eliminatingsynthetically cumbersome ligand alterations which can lead to differences inactive-site structure and reactivity or selectivity from the homogeneousanalogues. For example, we have shown sterically hindered surfaces to increaseselectivity toward *cis*-diol. Additionally, the high local concentrationsof surface carboxylates lead to increased reaction rates and shorter reactiontimes required for maximal productivity than is observed with analogoushomogeneous carboxylate co-catalysts. Recent work using other metal complexes (e.g.V), but the same general principles, are also described. Surface structures areunderstood through X-ray absorption fine structure, diffuse reflectanceUV-visible spectroscopy, and DFT modeling.

See more of this Session: Novel Catalytic Materials I
See more of this Group/Topical: Catalysis and Reaction Engineering Division

(41e) Unsupported NiMoW Sulfide Catalysts Prepared by Thermal Decomposition of Thiosalts for Hydrodesulfurization of Dibenzothiophene

Yanjiao Yi[1], Christopher T. Williams[2], Guang Xiong[1] and Changhai Liang[1], (1)School of Chemical Engineering, Dalian University of Technology, Dalian, China, (2)Chemical Engineering, University of South Carolina, Columbia, SC

More and more stringent environmental regulations and fuel specifications are demanding a significant reduction of sulfur and nitrogen containing compounds, as well as aromatics in fuels. Conventional hydrotreating processes are performed over alumina supported sulfide catalysts based on Mo or W and promoted with Ni or Co. These materials are produced by sulfidation of the corresponding oxides. It well known that incomplete sulfidation due to strong interaction between metal and support results in decreased hydrotreating activity [1]. Unsupported sulfide catalysts have attracted recent attention due to their high activity in the hydroteating reactions. The structures and catalytic properties of these sulfides obtained by comaceration [2], precipitation [3], and hydro/solvothermal route [4] depended strongly on their preparation atmosphere and thermal treatment conditions. Thiosalt decomposition provides a simple and reproducible method for obtaining unsupported sulfide materials with controlled stoichiometry and specific surface area [5]. Herein, a family of unsupported NiMoW sulfides with different ratios were prepared by thermal decomposition of tetramethylammonium thiomolybdotungstate and $Ni(NO_3)_2$. The catalytic materials were characterized by Fourier transformation infrared spectroscopy, ultraviolet-visible spectroscopy, thermogravimetric analysis, X-ray diffraction, N_2 physorption, scanning electron microscopy and transmission electron microscopy. The results show that structures and properties of unsupported NiMoW sulfides depend on nickel content and decomposition condition. The layering and stacking degree of the sulfide particles shows a decreasing trend as the nickel content increases in the NiMoW sulfides. This family of catalysts was evaluated for the hydrodesulfurization (HDS) of dibenzothiophene (DBT). Low nickel content NiMoW catalysts gave a positive effect on the HDS performance, while catalysts with high nickel content showed a negative effect. The $Ni_{0.5}MoWS_x$ catalyst showed the highest activity in the HDS reaction compared to other samples. The results suggest that unsupported NiMoW sulfide catalysts may be promising candidates for wide application in hydrotreating of fuel oil.

References

[1] R. Prins, *Adv. Catal.* 46 (2001) 399-464.

[2] G. Hagenbach, P. Courty, B. Delmon, *J. Catal.* 31 (1973) 264-273

[3] R. Candia, B. S. Clausen, H. Topsøe, *J. Catal.* 77 (1982) 564-566

[4] E. Devers, P. Afanasiev, B. Jouguet, M. Vrinat, *Catal. Lett.* 82 (2002) 13-17.

[5] X. Jin, C.L. Ma, Y.J. Yi, Q.M. Zhang, J.S. Qiu, C.H. Liang, J. Phys. Chem. Solids 71 (2010) 642-646.

See more of this Session: Novel Catalytic Materials I
See more of this Group/Topical: Catalysis and Reaction Engineering Division

(41f) Reactive Nanoparticles Immobilized In Hydrogel for Toxic Organics Degradation

Li Xiao and D.B. Bhattacharyya, Chemical and Materials Engineering, University of Kentucky, Lexington, KY

Nanoscale metallic particles have important applications in the environmentally important catalytic reactions due to the novel physical and chemical properties over bulk particles. Functionalized membrane supported metallic nanoparticles can be used to conduct reductive detoxification of chlorinated organics. In this research, a novel strategy was employed for stabilizing metalic nanoparticles with temperature responsive hydrogel (poly - N-isopropylacrylamide) as stabilizer. Hydrogel/Membrane immobilized nanoparticles possess many advantages: prevention of aggregation, tunable nanoparticle size and morphology, and potentially adjustable reactivity by the variance of hydrophobicity and hydrophilicity with the environmental change.

In this work, poly (N-isopropylacrylamide-co-acrylic acid) [p(NIPAAm-AA) immobilized metallic nanoparticles were synthesized with uniform size (around 30nm~50nm) and distribution immobilized in polymer domain. These iron nanoparticles containing hydrogel-composites were utilized as a reaction media for the reduction reactions of chlorinated organics. Also, this polymer network has tunable properties and it was found that reactivity toward TCE dechlorination with Fe/Pd (2 wt% Pd) increased three fold with a temperature change of only 4 ℃. On the other hand, experiments without any hydrogel showed an increase of only two fold, showing the importance of hydrogel tunable properties. Hydrogel supported nanoparticles can be regenerated and reused only with 30% reduction in the initial reduction rate. The dechlorination of dichlorobiphenyl was also conducted with the system of bimetallic nanoparticles Fe/Pd (4.6 wt% Pd) nanoparticles immobilized in hydrogel. The batch reaction results showed the formation of biphenyl and complete dechlorination (as measured by chloride analysis) at 25 ℃. Support of this research has been provided by NSF-IGERT and NIEHS-SRP.

See more of this Session: Novel Catalytic Materials I
See more of this Group/Topical: Catalysis and Reaction Engineering Division

(45a) Atmospheric Pressure Plasma Effects On the Composition and Adhesive Bonding Properties of Titanium and Titanium Alloy

Edward W. Harris[1], Justin Massey[1], **Thomas Williams**[2], S. F. (Dick) Cheng[3] and Robert F. Hicks[4], (1) Materials Engineering Labratory, NAVAIR North Island, San Diego, CA, (2)Department of Chemical and Biomolecular Engineering, University of California Los Angeles, Los Angeles, CA, (3)Chemical and Biomolecular Engineering, University of California, Los Angeles, Los Angeles, CA, (4)Chemical & Biomolecular Engineering, University of California Los Angeles, Los Angeles, CA

Atmosphericpressure plasma effects on the composition and adhesive bonding properties oftitanium and titanium alloy

Edward W. Harris[1],Justin T. Massey[1], Dick Cheng[2], Thomas Williams[3](speaker), and Robert F. Hicks[2]

[1]NAVAIR NorthIsland, San Diego, CA 92135

[2]SurfxTechnologies LLC, Culver City, CA 90232

[3]Chemical andBiomolecular Engineering Department, University of California, Los Angeles, CA 90095

Titaniumalloy is a lightweight, high-strength material that has many structuralapplications in aircraft, spacecraft, medical implants and dentistry.Preparing titanium surfaces for adhesive bonding is a complicated task thatrequires specialized chemical formulations and procedures. In this study, wehave examined the use of atmospheric pressure plasmas for surface preparationof titanium and Ti-6AL-4V alloy. Oxygen, nitrogen and hydrogen plasmachemistries were used to clean the metal surface after it had been mechanicallydeoxidized by sanding. Following sanding, the samples were coated with a sol-gel (AC technologies, AC-130-2), coated with a bond primer (Cytec BR6747-1), cured,and joined together with an epoxy film adhesive (3M FM300-2). Wedge crack-extensiontests (ASTM D3762) showed that the bonded titanium coupons exhibited 98% or 99%cohesive failure (i.e., failed within the adhesive and not at the interface),provided the metal surface was exposed to nitrogen-helium or hydrogen-heliumplasmas. Oxygen-helium plasmas yielded only 77% cohesive failure, whereasomitting the plasma process altogether yielded poorer results. X-rayphotoemission spectroscopy indicated that the titanium alloy surface wascovered with a layer of carbonaceous material after sanding, and that 74% ofthe atomic oxygen was present as $(OH)_3^{-3}$ species. Thenitrogen and hydrogen plasmas removed the carbonaceous deposit and the $(OH)_3^{-3}$species. These results demonstrate the potential of atmospheric pressureplasmas to improve the surface preparation of metal parts and achieve durable structuralbonds.

See more of this Session: Plasma Science and Technology
See more of this Group/Topical: Materials Engineering and Sciences Division

(45b) Engineering Antimicrobial Skin: In Situ Tissue Functionalization Using Atmospheric Pressure Plasmas

Matthew J. Pavlovich, Yukinori Sakiyama, Douglas S. Clark and David B. Graves, Department of Chemical Engineering, University of California - Berkeley, Berkeley, CA

Ambient-condition ionized gas plasmas are being used for an increasing variety of biological and biomedical applications. Prominent areas of research include microbial disinfection and surface antisepsis, wound healing, cancer treatment, and modification and functionalization of biomaterials. Plasma-initiated polymerization and surface modification has been widely studied as a method of imparting biological functionality, for example, to increase cell adhesion, decrease protein fouling, and enhance antimicrobial activity. However, the majority of surfaces modified by plasma have been inorganic surfaces, such as medical implants, modified ex situ. We have developed a proof-of-principle method for in situ functionalization of living tissues using an ambient-condition gas plasma jet, and we describe potential applications for this functionalization including imparting antimicrobial activity to skin tissue among other potential applications.

See more of this Session: Plasma Science and Technology
See more of this Group/Topical: Materials Engineering and Sciences Division

(45c) High Efficiency Chemical Synthesis Using Pulsed Plasma Gliding Arc Reactors with Water Spray

Bruce R. Locke[1], Kevin Hsieh[2], Robert Wandell[2], Wright Finney[2] and Radu Burlica[3], (1)Department of Chemical and Biomedical Engineering, Florida State University, FAMU-FSU College of Engineering, Tallahassee, FL, (2)Chemical and Biomedical Engineering, Florida State University, FAMU-FSU College of Engineering, Tallahassee, FL, (3)Electrical Engineering, Technical University "Gh.Asachi" Iasi, Iasi, Romania

Pulsed plasma gliding arc reactors utilizing liquid spray have been demonstrated to lead to very high efficiency production of hydrogen peroxide from water with argon carrier gases, nitrates from water with air carrier gases, and molecular hydrogen from organic liquids, e.g. methanol, using argon carrier gases. A gliding arc electrical discharge plasma is generated using two knife-edge electrodes separated by a small gap distance. The carrier gas (2 lpm) containing micrometer size liquid droplets (20 ml/min) flows between the electrodes where a pulse forming circuit delivers relatively low power in the range 200 – 300 mW. In the case of hydrogen peroxide generation, an energy yield of approximately 80 g/kwh has been obtained and this is close to 25% of the thermodynamic limits. Survey of all plasma processes that have been reported to generate hydrogen peroxide show this method to be highly competitive. This method has a major advantage over other chemical methods because only pure water and carrier gas are required in the feed. The generation of nitrates from air is also potentially useful for fertilizer and disinfection processes. The high solubility of products such as hydrogen peroxide and nitrate in the liquid droplets leads to rapid transfer of the plasma-formed products into the liquid phase, whereby they are protected from plasma degradation reactions and where they accumulate in significant amounts. The water droplets also provide a significant source of the needed water reactant into the plasma. The relatively low plasma and thermal reactor temperatures dramatically enhance the efficiency for generation of these species. In the case of hydrogen generation from organic liquids, high efficiency, energy yields, have been demonstrated. In all cases, the small liquid droplets lead to high efficiency due to rapid rates of mass transfer and large surface area contact with the surrounding plasma and gas phase. Such plasma reactors have significant potential for high efficiency chemical synthesis in gas-liquid environments.

See more of this Session: Plasma Science and Technology
See more of this Group/Topical: Materials Engineering and Sciences Division

(45d) Coupling of Surface Mixed-Layer Kinetics and Monte Carlo Modeling for Profile Evolution In Patterning Complex Oxides

Nathan P. Marchack, Calvin Pham and Jane P. Chang, Chemical and Biomolecular Engineering, University of California, Los Angeles, Los Angeles, CA

As the downscaling of integrated circuit devices continues, minute variations in the feature profiles from processing techniques such as plasma etching significantly affect device performance. With the increasing introduction of novel materials into integrated circuits, the need to predict surface response during etching of these materials, such as complex oxides, becomes critical to attainable device performance. In this work, a phenomenological model[1] based on high-k oxide etching in chlorine based plasmas is adapted into a translated mixed layer (TML)[2] kinetics-based format to be used a Monte Carlo-based feature profile simulator. To accurately represent the kinetics involved, experiments are conducted in this work in an inductively coupled plasma (ICP) reactor equipped with a quadrupole mass spectrometer (QMS) for analyzing etch products and a quartz crystal microbalance (QCM) for measuring the etch rate *in situ*. This reactor is connected to a UHV transfer tube which allows the surface composition to be studied via x-ray photoelectron spectroscopy (XPS) without exposure to ambient conditions. In the TML model, surface reactions such as ion impingement, neutral adsorption, physical sputtering and chemically enhanced ion etching are accounted for, and reaction parameters are either measured directly or extracted by comparing the model to etch yield data. The MC model used ion incident angle dependence and an elliptical energy deposition model to capture the effects of surface morphology on the profile evolution under the bombardment of energetic and directional ions. The material systems studied include HfLaO and HfSiON etched in Cl_2/BCl_3 plasmas, for both blanket films and trenches patterned by e-beam lithography. Very good agreement was demonstrated between the phenomenological and TML models, as well as between simulated profiles and cross-sectional SEM images of the patterned material systems.

[1] Martin et al. Journal of Vacuum Science and Technology A 27(2) 2009

[2] Kwon et al. Journal of Vacuum Science and Technology A. 24(5) 2006

See more of this Session: Plasma Science and Technology
See more of this Group/Topical: Materials Engineering and Sciences Division

(45e) Multiphysics Modeling of Ambient Gas Plasma-Based Wound Healing Process

Yukinori Sakiyama, Marat Orazov and David Graves, Chemical and Biomolecular Engineering, University of California, Berkeley, Berkeley, CA

Wound healing is one of the promising applications of atmospheric pressure gas plasmas in medicine. A recent clinical study shows a significant reduction of bacterial load in treated wounds without side-effects. Plasma-generated reactive oxygen/nitrogen species (ROS/RNS) are thought to be key factors. But the fundamental mechanisms are only poorly understood. We have been developing a multiphysics model of plasma-based wound healing process. Our model consists of three parts: plasma generation region, ROS/RNS transportation in the afterglow, and gas plasma-stimulated wound healing. In the presentation, we will especially focus on the role of nitric oxide (NO) during the healing process.

See more of this Session: Plasma Science and Technology
See more of this Group/Topical: Materials Engineering and Sciences Division

(45f) Combinatorial Chemistries for Reducing Protein Adhesion with Plasma Grafting

Minghao Gu[1], Arturo Vegas[2], James Kilduff[3], Daniel G. Anderson[4], Robert Langer[2] and Georges Belfort[5], (1)Howard P. Isermann Deptt of Chemical & Biological Engineering and The Center for Biotechnology & Interdisciplinary Studies, Rensselaer Polytechnic Institute, Troy, NY, (2)Department of Chemical Engineering, Massachusetts Institute of Technology, Cambridge, MA, (3)Civil and Environmental Engineering, Rensselaer Polytechnic Institute, Troy, NY, (4)Chemical Engineering, Massachusetts Institute of Technology, Cambridge, MA, (5)Department of Chemical and Biological Engineering, Rensselaer Polytechnic Institute, Troy, NY

We describe here for the first time a new combinatorial approach for synthesizing a wide range of surface chemistries with the aim of searching for reduced protein adhesion. In order to speed up the search, screening and selection procedure, we have utilized our novel high throughput platform (HTP) with atmospheric pressure plasma-induced graft polymerization (APPGP) to modify synthetic commercial membranes (poly(ether sulfone) PES). Monomers are first added to the porous polymeric film and then grafted polymerization was performed. Previously discovered monomers (poly(ethylene glycol) (PEG) and zwitterionic (Zwit) that exhibit protein-resistance were selected and so were new ones discovered. Many of these newly discovered monomers appear to have potential for reducing protein adhesion.

See more of this Session: Plasma Science and Technology
See more of this Group/Topical: Materials Engineering and Sciences Division

(45g) High Performance Thin Film Based Coatings In Aircraft Applications

Marvi A. Matos, Chemical Technologies, BR&T, The Boeing Company, Seattle, WA

Conference: **American Institute of Chemical EngineersAnnual Meeting**

Date of theconference: October 16-21st, 2011

Title: **HighPerformance Thin Film Based Coatings in Aircraft Applications**

Author: **Marvi A. Matos, PhD**

Organization:**The Boeing Company**

Abstract

The use polymer-based materialsand coatings in aircraft components is two-fold, first, the weight reductionsare highly desirable to decrease fuel consumption and second, polymer-basedmaterials are usually less expensive and convenient to manufacture. Howeverwhen used in applications with strict optical requirements, these polymericsurfaces can be susceptible to erosion which results in increased haze anddecreased clarity. For example, when staple materials such as polycarbonate orstretched acrylic are used in windows, windshields and canopies, one of thedrawbacks is the tendency to scratch and craze. Polymeric windows have beenhistorically coated with polysiloxane or polyurethanebased coatings to overcome this limitation by improving the surface resistanceto scratches. Still, improvements to the processes involved can decrease therequired long drying times and can offer long term solutions in which theresistance to erosion is maintained overtime. Advanced thin film coatings basedon Plasma Deposition Technologies can improve the durability of many componentson aircrafts. These technologies can be exploited to generate materials withhigh performance, which are also environmentally friendly and produced withwaste free processes. We are currently focusing our efforts in the developmentand study of thin films to improve the resistance to material erosion onpolymeric substrates with applications in the aerospace industry anddefense.

See more of this Session: Plasma Science and Technology
See more of this Group/Topical: Materials Engineering and Sciences Division

(57a) Increasing the Rigid Amorphous Fraction In Semi-Crystalline Homopolymers by Solid-State Shear Pulverization Leads to Major Reductions In Oxygen Permeability

Philip J. Brunner, Materials Science and Engineering, Northwestern University, Evanston, IL and John M. Torkelson, Departments of Chemical and Biological Engineering, Materials Science and Engineering, Northwestern University, Evanston, IL

Abstract

In semi-crystallinepolymers, a rigid amorphous fraction (RAF) exists at the interface of crystaland amorphous phases as a result of the immobilization of the polymer chainsdue to the crystal. Previous studies have shown that increasing RAF fraction inthe homopolymer can have a significant impact on physical properties. Here, weshow for the first time that the RAF content in semi-crystalline homopolymerscan be significantly increased by a process called solid-state shear pulverization(SSSP), which employs a modified twin-screw melt extruder that is cooled ratherthan heated, and thus results in solid-state processing. Following SSSP, overallcrystallinity remains constant, while RAF content increases, leading to majorreductions in the oxygen permeability coefficient. For example, the RAF contentof nylon 11 compression molded films increases from ~ 16% for the pellet (noexposure to SSSP) to ~ 37% for the pulverized sample, which leads to a ~ 40%reduction in the oxygen permeability coefficient after SSSP processing. Similarresults have been demonstrated with several other semi-crystalline homopolymers.The origin for these large decreases in oxygen permeability is due to the majorreduction in crystallite size of the SSSP processed material. The formation ofsmaller crystallites allows for a larger interface between the crystal andamorphous phases, ultimately resulting in greater immobilization of the amorphouspolymer chains. Polarized optical microscopy images demonstrate that crystallitesizes decrease ~ 10 fold following pulverization. These results demonstratethat SSSP processing of homopolymer prior to melt-processing can be verybeneficial for creating high oxygen barrier films, which is important forpackaging applications.

See more of this Session: Structure and Properties In Polymers I
See more of this Group/Topical: Materials Engineering and Sciences Division

(57b) Relaxation Characteristics of Thermally-Modified Aromatic Polyimides and Copolymers Designed for Selective Separations

Anthony C. Comer[1], Claudio P. Ribeiro Jr.[2], Benny D. Freeman[2], Sumod Kalakkunnath[3] and **Douglass S. Kalika**[1], (1)Department of Chemical and Materials Engineering, University of Kentucky, Lexington, KY, (2)Department of Chemical Engineering, University of Texas at Austin, Austin, TX, (3) ConocoPhillips Company, Bartlesville, OK

A new class of thermally-modified aromatic polyimides and copolymers has been identified as a viable membrane material for separation processes requiring high permeability and selectivity in combination with intrinsic thermal and chemical resistance properties. The membranes are cast from soluble aromatic polyimides and copolymers containing *ortho*-positioned functional groups on the diamine moiety; exposure of the polyimides to thermal rearrangement (TR) at temperatures above the glass transition (*i.e.,* > 350°C) leads to the formation of fully-aromatic polybenzoxazoles (PBO's) with exceptional thermal and chemical resistance. The thermal exposure step produces fundamental changes in molecular connectivity that alter chain packing, resulting in a unique distribution of free volume elements at the angstrom scale. It is the distinctive shape and distribution of these elements that appear to be responsible for the unprecedented gas separation performance reported by Park et al [1].

A series of aromatic polyimides (API) was synthesized based on the reaction of *ortho*-functionalized diamine-dianhydride pairs intentionally selected to foster high permeability; conversion of the API precursors was achieved by either chemical or thermal imidization. In select cases, the extent of functionalization (and the ultimate degree of thermal rearrangement) was controlled by the introduction of varying amounts of non-functional diamine co-monomer. API test specimens were cast from solution and then subject to thermal exposure at discrete temperatures in the range of 350°C to 450°C in order to achieve thermal rearrangement to polybenzoxazoles.

In this study, we examine the dynamic relaxation properties of the API polymers as a function of backbone structure and degree of thermal rearrangement. Specifically, dynamic mechanical analysis and broadband dielectric spectroscopy are used to elucidate the sub-glass and glass-rubber relaxation characteristics of the polymers as related to their structural details and thermal exposure history; the information obtained through these methods provides insight as to the relative flexibility of the membranes, their local relaxation environment and corresponding free volume, and the influence of thermal rearrangement on segmental mobility and ultimate separation performance. Further, the contrasting nature of the dynamic mechanical and dielectric probes can be used to more precisely establish the structural origins and mechanisms associated with each relaxation process.

The polymers display thermomechanical relaxation characteristics common to most polyimides [2].

Three relaxations are observed with increasing temperature: γ and β sub-glass relaxations, and the glass-rubber (α) relaxation. The ability of the API materials to undergo thermal rearrangement is governed by the glass transition temperature of the as-synthesized polymer, and dynamic mechanical sweeps beyond T_g can be used as a direct and sensitive means to monitor the progress of the TR conversion *in-situ* as a function of time and temperature. For those polymers exposed to conditions consistent with a high degree of thermal rearrangement, nearly complete suppression of the glass transition process is observed, with the resulting PBO structures showing high-modulus, glassy behavior up to the point of degradation.

Investigation of the sub-glass relaxations reveals a sensitivity to moisture, and shifts in relaxation position and intensity with thermal rearrangement. In all samples, the detection of the lowest-temperature γ relaxation is contingent upon the presence of ambient moisture in the sample; rigorous drying of the polymer prior to measurement leads to full removal of the γ relaxation response, which is restored with subsequent exposure to ambient conditions. This behavior can be correlated to the nature of the *ortho* functional groups, and the apparent disruption of intermolecular (*i.e.,* polymer-polymer) interactions upon adsorption of water molecules [3]. For the β relaxation, increasing extents of TR conversion result in a decrease in viscoelastic relaxation intensity and peak temperature, reflecting motions of a more compact, less cooperative character.

[1] H. B. Park, C. H. Jung, Y. M. Lee, A. J. Hill, S. J. Pas, S. T. Mudie, E. Van Wagner, B. D. Freeman and D. J. Cookson, *Science* 318 (2007) 254-258.

[2] A.C. Comer, D.S. Kalika, B.W. Rowe, B.D. Freeman and D.R. Paul, *Polymer* 50 (2009) 891-897.

[3] H. Ohya, V.V. Kudryavtsev, S.I. Semenova. *Polyimide Membranes: Applications, Fabrications and Properties (Chapter 3)*; Gordon and Breach Science Publishers: Amsterdam, 1996.

See more of this Session: Structure and Properties In Polymers I
See more of this Group/Topical: Materials Engineering and Sciences Division

(57c) High Strain Rate Performance of Thermoset Epoxy Resins

Daniel B. Knorr Jr., Chemical Engineering, University of Washington, Seattle, WA and Joseph Lenhart, Macromolecular Science and Technology Branch, US Army Research Laboratory, Aberdeen Proving Ground, MD

Fiber reinforced thermoset resins make up an important part of armor packages for ballistics applications. While a significant amount of work has been done understanding fiber and composite performance, relatively little has been done to understand or improve thermoset matrix energy absorption during high strain rate impact. Furthermore, the mechanical performance of epoxy resins is well understood at quasi-static to low impact velocity strain rates, but has not been extensively explored for regimes above about 10^2 s^{-1}. In the present work, we explore the high strain rate performance of diglycidyl ether of bisphenol A (DGEBA) and diglycidyl ether of bisphenol F (DGEBF) cured with a large variety of amine-based agents, representing a range in glass transition temperature of greater than 150°C. Samples were investigated using differential scanning calorimetry, dynamic mechanical analysis, rheological measurements and ballistic testing at temperatures below the glass transition with a view toward obtaining a fundamental understanding of the dissipative influence of molecular relaxations and, more broadly, molecular structure. Correlations between temperature and V50 (velocity at which 50% of samples are penetrated during the impact test) are presented and interpreted in light of dynamic mechanical analysis relaxation data.

See more of this Session: Structure and Properties In Polymers I
See more of this Group/Topical: Materials Engineering and Sciences Division

STRESS RELAXATION IN EPOXY NANOCOMPOSITES
AIChE 2011 FALL PROCEEDINGS

Suresh Ahuja, Xerox Corporation, Webster, NY

Introduction

Nano-composites are of great importance because of their application in polymer-based biomaterials, nano-particle drug delivery, mini-emulsion particles, fuel cell electrode polymer bound catalysts, layer-by-layer self-assembled polymer films, electrospun nano-fibers, imprint lithography, polymer blends and nano-composites. Nanocomposites are formed by the intercalation or exfoliation of nano inorganic fillers into polymers resulting in improvement of physical and mechanical properties relative to the unmodified polymer. In intercalated nanocomposite, the polymer is inserted between the silicate layers forming well-ordered multi-layers, where as in exfoliated nanocomposite the silicate layers break into single platelets and orient themselves forming a random pattern [1]. The elastic modulus of the nanocomposites increases with increase in nano filler concentration in the pure epoxy. The effect of nano filler on strength and toughness appears to depend on the polymer filler interaction, dispersion on the polymer matrix and aggregation of the filler. In terms of polymer chain mobility in a nanocomposite, glass transition has been reported to increase, decrease, or remain unaffected, in comparison with that of the pure polymer, and sometimes a second glass transition has as well been observed [2,3,4,5]. Nanocomposites can, in principle, be formed from clays and organo clays in a number of ways including various in situ polymerization, solution, high shear mixing, melt processing and latex methods. The degree of exfoliation depends on the structure of the clay, curing temperature and curing agent The commonly used techniques for processing clay-epoxy nanocomposites are by shearing/extension through the use of extruders, mixers and ultrasonification as well as solution mixing . The extent of exfoliation rather than intercalated/exfoliated composites depends on rates of deformation (shear rate /extension rate) and strain The molecular structure is modeled using a combination of the classical network theory and the elastic dumbbell model for polydispersed polymer segments and localized junctions.

Models and Simulations

Simple mechanical models are helpful for understanding the complex dynamical behavior of polymer molecules. The elastic dumbbell constitutes the simplest model that captures the effects of stretching and orientation of the polymer [6]. the polymer is represented by two beads which are connected by a spring. The classical Hookean dumbbell model is characterised by a linear spring force. Its relaxation times decrease like the inverse of the mode number and the relaxation modulus is of a single exponential form. The finite extensibility of the molecules has to be included in order to avoid unphysical behaviour such as infinite viscosities predicted by the Hookean model. The shear relaxation modulus for the Finite Extension Nonlinear Elastic Model (FENE model) [6] has been determined numerically [7].

A generalized Maxwell Model is given by

$$G(t) = \sum_{n=1}^{\infty} |\langle \sigma, f_{2n} \rangle|^2 e^{-t/\lambda_{2n}}, \qquad (1)$$

where G(t) is the relaxation modulus, σ is the stress and f is the eigenfunction.
The results are very close to Molecular dynamics (MD) simulations of bead-springs used for polymer liquids

and melts. Each chain is represented as a sequence of beads connected by finite-extensible, nonlinear (FENE) springs and interacting via the repulsive part of the Lennard-Jones 12-6 potential (LJ).

In nanocomposites, reinforcement takes effect once the filler concentration is above the percolation threshold. Experimental observation is that the filled polymers are reinforced once the good dispersion is achieved. In comparison, another explanation suggests that in addition to entanglement network, fillers are connected by a network of adsorbed chains forming a second network. The latter was used satisfactorily to qualitatively explain the nonlinear viscoelastic properties such as the Payne effect observed in nanocomposites. Investigation of the structural and dynamic properties of nano filled elastomer on the basis of the idealized model of elastomer and nanoparticle show that that the introduced nanoparticles induce more efficient chain packing. The decay of the bond orientation of polymer chains during the relaxation process is studied by changing the draw ratio, the filler loadings and the affinity between the nanoparticle and polymer. In order to equivalently investigate the stress relxation, we propose a new approach named conformational relaxation. It is observed that the conformational relaxation can be approximately fitted well by an exponential function for the pure and filled systems, similar to the stress relaxation [8].

Experimental Details

The matrix used in the present study was a two-component epoxy resin system where epoxy resin DGEBA was mixed with Cloisite, commercially available from Southern Clay Products, Inc., USA. It is a natural montmorillonite (MMT) that is organically modified. Materials and their composition are given in Table 1.Before mixing with epoxy, Cloisite 10A was washed with ethanol and water to remove excess of benzyl dimethyl stearyl ammonium chloride and benzyl dimethyl cetyl ammonium chloride from 7.5% to 0.4% and from 4.0 to 0.2%, respectively. Similarly, Cloisite 30A was washed with ethanol and water to remove bis(2-hydroxyethyl) methyl stearyl ammonium chloride from 7.4 % to 0.4%. Epoxy was crosslinked at different temperature of 90°C,110°C and 130°C using a crosslinker, hexahydrophthalic anhydride (75%) and Polyether polyol (25%) with 50/50 proportion of Crosslinked to epoxy.From dynamic viscoelastic measurements at various temperatures, elastic and viscoelastic modulus as function of frequency was used to get relaxation modulus by using transformation functions.

Relaxation modulus as function of time at different temperatures was used to get power law fits. From time-temperature superposition and Fourier transformation function and dynamic viscoelastic measurements of elastic and viscous modulus at various frequencies and temperatures, relaxation modulus was obtained [16]. Relaxation modulus as function of time at different temperatures was used to get power law fits. The pre-exponent of relaxation modulus as function of temperature was fitted with an exponential fit and the result is shown in Figure 1. The nanocomposite containing organoclay 30A is more temperature dependent relaxation modulus than the one containing organoclay Cloisite 10A may possibly due to differences in hydrogen bonding between the organic constituent bishydroxy ethyl stearyl ammonium chloride ion and the montmorillonite clay compared to benzyl dimethyl stearyl ammonium chloride ion and the montmorillonite clay.

Model and simulation analysis of nanocomposites is investigated by using the polymer as a bead spring with interactions between polymer segments and smooth nanoparticles, nanoparticles and nanoparticles are modeled using modified Lennard- Jones functions, which account for the excluded volume of the beads and particles by offsetting the interaction range. The relaxation time is extracted by fitting the conformational relaxation curves with an exponential function at different temperatures, exhibits an Arrhenius relationship with the temperature for the pure and filled systems [8]. For stronger interactions, characteristic relaxation timeτ increases. This can be explained in terms of an increase of the activation energy, which influences the relaxation time.

$$\tau = \tau_0 e^{\Delta U/RT} \qquad \text{---(2)}$$

In Figure 2 relaxation modulus as function of time is compared with crosslinked and uncrosslinked nanocomposites. From power law dependence, G(0) increases with nanofiller and crosslink temperature. Relaxation time,τ is significantly higher with nano filler while temperature effect is masked by the crosslinked structure.

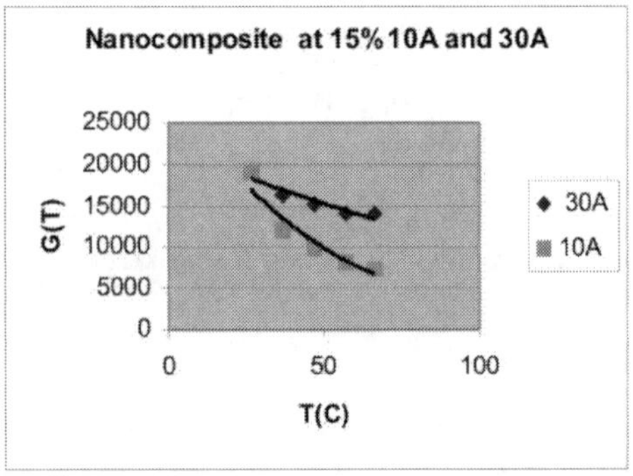

Figure 1

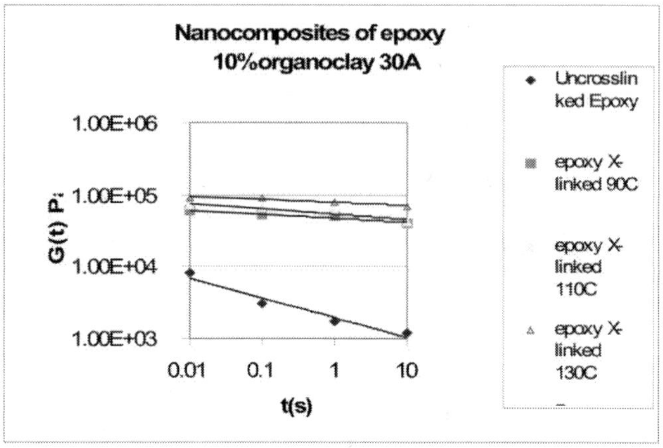

Temperature dependence of G(T) of nanocomposites

Figure 2Relaxation modulus as function of time

Table 1. Nanocomposites of Epoxy and 10%Cloisite 30A

Sample Description	G (0),Pa	t/τ
Uncrosslinked epoxy	1934	0.2718
Epoxy Crosslinked at 90C	47118	0.0554
Epoxy Crosslinked at 110C	53835	0.0737
Epoxy Crosslinked at 130C	78567	0.0379

Conclusion

Stress relaxation modulus of epoxy nanocomposites is analyzed. Qualitatively, FENE models can adequately explain experimental results.

References

1 T. J. Pinnavaia and G. W. Beall,(2000) "Polymer-Clay Nanocomposites", (Wiley, Chichester,) pp. 127-149

2 Kotsilkova, R.; Fragiadakis, D.; Pissis, P. (2005) J. Polym. Sci., Part B: Polym. Phys., 43, 522

3 Fragiadakis, D.; Pissis, P. (2007) J. Non-Cryst. Solids, *353*, 4344
4 Ash, B. J.; Siegel, R. W.; Schadler, L. S. *(*2004) *J.* Polym. Sci., Part B: Polym. Phys., 42, 4371
5 Tsagaropoulos, G.; Eisenberg, A. *(*1995) Macromolecules, *28*, 6067.
6 Bird R. B., Curtiss C. F., Armstrong R. C. and Hassager O. (1987)," Dynamics of Polymeric Liquids",Vol. **2** Kinetic Theory, 2nd ed. (Wiley, New York).
7 Ilg, P. Karlin1 I. V.and Succi, S. (2000)" Supersymmetry solution for finitely extensible dumbbell model", Europhys. Lett., **51** (3), pp. 355–360.
8 Liu, J., Cao, D. , , Zhang, L, and Wang W.,(2009)Macromolecules **9,** *42*, pp2831-2842

(57e) Relaxation Dynamics of Tethered Polymers

Praveen Agarwal, Chemical and Biomolecular Engineering, Cornell University, Ithaca, NY and Lynden A. Archer, Chemical & Biomolecular Engineering, Cornell University, Ithaca, NY

Dynamics of tethered polymers and polymers in confinements remains a debated issue till date. Most of the experiments reported in literature are for thin films where polymers are attached to a flat surface. We have created a bulk system to study the dynamics of tethered polymers by tethering Polyisoprene (PI) to the silica nanoparticle surface. Cis 1-4 PI is a type A dielectric implying the dipole moment of the chain is parallel to the chain backbone and relaxation of the normal mode provides information about the whole chain relaxation. We have explored the relaxation dynamics of the tethered polymer with the help of dielectric broadband spectroscopy. We find that the relaxation of tethered polymer is significantly slower than the free polymer. We have carried out this study for different molecular weight polymer chains and varying chain spacing by changing the grafting density of polymer chains.

See more of this Session: Structure and Properties In Polymers I
See more of this Group/Topical: Materials Engineering and Sciences Division

(57f) Tacticity Control Over Hyperbranched Poly(N-isopropylacrylamide) and Its Effects On Thermal Transition Temperature

Kai Chang, Nathan Rubright and Lakeshia Taite, Chemical and Biomolecular Engineering, Georgia Institute of Technology, Atlanta, GA

Tacticity control over hyperbranched poly(N-isopropylacrylamide) and its effects on thermal transitiontemperature

Kai Chang, Nathan C. Rubright, Lakeshia J. Taite.

Chemicaland Biomolecular Engineering, Georgia Institute ofTechnology.

Background: Hyperbranched polymers are aclass of polymers with very interesting properties, especially in the field ofdrug delivery. They have many similarproperties with dendrimers including globularstructure with many end groups as well as internal cavities that can be used toencapsulate drugs. Hyperbranched systems are usuallynot as well defined as dendritic systems; however,they are far less labor intensive to synthesize. Hyperbranched poly(*N*-isopropylacrylamide) (pNIPAAm)synthesis has been reported previously by Vogt et. al and Carter et. al.;however, the effect of branching on the ability to control tacticityis as yet unknown.

pNIPAAm is awell-studied thermally responsive polymer that is often associated withbiological applications due to its sharp aqueous lower critical solutiontemperature (LCST) of ~32°C.In a hyperbranched system, the additionalcomplexity of the architecture naturally decreases the LCST, a phenomenon thatis likely due to the steric hindrance of theformation of water structures resulting from the hydrogen bonding of the watermolecules to the acrylamide sidechain. Ordering the polymer in its tacticity, therefore, may have a synergistic effect inallowing additional order in the system and reducing the amount of LCSTreduction.

In previous studies, ReversibleAddition-Fragmentation Chain Transfer (RAFT) polymerization, a controlledliving radical polymerization scheme was used with hyperbranchinducing chain transfer agents (CTAs) in order to create these hyperbranched polymers.RAFT polymerization lends itself easily to copolymerization; therefore,the inclusion of copolymers was also studied; specifically copolymerizationwith acrylic acid (AAc). pNIPAAm-AAcis often used in biological systems due to the increased LCST of such a systemand a hyperbranched pNIPAAm-AAcwith a sufficiently high LCST (~45°C) has potentialdrug delivery applications.

Methods: Synthesis of 4-Vinylbenzyl-imidazole Dithioate (**1**)was done according to the procedure set forth by Carter et al.

Polymerization was conducted as follows. A 90:10:2:1 ratio of NIPAAm:AAc:**1**: Azobisisobutyronitrile was placed in asealed 100 mL roundbottomflask equipped with a magnetic stir bar.In the case of syndiotactic polymerization,3-methyl-3-pentanol was included in the reaction flask. The mixture was purged with Nitrogen andNitrogen purged Dioxane was added. The solution was reacted at 65°Cfor 48 hours and was quenched by exposure to air. The pNIPAAm-co-AAc was precipitated into ether, filtered and dried undervacuum. The resulting solid was resuspended in nanopure water (18MΩ) and dialyzed using 2500 MWCO cassettes (Pierce) overnight. The solution was then lyophilized to obtainthe product.

Results: Characterizationof polymer was conducted using NMR spectrometry and gel permeationchromatography. LCST was determinedusing UV-Visible Spectrometry.

As expectedfor RAFT polymerization, the polydispersity index forthe polymers were low, on the order of 1.1, and indicative of livingpolymerization. LCST data is shown inFig. 1.

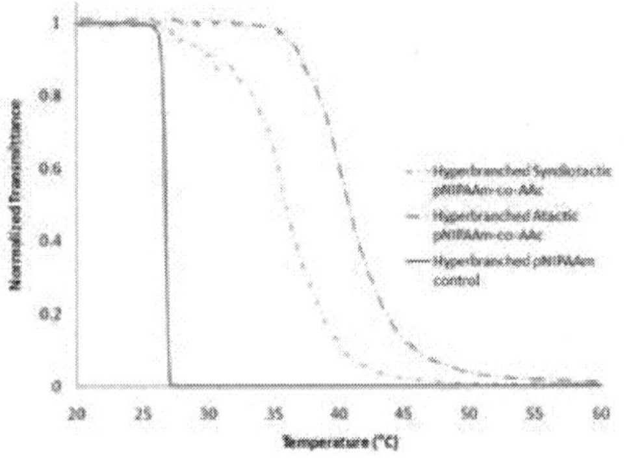

Fig.1: LCST curves of pNIPAAm-co-AAc at 10% AAc. Syndiotactic and atactic data are shown as well as a hyperbranchedpNIPAAm control.

While the hyperbranched pNIPAAmcontrol point and the atactic pNIPAAm-*co*-AAc showedexpected results of a below 32°C and far above 32°C LCSTrespectively, the syndiotactic data shows a contraryeffect. Previous studies into theeffects of tacticity on LCST linear pNIPAAm demonstrated that increased syndiotacticityincreases the LCST. The opposite trendseen in this case indicates that increased order in tacticityindeed has an effect on the LCST of hyperbranched pNIPAAm; however, the effect is not to allow for morehydrogen bonding of water, but rather less, possibly as a result of increasedpacking density of the ordered polymer.

Conclusions: Highlybranched systems behave differently from their linear counterparts in manyways. In the case of hyperbranchedpNIPAAm, increasing syndiotacticityhas the opposite effect from its linear counterpart. Further investigation is necessary todefinitively provide an explanation.

References:

Vogt AP. Macromolecules. 2008;41:7368-7373

Carter S. Macromolecular Bioscience. 2005;5:373-378

HiranoT. J. Polym Sci.Part A: Polym. Chem. 2006;44:4450-4460

See more of this Session: Structure and Properties In Polymers I
See more of this Group/Topical: Materials Engineering and Sciences Division

(57g) Anisotropic Thermal Conduction In Polymers and Its Molecular Origins

David Venerus, Jay D. Schieber and Sahil Gupta, IIT, Chicago, IL

The strong coupling of mechanical and thermal effects in polymer flows have a significant impact on both the processing and final properties of the material. Simple molecular arguments suggest that Fourier's law must be generalized to allow for anisotropic thermal conductivity in polymers subjected to deformation. In our laboratory we have developed a novel application of the optical technique known as Forced Rayleigh Scattering to obtain quantitative measurements of components of the thermal diffusivity (conductivity) tensor in polymers subjected to deformations. We report measurements of anisotropic thermal diffusivity and stress in molten, cross-linked and solid polymers subjected to several types of flows. The deformed samples have significant anisotropy in polymer chain orientation that results in significant anisotropy in thermal conductivity. Stress and thermal conductivity data support the validity of the stress-thermal rule, which is analogous to the well-known stress-optic rule. However, results for a semi-crystalline polymer do not obey the stress-thermal rule when significant crystallization occurs. We also report measurements on solid polymers with isotropic polymer chain orientation that are under stress, which display rather unexpected behavior. These measurements are used to develop an understanding of the molecular origins of anisotropic thermal conduction in polymeric materials.

See more of this Session: Structure and Properties In Polymers I
See more of this Group/Topical: Materials Engineering and Sciences Division

(74a) A Smart Material with Both Sticky and Non-Sticky Properties

Zhiqiang Cao, Department of Chemical Engineering, University of Washington, Seattle, WA and Shaoyi Jiang, Chemical Engineering, University of Washington, Seattle, WA

"Non-sticky" and "sticky" properties are two highly desired material characteristics. The presence of both is required for numerous applications. The former refers to the ability to resist non-specific biomolecular (or cellular) adhesion while the latter refers to the chemical reactivity, allowing for covalent attachment of recognition elements (e.g. antibodies). However, from conventional wisdom, these two distinct properties cannot exist within a "single" material. For example, most non-fouling materials must either be reacted to introduce functionalizable groups (e.g. carboxylate moieties) or reacted with coupling agents (e.g. carbodiimides) for conjugation to targeting ligands. For the first time, we propose an answer to this challenge.

By mimicking the drug camptothecin, we introduced a new zwitterionic material capable of switching between two equilibrium states, each with its own unique property: a lactone-ring structure (CB-Ring) and an acidic open-ring structure (CB-OH). We found that the zwitterionic nature of CB-OH make the material ultra-low fouling to complex media while the CB-Ring structure allow for reactivity with nucleophiles (e.g. amines). Control over these two material states (or properties) can be obtained via an acid/base driven equilibrium. In addition, we demonstrated a proof-of-concept experiment to realize a high throughput antibody array for early cancer diagnostics, in a simple way, by taking advantage of this novel smart material.

See more of this Session: Biomaterials I
See more of this Group/Topical: Materials Engineering and Sciences Division

(74b) Controlling the Dependency Between Hydrogel Rigidity and Permeability with the Inflexibility of a Polymer Cross-Linker

John J. Schmidt[1], Jae Hyun Jeong[2], Chaenyung Cha[3] and Hyunjoon Kong[2], (1)Chemical and Biomolecular Engineering, University of Illinois at Urbana-Champaign, Urbana, IL, (2)Chemical and Biomolecular Engineering, University of Illinois, Urbana, IL, (3)Chemistry, University of Illinois Urbana-Champaign, Urbana, IL

Hydrogels are extensively studied and engineered for use in many commercial products including sensors, membranes, drug delivery devices and synthetic extracellular matrices. The further success of hydrogels in commercial applications relies on the ability to control several physical properties in an independent manner. However, current hydrogel designs often encounter difficulty in the independent control of mechanics and permeability. For example, efforts to make a softer hydrogel often encounter the significant increase of the swelling ratio of the hydrogel, thus leading to undesirable disruption of micro-scale patterns and porous architecture. This inverse dependency between hydrogel mechanics and transport can cause significant limitations to hydrogel functionality.

In this study, we hypothesized that the chain inflexibility of a polymeric cross-linker is an important factor to tune the dependency between stiffness and permeability of a hydrogel. This hypothesis was examined using polymeric cross-linkers, which had a sugar-based backbone (alginate) or an alkyl chain backbone (acrylic acid), to represent various degrees of chain flexibility. These polymers were modified with varying degrees of substitution of methacrylic groups to control the number of cross-links of the hydrogel. Interestingly, the poly(acrylamide) hydrogels cross-linked by the more inflexible alginate methacrylates exhibited the significantly less dependency between swelling ratio and elastic modulus than those cross-linked by the flexible poly(acrylic acid) methacrylates. The critical role of these cross-linkers in mediating the mechanical and permeation properties of the hydrogel were further related to the ability of the cross-linkers to self-organize, by using a pyrene assay. Finally, the hydrogel cross-linked by alginate methacrylates maintained its pre-defined micropatterns of cell adhesion proteins independent of the hydrogel stiffness, such that the cells could respond to solely the substrate mechanics and not to changes in swelling of the hydrogel. Overall, the results of this study serve to advance the controllability of hydrogel properties, in order to greatly enhance the function and performance of the hydrogels in various applications.

See more of this Session: Biomaterials I
See more of this Group/Topical: Materials Engineering and Sciences Division

(74c) Isolating the Effects of Material Stiffness: Fabrication and Characterization of Univarient Synthetic Cell Culture Substrates

Alexander T. Leonard, Chemical and Nuclear Engineering, University of New Mexico, Center for Biomedical Engineering, Albuquerque, NM, Kirsten N. Cicotte, Biomedical Engineering, University of New Mexico, Center for Biomedical Engineering, Albuquerque, NM, Jared R. Funston, Chemical and Nuclear Engineering, University of New Mexico, Albuquerque, NM, Matthew N. Rush, Nanoscience and Microsystems, University of New Mexico, Albuquerque, NM and **Elizabeth L. Hedberg-Dirk**, Department of Chemical and Nuclear Engineering, University of New Mexico, Center for Biomedical Engineering, Albuquerque, NM

There are strong indications that the mechanical properties of a substrate plays an important role in many cellular functions. Native human soft tissue has elastic modulus in the range of 0.01 to 1000 kPa. The synthetic materials commonly used as cell culture platforms with varied moduli, including polyacrylamide (PA) and polydimethylsiloxane (PDMS), are limited in applicability due to a restricted range of achievable moduli and/or surface chemistry instabilities. The copolymer network of n-octyl methacrylate (nOM) and diethylene gycol dimethacrylate (DEGDMA) offers attractive material properties that overcome these limitations. In our laboratory, co-polymer networks were fabricated with 3 to 33 wt% DEGDMA. The compressive modulus was 25 ± 2 kPA for the 3% DEGDMA network and increased to 4700 ± 300 kPa at 33 wt% DEGDMA fraction. The networks demonstrated consistent surface wettability over the range of substrate formulations examined as determined by static Sessile drop contact angle. Surface interrogation with x-ray photoelectric spectroscopy (XPS) at the two extremes of formulations, 3 and 33 wt.% DEGDMA, showed similar elemental and chemical bond compositions. The 3% DEGDMA formulation had an elemental composition ratio (carbon to oxygen) of 4.7. The higher DEGDMA composition of 33 wt% DEGDMA showed a carbon oxygen ratio of 4.6. High resolution carbon XPS indicated similar ratios of ether, ester and aliphatic groups at the two extremes of DEGDMA compositions. Quantification of protein adsorption showed no statistical differences between formulations after incubation in a solution of the model protein bovine serum albumin (BSA). Atomic force microscopy was used to map as well as quantify the surface roughness for each of the formulations, obtaining a maximum surface roughness of R_{rms} 17 ± 6 nm. The murine osteoblastic cell line MC-3T3 E1 was used as a model for cell attachment and viability at six and 96 hours, respectfully. Our results indicate nOM/DEGDMA substrates can vary in modulus over three orders of magnitude while maintaining comparable chemical and topographical surface features. These networks are the first that allow for the study of the effects of material mechanics without the interference of other material properties.

See more of this Session: Biomaterials I
See more of this Group/Topical: Materials Engineering and Sciences Division

(74d) Stable Microwell Arrays As Platforms for Long-Term Tissue Culture

Grinia M. Nogueira[1], Michael Beste[2], Nicole Doyle[2], Tuyen Phung[1], Linda Griffith[2] and Paula Hammond[1], (1)Chemical Engineering, MIT, Cambridge, MA, (2)Bioengineering, MIT, Cambridge, MA

Microtechnology tools have been widely used to fabricate microenvironments for high throughput analyses of a large variety of cell types. Our group has been using microtechnology tools to create microwell arrays to assay the prevalence of rare cells in complex cell mixtures and subsequently study their proliferation and migration behavior under specific biological conditions. The microwell arrays are formed via soft lithography, using polydimethylsiloxane (PDMS) molds. Poly(ethylene)glycol (PEG) hydrogel composes the walls of square microwell arrays which are bonded to a silane-functionalized glass substrate. PEG hydrogel was chosen due to its resistance to cell and protein adhesion, which prevents cell migration outside of its original microenvironment. Our preliminary studies have shown that primary human endometrial stromal cells (ESCs) can be successfully isolated inside these microenvironments with relatively high platting efficiency. The dimensions of the microwells used in this study were 50x50um or 100x100um wells, 20um wall thickness and 40um height. We show that ESCs can proliferate in 100x100um microwells when cultured for up to 7 days. However, for long-term cell culture (> 3 days), optimization of the silanization protocol was necessary to ensure stable adhesion of the wells to the surface. We tested 3 types of silane coupling agents and several solvents' systems in order to achieve maximum durability of microwells under tissue culture conditions while maintaining high plating efficiency of endometrial cells. We also investigated the potential of varying mesh size of the PEG hydrogel, by varying synthetic parameters of the gel, for controlled molecule diffusion between wells.

Atomic Force microscopy (AFM), contact angle measurements, optical microscopy, and cell adhesion assays were performed to characterize the substrates. Average mesh size of PEG hydrogels with various molecular weights was measured via a correlation with the equilibrium degree of swelling of the polymeric networks. Stability of the microwell arrays were verified every 2 days by imaging analysis, and the average of the delaminated area was calculated. We demonstrate how we can use these arrays to perform long-term cell culture while isolating individual cells or a group of cells in order to study cell-cell interactions, cell migration and proliferation and several other facets of cell biology.

See more of this Session: Biomaterials I
See more of this Group/Topical: Materials Engineering and Sciences Division

(74e) High Throughput Screening of Tissue Specific Biomaterials

Courtney Dumont[1], **Pankaj Karande**[2] and Deanna M. Thompson[1], (1)Biomedical Engineering, Rensselaer Polytechnic Institute, Troy, NY, (2)Chemical and Biological Engineering, Rensselaer Polytechnic Institute, Troy, NY

High-throughput screening platforms have been used to investigate protein-protein binding interactions, drug toxicity, and gene expression to quantify cell viability, proliferation and differentiation in response to toxins, drugs and soluble factors. Cell behavior in 3D has largely been confined to a single scaffold type, and these assays have not been used to screen for more complex cell processes, such as migration. We propose the development of a flexible platform to investigate changes in number, morphology and migration for the rapid screening of biomaterials using a common flatbed scanner. We assert that this type of rapid screening may be valuable in the rational design of tissue-specific biomaterials. To illustrate this point, Schwann cell (SC) migration and repopulation have been found to be essential for axonal re-growth through the injury site in peripheral nerve injuries.It is known that SC migration and spreading are not supported in type I Collagen, while the incorporation of laminin and Collagen IV (found in Matrigel) promotes cell spreading and migration. We demonstrate sensitivity to changes in spot size, seeding density, composition and migration in a 3D biomaterial using a flatbed scanner and have confirmed these measurements using traditional microscopy. This type of screen will be vital in the rapid development and efficient screening of novel scaffolds in the development of novel cell and tissue optimized biomaterials. The proposed work evaluates a new assay to rapidly assess cell number, response to material composition and migration in 3D. Rapid assessment of biomaterials will facilitate the design of an optimized biomaterial supportive of SC migration for neural tissue engineering applications, which is currently identified to be a rate-limiting factor in vivo. Examination of neural and non-neural response will facilitate the development of a tissue-specific biomaterial for neural engineering applications and can be further extended to other tissue engineering applications for cell-specific biomaterials.

See more of this Session: Biomaterials I
See more of this Group/Topical: Materials Engineering and Sciences Division

(74f) Fabrication of Amine-Reactive Polymer Multilayers On Microwell Cell Culture Arrays: Combining Methods for the Topographic Patterning of Cell Substrates with Approaches to Facile Surface Functionalization

Adam H. Broderick[1], Samira M. Azarin[1], Maren E. Buck[2], Sean P. Palecek[1] and David M. Lynn[1], (1) Department of Chemical and Biological Engineering, University of Wisconsin-Madison, Madison, WI, (2)Department of Chemistry, University of Wisconsin-Madison

Numerous past studies have established the important and diverse roles that surface chemistry and substrate topography can play, independently, in directing the functions and fates of cells. The development of methods for the fabrication of topographically patterned surfaces that also display well-defined and tunable surface chemistry could (i) play an important role in understanding the combinatorial effects of these two environmental factors and (ii) provide a platform for the design of new materials and cell culture substrates that provide sophisticated control over a variety of cell behaviors.

While many different lithographic approaches can be used to fabricate topographically patterned surfaces, modification of the surface chemistry of the resulting 3-D substrates for use in biological studies remains a challenge. This problem arises from the fact that many polymers (e.g., polyurethanes) used to create 3-D substrates for cell-based studies are relatively inert and therefore limit opportunities for post-fabrication surface modification. Past studies have demonstrated approaches to spatially resolved functionalization of topographically patterned polymer surfaces by (i) physical adsorption of biological or synthetic macromolecules or (ii) the formation of self-assembled monolayers (SAMs) on thin films of metals (e.g. alkanethiols on gold). While these two approaches are useful for defining surface chemistry in several important contexts, the development of more general methods for the immobilization of chemical and biological functionality on a broader range of topographically patterned surfaces and materials could open the door to new biomedical applications and the development of new research tools.

Here, we present a 'reactive' layer-by-layer assembly approach to the fabrication of amine-reactive polymer thin films on the surfaces of topographically patterned polyurethane microwell cell culture arrays used previously for 3-D cell culture. Reactive assembly of branched polyethyleneimine (BPEI) and the azlactone-functionalized polymer poly(2-vinyl-4,4-dimethylazlactone) (PVDMA) yielded film-coated microwell arrays that could be chemically functionalized post-fabrication by treatment with different amine-functionalized molecules. Characterization using fluorescence microscopy revealed the presence of a uniform and conformal coating on the topographically patterned surface of the arrays.

Additional experiments demonstrated that it was possible to functionalize the films post-fabrication by treatment with amine-functionalized small molecules, and that this approach could be used to modify the surface properties of the arrays and their interactions with cells. For example, treatment of film-coated arrays with the small molecule amine D-glucamine resulted in microwell surfaces that resisted the adhesion and proliferation of mammalian fibroblast cells *in vitro*. The results of other experiments demonstrated that it was possible to functionalize different structural features of these arrays in a spatially resolved manner to create dual-functionalized substrates (e.g., to create arrays having either (i) azlactone-functionalized wells, with regions between the wells functionalized with glucamine, or (ii) substrates with spatially resolved regions of two different cationic polymers). In particular, spatial control over glucamine functionalization yielded 3-D substrates that could be used to confine and control populations of cells within the wells of these arrays for periods of up to four weeks and support the 3-D culture of cubiodal cell clusters.

This approach to the chemical functionalization of topographically patterned 3-D cell culture microarrays could prove useful for the long-term culture and maintenance of cell types (e.g., stem cells) for which the presentation of specific and chemically well-defined 3-D culture environments is believed to be required to control cell growth, differentiation, and other important behaviors. The results of additional studies demonstrating the patterning of arrays with peptide motifs and the potential of this approach to influence the culture of human plutipotent stem cells will be presented. More generally, this approach to the fabrication of reactive thin films provides methods for the straightforward chemical functionalization of otherwise unreactive topographically patterned substrates that could prove useful in a range of other fundamental and applied contexts.

See more of this Session: Biomaterials I
See more of this Group/Topical: Materials Engineering and Sciences Division

(74g) Nanoporous, Optically Transparent, and Free-Standing Polyelectrolyte Multilayer Films for Applications In Tissue Engineering

Adam L. Larkin[1], Reisha M. Parham[1] and Padma Rajagopalan[2], (1)Department of Chemical Engineering, Virginia Polytechnic Institute and State University, Blacksburg, VA, (2)Department of Chemical Engineering, Virginia Tech, Blacksburg, VA

The layer-by-layer assembly of alternately charged polyelectrolytes (PEs) is a simple and versatile technique to assemble polymer films with tunable properties. The resulting polyelectrolyte multilayer (PEM) films have been used in a variety of applications including anti-bacterial coatings, drug delivery, biomaterials and tissue engineering. The majority of the reports in the literature are on PEMs deposited and adhered to the underlying substrate. By modulating the deposition properties of the PE and through assembly on an inert, hydrophobic substrate, we have assembled detachable PEMs for applications in tissue engineering.

Chitosan and hyaluronic acid (HA) PEMs were assembled on hydrophobic polytetrafluoroethylene substrates. To achieve detachability, PEM assembly solution concentrations were varied from 1 to 5 mM and deposition times were varied from 20 to 80 min. PEMs ranging from 10 to 50 bilayers were detachable. The free-standing PEM thicknesses ranged from 0.8 to 3.5 μ m and were found to be optically transparent and stable in aqueous media. Glutaraldehyde, a crosslinker, was used to promote PEM stability and increase the Young's modulus. The PEMs exhibited Young's moduli ranging from 90 to 400 MPa depending upon exposure time to the cross-linker. PEMs were exposed to acidic solutions to introduce porosity in the film. The introduction of porosity may be due to phase segregation, reorganization, or degradation within the PEM. Exposure to acidic solutions with pH values of 3.0 and 3.5, resulted in sub-micron porous PEMs with an average pore size of 300 to 400 nm.

Free-standing, nanoporous PEMs are well suited to mimic basal membranes and the extracellular matrix due to their tunable thickness, porosity, and mechanical properties. One such application of these free-standing PEMs is in hepatic tissue engineering. In our research group, we have created a functional 3D liver model comprised of hepatocytes (the principal cell type of the liver) and liver sinusoidal endothelial cells (LSECs) separated by a HA and chitosan PEM interface.[1,2] The PEM mimics the Space of Disse found *in vivo*, a protein-rich interface between the hepatocytes and LSECs. The thickness of the Space of Disse ranges from 0.5 to 1 μ m and has an average pore size of 100 to 200 nm. PEMs with an average thickness of 0.8 μ m and porosity of 300 nm are ideal candidates to mimic the Space of Disse *in vitro*. Our current studies are focused upon tuning the properties to more closely mimic the Space of Disse and incorporating the detachable PEMs into the 3D liver model and monitoring hepatic function.

References:

1. Rajagopalan, P. et al. Tissue Eng Pt A, 2006, 12, 1553-1563.

2. Kim, Y. et al. Tissue Eng Pt A, 2010, 16, 2731-2741.

See more of this Session: Biomaterials I
See more of this Group/Topical: Materials Engineering and Sciences Division

(74h) A Novel Epoxy-Based, Injectable Hydrogel Scaffold for Tissue Regeneration

Adam K. Ekenseair, Kristel W. M. Boere, F. Kurtis Kasper and Antonios G. Mikos, Department of Bioengineering, Rice University, Houston, TX

While the majority of effort in tissue regeneration research has been focused on implantable scaffolds, there are many applications that would be better served with injectable, *in situ* forming materials capable of co-delivering cells and growth factors to optimize tissue regeneration. Injectable scaffolds are minimally invasive and can easily fill complex tissue defects or voids often found in applications such as craniofacial bone regeneration after trauma, tumor resection, or birth defects. Thermogelling polymers, such as poly(N-isopropylacrylamide), which pass through a lower critical solution temperature upon injection into the body, are promising candidates as scaffold backbones. Concomitant chemical crosslinking during thermogellation further enhances the stability and mechanical properties of such materials; however these networks must be made biodegradable on an appropriate timescale for tissue regeneration. This is a major issue for injectable materials, since many commonly used biodegradable polymers, such as polycaprolactone and poly(lactic acid), are not water soluble. Polyamidoamines, an emerging class of polymers, offer a biocompatible, biodegradable, and water-soluble option for the creation of such crosslinked networks; and their efficacy is herein evaluated.

To this end, a novel two-component system was developed. Poly(NiPAAm-co-glycidyl methacrylate) was synthesized by a free radical pathway to introduce functional epoxy rings into the thermogelling polymer. Diamine crosslinking macromers were formed from piperazine and methylene bisacrylamide with varying molecular weights through a step polymerization mechanism. The secondary amines present at the ends of these polyamidoamine macromers reacted readily with epoxy rings, even across phases. Thus, chemically and thermally gelling hydrogel scaffolds were formed and the effects of polymer wt %, time between solution mixing and injection, degree of crosslinking, and crosslinker length on the formation, equilibrium swelling, degradation, and mechanical properties of the scaffold were evaluated. This system is shown to be a promising candidate for injectable tissue engineering scaffolds with little or no syneresis upon gellation and relevant degradation rates.

See more of this Session: Biomaterials I
See more of this Group/Topical: Materials Engineering and Sciences Division

(80a) Near-Infrared Light Curable Composites for Advanced Dental Applications

Alexander Stepuk, Dirk Mohn and Wendelin J. Stark, Institute for Chemical and Bioengineering, ETH Zurich, 8093 Zurich, Switzerland

Light-curable dental polymers (resins) are commonly usedin restorative surgery, prosthodontics and surgical procedures. Despite thefact of wide application, there are clinical problems due to limitations ofblue light penetration: application is restricted to defects exposed to thelight source; layered filling of defect is required (see figure) [1].

In the present work, we therefore investigate a new wayto avoid problems inherent to conventional light-curing dental resins. We willdiscuss the innovative approach dealing with deep tissue penetrating nearinfrared light (NIR) and upconversion (UC) phosphors. The necessary light tohomogeneously harden a restorative resin inside a tooth can thus be generated"on spot" inside the polymer and thereby avoids the absorbance loss problem ofcurrent dental treatments (see figure).

We will focus on a novel technique for thepolymerization of light-curing resins: blending the resins with UC phosphorparticles and curing the material with blue UC emission upon NIR excitation. Theprepared composites could restore the cavities up to 7-10 mm deep in one step,even positioned behind thin enamel or dentin obstacles. Potential substitution of commonly used light-curing polymers, providingadditional advantages to be discussed.

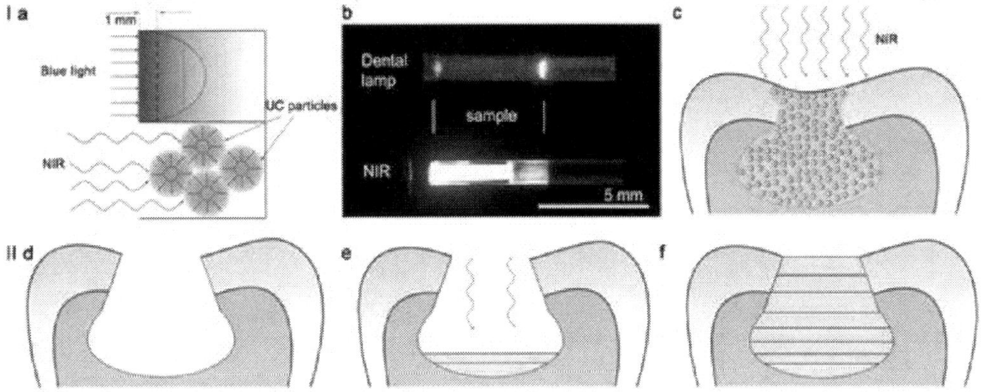

Figure 1: Approach scheme:
(I) Treatment with a composite: (a) light-curing of the polymer vs. blue light emission by upconversion of $NaYF_4$:Yb,Tm particles;
(b) photograph of the blue light emission/propagation for pure polymer and composite doped with UC phosphors;
(c) one-step filling and curing with NIR light
(II) Current treatment: (d) preparation and cleaning of a cavity; (e) layer-by-layer filling followed by curing with blue light;
(f) final filling consisting of several layers.

Reference: [1] R.G. Craig, in *RestoraiveDental Materials*, Mosby, St. Louis, 2002.

See more of this Session: Composites II
See more of this Group/Topical: Materials Engineering and Sciences Division

(80b) Comparison of Uniform and Structured Particles for the Protection of Ultraviolet Sensitive Molecules

Kristin Gilida Steeley[1], Kenneth Morabito[2], Dapeng Li[3], Paul Calvert[3], Anubhav Tripathi[4] and Nina C. Shapley[5], (1)Chemical and Biochemical Engineering, Rutgers University, Piscataway, NJ, (2)Division of Engineering and Warren Alpert School of Medicine, Brown University, Providence, RI, (3) Department of Bioengineering, University of Massachusetts Dartmouth, North Dartmouth, MA, (4) School of Engineering, Brown University, Providence, RI, (5)Chemical and Biochemical Engineering, Rutgers, the State University of New Jersey, Piscataway, NJ

Photodegradation is a problem observed in a wide array of systems including pharmaceuticals, paints, textiles, and even human skin. While many protection methods focus on protecting an entire system (such as applying sunscreen to skin), another approach is to protect the UV sensitive molecules themselves (such as a pharmaceutical product). Beta-carotene, the pigment that gives carrots their orange color and is also present in human skin, has a tendency to photodegrade and its degradation mechanism has been the subject of many studies. The wealth of information available on beta-carotene photodegradation along with its prominent absorption peaks which allow degradation to be quantified makes it an ideal molecule to compare UV protection mechanisms.

Particles comprised of a polymer, beta-carotene, and a UV absorber were synthesized. Uniform particles were produced in the range of 800-900 nanometers in diameter and contained a uniform distribution of the components. Structured particles were microcapsules with a shell of polymer and absorber and a core comprised of an aqueous suspension of beta-carotene and polymer uniform particles. Microcapsules were approximately 200 micrometers in diameter. The microcapsule geometry allows a much larger separation distance between the beta-carotene molecules and protective molecules than is typically found in uniform particles.

Both types of particles were exposed to UV radiation for various intervals and the amount of beta-carotene left undegraded was determined by use of an ultraviolet visible spectrophotometer since absorption is proportional to concentration. UV protection was measured by beta-carotene decay curves of normalized absorption over increased exposure time and quantified by decay rate constants. These tools make it possible to determine which polymer structure offers superior protection from photo-oxidation.

See more of this Session: Composites II
See more of this Group/Topical: Materials Engineering and Sciences Division

MECHANICAL AND RHEOLOGICAL PROPERTIES OF POLY(BUTADIENE ADIPATE-CO-TEREPHTHALATE) (PBAT) NANOCOMPOSITES ABOVE AND BELOW THE PERCOLATION THRESHOLD

Alexandria Niemoeller, Missouri University of Science and Technology, Rolla, MO
Mahin Shahlari, Missouri University of Science and Technology, Rolla, MO
Sunggyu Lee, Ohio University, Athens, OH

Abstract

Due to its elastomer-like properties, poly(butadiene adipate-co-terephthalate) (PBAT) is a biodegradable polymer with a wide range of potential applications. As nanofillers are used increasingly to alter the morphology, improve mechanical and barrier properties, and expand the applications of polymers and polymer blends, it is crucial to study the properties of nanocomposites at various filler loadings. The mechanical and rheological properties of composites are affected significantly as filler concentration is increased, particularly at and above the percolation threshold. For composites with a thermoplastic matrix, it has been demonstrated that the elasticity decreases considerably above the percolation threshold as a result of solid-like behavior. The aim of this work was to examine the mechanical and rheological properties of an elastomer-like matrix prepared with various percentages of nanoparticles, in order to study the deformation mechanism of PBAT-clay nanocomposites both below and above the percolation threshold.

Introduction and Background

Poly(butadiene adipate-co-terephthalate), or PBAT, is a biodegradable aliphatic/aromatic copolyester that is used primarily in film applications and as an extrusion coating material. It is thermoplastic in nature but has elastomer-like elongational properties and a relatively low modulus that can be attributed to the low level of crystallization found in the random copolymer. As one of the few biodegradable polymers with similar properties to low density polyethylene, it is critical to develop and characterize PBAT nanocomposites with improved mechanical, thermal, and barrier properties which also remain readily compostable.

Nanoclays are relatively inexpensive and widely available nanofillers with high aspect ratios of 100-1000. Modifying the surface of these silicate layers helps to further tailor their compatibility with polymeric materials and increase intercalation between and exfoliation of individual platelet layers. Organically-modified layered silicates are used in an increasing number of applications to improve the mechanical and thermal properties of polymers. In some cases, nanoclays can improve gas barrier properties and the flame retardancy of polymeric materials while also increasing the modulus and lowering the linear thermal expansion of the composite.

The use of any nanofiller necessitates the discussion of the percolation threshold of a nanocomposite: the filler percentage at which the ultimate property enhancement is achieved and beyond which property degradation typically occurs. In the short-range structure of a polymer clay nanocomposite, the optimal property improvement occurs when the clay tactoids are well dispersed throughout a polymer matrix and clay layers are fully exfoliated. This results in a larger aspect ratio of the dispersed phase, and a larger surface area of clay in contact with the polymer matrix.[1] On a more global scale, however, longer-range structures can be identified such as tactoid and individual platelet orientation or interactions with neighboring clay platelets. A secondary structure can thus be formed, and the interactions between the polymer matrix and the nanofiller structure can be related to global mechanical and viscoelastic properties.

Many opposing views of the mechanism of nanocomposite percolation have been popularized in recent publications.[2-5] One theory suggests that a secondary network exists between the clay tactoids and platelets and that these interactions affect the overall properties of the nanocomposite. Alternatively, it has been proposed that the global properties are governed mainly by polymer entanglements or trapped polymer chains between clay layers. Therefore if the clay volume percentage is high enough, well dispersed and exfoliated layers constrain a larger number of polymer entanglements. Further, it has been theorized that the clay exfoliation is driven mainly by polymer interactions with clay surface modifications, which essentially peel individual clay platelets from each tactoid. Viscoelastic measurements of a nanocomposite provides the most comprehensive method in which polymer chain confinement, polymer-clay interaction, and interparticle interaction effects can be studied.[2] Measurements such as the low frequency storage modulus provide a larger-scale analysis of the structure-property relationship as compared to localized measurements such as microscopy.

This work attempts to identify the deformation mechanism of a nanocomposite of PBAT and Cloisite 30B, and further elucidate the importance of the formation of a secondary clay network and/or polymer entanglements between the exfoliated layers. Viscoelastic response measurements will serve as the primary verification of the formation of a percolated clay structure and consequent property improvement, with transmission electron microscopy, x-ray diffraction and mechanical property testing as further clarification of the deformation mechanism.

Materials and Equipment

The PBAT used in this work was produced by BASF as Ecoflex® F BX 7011, which has a density of 1.25 g/cm^3 and a melt flow rate of 2.70 to 4.90 g per 10 minutes at 190°C.

Southern Clay supplied the Cloisite 30B, a natural montmorillonite which has been modified through an ion-exchange reaction with a methyl, tallow, bis-2-hydroxyethyl, quarternary ammonium, shown below:

$$CH_2CH_2OH$$
$$|$$
$$CH_3 — N^+ — T$$
$$|$$
$$CH_2CH_2OH$$

This particular clay modifier was chosen due to the hydroxyl groups present, which have been shown by our group's previous work to improve the compatibility between the clay and polar polymers such as PBAT and PLA, and ultimately improve dispersion of clay platelets and intercalation of polymer chains between the clay layers.[6]

Melt compounding PBAT with 0-7 wt% nanoclay was carried out in a Brabender conical counter-rotating twin screw extruder with an L/D of 8/1 and a screw diameter of 42 mm. Both materials were vacuum-dried at 60°C for 18 hours and pre-mixed prior to extrusion under starve-feed conditions. The two heating zones of the extruder were held at 145°C and the die was maintained at 150°C during fiber extrusion. Testing specimens were then prepared from a 1 mm thick compression molding apparatus with a Morgan Press.

Characterization and Analysis

All rheometric measurements were recorded with an ARES Rheometer in 25 mm diameter parallel-plate configuration under dynamic oscillatory shear and a strain rate of 3%. A dynamic strain sweep was carried out initially to ensure that dynamic moduli were independent of the particular strain rate used. X-Ray Diffraction patterns were obtained using a Rigaku Ultima IV X-Ray Diffractometer. Tensile properties of the composites were measured using a TA.XT Plus from Stable Micro Systems. Analysis of the clay dispersion and nano-scale structure was carried out with a JEOL JEM-2100F Transmission Electron Microscope.

Results and Discussion

First, a dynamic strain sweep was carried out to ensure that dynamic moduli were independent of any strain from 1-10% used during testing. Figure 1 below verifies that the storage and loss modulus are independent of the strain at which the viscoelastic response of the nanocomposites was recorded, 3%.

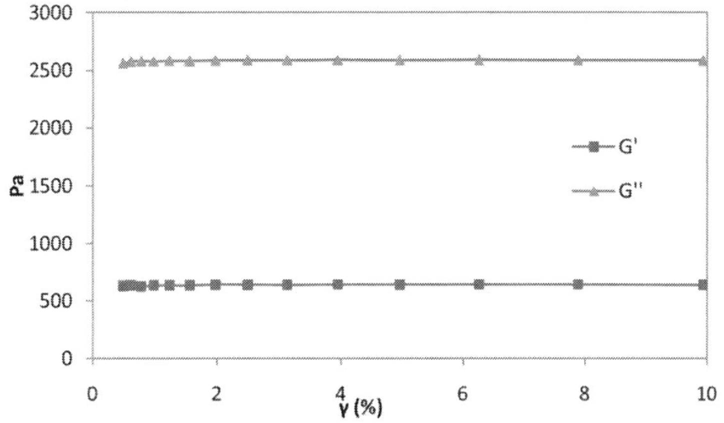

Figure 1. Strain dependence of the storage and loss modulus with 2 wt% clay

As can be seen in Figures 2 and 3, the low frequency storage and loss modulus (G′ and G″) are both improved as the clay loading is increased. This modulus improvement is said to be due to both confinement and interparticle effects. Once the clay loading has reached the percolation threshold, the distance between clay layers is on the same order as the size of the polymer entanglements and this can affect the relaxation dynamics of the polymer matrix.[2] Secondly, the higher clay loading increases the amount of interaction between individual clay tactoids and platelets. The friction between exfoliated clay layers can also affect the rheological properties and result in a more solid-like response at low frequencies. This also results in a reduced dependency of G′ and G″ on frequency in the terminal zone. The loss modulus of the 4, 5, and 7 wt% nanocomposites show this plateau particularly in the low frequency region.

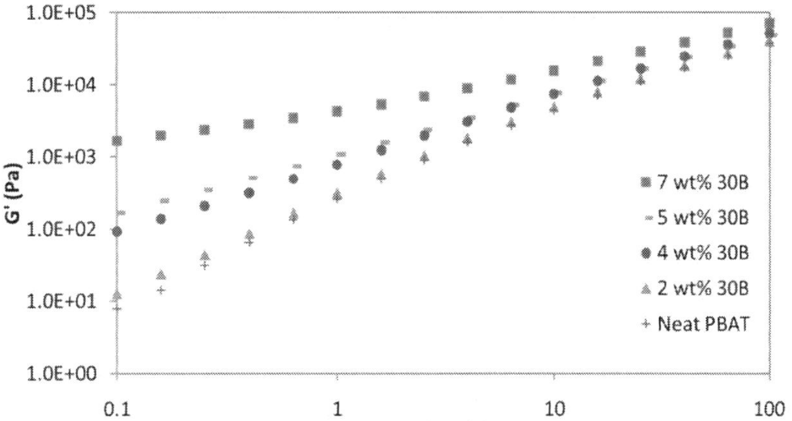

Figure 2. Comparison of storage modulus (G′) for PBAT nanocomposites at 3% strain

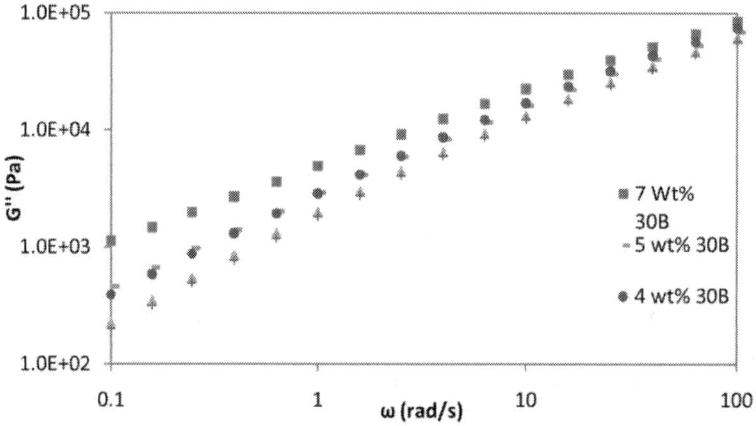

Figure 3. Comparison of loss modulus (G") for PBAT nanocomposites at 3% strain

From the low frequency region, the storage modulus of each prepared composite was calculated. Figure 4 shows the rapid increase in the storage modulus above the theoretical percolation threshold of 3 wt%. This signifies that a secondary network of rigid clay platelets has been formed, and the solid-like response of this network is evident most clearly in the low frequency measurements. Figure 5 also shows the cross-over frequency of the composite prepared with the highest clay loading. This marks the transition from the terminal, or viscous region, to the elastic region.

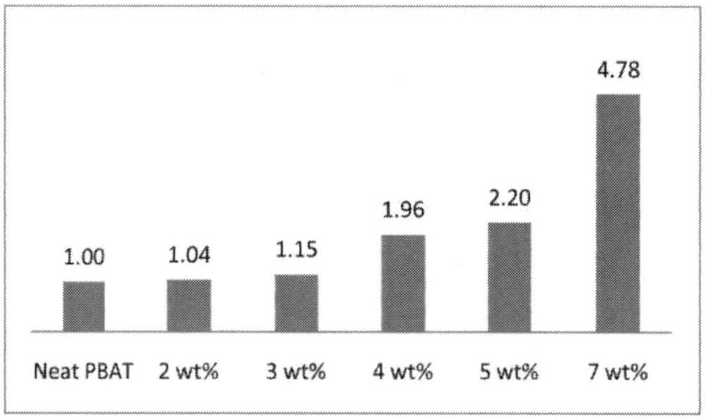

Figure 4. Increased relative storage modulus with increased clay loading at 3% strain

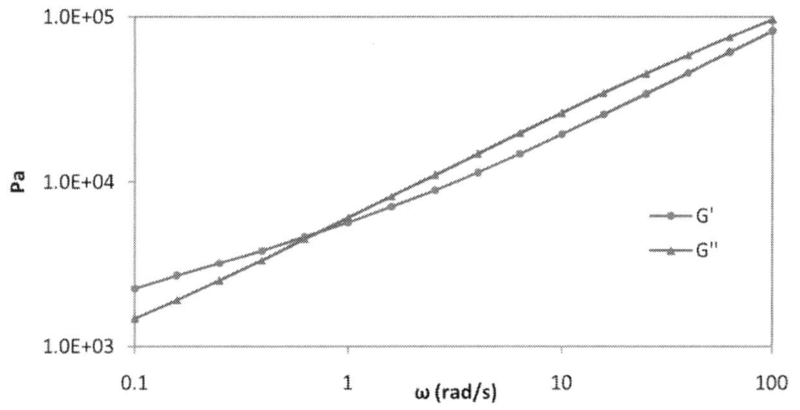

Figure 5. Crossover of G' and G" at high clay loading (7 wt%) and 3% strain

To further elucidate the deformation mechanism and structure-property relationships, TEM images, XRD patterns and tensile properties will be reported in the presentation of this work.

References

1. Kawasumi, M. J Polym Sci Part A Polym Chem 2004, 42, 819.
2. Li, J.; Zhou, C.; Wang G.; Zhao, D. J. Appl. Polym. Sci. 2003, 89, 3609.
3. Wu, D.; Zhou, C.; Zhang, M. J Polym. Sci. Part B: Polymer Physics. 2007, 45, 229.
4. Wu, D.; Wu, L.; Wu, L.; Zhang, M. Polym. Degr. Stabil. 2006, 91, 3149.
5. Paul, D.R.; Robeson, L.M. Polymer 2008, 49, 3187.
6. Shahlari, Mahin. *Biodegradable Polymer/Clay Nanocomposites Based on Poly(Butylene Adipate-co-Terephthalate and Polylactide.* Missouri University of Science and Technology, 2011.

(80d) Magnetic Carbon Nanocomposites

Jiahua Zhu[1], Suying Wei[2] and Zhanhu Guo[1], (1)Integrated Composites Laboratory (ICL), Dan F Smith Department of Chemical Engineering, Lamar University, Beaumont, TX, (2)Chemistry and Biochemistry, Lamar University, Beaumont, TX

A synergistic process for recycling polystyrene (PS)/Ni@NiO polymer nanocomposites (PNCs) and fabricating magnetic core@shell Ni@C nanoparticles is introduced. Different pyrolysis pathways of the PNCs have been suggested basing on the analysis of both gas and liquid pyrolysis products from gas chromatography/mass spectrometry (GC/MS). The pyrolysis pathway follows radical generation, recombination and hydrogenation of the PNCs, which generate more saturated structures. However, pure PS undergoes a radical generation and β-scission to form more unsaturated structures. The coke formed during the PS pyrolysis was used as a carbon resource to in-situ synthesize a graphite shell surrounding Ni nanoparticle surface. This Ni@C nanoparticle shows a significantly larger specific area (236.68 m^2/g) as compared to bare Ni@NiO nanoparticles (119.62 m^2/g) and can be stabilized in 1 M HCl for hours. A large saturated magnetization was also observed in the Ni@C nanoparticles. All these characters make it very useful in environmental remediation applications, with the advantages of higher absorbance capacity and easy magnetic separation process.[1, 2]

References

1. Zhang, D.; Wei, S.; Kaila, C.; Su, X.; Wu, J.; Karki, A. B.; Young, D. P.; Guo, Z. *Nanoscale* **2010,** 2, (6), 917-919.

2. Schartl, W. *Nanoscale* **2010,** 2, (6), 829-843.

See more of this Session: Composites II
See more of this Group/Topical: Materials Engineering and Sciences Division

(80e) Controllable Preparation of Nanoparticle-Coated Chitosan Microspheres In a Co-Axial Microfluidic Device

Wenjie Lan, The State Key Lab of Chemical Engineering, Department of Chemical Engineering, Tsinghua University, Beijing, China, Shaowei Li, Tsinghua University, Beijing, China, Jianhong Xu, Department of Chemical Engineering, The State Key Lab of Chemical Engineering, Beijing, China and Guangsheng Luo, Department of Chemical Engineering, The State Key Laboratory of Chemical Engineering, Beijing, China

In this work, we describe a novel, simple microfluidic method for fabricating nanoparticle-coated chitosan microspheres. Uniform droplets of aqueous chitosan solution were dispersed into an oil phase containing partially hydrophilic nanoparticles via a co-axial microfluidic device. Recirculating flow in the continuous phase in the area between drops enhanced mixing and allowed the nanoparticles to coat the surface of the droplets as they passed through the channel. The chitosan droplets were then crosslinked with glutaraldehyde and nanoparticle-coated microspheres were obtained. The SEM characterization shows that the microspheres are monodispersed with uniform nanoparticle distribution on the surface. The dispersity, size and composition of the microspheres could all easily be controlled by changing the microfluidic flow parameters and three different types of nanoparticles were successfully used to synthesize hybrid microspheres to demonstrate the method's versatility.

See more of this Session: Composites II
See more of this Group/Topical: Materials Engineering and Sciences Division

(80f) Influence of Wall Number and Surface Functionalization of Carbon Nanotube On Its Antioxidant Behavior In High Density Polyethylene

Xiaomei Shi, **Binbo Jiang**, **Jingdai Wang** and Yongrong Yang, Department of Chemical and Biological Engineering, Zhejiang University, State Key Laboratory of Chemical Engineering, Hangzhou, China

Carbon nanotubes (CNTs) are cylindrical allotropes of carbon, which can be classified into single-wall carbon nanotubes (SWNTs), double-wall carbon nanotubes (DWMTs) and multi-wall carbon nanotubes (MWNTs) according to the number of graphite cylinders. Given the superior physical and chemical properties of CNTs, many researchers have focused on finding the promising applications for them since their discovery in 1991. The use of CNTs as reinforcements for polymer composites is one of the suggested applications and has received considerable attention, in respect that they can greatly enhance mechanical, electrical, thermal and flame retardant properties of the polymer materials.

Recent research found that CNTs can also affect the thermal oxidative stability of polymers. However, the influence of wall number and surface functionalization of CNT on its antioxidant ability and its antioxidant mechanism in polymer matrix has not been well-documented.

In order to figure out the above issues, the antioxidant effect of SWNT, MWNT, and hydroxylated multi-walled carbon nanotube (MWNT-OH) on high density polyethylene (HDPE) are studied. Based on the measurement of the oxidation induction temperature (OIT) and oxidation induction time (OIt) of HDPE nanocomposites, it is found that the antioxidant efficiencies of the three CNTs are in the following order: MWNT-OH > MWNT > SWNT. The antioxidant ability and mechanism of CNTs are further illustrated by electron spin resonance (ESR) spectra and Raman spectra. It is observed that the antioxidant behaviors of CNTs in polyolefin follow free radical scavenging mechanism. The order of the defect concentration and the radical scavenging efficiency for CNTs are in good agreement with the order of their antioxidant efficiencies in HDPE. It suggests that the lattice defects in CNTs which caused acceptor-like localized states are responsible for their radical scavenging ability as well as antioxidant ability in HDPE.

See more of this Session: Composites II
See more of this Group/Topical: Materials Engineering and Sciences Division

(80g) Effect of Gas Phase Chemical Modification On Glass Fiber Reinforced Polymer Composite In Situ Joint Strength

Shishir Sinha, Department of Chemical Engineering, Indian Institute of Technology, Roorkee, Roorkee, India and Amit Kumar, Chemical Engineering, Indian Institute of Technology, Roorkee, Roorkee, India

In engineering materials; composites are becoming more and more prominent due to their high strength and stiffness to weight ratios. Composites are being used in the structural and aerospace applications. In fabrication of complex parts joining of composites is very important. Joint is the point of discontinuity which bears all the load applied to the structure, it means reliability of the mechanical structure entirely depends upon the strength of the joint. Researchers have been working in the field of process development to increase the joint strength . Therefore, in this study methods to increase the joint strength by chemical means has been investigated. Gas phase nitration (NO_x) and Amination have been employed to the joint in the present investigation and functional groups attached to the surface of glass fiber reinforced polymer (GFRP) by the used modification techniques have been verified by the Fourier Transform Infrared Spectroscopy(FTIR). It has been found that the strength of the single lap joint was increased by 47% under tensile shear loading when treated with NO_x and compared with the parent joint strength without any chemical modification and further increase in strength was observed by Amination treatment over the joint surface.

See more of this Session: Composites II
See more of this Group/Topical: Materials Engineering and Sciences Division

(87a) Electrodeposition of CuGaSe2 From Thiocyanate-Containing Electrolytes

Ian Ivar Suni, Department of Chemical and Biomolecular Engineering, Clarkson University, Potsdam, NY and Josiah Jebaraj, Dept. of Chemical and Biomolecular Engineering, Clarkson University, Potsdam, NY

We report electrodeposition of $CuGaSe_2$ from electrolytes that contain thiocyanate (SCN) as a complexing agent to cathodically shift the reduction potential of Cu(II). In the presence of thiocyanate, the peak reduction currents for Cu(II) and Ga(III) during cyclic voltammetry occur at potentials that differ by only 80 mV, even though the standard reduction potentials of the uncomplexed species differ by about 870 mV. Stoichiometric CuGaSe2 deposits are obtained by the induced codeposition mechanism at −300 mV vs SCE and *p*H 2.75 from electrolytes with Cu:Ga ratios ranging from 0.5 to 1.5. The stoichiometry is controlled by the thermodynamic driving force associated with stable compound formation, as illustrated by the anodically shifted reduction peak obtained in cyclic voltammograms of electrolytes containing all three elemental species. Oxygen incorporation into the electrodeposit will also be discussed

See more of this Session: Electrochemistry and Interfacial Phenomena In Electronic and Photonic Materials
See more of this Group/Topical: Materials Engineering and Sciences Division

(87b) Electrodeposition of AuFe Nanowire-Titania Composites and Their Photoelectrochemical Behavior

Savidra Lucatero and **Elizabeth J. Podlaha**, Chemical Engineering, Northeastern University, Boston, MA

The electrochemical fabrication of FeAu/Au porous, segmented nanowires were developed and integrated with TiO_2 to create a unique photoelectrochemical catalyst. Recent results of decorating Au nanoparticles with TiO_2, fabricated with a variety of different techniques, have resulted in an enhancement of its photoelectrochemical performance over that of TiO_2. We will show that a similar enhancement of the photocatalytic property occurs with our FeAu/Au porous, segmented nanowires as a photoanode in the UV photoelectrolysis of water, and that the nanowire alloy suppresses the loss of activity due to Au photo-enhanced oxidation. The nanowires were prepared with pulsed current in polycarbonate nanostructured templates from citrate and ascorbic acid electrolytes. They were then removed and transferred to the surface of an electrosynthesized TiO_2 film, creating a composite. The electrodeposition process is modeled on a microscopic, continuum scale capturing the kinetic and diffusion effects inherent in nanowire electrodeposition with an excess of Fe ions compared to Au ions. The titania film was deposited with a cathodic electrosynthesis route, induced by a local pH change at the electrode surface, modeled via diffusion of the hydroxyl ion near the electrode surface. The resulting photocurrent response was tested in 0.05 M NaOH with using a step, constant potential, constant illumination, experiment and impedance modulated photocurrent spectroscopy (IMPS). These results are fitted to a solid state model in order to show that the presence of Fe helps to reduce the drop in the flux of holes to the electrode surface over time, but has a deleterious effect of aiding recombination.

See more of this Session: Electrochemistry and Interfacial Phenomena In Electronic and Photonic Materials
See more of this Group/Topical: Materials Engineering and Sciences Division

(87c) Three Dimensionally Structured Electrodeposited CdS/CdTe Solar Cells

Carlos Hangarter[1], Behrang Hamadani[2], Suyung Jung[2], John Guyer[1], Carlos Beauchamp[2] and Daniel Josell[2], (1)Metallurgy Division, NIST, Gaithersburg, MD, (2)NIST, Gaithersburg, MD

Three dimensionally structured thin film photovoltaic devices based on interdigitated arrays of microscale electrodes are examined by electrodeposited CdS/CdTe heterojunctions and supported with EQE simulations, indicating considerable J_{SC} enhancement is possible through elimination of the front contact and window layer required in planar geometry devices. Electrode parameters including, pitch, width, height, and material are modeled and experimentally probed, demonstrating experimentally and capturing in models dependence on intrinsic material properties and electrode dimensions. In contrast to analogous silicon wafer back contact solar cells where the electrodes are placed on the silicon absorber at the end of processing, in this design the semiconductor is deposited on the electrodes, taking advantage of the thin film processing already required. CdS/CdTe thin film and three dimensionally structured heterojunction devices are electrodeposited from aqueous solutions with CdTe grain size and stress displaying strong dependence on [Cd^{2+}]. Typical annealing temperatures required for absorber optimization in thin films allow for only limited improvement in back contacted devices due to Schottky barrier formation on the nonoptimal electrode materials. Electrodeposited contact materials, ZnO and SbTe, are also examined and show promise of reduced reactivity and improved performance.

See more of this Session: Electrochemistry and Interfacial Phenomena In Electronic and Photonic Materials
See more of this Group/Topical: Materials Engineering and Sciences Division

(87d) Copper-to-Copper Electroless Bonding for High-Density off-Chip Interconnects

Rajarshi Saha, **Hyochol Koo** and **Paul Kohl**, Department of Chemical and Biomolecular Engineering, Georgia Institute of Technology, Atlanta, GA

Over the past several years, there has been an increasing demand for chip-to-chip and chip-to-package interconnects that can offer higher I/O density, better conductivity, and longer lifetimes for high speed integrated circuits. Copper-copper direct bonding is expected to be a potential candidate for high speed communications. Copper pillars electroplated on either chip or chip/substrate could be flip-chip bonded through electroless plating to produce a robust electrical/mechanical interconnect. Electroless bonding between Cu pillars could offer potential advantages due to its simple and fast process, and its ability to accommodate various process environments such as non-uniform offset between terminals [1]. In this presentation, fabrication of a void-free electroless Cu connection with control of electroless plating bath formulation will be discussed. The effect of several additives in the bath will also be studied. Thermomechanical reliability testing of chip-to-chip and chip-to-substrate bonding will also be discussed

The initial chip fabrication method is based on photo-lithography of high-aspect ratio polymers as a mold for through-hole Cu electroplating [2]. A variety of pillar diameters (300, 150, 50, 25, 15 um) and heights (0 and 20 μm) were tested for electroless bonding. High density and high aspect ratio interconnections could be fabricated with this technique. Addition of a leveler induced dome shaped top during electroplating which reduced void formation during electroless bonding. The chips containing pre-fabricated copper pillars were diced; flip-chip aligned using a flip-chip bonder and then dipped in electroless plating solution for bonding.

A wide range of accelerators and suppressor additives were investigated for electroless plating. A narrow gap between pillars hinders fluid flow during electroless deposition leading to trapped voids which could affect the mechanical and electrical integrity of the interconnects. A local suppressor is necessary to avoid the closing of the gap before deposition at the center. SPS (bis-(3-sulfopropyl)-disulfide) and similar molecules have been reported as diffusion-limited suppressors in electroless deposition [3], and their behavior was strongly dependent on the mass-transfer of its molecules, whereas polymer-based surface active suppressor (Polyethylene glycol, PEG) showed no strong dependence on the mass-transfer. Based on this result, it was shown that the addition of SPS could induce preferential Cu growth from the center to the outside of the gap between the two pillars. The effect of convection-currents on electroless deposition was also examined.

Smaller diameter pillars could be advantageous due to a small aspect ratio as the adequate electrolyte filling between small pillars could mitigate void-formation. In the case of 50 um pillars, the void trapping was not observed in FIB (Focused Ion Beam) analysis even without local suppressor.

Thermomechanical reliability tests were carried out in order to analyze structure/electrical reliability of the interconnect structures. The samples were loaded in temperature cycling chamber and the mechanical integrity was analyzed after 100~500 times of temperature cycling (-40°C~ 125°C). As shown in Fig. 3, most of the failure was observed in the interface between pillar and substrates and not within the electroless region. This implies that the bonding strength of electroless Cu region is stronger than the adhesion strength of Cu pillar-to-substrate. The fracture strength (50-110 MPa) (Table 1) of electroless bonded copper pillars with/without thermocycling was observed to be substantially higher than conventional solder-based bonding. Electrical continuity was measured through daisy-chain Cu-Cu structures fabricated on silicon. Integrity of the structures was retained even after high thermal cycling.

REFERENCES

[1] T. Osborn, N. Galiba, and P. A. Kohl, *J. Electrochem. Soc.*, **156** (7), D226 (2009).

[2] H.-C. Koo, C. H. Lightsey, and P. A. Kohl, *IEEE trans. Compon. Packag. Manuf. Tech.*, accepted.

[3] S. Shingubara, Z. Wang, O. Yaegashi, R. Obata, H. Sakaue, and T. Takahagi, *Electrochem. Solid St. Lett.*, 7(6), C78 (2004).

[4] C. H. Lee, S. C. Lee, and J. J. Kim, *Electrochimica Acta*, **50** (2005), 3563.

Cycles	Fracture Stress (MPa)
0	55
100	105
200	120
300	102

Table 1. Fracture stress of electroless copper bonds

(~ 50 μm) with/without thermal cycling.

See more of this Session: Electrochemistry and Interfacial Phenomena In Electronic and Photonic Materials
See more of this Group/Topical: Materials Engineering and Sciences Division

(87e) Vertically Aligned Ag Nanostructure and Its Application In SERS

Liang Su[1], Wenzhao Jia[2], Dan Manuzzi[1], Xiaopeng Li[3], Zhiyong Gu[3] and Yu Lei[2], (1)Chemical Engineering, University of Connecticut, Storrs, CT, (2)Department of Chemical, Materials & Biomolecular Engineering, University of Connecticut, Storrs, CT, (3)Chemical Engineering, University of Massachusetts Lowell, Lowell, MA

Surface-enhanced Ramen scattering (SERS) has intrigued considerable interest due to its extremely high sensitivity which enables the detection of single molecule. Silver or gold-based substrates have been widely applied in this area. However, the synthesis of such "hot spots" usually involves complicated procedures, harsh experimental conditions, or toxic chemicals. In this work, surface-engineered silver nanostructure with uniform and controllable morphology is prepared by a facile and green route, which is also applicable in large-scale production. Scanning electron microscopy, transmission electron microscopy, and X-ray diffraction are used to characterize the sample. Furthermore, SERS of crystal violet on this surface-engineered silver substrate is also demonstrated.

See more of this Session: Electrochemistry and Interfacial Phenomena In Electronic and Photonic Materials
See more of this Group/Topical: Materials Engineering and Sciences Division

(87f) Molecular Order of Mixed Self-Assembled Monolayers Studied by Surface Vibrational Spectroscopy

Armin Rumpel[1], Michael Novak[2], Bjorn Braunschweig[1], Marcus Halik[2] and Wolfgang Peukert[1], (1) Institute of Particle Technology, University of Erlangen-Nuremberg, Erlangen, Germany, (2)Organic Materials & Devices, University of Erlangen-Nuremberg, Erlangen, Germany

Self-organization of molecules into supra-molecular structures is a promising route to fabricate high performance, low cost and flexible electronic devices such as organic transistors. Here, we present a study of mixed self-assembled monolayers (SAM) of phosphonic acid (PA) and C60 modified PA on alumina, which are possible candidates for the active channel in thin-film field effect transistors. For these devices a detailed understanding and control of the interfacial properties and molecular composition is of great importance. However, studies of the interfacial molecular structure under technically relevant conditions represent a significant experimental challenge. For that reason analytical techniques that are highly interfacial specific and applicable under non-vacuum conditions need to be developed.

Vibrational sum-frequency generation (SFG) is an inherently interface sensitive optical technique which provides information on the composition and conformation of the SAM. Detailed information on mixed SAMs is revealed by a combination of SFG, ellipsometry and electronic measurements. SFG measurements show that the molecular order of pure PA SAMs is determined by the alkyl chain length. SAMs with short alkyl chains are in an amorphous state, while a quasi-crystalline state can be achieved for longer alkyl chains [1]. Evaluation of mixed SAMs is challenging since a distinction between CH stretching bands of PA and C60 modified PA cannot be made unambiguously. However, mixed monolayers that are composed of fluorinated and C60 modified PA offer a way to determine surface coverage and molecular properties of the two species independently. CF stretching vibrations of fluorinated C12-phosphonic acids give rise to vibrational bands at 1344 cm^{-1} and 1376 cm^{-1} in SFG spectra while a pronounced vibrational band centered at 1465 cm^{-1} is attributable to C60 vibrations of C60 modified C18-PA. The coexistence of CF stretching and C60 vibrational bands is indicative for assembly of a mixed monolayer that comprises both species. Furthermore, a pronounced maximum in SFG intensity of the C60 band is observed for coverages around 75%, which shows that C60 modified alkyl chains (C18) can be stabilized by C12-phosphonic acids. As a result, the molecular order and lateral conductivity of the mixed SAM is significantly improved.

[1] M. Novak M.; Jäger, C. M.; Rumpel, A.; Kropp, H.; Peukert, W.; Clark, T. and Halik, M. Organic Electronics 11 **2010** (1476–1482).

See more of this Session: Electrochemistry and Interfacial Phenomena In Electronic and Photonic

(95a) Silicate Prodrug-Loaded Nanoparticles

Adam R. Wohl[1], Jing Han[2], Barath R. Guru[3], Jayanth Panyam[3], Thomas R. Hoye[1] and **Christopher W. Macosko**[2], (1)Department of Chemistry, University of Minnesota, Minneapolis, MN, (2)Department of Chemical Engineering and Materials Science, University of Minnesota, Minneapolis, MN, (3) Department of Pharmaceutics, University of Minnesota, Minneapolis, MN

Nanoparticles of highly insoluble drugs can be created by flash nanoprecipitation, FNP. Turbulent mixing combined with a polymer stabilizer produces particles in the 100 nm size range, ideal for passive targeting of tumors and advantageous to increase solubility. We have used FNP to create nanoparticles containing a series of labile silicate prodrugs having customized hydrophobic character and hydrolysis rates. These are stabilized with poly(ethyleneglycol)-*b*-poly(lactic-co-glycolic acid) with well controlled molecular weights and narrow polydispersities. The particles are highly loaded and release studies indicate that hydrolysis occurs inside the particles with the drug released at rates controlled by the stereoelectronic character of the silicate. A 4T1 (breast cancer) cell culture assay demonstrated *in vitro* efficacy. *In vivo* studies will be reported.

See more of this Session: In Honor of Matthew Tirrell's 60th Birthday II
See more of this Group/Topical: Materials Engineering and Sciences Division

(95b) Fun and Profit with Janus Colloids

Steve Granick, Departments of Materials Science, Chemical Engineering, Chemistry, and Physics, University of Illinois at Urbana-Champaign, Urbana, IL

The colloidal state is the stage on which many of the great themes of Matt Tirrell's career find their natural meeting place. In this talk, I will consider Janus colloids, which violate the textbook notion that particle-particle interactions are isotropic and depend solely on separation. Instead, in the laboratory it is now straightforward to create diblocks, which are the colloidal analogue of molecular surfactants, and triblocks, which are the colloidal analogue of ABA and ABC triblock copolymers. Amazing and potentially useful superstructures result from their directed self-assembly. Powerful design rules to produce targeted supracolloidal structures will be described. Unsolved problems for the future will be emphasized.

See more of this Session: In Honor of Matthew Tirrell's 60th Birthday II
See more of this Group/Topical: Materials Engineering and Sciences Division

(95c) Biomimetic Peptide-Amphiphiles for Receptor-Targeted Therapeutics

Efrosini Kokkoli, Department of Chemical Engineering and Materials Science, University of Minnesota, Minneapolis, MN

Drug delivery systems that include site-specific surface ligands could further enhance selective targeting. We proposed to enhance selectivity by targeting vesicles (stealth liposomes or polymersomes) to the alpha(5)beta(1) integrin. The alpha(5)beta(1) integrin is an adhesion receptor that is expressed on the cell surface and has been implicated in dynamic processes such as suppressing tumor recurrence and metastasis in certain cases of cancer. Alpha(5)beta(1) is up-regulated by proliferating endothelial cells of the tumor vasculature and also found on different tumors. Colon cancer cells were chosen as the model target system in our study. A fibronectin-mimetic peptide-amphiphile (PR_b) was recently synthesized by our group. Surfaces functionalized with the PR_b sequence were shown to outperform fibronectin covered surfaces (fibronectin is the natural protein ligand for alpha(5)beta(1)) and other peptides in terms of various cellular phenomena, such as cell adhesion, spreading, cytoskeleton formation, and extracellular matrix production (experiments were performed with endothelial cells). The objective of our work is to engineer alpha(5)beta(1)-targeted delivery systems that reach their target and effectively release the encapsulated payload (DNA, chemotherapy agents, or proteins) intracellularly. In this talk I will discuss our results from PR_b-functionalized nanoparticles targeted to colon cancer cells in vitro and in vivo.

See more of this Session: In Honor of Matthew Tirrell's 60th Birthday II
See more of this Group/Topical: Materials Engineering and Sciences Division

(95d) Design Rules for Thermoresponsive Polymer Brushes

Deborah Leckband, Department of Chemical and Biomolecular Engineering, University of Illinois at Urbana-Champaign, Urbana, IL

...

See more of this Session: In Honor of Matthew Tirrell's 60th Birthday II
See more of this Group/Topical: Materials Engineering and Sciences Division

(103a) Coarse-Graining of Nanocoating Using MARTINI Force Field for Mechanical Property Evaluation

Rohan Uttarwar, Jeffrey Potoff and Yinlun Huang, Department of Chemical Engineering and Materials Science, Wayne State University, Detroit, MI

Nanocomposite coatings, either thermoplastics or thermosets, have been proven to possess significantly improved performance or novel functionalities, such as self cleaning, self healing, etc. Dramatically increased surface area of nanosize fillers inside the coating matrix is the main reason for achieving superior mechanical properties. Molecular simulation can be performed to identify the root cause behind enhancement in properties of nanocomposite coatings, which is increase in polymer number density around nanoparticles.[1] But these results of coarse-grained simulations can be difficult to validate with real experimental results due to loss of physical meaning and degrees of freedom of the system. Hence, effective coarse graining methodology needs to be developed to study molecular behavior of polymer systems and model with experimental validation.

Recently, a MARTINI based approach has been introduced to develop successfully coarse grained models for biomolecular systems.[2] A coarse grained model is generated from atomistic models, which should be more reliable in model accuracy and model-based prediction, and thus it can be compared with experimental data. In this paper, the MARTINI technique is adopted for coarse grained simulations of polymer nanocomposite material. Due to increasing environmental concerns, water based resin systems, such as acrylics, are used tremendously for nanocoating syntheses. Thus polymethyl methacrylate (PMMA) resin with spherical silica nanoparticles dispersed in it is modeled for simulating a nanocoating. The volume fraction of nanofillers is kept at 5%. All atomistic simulations are performed using the CHARMM force field. The verification to this model is obtained by obtaining density and glass transition temperature consistent with experimental data. The equilibrated structures of a polymer from all-atom model are used to generate the coarse grained model by mapping onto coarse-grained beads. The parameters from the MARTINI model are modified, and a new set of parameters are generated to obtain a coarse grained model with comparable densities and radius of gyration with that of atomistic structures. All molecular dynamics simulations are performed with NAMD, version 2.7.

The MARTINI models for pure polymer samples and nanocoating samples of different nanoparticle sizes are deformed with external force at a fixed strain rate in longitudinal and transverse directions to evaluate its mechanical properties. The Young's modulus of nanocomposite is observed to increase at lower temperatures. The relative change in stiffness of nanocomposite is lower below the glass transition (T_g) compared to above T_g due to increased mobility of polymer chains. It is also seen that the Young's modulus increases with decreases in the size of the nanoparticle. This justifies the increase in the surface area leading to improved mechanical properties. During the deformation testing, the externally applied stress reaches a peak and goes down, which dictates the onset of plastic flow of the nanocomposite. The behavior of polymer models under applied stress is in agreement with experimental results. This demonstrates the applicability of the use of coarse-grained potentials in the design of nanocoating formulations with superior mechanical properties.

References

1. Xiao J., Y. L. Huang, and C. Manke, "Computational Design of Thermoset Nanocomposite Coatings: Methodological Study on Coating Development and Testing," Chemical Engineering Science, 65, 753-771, 2010.
2. Marrink S. J., Risselada H. J., Yefimov S., Tieleman P. D. and Vries A. H., "The MARTINI Force Field: Coarse Grained Model for Biomolecular Simulations", Journal of Physical Chemistry B, 111, 7812-7824, 2007.

See more of this Session: Modeling and Simulation of Polymers I
See more of this Group/Topical: Materials Engineering and Sciences Division

(103b) Morphology Dependent Hydrophobic Drug Partitioning In PEO-PCL Micelles Investigated with Coarse-Grained Molecular Dynamics

Sharon M. Loverde[1], Michael L. Klein[2] and Dennis E. Discher[1], (1)Chemical & Biomolecular Engineering, University of Pennsylvania, Philadelphia, PA, (2)Institute for Computational Molecular Science and Department of Chemistry, Temple University, Philadelphia, PA

Atomistically accurate, coarse-grained (CG) polymer models have been developed and studied for poly (ethylene oxide) -poly(caprolactone) (PEO-PCL), with molecular dynamics simulations comparing well with experimental phase behavior. The anti-cancer drug paclitaxel is also being CG'd with intramolecular interactions again obtained from all-atomistic molecular dynamics. Solubility of paclitaxel indicates that partitioning of the drug is dependent on micellar morphology. Utilizing free energy techniques, it is found that the hydrophobic drug possesses a greater partitioning in a worm micelle morphology of the same diblock weight than a spherical micelle morphology. These findings are consistent with previously found experimental results from the Discher laboratory. Moreover, at larger loading concentrations, the taxol-PCL and taxol-taxol interactions significantly shift the micellar density profile, showing increased interactions with the PEO, which may be an explanation for the phenomena of 'burst release.'

See more of this Session: Modeling and Simulation of Polymers I
See more of this Group/Topical: Materials Engineering and Sciences Division

(103c) A Comparison Between Polymer Salt and Ionomer for Battery Applications

Kan-Ju Lin[1], Katherine Li[2] and Janna Maranas[1], (1)Chemical Engineering, Pennsylvania State University, University Park, PA, (2)Chemical and Biological Engineering, Princeton University

Salts solvated in a Polyethylene oxide (PEO) matrix are considered to be the most promising candidates for solid state polymer electrolytes in battery applications. Since the anions have high mobility, and do not participate in chemical reactions, they accumulate at the electrode/electrolyte interface, causing a concentration gradient. This degrades battery performance after several charge/discharge cycles. An electrolyte in which the cation is the only conducting species is desired. PEO based ionomers solvate the cations, and reduce anion motion because the anions are covalently bonded to the polymer backbone. Like their PEO-salt counterparts, these single ion conductors suffer from low conductivity.

In this study, we examine a PEO-ionomer system, and the corresponding PEO-salt system using molecular dynamics simulation. The PEO-ionomer system has been experimentally synthesized and studied at Penn State. To provide a clean comparison, we consider a hypothetical PEO-salt system, in which the cation and anion are the same, but the anion is no longer incorporated on the polymer backbone. Thus, the cation identity, the cation-anion, and the cation-PEO interactions are the same as in the PEO-ionomer system. Comparison of the two systems isolates the effect of attaching the anion to the polymer backbone by covalent bonding. A wide range of ion environments is observed in both systems. The number of ion crosslinks in the two two systems are similar, but 1.5 times more cations are solvated by the PEO matrix in the salt compared to the ionomer. As ion concentration increases in a PEO salt, the PEO chains slow significantly. In the ionomer, this mechanism is active, but ions also slow the polymer through ionic cross-linking. Thus, covalently bonding the anion to the PEO backbone slows its mobility in comparison to the PEO-salt system. This in turn impacts the mobility of the cations and anions. We also find that the cations in the PEO-salt system change their coordination more frequently than in the ionomer. We conclude that the use of single ion conductors requires a shift in mechanism, such as ion motion facilitated by string-like clusters. We show examples of such movement in the ionomer and suggest ways to increase this type of ion motion.

See more of this Session: Modeling and Simulation of Polymers I
See more of this Group/Topical: Materials Engineering and Sciences Division

(103d) Simulation of Gas Diffusion In Polymer Nanocomposites

Ben Hanson, Chemical Engineering, University of Texas, Austin, TX and Venkat Ganesan, Department of Chemical Engineering, The University of Texas at Austin, Austin, TX

ABSTRACT: The effectsof nanoparticles on the rates of gas diffusion through glassy polymers werestudied by a combination of molecular dynamics and Monte Carlo techniques. These techniques were used to overcome thelarge amount of computational time required to obtain gas trajectories in thediffusive regime for glassy systems.Polystyrene was simulated using molecular dynamics with a number of gasmolecules to obtain a probability profile of gas motion within the polymermatrix at a range of temperatures above and below glass transition temperature. These results were used in conjunction withMD simulations to examine the types of gas particle motion as a function of thesetemperatures. The use of the a timeextending kinetic Monte Carlo allowed simulation of a sufficient time frame tocalculate gas motion in the diffusive regime. We studied the effect of Fullerenenanoparticles added to the polymer upon the gas diffusivity in such polymers. These results will be discussed in this talkalong with a mechanistic explanation of our observations.

See more of this Session: Modeling and Simulation of Polymers I
See more of this Group/Topical: Materials Engineering and Sciences Division

(103e) Analysis of AFM Results with MD Calculations On Single-Molecule Stretching of Poly(ethylene oxide) In Water and Hexane

Mangesh Chaudhari, Department of Chemical and Biomolecular Engineering, Tulane University, New Orleans, LA and Lawrence R. Pratt, Chemical and Biomolecular Engineering, Tulane University, New Orleans, LA

Analysisof AFM results with MD calculations on single-molecule stretching of poly(ethylene oxide) in water and hexane

Mangesh I. Chaudhari and LawrenceR. Pratt

Department of Chemical and Biomolecular Engineering

Tulane University, New Orleans,LA, 70118, USA.

ABSTRACT :

Elasticityof a single polymer in solution is important for the modeling of foams/gels andcan be directly measurable by atomic force microscopy (AFM). The work comparesstretching behavior of poly (ethylene oxide) polymer in water and hexane.Parallel tempering molecular dynamic simulations on aqueous 11-, 21-, and31-mer poly (ethylene oxide) solutions differentiates low-, moderate-, andhigh- extension regions (FIG.1) and explains that water swells these polymers. The low extension (low force) behaviorobtained from experimental and simulation data is often complex. Our previousresults, which combined parallel tempering and window sampling, clarified thelow-extension behavior. Here we return to examine the measured high extension behaviorin detail. We compare simulation results for high polymer extensions withavailable experimental data, and we compare simulations for different solventswith one another. We analyzesimulation results on aqueous solutions from the perspective of cooperativeH-bonding of water molecules to the ether oxygens inorder to clarify the role of solvent H-bond bridges in the conformationalstructures of these polymers.

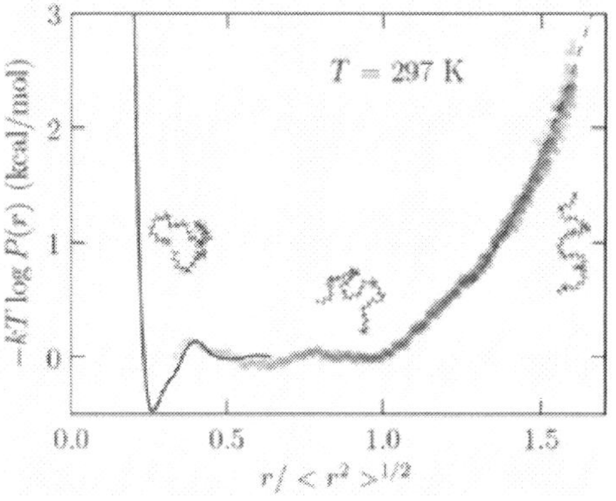

FIG. 1: Parallel tempering results forT=297 K distinguish low-, moderate-, and high-extension regimes. Thelow-extension regime includes the contact peak on the left of these graphs, andis associated with the analogous contact-minimum in the CH_4°CH_4potential-mean-force (pmf). That pmfhas been studied in numerous previous simulation calculations over severaldecades. The solid curve represents short-distance 'hydrophobic bonds' withwindow sampling that is consistent with the direct simulation results. X-axisis normalized with RMS end-to-end distance and emphasizes that the high extensionregion starts after this value. Parabola fit through high extensions iscomparable with experimental data available on stretching of these polymers.

See more of this Session: Modeling and Simulation of Polymers I
See more of this Group/Topical: Materials Engineering and Sciences Division

(103f) Molecular Dynamics Simulations of An Oxidized Vapor-Grown Carbon Nanofiber and Vinyl Ester Resin Interactions Leading to a Possible Interphase Formation In the Cured Nanocomposite

Changwoon Jang[1], **Sasan Nouranian**[2], Thomas E. Lacy[1], Steven R. Gwaltney[3], Hossein Toghiani[4] and Charles U. Pittman Jr.[3], (1)Department of Aerospace Engineering, Mississippi State University, Mississippi State, MS, (2)Center for Advanced Vehicular Systems (CAVS), Mississippi State University, Mississippi State, MS, (3)Department of Chemistry, Mississippi State University, Mississippi State, MS, (4)Dave C. Swalm School of Chemical Engineering, Mississippi State University, Mississippi State, MS

Molecular dynamics simulations were used to study the effect of vapor-grown carbon nanofiber (VGCNF) surface oxidation on VGCNF-vinyl ester (VE) resin monomer interactions. It was anticipated that the interfacial interactions between resin monomers and the oxidized carbon nanofiber surface would result in different local molar ratios of the resin monomers compared to the bulk resin. The resin was comprised of styrene and two VE monomers with either one or two bisphenol A groups in their backbones (designated as VE1 and VE2, respectively). Differences in local molar ratios of these resin monomers may result in the formation of an interphase region in the final cured nanocomposite. The time-averaged relative concentration profiles were used to monitor the temporal and spatial distributions of the resin monomers in a simulation cell of size 60×50×60 Å^3. Initially the cell was filled with monomers in a ratio consistent with a 33 wt% commercial epoxy vinyl ester resin (Derakane 441-400, Ashland Co.). The idealized VGCNF surface was represented by two overlapped (shingled) graphene sheets, which were in contact with the resin. Oxidation of the graphene sheets was represented by the introduction of functional groups on the edges and the surfaces of the graphene sheets. The edges were more highly oxidized than the surfaces, containing of hydroxyl, phenolic, lactone, quinone, hydroquinone, and anhydride functional groups. Hydroxyl, epoxide, aromatic, ketone and carboxylic acid groups were attached to the surfaces (e.g. basal planes). These functional groups have been previously identified in our laboratory [1]. The NVT (constant number of atoms, N; constant volume, V; constant temperature, T) ensemble was used for simulations. All simulations were conducted with a simulation time of 10 ns at 1000 K for monomer distribution equilibration followed by another 5 ns simulation at 300 K. The relative concentration profiles were generated for the direction roughly perpendicular to the graphene sheet surfaces. The VE1 monomer concentration was higher near the oxidized graphene surface, compared to a simulation performed using a pristine graphene [2], where VE1 was depleted near the graphene surface. Styrene accumulated in the interfacial region in both cases, but more so for the pristine surface. Overall, a ~5 Å thick interfacial region was observed near each surface of the graphene sheet. In these regions, the monomer molar ratios differed from those in the bulk resin. The increased concentration of VE1 near the oxidized carbon nanofiber surface may lead to a stiffer interphase compared to the bulk matrix. In addition, favorable VGCNF-matrix polar interfacial interactions may lead to increased interfacial adhesion and subsequently, increase the interfacial shear strength in the final cured nanocomposite. The results of this study can be extended to other resins and surfaces.

[1] Lakshminarayanan, PV, Toghiani, H, Pittman, CU. Nitric acid oxidation of vapor grown carbon nanofibers. Carbon 2004; 42: 2422-2433.

[2] Nouranian, S, Jang, C, Lacy, TE, Gwaltney, SR, Toghiani, H, Pittman Jr, CU. Molecular dynamics simulations of vinyl ester resin monomer interactions with a pristine vapor-grown carbon nanofiber and their implications for composite interphase formation. Carbon (In Press).

See more of this Session: Modeling and Simulation of Polymers I
See more of this Group/Topical: Materials Engineering and Sciences Division

(122a) Ultra-Stable Structured Polymeric Glasses

Rodney D. Priestley, Yunlong Guo and Kimberly B. Shepard, Chemical and Biological Engineering, Princeton University, Princeton, NJ

The density and stability of a glass are strongly correlated with the principal viewpoint being that low-density glasses are unstable. Upon heating, the structure of a low-density glass can abruptly collapse into a more dense state. This collapse can be dramatic, *e.g.*, resulting in glasses with negative thermal expansion coefficients upon heating within a certain temperature interval. The distinctive advantage of low-density glasses in technologies, *i.e.*, weight reduction, is obviated by their structural instability. A fundamental challenge is thus the development of strategies to manipulate – then arrest – the structure of low-density glasses to improve their stability. Here we demonstrate that Matrix Assisted Pulsed Laser Evaporation (MAPLE) can be used to fabricate polymer glasses with extraordinarily low densities and simultaneously exceptional thermal and kinetic stability. In addition, these glasses have significantly enhanced glass transition temperatures. This unique (and unprecedented) combination of properties is a result of the glass morphology, i.e., the glasses are nano-structured. The unique set of properties of MAPLE-deposited glasses may make them attractive in technologies where weight and stability are central design issues.

See more of this Session: Structure and Properties In Polymers II
See more of this Group/Topical: Materials Engineering and Sciences Division

(122b) Effects of Monomer Structure and Chain Architecture On Glass Transition Breadth In Homopolymers, Random Copolymers, Gradient Copolymers and Copolymer Blends

Stephen R. Marrou, Sean M. Wundrow and John M. Torkelson, Department of Chemical and Biological Engineering, Northwestern University, Evanston, IL

Recently, gradient copolymers have attracted significant interest as potential vibration or acoustic damping materials due to the extremely broad glass transition (T_g) response of these materials, in some cases approaching 100 K in breadth. This unique behavior was attributed to the large degree of compositional heterogeneity present in the material as a result of a sinusoidal composition profile in nanophase separated systems. However, we have discovered that specific homopolymers and random copolymers also exhibit large T_g breadths due to heterogeneities present in the repeat unit structure. We also determined that blending a strongly-segregating styrene/4-vinyl pyridine (S/4VP) gradient copolymer with a selective plasticizer significantly increases the T_g breadth of the resulting material, up to approximately 100 K.

Here, we present the T_g breadth of a range of homopolymers, primarily focused on the series of n-alkyl methacrylates. We determined that T_g breadth increases with increasing side chain length in the series of n-alkyl methacrylates as measured by differential scanning calorimetry (DSC), ellipsometry, and dynamic mechanical analysis. We conclude that these repeat unit structures exhibit significantly broad T_gs, up to 42 K for poly(n-hexyl methacrylate), due to heterogeneities corresponding to incompatible main and side chain interactions. We also incorporated these repeat units in random copolymers and determined that random copolymer T_g breadth is a strong function of copolymer composition, with upper and lower bounds dictated by breadths of the corresponding homopolymers.

Additionally, we show that blending oligomeric polystyrene into S/4VP gradient copolymers results in a dramatic shift in the T_g response of the bulk material. Oligomeric polystyrene is expected to selectively plasticize the S-rich regions of the copolymer, shifting the glass transition response of these regions to lower temperatures while leaving the 4VP-rich regions relatively unchanged. For a S/4VP gradient copolymer initially exhibiting T_g breadth of 35 K, addition of 20 weight percent oligomeric polystyrene yielded a T_g breadth of roughly 100 K.

We also blended weakly-segregating styrene/n-butyl acrylate (S/nBA) random copolymers of different compositions in an effort to approximate a sinusoidal composition profile as found in gradient copolymers. We demonstrate the range of composition differences in random copolymer blends to maintain a single continuous glass transition, and the resulting increase in T_g breadth. We also show that S/nBA gradient copolymer blends are able to maintain a single, broad, and continuous T_g with greater composition differences than that of random copolymer blends. We conclude that the presence of the full range of compositions in each gradient copolymer sample is extremely important in determining the compatibility of two copolymers.

See more of this Session: Structure and Properties In Polymers II
See more of this Group/Topical: Materials Engineering and Sciences Division

(122c) Physical Aging of Glassy Polymers In Confined Environments

Thomas Matthew Murphy, Chemical Engineering, University of Texas at Austin, Austin, TX, Benny Freeman, University of Texas at Austin and Don R. Paul, Department of Chemical Engineering, The University of Texas at Austin, Austin, TX

Amorphous polymers below their glass transition temperature (Tg) are almost always in a non-equilibrium state. They spontaneously undergo a densification process known as physical aging, which changes many of their physical properties, including gas permeability, modulus, refractive index, and enthalpy. Over the past decade, repeated observations have shown that thin, free-standing polymer films (<1 μ m in thickness) can undergo physical aging much more rapidly than thicker "bulk" samples. This is of great importance in many areas, including gas separation membranes, which often utilize a glassy skin layer of ~50-100 nm in thickness to effect the separation. Other studies of physical aging in confinement have shown both accelerated and retarded aging in confined polymers depending on the nature of the polymer-substrate interactions. The fundamental reasons why aging can be dependent on thickness are not completely understood.

A novel co-extrusion process, using an extruder with two feed streams and several layer-multiplying dies, was used to produce films of polysulfone (PSF) and a rubbery co-layering material. The co-layering material does not undergo physical aging (since it is above its Tg) and merely provides confinement for the PSF layers. Two sets of films were made: Dow Engage 8100 (EO, an ethylene-1-octene copolymer) was used as the rubbery co-layering material in the first set, and Dow Infuse 9007 (OBC, an olefin block copolymer) was used in the second set. The films have PSF layer thicknesses ranging from ~180-500 nm, but have overall thicknesses in the range of 50-100 μ m. The O2, N2, and He permeabilities of these multilayered films were measured over time to determine the effect of layered confinement on physical aging. Prior to study, the films were annealed for 10-15 minutes at 10-15 °C above the Tg of PSF (PSF Tg~186 °C) and then quenched to room temperature to define a starting point for the aging studies.

The results of experiments with these films show that the rate of decline in permeability for all layered films is similar to that of bulk films, regardless of PSF layer thickness, co-layering material, or penetrant gas. For free-standing PSF films with overall thicknesses similar to the PSF layer thicknesses in our films, physical aging would have been expected to proceed much more rapidly. We believe that the lack of free surfaces and possible chain entanglement at the PSF-rubber interfaces in layered films restricts the rapid, near-surface relaxation that can occur in thin polymer films.

See more of this Session: Structure and Properties In Polymers II
See more of this Group/Topical: Materials Engineering and Sciences Division

(122d) Developing High Impact Poly(lactic acid) (PLA) Blends Through Reactive Blending

Jinwen Zhang, Composite Materials and Engineering Center, college of Engineering and Architecture, Washington State University, Pullman, WA

Poly(lactic acid) (PLA) is one of a few commercially available biobased thermoplastic polymers and is considered the most promising one to substitute petroleum-based polymers for broad application because of its high strength, modulus and competitive price. However, the brittleness of PLA has prevented it from wide applications. To improve its toughness, PLA has been blended with various rubbery or flexible polymers. While most of these blends exhibit very high ductility compared with neat PLA, they only yield very limited (a few times higher) improvement on impact strength which is more important for many applications. Given the very low impact strength of PLA (ca.10-30 J/m), a few folds of increase is far from sufficient to satisfy most engineering applications. Because in most blends the polymer pair is immiscible, creation of sufficiently strong interphase interaction is a prerequisite for achieving high mechanical properties, particularly impact strength. Reactive blending may provide a crucial solution to realize the strong interfacial adhesion.

In this work, improving the toughness of PLA materials through blending is discussed in detail. In particular, the discussion is focused on our recent investigation of improving impact properties of PLA blends through reactive blending. Our novel PLA ternary blend system consisted of PLA, an epoxy-containing elastomer and a zinc ion ionomer of the ethylene-methyacrylic acid copolymer. The blending was performed in a twin-screw extruder. Mechanical, thermal dynamic mechanical properties and morphology of the blends were thoroughly studied. Blending temperature, elastomer/ionomer ratio, type of cations and neutralization degree of the ionomer all displayed significant effects on impact toughness of the ternary blends. Super toughness was successfully obtained for some blends compounded at 240° C. The curing of the elastomer and reactive interfacial compatibilization between the cured elastomer phase and the PLA matrix were noted to occur simultaneously during the blending. Zinc ionomer initiated and catalyzed the dynamic vulcanization of epoxy-containing elastomer and also promoted the reactive compatibilization at the interface between PLA and elastomer. TEM observation revealed that the 'salami'-like phase structure was formed in the ternary systems.

See more of this Session: Structure and Properties In Polymers II
See more of this Group/Topical: Materials Engineering and Sciences Division

(122e) Synthesis and Characterization of Multiblock Copolymers with Urethane Links

Intaek Lee and Frank S. Bates, Chemical Engineering & Materials Science, University of Minnesota, Minneapolis, MN

Multiblock copolymers containing 10 or more blocks are expected to have unique morphologies and mechanical properties relative to conventional triblocks. Poly(styrene-b-butadiene) (PS-PB) multiblock copolymers were synthesized using a combination of anionic polymerization and urethane based condensation reactions of dihydroxyl PS and PB. Two types of block sequences were prepared, one with random and the second with alternating PS and PB block architectures. These block copolymers were characterized with ^{1}H-NMR, gel permeation chromatography (GPC), and differential scanning calorimetry (DSC), demonstrating efficient block coupling of blocks and overall polydispersities close to the anticipated most probable distribution. Morphologies were examined using small-angle X-ray scattering (SAXS) and transmission electron microscopy (TEM). In all instances the multiblock copolymers contained a microphase separated bicontinuous structure regardless of composition or sequencing. Tensile tests revealed that the PS–PB multiblock copolymers are tougher than the corresponding linear PS–PB–PS triblock copolymers at similar molecular weight and composition.

See more of this Session: Structure and Properties In Polymers II
See more of this Group/Topical: Materials Engineering and Sciences Division

(122f) Role of Triblock Copolymer On the Molecular Structure and Compatibilization of PLA/PP Blends and Core/Sheath Fibers

Sara A. Arvidson[1], Kristen E. Roskov[2], Richard J. Spontak[2], Saad A. Khan[2] and Russell E. Gorga[3], (1) Department of Chemical and Biomolecular Engineering, North Carolina State University, Raleigh, NC, (2)Chemical and Biomolecular Engineering, North Carolina State University, Raleigh, NC, (3)Textiles Engineering Chemistry, North Carolina State University, Raleigh, NC

Bicomponent fiber spinning involves co-extruding two polymers or polymer blends from the same spinneret into a filament that contains both. Possible benefits of co-spinning include the lower costs associated with single-step processes, the performance of fibers based on multiple components, and the suppression of an unfavorable rheological behavior (i.e., Newtonian) of one polymer by spinning with a small fraction of another more advantageous (i.e., Maxwell-fluid-like) polymer [Park et. al (1990)]. While polymer blends generally result in properties intermediate to the individual components and requires trading off one quality for another, bicomponent spinning has the potential to retain the individual component properties.[1] However, bicomponent spinning can lead to fibers that split when subjected to mechanical force.[2] Maintaining adhesion between the individual components retains the mechanical properties of fibers and can direct the morphology of molecules within fibers.

At the polymer-polymer interface, immiscible polymers exhibit reduced entanglements which can lead to rupture of the interface at temperatures below the Tm of the components. Above the Tm of both components, subjecting the material to external forces in the plane of the interface may result in "slip" of one polymer across the other, also due to reduced polymer chain entanglements. Analogous to surfactants in aqueous-oil emulsions, block copolymers may be used to lower the interfacial tension at polymer-polymer interfaces which can improve adhesion between the polymers and reduce slip during extrusion, a concept which is commonly referred to as "compatibilization". Block copolymers make excellent surfactants for multiple-polymer systems and are judged to be effective if they reduce the domain size(s) of the individual phases in blends, reduce the interfacial tension of a polymer pairs, and/or increase the interfacial adhesion and mechanical properties of the composite.[1,3-4] The concept of compatibilizing polymer-polymer interfaces that are not blended but rather contacted, such as those involved in core-sheath fibers or laminates, however, has largely been ignored. To our knowledge there have been no reports on the compatibilization of the core and sheath of bicomponent fibers by the addition of a block copolymer to one component.

We found that during large scale mechanical deformations, PP/PLA co-spun fibers split at the interface when PP was in the core and PLA the sheath.[2] We explore the concept of compatibilization of two non-blended polymers by incorporating a triblock copolymer to influence the mechanical properties of melt spun bicomponent core/sheath fibers of PP and PLA. The morphology of the block copolymer in fibers is compared to that formed during melt mixing without extrusion and from solvent casting to elucidate the effect of elongational forces on block copolymer morphology. We discuss the application of steady shear rheological methods to evaluate the existence of slip at polymer-polymer interfaces and the efficacy of block copolymers in compatibilizing two polymer melts.

[1]R. Holsti-Miettinen and J. Seppälä. *Polym. Eng. Sci.* **32**(13), 868 (1992).

[2]S. Arvidson, et al. *ACS Appl. Mater. Interfaces* (under review).

[3]T. Del Castillo-Castro et al. J. *Appl. Polym. Sci.*, **119**, 2895 (2011).

[4]D. Wang, et al. *Polymer* **52**, 191 (2011).

See more of this Session: Structure and Properties In Polymers II
See more of this Group/Topical: Materials Engineering and Sciences Division

(122g) Study of Gas Permeabilities In Styrene Ethyleneoxide (SEO) Block Copolymers

Matteo Minelli[1], Marco Giacinti Baschetti[1], Daniel T. Hallinan Jr.[2] and Nitash Balsara[2], (1)University of Bologna, Bologna, Italy, (2)Chemical Engineering, University of California, Berkeley, Berkeley, CA

Study of Gas Permeabilitiesin Styrene Ethyleneoxide (SEO) Block Copolymers

MatteoMinelli[1], Marco Giacinti Baschetti[1], Daniel T. HallinanJr.[2], Nitash P. Balsara[2]

1 Dipartimento di Ingegneria Chimica, Mineraria e delle TecnologieAmbientali (DICMA) Alma Mater Studiorum - University di Bologna, via Terracini 28, 40131, Bologna,Italy.

2 Materials Sciences Division and Environmental EnergyTechnologies Division, Lawrence Berkeley National Laboratory, University of California, Berkeley, CA 94720, United States

Abstract

Transportproperties of different gases such as oxygen, nitrogen, methane and carbondioxide are characterized for a styrene-ethyleneoxide (SEO) triblock copolymer membraneat different temperatures ranging from 20 to 80°C.

Thematerial consists of a PEO fraction of about 50% by weight and a crystallinepart of about 10% at room temperature. DSC analysis allowed the evaluation ofmelting point at about 40°C whereas a lamellar structure with a domain spacingof 95 nm was observed through SAXS analyses.

Permeationexperiments were carried out by means of a manometric [1]device and were performed in a range of temperature covering the melting regionof PEO segment [2] allowing for the study of theeffect of crystallinity on the mass transport behaviour. This material, indeed,is suitable for different applications such as membrane for gas separation aswell as solid electrolyte in polymeric batteries [3, 4a&b].

Theresults of CO_2 permeability, reported in Figure 1, showed aremarkable increases of both permeability and diffusivity (calculated throughtime lag method) when temperature is raised, and the behaviour deviates significantlyfrom the Arrhenius law in the temperature range in which PEO melting takesplace. Interestingly, data show very similar activation energies both at high andlow temperature, but the calculated values are lower than those calculated byother authors [4-6], and reported in theliterature, for amorphous PEO, which is expected to be the phase involved inthe permeation phenomenon.

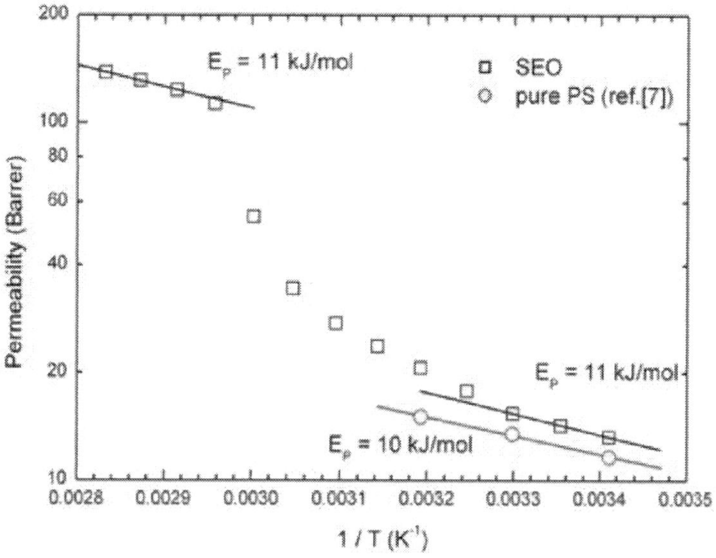

Figure 1, CO_2 Permeability inSEO block copolymer at different temperatures.

Thepermeability data were analyzed by considering the block copolymer as amultiphase system in which the two moieties are endowed with different permeabilities.In this concern, the results show that at low temperature, the gas diffusion ismainly located in the PS [7] phase due to crystallinityof the PEO. On the contrary, at temperature above 50-60°C, the PEO (melt) is remarkably more permeable than the styrenic part and the observed *P*for SEO copolymer is one order of magnitude higher than the value at roomtemperature.

Thesame behaviour of permeability with respect to temperature above described forCO_2 was also observed for the case of the other gases (O_2,N_2 and CH_4); the steep increase of *P* occurs at thesame temperature (between 40 and 60°C) and the trends at high and low temperature are once again comparable (similar activation energies) confirming theprevalent role of crystallinity change in the materials behaviour.

References:

[1] M. Minelli, M.G. De Angelis, F. Doghieri, M.Marini, M. Toselli, F. Pilati, Oxygen permeability of novel organic-inorganiccoatings: I. Effects of organic-inorganic ratio and molecular weight of the organiccomponent, Eur. Polym. Jou. l44 (2008) 2581-2588.

[2] H. Lin, B.D. Freeman, Gas solubility,diffusivity and permeability in poly(ethylene oxide), J. Membr. Sci. 239 (2004)105117.

[3] H. Lin, B.D. Freeman, Materials selectionguidelines for membranes that remove CO_2 from gas mixtures, J. Mol.Struct. 739 (2005) 57-74.

[4] a) Panday, A.; Mullin, S.; Gomez, E. D.;Wanakule, N.; Chen, V. L.; Hexemer, A.; Pople, J.; Balsara, N. P. *Macromolecules***2009**, *42*, 4632-4637.

b) Singh, M.; Odusanya, O.; Wilmes, G. M.;Eitouni, H. B.; Gomez, E. D.; Patel, A. J.; Chen, V. L.; Park, M. J.; Fragouli,P.; Iatrou, H.; Hadjichristidis, N.; Cookson, D.; Balsara, N. P. *Macromolecules***2007**, *40*, 4578-4585.

[5] S.J. Metz, M.H.V. Mulder, M. Wessling, Gas-PermeationProperties of Poly(ethylene oxide) Poly(butylene terephthalate) BlockCopolymers, Macromolecules 37 (2004) 4590-4597.

[6] D. Husken, T. Visser, M. Wessling, R.J.Gaymans,CO_2 permeation properties of poly(ethyleneoxide)-based segmented block copolymers, J. Membr. Sci. 346 (2010) 194-201.

[7] E. Sada, H. Kumazawa, H. Yakushiji, Y. Bamba,K. Sakata, S.T. Wang, Sorption and diffusion of gases in glassy polymers. Ind. Eng. Chem. Res. 26 (1987) 433-438.

See more of this Session: Structure and Properties In Polymers II
See more of this Group/Topical: Materials Engineering and Sciences Division

(123a) Surface Modified Electrospun Fibers for Investigating the Relative Influence of Chemical and Topographical Cues On Tumor Cell Migration

Shreyas S. Rao[1], Alex Hissong[2], Jed Johnson[3], John J. Lannutti[3], Atom Sarkar[4] and Jessica O. Winter[5], (1)Chemical and Biomolecular Engineering, The Ohio State University, Columbus, OH, (2)Department of Biomedical Engineering, The Ohio State University, Columbus, OH, (3)Material Science and Engineering, The Ohio State University, Columbus, OH, (4)Department of Neurosurgery and Laboratory for Nanomedicine, Geisinger Health System, Danville, PA, (5)Chemical and Biomolecular Engineering and Biomedical Engineering, the Ohio State University, Columbus, OH

Tumor cells are known to respond to the extracellular matrix (ECM) environment by altering their morphology and migration rate. However, there are few *in vitro* models that permit investigation of the relative effects of chemistry vs. topographical/mechanical cues in this process. Here, we present brain mimetic materials consisting of surface modified electrospun fibers (EFs) that maintain similar mechanical and topographical features, while displaying differing surface chemical functionalization.

As a model tumor system, we are investigating glioblastoma multiforme (GBM), a disease afflicting more than 22,500 individuals in the United States annually. These cancers are characterized by their infiltrative capacity and high degree of mobility; median patient survival is extremely low (~12-15 months) [1]. Clinical observations suggest that migration often occurs as single cells along white matter tracks, glial limitans externa and blood vessel periphery [2]. Aligned poly (ε-caprolactone) (PCL) electrospun fibers (EFs), *in vitro* biomaterials used in our study, closely mimic *in vivo* topographical features (i.e., blood vessels and white matter) and have diameters of ~ 700 nm, similar to physiological values ranging from 500 nm to 3 μm corresponding to fiber densities of ~10,000-30,000/mm^2 [3]. To further recapitulate *in vivo* environments in our *in vitro* models, we have modified PCL EFs with collagen-I (Col-I), collagen-IV (Col-IV), Hyaluronic acid (HA), and myelin basic protein (MBP)), representative of the blood vessel ECM, native brain ECM, and white matter ECM, respectively [4]. In addition, we have also created a model of white matter consisting of fibers myelinated by oligodendrocyte cells (myelinating cells of the central nervous system).

Aligned nanofiber multi-well plates were obtained from NanoFiber Solutions (Columbus, OH). Protein solutions of Col-I, IV and MBP (100 μg/mL) and HA (5 mg/ mL) were deposited on aligned fibers by simple adsorption. The presence of proteins on aligned fibers was quantified using immunofluorescence and that of HA confirmed using FITC-labeled HA. To complement protein adsorption, we also investigated EDC-mediated chemical conjugation techniques and quantified protein presence using micro-BCA assays. We are currently investigating GBM migration rates on surface modified aligned fibers using both techniques and examining specific chemistries that promote/hinder migration. In addition to protein modification of aligned fibers, we are also culturing oligodendrocytes on aligned fibers to create myelinated EFs as white matter mimetics. We have shown that oligodendrocytes align to PCL EFs and display myelination behaviors, including production of myelin basic protein (MBP). Myelinated fibers display remarkable similarities to native white matter. Ultimately, by using these biomimetic substrates, we hope to gain a better fundamental understanding of GBM migration mechanisms, eventually leading to discovery of better therapeutics.

References:

[1] P.Y. Wen et al., N Engl J Med. 2008. 359 (5); 492-507.

[2] A.C. Bellail et al., Int J Biochem Cell Biol. 2004. 36 (6); 1046-1069.

[3] M.K. Makino et al., Spine. 1996. 21(9): p. 1010-1016.

[4] A Giese and M Westphal., Neurosurgery. 1996;39(2):235-50; discussion 50-2.

See more of this Session: Surface Modification and Characterization of Biomaterials
See more of this Group/Topical: Materials Engineering and Sciences Division

(123b) Synthetic Polypeptide Macromers with Tunable Secondary Structure: Components of a Hydrogel Toolkit for Modeling Cell-Matrix Interactions

Abigail M. Oelker[1], Linda G. Griffith[2] and Paula T. Hammond[1], (1)Chemical Engineering, Massachusetts Institute of Technology, Cambridge, MA, (2)Biological Engineering, Mechanical Engineering, Massachusetts Institute of Technology, Cambridge, MA

Cellular functions including DNA synthesis, migration, and differentiation are co-regulated by both the stiffness of the extracellular matrix (ECM) and the presence of soluble signaling molecules (growth factors, cytokines, hormones, etc.). The purpose of this research is to synthesize and characterize a set of readily functionalizable synthetic polypeptides with tunable size and secondary structure as part of a hydrogel toolkit for modeling cell-matrix interactions. The polypeptides of interest were synthesized *via* ring-opening polymerization of N-carboxyanhydride derivatives of γ-propargyl L-and/or D-glutamate. These macromers exhibit low polydispersity and become water soluble after a highly-efficient "click" cycloaddition reaction between the alkyne moieties along the polymer backbone and azide-terminated poly(ethylene glycol) (PEG) chains. Circular dichroism and FTIR spectroscopy data demonstrate that stereoregular polyglutamate (containing only L-glutamate monomers) exhibits α-helix secondary structure in solution, whereas polyglutamate composed of both D and L monomers adopts a random coil conformation. This level of control allows us to design hydrogels with stiff rod-like segments or flexible coil segments, thus possibly impacting mechanical and other physicochemical properties of the gel. The grafted polypeptides form hydrogels upon reaction with a hetero-bifunctional cross-linker (PMPI) and thiol-terminated PEG. We are currently investigating the relationships between polypeptide structure and composition, as well as hydrogel formulation (i.e. mass fraction, cross-link density) on the physicochemical properties of the resulting hydrogels. These characterization data will be used to create a library of hydrogel substrates with systematically and independently varied stiffness, permeability, and ligand presentation for *in vitro* study of the fundamental parameters of cell-matrix interactions.

See more of this Session: Surface Modification and Characterization of Biomaterials
See more of this Group/Topical: Materials Engineering and Sciences Division

(123c) Development of Biocompatible Polymers for Modulating Lipid Membrane Integrity

Jia-Yu Wang[1], Brent Hammer[2], Sangram Parelkar[2], Jeremy Marks[3], Todd Emrick[2] and Ka Yee Lee[1], (1)Chemistry, The University of Chicago, Chicago, IL, (2)Department of Polymer Science and Engineering, University of Massachusetts, Amherst, IL, (3)Pediatrics, The University of Chicago, Chicago, IL

The cell membrane is in essence an integral lipid bilayer decorated by proteins and carbohydrates. It separates the interior of a cell from the exterior environment and serves as a permeable barrier by control over the passage of large molecules and ions in and out of the cell. Thus, lipid membrane integrity is essential for the function of the cell. Research studies have shown that a family of PEO-PPO-PEO triblock copolymers, also known as poloxamers, actively interact with lipid membranes but can have opposite effects on membrane integrity - depending on their architecture, they can behave either as membrane sealants in the sealing of injured cell membranes or as permeabilizers to promote drug intake for multidrug-resistant cancer cells. It is intriguing that these seemingly opposite biological effects can come from the interaction between biological cell membranes and the same family of triblock copolymers. However, no clear relationship between architectures and biomedical functionality of the copolymers has been established due to the lack of thorough understanding of the mechanism behind poloxamer-cell membrane interactions at the molecular level. In this work, we elucidate the origin of the poloxamer's biomedical functionality by investigating interactions between PEO-PPO-PEO triblock copolymers and lipid membranes using a suite of biophysical techniques. Our results indicate that the overall hydrophobicity of the triblock copolymer plays an important role in determining its ability to disrupt or preserve the integrity of lipid membranes. To test the universal effect of polymer hydrophobicity, we have further examined polymer-lipid membrane interactions using a new series of poloxamer-mimetic triblock copolymers, PMPC-PPO-PMPC. We discovered that this family of copolymers exhibits a similar interaction trend to poloxamers. As PMPC does not share any common chemical feature with PEO except that it is also highly hydrophilic - in fact, it is much more hydrophilic than PEO, we propose that it is the overall hydrophobicity of the polymer, rather than the exact chemical structure of both blocks, drives the transition of the polymer from being a membrane permeabilizer to a membrane sealant. Insights gained from this study should aid the design and development of novel copolymers as therapeutics.

See more of this Session: Surface Modification and Characterization of Biomaterials
See more of this Group/Topical: Materials Engineering and Sciences Division

(123d) Investigation of Stem Cell Responses to Dynamic Surface Topography

Murat Guvendiren and Jason A. Burdick, Bioengineering, University of Pennsylvania, Philadelphia, PA

Chemistry, mechanics, and topography of a biomaterial strongly regulate the interaction of cells such as attachment and proliferation. Specifically, topographical cues are shown to strongly influence nuclear organization/focal adhesions and alter cell alignment and hence function. However, the majority of reports on surface topography have focused on static micro/nano-patterns and the use of dynamic patterns is limited, potentially due to limited material systems for such studies. However, the native tissue environment is dynamic and dynamic changes in the microenvironment play critical roles in many biological events such as tissue regeneration, wound healing, fibrosis and tumor progression. In this work, we developed a dynamic PDMS patterning system and investigated the effect of pattern switching on nuclear orientation and stem cell behavior. PDMS surface patterns were prepared via strain induced buckling by stretching a surface and exposing to UV ozone for 30 min. In the stretched state the PDMS was flat, whereas lamellar patterns formed when the strain was released. hMSCs attached and spread randomly on the pre-stretched films; however, cells attached and aligned themselves to track the pattern shape on the pre-released substrates. After 1 day, cells had a random range of nuclei orientation from 10° to 90° for pre-stretched substrates whereas for pre-released substrates almost 80% of the cells had nuclei orientation $\leq 10^{o}$, which remained constant after 7 days. Dynamic studies (switching after 1, 3, and 7 days of culture) showed that the nuclear orientation of the cells on the pre-released films changed significantly from aligned to completely random when the film was stretched and that nuclear orientation went from random to aligned when stretched films were released. This study demonstrates that the strain responsive PDMS surface patterns can be used as a powerful tool to dynamically control the cell and nuclear alignment. This dynamic control over cell nuclei provides a novel platform to investigate the relationships between cellular differentiation and nuclear morphology. Ongoing long term studies are enabling us to study the influence of matrix generation of aligned and random cells, and the effect of switching on pre-formed matrix.

See more of this Session: Surface Modification and Characterization of Biomaterials
See more of this Group/Topical: Materials Engineering and Sciences Division

(123e) *In Vitro* Effects of Polyurethane Composites with Surface Modified Fillers On the Differentiation of Osteoclastic Cells

Edna Margarita Prieto, Department of Chemical and Biomolecular Engineering, Vanderbilt University, Nashville, TN, Erica Leah Von Stein, Department of Biomedical Engineering, Vanderbilt University, Nashville, TN and Scott A. Guelcher, Chemical and Biomolecular Engineering, Vanderbilt University, Nashville, TN

Injectable polyurethane(PU)/mineralized-filler composites offer a minimally invasive procedure for treating bone defects. In order to remodel effectively, these materials must undergo coupled cellular resorption and new bone deposition after implantation. PU/mineralized-filler composites have been shown to support new bone deposition. However, since the first phase of remodeling involves osteoclastic resorption, the objective of the present work was to study the effect of PU composites with surface modified fillers on the differentiation and function of osteoclastic cells *in vitro*. It was hypothesized that differences in the scaffold's characteristics (partly determined by the interactions between the filler and the matrix) could affect the cellular behavior. Allograft bovine bone and b-tricalcium phosphate (b-TCP) were surface modified to promote interactions with the PU phase. Allograft bone was surface demineralized, and polycaprolactone (PCL) was grafted on the surface of b-TCP particles. No significant differences were observed between the mechanical behavior of the bone composites, while the grafted PCL on b-TCP significantly improved the mechanical properties of the composites compared to the unmodified b-TCP. Osteoclast precursor cells were purified from mouse bone marrow cells and seeded on the composites and dentin (control). The cells were fed with macrophage colony stimulating factor (MCSF) and receptor activator of nuclear factor kB ligand (RANKL) and cultured up to 21 days. Differentiation rates were monitored by staining the samples for tartrate resistant acid phosphatase (TRAP) at days 5,13, and 21. Differences between the differentiation rates of the osteoclastic-like cells seeded on composites with un-modified vs. modified fillers were identified suggesting that the surface chemistry of the fillers impacts the cellular behavior. This information provides insight into the formulation of bioactive PU composites that actively participate in the bone healing process.

See more of this Session: Surface Modification and Characterization of Biomaterials
See more of this Group/Topical: Materials Engineering and Sciences Division

(123f) Targeting of Antioxidant Polymer Nanoparticles for Inhibition of Metal Nanoparticle Toxicity

David Cochran, Dr. Kimberly Anderson, Richard Eitel and Dr. Thomas Dziubla, Chemical and Materials Engineering, University of Kentucky, Lexington, KY

Vascular oxidative stress is a key pathological process in a variety of diseases (e.g., ischemia – reperfusion injury and hypoxia). Oxidative stress is characterized by the formation of a wide range of reactive oxygen species (e.g., super oxide, hydrogen peroxide, and peroxy lipids), which can cause severe DNA, protein, and lipid damage leading to cellular dysfunction and death. It is possible to suppress this injury through the addition of free radical scavengers, which can intercept oxidation of cellular components and thereby attenuate this damage. We have previously presented on the ability to formulate antioxidant nanoparticles with a platelet endothelial cell adhesion molecule (α-PECAM-1) antibody coating that can suppress oxidative stress in a static environment. In this work, we further characterize this active targeting drug delivery carrier. We studied the ability of these nanoparticles to suppress oxidative stress in an injury model (Nano-Co induced stress), along with the effective duration of protection post trauma. We have also characterized the ability of cells to internalize these antioxidant nanoparticles through the use of fluorescent antibody staining. The results obtained demonstrate the capability of this targeted antioxidant drug delivery system to suppress the formation of reactive oxygen species in injuries.

See more of this Session: Surface Modification and Characterization of Biomaterials
See more of this Group/Topical: Materials Engineering and Sciences Division

(123g) Synthesis and Characterization of Elastin-Like Polypeptide and Polyelectrolyte Conjugates for Tissue Engineering

Paul A. Turner and **Amol V. Janorkar**, Biomedical Materials Science, University of Mississippi Medical Center, Jackson, MS

Steatohepatitis and other metabolic disorders are a growing problem across the United States, contributing to a wide range of pathologies such as stroke, atherosclerosis, and progressive liver failure. Attempts to construct an adequate *in vitro* model simulating metabolic dysfunction have been inadequate due to brief survival of explanted hepatic and adipose cells and lack of physiologically-relevant behavior believed to be related to lack of appropriate cellular organization. It is hypothesized that cells arranged into three-dimensional spheroids will show enhanced activity and survival versus those grown on conventional two-dimensional cultures. In the current study, elastin-like polypeptide (ELP) was used as a substrate material due to its biocompatibility, tailored physiochemical properties, and ease of chemical modification. Polyelectrolytes including polyethyleneimine (PEI), poly-L-lysine, and poly-L-arginine were covalently bound to ELP by amide coupling reactions and characterized using a variety of qualitative and quantitative techniques such as DSC, FT-IR, HPLC, SDS-PAGE, and ninhydrin assay. Tissue culture polystyrene plates were coated with the ELP-PEs and seeded with 3T3-L1 and H35 rat hepatoma cells to assess the cell response to the modified surfaces. Modifying the charge of the ELP molecules via conjugated polyelectrolytes was found to induce cultured hepatic and adipose cells to form three-dimensional spheroids relative to the histological organization found *in vivo*.

See more of this Session: Surface Modification and Characterization of Biomaterials
See more of this Group/Topical: Materials Engineering and Sciences Division

(142a) Enhancement of Electrospun Polysulfone Mats Using Biocidal Nanomaterials

Jessica D. Schiffman[1], Yue Wang[2], Emmanuel P. Giannelis[2] and Menachem Elimelech[1], (1)Chemical and Environmental Engineering, Yale University, New Haven, CT, (2)Materials Science and Engineering, Cornell University, Ithaca, NY

Contamination of synthetic polymer surfaces by microorganisms is of great concern, especially for wound healing scaffolds, medical devices, and hospital equipment. Hence, there exists a high demand for antibacterial coatings that are flexible and highly porous. Single-walled carbon nanotubes (SWNTs) and silver (Ag) nanoparticles have demonstrated toxicity to a variety of microorganisms. In this work, by utilizing the electrospinning process, various concentrations of purified SWNTs have been immobilized in polysulfone (PSf) non-woven mats. The development of a second system wherein pure PSf fiber mats are plasma functionalized and subsequently surface modified with synthesized Ag nanoparticles will also be presented. The as-received and purified SWNTs, synthesized Ag nanoparticles, as well as the composite mats have been well characterized. Results on the interaction of the composite mats with bacterial cells will be presented directly linking bacterial cytotoxicity to nanomaterial availability.

See more of this Session: Biomaterials II
See more of this Group/Topical: Materials Engineering and Sciences Division

(142b) Materials-Based Approach to Implant Infection Mitigation

Ann O'Toole and Eric Nuxoll, Chemical and Biochemical Engineering, University of Iowa, Iowa City, IA

Bacterial biofilm infection is a common (~ 2 to 4%) complication for recipients of surgically implanted medical devices. Due to the tremendous chemical resistance of these biofilms, present treatment is a combination of antibiotics and replacement of the device, often requiring multiple surgeries. The objective of the current study is to determine the feasibility of using heat to deactivate biofilm as an alternative therapy. It is hypothesized that heat can be used to thermally eradicate the biofilm. In this study, Psuedomonas aeruginosa biofilm was cultured (37°C, 5% CO2 for 72 hours using glucose-enhanced media) and subjected to heat shocks. The live cell count per area was quantified for both control and heat shocked samples using confocal fluorescent microscopy. Decreased live cell count per area correlated to biofilm deactivation due to the heat shock and established the feasibility of thermal deactivation of infectious biofilms. Nanoparticle composite implant coatings for precise delivery of this heat via magnetic induction will be discussed.

See more of this Session: Biomaterials II
See more of this Group/Topical: Materials Engineering and Sciences Division

(142c) On the Design of Biocompatible Zwitterionic Surfaces: Does Chaotrope Have Lower Friction Than Kosmotrope?

Yi He, Qing Shao and Shaoyi Jiang, Chemical Engineering, University of Washington, Seattle, WA

We examine the frictional properties of zwitterionic surfaces and explore whether chaotropic or kosmotropic charged groups are preferred to achieve low friction. Carboxybetaine self-assembled monolayers (CB-SAMs) and sulfurbetaine (SB)-SAMs are used as model surfaces as they contain the same positively charged group, but different negatively charged ones. The negatively charged groups are kosmotropic carboxylates in the CB-SAM surfaces and chaotropic sulfonate groups in the SB-SAM surfaces, respectively. The results show that the friction of the SB-SAM surfaces is lower than that of the CB-SAM surfaces. This suggests that chaotropic charged groups are better in reducing friction than kosmotropic groups. The lower friction of the SB-SAM surfaces over the CB-SAM can be explained by the higher mobility of water near the SB-SAM surfaces, as shown in the survival autocorrelation function and the dipole autocorrelation function of hydration water molecules.

See more of this Session: Biomaterials II
See more of this Group/Topical: Materials Engineering and Sciences Division

(142d) Impact of NaCl Stress On the Mechanical and Material Properties of Staphylococcus Epidermidis Biofilms

Michael J. Solomon[1], Mahesh Ganesan[1], Leonid Pavolvsky[1], Elizabeth J. Stewart[2] and John G. Younger[3], (1)Chemical Engineering, University of Michigan, Ann Arbor, Ann Arbor, MI, (2)Chemical Engineering, University of Michigan, Ann Arbor, MI, (3)Department of Emergency Medicine, University of Michigan, Ann Arbor, MI

Biofilms are aggregates of bacteria encapsulated with a protective layer of extracellular polysaccharide (EPS). *Staphylococcus epidermidis* biofilms are frequent in hospital-related infections associated with contaminated medical devices. Studies have shown that NaCl can contribute to the overexpression of the ica operon, which plays a role in the production of EPS (Conlon, K.M., J. Bacteriology, 2002). This study determined the effect of osmotic stress imparted by high concentrations of NaCl on material and mechanical properties of *S. epidermidis* RP62A bacterial biofilms.

We probed the intercellular microstructure by analysis of 3D image volumes acquired by confocal laser scanning microscopy. We observed that increasing NaCl concentration significantly decreased the cellular density of biofilms grown in a 0.011 Pa shear environment. Parallel plate rheology was used to determine the mechanical properties of *S. epidermidis* biofilms under conditions of both linear and non-linear deformations. A non-monotonic effect of NaCl on the magnitude of the elastic and vicsous moduli of the biofilms was observed. Size-exclusion chromatography coupled with multi-angle laser light scattering and refractive index detection was used to measure the molar mass distribution of the EPS from biofilms grown under different NaCl concentrations. The relationship among the EPS molar mass, the intercellular structure, and the biofilm rheology is discussed so as to determine the structural and polymeric origins of the observed salt stress effects on biofilm mechanical properties.

See more of this Session: Biomaterials II
See more of this Group/Topical: Materials Engineering and Sciences Division

(142e) Investigations of the Strengths of Elliptical Nanoparticles In Drug Delivery Application

Mei-Hsiu Lai, Chemical and Biochemical Engineering, U of Illinois, Urbana-Champaign, Urbana, IL, Jae Hyun Jeong, Chemical and Biomolecular Engineering, University of Illinois, Urbana, IL and Hyun Joon Kong, Chemical and Biomolecular Engineering, University of Illinois Urbana-Champaign, Urbana, IL

Nanoparticles modified with bioactive motifs binding with specific proteins are becoming advanced drug carriers enabling targeted drug delivery. However, it is still challenging to enhance the mobility of nanoparticles in circulation, so the particles can quickly reach target sites. Previous studies reported that microparticles with an elliptical shape presents the better targeting efficiency than spherical microparticles because of the enhanced mobility. This important role of particle morphology can be extended to nanoparticles, but this hypothesis has not been examined to dates, because of the lack of technique to prepare nanoparticles with elliptical shape.

This study therefore presents a new method to control the morphology of nanoparticles using polymers self-assembled to form vesicles. The poly(N-hydroxyethyl-DL-aspartamide) (PHEA) was modified with pedant alkyl chains and poly(ethylene glycol) molecules to tune the aspect ratio of the nanoparticle, as confirmed with electron microscope. As expected, the resulting elliptical nanoparticles presented the higher diffusivity than the spherical ones. The elliptical PHEA nanoparticles further modified with oligopeptides containing Arg-Gly-Asp (RGD) sequence exhibited the enhanced binding efficiency to target cells, because of the increase of association rate and decrease of dissociation rate, according to surface plasmon resonance (SPR) spectroscope. Overall, this study serves to provide an invaluable biomedical tool to improve the efficiency of targeted delivery of a wide array of drug molecules.

See more of this Session: Biomaterials II
See more of this Group/Topical: Materials Engineering and Sciences Division

(142f) Injectable Thermosensitive Hydrogels Containing Cytokines for Modulating Dendritic Cell-Derived Immune Responses

Kye Il Joo, Liang Xiao and Pin Wang, Chemical Engineering and Materials Science, University of Southern California, Los Angeles, CA

Development of an effective vaccine and immunotherapy technology for cancer involves eliciting a potent tumor antigen-specific immune response by introduction of tumor antigens and generating systemic immunity against the tumors. Dendritic cells (DCs) are the most powerful antigen-presenting cells, capable of initiating immune response by stimulating both T cells and B cells, and much effort has therefore been devoted to methods for delivering appropriate antigens to DCs to develop effective vaccines. Immunization by adoptive transfer of autologous DCs that are loaded with antigens in vitro is, however, labor-intensive and requires significant regulatory concerns. In addition, most transplanted DCs remain non-functional, and few home to the lymph nodes where the subsequent activation of T-cell immunity is induced. More cost-effective and efficient protocols involve delivering antigen to host DCs in vivo by direct injection of vaccine vectors to program DCs in situ. Thus, the ability to deliver antigens to host DCs in vivo is a key step towards generating a potent antitumor immunity.

In this study, we demonstrate a general strategy to enhance antigen-uptake by DCs in vivo. It depends on recruiting and housing host DCs at the immunization injection site using functionalized biomaterials incorporating inflammatory cytokines. This approach offers the sustained and localized release of the cytokine granulocyte-macrophage colony-stimulating factor (GM-CSF), which has been identified as a potent stimulator of DC recruitment and proliferation, from the subcutaneously injected hydrogels. The hydrogel materials provide a controllable subcutaneous microenvironment where host DCs could be recruited and further educated in situ by following injection of immunogens (antigens) at the recruited site. Thermo-sensitive mPEG-PLGA hydrogels capable of solution-gel transition through changes in temperature enable greater efficiency of drug/cytokine loading and also exhibit a sustained and localized release of GM-CSF in the target site. The enhanced antigen-specific immune response was observed when viral vector-based vaccine carriers were administered at the site implanted with these engineered hydrogel biomaterials.

See more of this Session: Biomaterials II
See more of this Group/Topical: Materials Engineering and Sciences Division

(162a) Dramatic Effects of Nanoscale Confinement On Glass Transition Temperature (Tg) In Polymers: Perturbing the Tg by 50-100 K In Supported and Free-Standing Films and by the Presence of Neighboring Layers of Other Polymers

John M. Torkelson, Departments of Chemical and Biological Engineering, Materials Science and Engineering, Northwestern University, Evanston, IL

Using the temperature dependence of fluorescence intensity intrinsic to a polymer of interest or from dyes labeled at trace levels to a polymer, we show that Tg can be altered from bulk response by as much as 50-100 K via nanoconfinement in substrate-supported and freely standing films. Similarly large perturbations to polymer Tg can be observed in nanoconfined multilayer films of different polymers or blends with nanospheres of one polymer dispersed in a matrix of a second polymer.

In particular, we shall describe how supported homopolymer films may exhibit positive or negative deviations from bulk Tg response depending on the relative strengths of the effects of the polymer-air interfaces, which typically reduce Tg, and attractive interactions at polymer-substrate interfaces, which increase Tg. We shall also show how the Tg perturbations in free-standing homopolymer films (without a substrate) can be understood, at least in part, from the free-surface effects characterized in substrate-supported films. We shall further demonstrate how the perturbations that occur at interfaces lead to a gradient of Tg response some tens to hundreds of nanometers into the film interior. Additionally, we shall describe how fragility (as a glass former) may be important in the strength of the Tg-confinement effect, both in terms of the Tg change exhibited at a particular thickness as well as the thickness at which the average film Tg deviates from bulk response.

We shall also denote behavior that has resulted in new questions related to the fundamental nature of glass-forming materials. We have found via multilayer film studies that an ultrathin layer of one polymer species can have its glass transition behavior strongly modified by or even “slaved” to that of neighboring domains of other polymers. Similar results have been obtained in dilute blends of an A-B diblock copolymer in a B homopolymer, leading to nanospheres of the A block being dispersed in B-phase matrix. This indicates that multilayer films and nanodispersions of different, immiscible polymers can yield properties that are not possible with conventional polymer blends.

See more of this Session: In Honor of Matthew Tirrell's 60th Birthday III

(162b) Non-Equilibrium (rate, time, hysteretic, and history-dependent) Interactions In Polymer and Biopolymer Systems

Jacob Israelachvili, Department of Chemical Engineering, University of California at Santa Barbara, Santa Barbara, CA

...

See more of this Session: In Honor of Matthew Tirrell's 60th Birthday III
See more of this Group/Topical: Materials Engineering and Sciences Division

Unavailable Paper

(162c) How to Hit HIV Where It Hurts

Arup K. Chakraborty, Departments of Chemical Engineering, Chemistry, and Biological Engineering, Massachusetts Institute of Technology, Cambridge, MA

HIV is a scourge on the planet, particularly in poor countries. A vaccine is urgently needed to address this challenge. I will describe work that brings together methods from statistical physics with biological experiments and human clinical data that aims to identify targets for vaccine-induced immune pressure to which HIV may be most vulnerable.

See more of this Session: In Honor of Matthew Tirrell's 60th Birthday III
See more of this Group/Topical: Materials Engineering and Sciences Division

(162d) Protein Analogous Micelles

Matthew Tirrell, Institute for Molecular Engineering, University of Chicago, Chicago, IL

Peptides are functional modules of protein macromolecules that can be displayed apart from the whole protein to create biofunctional surfaces and interfaces, or can be re-assembled in new ways to create synthetic mimics of protein structures. Each of these routes are being employed to gain new insight into protein folding and to develop new, functional, biomolecular materials. Examples of work from our laboratory in this area using peptide-lipid conjugate molecules (peptide amphiphiles) will be discussed relating to multi-functional interfaces, DNA-binding peptide assemblies, and protein analogous micelles for cancer and cardiovascular therapeutics. This work has been pursued in collaboration with numerous people at the University of Minnesota, UC Santa Barbara and UC Berkeley, including: Gregg Fields, Erkki Ruoslahti, Frank Bates, Ishi Talmon, Peter Berndt, Efie Kokkoli, Angie Dillow, Havazelet Bianco-Peled, Shiv Chiruvolu, Markus Biesalski, Jim Schneider, Yoav Dori, Sanjay Kumar, Kraig Haverstick, Sarah Ochsenhirt, Tushar Gore, Teika Pakalns, Ray Tu, Alex Chu-Kung, Nathan Lockwood, Dimitris Stroumpoulis, Tomoko Shimada, Mark Kastantin, Badri Ananthanarayanan, Dimitris Missirlis, Won Huh, Matt Kade, Matt Black, Rachel Marullo, Katie Megley, Amanda Trent, Dan Krogstad and Laurie Drews.

See more of this Session: In Honor of Matthew Tirrell's 60th Birthday III
See more of this Group/Topical: Materials Engineering and Sciences Division

(171a) Drug Loading Into and Drug Release From pH- and Temperature-Responsive Cylindrical Hydrogels

Pravin Ninawe and **Satish J. Parulekar**, Chemical and Biological Engineering, Illinois Institute of Technology, Chicago, IL

Hydrogels that undergo deformation upon appropriate changes in pH or temperature have considerable promise as drug delivery vehicles. Drug uptake in swelling and non-swelling cylindrical hydrogels and drug release from these into a target fluid are investigated here. A mathematical model for hydrogel-solution composite, a composite of a distributed parameter system (cylindrical hydrogel) and a lumped parameter system (surrounding solution), is developed. The polymer network displacement in a swelling/deswelling hydrogel is described by a stress diffusion coupling model. The analytical solution for network displacement is used to predict solvent intake by swelling hydrogels, solvent efflux from deswelling hydrogels, and changes in pressure, porosity and effective drug diffusivity. These in turn influence drug uptake during and after hydrogel swelling and drug release from hydrogel during and after deswelling. Numerical results illustrate benefits of hydrogel swelling for drug loading and merits of different modes of drug release. Drug uptake and drug release by temperature-responsive hydrogels are compared to those by hydrogels not subject to deformation.

See more of this Session: Modeling and Simulation of Polymers II
See more of this Group/Topical: Materials Engineering and Sciences Division

(171b) Solvent Effects On Hysteresis In the Coil-Stretch Transition

Rangarajan Radhakrishnan and **Patrick T. Underhill**, The Isermann Department of Chemical and Biological Engineering, Rensselaer Polytechnic Institute, Troy, NY

The response of polymer solutions in elongational flow is important for many applications. One key feature of flexible polymers in elongational flow is the sharp transition from a coiled state to stretched state at a critical strain rate. It is important to understand how this transition is altered by polymer-solvent interactions and conformation dependent drag. In theta solvent conditions, conformation dependent drag leads to hysteresis in the transition.

We have recently developed a model of polymer elasticity that includes solvent effects in a new way. Using Brownian Dynamics simulations of this model, we have predicted that some polymers could have dramatically different behavior due to solvent effects; the coil-stretch transition can be eliminated under some conditions. This new model allows us to better understand the importance of molecular flexibility and entropic elasticity in determining the flow response. We will discuss how our model can be used to understand how conformation dependent drag affects the coil-stretch transition in good solvents, comparing a dumbbell model and a bead spring chain model including hydrodynamic interactions between the beads.

See more of this Session: Modeling and Simulation of Polymers II
See more of this Group/Topical: Materials Engineering and Sciences Division

(171c) Structure and Transport Properties of Nohms

Sushmit S. K. Goyal, Chemical Engineering, Cornell University, Ithaca, NY, Tuvalu and Fernando Escobedo, Department of Chemical and Biomolecular Engineering, Cornell University, Ithaca, NY

Structure and transport propertiesof Nanoparticle organic hybrid materials

Polymer nanocomposites have been a topic of interest in recentyears for their potential in such applications as water desalination, CO_2capture, photovoltaics and immersion lithiography [1-3]. Unlike colloids which tendto agglomerate irreversibly, polymer grafted colloids are stabilized bypolymer-polymer steric interactions. Nanoscale organic hybrid materials (Nohms)are a class of such materials which consist of an inorganic nanoparticlecore, functionalized with a corona of organic oligomers.These differ from common nanocomposites in that thetethered corona is the suspending medium for the cores. [1] Thehybrid nature of the suspension allows the fabrication of materials with tunableproperties by varying parameters such as nanoparticlechemistry, shape and size, as well as the polymer molecular weight, graftingdensity and chemistry. The range ofproperties exhibited by these composites vary from solids, stiff waxes, andgels for high core content to single component solvent free fluids for low corecontent.

Molecular simulations can help elucidate howthe properties of Nohms are affected (and can beoptimized for specific applications), by changes in molecular design. To thisend, we have used Molecular Dynamics (MD) to get a better understanding of thestructure and transport properties of these systems.

In this work, the translational and rotationaldiffusivity for pure Nohms was simulated viaequilibrium MD. Diffusivities are normalized by the corresponding values fromStokes-Einstein (translation) and Stokes-Einstein-Debye (rotation) relationsfor naked nanoparticles suspended in a melt of freechains. The simulated systems were chosen to mimic experimental systems using silicacores and PEO chains (having high grafting density and short chains[2])and to isolate the contributions of core and corona on Nohmsdynamics by varying the core volume fraction (f=*volume of nanoparticletotal volume* $\phi = \frac{volume\ of\ nanoparticle}{total\ volume}$) and polymer length.Non-equilibrium MD methods were implemented to obtain viscosities (by imposinga homogeneous steady state shear rate) and yield stresses. A non-Newtonianshear thinning behavior is observed in all cases along with reduction in yieldstress with decreasing f, a behavior partially consistent with experimental data.[2] Corevolume fraction affects therelative viscosity significantly; e.g., for f =0.2, theviscosities are significantly higher than that of free chains (pentamers) and those of the f=0.1 Nohms. Comparing systems with identical f=0.1, the Nohms with longer chain length have lower viscosities(closer to that of the melt of free with has the lowest viscosities).Altogether these observations show that the dynamics of the corona grows inimportance (relative to that of the cores) in Nohmsas f decreases andchain length increases. The diffusivity results indicate that the system withthe highest f (=0.2) has the smallest translational mobility, andfor the systems with equalf (=0.1) the Nohms with thelonger chains (20-mers) has higher translational motion and less structuredcores. Changes in the rotational diffusivity are less significant among the allcases examined, which could be due to a comparable effective friction resistingrotation.

We also observe a lower shear thinning slope for Nohms+polymer blends than for pure Nohms.Structural analyses reveal that preferential alignment of grafted polymers inthe direction of shear is one of the dominant phenomena associated with theshear thinning behavior. Longer chains tend to align more readily causing amore pronounced reduction in viscosity, consistent with the observed trends inviscosity. Free polymers are found to align the least, partly explaining theweaker shear thinning seen for blends (compared with pure Nohms). Ongoing work aims to provide a more completecharacterization of how *f*, graftingdensity, and polymer length can be tailored to produce Nohmswith a wide range of themophysical properties.

References

1. Bourlinos,A. B., Chowdhury, S. R., Herrera, R., Jiang, D. D., Zhang, Q., Archer, L. A.,et al. (2005). FUnctionalized Nanostructures with Liquid-Like Behavior :Expanding the Gallery of Available Nanostructures. *Advanced FunctionalMaterials* , *15*, 1285-1290.

2. Agarwal,P., Qi, H., & Archer, L. A. (2010). The Ages in a Self-SuspendedNanoparticle Liquid. *Nanoletters* , *10*, 111-115.

3. Rodriguez, R., Herrera, R., Lynden,A. A., & Giannelis, E. P. (2008). Nanoscale Ionic Materials. *Advanced Materials* , *20*,4353-4358.

See more of this Session: Modeling and Simulation of Polymers II
See more of this Group/Topical: Materials Engineering and Sciences Division

(171d) Dynamics of the Directed Assembly of Block Copolymers: The Influence of Entanglements

Abelardo Ramirez-Hernandez and Juan J de Pablo, Department of Chemical and Biological Engineering, UW-Madison, Madison, WI

The understanding of the kinetics of microphase ordering of block copolymers is important for controlling the morphology of these materials. In high molecular weight homopolymer melts, the entanglements have a strong influence on the dynamical properties of the melts, thus their influence is expected to be important on the kinetics of ordering of multipolymeric systems. Much of our current understanding of the equilibrium morphologies of block copolymers has emerged from studies using Self-Consistent Field Theory (SCFT), in which the effect of non-crossability of chains is not taken into account. In this work, we use a particle-based coarse grained model of block copolymers, and introduced slip-springs to model the effect of entanglements on the dynamics of the melts. The model is able to reproduce both Rouse and Entangled dynamical behavior of homopolymeric melts. We apply our computational approach to study the influence of entangled dynamics on the the dynamics of the directed assembly of block copolymers driven by chemically patterned surfaces.

See more of this Session: Modeling and Simulation of Polymers II
See more of this Group/Topical: Materials Engineering and Sciences Division

(171e) Dissipative Particle Dynamics Studies of Surface Dynamics of Thin Polymer Films

Hsin-Lun Wu, Chemical Engineering, National Taiwan University, Taipei, Taiwan, Yu-Jane Sheng, Chemical Engineering Department, National Taiwan University, Taipei, Taiwan and **David T. Wu**, Chemical Engineering and Chemistry Departments, Colorado School Of Mines, Golden, CO

Thin films of polymers show capillary wave fluctuations that can be measured experimentally by X-ray Photon Correlation Spectroscopy, and which reveal the dynamics at the surface. These surface dynamics have been shown to be well-modeled by a hydrodynamic continuum theory (HCT) considering the film to act as a thin film of the bulk continuum fluid when the polymer is a linear chain. However, experiments have also suggested that chain tethering and non-linear chain architecture can modify the surface dynamics to differ from those expected from the HCT. We have performed Dissipative Particle Dynamics simulations to investigate the surface dynamics of thin films of both linear and ring polymers. By time-correlating the surface fluctuation amplitudes, we compare the relaxation times to the HCT, and demonstrate the role of non-linear architecture on the surface relaxation.

See more of this Session: Modeling and Simulation of Polymers II
See more of this Group/Topical: Materials Engineering and Sciences Division

(171f) Self-Consistent Field Theory of Correlations In Linear/Branched Polymer Blends

Renfeng Hu, Chemical Engineering Department, Colorado School Of Mines, Golden, CO and David T. Wu, Chemical Engineering and Chemistry Departments, Colorado School Of Mines, Golden, CO

Blends of a linear polymer and its branched equivalent have been shown experimentally to exhibit a tendency for the two species to avoid each other on a molecular scale. This tendency to segregate in the bulk results in density correlations that can be measured by scattering and also produces a demixing contribution to the blend free energy. We present studies using a self-consistent field (SCF) theory that allows calculation of these correlations, normally neglected in SCF theories, by computing the average density of species around a specified monomer. We compute intra- and intermolecular radial distribution functions, and demonstrate the role of swelling at crowded branched junctions. In a linear and star blend, the intramolecular monomer pair distribution function shows a crossover from self-avoiding walk to random walk that depends on the monomer density and excluded volume. As an extension of SCF calculation, the structure factors are compared with the random phase approximation, and can provided a reference for neutron scattering data analysis.

See more of this Session: Modeling and Simulation of Polymers II
See more of this Group/Topical: Materials Engineering and Sciences Division

(172a) High Density Memory Devices Using Self-Assembled Gold Nanoparticle Arrays As Floating Gates

Girish Muralidharan[1], Navakanta Bhat[2] and Venugopal Santhanam[1], (1)Department of Chemical Engineering, Indian Institute of Science, Bangalore, India, (2)Department of Electrical Communication Engineering and Centre for Excellence in Nanoelectronics, Indian Institute of Science, Bangalore

Floating gate memory devices find applications in many portable electronic devices such as mobile phones, cameras, laptops etc. Floating gate memory devices consist of field effect transistor with dual gate architecture, namely, floating gate and control gate for its memory characteristics. Depending on the presence or absence of electrons in floating gate, memory state can be read as 0 or 1. Programming or erasing (P/E) operation is performed by controlling the flow of electrons from or to the substrate through the tunneling oxide. Until now, miniaturization was accomplished by the reduction of feature sizes with concomitant reduction of tunneling oxide thickness. However, with the feature size reaching sub-25 nm, further device scaling is hampered by: (1) significant cell-to-cell interference as the spacing between the floating gates has now become comparable to memory cell size, and (2) non-scalability of tunneling oxide thickness due to unreliable charge retention after several program/erase cycles. Use of nanoparticles as charge-trapping nodes will aid in both minimizing interference and enhancing reliability through compartmentalization of charges. Metal nanoparticles, in particular, gold nanoparticle arrays as floating gate will result in realizing superior endurance and retention characteristics, as it offers deeper potential well in comparison to its semiconductor counterparts. Thus, manufacture of next generation floating gate memory devices entails process development for fabricating large scale (wafer level), high density (~ 10^{12} particle per cm^{-2}) ordered gold nanoparticle arrays.

Here, we present fabrication and electrical characterization of metal-oxide-semiconductor (MOS) capacitor using self-assembled gold nanoparticle arrays as floating gates. We will also show a simple and cost-effective approach for fabricating wafer scale (~ 3 inch), close-packed arrays of gold nanoparticle by drop-casting gold colloids containing small amounts of excess ligand molecules on water surface. We will present the device charging, endurance and retention characteristics using both capacitance-voltage and electrostatic force microscopic measurements. Salient features of the MOS capacitor include: (a) less device-to-device variability over cm scale (b) reduced power consumption (c) good endurance characteristics and (d) reliable charge retention characteristics. This approach can be readily integrated into existing fabricating facilities for manufacturing next generation non-volatile memory devices.

See more of this Session: Nanoelectronic Materials
See more of this Group/Topical: Nanoscale Science and Engineering Forum

EFFECTS OF VARYING SURFACE FILM THICKNESS ON PARTICLE ADHESION IN SEMICONDUCTOR MATERIAL-BASED SYSTEMS

Katie M. Smith, Purdue University, West Lafayette, Indiana, USA
Jeffery W. Butterbaugh, FSI International, Chaska, Minnesota, USA
Stephen P. Beaudoin, Purdue University, West Lafayette, Indiana, USA

Abstract

The effects of surface film thickness on particle adhesion are examined for a number of common microelectronics device materials. These effects are investigated for a laboratory-scale model system, using atomic force microscopy, as well as an industrial-scale system, via particle removal efficiency experiments. Our results show that surface films below a critical thickness δ^* are effectively invisible, from an adhesion perspective, while those having thicknesses greater than δ^* exhibit bulk material behavior, essentially screening any influence from the underlying substrate. In systems with surface films of thicknesses (approximately) equal to δ^*, the net adhesion results from a combination of film and substrate contributions.

Introduction

The adhesion of contaminant particles to wafer and photomask surfaces is a problem of critical importance to the manufacture of microelectronics (1–10). As device dimensions continue to shrink and restrictions on material loss become more stringent, many are approaching operating regimes where the dimensions of the devices in one or more spatial directions are on the same length scale as van der Waals-London (vdW-L) forces, the primary forces that control particle adhesion. There is no well-accepted methodology for describing adhesion forces in systems with compositional variations at these length scales. As a result, the microelectronics industry does not have a theoretical framework to guide the development of new surface processing or cleaning technology. Research elucidating the fundamental origins and mechanisms of particle adhesion is essential, particularly in the context of wafer contamination prevention and removal.

Of particular interest here is the effect of varying surface film thickness on particle adhesion to layered substrates. The interaction phenomena arising between bodies in contact (e.g., a contaminant particle and a wafer surface) are governed predominantly by vdW-L dispersion forces (11–15). These forces have their origins in the electrostatic interactions between the dipoles within materials (14). A variety of approaches has been developed to describe the vdW-L forces between macroscopic bodies; however some of these appear in the literature without considering the appropriate regions of validity for the systems of interest. Furthermore, limited studies are available that focus on the adhesion of *inhomogeneous, layered* particles or substrates (16). In this work, we develop theory supporting the hypothesis that the thickness of the surface film material will affect the resultant interaction force between particle and substrate. We then examine a model system consisting of a pre-manufactured Si_3N_4 probe—representing a contaminant particle—and a Si substrate onto which GaN films had been deposited via atomic layer deposition (ALD). Finally, we present results from industrial-scale experiments involving the removal of model contaminant particles of SiO_2 from the surface of 300mm-diameter blanket Si wafers with surface films comprised of nano-scale TiN of varying thickness deposited over SiO_2. These results demonstrate the significant effect of the thickness of the nano-scale surface film on the resulting particle adhesion.

Theory

Model Development

In this work, we have used Vold's approach (17), with minor modifications in order to account for the non-additivity in multi-layered, highly polarized systems at small separation distances (i.e. <25 nm). Consider two multi-layered, semi-infinite slabs (named **L** and **R**) consisting of any number of different materials; a schematic of such a system appears in Figure 1.

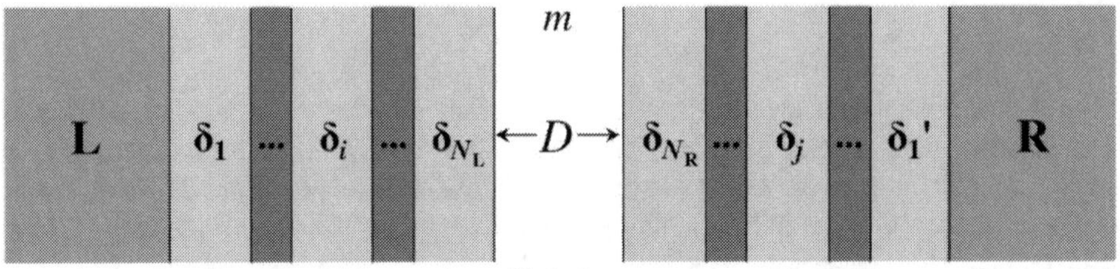

Figure 1. A schematic of two multi-layered parallel plates, separated by a distance *D*, interacting across a medium *m*.

The general expression for the vdW-L interaction energy between **L** and **R** is

$$E = -\sum_{i=1}^{N_L}\sum_{j=1}^{N_R} A(i,j)H_{ij} \qquad [1]$$

where N_L and N_R are the number of layers in the **L**- and **R**-side bodies; $A(i,j)$ is the 'effective' Hamaker constant for the interaction between any two layers *i* and *j*; and H_{ij} is a function that depends on the geometry of the layers as well as the distance between them. Derivations of H_{ij} for various ideal-geometry systems are available elsewhere (18). Modifying Equation [1] to correct for non-additivity and polarization effects, we arrive at

$$E_{\mathbf{L}\text{-}\delta_i\text{-}m\text{-}\delta_j\text{-}\mathbf{R}} = -\sum_{i=1}^{N_L}\sum_{j=1}^{N_R} A^{\text{eff}}(i,j)H_{ij} \qquad [2]$$

where $A^{\text{eff}}(i,j) = \alpha A(i,j)$ and α is a correction factor (19). For a system of layered materials, $A^{\text{eff}}(i,j)$ can be measured experimentally using instruments such as the surface forces apparatus (SFA) or atomic force microscope (AFM).

In this work, the idealized model system consists of a sphere {**1**} of radius *R* separated by a distance *D* across a medium {**0**} from a semi-infinite slab {**2**} having a surface film {**2'**} of thickness δ. See Figure 2 for a diagram of the idealized model system. The expression for the vdW-L interaction energy of this system is

$$E_{\{1\}\text{-}\{0\}\text{-}\{2'|2\}} = -\sum_{i=1}^{1}\sum_{j=1}^{2} A^{\text{eff}}(i,j)H_{ij} \qquad [3]$$

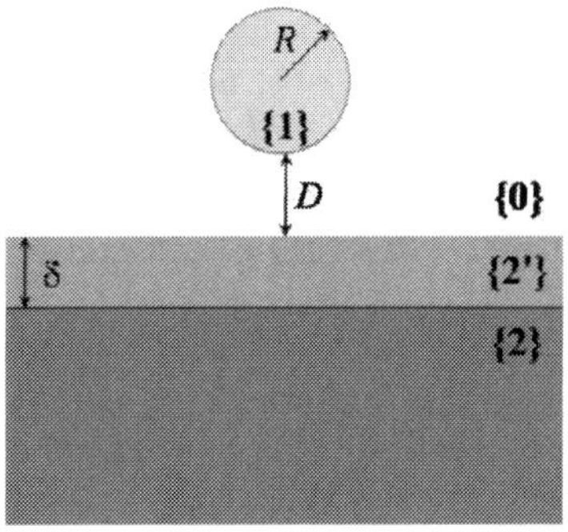

Figure 2. Diagram of the idealized sphere-coated flat plate system.

The vdW-L interaction force expression is obtained by differentiating Equation [3] with respect to separation distance D, which yields

$$F_{\{1\}\text{-}\{0\}\text{-}\{2'|2\}} = -\frac{\partial}{\partial D} E_{\{1\}\text{-}\{0\}\text{-}\{2'|2\}} = -\sum_{i=1}^{1}\sum_{j=1}^{2} A^{\text{eff}}(i,j)\frac{\partial}{\partial D}H_{ij} \qquad [4]$$

where $A^{\text{eff}}(i,j)$ is not affected by the differentiation as it is dependent only on the material properties of the multi-layer structure, which are taken to be unvarying. For the smooth sphere-flat plate system, it has been shown (18) that

$$\frac{\partial}{\partial D}H_{ij} = \frac{R}{6D_{ij}^2} \qquad [5]$$

where D_{ij} is the separation distance between the ith and j^{th} layers. Combining Equations [4] and [5], the final expression for the net vdW-L interaction force between a smooth sphere and a coated flat plate is given by

$$\frac{F_{\{1\}\text{-}\{0\}\text{-}\{2'|2\}}}{R} = \frac{1}{6}\left[A_{102'}^{\text{eff}}\left(\frac{1}{D^2} - \frac{1}{(D+\delta)^2}\right) + A_{102}^{\text{eff}}\left(\frac{1}{(D+\delta)^2}\right)\right] \qquad [6]$$

Critical Film Thickness

It is apparent from Equation [6] that in the limit of very small δ/D, the film will appear transparent to the particle such that as it approaches the substrate it will be interacting primarily with the underlying bulk material. However, as δ/D increases to the point where eventually $\delta >> D$, the dominant interaction transitions from that between the particle and the underlying substrate to that between the particle and the coating. This transition suggests the existence of three regimes of behavior: 1) the 'bulk substrate' region where the coating is so thin it is invisible to the approaching particle from an adhesion perspective; 2) the 'bulk coating' region where the coating is so thick that the adhesion is dominated by the properties of the coating; and 3) the 'transition region' where the adhesion is influenced by both the coating and the substrate. Coating thicknesses in the transition region are denoted by δ^*.

Consider a single spherical particle {**1**} approaching an isolated thin film {**2'**}, both of which are surrounded by a uniform medium {**0**}. The vdW-L interaction force between the particle and the film is obtained from Equation [6] by replacing material **2** with the medium **0** and recognizing that A_{100}^{eff} is equal to zero; this force expression is

$$\frac{F_{\{1\}\text{-}\{0\}\text{-}\{2'\}}}{R}=\frac{1}{6}A_{102'}^{eff}\left(\frac{1}{D^2}-\frac{1}{(D+\delta)^2}\right) \qquad [7]$$

Taking the limit of Equation [7] as $\delta \rightarrow \infty$ yields the formula for the adhesion force between the spherical particle and, what is now, a planar slab of bulk film material (13):

$$\frac{F_{\{1\}\text{-}\{0\}\text{-}\{2'_{bulk}\}}}{R}=A_{102'}^{eff}\frac{1}{6D^2} \qquad [8]$$

We define *critical film thickness* δ^* such that as $\delta \rightarrow \delta^*$, $F_{\{1\}\text{-}\{0\}\text{-}\{2'\}} \rightarrow F_{\{1\}\text{-}\{0\}\text{-}\{2'_{bulk}\}}$ i.e.

$$\frac{\Delta F}{F_{\{1\}\text{-}\{0\}\text{-}\{2'_{bulk}\}}}=\frac{F_{\{1\}\text{-}\{0\}\text{-}\{2'_{bulk}\}}-F_{\{1\}\text{-}\{0\}\text{-}\{2'\}}}{F_{\{1\}\text{-}\{0\}\text{-}\{2'_{bulk}\}}}\rightarrow \varepsilon \text{ as } \delta \rightarrow \delta^* \qquad [9]$$

where ε is termed the *permissible fractional error*. Substituting the force expressions from Equations [7] and [8] into Equation [9] yields the following formula for δ^*:

$$\delta^*=\left(\varepsilon^{-1/2}-1\right)D \qquad [10]$$

Experimental

Model System

The materials studied in these investigations include Si, SiO_2, Si_3N_4, and GaN. Sample substrates consisted of silicon coupons (~1 cm^2) having 5-, 11-, 12-, 16-, and 80nm thick GaN films, which were prepared via ALD, in the Anderson lab at the University of Florida (*Gainesville, FL*). Coupons of bare Si (with native oxide layer) and 'bulk' GaN (5μm on sapphire, prepared by University Wafer, *Boston, MA*) represented the limits of very thin coating/no coating (Si with native oxide), or very thick coating/'bulk' coating (5μm GaN). To simulate spherical contaminant particles, pre-manufactured AFM probed cantilevers, made of Si_3N_4 and having initial nominal radii of curvature (ROC) of ~15 nm, were employed (Veeco Probes, model OTR8-10, nominal spring constant 0.1 N/m). The exact ROC values of these probes were found using micrographs taken with a Hitachi S-4800 field emission scanning electron microscope (FE-SEM), in the Birck Nanotechnology Center at Purdue University.

Force measurements were made with a Veeco MultiMode® V Picoforce® scanning probe microscope (SPM) connected to a Veeco/Digital Instruments Nanoscope IV controller. All measurements were taken in a gaseous nitrogen ambient, maintained at a relative humidity of ~15%. A StaticMaster® ionizer (Amstat Industries, Inc.) was placed in the vicinity of the probe-substrate system to eliminate any electrostatic phenomena. For all samples, at least 1,000 approach-retract cycles were recorded in order to obtain statistically reliable average values for the adhesion force between probe and substrate.

Industrial-Scale System

This work was performed on six 300mm-diameter SiO_2-on-Si blanket wafers, onto which varying thicknesses (nominally: 5, 10, 15, and 20nm) of TiN had been deposited (prepared at SEMATECH, *Albany, NY*). Challenge particles were MSP dry-deposited SiO_2 particles having size distribution peaks at 55 and 110nm in diameter. After challenge particle deposition, the wafers were allowed to "age" for ~24 hours, and then analyzed using a KLA-Tencor SP1 to obtain pre-clean particle counts. Wafers were cleaned, at FSI International (*Chaska, MN*), using a less aggressive cryogenic aerosol particle removal process in order to clearly see the effects of TiN film thickness on particle removal efficiency (PRE).

Results

Model System: Force of Adhesion

The results from the AFM adhesion (pull-off) force experiments are shown in Figure 3. Note that the pull-off force values have been normalized with respect to the ROC of the Si_3N_4 probes.

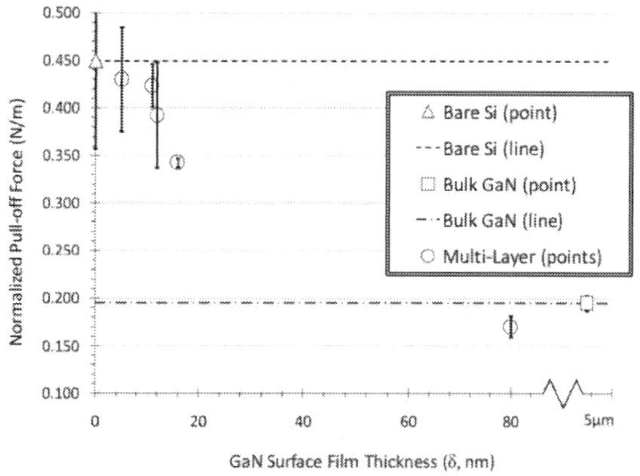

Figure 3. Variation in Si_3N_4 probe pull-off force with GaN surface film thickness.

We observe three regimes of behavior in Figure 3: Regime I, 'bulk substrate'; Regime II, 'transition'; and Regime III, 'bulk coating'. Regime I encompasses the bare Si data point (Δ) and the 5- and 11-nm thick GaN film data points. The pull-off force values for these two GaN film thicknesses vary from that of the bare Si by 4.2% (δ = 5nm) and 5.8% (δ = 11nm), both of which lie within the range of experimental error (~8%) for these measurements. Regime III includes the bulk GaN data point (□) and the 80-nm thick GaN film data point. At this GaN film thickness, the pull-off force value differs by 62% from the bare Si and by only 12% from the bulk GaN. Thus, an 80-nm thick GaN film on a Si substrate behaves much more like bulk GaN—since the contribution from the underlying Si is small—than bulk (bare) Si.

Regime II, the transition region, begins at δ = 12nm and extends beyond 20nm. For GaN film thicknesses of 12 and 16nm, the pull-off force values change by 12.5% and 23.6%, respectively, from the bare Si. This implies some combination of GaN film and underlying Si substrate contributions, both being significant to the determination of the net pull-off force. Regime II is also the region within which the greatest changes in pull-off force occur for the smallest changes in δ, which is seen clearly in Figure 3.

Industrial-Scale System: Particle Removal Efficiency

The PRE was measured when the cryogenic aerosol clean was applied. The results of the PRE experiments appear in Figure 4.

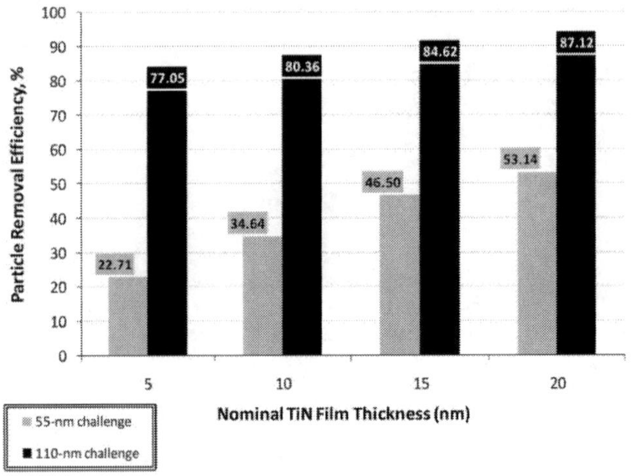

Figure 4. Variation in PRE with (nominal) TiN surface film thickness.

As seen in Figure 4, both 55- and 110-nm SiO_2 challenge PREs improve with TiN surface film thickness. Therefore, from a wafer cleaning perspective, a thin coating of TiN will result in enhanced wafer cleanliness (of SiO_2 particles) compared to that of the uncoated counterpart. The effect of particle size on PRE (for a cryogenic aerosol process) can also be seen in Figure 4: PRE increases as the contaminants become larger. This phenomenon is attributed to the fact that larger particles afford greater amounts of projected surface area for the transfer of momentum to occur from aerosolized ice crystals to contaminants thereby causing the latter's ejection from the wafer surface.

Figure 4 further elucidates the effect of particle size on the locations of the three regimes of behavior. For the 110nm challenge particles, a 15-nm change in TiN film thickness resulted in an improvement of the PRE just over 10%, while the same change improved the 55nm challenge PRE by more than 30%. We therefore conclude that nominal TiN film thicknesses of 5–20nm likely constitute some portion of the transition region (Regime II) for contaminants measuring 55nm in size, and for 110nm particles, comprise either of the bulk regions (Regime I or III, depending on the affinity of the contaminant particle material for TiN) instead.

Summary

The thickness of surface films on microelectronics devices has a significant impact on the magnitude of the forces with which particles will adhere to these surfaces, and therefore on the effectiveness of wafer cleaning processes to remove particulate contaminants. The range of influence imposed by the surface film thickness can be divided into three regions: at one extreme, the bulk substrate regime, in which the film is sufficiently thin to exhibit little to no contribution to the net force of adhesion between a particle and the surface; at the other extreme, the bulk coating regime, in which the film is so thick that any contribution from the underlying substrate to the adhesion force is small to effectively negligible; and a transition regime, lying between the two bulk regimes, in which the force of adhesion results from a combination of film and substrate contributions. The effects of surface film thickness are observed on the industrial scale when considering particle removal efficiency: depending on the affinity of the contaminant material for the film material and the size of the contaminant particles, the coating on a wafer can lead to considerable improvements in PRE.

Acknowledgments

The authors would like to thank Bill Taylor at SEMATECH for supplying the substrates for the PRE experiments, Murthy Peri and Greg Thomes, both of FSI International, for assistance with the PRE experiments, and Oh Hyun Kim, Joseph Revelli, and Tim Anderson, all of the University of Florida-Gainesville, for preparing the GaN thin film samples.

References

1. The International Technology Roadmap for Semiconductors: 2010, Yield Enhancement, Table ITWG14, pages 45–46.
2. J. W. Butterbaugh, S. Loper, G. Thomes, and D. Sheu, *MICRO*, **17**(6), 33 (1999).
3. J. W. Butterbaugh, S. Loper, and T. Wagener, in *Cleaning Technology in Semiconductor Device Manufacturing VIII/2003*, J. Ruzyllo, T. Hattori, R. L. Opila, and R. E. Novak, Editors, PV 2003-26, p. 116, The Electrochemical Society Proceedings Series, Pennington, NJ (2004).
4. T. J. Wagener and K. Kawaguchi, in *IEEE/SEMI Advanced Semiconductor Manufacturing Conference and Workshop/2004*, p. 467, IEEE ASMC Proceedings (2004).
5. T. J. Wagener, J. F. Weygand, and G. P. Thomes, *Solid State Phenomena*, **103–104**, 181 (2004).
6. J. M. Lauerhaas, J. F. Weygand, and G. P. Thomes, in *IEEE/SEMI Advanced Semiconductor Manufacturing Conference and Workshop/2005*, p. 11, IEEE ASMC Proceedings (2005).
7. J. M. Lauerhaas, C. Morote, and J. P. Plé, *Semicond. Int.*, **28**, 36 (2005).
8. J. W. Butterbaugh and C. K. Gould, *Semiconductor Manufacturing Magazine*, **7**(5), 33 (2006).
9. J. M. Lauerhaas, Y. S. Kok, A. Hamzah, and L. T. Tan, *ECS Trans.*, **11**(2), 33 (2007).
10. J. W. Butterbaugh, L. Larson, and R. Jammy, *Future Fab International*, **28**, 75 (2009).
11. F. London, *Zeit. Physik. Chem. B*, **11**, 222 (1931).
12. F. London, *Trans. Faraday Soc.*, **33**, 8b (1936).
13. H. C. Hamaker, *Physica*, **4**(10), 1058 (1937).
14. E. M. Lifshitz, *Soviet Physics JETP (English Trans.)*, **2**, 73 (1956).
15. I. E. Dzyaloshinskii, E. M. Lifshitz, and L. P. Pitaevskii, *Soviet Physics-Uspekhi (English Trans.)*, **4**(2), 153 (1961).
16. R. P. Jaiswal, *Adhesion Between Particles and Nano-structured Films (Doctoral Dissertation)*, p. 164, Purdue University Press, West Lafayette (2008).
17. M. J. Vold, *J. Coll. Sci.*, **16**, 1 (1961).
18. K. D. Danov, D. N. Petsev, N. D. Denkov, and R. Borwankar, *J. Chem. Phys.*, **99**, 7179 (1993).
19. R. P. Jaiswal, *Adhesion Between Particles and Nano-structured Films (Doctoral Dissertation)*, p. 164, Purdue University Press, West Lafayette (2008).

(172c) Electrical Characterization of Silicon Nanocrystal Films

Neema Rastgar[1], Dave Rowe[2], Lance M. Wheeler[2], Eray Aydil[1] and Uwe Kortshagen[2], (1)Chemical Engineering and Materials Science, University of Minnesota, Minneapolis, MN, (2)Mechanical Engineering, University of Minnesota, Minneapolis, MN

Electrical Characterization of Silicon Nanocrystal Films

Neema Rastgar, David Rowe, Lance Wheeler, Eray Aydil, Uwe Kortshagen

Thin films of semiconductor nanocrystals continue to receive attention as potential materials for making light-emitting diodes, photodiodes and solar cells. This approach to making optoelectronic devices may be promising because semiconductor nanocrystals are inexpensive to synthesize and their optoelectronic properties can be tuned by changing their size. However, devices based on thin films of nanocrystals typically show high electrical resistivity, and establishing control over electronic properties is difficult. The understanding of electronic transport in these nanocrystal films is in its infancy compared to bulk semiconductors. To improve this understanding and to learn how to manipulate charge carrier transport in semiconductor nanocrystal films, we study electronic transport in thin films of intrinsic and doped silicon nanocrystals.

Silicon nanocrystals with diameters ranging from 5-20 nm were synthesized through decomposition of silane in a radio-frequency plasma reactor. Thin films of these nanocrystals were deposited either through ballistic aerosol impaction onto substrates or through spin coating from colloidal dispersions of the nanocrystals. The former approach is in situ and the nanocrystals are deposited onto the substrate immediately after they leave the plasma. In the latter approach, the nanocrystals emerging from the plasma are collected, dispersed in a solvent and cast onto the substrate. In both cases, the nanocrystals are deposited between two 100 nm-thick thermally evaporated aluminum contacts to form thin films of randomly packed nanoparticles. Current-voltage characteristics of the nanocrystal films were measured as a function of doping and temperature between 100 and 300 K. Preliminary results show that the films exhibit space charge limited current above applied electric fields of 1000 V/cm, and Ohmic behavior at lower electric fields. The conductivity of annealed undoped films exhibits Arrhenius dependence on temperature, with an activation energy of 0.60 eV between room temperature and 225 K, indicative of conduction mediated by intrinsic carriers. Annealed boron-doped films, on the other hand, show moderate Arrhenius temperature dependence near room temperature, and weak temperature dependence below 225 K, characteristic of either dopant ionization or tunneling conduction.

Many films also demonstrate hysteresis in the current-voltage characteristics, ranging from insignificant to severe. The hysteresis is thought to arise from a parasitic capacitance in the film due to charging and appears to be most significant in films made of nanocrystals with ligands such as hexene. Resistance-capacitance (RC) time constants on the order of seconds describe the hysteresis. Conduction is sensitive to the surface conditions of the nanocrystals.

For example, the conductivity of nanocrystal films in vacuum increase by an order of magnitude when nitrogen, argon, and oxygen gases flow over the films and the chamber is purged continuously to reduce water partial pressure in the chamber. This change is reversible over a time scale of two hours and suggests that desorption of water from the film is the likely reason for improved conductivity.

This work was supported primarily by the MRSEC Program of the National Science Foundation under Award Number DMR-0819885.

See more of this Session: Nanoelectronic Materials
See more of this Group/Topical: Nanoscale Science and Engineering Forum

(172d) Electrostatic Coupling of Surface Charge to Bulk Defect Behavior In Metal Oxides

Prashun Gorai, Alice G. Hollister, Kristine Pangan-Okimoto and Edmund G. Seebauer, Department of Chemical and Biomolecular Engineering, University of Illinois, Urbana, IL

The technologically useful properties of semiconductor oxides such as titania and zinc oxide often depend on the concentration and diffusion of point defects. Near-surface effects are particularly important in nanoscale devices because most of the bulk is located in the vicinity of the surface. Past work in our laboratory with silicon and titania has shown that semiconductor surfaces serve as efficient pathways for generation and annihilation of point defects in the underlying bulk. Surfaces can, in addition, support electrically charged defects which create near-surface strong electric fields that can influence the local motion of charged defects resulting in the formation of space-charge layers. The electric field-driven accumulation or depletion of charged oxygen defects in such space-charge regions in metal oxides have direct implications on the performance of nanoscale devices such as gas sensors and memory resistors. Oxygen diffusion behavior was studied by exposing natural-abundance single-crystal rutile to isotopically labeled oxygen gas. The resulting profiles were measured by secondary ion mass spectrometry and subsequently modeled with continuum equations for the reaction, Fickian diffusion and electric field-driven diffusion of the key point defects. The degree of charge build-up at the surface can be quantified by an electric potential (V_i). The profiles calculated show a characteristic steep concentration gradient near the surface followed by a normal bulk diffusion profile. By identifying the charge-mediated field-driven diffusion mechanism as the controlling factor for the near-surface pile-up of oxygen, we demonstrate that this method allows the spatially resolved characterization of space charge regions near the surfaces of crystalline semiconductor metal oxides. The capability to predict oxygen pile-up in nanoscale metal oxide devices may be beneficial in device improvement via defect engineering of surfaces.

See more of this Session: Nanoelectronic Materials
See more of this Group/Topical: Nanoscale Science and Engineering Forum

(172e) Current-Driven Surface Morphological Stabilization of Coherently Strained Heteroepitaxial Thin Films

Georgios I. Sfyris, M. Rauf Gungor and Dimitrios Maroudas, Chemical Engineering, University of Massachusetts Amherst, Amhest, MA

The competition between elastic strain energy and surface energy is responsible for surface morphological instabilities in stressed elastic solids; these include the well-known Asaro-Tiller or Grinfeld (ATG) instability under conditions that promote surface diffusion and the Stanski-Krastanow (SK) morphological instability in the heteroepitaxial growth of thin films on solid substrates. In heteroepitaxial systems, the lattice mismatch between the thin film and the substrate materials induces a misfit strain, which is the source of elastic deformation. In recent studies, we have shown that applying on a stressed elastic conductor an external electric field of sufficient strength and proper direction can inhibit the ATG instability. Nevertheless, the role of surface electromigration in stabilizing the surface morphology of coherently strained epitaxial films remains largely unexplored.

In this presentation, we report the results of a morphological stability analysis for the driven morphological response of an epitaxial film surface. We have analyzed the surface morphological stability of a coherently strained thin film grown epitaxially on an elastic substrate and subjected simultaneously to an external electric field; both thick and thin (finite-thickness) elastic substrates have been examined. Due to its lattice mismatch with the substrate material, the film may undergo a SK instability, resulting in formation of islands on the film surface. We have developed a three-dimensional (3D) model for the surface morphological evolution of the thin film and conducted a linear stability analysis to examine the morphological stability of the epitaxial film's planar surface state. The analysis has shown that surface electromigration due to a properly applied and sufficiently strong electric field can inhibit SK-type instabilities, which can be used to control the onset of island formation on the film surface. We have also found that using a finite-thickness substrate can have the beneficial effect of reducing the critical strength of the electric field required to stabilize the planar surface morphology of the epitaxial film with respect to the field strength required in the case of an infinitely thick substrate.

We have determined the critical electric-field strength as a function of material properties and heteroepitaxial system parameters, as well as the electric field's optimal direction for the most efficient stabilization of the surface morphology. Furthermore, we have obtained detailed results for the effects of the finite-thickness substrate on the stabilization of the film's surface over a range of mechanical properties of the heteroepitaxial system's constituents. Perhaps more importantly, we have analyzed the surface morphological stabilization of heteroepitaxial films on substrates by combining the use of externally applied electric fields with well-known strain engineering techniques, such as the use of thin compliant substrates; such substrates provide some elastic accommodation of the lattice-mismatch strain in the epitaxial film due to the substrate's ability to relax parallel to its interface with the epitaxial film. It should be mentioned that employing substrate engineering techniques alone is not sufficient for stabilizing the planar morphology of the epitaxial films; for such stabilization, the simultaneous action of the externally applied electric field is required. Our analysis has demonstrated the benefits of the substrate compliance, as measured by the reduction in the critical electric-field strength required for stabilizing the epitaxial film surface; we have found this reduction in the critical field strength to be as high as two orders of magnitude with respect to that required for epitaxial films grown on conventional thick substrates. Our results generate experimentally testable hypotheses and motivate experimental measurements that can be compared directly with the theoretical predictions.

See more of this Session: Nanoelectronic Materials
See more of this Group/Topical: Nanoscale Science and Engineering Forum

(172f) Formation of Thin Films of IZO and ITO Nanoparticles

Stefan Schaefer, Michael Voigt, Wolfgang Peukert and Mahdi Mahajeri, Institute of Particle Technology, Friedrich-Alexander University of Erlangen Nuernberg, Erlangen, Germany

Nanoparticulate thin films are of high interest in various industries. In this contribution we focus on electronic applications as conducting layers in solar cells or displays [1]. The properties of these solution processed films are varying from highly porous to very densely packed films according to their field of application. We present our work on transparent conductive oxide films of indium tin oxide nanoparticles (ITO) [2] which were prepared in the gas phase and indium zinc oxide nanoparticles (IZO) produced by wet chemical sol-gel synthesis. Both are dispersed in alcoholic solutions in order to deposit the particles by spin- and dip-coating on silica wafers and glass substrates. The stability of the dispersions is verified by either dynamic light scattering (DLS) or UV-vis spectroscopy. Analytical tools for investigation of morphology, surface and conductivity of the thin films were AFM, XRD, SEM, HRTEM, four-point probe measurements and impedance spectroscopy. The effects of the indium concentration in the IZO films on the morphology, the optical and electrical properties of the thin films were investigated. Moreover, the influence of the layer thickness and annealing temperature on the film properties are discussed. The ITO and IZO films deposited on glass substrates showed a high transparency in the visible range of at least 80% to 95%. With the ITO thin nanoparticulate films we achieved conductivities up to 120 Scm^{-1} simply by varying preparation conditions [3] and up to 196 Scm^{-1} with a treatment in forming gas [2], the latter being the best known value of ITO particles produced in the gas phase. IZO dispersions were stable over at least two months. With IZO nanoparticulate thin films we achieved conductivities up to 104 S/cm^{-1} by forming gas treatment of the films.

[1] R. Marczak, F. Werner, J.-F. Gnichwitz, A. Hirsch, D.M. Guldi, W. Peukert, Journal of Physical Chemistry C, 113 (2009) 4669-4678.

[2] M. Mahajeri, M. Voigt, R.N. Klupp Taylor, A. Reindl, W. Peukert, Thin Solid Films, 518 3373-3381.

[3] A. Reindl, M. Mahajeri, J. Hanft, W. Peukert, Thin Solid Films, 517 (2009) 1624-1629.

See more of this Session: Nanoelectronic Materials
See more of this Group/Topical: Nanoscale Science and Engineering Forum

(172g) Formation of Fullerene Superlattices by Interlayer Bonding In Twisted Bilayer Graphene

Andre R. Muniz and **Dimitrios Maroudas**, Chemical Engineering, University of Massachusetts Amherst, Amherst, MA

Graphene, a two-dimensional material consisting of one-atom-thick sheets of sp^2-hybridized carbon atoms arranged in a honeycomb lattice, possesses an outstanding and unique set of electronic and mechanical properties, which has potential to enable a broad range of technological applications. Chemical modification of graphene, such as hydrogenation, has been used to properly tailor graphene properties aiming at specific applications. Exposure of graphitic single- and multi-layered carbon nanomaterials to atomic hydrogen leads to formation of sp^3-hybridized C-H and C-C bonds, which alter the materials structure remarkably. In this presentation, based on a systematic computational analysis, we report a class of novel carbon nanostructures formed from rotationally misoriented graphene bilayers (twisted bilayer graphene), upon creation of interlayer covalent C-C bonds. These interlayer C-C bonds are formed due to chemisorption of H atoms onto the outer surfaces of the graphene layers for certain H coverages and hydrogenation patterns. The analysis is based on a combination of first-principles density functional theory (DFT) calculations and classical molecular-dynamics (MD) simulations. The DFT calculations are used for relaxing the atomic structure and determining the electronic band structure of the relaxed atomic configurations, while the MD simulations are based on a realistic reactive bond order potential and are used to relax nanostructures that require computational supercells much larger than those employed in the DFT calculations.

The carbon nanostructures that we have discovered consist of superlattices of caged fullerene-like configurations embedded within the graphene bilayer and are generated from pairs of graphene layers rotated by angles $q \sim 30°$ with respect to each other; in this limit, a zigzag- and an armchair-oriented pair of planes are aligned. We have relaxed the atomic structures and computed the corresponding electronic structures of these materials for both non-commensurate bilayers at $q = 30^o$ and commensurate bilayers at an angle q deviating slightly from the perfect value of 30°. For certain such structures, we show that the linear dispersion around the K point in the Brillouin zone (Dirac cones), characteristic of single-layer graphene and twisted bilayer graphene, is preserved in spite of the introduction of sp^3 bonds due to hydrogenation and interlayer C-C bonding. We also demonstrate that an entire class of similar caged structures of various sizes can be generated in an analogous manner.

Our findings have far reaching implications: 2-dimensional carbon nanostructures can be functionalized by formation of superstructures through embedding of 0-dimensional carbon structures to tailor their electronic properties without deviating from those of single-layer graphene. This embedding of fullerene-like configurations can be accomplished by chemical modification through hydrogenation to form certain patterns of chemisorbed H. We have found these structures to be less stable than their hydrogenated non-bonded counterparts at low pressure; under higher pressure, however, these structures are expected to become thermodynamically favorable. The structures proposed in this study may be related to the intermediate carbon phases observed in laser-induced shock wave loading and high-temperature shock compression experiments on graphitic samples, which were likely to exhibit a turbostratic stacking of planes.

See more of this Session: Nanoelectronic Materials
See more of this Group/Topical: Nanoscale Science and Engineering Forum

(172h) Electrical Properties of Metallic Nanoparticle Films

David A. Walker, Yong Yan, Hideyuki Nakanishi and Bartosz Grzybowski, Chemical and Biological Engineering, Northwestern University, Evanston, IL

Nanoparticles (NPs) display a wide range of electrical behavior depending on the particle material, shape, and local environment. Particularly, the electronic properties of metallic NP films can be greatly influenced by the self-assembled monolayers (SAMs) which coat the particles' surfaces. This talk will focus on a class of nanoparticle-based materials whose conductivity can change drastically based on how the SAM is influenced by various external stimuli (i.e. irradiation with light causes photoconductance and inverse photoconductance). Such materials can be used to create elementary circuitry made entirely out of metallic NPs.

See more of this Session: Nanoelectronic Materials
See more of this Group/Topical: Nanoscale Science and Engineering Forum

(176a) Crystal Structure Engineering of Semiconductor Nanowires

Ildar Musin[1], Saujan Sivaram[2], Nae Chul Shin[2] and **Michael A. Filler**[3], (1)School of Chemical & Biomolecular Engineering, Georgia Institute of Technology, Atlanta, GA, (2)Chemical & Biomolecular Engineering, Georgia Institute of Technology, Atlanta, GA, (3)School of Chemial & Biomolecular Engineering, Georgia Institute of Technology, Atlanta, GA

Semiconductor nanowires provide a promising route to achieve next generation photonic, electronic, and energy conversion devices. Control of nanowire crystal structure (e.g. lattice, orientation, faceting) is critical to enable the appropriate function for a particular application. Unfortunately, this remains a challenging task with bottom-up syntheses. To this end, we demonstrate the ability to precisely engineer nanowire interface energetics via the addition of bifunctional precursors and, in doing so, rationally control Ge nanowire crystal growth orientation for the first time. More specifically, Ge nanowires are grown using the vapor-liquid-solid (VLS) technique with germane and a series of bifunctional alkylgermane species, including methylgermane, ethylgermane, and tert-butylgermane. Scanning electron microscopy (SEM) reveals dramatics change in growth direction as well as a reduction in tapering. In the case of methylgermane, transmission electron microscopy (TEM) confirms a transition from <111> to <110> oriented growth while maintaining single crystallinity throughout. Infrared and X-ray photoelectron spectroscopy (XPS) indicate that the nanowire surfaces are alkyl terminated. In addition to reducing the rate of oxidation, these surface species adjust the solid-vapor interface energy near the three-phase boundary, thereby influencing nucleation events and enabling changes in crystal orientation. The control of nuclei surface chemistry demonstrated by this work provides an important new handle for controlling nanowire properties for different applications. Furthermore the ability to effectively passivate sidewalls during growth is expected to enable more robust doping profiles by only permitting precursor incorporation through the catalyst tip.

See more of this Session: Nanostructured Materials for Electronic and Photonic Applications
See more of this Group/Topical: Materials Engineering and Sciences Division

(176b) Solvothermal Route to the Synthesis of Iron Sulfide Nanomaterials

Leize Zhu[1], Jessica Tanumihardja[2] and Qiuming Yu[2], (1)Dept. of Chemical Engineering, University of Washington, Seattle, WA, (2)Chemical Engineering, University of Washington, Seattle, WA

Recently, nanomaterials have employed as the new building blocks to construct solar energy harvesting devices. Of particular interest are the size dependent properties such as size quantization effects in semiconductor nanomaterials. Many semiconductor nanomaterials have been developed for solar cells application [1]. However, the use of toxic elements such as cadmium or lead remains a limitation as the environmental issues need to be addressed and these elements are not abundant [2]. Herein, we proposed a cheap, non-toxic and abundant semiconductor nanomaterials-pyrite nanowires, which has an indirect energy transition at 0.95 eV, a direct transition at 1.03 eV and an integrated absorption coefficient of 3.3×10^5 cm^{-1} for the energy spectrum of wavelength values (λ) between 300 and 750 nm [3]. With a high absorption coefficient from visible to near-infrared region, material consumption is significantly reduced in semiconductors for thin layer solar cells compared to conventional silicon-based solar cells. Pyrite has previously been prepared using several high temperature approaches including sulfurization of iron films, sulfurization of iron oxide films, reactive sputtering, and spray pyrolysis, yet at elevated temperatures, segregation of iron and sulfur species is unavoidable, which could change the stoichiometry and material phase of the deposited film. Thus, solvothermal process is a prefered method used for producing pyrite nanocrystals [4]. Despite the recent advancement of solvothermal routes to the synthesis of pyrite FeS_2 nanocrystals, to obtain pure pyrite phase and controllable shape and size of pyrite nanocrystals still remain challenges. In this work, we have synthesized iron sulfur nanowires, microflowers and microplates by using $FeCl_3$•$6H_2O$, $FeCl_2$ • $4H_2O$, $Fe(NO_3)_3$ • $9H_2O$ and $FeSO_4$ • $7H_2O$ as the iron source and thiourea, sulfur and $Na_2S_2O_3$ as the sulfur source. The solvent used is limited to ethylenediamine and ethylenediamine+ethyleneglycol mixture. Scanning electron microscope (SEM) was used to determine the morphology of the materials and energy dispersive x-ray analysis (EDAX) was used to determine the chemical composition of these materials. By tuning the concentration, we can obtain the length of wires ranging from 100-500 nanometers to 1-10 micrometers with 100-200 nm in diameter. X-ray photoelectron spectroscopy (XPS) results indicate that these materials are iron monosulfide along with some oxidized products. These wire-like materials are not well-crystallized which can be seen from high-resolution transmission electron microscopy (HRTEM) images. We found temperature is also an important factor in controlling the morphology of the iron sulfur materials. Microplates can be obtained at high temperature (210 • C). Only wires can be obtained at 180 • C . We also found that the counter-ion and valence state of the iron source play an important role in determining the morphology and phase of the final products. By changing the solvent from ethylenediamine to water, significant change in the phase of the product was observed (Pyrite along with macarsite), which is verified by X-Ray Diffraction (XRD). Further characterization of the magnetic and electronic property of these iron sulfur materials is carrying out.

References

[1] Kamat, P.V. J. Phys. Chem. C 2008, 112, 18737–18753.

[2] Alivisatos, A.P. et al. Chem. Mater. 2009, 21, 2568–2570.

[3] Chen, C.W. et al. Nanotechnology, 2009, 20, 405207-405212.

[4] Chaudhuri, S. et al. Chem. Phys. Lett. 2004, 398, 22–26.

See more of this Session: Nanostructured Materials for Electronic and Photonic Applications
See more of this Group/Topical: Materials Engineering and Sciences Division

(176c) Controllable Fabrication of Clustered Quantum Dots for Time-Correlated Hyperspectral Studies

Rajasekhar Anumolu[1], Hyeong Gon Kang[2], Matthew L. Clarke[2], Jeeseong Hwang[2] and Leonard F. Pease III[3], (1)Chemical Engineering, University of Utah, Salt Lake City, UT, (2)National Institute of Standards and Technology, (3)Chemical Engineering and Pharmaceutics & Pharmaceutical Chemistry, University of Utah, Salt Lake City, UT

Optical properties of clustered core(CdSe)-Shell(ZnS) quantum dots (QDs) are investigated. Monomers, dimers, and trimers of QDs are generated through a self-induced droplet assembly using an electrospray differential mobility analyzer. A fixed number of QDs in the clusters is achieved through controlled electrostatic deposition. QD clusters composed of both homoclusters and heteroclusters are generated. Multimodal time-correlated hyperspectral studies of these QD clusters are carried out, simultaneously measuring the dynamic photoluminescence (PL) intermittencies, PL life times, and spectral diffusion revealing the unique properties of interacting QDs in QD clusters under photooxidation. Controlled assembly of a fixed number of QDs into single clusters and analysis of their dynamic PL properties leads to better understanding of the excitonic charge transfer (photonic to electronic energy transfer) during photooxidation in single and clustered QDs. The clustered QDs will provide unique optical characteristics which will be beneficial for applications in energy devices such as solar cells and devices with a long-term photostability. Analysis of homoclusters and heteroclusters reveals a distinct variation in their optical properties which will lead to interesting aspects for QD based energy devices.

See more of this Session: Nanostructured Materials for Electronic and Photonic Applications
See more of this Group/Topical: Materials Engineering and Sciences Division

(176d) Multi-Scale Optical, Electrical, and Chemical Interrogation of Thiophene-Based Solar Cell Films

Chris Carach, Isaac Riisness and **Michael Gordon**, Chemical Engineering, University of California, Santa Barbara, Santa Barbara, CA

Understanding and controlling carrier transport in conjugated polymer films and composites is critical to the development and widespread application of plastic solar cells. Recent efforts have focused on "bulk heterojunction" (BHJ) structures where a conjugated polymer is mixed *at the nanoscale* with a fullerene acceptor to achieve large interfacial areas for exciton splitting. It has been seen that fabrication protocols drastically affect device efficiency and that charge transport is intimately tied to film morphology through local disorder, domain formation, and compositional heterogeneity. Since charge transport in these systems occurs over distances <100 nm, it is no surprise that *local* physicochemical properties have a tremendous impact on photovoltaic behavior. Unfortunately, our understanding of these issues is poor because conventional PV testing on devices tends to "average out" micro- and nanoscale heterogeneities. To this end, we combine device-level testing of BHJ films with confocal/near-field optical and electrical interrogation (conductive AFM and electrostatic force microscopy) to understand how nanoscale morphology, molecular alignment, chemical structure, and processing affect charge transport processes. In this work, we show that confocal Raman and photoluminescence (PL) imaging of polythiophene/fullerene blends, when combined with electrical AFM techniques, are highly descriptive means to study donor/acceptor mixing and nanoscale transport. For instance, spatially-correlated optical and electrical imaging are combined to differentiate between regions of phase separation, low donor concentration, and oxidative damage. The talk will also highlight how low temperature PL measurements and analysis of excitonic-vibronic coupling in polythiophene/blend films processed under different conditions (e.g. solvents, thermal annealing, etc.) can be used to probe local order and crystallinity. Preliminary results on near-field Raman investigations of P3HT-PCBM blends for localized chemical imaging at the nanoscale will be mentioned as well.

See more of this Session: Nanostructured Materials for Electronic and Photonic Applications
See more of this Group/Topical: Materials Engineering and Sciences Division

FULLY ORGANIC ITO REPLACEMENT THROUGH ACID DOPING OF DOUBLE-WALLED CARBON NANOTUBE THIN FILM ASSEMBLIES

Yong Tae Park[a] and Jaime C. Grunlan[a,b]
[a]Department of Mechanical Engineering and [b]Department of Chemical Engineering
Texas A&M University, 3123 TAMU, College Station, TX, 77840 (USA)
E-mail: yong.t.park@gmail.com, jgrunlan@tamu.edu

Abstract

Transparent electrodes made from metal oxides, of which indium tin oxide (ITO) is most common, suffer from poor mechanical flexibility and electrochemical stability. Highly transparent and electrically conductive thin films based on double-walled carbon nanotubes (DWNTs) were assembled layer-by-layer (LbL) as a potential replacement for ITO. The alternate deposition of positively charged poly(diallyldimethyl ammonium chloride) [PDDA] and DWNTs, stabilized with negatively charged deoxycholate (DOC), exhibit linear film growth. A five bilayer (BL) assembly exhibits a sheet resistance of 309 Ω/sq, and visible light transmittance of 84%. In an effort to further reduce sheet resistance, these thin films were exposed to HNO_3 vapor. Sheet resistance of the 5 BL film was reduced to 104 Ω/sq (4200 S/cm conductivity, 22.9 nm thickness), with no change in transparency. Single-walled carbon nanotube assemblies show an even greater change when exposed to HNO_3 vapor, with a 5 BL assembly (%T > 85) decreasing from 1236 to 237 Ω/sq. The DWNT-based assemblies maintained their low sheet resistance after repeated bending and also showed excellent electrochemical stability relative to ITO. This work demonstrates the excellent optoelectronic performance, mechanical flexibility, and electrochemical stability of CNT-based assemblies, which are potentially useful as transparent electrodes for a variety of flexible electronics.

INTRODUCTION

Transparent electrodes continue to be studied because of increasing demand for optoelectronics, such as photodiodes for solar cells and electronic displays. Various materials have achieved a combination of conductivity and transparency to satisfy the needs of various optoelectronic applications.[1,2] Carbon nanotubes (CNTs) have been intensively studied because of their impressive electrical, thermal, mechanical, structural and chemical properties. Recently, CNT thin films have been investigated as electronic devices including sensors,[3] solar cells,[4] integrated circuits,[5] and transparent conductive thin films.[6-8] These CNT thin films show good optoelectronic performance, with values in the middle of the range of commercial ITO-coated PET, which typically has 50–200 Ω/sq sheet resistance and ~ 83% T at 550 nm wavelength. An LbL technique is examined here as a means of tailoring the optical transmittance and electrical conductivity of CNT-based thin films. This method produces thin films by harnessing electrostatic interactions through alternately exposing a substrate to positively and negatively-charged aqueous solutions at room temperature. In the present work, our goal is to accomplish CNT-based LbL thin films as one candidate of ITO replacements. The increase in electrical conductivity is believed to be a combination of acid anion doping and removal of insulating materials. This film is highly flexible, transparent, and electrochemically stable, making it a potential alternative to ITO. After 100 bending cycles, these films exhibit the same sheet resistance, but ITO-coated PET increases two orders of magnitude. Cyclic voltammetry was used to demonstrate the electrochemical stability of these transparent electrodes. This combination of optoelectronic performance, mechanical flexibility, and electrochemical stability make this a suitable candidate for a variety of applications requiring a transparent electrode.

Experimental section

Purified electric arc SWNTs (average 0.5–3 μm length and 1.4 nm diameter, $C \geq 90$ wt%), synthesized using Ni/Y catalysts, from Carbon Solutions, Inc. were used as a base of comparison for HiPCO DWNTs (+1 μm length and ~3.0 nm diameter, $C \geq 90$ wt%), purchased from Continental Carbon Nanotechnologies Inc., used in this study. PDDA and DOC (≥98%) were purchased from Sigma-Aldrich. Fuming sulfuric acid (H_2SO_4, 20% free SO_3 basis), fuming nitric acid (HNO_3, 99.5%), and fuming hydrochloric acid (HCl, 37%) were also used for acid treatment. All chemicals were used as received. PET (0.175 mm thick) and PS (0.125 mm thick) films were used as flexible substrates for CNT assemblies. A 100 Ω/sq ITO-coated PET sheet was purchased from Sigma-Aldrich and cut to size to measure the optical transmittance and the change in sheet resistance during bending cycles.

A cationic 0.25 wt% PDDA solution was prepared with 18.2 MΩ deionized water. 0.05 wt% CNTs were dissolved in deionized water containing 1 wt% DOC, followed by 10 W tip sonication for 20 min. Each cycle of LbL assembly consists of substrate immersion into an aqueous mixture, with rinsing and drying after each deposition, beginning with the cationic solution. Cycles were repeated to deposit the desired number of BLs. All films were stored in a dry box for a minimum of 12 hours prior to testing or acid treatment. A schematic of this LbL deposition process is shown in Scheme 1. For acid treatments, assembled CNT films were held in a saturated acid vapor environment. Three types of concentrated acids (fuming sulfuric, nitric, and hydrochloric) in petri dishes were maintained at 70 °C with a water bath. After a 2 to 30 min exposure to acid vapor, the CNT films were rinsed with deionized water and dried with filtered air.

Thickness of assemblies was determined by averaging the values of at least five different positions using a PHE-101 Ellipsometer equipped with a 632.8 nm laser. Absorbance and transmittance were measured between 300 and 850 nm with a USB2000 UV-vis spectrometer. All absorbance values in the text are for one-sided coating (*i.e.*, absorbance from slides coated on both sides was halved). Sheet resistance of nanotube assemblies on PET was measured with a Signatone Pro4 four-point probe system with 0.4 mm probe tip diameter and 1.0 mm tip spacing. The Fourier-transform infrared (FT-IR) spectra were obtained with an ALPHA FT-IR spectrometer. Scanning electron microscopy was performed with a Quanta 600 FE-SEM at an operating voltage of 10 kV. Energy dispersive X-ray (EDX) analysis was carried out using an Oxford EDX system attached to the FE-SEM. Thin film cross-sections were imaged with a JEOL 1200 EX TEM with an operating voltage of 100 kV. The specimens were sectioned down to ~90 nm and placed on 300 mesh nickel grids to dry prior to imaging. An Epsilon 851 electrochemical workstation was used for cyclic voltammetry. A CNT- or ITO-coated PET electrode, platinum wire, and an Ag/AgCl (3M KCl) electrode were used as the working, counter, and reference electrodes, respectively. These measurements were performed at ambient temperature (22±2 °C) in 0.1 M Na_2SO_4 solution that had been bubbled with N_2 gas for more than 20 min prior to measurement. Cyclic voltammetry was performed between two potential limits, −0.2 and 0.8 V, at a scan rate of 100 mV/s.

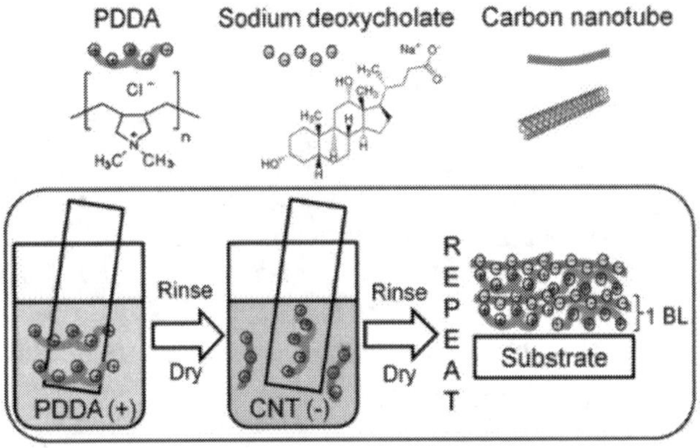

Scheme 1. Schematic of the LbL process. This recipe is represented by its own schematic: [PDDA/(CNT+DOC)]$_n$, where n is the number of cycles.

Results and discussion

Fig. 1 shows the linear growth of the following LbL assembly: cationic polyelectrolyte and anionic surfactant stabilized nanoparticles, with PDDA and DOC-stabilized DWNTs up to 10 BLs. Ellipsometric thickness of the CNT LbL thin film on a silicon wafer has the linear trend. The average thickness of one BL is 4.4 nm which is thicker than the DWNT thickness, 1.5–3 nm, suggesting that stabilizing surfactant and counterionic polymer molecules fully covered the DWNT network deposited at each BL step. Along with the thickness analysis, the graph of optical absorbance confirms that the CNT-based LbL assemblies of 10 BL grow in the same linear fashion as the graph of thickness does.

Also, Fig. 1 shows there is very little change in thickness and absorbance following acid vapor exposure. Previous studies indicate that swelling of CNT networks occurred with acid doping due to denser layers of acidic anions around the individual nanotubes.[9] In spite of this swelling effect, some CNT thin films have shown a significant decrease in thickness following acid doping.[6] The minimal reduction of thickness for the present LbL assemblies suggests a minimum amount of surfactant was associated with the DWNT during deposition. In some LbL systems, CNT assemblies with a polymeric stabilizer showed more than 30% thickness reduction after sulfuric acid treatment,[8] suggesting that these films contained excess polymer. No reduction of thickness for the LbL assembly indicates the minimum amount of surfactant associated with CNT during deposition.

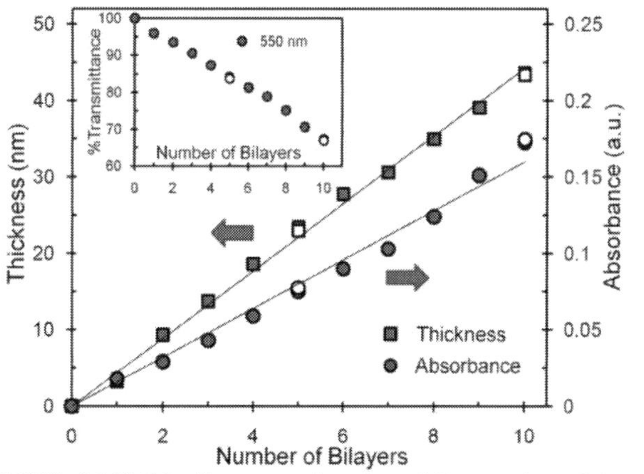

Fig. 1. Growth of PDDA/(DWNT+DOC) thin films, as a function of the number of BLs deposited. Transmittance was estimated from absorbance data. White ones with black maker lines are the values after 20 min nitric acid doping.

The morphology change in $[PDDA/(DWNT+DOC)]_6$ films before and after nitric acid treatment was analyzed using SEM, as shown in Fig. 2(A) and 2(B). DOC exfoliation of individual DWNTs in water results in a uniform distribution on the substrate.[9] Nanotubes, however, aggregate due to the high concentration in the deposited film, but they remain uniformly distributed, which creates a strong network that produces high electrical conductivity. Before treatment, the surface of the film was covered by a layer of insulators (polymer and surfactant) and perfectly exposed DWNTs could be hardly observed, while homogeneous DWNT networks were exposed due to removal of PDDA molecule after nitric acid treatment for 20 min. This removal of residual insulating materials enables direct contacts between conducting nanotubes. TEM cross-sectional micrographs (Fig. 2(C) and 2(D)) illustrate a highly inter-diffused nanostructure, in which the dark and light grey areas are DWNT-rich and PDDA-rich regions, respectively, and the darkest black dots are catalyst impurities from the HIPCO process. In addition to highlighting thin film structure, the TEM images verify the inner diameter of an individual DWNT (~3 nm) and the negligible change in the $[PDDA/(DWNT+DOC)]_{10}$ film thickness due to nitric acid treatment.

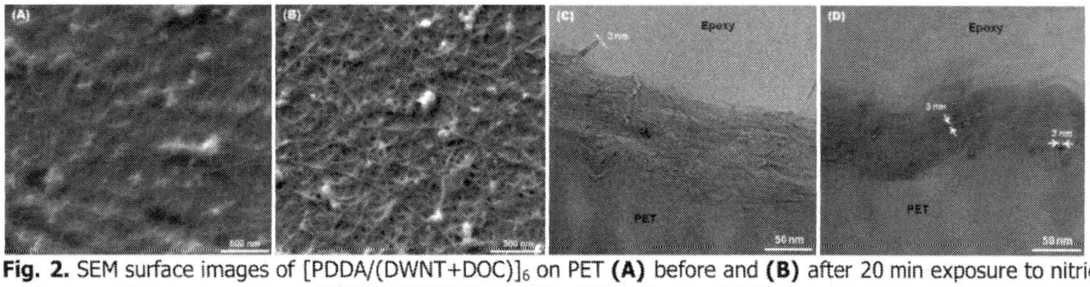

Fig. 2. SEM surface images of $[PDDA/(DWNT+DOC)]_6$ on PET **(A)** before and **(B)** after 20 min exposure to nitric acid vapor. TEM cross-sections of $[PDDA/(DWNT+DOC)]_{10}$ **(C)** before and **(D)** after 20 min treatment with nitric acid.

The sheet resistance and electrical conductivity of the [PDDA/(SWNT+DOC)]$_n$ and [PDDA/(SWNT+DOC)]$_n$ on PET substrates were measured as function of the optical transmittance according to the number of BLs deposited in Fig. 3. The sheet resistance of LbL films was shown to dramatically decrease with increasing number of BLs due to increased connectivity of CNT networks, along with gradual decrease in transmittance. A 10 BL DWNT thin film achieves sheet resistance of 127 Ω/sq with 67% T and 1180 Ω/sq with 94% T for just two BLs. The PDDA/(SWNT+DOC) system exhibits the same optoelectronic trend with higher sheet resistance and similar transmittance. A 5 BL DWNT-based film has the best optoelectronic performance of 309 Ω/sq with 84% T, but still higher sheet resistance than the requirement for ITO replacement.

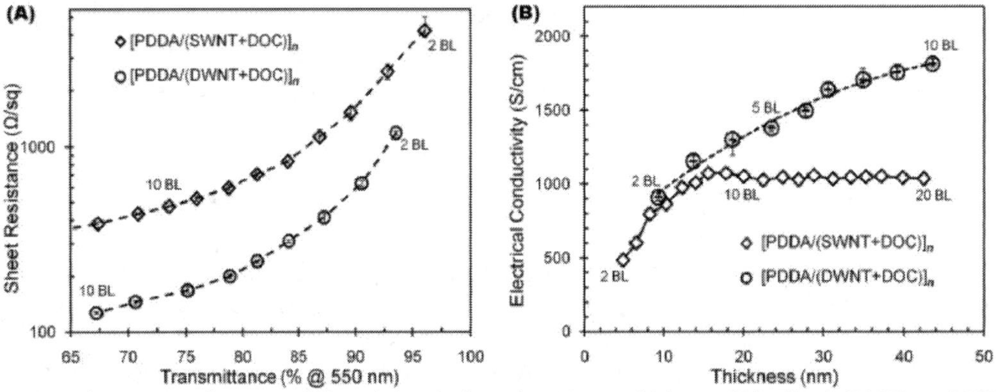

Fig. 3. Optoelectronic performance **(A)** and electrical conductivity vs. thickness **(B)** of both SWNT- and PDDA-base LbL thin films.

The sheet resistance of as-assembled films exhibit a higher value than the requirement for ITO replacement, due to the hindrance of the excessive polymer and surfactant to direct connections between CNTs. Strong acidic anions form a highly charge-transfer complex layer around individual CNTs and remove excessive polymer and surfactant.[6,9]

Most prior acid treatment studies have been performed with dip treatment (*i.e.*, immersing the films into a strong acid bath). Dip treating makes PET substrates brittle or even causes shrinkage. Moreover, dip treated CNT thin films readily de-adhere from substrates during rinsing. In the present study, films were exposed to acid vapor by heating the acid solution in a 70 °C water bath. Fig. 4 shows sheet resistance of DWNT assemblies after exposure to nitric, hydrochloric, or sulfuric acid vapor for increasing time periods. Nitric acid dramatically reduced the sheet resistance of as-assembled 5 BL PDDA/(DWNT + DOC) thin films, from 309 to 107 Ω/sq, after a 10 min exposure. Acid type plays a significant role in reducing sheet resistance. Nitric acid treatment provides the lowest resistance, while hydrochloric and sulfuric acids result in values that are nearly double that of HNO_3. This result suggests that PDDA and DOC molecules are more effectively removed by nitric acid vapor than the other two acids. The DWNT networks after each acid treatment were also characterized by SEM (not in this paper). 20 min nitric acid treatment exposes an extensive DWNT network, while more hazy DWNT networks are observed after 20 min hydrochloric and sulfuric acid treatments due to less removal of insulating molecules.

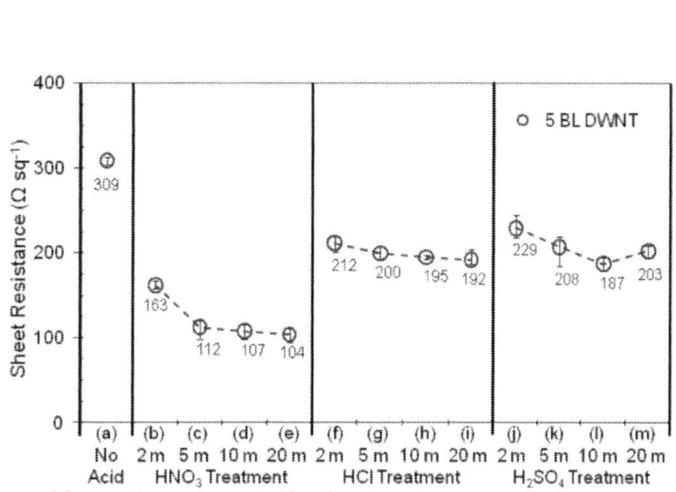

Fig. 4 Sheet resistance of [PDDA/(DWNT + DOC)]$_5$ after 2 to 20 min exposure to nitric acid, hydrochloric acid, and sulfuric acid. (a) is the sheet resistance of an as-assembled 5 BL DWNT thin film.

Fig. 5(A) shows the effect of nitric acid doping on CNT films according to the doping time for 2 to 20 min. The sheet resistance decreased significantly in the first few minutes when treated with 99.5% nitric acid. Upon the 2 min nitric acid doping to as-assembled CNT films, the sheet resistance of the 5 BL SWNT film exhibits an order of magnitude decrease, and that of the 5 BL DWNT film decreases moderately in half. However, further acid doping results in a very slight decrease, especially after 5 min. This may be because insulating materials on the surface of the film could be readily removed, while removal of DOC and PDDA inside the film took a longer time. The change in optoelectronic performances of several CNT-based LbL films due to acid doping is summarized in Fig. 5(B). [PDDA/(DWNT+DOC)]$_5$ films (107 Ω/sq and 84% T) had better optoelectronic performance than most of CNT-based films reported.[6-8]

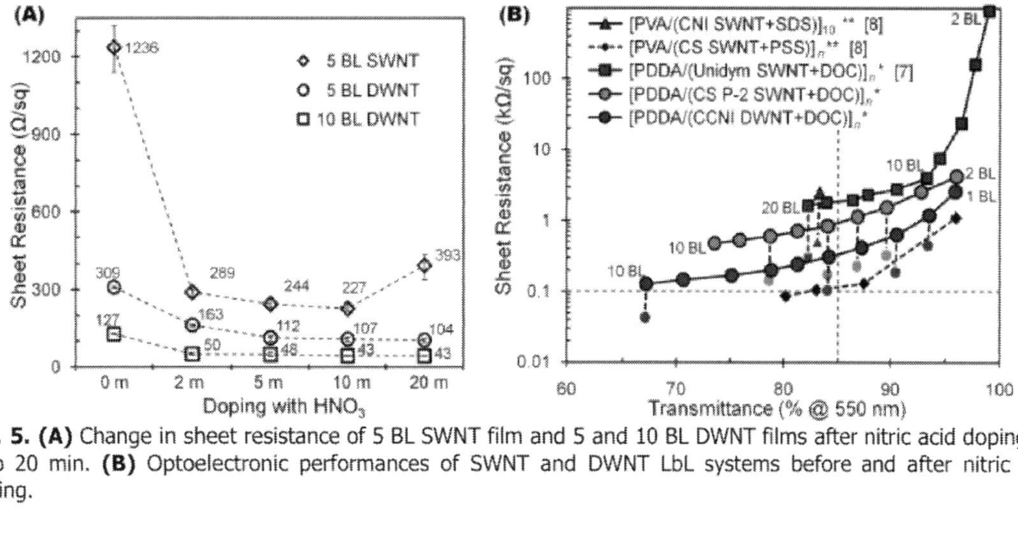

Fig. 5. (A) Change in sheet resistance of 5 BL SWNT film and 5 and 10 BL DWNT films after nitric acid doping for 2 to 20 min. **(B)** Optoelectronic performances of SWNT and DWNT LbL systems before and after nitric acid doping.

Recently, the bending performance of electrodes is very important to electronic applications. The bending property of CNT LbL films was compared to a commercial 100 Ω/sq ITO-coated PET sheet in Fig. 6(A). Their conductivity was maintained over 100 cycles of bending, while the conductivity of ITO coatings results permanent loss in conductivity. This outstanding feature of LbL films can be potentially used as a flexible electronics. Cyclic voltammograms (CV) of an HNO_3-treated 5 BL DWNT film and an ITO film (both on PET) were used to compare electrochemical stability, as shown in Fig. 6(B). The potential was cycled between -0.2 and 0.8 V vs. an Ag/AgCl reference electrode, with a scan rate of 100 mV/s, in 0.1 M Na_2SO_4 aqueous solution. ITO shows oxidation–reduction (redox) peaks at 0.1 and 0.3 V, respectively, indicating that ITO undergoes chemical changes under potentiodynamic stresses. A relatively large peak-to-peak separation of 0.2 V (between 0.1 and 0.3 V) is due to this transformation (irreversible change) that occurs in the ITO film during electrochemical cycling. The DWNT LbL films exhibit no redox peaks, suggesting they are electrochemically stable in this potential range.

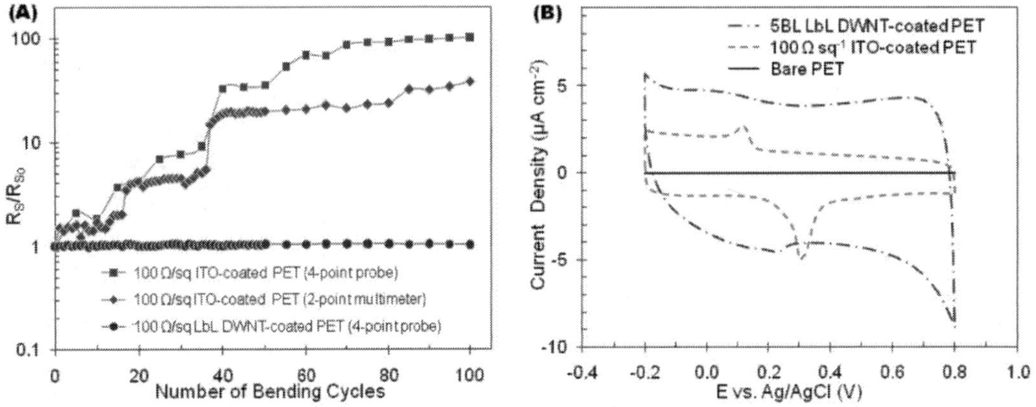

Fig. 6. (A) Sheet resistance as a function of bending cycles (to 1 cm radius of curvature) for 5 BL DWNT and 100 Ω/sq ITO-coated PET films. R_{So} indicates the resistance (~100 Ω/sq) prior to any bending. **(B)** Cyclic voltammograms of a $[PDDA/(DWNT + DOC)]_5$ thin film on PET and a commercial 100 Ω/sq ITO-coated PET.

CONCLUSIONS

Highly transparent and conductive thin films were prepared using LbL assembly with CNTs, DOC as a stabilizer and PDDA as polycations. This LbL adsorption of both species has linear increase independent on number of bilayers deposited. The self assembly of DOC-stabilized CNTs and PDDA was shown to realize the thinner and smoother films. Nitric acid doping was also treated for higher electrical conductivity. The optoelectronic performance of 5 BL DWNT LbL film here is much better than most other CNT thin films and capable of the ITO replacement. Additionally, the bending performance and electrochemical stability could suggest that these films potentially be used in a variety of electronic applications.

REFERENCES

1. Granqvist, C. G.; Hultaker, A. (2002) *Thin Solid Films*, 411, pp. 1-5.

2. Hu, L.; Hecht, D. S.; Gruuner, G. (2010) *Chem. Rev.*, 110, pp. 5790-5844.
3. Modi, A.; Koratkar, N.; Lass, E.; Wei, B. Q.; Ajayan, P. M. (2003) *Nature*, 424, pp. 171-174.
4. Lowman, G. M.; Hammond, P. T. (2005) *Small*, 1, pp. 1070-1073.
5. Friedman, R. S.; McAlpine, M. C.; Ricketts, D. S.; Ham, D.; Lieber, C. M. (2005) *Nature*, 434, pp. 1085.
6. Geng, H. Z.; Kim, K. K.; So, K. P.; Lee, Y. S.; Chang, Y.; Lee, Y. H. (2007) *J. Am. Chem. Soc.*, 129, pp. 7758-7759.
7. Park, Y. T.; Ham, A. Y.; Grunlan, J. C. (2011) *J. Mater. Chem.*, 21, pp. 363-368.
8. Shim, B. S.; Zhu, J.; Jan, E.; Critchley, K.; Kotov, N. A. (2010) *ACS Nano*, 4, pp. 3725-3734.
9. Ericson, L. M.; etc. (2004) *Science*, 305, pp. 1447-1450.
10. Wenseleers, W.; Vlasov, I. I.; Goovaerts, E.; Obraztsova, E. D.; Lobach, A. S.; Bouwen, A. (2004) *Adv. Funct. Mater.*, 14, pp. 1105-1112.

QUANTUM TUNNELING IN CARBON-NANOTUBE-BASED FIELD EMISSION DISPLAY AND BAND-TO-BAND TUNNELING

Meng-Mu Shih, University of Florida, Gainesville, FL

Abstract

In today's competitive electronics market, consumer products are required to have SWaP features and considerations: size, weight, power and price. Although the traditional display technology has been replaced by the flat display technology, the CNT-based display products, due to several unique properties, are still considered to be the next-generation display technology to enable more creative applications in our daily life.

This work uses a quantum mechanical model to explain the working principle of such CNT-based field emission display (FED). This model is modified by considering parameters such as the CNT diameter, tunneling probability, and electrical field. In the CNT-based band-to-band tunneling model, results show how the CNT diameter and applied electrical field affect the quantum tunneling probability. Numerical results can be used to improve the design and performance of applications.

Field Emission Display Model

CNT-based field emission display was proposed by Samsung in 1999 [1]. Figure 1 shows the schematic structure for the quantum tunneling at the metal-vacuum interface [2, 3]. The modified CNT-based quantum tunneling model has parameters: CNT diameter (d), effective mass (m_e) [4, 5], barrier height (ϕ_w), tunneling width (w), and electrical field (E).

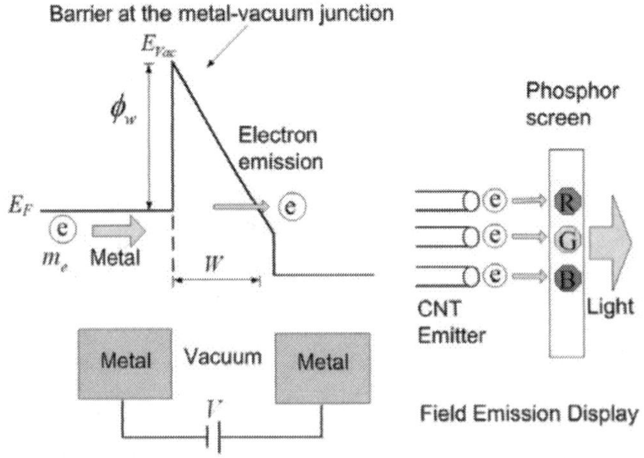

Figure 1 Schematic structure for quantum tunneling at the metal-vacuum interface

Band-to-Band Tunneling Model

CNT-based p-n junction with band-to-band tunneling in the transistor was proposed by IBM in 2004 [6]. Figure 2 shows the schematic structure for the band-to-band tunneling [3]. The modified CNT-based quantum tunneling model has parameters: CNT diameter (d), effective mass (m_e) [4, 5], CNT energy bandgap (Eg), tunneling width (w), and electrical field (E).

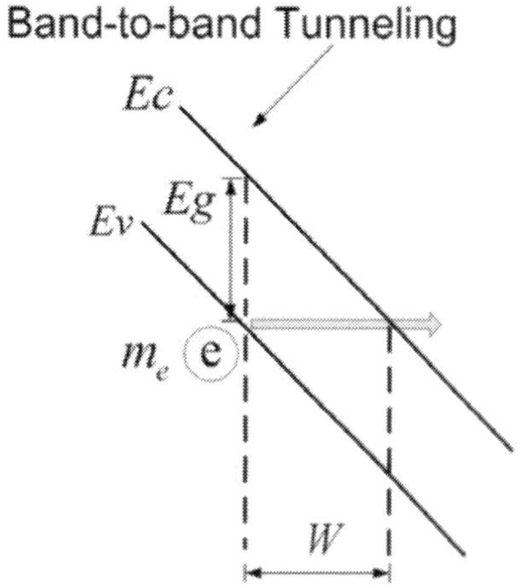

Figure 2 Schematic structure for quantum tunneling at the p-n junction

Numerical Results and Discussion

The calculation results for the field emission display model are based on the quantum tunneling model described in Figure 1. Figure 3 shows the electrical field (E) versus tunneling probability (T) for different materials. Since the metallic CNT with smaller diameter has larger work function as the tunneling barrier height, the tunneling is more difficult. The CNTs with diameters of 0.5 nm and 2 nm have the work function about 4.9 eV and 4.7 eV, respectively [7]. Tungsten has its work function of 4.55 eV and gold has its work function of 4.5 eV [3]. For a certain level of tunneling probability, the material with larger work function can have larger electrical field.

The calculation results for the band-to-band tunneling model are based on the quantum tunneling model described in Figure 2. Figure 4 shows the tunneling probability (T) versus CNT diameter (d) for electrical field E = 5×10^9 V/m and 10^9 V/m. Since the CNT with larger diameter has smaller energy bandgap as its tunneling barrier height and smaller effective mass, the tunneling is easier and the tunneling probability is higher than the CNT with smaller diameter. The larger electrical field can assist the tunneling and cause higher tunneling probability. Figure 5 shows the tunneling probability (T) versus electrical field (E) for CNT with diameter d = 6 nm and 2 nm. The CNT with larger diameter can reach tunneling probability of unity faster with smaller electrical field.

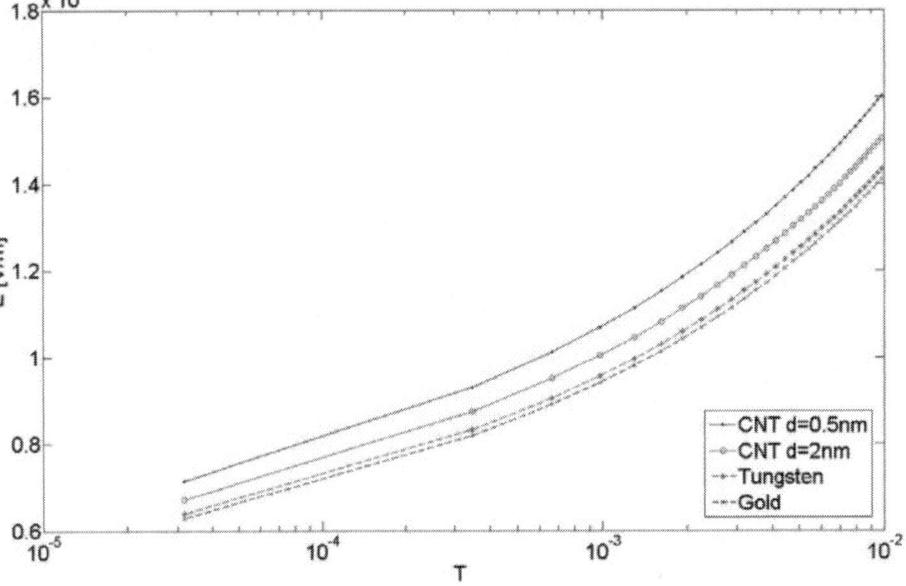

Figure 3 Electrical Field (E) versus tunneling probability (T) for different materials

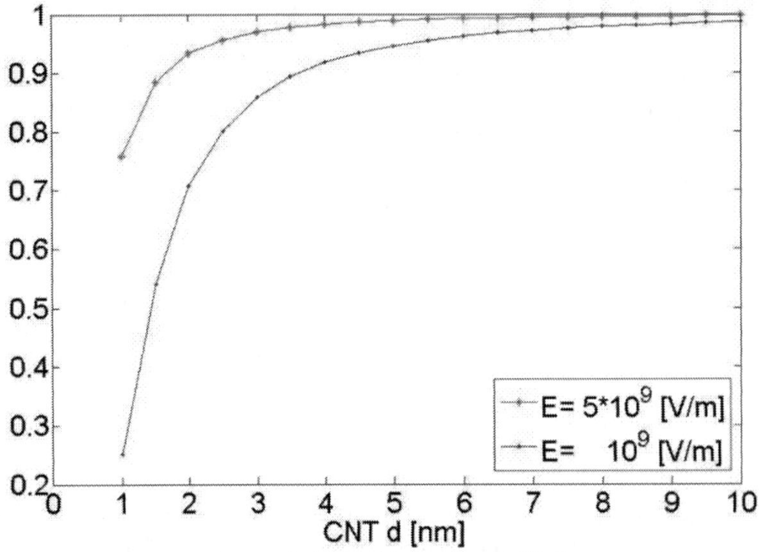

Figure 4 Tunneling probability (T) versus CNT diameter (d) for electrical field E = 5×10^9 V/m and 10^9 V/m

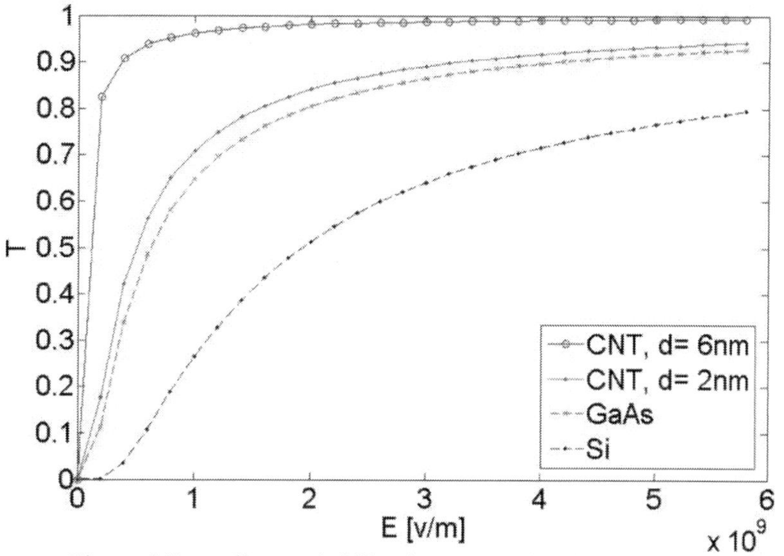

Figure 5 Tunneling probability (T) versus electrical field (E) for CNT with diameter d = 6 nm and 2 nm

References

1. Choi, W. B. et al. (1999), "Fully Sealed, High-brightness Carbon-nanotube Field-emission display," *Applied Physics Letters*, Vol. 75, No. 20, Nov., pp. 3129-3131.
2. Esaki, Leo (1973), "Long Journey to Tunneling," http://www.nobelprize.org/nobel_prizes/physics/laureates/1973/esaki-lecture.pdf
3. Singh, Jasprit (2004), "Modern Physics for Engineers," Weinheim, Germany, WILEY-VCH Verlag GmbH & KGaA.
4. Pennington, Gary and Goldsman, Neil (2003), "Theory and Design of Field-Effect Carbon Nanotube Transistor," in *International Conference on Simulation of Semiconductor Processes and Devices*, pp.167-170.
5. Marulanda, Jose M. and Srivastava, Ashok (2008), "Carrier Density and Effective Mass Calculations in Carbon Nanotubes," *Physica Status Solidi*, (b) Vol. 245, No. 11, pp. 2558–2562.
6. Appenzeller, J. et al. (2004), "Band-to-band tunneling in carbon nanotube field-effect transistors," *Physical Review Letters*, Vol. 93, No. 19, pp. 196805-1-3.
7. Shan, Bin and Cho, Kyeongjae (2005), "First Principles Study of Work Functions of Single Wall Carbon Nanotubes," *Physical Review Letters*, Vol. 94, 236602.

(176g) Palladium Nanowires Synthesized Via Templated Solid State Reduction for H2 Detection

Hector Mendez-Colberg and **Maria Martinez-Inesta**, Chemical Engineering Department, University of Puerto Rico, Mayaguez, PR

Pd nanowires have shown enhanced electrochemical detection of H_2 than the bulk due to higher surface to volume ratio. Here we show the results for ~5 nm Pd nanowires synthesized using zeolites as templates through a solid-state reduction method developed at our laboratory. After template removal, the Pd complex was characterized using TEM and hydrogen adsorption isotherms. Then it was deposited on a silicon based interdigitated electrode with platinum electrodes. The sensor showed the characteristic increase in resistance of bulk sensors, but with a mean response time of 10.12 seconds and sensitivity of 33% with exposure to 1% v/v H_2 that is orders of magnitude better than bulk sensors with recovery of the baseline. The hydrogen adsorption isotherm showed an α to β phase transition characteristic of nanoscale Pd.

See more of this Session: Nanostructured Materials for Electronic and Photonic Applications
See more of this Group/Topical: Materials Engineering and Sciences Division

(176h) Tunable Mirrors Made From Gold Nanoparticle Assembly At the Oil-Water Interface

Mingxiang Luo, Gloria Olivier and Joelle Frechette, Chemical and Biomolecular Engineering, Johns Hopkins University, Baltimore, MD

Nanoparticle self-assembly has opened the door to the design of materials with unique optical, electronic, and chemical properties. The interface between two immiscible liquids is a promising platform for nanoparticle assembly due to the rapid correction of defects afforded by the dynamic fluid interface, and optical applications afforded by the ability of light to access both sides of the interface. In particular, assembly of a two-dimensional array of metal nanoparticles at the oil-water interface enables the optical reflectance of the interface to be tuned according to the diameter of and interparticle spacing between the particles. Here we demonstrate the reversible or switchable assembly of reflective array of nanoparticles at a liquid interface. The films described herein possess both reversibility and mirror-like reflectance. Specifically, we show that 5 and 10 nm diameter gold nanoparticles, functionalized with an ion-pair, assemble spontaneously into pH-sensitive reflective sheets at the oil-water interface. The assembly occurs across a wide range of pH values (pH < 12) and more importantly, can be switched off and on reversibly by shifting the pH above and below pH 12, respectively. We address the role of the surface functionality on the reversible assembly and optical properties of the film at the oil-water interface. We also demonstrate control over the interfacial assembly based on an applied electrical potential.

See more of this Session: Nanostructured Materials for Electronic and Photonic Applications
See more of this Group/Topical: Materials Engineering and Sciences Division

(180a) Kinetically Trapping a Triblock Terpolymer Equilibrium Network Morphology for Use As a Polymer Electrolyte Membrane

Lucas D. McIntosh, Department of Chemical Engineering and Materials Science, University of Minnesota, Minneapolis, MN and Timothy P. Lodge, Department of Chemical Engineering and Materials Science and Department of Chemistry, University of Minnesota, Minneapolis, MN

We are exploring polymer electrolyte membranes exhibiting both high modulus and conductivity by kinetically trapping a triblock terpolymer equilibrium network morphology formed from poly[isoprene-*b*-(norbornenylethyl styrene-s-styrene)-b-ethylene oxide] (INSO). We show that INSO self-assembles into the O^{70} network morphology, an architecture in which both structural and conductive components are independently and three dimensionally continuous. The ring-opening metathesis polymerization of norbornene functionality incorporated into the styrene block is shown to kinetically trap the O^{70} morphology, preventing phase transitions induced by dissolving salts in the PEO phase. Dynamic mechanical spectroscopy is used to contrast mechanical properties of both uncrosslinked and crosslinked membranes; we find kinetic trapping slows segmental relaxation in glassy styrene by ca. one decade and prevents long-time relaxation of the O^{70} unit cell. Ionic conductivity through PEO in INSO membranes, as determined by impedance spectroscopy, is found to be largely unaffected by the kinetic trapping mechanism.

See more of this Session: Polymers for Energy Storage
See more of this Group/Topical: Materials Engineering and Sciences Division

(180b) Ionic Aggregation and Counterion Dynamics In Model Ionomers

Lisa M. Hall, Mark J. Stevens and Amalie L. Frischknecht, Sandia National Laboratories, Albuquerque, NM

Ionomers, polymers containing a small fraction of covalently bound ionic groups, have potential as solid, single ion conducting electrolytes in future batteries. However, the ions tend to form aggregates, making counterion diffusion unacceptably slow. A key materials design question is how molecular properties affect ionic aggregation and counterion dynamics. Recent experimental advances have allowed synthesis and extensive characterization of ionomers with a precise, constant spacing of charged groups. Because the molecular architecture is controlled and these materials show increased ionic aggregate ordering versus their randomly spaced analogs, this set of experiments is ideal for direct comparisons with molecular simulations.

We perform molecular dynamics simulations of coarse-grained ionomers with either periodically or randomly spaced charged beads. The charged beads are placed either in the polymer backbone (ionenes) or as pendants on the backbone. To understand the range of ionic aggregate morphologies possible in real materials, we vary the spacing of charges along the chain, degree of randomness (from periodic to random block to fully random), and dielectric constant. The well-known "ionomer peak" in the scattering is present in all cases. The peak is significantly more intense for pendant ions with a long periodic spacing of charged beads, which form roughly spherical aggregates. This morphology is in qualitative contrast to the extended aggregates of ionenes that show increased counterion diffusion. Depending on the degree of randomness in spacing of charged beads along the chain, counterion diffusion can increase or decrease versus that of the precisely spaced materials. Possible implications for ionomer electrolyte design will be discussed.

Sandia National Laboratories is a multi-program laboratory operated by Sandia Corporation, a wholly owned subsidiary of Lockheed Martin Corporation, for the U.S. Department of Energy's National Nuclear Security Administration under contract DE-AC04-94AL85000.

See more of this Session: Polymers for Energy Storage
See more of this Group/Topical: Materials Engineering and Sciences Division

(180c) Block°Copolymer Lithium Battery Electrolytes

Hany B. Eitouni, Seeo, Inc, Hayward, CA

With high energy density at low cost, Li ion has become the most prevalent portable rechargeable battery chemistry in the world. As demand for smaller and lighter batteries grows, the energy density limitation of Li ion batteries presents a significant hurdle. Pushing the existing Li ion platform to higher energy densities compromises lifetime and safety, and these have emerged as the most pressing challenges in today's industry. The weakest link in terms of safety and stability of Li ion batteries is the organic liquid electrolyte that facilitates ionic transport between the electrodes. The continuous electrochemical degradation of the electrolyte at the electrodes causes poor cycle life of the batteries, and in some cases, runaway reactions that lead to explosions.

Dry polymer electrolytes coupled to Li metal anodes had been considered a high energy alternative to liquid-based systems, as the solid-solid interface promised to alleviate the stability problems of the liquid electrolyte. However, repeated cycling of Li metal anodes leads to dendrite formation, reducing battery life and compromising safety. Recent theoretical work indicates that dendrite growth can be stopped if the shear modulus of current polymer electrolytes can be increased by three orders of magnitude without a significant decrease in ionic conductivity. Thus, the mechanical properties of polymer electrolytes are particularly important in rechargeable solid-state lithium batteries.

Because ion transport in polymers is coupled to the motion of the molecules that are solvating the ions, the presence of mobile molecules is essential to allow for a conductive medium. However, the same mobility of molecules is detrimental to the polymer's structural integrity. There is, thus, a clear need to develop methodologies for decoupling the conductive and mechanical properties of polymer electrolytes. Electrolytes comprised of self-assembled block-copolymer nanostructures overcome this principal constraint.

See more of this Session: Polymers for Energy Storage
See more of this Group/Topical: Materials Engineering and Sciences Division

(180d) Polyaniline/Vanadium Pentoxide Layer-by-Layer Battery Electrodes

Jodie Lutkenhaus, Department of Chemical Engineering, Texas A&M University, College Station, TX and Lin Shao, Department of Chemical Engineering, Yale University, New Haven, CT

As access to rare earths and other valuable transition metals narrows, the importance of identifying their potential replacements becomes a grand challenge. Cobalt is recognized as a material critical to the manufacture of Li-ion battery cathodes, but the supply of cobalt is limited. One potential alternative cathode material is polyaniline, which can be synthesized from domestic feedstock. Polyaniline is potentially an important material for organic energy storage applications because of polyaniline's high conductivity, specific energy, and cyclability. However, polyaniline's low specific power presents a major challenge to overcome. Here, we will present polyaniline/V_2O_5 cathodes made using layer-by-layer (LbL) assembly technique that may potentially address these challenges by carefully controlling (i) molecular structure via chemical synthesis, (ii) electrode topography, and (iii) mixing via LbL assembly. We will discuss *why* and *how* different types of polyaniline (commercial vs. custom-made) and different processing techniques (dip LbL vs. spray LbL) influence energy storage in the proposed hybrid battery electrodes. We explore these cathodes using non-aqueous electrochemistry, spectroscopy, and other materials characterization techniques. Results suggest that the electrochemical behavior is dominated by polyaniline and electrode thickness.

See more of this Session: Polymers for Energy Storage
See more of this Group/Topical: Materials Engineering and Sciences Division

(180e) Multilayer Assemblies of Polyaniline Nanofibers and Carbon Nanotubes for Electrochemical Applications

MD Nasim Hyder[1], Seung Woo Lee[2], Yang Shao-Horn[2] and Paula T. Hammond[1], (1)Chemical Engineering, Massachusetts Institute of Technology, Cambridge, MA, (2)Mechanical Engineering, Massachusetts Institute of Technology, Cambridge, MA

Polyaniline nanofibers (PANi) and acid functionalized multiwalled carbon nanotubes (MWNT) are used to fabricate nanostructured thin films for electrochemical capacitors using by layer-by-layer (LbL) assembly. The pH dependent surface charge of these components allow the system to grow in a linear but pH dependent fashion, much like weak polyelectrolytes, with controllable film thickness and morphology. The physicochemical properties and electrochemical behavior of the LbL films have been investigated by a variety of techniques including surface profilometry, infrared (IR) and Raman spectroscopy, scanning electron and atomic force microscopy and cyclic voltammetry. The experimental results reveal an interpenetrating network of PANi and MWNTs with a highly nanoporous morphology. The combination of the electrochemical redox behavior of PANi and the functionalized graphene surfaces of the MWNTs lead to high capacitance of 248 F/cm3 that are stable over 1000 cycles. Electrochemical measurements confirm the PANi/MWNT LbL film could be ideal for use in supercapacitor, fuel cell and battery electrode materials.

See more of this Session: Polymers for Energy Storage
See more of this Group/Topical: Materials Engineering and Sciences Division

(180f) Block Polymer Electrolyte Based Lithium-Air Batteries for Transportation

Daniel T. Hallinan Jr., Department of Chemical and Biomolecular Engineering, University of California, Berkeley, Berkeley, CA and Nitash P. Balsara, Department of Chemical Engineering, University of California, Berkeley, Berkeley, CA

The growing need for alternative energy and increased demand for mobile technology require higher density energy storage. Specifically, using batteries for transportation will enable improved efficiency and use of renewable energy. But, to maintain vehicle range and performance, batteries with an order of magnitude higher energy density are needed. New electrode chemistries that provide more energy with less mass, such as lithium-air, have the potential to fill this need. Lithium-air batteries are complex systems with many barriers to be overcome before they are competitive with current energy sources. Several of the challenges arise from the use of liquid electrolyte. In a system exposed to air, solvent in the electrolyte not only evaporates but is also flammable. Block polymer electrolyte mitigates this problem, while also serving multiple functions. In conventional systems the membrane separator and binder are inert, used only for their mechanical properties. Whereas a block polymer electrolyte can serve each of these functions and also conducts ions. Block polymer electrolyte battery cycling will be presented. The results compare with literature in terms of percent theoretical capacity accessed and fade. These batteries can be cycled at the same conditions and rate as conventional cathode chemistries. Improvements in fade and energy efficiency are being made by decreasing the over-voltage associated with charging by using better catalysts.

See more of this Session: Polymers for Energy Storage
See more of this Group/Topical: Materials Engineering and Sciences Division

POLYLACTIDE/CELLULOSE ACETATE/TRIETHYL CITRATE BLENDS AND THE EFFECT OF CELLULOSE ACETATE PERCENTAGE ON BLEND PROPERTIES

Barbara A. Wheelden, Amber R. Tupper, Leesha R. Blake, Mahin Shahlari[a], and Sunggyu Lee, Ohio University, Athens, OH

Abstract

Polylactide (PLA) is a bio-based, biodegradable polymer. Its relatively high tensile strength and stiffness make it a competitive option in the packaging industry. The greatest drawback to PLA which prevents its use in some applications is its brittle nature as evidenced by low impact strength and elongation at break. Efforts have been made to improve its toughness, including copolymerization of lactide with other monomers. Unfortunately, many methods used to toughen PLA can compromise its eco-friendly nature by increasing its carbon footprint and/or preventing eventual degradation. Cellulose acetate (CA) is another well-known bio-based and biodegradable polymer that is an organic acid ester of cellulose. Triethyl citrate (TEC) is an ester of citric acid. It is a natural compound that is nontoxic, acceptable for food packaging applications, and has been shown to plasticize both PLA and CA. The current work examines blends of PLA, CA, and TEC in varying percentages and examines the plasticizing effect of TEC on various properties of the blend including elongation at break, glass transition temperature, and modulus of elasticity. The goal of this work is to improve the properties of PLA without compromising PLA's novelty as an eco-friendly, petroleum-based polymer alternative.

Background

The replacement of conventional, petroleum-based polymer plastics with bio-based and/or biodegradable plastics has been a topic of much research and interest over the past decade. Historically, plastics are produced from nonrenewable fossil fuels, and once they are formed can exist for hundreds of years. The EPA estimates that 12.53 million tons of plastic packaging material were generated as waste in America in 2009[1]. Although many plastics are recyclable, only 7.1% of plastic products are recovered for recycling in the United States[1]. Even when plastics are recycled, the energy used to transport and recycle them contributes to increased energy usage and pollution.

Many promising alternative polymers have been developed to substitute traditional polymer packaging materials and are rapidly gaining popularity. Starch and cellulose based polymers are examples of bioplastics produced using familiar biological macromolecules[2,3]. Cellulose acetate is one such polymer. CA is made through the esterification of cellulose, which comes from a variety of natural resources including trees and cotton. It is used in various forms as a filler in polymer blends[4,5].

[a] Dr. Mahin Shahlari is currently with Northern Technologies International Corporation, Circle Pines, MN.

Bio-derived polyesters are produced from bio-based building blocks. Examples include polyhydroxyalkanoates (PHA) and polylactide (PLA) [6,7]. PLA is an aliphatic polyester that is produced from lactic acid (a fermentation product of sugar containing crops). This bioplastic has drawn much attention over the past two decades due to its comparable properties to many widely used petroleum-based plastics. PLA has a good tensile strength of 50-70MPa, but along with this desirable property comes the undesirable property of being brittle and stiff. Its elongation at break is a low 2-10% [7]. This hinders PLA's ability to be used in packaging applications where flexible films are often required. Plasticizing PLA by blending it with materials that enhance its flexibility and/or lower its glass transition temperature is a way to combat this drawback of the polymer. Much research has been conducted to find effective plasticizers of PLA[8]. Because of PLA's novelty as a bio-based polymer, it is important to maintain its eco-friendly nature even when processing it with different materials. Therefore, natural plasticizers for PLA have been found and studied[9].

Citrate esters, natural derivatives of citric acid, have been found to be effective and safe plasticizers of PLA[10,11]. Triethyl citrate is one such ester[12]. TEC has also been shown to effectively plasticize CA, which is difficult to process without plasticization owing to the fact that its degradation temperature is lower than its melting pressure when it is unmodified [13].

Blends of CA, TEC, and PLA can be produced by melt blending or solvent casting. Ultimately, melt blending in a batch mixer or an extruder is a more accurate representation of polymer processing in industry. However, since both PLA and CA are readily able to dissolve in a variety of organic solvents, and since small sample size is needed in order to verify that plasticization has occurred, solvent casting is an acceptable method to produce films simply and effectively. Melt blending and solvent casting techniques are explored in this work, and the benefits and drawbacks of each processing method discussed.

Materials

PLA 4042 was from Natureworks LLC. CA with a 39.8% acetyl content was from Eastman Chemical. TEC (99%) and chloroform (99+%) were from Acros Organics. Crown acetone was from Lowes.

Procedure

Solvent Casting Method

For samples produced via solvent casting, PLA in pellet form was ground in a bench top grinder from Cole-Parmer to expedite the dissolving process. It was then passed through a sieve to ensure a resulting size of less than 2.36mm diameter. CA was used as received in powder form.

PLA and CA were mixed in various weight percentages (wt%) with TEC accounting for a certain percentage of overall blend weight. Film samples were prepared via the solvent casting method by dissolving the polymer/plasticizer blend in an organic solvent. The solution was 95 wt% solvent. Solutions were placed in a round bottom flask which was then placed in a heated water bath. A condenser cooled by a recirculating water bath was used to prevent evaporation of solvent. Solutions were magnetically stirred until the polymer was fully dissolved, and additional solvent was added as needed. Solutions were then cast onto a glass dish, placed in a hood at room temperature, and the solvent allowed to evaporate. A film was produced on the bottom of the dish and removed.

Melt Blending Method

For samples prepared via melt blending, all materials were used as received. CA and PLA were dried at 60°C under vacuum for ~12 hours to remove excess moisture. After drying, the polymers were placed in a desiccator until use. For sample preparation, CA, TEC and PLA were mixed and added to a 300mL capacity CW Brabender three-piece counter-rotating mixer. The mixer operated at 32 rpm. A range of mixing sequences, temperatures and times were explored.

Sample Classification

Samples will be referred to as XXYYZZ where XX refers to the weight percentage of PLA in the blend with respect to total polymer content, YY refers to CA weight percentage with respect to total polymer percentage, and ZZ refers to overall weight percentage of TEC. For clarification, a 16.7g sample with 5g of PLA, 5g of CA, and 6.7 g of TEC will be referred to as 505040.

References

1. Environmental Protection Agency. Municipal Solid Waste in the United States. . 2009.

2. Cyras VP, Manfredi LB, Ton-That M, Vázquez A. Physical and mechanical properties of thermoplastic starch/montmorillonite nanocomposite films. *Carbohydr Polym.* 2008;73:55-63.

3. Simon J, Müller HP, Koch R, Müller V. Thermoplastic and biodegradable polymers of cellulose. *Polym Degrad Stab.* 1998;59:107-115.

4. Pandey JK, Lee CS, Ahn S. Preparation and properties of bio-nanoreinforced composites from biodegradable polymer matrix and cellulose whiskers. *J Appl Polym Sci.* 2010;115:2493-2501.

5. Alam MM, Ahmed T, Haque MM, Gafur MA, Kabir ANMH. Mechanical Properties of Natural Fiber Containing Polymer Composites. *Polymer-Plastics Technology & Engineering.* 2009;48:110-113.

6. Siracusa V, Rocculi P, Romani S, Rosa MD. Biodegradable polymers for food packaging: a review. *Trends Food Sci Technol.* 2008;19:634-643.

7. *Poly(Lactic Acid). Synthesis, Structures, Properties, Processing, and Applications.* . Hoboken, New Jersey: John Wiley & Sons, Inc., 2010.

8. Jacobsen S, Fritz HG. Plasticizing polylactide--the effect of different plasticizers on the mechanical properties. *Polymer Engineering & Science.* 1999;39:1303-1310.

9. Martin O, Avérous L. Poly(lactic acid): plasticization and properties of biodegradable multiphase systems. *Polymer.* 2001;42:6209-6219.

10. Labrecque LV, Kumar RA, Davé V, Gross RA, McCarthy SP. Citrate esters as plasticizers for poly(lactic acid). *J Appl Polym Sci.* 1997;66:1507-1513.

11. Ljungberg N, Wesslén B. Tributyl citrate oligomers as plasticizers for poly (lactic acid): thermo-mechanical film properties and aging. *Polymer.* 2003;44:7679-7688.

12. Ljungberg N, Wesslén B. The effects of plasticizers on the dynamic mechanical and thermal properties of poly(lactic acid). *J Appl Polym Sci.* 2002;86:1227-1234.

13. Park H, Misra M, Drzal LT, Mohanty AK. "Green" Nanocomposites from Cellulose Acetate Bioplastic and Clay: Effect of Eco-Friendly Triethyl Citrate Plasticizer. *Biomacromolecules.* 2004;5:2281-2288.

FORMATION OF POLYURETHANE FROM BIO-BASED MATERIALS

Amber R. Loftis, Barbara A. Wheelden, Leesha R. Blake, Mahin Shahlari[a], Sunggyu Lee
Ohio University, Athens, OH

Abstract

Polyurethanes have a wide range of applications in automobile seating, bedding, footwear, thermal insulation, and hard structural parts. Bio-based polyurethane is desirable as a renewable material. Vegetable oils have been considered for making polyurethanes since they are readily available, biodegradable, and have a low toxicity. Vegetable oils are triglycerides with different fatty acid chains which depend on the plant and the growing conditions. The chemical and physical properties of the oil depend on the degree of unsaturation. Oil derived from algae is a viable option as a polyurethane precursor because algae have a high yield compared to other oil sources and can grow in a variety of conditions. Different algae oils can be considered based on the desired fatty acids and the degree of unsaturation. In this work, castor oil was used as a surrogate for algae oil as it is more readily available and the same process can be used in polyurethane formation. The polyurethane was formed by the creation of a polyol through the ring opening of epoxidized castor oil which was reacted with diphenylmethane diisocyanate.

Introduction

Traditionally, polyurethanes are made from petroleum based polyols. The replacement of conventional, petroleum based polymer plastics with bio-based plastics has been a topic of much research and interest over the past decade. The EPA estimates that 12.53 million tons of plastic packaging material was generated as waste in America in 2009.[1] Roughly 7% of the world's oil and gas consumption goes to the manufacturing of polymers.[2] Polyols from vegetable oils are suitable bio-based alternatives to petroleum polyols.

Petrović has shown that the formation of polyurethanes from vegetable oils involves the reaction of a polyol with an isocyanate. One method to create polyols is through epoxidation and oxirane ring opening of vegetable oils. This converts the double bonds in the triglycerides into a hydroxyl group. The hydroxyl groups then react with the isocyanate. The resulting polyurethane is crosslinked due to the multiple hydroxyl groups per triglyceride. The hydroxyl groups are not terminal, meaning the remaining section of the fatty acid chain is a dangling chain which will plasticize the polyurethane and lower the elastic properties.[3] Crosslinking density, which increases with an increase in hydroxyl groups, has a greater effect on polyurethane properties than the position of the reactive sites does.[2,4,5]

Algae, specifically microalgae, is a promising source of oil because, compared to other crops, they have fast growth rates, potential for higher yield rates, and the ability to grow in a wide range of conditions.[6] The yield of oil per unit area of algae is at least seven times greater than that of palm oil, the second highest yielding crop.[7] Another benefit of using algae is that

[a] Dr. Mahin Shahlari is currently with Northern Technologies International Corporation, Circle Pines, MN

it overcomes the food versus fuel issue of other vegetables since algae is not a food crop and it does not take arable land from other crops.[6]

Fatty acids are named based on the chain length and number of double bonds; for example, linoleic acid (18:2) is an 18 carbon chain fatty acid with 2 double bonds. The placement of the first double bond further determines the name; for example, in ω3 fatty acid chains, the first double bond is on the third carbon from the chain end. The typical fatty acids in vegetable oils include palmitic (16:0), stearic (18:0), oleic (18:1), linoleic (18:2) and linolenic (18:3), while oil from algae can have significant percentages of long chain polyunsaturated fatty acid chains.[2] Examples of long chain fatty acids abundant in some algae are arachidonic acid (20:4), a ω6 fatty acid, and two ω3 fatty acid chains, eicosapentaenoic acid (20:5) and docosahexaenoic acid (22:6).[8]

Castor oil was used rather than algae oil as it is more readily available and can follow the same process to from polyurethanes. Castor oil is a good candidate for polyurethane production because its fatty acids consist of up to 90% ricinoleic acid.[9] Castor oil by itself is a polyol due to the hydroxyl group in these ricinoleic acid chains which means polyurethane can be formed directly from castor oil. There are about 2.7 hydroxyl groups per triglyceride in castor oil which are evenly distributed throughout the triglyceride resulting in a uniform crosslinked structure leading to good mechanical properties.[2] The two main disadvantages to using unmodified castor oil polyols are that it has a low hydroxyl number and it has a slow cure rate due to the secondary hydroxyls.[10] However, there are other double bonds that can be converted to hydroxyl groups which would result in a higher crosslink density.

The process of making vegetable oil based polyols has been the focus of many studies.[4,11] Vegetable oils used include soybean, canola, sunflower, corn and linseed oil. Many studies create polyurethane from unmodified castor oil polyols or castor oil modified by either a transesterfication reaction with another polyol or conversion to an amide.[2,9] One study synthesized epoxidized castor oil and found its thermal properties.[12]

Materials

Castor oil (Eur. Pharmaceutical Grade), acetic acid (99.8% for analysis), hydrogen peroxide (ACS 30%), methanol (99+%), fluoroboric acid (50 wt. % in water) were purchased from Acros Organics. Toluene (Histological Grade) was purchased from Fisher Scientific. Amberlite IR-120 (H) (Analytical) was purchased from Supelco. Lewatit MP-64 (Chloride Form) was purchased from Fluka. 4,4′-Methylenebis(phenyl isocyanate) (MDI) (98%) was purchased from Aldrich.

Procedure

The procedure to form a polyurethane by the reaction of the polyol, created though the ring opening of epoxidized castor oil, with MDI is similar to that done for other oils and triglycerides.[4,11,13]

Epoxidation of Castor Oil

The epoxidation of castor oil occurs when the oil reacts with peracetic acid which is formed by acetic acid and hydrogen peroxide. Toluene, castor oil, acetic acid and Amberlite IR-120 were added to a 3-neck round bottom flask with a thermometer, drop funnel, condenser and magnetic stirrer. The mixture was heated to 55°C. Hydrogen peroxide was added drop by drop so as to not exceed 60°C and reacted for 7 hours. The Amberlite was filtered off.

Oxirane Ring Opening

The epoxidized castor oil went through oxirane ring opening by reacting with methanol using fluoroboric acid as a catalyst. Methanol and fluoroboric acid were boiled in a 3-neck round bottom flask. The epoxidized castor oil was added drop by drop into the flask. The mixture was kept boiling for 35 minutes. Lewatit MP-64 was added to neutralize the catalyst and then removed by filtration. The solvent was removed by vacuum distillation followed by high vacuum at 60°C.

Polyurethane Formation

The polyurethane was synthesized by reacting the resulting castor oil based polyols with MDI. The polyols and MDI were heated to 60°C separately then mixed and poured into molds preheated to 100°C. The polyurethane was cured overnight at 100°C.

References

1. Municipal Solid Waste Generation, Recycling, and Disposal in the United States: Facts and Figures for 2009. *United States Environmental Protection Agency*. 2009.
2. Pfister DP, Xia Y, Larock RC. Recent Advances in Vegetable Oil-Based Polyurethanes. *ChemSusChem*. 2011;4:703-717.
3. Petrović ZS. Polyurethanes from Vegetable Oils. *Polymer Reviews*. 2008;48:109-155.
4. Zlatanić A, Lava C, Zhang W, Petrović ZS. Effect of structure on properties of polyols and polyurethanes based on different vegetable oils. *Journal of Polymer Science Part B: Polymer Physics*. 2004;42:809-819.
5. Xia Y, Larock RC. Vegetable oil-based polymeric materials: synthesis, properties, and applications. *Green Chem*. 2010;12:1893-1909.
6. Demirbas A, Fatih Demirbas M. Importance of algae oil as a source of biodiesel. *Energy Conversion and Management*. 2011;52:163-170.
7. Kent MS, Andrews KM. Biological research survey for the efficient conversion of biomass to biofuels. . 2007;SAND2006-7221.
8. Bigogno C, Khozin-Goldberg I, Boussiba S, Vonshak A, Cohen Z. Lipid and fatty acid composition of the green oleaginous alga *Parietochloris incisa*, the richest plant source of arachidonic acid. *Phytochemistry*. 2002;60:497-503.
9. Mutlu H, Meier MAR. Castor oil as a renewable resource for the chemical industry. *European Journal of Lipid Science and Technology*. 2010;112:10-30.
10. Ferrer MCC, Babb D, Ryan AJ. Characterisation of polyurethane networks based on vegetable derived polyol. *Polymer*. 2008;49:3279-3287.

11. Zlatanić A, Petrović ZS, Dušek K. Structure and Properties of Triolein-Based Polyurethane Networks. *Biomacromolecules.* 2002;3:1048-1056.
12. Park S, Jin F, Lee J. Synthesis and Thermal Properties of Epoxidized Vegetable Oil. *Macromolecular Rapid Communications.* 2004;25:724-727.
13. Petrovic Z, Guo A, Javni I. Process for the preparation of vegetable oil-based polyols and electroninsulating casting compounds created from vegetable oil-based polyols. . 2000;09/187,992.

(186c) Properties of Self-Assembled Peptide Fibers

Devin Ridgley and Justin R. Barone, Biological Systems Engineering, Virginia Tech, Blacksburg, VA

Micron-sized peptide fibers can be formed from the multi-scale assembly of combinations of peptides. The peptides required are a short, hydrophobic "template" and a longer, alpha-helix "adder". Nanometer-sized cross-beta fibrils assemble and then helically wind around each other to build the large fiber. The helical pitch is determined by the inter-fibril interaction energy and defines the modulus of the fiber. Modulus can then be controlled by choice of peptides and relative concentration. The fibers resemble silk and keratin in morphology and modulus can vary from a fraction of a GPa to several GPa. The peptide systems described are obtained from economical, abundant plant sources and show a unique way to easily build high performance biomaterials that rival more expensive, harder to obtain materials.

See more of this Session: Sustainable Biomaterials
See more of this Group/Topical: Materials Engineering and Sciences Division

(186d) Using a Natural Material to Enhance the Aggregation of Sediments and Bacteria: Implications In Water Treatment

Audrey Buttice, Chemical and Biomedical Engineering, University of South Florida, Tampa, FL and Norma A. Alcantar, Chemical & Biomedical Engineering, University of South Florida, Tampa, FL

Coagulant applications ranging from drinking water production to industrial waste treatment currently use inorganic chemicals and/or polymers. Costs and unwanted side effects from these chemicals have resulted in the study of naturally occurring, organic, inexpensive materials such as bacterial byproducts and plants. Indigenous knowledge and scientific studies suggests that the *Opuntia* genus of cactus may serve as a natural based coagulant for kaolin particles suspended in water. This project focuses on the use of the *Opuntia ficus-indica* as a potential coagulation tool for the separation of suspended bacteria (particularly *Bacillus cereus*, *Bacillus anthracis*, and *Escherichia coli*) and kaolin at high and low concentrations. Two distinct fractions of mucilage, a gelling (GE) and non-gelling (NE) were extracted from the cactus for studies. Through microscopy techniques, the surface characteristics of the extracted mucilage were studied and compared. Settling tests using kaolin and bacteria suspended in surrogate hard and soft waters demonstrated aggregation and increased settling times in the presence of mucilage. Ion concentration in the water was also observed to affect the settling and removal rates. Kaolin settling rate was observed to increase from 0.5 to 13.2 cm/min with mucilage treatments and flocculation of bacteria was observed in less than 10 minutes with high removal rates. Tests demonstrate that the two fractions of mucilage react differently under varying circumstances and suspension types. By testing both fractions of mucilage from the *Opuntia ficus-indica* the mechanism used in particle aggregation can be studied and the resulting knowledge can allow the mucilage to be applied according to its abilities. Mucilage is an ideal biomaterial for use as a coagulant because it is easy to grow, process, is biodegradable and avoids unwanted harmful side effects.

See more of this Session: Sustainable Biomaterials
See more of this Group/Topical: Materials Engineering and Sciences Division

(186e) Cactus Mucilage As An Emergency Response Biomaterial to Provide Clean Drinking Water

Daniela M. L. Stebbins[1], Dawn Fox[2], Audrey Buttice[2] and Norma Alcantar[2], (1)Department of Chemical & Biomedical Engineering, University of South Florida, Tampa, FL, (2)Chemical & Biomedical Engineering, University of South Florida, Tampa, FL

The production of potable water is usually very expensive, and involves chemical process such as flocculation and chlorination. The conventional flocculation process includes the use of inorganic salts and organic synthetic polymers as flocculants. Cost and environmental side effects of these flocculants are a public concern. Therefore, more than ever, the demand for environmentally friendly technologies and renewal bioproducts to replace synthetic flocculants in water treatment has greatly increased. Earthquakes or anthropogenic event have the potential to cause direct and indirect detrimental effects on the drinking water quality. The purpose of this study is to investigate cactus mucilage-based separation as a viable alternative treatment for water purification in emergency scenarios. The mucilage is an extract from the *Opuntia ficus-indica,* commonly known as Nopal or Prickly pear cactus. This readily available and inexpensive natural extract has been shown to remove turbidity, bacteria and arsenic from synthetic water in laboratory trials. It is hypothesized that the mucilage will be able to improve the quality of event-impacted potable water sources by lowering the suspended solids, bacteria and heavy metal content of the water. Samples of tap water (distributed from treatment center), well water, surface water and distributed water (bottled water and from tanker trucks) were collected from several places at a recently affected area: Port-au-Prince, Haiti. Turbidity, total suspended solids, fecal coliforms, biochemical oxygen demand, heterotrophic plate count, conductivity, pH, and trace metals were analyzed to establish the water quality and type of contamination. Up to this point, fifty percent of the evaluated samples have positive presence of coliforms. Water treated with cactus mucilage has lower levels of heavy metals. We have also tested the interactions and efficiency to clean water with surrogated bacteria for cholera, which is commonly seen after natural disasters when the anthropogenic conditions are susceptible. Samples from the top of water columns treated with the mucilage reveal that high bacterial removal rates were achieved. The cactus mucilage is a biomaterial with great potential to be used in water purification because of its long shelf life time, low cost and readily availability through sustainable agriculture.

See more of this Session: Sustainable Biomaterials
See more of this Group/Topical: Materials Engineering and Sciences Division

(186f) Arsenic Removal From Drinking Water Using Cactus Mucilage and Iron

Dawn I. Fox[1], Thomas Pichler[2], Daniel H. Yeh[3] and Norma A. Alcantar[1], (1)Chemical & Biomedical Engineering, University of South Florida, Tampa, FL, (2)Geosciences, University of Bremen, Bremen, Germany, (3)Civil & Environmental Engineering, University of South Florida, Tampa, FL

Consumption of arsenic contaminated groundwater poses a public health threat to millions of people globally, particularly those in developing communities without conventional, centralized water treatment facilities. Accessible technologies are needed which are easy to implement, operate and maintain and require little or no fossil fuel energy to work.

Cactus mucilage, extracted from the *Opuntia ficus-indica* (also known as Nopal and Prickly Pear cactus), is a natural flocculant shown to interact with dissolved arsenate. Ferric ions were introduced to the system to capitalize on the strong affinity of arsenate for ferric hydroxides. In batch experiments, ferric (Fe(III)) ions were contacted with arsenate (As(V)), then treated with cactus mucilage in a cylindrical column. After equilibration, sample aliquots were taken from the top and bottom of the column and tested for As and Fe. The mucilage accelerated precipitate formation and settling within 15 min of addition, achieving 90% arsenic removal as indicated by difference between initial and final dissolved As concentration in the column. Arsenic removal improved with increasing both mucilage concentration and Fe concentration. The role of the mucilage was demonstrated by untreated solutions showing no concentration difference. The mucilage is thought to assist the precipitation by providing a matrix for stable floc formation which leads to the removal of both iron and arsenic simultaneously, providing safe drinking water. This mucilage-based technology has the potential to be relatively inexpensive, and an environmentally sustainable alternative to synthetic polymer flocculants for removing arsenic from drinking water.

See more of this Session: Sustainable Biomaterials
See more of this Group/Topical: Materials Engineering and Sciences Division

(203a) Dual-Functional Electrospun Poly(2-Hydroxyethyl Methacrylate)

Lingyun Liu[1], Bo Zhang[1] and Reza Lalani[2], (1)Chemical and Biomolecular Engineering, University of Akron, Akron, OH, (2)Biology, University of Akron, Akron, OH

Poly(2-hydroxyethyl methacrylate) (pHEMA) has been widely used in many biomedical applications due to its well-known biocompatibility. For tissue engineering applications, porous scaffolds that mimic fibrous structures of natural extracellular matrix and possess high surface-area-to-volume ratios are highly desirable. So far a systematic approach to control diameter and morphology of pHEMA fibers has not been reported and potential applications of pHEMA fibers have barely been explored. In this work, pHEMA was synthesized and processed into fibrous scaffolds using an electrospinning approach. Fiber diameters ranging from 270 nm to 3.6 μm were realized. The pHEMA microfibrous membranes exhibited water absorption up to 280% (w/w) whereas the pHEMA hydrogel only absorbed 70% water. Fibrinogen adsorption experiments demonstrate that the electrospun pHEMA fibers highly resist nonspecific protein adsorption. Hydroxyl groups on electrospun pHEMA fibers were further activated for protein immobilization. The pHEMA fibrous scaffolds functionalized with collagen I significantly promoted fibroblast adhesion, spreading, and proliferation. It is concluded that the electrospun pHEMA fibers are dual-functional, i.e., they resist nonspecific protein adsorption meanwhile abundant hydroxyl groups on fibers allow effective conjugation of biomolecules in a non-fouling background. High water absorption and dual functionality of the electrospun pHEMA fibers may lead to a number of potential applications such as wound dressings, tissue scaffolds, and affinity membranes.

See more of this Session: Biomaterial Scaffolds for Tissue Engineering I
See more of this Group/Topical: Materials Engineering and Sciences Division

(203b) Injectable Polyurethane Composite Scaffolds Delay Wound Closure and Support Cellular Infiltration and Remodeling In Rat Excisional Wounds

Elizabeth J. Adolph[1], Andrea E. Hafeman[1], Lillian B. Nanney[2], Jeffrey M. Davidson[3] and Scott A. Guelcher[1], (1)Chemical and Biomolecular Engineering, Vanderbilt University, Nashville, TN, (2) Pathology, Cell & Developmental Biology, Vanderbilt University, Nashville, TN, (3)Pathology, Vanderbilt University, Nashville, TN

Injectable scaffolds present compelling opportunities for woundrepair and regeneration due to their abilities to fill irregularly shapeddefects and deliver biologics such as growth factors. However, there areseveral challenges that must be overcome, including toxicity of reactants orintermediates, curing in a reasonable amount of time, forming a sufficient porestructure, and achieving robust mechanical properties. In this study, we investigated the propertiesof injectable polyurethane (PUR) biocomposite scaffolds and their applicationin cutaneous wound repair using a rat excisional model. PUR scaffolds weresynthesized by reacting a lysine triisocyanate-poly(ethyleneglycol) prepolymer with a polysaccharide filler (hyaluronic acid [HA] orcarboxymethyl cellulose [CMC]) and a polyester triol with a backbone comprising60% caprolactone, 30% glycolide, and 10% lactide.°Cure profiles of the scaffolds weremeasured using a rheometer, thermal transitions were evaluated by differentialscanning calorimetry, and mechanical properties were measured using a dynamicmechanical analyzer. The reactivities of polyester triol, HA, CMC, and waterwith LTI-PEG were determined using attenuated total reflectance-fouriertransform infrared spectroscopy. The capacity of the scaffolds to facilitatedermal wound healing was evaluated in an excisional wound model inSprague-Dawley rats. The groups investigated were blank wounds, PUR scaffoldswith 15 wt% HA, and PUR scaffolds with 15% CMC. Wounds were harvested at days 7, 17, 26, and 35 (n = 4) and processedfor histological evaluation. Hematoxylin& eosin, Gomori's trichrome, picrosirius red, TUNEL, Ki67, a-SMA, and procollagen I immunostaining wereperformed on the tissue sections. The scaffolds have a minimal reaction exotherm (<10°C), a working time of 6-7 min, and a setting time of 16-19 min. Moreover, the compressive Young's modulus ofthe scaffolds under physiologic conditions ranges from 30-60 kPa, which iscomparable to that of human skin. In the ratexcisional wound model, injection of settable biocomposite scaffolds stentedthe wounds at early time points, resulting in a regenerative rather than ascarring phenotype at later time points. Measurements of wound length andthickness revealed that the treated wounds were less contracted at day 7compared to blank wounds. Analysis of cell proliferation and apoptosis showedthat the scaffolds were biocompatible and supported tissue ingrowth. Furthermore,the number of Ki67+ cells in the PUR scaffolds was significantly higher than inthe blank wounds from days 17 - 26. Myofibroblastformation and collagen organization provided evidence that the scaffolds have apositive effect on extracellular matrix remodeling by disrupting the formationof an aligned matrix under elevated tension. In summary, we have developed aninjectable biodegradable polyurethane biocomposite scaffold that enhancescutaneous wound healing in a rat model.

See more of this Session: Biomaterial Scaffolds for Tissue Engineering I
See more of this Group/Topical: Materials Engineering and Sciences Division

(203c) Characterizing Collagen Hydrogels for Tissue Engineering Applications

Yu Jer Hwang[1], Joseph Granelli[2], Jillian Larsen[2] and **Julia Lyubovitsky**[2], (1)Cell Molecular Biology Program, UC Riverside, Riverside, CA, (2)Bioengineering, UC Riverside, Riverside, CA

Developing analytical methods to report on collagen structures assembled *in vitro* would accelerate the pace of utilizing it in biomedical applications. This presentation will focus on our recent advances in characterizing fibrillar collagen prototype biomaterial systems prepared under the initial conditions commonly employed in contemporary biomedical, tissue engineering and medical device research. We examined the collagen nano- and micro-structures in the 3D collagen hydrogels formed at initial concentrations of 1 mg ml^{-1} to about 5 mg ml^{-1} , several incubation temperatures and those cross-linked with glyceraldehyde, ribose, glucose or genipin. This talk will describe the novel insights into the collagen nano-structures formed and origins of second harmonic generation signals within these hydrated scattering systems. The talk will also explain the effects of the cross-linking reagents employed to increase stiffness and strength of collagen-based hydrogels.

See more of this Session: Biomaterial Scaffolds for Tissue Engineering I
See more of this Group/Topical: Materials Engineering and Sciences Division

(203d) Assessing Viscoelastic Properties of Chitosan Scaffolds and Unifying with the Cyclical Tests

Swapnika Ratakonda, Upasana Manimagal Sridhar, R. Russell Rhinehart and Sundararajan. V. Madihally, School of Chemical Engineering, Oklahoma State University, Stillwater, OK

Assessing Viscoelastic Properties of Chitosan Scaffolds and Unifying with the Cyclical Tests

Abstract

With the availability of different natural and synthetic materials that can be introduced to human body, regenerating defective tissues outside the body has attracted significant interest. Three dimensional scaffolds can be prepared from both synthetic and natural materials that are i) compatible with the human body, ii) bio-degradable and iii) supportive of reparative cell colonization. Apart from being bio-compatible, tissue engineering scaffolds should have highly porous structures in order to aid biological activities and be mechanically strong to withstand the stresses and strains in the body. Since many tissues in the body exhibit viscous (like fluids) and elastic (like solids) behavior, prepared materials should have similar characteristics. Chitosan-based scaffolds have garnered significant consideration due to various advantages including cost, availability and biocompatibility. However, the viscoelastic characteristics of chitosan scaffolds have not been completely understood.

The objective of this study was to evaluate and model the viscoelastic characteristics of chitosan and chitosan-gelatin scaffolds prepared using freeze-drying. To validate the model, cyclical properties were predicted using the model and experiments were performed on the scaffolds. Scaffolds were frozen at different temperatures and freeze dried, similar to our previous publications. Using the scaffolds, uniaxial tensile properties were evaluated under physiological conditions (hydrated in Phosphate Buffered Saline (PBS) at 37°C) at a constant strain rate of 2.5 percent/s. Based on the linear region of tensile stress vs. strain plot, ramp and hold type of stress relaxation tests were performed for four successive cycles. Experiments were also performed by changing the ramp rates. Changes in the microstructure of the tested samples were evaluated using a stereo microscope. The micrographs of scaffolds after relaxation experiments showed orientation of pores suggesting the retention of the stretched state even after many hours of relaxation. Based on these observations, a model containing i) an hyperelastic spring (containing two parameters) and ii) retain pseudo components (containing three parameters) were developed in Visual Basic accessed through MS Excel. Also, leap-frogging method of optimization of parameters was used. Since one retain pseudo component could not predict the experimental stress relaxation, another component was added.

The model fitted the experimental stress-strain behavior of chitosan with an SSD of 1.792 and that of chitosan-gelatin with SSD of 0.933. Using the model parameters, cyclical behavior of chitosan and chitosan-gelatin were developed. Further, cyclical experiments were performed and compared to the model predictions. These results showed the model could be used to predict the cyclical behavior under the tested strain rates. When experiments were performed at very slow strain rates (5 %/min) than the model predictions (150%/min), larger discrepancy was observed. In summary, the pseudo-component modeling approach can accept any strain history, including sequential strain-and-hold stage, and provide good fits to experimental data. This approach can be extended to other biomaterial scaffolds to characterize the viscoelastic behavior.

See more of this Session: Biomaterial Scaffolds for Tissue Engineering I
See more of this Group/Topical: Materials Engineering and Sciences Division

(203e) Stiffness Tunable Peptide Amphiphile Hydrogels As Three Dimensional Cell Scaffolds

Brian Lin[1], Katie Megley[2], Nickesh Viswanathan[2], Dan Krogstad[3], Yichun Qian[2] and Matthew Tirrell[4], (1)Chemistry, University of California, Santa Barbara, Santa Barbara, CA, (2)Bioengineering, University of California, Berkeley, Berkeley, CA, (3)Materials, University of California, Santa Barbara, Santa Barbara, CA, (4)Institute for Molecular Engineering, University of Chicago, Chicago, IL

Peptide amphiphiles (PA) are an emerging class of materials that have been sought after for a wide breadth of applications including tissue engineering, drug delivery, vaccine, and biocatalysis. The bulk properties and modularity of PA-based hydrogels have made them particularly successful as cell scaffolds. The surface functionality of the gels can be easily modified by incorporating different PAs. However, little effort has been directed towards the tunability of gel stiffness, which has been demonstrated as an important cue in applications such as stem cell differentiation or regenerative medicine. In this report, we present our recent design of a pH responsive PA hydrogel as a three dimensional cell scaffold. The hydrogels exhibit pH responsiveness in the physiological range, where the stiffness is modulated by controlling peptide sidechain interactions. Below pH 6 the PAs are weak gels or viscous solutions and above pH 7, they form gels with storage moduli that range from 100 to 10,000 Pa. The applicability of this material as a three dimensional scaffold was examined with 3T3 fibroblasts, where cell toxicity was not observed over 72 hours. The design, physiochemical properties, and applications of this PA hydrogel will be discussed.

See more of this Session: Biomaterial Scaffolds for Tissue Engineering I
See more of this Group/Topical: Materials Engineering and Sciences Division

(203f) The Effect of Growth Factor Supplementation On Tendon Cell Migration, Viability, and Gene Expression In Anisotropic Collagen-Glycosaminoglycan Scaffolds

Steven R. Caliari, Lindsey S. Beyer and Brendan A. Harley, Chemical & Biomolecular Engineering, University of Illinois at Urbana-Champaign, Urbana, IL

It is estimated that over 62 million tendon and ligament injuries occur annually. The current clinical gold-standard for the treatment of tendon injuries is the use of autograft or allograft tissue. However, these approaches have limitations including donor site morbidity for the former and limited tissue availability for the latter. Optimized biomaterial constructs such as tissue engineering scaffolds could serve as potential alternatives. Collagen-glycosaminoglycan (CG) scaffolds have been used as regenerative templates for soft tissues like dermis and more recently bone and osteochondral tissue [1]. We have recently developed anisotropic CG scaffolds with aligned pore tracks for tendon tissue engineering [2]. Here we use these scaffolds to explore the effects of growth factor supplementation through both soluble and insoluble modes of presentation on tendon cell (TC) behavior.

Platelet-derived growth factor-BB (PDGF-BB) and insulin-like growth factor-1 (IGF-1) have both demonstrated dose-dependent effects on TC migratory capacity and viability in anisotropic CG scaffolds [2]. In this study two additional factors (basic fibroblast growth factor (bFGF) and stromal cell-derived factor-1α (SDF-1α)) were investigated as potential stimulators of TC bioactivity. Transwell assays were used to assess the effect of different growth factor doses and combinations on TC chemotaxis into CG scaffolds. Studies also compared the efficacy of growth factors presented freely in culture media (soluble) versus immobilized in CG scaffolds via carbodiimide chemistry [3] or direct photolithography [4] (insoluble). Ongoing work is exploring the effects of PDGF-BB, IGF-1, and bFGF, and SDF-1α on TC alignment, viability, metabolic activity, soluble collagen synthesis, and expression of tendon markers (collagen I and III, COMP, and tenomodulin).

References

1. Harley B.A.C. and L.J. Gibson. Chemical Engineering Journal, 2008. **137**(1): p. 102-121.
2. Caliari S.R. and B.A.C. Harley. Biomaterials, in press.
3. Shen Y.H. et al. Acta Biomater, 2008. **4**(3): p. 477-89.
4. Martin T.A. et al. Biomaterials, 2011. **32**(16): p. 3949-3957.

See more of this Session: Biomaterial Scaffolds for Tissue Engineering I
See more of this Group/Topical: Materials Engineering and Sciences Division

(210a) Carbon Fiber/Polyhedral Oligomeric Silsesquioxane/Carbon Nanotube (CF-POSS-CNT) Hybrid Reinforcement

Yudong Huang, School of Chemical Engineering and Technology, State Key Laboratory of Urban Water Resource and Environment, Harbin Institute of Technology, Harbin, China

A carbon fiber/polyhedral oligomeric silsesquioxane/carbon nanotube (CF–POSS–CNT) hybrid reinforcement was prepared by grafting CNTs onto the carbon fiber surface using octaglycidyldimethylsilyl POSS as the linkage in an attempt to improve the interfacial properties between carbon fibers and an epoxy matrix. X-ray photoelectron spectroscopy, scanning electron microscopy, dynamic contact angle analysis and single fiber tensile testing were performed to characterize the hybrid reinforcements. Interlaminar shear strength (ILSS), impact toughness, dynamic mechanical analysis and force modulation atomic force microscopy were carried out to investigate the interfacial properties of the composites. Experimental results show that POSS and CNTs are grafted uniformly on the fiber surface and significantly increase the fiber surface roughness. The polar functional groups and surface energy of carbon fibers are obviously increased after the modification. Single fiber tensile testing results demonstrate that the functionalization does not lead to any discernable decrease in the fiber tensile strength. Mechanical property test results indicate the ILSS and impact toughness are enhanced. The storage modulus and service temperature increase by 11 GPa and 17 °C, respectively. POSS and CNTs effectively enhance the interfacial adhesion of the composites by improving resin wettability, increasing chemical bonding and mechanical interlocking.

See more of this Session: Composites Interfaces
See more of this Group/Topical: Materials Engineering and Sciences Division

(210b) Magnetic Graphene Nanocomposites for Capturing Heavy Metal Ions From Polluted Water

Jiahua Zhu[1], Sowjanya B. Rapole[1], Suying Wei[2] and **Zhanhu Guo**[1], (1)Integrated Composites Laboratory (ICL), Dan F Smith Department of Chemical Engineering, Lamar University, Beaumont, TX, (2)Chemistry and Biochemistry, Lamar University, Beaumont, TX

A facile method to fabricate magnetic graphene nanocomposites decorated with monodispersed core@double-shell nanoparticles is developed in this work. High-resolution transmission electron microscopy and energy filtered elemental mapping revealed a core@double-shell structure of the nanoparticles with the core and inner shell being crystalline iron and iron oxide, and the outer shell of amorphous Si-S-O compound. This magnetic graphene nanocomposites shows an extremely fast chromium (VI) removal efficiency (completely removed in 5 min) from the waste water and a significantly high adsorption capacity (20.9 mg/g from 10 wt% nanoparticle loading). This material shows great advantages as compared to traditional adsorbents, such as active carbon, biomass, iron oxide that often require hours even days to remove the heavy metal ions. Also, we are able to separate the magnetic graphene nanocomposites from waste water easily by a permanent magnet within a few minutes.

See more of this Session: Composites Interfaces
See more of this Group/Topical: Materials Engineering and Sciences Division

(210c) Remendable Interfaces for Fiber-Reinforced Polymer Composites

Amy M. Peterson, Department of Chemical & Biological Engineering, Drexel University, Philadelphia, PA, Giuseppe R. Palmese, Drexel University, Philadelphia, PA and Robert Jensen, Materials Division, Army Research Laboratory, Aberdeen Proving Ground, MD

Interfacial damage is a major source of failure in fiber-reinforced composites. We have developed a method of healing at the resin-reinforcement interface based on the Diels-Alder reaction of furan and maleimide. The polymer network is a furan-functionalized epoxy-amine thermoset, while the reinforcement material is maleimide-functionalized glass fiber. Compatible functionalization of the polymer matrix and glass fiber results in a thermoreversible and remendable interface. Healing is induced by heating and subsequent cooling, allowing for Diels-Alder bonds to degrade and subsequently reform. Additionally, we have created a novel method of healing efficiency calculation in which any frictional or inertial forces are discounted. Therefore, any positive healing efficiency value is the result of the healing procedure. Based on this method of determining healing efficiency, the system reported demonstrates 42% healing efficiency with healing possible for up to five cycles. Strategies for characterizing Diels-Alder reaction kinetics at this interface and the development and characterization of alternate remendable interfaces will also be presented.

See more of this Session: Composites Interfaces
See more of this Group/Topical: Materials Engineering and Sciences Division

Morphological Study and Quantitative Analysis on Wear Debris of Carbon Nanofiber/High Density Polyethylene Composites

Tian Liu, Weston Wood, Bin Li, Brooks Lively and Wei-Hong Zhong[*]
Washington State University, Pullman, WA

1. Introduction

The polymeric nanocomposites as load-bearing materials have been investigated for many years [1-4]. These studies were mainly focused on the wear resistance and the wear surface analysis for the bulk materials. To date, very few studies on the characterization of wear debris of nanocomposites have been carried out [5]. However, the characterization of wear debris generated from sliding process is particularly important. The existence of debris with different morphologies and sizes may have a great impact on not only the wear surfaces between sliding pairs, but also the working environment around sliding components. In this case, the tribological behavior of a material could be changed during the continued wear process. The wear debris analysis can significantly affect mechanical systems with the multi-component composite system.

The main objective of this paper is to examine the influences of nanofillers on the wear debris morphologies and particle size distributions of high-density polyethylene (HDPE) nanocomposites during the continued wear testing. Morphology and particle size distribution of the debris of pure HDPE and its nanocomposites with 3 wt% filler loadings were analyzed. The analysis results indicated that polymeric composite system with nanofiller reinforcements exhibited the improved tribological behavior in comparison to pure polymer on these from two contributions: less mechanical damage and a reduction in the amount of long PE fibrils. In order to make a comparative study of the influences of different interfaces between the polymer matrix and the nanofibers on the composites wear debris, pristine CNF and our previously silane treated CNFs (thin and thick coatings) were selected as the reinforcements of HDPE [6]. Study results suggested that the debris size distribution and the morphology of the wear debris are related to the interaction status between the nanofibers and the polymer matrix. The experimental results remarkably revealed that heavily silanized CNF composites yielded smooth wear surfaces and a very small proportion of large wear debris.

2. Experimental procedures

2.1 Sample preparation

The preoxidized CNFs (oxCNFs) were modified under subsequent organosilane treatment. The coating thicknesses determined using TGA data were shown in our previous studies, which are heavily silanized CNF-h (about 46 nm) and lightly silanized CNF-l (about 2.8 nm) [6].

[*] Corresponding author. Tel.: +1 509 335 7658; fax:+1 509 335 4662
E-mail address: katie_zhong@wsu.edu

The 3 wt% fiber contents for both pristine CNFs (Comp-p) and silanized CNFs (Comp-l with light coating; Comp-h with heavy coating), were mixed with HDPE by a Haake Torque Rheometer. A neat HDPE reference was also prepared under identical conditions. Finally, HDPE and mixed polymeric nanocomposites were processed by Hot-press.

2.2 Wear debris treatment for separation

The wear tests were performed in a custom build rig for the samples, which was used in our pervious study [7]. All debris of each specimen that generated from the wear tests was collected from the surface of the disk after 24 hours. The detailed descriptions of each sample were shown in Table-1. The debris particles tended to be gathered through electrostatic attraction produced by the wear process. To separate the aggregated wear debris, the Nine-step Debris Treatment (NDT) was applied. All collected wear debris of each sample was mixed with NaOH and bath sonicated at room temperature. Then, glycerol was added to each sample. The mixtures were centrifuged overnight. The top layer of each sample was siphoned and placed in clean test tubes. All specimens were washed with deionized water. Finally, the treated samples were extracted randomly and put on different microscope glass slides for the particle length measurement and then quantitative analysis.

Table 1. Detailed descriptions for each sample

Samples	Reinforcement	Fiber Treatment	Coating Average Thickness (nm)	Fiber Content	Debris Collection
HDPE	-	-	-	-	24 hrs-Wear Testing
Comp-p	Pristine CNFs	-	-	3 wt%	
Comp-l	Silanized CNFs	lightly silane treated	2.8		
Comp-h	Silanized CNFs	heavily silane treated	46		

2.4 Characterizations

The particle sizes of separated wear debris after NDT were determined by using an optical microscope (Olympus BX51TRF). The quantitative analysis of wear debris size distribution was carried out through the length of each separated debris particle by optical image analysis. The experimentally determined particle number percentage provided the number proportion in each size range, and then two-dimensional particle size distributions of the wear debris were plotted.

To investigate the interaction between nanofibers and polymer matrix, the microstructure features on the fracture interface of HDPE nanocomposites before wear testing were characterized with field emission scanning electron microscope (FESEM type Quanta 200F). The overall fiber dispersion status in the nanocomposites were observed. In addition, the morphologies of separated wear debris of all specimens after NDT were also observed by FESEM.

3. Results and discussion

3.1 Morphology of the nanocomposites and amount of debris

The fracture surface morphologies of all nanocomposites with 3 wt% concentration of fibers were observed through SEM images (Fig.1 A-C). As we can see, no obvious big agglomerates were found in the nanocomposites with 3 wt% fiber-loading, indicating good dispersion of the nanofibers. Compared with Comp-p (Fig. 1A), large amounts of short CNFs were observed on the fracture surface of Comp-l (Fig. 1B) and Comp-h (Fig. 1C). Moreover, very few fibers were pulled out and there were few voids between the nanofibers and the matrix in these two types of nanocomposites. These phenomena are more obvious in Comp-h (Fig. 1C), revealing that sufficient entanglement of nanofibers with the polymer chains in heavily silanized CNF composites can induce an increased interaction between the fiber coating and the polymer matrix.

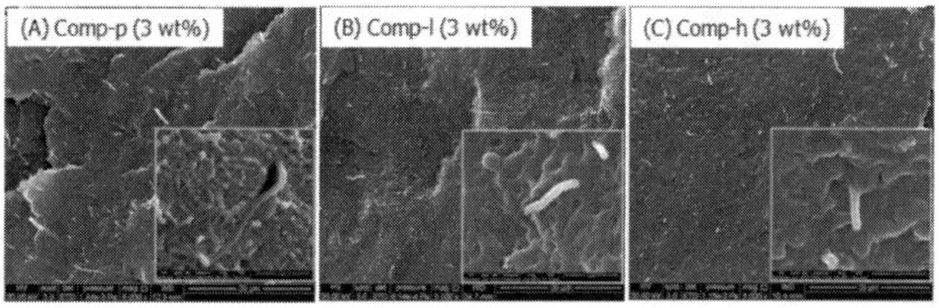

Figure 1 FESEM micrographs (scale bar: 20 μm; insert: 1 μm) of the fracture surfaces of HDPE nanocomposites reinforced by pristine CNFs (A), lightly silanized CNFs (B) and heavily silanized CNFs (C). The fiber concentrations of the nanocomposites were 3 wt%.

The debris amount was measured after 24 hours wear testing for the materials. It can be seen from Fig. 2 that all the nanocomposites exhibited lower wear debris amount than pure HDPE, indicating better wear resistance of the composite materials. Among three types of nanocomposites, Comp-h displayed the least amount of wear debris, reflecting the best wear resistance. Hence, it revealed that the organosilane coating, particularly for the thicker one, enhanced wear resistance of HDPE/CNF system effectively.

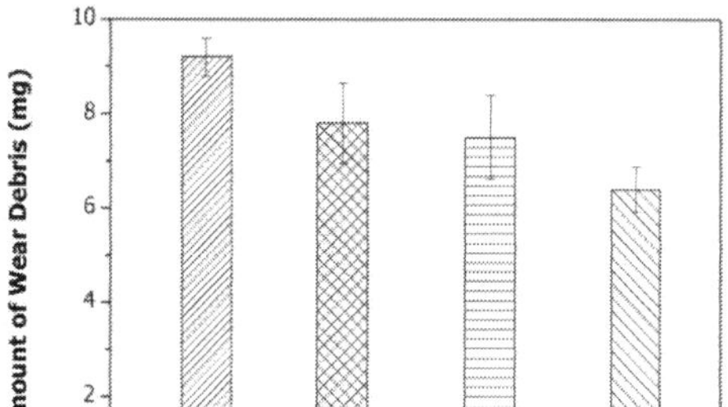

Figure 2 Amount of wear debris of pure HDPE and HDPE nanocomposites with different fiber loadings that collected after 24 hours pin-on-disk wear testing

3.2 Morphology of wear debris treated by Nine-step Debris Treatment

In our investigation of wear debris, some configurations of the separated pure HDPE wear debris after Nine-step Debris Treatment (NDT) were found. From Fig. 3 (A and B), there existed some cracks, ripples and micro-voids on debris surface. We deduced that they were attributed to the mechanical damage caused by the wear process.

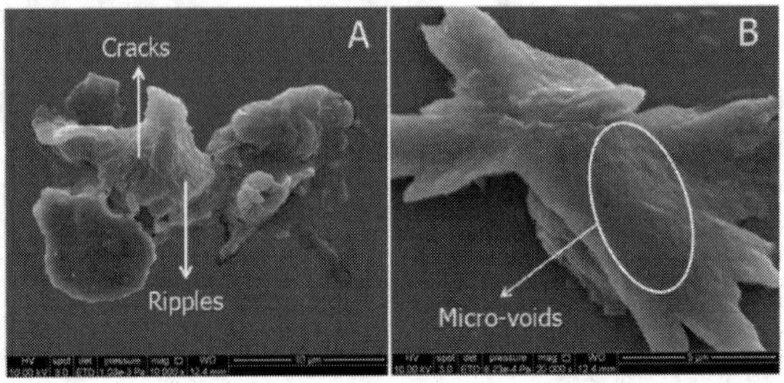

Figure 3 FESEM micrographs examples (scale bar: 10 μm for A; 5 μm for B) of separated pure HDPE wear debris

In Fig. 4 (A-C), the discrepancy of the wear debris morphologies among HDPE nanocomposites with various CNFs and different fiber concentrations can be observed. Compared with pure HDPE wear debris, there was fewer mechanical damage found in the nanocomposite debris, indicating the improved tribological behavior in the nanocomposites. This was more obviously seen in the two types of silanized CNF composites, especially for the samples of Comp-h (Fig. 4C). The wear debris surfaces of Comp-h were quite smooth. The distinction of the morphology suggests that the sufficient silane coating on the fiber surface could induce an increased strengthening in the interfacial zone between HDPE and CNF.

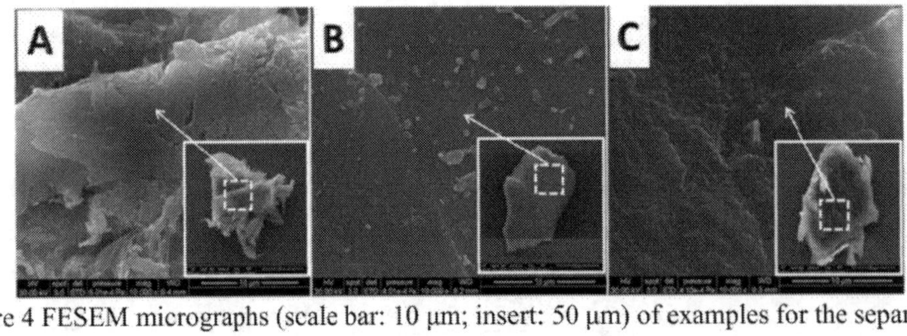

Figure 4 FESEM micrographs (scale bar: 10 μm; insert: 50 μm) of examples for the separated wear debris of HDPE nanocomposites: (A) Comp-p; (B) Comp-l; (C) Comp-h

3.3 Quantitative analysis of the wear debris

The individual wear debris treated by Nine-step Debris Treatment (NDT) was characterized using the maximum length of each particle observed from optical images. Some typical examples of the obtained optical images for the debris observation were shown in Fig. 5 (A-D). The measured maximum length of every separated debris particle was recorded for the further quantitative particle size distribution analysis.

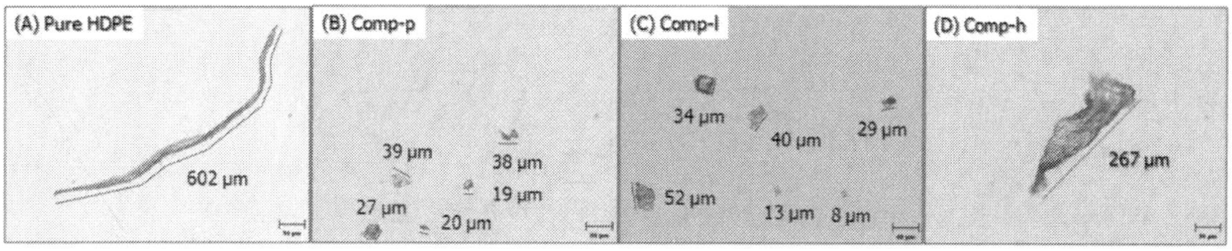

Figure 5 Examples of optical images (scale bar: 50 µm) for the separated wear debris of pure HDPE and its nanocomposites

Through statistical analysis, the percentage distributions of debris lengths below 1200 micron-meters for pure HDPE and HDPE nanocomposites with 3 wt% fiber concentration were displayed in Fig. 6 and Fig. 7 (A-C). It is clear that the proportions of small particles with the maximum lengths below 40 µm and large particles (length range: 300-1200 µm) in Comp-h wear debris were the lowest. For the specimens of Comp-p (Fig. 7A) and Comp-l (Fig. 7B), the proportions of small particles (below 40 µm) were higher than that of pure HDPE (Fig. 6).

The detailed situations of small (particle lengths were below 300 µm) and large (particle length range: 300-1200 µm) debris particle size distributions for pure HDPE and its composites can be seen clearly in the insert charts of Fig. 6 and Fig. 7 (A-C). According to the results of small particle size distribution, most of debris lengths were below 30 µm. For two types of silanized CNF composites (Fig. 7B and C), the small particles with the lengths below 10 µm constituted of high proportions of wear debris. From the other insert chart in Fig. 6 and Fig. 7 (large debris size distribution), it should be noticed that the maximum length of pure HDPE wear debris existed in the range of 750-1200 µm. Very large debris with the length above 1050 µm was attributed to the existence of long PE fibrils. Among the debris of three nanocomposites, there were no such very large particles and the size ranges were shifted to 300-900 µm. In particular, the maximum length s of large debris particles of Comp-h was fairly small (around 450 µm).

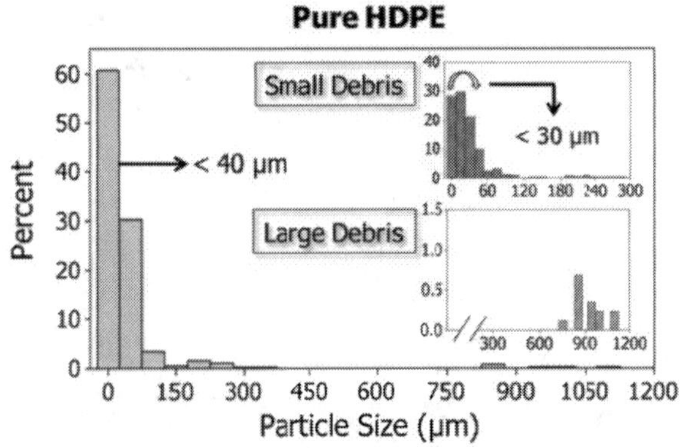

Fig.6 Particle size distributions (particle lengths were below 1200 μm) for the wear debris of pure HDPE. Two insert bar charts were the detailed particle size distributions for small debris with the sizes below 300 μm and large debris with the size range of 300-1200 μm.

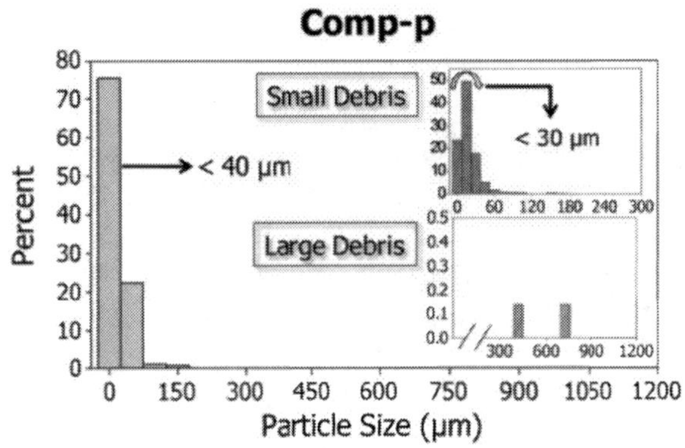

(A)

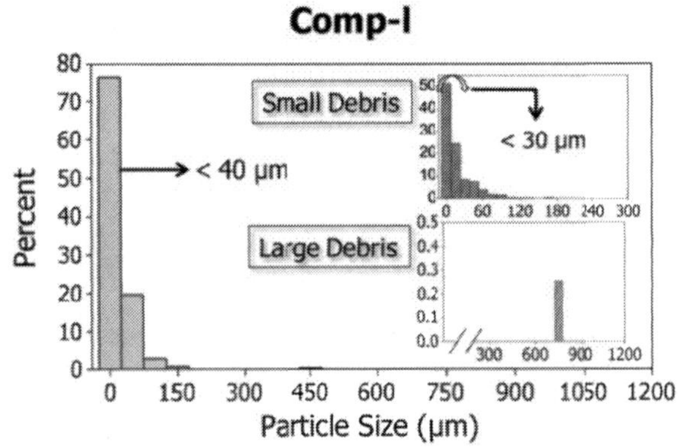

(B)

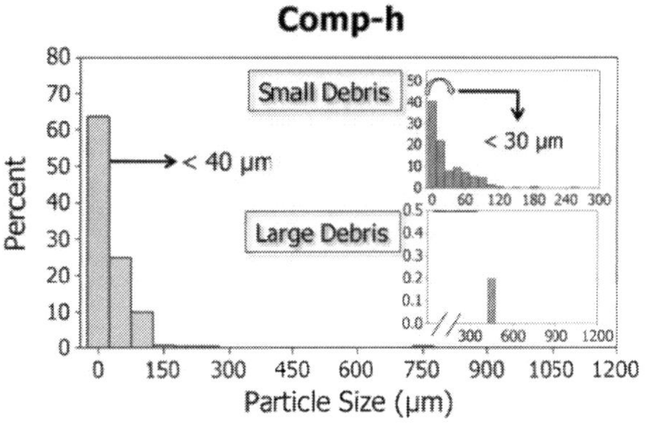

(C)

Fig.7 Particle size distributions (particle lengths were below 1200 μm) for the wear debris of HDPE nanocomposites specimens. (A) Comp-p; (B) Comp-l; (C) Comp-h. Two insert bar charts in each figure were the detailed particle size distributions for small debris with the sizes below 300 μm and large debris with the size range of 300-1200 μm.

4. Conclusions

The investigation on morphology characterization and quantitative analysis for the wear debris of HDPE/CNF multi-component systems were reported in this paper. The results revealed that less wear debris were generated during the wear process for the organosilane treated CNF composites with a thick coating on the fiber surface, indicating the best wear resistance. The morphology studies and quantitative analysis for the wear debris showed that the smooth surfaces with a very small amount of mechanical damage, and the lowest proportion of large debris existed in this modified CNF composites, which was attributed to the enhanced interaction and adhesion between the polymer matrix and the nanofibers.

References

1. K. Li, S.C. Tjong (2011), "Preparation and Mechanical and Tribological Properties of High-Density Polyethylene/Hydroxyapatite Nanocomposites", Journal of Macromolecular Science B, 50:1325-1337.
2. Y.S. Zoo, J.W. An, D.H. Lim, D.S. Lim (2004), "Effect of Carbon Nanotube Addition on Tribological Behavior of UHMWPE", Tribology Letters, 16:305-309.
3. S.Y. Li, H. Chen, D.M. Cui, J.Z. Li, Z.J. Zhang, Y.H. Wang, T. Tang (2009), "Structure and Properties of Multi-walled Carbon Nanotubes/Polyethylene Nanocomposites Synthesized by in situ polymerization with supported Cp_2ZrCl_2 Catalyst", *Polymer Composites*, 31:507-515.

4. G. Sui, W. H. Zhong, X. Ren, X.Q. Wang, X.P. Yang, M.A. Fuqua, C.A. Ulven (2009), "Structure, Mechanical Properties and Friction Behavior of UHMWPE/HDPE/Carbon Nanofibers", *Materials Chemistry and Physics*, 115:404–412.
5. J. Lu, M.Q. Zhang, M.Z. Rong, S.L. Yu, B. Wetzel, K. Friedrich (2004), "Thermal Stability of Frictional Surface Layer and Wear Debris of Epoxy Nanocomposites in Relation to the Mechanism of Tribological Performance Improvement", *Journal of Materials Science*, 39:3817-3820.
6. W. Wood, S. Kumar, W.H. Zhong (2010), "Synthesis of Organosilane-Modified Carbon Nanofibers and Influence of Silane Coating Thickness on the Performance of Polyethylene Nanocomposites", *Macromolecular Materials and Engineering*, 295:1125-1135.
7. T. Liu, W. Wood, W.H. Zhong (2010), "Sensitivity of Dielectric Properties to Wear Process on Carbon Nanofiber/High-Density Polyethylene Composites", *Nanoscale Research Letters*, 6:7, doi: 10.1007/s11671-010-9748-1.

CREASE RESISTANCE PROPERTY OF TEXTILE FABRIC / NANO CHEMICAL COMPOSITE STRUCTURE

Bilgen Çeliktürk, Cem Güneşoğlu

Textile Engineering Department, Gaziantep University, Gaziantep, 27310, Turkey

Textile fabrics are subjected to various treatments for required finishing effects by specific chemicals. Water / oil repellency and crease recovery are through the most applied treatments which are also used for products defined under "easy-care". Regarding to improve crease recovery, mostly applied approach has been to introduce crosslinks between individual fibre chains. Crosslinking agents are usually small molecules containing several functional groups capable of reacting with some active groups in the polymer, such as hydroxyl groups in cellulose. N-methylol reagents such as dimethyloldihydroxylethyleneurea (DMDHEU) have long been used as durable press finishes producing crease-resistant fabrics. The drawbacks of DMDHEU application are the well-known formaldehyde problems and the drop in some mechanical problems; thus there have been efforts to achieve nonformaldehyde alternative agents or reduce the amount of DMDHEU within treating recipes. Additionally, nano-sized cross-linking chemicals are claimed to emit less formaldehyde by higher level of fixation. This study evaluates the crease recovery properties of 100% cotton woven fabric composites produced by embedment of commercially available nano cross-linking chemicals via novel foam application system CFS by Gaston Systems to utilize from both reduced particle size and less chemical consuming effects. The production recipes had same curing conditions but varied at different pH levels. Crease recovery angle, abrasion resistance, tear and tensile strength, whiteness and yellowness index values of the samples were instrumentally evaluated.

(210f) Atmospheric Pressure Radio Frequency Plasma Activation of Composites for Adhesive Bonding

Thomas S. Williams[1], **Hang Yu**[2] and Robert F. Hicks[1], (1)Chemical & Biomolecular Engineering Department, University of California Los Angeles, Los Angeles, CA, (2)Chemical and Biomolecular Engineering Department, University of California Los Angeles, Los Angeles, CA

Atmospheric pressure radiofrequency plasma activation of composites for adhesive bonding

Thomas S. Williams, Hang Yu andRobert F. Hicks

Universityof California at Los Angeles, Department of Chemical andBiomolecular Engineering,

LosAngeles, California 90095-1592

The adhesion of bondedcomposite joints is of great concern to the aerospace, automotive, medicaldevice, and electronics industries. Recently, our group has developed methodsfor surface preparation of polymers and composites using low-temperature,atmospheric pressure radio frequency helium and oxygen plasmas. This is a promisingtechnique for replacing traditional methods of surface preparation byabrasion. The plasma source generates an afterglow containing ~$1x10^{17}cm^{-3}$ oxygen atoms, ~$1x10^{17}$ cm^{-3} metastableoxygen molecules ($^1\Delta_g$), and ~$4x10^{14}$ cm^{-3}ozone molecules. With sufficient exposure to the afterglow, polymer andcomposite surfaces are fully activated such that when bonded and cured withepoxy adhesives, they undergo 100% cohesive failure. Depending on thematerial, lap shear strength and delamination resistance can be increased over50% from that activated by solvent wiping or abrasion. Bond strength does notcorrelate with water contact angle or surface roughness. Instead it correlateswith the fraction of the polymer surface sites that are oxidized and convertedinto carboxylic acid groups, as determined by x-ray photoemission and infraredspectroscopy. Polymers with aromatic groups in the backbone are especially suitablefor activation with the atmospheric pressure plasma. The activation mechanism,surface characterization results, and bonding strength improvement will bediscussed in detail at the conference.

See more of this Session: Composites Interfaces
See more of this Group/Topical: Materials Engineering and Sciences Division

(210g) Polymer Nanocomposites Comprised of Diphenylamine End-Group Polymer and Quantum Dots

Yunfeng Li[1], Jiahua Zhu[1], Xi Zhang[1], Rahul Patil[1], Suying Wei[2] and Zhanhu Guo[1], (1)Integrated Composites Laboratory (ICL), Dan F Smith Department of Chemical Engineering, Lamar University, Beaumont, TX, (2)Chemistry and Biochemistry, Lamar University, Beaumont, TX

Nanocomposite films comprised of diphenylamine end-group polymers and CdSe/ZnS quantum dots (QDs) were fabricated electrochemically from the monomers and QDs solution on an Indium Tin Oxide (ITO) coated glass. The growth thermodynamics was investigated by the electrochemical methods such as cyclic voltammetry (CV). These films were also characterized by fluorescence microscopy, spectroscopies (FTIR, UV-Vis, photoluminescence), and SEM. These nanocomposite films show very strong fluorescence. The QDs were distributed as circles in the polymeric matrix as observed by fluorescence microscopy and SEM. Air-stable electrochromical window was also assembled and operated without obvious degradation even after 100 times scan. The UV-Visible incorporated electrochemical spectroscopies were also studied on these novel nanocomposite thin films.

See more of this Session: Composites Interfaces
See more of this Group/Topical: Materials Engineering and Sciences Division

(212a) Time-Dependent, Thermal-Capillary Analyses of the Micro-Pull-Down Crystal Growth System

Gaurab Samanta, Andrew Yeckel and Jeffrey J. Derby, Chemical Engineering and Materials Science, University of Minnesota, Minneapolis, MN

The micro-pull-down crystal growth technique is becoming increasingly popular to grow small-diameter, fiber crystals. Recently, it is also being developed for quick screening of promising scintillator crystals for radiation detectors in the crystal growth facility of E.D. Bourret-Courchesne at Lawrence Berkeley National Laboratory (LBNL). In this use, time delays are often associated with establishing suitable growth conditions for micro-pull-down experiments for new materials that have never been grown before.

The goal of our research is to develop computational models that can address the determination of suitable growth conditions in the micro-pull-down systems using a minimal amount of information. In keeping with this goal, we first developed a two- dimensional, quasi-steady-state (QSS), thermal-capillary model of the micro-pull-down system to grow sapphire crystals. We used this model to identify growth windows with respect to process parameters representing heater input, ambient temperature, pull rate, and melt static head. However, we did not establish the stability of the QSS solution branches.

In this presentation, we focus on results from transient simulations of the thermal- capillary model used to establish the stability behavior of those solution branches. For this purpose, starting with a quasi-steady-state solution, we provide step changes to any of the process parameters stated above and evolve the thermal-capillary system in time. By following the crystal radius evolution over time, we conclude whether a particular solution branch is stable or not. We also conduct transient simulations of a batchwise set-up of the micro-pull-down process and, for the first time, present temporal predictions of system evolution.

See more of this Session: Computational Studies of Electronic and Photonic Materials I
See more of this Group/Topical: Materials Engineering and Sciences Division

(212b) Simulation of Heat Transfer and Convection During Sapphire Crystal Growth by the Heat Exchanger Method

Hyun Gyoon Park[1], **Nan Zhang**[2] and Jeffrey J. Derby[2], (1)Advanced Materials Engineering, Korea Polytechnic University, Shiheung, Gyeonggi, South Korea, (2)Chemical Engineering and Materials Science, University of Minnesota, Minneapolis, MN

Sapphire single crystals are an important substrate for the fabrication of gallium-nitride light-emitting-diodes (LEDs) that emit white light. Theses solid-state devices promise great energy savings over both incandescent and compact fluorescent lighting technologies, and their market is expanding at double-digit growth rates. Key to continuing cost reduction for such LED-based lighting is reducing the cost of sapphire without compromising quality. The heat exchanger method (HEM) represents a possible yield improvement over existing growth technologies (primarily Kyropoulos growth) and is the subject of the study presented here.

We have performed quasi-steady state analysis using the commercial code CrysMAS to investigate the effects of furnace geometry and processing parameters on the temperature distribution, convection field, and melt-crystal interface shape during the crystal growth of sapphire by heat exchanger method (HEM). The temperature profile in the system is calculated to have lower temperatures both at the base of crucible and at the center top of the melt; however, the presence of convection flow arising from Marangoni force and buoyancy can ameliorate the latter effect. The interface shape is also predicted to change from flat to convex as crystal growth proceeds. Changes in geometry of furnace, such as segmentation of heater, changing the crucible position relative to heater center, and altering the width of crucible support, are shown to affect the temperature profile significantly. Based upon such results, optimization of furnace design will be presented.

See more of this Session: Computational Studies of Electronic and Photonic Materials I
See more of this Group/Topical: Materials Engineering and Sciences Division

(212c) Analysis of Current-Driven Morphological Evolution of Monolayer-Thick Coherently Strained Heteroepitaxial Islands On Substrates

Georgios I. Sfyris, Dwaipayan Dasgupta, M. Rauf Gungor and Dimitrios Maroudas, Chemical Engineering, University of Massachusetts Amherst, Amhest, MA

Surface morphological evolution under the action of external fields has attracted considerable attention over the past two decades. In addition to the interest in a fundamental understanding of field-induced nonlinear response and stability of surface morphology, the problem has been technologically significant in various engineering applications such as microelectronics and nanofabrication. Of particular interest is the driven assembly of confined quantum structures, such as islands grown epitaxially on substrates of different chemical composition than that of the islands. The lattice mismatch between the island and substrate materials induces a misfit strain, which is the source of elastic deformation of the islands. The surface morphological dynamics induced by an externally applied field, such as an electric field, under the simultaneous action of the misfit strain can give rise to morphological patterns, the stability and control of which may have significant impact on fabrication technologies.

In this presentation, we report the results of a theoretical and computational analysis of the current-driven morphological response of coherently strained heteroepitaxial islands on elastic substrates. We have developed a fully nonlinear model for studying the current-driven morphological evolution of one-monolayer-thick coherently strained heteroepitaxial islands on elastically deformable substrates. The model follows a two-dimensional (2D) approximation of island evolution, according to which diffusional mass transport is limited to the island circumference. Within this approximation (i.e., for this mass transport regime), we have carried out a theoretical analysis of the morphological stability of such isolated islands and of the current-driven migration of morphologically stable islands; the theory predicts the dependence of the stable island migration speed on the island size and on the heteroepitaxial system parameters. We have also developed self-consistent dynamical simulators of the driven morphological evolution of such islands, combining front tracking methods with an analytical solution for the corresponding electric-field component tangential to the island's circumference and a numerical computation of the corresponding elastic displacement field. Simulation results are reported together with systematic parametric studies to investigate thermal and size effects on the island's migration and morphological evolution, as well as effects on the island driven dynamical response of elastic properties of the substrate and island materials (such as elastic moduli and misfit strain). We demonstrate a variety of stable asymptotic states in the driven dynamical response of such heteroepitaxial islands and characterize in detail the resulting morphological patterns.

The simulation results validate our theory for stable heteroepitaxial island driven migration. The results also generate experimentally testable hypotheses and motivate experimental measurements that can be compared directly with the theoretical predictions. Finally, the present modeling study sets the stage for designing simulations of the current-driven evolution of populations of heteroepitaxial islands and examination of collective phenomena, such as island coalescence.

See more of this Session: Computational Studies of Electronic and Photonic Materials I
See more of this Group/Topical: Materials Engineering and Sciences Division

(212d) Dielectric Properties of Bismuth Pyrochlores From First Principles Calculations

Beverly Brooks Hinojosa[1], Juan C. Nino[1] and Aravind Asthagiri[2], (1)Department of Materials Science and Engineering, University of Florida, Gainesville, FL, (2)Department of Chemical & Biomolecular Engineering, The Ohio State University, Columbus, OH

There is considerable interest in pyrochlore systems ($A_2B_2O_7$) for use in high-permittivity dielectrics, capacitors, and high-frequency filter applications. The properties of these materials can be tuned through substitutions on the A and B cation sites, resulting in an extensive parameter space. Better understanding of the role of the local atomic structure and dynamics on the macroscopic properties will enable rational design within the vast number of possibilities. For example, the dielectric relaxation in Bi-based pyrochlores were linked to local atomic hopping events and it has been proposed that a highly polarizable cation, chemical substitution, atomic displacement, and the resulting fractional occupancy of equivalent sites, are necessary for dielectric relaxation in pyrochlores. Based on density functional theory (DFT) calculations, we have shown $Bi_2Ti_2O_7$ is expected to have ionic displacements. Being a bismuth based pyrochlore without chemical substitution of multi-valence cations, $Bi_2Ti_2O_7$ is an ideal compound to de-convolute these features and better understand the nature of the dielectric relaxation in pyrochlores. Quantum mechanical calculations were performed on $Bi_2Ti_2O_7$ and other pyrochlores ($Bi_{1.5}ZnNb_{1.5}O_7$, $Ca_{1.5}Ti_{1.5}NbO_7$) with the Fd-3m (No. 227) space group to determine the role of a highly polarizable cation, chemical substitution, and atomic displacement. We will present DFT results on the structural properties of $Bi_2Ti_2O_7$ and other Bi-containing pyrochlores including lattice stability, atomic displacement patterns, atomic hopping events, and the static dielectric permittivity. Based on these results, predictions will be made regarding the likelihood of dielectric relaxation in $Bi_2Ti_2O_7$ which further illustrate the need to synthesize the $Bi_2Ti_2O_7$ pyrochlore without the ferroelectric $Bi_4Ti_3O_{12}$ secondary phase, which is expected to mask the intrinsic dielectric behavior of the pyrochlore. Where possible, these predictions are compared with available experimental results.

See more of this Session: Computational Studies of Electronic and Photonic Materials I
See more of this Group/Topical: Materials Engineering and Sciences Division

(212e) SrTiO3 Based Anode Materials for Solid Oxide Fuel Cells: A Computational Attempt to Understanding and Improving Performance

Andreas Heyden and Suwit Suthirakun, Department of Chemical Engineering, University of South Carolina, Columbia, SC

In recent years, there has been a significant interest in alternative anode materials for solid oxide fuel cells (SOFCs) that are active and stable under hydrocarbon fuel conditions. The requirements for promising SOFC anode materials are good chemical and physical stability under operating conditions, high ionic and electronic conductivity, and good catalytic activity for the anode reaction. Furthermore, the material should not be susceptible to deactivation by both carbon formation and sulfur poisoning. Several alternative anode materials have been proposed that possess fluorite, rutile, tungsten bronze, pyrochlore, perovskite, and spinel structures. Among these anode materials, perovskite and double perovskite oxides are the most promising SOFC anode materials since they exhibit high ionic/electronic conductivity and high catalytic activity for various electrochemical reactions. It is the hypothesis of this research that modern computational tools permit the systematic investigation of the effect of dopants in the perovskite structure on various anode properties including electronic and ionic conductivity, catalytic activity, and resistance to coking and sulfur poisoning. In particular, we are interested here in understanding recently reported conductivity measurements of Ga and Nb doped A-site deficient $SrTiO_3$ [1].

Various configurations of Ga doped $Sr_{0.9}Ti_{0.8}Nb_{0.2}O_{3-x}$ at various concentrations of Ga were investigated to gain a better understanding of the electronic conductivity behavior of the material under SOFC operating conditions. Experimental results suggest that Ga doped $Sr_{0.9}Ti_{0.8}Nb_{0.2}O_3$ displays metallic behavior with a conductivity maximum at 10% Ga doping. Calculated densities of states of these materials suggest that 10% Ga doped $Sr_{0.9}Ti_{0.8}Nb_{0.2}O_3$ exhibits p-type doping behavior which would reduce the electronic conductivity of Nb (n-type) doped $Sr_{0.9}Ti_{0.8}Nb_{0.2}O_3$. On the other hand, in its reduced form $Sr_{0.9}Ti_{0.7}Nb_{0.2}Ga_{0.1}O_{2.9}$ displays increased n-type doping behavior due to the presence of an oxygen vacancy. This effect disappears whenever two Ga atoms are in close proximity.

Ab initio thermodynamic calculations were performed to determine the relative thermodynamic stability of various doped $SrTiO_3$ systems under experimental anode conditions ($P_{O2}= 10^{-20}$ bar). 10% Ga doped systems tend to form oxygen vacancies at elevated temperatures. As a result, the most stable structure of 10% Ga doped $Sr_{0.9}Ti_{0.8}Nb_{0.2}O_3$ is in the reduced form ($Sr_{0.9}Ti_{0.7}Nb_{0.2}Ga_{0.1}O_{2.9}$) and displays n-type doping behavior. To summarize, the experimentally observed conductivity maximum of A-site deficient

10% Ga-doped $Sr_{0.9}Ti_{0.7}Nb_{0.2}Ga_{0.1}O_{3-x}$ can be explained by small amounts of Ga promoting the reduction of the material which increases conductivity. Introducing larger amounts of Ga results in (reduced) Ga-Ga sites that do not increase n-type doping behavior and conductivity.

[1] G. Xiao, et al., *Synthesis and Characterization of A-site Deficient Perovskite* $Sr_{0.9}Ti_{0.8-x}Ga_xNb_{0.2}O_3$, Mater.Res.Bull. (2010), doi:10.1016/j.materresbull.2010.09.044

See more of this Session: Computational Studies of Electronic and Photonic Materials I
See more of this Group/Topical: Materials Engineering and Sciences Division

(212f) The Formation of Polytetrahedral Structures In Elongated Gold Nanowires

Christopher R. Iacovella[1], William R. French[1], Paul R.C. Kent[2] and Peter T. Cummings[3], (1)Chemical and Biomolecular Engineering, Vanderbilt University, Nashville, TN, (2)Oak Ridge National Laboratory, Oak Ridge, (3)Center of Nanophase Materials Sciences, Oak Ridge National Laboratory, Oak Ridge, TN

The observation of quantized conductance in single-atom wide metallic junctions has fueled much interest into understanding the behavior of nanowires undergoing tensile elongation [1]. Quantifying the structural changes that occur as a nanowire elongates is important to elucidating the origin of changes to nanowire properties, such as conductance or the opening of a bandgap, and has additional implications regarding the appropriateness of the assumed geometries used in high level theoretical calculations of nano- and molecular-electronics, e.g., the conductance through single atom wide chains or bridged single molecules.

In this work, we use molecular simulation to study the structural evolution of Au nanowires undergoing tensile elongation. While most of the previous work on gold nanowires has focused on the crystalline nature of the wire, here we focus on identifying the non-crystalline, "amorphous" structures that often form within the necks of the nanowires. To study this under conditions matching experiment, we perform molecular dynamics simulations with two different interatomic Au potentials. We use the tight-binding second-moment approximation (TB-SMA) potential, a common semi-emperical potential used to study Au nanowire elongation [2], as well as the more sophisticated ReaxFF bond-order potential with fitting parameters derived by Keith, et al. [3].

We report the widespread formation of polytetrahedral Z12 (i.e., icosahedron) and Z14 structures with both potentials, quantifying their presence using a shape-matching order parameter based on spherical harmonics [4,5]. We combine these results with density functional theory (DFT) calculations to investigate the energetic description of the polytetrahedral structures observed with TB-SMA and ReaxFF, finding that ReaxFF better captures the energetics predicted by DFT. While both potentials form polytetrahedra, our study suggests that TB-SMA based simulations over-predict their formation, most notably resulting in increased mechanical stability for nanowires with diameters less than 1.5 nm. We further perform calculations of the conductance [6] of the elongated nanowires configurations generated using ReaxFF, which show close agreement with experimental measurements [1], further validating the structural evolution predicted by our simulation results.

[1] J.M. Krans, J.M. van Rutlenbeek, V.V. Fisun, I.K. Yanson, and L.J. de Jongh, *Nature* 375, 767 (1995).
[2] Q. Pu, Y.S. Leng, P.T. Cummings, Rate-Dependent Energy Release Mechanism of Gold Nanowires under Elongation, *JACS* 130, 52, (2008)

[3] J.A. Keith, D. Fantauzzi, T. Jacob, A.C.T. van Duin, Reactive Forcefield for Simulating Gold Surfaces and Nanoparticles, *PRB*, 81, 235404, (2010)
[4] C.R. Iacovella, A.S. Keys, M.A. Horsch, S.C. Glotzer, Icosahedral Packing of Polymer-tethered Nanospheres and Stabilization of the Gyroid Phase, *PRE* 75, 4, (2007)
[5] A.S. Keys, C.R. Iacovella, S.C. Glotzer, Characterizing Structure Through Shape Matching and Applications to Self-Assembly. *Annual Review of Condensed Matter Physics*, 2, (2011)
[6] K. Varga, S.T. Pantelides, Quantum Transport in Molecules and Nanotube Devices, *PRL* 98, 076804 (2007)

See more of this Session: Computational Studies of Electronic and Photonic Materials I
See more of this Group/Topical: Materials Engineering and Sciences Division

(212g) Computational Study of Electronic and Optical Properties of Materials for Energy Applications

Maria Stoica, Energy Environmental and Chemical Engineering, Washington University in St Louis, St Louis, MO and Cynthia S. Lo, Energy, Environmental and Chemical Engineering, Washington University in St. Louis, Saint Louis, MO

We use *ab initio* atomistic DFT modeling as implemented in VASP to determine theoretical values of electronic and optical properties for skutterudites and zinc-based oxides. Among the properties we report are electrical conductivity and Seebeck coefficient. For skutterudites, we study the effect of fractionally filling both CoSb3 and CoAs3 with Ag and Au particles. For zinc oxide, we study the effects of doping ZnO with F and Al, both by direct replacement and interstitial insertion. We compare computationally derived values to experimentally observed values of the unfilled and undoped materials and make recommendations for improvements in estimation methods. We ultimately will use these computational methods to provide predictive insight into experimental materials behavior and guide the design of advanced thermoelectrics and transparent conducting oxides.

See more of this Session: Computational Studies of Electronic and Photonic Materials I
See more of this Group/Topical: Materials Engineering and Sciences Division

(229a) Evaluation of Alternative Fuels to Replace Coal In Cement Manufacturing

Steve R. Duke, Chemical Engineering, Auburn University, Auburn, AL, Anton Schindler, Civil Engineering, Auburn University, Auburn University, AL and Don Stafford, US East Business Unit, Lafarge North America, Alpharetta, GA

A multiyear DOE-supported study was completed to identify, determine, and understand fundamental burn characteristics and properties of alternative fuel sources to replace coal for energy generation, with emphasis on impacts in cement processing. Portland cement manufacturing involves the combustion of solid fuels along with raw materials to produce clinker. Temperatures exceeding 1,400°C are needed to carry out reactions for cement chemistry. Energy needs for a typical manufacturing plant are 12,000 MMBTU/day. Potential alternatives to fossil fuels must provide adequate and economical energy but they must also result in products of combustion that have negligible impact on cement quality and emissions.

An operating Portland cement manufacturing plant (Lafarge North America's Roberta Plant in Calera, Alabama) served as facility for a series of trial burns to replace coal at various substitution rates with bio-based and waste fuels. More than 20 three-day trial burns were conducted from 2007 to 2010 replacing a portion of the coal with scrap tires, waste plastics, woodchips, switch grass, poultry litter, forests trimmings, scrap railroad ties, soy waste products, or waste liquids. Each trial burn contained one or more of the alternative fuels. An elaborate sampling and analysis plan was carried out for each trial burn with close cooperation between the Lafarge plant, the Lafarge testing laboratory, Auburn University, and outside laboratories.

Results will be presented from a comprehensive evaluation of the raw materials, fuels, clinker, cement, emissions, and plant operations from the test burns. The technically viability for each fuel as supplements for coal for direct burn in a manufacturing plant will be discussed. Alternative Solid Fuel (ASF) handling and process operations were established and evaluated for the use of each of the fuels. Replacement ratios (on a BTU basis) were up to 18% for the waste and bio-based fuels and the synergistic burns where two or more of the alternative materials were used together. Impacts on product quality and emissions were negligible and in some cases advantageous.

See more of this Session: Innovations for Energy Intensive Industrial Processes and Building Technologies
See more of this Group/Topical: International Congress on Energy 2011

(229b) Kinetics Study of Calcium Carbonate Decomposition At High CO2 Environment

Yong-Hua Duan[1], **Jian Zhang**[1], Delong Xu[2], Hui Li[3], Le-le Yang[1], Xiao Fan[1], Wen-bin Yang[1], Yongxiang Ren[3] and **Yong Min**[3], (1)Material Institute, Xi'an University of Architecture and Technology,, Xi'an, China, (2)Institute for Powder Eng, Univ. of Archi. & Tech., Xi'an, China, (3)Xi'an University of Architecture &Technology, Institute of Powder Engineering, Xi'an, China

With the concerns of "global warming", CO2 sequestration from cement industry are becoming significant important for the "green processing" for the environments. This is because the cement processing released about 36% of CO2 comparing to the 12% from regular coal-power plant. CaO/ CaCO3, a major component and raw material from cement, was investigated in order to design suitable processing to sequestrate the CO2 during the cement production. It was found that the raw material particle size, surface area as well as chemical components were important factors, which will effect on the CO2 absorption efficiency, cycle times, particle activities, and so on. Various analytical equipments such as SEM, XRD, FTIR, TGA, were used to monitor the processing and analysis materials during the processing.

It was found that there are two competitive mechanisms involved during the processing i.e., chemical reaction and physical diffusion. In order to investigate the dominated effect, we selected the different particles as model compounds for the analysis. As a result, we found the small particle (particle size less than 5 micron) is dominated by chemical reaction mechanism, i.e., CO2 can completely diffusion into the core of the particle and result in a complete chemical reaction to form CaCO3; and the larger particle is dominated by the physical diffusion mechanism, i.e., only partial CO2 can diffuse into the CaO particles and formed a "core-shell" particles, where the centre core is CaO and out-shell is CaCO3. In addition, the small particle always accomplished with a high cycle times and more complete chemical reaction, which was recommended to use in the industry.

Based on the cement processing, various chemical looping processes were used to demonstrate the CaCO3 Calcinations / CaO Carbonation Cycle such as, vacuum processing, solids flow processing, fluidization processing, and cocking gas recycling processing. The energy efficiency, raw material activities, and corresponding thermodynamic date will be presented during the presentation.

See more of this Session: Innovations for Energy Intensive Industrial Processes and Building Technologies
See more of this Group/Topical: International Congress on Energy 2011

(229c) Application of Ionic Liquids In the Aluminum Electrorefining

Zheng Yong, Lu Xingmei and Zhang Suojiang, State Key Laboratory of Multiphase Complex Systems, Institute of Process Engineering, Chinese Academy of Sciences, Beijing, China

Application of Ionic Liquids in the Aluminum Electrorefining

Yong Zheng, Xingmei Lu *, Suojiang Zhang *

State Key Laboratory of MultiphaseComplex Systems, Institute of ProcessEngineering, Chinese Academy of Sciences, 100190, Beijing, P. R. China

**Corresponding author: xmlu@home.ipe.ac.cn,sjzhang@home.ipe.ac.cn*

The Hall-Heroult process for aluminum production has beenwidely employed in the past 100 years. According to this technique, the purityof primary aluminum is usually lower than 99.8% and needs to be improved further.At present, there are three main industrial methods applied in the aluminumrefining: zone melting, three-layer electrolysis and segregation process. However,these methods suffer from some inherent problems, such as high temperature and energyconsumption. Labeled as novel green solvents, ionic liquids have exhibited aseries of attracting properties and excellent performance in low-temperature aluminumelectrodeposition [1,2].The application of ionic liquids in aluminum electrorefiningis worth of investigation.

In our work, ionic liquid 1-butyl-3-methylimidazoliumheptachloroaluminate ([Bmim][Al_2Cl_7]) was prepared and used aselectrolyte in the aluminum electrorefining. The electrorefining processes were examined under different temperatures,potentials and current densities. Experimental results shownthat the main impurity elements, including Si, Fe, Tiand Na could be efficiently separated from primary aluminum in [Bmim][Al_2Cl_7]. Compact and smoothaluminum deposits with high purity (>99.9%) were obtained at 100 ^{0}C, -0.5~-1 V and 10~15 mA/cm^2(see Figure 1 and 2). Furthermore, the current efficiency of aluminum electrorefining increased to 98%. And the energyconsumption was lower than 5 kWh/kg-Al. As for [Bmim][Al_2Cl_7],it could be recycled and reused after purification. In general, ionic liquids area class of promising low-temperature electrolytes which open a new way to thealuminum refining.

Figure 1. SEM image of the Al deposits

Figure 2. EDAX profile for the Aldeposits

Reference

[1] A. Campion and P. Kambhampati, *Chem. Soc. Rev.*, 1998, **27**, 241.

[2] M. M. Zhang, V. Kamavaram andR. G. Reddy, *JOM*, 2003, **11**, 54.

See more of this Session: Innovations for Energy Intensive Industrial Processes and Building Technologies
See more of this Group/Topical: International Congress on Energy 2011

(229d) Novel Zeolite Membranes for Energy-Efficient Air Dehumidification and Conditioning

Rong Xing[1], Yuxiang Rao[1] and Wei Liu[2], (1)Energy and Environment Directorate, Pacific Northwest National Laboratory, Richland, WA, (2)Pacific Northwest National Laboratory, Richland, WA

Traditional vapor compression air conditioners have very high power consumption for cooling of hot and humid air due to condensation of a large amount of water inside the cooler. We have developed a novel zeolite membrane supported on a thin (~50 micrometer thick) porous metal sheet to remove water vapor from an incoming humid air stream without changing its temperature, i.e., through vapor-phase membrane separation. The membrane dehumidifier reduces air humidity levels well below its due point under the cooling conditions so that only sensible cooling occurs. In this way, the energy consumption for air conditioning in hot and humid climate can be reduced by 50%. The high permeance and high H2O/air selectivity are critical performance parameters. Compared to previous membranes reported in the literature, our membrane exhibits a high H2O/air separation factor (> 100), a permeation flux about 2 or 3 orders of magnitude higher than the literature result, and excellent stability. This presentation will provide systematic discussion on impacts of membrane separation conditions, such as temperature, feed air humidity, and permeate pressure, on the membrane separation performances. The new membrane technology has great potential to be developed as a commercially-viable membrane dehumidifier by obtaining exceptional membrane separation performances and substantially reducing the zeolite membrane fabrication cost at the same time.

See more of this Session: Innovations for Energy Intensive Industrial Processes and Building Technologies
See more of this Group/Topical: International Congress on Energy 2011

An Osmotic Membrane Dehumidifier

Arthur S. Kesten and Jack N. Blechner, Nanocap Technologies LLC, Longboat Key, FL

OSMOTIC MEMBRANE DEHUMIDIFIER

It has long been known that relying on a vapor compression system to cool air down to the dew point in order to dehumidify is expensive to build and uses far too much energy to operate. The ability to dehumidify independently of cooling provides substantial advantage, provided that it can be done using inexpensive materials, uses little energy to operate, and is easily packaged with a separate cooling system.

An osmotic membrane dehumidifier has been developed that uses a flexible, semi-permeable membrane to facilitate capillary condensation of water vapor and the transport of condensed water through the membrane into a salt solution by osmosis. Here a humid gas stream is brought into contact with a semi-permeable membrane, which separates the gas stream from an osmotic (e.g., salt) solution. Some of the pores of the membrane are small enough to permit capillary condensation. Liquid formed within these pores connects with liquid formed in adjacent pores, collectively forming continuous paths of liquid. These 'liquid bridges' extend across the thickness of the semi-permeable membrane and provide paths by which water can travel across the membrane. Because the membrane is so thin, water concentration gradients across the membrane can be large. This can provide a large driving force for water transport between the humid air and the osmotic fluid. The flexibility of the polymeric membrane allows for considerable design flexibility that enhances the potential for retrofit with any cooling system.

BACKGROUND

Conditioning of a gas stream, such as air, in warm weather generally involves removal of moisture and decrease of temperature to make the gas stream suitable for its intended environment. Current dehumidification technology is commonly based on the conventional, refrigerant vapor compression technology (DX technology) or on desiccant substrate capture (DS technology). DX technology requires cooling humid supply air, such as the air within a room and/or outside air, to the water vapor condensation point, with external heat rejection on the compression side. This usually requires the supply air to be cooled below comfortable temperatures, and, thereafter, either reheated or mixed with warmer dry air to raise its temperature to an acceptable level before directing it into the space being dehumidified. Twenty to thirty-five percent of the energy expended in cooling the high humidity air is utilized to remove the latent heat of condensation from the air. Cooling and dehumidification of the air are thus coupled. That makes it impossible to control comfort parameters independently, making the DX cycle less efficient, from an overall system perspective, than a technology that would allow independent control of sensible and latent heat.

In applications where the outside air has both high humidity and temperature and the functional use of the interior space generates high water vapor levels (e.g., populated convention halls, exercise rooms, school buildings, etc.), it may not be possible for the DX technology to maintain the air introduced into the interior space at the correct humidity and temperature for maintaining comfort. The air delivered is cool but 'muggy', since further cooling to remove additional water would result in the air being uncomfortably cool.

In stand-alone dehumidification using a conventional compression cycle, heat reject is in direct contact with room air. As a consequence, the room air becomes more comfortable from a humidity side, but may be too warm. Again, the comfort parameters are coupled.

DS systems are generally applied in central air, ducted systems. Water vapor capture on a solid phase substrate is efficient and rapid. However, removal of the water vapor from the pores requires energy input. It also requires removing the substrate from the high humidity air stream and placing it in an exhaust, water-reject stream, before adding the re-evaporation energy. Alternatively, the substrate remains fixed and the treated air and exhaust streams flow directions interchanged as is done in a parallel bed, desiccant dryer system. DS technology requires, in steady state operation, the addition of this energy at a rate equal to or greater than the latent heat of condensation of water in the desiccant substrate.

The advantage of DS technology is that humidity levels in the outside air and/or re-circulated air can be adjusted independently of the cooling step. The disadvantage is the requirement to move the substrate and treated air stream relative to each other for capture and rejection of the water vapor.

In a liquid desiccant system, direct contact equipment, such as packed-bed towers, allows for simultaneous heat and mass transfer between the air and the liquid. The essential issue here is the requirement for intimate contact of desiccant and air. The conventional technique of using a packed-bed adsorption tower is not energy efficient due to its low contact surface area per unit volume. Moreover, evaporation of some liquid desiccant in a packed-bed tower may cause contamination of the conditioned air stream.

Osmotic dehumidification provides a more compact and efficient option. The process is described in Refs. 1 and 2. In the earlier work, a capillary condenser with pore sizes around 2 to 3 nanometers was used to condense water vapor in air at moderately high relative humidity. The condenser (originally designed and fabricated by Mobil Oil) had been fabricated from a ceramic substrate using a molecular template which was subsequently calcined at high temperature to leave behind a fairly uniform porous structure. A semi-permeable membrane was used to separate the capillary condenser from an osmotic fluid. An osmotic driving force, resulting from a water concentration gradient, transports the condensed water through the condenser and into the osmotic fluid. This type of device is effective in promoting water transport at fluxes in excess of 1 liter/square meter-hour. Under some circumstances, however, the pore structure of the capillary condenser can become unstable and degrade over a relatively short period of time in a humid environment. In addition, the capillary condenser is typically made from a rigid material and is therefore limited to those applications where a rigid body is acceptable.

DEMONSTRATION – PROOF OF CONCEPT

The feasibility of using commercially available membranes for capillary condensation and subsequent transport of condensate into a salt solution was demonstrated first in a laboratory scale experiment. A glass container, filled to the top with a salt solution, was covered with a membrane. A narrow tube attached to the side of the container extends up above the container. Humid air was blown across the top of the membrane and the rate of removal of water from the air was measured by the change in liquid level in the tube. Membranes tested were those used commercially for water purification. Salt solutions that were very effective were lithium chloride and magnesium chloride. The latter salt is inexpensive, benign and readily available. Results suggest that 1 square meter of the membranes that worked most effectively removes about 0.7 liters/hr of water vapor at relative humidity levels between 80 and 90%. The rate of condensation per unit of surface area is considerably smaller than for a uniform pore size capillary condenser. However, the polymer membranes are stable, inexpensive, flexible, and offer simple methods for providing large surface area interfaces between humid air and films of liquid salt solution.

VALIDATION AND SCALABILITY – BENCH TESTS

Subsequent experiments were performed using a larger apparatus in which shallow trays were filled with a salt solution and each covered with a membrane. Trays were arranged in a cafeteria style with narrow spaces between the trays for air flow over the membranes. Air was blown over the membranes and then through an exhaust manifold. An anemometer was used to measure air flow and humidistats were placed upstream and downstream of the trays to measure temperature and humidity. A circulating pump was used to move the salt solution through the trays from inlet to outlet manifolds.

Tests were performed on arrays of trays and on single trays with various spacing and air flow. A single tray, for example, covered with a membrane of 0.13 square meters reduced relative humidity of an air stream flowing at 15 cfm (0.4 cubic meters per minute) by 7%. At 20 cfm, the relative humidity was reduced by 5%. An array of 12 of these trays (occupying a volume of less than 0.02 cubic meters) would remove about 0.5 liters/hr of water from a humid air stream. Reducing volumetric flow rates results in increased relative humidity reduction. However, capillary condensation with current membranes is limited by pore size and relative humidity is generally not reduced below 65-70%.

A mathematical model of the transport of air and removal of water vapor at membrane surfaces has been developed to help to optimize the design of the system. The tray system is but one of many potential configurations.

Ref. 1: Kesten, A.S. and Blechner, J.N., Versatile Dehumidification Process and Apparatus, U.S. Patent 7,758,671 B2, July 20, 2010

Ref. 2: Kesten, et al, Dehumidification Process and Apparatus, U.S. Patent 6,539,731 B2, April 1, 2003

(229f) Ethanol/Water Biomass Feedstock Separation Through Inorganic A-Type Zeolite Membrane on Thin Porous Ni Sheet Support

Yuxiang Rao[1], Rong Xing[1] and Wei Liu[2], (1)Energy and Environment Directorate, Pacific Northwest National Laboratory, Richland, WA, (2)Pacific Northwest National Laboratory, Richland, WA

Bio-Ethanol/water separation by fractional distillation becomes a popular industrial process to remove water from ethanol-water mixture. However, the mixture of 95.6wt% ethanol and 4.4 wt% water is an azeotrope and cannot be further purified by distillation. Alternatively, this can be achieved by membrane separation technology since it is not works on vapor-liquid equilibrium but on pervaporation mode.

In this work, real ethanol feeds (rectifier bottom, 40 wt%, PH=3.99 and rectifier overhead, 95wt%, PH=4.54) from Pacific Ethanol Inc. were chosen to test the performance and stability of A type zeolite membrane (2~3um), which supported on novel thin porous Ni sheet substrate (~50um). The preliminary results shows that NaA zeolite membrane was unstable at operation condition (~100C), mainly due to the strong acidity of real ethanol-water biomass feedstock (PH≈4).

However, after neutralized with 0.1M NaOH solution to PH=8.5 or by pass a guard bed filled with weak base ion exchange resin (Amberlite@IRA-93), this novel membrane shows high performance (Water flux >2 kg/m^2 h, water/ethanol selectivity > 5000) and stability (>100hrs). ICP analysis result also shows that sodium and sulfur ions will remains in the feed side during the separation process, mainly due to their large hydrated diameter (~7.2 Å for Na^+ and ~8.6 Å for SO_4^{2-}) comparing with the pore size of NaA zeolite membrane. (~4Å)

See more of this Session: Innovations for Energy Intensive Industrial Processes and Building Technologies
See more of this Group/Topical: International Congress on Energy 2011

(246a) Using Interfacial Manipulations to Control Ordering In Tapered Block Copolymers

Thomas H. Epps, Raghunath Roy, Jong Keun Park, Wei-Fan Kuan and Bin Wei, Chemical Engineering, University of Delaware, Newark, DE

Block copolymers permit the controlled fabrication of nanoscale materials for membrane and templating applications. To enable the increased utilization of block copolymer systems in commercial devices, additional understanding of the influence of interfacial manipulations on their bulk self-assembly is necessary. In this work, well-controlled tapered regions are inserted between blocks in a diblock copolymer to probe the effect of interfacial composition on block copolymer self-assembly. We systematically manipulate the interfacial region in tapered poly(isoprene-*b*-styrene) [P(I-S)] diblock copolymers to control ordering transitions. In addition to locating non-lamellar-forming materials, we find that normal and inverse tapered diblock copolymers show measurable decreases in the order-disorder transition temperature (T_{ODT}) relative to the corresponding non-tapered diblock copolymers; with the inverse tapered materials showing the greatest deviation in T_{ODT}. Additionally we have recently found the double-gyroid network morphology in several tapered diblock copolymers and show the ability to significantly reduce T_{ODT}'s while keeping mechanical properties nearly constant.

See more of this Session: Nanoscale Structure In Polymers I
See more of this Group/Topical: Materials Engineering and Sciences Division

(246b) Self Assembly of Soft Matter Quasicrystals and Their Approximants

Christopher R. Iacovella[1], Aaron S. Keys[2] and Sharon C. Glotzer[2], (1)Chemical and Biomolecular Engineering, Vanderbilt University, Nashville, TN, (2)Department of Chemical Engineering, University of Michigan, Ann Arbor, MI

The discovery of soft-matter quasicrystals and related periodic approximants in dendrimer [1-2] and block copolymer systems [3-4] hints at a unique thermodynamic mechanism that gives rise to their stability. In the past, specific interaction potentials have been contrived to stabilize quasicrystals and their approximants in computer simulations. However, such complex interactions are seemingly incompatible with the known properties of such soft-matter systems. Here, we use molecular simulation to demonstrate an alternative approach for assembling dodecagonal quasicrystals and their approximants based solely on particle functionalization and shape [5]. These effects mimic the characteristics of the specialized potentials, thereby replacing complex energetic interactions with simpler-to-achieve bonded and excluded-volume interactions. Our model consists of spherical building blocks functionalized with mobile surface entities to encourage the formation of structures with low surface contact area, including non-close-packed and polytetrahedral structures. The building blocks also posses shape polydispersity, where a subset of the building blocks deviate from the ideal spherical shape, which discourages the formation of close-packed crystals. We show that a model system possessing both mobile surface entities and shape polydispersity consistently assembles complex quasicrystal-like structures. We also report the spontaneous assembly of quasicrystal approximants in two micelle-forming systems of tethered nanoparticle building blocks that possess these key features. We argue that this mechanism can be widely exploited to assemble quasicrystals and approximants on the nano and microscales, and may further elucidate the recent formation of soft matter quasicrystals in experiment [1-4].

[1] G. Ungar, et al., Giant supramolecular liquid crystal lattice, *Science* 299 (2003)
[2] X. Zeng, et al., Supramolecular dendritic liquid quasicrystals, *Nature* 428, (2004)
[3] S. Lee, M.J. Bluemle, F.S. Bates, Discovery of a frank-kasper sigma phase in sphere-forming block copolymer melts, *Science*, 330 (2010)
[4] S. Fischer, et al. Colloidal quasicrystals with 12-fold and 18-fold diffraction symmetry, *Proc. Natl. Acad. Sci.*, 108, (2011)
[5] C.R. Iacovella, A.S. Keys, S.C. Glotzer, Self assembly of soft matter quasicrystals and their approximants, preprint, http://arxiv.org/abs/1102.5589

See more of this Session: Nanoscale Structure In Polymers I
See more of this Group/Topical: Materials Engineering and Sciences Division

(246c) Oligosaccharide/Silicon-Containing Block Copolymers for Lithography Applications

Julia D. Cushen[1], Issei Otsuka[2], Sami Halila[2], Sbastien Fort[2], Redouane Borsali[2], Erica Rausch[1], C. Grant Willson[3] and **Christopher J. Ellison**[1], (1)Department of Chemical Engineering, The University of Texas at Austin, Austin, TX, (2)Centre de Recherche sur les Macromolcules Vgtales, CERMAV, CNRS UPR 5301-ICMG affiliated with Joseph Fourier University, Grenoble, France, (3)Departments of Chemical Engineering and Chemistry, The University of Texas at Austin, Austin, TX

The need to overcome feature-size limitations in conventional lithography has led to the development of new patterning techniques using block copolymer templates. Ideal block copolymer systems for these applications have high etch contrast between blocks to promote good feature resolution and high chi-parameters to achieve small features. An additional desireable attribute is polymers with high silicon content such that they form a robust oxide mask during reactive ion etching with oxygen. To achieve etch contrast, these silicon-containing polymers can be incorporated into a block copolymer where the adjacent block is organic and etches easily. It is also observed that, in some cases, incorporating silicon into one of the blocks increases chi compared to similar silicon-deficient block copolymers. Here we will present three new block copolymer systems exhibiting cylindrical morphologies that incorporate fast-etching oxygen-rich oligosaccharides coupled to a silicon-containing polymer. The silicon-containing block provides sufficient etch resistance to achieve robust patterns in addition to promoting high chi parameters which allows access to cylinder diameters between 2 and 5nm.

See more of this Session: Nanoscale Structure In Polymers I
See more of this Group/Topical: Materials Engineering and Sciences Division

(246d) Photopolymerization In Lyotropic Liquid Crystal Templates for Improved Mechanical and Transport Properties

Bradley S. Forney and C. Allan Guymon, Chemical and Biochemical Engineering, University of Iowa, Iowa City, IA

A promising method of synthesizing polymers with useful property relationships utilizes self-assembling lyotropic liquid crystals (LLCs) as polymerization templates to direct polymer structure on the nanometer scale. Of particular importance is the potential to improve and tailor properties dependent on local order including mechanical strength and molecular transport by controlling polymer nanostructure. Herein, ordered nanostructures are generated in organic polymers through photopolymerization in LLC templates. Properties including solvent uptake, mechanical strength, solute release, and stimuli-response are significantly improved in polymer systems containing LLC nanostructure. These results could allow development of polymers with particular properties required for successful performance in a growing number of advanced biological and industrial applications.

See more of this Session: Nanoscale Structure In Polymers I
See more of this Group/Topical: Materials Engineering and Sciences Division

(246e) Janus Double Brush Copolymers

Yukun Li, Leela Christian-Tabak, Jiong Zou and **Chong Cheng**, Department of Chemical and Biological Engineering, SUNY-Buffalo, Buffalo, NY

Brush polymers have emerged as a class of nanostructured polymers. Representing a novel type of brush polymers, double brush copolymers (DBCs) have well-defined architectures with every grafting site carrying two different grafts. DBCs have been synthesized by tandem living polymerizations. With a variety of interesting nanostructures, DBCs have promising application potentials. For instance, they can serve as nano-surfactant for miniemulsions with remarkable stability.

See more of this Session: Nanoscale Structure In Polymers I
See more of this Group/Topical: Materials Engineering and Sciences Division

(246f) Self Assembly of POSS and Sorbitol and Their Effects On the Reinforcement of Polypropylene Spun Fiber

Sayantan Roy and Sadhan C. Jana, Department of Polymer Engineering, University of Akron, Akron, OH

Abstract:

Thisstudy developed fundamental understanding of nanocomposite formation by "bottom-up"self-assembly of inorganic-organic hybrid molecules (POSS) in isotactic polypropylene(iPP) matrix. A sorbitol nucleating agent (DBS) compatible with POSS withsilanol (Si-OH)functionalities promoted dispersion of POSS molecules andprovided templates for self-assembly into cylindrical nanoparticles in the spunfibers. It was found that at the optimum combination of phenyl-POSS and DBS, iPPcomposites can be spun into fibers at higher draw-ratios compared to unfilled polymerand provided superior mechanical properties. Rheological studies revealedphenyl silanol-POSS prevented fibrillation of DBS. The interactions between DBSand silanol POSS with various functionalities were investigated using infra-redspectroscopy (FT-IR), differential scanning calorimetry (DSC), wide-angle X-raydiffraction (WAXD), nuclear magnetic resonance spectroscopy (NMR), massspectrometry (MS), and their synergistic morphology by scanning electronmicroscopy (SEM). The mixtures of phenyl silanol-POSS with DBS when heated atthe processing temperature of iPP (200°C) formed a low viscous transparentliquid. The cylindrical shaped nanoparticles as found in spun fibers werebelieved to have resulted from the stretching of this liquid complex. Nochemical reaction occurred between DBS and POSS as identified from the solidstate NMR and complexes of different molecular ratios were identified from massspectrometry study. It was inferred that a combination of hydrogen bondinginteractions and intermolecular π-π stacking between the phenyl ringsplayed a major role in the formation of supramolecular complexes. Higher orientationof polymer chains at higher draw-ratios was identified to be the possible causeof mechanical strength improvement.

1. Introduction

Our study focused on dispersion of polar, crystalline polyhedraloligomeric silsesquioxane (POSS) molecules of 1-2 nm size in non-polar polyolefins to obtainlarge enhancements in mechanical properties. The incompatible POSS molecules do not dispersewell and remain as agglomerates of 10-20 mm size in polyolefins.[1] Such large size POSS particles act as defects anddeteriorate mechanical properties.[2] In this context, POSS with non-polar side groupsdisperses well in polyolefins, but do not contribute much to the properties ofthe molded composites due to greater plasticization effects.[2, 3] So the scientific challenge of the work is to find a dispersionaid that can produce nano-scale dispersion of an incompatible POSS in non-polar polyolefins andhaving a subsequent improvement in mechanical properties.

Thisresearch exploited the fundamentals of nanocomposite formation by a "bottom-up"self-assembly approach from dispersions of POSS molecules in polyolefins. Thesecomposite materials had the processing ease of unfilled polymers and weresuitable for manufacturing of articles with micro- and nano-scale features byhigh-speed fiber spinning. The fundamentals developed in this work offeredsuperior alternatives to most "top-down" polymer nanocomposites, where ordersof magnitude increase in viscosity over that of the host polymers is a norm andachieving nanoscale dispersion is a challenge. The degree of nanofiller/polymerand nanofiller/nanofiller interactions was governed by the choice of polymer andnature of the side groups of POSS.[1, 4, 5]

In our work, non-reactive POSS containingmultiple silanol groups were dispersed in iPP with the aid of sorbitol-typenucleating agent, di-benzylidene sorbitol (DBS). Theunderlying chemistry between DBS and POSSramified some unique molecular structures in molten PP, which in turn led tonanoparticle formation in the final stages of fiber spinning and providedmechanical property improvement.

2. Materials and methods

Variousgrades of POSS were chosen tetra-silanol phenyl-POSS (tetra-POSS),tri-silanol phenyl-POSS (tri-POSS), tri-silanol cyclopentyl-POSS (cyclo-POSS),and tri-silanol isobutyl POSS (iso- POSS). These were obtained from HybridPlastics. The sorbitol derivative chosen for this study wasdi(benzylidene)sorbitol (DBS) and was obtained from Milliken Chemicals. Forcomposite preparation, isotactic polypropylene (iPP), P4G2Z-159, was obtainedfrom Flint Hill Resources (MFI 1.95 g/10 min, T_m 165°C). The materials were mixed in a batch mixer at 200°C for 5minutes to obtain compounds. In some compounds the nucleating agent DBS wasnot used to determine the potential of nanoparticle formation by POSS molecules alone. Anoptimum ratio of DBS and POSS were determined using rheological study for furtherprocessing. The optimized polymer composites were subsequently melt-spun into monofilamentfibers using a capillary rheometer (Instron) and a take-up device. Thevalues of tensile modulus, tensile strength and yield strength of the fibers were determined. The fibers were characterized by DSC andWAXD and morphology by SEM.

The synergistic interactions between POSS and DBS moleculeswere studied using FT-IR, DSC, and MS and any possible reaction by solid stateNMR. DBS/POSSself-assembled structures interdependent of iPP were observed by SEM.

3. Results and discussion

The POSS molecules with multiple silanol groups were selected sothat non-covalent hydrogen bonding interactions can occur with the freehydroxyl groups of DBS which was identified by FT-IR. DBS is known toform 3-D fibrillar networks in PP melt.[6] However, the supramolecular complexes between phenyl silanol-POSS and DBS suppressed such fibrillar network formation as revealed from oscillatory shear rheology.[7] For cyclo- and iso-butyl POSS, the DBS fibrillation was still visible which suggests insufficient interactions between non-phenyl POSS and DBS. This trend was interpreted by invoking the possibility of π-π stacking between the phenyl rings. It was confirmed by molecular dynamics simulation that DBS/tri-POSS and DBS/tetra-POSS molecules first engaged in π-π stacking and then formed hydrogen bonding leading to the formation of mutlimolecular complexes, which in turn subdued DBS fibrillation completely. Such π-π stacking was not possible in the case of cyclo-POSS and iso-POSS with DBS system. In corroboration with the rheology data, the optimum combinations of DBS with tri- and tetra-POSS in PP also produced lowest crystallinity. In this case, the nucleating function of DBS[8] was arrested due to participation of DBS in DBS-POSS multi-molecular complexes. Insufficient interaction of cyclo-POSS and iso-POSS with DBS also reflected in no such changes in crystallinity with increasing POSS content.

Tri-POSSand tetra-POSS formed crystalline lamella structures through self-aggregation viaintermolecular hydrogen bonding between the silanol POSS molecules. In the presenceof DBS, the phenyl silanol-POSS formed much smaller crystalline aggregates anddistributed uniformly into the DBS network (Figure 1). On the other hand, cyclo-POSS and iso-POSS retained their circular and lamellar domain aggregates respectively, when mixed with DBS. The nature of assembly of cyclo-POSS and iso-POSS indicates greater POSS-POSS affinity,unlike phenyl silanol-POSS, which showed greater affinity towards DBS.

Theinteractions towards the formation of supramolecular complexes between POSS andsorbitol derivatives at various molecular ratios were also systematicallyinvestigated using WAXD, NMR and MS. Nochemical reaction occurred between them as identified from the solid state ^{29}Siand ^{13}C NMR and supramolecular complexes of different molecularratios were identified from MS - Electrosprayionization study. Only complexes with 1:1 molecular ratio wereidentified between cyclo-POSS and DBS, whereas higher molecular ratio complexeswere observed for tri-POSS and tetra-POSS with sorbitol derivatives.

Finefibers were obtained with DBS/tri-POSS and DBS/tetra-POSS compounds. PPcomposite formulations with 2-5 wt% of POSS were spun into fibers with close to40% reduction in diameter. These fibers offered 60-80% increase in modulus,50-60% increase in tensile strength, and 100% increase in yield strengthcompared to unfilled iPP. On the other hand, in spun fibers, the domain size of POSS particles were found to be cylindrical with close to 100 nm in diameterand 200-300 nm in length (Figure 2). The complex materials formed between DBS and phenyl silanol-POSS were liquids with low viscosity which enhanced the melt-spun fiber to be drawn at very high draw-ratio, where the liquid droplets stretched and deformedto form the cylindrical nanoparticles. Maximum tensile properties were identifieddue to increased draw-ratio of the fibers, which in turn produced finer fibers.

4. Conclusions

The following conclusions can be drawn from the resultsobtained in this study.

PP/DBS/POSS ternary blend system provided the best reinforcementof PP at its optimum combination with maximum draw-ratio of spun fiber.

The spun fiber allowed formation of self-assembled cylindrical nanoparticlesof POSS/DBS complex.

A combination of hydrogen bonding interactions and intermolecularπ-π stacking between the phenyl rings played a major role in theformation of supramolecular complexes between DBS and POSS molecules.

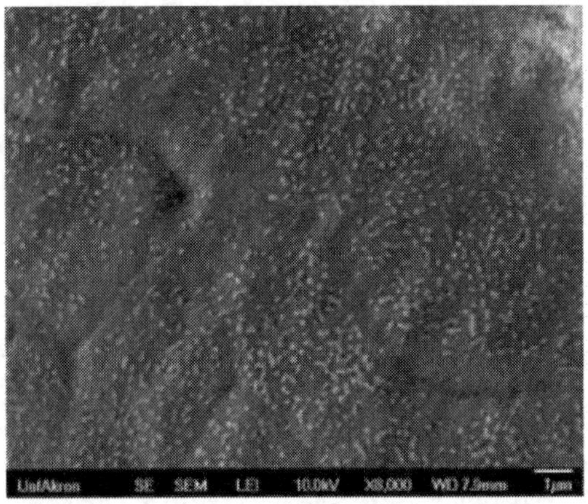

Figure 1. Uniformly dispersed POSSnanoparticles on DBS fibrils (solution cast)

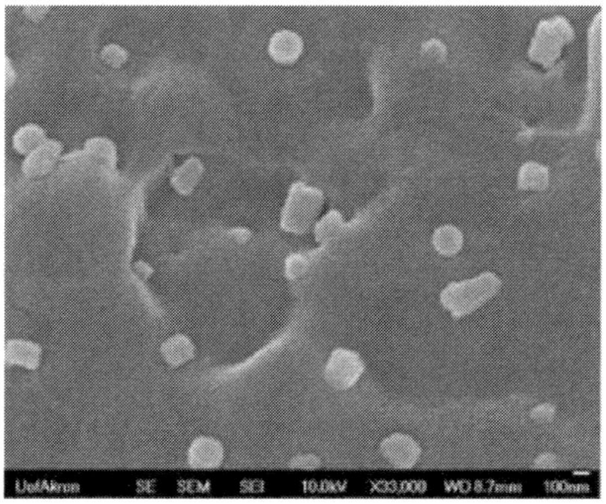

Figure 2. Cylindricalnanoparticle formation in iPP spun fiber

5.Acknowledgements:

Financialsupport from National Science Foundation in the form of award CMMI-0727231 isgratefully acknowledged.

6.>References:

1. Fina A, Tabuani D, Frache A, andCamino G. Polymer 2005;46(19):7855-7866.

2. Baldi F, Bignotti F, Fina A, TabuaniD, and Ricco T. Journal of Applied Polymer Science 2007;105(2):935-943.

3. Misra R, Fu BX, and Morgan SE.Journal of Polymer Science, Part B Polymer Physics 2007;45(17):2441-2455.

4. Jeon HG, Mather PT, and Haddad TS.Polymer International 2000;49(5):453-457.

5. Zheng L, Farris RJ, and CoughlinEB. Macromolecules 2001;34(23):8034-8039.

6. Kristiansen M,Werner M, Tervoort T, Smith P, Blomenhofer M, and Schmidt H-W. Macromolecules2003;36(14):5150-5156.

7. Perilla JE, Lee B-J, and Jana SC.Journal of Rheology (Melville, NY, United States) 2010;54(4):761-779.

8. Thierry A, Fillon B, Straupe C,Lotz B, and Wittmann JC. Progress in Colloid & Polymer Science1992;87(Solidif. Processes Polym.):28-31.

See more of this Session: Nanoscale Structure In Polymers I
See more of this Group/Topical: Materials Engineering and Sciences Division

(253a) Effect of Nanoparticles Location On the Orientation of Block Copolymer Thin Films

Joona Bang, Dept. Of Chemical and Biological Engineering, Korea University, Seoul, CA, South Korea

The nanocomposites which consist of polymers and inorganic nanoparticles (NPs) have attracted great interest due to their infinite potentials to diverse applications such as solar cell, sensors, and ferroelectric devices. Previously, we successfully synthesized thermally-stable Au NPs using UV-crosslinkable polymeric ligands and applied Au NPs as compatibilizers in PS/PMMA blends. In this work, we investigate the effect of Au NPs on the orientation of PS-b-PMMA thin films. To mediate the interfacial interaction between PS-b-PMMA block copolymer and Au nanoparticles, we adjusted the surface properties of Au NPs by varying the PS/PMMA composition in the ligands and thus controlled the NPs location in PS-b-PMMA. As a result, the orientations of block copolymer thin film are remarkably affected by the NPs location.

See more of this Session: Polymer Thin Films and Interfaces I
See more of this Group/Topical: Materials Engineering and Sciences Division

(253b) Tuning Structure and Properties of Graded Triblock Terpolymer-Based Mesoporous Films

William A. Phillip[1], Rachel Mika Dorin[2], Joerg Werner[2], Eric M.V. Hoek[3], Ulrich Wiesner[2] and Menachem Elimelech[4], (1)Department of Chemical and Biomolecular Engineering, University of Notre Dame, Notre Dame, IN, (2)Department of Materials Science and Engineering, Cornell University, Ithaca, NY, (3)Dept. Civil & Environ. Eng., University of California, Los Angeles, Los Angeles, CA, (4) Chemical & Environmental Engineering, Yale University, New Haven, CT

This talk will focus on the fabrication and performance of ultrafiltration membranes made from multiblock polymers, whose self-assembled structure promises superior selectivity for separating viruses and other undesirable solutes from solution. Despite considerable efforts towards fabricating ordered, water-permeable, mesoporous films from block copolymers, fine control over pore dimensions, structural characteristics, and mechanical behavior remains a major challenge. To this end, we describe the fabrication and performance characteristics of graded mesoporous and hybrid films derived from the newly synthesized triblock terpolymer, poly(isoprene-*b*-styrene-*b*-4-vinylpyridine). The film structure comprises a thin selective layer containing vertically aligned and nearly monodisperse mesopores at a density of more than 10^{14} per m^2 above a macroporous support layer. Solvent flow and solute separation experiments demonstrate that the terpolymer films have permeabilities comparable to commercial membranes, are stimuli-responsive, and contain pores with a nearly monodisperse diameter. These results suggest that moving to multiblock polymers and their hybrids may open new paths to produce high performance graded membranes for a myriad of applications.

See more of this Session: Polymer Thin Films and Interfaces I
See more of this Group/Topical: Materials Engineering and Sciences Division

(253c) Defect Probability In Block Copolymer Directed Assembly

Umang Nagpal, Chemical Engineering, University of Wisconsin - Madison, Madison, WI and Jaun J de Pablo, Chemical Engg., University of Wisconsin - Madison, Madison, WI

Block copolymers (BCP) find applications in a wide variety of technologies, ranging from adhesives to nanolithography. Device fabrication relies on the ability to create precisely ordered nanoscale structures. Recent works from our group and others have shown that BCPs can be used to create such structures. A central challenge in the nano-fabrication area is to determine the origin and frequency of defects. Although these defects are thermodynamically unstable, they are formed when BCP's are trapped kinetically due to their low mobility. Our research uses Theoretically Informed Monte Carlo (MC) simulations to study some of the more commonly occurring defects in BCPs. In particular, precise calculations of free energy enable concrete predictions of the probability of forming distinct defects over large surface areas. In our work, we observe that the defect occurrence becomes less probable when the strength of the pattern on the bottom surface is increased. When the defects occur on the top surface (rather than through the film); their occurrence is more probable when the pattern is stretched or compressed in comparison to pattern at its natural periodicity.

See more of this Session: Polymer Thin Films and Interfaces I
See more of this Group/Topical: Materials Engineering and Sciences Division

(253d) Diblock Copolymer Thin Films On Substrates Modified by Random Copolymer Brushes

David M. Trombly[1], Victor Pryamitsyn[1] and Venkat Ganesan[2], (1)Chemical Engineering, University of Texas at Austin, Austin, TX, (2)Department of Chemical Engineering, The University of Texas at Austin, Austin, TX

The self assembly of diblock copolymer thin films into lamellar and cylindrical structures has been proposed as a method of creating small patterns in polymer thin films for electronic materials applications. A key obstacle, however, is inducing the structures to align perpendicularly, rather than parallel, to the substrate. To achieve this goal, a "neutral" substrate may be created using a random copolymer brush composed of the species present in the diblock copolymer. We use self consistent field theory to study the system of diblock copolymer films on random copolymer brushes. Our study focuses on the "neutral window" and examines its dependencies on grafting density, relative chain length between the free and grafted polymers, and blockiness of the random copolymer chains.

See more of this Session: Polymer Thin Films and Interfaces I
See more of this Group/Topical: Materials Engineering and Sciences Division

(253e) Composite Fluorocarbon Membranes by Surface-Initiated Polymerization

Carlos A. Escobar, A. Ridzanoel Zulkifli and G. Kane Jennings, Chemical and Biomolecular Engineering, Vanderbilt University, Nashville, TN

This presentation describes the fabrication and characterization of a novel composite membrane that consists of two types of nanoporous materials, namely, nanoporous gold leaf (NPGL) and nanoporous alumina, and a selective poly(perfluorohexyl norbornene) (pNBF6) polymer. Integration of the three materials is achieved by means of silane and thiol chemistry, and the use of surface-initiated ring-opening metathesis polymerization (SI–ROMP). The former two provide functionalization of the nanoporous substrates, and the latter promotes the generation of the polymer film within and atop of the alumina-NPGL membrane. The synthetic process is versatile in that simultaneous or selective growth of the polymer film in each nanoporous layer is straightforward. The use of SI–ROMP allows tailoring of the extent of polymerization of pNBF6 throughout the structure by varying polymerization time. Advancing contact angle measurements show that the surface of these composite membranes exhibits both hydrophobic and oleophobic behavior. Scanning electron microscopy (SEM) images indicate that the thin polymer films cover the porous substrates entirely. Results from electrochemical impedance spectroscopy (EIS) confirm that the membranes provide effective barriers to aqueous ions and that sulfonation of the polymer backbone substantially enhances ion transport through the composite membrane. Ion transport and selectivity of the membrane change by regulating the polymerization time. The fluorinated nature of the polymer thin film renders the membrane selective towards molecules with similar chemical characteristics.

See more of this Session: Polymer Thin Films and Interfaces I
See more of this Group/Topical: Materials Engineering and Sciences Division

(253f) Understanding Polymer Properties That Impact Thin Film Behavior

Jessica M. Torres, Chemical Engingeering, Arizona State University, Tempe, AZ, Christopher M. Stafford, Polymers Division, National Institute of Standards and Technology, Gaithersburg, MD and **Bryan D. Vogt**, Department of Polymer Enigneering, University of Akron, Akron, OH

The physical properties of polymers when confined to nanoscale dimensions are a source of major disagreement. Depending upon the techniques and processing conditions, apparent contradictions in the effect of confinement have been reported for the same polymer system. In particular, the glass transition temperature (T_g) in polymer thin films has been the focus of many researchers. Moreover, polystyrene is the most examined system, but there is still no consensus on its thin film T_g despite nearly 20 years of experiments. Here we take an alternative approach by utilizing two techniques (wrinkling and ellipsometry) to examine the elastic moduli and T_g of systematic series of glassy polymers. We find several simple relationships between bulk polymer properties and deviations in thin film T_g and moduli. There appear to be two important contributions: the free surface of the polymer and the relative flexibility of the polymer chain. From this proposed mechanism, we can reconcile some of the apparent contradictions in the literature.

See more of this Session: Polymer Thin Films and Interfaces I
See more of this Group/Topical: Materials Engineering and Sciences Division

(254a) Solar Grade Silicon Production In Fluidized Bed Reactor

Juan Du, Soham Dutta and B. Erik Ydstie, Department of Chemical Engineering, Carnegie Mellon University, Pittsburgh, PA

In this work, we present a multi-scale modeling approach for solar grade silicon production in fluidized bed reactor (FBR) by thermal decomposition of silane. The FBR process has been commercialized by several companies and promises to deliver poly-silicon at reasonable and stable price. Silane decomposes to form hydrogen and silicon and seed particles grow by heterogeneous chemical vapor deposition and by scavenging silicon powder produced in homogeneous gas phase reaction. The subsequent particle growth process is represented by a discretized version of population balance. The model uses ordinary differential and algebraic equations to track particle movement through discrete size intervals to simulate changes in the size distribution. The dynamics of gas phase is modeled Computational Fluid Dynamics (CFD) model. The CFD module shows the hydrodynamics via momentum balance, mass and heat transfer between different phases. The volume fraction of phase, concentration of each components and temperature of the bed is generated and are imported into the population balance. The particle size distribution calculated by population balance is used as the input of the CFD module and thus the complex interplay between particulate phase and continuous phase is captured by integrating those two modules.

Based on the multi-scale model, a novel method is proposed to derive the global stability condition to guarantee the stable operation of FBR. The resulting analytical stability condition will be verified by numerical simulations. The particulate process in FBR is complex and typically has very few measurements, inventory control is a simple method for control of complex systems and thus has potential for industrial application. We apply inventory control strategy to control particle size distribution to improve the performance of the silicon production process. Due to the lack of knowledge of the reaction coefficient, which is used in population balance to describe the mass transfer rate from gas phase to dispersed phase, an on-line estimator based on adaptive mechanism will be developed to track experimental behavior more accurately and in addition to enhance robustness of the inventory control system.

See more of this Session: Process Development and Design for Solar-Grade Silicon Production
See more of this Group/Topical: International Congress on Energy 2011

(254b) Solar Thermochemical Production of Solar-Grade Silicon: Thermodynamic and Economic Analyses

Ronald Michalsky[1], Peter Pfromm[1], Bryon Parman[2] and Vincent Amanor-Boadu[2], (1)Department of Chemical Engineering, Kansas State University, Manhattan, KS, (2)Department of Agricultural Economics, Kansas State University, Manhattan, KS

Solar thermochemical production of solar-grade silicon:thermodynamic and economic analyses

RonaldMichalsky [1], Peter H. Pfromm [1], Bryon Parman[2], Vincent Amanor-Boadu [2]

[1] Department of Chemical Engineering, [2]Department of Agricultural Economics,

Kansas StateUniversity, Manhattan, Kansas, USA

Availability oflow cost solar-grade silicon will improve the sustainability of electricitygeneration from photovoltaic thin-film-Si cells. A precursor for manufacturingsolar-grade Si, polycrystalline Si is conventionally produced via carbothermalreduction of silica above 1700 C in an electric arc furnace with carbonelectrodes. Besides the use of electricity as a costly energy source, theprocess emits fossil CO_2 at least equimolar to the amount of Siproduced. Sustainability would be greatly improved if the release of fossilcarbon as CO_2 indirectly due to electricity generation and directlyfrom silicon reduction could be reduced or eliminated.

Solar-thermal H_2production via H_2O hydrolysis oxidizing a Zn reactant and sequentialZnO thermal dissociation at up to 2000 C using concentrated solar radiation isa well studied solar thermochemical approach. Materials such as Al, $Si_3N_4$and SiC have been produced from their metal ores via solar carbothermalreduction.

To reduce theenergy expenses during Si production and to avoid associated CO_2emissions, this work proposes solar thermochemical reduction of SiO_2at high temperatures using sulfur as reducing agent. Production of elementalsulfur is outpacing demand in the marketplace raising issues with sulfurdisposal. Based on Gibbs free energy computations, a process sequence forproducing polycrystalline Si and sulfuric acid from SiO_2, elementalsulfur, H_2, air and water at near ambient pressure and hightemperatures is proposed here. Technical considerations such as temperature requirementsand the avoidance of the cumbersome handling and mixing of two solids will bediscussed. To assess sustainability and economic competitiveness the mass andenergy balance for the conceptual process will be presented along with an economicanalysis determining the price of sulfur-based solar-grade Si required forbreaking even with plant investment and operational costs.

See more of this Session: Process Development and Design for Solar-Grade Silicon Production
See more of this Group/Topical: International Congress on Energy 2011

(254c) The Horizontal Ribbon Growth of Solar Silicon Crystals: Process Analysis, Stability, and Control

Parthiv Daggolu[1], Andrew Yeckel[1], Carl Bleil[2] and Jeffrey J. Derby[1], (1)Chemical Engineering and Materials Science, University of Minnesota, Minneapolis, MN, (2)Ribbon Technology International

Present-day photovoltaic technology is predominantly based on the fabrication of solar cells from silicon. The cost and efficiency of silicon-based solar cells depend primarily on the growth rate and degree of single-crystallinity attained by the growth process. Traditional silicon crystal growth methods are either very expensive or compromise on efficiency. The horizontal ribbon growth (HRG) process is a promising crystal growth technology that has the potential to overcome these limitations. William Shockley first conceived of the HRG process in late 1950's, with subsequent efforts by Kudo in Japan in the late 1970's and by Energy Materials Corporation in the US in the early 1980's. After encouraging initial development, these efforts stalled owing to various technical difficulties.

We are applying a comprehensive thermal-capillary model to study the coupled phenomena of heat transfer and interfacial phenomena (solidification and capillarity) in the HRG process. This model accounts for heat transfer in the melt-crystal-crucible domains with radiative heat loss from high temperature surfaces, melt convection due to buoyancy and surface-tension forces, and the self-consistent determination of melt-crystal, melt-ambient and crystal-ambient interface shapes. This moving-boundary problem is solved numerically by the Galerkin finite element method, using elliptic mesh generation.

Extensive parametric sensitivity studies are conducted to identify the variables that impact the process. The effect of pull rate, pulling angle, melt height, crucible geometry and furnace heat transfer are investigated. Transient simulations are performed to analyze the dynamics of the growth process and to determine its stability. Preliminary results indicate that quasi-steady-states exist for this system but that most are unstable. In this light, the dynamic analysis of this system will be require to identify attainable stable states and possible strategies to stabilize unstable ones.

See more of this Session: Process Development and Design for Solar-Grade Silicon Production
See more of this Group/Topical: International Congress on Energy 2011

(254d) The Impurities Removal In Polysilicon

Guoqiang Huang and Qiuling Shi, Department of Chemical Engineering, Tianjin University, Tianjin, China

Polysilicon, as a base material, is widely used in the electronic information and photovoltaic industry. The Japan earthquake, happened on March 11, and the subsequent nuclear leakage caused a rise in polysilicon price. An American market research organization reported that about 25% of the global silicon was suspended due to the earthquake. At the same time, nuclear leakage made it understandable that solar energy has found favor in clean energy. Our work aims at the purification of polysilicon in Modified Simens Process and the methods adopted to remove boron and phosphorus impurities affect the product quality most, which are important problems hamper the industry's development.

At present, Modified Simens Process which accounts for about 85% of the total output is the best way to produce polysilicon. The feedstock trichlorosilane(TCS) used here requires that the purity is above 99.999%, the content of boron and phosphorus elements is no more than 10ppb, polymers 10ppm, total metal elements is less than 10ppb. It is of great importance to control the content of impurities. Here 26 metallic impurities, 5 non-metallic impurities, polysilanes, organosilane compounds et al. were summarized.

New ways to removal boron and phosphorus impurities contain: 1) Reactive distillation; 2) Distillation – Absorption; 3) Wet nitrogen bubble method; 4) Adsorptive distillation. Our work is to exploit new purification technology, integrate the existing technological process and improve purification efficiency. Energy saving and environmental protection are also discussed. Reactive distillation and adsorptive distillation are two methods we use.

The key thing in reactive distillation is the choice of reactant. Reactant suitable for industrial application should contain these features: 1) It will react with boron or phosphorus compounds to form stable complex, quickly and efficiently. 2) The boiling point of it and its complex must be higher than the components of chlorosilanes, only in this way will the excess reactant and the complex formed be removed easily. 3) It will not react with the component of chlorosilanes, unless there is a special need. 4) It never reacts with components of the reactor, vessel, column and their accessory devices to introduce new impurities into the chlorosilane system. 5) The price and amount should be suitable for large-scale production. Some kinds of reactants are recommended.

Accordingly, choosing the adsorbents in adsorptive distillation, we should regard physical and chemical properties of the components and impurities in the system, e.g. boiling point, polarity. The adsorbent should be of the following characters. 1) It should have bigger adsorption amounts and adsorb impurities effectively at appropriate temperature and pressure. 2) It will not release impurities into the system.3) It has a uniform particle size and large surface area. 4) Regeneration is easy. Activated charcoal, silica gel and some special adsorbents are good choices.

We also proposed two technological processes using the two methods mentioned above to reduce the boron and phosphorus in polysilicon. The simulation result is good.

See more of this Session: Process Development and Design for Solar-Grade Silicon Production
See more of this Group/Topical: International Congress on Energy 2011

(254e) Highly Active Oxide Photocathode for Photoelectrochemical Water Reduction

Elijah Thimsen[1], Adriana Paracchino[2], Vincent Laporte[3], Kevin Sivula[2] and Michael Graetzel[2], (1) Energy, Environmental and Chemical Engineering, Washington University in St. Louis, St. Louis, MO, (2)Institute of Chemical Sciences and Engineering, Ecole Polytechnique Fdrale de Lausanne, Lausanne (VD), Switzerland, (3)Interdisciplinary Centre for Electron Microscopy, Ecole Polytechnique Fdrale de Lausanne, Lasuanne (VD), Switzerland

A clean and efficient way to overcome the limited supply of fossil fuels and the greenhouse effect is the production of hydrogen fuel from sunlight and water through the semiconductor/water junction of a photoelectrochemical (PEC) cell, where energy collection and water electrolysis are combined into a single semiconductor electrode. We present a highly active photocathode for solar H_2 production, consisting of electrodeposited cuprous oxide, which was protected against photocathodic decomposition in water by nanolayers of Al-doped zinc oxide and titanium oxide deposited by atomic layer deposition (ALD) and activated for hydrogen evolution with electrodeposited Pt nanoparticles. The roles of the different surface protection components were investigated, and in the best case electrodes showed photocurrents up to –7.6 mA cm^{-2} at a potential of 0 V vs. RHE (Reversible Hydrogen Electrode) at mild pH. The electrodes remained active after 1h of testing, cuprous oxide was found to be stable during the water reduction reaction and the Faradaic efficiency was estimated to be close to 100%.

See more of this Session: Process Development and Design for Solar-Grade Silicon Production
See more of this Group/Topical: International Congress on Energy 2011

(262a) Insights Into Crosslinking Nitroxide-Mediated Radical Copolymerization of Styrene and Divinylbenzene with a Unimolecular Initiator

Afsaneh Nabifar[1], Neil T. McManus[1], Eduardo Vivaldo-Lima[2] and Alexander Penlidis[1], (1) Department of Chemical Engineering, Institute for Polymer Research, University of Waterloo, Waterloo, ON, Canada, (2)Facultad de Qumica, Universidad Nacional Autnoma de Mxico, Mexico City, Mexico

Crosslinked polymer networks are very important in technology, medicine, biotechnology, agriculture and other areas. They find uses over a wide variety of applications, including construction materials, paints and coatings, polymer glasses with high mechanical strength and thermal stability, rubbers, ion-exchange resins and sorbents, insoluble polymer supported reagents, controlled drug-release matrices, electronics and cables, food packaging, sensors, "smart" materials, artificial organs, implants, superabsorbent materials, etc. Most of these applications of polymer networks require a homogeneous structure (morphology) to obtain optimal performance. However, polymer networks obtained by regular free radical copolymerization are rather heterogeneous in nature. The non-uniformity of network morphology within crosslinked polymers makes consistent production challenging and limits their marketability. Hence, it would be desirable to have a synthetic route to produce homogeneous polymer networks. Copolymerizing vinyl and divinyl monomers in the presence of controlled radical polymerization (CRP) controllers might result in a more homogeneous network, therefore, synthesis, characterization and modelling of polymer networks by CRP processes have received considerable attention in the last few years.

The crosslinked copolymer of styrene (STY) and divinyl benzene (DVB) is used for chromatographic applications and as a precursor for ion-exchange resins. It is also a system relatively well studied under regular radical polymerization conditions, hence a good choice for fundamental comparisons between CRP and regular crosslinking polymerizations. In regular crosslinked copolymerization of STY/ DVB, the rate of polymerization (and the corresponding evolution of conversion versus time) is very fast and high (theoretically, infinite) molecular weight crosslinked network material forms very early in the reaction. The (complex, multi-radical) gel portion grows very quickly like a 'sponge', consuming rapidly the adjacent sol molecules, and hence the polymerizing mixture becomes viscous from the very beginning. This greatly affects the nature of the crosslinked polymer network, resulting in the formation of a heterogeneous one.

Recently, the claim has been made that several types of controlled radical polymerization may lead to the formation of much more (almost, ideally) homogeneous crosslinked networks (Ide and Fukuda, Macromol., vol 32, 95 (1999)). The ultimate target for characterizing a polymer network is its crosslink density distribution, albeit a very difficult distribution to obtain (and even predict via a mathematical model, due to many unknowns in the involved parameters). Hence, several other indicators are used that give some idea about the degree and kinetics of crosslinking. Essentially, researchers use these indirect macro-indicators in order to make claims about the homogeneity of the polymer network, i.e., the fact that the crosslink density distribution is spatially the same within the network. However, this jump from, essentially, kinetic data and gel average characteristics is based on very speculative arguments. These are at best vague and often based only on indirect, theoretical (and usually not well founded modelling) studies. Perhaps, a more formal, direct, and reliable way (if such a way exists) of characterizing the crosslinked polymer network with respect to its crosslink density distribution could clarify many existing contradictory statements encountered in the literature. In fact, there is not even a widely accepted definition for the concept of homogeneity in a polymer network. These observations motivated our study on polymer networks in controlled radical polymerization, given our prior experience with both regular and CRP-type free radical crosslinking copolymerizations, both experimentally and via mathematical modelling approaches (Vivaldo-Lima et al., Polym. React. Eng. J., vol 2, 97 (1994); Tuinman et al., J. Macromol. Sci., Part A: Pure & Appl. Chem., vol 43, 995 (2006); Hernandez-Ortiz et al., Macromol. React. Eng., vol 3, 288 (2009)).

Our current work examines crosslinking nitroxide-mediated radical polymerization (NMRP) of STY in the presence of a small amount of DVB as the crosslinker, at elevated temperatures (above 100 °C). N-tert-butyl-N-(2-methyl)-1-phenylpropyl)-O-(1-phenylethyl) hydroxylamine (TIPNO) is used as the unimolecular initiator, which acts as both initiator and nitroxide-type controller. The reasons for choosing a unimolecular initiator (instead of a bimolecular system with a peroxide) were simplicity and prior reliable kinetic information on the novel TIPNO (Drache et al., Polymer, vol 48, 1875 (2007)). Results are contrasted with regular free radical copolymerization of STY/DVB and nitroxide-mediated radical homopolymerization of STY in the presence of TIPNO, as reference systems. The effects of TIPNO concentration and DVB level are investigated on polymerization rate (conversion versus time profiles), average molecular weights (number- and weight-average), gel content and swelling index.

Our investigations show that the presence of TIPNO as a mediator in the copolymerization of STY/DVB slows down the rate of polymerization, delays the onset of gelation, yields lower average molecular weights and polydispersities, and produces a 'looser' polymer network compared to the regular free radical copolymerization of the two monomers. The copolymerization reaction exhibits controlled behaviour up to the vicinity of the gelation point (i.e., linear increase of average molecular weights with conversion, and low polydispersity values (well below the typical PDI values in regular free radical polymerization, i.e., much less than 2). However, after the gelation point, and due to the presence of the crosslinker (DVB), the polymerization mixture becomes very viscous, and this leads to the loss of 'livingness' of the NMRP process at higher conversion levels.

In our designed experiments (with careful independent replication, in order to build more confidence in the observed data from such complex, noisy and experimentally uncertain systems), two concentration levels were used for TIPNO and DVB, which resulted in four different concentration ([DVB]/ [TIPNO]) ratios. Results show that TIPNO and DVB concentrations do not affect the rate of polymerization (at least at the DVB levels employed; the non-significant effect of TIPNO concentration was as expected). On the other hand, their effect on average molecular weights, gel content and network morphology are noticeable. Using a higher TIPNO concentration, while keeping the same DVB concentration, will delay the gel point. In contrast, using a higher DVB concentration, while keeping the same TIPNO concentration, will accelerate the formation of gel. All in all, our studies reveal that combinations of high DVB concentration and low TIPNO concentration (i.e., maximizing the [DVB]/ [TIPNO] ratio) result in the earliest gelation point and hence, fastest loss of 'livingness'. As the [DVB]/ [TIPNO] ratio decreased, the gel point was observed at higher conversion levels. These observations were confirmed both indirectly, based on the jump in average molecular weight values versus conversion data, and also through more direct steps measuring the gel content and swelling index using a Soxhlet extraction set up. Furthermore, our Soxhlet results indicated that the lower the [DVB]/ [TIPNO] ratio, the 'looser' the polymer network was at higher conversions (for instance, at 85% conversion, highest swelling index and lowest gel content were obtained for the lowest [DVB]/ [TIPNO] ratio).

In parallel to our experimental investigations, a detailed mathematical model for the copolymerization kinetics of STY/ DVB in the presence of different nitrodixe-type controllers has been developed. The predicted profiles for polymerization rate, molecular weight averages and gel content (or swelling index) have been validated with the respective experimental data for NMRP copolymerization of STY/ DVB in the presence of TIPNO. These validations took place independently (between two different research groups in two universities in two countries), without additional parameter fitting (thus indirectly confirming the validity and generality of the developed mathematical model and accompanying database of kinetic and physical/ chemical characteristics and parameters). Not only did model predictions follow the general experimental data trends but also were in good agreement with experimental observations. As our investigations for a more reliable and comprehensive indicator for network homogeneity continue, more detailed discussions on the results of our studies, both modelling and experimental, will be presented at the time of the conference.

See more of this Session: Structure and Properties In Polymers III: Networks and Gels I
See more of this Group/Topical: Materials Engineering and Sciences Division

(262b) Side-Chain Metallo-Supramolecular Polymers: Synthesis and Characterization

Robert H. Lambeth, Berend Christopher Rinderspacher, Kenneth E. Strawhecker, Jan W. Andzelm and Adam M. Rawlett, U.S. Army Research Laboratory, Aberdeen Proving Ground, MD

Metallo-supramolecular polymers represent an important class of self-assembled materials. The metal-ligand bond is highly directional and specific and the strength of the interaction can be easily tuned by changing the metal or manipulating the ligand structure. These properties offer the possibility to readily prepare materials whose macroscopic properties can be easily modified for a particular property or application. To this end, we have initiated a program to incorporate non-covalent metallo-supramolecular bonds into polymer systems in an effort to prepare materials with enhanced, highly tunable properties. Initial studies focus on a model system consisting of a poly(n-butyl acrylate) copolymer with 2,6-bis-(1' methyl benzimidazoyl) pyridine tethered along the backbone. Addition of metal ions cross-links the polymer forming a network structure. The kinetics and thermodynamics of the self-assembly process of the ligand with various metal ions was studied with UV/Vis, ^{1}H NMR, and capillary viscometry. The impact of changing the metal-ligand combination on material properties was investigated with dynamic mechanical analysis. Furthermore, the metal-ligand interaction was directly probed via atomic force spectroscopy based single molecule force spectroscopy and compared with computationally derived force data.

See more of this Session: Structure and Properties In Polymers III: Networks and Gels I
See more of this Group/Topical: Materials Engineering and Sciences Division

(262c) Swelling, Shrinking, and Mechanical Relaxations In Glucose-Sensitive Hydrogels

Ronald A. Siegel[1], Siddharthya K. Mujumdar[2] and Arum Kim[2], (1)Pharmaceutics/Biomedical Engineering, University of Minnesota, Minneapolis, MN, (2)Biomedical Engineering, University of Minnesota, Minneapolis, MN

Hydrogels containing phenylboronic acid (PBA) sidechains bind glucose and other sugar molecules. Such binding stabilizes the ionized Lewis base form of PBA, giving rise to an osmotic swelling force. Binding occurs by formation of bisbidentate condensation complexes between cis-diols on PBA and tetragonal hydroxyls of the PBA Lewis base. Unlike other sugars, glucose possesses two cis-diols, which provides the opportunity to form reversible crosslinks between polymer chains, leading to shrinkage of the hydrogel. These properties enable PBA-hydrogels to be utilized in microfabricated implantable wireless glucose sensors. In this presentation, data concerning swelling and mechanical stress relaxation of PBA-based hydrogels is reported as a function of pH and concentrations of fructose or glucose. Hydrogels contained methacrylamidophenylboronic acid (MPBA), acrylamide (AAm), and dimethyaminopropylmethacrylamide (DMP).

Hydrogels containing MPBA and AAM swelled in the presence of fructose at all pH values, whereas glucose had a swelling effect at lower pH values but cause shrinking at higher pH values, reflecting the effect of pH on the fraction of PBA moeities that are in the Lewis base form. Swelling data was well fit by a modified Flory-Rehner-Donnan-Langmuir model, with the interaction parameter depending linearly on the proportion of PBA units that are ionized. Mechanical stress relaxation studies demonstrated that glucose-mediated crosslinks are reversible, with a wide spectrum or relaxation times observed, reflecting kinetics of crosslink breakage and reformation, network reorganization, and collective network diffision.

Hydrogels containing MPBA, DMP, and AAm deswelled with increasing glucose concentration near physiologic pH, reflecting the polyampholytic character of these hydrogels in this pH range, and the Lewis base action of DMP on MPBA.

See more of this Session: Structure and Properties In Polymers III: Networks and Gels I
See more of this Group/Topical: Materials Engineering and Sciences Division

(262d) High Throughput Screening of Polyurethane Coating Mechanical Properties

Ismael J. Gomez[1], Johannes E. Leisen[2], Haskell W. Beckham[3] and J. Carson Meredith[1], (1)Chemical and Biomolecular Engineering, Georgia Institute of Technology, Atlanta, GA, (2)Chemistry and Biochemistry, Georgia Institute of Technology, Atlanta, GA, (3)Materials Science and Engineering, Georgia Institute of Technology, Atlanta, GA

High throughput screening (HTS) is of particular interest to many applications that require rapid screening of limited quantities of material. HTS is significantly faster and requires significantly less material than conventional assays. To further extend the functionality of screening available for coatings, we report on the development of a high-throughput mar (HT-Mar) characterization instrument for quantitative measurement of scratch damage to coatings. We apply this new technique, along with high-throughput mechanical (HT-MECH) characterization, ATR-FTIR, and solid-state NMR, to a comprehensive screening of polyurethane films composed of blended isocyanates (Desmodur-N-3390, Desmodur Z-4470, or ADI Trimer) and the polyols Desmophen A-365 or Joncryl 920. We report response of tensile strength, elongation at break, and scratch depth to differences in composition (3 levels), chemistry (7 systems), and curing conditions (3 conditions), and how these responses are correlated to structure from FTIR and NMR measurements.

See more of this Session: Structure and Properties In Polymers III: Networks and Gels I
See more of this Group/Topical: Materials Engineering and Sciences Division

(262e) Poly(N-Vinyl Formamide) Hydrogels As An Alternative to Polyacrylamide

Tiffany C. Suekama[1], Anthony Livengood[1], Vara Aziz[1], Zahra Mohammadi[1], Cory Berkland[2] and Stevin H. Gehrke[2], (1)Department of Chemical and Petroleum Engineering, University of Kansas, Lawrence, KS, (2)Departments of Chemical & Petroleum Engineering and Pharmaceutical Chemistry, University of Kansas, Lawrence, KS

A novel hydrogel, crosslinked poly (N-vinyl formamide) (PNVF), was synthesized and its physical properties were comprehensively evaluated when swollen to equilibrium in water. N-Vinyl formamide (NVF) is a water soluble monomer that has been produced on the commercial scale by several major chemical companies in the US, Europe and Japan though with limited commercial success to date. NVF is an isomer of the technologically important hydrogel-forming monomer acrylamide, but NVF has a number of advantages over acrylamide, including lower toxicity and that it is more easily processed because NVF monomer is a liquid at room temperature. Another useful feature of PNVF gel is that it can be hydrolyzed in acid or base to form a cationic hydrogel based on polyvinylamine, which has potential utility in drug delivery and chemical separations. Its synthesis as a molded hydrogel has not been previously reported.

PNVF gel was synthesized through free radical copolymerization/cross-linking with a novel crosslinker, initiated by the thermally-activated VAZO-44 at 50° C. Transparent gels of reproducible, uniform properties were obtained. The monomer concentration was varied from 10 to 30 wt% in water and the crosslinker concentration was varied from 0.5% to 5% on a mass basis, to study the effect on swelling degree and mechanical properties; however the crosslinker concentration was compared against PAAm on a mole basis to accurately compare the properties of the two different crosslinkers with different molecular weights. PNVF swelling is up to twice that PAAm for a given formulation. From swelling and compression measurements, PNVF gels were found to be more hydrophilic than PAAm gels such that Flory-Huggins solubility parameters were found to be $X=0.35\Phi_2+0.47$ for PNVF and $\chi=0.26\Phi2+0.49$ for PAAm, where Φ_2 = polymer volume fraction. When correlated against Φ_2, similarity of mechanical properties suggest that these PNVF gels have a network structure to comparable PAAm gels. The fracture strains of both gels declined with Φ_2 by the same linear function while fracture stresses were all around 600 kPa regardless of gel formulation. The shear modulus for PNVF scales with $\Phi_2^{2.0}$ and $\Phi_2^{2.5}$ for PAAm, consistent good solvent behavior for both gels, with PNVF somewhat more hydrophilic. PNVF gel properties were found to be generally comparable to polyacrylamide, suggesting that PNVF gels crosslinked with a novel crosslinker could substitute for PAAm gels. We are also developing strategies for making interpenetrating networks and nanostructured gels for the improvement of fracture properties of this class of hydrogels, since poor ultimate properties limit the utility of such materials in applications as diverse as tissue engineering, biomedical devices, and separations media.

This work was supported by NSF DMR 0805264.

See more of this Session: Structure and Properties In Polymers III: Networks and Gels I
See more of this Group/Topical: Materials Engineering and Sciences Division

(262f) Modeling of Crosslinking In Free-Radical Polymerization

Stefano Lazzari, David Pfister, Giuseppe Storti and Massimo Morbidelli, Institute for Chemical and Bioengineering, ETH Zurich, Zurich, Switzerland

Since the pioneering studies of Flory and Stockmayer many different models have been introduced to describe crosslinking in free-radical polymerization. Purely kinetical, statistical, Monte Carlo and combinations of these models have been proposed.

In the present work, a mixed statistical/kinetic model, based on Flory's gelation theory and on Tobita's extension of it, has been developed. The most important cyclization reactions have been implemented in a classical population balance equation scheme and a thorough sensitivity analysis, focusing on the cyclization rate constants, has been performed. The influence of these rate constants has been evaluated in terms of their effect on system properties such as gelation time, crosslinking density distribution, elasticity and molecular weight distribution between crosslinks. The birth time of the chains and the "actual" time, at which the polymerization occurs, had to be distinguished as the crosslinking distribution has been proven to be non-random in time. The present model describes bulk and heterogeneous crosslinking copolymerization and has been validated by comparison with Monte Carlo simulations.

See more of this Session: Structure and Properties In Polymers III: Networks and Gels I
See more of this Group/Topical: Materials Engineering and Sciences Division

(262g) Impact of Sol Molecular Weight and Network Architecture On the Strain Rate-Dependent Mechanical Properties and Fracture Behavior of Elastomeric Polysiloxanes

Randy A. Mrozek[1], Kathryn Otim[2], Kenneth R. Shull[2] and Joseph Lenhart[1], (1)Macromolecular Science and Technology Branch, US Army Research Laboratory, Aberdeen Proving Ground, MD, (2) Northwestern University, Evanston, IL

Polymer gels have the potential to be utilized in a wide variety of applications including coatings, robotics, and sensors. However, widespread implementation of polymer gels has been limited due to their relatively poor mechanical performance and toughness. In this report, we investigate the impact of the solvent molecular weight and network architecture on the mechanical properties and fracture behavior of model polydimethylsiloxane gels. When the solvent molecular weight is increased above the molecular weight of entanglement (MW_{ENT}) the resulting gels exhibited a decreased impact on the modulus relative to a low molecular weight solvent. The decreased impact on the modulus is attributed to the solvent entangling with the polymer network and contributing to the stiffness. In addition, gels containing solvents above MW_{ENT} exhibited enhanced strain rate dependent mechanical properties on timescales consistent with the solvent reptation time. Gels containing solvent above MW_{ENT} have also exhibited higher energy release rates during tear testing. We are now coupling the solvent molecular weight with the network architecture to further enhance the mechanical response and toughness.

See more of this Session: Structure and Properties In Polymers III: Networks and Gels I
See more of this Group/Topical: Materials Engineering and Sciences Division

(275a) Bone Tissue Regeneration Using Mesenchymal Stem Cell-Encapsulated Acidic Microparticle-Filled Collagen Gel

Ross J. DeVolder, Chemical and Biomolecular Engineering, University of Illinois, Urbana, IL and Hyun Joon Kong, Chemical and Biomolecular Engineering, University of Illinois Urbana-Champaign, Urbana, IL

Bone regenerative therapies have been extensively studied to improve the quality of clinical treatments for various bone defects. These therapies integrate an assortment of bioactive molecules, stem and progenitor cells, and biomaterials to orchestrate osteogenesis. These approaches widely use collagen gels to sequester transplanted cells in bone defects and further stimulate cellular osteogenic activities. However, collagen gels are often plagued by a limited control of the mechanical rigidity and mineral deposition, which leads to an unsatisfactory regeneration of mineralized bone with the desired osteointegration. This study presents a novel method to enhance both the mechanical rigidity and mineral deposition of collagen gels by introducing poly(lactic-co-glycolic acid) (PLGA) microparticles in the gels. We found that incorporation PLGA particles at a concentration of 10 mg/mL increased the storage modulus of the collagen gels from 4 to 36 Pa. The osteogenic differentiation of mesenchymal stem cells (MSCs) within the soft PLGA particle-filled collagen gels generated a hard mineralized matrix, of mostly hydroxyapatite. Finally, implantation of MSC-encapsulating PLGA microparticle-filled collagen gels on chick chorioallantoic membranes (CAMs) generated mineralized tissues similar to woven bone tissue. Overall, the material system designed in this study displays a strong potential to significantly improve clinical treatments of critical bone tissue defects.

See more of this Session: Biomaterial Scaffolds for Tissue Engineering II
See more of this Group/Topical: Materials Engineering and Sciences Division

PDMSstar-PEG Hydrogels for Directed MSC Differentiation

Statement of Purpose: The use of mesenchymal stem cells (MSC) for bone regeneration requires the identification of osteoinductive scaffolds. Hydrophobic biomaterials such as poly(propylene fumarates) and polyanhydrides have shown promise as osteoinductive matrices, and recently, hydrophilic polyethylene glycol (PEG) hydrogels were employed as bone regeneration scaffolds. Thus, tuning scaffold hydrophobicity may be a potent method for optimizing the formulation of bone tissue engineering matrices. Furthermore, an extensive body of literature indicates that appropriately tailoring scaffold inorganic composition (e.g., bioactive glass) can enhance its osteoconductivity. In the present study, we therefore examined the impact of tailoring scaffold inorganic content and hydrophobicity through the photo-cure of star polydimethylsiloxane (PDMSstar) (hydrophobic inorganic polymer) and linear hydrophilic PEG on MSC differentiation. Results indicated a strong dependence of MSC fate on PDMSstar Mn at relatively high PDMSstar levels.

Methods: *Materials.* Pt-divinyltetramethyldisiloxane complex, tetrakis(dimethylsiloxy)silane (tetra-SiH), and octamethylcyclotetrasiloxane (D4) were obtained from Gelest. Other reagents were obtained from Sigma Aldrich. *Synthesis of Macromers.* PDMSstar-MA was prepared in two synthetic steps. First, PDMSstarSiH was synthesized by the acid-catalyzed equilibration of D4 with tetra-SiH. Mn was controlled by reaction stoichiometry. Allyl methacrylate was then added to each arm. PEG-DA was prepared by acrylation of corresponding PEG.4 *Hydrogel Preparation.* Gels containing 1 mM acrylate-derivatized RGDS were prepared by the photo-polymerization of aqueous mixtures of PDMSstar-MA (2k, 7k) and PEG-DA (3.4 k) macromers. Aqueous precursor solutions were prepared at 10 wt% with the following wt% ratios of PDMSstar to PEG-DA: 0:100, 5:95, and 20:80. *Hydrogel Characterization.* Gel composition was characterized by XPS, and gel material properties were assessed by DMA, swelling, and serum protein adsorption analyses. Cell Characterization. Following three weeks of culture, the differentiation status of MSCs encapsulated within each hydrogel was examined using competitive ELISA.

Results: MSC differentiation toward osteoblast, adipocyte, and smooth muscle cell fates was evaluated by quantifying the levels of osteocalcin, A-FABP, and SM22α, respectively, across hydrogel formulations. Increasing the ratio of PDMSstar to PEG from 0:100 to 5:95 appeared to have limited impact on MSC differentiation for both 2k and 7k PDMSstar (Figure 1). In contrast, further increasing PDMSstar levels to 20:80 resulted in marked shifts in MSC differentiation for both PDMSstar Mn. However, the directionality of these shifts varied with PDMSstar Mn. Specifically. 20:80 PDMSstar (2k)-PEG gels were enriched in osteoblast-like cells, whereas MSC differentiation toward a smooth muscle cell phenotype was enhanced in 20:80 PDMSstar (7k)-PEG gels. To gain insight into material-based stimuli underlying observed differences in MSC differentiation, the modulus, swelling, and protein adsorption profiles (an indicator of hydrophobicity) of each hydrogel formulation was assessed. Comparison of gel mechanical properties demonstrated that the modulus of each PDMSstar (7k)-PEG gel was statistically indistinguishable from the PDMSstar (2k)-PEG gel of the corresponding weight ratio. In contrast, gel protein adsorption profiles, and therefore the biochemical stimuli presented to cells, varied markedly with PDMSstar Mn.

Conclusions: Results from the present study indicate that tuning scaffold inorganic content and hydrophobicity can serve as a powerful tool for directing MSC differentiation. Hydrogel characterization studies suggested that alterations in adsorbed serum proteins and/or calcium deposits with increasing PDMSstar levels and Mn may underlie the present differentiation results. Future studies will explore the influence of PDMSstar on MSC behavior over a broader Mn range toward identifying osteoinductive scaffold formulations.

(275c) Modulating the Degradation Rate of Fumarate-Based Polymers for Bone Tissue Engineering

Kirsten N. Cicotte, Biomedical Engineering, University of New Mexico, Center for Biomedical Engineering, Albuquerque, NM, Shawn M. Dirk, Organic Materials, Sandia National Laboratories, Albuquerque, NM and **Elizabeth L. Hedberg-Dirk**, Department of Chemical and Nuclear Engineering, University of New Mexico, Center for Biomedical Engineering, Albuquerque, NM

Poly(propylene fumerate) (PPF) has been developed and used widely in the field of tissue engineering, specifically bone tissue engineering. Currently, the rate of degradation of PPF is less than the rate of bone formation. Due to this limitation, work in our laboratory focuses on new synthetic routes to increase the ester hydrolysis of fumarate-based polymers over PPF. Poly(butylene fumarate) (PBF), an unsaturated linear polyester which differs from PPF by a methylene group, has been synthesized through a ring opening polycondensation reation from maleic anhydride (MA) and butylene glycol (BG). Both fumarate polymers, PPF and PBF, were synthesized to have similar molecular weights (1301 and 1573, respectively). Upon synthesis samples were subjected to accelerated degradation conditions (0.1N NaOH, 60 C with gentle agitation at 65 rpm), evaluating mass loss as well as change in compressive moduli up to 48 hrs. PPF and PBF both exhibited mass loss of ~20% in the first 24 hrs, with PPF remaining relatively constant throughout the rest of the incubation and PBF continuing to undergo mass loss. Compression testing results show a similar trend as the mass loss, with the compressive moduli continuing to decrease for PBF after 24 hrs, as PPF again shows little change after 24 hrs. These results indicate that with the addition of a methylene in the polymer backbone in turn increases the rate of degradation. Synthesis, substrate fabrication and degradation of the bulk material will be discussed. Additionally, fabrication and characterization of three-dimensional non-woven fibrous scaffolds using a novel *in situ* electrospinning will be presented.

See more of this Session: Biomaterial Scaffolds for Tissue Engineering II
See more of this Group/Topical: Materials Engineering and Sciences Division

(275d) Composite Hydrogel From Chitosan and Hemicellulose for Musculoskeletal Tissue Engineering

Joshua R. Bush and Hai xiang Liang, Department of Orthopaedic Surgery, University of Virginia, Charlottesville, VA

A thermally-responsive compositehydrogel has been developed and synthesized from the natural polymers chitosan and xylan. The new material is a viscous liquid at room temperature, but turns to a solid gelat physiological temperature (37 C).Rate of gellation is controlled with addition of a salt solution. Applications are for tissue engineering and local delivery of therapeutic agents including proteins and other drugs. The specific application that has been tested is for delivery of the protein BMP-2 for creation of new bone and 2D cell culture for osteogenic stimulation of mesenchymal cells.

The polymer composite remains a viscous liquid for more than 2 hours at room temperature after the salt solution is added. This allows for addition of the desired mass of growth factors and ample time to complete the surgical procedures. Once the polymer solution reaches physiological temperature it undergoes a phase change to become a solid gel in less than 10 minutes. This system allows for a known mass of therapeutic agent(s) to be accurately delivered to injuries of complex shape. *In situ* gelling lets the material match the complex geometries of injuries that prefabricated scaffolds and carriers are unlikely to match. The material is delivered with commonly used techniques, requiring only a common syringe to administer the liquid solution to the correct target area. This gelling material can be used in combination with current surgical techniques without disrupting the procedure. For example, anon-union fracture is usually fixed in place with titanium rods or fixation plates. This thermally-responsive composite hydrogel can be injected after placement of the more traditional healing aids, filling in the areas of likely non-union to accurately provide a precise dose of growth factor and medium in which the cells can move and bridge the defect.

A thermally-sensitive hydrogel (thermogel) graft substitute made from ultrapure chitosan has been previously used to treat critical sized unicortical defects in a rat femur.[1] The thermogel successfully delivered osteogenic peptides that induced healing over an 8 week time period. The pure chitosan thermogel was successful at delivery of osteogenic factors, however it did not promote in growth of tissue or cells before complete degradation. Thehydrogel material did not allow cells to penetrate into the interior or allow tissue growth into the area until the chitosan was completely removed from the defect site. One reason for this is that he chitosan only displays positive charges in the thermogel matrix. This has a large potential to impede the healing at an injury site as cell movement is blocked at the boundary of the hydrogel.

A multi-ion environment is moreconducive to cell migration within biomaterials. To address this issue, chitosan was blendedwith another natural polymer, xylan, which is ahemicellulose that displays negative charges on its side chains. We have developed this hydrogel to behave asa thermally-sensitive hydrogel, or thermogel, basedon the results from the pure chitosan thermogel. Natural polymers were chosen for thishydrogel composite because of a large emphasis in engineering circles toincrease sustainability in all areas of research from biofuels tobiomaterials. Chitosan is already arenewable material as it is found in crustacean shells. Hemicellulose can be up to 90% of the biomassmaterial from plants used as feed stock in other applications (sugarcane, switchgrass, wood, algae, etc.). However, as hemicellulose is only efficientlydegraded by enzymes, it is not an easy product to use for furtherconversion. Use of this material inbiomedical technology applications provides added value for the hard to digestbyproducts of biomass production as well as provides tissue engineeringapplications with an abundant source of renewable and biocompatible materialreadily available in most areas of the world.

The success of this chitosan/xylan composite hydrogel has been seen both *in vivo* and *in vitro*. This new chitosan/xylan composite hydrogel has been used to deliver 0.25 g/L BMP-2 to an ectopic site in a rat where bone was seen toform. With a chitosan/xylan ratio of 3/1, BMP-2 was successfully delivered to thesurrounding tissue in rat muscle and created mineralized bone (Figure 1). PLGA microsphere scaffolds were filled witheither the pure chitosan thermogel or the newchitosan/xylan composite. The scaffolds were placed into the thighmuscle of a rat without any other stimulating factors. After 2 weeks the scaffolds were removed, cutinto 8 micron sections and stained with H&E. When compared to the pure chitosan thermogel, the composite implants showed greaterintegration of the surrounding tissue into the scaffold (Figure 2).

Invitro cell culture experiments have also revealed benefits of the compositematerial over a pure chitosan hydrogel. Realtime PCR measurement of the 18s gene expression in D1 mouse mesenchymalstem cells (a common gene used as a control that can reveal the difference inrelative cell numbers between cultures) showed that cells cultured on thechitosan/xylan composite had much higher expressionof this gene compared to the cultured on the pure chitosan material (real timePCR results: composite 18s: 1.08x10-4 7.42x10-6; pure chitosan 18s: 3.54x10-8 8.89x10-9). This indicates that thereare many more cells living on the chitosan/xylanhydrogel after 3 days than on the pure chitosan hydrogel.

Real Time PCR measurement of theexpression of the osterix and runx2 genes in D1 stemcells exposed to chitosan/xylan for 3 days, or cells inmonolayer with differentiation medium for 7 days, or cells in monolayer withgrowth medium for 7 days (Figure 3).Gene expression levels are normalized with respect to cell numbers ineach culture. The results show thatafter only 3 days (chitosan/xylan 3 day), the cellsgrown on the xylan/chitosan composite match the bonerelated gene expression level of stem cells grown in osteogenicmedia (ODM) for an entire week (monolayer/ODM 7 day). DMEM is normal cell culture media which doesnot promote any specific cell changes.The third group (labeled monolayer/DMEM 7 day) which shows very lowlevels of osteogenic markers indicates the results ofcells in normal media grown as a monolayer on tissue culture plastic. The left bar of each of the three groups(black) represents osterix expression, and the rightbar represents runx2 expression. Theordinate represents normalized gene expression, based on gene/18s.

The results of these experimentsshow this chitosan/xylan composite hydrogel to be abiocompatible material that provides marked improvement in host tissue responsewhen compared to pure chitosan hydrogels.

Reference:

1. Bush JR, Madhu V, BalianG, Laurencin CT, Nair LS. Thermosensitivechitosan as a matrix for the controlled delivery of biologically activemolecules for accelerated bone repair. American Institute of ChemicalEngineers Annual Meeting, Salt Lake City, Utah, November 4-9, 2007

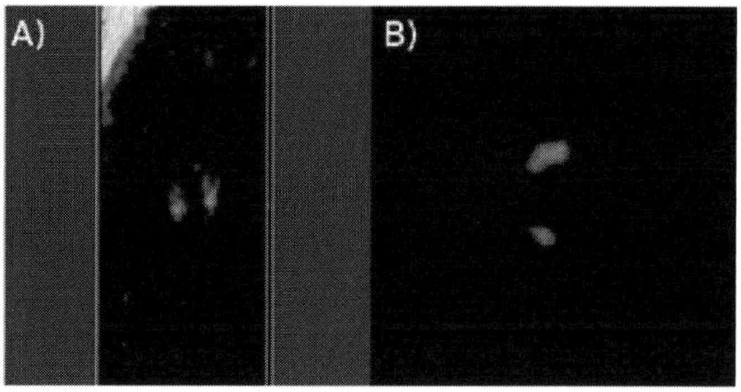

Figure 1. Mineralizedbone formed in the thigh muscle of a rat images with CT at 4 weeks. BMP-2 was delivered from the chitosan/xylan composite thermogel at thisectopic site. 0.25 g/Lwas delivered in a 40 L injection of the thermogel. A) ectopic bone inthigh muscle, B) enlarged image with higher threshold value to focus on hardtissue.

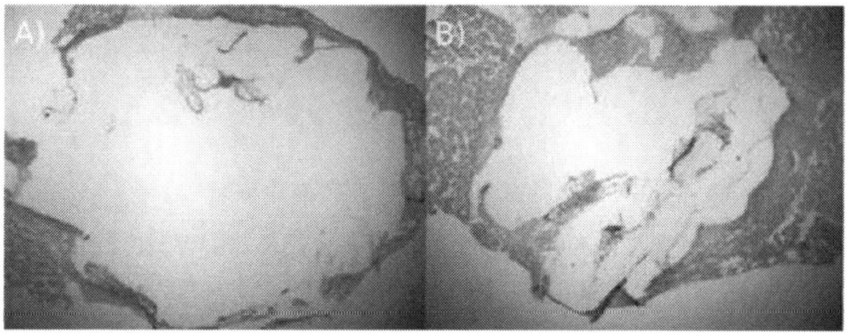

Figure 2. PLGAmicrosphere scaffolds were filled with A) pure chitosan thermogel, and B) the chitosan/xylanthermogel and placed in the thigh muscle of arat. The chitosan/xylancomposite allowed much more tissue penetration (B) than the pure chitosan thermogel (A). Bothimages are magnified at 4x.

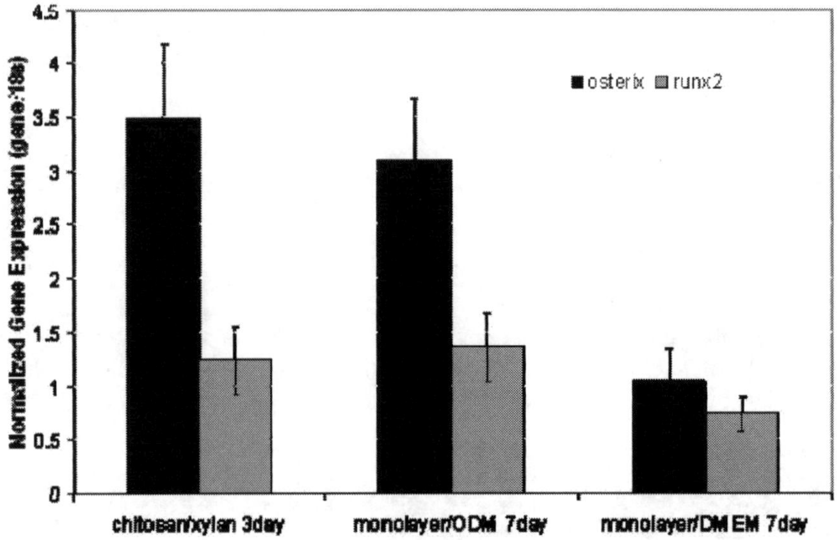

Figure 3. Real Time PCR measurement of the expression of the osterix and runx2 genes. Gene expression levels are normalized withrespect to cell numbers in each culture.The results show that after only 3 days, the cells grown on the xylan/chitosan composite match the bone related geneexpression level of stem cells grown in osteogenicmedia (ODM) for an entire week. DMEM isnormal cell culture media which should not promote any specific cell changes.

See more of this Session: Biomaterial Scaffolds for Tissue Engineering II
See more of this Group/Topical: Materials Engineering and Sciences Division

Multicomponent Polymer Networks Enhance Mechanical Properties and Cellular Response of Hydrogel Scaffolds for Cartilage Tissue Engineering

Ganesh C. Ingavle, Anahita Khanlari, Deena Rennerfeldt, Tiffany C. Suekama, Michael S. Detamore and Stevin H. Gehrke.
Department of Chemical and Petroleum Engineering, University of Kansas, Lawrence, KS 66045

Introduction

Hydrogels made of single components typically do not have either the mechanical properties or cellular interaction that adequately mimics the native extra cellular matrix (ECM) for tissue engineering application. ECM-mimetic modification of synthetic hydrogels has emerged as an important strategy to modulate specific cellular responses. Therefore, developing a biomaterial with both biological and synthetic components with tissue adhesive properties is of significant interest. The approach taken in this work is to use multiple polymeric components to more closely capture the key features of the ECM necessary to provide the mechanical response and cytocompatibility needed to improve cellular response for cartilage tissue engineering. In this work, synthetic and biological polymers of established utility in tissue engineering are combined in different fashions to obtain desirable combinations of mechanical and biological responses. The primary approach taken is to use interpenetrating networks (IPNs) of agarose and poly(ethylene glycol) diacrylate (PEGDA) to provide mechanical strength while supplementing the network with methacrylated chondroitin sulfate (MCS) or other glycosaminoglycans to improve the biological response. In a related approach, PEGDA is used as a crosslinker for MCS to create a strong yet biodegradable scaffold in a single synthesis step. The overall objective of this study was to develop a promising new class of bio-functionalized engineered IPN hydrogel scaffold materials with a unique combination of high mechanical strength and cyto-compatibility to improve cellular response for cartilage tissue engineering.

Materials and Methods

In this study, two biocompatible materials, agarose and poly(ethylene glycol diacrylate (PEGDA), were combined to create a new IPN hydrogel with greatly enhanced mechanical performance, without sacrificing cell viability.[1] Bioactive IPNs are created by encapsulating porcine chondrocytes in thermally gelling agarose, followed by soaking for 2.5 h in a 15% solution of PEGDA containing 40 μg/mL MCS and 0.1wt% photoinitiator Irgacure 2959 which is subsequently photo-polymerized (10 min at 312 nm) to form a copolymer network which interpenetrates the agarose constructs. Crosslinked MCS gels are also made in a similar fashion but by photopolymerizing a solution of MCS with PEGDA in a single step. The viability of encapsulated chondrocytes was examined with culture time and as a function of the gel chemistry to examine the role of cell–matrix interactions in promoting long-term viability and biosynthesis.

Results and Discussion

The IPN displayed a 4-fold increase in shear modulus relative to a pure PEG-DA network (39.9 vs. 9.9 kPa) and a 4.9-fold increase relative to a pure agarose network (8.2 kPa). PEGDA and IPN compressive failure strains were found to be 71 ± 17% and 74 ± 17 %, respectively, while pure agarose gels failed around 15 % strain.[1] Similar mechanical property improvements were seen in

IPN gels with encapsulated chondrocytes. Live/Dead assays demonstrated that the viability of IPN-encapsulated chondrocytes was over 90% initially, but viability while in culture drops to 45% 3 weeks after encapsulation. However, by incorporating chondroitin sulfate in a small concentration (40 μg/mL) within the IPN, this lower viability was rescued to 70 ± 10 % at 3 weeks post-encapsulation. A continuous increase in glycosaminoglycan (GAG) content was observed during the culture period. CS may have served in a role of an adhesion molecule and therefore provided an overall positive impact on cell a) viability, b) cell proliferation, and c) biosynthesis. Crosslinked CS molecules with PEGDA networks in the IPN provide a host tissue mimetic microenvironment in which cells grow and proliferate. Therefore, to promote cellular functions, it is highly desirable that the scaffolds have cell specific adhesion and the ability to carry signaling biomolecules. The combination of MCS (10-15%) and PEGDA (1-3%) substantially increased the Young's modulus and fracture stress of MCS gel to levels comparable to the IPNs.

Conclusions

We successfully designed and synthesized a promising new class of *bio-functionalized* engineered IPN hydrogel scaffolds with a unique combination of high mechanical strength and cyto-compatibility for the first time in cartilage tissue engineering. The method introduced in this study may provide a platform for developing suitable engineered scaffolds for cartilage regeneration that provide not only a superior mechanical performance, but also harness a biological cue to promote new tissue growth for cartilage defect repair.

Acknowledgment

The present work was supported by the NIH (1 R21 EB008783-0, 5 P20 RR 16475-08), and the NSF (IOS 0726425, DMR 0805264).

References

1) DeKosky BJ, Dormer NH, Ingavle GC, Roatch CH, Lomakin J, Detamore MS, Gehrke SH. Tissue Engineering C.**16**, 1533-1542, 2010.

(275f) Optimization of Thermogelling, Injectable Scaffolds for Craniofacial Tissue Regeneration

Adam K. Ekenseair, Nicholas Zuiker, F. Kurtis Kasper and Antonios G. Mikos, Department of Bioengineering, Rice University, Houston, TX

The development of a suitable replacement for autogenous bone grafts in the treatment of craniofacial bone defects resulting from trauma, deformities, degenerative diseases, and tumor resection is a major scientific goal. An injectable, *in situ* forming synthetic scaffold capable of co-delivering cells and growth factors to enhance tissue regeneration would enable a non-invasive procedure capable of filling complex bone defects with no donor-site morbidity. A promising approach is to utilize organic, thermogelling polymers, such as poly(N-isopropylacrylamide, which pass through a lower critical solution temperature (LCST) upon injection into the body. Simultaneous chemical crosslinking of the polymer matrix has been used to enhance the mechanical properties of the formed gels and reduce syneresis. This material must also be made biocompatible and biodegradable with non-toxic and soluble degradation products.

In this paper, a recently developed system is presented that is capable of eliminating the commonly encountered problem of significant syneresis as a result of the thermogellation mechanism. Modest incorporation of 3-sulfopropyl acrylate potassium salt into poly(NiPAAm-co-acrylamide-co-2-hydroxyethyl acrylate) progressively increased the equilibrium degree of swelling to create non-shrinking, injectable scaffolds. Additionally, a novel method to produce soluble degradation products from a thermogelling scaffold is presented. By modifying the base polymer with methacryloyl chloride to introduce crosslinkable moieties, the LCST of the polymer was reduced below body temperature. After injection and crosslinking through free radical pathways, the hydrolytic degradation of the scaffold resulted in soluble polymer chains. Thus, a promising system for injectable tissue engineering scaffolds has been developed through improvement and optimization of previous research efforts with similar polymers.

See more of this Session: Biomaterial Scaffolds for Tissue Engineering II
See more of this Group/Topical: Materials Engineering and Sciences Division

(283a) Segregated Network Polymer Nanocomposites for Thermoelectric Energy Conversion

Jaime C. Grunlan, Department of Mechanical Engineering, Texas A&M University, College Station, TX, Choongho Yu, Mechanical Engineering, Texas A&M University, College Station, TX and Greg Moriarty, Materials Science and Engineering, Texas A&M University, College Station, TX

Segregated network (latex-based) composites containing carbon nanotubes can produce electricity from a thermal gradient. These thermoelectric materials harvest electricity from waste heat or any temperature gradient in the environment. The present work demonstrates that nanotube-filled polymer composites can be viable for light-weight and economical thermoelectric energy conversion. By combining double-walled carbon nanotubes (DWNTs), stabilized with poly(3,4-ethylenedioxythiophene): poly(styrene sulfonate) [PEDOT:PSS] in water, an electrical conductivity (σ) near 2000 S/cm is achieved in a poly(vinyl acetate) [PVAc] latex-based matrix. When this conductivity is combined with a Seebeck coefficient (S) above 40 mV/K, a power factor ($S^2\sigma$) above 370 mW/m·K^2 is achieved at room temperature. Thermal conductivity (k) of these composites can be tailored between 0.3 and 10 W/m•K, suggesting the thermoelectric figure of merit ($ZT = S^2\sigma T/k$) could be as high as 0.4 at room temperature, which would make these fully organic composites competitive with bismuth telluride (i.e., the most efficient thermoelectric material) at room temperature.

See more of this Session: Composites for Energy Applications
See more of this Group/Topical: Materials Engineering and Sciences Division

(283b) Engineered Si-Graphene Composite Papers with Enhanced Lithium Ion Transport for High-Power Batteries

Xin Zhao, Cary M. Hayner, Mayfair C. Kung and Harold H. Kung, Chemical and Biological Engineering, Northwestern University, Evanston, IL

The development of more efficient energy conversion and storage devices remains a pressing need in attempting to achieve a high level of market penetration of electric vehicles and use of intermittent renewable energy sources such as wind and solar energy. Current lithium-ion batteries, although an attractive energy storage device, have limited charge capacity that diminishes rapidly at high charge/discharge rates. We have shown previously that a Si-graphene composite integrated into a 3-dimensional, conducting graphitic network offers high structural and cycling stability and substantially higher charge capacity as an anode material than graphite. However, the composite suffers from the problem of diminishing charge capacity during high current operations. Here, we report a facile microscopic engineering strategy to enhance the Li ion transport in graphene sheets, which could be utilized as a platform to create high-power battery materials. When used with Si nanoparticles to form a Si-graphene composite and integrating the composite into a 3-dimensional graphitic network to form a self-supporting electrode paper, the electrode exhibits an outstanding energy density-power density combination, without compromising the usable lifetime and mechanical flexibility. The power capability was found to be comparable to supercapacitors but with a much higher energy density. For example, a charge capacity of over 1000 mAh/g was achieved at a discharge rate of 8 A/g. Structural characterization by high resolution electron microscopy and impedance measurements provided evidence that relates the structural properties to the Li ion diffusivity for this material, which can bridge the performance disparity between batteries with high energy density and capacitors with high power density.

See more of this Session: Composites for Energy Applications
See more of this Group/Topical: Materials Engineering and Sciences Division

(283c) Functionalized Porous Materials As Liquid Springs for Energy Conversion

Yu Qiao[1], Xi Chen[2], Xiaoyue Li[3] and **Aijie Han**[3], (1)Department of Structural Engineering, University of California-San Diego, La Jolla, CA, (2)Department of Earth and Environmental Engineering, Columbia University, New York, NY, (3)Department of Chemistry, University of Texas-Pan American, Edinburg, TX

For a few decades, shape-memory materials, also known as smart materials, have drawn increasing attention. In general, any materials or structures that change their dimensions or configurations as temperature varies can be potentially used to develop thermally controllable intelligent systems. The functionalized porous materials lead to a high degree of hydrophobicity. When the external pressure is low, water does not enter the nanopores, due to the well known capillary effect. When the pressure is relatively high, it becomes energetically favorable for the liquid molecules to enter the nanopores, and pronounced pressure-induced infiltration occurs. Since the infiltration and defiltration of water can be controlled by adjusting external pressure and the hysteresis of sorption isotherm is negligible, the system behaves as a 'liquid spring' where mechanical energy is converted to the interfacial energy. With the addition of sodium chloride, the temperature sensitivity of the system increases with the electrolyte concentration, beneficial to improving the output energy density. Due to the large surface area, its energy density and the deformability are much higher than that of conventional shape-memory solids.

See more of this Session: Composites for Energy Applications
See more of this Group/Topical: Materials Engineering and Sciences Division

Simultaneous Improvement in Ionic Conductivity and Mechanical Properties of Multi-functional Block-copolymer Modified Solid Polymer Electrolytes for Lithium Ion Batteries

Jianying Ji, John Keen and Wei-Hong Zhong[*]
Washington State University, Pullman, WA 99164

1. Introduction

Solid polymer electrolytes (SPEs) have attracted world-wide attention due to their unique characteristic properties wherein ions can be transported inside a polymer matrix under an external electrical field. Thus, SPEs will make it possible to fabricate thin film lithium batteries with high specific energy and specific power, safe operation, flexibility in packaging, and low cost [1-3]. However, existing SPEs have inherently lower conductivities due to the more restricted motion of the polymer molecules [4], and thus, make them unacceptable for practical use. Many efforts have been made to increase the ionic conductivity of polymers, such as blending, plasticization, blocking, cross-linking, and inorganic composite [5-9]. Some of these methods have improved the ionic conductivity to some extent. However, the mechanical strength is often sacrificed in order to increase ionic conductivity so that the SPE film preparation can become more difficult.

Enhancements in both ionic conductivity and mechanical properties are usually in conflict because both these two types of properties are contrary with respect to the movements of the polymer chains. The high chain mobility can lead to high ion conductivity but at the same time can lead to lower mechanical properties [10-11]. Many researchers have attempted to find an optimized balance between the ionic conductivity and mechanical properties, but the outcomes are not yet satisfactory. Our previous work revealed that poly(ethylene oxide) -*block*- polyethylene (PEO-*b*-PE) possesses a capability for enhancing the multiple performances of poly(ethylene oxide)(PEO)-based polymer electrolyte, in which the conducting pathways, provided by polar groups contained within PEO parts (PEO parts include PEO matrix and PEO block in the copolymer), are called conductivity parts; while a non-conducting block (PE-*b*, PE block in the copolymer) imparting the desired mechanical properties, is called a reinforcement block [12,13]. In this paper, a series of diblock copolymer with different composition and molecular weight are studied. The effects of the composition and molecular weight of the copolymer on both ionic conductivity and mechanical properties of the resulting SPEs are examined.

2. Experimental procedures

Poly(ethylene oxide) (PEO) (Mw = 4,000,000 g/mol), pure polyethylene glycol (PEG, low molecular weight of PEO), poly(ethylene oxide)-block-polyethylene (PEO-*b*-PE) (875 with 80 wt% PE, 920 with 50 wt% PE, and 1400 with 50 wt% PE) and lithium perchlorate ($LiClO_4$) were obtained from Aldrich. The $LiClO_4$ was dried in a vacuum oven at 70°C for 24 hrs. The concentration of $LiClO_4$ was fixed at 15 wt% for all samples. The amount of copolymer is same as the weight of PEO. Calculated amounts of PEO, $LiClO_4$, and PEO-b-PE were dissolved in acetonitrile and then stirred by magnetic

* Corresponding author. Tel.: +1 509 335 7658; fax:+1 509 335 4662
E-mail address: Katie_zhong@wsu.edu (W. H. Zhong).

stirring. A homogeneous solution was casted onto a glass plate after 3hr ultrasonic treatment for de-aerating. The solvent was allowed to slowly evaporate at room temperature in a hood. All the films were put in vacuum oven at 40 °C before any tests in order to wipe off residue solvent traces.

The microstructures of the films were analyzed using a polarized optical light microscope (Olympus BX51). The ionic conductivity was measured using a Universal Dielectric Spectrometer BDS 20 in the frequency range from 10^{-2} Hz to 10^{6} Hz. The film was sandwiched between two gold electrodes of 2 cm diameter. The input voltage (Vrms) was 1V. Tensile tests were performed using an Instron 4466 with a 500N load cell and a laser extensometer.

3. Results and Discussion

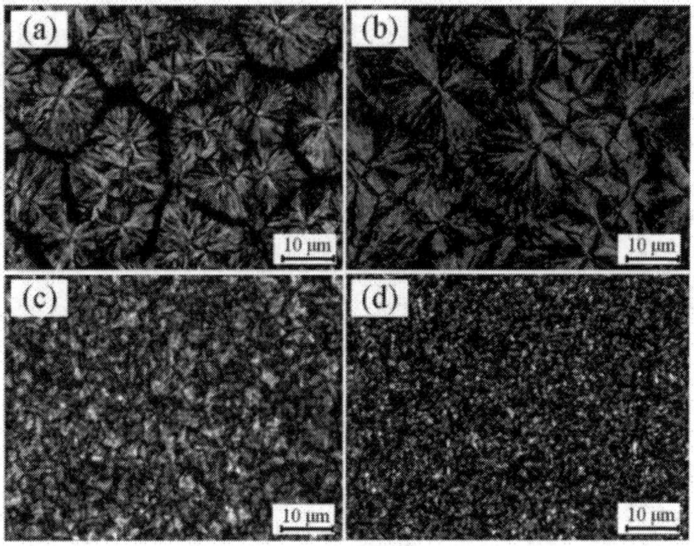

Figure 1 Optical micrograph of (a) SPE-0 (pure PEG contained), (b) SPE_h-50 (high molecular weight PEO-*b*-50 wt% PE contained), (c) SPE_l– 50 (low molecular weight PEO-*b*-50 wt% PE contained) and (d) SPE-80 (PEO-*b*-80 wt% PE contained) films at the magnification of 50×

Fig. 1 shows optical micrographs of four kinds of polymer electrolyte films. Distinct spherulites are visualized in the micrographs of the pure PEG included film (Fig. 1a). The boundary between the spherulites shows the existence of an amorphous phase. After addition of the copolymer, the microstructure of the electrolyte film was dramatically changed as shown in Fig. 1(b, c and d). It indicates the addition of the copolymers indeed changed the structure of the electrolytes that are believed to be related to conductivity properties for the materials. The characteristic morphology of the SPE materials is controlled by the composition and molecular weight of the block copolymer chains. Comparing the morphology of SPE_l-50 and SPE-80 (Fig. 1 c and d), the size of the crystals in SPE_l-50 is larger than in SPE-80. In SPE-80, due to the high disturbance of the pre-existing PE crystal parts, the crystal structure of PEO block is not visible. While in SPE_l-50, the crystallinity of the PE-*b* is much lower (shown in the following DSC results), and the imperfect crystallization of PE-*b* cannot disturb the order of the PEO as much as the PE-*b* did in SPE-80. As for the molecular weight, for example in SPE_h-

50 and SPE_l – 50 (Fig. 1b and c), the higher molecular weight resulted in more well-shaped crystal structure.

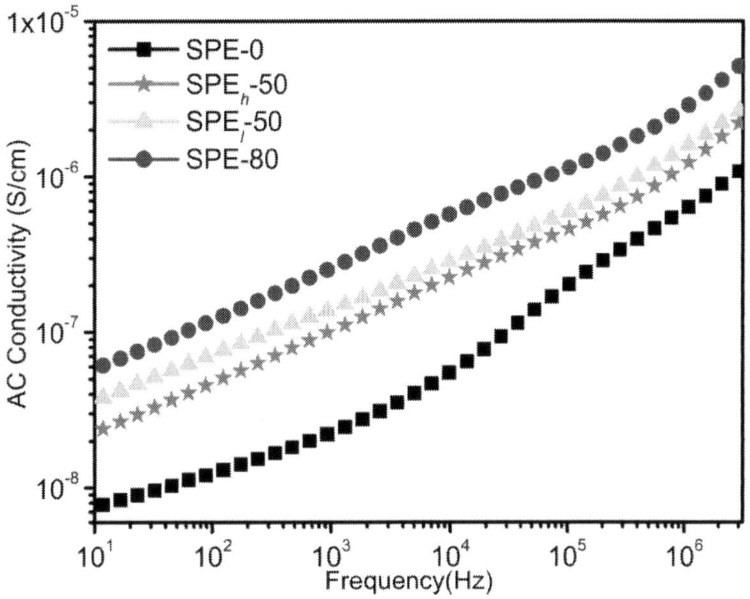

Figure 2 Frequency dependence of AC conductivity of the SPE films

In order to understand the ion dynamics in the polymer electrolytes, the frequency dependence of the AC conductivity has been analyzed for all samples [14]. Fig. 2 shows the AC conductivities acquired from all four electrolyte films, SPE-0 (only with pure PEG), SPE_h-50 (with high molecular weight PEO-*b*-50 wt% PE), SPE_l-50 (with low molecular weight PEO-*b*-50 wt% PE), and SPE-80 (with PEO-*b*-80 wt% PE), over the frequency range of 10^1 Hz to 10^6 Hz. The plots consist of two distinct regions: a weak frequency-dependent region in the low frequency range and a strong frequency-dependent region in the high frequency range where AC conductivity varies significantly with frequency. It is believed that the variation of AC conductivity is reflected in the mechanism of charge transport and interactions among charge carriers. According to the jump relaxation model [14, 15], at low frequencies, an ion can jump from one site to its neighboring vacant site successfully. While at higher frequencies, due to the short time periods, the probability for the ions to go back to their original sites increases. This high probability is responsible for the higher conductivity in higher frequencies. With the copolymer addition, the switchover (the vertical dotted line) form the frequency independent to frequency dependent region that shifts to higher frequencies, which indicates a higher probability for the ions to return to their original sites. This increased probability contributes to the improvement in ion conductivity.

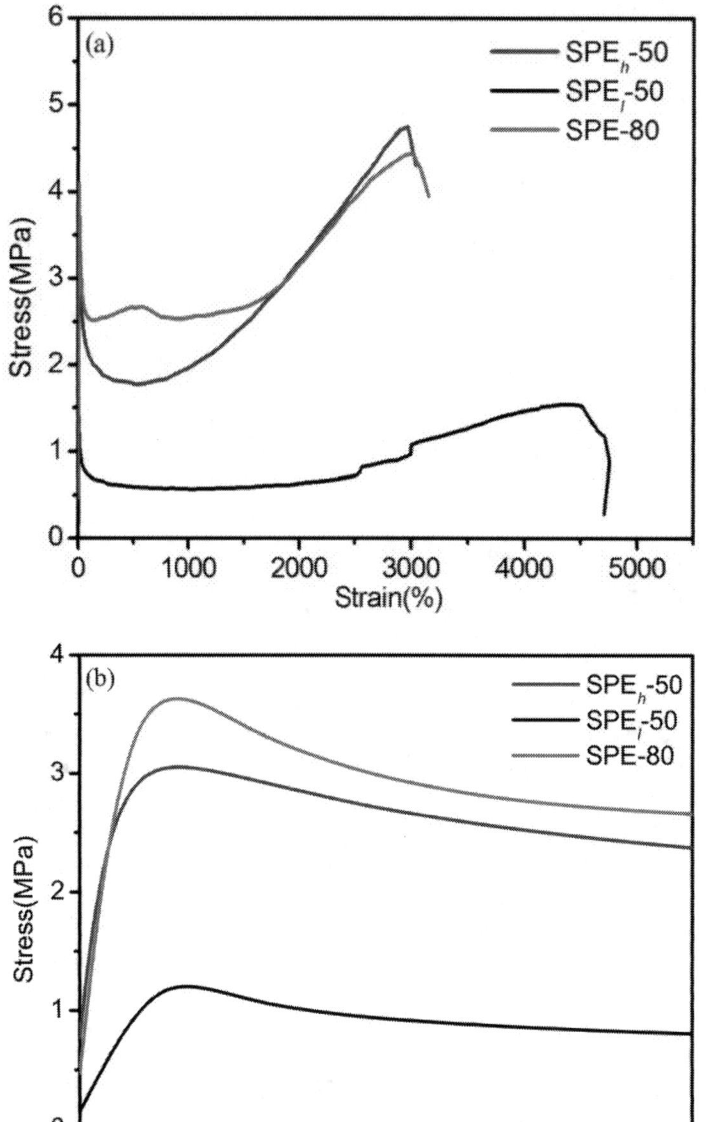

Figure 3 Stress-/strain curves for the block copolymer contained SPE films: (a) complete curves and (b) enlarged portion of strain less than 50%

Fig. 3a shows the stress-strain curves of the SPE films. The elastic region of the stress strain plots are magnified in Fig. 3b and the modulus was calculated by the slope of the curves in linear portion of Fig. 3b.

As seen in Fig. 3, the mechanical properties of these films depend on the composition and the molecular weight of the copolymer added in the SPE. The mechanical property of the SPE-0 is too weak to be tested, and failed to obtain mechanical property values. It can be seen that the presence of PE-*b* in the copolymers helps enhance the mechanical properties that can be observed in the stress-strain behavior of the films. The SPE films show higher ultimate stress when the PE-*b* weight percent is higher. When the weight percent of the PE-*b* in copolymer increased from 50 wt% to 80wt%, the ultimate tensile strength and modulus increased ~97% and ~359%, respectively.

It is apparent from Fig. 3 that molecular weight also plays a big role with respect to the mechanical properties. Comparing the SPE_h-50 and SPE_l-50, ~350% increase in tensile modulus and ~130% increase in ultimate tensile strength is found when the molecular weight of the copolymer increased from 920 to 1400. This is possibly due to the fact that the structural block formed from PE chains in the lower molecular weight copolymers are shorter and are not able to rearrange and organize into an ordered crystal structure, which is confirmed by the following DSC results.

4. Conclusions

From this study we can conclude that the composition and molecular weight of the multi-functional copolymer, PEO-*b*-PE, have significant effects on the properties of PEO-based SPE films. The best ionic conductivity and mechanical properties are obtained in 80 wt% PE-*b* solid polymer electrolytes (SPE-80) by comparing the properties with the pure PEG (SPE-0) and 50 wt% PE-*b* (SPE_l-50) contained electrolytes. Molecular weight did not seem to have much effect on the ionic conductivity. However, for the mechanical properties, tensile tests showed that increasing the molecular weight of the block copolymer resulted in ~350% increase in tensile modulus and ~130% increase in ultimate tensile stress.

References

1. N. Srivastava, T.Tiwari (2009), *e-polymer*, 146:1-17.
2. C. A. Angell (1992), *Annu. Rev. Phys. Chem.*, 43:693-717.
3. P. Mustarelli, C. Capiglia, E. Quartarone, C. Tomasi, P. Ferloni (1999), *Physical Review B*, 60:7228-7233.
4. D. Saikiaa, H. Wub, Y. C. Pana, C. P. Lina, K. P. Huangc, K. N Chend, G. T. K. Feyc, H. M. Kaoa (2011), *Journal of Power Sources*, 196:2826–2834.
5. J. Y. Song, Y. Y. Wang, C. C. Wan (1999), *Journal of Power Sources*, 77:183-19.
6. P. G. Bruce (2008), *Solid State Ionics*, 179:752-760.
7. T. Ren, X. B. Huang, X. Zhao, X. Z. Tang (2003), *Journal of Materials Science*, 38:3007-3011.
8. J. Ding, C. Chuy, S. Holdcroft (2002), *Adv. Funct. Mater.* 12:389-394.
9. S. Liua, N. Imanishi, T. Zhang, A. Hirano, Y. Takeda, O. Yamamoto, J. Yang (2010), *Journal of Power Sources*, 195:6847–6853.
10. L. Fan, Z. Dang, C. Nan, M. Li (2002), *Electrochimica Acta*, 48:205-209.
11. M. Singh, O. Odusanya, G. M. Wilmes, H. B. Eitouni, E. D. Gomez, A. J. Patel, V. L. Chen, M. J. Park, P. Fragouli, H. Iatrou, N. Hadjichristidis, D. Cookson, N. P. Balsara (2007), *Macromolecules*, 40:4578-4585.
12. J. Y. Ji, B. Li, W. H. Zhong (2010), *J. Phys. Chem. B*, 114:13637–13643.
13. J. Y. Ji, B. Li, W. H. Zhong (2010), *Electrochimica Acta*, 55:9075–9082.
14. B. Natesan, N. K. Karan, and R. S. Katiyar (2006), *Physical Review E*, 74:042801-1-042801-4.

15. P. Basak, S. V. Manorama (2005), *J. Phys. Chem. B*, 1091174-1182.

(283e) Electrocatalysis of Sandwich-Structured Pd/Polypyrrole/Pd Composites towards Formic Acid Oxidation

Suying Wei[1], Haitao Jia[2], Keqiang Ding[2] and Zhanhu Guo[3], (1)Chemistry and Biochemistry, Lamar University, Beaumont, TX, (2)College of Chemistry and Materials Science, Hebei Normal University, Shijiazhuang, China, (3)Integrated Composites Laboratory (ICL), Dan F Smith Department of Chemical Engineering, Lamar University, Beaumont, TX

Palladium (Pd) nanoparticle catalysts were electrodeposited onto a graphite electrode by cyclic voltammetry (CV) technique from a 0.5 M HCl solution containing 0.2 M $PdCl_2$, and then a film of polypyrrole (PPy) was immobilized onto the as-prepared Pd-coated graphite electrode, leading to a PPy/Pd-coated electrode. Pd particles were repeatedly electrodeposited on the PPy/Pd-coated electrode to generate a Pd/PPy/Pd sandwich structured composite coated electrode. This developed composite electrode showed a satisfactory electrocatalysis toward the formic acid oxidation (FAO). The catalysis mechanism of the sandwich structured Pd/PPy/Pd composite catalyst towards FAO was also discussed based on the obtained cyclic voltammograms and FT-IR spectra.

See more of this Session: Composites for Energy Applications
See more of this Group/Topical: Materials Engineering and Sciences Division

(283f) Hierarchical Ternary Nanocomposites for High Performance Electrochemical Energy Storage

Yuanbing Mao[1], Elizabeth M. Martinez[2] and Suresh B. Alaparthi[2], (1)Department of Chemistry, University of Texas - Pan American, Edinburg, TX, (2)Chemistry, University of Texas - Pan American, Edinburg, TX

Nanomaterials have significantly better potentials for applications in the field of energy store. To achieve high energy densities at high rate capabilities, it is necessary to use innovative electrode materials with architecturally tailored nanostructures for these devices, such as batteries and supercapacitors. Heterogeneous nanostructured materials with multi-components can be tailored to address different demands for these types of novel electrodes, which are expected to exhibit synergic properties. In this presentation, the design and synthesis of hierarchical carbon nanotubes/conductive polymer/metal oxide nanoparticles ternary nanocomposites for high performance electrochemical energy storage will be reported. The performance, i.e. specific capacitance, of the fabricated supercapacitors will be shown based on cyclic voltammetry and charge-discharge measurements and compared with literature results. It is expected that the nanocomposites will effectively utilize the full potential of all the desired functions of each component.

See more of this Session: Composites for Energy Applications
See more of this Group/Topical: Materials Engineering and Sciences Division

NANO-SIZED TIN PARTICLE/CARBON NANOFIBER COMPOSITES AS ANODE MATERIALS FOR LITHIUM-ION BATTERIES

Yun-Hua Yu [a], *Jie Chang* [a], *Xiao-Ping Yang* [a*], *Wei-Hong Zhong* [b]

[a] *School of Materials Science and Engineering, Beijing University of Chemical Technology, Beijing (China)*; [b] *School of Mechanical and Materials Engineering, Washington State University, Pullman, WA (USA)*

Abstract

Nano-sized Sn particle dispersed in carbon nanofiber (Sn/CNF) composites as anodes for lithium-ion batteries (LIBs) were prepared by using electrospinning techniques and heating treatment. The Sn/CNF composites with amorphous and crystalline structure were obtained by changing heating temperature or time. Among the products carbonized at 850℃ for different time, the Sn/CNF composite carbonized at 850℃ for 1h presents an amorphous structure and delivers the most superior electrochemical performance.

Keyword: anode, carbon nanofiber, electrospinning, lithium-ion batteries, tin

Introduction

Metallic tin (Sn) has been recently considered as a promising anode material for lithium-ion batteries (LIBs) due to its much higher specific capacity than the commercial graphite electrode [1-4]. However, its practical usage in LIBs has been greatly hampered by the extremely large volume change of about 260% when Sn alloys with Li to form $Li_{4.4}Sn$ during Li^+ insertion/extraction cycles, which leads to pulverization of the electrode and very rapid capacity fading [5]. In this work, in order to overcome this problem, a profound new strategy of dispersing nano-sized Sn particles in carbon nanofibers (CNFs) by using electrospinning techniques and heating treatment was proposed to prepare a novel nano-sized tin particle/carbon nanofiber (Sn/CNF) composite as an anode material for LIBs.

Experimental

The product was prepared by using the electrospinning techniques and heating treatment. The preparation flowchart is shown in Figure 1. Field emission scanning electron microscopy (FESEM, Zeiss, supra55), high-resolution transmission electron microscopy (HRTEM, Hitachi H-7650B), X-ray powder diffraction (XRD, Rigaku D/max 2500VB2+/PC diffractometer with Cu Kα, λ=0.1542 nm, 40 kV, 200mA) were used to characterize the microstructure of the products. The electrochemical properties of the as-prepared products were measured using 2025 coin cells assembled following our previous study. [4]

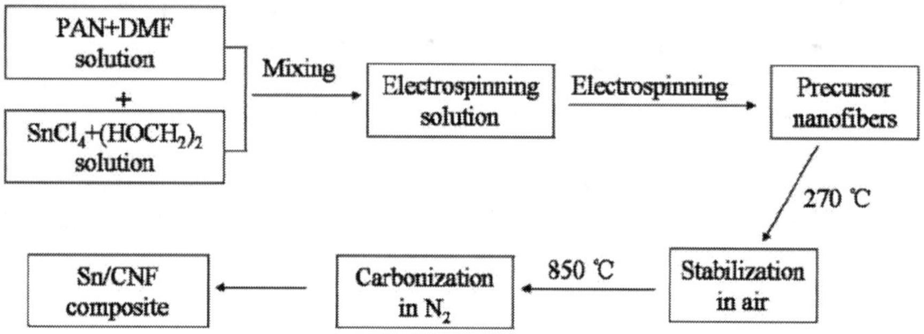

Figure 1. Preparation flowchart of Sn/CNF composite

Result and Discussion

Figure 2 shows the optical, FESEM and HRTEM images of the as-prepared Sn/CNF composite carbonized at 850℃ for 1h. The Sn/CNF product exists in the form of flexible and thin film (Figure 2a), which can be directly used as electrode without adding any binder and conductor. From the SEM image as shown in Figure 2b, it can be found that the product consists of continuous, uniform and partly aligned nanofibers. The surfaces of the fibers are smooth, on which no Sn particles can be found. The HRTEM image in Figure 2c shows that the Sn particles are very fine and uniformly dispersed in the carbon nanofibers.

Figure 2. Optical, FESEM and HRTEM images of Sn/CNF composite carbonized at 850℃ for 1h

Figure 3 shows the XRD patterns of the Sn/CNF composites carbonized at 850℃ for different time. The Sn/CNF composites carbonized at 850℃ for less than 2 hours present an amorphous structure. While increasing the carbonization time, the Sn particles are transformed into crystalline form like that at the higher carbonization temperature of 1000℃.

Figure 4 shows the voltage profiles and capacity-cycle number curves of the Sn/CNF composite anodes carbonized at 850℃ for different time. From the Figure 4a, it can be seen that the Sn/CNF anode carbonized at 850℃ for 1h delivers the largest reversible capacity of about 796 mAh g^{-1} and the highest coulombic efficiency of 74% although there is also an irreversible capacity loss during the first cycle due to the formation of solid-electrolyte interfacial (SEI) layers on the CNFs. [3] Moreover, as shown in Figure 4b, the Sn/CNF anode carbonized at 850℃ for 1h presents the most superior cycling performance with a high reversible capacity of over 650 mAh g^{-1} after 30 cycles.

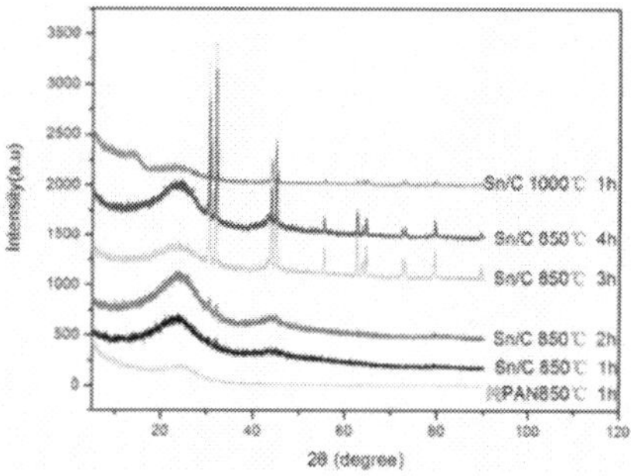

Figure 3. XRD patterns of Sn/CNF composites carbonized at 850℃ for different time

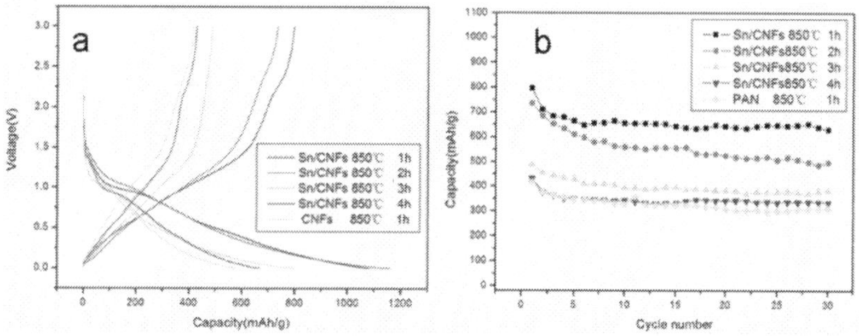

Figure 4. Voltage profiles and capacity-cycle number curves of Sn/CNF composites carbonized at 850℃ for different time

Figure 5 shows the FESEM image of the Sn/CNF anode (carbonized at 850℃ for 1h) after 30 charge-discharge cycles. The nanofibers become more uneven because of the SEI film formation and the large volume expansion during the cycling. However, the integral reticular structure of the Sn/CNF electrode still maintains well after 30 cycles.

Figure 5. FESEM image of Sn/CNF electrode after 30 charge-discharge cycles

Conclusions

A novel Sn/CNF composite anode for LIBs with a stable reticular structure, high specific capacity and good cycling performance can be prepared by using the facile electrospinning technique and heating treatment. Heating time and temperature influence the microstructure and the electrochemical performance of the product. The Sn/CNF composites carbonized at 850°C for less than 2 hours present an amorphous structure and show more superior electrochemical performance than the ones with crystalline structures. Such a Sn/CNF composite with an amorphous structure opens up a possibility of fabricating higher energy-density lithium-ion batteries.

References

1. Gaelle Derrien, Jusef Hassoun, Stefania Panero, and Bruno Scrosati (2007), "Nanostructured Sn–C Composite as an Advanced Anode Material in high-Performance Lithium-Ion Batteries," *Adv. Mater.*,19, PP. 2336–2340.
2. Wei-Ming Zhang, Jin-Song Hu, Yu-Guo Guo, Shu-Fa Zheng, Liang-Shu Zhong, Wei-Guo Song, and Li-Jun Wan (2008), "Tin-Nanoparticles Encapsulated in Elastic Hollow Carbon Spheres for High-Performance Anode Material in Lithium-Ion Batteries," *Adv. Mater.*, 20, PP. 1160–1165.
3. Yan Yu, Lin Gu, Changbao Zhu, Peter A. van Aken, and Joachim Maier (2009), "Tin Nanoparticles Encapsulated in Porous Multichannel Carbon Microtubes: Preparation by Single-Nozzle Electrospinning and Application as Anode Material for High-Performance Li-Based Batteries," J. AM. CHEM. SOC., 131, PP. 15984–15985.
4. Yunhua Yu, Qing Yang, Donghua Teng, Xiaoping Yang, Seungkon Ryu (2010), "Reticular Sn nanoparticle-dispersed PAN-based carbon nanofibers for anode material in rechargeable lithium-ion batteries," *Electrochemistry Communications* , 12, PP. 1187–1190.
5. Kyu T. Lee, Yoon S. Jung, and Seung M. Oh (2003), "Synthesis of Tin-Encapsulated Spherical Hollow Carbon for Anode Material in Lithium Secondary Batteries," *J. AM. CHEM. SOC.*, 125, PP. 5653.

Acknowledgements

The authors acknowledge the financial support from National Natural Science Foundation of China (No. 51072013).

(285a) Segregation and Interface Shape Control During EDG Growth of CZT Crystals

Nan Zhang, Andrew Yeckel and Jeffrey J. Derby, Chemical Engineering and Materials Science, University of Minnesota, Minneapolis, MN

The availability of large, single crystals of cadmium zinc telluride (CZT) with uniform properties is key to improving the performance of gamma radiation detectors fabricated from them. In this presentation, we discuss the validation of our computational models by comparison with experimental zinc distributions via photoluminescence measurements of Burger (Fisk University) and Lynn (Washington State University) for a CZT crystal grown using the electrodynamic gradient-freeze (EDG) method. Furthermore, we predict non-classical zinc distributions for crystals grown in larger-scale EDG systems using classical thermal growth profiles. This surprising result is consistent with many prior growth outcomes where "anomalous" zinc distributions have been observed, quite different from the segregation behavior exhibited by traditional Bridgman growth systems.

We follow with a novel analysis to optimize EDG growth profiles to promote the growth of CZT crystals with a uniformly convex interface shape. Such interface shapes may be very beneficial toward reducing the probability of new grain formation, thus improving the single-crystal yield for a growth run. We also demonstrate that growth under these optimized thermal profiles results in zinc composition that is much more uniform than achieved using classical profiles. We make the case that these results can be put directly into practice using existing EDG growth systems.

See more of this Session: Computational Studies of Electronic and Photonic Materials II
See more of this Group/Topical: Materials Engineering and Sciences Division

(285b) Analysis of a Traveling Magnetic Field (TMF) for Active Control of the Bridgman Growth of CZT Crystals

Gaurab Samanta, Andrew Yeckel and Jeffrey J. Derby, Chemical Engineering and Materials Science, University of Minnesota, Minneapolis, MN

Large, single crystals of cadmium zinc telluride (CZT) are needed for portable, low-cost, and sensitive devices to monitor radioactive materials. However, while CZT improvements are needed, the growth of these crystals is not well understood. We aim to advance the practice of crystal growth by applying physics-based, computational process models along with modern process control theory. Specifically, we are developing model-based control for the growth of cadmium zinc telluride crystals via the Bridgman method.

In this presentation, we describe a novel approach that may be employed to actively control features of CZT growth in Bridgman systems, namely the application of a traveling magnetic field (TMF). The application of TMF produces Lorentz forces through the molten phase, and these forces can be tuned in magnitude and spatial orientation to profoundly impact melt flows. In turn, these flows modify heat and mass transfer, influencing the shape of the melt-solid interface and segregation along the interface. Since the applied TMF can be instantaneously changed, it provides for a potentially ideal control action, if its complicated effects can be predicted and understood. This is the immediate goal of our work; we discuss the mathematical formulation of our TMF analysis and, more importantly, our initial results on the effects of TMF applied to a prototypical CZT growth system.

See more of this Session: Computational Studies of Electronic and Photonic Materials II
See more of this Group/Topical: Materials Engineering and Sciences Division

(285c) First Principles Calculation of the Raman Spectra of Cu2ZnSnS4, a Promising New Photovoltaic Material

Ankur Khare, Burak Himmetoglu, David J. Norris, Eray S. Aydil and Matteo Cococcioni, Chemical engineering and materials science, University of Minnesota, Minneapolis, MN

Thin film solar cells based on copper zinc tin sulfide (Cu_2ZnSnS_4 or CZTS) are emerging as an alternative to cadmium telluride (CdTe) and copper indium gallium diselenide (CIGS) solar cells. Although CIGS and CdTe solar cells have already achieved impressive power conversion efficiencies (15-20%) the elements commonly used for their production are either toxic (e.g., cadmium) or rare in the earth's crust (e.g., indium, tellurium). CZTS is made of nontoxic and abundant elements, has a band gap of ~1.4 eV and absorbs light strongly in the visible region of the electromagnetic spectrum. CZTS crystallizes in a tetragonal crystal structure where $c \approx 2a$. The structure can be thought of two face centered cubic sulfur anion lattices stacked on top of each other with tetrahedral voids filled with metal cations. Half the tetrahedral voids within the sulfur anion FCC lattice are occupied by copper, zinc and tin cations in the ratio 2:1:1. Depending on the local arrangement of copper, zinc and tin within the anion lattice, CZTS can exist in three different phases: kesterite, stannite and the pre-mixed Cu-Au structure (PMCA). The unit cell for CZTS can also be viewed as two zinc blende lattices stacked on top of one another with copper and tin cations replacing some of the zinc cations until the CZTS metal cation stoichiometry is obtained. The structures of the three phases of CZTS and other metal sulfides such as ZnS and Cu_2SnS_3 (CTS) are very similar and they are practically impossible to differentiate based on X-ray diffraction (XRD). This problem becomes acute especially in the case of nanometer size crystals where the XRD peaks broaden with decreasing crystal size. Substitution of S with Se is used to make $Cu_2ZnSnS_{4-x}Se_x$ (CZTSSe) and vary the band gap. Raman spectroscopy may help differentiate between CZTS, CZTSe, CTS and ZnS. However, it is still not known if Raman spectroscopy can be used to differentiate between the three phases of CZTS primarily because the three phases have not been isolated in the laboratory yet.

We have used density functional theory calculations within the generalized gradient approximation (GGA) as implemented in QUANTUM ESPRESSO to calculate the phonon frequencies and vibrational densities of states in CZTS and related compounds and compared the predictions to experimental measurements. Specifically, we compared our calculations to experimental Raman scattering due to characteristic phonon vibrations at the gamma point. We obtain excellent agreement between calculated and measured Raman scattering frequencies for CZTS. Having validated the calculation approach and the pseudo potentials, we extended the predictions to compositions and CZTS related compounds where data is either not available or scarce. We find that the three different phases of CZTS give Raman scattering peaks which are shifted from each other by a few wavenumbers. We have also studied the

Raman spectra for CZTSe and CZTSSe and found significant variations in the spectra with changes in the S:Se ratio. Raman scattering peaks also shift with changes in the arrangements of S:Se within the lattice. These results will be of practical use and allow one to gain deeper insights from the Raman spectra during synthesis of CZTS, CZTSe and related materials.

See more of this Session: Computational Studies of Electronic and Photonic Materials II
See more of this Group/Topical: Materials Engineering and Sciences Division

(285d) Quantifying the Surface Generation Rate for Bulk Point Defects In TiO_2

Kristine Pangan-Okimoto, Alice Hollister, Prashun Gorai and Edmund G. Seebauer, Department of Chemical and Biomolecular Engineering, University of Illinois, Urbana, IL

The defect and surface properties of titania have been well-studied for its potential use in memory resistors and nanoelectronic gas sensors but the control of these properties in such fields has still not been fully realized. Such control of defect mobility and concentration in titania is essential for improving the operation of these technologies. Previous work in our research group has found that controlling the rutile (110) surface has opened up a new pathway for oxygen self-diffusion via an oxygen-interstitial mechanism in conditions where oxygen vacancies and titanium interstitials are thought to dominate. The present work models the detailed diffusion-reaction network of point defects in rutile to simulate interstitial-mediated self-diffusion of oxygen and to calculate a generation rate of bulk point defects at the surface. The model explicitly incorporates gas adsorption onto the (110) surface, interstitial generation at the surface/near surface, and diffusion of the oxygen in an interstitial-mediated mechanism through the bulk. In addition, the model employs techniques drawn from systems engineering to estimate the key kinetic parameters. Simulated diffusion profiles were compared to experimental profiles obtained by exposing annealing single-crystal rutile titania to isotopically labeled oxygen gas and then measuring isotopic oxygen concentration with secondary ion mass spectrometry. The simulated diffusion profiles matched experimental results, verifying that oxygen interstitials do indeed mediate oxygen self-diffusion. More importantly, our model allows us to calculate the oxygen interstitial generation rate at the (110) surface, a first for titania.

See more of this Session: Computational Studies of Electronic and Photonic Materials II
See more of this Group/Topical: Materials Engineering and Sciences Division

(285e) Multiple Exciton Generation In Organic Materials Through Singlet Fission: a Theoretical Perspective

Paul M. Zimmerman, Franziska Bell and Martin Head-Gordon, Dept. of Chemistry, University of California at Berkeley, Berkeley, CA

Future generations of solar cells may break theoretical efficiency limits by simultaneously harvesting high and low energy photons. This possibility has sparked interest in singlet fission, the process where one absorbed photon generates two triplet charge carriers. A detailed understanding of the mechanism is required, however, to exploit the efficiency gain promised by singlet fission. Theoretical study of singlet fission is especially challenging and requires characterization of multi-exciton (multi-electron/hole pair) states, which are inaccessible to most simulation techniques. In this talk, we detail our efforts to characterize multi-exciton states using restricted active space double spin-flip configuration interaction (RAS-2SF). Multi-exciton states cannot form directly by photon absorption, but instead form via coupling to the optically allowed excited states. RAS-2SF simulations show this coupling proceeds through pi-pi interactions caused by intermolecular motion. This analysis shows how singlet fission proceeds through molecular excited states in acene crystals, and this process is distinct from multiple exciton generation mechanisms occurring in inorganic materials.

See more of this Session: Computational Studies of Electronic and Photonic Materials II
See more of this Group/Topical: Materials Engineering and Sciences Division

(285f) Prediction of the Effective Medium Properties of Concentrated Multiple Species Plasmonic Composites

Satvik Wani[1], Tao Cong[1], Ashok S. Sangani[2] and Radhakrishna Sureshkumar[1], (1)Department of Biomedical and Chemical Engineering, Syracuse University, Syracuse, NY, (2)Division of Chemical, Bioengineering, Environmental and Transport Systems, National Science Foundation, Arlington, VA

Plasmonic materials have been actively considered for light trapping in photovoltaic devices. A metal nanoparticle can sustain a localized plasmon resonance that leads to enhanced absorption and storage of energy. The simplest form of a plasmonic material is a random particulate composite. However, the modeling of optical response of such composites under excitation by an electromagnetic wave is not straightforward. Here, we have developed a new technique for calculating the effective medium properties of concentrated linear plasmonic composites containing multiple metallic (Ag, Au, Al and Cu) monodisperse spherical inclusions in a homogenous absorbing matrix. The underlying physical motif was the separation of the space surrounding any inclusion into two regions, one immediately surrounding the particle with the properties of the matrix (the size of this region depends on the static structure factor) and an effective medium. Self consistent closure relations were found for the conditionally averaged fields by solving the vector Helmholtz equations for a layered sphere in an infinite matrix by utilizing a multipole expansion technique in conjunctions with a Newton-Raphson algorithm for the evaluation of the spectral coefficients. In the quasi-static limit, the permittivity was in an excellent agreement with the classical Maxwell-Garnett theory. For large particle sizes and relatively high concentrations (~2 %), we found that the relationship between the concentration and the attenuation constant is nonlinear and is evidenced by experimental measurements on Ag nanoparticle/ethylene glycol suspensions. Our computations also suggest that at high concentrations (~4%), the plasmonic composites become Fano resonant media with an asymmetric absorption peak. We support this by experimental evidence.

Acknowledgement: NSF CMMI 0757589

See more of this Session: Computational Studies of Electronic and Photonic Materials II
See more of this Group/Topical: Materials Engineering and Sciences Division

MODELING AND DESIGN OF METAL-SEMICONDUCTOR PHOTONIC WAVEGUIDES

Meng-Mu Shih, University of Florida, Gainesville, FL

Introduction

Optical waveguides are common structures in photonic devices. To have better light-emitting performance and efficient use-of-energy, specific periodic structures with proper materials can generate stable wavelengths required in scientific research and engineering applications.

How to define the geometry and select proper materials for this model will be described first. This waveguide structure generally has three types of semiconductor layers: an active layer, a cladding layer, a buffer layer, plus a corrugated metal-grating layer. The interface between the metallic layer and its adjacent semiconductor layer has sinusoidal corrugation. To have a stable wavelength, a specific grating period is determined. The corrugation depth of this grating can greatly affect the magnitude of this backward wave mode-coupling. The light beam is emitted from the active layer. However, the active layer thickness can affect this coupling in a more complicated way. The improper design and manufacturing of the active layer thickness can cause this coupling to become smaller or even vanish. Therefore, an optimal range of the active layer thickness is important to make a photonic device function properly.

This work demonstrates a multi-parametric model to compute the mode-coupling, which is important to the analysis and integration of electronic and photonic devices. For such a hybrid metal-semiconductor structure with built-in corrugation, a proper technique, by considering the electronics and photonics, needs to be utilized to solve this computational complexity. This work modifies a prior model which only can work for ideal perfect-metal gratings. Now, this modified model can deal with the practical metal, gold, which can be used in the real fabrication process. Numerical results show comparisons between the ideal model and modified practical model. Two computational methods are used to verify the accuracy of the above results for both models. Further physical interpretation and discussion can support and explain the above results.

Model Description

This work will describe the laser device model, operating at the wavelength of 850 nm in the near infrared range. The lasers without gratings have poor performance. To improve the performance, GaAs/AlGaAs semiconductor laser with metal-gratings were discussed in the 1970s [1]. The performance can be improved for some applications, for example, low beam divergence for targeting [1], narrow linewidth for sensing [1] and stable wavelength for optical communication [2]. Direct band-gap materials are widely used for photonic devices. The metal materials used in the gratings can have the advantages [3] by serving as an electrical contact, heat sink, and a shiny metal with high diffraction efficiency.

Modified Model

Figure 1 shows the schematic structure represented by the cross-section view of a quantum well semiconductor lasers. Two major types of parameters in this model are optical and geometric parameters.

The refractive index is the most important optical parameter to determine the mode distribution inside the optical waveguide and affect the mode coupling. This index can combine many other parameters in material properties, fabrication processes, and physical effects. This multi-layer structure has the following four layers with refractive indices [3-4]: active layer or called active region (na=3.6 for GaAs), buffer layer (nb=3.4 for AlGaAs), cladding layer (nd=3.4 for AlGaAs) and metal contact layer (nc=0.15-5.3i for gold at 850 nm). The geometric parameters for the above layers are: active layer (d), buffer layer thickness (t), grating depth (2a) with grating period (Λ). The grating period Λ is determined by the phase matching condition and related to the wavelength and effective refractive index (ne) of the waveguide. The coordinate system chooses the reference plane (x=0) at the imaginary planar interface between the corrugated buffer layer and metal-grating layer in order to simplify the mathematical derivation.

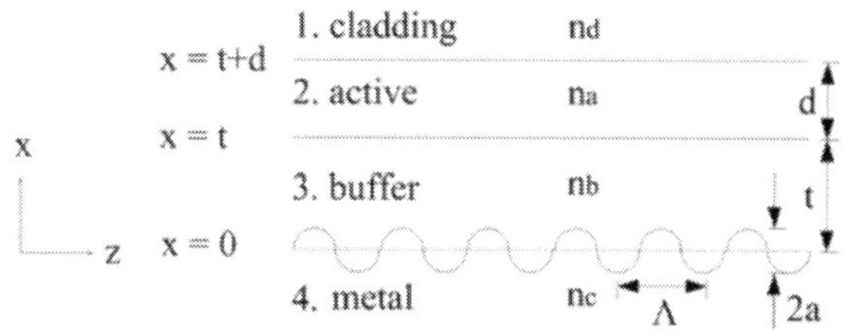

Figure 1 Schematic structure for four-layer quantum-well metal-semiconductor lasers

Computational Techniques

This model calculates the key parameter, backward coupling coefficients (*K*), for analyzing the laser performance such as frequency selectivity and threshold mode gain [5]. The conventional technique of calculating the coupling coefficients can not work well in this corrugated metal-semiconductor structure due to the huge difference of refractive indices at this corrugated interface between the metal and semiconductor [6]. Two calculation techniques [3], ray-optics (RO) and Floquet-Bloch (FB), are used to calculate the coupling coefficients. Floquet-Bloch technique applies the stop-band concept and Bloch theorem in semiconductor electronics to the dispersion diagram in photonics [7]. However, the prior model [3] used Floquet-Bloch technique to calculate the coupling coefficients only in the perfect-metal/semiconductor devices. A modified model by FB technique [8] was demonstrated to calculate the coupling coefficients versus the grating amplitude (a) of the sinusoidal corrugation in shiny-metal/semiconductor devices. This work will use these two techniques to calculate the coupling coefficients versus the active layer thickness to verify the accuracy of the modified model for shiny-metal/semiconductor devices.

Numerical Results and Discussion

Figure 2 and 3 shows the coupling coefficients versus the active layer thickness when waveguides have a=67nm and buffer layer thickness t=300 nm. Both figures show the RO and FB techniques have close numerical results. When the active layer thickness decreases, the electric field interaction at the corrugated interface becomes larger, and the coupling coefficient increases to a peak value. After this peak value, because the thin active layer causes the mode to be more far away from the corrugated interface [3], the coupling coefficient drops quickly and finally reaches to a left-end on each curve. This left-end has the cut-off situation with the effective refractive index close to 3.4. This "cliff region" between the peak and cut-off should be avoided in order to make such photonic devices function properly.

Figure 2 and 3 show the coupling coefficients for the perfect-metal gratings and shiny-metal gratings, respectively. For both RO and FB techniques, Figure 3 shows the coupling coefficients are smaller than those in Figure 2 because the shiny-metal has diffraction efficiency less than ideal value of 1.

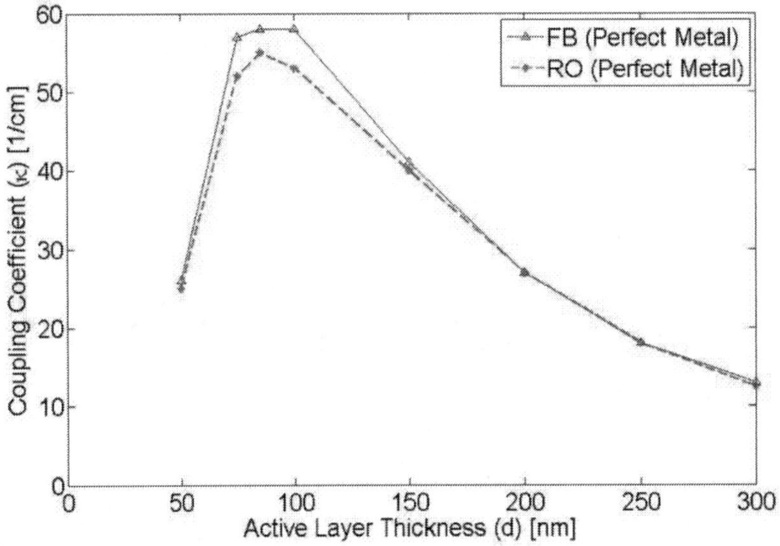

Figure 2 Coupling coefficients versus active layer thickness for perfect-metal gratings

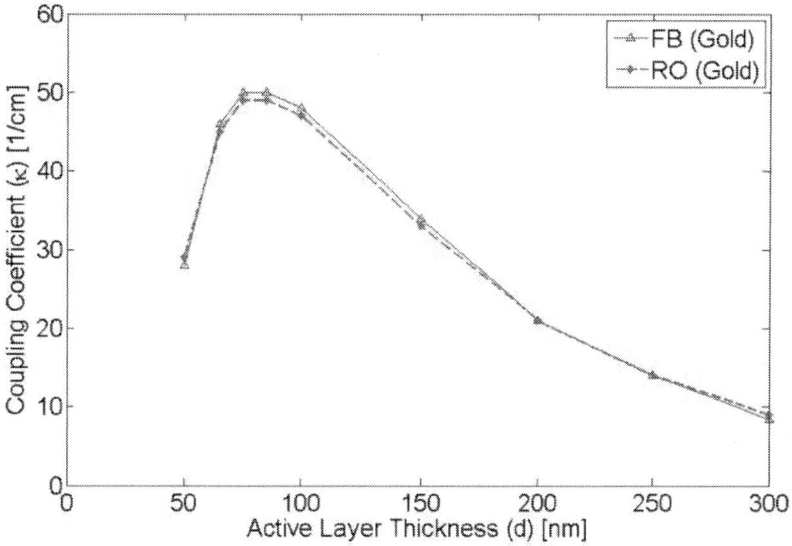

Figure 3 Coupling coefficients versus active layer thickness for gold gratings

Conclusion

This work shows the modified model by using Floquet-Bloch technique can compute the coupling coefficients versus geometric parameters such as grating corrugation amplitude and active layer thickness. The RO and FB techniques have close numerical results.

References

1. Zory, P. S. and Comerford L. D. (1975), "Grating-coupled Double-heterostructure AlGaAs Diode Lasers," *IEEE J. Quantum Electron.*, 11(7), pp. 451-457.
2. Hecht, J. (2010), "Short History of Laser Development," *Opt. Eng.*, 49(9), 091002-15.
3. Luo, H. J. and Zory, P. S. (1990), "Distributed Feedback Coupling Coefficient in Diode Lasers with Metalized Gratings," *IEEE Photonics Tech. Lett.*, 2(9), pp. 614-616.
4. Johnson, P. B. and Christy, R. W. (1972), "Optical Constants of the Noble Metals," *Phys. Rev. B*, 6, pp. 4370-4379.
5. Kogelnik, H. and Shank, C. V. (1972), "Coupled-wave Theory of Distributed Feedback Lasers," *J. Appl. Phys.*, 43, pp. 2327-2335.
6. Luo, H. J. and Zory, P. S. (1994), "Ray Optics Determination of the DFB Coupling Coefficient in Separate Confinement and Multiquantum-well Laser Structures," *IEEE J. Quantum Electron.*, 30(1), pp. 10-14.
7. Yariv, A and Yeh, P. (2007), "Photonics: Optics Electronics in Modern Communication", New York, Oxford University Press, pp. 575-579.
8. Shih, M. (2011), "Modeling First-order Distributed Feedback Semiconductor Lasers at NIR," *Laser Technology for Defense and Security VII, 8039 of SPIE Proceedings*, 80391B, pp. 01-06.

(285h) Modeling and Simulation of Hydrogen Diffusion and Impurity Passivation In n- and p-Doped III-V Semiconductor Photonic Materials

Joshua A. Levinson[1], Kenneth G. Glogovsky[2] and Prajesh Adhikari[1], (1)Chemical and Biomolecular Engineering, Lafayette College, Easton, PA, (2)CyOptics, Inc., Breinigsville, PA

Hydrogen has been identified as an effective passivating species for n- and p-type III-V semiconductor photonic materials. Briefly, upon exposure to hydrogen plasma, hydrogen atoms penetrate the surface of a semiconductor substrate and diffuse through the various thin-film layers. These atoms then react with n- and/or p-type impurities to form a passivated complex, thereby causing electrical or photonic isolation of these regions. As this technique is presently used in photonic device fabrication, it is desired to have a robust, predictive reaction-diffusion model to more efficiently develop passivation techniques with respect to time and cost. The purpose of this work is to develop a fundamental model of the relevant kinetic and transport mechanisms involved in the passivation of photonic materials (with particular emphasis on Zn-doped, p-type InP), to create a predictive computer simulation, and to validate it against experimental data. To meet these objectives, a theoretical model was developed that incorporates concentration-based and potential field-enhanced diffusion of hydrogen in n- and p-type materials while also considering the passivation reactions between the diffusing hydrogen and photonically active impurities. Experiments have been performed to follow the diffusion and/or reaction of deuterium (as a model for hydrogen) in undoped and Zn-doped InP substrates and heterostructures, with deuterium and impurity concentrations determined using SIMS and with the optoelectonic activity of Zn determined using Polaron measurements. This data has been used to determine key parameter values included in the model as a function of temperature, which are subsequently used to predict diffusion-passivation profiles in other thin-film substrate configurations. Computational results have shown excellent agreement between experiment and theory, predicting the passivation profiles and penetration depths for various heterostructures. Present work is focused on expanding the simulation to another computational platform, with development of n-type, p-type, and merged n/p-type versions to address different impurities and their positions in the crystal structure of the substrate. Future work will seek to validate these simulations and to more fully explore the behavior of this reaction-diffusion system under a wider array of thin-film structures.

See more of this Session: Computational Studies of Electronic and Photonic Materials II
See more of this Group/Topical: Materials Engineering and Sciences Division

(286a) Liquid Fuel Production Using Solar-Thermal Energy: Process Development and Technoeconomic Evaluation

Jiyong Kim[1], Terry A. Johnson[2], James E. Miller[3], Ellen B. Stechel[4] and **Christos Maravelias**[1], (1) University of Wisconsin, Madison, WI, (2)Thermal Fluids Science and Engineering, Sandia National Labs, Livermore, CA, (3)Ceramic Processing and Inorganic Materials, Sandia National Labs, Albuquerque, NM, (4)Fuels and Energy Transitions, Sandia National Laboratories, Albuquerque, NM

We have previously described a novel solar-based concept for the production of methanol from carbon dioxide and water. This system applies concentrated solar energy to thermochemically reenergize CO_2 into CO, and then water-gas shift (WGS) to produce syngas (a mixture of CO and H_2) that is fed to a methanol synthesis reactor. Aside from the thermochemical reactor, which is currently under development at Sandia National Laboratories, the approach is based on well-established industrial processes and component designs. Our previous work included an assessment of energy efficiency and economic feasibility of this baseline configuration for an industrial-scale methanol plant. We determined that an overall process energy efficiency (solar-to-fuel) of 7.1% could be achieved and a break-even price of the methanol produced using this approach would be 1.22 USD/kg. Our current work focuses on improvements to the baseline system. We propose several avenues for improvement: 1) alternative pathways to syngas including water splitting and combined CO_2 and H_2O splitting, 2) improved separations, and 3) alternative hydrocarbon products.

For alternative pathways and products, we studied three systems in addition to the baseline system previously reported:
(i) H_2O water splitting, followed by reverse water-gas shift (RWG) reaction, followed by methanol synthesis (MS)
(ii) Simultaneous CO_2 and H_2O splitting (mixed pathway) followed by MS
(iii) Simultaneous CO_2 and H_2O splitting (mixed pathway) followed by Fischer-Tropsch (FT) synthesis

In all designs, we employed amine-based separation systems to remove CO_2 from CO/CO_2 or H_2/CO/CO_2 mixtures. However, we also studied the effect of replacing the CO_2 separations system with an analogous CO separations system. We found that CO separations could provide a significant improvement in energy efficiency and utility costs for the CO_2 splitting and mixed pathways.

To compare the improved systems to our baseline, we performed rigorous process simulations to determine equipment sizing and costing, as well as operational expenses (e.g., raw material and utility costs). We then performed detailed profitability calculations. We determined that the mixed pathway (i. e., simultaneous splitting of CO_2 and H_2O) has a number of advantages over the CO_2-splitting or H_2O-splitting pathways. Also, the production of methanol is slightly more attractive than the production of FT diesel.

Using the mixed pathway with MS as our new baseline, we then carried out sensitivity analysis studies to identify the main cost drivers. We found that the primary economic drivers are the high capital investment associated with the solar concentrator/reactor sub-system which accounts for more than 90% of the capital expenditure, and the high utility consumption for CO/CO_2 separation. Economic parameters such as tax rate and interest rate are also important.

Rather than use a cost of CO_2 as had been done previously, we expanded the boundaries of our system to include CO_2 capture at a power plant. We determined that the effect of CO_2 capture leads to small increases in the minimum selling price (i.e., break-even price) of methanol. We also studied how the minimum selling price (MSP) of methanol increases as a function of the distance between the power plant and the solar facility and the transportation technology.

The energy efficiencies of the different systems were also analyzed. Along with lower cost, the mixed pathway with MS demonstrated the highest efficiency. The primary energy efficiency (= chemical energy out (HHV) over total primary energy in) was calculated as 12.7% if process heat and electricity are provided from fossil fuels and 10.7% if solar power is the sole primary energy source. This 10.7% system efficiency is significantly higher than what can currently be achieved with photosynthesis-based processes, and illustrates the potential for solar thermochemical based strategies to overcome the resource limitations that arise for low-efficiency approaches.

Finally, the analysis indicates that a solar-thermochemical pathway to fuels has significant potential, and points towards future research opportunities to increase efficiency and reduce cost. Particularly, it is evident that there is much room for improvement in the development of a less expensive and more efficient solar concentrator/reactor sub-system, an opportunity that will benefit from the increasing deployment of concentrated solar power. In addition, significant advances are achievable through improved separations, different end products, and different degrees of process integration and distribution. Our investigation establishes a methodology for comparison and assessment of impact on both the economics and the efficiency for advanced system designs.

"This work was supported by the Laboratory Directed Research and Development program at Sandia National Laboratories, in the form of a Grand Challenge project entitled Reimagining Liquid Transportation Fuels: Sunshine to Petrol. Sandia is a multiprogram laboratory operated by Sandia Corporation, a Lockheed Martin Company, for the United States Department of Energy's National Nuclear Security Administration under Contract DE-AC04-94AL85000. "

(286b) Solar Thermochemical Recycling of CO2 Using ALD Deposited CoFe2O4 On Alumina Supports

Paul Lichty, Department of Chemical and Biological Engineering, University of Colorado, Boulder, CO, Xinhua Liang, Chemical and Biological Engineering, University of Colorado at Boulder, Boulder, CO, Carl Bingham, National Renewable Energy Laboratory, Golden, CO and Alan W. Weimer, Department of Chemical and Biological Engineering, University of Colorado at Boulder, Boulder, CO

Fuels production through metal oxide redox cycling can be completely renewable if reduction temperatures are achieved using concentrated solar energy. Many combinations of metal oxides have been investigated over the years and cobalt ferrites show much promise as they have good thermodynamic properties and high melting points. However, bulk ferrites degrade when cycled due to sintering of starting materials and resulting diffusion limitations. Recent work has found that coating the ferrites on rigid supports helps to reduce sintering. When using alumina as a support several Co/Fe/Al spinel species are formed which are capable of being cycled at lower temperatures. Cobalt ferrite has been coated on alumina nano powder and high surface area alumina shells. Large, high surface area polymer particles (~600μm, 43.5 m^2/g) were coated via atomic layer deposition (ALD) with several layers of alumina. The alumina layers were applied using TMA/water ALD chemistry. The particles were then heated to 800°C in air to burn out the polymer substrate. The alumina substrates were then coated with alternating layers of Fe_2O_3 and Co-O using ferrocene and cobaltocene ALD chemistries.

ICP results also show metal oxide mass loadings up to 27%. Thermo-chemical cycling was performed in a Netzch DIL 402 thermogravimetric analyzer with CO production measured using an in-situ Netzch QMS 403C mass spectrometer. Particles with alumina supports maintain cycling efficiency as compared to unsupported particles. Testing has been conducted on sun in a multi-tube reflective cavity reactor at the National Renewable Energy Laboratory in Golden, Colorado. Cycling results help to estimate reactor efficiencies.

See more of this Session: Concentrated Solar for Power Generation and Chemical Processing I
See more of this Group/Topical: International Congress on Energy 2011

(286c) Computational Modeling and On-Sun Model Validation for a Multiple Tube Solar Reactor with Reflective Cavity Walls

Janna Martinek[1], Carl Bingham[2] and Alan W. Weimer[1], (1)Department of Chemical and Biological Engineering, University of Colorado, Boulder, CO, (2)National Renewable Energy Laboratory, Golden, CO

A three-dimensional steady state computational fluid dynamics model is developed for a solar receiver consisting of a reflective cavity with a windowed aperture enclosing an array of five tubes. The computational model couples radiative transfer with fluid flow, heat transfer, mass transfer, and chemical reaction kinetics. The accuracy of the finite volume (FV) method for the approximate solution of the radiative transfer equation is examined for a 2D cross-section of the receiver for the solar and emitted energy components on each of four unstructured grids with three sets of control angles and constant pixelization. FV model accuracy was judged by comparison with Monte Carlo solutions. FV results for the solar component either failed to approach the MC solution or required highly refined spatial and angular grids to achieve accurate, non-oscillatory solutions whereas solutions for the emitted component showed only a weak dependence on the angular and spatial grids.

Based on these results, radiation heat transfer is included in the full 3D model via a combination of ray tracing, Monte Carlo, and finite volume methods. Ray trace modeling of the concentrating system provides the magnitude and direction of the solar energy incident on the receiver window surface. Transport of solar radiation in the cavity is decoupled from all other transport processes occurring in the receiver and profiles of the absorbed solar radiation are determined via a Monte Carlo model using only the receiver geometry, magnitude/direction of radiation incident on the window, and spectral/directional optical properties. A finite volume model is implemented in conjunction with the overall CFD model to account for thermal radiation emitted by heated surfaces in the receiver. Absorption coefficients, scattering coefficients, and the scattering phase function for the cloud of particles in each tube are determined via Mie theory. Transport of the aerosol particles is modeled using a single fluid mixture model with a population balance including transport by convection, Brownian motion, and thermophoretic diffusion. Gasification of 40nm carbon particles is used as a test reaction with Arrhenius kinetic parameters from the literature. Temperature distributions and, correspondingly, reaction conversions among the five tubes are found to be considerably non-uniform, with large temperature gradients developing between the front and back surfaces of each tube. Validation of the base heat transfer and reaction models is accomplished using experimental data taken on-sun at the High Flux Solar Furnace (HFSF) at the National Renewable Energy Laboratory (NREL) with both inert and reacting materials reaching temperatures up to 1400°C.

See more of this Session: Concentrated Solar for Power Generation and Chemical Processing I

(286d) Design and Optimization of Solar Thermal Systems

Amin Ghobeity, Corey J Noone, Enrique Lizarraga-Garcia and **Alexander Mitsos**, Mechanical Engineering, Massachusetts Institute of Technology, Cambridge, MA

Solar-thermal systems based on Concentrated Solar Power (CSP) are considered a promising method to generate electricity from solar energy. CSP systems have been considered for several decades and in recent years generate renewed interest in research, development and commercialization. Several difficulties associated with CSP systems have to be overcome before a successful wide-scale commercial adoption.
The focus of this talk is the design and optimization of CSP with a central receiver, in particular integrated with thermal energy storage. Such technologies are still at a research/development stage.

Central receiver systems are typically employed in flat land using dozens to thousands of heliostats for a single receiver. In the first part of the talk we describe a recent proposal to use hilly terrain [1,3]. Hilly terrain is often significantly cheaper than flat lands which is suitable for many uses. A tool for site selection is presented for both beam-up and beam-down configurations [1] and used to demonstrate the abundance of suitable hills. Additionally a computationally efficient tool for heliostat placement in both planar and hilly terrains is described and used to identify novel heliostat patterns that are significantly better than any of the existing ones [2].

An interdisciplinary team at MIT proposed a new concept, termed CSPonD, that collocates the receiver and storage system [3]. CSPonD has the potential to overcome some of the operational challenges of power towers and reduce capital cost of the receiver/storage system without compromising the efficiency. On the other hand, identifying an optimal design and operation is very challenging. The second part of the talk describes a methodology based on dynamic optimization [4,5,6]. Both single-purpose (electricity only) and dual-purpose plants (electricity and seawater desalination) are discussed. The effect of feed-in tariff (FIT) on optimal design and operation is demonstrated and the influence of seasonal variations on operability discussed. Finally, different energy storage systems (both conventional and novel) are compared in terms of thermodynamic efficiency and cost.

[1] C. J. Noone, A. Ghobeity, A. H. Slocum, G. Tzamtzis and A. Mitsos. Site Selection for Hillside Central Receiver Solar Thermal Plants, Solar Energy, 85 (5):839-848 2011.

[2] C. J. Noone and A. Mitsos, Heliostat Layout Optimization via Automatic Differentiation: Adjoints and Convex Relaxations. SIAM Computational Science and Engineering Conference, Reno, NV, 2011.

[3] A. H. Slocum, D. S. Codd, J. Buongiorno, C. Forsberg, T. McKrell, J.-C. Nave, C. N. Papanicolas, A. Ghobeity, C. J. Noone, S. Passerini, F. Rojas and A. Mitsos. Concentrated Solar Power on Demand,

In Press: Solar Energy, April 11, 2011.

[4] A. Ghobeity and A. Mitsos, Optimal Use of Solar Thermal Energy for Combined Power Generation and Water Desalination (Plenary Talk, Paper #174), Nicosia, Cyprus. Conference on the Promotion of Distributed Renewable Energy Sources in the Mediterranean Region (DISTRES), 2009.

[5] A. Ghobeity and A. Mitsos. Optimal Operation of a Concentrated Solar Thermal Cogeneration Plant. 21st European Symposium on Computer Aided Process Engineering ESCAPE 21. Chalkidiki, Greece. 2011

[6] A. Ghobeity, E. Lizarraga-Garcia and A. Mitsos. Optimal Design and Operation of a Volumetric Solar-Thermal Energy Receiver and Storage. ECOS 2011, The 24th International Conference on Efficiency, Cost, Optimization, Simulation and Environmental Impact of Energy Systems, Novi Sad, Serbia, 2011.

See more of this Session: Concentrated Solar for Power Generation and Chemical Processing I
See more of this Group/Topical: International Congress on Energy 2011

(286e) Techno-Economical Analysis of Solar Thermochemical Ammonia Production At near Atmospheric Pressure

Ronald Michalsky[1], Peter Pfromm[1], Bryon Parman[2] and Vincent Amanor-Boadu[2], (1)Department of Chemical Engineering, Kansas State University, Manhattan, KS, (2)Department of Agricultural Economics, Kansas State University, Manhattan, KS

Techno-economical analysis of solar thermochemicalammonia production at near atmospheric pressure

RonaldMichalsky [1], Peter H. Pfromm [1], Bryon Parman[2], Vincent Amanor-Boadu [2]

[1] Department of Chemical Engineering, [2]Department of Agricultural Economics,

KansasState University, Manhattan, Kansas, USA

Ammonia is thebasis for modern agriculture and ensures the food supply for a growing worldpopulation. It is used widely as base chemical for the chemical industry. Over100 million metric tons NH_3 are produced world wide per year. NH_3has been proposed recently as sustainable transportation fuel and may enableefficient H_2 storage.

At industrialscale the Haber-Bosch process synthesizes NH_3 catalytically at high pressure and elevated temperature with naturalgas and nitrogen as main inputs. Globally this process consumes up to 5%of all natural gas produced and 2% of the total world energy production, withsignificant fossil-based CO_2 emissions. Aiming at reactive NH_3synthesis without natural gas from air and steam at ambient pressure, this workstudies a solar thermochemical cycle employing a molybdenum-based reactant.

Based on Gibbsfree energy computations, a process sequence of Mo_2N formation fromits elements, nitride hydrolysis forming NH_3 and MoO_2 atambient pressure, and metal oxide reduction using solar heat and H_2as chemical reducing agent below 1200 C, is proposed. Experimental data on aMo-based transition element reactant utilized for the fixation of $N_2$and formation of NH_3 from the solid nitride hydrolyzed with steamwill point out process limiting steps such as diffusion in the solid state andionicity of an interstitial nitride reactant. To assess sustainability andeconomic competitiveness compared to the state of the art Haber-Boschprocess, the mass and energy balance for Mo-based solar thermochemical NH_3synthesis including on-site production of H_2 used as reducing agentwill be presented along with an economic analysis determining the price of NH_3required for breaking even with plant investment and operational costs.

See more of this Session: Concentrated Solar for Power Generation and Chemical Processing I
See more of this Group/Topical: International Congress on Energy 2011

(286f) A Sulfur-Sulfur Thermochemical Water Splitting Cycle for Thermal-to-Chemical Energy Conversion

Nick AuYeung, Malachi D. Bunn and Alexandre F.T. Yokochi, School of Chemical, Biological and Environmental Engineering, Oregon State University, Corvallis, OR

Thermochemical water splitting presents a way to store thermal energy in the chemical bonds of a H_2 molecule. Besides becoming an increasingly important energy carrier, hydrogen already has vast importance in petroleum upgrading and ammonia synthesis. In an effort to improve the Sulfur-Iodine Thermochemical Water Splitting Cycle, a novel Sulfur Sulfur Thermochemical Cycle may offer an easier, all fluid route to hydrogen than its predecessor:

Step 1A, liquid phase: $4I_2+4SO_2+8H_2O \leftrightarrow 4H_2SO4 + 8HI$

Step 1B, liquid phase: $8HI+H_2SO_4 \leftrightarrow H_2S+4H_2O+4I_2$

Step 2, gas phase: $3H_2SO_4 \leftrightarrow 3H_2O+3SO_2+1.5O_2$

Step 3, gas phase: $H_2S+2H_2O \leftrightarrow SO_2+3H_2$

An overall theoretical thermal efficiency assuming a maximum temperature of 1200 K is estimated to be 55%. Here, the I_2 consumed in Step 1A is conveniently regenerated in situ via Step 1B. Furthermore, H_2S easily desorbs from the liquid reaction medium and can then be steam reformed to produce H_2 and SO_2, (Step 3) the latter of which is recycled. The H_2S generation and steam reformation steps (Step 1 and Step 3 respectively) have been demonstrated in our laboratory.[1]

This presentation will focus on the parametric dependence of this highly endothermic, high temperature steam reformation reaction on reactant concentrations, temperature, and residence time in tubular reactor.

[1] Yokochi, A., AuYeung, N. In *A Proposed New Sulfur-Sulfur Thermochemical Cycle*, AIChE National Meeting, Nashville, Tennessee, Nashville, Tennessee, 2009.

See more of this Session: Concentrated Solar for Power Generation and Chemical Processing I

(302a) Bringing Broader Understanding of Step Growth Polymerization

W. Harmon Ray, Chemical and Biological Engineering, University of Wisconsin - Madison, Madison, WI

This paper will discuss how Santosh Gupta's work brought an understanding of Step Growth Polymerization to a wide audience.

See more of this Session: In Honor of Professor Santosh K Gupta I
See more of this Group/Topical: Catalysis and Reaction Engineering Division

Unavailable Paper

(302b) Polyelectrolyte Brushes In Multi-Valent Ionic Media

Matthew Tirrell, Institute for Molecular Engineering, University of Chicago, Chicago, IL

Multi-valent ionic media produce attractions between polyelectrolyte chains, in contrast to mono-valent media in which interactions are always repulsive. Numerous examples from our lab of this interesting and important realm of polymer science will be presented and explained drawing on data from surface force measurement, electrochemical measurement, x-ray scattering, rheology and cryo-transmission electron microscopy. Such electro-statically generated attraction offer new possibilities for self-assembled materials in polymer science.

See more of this Session: In Honor of Professor Santosh K Gupta I
See more of this Group/Topical: Catalysis and Reaction Engineering Division

(302c) Adventures with and Insights Into the Bayesian Design of Experiments In Complex Polymerizations

Alexander Penlidis, Department of Chemical Engineering, Institute for Polymer Research, Waterloo, ON, Canada and Afsaneh Nabifar, Department of Chemical Engineerign, Institute for Polymer Research, Waterloo, ON, Canada

For many polymerizations, experimental data may be available either from industrial or exploratory laboratory work. In addition, mathematical models usually do exist, albeit often with unreliable and/or highly correlated parameters, and sometimes even unverified mechanistic bases. In complex polymerization systems, these problems are even more prominent. Hence, ideas from the (statistical) design of experiments applied as early as possible can be very beneficial for the clarification of polymerization kinetics. Classical experimental design methods (for example, (fractional) factorial designs) have been used extensively and are useful in optimizing a wide variety of systems. However, these designs usually ignore the prior knowledge available and are often limited when the system digresses from linear behavior. Using more efficient experimental designs, which can accommodate both prior information and nonlinearities, could (hopefully) lead to optimal performance in fewer trials.

Bayesian design is a powerful and largely unstudied (in the polymerization area) experimental design methodology, which can accommodate practical limitations encountered in classical (fractional) factorial designs. In addition, Bayesian design allows the use of a nonlinear (fully mechanistic) model along with experimental information (hence, it is essentially an optimal model-based design of experiments). The approach can shed light on the most uncertain parts of our process understanding, identify the least reliable (less well known) parameters (e.g., uncertain values of kinetic rate constants), and further guide sensitivity analysis studies focusing on key uncertain parameters in one's model.

The Bayesian design approach is an experimental design technique which has many advantages over standard experimental designs. The approach incorporates prior knowledge about the process (or product) into the design to suggest a set of future experiments in an optimal, sequential and iterative fashion. Since for many complex polymerizations prior information is available, either in the form of experimental data or mathematical models, use of a Bayesian design methodology could be highly beneficial. Hence, exploiting this technique could hopefully lead to optimal performance in fewer trials, thus saving time and money. Advantages of the Bayesian design approach will be illustrated via case studies drawn from different complex polymerization systems. However, since this technique is perfectly general, and exploiting the duality between many systems in science and engineering, it can potentially be applied to other chemical engineering processes.

Why Bayesian design? As mentioned above, standard experimental design methods (for example, (fractional) factorial designs) have been employed extensively and are useful in optimizing a wide variety of systems. However, these designs usually suffer from several limitations and cannot handle certain situations, as listed in Table 1. In addition, these approaches do not take direct advantage of the considerable prior knowledge that is available about the reaction system to design experiments. As prior information is already available within existing data, it is logical that it should be used in order to contribute to the optimality of the designed experiments, and hence to improved models and performance of the process in question. Using more efficient experimental designs, like the family of Bayesian designs, can accommodate these restrictions and could lead to optimal performance in fewer trials.

Table 1. Typical Limitations in Standard Designs

Available resources do not match number of trials that can be designed
Handling of impractical treatment combinations
Handling of situations with missing observations
Factor levels change in the middle of experimentation
Factors with several (or combination of) levels
Dropping/adding factors
Not a sequential and iterative approach
Nonlinearities
Incorporation of prior knowledge

The Bayesian experimental design is based on Bayes' theorem which is well established, especially among statisticians. Recent applications more relevant to chemical and process engineering include heat transfer in packed beds [1], emulsion terpolymerization kinetics [2], drug and cell transport kinetics [3], particle size distribution studies in suspension polymerization [4], pharmaceutical kinetics [5], and analyses of catalytic systems [6]. The common characteristic of all these Bayesian applications is that they are concerned with, admittedly very important, (multi)parameter estimation questions (of kinetic rate constants), but not with the issue of the design of experiments (with the exception of [2], [4], and [6]). In other words, looking at the last 15 years or so, whereby many advances have taken place in polymerization, Bayesian design of experiments has not been exploited extensively nor frequently in complex polymerization systems, which could benefit tremendously from its important traits.

What could one gain from the Bayesian design? In our presentation, principles and capabilities/benefits of the Bayesian design approach will be illustrated and its superiority to the currently practised (standard) design of experiments will be presented via case studies drawn from representative complex polymerization processes. These include examples from nitroxide-mediated controlled radical polymerization (both bimolecular and unimolecular), emulsion copolymerization of acrylonitrile/butadiene (nitrile or NBR) and styrene/butadiene (SBR) rubber, and crosslinking copolymerizations (of styrene and divinyl benzene under controlled radical polymerization conditions). All these case studies address important, yet practical, issues in not only the study of polymerization kinetics but also, in general, in process engineering and improvement. A preview of these important process issues and, accordingly, topics that can be handled efficiently by the Bayesian design approach are listed in Table 2.

Table 2. Overview of Gains in Bayesian Design Approach

Flexible wrt number of trials that can be designed
Changing factor level/range in the middle of experimentation
Accommodating extra trial(s) mid-way through experimentation
Sequential nature (n-trials versus sequences of fewer trials)
Process constraints (and impractical treatment combinations)
Situations with missing observation(s)
Increase of information content (flexibility and cost-effectiveness)
Incorporation of prior knowledge (informative versus non-informative priors)
Flexible wrt source and quality of prior knowledge (screening experiments versus models and/or combinations)
Factors with several (or combination of) levels
Dropping/adding factors
Single versus multi-response scenarios

In addition, the analysis is amenable to a series of statistical diagnostic tests that one can carry out in parallel. These diagnostic tests serve to quantify the relative importance of the parameters (intimately related to the significance of the estimated factor effects) and their interactions, as well as the quality of prior knowledge (in other words, the adequacy of the model or the expert's opinions used to generate the prior information, as the case might be).

The case studies will clearly show that what is novel in the Bayesian approach is the simplicity and the natural way with which it follows the logic of sequential model building paradigm, taking full advantage of the researcher's expertise and information (knowledge about the process or product) prior to the design, and invoking enhanced information content measures (the Fisher Information matrix is maximized, which corresponds to minimizing the variances and reducing the 95% joint confidence regions, hence improving the precision of the parameter estimates).

1. Duran, M. A.; White, B. S. *Chem. Eng. Sci.* **1995**, *50*, 495.
2. Dube, M. A.; Penlidis, A.; Reilly, P. M. *J. Polym. Sci. Part A-Polym. Chem.* **1996**, *34*, 811.
3. Murphy, E. F.; Gilmour, S. G.; Crabbe, M. J. C. *Febs Letters* **2004**, *556*, 193.
4. Vivaldo-Lima, E.; Penlidis, A.; Wood, P. E.; Hamielec, A. E. *J. Appl. Polym. Sci.* **2006**, *102*, 5577.
5. Hermanto, M. W.; Kee, N. C.; Tan, R. B. H.; Chiu, M. S.; Braatz, R. D. *AIChE J.* **2008**, *54*, 3248.
6. Hsu, S. H.; Stamatis, S. D.; Caruthers, J. M.; Delgass, W. N.; Venkatasubramanian, V.; Blau, G. E.; Lasinski, M.; Orcun, S. *Ind. Eng. Chem. Res.* **2009**, *48*, 4768.

(302d) Novel Pathways for Biomass-to-Liquid Fuel Production

Rakesh Agrawal, Dharik S. Mallapragada, Fabio H. Ribeiro and W. Nicholas Delgass, School of Chemical Engineering, Purdue University, West Lafayette, IN

In a limited fossil fuel world, liquid fuels from biomass offer a renewable solution for meeting the transportation sector's fuel needs. In this context, for a given liquid fuel demand, the role of increasing the liquid fuel output per unit of biomass input is critical to reducing the land area requirement for harvesting the feedstock. A grand challenge for chemical engineers is to develop novel processes for converting biomass to high energy density liquid fuels while requiring minimum additional energy input. Moreover, when the additional energy input is derived from solar, increased process energy efficiency would also lead to reduced corresponding land area requirements. Through a systems approach, this presentation will identify some novel biomass-to-liquid fuel pathways based on the efficient use of different forms of supplemental solar energy. In addition, experimental findings pertaining to one particular pathway will also be presented.

See more of this Session: In Honor of Professor Santosh K Gupta I
See more of this Group/Topical: Catalysis and Reaction Engineering Division

(302e) Application of Multi-Objective Optimization In Polymer Reaction Engineering

Ajay K. RAY, Chemical and Biochemical Engineering, University of Western Ontario, London, ON, Canada

Multi-objective optimization involves the simultaneous optimization of more than one objective function. This is quite commonly encountered in Chemical Engineering, particularly in polymer reactors. A considerable amount of research has been reported in this area over the years. The general background of this area will be presented at the beginning, followed by a description of how the results can be described in terms of Pareto sets. We will then present the several methods available for generating these optimal solutions. Applications of optimization in Polymer Reaction Engineering, wherein multiple objectives are encountered, as well as special adaptations of the basic algorithms required to solve these problems, will then be discussed. Most importantly, major contributions of Dr. Santosh K. Gupta in this field will be presented.

See more of this Session: In Honor of Professor Santosh K Gupta I
See more of this Group/Topical: Catalysis and Reaction Engineering Division

(302f) Self-Organized Nano-Lens Arrays by Intensified Dewetting of Polymer Thin-Film

Ankur Verma, Department of Chemical Engineering and DST Unit on Nanosciences, Indian Institute of Technology Kanpur, Kanpur, India and **Ashutosh Sharma**, Chemical Engineering, Indian Institute of Technology Kanpur, Kanpur, India

Dynamics of instabilities and structure formation in thin polymer films havebeen widely studied because of their increasing applications in the areas ofcoatings (thermal and electrical insulation, optical, protective, biologicaletc.), plastic electronics, organic light emitting diodes (OLEDs), membranes,lubricants and adhesives. In many such applications instabilities are undesiredand have to be suppressed for the product to remain functional. However, thestudy of instabilities and their kinetics in ultrathin films (<100 nm) canprovide some important insights into the inter-surface potentials that cannotbe directly measured otherwise. It also can be used to generate useful meso-and nano-scale patterns over large areas (~cm^2). Therefore, thespontaneous dewetting of ultrathin polymer films on a non-wetting substrate hasbeen a subject of great interest owing to its scientific and technologicalimportance.

Dewetting on a flat homogeneous surface starts with the nucleation ofrandomly placed holes at a certain mean separation (λ), which grow andcoalesce with time and result in randomly placed droplets of the polymer.Average diameter as well as the mean separation of dewetted structures are afunction of initial thickness (h) and the interfacial tension of thepolymer film. However, randomness of the dewetted structure limits theusefulness of this method in some applications as a potential soft patterningtechnique. Various strategies have been explored to impose a long range-orderin the dewetted structures. One strategy for the alignment of the dewettedstructure is to combine it with other top-down lithographic approaches suchascontrolled dewetting on topographically or chemically patterned substrates.Dewetting of ultrathin polymers films on physico-chemically patternedsubstrates has also been extensively studied both theoretically andexperimentally. However, despite its promising scientific and technologicaladvantages, the feature-size generated by the self-organized dewetting has twomajor limitations on the pattern resolution and its aspect ratio. The firstlimitation arises from the weak van der Waals destabilizing force and highsurface tension (γ), both of which impose a severe limit on minimumfeature size, which is related to the wavelength, (λ) of the long-waveinstability in spinodal dewetting:

$$\lambda = [-8\pi^2\gamma/(\partial\varphi/\partial h)]^{1/2}$$

Where h is the film thickness and f is the destabilizingintermolecular potential (~ h^{-3} for van der Waalsinteraction). In a model system of polystyrene (PS) thin-film on siliconsubstrate, this limits the spot size to >1m even in case of films as thinas 10 nm.

Other limitation is the very small contact angle (< 10°) and thusthe aspect ratio of dewetted structures in air is rather small.

We have recently developed a novel method which overcomes the limitations onthe feature size and aspect ratio to a great extent. The PS thin (< 100 nm)films were destabilized by immersing into an optimal mixture of solvent(methyl-ethyl ketone and acetone) and water at room temperature. Selectivediffusion of solvent molecules into the polymer matrix brings down its glasstransition temperature (T_g) below room temperature and thuscauses the instability and dewetting of the film. However, water being themajority phase in the bounding media inhibits the dissolution of PS. Roomtemperature dewetting in the liquid media provides the advantage of cleanerenvironment, greater control over the shape of structure and better defectcontrol as compared to thermal annealing or dewetting in air. Further itprevents the sensitive materials from harsh conditions (UV, laser, electronbeam, heating high vacuum etc.) of patterning. Interestingly, dewetting underthe water-solvent mix reduces the droplet diameter by more than one order ofmagnitude compared to dewetting in air. We have also investigated theunderlying mechanisms of intensification of surface instability underwater-solvent mix including the length scale of instability beforefragmentation of the film into droplets and compared the conditions for theformation of ordered patterns on topographically structured substrates in airand under the water-organic mix. Toward these ends, we first systematicallyinvestigate the length scale of instability, droplet diameters andinter-droplet spacing as a function of film thickness in order to clearlydifferentiate these from the well-known aspects of spontaneous dewetting inair. Finally, we explore the conditions for the formation of ordered patternson topographically structured 1-D and 2-D substrates where dewetting occurs inhighly confined but structured spaces. We demonstrate the use this techniquefor fabrication of ordered arrays of polymeric nano-lenses of tunable curvatureby controlled dewetting under the water-organic mix.

We have achieved more than an order of magnitude reduction in the featuresize of the dewetted structures and thus fabricate structures as small as ~60nm. The mean separation of these structures is also reduced close to ~200 nm.Further, the mechanisms of miniaturization of the interfacial instability wasexplored and it was found that the thickness dependence of the characteristiclength scales of dewetted structures is weaker ($\lambda \sim h^{1.51}$)as compared to the instability caused by van der Waals attraction ($\lambda \sim h^2$).Moreover, a far greater control over the shape and aspect ratio of the dewettedstructures is demonstrated as the contact angles can be tuned in the range of40-140°. The ability to tune the curvature of the dewetted structures can beexploited in the fabrication of polymeric nano-lenses. Further, the controlleddewetting under the water-solvent mix on topographically patterned substrateswas also employed for fabrication of sub-micron ordered structures. It was alsoshown that under the same conditions, dewetting in air produces no dewetting,incomplete dewetting or is incapable of producing ordered structures owing to alarge length scale of instability in air. A further miniaturization of lengthscales because of the 2-D confinement is also demonstrated.

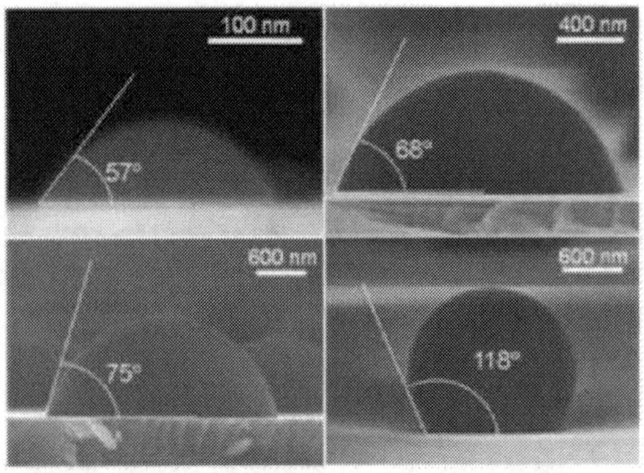

Figure 1. PS droplets of different contact anglesand size visualized in the scanning electron microscope in transverse view.

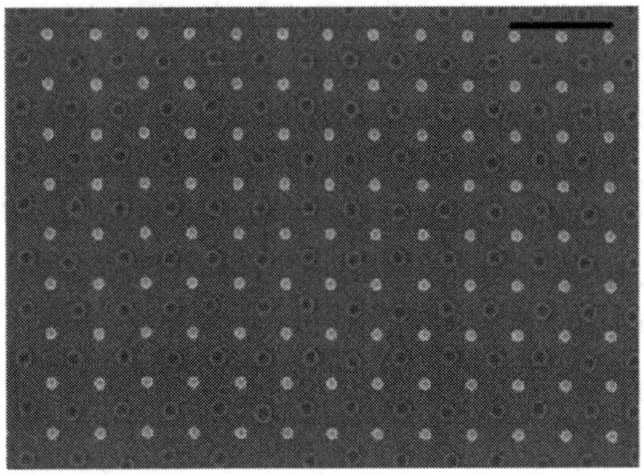

Figure 2. 2-D array of polymeric nano-lenses obtained from the dewettingof PS thin film on the topographically patterned substrate. (Scale bar: 2 mm)

Field-induced self-organized patterning in ultra-thin polymer films producesstructures that are limited to few microns to tens of microns in size becauseof the high energy penalty for the surface deformations on small scales andbecause of weak destabilizing van der Waals forces. We have resolved this longstanding problem of miniaturization of length scales in thin filmself-organization to sub-100 nm scale by reducing the interfacial tension andintensifying the field using a water-solvent mixture for dewetting. This provesto be a simple, powerful, flexible and inexpensive technique for the roomtemperature fabrication of sub-micron polymeric structures such as nano-lensesand their ordered arrays.

In conclusion, dewetting under water-solvent mixture takes the physicalself-assembly of polymer thin-films to its limits by producing sub-micronstructures of various degrees of ordering and tunable shapes. Among otherthings, such patterns have the potential to be used as tunable polymericmicro-lenses and lens-arrays, modifiers for the optical properties, and asdelivery and positioning tools for a variety of functional materials that canbe mixed with the polymer.

References

1. Ankur Verma andAshutosh Sharma, "Enhanced Self-organized Dewetting of Ultrathin Polymer Filmsunder Water-organic Solutions: Fabrication of Submicrometer Spherical LensArrays", *Advanced Materials* **2010**, *22*, 5306-5309.

2. Ankur Vermaand Ashutosh Sharma, "Submicrometer Pattern Fabrication by Intensification ofInstability in Ultrathin Polymer Films under a Water-Solvent Mix", *Macromolecules***2011**, *communicated.*

See more of this Session: In Honor of Professor Santosh K Gupta I
See more of this Group/Topical: Catalysis and Reaction Engineering Division

(321a) Melt and Solid-State Structures of Semicrystalline Linear ABC "Block-Random" Copolymers

Bryan S. Beckingham and Richard A. Register, Chemical and Biological Engineering, Princeton University, Princeton, NJ

The solid-state structure ofsemicrystalline block copolymers is set either by block incompatibility or bycrystallization of one or more blocks. A variety of solid-state morphologiesmay be observed depending on the block interaction strength, ranging fromspherulitic to confined crystallization within preexisting microphase-separateddomains[1]. We aim to explore crystallization from both homogeneousand microphase-separated melts and to characterize the resulting solid-statestructure for linear ABC "block-random" copolymers that incorporate asemicrystalline polyethylene endblock.

"Block-random"copolymers-represented generally by the structure (A_xB_{1-x})-(A_yB_{1-y}),where each of the two blocks is a random copolymer of monomers A and B, simplywith different fractions of A (x and y) present a convenient and usefulvariation on the typical block copolymer architecture, as the interblockinteractions and physical properties can be tuned continuously, via x and y,through the random block's composition. The ability to tune the effectiveinteraction parameter between the blocks continuously, allows for theorder-disorder transition temperature (T_{ODT}) to be tunedindependently of molecular weight using only two monomers, via the difference[2]between x and y. This flexibility makes block-random copolymers a versatileplatform for the exploration of polymer phase behavior and structure-propertyrelationships.

In previouswork we showed that in a cyclohexane/triethylamine mixture, narrow-distributioncopolymers of styrene (S) and isoprene (I) of any desired composition, with nomeasurable down-chain gradient could be synthesized. These random copolymers (S*r*I)have been successfully incorporated into well-defined symmetric blockcopolymers (I-S*r*I diblocks) and subsequent isoprene-selectivehydrogenation yields thermally stable hI-S*r*hI diblocks, whichself-assemble into well-defined lamellar morphologies with sharply-definedorder-disorder transitions, whose temperatures scale predictably with diblockmolecular weight. The use of S*r*hI in lieu of a styrene homopolymer blockeffectively dilutes the unfavorable contacts between the two blocks in thehomogeneous phase, reducing the effective χ; this reductionwas generally consistent with the mean-field °Copolymer equation[3].By studying these block-random diblocks we were able to extract solubilityparameters for both hI and S*r*hI. By comparison with the saturatedhydrocarbon polymers studied extensively by Graessley and coworkers,[4,5]we find that our determined solubility parameter for S*r*hI matched wellwith that calculated from homopolymer solubility parameters, for hI and S,using a volume-fraction-weighted average and, interestingly, that thedetermined value of dS*r*hI is close to that forpolyethylene,[5] suggesting potential miscibility between polyethyleneand S*r*I which would facilitate the formation of homogeneous melts. Here,we extend these hI-S*r*hI studies by adding a semicrystalline endblock,polyethylene (E), via hydrogenation of high-1,4-polybutadiene (B), to formlinear ABC (E-hI-S*r*hI) and ACB (E-S*r*hI-hI) "block-random"copolymers. Such ABC triblock copolymers are more complex than more typical ABdiblock copolymers, with two additional independent interaction parameters andtwo independent volume fractions.

We synthesize linear triblockcopolymers, B-I-S*r*I and B-S*r*I-I, via organolithium-initiatedanionic polymerization in cyclohexane. After polymerization of the butadieneblock, triethylamine is added facilitating the random copolymerization ofstyrene and isoprene while also increasing the vinyl content of the isopreneblock. Selective hydrogenation of the diene units with a Ni-Al catalyst yieldsa semicrystalline polyethylene endblock. For E-hI-S*r*hI triblock-randomcopolymer with block molecular weights of 30-14-14 kg/mol, small-angle x-rayscattering has revealed the formation of a well-ordered three-domain lamellarmelt from which crystallization of the hydrogenated polybutadiene(polyethylene) block proceeds. Currently this polymer is being studied afterorientation, compression in a channel die[6] above T_m(E),to further examine the solid state structure, to confirm confinement ofcrystallization into the melt morphology, and to reveal the orientation of thecrystal stems relative to the pre-existing lamellar interfaces formed byself-assembly in the melt. We are currently synthesizing, hydrogenating andcharacterizing additional E-hI-S*r*hI and E-S*r*hI-hI triblock-randomcopolymers to explore the resulting morphologies.

This work was generously supported by the National ScienceFoundation, Polymers Program (DMR-1003942).

References

(1) Loo,Y. L.; Register, R. A.; Ryan, A. J. *Macromolecules*, **2002,** *35*,2365-2374.

(2) Smith,S. D.; Ashraf, A.; Satkowski, M. M.; Spontak, R. J. *Polym. Prepr. (Am. Chem.Soc., Div. Polym. Chem.)* **1994**, *35(1)*, 651-652.

(3) Graessley,W. W. *Polymeric Liquids & Networks: Structure and Properties*;Garland Science: New York, 2004, pp. 341-407.

(4) Graessley,W.W.; Krishnamoorti, R.; Reichart, G. C.; Balsara, N. P.; Fetters, L. J.;Lohse, D. J. *Macromolecules* **1995**, *28*, 1260-1270.

(5) Krishnamoorti,R.; Graessley, W. W.; Dee, G. T.; Walsh, D. J.; Fetters, L. J.; Lohse, D. J. *Macromolecules***1996**, *29*, 367-376.

(6) Lee,H. H.; Register, R. A.; Hajduk, D. A.; Gruner, S. M; *Polymer Engineering andScience*, **1996,** *36*, 1414-1424.

(321b) Characterization of Glass Transition Temperatures In Block Copolymer/Ionic Liquid Micelle Cores

Michelle M. Mok, Chemistry, University of Minnesota, Minneapolis, MN and Timothy P. Lodge, Department of Chemistry and Department of Chemical Engineering and Materials Science, University of Minnesota

Achieving equilibrium structures in block copolymer micelles is significantly more difficult than in small-molecule surfactant micelles, due to the much higher thermodynamic barriers and slower kinetics associated with the long-chain structures of their unimers. Polymer chains can also become trapped or frozen within micelles in cases when the core-forming block has a glass transition temperature (Tg) at the temperature ranges of interest. Measurements of micelle core glass transitions are challenging, given the nanoscale volumes and small volume fractions being probed. Such measurements would not only provide insight into block copolymer micellar equilibration, but also have the potential to yield contributions to the field of polymer systems confined at the nanoscale. The physical properties of such systems can be significantly different from those of bulk systems; this has been demonstrated in a number of cases in terms of changes to the Tg of polymer thin films, but few studies have been carried out in other geometries. Frozen micelles present an important opportunity for expanding the range of studies in low-dimensional structures

We discuss the development of a fluorescence-based technique for the investigation of Tgs in polystyrene (PS) micelles cores in ionic liquid solvents. Specifically, we demonstrate the ability to detect the transition from their fully glassy state to their fully rubbery state in micelles of poly(styrene-b-ethylene oxide) and poly(styrene-b-methylmethacrylate). The use of non-volatile ionic liquids as the solvent allowed for studies over extended temperature ranges while remaining in a stable, liquid state. The systems are explored as a function of copolymer block molecular weight and composition and micelle preparation protocol. We find that the PS core Tgs are shifted 20–45 °C below reported bulk PS homopolymer Tgs of equivalent molecular weight, and that they can also be significantly broader (up to 45 °C in breadth).

See more of this Session: Nanoscale Structure In Polymers II
See more of this Group/Topical: Materials Engineering and Sciences Division

(321c) Polystyrene Nanoparticles As a Model System to Investigate 3-Dimensional Confinement Effects On the Glass Transition Temperature

Chuan Zhang, Yunlong Guo and Rodney D. Priestley, Chemical and Biological Engineering, Princeton University, Princeton, NJ

When confined to the nanoscale, the glass transition temperature (T_g) of polymers can deviate substantially from the bulk, in what is commonly referred to as the T_g-confinement effect. Due to ease of processing, most studies have focused on the size-dependent T_g of thin films, while few have extended investigations to other geometries. As polymers confined to higher geometrical dimensions become the enabling material in technologies ranging from drug delivery to plastic electronics, a greater understanding of size effects on the T_g is warranted. Here, we investigate the effect of soft and hard three-dimensional confinement on the T_g of polymer nanoparticles. Via modulated differential scanning calorimetry, we show that the T_g can decrease significantly with size for aqueous suspended polymer nanoparticles, i.e., the case of soft confinement. Furthermore, capping of the nanoparticles with a silica shell leads to a size-invariant T_g, i.e., the case of hard confinement. Lastly, using a unique flash nanoprecipitation method to generate polymer nanoparticles of the same size but different molecular weights, we examine the impact of polymer molecular weight on the T_g of 3-dimensional confined polystyrene nanoparticles.

See more of this Session: Nanoscale Structure In Polymers II
See more of this Group/Topical: Materials Engineering and Sciences Division

(321d) Characterization of Oxygen Scavenging Films Based On Butadiene-Containing Polymers

Kevin K. Tung, Richard H. Li, Benny D. Freeman and Don R. Paul, Department of Chemical Engineering, The University of Texas at Austin, Austin, TX

Butadiene-containing polymers, such as styrene-butadiene-styrene block copolymers and 1,4 polybutadiene are of use as oxygen scavenging polymers (OSP) in barrier applications. As part of an evaluation of this use in such applications, films of these polymers were oxidized catalytically at four catalyst loadings and various thicknesses. Oxidation led to formation of crust layers at the film outer surfaces. Gas permeabilities of the crust layers are 2-3 orders of magnitude lower than those of neat, un-oxidized polymers. The diffusion-controlled oxidation creates a transient moving boundary that separates the oxidized and un-oxidized regions. Maximum oxygen uptake was measured to understand the kinetics and oxidation capacities of OSP systems and extracted from thin film experiments.

The thickness of the oxidized regions at long oxidation times is called the critical thickness, and it reaches a maximum at a certain catalyst loading. Films that are thinner than the critical thicknesses will oxidize fully and homogeneously. Results from three methods of analysis: 1) Thickness Model Analysis, 2) Inverse Thickness Analysis, and 3) SEM Imaging Analysis, all indicate a similar critical thickness value in each OSP system. Critical thicknesses appear related to the values of maximum uptake. Critical thicknesses set the upper OSP thickness limit in future diffusion-controlled oxidation systems, such as multi-layered and blend films.

See more of this Session: Nanoscale Structure In Polymers II
See more of this Group/Topical: Materials Engineering and Sciences Division

(321e) Exploring Molecular Architecture Effects On the Microstructures of Block Copolymer Liquid Crystals

James A. Bergman and Jennifer M. O'Donnell, Chemical and Biological Engineering, Iowa State University, Ames, IA

Many amphiphilic block copolymers self-assemble in water to form periodic structures or liquid crystals. Liquid crystal microstructure greatly impacts the bulk material properties. For example, the viscosity of bicontinuous cubic microstructures is significantly greater than the viscosity of lamellar or hexagonal-close packed microstructures. In addition, bicontinuous cubic microstructures have superior transport properties relative to other microstructures. Therefore, certain microstructures are optimal for specific applications. Unfortunately, block copolymer chemistry and microstructure are coupled, and the optimal chemistry for a particular application often does not produce the optimal microstructure, so current applications require a compromise between chemistry and microstructure to optimize product performance. This work investigates the effect the copolyomer architecture at the hydrophilic – hydrophobic interface has on the self-assembled microstructure for a series of polymers with identical chemical composition.

Five polymer architectures, which differ only in the number of hydrophilic branches, have been synthesized by a combination of reversible addition-fragmentation chain transfer (RAFT) polymerization, and click chemistry. These methacrylic polymers contain the same number of hydrophobic butyl methacrylate (BMA) monomer units and the same number of hydrophilic hydroxyethyl methacrylate (HEMA) monomer units. The BMA block is grown as a straight polymer chain, and the HEMA block is grown as either a straight chain, or as two, three, four, or five branches. Branches are added via a 'graft from' scheme by clicking additional RAFT agents to the BMA backbone, then growing the HEMA block accordingly. The microstructure of each polymer architecture as a function of concentration in aqueous solution has been studied using small angle x-ray scattering (SAXS) and cryogenic transmission electron microscopy (cryo-TEM). The results indicate that altering the architecture of the copolymer without altering the chemical composition induces a change in the thermodynamically favorable microstructure.

See more of this Session: Nanoscale Structure In Polymers II
See more of this Group/Topical: Materials Engineering and Sciences Division

(321f) Structures of Polyelectrolytes In Differently Charged Colloidal Solutions

Chongli Yuan, School of Chemical Engineering, Purdue University, West Lafayette, IN and Ian Smith, Chemical Engineerign, Purdue University, West Lafayette, IN

Polyelectrolytes are commonly used as a stabilizing agent in colloidal solutions, e.g., oil, lotion and etc. In addition to the electrostatic contribution, polyelectrolytes are capable of rearranging the nanoparticles into distinctive spatial organizations due to their unique conformational features, i.e., length, geometry and flexibility. This study explores the conformational changes of different types of model polyelectrolytes, e.g., flexible, semi-flexible and rigid polymers, when mixed with distinctively charged nanoparticles. The conformation and the dynamics of the model polymers will be characterized using fluorescence spectroscopy approach. This knowledge will help us to elucidate the interactions between polyelectrolytes and colloidal particles and lead to an unambiguous identification of an energetic term accounting for the effects of polyelectrolytes on colloidal suspensions.

See more of this Session: Nanoscale Structure In Polymers II
See more of this Group/Topical: Materials Engineering and Sciences Division

(327a) Humidity Swelling Hysteresis In Electrostatically Crosslinked Polyelectrolyte Multilayer Films

Adam J. Nolte, Chemical Engineering, Rose-Hulman Institute of Technology, Terre Haute, IN and Kimberly E. Secrist, Chemistry, Rose-Hulman Institute of Technology, Terre Haute, IN

Polyelectrolyte multilayers (PEMs) are thin polymer films created via the alternating adsorption of positively and negatively charged polyelectrolytes from aqueous solutions. Although PEMs are generally insoluble due to internal electrostatic crosslinks formed during the deposition process, the films remain moisture-responsive and hence are good candidates as humidity sensors and stimuli-responsive coatings. In this talk, we demonstrate that PEMs assembled from poly(allylamine hydrochloride) and poly(acrylic acid) demonstrate humidity swelling/deswelling hysteresis, which ultimately can be attributed to the internal crosslink structure of the PEM. The hysteresis phenomenon is remarkably persistent, with a characteristic relaxation time on the order of hours to days. At even longer times, films retain a swelling difference of a few percent, demonstrating that a PEM can maintain a "memory" of its drying history. These results suggest that careful control over PEM drying and storage may be warranted, as such treatments may affect subsequently measured film properties.

See more of this Session: Polymer Thin Films and Interfaces II
See more of this Group/Topical: Materials Engineering and Sciences Division

(327b) Poly(ethylene oxide) (PEO) Chains Are Not "Hydrophilic" When They Exist As Polymer Brush Chains

Hoyoung Lee[1], Kimberly Ohn[1], Kevin N. Witte[1], Dae Hwan Kim[1], Bulent Akgun[2], Sushil Satija[2] and **You-Yeon Won**[1], (1)School of Chemical Engineering, Purdue University, West Lafayette, IN, (2)NIST Center for Neutron Research, National Institute of Standards and Technology (NIST), Gaithersburg, MD

A model poly(ethylene oxide) (PEO) brush system was prepared by spreading a poly(ethylene oxide)-poly(n-butyl acrylate) (PEO-PnBA) amphiphilic diblock copolymer on to an air-water interface. The polymer segment density profiles of the PEO brush in the direction normal to the air-water interface under various grafting density conditions were determined by using the neutron reflectivity (NR) measurement technique. In order to achieve a theoretically sound analysis of the reflectivity data, we developed a new data analysis method that uses the self-consistent field (SCF) theoretical modeling as a tool for predicting expected reflectivity results for comparison with the experimental data. Using this new data analysis technique, we discovered, for the first time, that contrary to what is known for PEO in normal situations, the PEO brush chains are not "hydrophilic" (i.e., $\chi_{PEO/water} > 0.5$) because of the many body interactions that are forced to be effective in the brush situation, in particular, in the region close to the grafting surface. We suspect that this effect is key to the mechanism by which a PEO brush system, a coating structure commonly used in biomedical materials for protein resistance and biocompatibilization, is able to resist, so effectively, the adsorption of proteins in physiological conditions.

See more of this Session: Polymer Thin Films and Interfaces II
See more of this Group/Topical: Materials Engineering and Sciences Division

(327c) Effects of the Salt Concentration On Charge Regulation In Tethered Polyacid Monolayers Designed to Model ssDNA Monolayers

Mark J. Uline, Department of Biomedical Engineering, Northwestern University, Evanston, IL and Igal Szleifer, Biomedical Engineering, Northwestern University, Evanston, IL

Charge regulation in polyacid monolayers attached at one end to a planar surface is studied theoretically. The polyacid layers are designed to mimic single-stranded DNA monolayers. The effects of the local pH and salt concentration on the protonation states of the polyacid layer are studied using a molecular mean-field theory that includes a microscopic description of the conformations of the polyacid molecule along with electrostatic interactions, acid–base equilibrium, and excluded volume interactions. We predict that, in the case of a monovalent salt, NaCl, the amount of proton binding increases dramatically for high surface coverage of polyacid and low bulk salt concentration. When the polyelectrolyte is almost completely charge neutralized by bound protons, there is an expulsion of sodium from the layer. We show that the degree of protonation can go all the way from 0% to 100% when the bulk pH is kept fixed at 7 by changing the surface coverage of polyacid and the bulk salt concentration. The effects of increasing protonation and the expulsion of the cations from the monolayer are reduced when sodium ions are replaced by divalent magnesium ions. Our theoretical results concur with X-ray photoelectron spectroscopy studies of ssDNA monolayers on gold.

See more of this Session: Polymer Thin Films and Interfaces II
See more of this Group/Topical: Materials Engineering and Sciences Division

(327d) Directed Deposition of Functional Polymers Onto Porous Substrates Using Metal Salt Inhibitors

Philip Kwong, Cristofer Flowers and Malancha Gupta, Mork Family Department of Chemical Engineering and Materials Science, University of Southern California, Los Angeles, CA

In this talk, we will demonstrate the ability to control the location of polymer deposition onto porous substrates using vapor phase polymerization in combination with metal salt inhibitors. Functional polymers such as hydrophobic poly(1H,1H,2H,2H-perfluorodecyl acrylate), click-active poly (pentafluorophenyl methacrylate), and light-responsive poly(*ortho*-nitrobenzyl methacrylate) were patterned onto porous hydrophilic substrates using metal salts such as $FeCl_3$ and $CuCl_2$. A combinatorial screening approach was used to determine the effects of different transition metal salts and reaction parameters on the patterning process. Our study offers a new and convenient method to selectively deposit a wide variety of functional polymers onto porous materials and will enable the production of next-generation multifunctional paper-based microfluidic devices, polymeric photonic crystals, and filtration membranes.

See more of this Session: Polymer Thin Films and Interfaces II
See more of this Group/Topical: Materials Engineering and Sciences Division

(327e) Manipulating the Growth and Disassembly of Layer-by-Layer Polymer Thin Films

Biswa P. Das, Department of Chemical and Biological Engineering, University at Buffalo - The State University of New York, Buffalo, NY and **Marina Tsianou**, Department of Chemical and Biological Engineering, University at Buffalo - The State University of New York (SUNY), Buffalo, NY

Multilayered thin films have been the focus of intense research because of their vast array of applications. We investigate the growth and disassembly under different pH conditions of layer-by-layer (LbL) films composed of oppositely charged polyelectrolytes (including poly (diallyldimethylammonium chloride) (PDDA), poly(styrene sulfonate) (PSS), and poly(acrylic acid) (PAA)) and synthetic clays. We focus on the fundamentals governing the LbL assembly and triggered disassembly in order to get a better understanding of the mechanisms that control the internal ordering and organization in these films. We have confirmed the presence of layering in these nanostructures by scanning electron microscopy (SEM), obtained the surface roughness via atomic force microscopy (AFM), determined the regular spacing in these layers by X-ray diffraction (XRD), and studied the layer growth and disassembly using UV-vis absorbance spectroscopy. We evaluate the structural characteristics and properties of these multilayers, and we are especially interested in investigating their sensitivity and stability to external stimuli such as pH, temperature, humidity, and ionic strength.

See more of this Session: Polymer Thin Films and Interfaces II
See more of this Group/Topical: Materials Engineering and Sciences Division

(327f) Degradation of Organic Coatings In Deionized Water and NaCl Solution

Qixin Zhou[1], Yechun Wang[1] and Gordon P. Bierwagen[2], (1)Mechanical Engineering, North Dakota State University, Fargo, ND, (2)Coatings and Polymeric Materials, North Dakota State University, Fargo, ND

Water percolation into the metal-coating interface is usually the main cause of the loss of barrier properties of coatings and leads to coating delamination and underfilm corrosion. Previous studies focus on coating degradation behaviors under electrolyte solution immersions. Reported studies using pure water (e.g. deionized or distilled water) are only for metallic materials. Traditionally, investigations on the degradation of organic coatings are performed for organic coatings immersed in stationary solutions. Recently, flowing fluid accelerated coating degradation has been received more attention since flowing fluid may enhance the transport of ions and oxygen through the coating film, abrade the coating surface with fluid shear, and degrade the barrier properties of organic coatings.

In this study, both deionized water and 3.5 wt% sodium chloride solution are chosen as the working fluids. To investigate the degradation behavior of coatings under a variety of flow rates, the organic coatings are exposed to stationary immersion as well as the laminar flow. A variety of coatings have been investigated including epoxy and polyurethane based clear coating, commercially available air force coating, and marine coating. The barrier properties of coatings are monitored inline by electrochemical impedance spectroscopy (EIS) measurements. Moreover, equivalent circuit models are developed to interpret EIS spectra and to analyze the physical behavior of coatings as they are degrading. We find that the impedance spectra of coatings decrease with the immersion time and the decrease is more substantial for flowing fluids at higher flow rate disregard of the fluid type. Different types of coatings respond differently to DI water and 3.5 wt% sodium chloride solution for their degradation rates.

See more of this Session: Polymer Thin Films and Interfaces II
See more of this Group/Topical: Materials Engineering and Sciences Division

(337a) Monodisperse Core-Shell Chitosan Microcapsules for pH-Responsive Burst Release of Hydrophobic Drugs

Li Liu, Jian-Ping Yang, Xiao-Jie Ju, Rui Xie, Ying-Mei Liu, Wei Wang and Liang-Yin Chu, School of Chemical Engineering, Sichuan University, Chengdu, China

Microencapsulationof drugs has attracted great attention because of its capability of maskingunpleasant drug tastes and odors, in controlled release of drugs and inprotecting drugs from undesirable degradation.[1] Due to a largeinterior space, core-shell microcapsules can enhance loading efficiencycompared with homogeneous microspheres. Stimuli-responsive microcapsules showeven greater advantages in controlling the release rate and targeting therelease site, because the release can be triggered by environmentaltemperature, pH, ionic concentration and/or magnetic field.[2] Amongthose microcapsules, pH-responsive microcapsules have been paid much attentionbecause of large variations in physiological pH at different body sites innormal as well as pathological conditions.[3] For example, the pH valueof the gastric content can vary from a lower value of 1.2 in the fasted stateto a higher value of 5.0 under fed state.[4] Gastric acid secretiondisorder can cause diseases such as gastroesophageal reflux disease and pepticulcer disease. Acid-suppression therapy has been adopted to treat thosediseases for decades. This treatment demands agents to reach its maximalefficacy in a short time to alleviate patients' symptom as soon as possible.[5]Thus, the development of acid-triggered core-shell microcapsules for gastricdelivery with prompt onset and complete release characteristics in acontrollable manner is of both scientific and therapeutic interest.

Here, we reporton a novel strategy to fabricate core-shell chitosan microcapsules forstomach-targeted rapid and complete drug release in a controllable manner bydesigning an acid-triggered burst release mode for the capsules. The core-shellchitosan microcapsule membrane is composed of terephthalaldehyde-crosslinkedchitosan hydrogel. In this study, our strategy for microcapsule preparation isto use uniform-sized oil-in-water-in-oil (O/W/O) emulsions fabricated bycapillary microfluidic technique[6] as templates and convert these emulsionsinto core-shell microcapsules via interfacial crosslinking reaction. Inner oilphase, middle water phase, and outer oil phase solutions were separately pumpedinto the injection tube, the transition tube, and the collection tube throughpolyethylene tubing attached to disposable syringes. Based on the coaxialco-flow geometry, monodisperse O/W single emulsions were generated in thetransition tube and monodisperse O/W/O double emulsions were generated in the collectiontube. In our O/W/O double emulsion templates, the chitosan is in the middle waterlayer and the inner oil phase contains oil-soluble terephthalaldehyde, whichacts as crosslinker in the subsequent interfacial crosslinking reaction. Thecrosslinking reaction occurs at the inner O/W interface of the double emulsiontemplate as soon as the inner oil fluid contacts the middle water fluid in the transitiontube of the microfluidic device. The obtained O/W/O double emulsions werecollected in a container, and left to stand overnight to make sure the chitosanin the water phase was completely crosslinked. The resultant microcapsules werewashed using a mixture of ethyl acetate and isopropanol (1U5 v/v) to removethe inner and outer oil solutions, and finally dispersed into water. Suchone-pot method presented here has competitive advantages for preparingchitosan-based core-shell microcapsules with more controllable structure andsimpler procedure. Furthermore, lipophilic substances can be easilyencapsulated into the proposed chitosan microcapsules through the microfluidicapproach.

The microcapsuleshave been found to display autofluorescent properties because of the formationof Schiff's bases.[7] The Fourier transform infrared (FTIR) analysisalso confirms the crosslinking reaction between chitosan and terephthalaldehyde

The averageouter diameters of the double emulsions and microcapsules are 292 μm and 224 μm,respectively. The coefficient of variation (*CV*), which is defined as theratio of the standard deviation of the size distribution to its arithmeticmean, is used to characterize the size monodispersity of particles. The *CV*values for the double emulsions and the resultant microcapsules are 0.94% and2.3% respectively, which indicate their narrow size distributions. Thethroughput rate of the chitosan microcapsules prepared with a singlemicrofluidic device is $1.25\ 10^5$ per hour.

In neutralmedium with pH 7.1, microcapsules shrink considerably during the initial periodof 24 h and afterwards the size of microcapsules shows nearly no change. Fivedays later, the average outer diameter of microcapsules is reduced from 253 μm to201 μm and the thickness of capsule membrane decreases by 23%. Although aslight volume change is observed for the crosslinked chitosan microcapsules,the microcapsules maintain good spherical shape and structural integrity inneutral medium.

To estimate thecapability of acid-triggered burst release from microcapsules, thedecomposition processes of chitosan microcapsule membranes in the pH range of1.54.7were studied systematically. Microcapsules were firstly immersed in deionized water.Then, we introduced a sudden change to the pH value of their environmentalsolution by quickly adding HCl or phosphate buffer solution with different pHvalues.Allthe microcapsules swell first, and then gradually collapse and finallydecompose. The lower the environmental pH value, the faster the acid-triggeredswelling and the faster the decomposition of microcapsule membrane. When the environmentalpH value is 4.7, it takes 22 min for the microcapsules to completely decompose;whereas, when the environmental pH value decreases to 1.5, the microcapsulesdecompose rapidly in 39 s. This pH-dependent decomposition manner can be utilizedto develop smart gastric delivery systems, from which anti-acid agents can bereleased at a rate that depends on the pH value of the gastric juice.

To demonstratethe feasibility to encapsulate lipophilic drugs using our technique and theacid-triggered burst release behavior of the prepared microcapsules, LR300, ared fluorescent dye, is successfully encapsulated in the crosslinked chitosanmicrocapsules as a lipophilic model drug (Fig. 1).

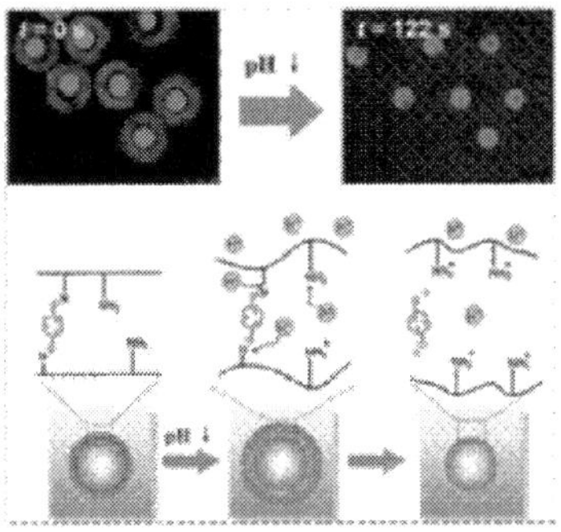

Fig. 1 The acid-triggeredburst release process of chitosan microcapsules.

Theburst and complete release pattern from the prepared chitosan microcapsules mayenable them to be promising stomach-specific drug carrier candidates forquick-release.

References

[1] (a) W. E. Yuan, F. Wu and T. Jin, Polym. Adv.Technol., 2009, 20, 834; (b) A. S. Pedro, E. Cabral-Albuquerque, D. Ferreiraand B. Sarmento, Carbohydr. Polym., 2009, 76, 501.

[2] W. Wei, C. L. Zhang, S. J. Ding, X. Z. Qu, J. G.Liu and Z. Z. Yang, Colloid Polym. Sci., 2008, 286, 881.

[3] P. Gupta, K.Vermani and S. Garg, Drug Discovery Today, 2002, 7, 569.

[4] R. Hejazi and M.Amiji, Int. J. Pharm., 2002, 235, 87.

[5] K. R. DeVault andN. J. Talley, Nat. Rev. Gastroenterol. Hepatol., 2009, 6, 524.

[6] (a) A. S. Utada,E. Lorenceau, D. R. Link, P. D. Kaplan, H. A. Stone and D. A. Weitz, Science,2005, 308, 537; (b) A. S. Utada, L. Y. Chu, A. Fernandez-Nieves, D. R. Link, C.Holtze and D. A. Weitz, MRS Bull., 2007, 32, 702.

[7] (a) W. Wei, L. Y. Wang, L. Yuan, Q. Wei, X. D.Yang, Z. G. Su and G. H. Ma, Adv. Funct. Mater., 2007, 17, 3153; (b) W. Wei, L.Yuan, G. Hu, L. Y. Wang, H. Wu, X. Hu, Z. G. Su andG. H. Ma, Adv. Mater., 2008,20, 2292.

See more of this Session: Structure and Properties In Polymers III: Networks and Gels II
See more of this Group/Topical: Materials Engineering and Sciences Division

(337b) Dual Photo and Thermally Polymerizable Monomers Derived From Methyl Esters of Vegetable Oils

Brian Dillman, Chemical and Biochemical Engineering, Univerisity of Iowa, Iowa City, IA and Julie L. P. Jessop, Chemical & Biochemical Engineering, University of Iowa, Iowa City, IA

DualPhoto and Thermally Polymerizable Monomers Derived from Methyl Esters of VegetableOils

BrianDillman and Julie L.P. Jessop

Photo-polymerizationor polymerization initiated by light is a facile material synthesis techniqueused widely in coating, adhesive, lithographic, and dental restorativeapplications. A wide variety of photo-polymerizable monomers are available thatresult in a range of polymer properties. Of the monomers available, few arederived from renewable resources. In this study, methyl esters of soybean andlinseed oil, produced using standard biodiesel synthetic procedures, were usedto synthesize hybrid monomers (i. e., monomers bearing two types of polymerizablefunctional groups). The methyl esters were amidated with ethanol amine,resulting in α-hydroxy amides of vegetableoil. In one case, the double bonds in the fatty acid residues were epoxidizedusing H_2O_2 and acetic acid. In another case, the doublebonds were not functionalized further. The terminal hydroxyl group was thenreacted with an isocyanate acrylate monomer, resulting in acrylated amides ofvegetable oil or acrylated epoxidized amides of vegetable oil. Both monomertypes were formulated with photo-initiators, and the rapid acrylatephoto-polymerization was observed by real-time vibrational spectroscopy. Afterphotopolymerization, the materials were annealed at higher temperatures topromote cross-linking through the epoxide moieties or double bonds in the fattyacid residues by ring-opening or alkyd-type reactions, respectively. The degreeof cross-linking for each monomer type was evaluated by measuring thecross-link density of the polymer before and after annealing using dynamicmechanical analysis. The materials developed in this study incorporate approximately60 wt% renewable materials using green synthetic chemistry and low energydemand processing methods. The high efficiency of monomer and polymer synthesiswas accomplished by minimizing solvent use, selecting reactions that proceed tohigh yields, and utilizing photopolymerization. The polymers prepared haveunique properties which allow for simple processing of polymers to complexshapes.

See more of this Session: Structure and Properties In Polymers III: Networks and Gels II
See more of this Group/Topical: Materials Engineering and Sciences Division

(337c) Analyzing the Interdependence of Polymer Chain Growth and Macromolecular Memory Creation In Molecularly Imprinted Polymers

Vishal D. Salian and Mark E. Byrne, Biomimetic & Biohybrid Materials, Biomedical Devices, & Drug Delivery Laboratories, Department of Chemical Engineering, Auburn University, Auburn, AL

Molecular imprinting is an effective technique to create macromolecular memory within a polymer. It is generally accepted that polymer network formation affects macromolecular memory creation and subsequently, template binding and transport within the polymer network. In this work, we demonstrate that the converse is also true; that the presence of the template during polymerization has a significant impact on polymer chain growth. Polymerization reactions for poly(HEMA-co-DEAEM-co-PEG200DMA) networks were examined as the template (diclofenac) concentration in the pre-polymerization mixture was increased. GPC analysis of the polymer chains formed showed that the presence of template leads to a 29% decrease in the kinetic chain length (KCL) and nearly three fold increase in the polydispersity index (PDI). In addition we demonstrate that the living polymerization process not only negates the retardation of polymer chain growth due to template but also results in an improved memorization of template within the network. Thus, we investigate the interdependent relationship between polymer chain growth and macromolecular memory creation and offer a means of improving the network architecture. This work could have impact beyond the field of molecular imprinting and result in significant advances in the formation of better polymer networks.

See more of this Session: Structure and Properties In Polymers III: Networks and Gels II
See more of this Group/Topical: Materials Engineering and Sciences Division

PREPARATION OF GEL STRUCTURED POLY STYRENE-ETHYLENE GLYCOL DIMETHACRYALATE (PS-EGDM) BASED HIGH CAPACITY ION EXCHANGE RESIN AND ITS CHARACTERIZATION

Gagnesh Sharma[1], Shishir Sinha[1], Shashi Kumar[1]

[1]Department of Chemical Engineering, Indian Institute of Technology Roorkee

Roorkee- 247667, Uttrakhand, India

EXTENDED ABSTRACT

There are so many unit operations in which application of ion exchangers employed such as de-ionization and de-minralization of impure water as packed bed for exchange of ions. These ion exchangers or ionic polymers either cationic, anionic or both are chemically bonded materials have exchangable functional groups within their structure. Most common ion exchangers in this class are polystyrene-divinyl benzene (PS-DVB), Poly (methyl methacrylate)-ethyleneglycol dimethacrylate (PMMA-EGDM) based high capacity anion exchange resin. These ion exchangers have different network structure which depends upon the crosslinking by crosslinking media. The choice of crosslinking media has been made on the basis of degree of crosslinking. The exchanging property of ion exchangers is because of these functional groups in it, and functional groups are created on the crosslinked polymer gel by various liquid and gas phase modification[1-2]. The PS-DVB resin is very popular as ion exchange resin during last 15 years study due to good mechanical strength and stability in the entire range of pH [3]. These resins can be prepared in the form of beads by suspension polymerization or alternatively emulsion polymerization. Mostly different copolymers based ion exchange resin were prepared using suspension polymerization technique using different monomers, crosslinking agents, porogen and different parameters [4-7]. Another polymerization technique called emulsion polymerization in which polymer formation occur in an inert medium in which the monomer is partially soluble[8-10]. In the present study preparation of high capacity polystyrene-ethyleneglycol dimethacrylate (PS-EGDM) based ion exchange resin by suspension polymerization technique which is having higher mechanical strength, stable and hydrophilic in nature. Suspension polymerization of polystyrene-ethyleneglycol dimethacrylate (PS-EGDM) is carried out in the presence of toluene as porogen and gelatin as stabilizer. This resin is subsequently functionalized by gas phase nitration. Weak base groups NO_2 have been introduced into crosslinked polymethyl methacryalate divinyl benzene (PS-DVB) by nitration (gas phase modification) using NO_X and then resin undergo subsequent chemical reaction with amination route by hydrazine hydrate for further functionlization i.e. reduction of the NO_2

into NH_2 groups which is a wet chemical modification scheme. Resins of this type carry amino groups directly attached to the phenyl rings, have very low base strength. The liquid phase nitration using nitric acid degrades the resin and it loses its mechanical strength [11].

EXPERIMENTAL

Material and method of preparation of Macroporous PS-EGDM resin:

The macroporous polystyrene-ethyleneglycol dimethacrylate (PS-EGDM) resin was prepared by suspension polymerization technique [12-13]. The recipe consist of organic phase and inorganic phase. Organic phase consist of Styrene, Ethyleneglycol Dimethacrylate (EGDM) as crosslinking media, Toluene as diluent (porogen) and Azobisisobutyronitrile (AIBN) as initiator. This organic phase is suspended in inorganic phase which consists of Calcium Carbonate , Sodium Sulphate, as suspension stabilizers, and distilled water. The reaction was carried out using suspension polymerization technique by mixing the organic phase and inorganic phase in a polymerization reactor, under constant stirring and at atmospheric pressure for 8 hr at $90^{o}C$ as shown in Figure 1. The washed resin is then dried in oven maintained at $100\pm5^{O}C$ for 24 hr. Dried resin now have considerable mechanical strength and it becomes very hard and size was found to be uneven.

Modification of PS-EGDM resin by Gas Phase Nitration using NO_X

NO_X, a mixture of NO_2 and small amount NO, is used to carry out direct gas phase nitration of polystyrene-ethyleneglycol dimethacrylate (PS-EGDM) efficiently. This reaction is carried out in a 500ml conical flask which is closed with a silicon cork at the opening of flask. NO_X is generated by reacting sodium nitrite with sulfuric acid in the presence of ferrous sulfate $FeSO_4$ at room temperature in a 2 lit 3-neck round bottom flask. This flask is equipped with silicon cork for withdrawing NO_X with help of a 50 mL syringe. Gas phase nitration as reported was carried out in reaction vessel (500 mL conical flask) by introducing approximately 1.5 g of PS-EGDM resin into vessel. After injecting required amount of NO_X gas mixture the reaction vessel is kept in oven at desired reaction temperature for a certain period of time. After completion of reaction there is peculiar change in color from white to pale yellow was observed for nitrated resin. If one repeated unit of unmodified PS-EGDM resin is represented with unmodified resin 1 then the overall nitration reaction is as below:

$$\boxed{P} - CH_2CHC_6H_5 \overset{NO_x}{\longrightarrow} \boxed{P} - CH_2CHC_6H_4NO_2 \qquad (2)$$

Resin 1 *Resin 2*

Amination of Nitrated PS-EGDMresin

The nitrated PS-EGDM resin is aminated using hydrazine hydrate to reduce the NO_2 functional group to amine group on the surface of resin. For this approximately 1.5 g of nitrated resin was placed in a 250mL single neck round bottom flask with 50mL of hydrazine hydrate. The reaction system was refluxed at about constant temperature of $60^{O}C$ maintained in an oil bath. The resin's colour was changed from pale yellow to light golden yellow. After completion of reaction the resin is repeatedly washed with distilled water and dried in oven maintained at $70^{O}C$. The amination reaction can be represented as below

$$\boxed{P} - CH_2CHC_6H_4NO_2 \xrightarrow{\text{Hydrazine Hydrate}} \boxed{P} - CH_2CHC_6H_4NH_2 \quad (3)$$

Resin 2 *Resin 3*

Determination of consumption of NO_X during Nitration of PS-EGDMresin

The consumption of NO_X during nitration reaction was calculated by determining the amount of unconsumed NO_X.The unreacted NO_X was determined by dissolving 50 mL distilled water in the reaction vessel after completion of nitration reaction. Then after vigrous shaking unconsmed NO_X with water form HNO_3 which is light pale yellow in color. The solution is filtered and resin is dried in oven. The filtrate, contains HNO_3, was titrated with 0.1 N NaOH to determine the strength of HNO_3 in water. In this way we can determine the amount of NO_X consumed during the nitration reaction for different temperature and time of nitration. The effect of time of nitartion on NO_X consumption at different temperatures is shown in figure 3.

Determination of exchange capacity of aminated resin by Titration Method

This method is easy and less time consuming as reported[16]. It is based on the fact that all amine groups of aminated would react with HCl to form ammonium salt. In this method about 1.5 g of aminated resin is placed in a 250mL single neck round bottom flask and refluxed with 0.1N HCl for 4 hr in an oil bath at $50^{O}C$. The reaction mass is then filtered and filtrate is titrated with 0.1 N NaOH which contains unconsumed HCl. The titration determines the chlorine intake by resin 3 and carried out using phenolphthalein indicator. The filtered resin 3 is then washed carefully and dried. The dried aminated resin is again refluxed with 0.1 N NaOH for 4 hr at $50^{O}C$ in oil bath. Remaining NaOH concentration in aminated resin was determined by titrating the filtrate with 0.1 N HCl using phenolphthalein indicator and the capacity of aminated resin is calculated for different time of nitration and at different temperature. The effect of nitration time and temperature on exchange capacity has been shown in figure 3.

FTIR analysis of resins

Fourier Transform Infrared Spectroscopy (FTIR Spectroscopy) is used to identify the functional groups present on the resins 1, 2, and 3 after the reactions were determined by FTIR instrument (Thermoscientific NICOLET 6700) were used to determine FTIR spectra [17] as shown in figure 4. There was a characteristic sharp peak at 709cm^{-1} and 1384 cm^{-1} confirms the presence of NO_2 group for nitrated resin. In aminated resin strong absorption near 3100–3200 cm^{-1}, and simultaneous disappearance of the peak at 1384 cm^{-1} which belongs to NO_2 group, confirms the presence of NH_2 group.

RESULT AND DISCUSSION

Suspension polymerization of PS-EGDM resin

In this work, crosslinked PS–EGDM copolymer resin in the form of macroporous beads were prepared with a size range of 500μm to 1.0mm by suspension polymerization. As crosslinked polymer PS was preferable due to its ease of chain length extension because of presence of methyl groups onto its surface structure. This results in the increase in surface area of crosslinked PS-EGDM resin and thereby increasing the free active centers onto the surface if it. Due to increase in these active centers the chemical modification becomes easier. Gelatin content plays a vital role in controlling the particle size.

Nitartion of PS-EGDM resin

Macroporous crosslinked PS-EGDM resin is prepared using procedure given earlier was subjected to gas phase nitration scheme. Same size resin particles were separated using sieve shaker and nitrated in a 500 mL conical flask to accommodate NO_X without increasing the inside pressure significantaly at different reaction temperatures. The change of color from white to pale yellow was observed for nitrated resin. This indicates the extent of nitration over the surface of nitrated resin. Nitration was done for different time of nitration and at different temperatures. The effect of nitration with time and temperature has been shown in figure 2.

Amination of Nitarted resin

Nitrated resin was reduced using hydrazine hydrate to introduce amine group (NH_2) by replacing NO_2 group from the surface of the nitrated resin. This was confirmed in the Laissagne spot test and

FTIR analysis. The color change from pale yellow to light golden yellow was observed for aminated resin. The exchange capacity was calculated for aminated resin for different nitration temperature and time by titration method shown in figure 3. The ion exchange capacity of resin was plotted with time shown in figure.

Exchange Capacity

The exchange capacity aminated resin 3 was determined by titration method reported earlier. One time exchange capacity was determined for weak base aminated resin. The values for exchange capacity (meq/g) using titration method of aminated PS-EGDM resin for nitartion temperature range of 70-120°C and for nitartion time of 4-9hr are given in table 2. The results indicate that the exchange capacity of PS-EGDM resin increases with increase in temperature upto 100°C and then becomes almost stagnant there after 100°C to 120°C. This stagnancy was attributed to the exhausting of the active centers on the surafce of resin. Also it increases with increase in time of nitration upto 6 hr and almost stagnant after 6hr upto 9hr. At 100°C and 6hr. of nitration the exchange capacity value is optimized. The effect of time of nitration on exchange capacity of aminated resin at different temperatures is illustrated in figure 3. It shows that at 100 ^{O}C the value of exchnage capacity is highest in correspondance with the subsequent higher temperatures.

CONCLUSION

The unmodified PS-EGDM resin has been prepared by suspension polymerization technique using gelatin as the stabilizer, toluene as porogen and AIBN as initiator followed by the gas phase nitration in order to introduce weak base group NO_2 onto the surface. The nitration of these resins using NO_X is an extremely simple reaction, and the bond formation in the nitrated resin has been confirmed using the Laissagne spot test, and FTIR. The resin obtained was observed to have a good mechanical strength, to be more hydrophilic than PS-DVB copolymer, macro porous in nature, and with a higher exchange capacity. These exchangers have been applied in the wide range of applications like de ionization, de minlarization of water in DM plant.

References

1. Kumar, A.; Gupta, R., Fundamental of Polymers; McGraw Hill: New York, (1996).
2. Helfferich, F. Ion Exchange; 1st ed. McGraw Hill: New York, (1962).
3. Lewandowski, K.; Svec, F.; Frechet, M.J.J. J Appl Polym Sci (1992), 67, pp.2443.

4. Balakrishnan, T.; and Ford, W.T. J Appl Polym Sci (1982), 27, pp.133.
5. Kabay, N.; Egawa, H. J Appl Polym Sci, (1994), 51, pp.381.
6. Zhang, B.W.; Zhang, Y.; Grote, M.; Kettrup, A. Reac Polym (1994), 22, pp.115.
7. Trochimczuk, A.W.; Alexandratos, S.D. J Appl Polym Sci (1994), 52, pp.1273.
8. Matynia, T.; Gawdizik, B. J Appl Polym Sci (1995), 58, pp.861.
9. Tamai, H.; Sakurai, H.; Hirota, Y.; Nishiyama, F.; Yasuda, H. J Appl Polym Sci (1995), 56, pp.441.
10. Zou, D.; Aklonis, J.J.; Salovey, S. J Polym Sci Part A: Polym Chem (1992), 30, pp.2443.
11. Sakaguchi, S.; Nishiwaki, Y.; Kitamura, T.; Ishii, Y. Angew Chem Ind Ed (2001), 40, pp.222.
12. Sinha, S.; Jayaswal, N.; Kumar, A. J Appl Polym Sci (2003), 89, pp.1991.
13. Jayaswal, N.; Sinha, S.; Kumar, A. J Appl Polym Sci (2001), 79, pp.1735.
14. Vogel, A. I. Text Book of Macro and Semi Macro Qualitative Inorganic Analysis; (1953); pp.338-340.
15. Cotton, F. A.; Wilkinson, G. Advanced Inorganic Chemistry, 5th ed.; Wiley: New York, (1992); pp.320-325.
16. Shishir, S; Kumar, A. Sep Sci Technol (2002), 37, pp.895.
17. Silverstein, R.M. Spectrometric Identification of Organic Compounds; Wiley: New York, (1991); pp.123-127.

Figure 1. Experimental Setup for Preparation of PS-EGDM resin

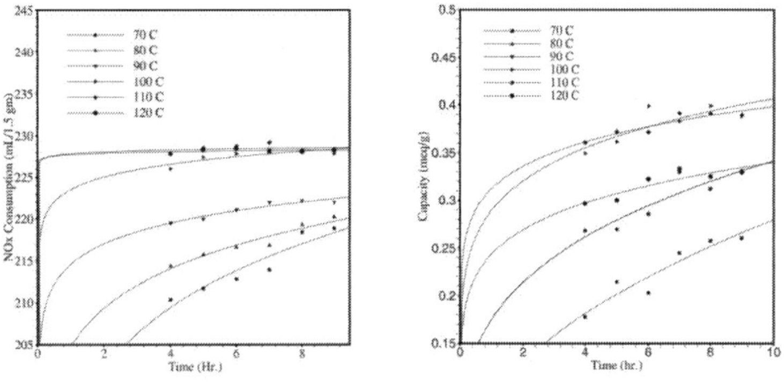

Figure 2 Effect of nitration time and temperature on consumption and one time exchange capacity of aminated resin

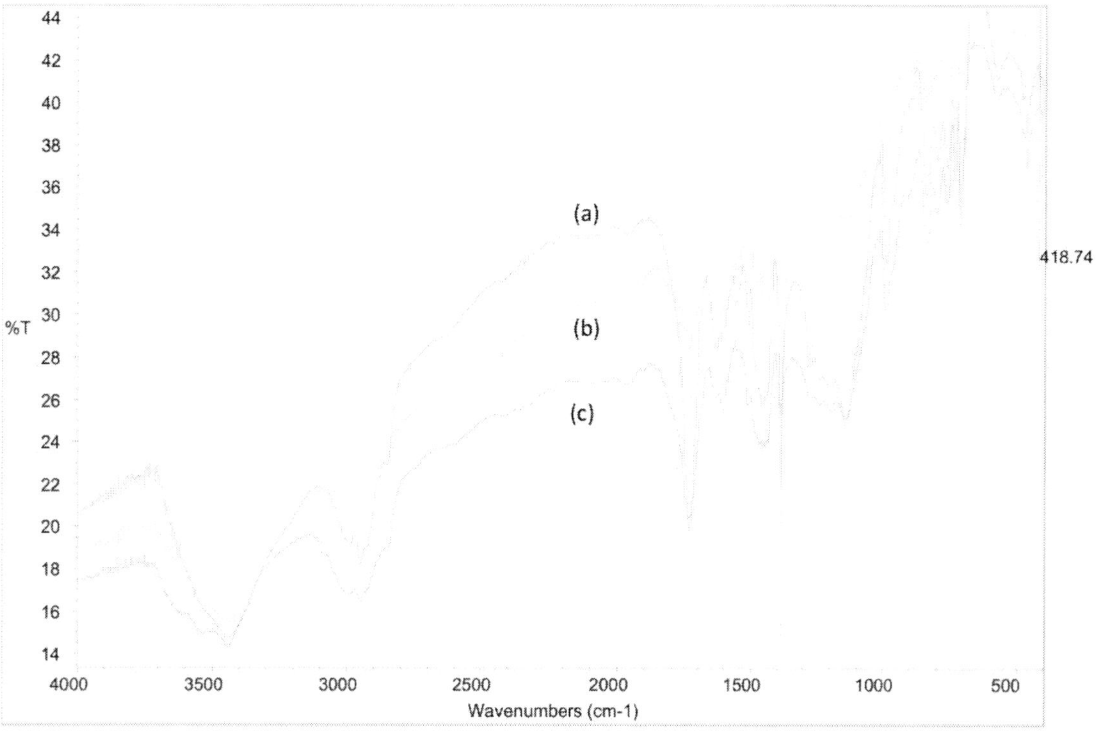

Figure 4 FTIR analysis of PS-EGDM resin: (a) unmodified resin (b) nitrated resin (c) aminated resin

(337e) Recent Developments In APT-Based Polymerization Reactions: The Iodo-Ene Reaction

Christopher J. Kloxin, Chemical Engineering and Materials Science and Engineering, University of Delaware, Newark, DE and Timothy F. Scott, Chemical Engineering, University of Michigan, Ann Arbor, MI

The thiol-ene 'click' reaction has been described as the only radical-mediated, step-growth polymerization mechanism;[1] however, our recent account of radical-mediated thiol-yne polymerizations belies this premise.[2] Rather, thiol-ene reactions constitute only one category of a greater class of radical-mediated, step-growth polymerizations, designated here as "Alternating Propagation-chain Transfer" (APT) polymerizations. In this presentation, we will discuss photoinitiated APT polymerizations, which afford several distinct and advantageous attributes, such as low shrinkage and shrinkage stress, delayed gelation, low extractable content, and the formation of homogeneous polymer networks with very narrow glass transition regions. Additionally, we will introduce a new radical-mediated APT polymerization: the iodo-ene polymerization. While possessing all the advantages of thiol-ene APT polymerizations, the iodo-ene reaction utilizes iodinated perfluoroalkane comonomers, assuring the covalent incorporation of both iodine and fluorine in high concentrations throughout the polymer matrix. This imparts the iodo-ene polymer product with intrinsic radiopacity, an inert, low energy surface, and chemical stability, which are desirable in a number of biomedical applications. Moreover, this material has the capacity for facile nucleophilic substitution for surface passivation while the bulk of the material remains iodinated but inert.

1. H. Lu, J. A. Carioscia, J. W. Stansbury, C. N. Bowman, *Dent. Mater.* **21**, 1129 (Dec, 2005).

2. B. D. Fairbanks, T. F. Scott, C. J. Kloxin, K. S. Anseth, C. N. Bowman, *Macromolecules* **42**, 211 (Jan, 2009).

See more of this Session: Structure and Properties In Polymers III: Networks and Gels II
See more of this Group/Topical: Materials Engineering and Sciences Division

Lipophilic Super-Absorbent Polymer Gels as Surface Cleaners for Oil and Grease

Christopher Myers[1], Veera M. Boddu[1*], Minori Uchimiya[2]

[1]US Army Engineer Research and Development Center Construction Engineering Research Laboratory, 2902 Newmark Drive, Champaign, IL 61822
[2]USDA-ARS Southern Regional Research Center, 1100 Robert E. Lee Blvd., New Orleans, LA 70124

*Veera.Boddu@usace.army.mil

ABSTRACT

Increasingly stringent environmental regulations on volatile organic compounds (VOCs) and hazardous air pollutants (HAPs) demand the development of disruptive technologies for cleaning weapons systems and platforms. Currently employed techniques such as vapor degreasing, solvent, aqueous, or blast cleaning processes suffer from shortcomings in environmental friendliness, personnel health and safety, cleaning efficiency, cost-effectiveness, management of contaminated cleaning media, or in maintaining the integrity of equipment material surfaces. Lipophilic super-absorbent swelling gels have been developed that will not only absorb the oil and grease from these machine parts, but will also act as an automated sweeper due to the self-generating mechanical force of the gel. An octadecylacrylate-co-ethylene glycol dimethacrylate (ODA-co-EGDMA) lipophilic gel (G-1) and commercially available Alkylstyrene Copolymer (imbiber beads) were evaluated for swelling and oil sorption capacity. Along with the imbiber beads and G-1 gel, poly stearylacrylate-co-ethyleneglycol dimethacrylate neutral gel (NG-18), another lipophilic super-absorbent gel, was synthesized and tested. For each gel, the swelling degree and the oil absorption capacity were quantitatively investigated at 0-60°C using a variety of polar and nonpolar solvents. The mechanical strength of each gel was studied by measuring compression strength, and cross-linking of the polymers was studied using infrared spectroscopy. The cleaning tests were performed using metal coupons. Details of gel preparation and evaluation of oil sorption capacity are discussed here.

BACKGROUND

Environmentally benign VOC-exempt, and HAP-free surface cleaning technology, will support ongoing Department of Defense (DoD) programs such as the Sustainable Painting Operations for the Total Army (SPOTA). New technologies developed are anticipated result in overall reductions of VOCs and HAPs emissions from DoD surface cleaning operations. Polyelectrolyte gels are ionic polymer networks composed of charged polymer chains and freely mobile counter-ions. Polyelectrolyte super-absorbent wet-swelling hydrogels are known to undergo a dramatic but reversible volume change by absorbing large quantities of water. The polyelectrolyte hydrogels swell in water because of (1) osmotic pressure induced by freely mobile counter-ions within the polyelectrolyte, (2) increased entropy arising from the solvation of polymer ions and counter-ions, (3) electrostatic repulsion between the oppositely charged ions within the polyelectrolyte gel, and (4) stretching of polymer chains between crosslinks caused by the increase in entropy associated with mixing polymer with solvent [1,2]. Polyelectrolyte hydrogels have found a wide range of applications in diapers, inks and display devices, separation media, and clean up of aqueous spills. Polyelectrolyte hydrogels are particularly useful for a wide range of environmental applications, because expansion and contraction of the gels can be engineered to be triggered by small changes in environmental parameters such as temperature, pH, and ionic strength. However, until recently, reports on gels that will swell by absorbing large quantities of nonpolar organic solvents were nearly nonexistent. In nonpolar solvents, most polyelectrolyte gels collapse, because the oppositely charged ions within the gel form ion pairs that then aggregate, rather than becoming solvated.

In 2007, Sada and colleagues [3] reported, for the first time, a novel class of lipophilic polyelectrolyte gels bearing positively charged repeating units (substituted tetraalkylammounium with long alkyl chains) and negatively charged counter-ions (substituted tetraphenylborate; $TFPB^-$) that swell dramatically but reversibly by absorbing organic solvents having various polarities (ε = 1.9-46; the lower the dielectric constant (ε), the less polar the solvent). Superior swelling ability in nonpolar solvents is enabled by making both the polymer chains and the counter-ions lipophilic, preventing counter-ions from forming ion pairs, thereby enabling the solvation of ionic gel components in solvents. Lipophilic polyelectrolyte gels, hereby termed EG18 and G1 can serve as cleaners in their swollen state. A neutral analogue of 1G that does not contain the ionic tetraalkylammonium tetraphenylborate unit is NG18. As shown in Figure 1, neutral gel NG18 swells to a much lesser extent than the ionic EG18 gel. Neutral polymer gels swell in organic solvents because of the stretching of polymer chains between crosslinks caused by the increase in entropy associated with mixing polymer with solvent.

A promising candidate is the lipophilic super-absorbent gel that will swell by absorbing nonpolar organic solvents (e.g., hydrocarbon oils, VOCs) several hundred times their dry weight. This study aims to utilize intricate designing of lipophilic super-absorbent swelling gels through careful selection of polymer backbone and ionic components, and the cross linking density. Designed swelling gels will function as the cleaner of the metal and non-metal surfaces by (1) absorption of oil and grease and (2) removal of particulate contaminants by self-generated mechanical forces obtained from swelling in (1). After the cleaning operation, cleaning media can be safely collected and recycled or used in fuel blends.

In order to develop a new cleaning technology based on lipophilic super-absorbent swelling gels for the removal of oil, grease and particulate matter from metal and non-metal surfaces, specific tasks are formulated to maximize the cleaning efficiency of lipophilic gels by (1) testing cleaning ability of candidates EG-18 and NG-18 gels and (2) designing lipophilic gels with improved cleaning ability by appropriate selection of lipophilic polymer backbone, weakly coordinating anions, and enhancement of mechanical strength. Elimination of fluorinated compounds in the gel synthesis will be carried out during the subsequent phases of this research.

TECHNICAL APPROACH

The lipophilic polyelectrolyte gel, EG-18, and the neutral gel, NG-18 are synthesized following solution polymerization methods described in detail elsewhere [3-5]. These gels are evaluated following the ASTM G122-96(2008) and MIL-PRF-680B protocols. The synthesis, characterization, and oil sorption studies are briefly described here.

Materials and Methods

All chemical reagents were used as received. Stearylacrylate (SA), ethylene glycol dimethacrylate (EGDMA), benzene, azobisisobutylonitrile (AIBN), methanol, ethanol, diethylether, and carbon tetrachloride were purchased from Sigma-Aldrich (Milwaukee, WI). Tetrahydrofuran (THF), isopropanol, acetonitrile, and dichloromethane were obtained from Acros Organics (Morris Plains, NJ). Dimethylsulfoxide (DMSO) and 1-octanol were supplied from Alfa Aesar (Ward Hill, MA). Chloroform and cyclohexane were distributed from VWR International (West Chester, PA). Acetone, methylisobutylketone (MIBK), toluene, and *n*-hexane were supplied from Fisher Scientific (Pittsburgh, PA). BioAct MSO was suppled by Petroferm Cleaning Products (Gurnee, IL). An Instron 5900 Electromechanical System was used for the compression of the gels. A Jasco FT/IR-4100 Fourier Transform Infrared Spectrometer was used for infrared spectroscopy.

Synthesis

A typical protocol for the synthesis of NG-18 gel is as follows: 10.0 g (30.8 mmol) of SA (monomer) and 61 mg (0.31 mmol, the case of x=1) of EGDMA (crosslinker) as initiator were placed in a vial tube and dissolved in 2 mL of benzene by heating at 50 °C. Oxygen in the solution was excluded by bubbling nitrogen gas for 45 min then 101 mg (0.62 mmol) of AIBN was added. The vial tube was sealed tightly and heated at 65 °C for 24 h for polymerization. Gels with low crosslinking densities were prepared in a similar way by reducing the feed ratios of EGDMA. The synthesized gels were washed by swelling in hexane repeatedly, air-dried for 2 days, and dried in vacuum overnight. SA-*co*-EGDMA with two crosslinker ratios were prepared by radical copolymerization, which are represented as NG-18-x% (x= 1 or 0.5; x denotes the mole ratio of crosslinker to SA). A typical protocol for the synthesis of EG-18 gel is as follows: 125 mg (0.1 mmol) of $TFPB^-$ and 617 mg (1.9 mmol) of ODA, 3.96 mg (0.02 mmol) of EGDMA, and 6.57 mg (0.04 mmol) of AIBN were placed in a capillary of 7.0 mm in diameter and dissolved in benzene adjusted to 1.0 mL. The solution was degassed and polymerized by heating at 60°C for 24 h. The feed ratio was adjusted to $TFPB^-$:ODA:EGDMA = 5:95:1. Gels with low crosslinking densities were prepared in a similar manner by reducing the feed ratios of EGDMA. The formed gels were washed by swelling in benzene for 10 h, and then air-dried at room temperature. The sample was cut into cylinders of about 1.0 mm in length, and the cylinders were dried *in vacuo* at 40°C. Synthesis and swelling properties of EG-18 was reported by Sada and co-workers[3].

Characterization

The Fourier transform infrared (FTIR) spectra were performed using Jasco FT/IR-4100. Compression strength was measured with Instron 5900 electromechanical system and the compression speed was 0.25 mm/min. UV-vis spectra were colleceted using a ThermoSpectronic Aquamate 100 UV-vis Spectrometer. Experimental details of the swelling behavior, critical temperature studies, and other surface cleaning properties are presented in a technical report [6]. Swelling behavior of commercially available imbiber beads were also studied.

Compression Strength, oil sorption, and grease cleaning studies are also included in the report.

The cleaning studies were conducted using stainless steel metal coupons. Clean coupons were soaked in SAE-30 oil and allowed to drip excess oil for 30 minutes. The percentage of oil absorbed by the gel (removed from the coupon) was then obtained by comparing the weight of each coupon before and after immersion in the swollen gel. Grease (MIL-PRF-10924 grease) cleaning power of the swollen gel was obtained following a described in a protocol based on ASTM standard. The percent of grease absorbed was determined by comparing the weight of each coupon before and after immersion. The cleaning power was also determined by the following equation:

where A is the average cleaning time of the three tested runs.

RESULTS AND DISCUSSION

Characterization of NG-18 gels

FT-IR spectra of SA monomer and NG-18-1%, -0.5% are provided in Figure 1. Compared to the spectrum of stearylacrylate monomer, NG-18 gels showed the disappearance of peaks in four regions. Each peak was identified as follows: 1634 cm^{-1} is C=C bond vibration, 1410 cm^{-1} is C-H of $C=CH_2$ in-plane scissoring, 1297 cm^{-1} is C-H of C=CH in-plane vibration, and 997 and 893 cm^{-1} are C=CH out-plane vibration. The disappearance of these peaks of the vinyl group indicates that NG-18 gels include a little non-reacted SA. Figure 2 shows the compression strength of swollen NG-18-1% gel in THF. The first breaking point is the stress of 0.371 MPa and the fracture strain of the gel is λ=67 %. The NG-18-1% gel could withstand a similar degree of compression as reported by single network poly acrylamide gel prepared by Gong et al. [6].

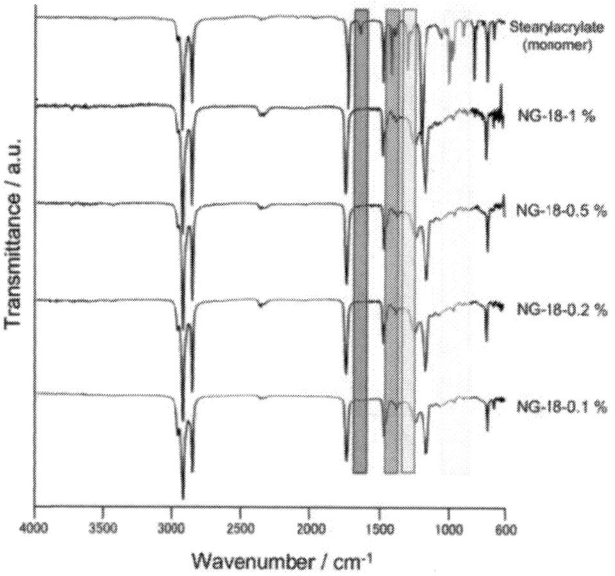

Figure 1. FT-IR spectra of NG-18 gels

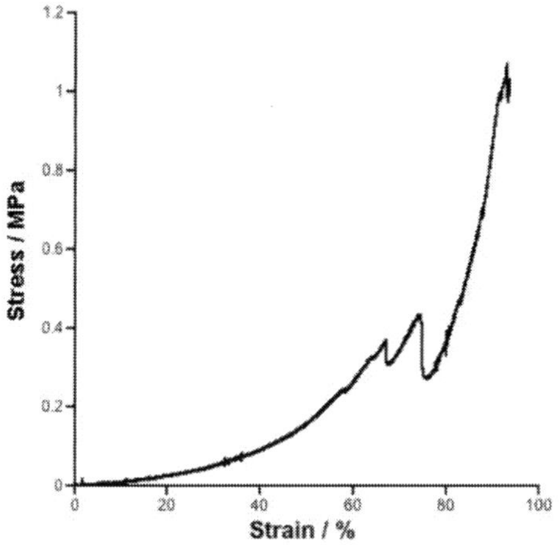

Figure 2. Compression strength of swollen NG-18-1%, THF has a breaking point around 0.371 MPa.

Swelling behavior

The swelling behavior of NG-18 gels (NG-18-1% and -0.5%) in the solvents with various polarities from cyclohexane to water at 25 °C was investigated (Figure 3). The swelling degree Q increased with increasing polarity from cyclohexane, and the maximum value was observed in chloroform. On the other hand, the gels collapsed in the more polar solvents (dielectric constant $\varepsilon > 10$). Particularly, NG-18 gels absorbed large amounts of chlorinated solvents such as chloroform (Q=35 on NG-18-1%) and carbon tetrachloride (Q=36). Moreover, NG-18 swelled in a moderate amount of solvents such as ether (diethylether: Q=12, THF: Q=17), aromatic compounds (benzene: Q=21, toluene: Q=22), and aliphatic reagents (hexane: Q=14, cyclohexane: Q=20). In more polar solvents, such as water, DMSO, alcohols (methanol, ethanol, isopropanol, and 1-octanol), ketones (acetone and MIBK), and acetonitrile, NG-18 didn't swell at all (Q<1). Also, enhancing swelling ability by reducing the cross linker density was attempted. Reducing the feed ratio of the crosslinker to the monomer from 1 mol% to 0.1 mol% enhanced the swelling ratio. However, the gels less than 0.2 mol% crosslinker density were too soft to separate excess solvent and swelling degree could not be accurately measured. NG-18-0.5% indicated the same tendency as NG-18-1% and had a higher swelling degree than NG-18-1%. These swelling behavior of NG-18 gels essentially depend on the compatibility of the polymer chain with the media. NG-18 didn't allow penetration of the highly polar molecules into the polymer networks, while non-polar solvents were absorbed.

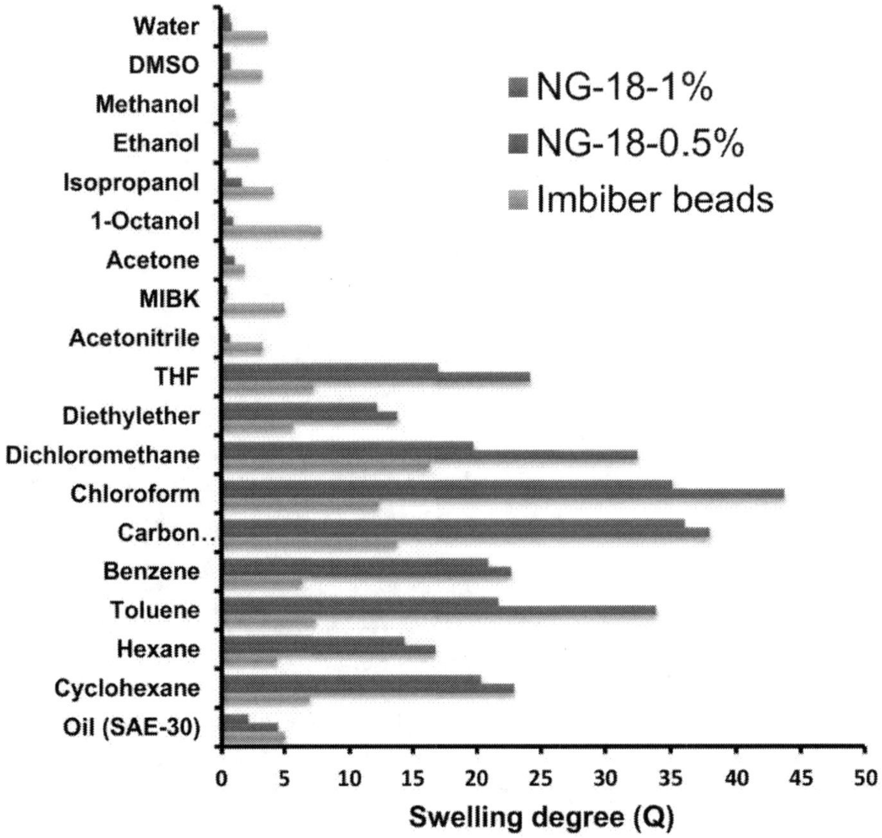

Figure 3. Swelling degree (Q) of NG-18-1%, -0.5% and imbiber beads in various solvents.

Oil Absorption

Oil absorption was tested with both NG-18 gels in toluene and in THF, the results of which are shown in Figures 13-14. It was determined that NG-18 1% in THF was the best performer in this category because of its high swelling degree, good recyclability, the high oil and alumina absorption properties. THF is also a low- or non-toxic and environmentally friendly solvent. NG-18 0.5% gel was not able to clean all of the alumina powder and oil. Along with metal coupons, NG-18 gels were tested on painted coupons and stainless steel parts with bolts, the results of which are shown in the appendix. The painted coupons did not show any signs of peeling, and the bolts and flat coupons were cleaned almost as well as using solvent by itself. From these data an oil recycling system (Figure 16) was made, which can recycle up to 25% of solvent when cooled to 0°C.

Temperature did change the swelling degree of NG-18-1% in 1-Octanol, methylisobutylketone, and SAE-30 oil when heating from 20 °C to 40 or 60 °C (Figure 7). Even if they were cooled to 0 °C, however, they did not collapse. In other words, these changes were irreversible.

Grease Cleaning

The grease absorption capabilities of NG-18 gels were nearly as good as trichloroethylene all showing greater than 99% absorption capabilities. As shown in Figures 21-22, NG-18 0.5% outperformed the TCE in two of three trials. However, the NG-18 0.5% also had an ability to slightly damage the metal as in Trial 1. NG-18 1% was the worst of the three tested, though it showed a higher cleaning power than NG-18 0.5%. This could be due to the tendency of the NG-18 1% to stay on the metal after removing the grease. This could be prevented by washing with ethanol immediately after cleaning.

SUMMARY

In this study, we demonstrated the synthesis and characterization of poymer gels that can absorb oil and grease. The swelling characteristics of the gels were studied as a function of the solvent polarity and temperature. Swelling kinetics of the gels in various solvents were also studied. The gels were evaluated for their oil and grease sorption capacities. The gel system uses THF for swelling, however, cyclohexane or other benign solvents with similar swelling properties may also be used as a swelling agent. In particular cyclohexane is a good possibility with swelling ratio (Q) of 20. Other solvents such as toluene or other non-polar solvents may also be used. We have not tested several other potential solvents during this phase of the research. The gel cleaning process has the advantage of avoiding emissions of hazardous air pollutants (HAPs) and volatile organic compounds (VOCs).

REFERENCS

(1) Ohmine, I.; Tanaka, T. Salt effects on the phase transition of ionic gels. *J. Chem. Phys.* **1982**, *77*, 5725-5729.

(2) Rubinstein, M.; Colby, R. H.; Dobrynin, A. V.; Joanny, J. F. Elastic modulus and equilibrium swelling of polyelectrolyte gels. *Macromolecules* **1996**, *29*, 398-406.

(3) Ono, T.; Sugimoto, T.; Shinkai, S.; Sada, K. Lipophilic polyelectrolyte gels as super-absorbent polymers for nonpolar organic solvents. *Nat. Mater.* **2007**, *6*, 429-433.

(4) Ohta, M.m Boddu, V. M., M. Uchimiya, Sada, K. Thermal response and recyclability of poly(sterylacrylate-*co*-ethylene glycol dimethylacrylate) gel as a VOCs absorbent, *Polym. Bull.* **2011**, *67*, 915-926.

(5) Boddu, V. M., Uchimiya, M., Ohta, M., Sada, K., Myers, C., Ziegler, W., Torres, T. Lipophilic Super-Absorbent Swelling Gels as Cleaners for Use on Weapons Systems and Platforms, Draft Annual Report to SERDP, August 2011.

(6) Gong, J. P.; Katsuyama, Y.; Kurokawa, T.; Osada, Y. Double-network hydrogels with extremely high mechanical strength. *Adv. Mater.* **2003**, *15*, 1155-1158.

(337g) Mechanical Response of Polydomain Smectic Elastomers: Influence of Thermal History

Ronald C. Hedden[1], Huipeng Chen[1], Ziniu Yu[1] and Daniel M. Lentz[2], (1)Chemical Engineering, Texas Tech University, Lubbock, TX, (2)3M Corporate R&D, Saint Paul, MN

Smecticmain-chain liquid crystalline elastomers (S-MCLCE) with polydomain morphologyare rare examples of elastomers that can yield and undergo cold drawing undertension.[1,2] Yielding andformation of a "neck" are observed only under certain conditions of strainrate and temperature. Inhomogeneousdeformation is increasingly favored as the strain rate increases at constanttemperature, or as the temperature decreases toward T_g.[2] As cold drawing proceeds, significant creepoccurs continuously within the neck.Thermal imaging of the MCLCE during elongation corroborates thisobservation, yet illustrates that viscous heating is not a prerequisite forneck formation.[2] Rather, inherent softening of the material during yielding due to morphological changesleads to an enhanced rate of deformation and contraction at the neck.

Thephysical basis for the observed mechanical instability is of considerableinterest, given the rarity of necking phenomena in elastomers. Recent studies revealed that thermal historyof the material plays a significant role in promoting mechanical instability,which is enhanced in elastomers having large, stable microdomains. Polydomain S-MCLCE cooled quickly from abovethe clearing temperature ($T > T_{si}$) to room temperature exhibit alow nominal yield stress, a lessened tendency to form a neck during elongation,and high extensibility. S-MCLCE cooledslowly from $T > T_{si}$ and annealed for an extended period of timeat T slightly less than T_{si} exhibit increased yield stress,well-defined necking, and in some cases, lower elongation at break. Analysis of the lineshape of the low-angleX-ray reflection (associated with smectic layering) reveals a significantgrowth in domain size with increased annealing time. In addition, very long annealing times caninduce formation of a second, higher-order smectic phase. The increase in the yield stress duringannealing is initially due to the growth of domain size, which increases theenergetic penalty for yielding (via transient disordering of smecticmicrodomains). After a long annealingtime, the material can become brittle due to the formation of the higher-orderphase.

References

[1] H.P. Patil, D.M. Lentz, R.C.Hedden, *Macromolecules* **42**, 3525-3531 (2009).

[2] D.M. Lentz, H.P. Chen, Z.Y. Yu, H.P. Patil, C.A. Crane, R.C. Hedden, *J Polym Sci Pol Phys* **49(8)**, 591-598 (2011).

See more of this Session: Structure and Properties In Polymers III: Networks and Gels II

(338a) Comparing Resource Consumption and Sustainability of Biobased Materials Using Eco-LCA

Erin F. Landers and Bhavik R. Bakshi, William G. Lowrie Department of Chemical and Biomolecular Engineering, The Ohio State University, Columbus, OH

Ecosystem goods and services are essential to all human activity. In spite of this importance, the Millennium Ecosystem Assessment has found that many ecosystem services are highly degraded. In many engineering analyses, these necessary goods and services are undervalued, or even ignored completely. It is therefore crucial to develop and modify existing methods to allow these services to be valued and accounted for in decision making. Conventional life cycle assessment (LCA) tends to only look at energy consumption and emissions. However, Ecologically Based LCA (Eco-LCA) is a thermodynamically based LCA that allows for many ecosystem goods and services used throughout our economy to be quantified, in addition to the traditional LCA results of energy consumption and emissions.

To create the Eco-LCA model, a nationally based economic model must first be obtained, to which the resources will then be added. Eco-LCA uses the national input-output economic model that the Bureau of Economic Analysis provides every five years. This current model, based on the 2002 national economy, shows the interactions between different industries and is used to track the use of ecosystem goods and services through the economy, which will be compared with previous research done to create an Eco-LCA inventory for the 1997 economic model. Resource use must then be added to the model; this is done by finding values for resource use from government agencies and various research groups. The thermodynamic properties of exergy and emergy allow for the many different units of ecosystem goods and services to be converted to the common basis of thermodynamic work. This common basis means the results of Eco-LCA analyses can be aggregated and directly compared in metrics such as Cumulative Exergy Consumed and Percent Renewable. After adding the ecosystem goods and services to the national economic model, resource intensities are calculated for every resource and economic industry by taking into account the resources entering the economy, the interactions between industries, and the total monetary activity of every industry. This then provides a measure of individual and total resource use per dollar of economic activity in each industry. These new resource intensities can be directly compared to the resource intensities from the 1997 Eco-LCA model and provide insight as to changes in both resource use and interactions between industries from 1997 to 2002.

A case study will be presented using Eco-LCA to evaluate biobased materials such as biofuels and biopolymers. The sustainability of biofuels will be discussed by investigating the resources required and emissions produced for various feedstocks, including changes since 1997. This will allow for a more dynamic view of the progress towards the most viable, sustainable biofuels. Resource consumption to make biopolymers will be compared to traditional polymers, and conclusions will be drawn as to the most sustainable polymer materials.

See more of this Session: Sustainability and Biobased Materials
See more of this Group/Topical: Forest and Plant Bioproducts Division

(338b) Sustainable Green Composites

Amar K. Mohanty, Department of Plant Agriculture & School of Engineering, Guelph, ON, Canada and Manju Misra, School of Engineering and the Department of Plant Agriculture, University of Guelph, Guelph, ON, Canada

In progress

See more of this Session: Sustainability and Biobased Materials
See more of this Group/Topical: Forest and Plant Bioproducts Division

Unavailable Paper

(338c) Integrated Biorefinery Operations: Biomass Pretreatment As It Relates to Biochemical and Thermochemical Products

Brian L. Cooper, Jeff London and Robert Mellon, Hazen Research, Inc., Golden, CO

Biomass based chemical refining represents many new opportunities in engineering and process design. The population of companies seeking to develop their technology is large and diverse, from pure liquid fuel producers to integrated multi-product refineries. A common flow sheet among many of these organizations shows a hybrid biochemical/thermochemical process, whereupon the carbohydrate values in a biomass are removed and the resultant lignin portion is utilized for co-products or process heat. As a company with capabilities in both areas, Hazen Research is conducting a series of experimental programs that generate data for a hypothetical process for the biorefining of Lodgepole Pine (Pinus contorta) and Ponderosa Pine (Pinus ponderosa) into chemical values and thermal chemical products. Initial studies demonstrated a clear correlation between biomass pretreatment conditions and the resultant pyrolysis/gasification products generated from the hydrolysis residue. The presentation will focus on new data generated in a second round of experiments, which include optimizing carbohydrate yields in hydrolysis as well as running a continuous thermal processing in place of batch experiments.

See more of this Session: Sustainability and Biobased Materials
See more of this Group/Topical: Forest and Plant Bioproducts Division

Green Poly (lactic acid) Composites Reinforced with Sugarcane Bagasse Residues

Letian Wang, Zhaohui Tong

Agricultural and Biological Engineering Department, IFAS, University of Florida,

Rogers Hall Museum Road, PO Box110570, Gainesville, FL, 32611-0570

Email: ztong@ufl.edu

Poly (lactic acid) (PLA) composites reinforced by sugarcane bagasse residues were prepared using twin-extrusion. By using near zero-value bio-based residues, these composites are not only 100% renewable and biodegradable, but also have the competitive cost in comparison with petroleum-based composites. The sugarcane bagasse residues (10-30 wt.%), containing different content of fibers and all the remaining lignin, were used to prepare lignocelluloses reinforced PLA composite with desirable physical and chemical properties. In this study, the sugarcane bagasse residues include the residues originally from the sugar extraction process, after pretreatment, and after fermentation of biorefinery processes. The varieties of different bagasse residues include the composition, particle size and its surface reactivity during the chemical and biological (enzymes, microbes) treatments that the biomass is subjected to. The effect of these varieties on the morphologies, thermal properties, mechanical properties and the molecular weight of PLA composites was studied. Furthermore, a coupling agent (Desmodur® VKS 20) was used to increase the interfacial bonding of PLA with lignocelluloses. Desmodur® VKS 20 is a mixture of diphenylmethane-4, 4'-diisocyanate (MDI) with isomers and higher functional homologues (PMDI). The microstructure and the properties of the composites in the presence of couple agents were also investigated in this study. It was found that this coupling agent could extend the polymer chain and therefore improved the mechanical properties of the composites.

ALTERNATIVE USES OF SUGARCANE MOLASSES AND JAGERRY FOR PRODUCTION OF VALUE ADDED FATTY ACIDS, ESPECIALLY OXALIC ACID

P.Sanyal*, S. Gupta, S. Chattopadhyay and A.K. Ray
Department of Paper Technology
Indian institute of technology roorkee
Saharanpur campus, Saharanpur-247001, India
*National Sugar Institute, Ministry of Food and Civil Supplies, Govt. of India, India

ABSTRACT

Sugarcane molasses and jaggery provide a unique source for large number of downstream products which include livestock and human food sources, potable alcohol and absolute alcohol, alcohol based chemicals like oxalic acid, citric acid, tartaric acid, lactic acids, acetic acid, diethyl oxalate, ethyl acetate, acetic anhydride etc., which are raw materials for manufacture of pharmaceuticals, plastics, synthetic rubber, base for yeast, lysine, protein and innumerable other products. In this present study, status and economics of Indian industries who utilize molasses to produce organic acids through both chemical and biochemical routes like lactic, citric, tartaric, aconitric and oxalic acid etc. are reviewed with particular reference to oxalic acid production and updated research findings. The oxalic acid or its derivatives have widespread industrial applications such as textiles, leather, oil refinery, metallurgy ,nuclear power plant, dyes, catalysts, explosives, straw bleaching, printing, marble polishing, metal and cloth cleaning, pharmaceuticals and antibiotics. It is also a very important chemical in petroleum, rare-earth, ink, rust, corrosion inhibitor, and dental adhesive processing.

Oxalic acid can be produced from a number of sources from sugar itself, from sugarcane or sugar industry wastes, sugarcane trash, wastes from paper industry by oxidation of their main components of carbohydrates e.g. sucrose, glucose, and other sugar bearing raw materials, from dry powder of parthenium hysterophorus L., and other cellulosic wastes, wood dust, wood cellulose etc. Sugar cane or sugar beet molasses, sugarcane jaggery, deionized cane juice can be a potential source from sugar industry wastes. Both chemical as well as biochemical process may be followed to produce this important chemical. In chemical route generally catalytic oxidation of the above ingredients by nitric acid or by formed nitrogen oxides in batch or continuous stirred tank reactors produces oxalic acid, acid sachharate, and tartaric acid. Production of oxalic acid by fermentations of carbohydrates using *Aspergillus niger* and *Pseudomonas fluorescens* and radiation processing of carbonate solutions and molasses are reported. Production of tartaric acid and citric acid under certain conditions can also be produced. However, the majority of work involves oxidation of sugar bearing materials by nitric acid to produce oxalic acid. In this present study, an attempt has also been made to optimize the processes of producing oxalic acid from sugarcane molasses and sugar cane jaggery by oxidation with nitric acid in presence of dehydrant and catalyst with or without the presence of catalyst promoters. Parameters affecting the reaction were: air flow rate, temperature, the amount of V_2O_5, catalyst promoters and the concentrations of reactants and dehydrant. The maximum yields obtained by different processes were compared. Treatments of NO_x vapour by absorbing in a suitable medium (in soda ash or sulphuric acid), to avoid air pollution (and also to produce value added by-products, glauber salt and sodium nitrate), mother liquor and catalyst recycling were attempted to enhance the economy of the plant.

INTRODUCTION:

India is the largest producer of sugar in the world [1-4]. In terms of sugarcane production, India and Brazil are almost equally placed now a days. Two to three years' before it was the second largest producer of sugar in the world next to Brazil. At present, total world production is of the order of 110-116 million tons from both beet and cane sugar.India is expected to produce round 27.5 million metric tonnes (MMt)and 30.5MMt in 2011-12 and 2016-17 respectively.

It is wellknown that one of the important by- product of the Sugar Industry is the cane molasses, which is a strong building block for many chemical industries. In spite of producing cane molasses in large quantity in the country, several molasses based products are being imported in substantial quantity[4]. Department of Science and technology in India submitted a report which detailed the demand supply scenario and status of the Indian molasses based industry concerning the five products namely Lactic Acid, L-Lysine, Gluconic Acid, Glutamic Acid/Monosodium Glutamate and Itaconic Acid. It is necessary that India should set up facility for the production of the above molasses based products for the reasons [4]such as to utilize the molasses availability, to avoid the import of the molasses based downstream products, to export the molasses based downstream products in the global market. To achieve breakthrough in the fermentation technology for the molasses based products, as molasses is the byproduct of the sugarcane industry it is important to highlight regarding the status of this industry in this country.

STATUS OF SUGAR CANE INDUSTRY IN INDIA:

Sugar industry, worth Rs 60,0000 millions is also one of the largest agro-based industries in India-a country that happens to be also the largest consumer of white plantation sugar.[1-2] It has manufactured 26 million MT in 2007-08 which has declined over the following years due to varieties of reasons, 14.7 million tons 2008-2009 and again increased to 16.0 million tons 2009-2010. As already indicated in the coming years, the production of sugar is likely to go up steadily in the country [1-3].With approximately 25 million tons of sugar production in 2010-11 and 24 million tons of consumption, both sugar production and consumption are bound to increase lineally. The growth rate of sugar production in India is shown in Fig.1 and the same for molasses is given in Fig.2.

Indian sugar consumption is growing at a CAGR of 3.75% .Sugar production in India has occupied a place of pride in the Indian economy being one of the five largest industries comprising of more than 571 units spread all over the country.

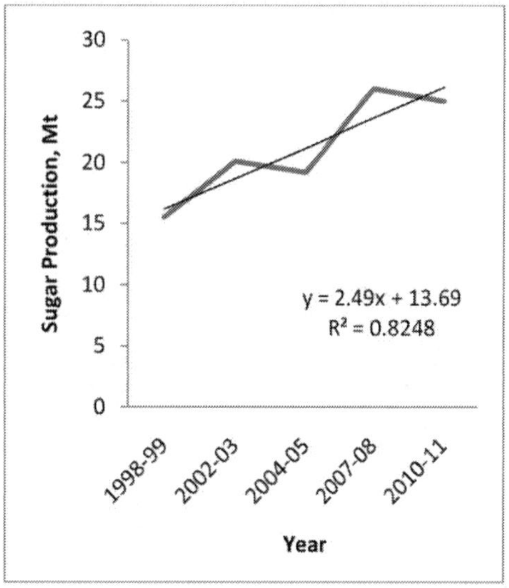

Fig.1 Production of white plantation sugar.

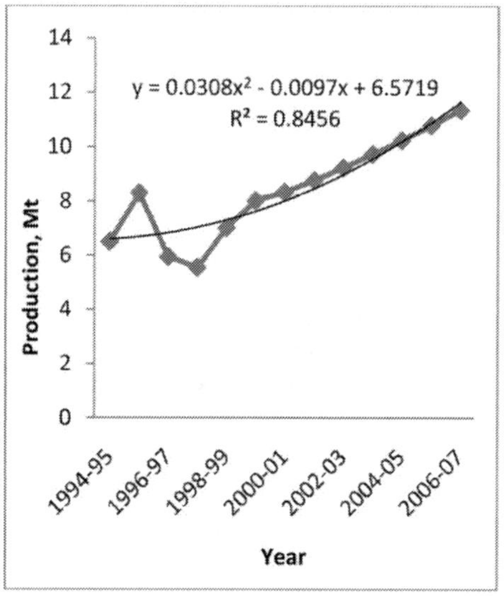

Fig. 2 Production of Molasses

The average yield of cane in India is around 70 -77 t/ha in tropical states to about 52t/ha in subtropical stases[1] though in UP(which produces 40% sugar) the yield has decreased from 55 MT/ha to 45 MT/ha in the recent past. High yielding new varieties, namely Co 238, Co 239 and Co 118 identified can be used to increase the productivity of sugar cane and sugar. The area under sugarcane is presently less than 2% of total cultivable area in the country and about 3% of the irrigated area. There is considerable scope for increasing the area under sugarcane considering the fact that it is more profitable compared to other crops.

INDIAN MOLASSES CHARACTERISTICS AND ITS UTILIZATION:

Molasses is the uncrystallizable syrup which remains as final residue in process of manufacture of sugar. Its composition varies due to a number of factors, including cane maturity, variety, agricultural conditions, industrial processes etc. A typical composition of final molasses(all in %) in Indian Sugar Mill is as under:

H_2O	-	17-25	
Sucrose	-	30-40	50% Saccharine matter
Glucose	-	4-9	
Fructose	-	5-12	
Other reducing sugars	-	1-4	
Total reducing sugar	-	10-25	
Carbohydrates, Gums, Starch	-	2-5	
Water soluble carbonates	-	7-5	
Nitrogen	-	2.5-4.5	
Others	-	0.5-1.5	

Waxes, sterols phosphorus	-	0.1-1.0
Vitamins	-	varying amounts

As molasses contains substantial quantity of sucrose as well as reducing sugars, it must produce many chemicals acids as pure sucrose does as shown in the following diagram[1]:

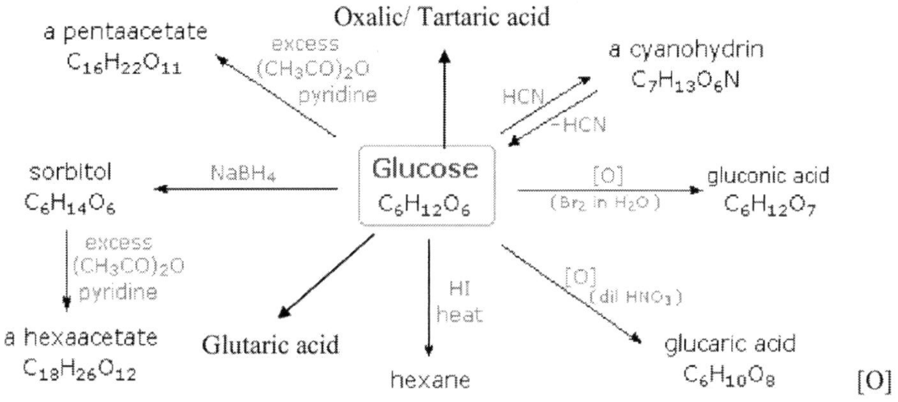

Most of the chemicals are however imported in India.

Use of Molasses

(a) Direct utilization: Fertilizer, animal feed
(b) Distilling industry: Ethyl alcohol, rectified spirit, power alcohol, alcohol derivatives, potable liquor, such as beverages
(c) Other fermentation industries: Vinegar and acetic acids, acetone and butanol, citric acid, lactic acid, glycerine, yeast,
(d) Other chemical industries: Oxalic acid, and tartaric acid
(e) Miscellaneous industries: Aconitic acid, monosodium glucamate, dextran,L-Lysine, Xanthenes gum, 2-keto gluconic, 5-keto gluconic , a-ketoglutaric, fumaric acids , glycolic acid,kojic acid and glucaric acid etc.

Besides the above, molasses can be used for many other value added products.Some of the important value added chemicals (mostly imported) are briefly described below:

Molasses is the only raw material available in large quantities in India for an economic production of alcohol until petrochemical source develops. Almost all the commercial molasses in the world is cane molasses because the beet-sugar producing countries usually consume it domestically. The most significant use of final molasses in many industrialized, countries is in animal feed. Other uses of molasses are as the raw material for alcohol production and as the base for yeast, lysine and citric acid production and may others. The chemicals derived from molasses could be produced cost effectively and be profitably exported [3,5-40].

Molasses contains small amount of Lysine, or L-lysine, an essential amino acid, 2,6-diaminohexanoic acid, building blocks of protein, required for proper growth but the body can't manufacture it; lysine has to be obtained from food. Lysine it plays an essential role in the production of carnitine, a nutrient responsible for converting fatty acids into energy and helping to lower cholesterol. Lysine appears to help the body absorb calcium, and plays an important role in the formation of collagen, a substance important for bones and connective tissues including skin, tendon, and cartilage. Glucaric acid is a member of a larger group of compounds known as sugar acids – and more specifically, aldaric acids. Glucaric acid has gained attention because it was identified as one of the top 12 renewable building block chemicals.

Fumaric acid

The principal industrial use of fumaric acid is as a mordant in textile dyeing. It appears that it could find application as a replacement for tartaric acid in leavening agents, as well as in the manufacture of resins and plastics for special purposes. Itaconic and fumaric acids, if sufficiently cheap, could flow into plastics. The same is true of citraconic acids which may be derived from citric acid. In addition, this group is outstanding as acidulating agents in foods. Citric, lactic, and gluconic acids are known to be safe ingredients of the diet and are extensively employed in foods, soft drinks, confectionary and pharmaceuticals.

Itaconic acid

Itaconic acid, 2-Ketogluconic acid is produced by the action of various species of bacteria of the genus *Pseudomonas* on glucose in aerated cultures in the presence of calcium carbonate. This is formed by *Aspergillus itaconicus* and is now available commercially. It has the possibilities in the preparation of resins and plastics because it is a substituted acrylic acid. It is now possible to get 37-percent weight yields in 12 days at about pH 2.2. With submerged fermentations, yields as high as 45% have been obtained in 5 days at a pH of 1.8.Weight yields of 90 percent can be obtained in a 30-hour submerged fermentation. The keto acids can be converted to d-araboascorbic acid which is a good antioxidant for food use. In this capacity, it can serve in the preservation of flavors in the canning industry and in preventing the development of oxidation flavors.

Kojic acid

It is very chemically reactive compound and many derivatives have been prepared from it. It can be prepared readily in yields of more than 50 percent by molds growing on a variety of carbon sources like glucose and xylose, with the use of members of the *Aspergillus flavus, A.oryzae* group.

5 keto gluconic acid is produced by various species of *Acetobacter* in weight yields exceeding 90 percent. When oxidized by air in the presence of vanadium pentoxide, it is converted to d-tartaric acid with good yield. A ketoglutaric acid can be produced in a 25 percent yield when fermentation of glucose by *Pseudomonas fluorescens* is allowed to proceed beyond the 2-ketogluconic acid stage to the point where no more reducing sugar is observed. Good yields of bionic acids such as maltobionic and lactobionic acids could be obtained by the action of *Pseumonas graveolens* on maltose and lactose. The pentonic acids such as d-arabonic, and d-

ribonic acids are formed with good yields during oxidation of d-arabinose , d-xylose and d-ribose by the genus *Pseudomonas*.

Lactic Acid:

Five tons of molasses would be needed to produce one ton of lactic acid.

Lactic acid is used primarily as an acidulant and preservative agent in the pharmaceutical, chemical, and food industries like beverages, beer, jelly, etc. Described as ``the chemical of the future" because of its use in the manufacture of polylactic acid, it is reckoned that polylactic acid could substantially replace plastic bags, since it is bio-degradable [9-10]. The current interest lies in making biodegradable plastics used for general and medical applications as a delivery vehicle for vaccinations.

Lactic acid production parameter optimization[9] was performed using cane molasses by the design of experiment(DOE).Eight factors with three levels studied were yeast extract, $CaCO_3$, $MnSO_4$, pH, temperature, molasses, urea, and Tween 80. Lactic acid production was significantly affected by interactions of two factors such as temperature-urea, pH-$CaCO_3$, temperature- Tween 80, but individually they have minimum impact on lactic acid production. Individually pH, yeast extract, molasses, and urea are the most significant factors in lactic acid production. Lower amount of $CaCO_3$ and $MnSO_4$ and Tween 80 enhanced lactic acid production. The expected yield of lactic acid under optimal condition was 98.66% and the actual yield was 96.8%. Overall 6% improvement in yield has been observed when compared with same (mutant) species grown on cane molasses.Lourdes R. Fragante [5] has conducted fifteen experiments using *Lactobacillus delbreukii* in aqueous molasses and molasses/coco water solutions containing 9, 10, 11 and 12% sugar as invert. The additives used were 2% nitrogen, 1% phosphate and 8.3% invertase per 35g. molasses. The incubation temperature was 45°C. Highest yields were obtained at 10% sugar level for all substrates. Coconut water was found to support lactic acid production. Substrate-Molasses in coco water solution with invertase further increased the yields but only up to 10% sugar level. The presence of higher glucose/fructose levels seemed to inhibit lactic acid production. All substrates gave decreasing yields at 11 and 12% sugar.

Citric Acid

Seven tons of molasses would be required to produce one ton of citric acid.

This chemical is mainly used as a preservative in food processing, pharmaceutical and cosmetic industries. It is also used to manufacture sodium citrate, used in the production of detergents. Industrial production of citric acid by fermentation is one of the most efficient examples of fungal fermentation technology[7-8].The food industry is thought to comprise as much as 70% of this market; the remainder being largely pharmaceutical produce, though the metal cleaning, agriculture and detergent industries would also be of notable importance.

Tartaric Acid:

Tartaric acid is used above all in many applications in the food and textile industries. For example, tartaric acid is used as a food additive, especially as an antioxidant and acidification agent. In addition, tartaric acid is used as a building material additive for modification of the setting characteristics of cement and serves as a starting substance for numerous chemical reactions, especially for chiral synthesis.

A method of preparing tartaric acid in which 5-ketogluconate is oxidized in a carbonate buffer at an alkaline pH preferably in the range of 8 to 10 using vanadate has been described [11-13]. One way of producing tartaric acid is to recover it from tartar, a substance which is produced during the production of wine.

However, the crude tartar which results from the production of wine is as a rule contaminated by organic materials so that expensive preparation is necessary to obtain the tartaric acid from the tartar.

It has been found surprisingly that under these reaction conditions, i.e. also without the addition of vanadate or another catalyst, a conversion of 5-ketogluconate to tartaric acid results whereby a conversion of 5-ketogluconic acid to tartaric acid of up to 55% is achieved. If vanadate is additionally supplied to the reaction mixture, conversions up to 75% of the 5-ketogluconate used to tartaric acid are obtained[11-12].

A further process for producing tartaric acid resides in the conversion of glucose to tartaric acid through microorganisms like *Acetobacter* or *Gluconobacter* in the presence of vanadate as trace element. It can be demonstrated that in these processes the tartaric acid production is carried out in two steps. In the first step 5-ketogluconate is excreted into the medium of microorganisms cultured in a glucose-containing medium, while in a second step the conversion of 5-ketogluconate to tartaric acid is carried out purely chemically with the vanadium salt as trace element, i.e. independently of the presence of microorganisms[13].

PRODUCTION STATISTICS AND ECONOMICS:

Although India is a major producer of molasses (fig.2). Sugar mills produce around 8 million tons (mt) of molasses, which is projected to increase to 10.5 mt over the next ten years. About 14 per cent of the molasses produced today is exported, without any value addition. There is a growing demand world over for molasses-based chemicals, which are presently imported [6]. The huge export opportunity is being lost, because, principally, the Indian industry is not fully aware of the business potential in this area. There is a possibility of manufacturing these chemicals with indigenous resources. Here again, a huge export opportunity exists manufacturing this chemical at competitive prices, further harnessing of overseas business opportunities. On the other hand, the country imports molasses-derived chemicals in very large amounts

Estimated demand for lactic acid (partly imported) is 2500 tons in the year 2015 A.D. Lactochem Ltd of the Malladi group has a 200 tpa plant in operation. Karnataka-based Munirabad Chemical Co, which has a 180 tpa plant, is another example of a successful plant. Orchem Ltd, an associate unit of Atul Drug House, attempted to put up a 300-tpa plant, but did not succeed. The company first attempted to manufacture the chemical synthetically on a laboratory scale, but ran into difficulties when getting into commercial scale operations. Another company called Sangita Bio-Chem of Orissa also attempted a venture, but is said to be held up for want of funds. Two new units are coming up; one is Prathista Industries in Andhra Pradesh, with 10000 tpa capacity and Interspice Biotech near Chennai. The latter is coming up with technology from GaLactic S.A. of Belgium. The study recommends setting up of a plant of 3,000 tons capacity, which would cost around Rs 27 crore.

In 2002, citric acid production by fermentation was estimated to be in the region of $7.0x10^5$ ton/year [8]. Such production levels are necessary to meet the high demands of the now massive market. From the design of the original fermentation process right up to the present day, citrate production has largely been overseen by Pfizer. Deviation from the market profile of the 60s and 70s has seen the once dominant pharmaceutical market demand overshadowed by that of the ever-growing food market, particularly the carbonated drinks industry. This shift has seen the price of citrate fall significantly, thus encouraging further improvements in the rate and efficiency of its production. Unlike oxalic acid, the technology for producing citric acid from molasses is established and available. The Indian industries worth mentioning are Bharat Starch Industries of the Thapar group, Citric India Ltd of Mumbai and Andhra Citrates of Hyderabad, Citurgia Biochemicals of Mumbai.

For the first time, a detailed study (commissioned by the Centre and conducted by a consultancy company called Nandini Chemical Consultancy[3] has identified some molasses-based products and the business potential for each of them. The study, which advocates setting up of plants for the identified molasses-based chemicals, makes two postulates for its recommendations. First, the price fluctuation of molasses in the country is immaterial because the value addition in the manufacture of these chemicals is so high that volatility in prices is inconsequential. Second, the target market is the world market and any entrepreneur who wants to first establish a ``good domestic base" might as well look for opportunities elsewhere.The following table indicates the approximate business statistics for three principal molasses derived acids.

Table Major Molasses based acids, their demand-Supply, growth rate, export potential

Acids	Global Demand, tons(year)/ Growth Rate,%	Demand/Installed and Production Capacity, tons	Export, tons
Oxalic acid	250,000; 3.0	30,000 /22,000	9,000
Citric acid	703,200(2001), $7.0x10^5$ (2002),1,040,900(2011)	12,000(2001)/ 38,475(2011)	4,610(0.4% of world market)
Lactic acid	70,000(2001)/100,000(2002) 181,560(2005),500,000 (2011)	560(2001)/ 1,980 (2011)	5,000(3% of world share)

Techno-Economic feasibility studies:

Department of Science and technology (DST), Government of India, has estimated the likely gap in supply by 2015 AD in India and recommended the Installed Capacity of the following chemicals. The detailed project report has been made based on number of 300 working days per annum and 24 numbers of working hours.

Lactic acid, is presently consumed 200 Mt, partially met by imports, likely to increase to 600 M t. Only M/s Lactochem India Ltd. is manufacturing it. L-Lysine, itaconic acid, gluconic acid and glutamic/ MSG acid all are imported, India will need 18,000, 1800, 1500 and 10,000 Mt respectively. Adequate electrochemical process technology has been developed for the production of Gluconates from Glucose by Central Electrochemical Research Institute, Karaikudi to produce the later two chemicals.

The Nandini Chemicals study has also said that huge export possibilities exist for other molasses-based products such as lysine, glutamic acid, itaconic acid, gluconic acid, ephedrine hydrochloride and yeast.

Oxalic Acid:

To produce one ton of oxalic acid, 2.5 tons of molasses would be required.
Oxalic acid and its derivatives have widespread industrial applications like textile, leather or tanning, oil refinery, catalysts, dyes, metallurgy, pharmaceutical, marble polishing, petroleum, nuclear engineering, explosives, detergent, plastic, film industries and antibiotics.
Other uses include dental adhesive processing, corrosion inhibition, straw bleaching, printing, precipitating agent in rare-earth mineral processing, bleaching agent in the textile activities, wood pulp bleaching, wood bleaching agent(used to lighten and brighten dingy gray wood), remover of cement residues, rust remover for metal treatment, commercial rust removers to remove rust stains from tubs and sinks, food and rust stains from kitchen countertops, grinding agent, such as marble polishing ,waste water treatment, removing calcium in water, cleaning and sterilizing home brewing equipment, as a reducing agent for photography and ink removal ,as purifying agent in pharmaceutical industry, clean rock collectors, clean mineral specimens, , plumbing fixtures and fabric.

Indian Statistics for oxalic acid manufacture:

Nandini Chemicals recommends a minimum economic size of 15,000 tons, which will call for an investment of about Rs 17 crore. In India, oxalic acid is produced from sugar, by a dozen-odd companies, most of them Mumbai-based. Two companies have attempted to produce oxalic acid from molasses instead of sugar - but both attempts failed. One was a Chennai-based company called Krebs India, which put up a 300 tpa plant at Pondicherry. The other was Punjab Chemicals and Pharmaceuticals Ltd, the largest producer of oxalic acid through the sugar route. The cost advantage can be gauged by the fact that while it takes 600 kg of sugar to produce one ton of oxalic acid, it takes 2.5 tons of molasses to produce the same.

Present cost of molasses based acids:

The present market cost of the following acids from molasses in Indian currency is as under:

Citric acid:Rs. 51.74,Oxalic acid, Rs.52,Fumaric acid(Food grade):60.72,Lactic acid:Rs.68.33,L-Tartaric acid:122.75,D-Tartaric acid :251.09,. These are in the ratio of 1:1.005:1.174:1.321:2.372 :4.853. It is reported that the ratio of prices of some important acids including the mentioned one are as under: Glycolic: Formic acid: Acetic Acid: Oxalic acid : Tartaric acid: Butyric Acid=1:1.44:1.5:1.61:3.56:3.78. From the current market study it is usual noticed that the imported chemicals are found to be cheaper than those indigenously manufactured.

R& D efforts for Oxalic acid production:

Literature Survey:

There are various methods for manufacturing oxalic acid [14-40]. Process depends upon the raw material selected. The conventional method of production generally consists of glucose or sucrose oxidation in presence of nitric acid and vanadium pentoxide is used as a catalyst. Attempts have been made to prepare oxalic acid from ethylene, ethylene glycol, propylene, lignin, jaggery, cane and beet molasses, sugarcane, sugars, flour, plant wastes, formate, carbonate and bicarbonate salts, cellulose and wood wastes by using different methods. These methods are: oxidation of pure sucrose or impure commercial sugars, cellulose and other carbohydrates by nitric acid, or oxides of nitrogen, fermentation of carbohydrates, fusion of sawdust with caustic soda, oxidation of olefins(ethylene , propylene) and glycols, radiation processing of carbonate solutions and molasses, and decomposition of formates.

Of the above the first two received much attention.Other methods such as thermal decomposition of sodium formate (Radulescu et al., 1993)have little bearing on economic production of oxalic acid.

Fermentations of carbohydrates to oxalic acid have been reported for *Aspergillus niger* (Rujiter et al., 1999) and *Pseudomonas fluorescens* (Hamel *et al*., 1999).

Number of attempts have made for the production of oxalic acid by the nitric acid oxidation of carbohydrates such as glucose, sucrose, cellulose and commercial waste materials (Metin Gürü,*et al*.,2001,Guo *et al*., 1997, Sattar *et al*., 1988, Mane *et al*., 1986; Mane *et al*., 1988, 1986, Nowak,1986 , Sullivan *et al*.,1983,Dube *et al*., 1982, Shukla *et al*., 1982 ,Sanyal *et al*. 1981,82,Gupta *et al*., 1980, Despande and Vyas ,1979,. Agrawal *et al*., 1979, Raha *et al*., 1976, Bose *et al*.,1971, Preen,1942)

Bretschneider *et. al,* [15] reported oxidation of sucrose by nitric acid and discussed the kinetics of the process for tartaric acid as the main product., not oxalic acid.
Berntsen and Degering(1949) oxidized glucose to produce potassium acid sacchharate with oxides of nitrogen and Chia and Lin(1962) oxidized sucrose to produce gluco-saccharic acid. Tanz and Kao (1944) reported work on the selection of catalyst and promoters for this oxidation reaction using oxides of Vanadium, Molybdenum and Tungsten. The oxidation of ethylene glycol to oxalic acid has been discussed by Vishwalingam *et al*., (1980) and kinetics was studied. The production of oxalic acid from other sugar bearing commercial by-product materials such as jaggery (gur) , beet molasses , cane molasses and wood cellulose have also been tried by a very few investigators [14].
Raha *et al*., [30] have described especially the method for producing oxalic acid from Gur. In most of these cases after pretreatment the raw material was oxidized by nitric acid in presence of vanadium pentoxide. Agarwal and Rao [14] oxidized cellulosic materials by nitric acid and found ammonium-meta-vanadate which is a cheaper material to be equally effective as the oxide of vanadium.
The attempts for producing oxalic acid using waste materials like final molasses have not been very economical. Bose *et al.,* [17] attempted the preparation of oxalic acid from sugarcane

molasses using nitric acid and vanadium pentoxide and obtained about 55 percent of the theoretical yield.

Deshpande and Vyas [19] have produced oxalic acid from sugar and the nitrogen oxides evolved were absorbed to produce sodium nitrite as a by- product . Gupta *et al*., [22] reported the production of oxalic acid from deionized sugar cane juice.

Optimum conditions of reaction as developed earlier were employed in the present investigation and the raw materials (sugarcane molasses) of three different compositions were utilized.

Metin Gürü,*et al*., [23]conducted exhaustive investigation for the production of oxalic acid from sugar beet molasses using a series of three reactors with formed nitrogen oxides in the following two reactors Parameters, air flow rate, temperature(45 °C to 58°C), the amount of V_2O_5 catalyst and the concentrations of molasses and H_2SO_4 were optimized. The process is however a noble one in respect of recovery of NOx, to prevent air pollution ,to reduce the unit cost of reactant for oxalic acid production and the equipment arrangement. The yield was reported of the order of a maximum 77.5% and 74.1% of theoretical yield obtained by using the absorption solutions with NOx.

Though the process is claimed to be Novel, definitely improvement over the earlier one, in India firstly the raw material used by Guru *et al*., beet molasses is scarcely available(almost negligible) whereas cane molasses are abundantly available. Besides, beet molasses from refinery goes for cattle feed, hence not available. It however contains more sucrose than that of molasses but lower than jaggery. The sugar beet molasses used in the experiment contained 48.1% sucrose, 3.2% invert sugar and 19.4% water. These are compared in the following table.

Investigators	**Yield of theoretical,%**	**Raw Materials employed**
Metin Guru *et al*.,[23]	78.9/ 77.5 /74.1	Sugar beet Molasses
Sanyal *et.al*., [33]	74	Jaggery (Gur)
Raha *et al*., [30]	74	Jaggery(Gur)
Shukla *et al*.,[35]	48.5	Cane Molasses
Agarwal *et al*.[14]	< 50	Cellulosic materials
Sattar *et* al.,[34]	48.5	Molasses
Mane *et al*.,[27]	51.6/60	Sugar cane trash/Saw dust
Sullivan [36]	80.2	Saw dust
Bose *et al*.,[17]	55	Sugar cane molasses

Therefore it can be ascertained if final cane molasses is at all allowed to use , manufacture of oxalic acid, and tartaric acid with the three CSTR as proposed by Metin Gürü,*et al*., with optimum process conditions of temperature, catalyst concentration etc can give the best yield.

Therefore the work done previously by Sanyal *et al*., [33] are reproduced here under for effect of two different raw materials , namely jaggery and final cane molasses for further critical analysis;

Raw Material and Chemicals Used:

1. Sugarcane Molasses: Khandsari and final sugar cane factory molasses were used and their analyses are as follows:

Specifications on Sugarcane molasses

Sl. No.	Item	Molasses Sample		
		A	B	C
(i)	Brixs (Bx^0)	70.30	80.30	84.00
(ii)	Pol.	33.00	34.90	-
(iii)	Purity	46.94	43.46	-
(iv)	Sucrose %	29.90	28.00	39.98
(v)	Sucrose Purity	42.53	34.80	41.80
(vi	Reducing Sugar %	21.42	18.36	20.80

2. Gur(Sugar cane Jaggery):

Large quantities of gur are produced in India mostly from small scale units and the properties are likely to change depending upon the sugar cane used. Sudies were made using a single batch product with the following composition:

Specifications of Sugarcane Jaggery (Gur)

Parameters	Percentage
Sucrose	76.3
Reducing sugars	9.5
Purity	84.65
Ash	3-6
Colour	Yellow Brown
pH	6.3

1. Acids: Commercial grade Nitric Acid (60 % HNO_3) and A.R. grade Dehydrant(H_2SO_4)
2. Catalyst: LR grade Vanadium Pentoxide (V_2O_5) in powder and pellet forms.
3. Promoters: LR grade oxides of Tungsten(WO_3) Molybdenuim(MoO_3) and Iron(Fe_2O_3)
4. Soda Ash: commercial grade Soda Ash (40% Na_2CO_3).

Experimental:

The experimental set-up for the batch reactor and other accessories used has been described elsewhere [33, 35]. The reactor consisted of long glass tube vessel of 1 liter capacity fitted with internal glass cooling coil through which constant temperature water was circulated. Three side tubes were attached for liquid feed, for air bubbling and for continuous removal of the reaction gases.

Procedure:

Predetermined amounts of gur/ molasses, distilled water and catalyst 0.08 % were charged in the reactor. The air compressor was started to introduce at a rate of 9-10 l/min and the acid mixture containing measured amount of nitric acid and sulfuric acid was added slowly. Reaction was allowed to proceed for a particular batch time and the temperature was maintained at 54.0 ± 1.0^oC.

After completion of the batch whole mass was transferred to a flask and was left overnight at a room temperature of 25.0±4.0°C. Needle shaped crystals of oxalic acid dehydrate were formed which were filtered, washed and dried. This product gave the yield for first crop. The filtrate was again cooled in the freezer at 2.0±1.0°C for about 12 hrs. which gave further crystals and the product gave the yield for the second crop. The two crystal yields were combined to get the total yield.

The repeated experiments were carried out for both jaggery as well as molasses at desired air flow rates, reaction time, temperature and catalyst concentration. Product yield was defined as the ratio of oxalic acid dehydrate produced to the amount of oxalic acid dehydrate that was expected by reaction stoichiometry from a particular amount of sucrose present in the jaggery sample. For further experiments on molasses, the filtered air flow rate was kept constant and the reactant to acid ratio was varied. Oxalic acid yield was computed after separating the crystals from mother liquor.

Results and Discussions:

Experimental results on sugar cane molasses were shown in Fig.3 indicating how reactant ratio(molasses : dehydrant: HNO_3).The molasses however, has been kept in the ration as 1.

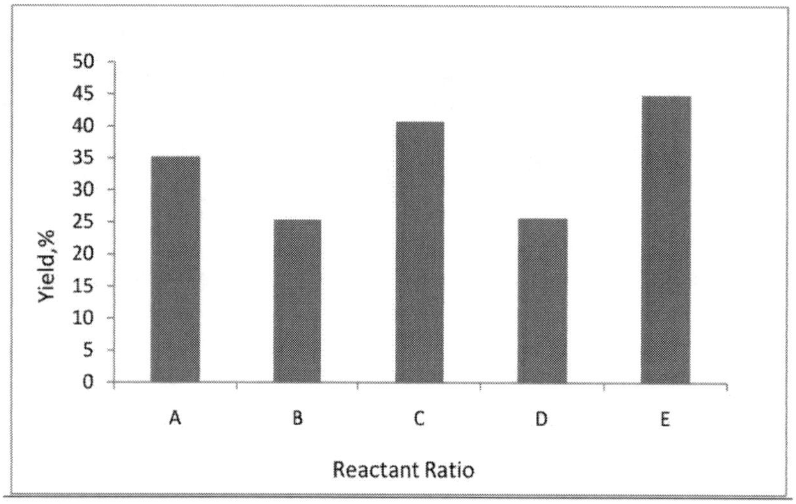

A=1:0.92:4.20, B=1:1.38:4.06, C=1:1.84:4.06, D=1: 1.895:3.60, E=1:2.134:5.41

Fig.3 Yield of oxalic acid as a function of reactant ratio

It is clear from the plot that slightly higher amount of dehydrant and nitric acid gave maximum yield of oxalic acid about 48% of theoretical yield. There was not much difference observed between C and E. considering the cost of dehydrant and nitric acid C might be chosen as the optimum yield.

Vapor Treatment: Reaction vapor products consisting of moisture and nitrogen oxides were allowed to pass through soda ash solution. Thus the gas absorption with chemical reaction took place in the absorber. After the batch completion, the solution was concentrated and cooled for crystal formation. The crystalline product was analyzed for carbonate and nitrate fractions. It is evident from the results that product ratio of $NaNO_3$ and Na_2CO_3 ranges between 1.69 -1.90 with the percentage conversion of soda ash between 60-65%.

Mother Liquor Treatment and recycling: After complete possible extraction of oxalic acid, the mother liquor was treated with soda ash, under constant stirring, till neutral solution was obtained. This neutralized solution was concentrated and cooled to get glauber salt (Na_2SO_4) mixed with unreacted soda ash (Na_2CO_3). The yields and analysis of results indicated that $CO_3^{//}$: $SO_4^{//}$ and yield of Na_2SO_4 (in percentage of theoretical) were between 1:2.70- 1:3.35 and 67.0 - 78.2 respectively. Analytical procedure given in the literature was employed [37- 38].Mother liquor recycling results gave yield (in % of theoretical yield) 18.01% -22.60%.

Product Purity: Oxalic acid crystals were tested chemically and physically using Bausch a Lomb Spectrophotometer and Erma Colorimeter. A comparison was made with I.S.I. grade sample and the product was found to be 99.5% pure.

Parametric relationship for jiggery utilization:

Effect of Reaction Time:

The effect of reaction time on yield was evaluated by varying the reaction time from 1 to 12 hours at constant 0.08% catalyst concentration as shown in Fig.4. The experimental results indicated that the yield of about 74.0% of oxalic acid was attained at a reaction time of 10.0 hrs, beyond which the yield started decreasing. This

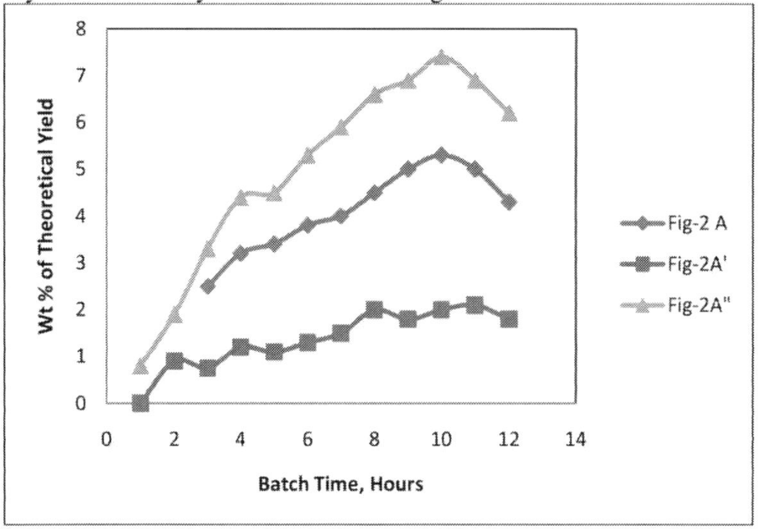

Fig.4 Percent of theoretical yield as a function of reaction time

Effect of Catalyst Concentration:

Vanadium pentoxide powder was used as catalyst. The effect of catalyst concentration on yield was studied at the reaction time of 10 hrs. Amount of catalyst in relation to the amount of gur was varied from 0.01 percent to 0.20 percent. The experimental results as shown in Figs.5 and 6 revealed that the maximum yield could be obtained at a catalyst concentration of 0.08 percent.

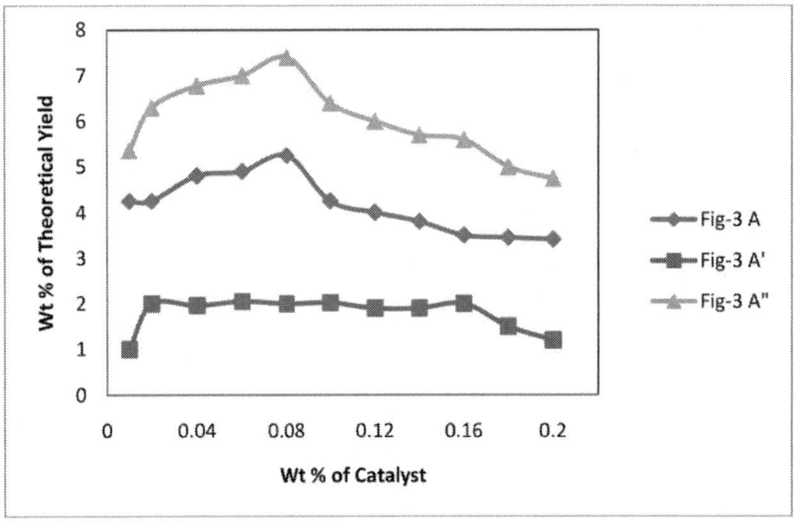

Fig. 5 Percent of theoretical yield of oxalic acid as a function of percentage of catalyst

Effect of Catalyst Promoters:

At the optimum amount of catalyst (0.08 wt. % V_2O_5), the effect of three promoters was studied using oxides of Tungsten (WO_3), Molybdenum (MoO_3) and Iron (Fe2O3). The amount was varied from 0.04 to 0.12 wt percent. The experimental results are shown in Fig. 6 reveals the following:

In case of Tungsten Oxide the yield gave a maximum at wt. 0.08% while in other two cases the yield decreased as the amount of promoters was increased. Furthermore, on comparing the results, it was observed that vanadium pentoxide alone used as catalyst gave better yield (74.4% of theoretical) than the maximum obtained in the case of Tungsten Oxide (66.0 wt% of the theoretical). It was also observed that the use of promoters added colour to the oxalic acid crystals such as reddish brown in the case of ferric oxide, and lost the aesthetic value of the product.

For sugarcane molasses the following additional experiments were performed.

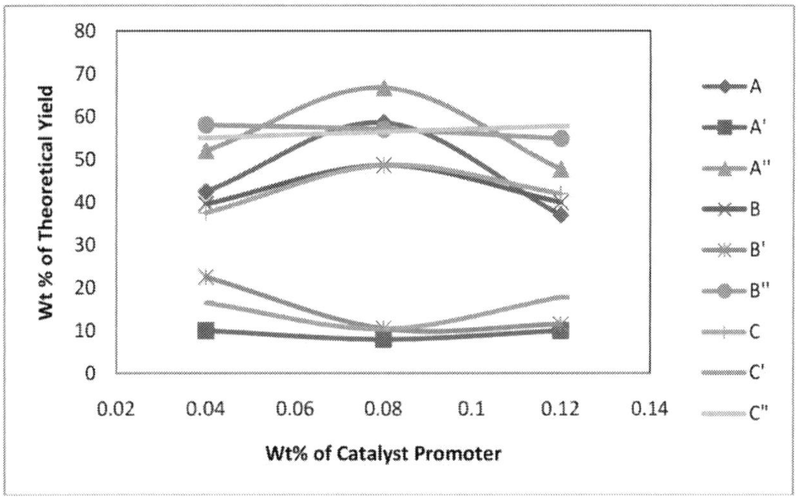

Fig.6 Percent of theoretical yield as a function of percentage of catalyst promoters.

Catalyst Recirculation: Vanadium pentoxide was impregrated in glass beads and alumina and used in place of the powder. It was observed that although at high acid concentrations there was no effect but at lower acid concentrations after the batch was over the oxide was soluble and hence it could not be recovered or recycled.

Gas Absorption: Nitrogen oxides evolved along with air and water vapor could be arrested only partially. As it is clear from the results, it is felt that a better absorber should be designed for complete removal of nitrogen oxides may cause air pollution and also offset the cost of oxalic acid production. The best way is to absorb it and recycle so as to reduce the nitric acid requirements.

Chemicals from Mother Liquor: Effluent liquid in the form of mother liquor contains unreacted acids, catalyst and other impurities originally present in the raw material (in the present case sugarcane molasses). Production of oxalic acid on industrial scale may pose the liquid effluent disposal problem and hence this treatment was tried. Effluent stream should be treated in such a manner that the pollution hazards are controlled as well as the useful chemicals are recovered to reduce the cost of production effluent treatment. Commercial grade soda ash was employed and the results are given in Table 3. It was noticed that no nitric or nitrates were present and about 73% conversion to glauber salt was obtained. In this way the left over sulfuric acid was recovered as well as the liquid effluent was treated.

Mother Liquor Recycling: Mother liquor analysis (after oxalic acid extraction) showed that a large portion contained unreacted dehydrant. Five batches were run using mother liquor. Make up nitric acid and fresh sample of molasses were added to start the batch. The results as discussed clearly show that an overall reduction in oxalic acid yield to an extant of 50% was obtained. Hence it is not instead it should be treated as discussed earlier.

CONCLUSIONS:

There are possibilities of establishing various molasses based acids in India such as lactic acid, oxalic acid, tartaric acid, ctric acid, aconitic acid, monosodium glucamate, dextran,L-Lysine, Xanthenes gum, glycerine, yeast, 2-keto gluconic, 5-keto gluconic , a-ketoglutaric, fumaric acids , glycolic acid,kojic acid and glucaric acid etc.These are mostly imported. Production are very scarce.
Projects regarding production of lactic acid, citric acid, tartaric acid and oxalic acid can be considered as competitive though there could be scope for further optimization of quality & cost standards. Also, more R&D initiatives are required to produce products of various grades, catering to the requirement of specific user. R& D work on the above acids are going on in various industries as well as research laboratories in India. Little R&D push, a unit would be able to produce oxalic acid from molasses. Producing the acid from cane molasses would be very cost-effective and India would be able to take a substantial share of the world market.

The following conclusions can be drawn from the results of investigation for producing oxalic acid dehydrates crystal using sugarcane jiggery and molasses as the raw materials.

1. The maximum yield, 74 wt percent of the theoretical could be obtained in 10 hrs. and increasing or decreasing the reaction time would decrease the yield.
2. Vanadium pentoxide at a concentration of 0.08wt% gave the maximum yield.
3. Addition of promoters like WO_3, MoO_3 and Fe_2O_3 decreased the yield and added colour to the final product, hence it is not recommended.
4. Maximum yield, 48.5 weight percent of theoretical could be obtained while in the case of gur (sugarcane jaggery) under similar condition it was 74.0 wt. percent.
5. Vanadium Pentoxide used as a catalyst has some solubility in dilute acids and hence cannot be re-cycled at this stage.
6. Nitrogen oxides can be absorbed in alkaline solution to give useful products (fertilizers) and control the air pollution.
7. Mother liquor re-cycling reduces the yield by 50% and is hence not recommended at this stage.
8. Mother liquor can be treated with alkalis to get useful products like glauber salt, ammonium sulphate etc.

REFERENCES:

1. http://search.conduit.com/?SearchSource=10&ctid=CT2776682
2. http://www.google.co.in/
3. *http://www.nandinichemical.com/publications/molassesbasedprojects.html*
4. Lourdes R. Fragante, Lactic acid from molasses, RDE abstract of completed research,Sugar and co-products utilization/molasses,ibid
5. TIFAC Report on "Technology for Production of Lactic Acid" 2001
6. Chemicals Weekly, July 5,2011 pp.239-273
7. Ali. S. et al (2002), Production of citric acid by *Aspergillus niger* using cane molasses in a stirred fermentor, Electonic Journal of Biotechnology, 5, Issue December 15th 2002

8. Ikram-ul-Haq et al (2004), Citric Acid production by selected mutants of *A. niger* from cane molasses, Bioresource Technology 93:125-130
9. Bhatt S.M. and S.K.Srivastava,Lactic Acid Production from Cane Molasses by *Lactobacillus delbrueckii* NCIM 2025 in Submerged Condition: Optimization of Medium Component by Taguchi DOE Methodology, Food Biotechnology, 22:115–139, 2008
10. Aksu Z. and T.Kutsal,Lactic acid production from molasses utilizing *Lactobacillus delbrueckii* and invertage together, Biotechnology Letters , Vol.8, No.3, pp.157-160,1986
11. James M. Church, Ruth Blumberg, Synthesis of Tartaric Acid, *Ind. Eng. Chem.*, 1951, *43* (8), pp 1780–1786
12. W. E. Barch, The Oxidation of 5-Keto-gluconic Acid with Nitric Acid in the Presence of Vanadium, J. Am. Chem. Soc., 1933, 55 (9), pp 3653–3658
13. R. Klasen, S. Bringer-Meyer and H. Sahm, Biotechnology and Bioengineering 40, 183, 1992.
14. Agrawal, Prasad, H. and Rao, M.B., 1979. "Influence of different processing conditions on the production of oxalic acid from cellulosic materials by nitric acid oxidation. Ind. J. Technol. 17, pp. 11–15.
15. Bretschneider, R.B.,Kopriva, J.. Markova and Z.Ondrackova, " Oxidation of Sucrose by nitric acid",Listy Cuker 1966-82,215-219,1967,36-39,and 227-232.
16. Bernsten,R.A.,and E.F. Degering," Oxidation of glucose with oxides of Nitrogen "Sugar Industry Abstracts,1949,11,87,258
17. Bose , S., S.Mukherjee,and A.N. Srivastava," Production of oxalic acid from sugarcane molasses", Proc.4th Joint Conv. S.T.A.I., D.S.T.A. and S.I.S.S.T.A.,G-17-21,1971
18. Chia, K.S. and C.C.Lin," Oxidation of Carbohydrates-synthesis of Glucosachharic Acid from sucrose",Chemistry, Taiwan,1962,98-104
19. Deshpande, S.D. and Vyas, S.N., 1979. Oxidation of sugar to oxalic acid and absorption of oxides of nitrogen to sodium nitrite. Ind. Eng. Chem. Prod. Res. Dev. 18 1, pp. 69–71.
20. Dube, S.K., Vasudevan, P. and Khandelwal, B., 1982. Oxalic Acid Manufacture. J. Chem. Tech. Biotechnol. 32, pp. 909–919.
21. Guo, W., Jiang, H. and Huaxue, Yu.N., 1997. Manufacture of oxalic acid from beet waste slag by hydrocarbonylation at atmospheric pressure. J. CA 45 4, pp. 240–241.
22. Gupta, K.C., Suri, S.K. and Bose, S., 1980. Oxalic acid from deionised sugarcane juice. Proc. STAI 44, pp. 119–125.
23. Gürü, M., et al., 1991. eker Pancarı Melasından Nitrik Asitle Okzalik Asite Yükseltgenmesinde Optimum artların Belirlenmesi. VII. Kimya ve Kimya Mühendisli i Sempozyumu, 2, 464, Gazima usa
24. Gürü, Metin, Ali Y. Bilgesüb and Vecihi Pamuka, Production of oxalic acid from sugar beet molasses by formed nitrogen oxides ,Bioresource Technology,Volume 77, Issue 1, March 2001, Pages 81-86
25. Hamel, R., Levasseur, R. and Appanna, V.D., 1999. Oxalic acid production and aluminum tolerance in *Pseudomonas fluorescens*. J. Inorg. Biochem. 76 2, pp. 99–104.
26. Mane, J.D., Jadhav, S.J. and Ramaiah, N.A., 1986. Production of oxalic acid from dry powder of parthenium hysterophorus L. J. Agricultural Food Chem. 34 6, pp. 989–990.
27. Mane, J.D., Modak, H.M., Ramaiah, N.A. and Jadhav, S.J., 1988. Utilisation of sugarcane trash and other cellulosic wastes for production of oxalic acid. Biological Wastes 25, pp. 171–176.

28. Nowak, P., 1986. Technology for obtaining oxalic acid from sugar by a pressure method. Przemysl Chemiczny 65 9, pp. 488–490.
29. Preen, F.H.E., "Oxalic acid from wood waste", Chemical age,1942-32,72,
30. Raha A.C., R.B.Nigam,and P.Sanyal : Production of oxalic acid from gur", Proc.S.T.A.I.,44(part 2),G-1-6,1976
31. Radulescu, G., Moise, M.-I., Cornos, L., Vapler, G., 1993. Preparation of oxalic acid by the thermal decomposition of sodium formate followed by cation exchange. Patent CA Section:45, Rom. RO 106878 B1 30: 4
32. Rujiter, G.J.G. et al., 1999. Oxalic acid production by *Aspergillus niger*: an oxalate-non-producing mutant produces citric acid at pH 5 and in the presence of manganese. Microbiology 145 9, pp. 2569–2576.
33. Sanyal, P., Ray, A.K. and Shukla, N.P., 1981. Oxalic acid from gur-effect of catalyst. Proc. STAI 45, pp. 21–26.
34. Sattar, A., Muhammad, D., Ashraf, M., Khan, S.A. and Bhatty, M.K., 1988. Oxidation of molasses to produce oxalic acid. Pakistan J. Sci. Ind. Res. 31 3, pp. 163–166.
35. Shukla, N.P., Ray, A.K., Sanyal, P., Oxalic acid from sugarcane molasses: process development and waste utilisation. Proc. 46th Convention STAI, pp. 93–101,1982
36. Sullivan, J.M., Williard, J.W., White, D.L. and Kim, Y.K., 1983. Production of oxalic acid via the nitric acid oxidation of hardwood (Red Oak) sawdust. Ind. Eng. Chem. Prod. Res. Dev. 22, pp. 699–709.
37. Scott., W.W.,Standard Methods of Chemical analysis, Fifth Ed., Vol.1, The Technical Press, London,644
38. Snell, F.D. and F.M.Biffen,Commercial Methods of analysis, First Ed., McGraw Hill Book Co., New York,1944,140
39. Tanz and Kao, Chemical Abstracts,1944,355466
40. Vishwalingam,K.,T.V.Suramaniam and B.Jagannadhaswamy," Process oxidation of ethylene glycol to oxalic acid in liquid in the presence of nitric acid" Ind. Chem.Engr.1980-22(4),11-13

(338f) Modeling the Hydrolytic Degradation and Distribution of Polymer Species In Biopolymers

Justin T. Kaffenberger, Huajiang Huang, Ulrike Tschirner, Ben Schroeder, Waleed Al-dajani and Shri Ramaswamy, Bioproducts and Biosystems Engineering, University of Minnesota, St Paul, MN

Modeling thehydrolytic degradation and distribution of polymer species in biopolymers

Justin Kaffenberger, Huajiang Huang, Ulrike Tschirner, BenSchroeder, Waleed Al-dajani, and Shri Ramaswamy

Department of Bioproducts and Biosystems Engineering,University of Minnesota, St. Paul, MN 55108

Abstract:

The US Department of Energy in its Plant/Crop-BasedRenewable Resources Report 2020 has put forth an initiative to have 10 percent ofchemical feedstock coming from plant-based renewable sources by 2020 and afurther increase to 50 percent by 2050.[1] Meeting these targets will require productionof current chemical feedstocks from renewable sources, such as the dehydrationof plant-based ethanol to produce ethylene.It will also require the development of alternative feedstocks that cansupplant the petroleum-based ones that currently predominate. One example of a chemical feedstock ofincreasing importance is lactic acid.

Lactic acid is a metabolite formed during glucosefermentation by many organisms. Due toits versatility, lactic acid has the potential to be a major renewable chemicalfeedstock. It can be esterified to formlactate esters, reduced to propylene glycol, or dehydrated to acrylic acid.[2] It can also be homopolymerized to formpolylactic acid (PLA) via condensation reaction. The resulting polymer exhibits mechanicalproperties similar to many petroleum-based polymers, including polyethyleneteraphthalate, polyethylene, and polystyrene.

The degradability of PLA is critical to its currentcommercial success. In disposable foodpackaging and service ware, the degradability and bio-absorptivity of thedegradation product of PLA makes it marketable as a green and environmentally-friendlyproduct. In biomedical applications, thedegradability and biocompatibility of PLA are essential. As PLA's ability to degrade is a major factorin its application functionality, understanding the fundamentals of thisprocess is critical to all stages of PLA manufacture and end-use application.

To this end, a mathematical model has been developed thatpredicts the time dependent change in the average molecular weight ofpolylactic acid (PLA) of varying crystallinity as it hydrolytically degrades ata given temperature. This model includes three possible hydrolysis reactionroutes: random chain scission, end scission (monomer removal from the acid endgroup, and monomer removal from the alcohol end group). Apparent rate constants for these reactionpathways were determined from experimental data. The resulting values suggest that,independent of the crystallinity of the starting PLA, acid end groups arehydrolyzed more rapidly than either alcohol end groups or random chainscission. This may help develop suitableapproaches to control the polymer properties and degradation and help developwider application for PLA products.

See more of this Session: Sustainability and Biobased Materials
See more of this Group/Topical: Forest and Plant Bioproducts Division

(338g) Experimental Study On Bast Fiber Reinforced Soy-Based Polyurethane

Shanshan Huo and Chad A. Ulven, Mechanical Engineering, North Dakota State University, Fargo, ND

Growing importance has been placed on the research of natural fibers and resins derived from natural resources as the need for reducing dependency on petroleum-derived products has increased. Natural fibers and biobased resins are potentially inexpensive, renewable resources capable of increasing the specific strength, specific stiffness, and damage tolerance of a variety of composites.

Polyurethane is one of the most versatile and intensively used industrial materials. Bio-based polyols developed from vegetable oils such as soybean oil, canola oil, palm oil and castor oil, are now being used in industry. Soy-based polyurethanes normally exhibit equivalent or improved physical and chemical properties due to the hydrophobic nature of triglycerides. Cellulosic fibers as an alternative for fiberglass have been studied by many researchers. Chemical modifications can be applied to reduce the amount of hydroxyl groups in cellulosic fiber to increase the hydrophobicity and improve the interfacial properties between fibers and matrices. Distillers dried grains with solubles (DDGS) is a by-product of corn-ethanol. Their potential use as reinforcing filler in biocomposites has been studied in thermoplastic composites, but not to a great extent in thermoset composites.

In this study, bast fiber reinforced soy-based polyurethane composites with DDGS filler are produced. The mechanical properties of soy-based PU matrix, soy-based PU matrix with DDGS, and cellulosic fiber reinforced soy-based PU composites with DDGS are investigated. The thermal properties of the matrix system and their composites are evaluated.

See more of this Session: Sustainability and Biobased Materials
See more of this Group/Topical: Forest and Plant Bioproducts Division

(338h) Plant Seed Oils and It's Derivatives Based Value-Added Products From Cenral India Region of the India

Bhalchandra Vibhute, Department of Oil Technology, Laxminarayan Institute of Technology,Nagpur (Maharashtra),India, Purna, India and Dr.Anand S. Kulkarni, Dept.of Oil Technology, Laxminarayan Institute of Technology,Nagpur(Maharashtra),India, Purna, India
See more of this Session: Sustainability and Biobased Materials
See more of this Group/Topical: Forest and Plant Bioproducts Division

Unavailable Paper

(345a) Patient-Specific Design of Polymeric Delivery Vehicles for Anti-HIV Microbicides

Thora W. Whitmore[1], Taylor Wilson[2], Amber Markey[1], Ichie Osaka[3], P. Scott Hefty[3], Kyle V. Camarda[2] and Sarah L. Kieweg[4], (1)Bioengineering, University of Kansas, Lawrence, KS, (2)Chemical and Petroleum Engineering, University of Kansas, Lawrence, KS, (3)Molecular Biosciences, University of Kansas, Lawrence, KS, (4)Mechanical Engineering, University of Kansas, Lawrence, KS

The purpose of this research is to design a novel microbicide delivery mechanism to prevent the transmission of HIV/AIDS to women in developing countries that will be both readily available and affordable. This delivery vehicle, either a gel or a cream, is usually polymeric. The design of these polymeric vehicles involves optimizing the molecular structure subject to specific property requirements. Computational Molecular Design (CMD) provides the methodology for this optimization problem. In the application of CMD, Quantitative Structural-Property Relationships (QSPRs) are generated and validated to relate the structure of possible polymers to their physicochemical properties. QSPRs are generated by regressing molecular topological indices, a quantitative method to represent the structure, to measured physiochemical properties. Such property data for 14 different polymeric liquids of different chain length prepared at the same concentration have been collected. The physiochemical properties considered for the QSPRs are power-law consistency, shear-thinning index, storage modulus, loss modulus, compressibility, anti-chlamydial activity, toxicity, and osmolality. Correlations were generated relating each of these properties to the structures of the 14 distinct polymeric liquids, and were then cross-validated statistically. These property prediction models were incorporated into the objective function of an MINLP formulation. Ideal values of these physicochemical properties were obtained from separate *in vitro* and *in vivo* experiments as well as review of the relevant literature. These data were then incorporated into the objective function, and the MINLP formulation was solved for local optima. This generated novel structures possessing physical property values near the ideal values. This problem was solved several times, with different ideal property values corresponding to differences in size and mechanical properties of vaginal tissue among different groups of women. Examples of these novel structures are provided.

See more of this Session: Advances In Biomaterial Evaluation
See more of this Group/Topical: Materials Engineering and Sciences Division

(345b) Multifunctional Zwitterionic Nanogels for Targeted Drug Delivery and Renal Clearance

Lei Zhang, Department of Chemical Engineering, University of Washington, Seattle, WA and Shaoyi Jiang, Chemical Engineering, University of Washington, Seattle, WA

Zwitterionic polymers have been shown the excellent ability to prepare non-fouling surface and stealth nanoparticles. Poly(carboxybetaine methacrylate) (pCBMA) is one important type of zwitterionic polymer, beside the non-fouling properties, it can also be easily functionalized, so it is an ideal type of material to develop long circulating nano-carriers for targeted drug delivery. In this work, degradable pCBMA nanogels were prepared, the nanogels could load different imaging reagents and model drugs by suitable methods. *In vitro* studies show the nanogels poccess extraordinary stability in different environment including saline solutions or undiluted blood serum, and they have very low interaction with immune cells. After conjugated with a targeting ligand, the cell uptake of the nanogels was greatly increased, but their stealth properties were not compromised. By using the degradable crosslinker, the nanogels can be degradable at certain situation, rendering the simultaneous release of the loaded model drug and the disassembling of the nanogels. The degraded products could be removed from the body via renal clearance. The above results suggest the pCBMA nanogels could be an ideal and safe platform for targeted drug delivery.

See more of this Session: Advances In Biomaterial Evaluation
See more of this Group/Topical: Materials Engineering and Sciences Division

In Vitro Three-Dimensional Fluorescent Mouse Embryonic Stem Cell-based Models for Embryotoxicity and Teratogenecity Screening

Abstract

Numerous animals are commonly used to analyze the mutagenic, embryotoxic or teratogenic potential of environmental factors, pharmaceuticals and xenobiotics *in vivo*. In an attempt to reduce animal tests, mouse embryonic stem cell-based models are highly desirable. ES cells possess two unique characteristics including pluripotency and unlimited proliferation. Both of the properties are associated with survivin, which belongs to IAP (inhibitors of apoptosis protein) family and specifically is expressed in cancer cells and embryonic stem cells. In this work, we developed an *in vitro* three-dimensional (3-D) dynamic embryonic stem (ES) cell-based model for indentifying potential chemicals associated with embryotoxiciy or teratogenecity. In order to achieve real-time quantification of survivin, a survivin-reporter gene system was constructed. RT-PCR and neuronal differentiation of ES cells results demonstrated that survivin could be estimated by measuring green fluorescence protein (GFP) expression using a commercial well-plate reader. In order to rule out false positive hits due to the cytotoxic effects of compounds, a constitutive cytomegalovirus (CMV)-reporter system was also developed. Furthermore, bioactivity of well-known teratogens including acrylamide and 4-tetracycline (4-TH) in survivin expression was studied using the ES cell-based model. A good *in vitro-in vivo* consistency was found regarding the toxicity of acrylamide. Acrylamide with a concentration of 25 mg/L significantly enhanced survivin expression, which is consistent with previous *in vivo* result that 20 mg/L was the lowest concentration when teratogenic effects were observed using animals. Besides, compared to controls and

4-TH-treated cells, acrylamide-treated cells exhibited different cell morphologies. Additionally, 4-TH at a concentration of 0.1 mg/L did not significantly influence cell growth in comparison with controls. However, 4-TH did reduce survivin expression, which may explain its embryotoxic effects on the development of fetuses. These results showed that the 3-D dynamic ES cell-based model could serve as a reliable, sensitive and robust method to assess embryotoxic and teratogenic potential of chemical compounds.

Keywords: Embryonic stem cells; Survivin; Teratogenecity; Embryotoxicity; Green fluorescence protein

(345d) In Vivo Analysis of Insulin Delivery Using CPEs Designed Via In Silico QSPR Approach

Ayman Grada, Khaled A. M. Gasem and Sundararajan. V Madihally, School of Chemical Engineering, Oklahoma State University, Stillwater, OK

The uneasiness and pain of multiple injection therapy courses for diabetic patients and the non-compliance of patients with poorly controlled type 2 diabetes have prompted the development of alternative, pain-free delivery methods. Transdermal drug delivery is gaining significance in the recent years due to its advantages over other invasive methods. An insulin transdermal patch placed on the skin may be able to deliver sustained physiological levels of basal insulin in a pain-free manner; however, since insulin is a relatively large molecule, it does not penetrate the skin easily. Altering the characteristics of the skin by using chemicals called penetration enhancers is one of the promising ways to breach the skin's barrier to drugs. Using this concept, we have reported on the methodology of predicting CPEs using improved algorithms for non-linear, quantitative structure-property relationship (QSPR) models based upon representative molecular properties [1-7]. The predicted molecules were extensively tested for cytotoxicity and *ex vivo* delivery. This study focuses in evaluating the *in vivo* efficacy delivering insulin across the skin with and without CPEs in a rodent model. Rats were starved eight hours before the application of the patch and then blood glucose levels were monitored. After five days, we evaluated the overall body weight change, hepatotoxicity and skin toxicity. Our findings through animal trials indicate proven efficacy of the newly designed CPEs.

References

[1] Yerramsetty KM, Rachakonda VK, Neely BJ, Madihally SV, Gasem KAM. Effect of Different Enhancers on the Transdermal Permeation of Insulin Analog. International Journal of Pharmaceutics. 398 (1-2): 83-92. 2010.

[2]. Yerramsetty KM, Neely BJ, Madihally SV, Gasem KAM. A Skin Permeability Model of Insulin in the Presence of Chemical Penetration Enhancer. International Journal of Pharmaceutics. 388(1-2):13-23. 2010.

[3] Godavarthy SS, Yerramsetty KM, Rachakonda VK, Neely BJ, Madihally SV, Robinson Jr RL, Gasem KAM. Design for Improved Permeation Enhancers for Transdermal Drug Delivery. Journal of Pharmaceutical Sciences. 98(11):4085-4099. 2009.

[4] Neely BJ, Golla S, Whitebay E, Madihally SV, Robinson Jr RL, Gasem KAM. Non-linear Quantitative Structure-Property Relationship Modeling of Permeation Coefficients. Journal of Pharmaceutical Sciences. 98(11):4069-4084. 2009.

[5] Golla S, Madihally SV, Robinson RL Jr., Gasem KAM. Quantitative Structure-Property Relationship Modeling of Skin Sensitization: A Quantitative Prediction. Toxicology In vitro. 23(3):454-65. 2009.

[6] Golla S, Madihally SV, Robinson RL Jr., Gasem KAM. Quantitative Structure-Property Relationship Modeling of Skin Irritation. Toxicology In vitro. 23(1):176-184. 2009.

[7] Rachakonda VK, Yerramsetty KM, Madihally SV, Robinson RL Jr, Gasem KAM. Screening of Chemical Penetration Enhancers for Transdermal Drug Delivery using Electrical Resistance of Skin. Pharmaceutical Research. 25(11):2697-704. 2008.

See more of this Session: Advances In Biomaterial Evaluation
See more of this Group/Topical: Materials Engineering and Sciences Division

(345e) Evaluating a Library of Modified Alginates to Reverse Diabetes Through Islet Encapsulation

Kaitlin M. Bratlie, Arturo J. Vegas, Thema M. Vietti, Alan Chiu, Nimit Dholakia, Robert Langer and Daniel G. Anderson, Chemical Engineering, MIT, Cambridge, MA

According to the American Diabetes Association, diabetes affects over 20.8 million people in the United States or about 7% of the population. Current treatments involve daily injection of insulin from recombinant human or animal sources. Controlling diabetes has proven difficult since regulation of insulin secreted by the β-cells of the pancreatic islets in response to blood glucose is a highly dynamic process. We are particularly interested in developing materials for reducing host-responses for polymer encapsulated insulin producing β-cells, which has potential as a type I diabetes therapeutic. Encapsulating β-cells in polymers to evade host-responses was first proposed by Lim & Sun[1] using alginate, a natural polymer derived from seaweed. Elliott *et al.*[2] found that alginate encapsulated porcine islets remained viable 9.5 years after transplantation. However, the alginate particles were encapsulated in fibrous tissue, resulting in poor diffusion of insulin into the patient's blood stream. Based on these findings, our approach has been to develop a library of modified alginates for encapsulating β-cells that are more biocompatible than existing alginate.

In assessing biocompatibility, our goal is to develop *in vivo* fluorescence imaging for assessing the host-response to libraries of implanted biomaterials to determine which chemical functionalities mitigate the foreign body response. We have developed a method for imaging the foreign body response through cathepsin activity imaging. Through this imaging platform, we are able to assess early time-point responses to implanted materials. We have examined a 1000 member library based on alginate for encapsulating β-cells. Histological analysis of these implanted materials complements early time point imaging in assessing host responses.

Novel alginates identified as biocompatible through *in vivo* imaging and histology were used to encapsulate β-cells isolated from rats and were implanted into diabetic mice. Several polymers were found to be as efficacious as unmodified alginate in stabilizing blood glucose levels for periods longer than one month. Future plans in this work include developing a more apt mouse model to differentiate between unmodified and modified alginates.

References

1. Lim, F.; Sun, A. M., Microencapsulated islets as bioartificial endocrine pancreas. *Science* **1980,** *210*, 908-910.

2. Elliott, R. B.; Escobar, L.; Tan, P. L. J.; Muzina, M.; Zwain, S.; Buchanan, C., Live encapsulated

porcine islets from a type 1 diabetic patient 9.5 yr after xenotransplantation. *Xenotransplantation* **2007,** *14* (2), 157-161.

See more of this Session: Advances In Biomaterial Evaluation
See more of this Group/Topical: Materials Engineering and Sciences Division

(345f) A Fluorescence Correlation Spectroscopy Study of Hindered Probe Diffusion In Complex Media

Silviya Petrova Zustiak[1], Ralph Nossal[2] and Dan Sackett[2], (1)Laboratory of Integrative and Medical Biophysics, National Institute of Health, Bethesda, MD, (2)National Institute of Health, Bethesda, MD

Fluorescence Correlation Spectroscopy (FCS) has been recognized as a promising technique for evaluating the movement of molecules within complex media. Because FCS is intrinsically a number fluctuation technique, the target material is studied at low concentrations and small volumes, in contrast to other physical techniques such as small angle neutron scattering (SANS) and nuclear magnetic resonance (NMR).

Here, we present application of FCS for the study of hindered probe diffusion in polymeric solutions. This study was mainly aimed at recapitulating the intracellular environment, where diffusion of molecules is impaired predominantly by crowding and charge-mediated binding. Intracellular diffusion of molecules has long been of interest in the area of targeted drug delivery as well as for describing basic cellular processes and yet, it is still poorly understood. In particular, the combined effect of crowding and non-specific binding has rarely been addressed, partly due to the inability to decouple their individual contributions in the heterogeneous cell environment.

In this work, we developed a model system in which both molecular crowding and charge-mediated binding were addressed independently in a controlled manner. In particular, using FCS, we obtained the translational diffusion coefficient of the positively charged protein, RNase, in solutions of dextrans of various charges (binding) and concentrations (crowding). In agreement with existing data, we observed an overall 5-fold decrease in RNase diffusivity at the highest concentration of dextran, where binding accounted for 75% and crowding for 25% of the decrease. Interestingly, binding decreased RNase diffusivity by 32% even at 0.4 μ M dextran. In contrast, crowding affected diffusivity only above a crowder concentration of 20 μ M. Further analysis revealed that 100 μ M crowder, as compared to 1 μ M binder, was needed to achieve equivalent reduction in RNase diffusivity. However, the data suggested that at a higher crowder concentration (e.g., similar to that in the cell), crowding could "mask" the effect of binding.

This is one of the first studies to highlight the relative contribution of non-specific binding and crowding to hindered diffusion in complex media and thus can facilitate future understanding of molecular transport implicated in key cellular processes. However, in order to translate this study to an *in-vivo* setting, we would need to take into account the turbidity of the media, since it can compromise the accuracy of the FCS measurements. Therefore, we further present the impact of multiple light scattering due to background turbidity on FCS measurements.

A Spinning Device for the Mechanical Characterization of Soft Tissues and Polymers

Matthew A. Reilly, University of Texas at San Antonio, San Antonio, TX, United States
Philipp Martius, University of Rostock, Rostock, Germany
Harvey J. Burd, University of Oxford, Oxford, United Kingdom
Oliver Stachs, University of Rostock, Rostock, Germany
Rudolf Guthoff, University of Rostock, Rostock, Germany

Abstract

Nondestructive evaluation of the elastic properties of soft tissues and polymers may be achieved by spinning a sample about an axis of symmetry. If the angular velocity is known, the centrifugal body force can be determined at each point in the sample. Sample deformation may be measured by comparing high resolution digital photographs at rest and at known velocities using image analysis algorithms. Here, we present an automated testing method which offers repeatable measurements by photographing the sample at eight fixed angular locations both at rest and at speed. The apparatus and imaging methods are validated using a rigid sample of known dimensions. A climate control system which allows modulation of both temperature and humidity is also detailed.

An analysis method was developed which allowed estimation of the linear elastic properties of general spheroidal elastic bodies using a separable response surface method. This method computed deformations in the sample as a function of nondimensionalized parameters related to the sample geometry and mechanical properties, apparatus geometry, and angular velocity.

Further validation of the device was achieved using a multi-level, full factorial experimental design and corresponding analysis. Here, we measured the effects of storage (duration and solution) and testing (temperature, humidity, and angular velocity) conditions on the geometry and mechanical properties of 99 porcine ocular lenses. Confocal scanning laser microscopy was used to determine the critical angular velocity above which structural damage to the lens occurred.

Introduction

Testing of soft tissues often requires the development of bespoke instrumentation tailored to the tissue's geometry and mechanical properties. A climate control system must also be employed to maintain the tissue's properties during testing. These systems may also be used to characterize potential prosthetic materials with similar mechanical properties.

One such tissue is the ocular lens. The lens is a soft (elastic modulus 10^2-10^4 Pa) collection of cells surrounded by a relatively stiff collagenous capsule. The mechanical properties of both the lens matrix and capsule vary with age, contributing to a breakdown in the overall function of the eye. Because of the complex structure, softness, and geometry of the lens, mechanical testing of the lens using traditional techniques is futile.

We therefore developed an automated instrument which combines modern control technology and digital photography to monitor the deformation of the lens due to centrifugal stresses exerted due to spinning the lens about its axis. An environmental control chamber was also designed to ensure that the temperature and humidity about the lens were maintained at the necessary level throughout.

Methods

Porcine lenses were dissected and stored for up to one week in the whole eye, in balanced salt solution, and in M-199 media. Lenses were then tested at both room and body temperature. Fresh lenses tested at body temperature were used as a baseline for comparison.

A brine solution was used to fix the equilibrium vapor pressure of the chamber near the vapor pressure of the lens. High-power resistors were used to heat the sample chamber to body temperature. An infrared thermometer was used to monitor the surface temperature of the lens.

A motor controller was used to spin each lens up to a known angular velocity, then triggering a digital camera and flash at known angular locations. Image analysis was applied after the experiment to objectively determine lens dimensions. Changes in these dimensions were measured at a variety of angular velocities.

Statistical analysis was performed to determine whether the lens' mechanical properties were affected by storage time, storage solution, and test temperature. Lenses were also weighed before and after testing to examine the efficacy of the environmental control chamber in maintaining the lens during testing.

Results and Discussion

Porcine lens deformation due to spinning was significantly impacted by storage time, storage solution, and test temperature. Lenses stored in M-199 media for up to 24 hours post mortem were most similar to the fresh lens in both initial geometry and mechanical properties. Temperature had a significant impact on the mechanical properties of the lens, indicating that previous tests conducted at room temperature may give misleading results.

These findings establish a best practice for storage of lenses post mortem and indicate that lens properties change significantly after less than 24 hours in the eye. Future testing of human cadaver tissues must account for post mortem changes to the mechanical properties of the lens which will inevitably arise after prolonged storage. Additional storage media should be investigated to examine whether they may maintain the lens in its native state through the

72-hour post mortem period usually encountered by researchers waiting for the arrival of human tissue.

Acknowledgements

This study was supported in part by the German Research Foundation DFG (Transregio 37, Micro- and Nanosystems in Medicine – Reconstruction of Biological Functions) and Federal Ministry of Education and Research BMBF (REMEDIS).

(350a) Light-Curable Bioactive Polymeric Composite Glues for Bone Defect Treatment

Alexander Stepuk, Oliver Schneider, Dirk Mohn and Wendelin J. Stark, Institute for Chemical and Bioengineering, ETH Zurich, 8093 Zurich, Switzerland

Recentresearch on bone implant materials targets combination of resorption andsufficient mechanical properties [1]. Photopolymerizabledegradable resins are particularly attractive to clinicians as they allowaccurate control on initiation of the hardening process. However a majordrawback is the difficulty to set them *in vivo*. We will demonstrate theintroduction of a reactive filler into a light-curable, clinically used polymerand its effect on adhesion to wet bone surfaces in a close to *in vivo*setup (see figure) [2].

Inthis present study, we combine an advanced light-curable clinical polymer mixtureconsisting of bisphenol A diglycidylmethacrylate and tri(ethylene glycol) dimethacrylate (BisGMA/TEGDMA) with most recently developed reactive amorphous calciumphosphate (a-CaP) nanoparticles [3] as filler.Here we will present an investigation of *in vitro* biomineralization, compressivestrength and preliminary tests on adhesion to bone. The hypothesis within thisstudy is that the performance of the adhesive polymer can be increased throughthe incorporation of the inorganic filler. The incorporation of 20 wt.% amorphous calcium phosphate (a-CaP) nanoparticles inthe adhesive polymer significantly improved the adhesion to wet bone.

Thiswork gives access to light-curable polymers with significantly improved boneadhesion. New frontiers and possibilities in complex bone defects treatmentusing light-curing polymeric biomaterials to be discussed.

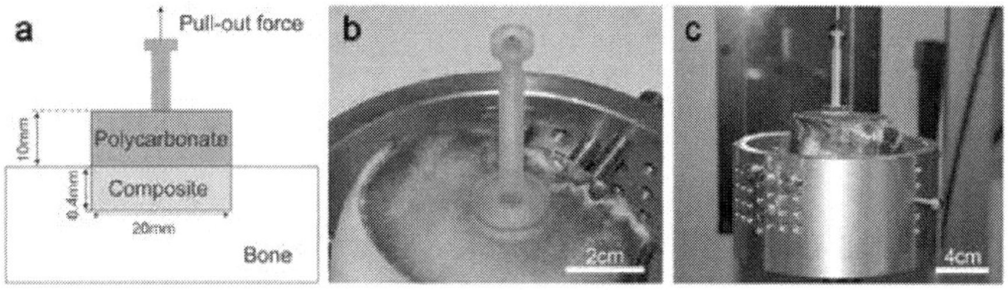

Figure 1: Adhesion to bone: (a) Schematictest setup. (b) Image of a glued polycarbonate rod on a fresh cut cowhip bone after light-curing of the sample. (c) Pull-out test in a conventionaltesting machine.

References: [1] J.L. Ifkovits, J.A.Burdick, *Tiss. Eng.*, **2007**, *13(10)*,2369.

[2] O. D. Schneider, A. Stepuk,D. Mohn, N. A. Luechinger, K. Feldman, W. J. Stark, *Act. Biomat.* **2010**,*6*, 2704.

[3] T.J. Brunner, M. Bohner, C. Dora, C. Gerber, W.J.Stark, J. Biomed. Mater. Res. Part B, **2007**, *83*, 400.

See more of this Session: Biomaterials at Interfaces
See more of this Group/Topical: Materials Engineering and Sciences Division

(350b) Model Myelin Membranes: Asymmetry, Domains, Miscibility and Equilibrium

Joesph A. Zasadzinski, Department of Chemical Engineering, University of California, Santa Barbara, CA, Younjin Min, Chemical Engineering, Massachusetts Institute of Technology, Cambridge, MA, Dong-woog Lee, Chemical engineering, University of California at Santa Barbara, Santa Barbara, CA, Prajnaparamita Dhar, Chemical Engineering, University of Kansas, Lawrence, KS, Arun Ramachandran, Department of Chemical Engineering and Applied Chemistry, University of Toronto, Toronto, ON, Canada and J. Israelachvili, Department of Chemical Engineering, University of California at Santa Barbara, Santa Barbara, CA

This myelin membrane bilayer is asymmetric, with different composition and phase behavior on either side of the bilayer. Both cytoplasmic and extracellular sides of the myelin phase separate into liquid ordered and liquid disordered domains that are miscible at low surface pressures. With increasing surface pressure the domains mix; for the cytoplasmic side, the transition is nearly immediate, the extracellular side transitions via a critical point at which the domains coalesce, elongate and composition fluctuations persist into the homogenous phase. We show that the domain distribution on the cytoplasmic side is consistent with the equilibrium distribution predicted through a balance of domain line tension, electostatic self-energy and the entropy of mixing. From the size distribution we extract the line tension and dipole density difference and show that the line tension is extremely low in complex lipid mixtures compared to single component monolayers. Subtle differences between normal myelin and myelin from animals with multiple sclerosis may lead to differences in the adhesion and fluidity of the membranes and correlate with the origination of this disease.

See more of this Session: Biomaterials at Interfaces
See more of this Group/Topical: Materials Engineering and Sciences Division

(350c) Hierarchical Biomaterials Based On the Self-Assembly of the Silk Protein Sericin

Nicholas Kurland[1], Subhas C. Kundu[2] and Vamsi K. Yadavalli[1], (1)Chemical and Life Science Engineering, Virginia Commonwealth University, Richmond, VA, (2)Biotechnology, Indian Institute of Technology, Kharagpur, India

Silk consists primarily of two proteins - a fibrous core protein, fibroin (~70%) and a glue protein, sericin (~25-30%), which envelops the fibroin fibers with sticky layers thus helping in the formation of cocoon, achieved by cementing the silk fibers together. Sericin, a water soluble protein has traditionally been discarded in silk processing, despite great potential for use as a biomaterial. Here we present results showing that this glue protein, sericin has the ability to form self-assembled nano and micro-structures with hierarchical self-similarity across length scales in the form of a diffusion-limited assembly. High-resolution atomic force microscopy was used to image and take nanomechanical measurements of the self-assembled protein. The self assembly of the sericin protein was analyzed based on the theory of diffusion limited aggregation to explain the fractal nature of these architectures observed. Using nanoindentation, the stiffness of these nanostructures was measured. The unique physical behavior, mechanical properties and fractal nature of this glue protein may represent a key step in understanding its biological relevance of sericin as well for directing the design and formation of silk-based biomaterials.

See more of this Session: Biomaterials at Interfaces
See more of this Group/Topical: Materials Engineering and Sciences Division

(350d) Efficiency of a Mixed Copolymer As a Non-Fouling Surface

Tapashree Tah and Matthew T Bernards, Chemical Engineering, University of Missouri, Columbia, Columbia, MO

Non-specific protein binding is a very important factor in biomaterial application. To address this issue different kinds of non-fouling surfaces have been investigated. In this paper we report on the nonfouling properties of a mixed-charge polyampholyte copolymer composed of positively charged [2-(acryloyloxy) ethyl] trimethyl ammonium chloride (TMA) and negatively charged 2-carboxy ethyl acrylate (CAA) monomers. The thickness of the polymer brush coating was controlled by varying the surface initiated atom transfer radical polymerization conditions. The variation of nonspecific protein adsorption with thickness was noted. Protein conjugation was then done on each of the surfaces. The conjugation experiments clearly demonstrate that TMA:CAA copolymer is suitable for protein conjugation and nonfouling. Fetal bovine serum (FBS) adsorption onto the surface was also characterized and indicated that nonspecific protein adsorption from complex media remains in the nonfouling range The results of this investigation indicate that TMA:CAA mixed charge surfaces show promise for biomedical application.

See more of this Session: Biomaterials at Interfaces
See more of this Group/Topical: Materials Engineering and Sciences Division

(350e) Phase Behavior of Lipid-Polymer Nanobubbles

Joseph V. Badami, Chemical Engineering, The City College of New York, New York, NY and Raymond Tu, Department of Chemical Engineering, The City College of City University of New York, New York, NY

The development of microbubble systems has allowed the clinical application of ultrasound to progress as a viable option for medical imaging purposes due to the echogenic properties they demonstrate. Simultaneously, these structures also serve as practical carriers for drug delivery purposes. These systems are typically limited however by overall size which hinders their ability to navigate throughout specific vasculature, such as that commonly seen in tumors. Size and stability of these systems comes as a direct consequence of interfacial interactions that take place amongst shell components and a greater understanding of these interactions is instrumental to the further development of the field. Polymeric materials show great promise as stabilizing agents since they can be finely controlled in both size and composition. In this work we investigate the fundamental aspects of bubble stabilization using polymeric materials to further enhance the lifetime associated with such bubbles. We explore the fundamental phase behavior demonstrated by these shell materials and aim to determine the particular effects they invoke on the elasticity and stability of bubbles at sub-micron scales. Lastly, the microstructure associated with such phase behavior is studied using microscopy techniques in order to gain a greater understanding of the interfacial ordering present.

See more of this Session: Biomaterials at Interfaces
See more of this Group/Topical: Materials Engineering and Sciences Division

(350f) Mixed-Charged, but Overall Neutral Hydrogels As Nonfouling, Mechanically Strong Materials

Sean C. Dobbins and Matthew Bernards, Chemical Engineering, University of Missouri-Columbia, Columbia, MO

A crucial predicament in the field of biomaterials is the often-undesirable, but immediate coating of implants with nonspecific proteins, which adsorb to the implant upon contact with bodily fluids. Nonfouling materials are defined as materials to which no proteins adsorb. Prior research has shown that overall-neutral materials containing a homologous arrangement of mixed-charges exhibit nonfouling properties. This has been shown for materials formed by either the creation of a hydrogel containing a cross-linked, equimolar mixture of positively and negatively charged monomers, or by use of zwitterionic compounds. The goal of this research is to develop a greater level of control over mixed-charge hydrogels through the selective choice of the individually charged monomers. In particular, hydrogels were formed from mixtures of [2-(methacryloyloxy)ethyl]trimethylammonium chloride (TM) and 3-sulfopropylmethacrylatepotassium salt (SA), using a triethyleneglycoldimethacrylate (TEGDMA) crosslinker. Enzyme Linked Immunosorbant Assay (ELISA) experiments were completed to study the impact of the crosslinker density on the nonfouling characteristics of the hydrogels. The results have shown promise, indicating that a variety of hydrogels with ultra-low to no quantifiable adsorption of negatively-charged fibrinogen or positively-charged lysozyme were developed. Swelling characterization and hydration studies were completed to gain insight into mechanical properties of the resulting hydrogels in order to quantitatively study the effects of the concentration of the crosslinker on the mechanical strength of the TM/SA hydrogel. Ongoing research efforts are focused on tailoring these properties based on the concentration of the TEGDMA crosslinker. The hydrogels being investigated in this work will be an important component of future research efforts to develop functionalizable hydrogel implant materials.

See more of this Session: Biomaterials at Interfaces
See more of this Group/Topical: Materials Engineering and Sciences Division

(350g) Tuning the Binding Kinetics of Angiogenic Factors for Revascularization Therapies

John J. Schmidt, Chemical and Biomolecular Engineering, University of Illinois at Urbana-Champaign, Urbana, IL and Hyunjoon Kong, Chemical and Biomolecular Engineering, University of Illinois, Urbana, IL

One common type of ischemic disease is Peripheral Arterial Disease (PAD), which affects nearly 10 million Americans every year. PAD results from the lack of blood supply to the peripheral tissues, typically caused by a blockage of in a peripheral artery. One promising approach to treat PAD involves the growth of new blood vessels from existing blood vessels in the body, termed as angiogenesis, to restore blood flow to the peripheral tissue. A variety of angiogenic growth factors including vascular endothelial growth factor (VEGF), have been extensively studied and shown to play crucial roles in the growth of new blood vessels. VEGF, along with other growth factors, stimulate an angiogenic response through their interaction with endothelial cell surface receptors on endothelial cells and subsequently causing a signaling cascade. Clinically, it is typical to extend the stimulatory effect of VEGF with its dose and duration of administration, but this approach often triggers several side-effects.

This study presents a new way to tune the efficacy of VEGF signaling at a given dose, by controlling binding kinetics of VEGF to cellular receptors. We hypothesized that a molecule which can bind with cellular integrins and growth factors will greatly improve the efficacy of VEGF to stimulating blood vessel formation. We examined this hypothesis using natural polymers (e.g., fibronectin and fibronectin fragments) and a fibronectin-mimicking synthetically-modified polymer, alginate. Interestingly, the co-administration of VEGF with the fibronectin molecules significantly increased the phosphorylation of cellular receptors. The role in mediating the increase in VEGF receptor activation was further related to the increased association between the VEGF and the fibronectin molecules and the cell's surface. Finally, the interaction of these molecules led to both improvements in blood vessel formation in vitro, using an angiogenesis assay, and in vivo, using a mouse model. Overall, this strategy will not only enhance efficacy of vascular medicine but also be readily extended to a wide array of molecular therapies.

See more of this Session: Biomaterials at Interfaces
See more of this Group/Topical: Materials Engineering and Sciences Division

(351a) Multilayer Drug Delivery Coatings for Trauma Relief

Anita Shukla and Paula Hammond, Chemical Engineering, Massachusetts Institute of Technology, Cambridge, MA

Current techniques used to address aspects of traumatic injury, including bleeding, infection, and inflammation, are plagued by a lack of efficacy due to difficulty of application, low drug loading, and lack of control over therapeutic release. In this work, we have focused on the design of efficient and highly therapeutic device coatings that target these conditions. We have engineered multilayer films using the layer-by-layer (LbL) assembly technique to exhibit a range of favorable drug release profiles and high loadings of antimicrobial peptides, antibiotics such as vancomycin, anti-inflammatory drugs such as diclofenac, and hemostatic agents such as thrombin. Both electrostatic and hydrogen bonding interactions were explored as methods of incorporating therapeutics into these film coatings. The drug loading and release properties of these films are strong functions of the film architecture and the LbL assembly technique utilized (spray or dip LbL), which both influence the formation of favorable secondary interactions between film components. All film released therapeutics retained complete functionality *in vitro*. Hemostatic films were also tested *in vivo* and found to be highly effective in promoting blood clotting.

See more of this Session: Biomaterials for Drug Delivery
See more of this Group/Topical: Materials Engineering and Sciences Division

(351b) Development of Biodegradable Hydrogels for the Controlled Release of Antimicrobial and Antioxidant Agents

Andrew L. Vasilakes[1], Dipti Biswal[2], Rebecca Peyyala[3], David A. Puleo[4], J. Zach Hilt[2] and Thomas D. Dziubla[2], (1)Chemical and Materials Engineering, University of Kentucky, Lexington, KY, (2) Chemical Engineering, University of Kentucky, Lexington, KY, (3)Dentistry Research, University of Kentucky, Lexington, KY, (4)Biomedical Engineering, University of Kentucky, Lexington, KY

Antibiotic resistance emergence is a major medical challengedue to the resulting reduced clinical effectiveness of antibiotics. Mitigation of these resistant bacteria inwounds would be greatly beneficial towards the reduction of infection andprognosis of patients. In this research,a multidrug loaded hydrogel is devised which can interfere with the ability of *Staphylococcus aureus* to developantibiotic resistance towards the commonly used antibiotic vancomycin. By acting upon the bacteria with two agentswhich function through independent mechanisms, it is possible to suppress theability of bacteria to propagate as well as evolve resistance via hydrogenperoxide signaling in formed biofilms.Sustained release of the drugs is important as eventually the hydrogelwill have a longer life-span *in vivo*versus a burst-release system, and these two components are shown totime-release by adding them to a tunable biodegradable poly(beta-amino ester)hydrogel. This time-release is shownthrough use of degradation studies in PBS as well as through zone of inhibitionvia hydrogel diffusion onto *S. aureus*seeded blood agar petri dishes. Preliminary data is presented upon the experimental design of showingthe reduction of antibiotic resistance emergence with multidrug loadedhydrogels. The synergistic effect ofvancomycin and catalase released from a hydrogel illustrates that this systemis potentially effective *in vivo* asan infection reducing and preventing medical device.

See more of this Session: Biomaterials for Drug Delivery
See more of this Group/Topical: Materials Engineering and Sciences Division

(351c) Injectable Modular Hydrogels for Drug Delivery

Mathew J. Patenaude and Todd R. Hoare, Chemical Engineering, McMaster University, Hamilton, ON, Canada

Injectable Modular Hydrogelsfor Drug Delivery

Mathew Patenaude and Todd Hoare

Department of Chemical Engineering, McMasterUniversity, Hamilton, Ontario, L8S 4L7

Introduction

The development of novel drug delivery systems is an essential step towardcontrolled site-specific administration of therapeutics within the body. It isdesirable for delivery vehicles to be introduced into the body throughminimally invasive means and, once present at their desired site, thesevehicles should be capable of releasing drug to their intended location at acontrolled rate. Furthermore, it is desirable to develop drug delivery vehiclesthat are capable of *in vivo* degradation once delivery of the drug iscomplete, avoiding the need to surgically remove the vehicle at the end of itsuseful lifetime. Hydrogels are of particular interestfor drug delivery drug delivery applications due to their ability to addressthese needs in addition to their generally good biocompatibility, tunablenetwork structure to control the diffusion of drug out of the network, andtunable affinity for specific drugs[1]. However, hydrogelsare limited for drug delivery applications due to the often quick elution ofdrug from their highly swollen polymer matrices as well as the difficultyinherent in injecting macroscopic hydrogels into thebody.

These challenges can in part be addressed by fabricating hydrogelsbased on poly(N-isopropylacrylamide)(polyNIPAM). These hydrogelsundergo a reversible thermal transition at ~32 °C, causing them to collapse andsubsequently expel water from their network when placed *in vivo*. Uponintroduction into the body, these vehicles are capable of entrapping drugfollowing their thermal collapse, allowing for the elution of a therapeuticinto the surrounding environment at a rate that depends on the diffusioncoefficient of the drug through the network and relative affinity of the drugtoward the hydrogel phase over its aqueous surroundings. To take advantage ofthese thermoresponsive properties, hydrogels must be designed such that they formspontaneously inside the body from liquid-like precursors outside the body.Hydrazone cross-links between aldehyde and hydrazide-functionalized polymer precursors provide a meansof rapid *in vivo* hydrogel formation following their co-injection intothe body. The hydrazone bond offers the additional advantage of being slowlyhydrolyzed at normal physiological pH, allowing for the excretion of polymerprecursors following hydrogel degradation provided that their molecular weightis kept below the renal cutoff (< 40 kDa)[2].Hydrogel degradation, swelling, and thermosensitivitycan further be modified by cross-linking the polyNIPAMoligomers with natural polysaccharides,functionalized to contain the same hydrazide and aldehyde groups. By mixing these synthetic andnatural polymer building blocks in any combination and at any ratio, "modular" hydrogelsmay be generated that possess targeted physical and biological properties aswell as finely tuned drug release characteristics.

Materials and Methods

Hydrogels were fabricated from hydrazide-functionalized poly(NIPAM)co-polymers and aldehyde-functionalizedpolysaccharides, such as carboxymethyl cellulose(CMC), hyaluronic acid (HA), and dextran(dex). Hydrazide-functionalizedpoly(NIPAM) was generated by copolymerizing NIPAM withacrylic acid and subsequently grafting the resulting copolymer with adipic acid dihydrazide usingEDC/NHS chemistry. Aldehyde-functionalizedpolysaccharides were generated using using periodate-mediated oxidation.

The molecular weight of precursor poly(NIPAM)copolymers was determined using aqueous gel permeation chromatography. Thedegree of hydrazide functionality of poly(NIPAM) copolymers was determined through conductometric titration. Polysaccharide aldehyde functionality was quantified with a carbazate-based aldehydedetection assay.

Hydrazide and aldehyde-functionalizedpolymers were mixed at various concentrations via co-injection through a needlealong with bupivacaine, a cationic local anesthetic. These polymers rapidlyform a hydrazone-crosslinked hydrogel network upon mixing. Hydrogel drugrelease properties were determined using a transwellplate technique and quantified using UV/VIS spectrophotometry.Cytotoxicity was assayed using an MTT assay.

Results

Poly(NIPAM)-only hydrogels weresuccessfully synthesized by mixing hydrazide and aldehyde-functionalized poly(NIPAM) oligomers,exhibiting thermal deswelling on the order of 70 %upon heating from 25 °C to 37 °C. Such hydrogels alsoexhibited the classical poly(NIPAM) burst drug releaseprofile, rapidly expelling 90 % of pre-loaded drug via convective mass transferfollowing the rapid collapse of the hydrogel network. This is, to ourknowledge, the first demonstration of an injectableand degradable poly(NIPAM)-basedhydrogel. The thermosensitivity, swelling, anddegradation of the hydrogels can be readily modulatedby replacing all or part of the aldehyde-functionalizedsynthetic polymer with one or more aldehyde-functionalizedcarbohydrates (see figures). The thermosensitivity ofthese hydrogels is shown to increase by increasingtheir dextran content, while it can be effectivelyturned off by increasing their CMC content. Completely replacing poly(NIPAM)with CMC resulted in a non-thermoresponsive hydrogelthat swelled by 75 % under physiological conditions (relative to 25 °C) while replacing poly(NIPAM) with dextranresulted in a thermoresponsive hydrogel thatcollapsed by 10 % more under physiological conditions than at

roomtemperature. Similarly, while CMC-crosslinked hydrogels degraded within two days, dextran-crosslinkedhydrogels degraded over the period of several weeks,providing tunable control over the lifetime of the hydrogel *invivo.*
By mixing CMC and dextran,hydrogels with any lifetime, any degree of thermosensitivity, and any degree ofswelling (including zero swelling) can be generated by simple co-injection andsubsequent spontaneous gelation of multiple, off-the-shelf modular polymerprecursors. Hydrogel composition and swelling can similarly be tunedindependently to generate hydrogels with more or lessaffinity for a target drug or faster or slower drug release kinetics, allowingcontrol over the drug release profiles achieved using the modular hydrogels.

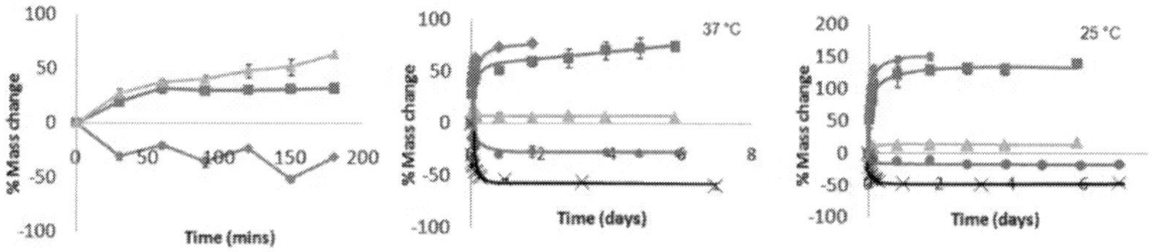

The dependence of hydrogel thermosensitivity on polymer composition:

▲poly(NIPAM):75%CMC/25%Dextran
■poly(NIPAM):50%CMC/50%Dextran
◆poly(NIPAM):25%CMC/75%Dextran

0-30 min = 37 °C, 31-60 min = 25 °C, 61-90 min. = 37 °C
91-120 min = 25 °C, 121-150 min = 37 °C, 151-180 min = 25 °C

The dependence of hydrogel swelling on polymer composition, shown above and below the LCST of poly(NIPAM)

◆ poly(NIPAM):100%CMC, ■poly(NIPAM):75%CMC/25%Dextran, ▲poly(NIPAM):50%CMC/50%Dextran
●poly(NIPAM):25%CMC/75%Dextran, X poly(NIPAM):100%Dextran

Conclusions

By mixing hydrazide and aldehyde-functionalizedpolymeric oligomers with various chemistries, injectable hydrogels withtargeted degradation rates, thermoresponsivity,swelling, and drug delivery properties can be generated by simple mixing. This "modular" approach to the synthesis of degradable hydrogels,by which hydrazide and aldehyde-functionalizednatural or synthetic polymers can be mixed in any combination and any ratio,permits the facile preparation of any desired hydrogel composition.

References

1 - Hoare,T.; Kohane, D. *Polymer*, 49, 1993 (2008)
2 - Lee, K.,*Macromolecules*, 33, 97 (2000)

Acknowledgments: Funding from the Natural Sciencesand Engineering Research Council of Canada (NSERC) and the Ontario Ministry ofResearch and Innovation (Early Researcher Award program) is gratefullyacknowledged.

See more of this Session: Biomaterials for Drug Delivery
See more of this Group/Topical: Materials Engineering and Sciences Division

Hydrogels Decorated with Hydrophobic Nanoparticles for the Oral Delivery of Chemotherapeutics

Cody A. Schoener[1], *Heather Hutson and Nicholas A. Peppas*[1,2,3]

[1]Department of Chemical Engineering, [2]Department of Biomedical Engineering, [3]Division of Pharmaceutics,
University of Texas at Austin, Austin, TX, 78712, USA
peppas@che.utexas.edu

ABSTRACT

A sequential interpenetrating network (IPN) composed of hydrophilic P(MAA-g-EG) and hydrophobic PBA has been developed for the oral delivery of chemotherapeutic agents. P(MAA-g-EG) provides pH responsive properties for protection and proper release in the upper small intestine while PBA provides the hydrophobicity necessary to preferentially load therapeutic levels of anticancer agents. A series of IPNs was synthesized with varying amounts of PBA and their loading and release properties were investigated against fluorescein, a cancer drug analogue.

INTRODUCTION

Post-surgical and early stage cancer patients are traditionally subjected to chemotherapeutic therapy for improved survival rates. Traditional chemotherapeutic regimes include intravenous treatments (I.V.) where potent drugs flow through the entire circulatory system, attach, and destroy cancerous cells, but consequently this non-targeted treatment also destroys healthy cells plaguing patients with side effects including elevated pain, hair loss, and significant energy loss.

New methods of delivering chemotherapeutic agents can improve patient survival, comfort, and compliance to treatment plans. One such method is developing polymer carriers for the oral delivery of anticancer agents which has shown promising results when compared to conventional intravenous treatments.[1] Oral delivery of chemotherapeutic agents provides a host of obstacles and challenges including protecting the therapeutic agent from the low pH of the stomach and degradative enzymes present, protecting the gastrointestinal (GI) tract from the cytotoxic agent, absorption and transport across the lumen wall to desired cancerous location, and loading hydrophobic chemotherapeutics in conventional hydrophilic polymer carriers. For these reasons the speed at which a cohort of cancer drugs such as paclitaxel, tomoxifen, 5-fluorouracil, and doxorubicin has been slow.

However, we recently developed a hybrid sequential interpenetrating network (IPN) composed of poly(methacrylic acid – *g* – ethylene glycol) (P(MAA-g-EG)) and poly(n-butyl acrylate) (PBA). IPNs are two networks intertwined and entangled with each other, but not covalently bound. In this way, individual characteristics of each polymer network may be combined and give rise to a new set of unique properties appropriate for certain applications, in this case an alternative to I.V. treatments.

P(MAA-g-EG) is a hydrophilic, anionic, pH-responsive hydrogel desirable for oral delivery due to its ability to respond to the change in the GI environment to properly deliver therapeutic agents to the upper small intestine. Our lab has shown previous success using these hydrogels to deliver hydrophilic compounds such as insulin.[2-37] By modifying these P(MAA-g-EG) hydrogels with PBA, a hydrophobic polymer, we can improve our ability to deliver anticancer agents at therapeutic levels necessary for eradicating cancerous cells.

EXPERIMENTAL METHODS

Hydrophobic-hydrophilic IPNs were prepared by sequential photo-polymerizable methods. An example preparation of hydrophilic P(MAA-g-EG) is as follows: a precursor solution composed of 3.55 g of distilled methacrylic acid, 2 g of poly(ethylene glycol monomethacrylate) (PEGMMA; MW ~ 1000 g/mol), 0.093 g of tetraethylene glycol dimethacrylate (TEGDMA; crosslinker), 0.028 g of 1-

hydroxycyclohexylbenzophenone (Irgacure 184; photocatalyst) where dissolved in a solvent composed of 50% (w/w) ethanol (EtOH) in water and sonicated for 15 min. The precursor solution was purged with N_2 for 30 min, pipetted between glass plates separated by 0.7 mm Teflon spacer, and exposed to UV light at an intensity of 16 mW/cm^2 for 30 min. The resulting film was removed from the glass plates and washed with deionized water for 7 days to remove unreacted components. The washed film was punched into 18 mm discs and dried in vacuum at 30 °C for 1 wk.

In previous work, dried P(MAA-g-EG) discs increased in the degree of swelling when the percentage of water was increased in a cosolvent of EtOH and water. Thus, using this parameter we could develop three unique IPNs with increasing hydrophobic PBA content as determined by the degree of swelling. An example of preparing an 80/20 IPN hydrophobic network with the P(MAA-g-EG) network is as follows: a soaking solution of 4.5 g of butyl acrylate, 0.075 g of TEGDMA, 0.0225 of Irgacure 184 was created by dissolving in a solvent composed of 80% (w/w) EtOH in water. A P(MAA-g-EG) disc was placed in this soaking solution and shaken for 24 hr, purged with N_2 for 45 min, placed on an open-faced glass plate, and exposed to UV light at an intensity of 16 mW/cm^2 for 45 min. The resulting IPN was washed in 30% (w/w) EtOH in water for 7 days to remove unreacted components and crushed with pastel and mortar or razor and dried in vacuum at 30 °C for 1 wk. In this way an 80/20 IPN was developed, similarly 75/25 IPN and 70/30 IPN were synthesized. Hypothetically the amount of PBA in the IPN should increase as follows 80/20 IPN < 75/25 IPN < 70/30 IPN.

Equilibrium swelling experiments were carried out in 0.1 M dimethylglutaric acid (DMGA) buffers between pH 3.2 and 7.6 at 37 °C on dry 10.2 mm IPN discs. Surface analysis was carried out on a Rame-Hart Goniometer (Model No. 100) and analyzed with DROPimage software. Loading and release properties were determined using a DISTEK Dissolution Apparatus 2100 B (100 rpm, 37 °C) and BIO-TEK SynergyHT UV/Vis plate reader. All experiments were run in triplicate.

RESULTS AND DISCUSSION

Contact angle analysis using a 5 μL sessile drop demonstrated contact angle measurements of 47 ± 1°, 62 ± 1°, 68 ± 1°, 106 ± 1° for P(MAA-g-EG), 80/20 IPN, 75/25 IPN, 70/30 IPN respectively. This confirms our hypothesis that PBA increases as a result of the degree of swelling during the second network photo-polymerization.

Equilibrium swelling studies (**Figure 1**) illustrated the influence of PBA on resulting IPNs as compared to pure P(MAA-g-EG) hydrogels. As the amount of PBA increases in the IPN there is a greater degree of entanglement between the polymer networks resulting in decrease swelling. Furthermore, all polymer carriers demonstrated a collapsed, non-swollen state at low pH's indicative of drug protection during residency in the stomach.

IPN microparticles (75 – 400 μm) were loaded with fluorescein, a fluorescent dye with similar hydrophobicity, chemical structure, and molecular weight as chemotherapeutics. Loading efficiency for P(MAA-g-EG), 80/20 IPN, 75/25 IPN, 70/30 IPN was 25 ± 4 %, 28 ± 3%, 45 ± 7%, 21 ± 7%, respectively. 75/25 IPN showed a balance of swelling and hydrophobic components to optimize loading levels. The release profiles (**Figure 2**) for the IPNs illustrated over 80% release of contained drug by 2 hr which would be sufficient for drug release in the upper small intestine.

Efforts are underway to characterize pure P(MAA-g-EG) loading and release profiles as well as loading and releasing doxorubicin, a chemotherapeutic used for a variety of cancers.

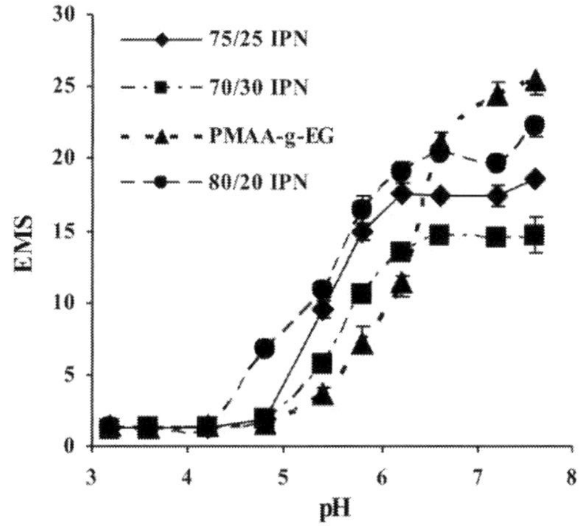

Figure 1.
Equilibrium mass swelling (EMS) of P(MAA-g-EG) and IPN discs.
Samples were soaked at 24 hr, blotted with filter paper and gravimetrically measured.
EMS = M_s/M_d ; M_s = swollen mass, M_d = dry mass.

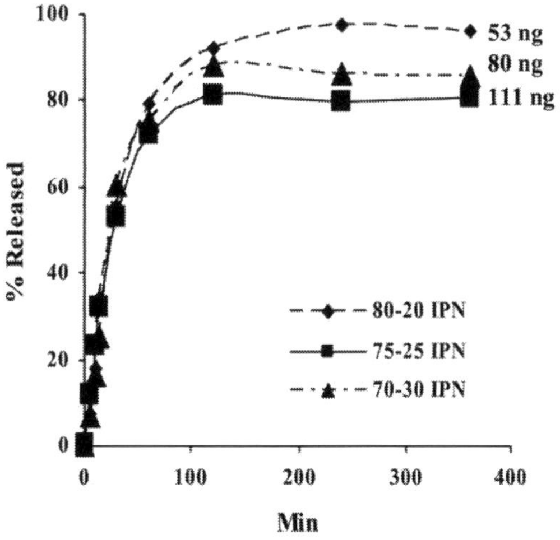

Figure 2.
Percent release profiles of all IPNs.
IPN microparticles were released in 1X PBS at 37 °C and
samples periodically taken over 6 hr time period.
UV/Vis plate reader confirmed amount released using calibration curve.

CONCLUSIONS

A polymer carrier device for the oral delivery of chemotherapeutics has been developed using novel sequential interpenetrating networks. These IPNs are composed of hydrophilic P(MAA-g-EG) and hydrophobic PBA and demonstrate the ability to load and release fluorescein, a cancer drug analogue..

REFERENCES

1. M. Beran, S. O'Brien, S. and D.A. Thomas. *Br. J. Cancer* 73, 1268 – 1271 (1996)
2. J. Zhang, and N. A. Peppas *J. Biomat. Sci., Polym. Ed.*, 13, 511-525 (2002)
3. J. Zhang and N. A. Peppas, in "Polymeric Drugs and Drug Delivery Systems," R.M. Ottenbrite and S.W. Kim, eds., 157-168 Technomic, Lancaster, PA, 2001.,
4. A.M. Lowman, N.A. Peppas, M. Morishita and T. Nagai, "Novel Bioadhesive Complexation Networks for Oral Protein Drug Delivery," in I. McCulloch and S.W. Shalaby, eds., Tailored Polymeric Materials for Controlled Delivery Systems, ACS Symposium Series, Vol. 709, 156-164, ACS, Washington, DC, 1999.
5. A.M. Lowman, M. Morishita, M. Kajita, T. Nagai and N.A. Peppas, "Oral Delivery of Insulin Using pH-Responsive Complexation Gels," *J. Pharm. Sci.*, *88*, 933-937 (1999).
6. M. Torres-Lugo and N.A. Peppas, "Molecular Design and in vitro Studies of Novel pH-Sensitive Hydrogels for the Oral Delivery of Calcitonin," *Macromolecules*, *32*, 6646-6651 (1999).
7. A.M. Lowman and N.A. Peppas, "Solute Transport Analysis in pH-Responsive, Complexing Hydrogels of Poly(Methacrylic Acid-g-Ethylene Glycol)," *J. Biomat. Sci., Polym. Ed., 10,* 999-1009 (1999).
8. M. Morishita, A.M. Lowman, K. Takayama, T. Nagai and N.A. Peppas, "Elucidation of the Mechanism of Incorporation of Insulin in Controlled Release Systems Based on Complexation Polymers," *J. Controlled Release*, *81*, 25-32 (2002).
9. M. Torres-Lugo, M. Garcia, R. Record and N.A. Peppas, "Physicochemical Behavior and Cytotoxic Effects of P(MAA-g-EG) Nanospheres for Oral Delivery of Proteins," *J. Controlled Release*, *80*, 197-205 (2002).
10. H. Ichikawa and N.A. Peppas, "Novel Complexation Hydrogels for Oral Peptide Delivery: In vitro Evaluation of their Cytocompatibility and Insulin-Transport Enhancing Effects Using Caco-2 Cell Monolayers," *J. Biomed. Mater. Res.*, *67*, 609-617 (2003).
11. A.C. Foss, T. Goto, M. Morishita and N.A. Peppas, "Development of Acrylic-Based Copolymers for Oral Insulin Delivery," *Europ. J. Pharm. Biopharmac.*, *57*, 163-169 (2004).
12. K. Nakamura, R.J. Murray, J.I. Joseph, N.A. Peppas, M. Morishita and A.M. Lowman: "Oral Insulin Delivery Using P(MAA-g-EG) Hydrogels: Effects of Network Morphology on Insulin Delivery Characteristics", *J. Controlled Release, 95,* 589-599 (2004).
13. J. O. Blanchette, N. Kavimandan and N.A. Peppas, "Principles of Transmucosal Delivery of Therapeutic Agents", *Biomedicine & Pharmacotherapy, 58,* 142-151 (2004).
14. A.C. Foss and N.A. Peppas, "Investigation of the Cytotoxicity and Insulin Transport of Acrylic-based Copolymer Protein Delivery Systems in Contact with Caco-2 Cultures," *Europ. J. Pharm. Biopharm.., 57,* 447-455 (2004).
15. J.E. López and N.A. Peppas, "Cellular Evaluation of Insulin Transmucosal Delivery," *J. Biomater. Sci., Polym. Ed., 15,* 385-396 (2004).
16. M. Morishita, T. Goto, N.A. Peppas, J.I. Joseph, M.C. Torjman, C. Munsick, K. Nakamura, K. Takayama and A.M. Lowman, "Mucosal Insulin Delivery Systems Based on Complexation Polymer Hydrogels: Effect of Particle Size on Insulin Enteral Absorption", *J. Controlled Release, 97,* 115-124 (2004).
17. N.A. Peppas, "Devices Based on Intelligent Biopolymers for Oral Protein Delivery," *Intern. J. Pharmac., 277,* 11-17 (2004).
18. N.A. Peppas, K.M. Wood and J.O. Blanchette, "Hydrogels for Oral Delivery of Therapeutic Proteins", *Expert Opin. Biol. Ther., 4,* 881 – 887 (2004).
19. J.O. Blanchette and N.A. Peppas, "Cellular and Molecular Evaluation of Novel Carriers for Oral Chemotherapy", *Proceed. Intern. Symp. Control. Rel. Bioact. Mater., 31,* 436 (2004).

20. J.E. López and N.A. Peppas, "Effect of PEG Molecular Weight and Microparticle Size on Oral Insulin Delivery from P(MAA-g-EG) Microparticles," *Drug Develop. Ind. Pharm, 30,* 497-504 (2004).
21. J. O. Blanchette and N. A. Peppas, "Cellular Evaluation of Oral Chemotherapy Carriers", *J. Biomed. Mater. Res., 72A,* 381-388 (2005).
22. J. O. Blanchette and N. A. Peppas, "Oral Chemotherapeutic Delivery: Design and Cellular Response", *Ann. Biomed. Eng., 33,* 142-149 (2005).
23. T. Goto, M. Morishita, N.J. Kavimandan, N.A. Peppas, and K. Takayama, "Gastrointestinal Transit and Mucoadhesive Characteristics of Complexation Hydrogels in Rats", *J. Pharm. Sci., 95,* 462-469 (2006).
24. A. Besheer, K. M. Wood, N. A. Peppas and K. Mäder, "Loading and Mobility of Spin-Labeled Insulin in Physiologically Responsive Complexation Hydrogels Intended for Oral Administration", *J. Controlled Release, 111,* 73-80 (2006).
25. M. Morishita, T. Goto, K. Takayama, and N. A. Peppas, "Oral Insulin Delivery Systems Based on Complexation Polymer Hydrogels", *J. Drug Deliv. Sci. Techn., 16,* 19-24 (2006).
26. M. Morishita, T. Goto, K. Nakamura, A. M. Lowman, K. Takayama and N.A. Peppas, "Novel Oral Insulin Delivery Systems Based on Complexation Polymer Hydrogels: Single and Multiple Administration Studies in Type 1 and 2 Diabetic Rats", *J. Controlled Release., 110,* 587-594 (2006).
27. T. Yamagata, M. Morishita, N.J. Kavimandan, K. Nakamura, Y. Fukuoka, K. Takayama and N.A. Peppas, "Characterization of Insulin Protection Properties of Complexation Hydrogels in Gastric and Intestinal Enzyme Fluids", *J. Controlled Release, 112.* 343-349 (2006).
28. N.A. Peppas, "Vecteurs de médicaments innovants et intelligents: Leur applications pharmaceutiques", *Ann. Pharm. Fr, 64,* 260-275 (2006).
29. M. Morishita and N. A. Peppas, "Is the Oral Route Possible for Peptide and Protein Drug Delivery?", *Drug Discovery Today, 11,* 905- 910 (2006).
30. N. A. Peppas and N. J. Kavimandan, "Nanoscale Analysis of Protein and Peptide Absorption: Insulin Absorption Using Complexation and pH-Sensitive Hydrogels as Delivery Vehicles", *Europ. J. Pharm. Sci., 29,* 183-197 (2006).
31. N. J. Kavimandan and N. A. Peppas, "Confocal Microscopic Analysis of Transport Mechanisms of Insulin Across the Cell Monolayer", *Intern. J. Pharm., 354,* 143-148 (2008).
32. T. Tuesca, K. Nakamura, M. Morishita, J. Joseph, N.A. Peppas and A.M. Lowman, "Complexation Hydrogels for Oral Insulin Delivery: Effects of Polymer Dosing on In Vivo Efficacy", *J. Pharm. Sci., 97,* 2607-2618 (2008).
33. N. Kamei, M. Morishita, H. Chiba, N.J. Kavimandan, N.A. Peppas, and K. Takayama, "Complexation Hydrogels for Intestinal Delivery of Interferon beta and Calcitonin", *J. Controlled Release, 134,* 98-102 (2009).
34. D.A. Carr and N.A. Peppas, "Molecular Structure of Physiologically-Responsive Hydrogels Controls Diffusive Behavior", *Macromol. Biosciences, 9,* 497-505 (2009).
35. K.M. Wood, G.M. Stone and NA Peppas, "The Effect of Complexation Hydrogels on Insulin Transport in Intestinal Epithelial Cell Models", *Acta Biomaterialia,* 6, 48-56 (2010).
36. D.A. Carr and N.A. Peppas, "Assessment of Poly(Methacrylic Acid-co-N-Vinyl Pyrrolidone) as a Carrier for the Oral Delivery of Therapeutic Proteins Using Caco-2 and HT29-MTX Cell Lines", *J Biomed. Mater Res, Part A,* 92A, 504-512 (2010).
37. M. A. Phillips, M. L. Gran and N.A. Peppas, "Targeted Nanodelivery of Drugs and Diagnostics", *NanoToday, 5,* 143-159 (2010).

ACKNOWLEDGMENTS

This work was supported in part by NIH grant U54-143837-02 (Center for Transport Oncophysics). Cody A. Schoener is a National Science Foundation Research Graduate Fellow.

(351e) Remotely-Triggered Nanocomposite Membranes for Drug Delivery

Brian P. Timko, Chemical Engineering, MIT / Children's Hospital Boston, Cambridge, MA, Robert Langer, Chemical Engineering, Massachusetts Institute of Technology, Cambridge, MA and Daniel S. Kohane, Laboratory for Biomaterials and Drug Delivery, Department of Anesthesiology, Division of Critical Care Medicine, Harvard Medical School/Children's Hospital Boston, Boston, MA

Sustained drug release technology has been applied to many medical fields. However, drug release kinetics from devices is usually monotonic over time. To this end, we synthesized externally-triggerable membranes based on the thermoreversible polymer poly(N-isopropylacrylamide)(PNIPAm) and remotely-triggered nanoparticles. Drug flux across the membrane can be turned “on” by applying an external stimulus, with continuous release rate and good reproducibility over multiple cycles. We examine several membrane design parameters, such as thickness and PNIPAm nanoparticle concentration, and evaluate how they affect drug release characteristics.

See more of this Session: Biomaterials for Drug Delivery
See more of this Group/Topical: Materials Engineering and Sciences Division

(351f) ATP Delivery to Insulin Producing β Cells Using Peptide Targeted Liposomes

Nicole Atchison[1], Klearchos K. Papas[2], Michael Tsapatsis[3] and Efrosini Kokkoli[3], (1)Department of Biomedical Engineering, University of Minnesota, Minneapolis, MN, (2)Department of Surgery and Schulze Diabetes Institute, University of Minnesota, Minneapolis, MN, (3)Chemical Engineering and Materials Science, University of Minnesota, Minneapolis, MN

Islet cell transplantation is becoming a promising treatment option for patients with type 1 diabetes. However, stresses imposed during periods of ischemia, including isolation, culture, and after transplantation, reduce islet survival. The isolation process destroys the dense vascular network within the islet requiring diffusion to transport nutrients and oxygen to the cells within the islet. The period of low oxygen availability, known as hypoxia, during isolation limits the cell's ability to produce adenosine triphosphate (ATP), the primary energy component within the cell, leading to loss of cell functions and eventually cell death.

In this work we demonstrate the synthesis, characterization, and application of targeted ATP containing liposomes (ATP-L) to insulin producing β cells. ATP is unable to cross the cell membrane but by encapsulating ATP within liposomes, intracellular delivery can be achieved. In our previous work, we have shown that functionalizing liposomes with the fibronectin-mimetic peptide, PR_b, increases binding and internalization of liposomes into pig islet cells that express the $\alpha_5\beta_1$ integrin compared to conventional liposomes. Here we demonstrate the ability of the PR_b functionalized liposomes to deliver ATP to β cells. We show that delivery of exogenous ATP protects the cells from hypoxic mediated loss of cell function and death. This work highlights the promising role for PR_b targeted ATP-L in protecting β cells during periods of ischemia, potentially leading to increases in cell survival after both isolation and transplantation.

See more of this Session: Biomaterials for Drug Delivery
See more of this Group/Topical: Materials Engineering and Sciences Division

(351g) Self-Assembled Triblock System to Study Drug Delivery Under Dynamic Strain

Murat Guvendiren and Jason A. Burdick, Bioengineering, University of Pennsylvania, Philadelphia, PA

Polymer systems that can self-assemble to form ordered structures such as micelles, vesicles and physically crosslinked gels are potentially attractive for drug delivery applications. These systems are also responsive to external stimuli such as temperature, pH, ion concentration, etc., which is highly valued in drug delivery applications. Although the drug delivery kinetics of many of these systems has been reported, comparably few of them are studied under a mechanically dynamic environment. In this study, we investigated the drug delivery properties of a self-assembled hydrogel under dynamic loading. For this purpose, triblock copolymers composed of poly(methyl methacrylate) (PMMA) endblocks and a poly(methacrylic acid) (PMAA) midblock were dissolved in a good solvent that is also miscible to water. When injected into an aqueous environment, the water-miscible solvent is replaced by water, and the polymer system self-organizes to form a physically associating hydrogel. When a model hydrophobic drug, doxorubicin (DOX), was introduced into to the system prior to self-assembly, our results showed that DOX was trapped either inside the hydrophobic aggregates during self-assembly or in the bulk network and was free to diffuse out during swelling or de-swelling. For 10wt% hydrogels, ~15% of the initial DOX (3wt%) was lost during injection (self-assembly) and 65% was released gradually for the following 14 days. The remaining ~20% was kinetically trapped in the PMMA domains. The amount of DOX released did not show any significant difference under 1% and 5% dynamic compressive strain applied at days 1, 2 and 3 for 6 hours, which was due to strong hydrophobic interactions within glassy PMMA domains. However, triethyl citrate (TC) was used (plasticizer for PMMA) to tune these interactions. The amount of DOX released increased almost 4 times with 20% (weight of PMMA) TC in steady state. This trend decreased significantly with increasing dynamic strain which was due to a significant increase in swelling under strain. For hydrogels with TC at day 3, static release was ~40% whereas it was reduced to ~35% for 1% strain and further reduced to ~20% for 5% strain. However, for gels without TC, ~10% was released for all conditions.

See more of this Session: Biomaterials for Drug Delivery
See more of this Group/Topical: Materials Engineering and Sciences Division

(352a) Invited Speaker: Influence of Uptake Pathway On Polymer-Mediated DNA and siRNA Delivery

Daniel W. Pack, Mark E. Hwang, Mihael Lazebnik and Nathan P. Gabrielson, Chemical & Biomolecular Engineering, University of Illinois Urbana-Champaign, Urbana, IL

Continued progress in human gene therapy requires the development of safe and effective delivery strategies. Polymeric gene delivery agents hold important advantages including safety, targetability and versatility, but are hindered by generally poor efficiency. Design of polymers capable of efficient delivery will be facilitated by understanding of the structure-function relationships by which polymers mediate cellular internalization, escape from endocytic vesicles, cytosolic transport, nuclear translocation, and release of nucleic acids. We have recently begun to investigate how the first step in this process, cellular uptake, affects the subsequent steps. In particular, we show that internalization may occur through pathways including clathrin-dependent endocytosis, caveolin-dependent endocytosis, macropinocytosis, etc. The subsequent processing of polyplexes can differ significantly for each pathway. Most importantly, we demonstrate that, depending on the polymer chemistry, the nucleic acid (DNA or siRNA), and the cell type, some uptake pathways lead to more effective transgene expression than other pathways. Thus we conclude that uptake pathway is a key parameter for gene delivery polymer design, especially for vectors that are targeted to cell-specific ligands.

See more of this Session: Biomaterials for Nucleic Acid Delivery
See more of this Group/Topical: Materials Engineering and Sciences Division

(352b) Intracellular Trafficking and Activity of Histone-Mimetic Gene Delivery Vehicles

Millicent O. Sullivan, John D. Larsen and Meghan J. Reilly, Chemical Engineering, University of Delaware, Newark, DE

Effective gene transfer strategies would fundamentally alter healthcare, and depend upon our ability to create materials that can chaperone DNA transport and release within a diverse biological environment. A critical problem in non-viral gene therapy is how to trigger nucleus-specific DNA release: current electrostatic DNA packaging techniques condense DNA to prohibit DNase access, but are believed to inhibit transcription in the process. We have created a new class of DNA packaging materials that are designed to promote efficient transcription within the nucleus by interacting with histone effector proteins. These materials contain peptides that are comprised of histone tail sequences known to signal nuclear import and transcriptional activation. In this work, we have explored the activity of polyplexes containing trimethylated H3 histone tail peptides ($H3K4Me_3$) that are known to interact with two nucleus-specific effector complexes that initiate transcriptional activation. We demonstrate that these materials initiate gene expression more rapidly than traditionally packaged polyplexes or polyplexes made from non-methylated H3 histone tail peptides when introduced directly into the nucleus. Furthermore, we have developed a simple mass action kinetic model that demonstrates that the enhanced activity of the $H3K4Me_3$-containing polyplexes results from their more rapid nuclear unpackaging/ activation. We have used co-immunoprecipitation to show that the $H3K4Me_3$–containing polyplexes interact with elements of the transcriptional machinery, presumably as a result of their specific interactions with resident nuclear proteins. Finally, when the $H3K4Me_3$ peptides are combined with traditional gene transfer polycations such as polyethylenimine (PEI), the hybrid polyplexes promote higher levels of gene transfer into cultured cells than PEI alone, suggesting that nuclear unpackaging and activation represent significant barriers to effective gene transfer.

See more of this Session: Biomaterials for Nucleic Acid Delivery
See more of this Group/Topical: Materials Engineering and Sciences Division

(352c) MicroRNA Delivery by Cationic Lipoplexes for Lung Cancer Therapy

Yun Wu[1], Melissa Crawford[2], Bo Yu[3], Yicheng Mao[4], Serge P. Nana-Sinkam[2] and Ly James Lee[5], (1) Nanoscale Science and Engineering Center, The Ohio State University, Columbus, OH, (2)Division of Pulmonary, Allergy, Critical Care and Sleep Medicine, The Ohio State University, Columbus, OH, (3) Nanoscale Science and Engineering Center, Ohio State University, Columbus, OH, (4)Division of Pharmaceutics, College of Pharmacy, the Ohio State University, Columbus, (5)Chemical and Biomolecular Engineering, The Ohio State University, Columbus, OH

Lung cancer is the leading cause of cancer deaths in western countries and carries a poor overall five year survival rate. Several studies demonstrate that microRNAs (miRNAs or miRs) are actively involved in tumor development by serving as tumor suppressors, oncogenes or both. In lung cancer, miRNAs may serve as both diagnostic and prognostic biomarkers as well as regulate *in vitro* and *in vivo* tumor progression. However, miRNA-based therapeutics are faced with several challenges including lack of tissue specificity, lack of optimal delivery systems, poor cellular uptake and risk of systemic toxicity. Here, we report a cationic lipid based miRNA delivery system to address some of these challenges. Among many lung cancer related miRNAs, miR-133b, a tumor suppressor, was selected as a therapeutic target because it directly targets the prosurvival gene MCL-1 thus regulating cell survival and sensitivity of lung cancer cells to chemotherapeutic agents. The efficacy of pre-miR-133b containing lipoplexes was evaluated in A549 non-small cell lung cancer (NSCLC) cells. Compared with siPORT™ NeoFX™ transfection agent, lipoplexes delivered pre-miR-133b in a more efficient manner with ~2.3 fold increase in mature miR-133b expression and ~1.8 fold difference in MCL-1 protein down-regulation *in vitro*. In the *in vivo* biodistribution study, lipoplexes achieved ~30% accumulation in lung tissue, which was ~50 fold higher than siPORT™ NeoFX™ transfection agent. Mice treated by pre-miR-133b containing lipoplexes had mature miR-133b expression in lung ~52 fold higher than untreated mice. Our results demonstrated that cationic lipoplexes are a promising carrier system for the development of miRNA-based therapeutics in lung cancer treatment.

See more of this Session: Biomaterials for Nucleic Acid Delivery
See more of this Group/Topical: Materials Engineering and Sciences Division

(352d) Invited Speaker: Directed Evolution of New Viruses for Enhanced Gene Delivery

David Schaffer, Chemical Engineering, University of California - Berkeley, Berkeley, CA

Viral vectors – composed of protein biopolymer capsids surrounding a recombinant genome – offer a number of advantageous properties including the potential for safe and efficient gene delivery in vivo. However, they also face a number of challenges – including inefficient delivery to some therapeutically valuable cell types, difficulty with targeted delivery, and an inability to overcome some tissue transport barriers in vivo – all which arguably stem from the fact that they did not naturally evolve to be utilized as human therapeutics. Rational design approaches have generated vectors with enhancements in efficiency and selectivity of gene delivery. However, in many cases there is insufficient knowledge of viral infection mechanisms and structure-function relationships that underlie gene transfer challenges to empower rational design efforts to improve the gene delivery properties of viral proteins. As a complementary approach, we have been developing directed evolution – the iterative generation of large libraries genetic mutants and selection for enhanced properties – as an approach to create new viruses with useful properties for gene therapy. For example, adeno-associated virus (AAV) vectors have enjoyed recent success in several clinical trials, and we have evolved AAV for the ability to transduce nonpermissive cells in vitro and in vivo, cross tissue barriers to gene delivery, target delivery to specific cell types, and evade pre-existing immunity. Furthermore, we have engineered retroviral vectors for both targeted delivery and safe harbor integration into the human genome. Directed evolution thus represents a powerful approach for engineering 'designer' viruses with enhanced gene delivery properties and therapeutic potential.

See more of this Session: Biomaterials for Nucleic Acid Delivery
See more of this Group/Topical: Materials Engineering and Sciences Division

(352e) Development of An Integrated Peptide for Gene Delivery

Qiong Tang, Bin Cao and Gang Cheng, Department of Chemical and Biomolecular Engineering, University of Akron, Akron, OH

To achieve an efficient delivery of therapeutics to physiological targets, the design and/or discovery of effective delivery vehicles play an important role. Polymer-based carriers have been intensively investigated and shown potential clinical applications due to their merits in safety including the avoidance of immunogenicity and the possibility of repeated administration. However, multi-step reaction and purification involved in the synthesis make the purity and fidelity of materials the most challenging issues for large scale preparation of synthetic drug delivery materials. Besides, long-term toxicity and biocompatibility are also concerns for synthetic materials. Peptide-based vectors are gaining prominence in the field of precisely designed and controlled biomaterials. However, these peptide-based materials only occupy partial functions compared to a fully functional drug delivery carrier which is supposed to have drug loading capability and specific targeting. In order to address these challenges in the field of gene delivery, we developed a peptide for DNA delivery. Evaluation of the peptide-based carrier is presented in details, focusing on size and zeta potential of the peptide/DNA complex, blood stability, cytotoxicity and gene transfection efficiency on target cells.

See more of this Session: Biomaterials for Nucleic Acid Delivery
See more of this Group/Topical: Materials Engineering and Sciences Division

(352f) Synergistic Silencing: Combinations of Lipid-Like Materials for Improved siRNA Delivery

Kathryn A. Whitehead[1], Gaurav Sahay[2], Kevin Love[3], Christopher Alabi[2], Robert Langer[3] and Daniel Anderson[3], (1)Department of Chemical Engineering, Massachusetts Institute of Technology, Cambridge, MA, (2)Koch Institute for Integrative Cancer Research, Massachusetts Institute of Technology, Cambridge, MA, (3)Chemical Engineering, Massachusetts Institute of Technology, Cambridge, MA

Despite the promise of RNA interference therapeutics, progress towards the clinic has been slowed by the difficulty of delivering short interfering RNA (siRNA) into cellular targets within the body. Nearly all siRNA delivery vehicles developed to date employ a single cationic or ionizable material. In order to increase the material space available for development of siRNA delivery therapeutics, this study examined the possibility of using binary combinations of ionizable lipid-like materials to synergistically achieve gene silencing. Interestingly, it was found that ineffective single lipid-like materials could be formulated together in a single delivery vehicle to induce near-complete knockdown both *in vitro* and *in vivo*. Synergistic liver delivery in mice was initially observed at doses of 5 mg/kg total siRNA, with the most efficacious binary combination demonstrating an IC_{50} value of 1.5 mg/kg. Microscopy experiments suggested that synergistic action resulted when combining materials that respectively mediated cellular uptake and endosomal escape, two important steps in the delivery process. Together, the data indicate that formulating lipid-like materials in combination can significantly improve siRNA delivery outcomes while increasing the material space available for therapeutic development.

See more of this Session: Biomaterials for Nucleic Acid Delivery
See more of this Group/Topical: Materials Engineering and Sciences Division

(352g) Invited Speaker: Targeted Nanoparticle Delivery of siRNA: From Concept to Clinic

Mark E. Davis, Chemical Engineering, California Institute of Technology, Pasadena, CA

One of the major challenges in the development of siRNA-based therapeutics for human use is their effective, systemic delivery. This issue is of particular relevance for cancer as there remains a lack of effective treatments for metastatic disease. We have been investigating the potential of targeted nanoparticles for the systemic delivery of siRNA in cancer, and have reported that transferrin targeted nanoparticles formulated with a cyclodextrin-containing polycation and anti-EWS-FLI1 siRNA can be effective anti-tumor agents in a mouse model of Ewing's Sarcoma [1]. Targeted nanoparticles show behaviors that provide advantages in the systemic delivery of siRNA. For example, they can protect and deliver non-chemically modified siRNA, they can deliver a large "packet" of siRNA, they can have tunable binding affinities to target cell surfaces and when correctly assembled can systemically deliver siRNA without immune stimulation. Nanoparticles in the size range of 50-100 nm can circulate and localize in tumors [1,2]. These particles can carry a large amount of siRNA as the polycation protects the RNA from degradation and transports it into cells where the kinetics of gene inhibition are a strong function of the cell doubling time [3]. Using a combination of PET and bioluminescent imaging, the biodistribution of the nanoparticles is shown to not be a strong function of the presence of the targeting ligand, while the uptake and function in tumor cells are critically dependent on the function of the targeting ligand [4]. We have confirmed this behavior with other targeted nanoparticles [5]. Repetitive dosing in monkeys with the cyclodextrin-containing polycation nanoparticles can be safely accomplished without eliciting complement activation or interferon and other immunostimulatory processes [6]. This delivery system entered a Phase I clinical trial in 2008 [7], and results from the clinical trial [8,9] and how they compare to results from animal studies will be discussed.

[1] Hu-Lieskovan S et al., Cancer Res., 65 (2005) 8984-8992.
[2] Pun SH et al., Cancer Biology & Therapy, 3 (2004) 641-650.
[3] Bartlett DW and Davis ME, Nucleic Acids Res., 34 (2006) 322-333.
[4] Bartlett DW et al., PNAS, 104 (2007) 15549-15554.
[5] Choi, CHJ et al., PNAS, 107 (2010) 1235-1240.
[6] Heidel JD et al., PNAS, 104 (2007) 5715-5721.
[7] Davis, ME, Mol. Pharm., 6 (2009) 659-668.
[8] Davis, ME et al., Nature, 464 (2010) 1067- 1070.
[9] Ribas, A et al, ASCO Abstract No. 3022 (2010).

See more of this Session: Biomaterials for Nucleic Acid Delivery
See more of this Group/Topical: Materials Engineering and Sciences Division

(361a) Molecular Simulation Studies On the Rheological Properties of Silica Nanoparticles Embedded In a Polyethylene Melt

Yangyang Shen and M. Silvina Tomassone, Chemical and Biochemical Engineering, Rutgers University, Piscataway, NJ

The addition of nanoparticles to polymer composites has been shown to significantly influence their mechanical, optical, and electrical properties. We use a combination of thermodynamics theory, Molecular Dynamics (a coarse grained approach) and atomistics simulations to develop a molecular model for Nanoparticle Composites. In our coarse-grained model, eight methylene groups of polyethylene are represented by one soft bead. The nanoparticles are modeled as spherical clusters of beads kept together by rigid harmonic bonds. The particles move in a simulation box of dimensions 30nmx30nmx16nm. Using this coarse-grained model, we explore the diffusion of particles in the melt and rheological properties of the polymeric fluids and nanoparticle composites under shear flow. Specifically we investigate the effects of polymer chain length, nanoparticle dimension and filling fraction on viscosities. Over the range of shear rates studied, the shear viscosity is strongly dependent on the shear rate. These trends are quantified by fitting the power-law behavior. For all the polymer matrices, the shear viscosity consistently increases with filler concentration, and increases up to 2 orders of magnitude with respect to the viscosity of pure polymer. The monotonic increase of viscosity with filler concentration is commonly observed in experimental systems. We also demonstrate that the addition of nanoparticle fillers leads to a more pronounced shear thinning behavior, as a result of increasing shear rates in the gaps between the filler particles. The results of our simulations are compared with an experimental analysis of a polyethylene filled with spherical silica nanoparticles of various sizes. The viscosity of polymer nanocomposite is sensitive to the size of the filler particles. We found that decreasing the size of spherical filler particles while keeping the filler volume fraction constant leads to an increase of the viscosity. This is mainly due to the aggregation of small nanoparticles reduces the chain movements and slows down the relaxation of polymer, as a result, less material that flows, leading the melt viscosity to increase. For these polymer composites where fillers are in the nanometer scale, Einstein and Batchelor equations greatly under predict the data. These findings qualitatively agree with the experimental studies reported in the literature.

See more of this Session: Composites: Modeling of Composites
See more of this Group/Topical: Materials Engineering and Sciences Division

(361b) Long Glass Fiber Orientation In Extensional Flow

Kevin J. Meyer, Kevin C. Ortman and D. G. Baird, Chemical Engineering, Virginia Tech, Blacksburg, VA

Long Glass Fiber Orientation in Extensional Flow

Kevin J. Meyer, Kevin C. Ortman andDonald G. Baird

Department of ChemicalEngineering, Virginia Tech, Blacksburg VA 24061

With the need for alternative lightweightmaterials growing daily, significantly more research is being conducted in thearea of reinforced composites. A cheap and effective way of increasing theproperties of a polymer is to embed both short (L < 1mm) and long glassfibers (L > 1mm) in the matrix. The bulk properties of the composite materialin question are highly governed by the orientation of the embedded fibers. Whilesignificant effort and emphasis has been placed on studying the orientation offibers in injection molding, which are shear-dominated flow fields, the flowsare usually very complicated with final fiber orientation being governedprimarily by the processing technique. In extensional flow (shear-free flow) wepresent a way to govern the fiber orientation kinetics completely independentof processing technique. Under biaxial deformations, due to isotropic stressesoccurring during squeezing, the fibers retain their initial orientation. Withthis knowledge, initial fiber orientation can be chosen a priori thus governingfinal material properties. Currently, modeling of fiber orientation inthermoplastic composites is done by way of the widely used Folgar-Tucker modelmodified for slow orientation kinetics. Due to the over prediction of fiberorientation in shear and extensional flow fields, a different modeling approachfirst provided by Strautins and Latz,is presented as an alternative to modeling thermoplastic composites. More specifically,we present a way to obtain model parameters used in the orientation equationsfrom rheological data only. The orientation parameters are obtained by usingthe Dinh-Armstrong variation of the Lipscomb stressmodel for concentrated suspension. This technique provides a way to model fiberorientation without any knowledge of transient orientation data to predictfinal orientation. It is the purpose of this paper to apply current theoreticalmodels and compare these models to experimentally obtained orientation data togain insight into the orientation kinetics of long glass fibers in shear-freeflow fields for predictive ability of composite properties of compressionmolded samples.

See more of this Session: Composites: Modeling of Composites
See more of this Group/Topical: Materials Engineering and Sciences Division

(361c) Hybrid Model and Application In the Study of Interfacial Properties of Nanocomposites

Jie Feng and Sadhan C. Jana, Polymer Engineering, The University of Akron, 44325, OH

All-atom model provides accurate results for simulation of large systems, but is time-consuming. Coarse-grained model, on the other hand, accelerates the simulation but loses accuracy in many situations. To alleviate this dilemma, we propose a "concern oriented" hybrid approach (COHA), which achieves a balance between accuracy and simulation time. In this approach, all atom resolution is only kept for the parts that are of utmost importance, while coarse-grained resolution is applied for the rest playing minor roles. The proposed model will be elaborated using a few examples of the binding phenomenon of polymers on the inorganic surfaces in solutions. In these cases, the atomistic details for polymers and surfaces are kept, but the structures of the solvent molecules are coarse-grained. The coarse-grained force field for solvents has been identified via verification with experimental values of physical quantities, such as density and evaporation energy.

See more of this Session: Composites: Modeling of Composites
See more of this Group/Topical: Materials Engineering and Sciences Division

(361d) A More Realistic Model for the Study of Thermal Conductivity of Nanocomposites

Khoa N. D. Bui and Dimitrios V. Papavassiliou, School of Chemical, Biological and Materials Engineering, The University of Oklahoma, Norman, OK

Calculating the thermal conductivity (K_{eff}) for carbon nanotube-based nanocomposites is a challenging task due to the presence of thermal boundary resistance (TBR) as well as the randomness of dispersion state and of the geometry of carbon nanotubes (CNTs). It has been found that the TBR of carbon nanotube to carbon nanotube ($TBR_{CNT-CNT}$) can be higher than the TBR between a nanotube and the surrounding matrix [1-3]. The effective medium theory is capable of calculating K_{eff} assuming that the CNTs are in a perfect dispersion state and excluding the $TBR_{CNT-CNT}$ effect [4]. However, our previous work with CNTs modeled as stiff cylinders showed that the tendency of CNTs to create bundles and the $TBR_{CNT-CNT}$ can dramatically suppress K_{eff} at high volume fraction of CNTs [5]. The CNTs are observed to have "worm-like" shape in reality [4] and no prior work has been reported in studying the effect of the curvature (expressed in terms of the persistence length of the CNTs) on the K_{eff} of nanocomposites.

In this work, we will report on the development of a new algorithm allowing the generation of CNTs with worm-like geometry in 3D, and with different persistence length. The use of these geometries in conjunction with off-lattice Monte Carlo simulations [3,5] in order to study the effective thermal properties of nanocomposites will be discussed.

References

[1] J.E. Peters, D.V. Papavassiliou, B.P. Grady, Macromolecules, 41 (2008) 7274-7277.

[2] S. Maruyama, Y. Igarashi, Y. Taniguchi, J. Shiomi, J. Therm. Sci. Tech., 1 (2006) 138-148.

[3] K. Bui, H. M. Duong, A. Striolo, and D. V. Papavassiliou, J. Phys. Chem. C, 115 (2011)10.

[4] C. W. Nan, G. Liu, Y. Lin, and M. Li, Applied Physics Letters, 85 (2004) 16.

[5] K. Bui, B. P. Grady and D. V. Papavassiliou, Chemical Physics Letters, doi:10.1016/j.cplett.2011.04.005 (2011).

(361e) New Insight Into the Reaction Mechanism of SiC Oxidation and Nitridation: A Density Functional Theory Study of β-SiC (001) Surface

Satyender Goel and Linda J. Broadbelt, Chemical and Biological Engineering, Northwestern University, Evanston, IL

Wide interest in detailed investigations of SiC can be divided into two major areas: exploring their potential as high-frequency materials in electronic applications and assessing their continuous stability when exposed to high temperature environments. One common factor in both types of applications is the importance of SiC oxidation. The mechanism of this process likely proceeds via a complex set of serial and parallel reaction steps, and while predicting all of them is challenging mapping a comprehensive set computationally would offer significant insight into the dominant adsorption and reaction chemistry of gaseous species interacting with SiC surfaces. In this work, reaction pathways leading to active/passive oxidation of SiC surfaces were investigated systematically using density functional theory. First, adsorption of gaseous species (small molecules containing O, C, H and N) interacting in different orientations and positions on the β-SiC (001) surface was studied. Based on the most favorable adsorption configurations, various reaction channels were explored by identifying transition state structures. Interactions with both the carbon and silicon layers of SiC were analyzed. Interactions with the carbon layer were found to be important for NO dissociation leading to Si-C bond breaking by insertion. The observations made for atomic and multiatomic species adsorption on the SiC surface were extended to form reaction pathways resulting in SiC oxidation products, including CO, SiO_2 and SiO.

Rate constants were calculated with transition state theory, and microkinetic modeling was carried out to identify the dominant pathways that were mapped. The thermodynamic data was interpreted to understand the role of temperature in the formation of SiO instead of SiO_2.

See more of this Session: Composites: Modeling of Composites
See more of this Group/Topical: Materials Engineering and Sciences Division

(374a) Encapsulation and Permeability Characteristics of Amorphous Hydrogenated Carbon Films Formed by Plasma Enhanced Chemical Vapor Deposition Technique

Anaram Shahravan and Themis Matsoukas, Department of Chemical Engineering, The Pennsylvania State University, University Park, PA

We have used plasma enhanced chemical vapor deposition technique to coat the materials and control the rate of their release. This powerful technique coat nanoparticles and nanopowders ranging from 15 nm to 1 micron and we can control the thickness (from 1nm to upwards of 100 nm) of the film by controlling the duration of the deposition time. We present results with two different core materials, silica nanopheres and KCl nanoparticles. By coating silica and etching the core in HF, we form hollow nanoshells whose dimensions depend on the size of the silica particle and the thickness of the shell. Permeation studies are conducted using KCl, whose release profile is measured easily by monitoring the conductivity of the solution. The thickness of the shell is one of the independent variables by which to control the rate of release. It has a strong effect on the rate of release, which increases from a six-hour release (shell thickness is 20 nm) to a long-term release over 30 days (shell thickness is 95 nm). The profile of release for any thickness follows a unique behavior: a fast release (35% of the final materials) during the first five minutes after the beginning of the dissolution, and a slower release till all of the core materials come out. In order to analyze this behavior a model based on an unsteady-state diffusion from sphere is developed. Very good agreement between experimental data and the model after half an hour from the beginning of the release is observed and based on that a diffusion coefficient for the shell is calculated. The diffusion coefficient of $(3.5 \pm 0.5) * 10^{-21}$ is obtained for four different samples with different shell thicknesses (20, 40, 75, 95 nm).

See more of this Session: Functional Nanoparticles and Nanocoatings on Particles I
See more of this Group/Topical: Particle Technology Forum

(374b) From Embedded to Supported Metal/Oxide Nanomaterials: Thermal Behavior and Modelling of Structural Evolution At Elevated Temperatures

Stephanie B. Bubenhofer, Wendelin J. Stark and Robert N. Grass, Institute for Chemical and Bioengineering, ETH Zurich, Zurich, Switzerland

The relation between structure and function is of major importance at material interfaces which often excel in terms of chemical or physical effects. Therefore, multi-phase nanocomposites (e.g. core-shell, janus-shaped and carrier-supported nanoparticles) attract attention due to their possibility to contact two materials at the nanometer scale profiting from their highly increased interaction area.

Especially at metal/oxide interfaces, accounting for a majority of heterogeneous catalysts, the structure-function relationship has been studied intensively. High temperature, dry aerosol processes (flame spray synthesis) are common for building such composites. But the formation of key structural elements in such multi-component systems, depending on coagulation-order and solid-solid diffusion, is not fully understood.

Here, we present the synthesis of silica coated crystalline Pd nanoparticles (Pd in silica) in a single step flame spray pyrolysis process. The nanometer scale transformation of this core-shell material to nanocrystalline Pd supported on the amorphous silica matrix (Pd on silica) was achieved at temperatures exceeding 700°C. A physical, population balance based aggregation/diffusion model gives an understanding of the influence of temperature, matrix viscosity, particle sizes and concentrations on this transformation process. It enables morphological predictions (core-shell vs. carrier-supported particles) in nanocomposite formation processes such as flame spray synthesis. It further allows to understand the performance of multi-phase materials during thermal cycling in catalysis or fuel cells.

Figure 1. Morphological transformation of noble metal encapsulated in oxide (core-shell, left TEM image) nanoparticles to supported on oxide particles (right TEM image).

Reference :

S. Bubenhofer et al., J. Phys. Chem. C, 115(4), 1269-76 (2011).

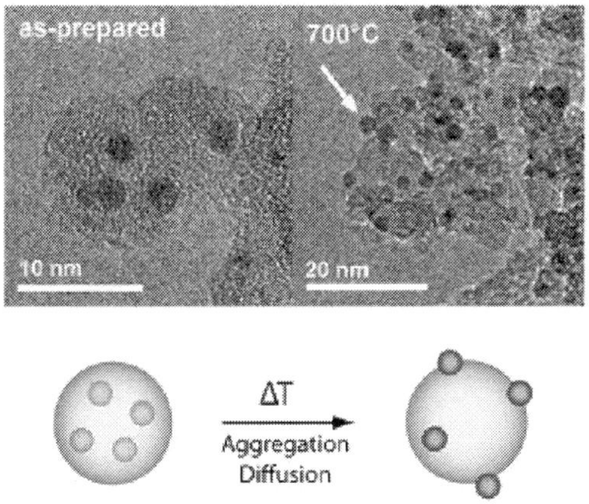

See more of this Session: Functional Nanoparticles and Nanocoatings on Particles I
See more of this Group/Topical: Particle Technology Forum

(374c) Atomic Layer Deposition As a Catalyst Synthesis Technique for Nickel Nanoparticles

Troy D. Gould[1], John L. Falconer[2], J. Will Medlin[3] and Alan W. Weimer[2], (1)Chemical and Biological Engineering, University of Colorado at Boulder, Boulder, CO, (2)Chemical & Biological Engineering, University of Colorado at Boulder, Boulder, CO, (3)Department of Chemical and Biological Engineering, College of Engineering and Applied Science, University of Colorado-Boulder, Colorado Center for Biorefining and Biofuels, Boulder, CO

Efforts to design new catalysts for renewable energy applications have revolved around increasing selectivity & activity by exploring advantages of catalysis at the nanoscale, mainly due to geometric & electronic effects. This work investigates the synthesis of Ni nanoparticle catalysts through the scalable gas phase synthesis route of Atomic Layer Deposition (ALD). Typically ALD deposits Ni as a NiO film, however; Ni nanoparticles were synthesized by using elevated reaction temperatures to produce particles with diameters <5nm. Smaller particles were created to have more catalytically active defects on the surface of the particles. The catalysts were characterized via Temperature Programed Reduction (TPR), chemisorption, Transmission Electron Microscopy (TEM), & Inductively Coupled Plasmon Mass Spectroscopy (ICP-MS) & compared to traditional incipient wetness Ni catalysts. The incipient wetness Ni catalysts were also used as a base comparison of catalytic activity for alkene hydrogenolysis reactions. We will also present the investigations of the controlled preparation of bimetallic catalysts using ALD.

See more of this Session: Functional Nanoparticles and Nanocoatings on Particles I
See more of this Group/Topical: Particle Technology Forum

(374d) Facile Synthesis and Optical Characterization of Hybrid Upconverting and Plasmonic $NaGdF_4$: Yb^{3+}, Er^{3+}/Silica/Gold Nanoparticles

Sha Liu[1], Guanying Chen[2], Tymish Y. Ohulchanskyy[2], Paras N. Prasad[2] and Mark T. Swihart[1], (1) Chemical and Biological Engineering, The University at Buffalo (SUNY), Buffalo, NY, (2)Institute for Lasers, Photonics, and Biophotonics, The University at Buffalo (SUNY), Buffalo, NY

We present a simple process for preparing water dispersable hybrid $NaGdF_4$: Yb^{3+}, Er^{3+}/silica/gold nanoparticles. Monodispersed upconversion nanoparticles of $NaGdF_4$:2% Er, 20% Yb capped by oleic acid were first synthesized via a thermal-decomposition method. The hydrophobic UC NPs were then capped with silica via a reverse-micelle method to make them hydrophilic. Amino-propyltriethoxysilane (APTES) was then used to functionalize the silica-capped UC NPs with amine groups. Finally, gold NPs capped with PVP were mixed with the silica-coated UC NPs. Amine groups on the UC NPs displaced the PVP from the AuNP surface, yielding AuNP-decorated UC NPs that were dispersible in water or ethanol. The emission intensity and color of the upconverting cores are modulated by the plasmonic absorbance and field enhancement of the gold nanoparticles. Energy transfer happened in the hybrid nanoparticles from UC core to the decorating gold nanoparticles, especially at the wavelength of 530 nm at which gold nanoparticles have strong absorption. The lifetime analysis shows that the plasmon resonance effected the emission of upconversion nanoparticles. The applicability of hybrid NPs for multi-modal bioimaging probes was also illustrated with confocal microscopy of living cancer cells in vitro.

See more of this Session: Functional Nanoparticles and Nanocoatings on Particles I
See more of this Group/Topical: Particle Technology Forum

(374e) Single-Walled Aluminosilicate Nanotubes with Organic-Modified Interiors

Dun-Yen Kang, Ji Zang, Christopher W. Jones and Sankar Nair, School of Chemical & Biomolecular Engineering, Georgia Institute of Technology, Atlanta, GA

This talk describes a methodology for modifying the interior of single-walled metal oxide (aluminosilicate) nanotubes by covalently immobilizing organic functional entities on the interior surface of the nanotube structure. Characterization of the modified nanotubes by a range of solid-state characterization techniques - including nitrogen physisorption, thermogravimetric analysis, transmission electron microscopy (TEM), powder X-ray diffraction (XRD), and solid-state NMR - strongly indicates that the organic entities are immobilized on the inner surface of the nanotubes by reaction with the silanol groups on the interior wall of the nanotube. The resulting organic-modified SWNTs show higher hydrophobicity than bare nanotubes based upon water adsorption measurements. Furthermore, a mechanistic understanding of water adsorption in the modified SWNTs is developed, by interpretation of the water adsorption data with a multilayer adsorption model. The degree of interior surface silanol substitution is estimated, with up to 35% of the silanols being substituted through the present modification chemistry. This methodology of immobilizing various functional entities at the inner wall of aluminosilicate nanotubes opens up a range of previously inaccessible "molecular recognition"-based applications for nanotube materials in areas such as catalysis, molecular encapsulation, sensing, and separation.

See more of this Session: Functional Nanoparticles and Nanocoatings on Particles I
See more of this Group/Topical: Particle Technology Forum

(374f) Phase Separation of Mixed Monolayers On Silica Nanoparticles Induced by Hydrogen Bonding

Daniel Sunday, Chemical Engineering, University of Virginia, Charlottesville, VA and **David L. Green**, Department of Chemical Engineering, University of Virginia, Charlottesville, VA

Controlling the self-assembly of mixed monolayers on silica nanoparticles is challenging due to irreversible ligand attachment. In the course of attempting to control the graft density of polystyrene brushes on silica nanoparticles, we synthesized both a surface atom transfer radical polymerization (ATRP) initiator along with an inert analog containing an amide group. The initiator and analog were reacted to the nanoparticles in methylisobutyl ketone to form a mixed monolayer from which polymers could grow. The inert analog reacted preferentially to the nanoparticle surface, which affected the polymerization of styrene with ATRP. In particular, analysis of the initiation efficiency, or the ratio between the ATRP and graft polymer surface coverage provides evidence for phase separation between the ATRP initiator and inert analog in the mixed monolayer. This is most likely due to hydrogen bonding primarily between the inert analog in solution and on the particle surface. Small-angle x-ray scattering (SAXS) and NOESY NMR were used to show that in solution, the analog and initiator form disc-shaped molecular aggregates about 1 nm thick and 1.6 nm in diameter through hydrogen bonding between amide groups. Based on kinetic deposition experiments and FTIR measurements, the molecular aggregates, composed primarily of the inert analog, adsorb strongly to the surface. This adsorption leads to phase separation between the inert analog the ATRP initiator in the monolayer as inferred from FTIR by tracking the non-linear shift in the methylene stretches of the carbons in the ligands, which was a function of the amount of the initiator and analog used to functionalize the nanoparticles. To our knowledge, this reaction mechanism in which the ligand solution behavior affects mixed monolayer structure on silica nanoparticles has not been previously observed.

See more of this Session: Functional Nanoparticles and Nanocoatings on Particles I
See more of this Group/Topical: Particle Technology Forum

(377a) Highly Efficient Electroluminescence From Hybrid Silicon Nanocrystal-Organic Light-Emitting Devices

Kai-Yuan Cheng[1], Rebecca Anthony[2], Uwe R. Kortshagen[2] and Russell J. Holmes[1], (1)Department of Chemical Engineering and Materials Science, University of Minnesota, Minneapolis, MN, (2) Department of Mechanical Engineering, University of Minnesota, Minneapolis, MN

Colloidal semiconductor nanocrystals (NCs) have received considerable attention for optoelectronic applications due to their high photoluminescence efficiency and broad spectral tunability. Semiconductor NCs can be solution processed permitting integration into hybrid light-emitting devices that use organic semiconductors as charge transport layers. While electroluminescence (EL) from group II-VI and III-V NCs has been well studied, emission from group IV NCs including silicon (Si) has not been characterized as extensively. Here, we demonstrate efficient EL from hybrid nanocrystal-organic light-emitting devices (NC-OLEDs) containing SiNCs passivated with ligands of 1-dodecene. In such hybrid NC-OLEDs, a forward-emitted external quantum efficiency of 8.6% at a wavelength of 853 nm is obtained for devices containing 5-nm SiNCs with emission originating solely from the SiNCs. Comparable high efficiency is also obtained for SiNCs emitting at a wavelength of 777 nm. The high device efficiency is noted to be very robust and can be observed with electron transport materials whose mobility varies across three orders of magnitude. In this study, the relative roles of electron mobility and charge/exciton confinement will be discussed in the content of realizing high efficiency.

See more of this Session: Hybrid Organic-Inorganic Materials for Electronics, Photonics, and Photovoltaics
See more of this Group/Topical: Materials Engineering and Sciences Division

Hybrid ZnO complexes from a simple thiol modification process: understanding electro-optical properties

[1]Jisun Im, [1]Jagdeep Singh, [1]James E. Whitten, [2]Diane M., Steeves, [2]Jason W. Soares *

[1]Department of Chemistry and Centers for Advanced Materials and High-Rate Nanomanufacturing, The University of Massachusetts Lowell, Lowell, Massachusetts 01854
[2]U.S. Army Natick Soldier Research, Development & Engineering Center, Natick, Massachusetts 01760

Extended Abstract

Metal oxide nanoparticles, nanotubes, nanorods, nanowires and whiskers are finding applications ranging from catalysis to optoelectronics. In particular, nano-crystalline zinc oxide (nano-ZnO) is an important electro-optical material due to its wide bandgap and bimodal photoluminescence (PL) spectrum. Nano-ZnO represents a promising semiconductor for future hybrid devices; however, use has been limited due to inefficient charge transport. Towards creating more efficient ZnO hybrid organic-inorganic complexes, a simple thiol modification approach has been developed. Organothiol surface modifiers are covalently attached to the surface of ZnO nanoparticles (rods and spheres) by stirring an ethanolic nanoparticle/organothiol mixture for one hour. Stirring under ambient conditions creates a thiol monolayer, while elevated temperatures result in complete encapsulation of the ZnO nanoparticles, as determined by transmission electron microscopy (TEM), X-ray photoelectron spectroscopy (XPS) and thermogravimetric analysis. While previous studies have demonstrated that alkanethiols adsorb on ZnO surfaces and nanoparticles, the present work is the first to demonstrate encapsulation of ZnO nanoparticles. The encapsulation layer consists of a thick (e.g., 100-500 Å) organic shell comprised of a 1:2 Zn:thiol complex. The thickness and morphology of the encapsulating layer is controllable by choice of thiol and preparation conditions. Furthermore, because a large selection of functionalized thiols is available, it is possible to surround ZnO nanoparticles with a variety of chemical functional groups for subsequent interactions. The effect of thiol monolayer and encapsulating layer formation on the inherent PL of nano-ZnO has also been analyzed, demonstrating the ability to induce unique changes in the PL, when compared to other surface modifiers.

For organic-inorganic hybrid devices, it is often desirable to covalently attach organic molecules to metal oxide surfaces such that electrons and holes may be transported across the organic-inorganic interface. Here, we discuss the creation of ZnO inorganic-organic and metal-metal oxide hybrid complexes via thiol linkages using the facile thiol modification approach described above. One such hybrid, formed from nano-ZnO and a potential photovoltaic dye, was created. A thiol-derivatized ruthenium-based dye was synthesized and directly attached through covalent interaction of the thiol groups and the surface of nano-ZnO under monolayer adsorption conditions while flushing with argon gas. Detailed understanding of the electronic structure of the hybrid complex, which has previously been lacking, was generated through XPS, ultraviolet photoelectron spectroscopy (UPS), and absorbance spectroscopy studies. An energy level diagram was constructed, indicating that the lowest unoccupied molecular orbital (LUMO) of this dye is lower in energy than the ZnO conduction band edge, providing minimal enthalpic driving force for photovoltaic electron injection.

Toward the goal of tailoring the optical properties of nano-ZnO, the influence of dye adsorption on the inherent PL emission of nano-ZnO was determined. Adsorption of the dye caused complete quenching of the nano-ZnO inherent visible emission intensity, while the excitonic UV emission intensity remained unaltered. Creation of this complex using the simple thiol modification approach confirms the possibility of using thiol-terminated dyes for ZnO-based DSSC devices, and understanding the electronic structure will facilitate future optimization of the ZnO:dye hybrid complex.

Another hybrid system, ZnO nanorods tethered to gold nanoparticles (AuNPs) via dithiol linkers, has also been synthesized. Although ZnO:AuNP complexes have previously been created, this is the first successful attempt to covalently attach AuNPs to the surface of ZnO using dithiol chemistry and to understand the influence on electronic transfer processes at the interface. AuNPs functionalized with octanethiol were tethered to the nanorod surface using the thiol monolayer absorption approach by mixing octanethiol-protected AuNPs and the nanorods in the presence of p-terphenyl dithiol. One end of the dithiol linker bonds to AuNPs via a ligand place-exchange reaction, and the other end attaches to ZnO via Zn-S bonding. Hybrid complexes of varying AuNP surface density were created by adjusting the AuNP:ZnO ratio and dithiol concentration. XPS and TEM confirm attachment of gold nanoparticles to the nanorod surface, and fluorescence spectroscopy was employed to measure the influence of the proximal metal on the photoluminescence of the ZnO. As the AuNP surface coverage increases, the visible emission band, centered at ~520nm, and the UV emission band decrease in intensity. Furthermore, the UV emission band is increasingly blue shifted as the AuNP surface density increases. The electronic processes that enable control of the nano-ZnO optical properties upon AuNP adsorption were also investigated in-depth. The information derived will facilitate formation of future ZnO:AuNP complexes with increasing control of the electro-optical properties, enabling design of a series of hybrid materials with specific optical responses. The creation of hybrid organic:inorganic and metal: metal oxide complexes through the described simple thiol modification approach could have significant implications on future chemical sensors, as well as photovoltaic devices.

(377c) Polymer-Nanocrystal Hybrid Solar Cells Using P3HT and Pyrite FeS2

Beau J. Richardson, John Bae, Leize Zhu and Qiuming Yu, Dept. of Chemical Engineering, University of Washington, Seattle, WA

Polymer-NanocrystalHybrid Solar Cells Using P3HT and Pyrite FeS_2

BeauRichardson, John Bae, Leize Zhu, and Qiuming Yu

Departmentof Chemical Engineering, University of Washington, Seattle, WA 98195

qyu@uw.edu

In order to reducedependence on exhaustible energy sources and reduce carbon emissions, muchscientific effort has been directed towards reducing the cost of renewableenergy sources to make them competitive with conventional sources. Solar cellshave great potential as a low-carbon, sustainable energy source but are stilltoo expensive to compete. Searching for new materials to design low-cost andmore efficient solar cells can reduce the cost per kilowatt-hour (kWh) ofenergy produced from these devices and bring them closer to grid parity. Specifically,pyrite FeS_2 nanocrystals have several advantages for making solarcells but its use in thin-film, bulk-heterojunction type devices has not beenthoroughly studied.

Recently, our grouphas demonstrated controllable synthesis of pyrite nanocrystals in octahedraland cubic shapes. We are incorporating these nanocrystals into a hybrid, bulkheterojunction device with the semiconducting polymer Poly (3-hexylthiophene-2,5-diyl)(P3HT) and studying how the different nanocrystal sizes andshapes affect charge transport, light absorption, and overall deviceperformance. In conjunction with the fabrication of these hybrid devices, weare also fabricating organic devices with P3HT:PCBM active layers under thesame conditions for direct comparison. The typical device is fabricated on anITO coated glass substrate with a ~40 nm PEDOT:PSS hole conducting layer,followed by a ~100-200 nm P3HT/pyrite active layer and an ~85 nm Al electrode. Thefilm thickness and the surface morphology of the hybrid thin films are measuredby AFM. The optical properties of the film are studied using ellipsometry andUV-vis-NIR absorption spectroscopy. The power conversion efficiency (PCE) isdetermined by measuring the I_{sc} and V_{oc} under AM 1.5Global Spectrum at 100 mW/cm^2. A time-correlated single photoncounting (TCSPC) system is used to analyze the relaxation of electrons in thesefilms from an excited state to a lower energy state. By employing TCSPC tostudy P3HT, pyrite nanocrystals, PCBM, P3HT:PCBM, and hybrid P3HT:pyritenanocrystal films, we can elucidate the charge transport characteristicsbetween the n- and p-type materials. The effects of the ratio of P3HT:

pyritenanocrystals, solvent, film thickness and annealing conditions on the deviceperformance will be discussed.

See more of this Session: Hybrid Organic-Inorganic Materials for Electronics, Photonics, and Photovoltaics
See more of this Group/Topical: Materials Engineering and Sciences Division

(377d) Correlating Interfacial Interactions and Macroscopic Photovoltaic Properties of Self-Assembled Hybrid Materials

Justin P. Jahnke[1], Shany Neyshtadt[2], Tamar Segal-Peretz[2], Mara Daz-Garca[3], Gitti L. Frey[2] and Bradley F. Chmelka[1], (1)Chemical Engineering, UC Santa Barbara, Santa Barbara, CA, (2)Materials Engineering, Technion, Israeli Institute of Technology, Haifa, Israel, (3)Applied Physics, University of Alicante, Alicante, Spain

Correlating interfacial interactions and macroscopic photovoltaicproperties of self-assembled hybrid materials

JustinP. Jahnke[1], Shany Neyshtadt[2], Tamar Segal-Peretz[2], Maria Diaz-Garcia[3],Gitti L. Frey[2], Brad F. Chmelka[1*]

[1] *Department of Chemical Engineering, University of California, Santa Barbara*

[2] *Department of Materials Engineering, Technion- Israel Institute of Technology*

[3] *Department of Applied Physics, University of Alicante, Spain*

The use of surfactant-directed mesostructured inorganic materials allows for the preparation of oxide frameworks ascontinuous transparent films with high inorganic-organic interfacial contactand the inclusion of diverse functional guest species. For example, mesostructured titania films (with ca. 10 nmchannel dimensions) containing light-absorbing conjugated guest species exhibitphotovoltaic properties that are influenced strongly by solution processingconditions. Solution compositions must be chosen for compatibility with the dissimilarmolecular components, which co-assemble into functional hybrid materials as thesolvent is removed. Subsequent integration of these materials into photovoltaicdevices leads to performances that depend on many factors, including thedegree of exciton dissociation at the conjugated guest-titania interface andcharge transport to the electrodes. The degree of exciton dissociation isstrongly influenced by the extent of interfacial contact between the conjugatedguest species and titania network, while charge transport may be enhanced bythe introduction of orientational order. Detailed molecular-level understandingof the interactions, mobilities, and proximities among the different functionalcomponents in the hybrid materials can be obtained through solid-state two-dimensionalNMR and other techniques and correlated with macroscopic photo-currentproperties. In particular, judicious selections of the synthesis andprocessing conditions are demonstrated to promote enhanced interfacial contactand charge carrier transport, leading to improved photovoltaic deviceperformance.

Resulting insights allow mesostructured titania-conjugatedpolymer hybrid materials and device properties to be controllablymodified and optimized.

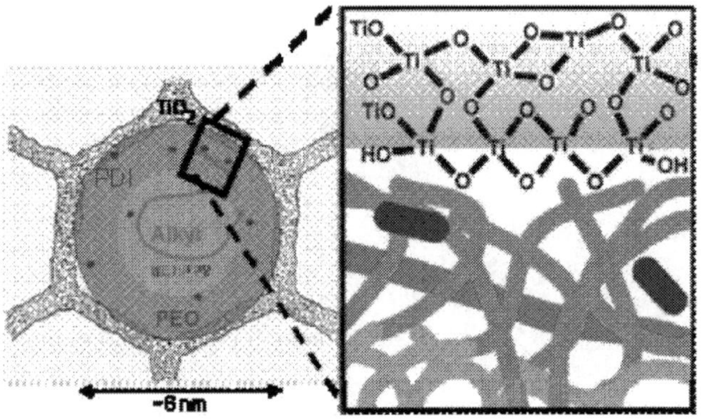

See more of this Session: Hybrid Organic-Inorganic Materials for Electronics, Photonics, and Photovoltaics
See more of this Group/Topical: Materials Engineering and Sciences Division

(377e) Interfacial Stability and Compositional Distribution Effects On Electronic Properties of ZnSe/ZnS Core/Shell Nanocrystals

Sumeet C. Pandey, Jun Wang, T. J. Mountziaris and Dimitrios Maroudas, Department of Chemical Engineering, University of Massachusetts, Amherst, MA

Ternary semiconductor nanocrystals exhibit size-dependent excitonic properties and allow for exceptional tunability of their band gaps by modulation of the nanocrystal composition. Band structure enegineering based on such controllable parameters enables the development of nanomaterials suitable for photovoltaic and light-emitting devices, as well as for highly luminescent biological labels. A similar class of hetero-nanocomposite materials with widespread technological interest is that of core/shell quantum dots (QDs). For example, the so-called type-I QDs, with wider band-gap shell material such as ZnSe/ZnS, exhibit an increased stability against photodegradation and enhanced quantum yields. Synthesis of such core/shell structures has followed various colloidal chemistry routes, including the coating of a narrower-band-gap semiconductor core with a shell of a wider-band-gap material in a two-step process. However, the thermodynamic stability of core/shell nanostructures synthesized from such processes and the resulting equilibrium compositional distributions have not been investigated systematically.

The nm-scale diffusion lengths in nanocrystals introduce an interesting interplay between the kinetic and thermodynamic stability of core/shell interfaces. Based on this concept, in this presentation, we report results of a systematic analysis of the thermodynamic stability of ZnSe/ZnS core/shell nanocrystals. The analysis is based on first-principles density functional theory (DFT) calculations and MC simulations combined with X-ray photoelectron spectroscopy (XPS) and determination of the photoluminescence (PL) spectra of QDs that we have synthesized. In this synergistic procedure, the MC simulations are based on classical valence force fields that we have parameterized based on our DFT calculations of surface segregation energies using slab supercells. Our findings indicate the possibility of degradation of electronic properties over time, which has far reaching implications for the application of such nanostructures in devices. Moreover, our electronic structure calculations elucidate compositional effects on the band gaps of $ZnSe_{1-x}S_x$ nanoclusters. The electronic structure dependence on the compositional distribution obtained for core/shell nanocrystals at various compositions provide an interpretation for the observed peak shifts in the PL spectra of the QDs.

See more of this Session: Hybrid Organic-Inorganic Materials for Electronics, Photonics, and Photovoltaics
See more of this Group/Topical: Materials Engineering and Sciences Division

(377f) Role of Quantized and Mid-Gap States In Charge Transport In Semiconductor Nanocrystal Thin Films

Prashant Nagpal, Physical Chemistry and Applied Spectroscopy, Los Alamos National Labratory, Los Alamos, NM

Colloidal semiconductor nanocrystals (NCs) have attracted significant interest for applications in solution-processable devices such as light-emitting diodes and solar cells. However, poor understanding of charge transport in NC assemblies, specifically the relation between electrical conductance in dark and under light illumination, hinders their technological applicability. Here, we simultaneously address the issues of "dark" transport and photoconductivity in films of PbS NCs by incorporating them into optical field-effect transistors (OFETs), in which the channel conductance is controlled by both gate voltage and incident radiation. Spectrally resolved photoresponses of OFETs reveal a weakly conductive mid-gap band (MGB), formed due to defects on NC surfaces, which is responsible for charge transport in dark. The mechanism for conductance, however, changes under illumination when it becomes dominated by band-edge quantized states. In this case, the MGB still plays an important role as its occupancy (tuned by the gate voltage) controls the dynamics of band-edge charges. This demonstrates the role of surface states in NC thin film conductance, and also highlights the role of defects in thin film photoconductance by modulation of recombination dynamics.

See more of this Session: Hybrid Organic-Inorganic Materials for Electronics, Photonics, and Photovoltaics
See more of this Group/Topical: Materials Engineering and Sciences Division

(394a) Integrating Nanocellulose Production with Biofuel

Junyong Zhu, USDA Forest Products Laboratory, Madison, WI

This study demonstrated the feasibility of integrating the production of nano-fibrillated cellulose (NFC), a potentially highly valuable biomaterial, with sugar/biofuel (ethanol) from wood fibers. Commercial cellulase enzymes were used to fractionate the less recalcitrant amorphous cellulose from a bleached kraft eucalyptus pulp resulting in a highly crystalline and recalcitrant cellulose (RC). The RC is difficult to hydrolyze to sugars but very suitable for producing biobased nanomaterials through mechanical homogenization. A range of fractionation yields of RC from 10–70% can be achieved by varying fractionation duration and enzyme dosage. The crystallinity of the RC was found to be as much as 24% higher than that of the original bleached pulp. The cellulase fractionation process facilitated mechanical homogenization by significantly reducing the degree of polymerization (DP) to about 400 and the length of the fibers to about 200 mm. The hydrolyzed sugars were found to be easily converted to ethanol through yeast fermentation with excellent efficiency of 91%. Films made from nano-fibrillated cellulose were found to be optically transparent with opacity as low as 12%. The NFC films were mechanically strong and stiff with tensile strengths and moduli approximately 10 and 6 times higher than film made from fibers that had not been nano-fibrillated.

See more of this Session: Nanocellulose-Based Materials and Composites
See more of this Group/Topical: Forest and Plant Bioproducts Division

SURFACE AND AGING CHARACTERIATICS OF PAPER COATED WITH NANO SIZE ZINC OXIDE PIGMENT

Sanjay Tyagi. Renu Tyagi* , B P Thapliyal* R M Mathur* and A K Ray***

**Central Pulp and Paper Research Institute, Saharanpur – 247001*
*** Indian Institute of Technology, (Saharanpur Campus), INDIA*

ABSTRACT

Quality of paper is improved on pigment coating in terms of surface, optical properties and printability characteristics. One of the major parameter for aqueous pigment coating is the type of pigment used and also its particle size and distribution. Zinc oxide is one of the pigment s employed for special coating application. It is offlet nano size particles are playing prominent roles in several areas ranging from medicine to electronics. In paper making there is no exception. Few specialty grades of paper are already using nano size calcium carbonates both as filler and also as coating pigments. Application of water based barrier coating in paper and board in food packaging industry have proven successful. It is also important due to substantial smaller size particles (nano range) compared to macro size solid particles (in micron range) can provide better functional properties of paper in some specialized applications. However, ZnO is present in various forms nano size ZnO particles can enhance the properties significantly to improve optical and surface properties and specially the effect of aging. In this present therefore an attempt has been made to study the potential technological applications of nano zinc oxide particles on paper surface made from hardwood furnish and bamboo using 100% nano ZnO and with admixture of coating grade China clay using laboratory rod coater. All the required properties were evaluated as per ISO standard and compared with the same base paper coated with micro size ZnO pigment particles. From the detailed study it is observed that paper coated with nano ZnO particles imparts better properties as compared to bulk ZnO in terms of brightness, whiteness, gloss, smoothness and printability. Aging properties were also thoroughly studied under artificial condition of varying temperature (50% – 120%)and humidity (10% - 90%). And also UV light, paper coated with nano ZnO showed better resistance to brightness reversion and UV protecting potential.

INTRODUCTION

Paper is pigment coated to improve the quality of paper in terms of surface, optical and printability characteristics. Paper coating ingredients which generally contains pigments, binders and additives is applied to the paper surface to improve the above properties of the paper. One of the major parameter for aqueous pigment coating is the type of pigment used and also its particle size distribution [(1,2,3)]. A commonly used pigment includes china clay, calcium carbonate, silica, zeolite, titanium dioxide, alumina trihydrate and polystyrene [(2)]. Zinc oxide is one of the pigments employed for special coating application [(4)]. Presently all above mentioned conventional pigments are available in the market have their particle size in the range of 1 – 10 micron [(5,6)]. The size of these particles is too big for the preparation of coating pigments to be used efficiently in coating machines. Evenness and uniformity in surface smoothness and thickness of coating layer which is one of the two prime requirements for producing good multi color print, generally lacks with conventional micron sized coating pigments [7,8,9]. Pore structure and contact angle wettability effect print quality by preventing ink spreading. In order to prevent wicking and feathering, it is important that the thickness of coating layer be homogeneous to a scale of few microns in depth which also helps in the absorption of successive droplets of ink at high delivery rates and any water present[9,10].

Application of nano size Zinc oxide in papermaking

While papermaking is normally a statistical process, nano technology makes use of the classic process where the position of every atom is specified and controlled. Yet despite this apparent discrepancy, between two technologies, the paper industry has been practicing nano particles over the conventional pigments long before it was marketed as latest technology trend [5,6].

Nano sized calcium carbonate has been used by paper industry since long as filler due to the advantage of high opacity, high brightness and whiteness [11]. At present, it is also used in special paper products such as baby diapers, sanitary napkins, etc.

Zinc oxide is also one of the pigments finding its use for special coating applications[4]. It is offlet nano size particles are playing prominent role in several areas ranging from medicine to electronics [12]. ZnO is essential for rubber and tier industry and also used as natural rubber, synthetic rubber and latex vulcanization active agents, reinforcing agent and coloring agents. In paint industry, ZnO in addition to color strength and hiding power outside but also a light coating of preservative and agents [13]. In medical and health and food industries, ZnO has a Sida, bleeding, hygienic convergence of functions. In electronics industry, ZnO varistor is the main raw material is magnetic, optical material such as the main additive. Prepared using ZnO varistor, not only has low sintering temperature, but also improve the performance of porous sensitive resistor such as the flux, the non linear coefficient [14,15].

In the present study, an attempt has been made to study the potential technological applications of nano zinc oxide particles on the paper surface made from the pulp of hardwood furnish and bamboo using 100% nano ZnO and with the admixture of coating grade china clay by taking trials on laboratory rod coater. The Zinc oxide soluble starch nano composite was prepared by wet chemical method using Zinc nitrate and sodium hydroxide as precursors and soluble starch as stabilizing and compared with the same base paper coated with micro size ZnO pigment particles.

EXPERIMENTAL

Coating formulation

The coating trials were carried out using Laboratory rod coater. The coated sheets were dried immediately in laboratory oven at the temperature of 105 C for maximum 2 minutes. The following two coating formulations were used for coating trials.

Ingredient	Composition I	Composition II
Pigment		
• Zinc Oxide (Nano or Ordinary)	500 parts	1000 parts
• China clay	500 parts	
Binder		

• Oxidized Starch	90 parts	90 parts
• Styrene Butadiene Latex (SBR Latex)	60 parts	60 parts
Dispersant		
• Anionic polymer	5 parts	5 parts
	Total solids maintained- 65±2%	

Preparation of coating colour

The preparation method for the coating slip is as under-

(a) Starch cooking

The oxidized starch solution was prepared separately by cooking under continuous agitation in hot water at temperature 90-95 ^{0}C, for about 15 minutes, i.e. upto the point when the whole solution became fully gelatinized.

(b) Pigment preparation

Pigment slurry was made in laboratory high-shear coating mixer. The dispersant was firstly added to desired quantity of water. When homogeneous solution was formed, weighed quantities of pigments were added, under constant stirring. A thick paste thus resulted, which became uniform at high shear. Time of agitation was, however kept at about 15 minutes.

(c) Addition of binders to pigments

The freshly prepared starch solution and synthetic binder SBR latex (in required amount) were added to the pigment paste in laboratory high-shear coating mixer, under constant agitation. The mixture was kept under constant agitation at high shear for about 30 minutes to ensure uniform mixing and dispersion of all the ingredients. Finally the slurry was taken out of the mixer.

(d) Screening

The coating slurry was next screened on a 300-mesh screen to eliminate foreign matters, coagulated coating and particles of dried coating slurry etc. Thus a uniform coating slurry; free from dirt, was ready for coating.

Calendering

Coated sheets after dried in laboratory oven were calendered using the laboratory calender with softnip mode, at 40 ^{0}C temperature & 30 bar pressure. The sheets were passed twice through the softnip.

Evaluation of optical characteristics of the pigments

The two pigments- Zinc Oxide (ordinary) and Zinc Oxide (nano particles), which were used in coating trials, were evaluated also for following optical characteristics; as an extension of the detailed study.

- Brightness (%)
- Whiteness (%)
- Yellowness (%)

Evaluation of base paper and coated paper

Base paper as well the coated & calendared papers were conditioned at 27±1 ^{0}C temperature and 65±2 % relative humidity prior to testing of following properties as per standard methods mentioned below.

Property	Standard methods
Grammage (g/m^2)	ISO 534
Brightness (% ISO)	ISO 2470

Gloss (%)	Tappi 480 om-99
Whiteness (%)	ISO 2470
PPS Roughness (micron)	ISO 8791/4
Picking velocity (cm/s)	ISO 3783
Print density (g/cm^3)	Using IGT printability tester Model AIC2-5
Print uniformity index	Using IGT printability tester Model AIC2-5
Stain length (mm), oil absorbency	Using IGT printability tester Model AIC2-5

RESULTS & DISCUSSION

Optical characteristics of base paper

The base paper used in the coating trials, was procured from commercial paper mill. The base paper was made from the pulp containing furnish of hardwood pulp and bamboo. Base paper was evaluated for different optical & printing properties for comparison purpose with coated & calendered paper. The result is tabulated in Table-I.

Optical characteristics of pigments

The optical characteristics of the pigments used in coating are given in Table-II. By comparing the values of the properties obtained for nano Zinc Oxide and ordinary Zinc Oxide, it was found that the starch-encapsulated nano Zinc Oxide particles had slightly lower brightness, whiteness and was relatively more yellowish in colour as compared to ordinary Zinc Oxide particles. This is probably due to the preparation process.

Characteristics of Coated & Calendered Paper samples

The coated paper samples, after softnip calendering, were evaluated for

(a) Optical characteristics
(b) Surface characteristics
(c) Printing characteristics
(d) Aging characteristics

(a) Optical characteristics

The effect of nano size zinc oxide particles on brightness of coated and calendered paper is given in Figure 1. Results show that nano zinc oxide gives better brightness as compared to conventional zinc oxide in both cases. i.e. when used alone and when used in combination with coating grade china clay. The same trend is true with whiteness value of coated paper (Fig. 3).

The value gloss of paper coated with nano ZnO also improved as compared to paper coated with conventional ZnO (Fig. 2). However, the improvement is low as compared to brightness and whiteness values of base paper, there is a fall in their values after coating with ZnO and China clay. The reduction in values is more in case when china clay is used in coating formulation. This is probably due to the lower brightness of china clay and binder additives which results in overall reduction of optical properties of coated paper.

(b) Surface characteristics

PPS Roughness

The surface smoothness of coated paper which is an important property in deciding the print performance of paper, improved with the application of nano ZnO. The PPS roughness in case of nano ZnO is relatively lower than conventional ZnO. The same trend was obtained when nano ZnO is used with china clay. A smooth surface was obtained when china clay is used in coating furnish. This is probably due to particle shape. China clay being platelet shape flattened more during the calendaring operation in compared to ZnO which is spherical in shape (Fig. 4).

(c) Printing characteristics

Picking velocity

Picking velocity of paper coated with nano ZnO was found to be higher as compared to conventional ZnO (Fig. 5) in both cases when used alone and in combination with china clay. This indicates that paper coated with nano ZnO can run easily on high speed printing presses or at the same printing speed, less binder is required in coating formulations with nano ZnO. The excess amount of binder always has negative effect on the printing characteristics.

Print Mottle (Print Non-Uniformity)

Print non uniformity leads to print mottle which is uneven appearance of colour density in a print. This is due to the surface factors to absorb ink evenly due to variation in base paper formation or coat weight. The non uniformity printing index is lower for paper coated with nano ZnO as compared to paper coated with conventional ZnO (Fig. 8).

The variation in sheet matrix formation formed by coating layer of conventional ZnO (due to non uniform size of ZnO particles) probably the reason for this. Paper coated with ZnO alone showed lower number as compared to paper coated with a mixture of china clay and ZnO since particle size of china clay is largely varied. The sheet matrix variation can lead to print mottle non uniform compression during calendering, resulting in variation in coating layer densification. Sheets that contain binder migration, coat weight variation, inferior coating profile and heterogeneous pigment blends can all cause mottle formation on the printed sheet.

Print density

Print density is affected by ink setting/ink holdout. For good density the vehicle of printing ink should stay on the top of paper together with ink pigment. The more the dried ink is retained on the outer surface of the paper, the higher will be the ink density, accordingly, the lower will be the ink demand. Paper with large pores in coating will have more penetration of printing ink leading the lower density. This is observed in case of papers coated with conventional ZnO coating (Fig. 7). The same is true with china clay in the coating formulation. This indicates that more ink will be required to print coated papers with conventional ZnO coatings.

Oil absorbency

Oil absorbency which is an indication of rate of penetration of oil was found to be more in the case of paper coated with nano size ZnO alone and in combination with china clay. (Fig 6)

(d) Effect of Aging

The effect of aging on the optical properties of coated and calendered paper is shown in the graph (Fig. 9-11) and tabulated in table (4-6). The results show that nano zinc oxide gives excellent UV protecting strength in both cases when used alone and when used in combination with coating grade china clay. The reduction in brightness and whiteness is minimum in the case of nano zinc oxide. The lower value of indicated the higher yellowness, which incase after aging maximum when china clay is used.

CONCLUSIONS

The use of nano zinc oxide as a coating pigment improves the paper properties in terms of optical, surface and printability as compared to conventional zinc oxide. This indicates that use of nano ZnO can be a good substitute for the replacement of conventional ZnO in the case of speciality grade papers. This opens a new scope for the use of nano Zinc oxide as a coating pigment for speciality grade of papers. The application of nano ZnO also exhibits excellent UV protection characteristics when tested at different condition of humidity and temperature. Paper represents good aging resistance properties which increases the longevity of paper. This opens a new scope of manufacturing paper with much longer life which lasts for longer years paper manufactured under connectional methods

The use of nano Zinc oxide as a coating pigment results in the paper with better properties as compared to conventional ZnO in terms of brightness, whiteness, gloss, smoothness, UV protection and printability characteristics i.e. picking velocity, print uniformity, print density, and oil absorbency .

Refrences:

1. 'Y.V.Sood, S.Tyagi. R.Tyagi, P.C.Pande, R.Tandon & T.K.Roy', 'Quality of indigenous coated paper and ways to improve it', *'J.IPPTA, 2008, 20 (4), pp'*
2. 'Coating pigments: from size and shape to nanotech', *'www.thefreelibrary.com/coating+pigments:+from+size+and+shape+to+nanotech.....'*
3. 'Surface characteristics of coating layers formed by coating pigments with different particle sizes and size distribution' *'J. Coat. Tech. and Res'*
4. 'N. Vatansever & S. Polat', 'Effect of zinc oxide type on aging properties of styrene butadiene rubber compounds', *'www.sciencedirect.com, 2009'*
5. Roxare, 'Nanotechnology in paper making', *'http://pulpandpapercircle.com/2008/07/nanotechnology –in-paper-making'*
6. 'J.Shen, Z.Song, X.Ojan, F.Yang & F.Kong', 'Nano fillers for papermaking wet end applications', *'www.bioresources.com, 2010, 5(3)'*
7. G. Chinga, 'Detailed characterization of paper surface structure for gloss assessment', *'JPPS, 2004, 30 (8), pp 222-227'*
8. G. Ström, S. E. Research, A. Englund & M. Karathanasis, 'Effect of coating structure on print gloss after sheet fed offset printing', *'J. Nord. Pulp & Paper, 2003, 18 (1), pp108-115'*
9. 'G. Chinga-Carrasco, M. Axelsson, Ø. Eriksen & S. Svensson', 'Structural characteristics of pore networks affecting print through', *'JPPS, 2008, 34 (1), pp 13-22'*
10. 'D.W. Donigian, J. N. Ishley & K.J. Wise, 'Coating pore structure and offset printed gloss', *'J.TAPPI, 1997, 80 (5), pp 163-172'*
11. T. Nypelö, M. Österberg & J. Laine, 'Ultra thin coating of paper by tailor made nanoparticles', *'TAPPI International Conf. on Nanotechnology for the Forest Product Industry, 2010'*
12. 'V. Bianchard & P. Blanchet', 'Color stability for wood products during use: Effects of inorganic nanoparticles', *'www.bioresources.com, 2011, 6(2)'*
13. 'B. Wang, N. Han, D. Meng, R. Yue, J. Yan & Y. Chen', 'Filter paper- templated preparation of ZnO thin films and examination of their gas sensing properties', *'www.sciencedirect.com, 2011'*
14. 'V. Prasad, A. J. Shaikh, A. A. Kathe, D. K. Bisoyi, A.K. Verma, N. Vigneshwaran', 'Functional behavior of paper coated with zinc oxide-soluble starch nanocomposites', *'J. Mat. Proc. Tech., 2010, 210 (14), pp 1962-1967'*
15. 'Surface characteristics of coating layers formed by coating pigments with different particle sizes and size distribution' *'J. Coat. Tech. and Res.'*

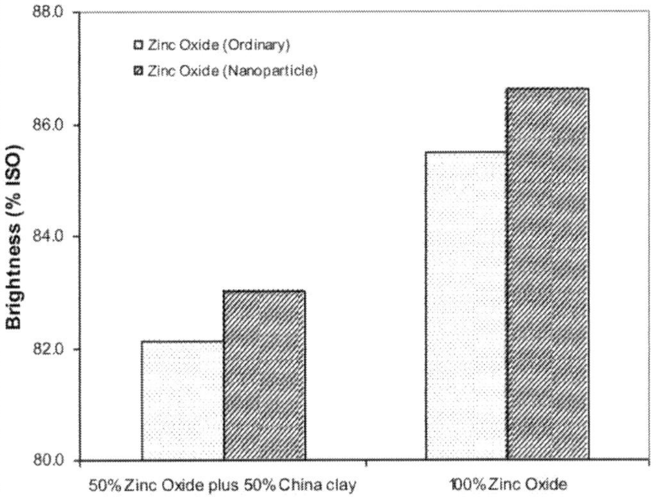

Fig. 1: Effect of Zinc Oxide nanoparticles on brightness of coated & calendered paper

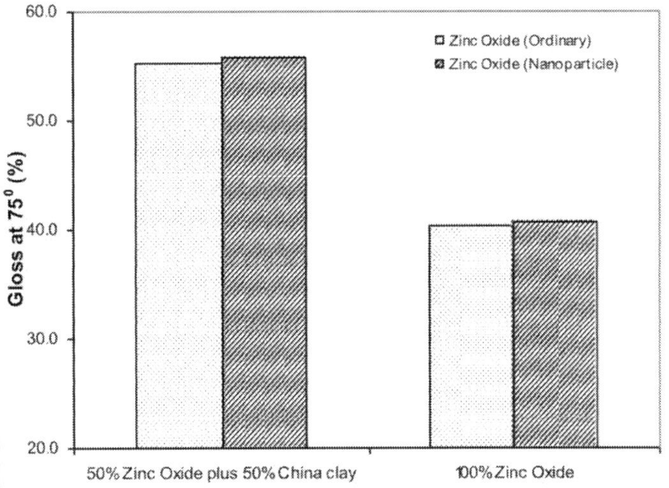

Fig. 2: Effect of Zinc Oxide nanoparticles on gloss of coated & calendered paper

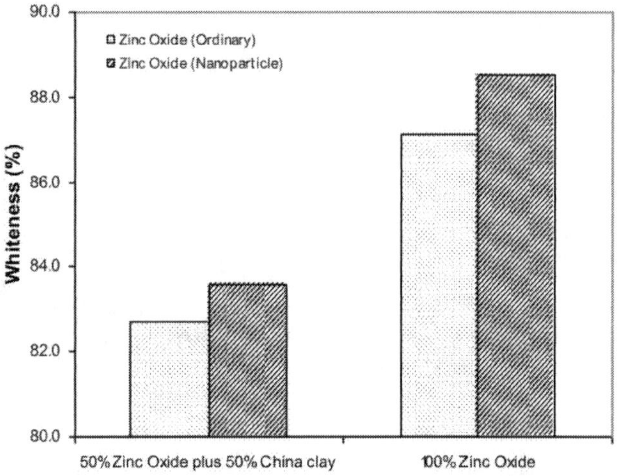

Fig. 3: Effect of Zinc Oxide nanoparticles on whiteness of coated & calendered paper

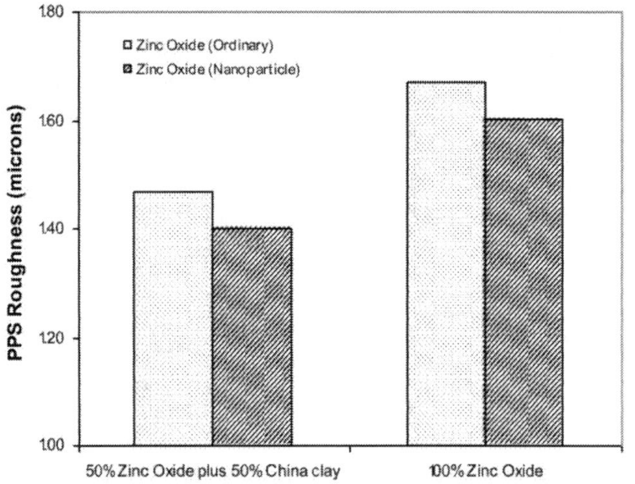

Fig. 4: Effect of Zinc Oxide nanoparticles on the roughness of coated & calendered paper

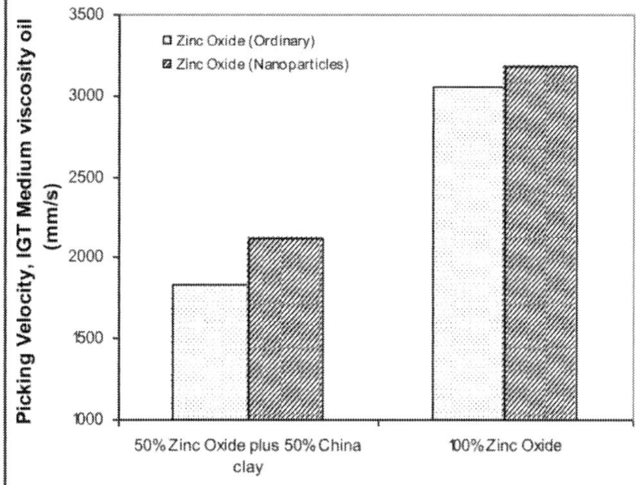

Fig. 5: Effect of Zinc Oxide nanoparticles on picking velocity of coated & calendered paper

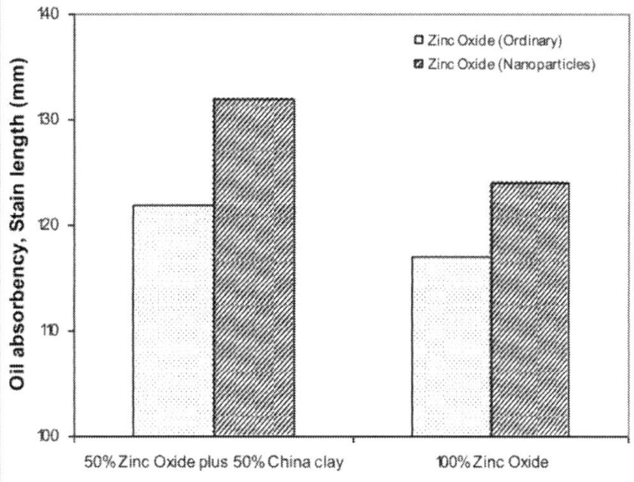

Fig. 6: Effect of Zinc Oxide nanoparticles on oil absorbency of coated & calendered paper

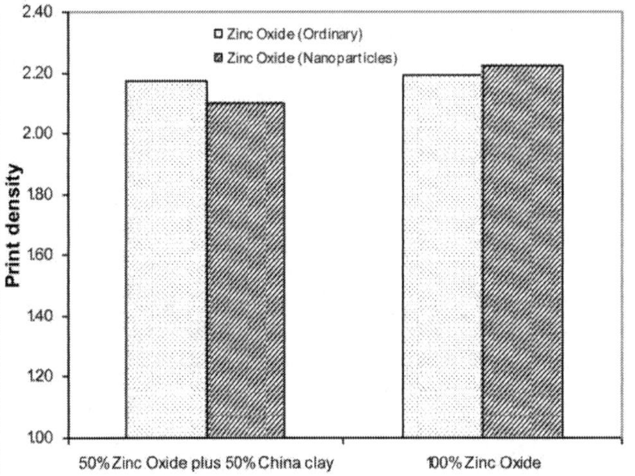

Fig. 7: Effect of Zinc Oxide nanoparticles on print density of coated & calendered paper

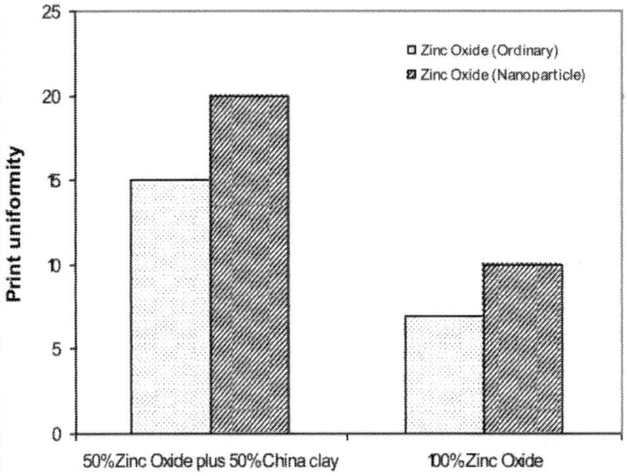

Fig. 8: Effect Zinc Oxide nanoparticles on print uniformity of coated & calendered paper

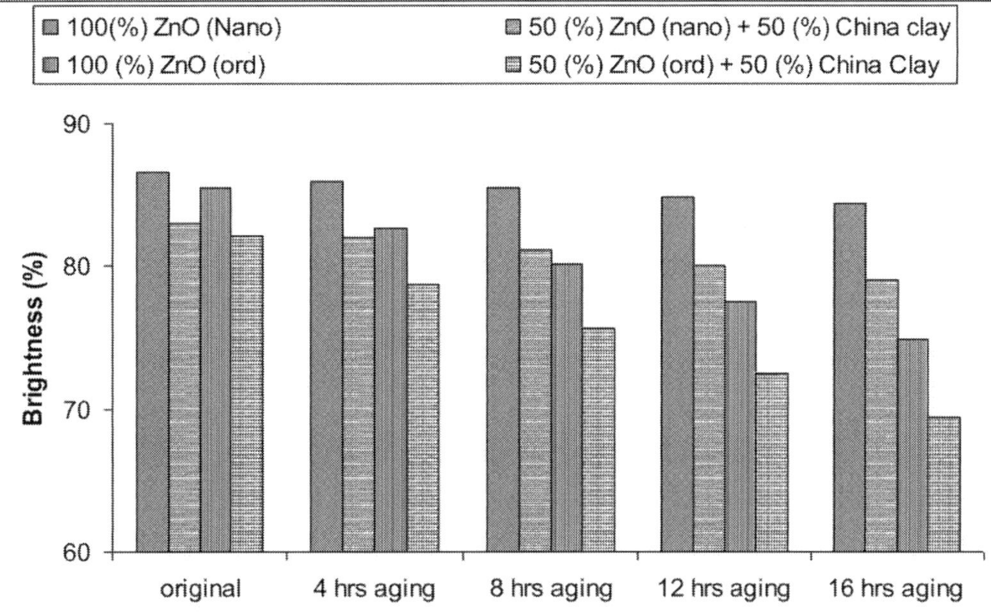

Fig. 9: Effect of zinc oxide nano particle on brightness reversion of coated and calendared paper

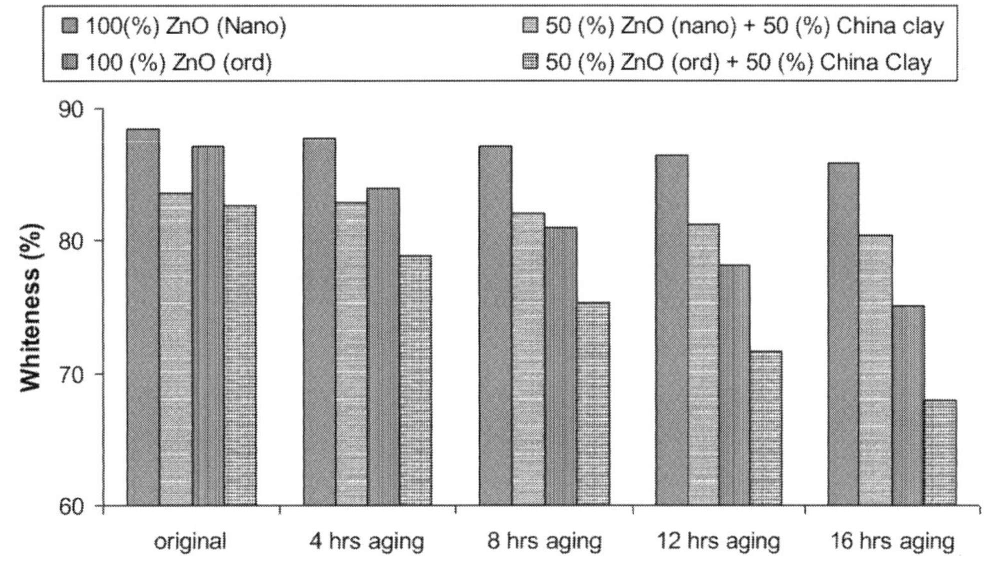

Fig. 10: Effect of zinc oxide nano particles on whiteness reversion of coated and calendared paper

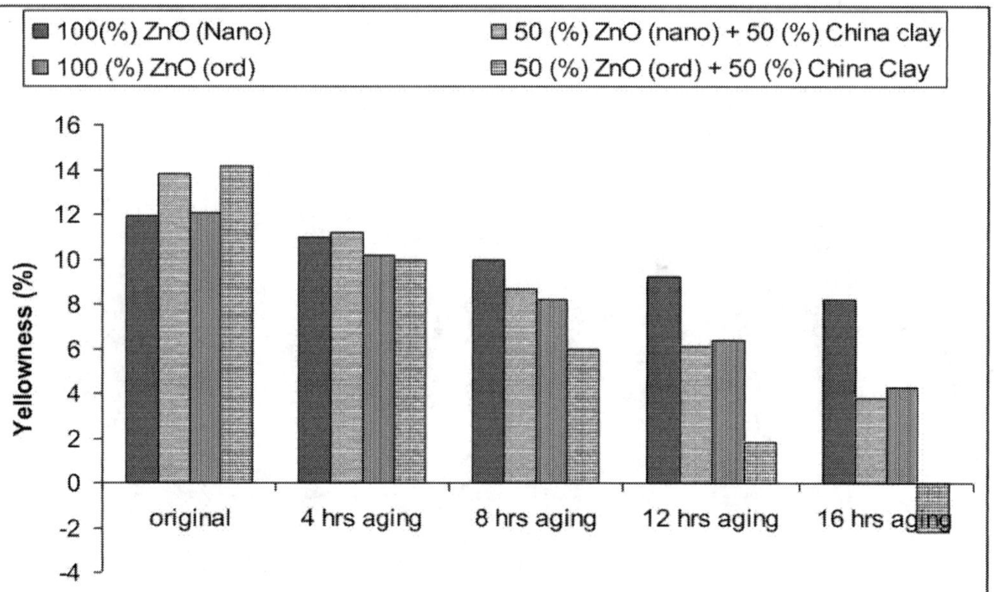

Fig.11: Effect of zinc oxide nano particle on yellowness reversion of coated and calendared paper

Table-I: Optical & printing characteristics of base paper used for coating

Property	Values obtained for the sample
Brightness (%)	88.5
Opacity (%)	80.0
Yellowness (%)	-5.13
Whiteness (%)	124.2
Gloss at 75^0 (%)	6.9
Porosity, Bendtsen (ml/min)	593

Table-II: Optical & printing characteristics of Zinc Oxide pigments used in coating

Property	Zinc Oxide (Ordinary)	Zinc Oxide (Nanoparticle)
Brightness (%)	97.8	96.2
Whiteness (%)	93.9	91.2
Yellowness (%)	1.64	1.70

Table-III: Different optical & printing characteristics of coated & calendared paper

Property	Pigment composition for coating			
	50% Zinc Oxide (Ord.) plus 50% China clay	50% Zinc Oxide (Nano) plus 50% China clay	100% Zinc Oxide (Ord.)	100% Zinc Oxide (Nano)
Coat weight (g/m^2)	12.7	12.6	12.8	12.4
Brightness (%)	82.1	83.0	85.5	86.6
Whiteness (%)	82.7	83.6	87.1	88.5
Gloss at 75°(%)	55.2	55.8	40.4	40.6
PPS Roughness (microns)	1.47	1.40	1.67	1.60
Oil absorbency, IGT Stain length (mm)	122	132	117	124

Picking velocity, IGT Medium viscosity oil (mm/s)	1830	2120	3060	3180
Picking Viscosity Velocity Product (VVP), IGT(Poise.m/s)	805	933	1346	1399
Print density	2.17	2.10	2.19	2.22
Print uniformity	15	20	7	10

Table-IV: Brightness reversion of coated and calendared paper after aging

Pigment Composition	Brightness (%)				
	Original	4 hrs aging	8 hrs aging	12 hrs aging	16 hrs aging
100(%) ZnO (nano)	86.6	86	85.5	84.9	84.4
50(%) ZnO (nano) + 50 (%) China Clay	83	82	81.1	80	79
100 (%) ZnO (ord)	85.5	82.7	80.2	77.5	74.9
50 (%) ZnO (ord) + 50 (%) China Clay	82.1	78.7	75.7	72.5	69.4

Table-V: Whiteness reversion of coated and calendared paper after aging

Pigment Composition	Whiteness (%)				
	Original	4 hrs aging	8 hrs aging	12 hrs aging	16 hrs aging
100 (%) ZnO (nano)	88.5	87.7	87.2	86.5	85.9
50 (%) ZnO (nano) + 50 (%) China Clay	83.6	82.9	82.1	81.2	80.4
100 (%) ZnO (ord)	87.1	84	81	78.1	75.1
50 (%) ZnO (ord) + 50 (%) China Clay	82.7	78.8	75.3	71.6	68

Table-VI: Yellowness reversion of coated and calendared paper after aging

Pigment Composition	Whiteness (%)				
	Original	4 hrs aging	8 hrs aging	12 hrs aging	16 hrs aging
100 (%) ZnO (nano)	11.9	11	10	9.2	8.2
50 (%) ZnO (nano) + 50 (%) China Clay	13.86	11.2	8.7	6.1	3.8
100 (%) ZnO (ord)	12.1	10.2	8.2	6.4	4.3
50 (%) ZnO (ord) + 50 (%) China Clay	14.2	10	6	1.8	-2.2

(394c) Investigation of Mass Transport Properties of Microfibrillated Cellulose (MFC) Films

Matteo Minelli[1], **Marco Giacinti Baschetti**[2], Ferruccio Doghieri[2], Mikael Ankerfors[3], Tom Lindstrm[3], David Plackett[4] and Istvn Sir[4], (1)Department of Chemical Engineering, Mining and Environment Technology, University of Bologna, Bologna, Italy, (2)Department of Chemical Engineering, Mining and Environmental technology, University of Bologna, Bologna, Italy, (3)Innventia AB, Stockholm, Sweden, (4)Solar Energy Programme, Ris National Laboratory for Sustainable Energy, Technical University of Denmark, Roskilde, Denmark

Investigation of masstransport properties of microfibrillated cellulose (MFC) films

M.Minelli, M. Giacinti Baschetti, F. Doghieri, M. Ankerfors, T. Lindstrm, I.Sir, D. Plackett

Cellulose-based nanocomposites are now emerging as promising newmaterials suitable for a variety of applications. One source for such newnanocomposites is microfibrillated cellulose (MFC), produced by delamination ofcellulosic fibers in high-pressure homogenizers. MFC is an organic andbiodegradable reinforcement in polymer nanocomposites because of its highaspect ratio, good mechanical properties and ability to form networks: promisingproperties that may make this material suitable for use in a number ofindustrial fields and products.

The structure and transport properties of a four different filmsbased on two different generations of microfibrillated cellulose (MFC), aloneor in combination with glycerol as plasticizer, were investigated throughFE-SEM analysis and sorption or permeation experiments. FE-SEM revealed the existenceof complex structures in the different samples. A porous, closely packed fibernetwork, more homogeneous in the samples containing glycerol, wascharacteristic of the surface of MFC films; while film cross-sections presenteda dense layered structure with no evidence of porosity. Water vapor sorptionexperiments confirmed the hydrophilic character of these cellulosic materialsand showed a dual effect of glycerol, which reduced the water uptake at lowwater activity while enhancing it at high relative humidity, as showed inFigure 1. The observed trends of water sorption can be described considering thatboth physical adsorption on fiber surfaces and absorption in the celluloseamorphous phase takes place and therefore a dual-model sorption model wasemployed.

The water diffusivity (in the panel in Figure 1) in dry samples wasremarkably low for a porous material (10^{-11}-10^{-12} cm^2/s), confirming the existence of complex structures below thefilm surface; however, when water is present in the system, *D* rapidlyincreases with an exponential trend. Diffusivity is also definitely affected byplasticization, being higher for glycerol-containing samples.

Similar behavior was observed in permeation experiments. Dry MFCfilms showed excellent oxygen barrier properties, comparable with those ofpolymers usually considered suitable for ultra-high barrier applications.Indeed, when dry, the cellulose network presents a particularly compact andstiff structure, causing the observed barrier effect. Furthermore, although MFCfilms are characterized by a certain porosity, FE-SEM images of sample surfacessuggests that the pores are not substantially interconnected

When the water content in the membrane is raised, however, adramatic decrease in these properties was observed (Figure 2).

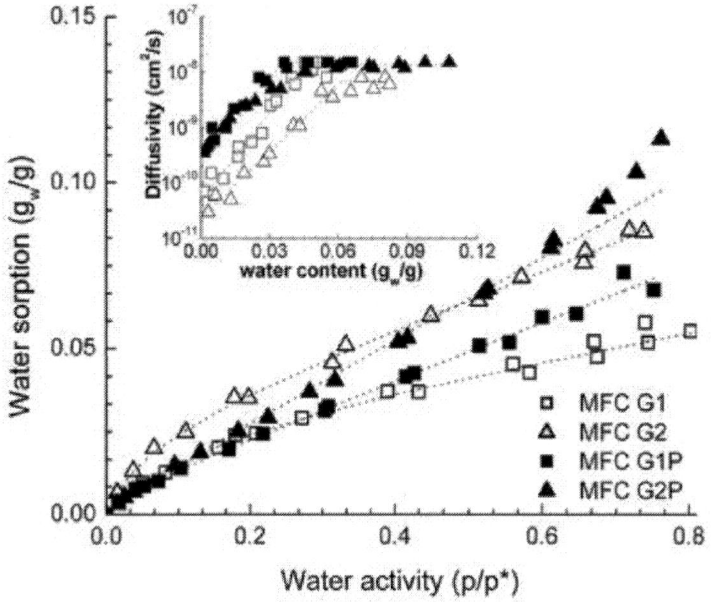

Figure 1. Water vapor sorption isothermsin MFC samples at 35°C, dotted lines are given by the best fitting of the data withthe dual-mode model equation. In the panel: water diffusivity at 35°C in the four different MFC samples as function of the average water concentration.

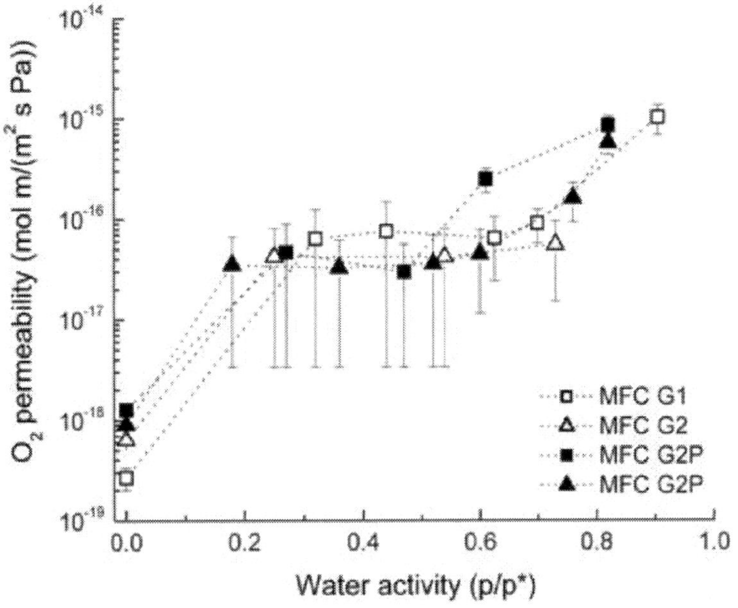

Figure 2. Oxygen permeability in thefour MFC samples at 35°C as a function of water activity.

See more of this Session: Nanocellulose-Based Materials and Composites
See more of this Group/Topical: Forest and Plant Bioproducts Division

(394d) Co-Products of Bioenergy System: Characterization of Saccharification Residuals

Han-Seung Yang, Shona Duncan, William T. Y. Tze and Jonathan Schilling, Department of Bioproducts and Biosystems Engineering, University of Minnesota, Saint Paul, MN

AIChE2011 Annual Meeting - Forest and PlantBioproducts Division

Co-Products of Bioenergy System: Characterizationof Saccharification Residuals

Han-Seung Yang[1], WilliamT. Y. Tze[1], Jonathan Schilling[1] and Shona Duncan[1]

[1]Departmentof Bioproducts and Biosystems Engineering

University of Minnesota

Saint Paul, MN 55108-6130 USA

Tel: 612-625-5200   Fax: 612-625-6286

e-mail: hsyang@umn.edu

The objective of this study is to investigate the effects ofdilute acid pretreatment and enzymatic saccharification on the reactivity,porosity, and crystallinity of solid residuals of biomass. Aspen wood flour was pre-treated with adilute acid treatment followed by enzymatic saccharification. The solid phase of the saccharificationprocess was analyzed using standard NREL protocols for amounts of lignin, pentosans,and hexosans. The functional groupsof solid residuals were analyzed using mid-infrared spectroscopy and hydroxylnumbers. The porosity andcrystallinity of the solids were also evaluated usingrespectively, nitrogen adsorption and X-ray diffraction. Cellulose nanocrystals were extractedfrom the solid residuals and verified using atomic force microscopy. The results will be discussed in thelight of reactivity, porosity, and crystallinity of saccharification residues topave the way for down-stream conversion into co-products of bioenergy(bio-alcohol) systems.

See more of this Session: Nanocellulose-Based Materials and Composites
See more of this Group/Topical: Forest and Plant Bioproducts Division

(394e) Bio-Nano Reinforcement of Polylactic Acid with Surface Modified Cellulose Nanocrystals

Jos Luis Orellana, Esteban E. Urea-Benavides and Christopher L. Kitchens, Department of Chemical and Biomolecular Engineering, Clemson University, Clemson, SC

Cellulose nanocrystals (CNC) have the potential to serve as a nanoscale reinforcement for polylactic acid (PLA), which has become a popular polymer but some of the deficiencies still need to be overcome. The main goal of this work is to improve the mechanical properties of the PLA films – especially toughness – by the addition of surface-modified CNC in the polymer matrix without sacrificing the thermal stability and processability.

CNC were isolated via acid hydrolysis from cotton and surface modified with an esterification process. Transmission Electron Microscopy (TEM) images demonstrated the isolation of the nanocrystals although aggregation can be evidently observed. CNC was further functionalized with different surfactants including decylamine (DA) and cetyltrimethylammonium bromide (CTAB). The dispersibility of the CNC was studied in different solvents such as THF, chloroform and ethyl acetate. Surface modified CNC were added in different concentrations to PLA. Tensile properties of the films were studied using an Instron universal testing machine.

It was found that the stability of the surface-modified CNC improved in organic solvents. DA greatly improved the stability of CNC in both THF and ethyl acetate; while CTAB did in chloroform. No settling was observed in these suspensions for at least three weeks. The toughness and elongation at break of PLA-CNC nanocomposites were improved by the addition of CNC by at least two-fold increase compared to plain PLA. Tensile strength and tensile modulus were not significantly affected by the addition of the crystals. The mechanical properties of the composites were compared and related to the CNC dispersion quality in the different solvents.

See more of this Session: Nanocellulose-Based Materials and Composites
See more of this Group/Topical: Forest and Plant Bioproducts Division

(395a) Template Directed Synthesis and Characterization of Tunable Mesoporous Polymer Resins

Manasa Sridhar, Krishna Reddy Gunugunuri, Panagiotis Smirniotis and Neville G. Pinto, Chemical Engineering Program, School of Energy, Environmental, Biological and Medicinal Engineering, University of Cincinnati, Cincinnati, OH

Ordered organic porous materials possess tremendous potential to serve as radically advanced adsorbents in the field of bio-separations. These biocompatible organic mesoporous polymer replicas are expected to offer unique selectivities based both on size and surface-biomolecule affinity, ultrahigh capacity, reversible adsorption and rapid adsorption/desorption kinetics. In our current study we have synthesized a broad spectrum of ordered mesoporous phenol formaldehyde polymer resins by a facile nanocasting method with siliceous Mesocellular Foams (MCFs) as hard templates. A systematic investigation into the textural, structural and framework properties has been carried out through BET, SAXS, TEM, ^{13}C NMR and FT-IR studies. Faithful replication of pore structure in the resultant mesoporous organic polymer materials by optimizing the resol precursor to template ratio and the thermopolymerization temperature has been achieved. Highly cross-linked phenolic resin frameworks with high surface areas (up to 1200 m^2/g) and large pore volumes (up to 1. 3 cm^3/g) have been synthesized. Well-defined gradation of pore characteristics in the replicas has been established by controlling the morphology of the inorganic hard template used. These interesting findings will be discussed in detail during the presentation.

See more of this Session: Nanoscale Structure In Polymers III
See more of this Group/Topical: Materials Engineering and Sciences Division

Nanoscale Dispersions of Polymers: Aramid Nanofibers

Ming Yang[1], Keqin Cao[2], Lang Sui[3], Ying Qi[4], Jian Zhu[1], Anthony M. Waas[5], Ellen M. Arruda[2], John Kieffer[3], M. D. Thouless[2] and Nicholas Kotov[1*], (1)Chemical Engineering, University of Michigan, Ann Arbor, MI, (2)Mechanical Engineering, University of Michigan, Ann Arbor, MI, (3)Materials Science & Engineering, University of Michigan, Ann Arbor, MI, (4)Electron Microbeam Analysis Laboratory, University of Michigan, Ann Arbor, MI, (5)Aerospace Engineering, University of Michigan, Ann Arbor, MI, *Corresponding author: kotov@umich.edu

Introduction: Kevlar is a well-known ultra-strong para-aramid synthetic macroscale fiber with a high tensile strength-to-weight ratio.[1,2] Typical threads and fibers of Kevlar consist of long molecular chains produced from PPTA (poly-paraphenylene terephthalamide) between which there are inter-chain bonds making the material extremely strong and stiff with a tensile strength 3.6 GPa and modulus of ~90 GPa,[3] comparable to the properties of CNT fibers.[4,5] Despite the fact that Kevlar processing might have involved the polymer in the nanoscale state, Kevlar fibers were never associated with nanoscale building blocks of composites. The formation of nanoscale versions of aramid fibers was never experimentally observed or theoretically predicted before.

Instead of the traditional synthesis from monomers used in many other methods of preparation of nanofibers, which can also be described as the bottom-to-top approach analogous to the terminology used in many nanoelectronic devices, alternative approaches must be explored for making polymeric structures that would satisfy the needs of both hybrid composites from INCs and multifunctional organic coatings. Importantly, the strength of polymeric components can come not only from the strength of individual polymeric chains but also from their assemblies, which, in turn, underscores the significance of the preparation method. Distribution of stress over multiple polymeric chains allows one to avoid stress concentration at particular bonds, which is the most daunting problem of molecular design of new ultrastrong polymers. Taking this further, we decided to use the so called top-down approach reminiscent to that being used for cellulose to prepare suitable polymeric components from macroscale fibers of synthetic polymers, which are known to have high mechanical strength but are processed in a form incompatible with INCs or other polymers.

Results and discussions: 1. Synthesis of ANFs: Inspired by Takayanagito's work on deprotonation of amide groups,[6,7] Burch *et al.* made aramid films by casting the PPTA solution followed by water immersion[8] or thermal decomposition.[9] In all these studies, the nature of the solution was not studied and it was assumed that the solution consisted of polyanions, which was probably only partially correct. Analyzing these works we realized that fine control over the deprotonation and dissolution process can lead to nanoscale fibers because the balance between electrostatic repulsion and attraction due to π-π and van der Waals interactions must lead to nanoscale version of this material. Notably in many ways the balance of forces and the process itself would be similar to the methods of preparation of graphite dispersions with intermediate step of charged graphene oxide. One could expect that the nanometer-scale forms of Kevlar fibers would possess equally high or better strength, stiffness, toughness and possibly other properties. Nanoscale fibers will allow the distribution of stress over many polymer chains and multiple bonds. This will avoid fast breakage of the weakest bond(s) taking place in non-assembled polymer chains.

High temperatures are known to induce the depolymerization of Kevlar,[8] so we decided to employ a different method of Kevlar dissolution than those used before, which allows for processing to be done at room temperature with minimal reduction of the molecular weight of aramid chains. It was found that the dissolution of Kevlar threads or fabric in KOH splits the bulk macroscale fibers into nanoscale fibers (Figure 1). This process is likely to occur by abstraction of mobile hydrogen from >NH groups and substantial reduction of the strength of hydrogen bonds between the polymer chains (Figures 1f,m), while increasing electrostatic repulsion. Importantly, the extent of this destruction is limited and does not appear to proceed down to the level of individual polymer chains. Complete disintegration of the fibers into individual chains is counteracted by hydrophobic attraction and π-π stacking in the polymer backbone. The diameters of ANFs after the dissolution of Kevlar fabric are 20-30 nm (Figure 1a,g). These diameters decrease with the addition of greater amounts of water (Figure 1b-1d, 1h-1j), reaching 5-10 nm when the volume ratio of water to DMSO is 1:40 (Figure 1d,j). From the materials perspective this is quite remarkable and very convenient for processing into composites although will require further studies. The length of these fibers appears to be in the range of 5 to 10 μm (Figure 1l). ANFs are thicker than single-walled CNTs, which are typically about 1 nm in diameter and several micrometers in length. ANFs are thinner and longer than the cellulose fibers, which are 5-15 nm in diameters and 200 nm to several micrometers in length.[10] Abundance

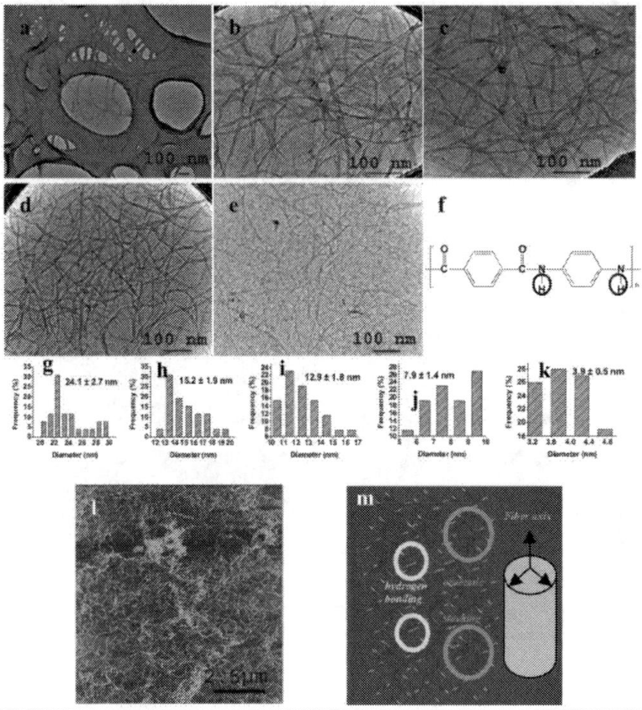

Figure 1. TEM images of ANFs in DMSO solution with different water to DMSO volume ratios: a) 0, b) 1/200, c) 1/100 and d) 1/40. e) TEM image of one bilayer PDDA/ANFs. f) The molecular structure of aramid; the H atoms removed by KOH are marked by circles. (g-k) Statistical analysis of diameter distribution for ANFs under same conditions as that shown in a-e, respectively. l) AFM image of one bilayer PDDA/ANFs. m) Molecular description of ANFs. Yellow circles mark the inter-molecular hydrogen bonds formed between the carbonyl groups and >NH groups; green circles mark aromatic stacking interactions between adjacent strands.

of polar functional groups on ANFs provides additional opportunities compared with CNTs to dramatically improve nanofiber-matrix stress transfer. Note that other potential nanofibers investigated so far, such as cellulose nanofibers, are quite rigid. While materials from them have remarkable properties, their successful utilization with INCs, for instance CNTs, to obtain effective stress transfer is impeded by insufficient flexibility as indicated by multiple recent studies.[11-14] Their sensitivity to humidity could also be problematic.[15]

2. ANF Multilayer Assembly: The availability of dispersions makes it possible to utilize a variety of different materials design techniques based on simple liquid processing. The layer-by-layer (LBL) assembly technique is one of the most versatile methods for the preparation of organic-inorganic composites with enhanced

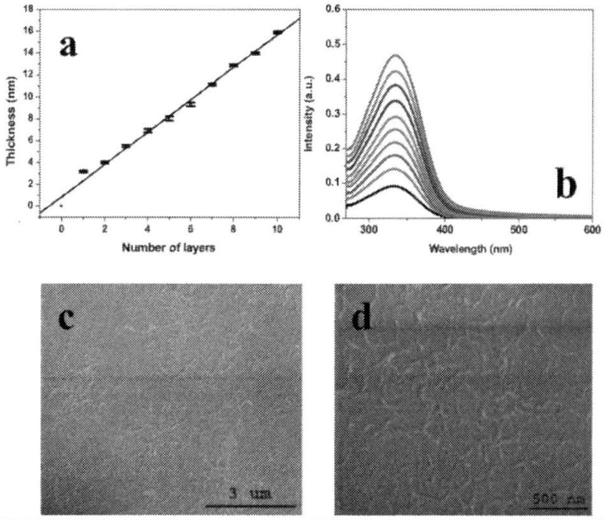

Figure 2. a) Ellipsometry results for LBL assembly of ANFs on silicon substrate. b) UV-Vis spectra of 1-10 layers of ANF films (from bottom to top: 1-10 layers of ANFs). c and d are SEM images of one bilayer PDDA/ANFs on a glass slide.

mechanical parameters. Among other methods of composite synthesis, LBL is advantageous due to intrinsically strong interactions between the components and great uniformity of the resulting materials. Successful examples of LBL composites with exceptional mechanical properties include several INCs.[16,17] It is essential to demonstrate that ANFs can be assembled in LBL systems. It was found that the ANF fibers were both dispersed and ionized in polar solutions and are therefore likely to be suitable for LBL processing. LBL assembly of ANFs was accomplished by using a traditional LBL dipping however only the first layer was PDDA and therefore will be denoted as PDDA/ANF$_n$ where n is the number of dipping cycles in ANF dispersion. The ANFs are fairly uniformly packed on the surface of glass slide during LBL assembly (Figure 2c,d). The gradual growth of ANF multilayers was monitored by ellipsometry and UV-Vis absorption spectra. The growth of PDDA/ANF$_n$ multilayers is nearly linear with the first 10 layers reaching a film thickness of about 16 nm (Figure 2a). This gives a thickness increment for each LBL bilayer of 1.6 nm, which corresponds to fairly loose two-dimensional packing of ANF in each adsorption cycle, as can be seen in the AFM image of Figure 1I.

As more LBL cycles were applied, the enhancement of an absorption peak at around 330 nm, which is characteristic of Kevlar, was observed in the UV-Vis spectrum (Figure 2b). Note that although the initial adsorption of negatively charged ANFs on positively charged PDDA can be attributed to electrostatic interactions, intermediate rinsing with water is expected to regenerate the H^+ on the aramid macromolecules removed by KOH (Figure 1f) and near-electroneutrality of the ANFs. At this moment, stability of ANF films outside of KOH is attributed primarily to

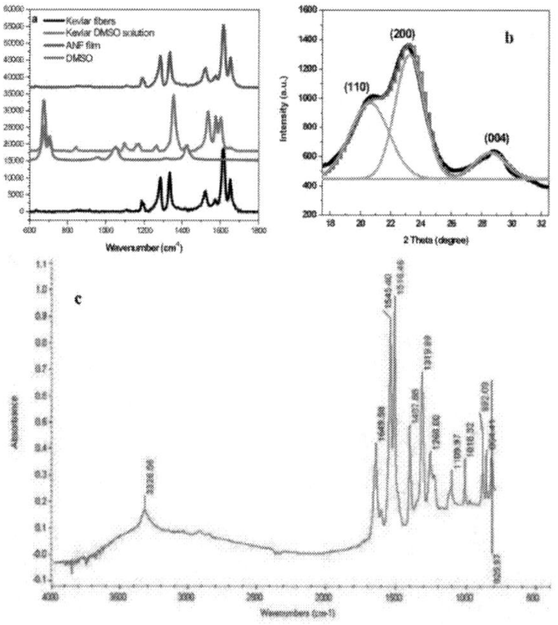

Figure 3. a) Raman scattering of Kevlar fibers (black line), Kevlar DMSO solution (red line) and ANF film (blue line). The Raman scattering of DMSO (green line) is also given near the spectra of Kevlar DMSO solution to clarify the solvent influence. b) XRD pattern of ANF film (black line) with Gaussian peak fitting (green and red lines). c) IR spectra of ANF film.

aromatic stacking interactions and hydrogen bonding between the multilayer components (Figure 1m), which are also believed to be the main interactions between individual nanofibers. Van der Waals interactions are also expected to contribute to the mechanical integrity of the fibrous network. The successful reconstruction of aramid films with the inclusion of ANFs during the LBL process can be confirmed by Raman scattering (Figure 3a). The Raman spectra of the original macroscale Kevlar material are similar to the results reported before,[18] but they are drastically changed after the dissolution in KOH DMSO solution (Figure 3a). These changes disappear after the construction of the ANF films and the Raman spectra of our final films are nearly the same as those of the original Kevlar fibers (Figure 3a). Identification of molecular structures of ANFs in the film was also confirmed by IR spectra (Figure 3c). Compared with Kevlar macrofiber,[19] the intensity and sharpness of IR peaks (for example, N-H stretching vibrations) is decreased due to broader distribution of bond lengths and surface states of the fibers. XRD measurements also show broadening of characteristic Kevlar peaks (Figure 3b), which can be assigned to (110),

(200) and (004) reflections, respectively.[20] Based on the Scherrer equation, $D = 0.9\lambda/\ \cos\theta$, where D is the size of crystallites, λ is the wavelength of X-ray beam used in diffraction experiments, is the line broadening at half the maximum intensity, θ is the Bragg angle, the diameter of the nanofiber can be calculated to be ca. 4.3 nm based on the broadening of (200) reflection obtained after Gaussian peak fitting. This size of crystalline domains matches

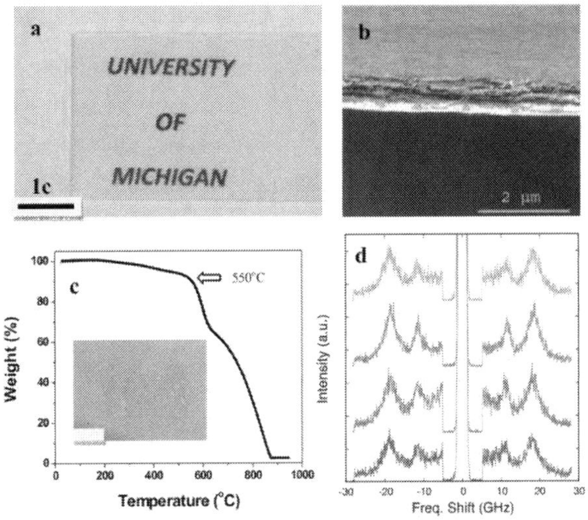

Figure 4. a) Optical image of 300 layers LBL ANF film on glass slide, b) cross-sectional SEM image of PDDA/ANF_{300} and c) thermogravimetric analysis of PDDA/ANF_{300}. Inset: free-standing PDDA/ANF_{300} film. d) Stokes and anti-Stokes peak of the Brillouin Spectra for PDDA/ANF_{300} taken at four different spots, the spectra are offset for the ease of viewing.

really well with the diameter of ANF from TEM observations (Figure 1e), and indicates that the nanoscale fibers retain substantial crystallinity despite reduction in diameter. This fact is also indicative of the retained mechanical properties characteristic of macroscale Kevlar fibers.

*3. **Physical Properties:*** The ANF LBL films are transparent allowing for potential use in ultrastrong surface coatings (Figure 4a). They can also be easily peeled from a surface by HF etching (Figure 4c inset). SEM images indicated that a PDDA/ANF_{300} film has a thickness of about 580 nm (Figure 4b), which is comparable to the results of ellipsometry measurements (Figure 2a) and suggests that the average ANF layer is about 1.9 nm thick. Thermogravimetric analysis (TGA) of PDDA/ANF_{300} (Figure 4c) is similar to the bulk Kevlar material.[1] This is very encouraging because heat-resistant, flexible, transparent films represent a bottleneck for many applications in flexible electronics and flexible photovoltaic cells based on semiconductors with high electron/hole mobility.

It is expected that PDDA/ANF_n films will be elastically orthotropic, with the in-plane and out-of-plane properties being different as is the case with clay and many other multilayers. Brillouin light scattering (BLS) is a convenient method to determine the orthotropic moduli. The in-plane and out-of-plane longitudinal moduli were determined to be 22 ± 1.2 GPa and 11.8 ± 0.6 GPa, respectively (Figure 4d , Table 1). A higher in-plane modulus is expected as ANFs are spread on the substrate with axes parallel to the plane. PDDA/ANF_n films outperform a well-known high-performance polymeric coating material, *i.e.*, polyimide thin film with 17.7 GPa in-plane longitudinal modulus and 7.7 GPa out-of-plane longitudinal modulus recently probed by a similar technique.[21] These mechanical properties of PDDA/ANF$_n$ films also exceed those for cellulose nanofibers with no effect of humidity that greatly reduces the performance of cellulose LBL films. Furthermore, the Young's modulus is 10-20 times higher than that of aligned cellulose fibers and CNTs.

The compression stiffness and hardness of the ANF LBL films were estimated using a Berkovich nanoindenter at various indentation depths. Care was taken to ensure that the depths of the indentations were shallow enough to avoid significant influence from the substrate, which was indeed detected at greater penetration depths (Figure 5). The tests revealed a hardness of 0.46 ± 0.04 GPa (Figure 5c, Table 1) for our film, which is higher than some high strength polymers, such as isotactic polypropylene (H = 0.125 GPa) and high-density polyethylene (H = 0.06 GPa). The Young's modulus was obtained using the continuous stiffness measurement technique of Oliver and Pharr during both the loading and unloading portions of the indentation. The film was found to have a

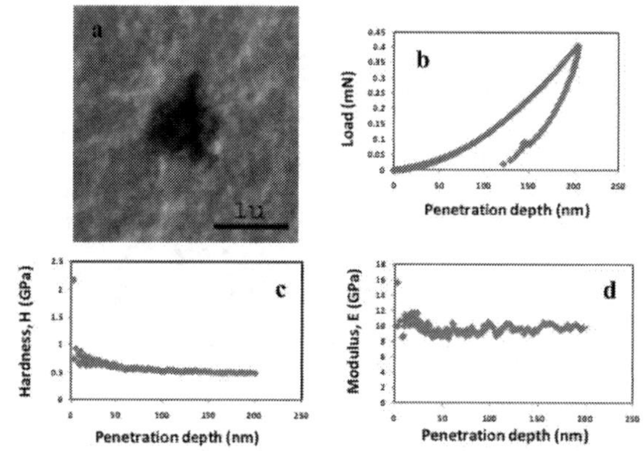

Figure 5. Typical results from nanoindentation experiments for PDDA/ANF$_{300}$ LBL film with maximum depth set at 200 nm. a) SEM images of an indent left by a Berkovich tip in PDDA/ANF$_{300}$ film. b) Represents load-displacement curve. c) Represents the hardness as a function of penetration depth and d) corresponding modulus. These results are from the loading part of the experiment.

modulus value of 9 ± 1 GPa (Table 1), and the value appeared not to be affected by the substrate for indentation depths of less than 200 nm (Figure 5d). The modulus of the films was higher than the 5.5 GPa reported for a cast film of pure Kevlar.[8] This indicates that the preparation of ANFs is indeed advantageous for preparation of high-performance materials. Solvent-based LBL process is particularly effective at producing reasonably dense and

A-nanoindentation	Loading	Unloading
Modulus (GPa)	9 ± 1	8 ± 1
Hardness (GPa)	0.46 ± 0.04	0.45 ± 0.05
B-Brillouin light scattering	In plane	Out of plane
Modulus (GPa)	22 ± 1.1	11.8 ± 0.6

Table 1. A) Summary of modulus and hardness derived from loading and unloading part. B) Summary of in-plane and out-of-plane modulus obtained from Brillouin light scattering.

better ordered arrays of the nanofibers, compared with the casting technique. The modulus is also higher than the reported elastic modulus obtained from nanoindentation of various uniaxial anisotropic polyimide films, which were in the range of 5 to 7 GPa.[22,23] Since BLS measures the longitudinal modulus, the corresponding Young's modulus can be extracted,[24] assuming a Poisson's ratio of 0.25,[25] resulting in a value of 9.8 ± 0.5 GPa, which compares remarkably well with the nanoindentation measurement of 9 ± 1 GPa.

Conclusions: Negatively-charged, uniformly-sized ANFs can be prepared in DMSO media by controlled deprotonation with KOH. The diameter of the fibers can be effectively controlled by addition of water. Such dispersions represent the first example of nanofiber dispersions of synthetic polymers. They are suitable for making nanoscale materials using many techniques and can be considered as new nanoscale building blocks for ultrastrong materials. They also reveal intriguing similarities to CNTs in respect to morphology and mechanical behavior in polymer composites. As a potential direction of future research, evaluation of the mechanical properties of individual ANF to compare those to carbon nanotubes would be quite interesting from both fundamental and practical perspectives. We believe that the method demonstrated here provides a way to prepare ANF-based high-performance films and to incorporate ANFs into multifunctional films lending its outstanding mechanical properties to the resulting hybrid material(s).

References:

1. Tanner, D.; Fitzgerald, J. A.; Phillips, B. R., The Kevlar Story - An Advanced Materials Case-Study. *Angew. Chem.-Int. Edit. Engl.* **1989**, *28*, 649-654.

2. Chae, H. G.; Kumar, S., Rigid-Rod Polymeric Fibers. *J. Appl. Polym. Sci.* **2006**, *100*, 791-802.

3. Vollrath, F.; Knight, D. P., Liquid Crystalline Spinning of Spider Silk. *Nature* **2001**, *410*, 541-548.

4. Koziol, K.; Vilatela, J.; Moisala, A.; Motta, M.; Cunniff, P.; Sennett, M.; Windle, A., High-Performance Carbon Nanotube Fiber. *Science* **2007**, *318*, 1892-1895.

5. O'Connor, I.; Hayden, H.; O'Connor, S.; Coleman, J. N.; Gun'ko, Y. K., Kevlar Coated Carbon Nanotubes for Reinforcement of Polyvinylchloride. *J. Mater. Chem.* **2008**, *18*, 5585-5588.

6. Takayanagi, M.; Katayose, T., N-Substituted Poly(p-phenylene terephthalamide). *J. Polym. Sci. Pol. Chem.* **1981**, *19*, 1133-1145.

7. Takayanagi, M.; Kajiyama, T.; Katayose, T., Surface-Modified Kevlar Fiber-Reinforced Polyethylene and Ionomer. *J. Appl. Polym. Sci.* **1982**, *27*, 3903-3917.

8. Burch, R. R.; Sweeny, W.; Schmidt, H. W.; Kim, Y. H., Preparation of Aromatic Polyamide Polyanions – A Novel Processing Strategy for Aromatic Polyamides. *Macromolecules* **1990**, *23*, 1065-1072.

9. Burch, R. R.; Manring, L. E., N-Alkylation and Hofmann Elimination from Thermal-Decomposition of R_4N^+ Salts of Aromatic Polyamide Polyanions – Synthesis and Stereochemistry of N-Alkylated Aromatic Polyamides. *Macromolecules* **1991**, *24*, 1731-1735.

10. Podsiadlo, P.; Choi, S. Y.; Shim, B.; Lee, J.; Cuddihy, M.; Kotov, N. A., Molecularly Engineered Nanocomposites: Layer-by-Layer Assembly of Cellulose Nanocrystals. *Biomacromolecules* **2005**, *6*, 2914-2918.

11. Chen, P.; Yun, Y. S.; Bak, H.; Cho, S. Y.; Jin, H.-J., Multiwalled Carbon Nanotubes-Embedded Electrospun Bacterial Cellulose Nanofibers. *Mol. Cryst. Liq. Cryst.* **2010**, *519*, 169-178.

12. Li, M.; Kim, I.-H.; Jeong, Y. G., Cellulose Acetate/Multiwalled Carbon Nanotube Nanocomposites with Improved Mechanical, Thermal, and Electrical Properties. *J. Appl. Polym. Sci.* **2010**, *118*, 2475-2481.

13. Miyauchi, M.; Miao, J.; Simmons, T. J.; Lee, J.-W.; Doherty, T. V.; Dordick, J. S.; Linhardt, R. J., Conductive Cable Fibers with Insulating Surface Prepared by Coaxial Electrospinning of Multiwalled Nanotubes and Cellulose. *Biomacromolecules* **2010**, *11*, 2440-2445.

14. Lu, P.; Hsieh, Y.-L., Multiwalled Carbon Nanotube (MWCNT) Reinforced Cellulose Fibers by Electrospinning. *ACS Appl. Mater. Interfaces* **2010**, *2*, 2413-2420.

15. Cranston, E. D.; Eita, M.; Johansson, E.; Netrval, J.; Salajkova, M.; Arwin, H.; Wagberg, L., Determination of Young's Modulus for Nanofibrillated Cellulose Multilayer Thin Films Using Buckling Mechanics. *Biomacromolecules* **2011**, *12*, 961-969.

16. Podsiadlo, P.; Kaushik, A. K.; Arruda, E. M.; Waas, A. M.; Shim, B. S.; Xu, J. D.; Nandivada, H.; Pumplin, B. G.; Lahann, J.; Ramamoorthy, A.; Kotov, N. A., Ultrastrong and Stiff Layered Polymer Nanocomposites. *Science* **2007**, *318*, 80-83.

17. Tang, Z. Y.; Kotov, N. A.; Magonov, S.; Ozturk, B., Nanostructured Artificial Nacre. *Nat. Mater.* **2003**, *2*, 413-418.

18. Washer, G.; Brooks, T.; Saulsberry, R., Investigating the Effects of Aging on the Raman Scattering of Kevlar Strands. *Res. Nondestruct. Eval.* **2008**, *19*, 144-163.

19. Crompton, T. R., *Polymer Reference Book* Smithers Rapra Technology, **2006**, p181-182.

20. Jain, A.; Vijayan, K., Kevlar 49 Fibres: Thermal Expansion Coefficients from High Temperature X-Ray Data. *Curr. Sci.* **2000,** *78*, 331-335.

21. Gomopoulos, N.; Saini, G.; Efremov, M.; Nealey, P. F.; Nelson, K.; Fytas, G., Nondestructive Probing of Mechanical Anisotropy in Polyimide Films on Nanoscale. *Macromolecules* **2010**, *43*, 1551-1555.

22. Lee, C.; Kwon, J.; Park, S.; Sundar, S.; Min, B.; Han, H., Nanoindentation Studies of Polyimide Thin Films with Various Internal Linkages in the Diamine Component. *J. Polym. Sci. Pt. B-Polym. Phys.* **2004**, *42*, 861-870.

23. Jang, W.; Seo, J.; Lee, C.; Paek, S. H.; Han, H., Residual Stress and Mechanical Properties of Polyimide Thin Films. *J. Appl. Polym. Sci.* **2009**, *113*, 976-983.

24. The longitudinal modulus, C_{11}, is related to the Young's modulus, E, and Poisson's ratio, ν, through the formula, $C_{11}=E(1-\nu)/(1+\nu)(1-2\nu)$.

25. Nakamae, K.; Nishino, T.; Xu, A. R., Poissons Ratio of the Crystal-Lattice of Poly(p-phenylene terephthalamide) by X-Ray-Diffraction. *Polymer* **1992,** *33*, 4898-4900.

(395c) Template-Directed Synthesis of Micro and Nano-Structures of Functional Conducting Copolymers by Oxidative Chemical Vapor Deposition

Dhiman Bhattacharyya and Karen K. Gleason, Chemical Engineering, Massachusetts Institute of Technology, Cambridge, MA

Template-directed synthesis of micro and nano-structures of functional conducting copolymers by oxidative chemical vapor deposition

Dhiman Bhattacharyya, and Karen K. Gleason*

Department of Chemical Engineering, Massachusetts Institute of Technology, Cambridge, MA 02139

Abstract:

Template-directed syntheses of inorganic and organic materials are becoming very popular because it is easy to control the desired structures in the products. Micro and nanostructures are important because they provide very high surface area to volume ratio. It increases the interfacial area of the surfaces. Thus synthesized increased surface areas have many applications, for example, to increase the sensitivities of sensors, performances of the photovoltaic cells. Previously, nano and micro-structures made of conductive polymers have been reported. However, they were free floating nanostructures of conductive polymers. In this work, we present a single step preparation of micro-tubules and nanoforests of functional conductive copolymers by oxidative chemical vapor deposition (oCVD) process at low temperature. Here, 3,4-ethylenedioxythiophene (EDOT) and 3-substituted thiophene monomers were employed to synthesize these structures. Iron chloride was used as an oxidant during the vapor phase polymerization step, as well as a dopant in the polymeric system in order to make it conductive. The copolymeric films were characterized by FT-IR, UV-Vis and X-ray photoelectron spectroscopies and 4-point probe measurements and SEM. The available functional groups in the copolymers were used for further derivatization reactions.

See more of this Session: Nanoscale Structure In Polymers III
See more of this Group/Topical: Materials Engineering and Sciences Division

(395d) Polymeric Nanofiber Braid Manufacturing and Characterization

Ji Wang, Engineering Science and Mechanics, Virginia Tech, Blacksburg, VA and **Amrinder S. Nain**, Mechanical Engineering, Virginia Tech, Blacksburg, VA

Using our previously reported Spinneret based tunable Engineering Parameters (STEP) technique, we demonstrate aligned deposition of high aspect ratio uniform diameter Polystyrene (PS) fibers (diameter: sub 50nm to 800nm, length: mm-cm). PS fibers made from STEP technique were deposited in well aligned configuration with parallelism ≤ 2.5 degrees which makes it possible to prepare unidirectional PS fiber bands composed of thousands of individual fibers. PS fiber bands having 600nm diameter uniform fibers of 22 mm in length were made from 10% wt PS(M_w=2000K) dissolved in xylene. PS fiber braids were made by twisting two parallel PS fiber bands using home-designed braiding device and the braiding angle was controlled by number of twists added to the braids. For 3 , 6, 9 and 12 twists braids, the braiding angles were measured to be 6°, 15°, 21° and 32 ° respectively. The break load of PS fiber braids increases as braiding angle becomes larger and peaks (around 700mN)at 21°. Further increase of braiding angle leads to decrease of break load. Compared to non braided PS fiber bands, PS fiber braids demonstrate higher ultimate elongation, higher yield strength and lower modulus. The ultimate elongation increases from 15% to 36% as braiding angle increases from 0° to 15°while even larger braiding angle do not contribute to elongation. SEM images of braids fracture reveals that sheath-core structure forms during the elongation of densely packed fiber braids, which causes fibers at inner portion of braids to fuse together. To further enhance mechanical properties of PS fiber braids, 2%wt multi wall carbon nanotubes(MWNTs) were embedded into polystyrene fiber, which causes the fiber diameter to increase from 600nm to 1000nm. The break load for non fiber bands and 9 twists braids of PS-CNT fibers are 431 mN and 1000mN respectively, both of which are much larger than corresponding pure PS fiber bands and braids. Besides, 9twists braids of PS-CNT fiber demonstrates ultimate elongation up to 82% which is twice of 9 twists braids of pure PS fibers. Ongoing efforts are devoted towards detecting reinforcing effect of CNTs with different aspect ratio and understanding the influence on CNTs on polymer chain orientation in PS fibers.

See more of this Session: Nanoscale Structure In Polymers III
See more of this Group/Topical: Materials Engineering and Sciences Division

(395e) Nanofiber Structure Influenced by Air Gap In a Collector Plate of Electrospinning

Jong Kyu Hong[1], Guan Xu[2], Daqing Piao[2] and Sundararajan. V. Madihally[1], (1)School of Chemical Engineering, Oklahoma State University, Stillwater, OK, (2)Department of Electrical and Computer Engineering, Oklahoma State University, Stillwater

Electrospinning has recently emerged as a technique due to the possibility of fabricating nanofibers. Electrospraying apparatus consisted of a syringe pump, syringe, needle tip, high voltage power supply, earth grounding, and an air gap having collector plate wrapped with aluminum foil. In this study, we address the influence of the shape and size of air gap in a collector plate on fiber density, fiber configuration, and pore size of fibers using synthetic and natural polymers.

SEM morphology showed that the shape of air gap affects the pattern of fibers. For the circular gap in the collector plate, the fibers were randomly deposited throughout the entire void of the collector plate. For the triangular gap, the fibers aligned perpendicular to the bisector of the vertices at the edge of the vertices. For the rectangular void, the fibers aligned from one short side to the other side perpendicular to the long sides of the rectangle in large scale. Quantification study revealed that when the size of air gap was increased from 0.9 cm via 1.4 cm to 1.9 cm, the average pore size of fibers increased from 14.71, 22.83, and 57.77 μ m since fiber density on air gap decreased when the gap size increased.

Simulation study using electrostatics module of COMSOL with passion equation revealed that the equi-potential surface collapses into the air gap. Since the dragging force generated by the electric field will be perpendicular to the equi-potential surfaces and the polymer fiber with be directed along the steepest potential decent direction, that is, to take the shorter path with respect to its initial launching direction and accumulate at the aluminum surfaces. Otherwise, fiber density is low at air gap due to the collapsed equi-potential surface. Also, the strong electric forces at the aluminum gap edges will stretch the polymer fiber and form aligned fiber at the edge of triangular gap.

See more of this Session: Nanoscale Structure In Polymers III
See more of this Group/Topical: Materials Engineering and Sciences Division

(395f) Controlling TiO_2 Nanoparticle Distribution within a Coating Film for Surface Mechanical Property Study

Rohan Uttarwar[1], Sunxi Wang[2], Guangzhao Mao[2] and Yinlun Huang[2], (1)Department of Chemical Engineering and Materials Science, Wayne State University, Detroit, MI, (2)Chemical Engineering and Materials Science, Wayne State University, Detroit, MI

Experimental studies have shown that nanocomposite coatings may show superior mechanical and barrier properties and new functionalities, such as self cleaning, self healing, super scratch resistance, super corrosion resistance. These smart functional coatings should have great application potentials in the automotive, semiconductor, defense, construction, aerospace, and other industries. In nanopaint design, a certain type of functional or non-functional nanoparticles needs to be dispersed in a coating matrix; a significantly increased surface area of these fillers at nanosize is the main reason behind improved mechanical and barrier properties of nanocoating films. It is known that layer-by-layer assembly of nanoparticles can be useful in incorporating self-cleaning phenomenon in coating thin films.[1],[2] The finding justifies the importance of nanoparticle size and its depth inside coating thin films to obtain superior performance properties. Hence, understanding and developing a correlation among them is critical to optimize nanocoating's performance properties.

In developing nanocoatings, the use of water-borne composites such as acrylics is increasing tremendously due to growing environmental concerns with solvent borne coatings. Some nanoparticles like titanium dioxide (TiO_2) can help in introducing smart functionality into coatings as they behave as oxidizing agents in the matrix. But formulating these nanocoatings with uniform dispersion without aggregation of nanoparticles is still a challenge.

This paper reports formation of nanocoating thin films of polymethyl methacrylate (PMMA) resin with dispersed TiO_2 nanoparticles. Thin films are prepared by sandwiching a nanoparticle film with known thickness and size of nanoparticles between two polymer nano-layers with varying thicknesses. The multilayer coat is characterized for its surface scratch resistance and hydrophilicity. A spin coating technique is employed to obtain all nanofilms of polymer and nanoparticles. A uniform dispersion of nanoparticles in selected solvent chloroform is achieved by modifying the surface of TiO_2.[3] The aggregation of particles on applied film is avoided by optimizing the TiO_2 solution concentration and spin coating parameters. The film thickness and surface smoothness are measured using X-ray diffraction studies of all polymer samples. A correlation between polymer solution concentration and nano-film thickness is then obtained. AFM images of TiO_2 films show uniform distribution of particles over the substrate. Various samples with nanoparticle film depth varying from substrate to top of the surface of the coating film are prepared by keeping the total coating film thickness constant at about 200nm. Surface mechanical properties are characterized by nanoindentation using round colloidal probe attached to AFM cantilever. The mechanical properties are found to be strongly affected by the depth of nanoparticles within the film and a correlation between them is obtained. The surface hydrophilicity is compared by measuring the change in contact angle of nanocoating films. The depth at which there is a maximum decrease in the contact angle due to TiO_2 incorporation is recorded. The experimental results are discussed in terms of betterment of nanocoating film performance properties by controlling the distribution of nanoparticles inside the film.

References

1. Wu Z, Lee D, Rubner MF, Cohen RE, Structural color in porous, superhydrophilic, and self-cleaning SiO_2/TiO_2 Bragg stacks, Small, 2008, 3(8), 1445-1451.
2. Zhang X, Sato O, Taguchi M, Einaga Y, Murakami T, Fujishima A, Self-Cleaning Particle Coating with Antireflection Properties, Chemistry of Materials, 2005, 17(3), 696-700.
3. Nakayama N, Hayashi T, Preparation of TiO_2 nanoparticles surface-modified by both carboxylic acid and amine: Dispersibility and stabilization in organic solvents, Colloids and Surfaces A, 2008, 317, 543-550.

See more of this Session: Nanoscale Structure In Polymers III
See more of this Group/Topical: Materials Engineering and Sciences Division

(395g) Impact Resistance and Adhesive Properties of a Functionally Graded Polymer Composite Interlayer

Michael K. Opoku[1], Robb M. Winter[1] and David R. Salem[2], (1)Department of Chemical and Biological Engineering, South Dakota School of Mines and Technology, Rapid City, SD, (2)Composites and Polymer Engineering Laboratory, South Dakota School of Mines and Technology, Rapid City, SD

There is growing interest in lightweight materials with high impact resistance for various applications. In the development of these materials, researchers have started to explore polymer composites containing very low concentrations of ceramics and/or metals with at least one dimension ranging from nanometers to a few microns. In this research, we attempt to synthesize an interlayer nanocomposite polymer which provides a gradual transition of elastic modulus between two laminates of dissimilar modulus. This is expected to minimize the stress concentration gradient at the interlayer boundaries and improve impact resistance.

Formulation of polyurea and polyurethane from disocynate and diamine and/or polyol prepolymers was investigated, and promising candidates were mixed with nanoparticles, nanowhiskers or nanofibers of alumina or silicon carbide at different loading levels and size/loading distributions. These nanoparticle/ nanofiber resin systems were designed to serve as a bonding layer between materials of dissimilar modulus (such as ceramics, metals and polymers) while creating a functionally graded composite with minimal interfacial stress and high energy absorbing properties. We will present details of the synthesis of the polyurethane/polyurea resins and the preparation of the nanocomposite interlayer material; characterization of the nanocomposite layer and interface regions; and results from mechanical property testing.

See more of this Session: Nanoscale Structure In Polymers III
See more of this Group/Topical: Materials Engineering and Sciences Division

(401a) Photonic Metamaterials: Challenges and Oppurtunities

Costas M. Soukoulis, Ames Laboratory and Iowa State University, Ames, IA

In the last decade, a new area of photonics research has emerged, that has given the ability to produce materials with entirely novel electromagnetic properties. Known as metamaterials for their ability to take beyond conventional materials. Clearly, the field of metamaterials can develop mould-breaking technologies for a plethora of applications, where control over light (or more generally electromagnetic radiation) is a prominent ingredient—among them telecommunications, solar energy harvesting, biological and THz imaging and sensing, optical isolators and polarizers. In this talk, I give an introduction into this emerging field, review recent progress, and highlight remaining challenges and opportunities.

See more of this Session: Plasmonic Nanomaterials for Energy Applications
See more of this Group/Topical: Topical 5: Nanomaterials for Energy Applications

(401b) Trapping Light with Dark Plasmons In Nanoparticle Arrays

Teri W. Odom, Department of Chemistry and Materials Science and Engineering, Northwestern University, Evanston, IL

Metal nanostructures concentrate optical fields into highly confined, nanoscale volumes, which is important for applications such as plasmonic nanolasers and white-light generation. However, the short lifetimes of the measured or "bright" plasmons preclude amplification of local fields because of rapid depletion of the plasmon energy. To overcome this problem, we will describe a new type of "dark" or subradiant plasmon that can be supported by 2D arrays of large (> 100 nm, all three dimensions) gold particles. This dark lattice plasmon can be tuned over a broad range of wavelengths simply by changing the nanoparticle height. At resonance, strong coupling between out-of-plane nanoparticle dipolar moments suppresses radiative decay, trapping light in the plane of the array and strongly localizing optical fields on each nanoparticle. This new mechanism can open applications for subradiant plasmons because height-controlled nanoparticle arrays can be manufactured over wafer-scale areas on a variety of substrates.

See more of this Session: Plasmonic Nanomaterials for Energy Applications
See more of this Group/Topical: Topical 5: Nanomaterials for Energy Applications

(401c) Tailoring Thermal Emission Using Photonic and Plasmonic Metal Nanostructures

Prashant Nagpal, Physical Chemistry and Applied Spectroscopy, Los Alamos National Labratory, Los Alamos, NM

Radiative thermal emission from hot bodies is broad in energy and is non-directional. This poses problems for efficient energy conversion of this emitted radiation from hot body sources (eg. from sun) into electricity using an infrared photocell. Metal nanostructures, patterned as three-dimensional photonic crystals, selectively tailor the light-matter interaction near the photonic bandgap and the increased photonic density of states is used to selectively emit light of particular frequency from a hot metallic nanostructure. Here I will present results from selective tailoring of thermal emission at low temperatures for development of efficient thermophotovolatic emitters. This can possibly serve as an efficient route for harnessing solar-thermal energy and waste heat, and converting it into electricity with an upper conversion limit of over 60 percent.

Another possible route towards tailoring thermal emission is utilizing surface plasmons. While these hybrid photon-electron waves on metal surfaces are typically excited using light or fast electrons, recently we have explored excitation of surface plasmon polaritons using heat. I will discuss recent results on fabricating ultrasmooth patterned surface with nanometer scale precision for low loss propagation of surface plasmons. These patterned surfaces will then be combined with designed nanostructures for modulating the frequency of emitted radiation, and also the directionality of the emitted glow. Such unprecedented control over hot body radiation opens up new opportunities and challenges for utilizing plasmonics, while also opening up alternate route towards efficient photovoltaic conversion.

See more of this Session: Plasmonic Nanomaterials for Energy Applications
See more of this Group/Topical: Topical 5: Nanomaterials for Energy Applications

(401d) Self-Assembled Plasmonic Electrodes for High-Performance Organic Photovoltaic Cells

Wade A. Luhman[1], Si Hoon Lee[2], Timothy W. Johnson[2], Russell J. Holmes[1] and **Sang-Hyun Oh**[2], (1) Chemical Engineering and Materials Science, University of Minnesota, Minneapolis, MN, (2)Electrical and Computer Engineering, University of Minnesota, Minneapolis, MN

Thin Ag films perforated with subwavelength hole arrays are used as transparent conducting electrodes for organic photovoltaic cells. Plasmonic electrodes are fabricated using nanosphere lithography to create hexagonal nanohole arrays. Organic photovoltaic cells constructed using a nanopatterned Ag anode show power conversion efficiencies that exceed those of devices constructed on conventional indium-tin-oxide, independent of light polarization. In comparison to cells constructed on unpatterned Ag, the power conversion efficiency is noted to double with patterning.

See more of this Session: Plasmonic Nanomaterials for Energy Applications
See more of this Group/Topical: Topical 5: Nanomaterials for Energy Applications

(401e) Coupling Plasmons In Metal Nanoparticles to Excitons In Semiconductor Nanocrystals

Matthew Pelton, Center for Nanoscale Materials, Argonne National Laboratory, Argonne, IL

Coupling plasmon resonances in metal nanoparticles to excitons in semiconductor nanocrystals, or quantum dots, opens up new possibilities for light harvesting and management of optical energy at the nanoscale. These nanoparticles can be synthesized and organized using wet-chemical methods, providing a bottom-up method for the development of nanomaterials with new optical functionality. Exploiting plasmon-exciton interactions will require a fundamental understanding of the relationship between nanometer-scale structure of hybrid metal-semiconductor materials and their optical properties, particularly in the limit of strong coupling where new optical phenomena can arise. I will present progress in this direction, including calculations showing that a single quantum dot can induce transparency in a plasmonic nanoresonator, a strategy to overcome radiative losses by coupling to dark plasmon modes, and measurements on a model system consisting of single semiconductor nanocrystals near a metal surface.

See more of this Session: Plasmonic Nanomaterials for Energy Applications
See more of this Group/Topical: Topical 5: Nanomaterials for Energy Applications

(401f) Influence of Plasmonic Au Nanoparticles On the Photoactivity of Fe2O3 Electrodes for Water Splitting

Elijah Thimsen[1], Florian Le Formal[2], Michael Graetzel[2] and Scott C. Warren[2], (1)Material Science Division, Argonne National Laboratory, Argonne, IL, (2)Institute of Chemical Sciences and Engineering, Ecole Polytechnique Fdrale de Lausanne, Lausanne (VD), Switzerland

An experimental study of the influence of gold nanoparticles on α-Fe_2O_3 photoanodes for photoelectrochemical water splitting is described. A relative enhancement in the water splitting efficiency at photon frequencies corresponding to the plasmon resonance in gold was observed. This relative enhancement was observed only for electrode geometries with metal particles that were localized at the semiconductor-electrolyte interface, consistent with the observation that minority carrier transport to the electrolyte is the most significant impediment to achieving high efficiencies in this system.

See more of this Session: Plasmonic Nanomaterials for Energy Applications
See more of this Group/Topical: Topical 5: Nanomaterials for Energy Applications

(401g) Nanomanufacturing of Multicomponent Plasmonic Nanogels and Interfaces with Broadband Solar Absorption Capability

Tao Cong, Satvik Wani, Peter Paynter and Radhakrishna Sureshkumar, Department of Biomedical and Chemical Engineering, Syracuse University, Syracuse, NY

Thin film photovoltaics (PVs) offer much potential in cost-effective harvesting of solar energy through efficient use of the semiconductor materials, typically in the form of thin films of thickness 1-2 microns. However, the efficiency of single junction thin film PVs is significantly lower compared to their wafer-based counterparts due to poor light trapping by such devices. Improving the efficiency of thin-film PVs using plasmonic interfaces is an area of active research. A promising approach is the incorporation of a light trapping layer that consists of noble metal nanoparticles (NPs) onto the PV device [1-3]. Nanostructured plasmonic interfaces for this purpose have been fabricated by using lithography, vapor deposition, dewetting of thin metal films by ns and fs pulsed lasers and wet chemistry using self-assembled monolayers [4-13]. However, economical scale up and adaptation of such processes to fabricate interfaces with multiple species/shapes/sizes in a controllable and repeatable fashion are not straightforward.

In this work, we show how a nanostructured interface consisting of multiple metals and/or multiple shapes (e.g. sphere, rod) of a given metal can serve as a "broadband antenna" that would trap light in the UV-visible range. Specifically, a network of wormlike surfactant micelles (WLMs) in an aqueous solution is used as a template for producing stable multicomponent suspensions of Au and/or Ag NPs with desired optical properties. Such suspensions, hereafter referred to as plasmonic nanogels (PNGs), exhibit a long shelf life (~ weeks) and remarkable color uniformity. The structure of the PNGs was studied by small angle X-ray scattering, cryogenic transmission electron microscopy (Cryo-TEM) and rheological experiments. These studies, together with Molecular Dynamics simulations, support a mechanism of self-assembly in which the nanoparticles bridge the micelle fragments to form a stable double network. As evidenced by UV-visible transmission spectroscopy, the shape, size and concentration of the NPs can be varied to tune the optical properties of the PNGs in a way that they absorb radiation over a broadband of wavelengths. Multicomponent PNGs reported in this work have relatively low viscosity and low elastic modulus. Hence they are processable by conventional coating techniques. We employed spin- and dip-coating to produce multicomponent plasmonic interfaces. The structure and optical properties of such interfaces and their integration into PV devices will be discussed.

Acknowledgements: We acknowledge National Science Foundation grant CBET-1049454 for partial support of this research. *Syracuse University has filed a provisional patent application based on the findings of this work.*

References

1. H. A. Atwater and A. Polman, Nat. Mater. **9** (3), 205 (2010).

2. S. Pillai, K. R. Catchpole, T. Trupke, and M. A. Green, J. Appl. Phys. **101**, 093105 (2007).

3. J. Trice, H. Garcia, R. Sureshkumar, and R. Kalyanaraman, Proc. SPIE **6648**, 66480L (2007).

4. E. M. Hicks, S. Zou, G. C. Schatz, K. G. Spears, R. P. Van Duyne, L. Gunnarsson, T. Rindzevicius, B. Kasemo, and M. Käll, Nano Lett. **5** (6), 1065 (2005).

5. L. M. Campos, I. Meinel, R. G. Guino, M. Schierhorn, N. Gupta, G. D. Stucky, and C. J. Hawker, Adv. Mater. **20** (19), 3728 (2008).

6. Y. H. Lanyon, G. De Marzi, Y. E. Watson, A. J. Quinn, J. P. Gleeson, G. Redmond, and D. W. M. Arrigan, Anal. Chem. **79** (8), 3048 (2007).

7. Y. Wu, J. D. Fowlkes, P. D. Rack, J. A. Diez, and L. Kondic, Langmuir, 739 (2010).

8. L. Kondic, J. A. Diez, P. D. Rack, Y. Guan, and J. D. Fowlkes, Phys. Rev. E **79** (2), 026302 (2009).

9. C. Favazza, J. Trice, R. Kalyanaraman, and R. Sureshkumar, Appl. Phys. Lett. **91**, 043105 (2007).

10. J. Trice, C. Favazza, D. Thomas, H. Garcia, R. Kalyanaraman, and R. Sureshkumar, Phys. Rev. Lett. **101** (1), 17802 (2008).

11. C. Favazza, R. Kalyanaraman, and R. Sureshkumar, Nanotechnology **17**, 4229 (2006).

12. A. I. Kuznetsov, J. Koch, and B. N. Chichkov, Appl. Phys. Mater. Sci. Process. **94** (2), 221 (2009).

13. L. A. Porter Jr, H. C. Choi, J. M. Schmeltzer, A. E. Ribbe, L. C. C. Elliott, and J. M. Buriak, Nano Lett. **2** (12), 1369 (2002).

See more of this Session: Plasmonic Nanomaterials for Energy Applications
See more of this Group/Topical: Topical 5: Nanomaterials for Energy Applications

(402a) In-Situ Monitoring of Block Copolymer Ordering In Solution-Cast Films During Solvent Removal

Michael J. Heinzer[1], D. G. Baird[1] and **Stephen M. Martin**[2], (1)Chemical Engineering, Virginia Tech, Blacksburg, VA, (2)Chemical Engineering, Virginia Polytechnic Institute and State University, Blacksburg, VA

The manner in which solvent is removed during the solution-casting of block copolymer films has a significant influence on the polymer morphology in the final film, which in turn governs film properties. These solution-cast films have various applications, including the processing of polymer electrolyte membranes for fuel cells and reverse osmosis water purification membranes. It is, therefore, desirable to understand the mechanism by which ordered structures develop during the drying process. We have developed a new small angle X-ray scattering method that allows for real-time tracking of the phase separation of block copolymers as solvent is continually extracted from a film. These experiments are capable of tracking changes in the ordering kinetics and domain dimensions as a function of concentration across the order-disorder transition and deeper into the phase diagram. The effect of drying temperature on the ordering kinetics and domain compression is explored. Removing solvent in the presence or absence of a sweep gas allows the effect of drying rate on the ordering process to be decoupled from the effect of drying temperature. In addition, iso-concentration measurements of the ordering kinetics at various concentrations along the drying path can be made using SAXS and compared to the phase kinetics measured using rheology. Trends in the ordering kinetics observed in these iso-concentration phase separation experiments will be compared to those observed in in-situ ordering experiments.

See more of this Session: Polymer Thin Films and Interfaces III
See more of this Group/Topical: Materials Engineering and Sciences Division

(402b) Shear Induced Alignment of Standing Lamellar Block Copolymer Thin Films

Saswati Pujari, Chemical & Biological Engineering, Princeton University, Princeton, NJ, Richard A. Register, Chemical and Biological Engineering, Princeton University, Princeton, NJ and Paul M. Chaikin, Department of Physics, New York University, New York, NY

Introduction.

Thin films formed from microphase-separated block copolymers are of interest because they form periodic structures on the order of tens of nanometers with domain size tunable by their molecular weight.[1, 2] This makes them ideal candidates for templates for nanopattering applications, which can be easily integrated into the existing manufacturing processes.[3] Cylinder forming block copolymers have been traditionally used as templates for making parallel stripes in nanopattering applications, like templates for nanowires and polarizers.[4] However, lamellae oriented perpendicular to the substrate have advantages in pattern transfer over parallel cylinders, because of the higher aspect ratio of the resulting templates and vertical side walls.[5] One of the ways to achieve perpendicular orientation of lamellar block copolymer domains is by using brush layers of random copolymers.[6-8] Chemically grafted random copolymers of the brush layers have the effect of modifying the chemical composition of the substrate and in turn tuning the interfacial or wetting behavior of the blocks of the copolymer film on the substrate.

Once the out-of-plane orientation of the block copolymer domains has been controlled, it is important to control their in-plane order, to make then useful for the intended applications. Previously our group has used mechanical shear as an easy alternative approach to orient spherical and cylindrical nanodomains over macroscopic areas.[9-12] Alignment can be achieved with a few minutes of shearing, over cm^2 regions of film, and with stresses (order 1000 Pa) small enough that they can be imparted by either a moving solid surface in contact with the film[9, 10] or simply a fluid flowing over the film's surface.[11, 12]

In this talk we will present recent results in which we have used shear to achieve long range order in a block copolymer thin film forming perpendicular lamellae. Lamellar polystyrene/ poly (methyl methacrylate) (PS/PMMA) block copolymers were made to orient perpendicular to Si substrate by grafting a layer of random terpolymer of poly (styrene/ (methyl methacrylate)/ 2-hydroxyethylmethactylate). To date, films less than one domain spacing thick show good alignment under shear, while thicker films switch back to parallel orientation. Further, we are investigating alignment as a function of stress, terpolymer thickness, terpolymer composition and domain spacing.

Experimental.

Terpolymers of styrene, methyl methacrylate, and 2-hydroxyethylmethacrylate were synthesized using classical free radical polymerization and AIBN initiator at 70 C, to a conversion of 20% to avoid compositional drift. Terpolymers with styrene content, F_S, varying from 0.5-0.8 were prepared and the HEMA content in the terpolymers was fixed at 0.01 mole fraction. ^{1}H NMR was used to determine the exact styrene fraction in each of the synthesized terpolymers. The PS/PMMA block copolymer used in this study was purchased from Polymer Laboratories Inc. and had block molecular weights of 28.5 kg/mol PS and 30 kg/mol PMMA. This PS/PMMA block copolymer has a lamellar repeat spacing of ~30 nm as determined from AFM images of thin films.

The brush layer was prepared by spinning a film of terpolymer on bare pieces of 3" silicon wafer from Silicon Quest International and annealing them at 140 °C for 12 hours. Unreacted terpolymer was removed by rinsing with toluene leaving behind a uniform brush layer of 4-6 nm. Dilute solutions (1-2 % in toluene) of the PS/PMMA block copolymer were then spin-coated on the modified Si substrate, at various spin speeds to obtain a range of film thicknesses, and annealed at 150-170 °C to induce phase separation.

Shear stress was applied to the film via PDMS pads (1 cm x 1 cm), prepared by curing a mixture of Dow Corning Sylgard 184/curing agent (10/1 w/w) at 60 °C for 2 h. While holding the film at a constant temperature of 150 °C via a digital hot plate, shear stresses of 10-40 kPa were applied for 30 min. The specimen was then cooled to below the glass transition temperature of both PS and PMMA (~ 100 °C), while maintaining the shear stress, so as to ''lock in'' the structure induced by shear. The block copolymer domains were then imaged using tapping mode AFM, to determine their orientation.

Results and Discussion.

Use of terpolymer brushes with a range of styrene content enabled the determination of a *window of neutrality* in styrene mole fraction, which reliably resulted in perpendicularly oriented block copolymer lamellae. As far as we know, the window of neutrality for the lamellae forming PS/PMMA block copolymer lies between a styrene fraction of 0.68-0.74 in the brush layer. Terpolymers with higher styrene fractions (up to 0.8) are still being investigated to determine more accurately this neutrality window. It is to be noted here that this window of neutrality is much higher in styrene content (~ 0.1 mole fraction higher) than that determined by Han et al. [7]. Han et al. prepared their terpolymers using a living radical polymerization, while we use conventional free radical polymerization. The difference in synthesis techniques can result in differences in the detailed chain architecture, resulting in a substantial shift of the window of neutrality. However, our terpolymers (just like those prepared by Han et al.) result in perpendicular lamellae to film thickness up to 2 times the domain spacing of the block copolymer.

Perpendicular domains of block copolymer films of thickness less than 1 domain (< 30 nm) spacing can be shear aligned using stresses of 30 kPa and higher. Stresses lower than 30 kPa do not result in significant alignment of the perpendicular lamellae. This suggests the occurrence of a critical stress for alignment to occur, similar to that seen in sphere and cylinder forming systems. However, shearing films with thickness greater than 1 domain spacing (> 30 nm) results in the lamellae switching back to parallel orientation. There are several possible reasons for this. First, in the current study we have used a brush layer with $F_s = 0.68$, which is at the edge of the neutral window for forming perpendicular lamellae. It is possible that the substrate is still PMMA preferential, such that mobilization of chains during shear can flip the lamellar orientation. A second possibility is that the film/ PDMS wetting conditions are PS selective (since PDMS has lower surface tension), resulting in the PS domain preferentially wetting the PDMS surface. To overcome these issues we are currently trying to better investigate the window of neutrality in this system, so that we can achieve truly neutral substrates. Also we are working towards neutralizing the PDMS pad used for shearing.

References

1. Fasolka, M.J. and A.M. Mayes, *Block copolymer thin films: Physics and applications.* Annual Review of Materials Research, 2001. **31**: p. 323-355.

2. Segalman, R.A., *Patterning with block copolymer thin films.* Materials Science and Engineering R - Reports, 2005. **48**(6): p. 191-226.

3. Black, C.T., *Polymer self-assembly as a novel extension to optical lithograph.* ACS Nano, 2007. **1**(3): p. 147-150.

4. Pelletier, V., et al., *Aluminum nanowire polarizing grids: Fabrication and analysis.* Applied Physics Letters, 2006. **88**(21): p. 211114

5. Stoykovich, M.P. and P.F. Nealey, *Block copolymers and conventional lithography.* Materials Today, 2006. **9**(9): p. 20-29.

6. In, I., et al., *Side-chain-grafted random copolymer brushes as neutral surfaces for controlling the orientation of block copolymer microdomains in thin films.* Langmuir, 2006. **22**(18): p. 7855-7860.

7. Han, E., et al., *Effect of Composition of Substrate-Modifying Random Copolymers on the Orientation of Symmetric and Asymmetric Diblock Copolymer Domains.* Macromolecules, 2008. **41**(23): p. 9090-9097.

8. Mansky, P., et al., *Controlling polymer-surface interactions with random copolymer brushes.* Science, 1997. **275**(5305): p. 1458-1460.

9. Angelescu, D.E., et al., *Macroscopic Orientation of Block Copolymer Cylinders in Single-Layer Films by Shearing.* Advanced Materials, 2004. **16**: p. 1736-1740.

10. Angelescu, D.E., et al., *Shear-Induced Alignment in Thin Films of Spherical Nanodomains.*

Advanced Materials, 2005. **17**: p. 1878-1881.

11. Pelletier, V., et al., *Writing Mesoscale Patterns in Block Copolymer Thin Films through Channel Flow of a Nonsolvent Fluid.* Applied Physics Letters, 2007. **90**: p. 163105.

12. Wu, M.W., R.A. Register, and P.M. Chaikin, *Mechanisms for Current Induced Conductivity Changes in a Conducting Polymer.* Physical Review E, 2006. **74**: p. 040801R.

See more of this Session: Polymer Thin Films and Interfaces III
See more of this Group/Topical: Materials Engineering and Sciences Division

(402c) Confinement of Elastomeric Block Copolymers Via Forced Assembly Co-Extrusion

Tiffani M. Burt, Eric Baer and LaShanda T.J. Korley, Macromolecular Science and Engineering, Case Western Reserve University, Cleveland, OH

Forced-assembly processing provides a unique opportunity to examine the effects of confinement on block copolymers via conventional melt processing techniques. The microlayering process was utilized to produce novel materials with enhanced mechanical properties through selective manipulation of layer thickness. Multilayer films consisting of an elastomeric, block copolymer confined between rigid PS and PMMA layers were produced with layer thicknesses ranging from 100 nm to 1 μm. Deformation studies of the confined BCP showed an increase in ductility as the layer thickness decreased to 200 nm due to a shift in the mode of deformation from crazing to shear yielding. Post-extrusion annealing was performed on the multilayer films to investigate the impact of a highly ordered morphology on the mechanical properties. The annealed multilayer films showed increased toughness with decreasing layer thickness and resulted in homogeneous deformation compared to the as-extruded films. Multilayer co-extrusion proved to be an advantageous method for producing continuous films with tunable mechanical response.

See more of this Session: Polymer Thin Films and Interfaces III
See more of this Group/Topical: Materials Engineering and Sciences Division

(402d) Controlling Domain Orientations In Thin Films of Lamellar Copolymers

Gila Stein, Thai Vu and Nikhila Mahadevapuram, Chemical and Biomolecular Engineering, University of Houston, Houston, TX

Thin films of block copolymers are widely investigated for low cost, large area nanopatterning. For applications in semiconductor manufacturing, the copolymer domains must be oriented perpendicular to the substrate through the film thickness. It is well known that a perpendicular orientation can be sustained in thin films of diblock copolymers if the following criteria are met: (i) The copolymer constituents have similar surface energies; (ii) The substrate is energetically neutral with respect to the copolymer constituents; and (iii) The film thickness is slightly incommensurate with the equilibrium domain periodicity. In this work, we demonstrate that copolymer architecture significantly affects the energetics that control domain orientations in thin films. Two copolymer architectures are compared: An AB diblock of poly(styrene-b-methyl methacrylate) and an ABA triblock of poly(methyl methacrylate-b-styrene-b-methyl methacrylate). Both copolymer systems form lamellar domains with an equilibrium periodicity of 50 nm. All films are cast on ``neutral'' substrates, and the resulting structures are evaluated with optical microscopy, scanning force microscopy, and grazing-incidence small-angle X-ray scattering. For AB diblock copolymers, domain orientations are very sensitive to film thickness, annealing temperature, and imperfections in the "neutral" substrate coating. The desired perpendicular orientation is most reliable obtained at high annealing temperatures where PS and PMMA have nearly identical melt surface energies. For ABA triblock copolymers, the perpendicular domain orientation is stable for all film thicknesses and annealing temperatures that were studied. These findings are consistent with recent works that consider architectural effects when calculating the copolymer surface tension.[1,2] Significantly, the data demonstrate that triblocks are easier to process for applications in nanopatterning - particularly when high-aspect-ratio nanostructures are required. However, both diblock and triblock films contain a high density of "tilted" or bent domains, and these kinetically-trapped defects should be minimized for most patterning applications.

[1] Khanna et al., Macromolecules 2006, 39, 9346–9356.

[2] Matsen, Macromolecules 2010, 43, 1671–1674.

See more of this Session: Polymer Thin Films and Interfaces III
See more of this Group/Topical: Materials Engineering and Sciences Division

(402e) Molecular Assembly of Ultrathin Polymer Films: An Application In Tribology

Raju K. Gupta[1], Sujeet K. Sinha[2], N. Satyanarayana[2] and **M.P. Srinivasan**[3], (1)Chemical and Biomolecular Engineering, National University of Singapore, Singapore, Singapore, (2)Department of Mechanical Engineering, National University of Singapore, Singapore, Singapore, (3)Department of Chemical and Biomolecular Engineering, National University of Singapore, Singapore, Singapore

We demonstrate the application of covalent molecular assembly in fabricating robust thin film structures. An anhydride functionalized polymer has been deposited over an amine-terminated silicon surface through covalent binding. The process conditions to achieve reproducible molecular thicknesses for the polymer film have been optimized. X-ray photoelectron spectroscopy, atomic force microscopy and ellipsometry were employed to study the interfacial chemistry, morphology and thickness of the assembled film. The films show excellent stability and mechanical strength attributable to the covalent interlayer linkage. The applicability of such polymeric film as an intermediate layer between Si and PFPE to improve the tribological properties of PFPE films has been demonstrated. Such films showed wear life of >100,000 cycles and can potentially be used as lubricants in several technological applications, such as information storage devices and micro-electro-mechanical systems (MEMS).

See more of this Session: Polymer Thin Films and Interfaces III
See more of this Group/Topical: Materials Engineering and Sciences Division

(402f) High Performance Thin Film Based Coatings for In Aircraft Applications

Marvi A. Matos, Chemical Technologies, BR&T, The Boeing Company, Seattle, WA

Conference: **American Institute of Chemical EngineersAnnual Meeting**

Date of theconference: October 16-21st, 2011

Title: **HighPerformance Thin Film Based Coatings in Aircraft Applications**

Author: **Marvi A. Matos, PhD**

Organization:**The Boeing Company**

Abstract

The use polymer-based materialsand coatings in aircraft components is two-fold, first, the weight reductionsare highly desirable to decrease fuel consumption and second, polymer-basedmaterials are usually less expensive and convenient to manufacture. Howeverwhen used in applications with strict optical requirements, these polymericsurfaces can be susceptible to erosion which results in increased haze anddecreased clarity. For example, when staple materials such as polycarbonate orstretched acrylic are used in windows, windshields and canopies, one of thedrawbacks is the tendency to scratch and craze. Polymeric windows have beenhistorically coated with polysiloxane or polyurethanebased coatings to overcome this limitation by improving the surface resistanceto scratches. Still, improvements to the processes involved can decrease therequired long drying times and can offer long term solutions in which theresistance to erosion is maintained overtime. Advanced thin film coatings basedon Open Air Plasma Deposition Technologies can improve the durability of manycomponents on aircrafts. These technologies can be exploited to generatematerials with high performance, which are also environmentally friendly andproduced with waste free processes. We are currently focusing our efforts inthe development and study of thin films to improve the resistance to materialerosion on polymeric substrates with applications in the aerospace industry anddefense.

See more of this Session: Polymer Thin Films and Interfaces III
See more of this Group/Topical: Materials Engineering and Sciences Division

(415a) Phase Behavior of "Block-Random" Copolymers

Bryan S. Beckingham and Richard A. Register, Chemical and Biological Engineering, Princeton University, Princeton, NJ

"Block-random" copolymers-represented generallyby the structure (A_xB_{1-x})-(A_yB_{1-y}),where each of the two blocks is a random copolymer of monomers A and B, simplywith different fractions of A (x and y)-present a convenient and usefulvariation on the typical block copolymer architecture, as the interblockinteractions and physical properties can be tuned continuously, via x and y,through the random block's composition. The ability to tune the effectiveinteraction parameter between the blocks continuously, allows for theorder-disorder transition temperature (T_{ODT}) to be tuned independentlyof molecular weight using only two monomers, via the difference[1]between x and y. This flexibility makes block-random copolymers a versatileplatform for the exploration of polymer phase behavior and structure-propertyrelationships.

Typical living or controlledpolymerizations produce compositional gradients along the random block, sincethe reactivity ratios are generally different from unity, which can in turninfluence the phase behavior. Living polymerizations which proceed with *random*monomer addition (no gradient) are consequently sought because they wouldpermit the synthesis of well-defined polymers, and block copolymers, of tunablecomposition (hence tunable properties) which are effectively homogeneous onlength scales larger than the monomer size.[2] It is desirable toachieve such random copolymerizations of styrene and isoprene (SI), to expandthe accessible range of polymer properties. In particular, hydrogenatedhigh-vinyl polyisoprene has an exceptionally low cohesive energy density,[3]lower than that for hydrogenated high-vinyl polybutadiene, meaning that a broadrange of solubility parameters (>2 $MPa^{1/2}$) could be accessed inrandom copolymers containing styrene and hydrogenated vinyl isoprene (hI)units. This increased range of solubility parameters translates directly intoan increased range of accessible values of the interblock Flory-Hugginsinteraction parameter χ in block-random copolymers.

The reactivity ratios for SI inhydrocarbon and ether solvents are qualitatively similar to those for SB,[4]but despite the seemingly minor difference between butadiene and isoprene, N,N,N,N-tetramethylethylenediamine,TMEDA is not an effective randomizer for SI copolymerizations.[5] Overhalf a century ago, Kelley and Tobolsky[6] showed that triethylamine(TEA) is an excellent randomizer for equimolar SI copolymerizations (~60 wt%S). Later, Annighfer and Gronski[7] used a TEA/benzene mixture (20/80v/v) to synthesize S-S*r*I-I triblock copolymers with styrene-*ran*-isoprene(S*r*I) midblocks (~50 wt% S); they reported reactivity ratios r_I= 1.0 and r_S = 0.8, implying that effectively randomcopolymers could be synthesized at essentially any S:I ratio with thisapproach.

Here, organolithium initiation in acyclohexane/triethylamine mixture yields narrow-distribution copolymers ofstyrene and isoprene of any desired composition, with no measurable down-chaingradient. These random copolymers (S*r*I) have been successfullyincorporated into well-defined symmetric block copolymers (I-S*r*I diblocks)and subsequent isoprene-selective hydrogenation yields thermally stable hI-S*r*hIdiblocks, which self-assemble into well-defined lamellar morphologies withsharply-defined order-disorder transitions, whose temperatures scalepredictably with diblock molecular weight. The use of S*r*hI in lieu of astyrene homopolymer block effectively dilutes the unfavorable contacts betweenthe two blocks in the homogeneous phase reducing the effective *X*, where *X*if the interaction energy density ($X = \chi(N\rho/M)RT = (\chi N)_{ODT}\rho RT_{ODT}/M$)by a factor of ~8, and allowing for greatly elevated molecular weights and *d*spacings at a given value of T_{ODT}. The measured interaction energydensity between hI and S*r*hI is consistent with the mean-field "copolymerequation"[8], predicts factor of ~5 reduction in *X*, providing afirst step towards the design of styrene-isoprene block-random copolymers ofdesired molecular weight and T_{ODT}.

Within the context of mean-field theory,as in the Flory-Huggins model, *X* is related to the difference insolubility parameters (d*X*=(Δd^2.Having measured values of *X* we can rank S*r*hI in solubilityparameter relative to other polymers, particularly the saturated hydrocarbonpolymers studied extensively by Graessley and coworkers.[3,9] We findthat our determined solubility parameter for SrhI matches well with what wouldbe calculated from homopolymer solubility parameters, for hI and S, using avolume fraction weighted average. Interestingly, the value of $dSrh_I$determined here is close to that for polyethylene,[9] suggestingpotential miscibility between polyethylene and S*r*I, an idea we are inthe process of testing.

This work was generouslysupported by the National Science Foundation, Polymers Program (DMR-1003942).

References

(1) Smith,S. D.; Ashraf, A.; Satkowski, M. M.; Spontak, R. J. *Polym. Prepr. (Am. Chem.Soc., Div. Polym. Chem.)* **1994**, *35(1)*, 651-652.

(2) Tierney,N.K.; Register, R. A. *Macromolecules* **2003**, *36*, 1170-1177.

(3) Graessley,W.W.; Krishnamoorti, R.; Reichart, G. C.; Balsara, N. P.; Fetters, L. J.;Lohse, D. J. *Macromolecules* **1995**, *28*, 1260-1270.

(4) Morton,M. *Anionic Polymerization: Principles and Practice*; Academic Press: NewYork, 1983, pp. 134-145.

(5) Chang, C. C.; Halasa, A. F.;Miller, Jr., J. W.; Hsu, W. L. *Polym. Int.* **1994**, 33, 151-159.

(6) Kelley, D. J.; Tobolsky, A. V. *J.Am. Chem. Soc.* **1959**, *81*, 1597-1600.

(7) Annighfer, F.; Gronski, W. *Coll. Polym. Sci.* **1983**,*261*, 15-25.

(8)Graessley, W. W. *Polymeric Liquids & Networks: Structure and Properties*;Garland Science: New York, **2004**, pp. 341-407.

(9) Krishnamoorti,R.; Graessley, W. W.; Dee, G. T.; Walsh, D. J.; Fetters, L. J.; Lohse, D. J. *Macromolecules***1996**, 29, 367-376.

See more of this Session: Thermodynamics of Polymers
See more of this Group/Topical: Materials Engineering and Sciences Division

(415b) The Phase Behavior and the Flory-Huggins Interaction Parameter of Blends Containing Amorphous Poly(resorcinol phthalate- block-carbonate), Poly(bisphenol-A carbonate) and Poly(ethylene terephthalate)

Philip Bell, Corporate Research, SABIC Innovative Plastics, Mt Vernon, IN, Sung Dug "Doug" Kim, Module R&D Group, SunPower Corporation, San Jose, CA, Jun Tian, Analytical Technology, Liquidia Technologies, Inc., Research Triangle Park, NC and Shreyas Chakravarti, Global Crystalline Polymer Technology, SABIC Innovative Plastics, Mt Vernon, IN

The phase behavior of the blends of poly(ethylene terephthalate) (PET) and poly(Resorcinol Phthalate-*block*-Carbonate) (RPC) and the blends of PET and poly(Bisphenol-A Carbonate) (PC) was investigated by dynamic mechanical analysis (DMA), differential scanning calorimetry (DSC), and transmission electron microscopy (TEM). Blends of high molecular weight PET and RPC copolymer with 20 mol% resorcinol phthalate (RPC20) showed two glass transition temperatures in DMA and DSC but the cold crystallization rate of PET phase was substantially lowered as compared to neat PET, indicating partial miscibility at all compositions. The RPC20 with M_w = 31,500 g/mol formed miscible blends with PET when PET has weight-average molecular weight <9500 g/mol. The Flory–Huggins interaction parameter between PET and RPC20 was calculated to be 0.029 ± 0.003 by using the Flory–Huggins equation at critical composition and molecular weight. PC with M_w = 30,000 g/mol formed miscible blends with PET only when PET had molecular weight <2800 g/mol, indicating PC/PET blends were much less miscible than RPC20/PET blends. Group contribution methods agreed well with the experimental results obtained both in the present study and a previous study [1], predicting that the addition of a resorcinol phthalate block to a PC backbone should increase the miscibility of PC and PET.

See more of this Session: Thermodynamics of Polymers
See more of this Group/Topical: Materials Engineering and Sciences Division

(415c) Behavior of Homopolymers and Block Copolymers In Ionic Liquids: Glass Transitions, Viscoelasticity and Critical Micelle Concentrations

Michelle M. Mok[1], Xingcheng Liu[1], Zhifeng Bai[1], Yu Lei[1] and Timothy P. Lodge[2], (1)Chemistry, University of Minnesota, Minneapolis, MN, (2)Department of Chemistry and Department of Chemical Engineering and Materials Science, University of Minnesota

Ionic liquids (ILs) are a relatively new class of solvents composed exclusively of ions. Specific advantages that they offer over traditional solvents include negligible volatility, tailorable solubilization, and high thermal and chemical stability, allowing them to be used as a liquid phase over very large temperatures ranges. Initially, polymer scientists took advantage of their solvating properties for use as polymerization media, and more recently, for their potential to create composite materials for diverse applications such as dielectric materials for thin-film transistors and solid-state polymer electrolytes for electrochemical devices. In terms of more dilute systems, homopolymers in ILs have been proposed as designer solvents, while block copolymers in ionic liquids have been studied for their micelle- and vesicle-forming capabilities with potential applications as vessels for transport and extraction within biphasic systems. With this increasing focus on their materials applications, it is important to gain a complete understanding of the basic phase behavior of polymers and IL solvent systems.

First, we examine the behavior of a model polymer-IL system over the range of compositions from dilute polymer to pure polymer. Specifically, we study the glass transition and viscoelastic properties of poly(methyl methacrylate) in the ionic liquid 1-ethyl-3-methylimidazolium bis(trifluoromethylsulfonyl) amide over the temperature range from -180 to 230 °C. Secondly, we discuss the critical micelle concentrations (CMCs) of block copolymers in IL systems as functions of block copolymer and IL composition. These results are related to the balance of solvophobic and solvophilic interactions induced by the different copolymer and cation/anion combinations. Overall CMC values were found to be over four orders of magnitude greater than those measured in aqueous systems, in accord with the lower solvophobicity of ILs. Finally, we explore the effect of temperature on these systems at concentrations near the CMC.

See more of this Session: Thermodynamics of Polymers
See more of this Group/Topical: Materials Engineering and Sciences Division

(415d) Cosurfactant Effects Affecting the Phase Diagram for a Binary Blend of Diblock Copolymers

Poornima Padmanabhan, Juan C. Araque and Fernando Escobedo, School of Chemical and Biomolecular Engineering, Cornell University, Ithaca, NY

Blends of diblock copolymers present enormous possibilities of designing bicontinuous and monocontinuous microphases, which are otherwise unstable in a pure diblock melt due to packing frustration. We present a study of the phase behavior of a binary blend of a long, asymmetric diblock with a shorter, symmetric diblock through self-consistent field theory and dissipative particle dynamics simulations. Such a system has been studied previously by experiments which mapped out a phase diagram that included a narrow region where a bicontinuous phase existed of undetermined identity. Theory is shown to predict the gyroid phase in a narrow region of the phase diagram. However, detailed investigation through particle-based simulations of the same region shows that the cylinder phase is stable, and the gyroid phase is metastable (though very nearly stable). These results indicate that the previously unidentified bicontinuous phases in experiments can be attributed to a defective gyroid phase. We found several other phases competing with the cylinder phase in the phase diagram – cocontinuous phases, perforated lamellar, and the woodpile phase (never before been observed in diblock systems), which indicate that a rich phase diagram is possible with binary blends of diblock copolymers. Analysis of the simulated structures shows that the spatial distribution of the short, symmetric diblocks (cosurfactant effect) provides a key to understanding the limited stability observed in the gyroid phase. Some of the ongoing work entails understanding the effect of uniform shear on the structure of selected phases.

See more of this Session: Thermodynamics of Polymers
See more of this Group/Topical: Materials Engineering and Sciences Division

(415e) Influence of Finite Size In Charge Reversal of Polyelectrolytes: A Simulation Study

Maria Sammalkorpi, Department of Chemistry, Aalto University, Helsinki, Finland and Paul R. Van Tassel, Chemical Engineering, Yale University, New Haven, CT

In recent experiments, we have uncovered conditions where polyelectrolyte adsorption to a conducting surface may become continuous in the sense of scaling linearly with time over hours [1,2]. This discovery of continuous layer growth offers an enticing possibility of nanoscale thin film growth in a single step process, but also brings forth questions of the underlying mechanisms.

We present a molecular Monte Carlo simulation study aimed at understanding mechanistically the like-charge attraction involved in the continuous adsorption process and, more broadly, polyelectrolyte adsorption in general. Our system consists of two parallel polymer chains composed of charged tangent spheres above a surface of variable dielectric discontinuity between the substrate and the solution, and spherical counterions and salt ions. We find that counter ion correlations act to enable the formation of stable polymer-polymer binding and aggregation. We discuss the sensitivity of the attractive regime to a Coulombic coupling parameter and to finite ion size, and the implications of finite size effects and charge distribution both in the polyelectrolytes and in the ions. We also explore implications of these results to our ongoing experiments.

[1] A. P. Ngankam and P. R. Van Tassel, Proc. Nac. Acad. Sci. **104**, 1140-1145 (2007).

[2] C. Olsen and P. R. Van Tassel, J. Colloid and Interface Science **329,** 222-227 (2009).

See more of this Session: Thermodynamics of Polymers
See more of this Group/Topical: Materials Engineering and Sciences Division

(415f) Quantifying Fluctuation Effects On the Order-Disorder Transition of Symmetric Diblock Copolymers

Jing Zong and Qiang Wang, Department of Chemical and Biological Engineering, Colorado State University, Fort Collins, CO

Howfluctuations change the order-disorder transition (ODT) of symmetric diblockcopolymers is a classic yet unsolved problem in polymer physics.[1]While it is qualitatively known that fluctuations change ODT to a weakfirst-order phase transition and increase the ODT temperature, their effectshave not been ambiguously quantified as a function of the invariant degree ofpolymerization □ controlling the system fluctuations.Although many groups addressed this problem with molecular simulations, theywere hindered by the difficulty of accurately locating ODT in the simulationsand the use of different model systems in the simulations and mean-fieldtheory.

Here weunambiguously quantify the fluctuation effects by direct comparisons betweenfast off-lattice Monte Carlo (FOMC) simulations[2] and mean-fieldtheory using exactly the same model system (Hamiltonian), thus without any parameter-fitting.The symmetric diblock copolymers are modeled as discrete Gaussian chains withsoft, finite-range repulsions as commonly used in dissipative-particle dynamicssimulations. Such soft potentials give much better sampling of configuration spaceby allowing particle overlapping, and further enable the simulations to beperformed at experimentally realistic □-valuesnot accessible by conventional molecular simulations using hard-core repulsions(such as in the Lennard-Jones potential or the self- and mutual-avoiding walk).The effects of chain discretization and finite-range interactions on ODT areproperly accounted for in our mean-field theory.[3] Our FOMCsimulations are performed in a canonical ensemble with variable box lengths toeliminate the adverse effects of fixed box sizes on ODT.[4]Furthermore, with a new order parameter for the lamellar phase, we use replicaexchange and multiple histogram reweighting to accurately locate ODT in oursimulations.

References:

[1] *L. Leibler*, **Macromolecules, 13**, 1602 (1980); *G. H. Fredrickson and E. Helfand*, **J. Chem.Phys., 87**, 697 (1987).

[2] *Q. Wang and Y. Yin*, **J. Chem.Phys., 130**, 104903 (2009).

[3] *Q. Wang*, **J. Chem. Phys.,129**, 054904 (2008); **131**, 234903 (2009).

[4] *Q.Wang et al.*, **J. Chem. Phys., 112**, 450 (2000).

See more of this Session: Thermodynamics of Polymers
See more of this Group/Topical: Materials Engineering and Sciences Division

(415g) Self-Assembly of Globular Protein-Polymer Diblock Copolymers

Carla S. Thomas, Christopher N. Lam, Liza Xu and **Bradley D. Olsen**, Department of Chemical Engineering, Massachusetts Institute of Technology, Cambridge, MA

Protein-based materials show a great deal ofpotential as catalysts and optoelectronics, where the unique efficiency,selectivity, or activity of enzymes can be captured to improve materialperformance. The self-assembly ofglobular protein-polymer diblock copolymers into nanostructured phases isdemonstrated as an elegant and simple method for structural control inprotein-based biocatalysts or bioelectronics.In order to fundamentally investigate self-assembly in these complexblock copolymer systems, a mutant of the red fluorescent protein mCherry is expressed in *E.coli* and site-specifically conjugated to a low polydispersitypoly(N-isopropyl acrylamide) (PNIPAM) block.Using thiol-maleimide coupling, a well-definedmodel globular protein-polymer diblock copolymer isformed. Nanostructured protein materialsare self-assembled by the evaporation of water from concentrated solutions ofthe diblock copolymer. Through changes in the temperature of thesolution or use of volatile acids and bases, different pathways towardself-assembly may be accessed. Smallangle X-ray scattering and transmission electron microscopy are used to explorethe dependence of nanostructure formation on the processing pathway, andsolvent annealing was used to achieve more fully equilibrated structures. The effect of the molecular weight of thePNIPAM block was also investigated, demonstrating the ability to self-assemblea variety of different nanostructured morphologies from globular proteins. Wide angle X-ray scattering illustrates thatdiblock copolymer self-assembly results in a noncrystallinestructure within the protein nanodomains.Circular dichroism , UV/Vis spectroscopy, and Fourier transforminfra-red (FTIR) spectroscopy show that a large fraction of the protein remainsin its folded state after conjugation, and the activity of the protein isexplored as a function of processing conditions to identify conditions that maximizethe preservation of protein function.

See more of this Session: Thermodynamics of Polymers
See more of this Group/Topical: Materials Engineering and Sciences Division

(433a) Developing a Novel Preparation Method for Residual Biomass Polymer Blends

Jinwen Zhang, Composite Materials and Engineering Center, college of Engineering and Architecture, Washington State University, Pullman, WA

There is a growing interest in polymeric materials from renewable resources. Nevertheless, in facing a society where petroleum-based plastic products are almost ubiquitous, the challenge for the emerging biobased polymer materials is twofold: being cost effective and performance competitive while lowering the environmental footprint.

Soy protein meal, a residue after oil crush, has being explored as a new resource for plastics. However, owing to some obstacles such as water/moisture sensitivity, high melt viscosity, narrow process window and brittleness, neat soy protein (SP) plastics have little practical values. To overcome these disadvantages, blending SP with other biodegradable polymers has received increasing attention. Sugar beet pulp (SBP) is another agricultural residue which has received similar attention for plastic applications as SP. In the current study of SP or SBP polymer blends, SP and SBP are merely used as organic fillers in most cases. Although certain processing and end use properties can be improved with respect to that of neat SP and SBP plastics, the mechanical properties of the blends are often not satisfactory and even lower than that of both the neat polymer and the neat SP (or SBP) plastics.

In this work, we introduced a novel method to prepare polymer blends of residual biomass and biodegradable thermoplastic polymer. SP and SBP were used as plastic components during the compounding of the blends. The polymers used were poly(lactic acid) (PLA) and poly(butylene adipate-*co*-terephthalate) (PBAT), respectively. The mixing of the blends was carried out in a twin screw extruder and the test specimens were prepared by injection molding. Plasticization of SP and SBP by water during compounding played a critical role in determining SP (or SBP) to be a plastic component or filler in the mixing and hence the morphological structure of the resulting blends. When extra water was added, SP (or SBP) behaved like a melt during the compounding, consequently the resulting blends demonstrated significant different properties and morphological structures from those made of dry SP (or SBP). Because water evaporated during mixing and after drying, the resulting blends actually became in situ formed thermoplastic composites. Percolation network structure of SP (or SBP) was noted in some of these blends. Different effects of SP (or SBP) as a plastic component and filler on phase structure and mechanical properties of the resulting blends were investigated. Molecular weight changes of the thermoplastic PLA and PBAT after processing were examined. The change of SP molecular structure caused by melt processing was also evaluated by protein solubility test. Effects of various compatibilizers on the properties of the blends were also studied. The results suggest that when SP and SBP were used as plastic components in mixing, the resulting blends demonstrated an overall superior performance than when they are used as fillers.

(433b) Air-Templated Nanocellulose Reinforced Macroporous Foams

Alexander Bismarck, ICL, London, United Kingdom

In progres

See more of this Session: Biobased Materials I: General
See more of this Group/Topical: Forest and Plant Bioproducts Division

Unavailable Paper

(433c) Polymerization of Lactide to Polylactic Acid and Co-Polymers of Polylactic Acid Using High Viscosity Kneader Reactors

Boyd T. Safrit and George E. Schlager, LIST USA, Inc., Charlotte, NC

Polymerization of lactide to polylactic acid (PLA) can be performed using conventional reactor technology such as stirred tank reactors, but the conversion and/or final molecular weight must be controlled to a lower level. At higher conversion and/or molecular weight, the reaction mass will become very viscous , which limits the ability of conventional reactor technology to provide adequate mixing, minimize mass transfer effects on reaction kinetics, remove exothermic heat of reaction and ensure proper heat transfer in order to eliminate hotspots/thermal degradation.

Kneader reactor technology has been used over 60 years in many high viscosity applications such as reactions and polymerization, devolatilization, and drying. This technology can handle the higher conversion and molecular weight polymerizations of lactide and other copolymers of lactide, while also providing the heat transfer required for proper temperature control. Using model kinetics and rheology data, a study was performed that shows the capability of kneader reactor technology for lactide polymerizations as well as other copolymers. Kneader reactor technology can also be used to remove the uncovered monomers from the polymer and expected results from the continuous operation of a polymerizer and finisher will be shown.

See more of this Session: Biobased Materials I: General
See more of this Group/Topical: Forest and Plant Bioproducts Division

(433d) Biodegradable Poly(hydroxyalkanoate) Foams

Amy Tsui, Qi Liao and Curtis W. Frank, Chemical Engineering, Stanford University, Stanford, CA

Poly(3-hydroxybutyrate-*co*-3-hydroxyvalerate) (PHBV) is a member of the bacterial polyester family of poly(3-hydroxyalkanoates) (PHAs) known for its ability for enzymatic degradation and good mechanical properties. The use of biodegradable polymers for producing sustainable materials has received increasing attention, and PHBV is an excellent candidate, shown to have similar properties to polypropylene (PP), known to be a tough and robust material. Specifically, PHBV could be a useful material for construction materials such as insulation foam. Current insulation foams are made primarily of polyurethane, polystyrene and polyethylene which are derived from non-renewable (petroleum) feedstock. Conventionally, foam structures in nonaqueous systems, such as polymers are stabilized simply by solidification due to vitrification, cross-linking, or crystallization (1). This requires that the polymer matrix have a high enough melt strength to retain cell structures until solidification occurs. However, processing of the PHBV into foam is limited by its low melt elongational viscosity and narrow thermal processing window. Low melt elongational viscosity contributes to cell coalescence and collapse, resulting in undesirable high density foam. The poor foamability of PHBV can be addressed via the addition of a nucleating agent, orotic acid.

Orotic acid (OA), a biodegradable and nontoxic chemical, was shown to be a very effective nucleating agent (NA) for crystallization in poly(3-hydroxybutyrate) (PHB) (2,3). Nucleating agents can also be used to promote bubble formation (4). With NAs, more nucleation sites are created for gas to diffuse into so bubbles are limited to a smaller, more uniform cell size (5), while also increasing cell density (6). Additional benefit to using nucleating agents is that they often act as plasticizers which reduce the processing temperature needed for melt extrusion (4). As NAs also promote crystallization, the addition of NAs may also reduce the time required for stabilization of polymeric foam structure.

To first evaluate the nucleating efficiency of OA in PHBV, films of PHBV and OA were made using the solvent casting method. The compositions ranged from 0 to 10 wt% OA. Polarized light optical microscopy (PLOM) showed that the OA particles were well-dispersed throughout the films, which is optimal for homogeneity of nucleation sites. Differential scanning calorimetry (DSC) indicated that 1 or 2 wt% OA resulted in the highest percent crystallinity. It was also seen that presence of OA resulted in more uniform crystal size, as indicated by the narrower full-width-half-maximum (FWHM) values of the melting endotherm. Further studies are being performed to evaluate the effect of OA content on crystallization kinetics.

Foaming of poly(3-hydroxybutyrate-co-3-hydroxyvalerate) (PHBV) with 5% hydroxyvalerate content (Tianan Biologic Materials Company, China) was performed using an activated azodicarbonamide (AZ) (Actafoam 780, Chemtura) chemical blowing agent that decomposed at 140-150°C, generating approximately 150 cc of nitrogen per gram of solid. After blending and drying the PHBV and blowing agent, materials were extruded through a ¾-inch single-screw extruder (L/D ratio 25:1, compression ratio 3:1, C.W. Brabender) equipped with a 2-inch horizontal flex-lip ribbon die. The extruder temperature profile was chosen based on the temperature requirements for melting the PHBV and decomposing the blowing agent, while preventing PHBV decomposition (7-9). Foaming occurred upon leaving the die. SEM images obtained on cryo-fractured surfaces of extruded foam samples showed varying cellular morphologies with changes in AZ content. Cell coalescence and collapse occurred when the amount of blowing agent AZ exceeded 1.5 parts per hundred (phr), suggesting that the PHBV melt strength does not withstand more increase in pressure posed by gas generation and expansion (10). Using the same procedure as for the PHBV foams, results from the addition of 1 or 2 wt% OA will be shown.

The biodegradable polyester PHBV is a promising material for replacing conventional petroleum-based polymers. However, the melt strength and thermal processing window of PHBV must be improved in order to develop an alternative insulation foam. Adding orotic acid as a nucleating agent for both crystallization and bubble formation could be a promising method for making such enhancements to the foamability of PHBV.

References

1. Thareja P; Ising BP; Kingston SJ; Velankar SS. Macromol. Rap. Comm. 2008, 29, 1329

2. Jacquel N; Tajima K; Nakamura N; Miyagawa T; Pan P; Inoue Y J. App. Polym. Sci. 2009, 114, 1287

3. Jacquel N; Tajima K; Nakamura N; Kawachi H; Pan P; Inoue Y J. App. Polym. Sci. 2010, 115, 709

4. Kim SG; Lee JWS; Park CB; Sain M J. App Polym. Sci. 118, 1691

5. Mills N Polymer Foams Handbook 2007, Chapter 3, 40

6. Rodrigue D; Gosselin R. The Effect of Nucleating Agents on PP Foam Morphology 2003, 169

7. G. Schmack; D. Jehnichen; R. Vogel; B. Tandler J. Polym. Sci. 2000, 38, 2841

8. M. Yamaguchi; K. Arakawa Euro. Polym. J. 2006, 42, 1479

9. R. Vogel; B. Tandler; D. Voigt; D. Jehnichen; L. Hausler; L. Peizsch; H. Brunig Macrom. Biosci. 2007, 7, 820

10. Q. Liao 2010 Ph.D. Thesis, Stanford University, Stanford, CA.

(433e) Carbon Fiber Reinforced BioPlastics: Present Status and Future Directions

Manju Misra, School of Engineering and the Department of Plant Agriculture, University of Guelph, Guelph, ON, Canada and **Amar K. Mohanty**, Department of Plant Agriculture & School of Engineering, Guelph, ON, Canada

In Progress

See more of this Session: Biobased Materials I: General
See more of this Group/Topical: Forest and Plant Bioproducts Division

(433f) Modification of Biomass Rheology: The Influence of Physical Chemistry

Joseph R. Samaniuk[1], C. Tim Scott[2], Thatcher W. Root[1] and Daniel J. Klingenberg[1], (1)Chemical and Biological Engineering, University of Wisconsin-Madison, Madison, WI, (2)USDA Forest Products Laboratory, Madison, WI

Chemicals derived from lignocellulosic biomass, including liquid fuels, are a sustainable alternative to petroleum derivatives. Many different processes have been proposed to convert lignocellulosic materials to liquid fuels but a common operation required in all of them is heating to high temperatures (110-230 ^{o}C). Reducing the concentration of water during this step decreases utility costs and decreases overall capital costs by reducing equipment size. Increasing the solids concentration however, results in rheological changes like a rapidly increasing yield stress. Mixing, pumping and pouring biomass slurries becomes difficult, if not impossible, above 20-30 wt% solids.

We have shown previously that the rheology of biomass can be controlled with the use of certain water soluble polymers (WSPs). AFM experiments have revealed that WSPs alter stresses in cellulose fiber suspensions by lubricating the surface of individual fibers. Using these additives, we have been able to mix and convey untreated biomass at solids concentrations up to 50 wt%. However, the efficacy of WSPs is reduced after certain treatments such as acid hydrolysis. How changes in the physical chemistry of biomass during treatment affect rheology modification with WSPs is unknown.

In this presentation our focus will be on the influence of the physical chemistry of both biomass and WSPs on rheological modifications. The physical chemistry of the biomass is altered by the treatment type, e.g. acid hydrolysis, and the treatment severity. Specific changes to the physical chemistry of the biomass are probed with fiber suspensions containing known chemistries, e.g. cellulose fibers vs. hemicellulose-rich fibers. The physical chemistry of the WSPs are altered by varying the type of polymer, the molecular weight, the degree of substitution, and by the addition of chemical additives. We will show how these variables influence not only the instantaneous rheological modification but also its behavior over time.

See more of this Session: Biobased Materials I: General
See more of this Group/Topical: Forest and Plant Bioproducts Division

Torrefied wood pellets production using natural biomass compounds

Venkata K Penmetsa and Philip H Steele*
Forest Products Department, Mississippi State University,
100 Blackjack Rd. Starkville, MS 39759. Tel. no: 662-325-8083
kpenmetsa@cfr.msstate.edu

The U.S. coal industry is highly interested in the potential for combusting torrefied wood as a co-feed with coal. Unlike biomass torrefied wood can be cofed with coal at a high percentage. Biomass combustion facilities cannot use the existing infrastructure and new biomass fed plants cost from $140 to $200 million to construct.

Torrefaction of biomass has attracted considerable attention as a means to reduce biomass weight by 70% through removal of water and the hemicellulosic fractions while retaining most of the original high-energy cellulose and lignin content. A further advantage of torrefied wood is that it has anti-microbial properties as well as limited water resistance such that microbial degradation is eliminated.

Unlike biomass torrefied wood can be co-fed with coal at a high percentage and requires no major infrastructure investment. One issue is the current inability to ship torrefied wood as coal is currently shipped. A waterproof pelletized product is required to allow bulk shipping via open railcar or by barge. Production of a waterproof torrefied wood pellets or briquettes has not been achieved to date.

Mississippi State University has developed a low-cost binder to produce a high-density pellet with high water resistance. The binder is produced from biomass and has a neutral pH of 6.2 such that the pellets do not produce corrosive gases during their combustion. No chemical cross linking agents, such as formaldehyde, are required making this binder very environmentally friendly. The binder is suitable to produce densified water repellent pellets with many different biomass feed stocks. The pellets have been subjected to crushing strength and water repellant test.

Pellet hardness was determined for each test specimen as described in ASTM standard D143 for small clear specimens. Each specimen was weighed immediately before and after the test and the density values were computed. All test results are given in Table 1.

The pellets were also tested in a water absorption test. The pellets were weighed before immersing in water and were fully immersed in water for 14 days. The pellets were weighed every 24 hours to monitor any increase in the weight of each pellet due to water absorption. The weight difference quantifies the amount of water absorbed inside the pellet pores. Figure 1 demonstrates the behavior of the pellets immersed in water over 14 days at various binder loadings as the weight percent of the total pellet weight. The 30%, 20% and 15% binder proportions showed very minor weight gains of water ranging from 0.05% to 0.6%.

These results indicate that at these high binder ratios the pellets were so densified that limited pore spaces are available for water to penetrate over time. These results on water weight gain over time indicate that binder loading magnitude was directly linked to water absorbance. The pellets with the lowest binder proportion had the highest water absorption values. Over 14 days the pellet produced with 2% binder had maximum weight gain of about 11%; the 5% binder loaded pellet had a water weight gain of 9.6%; that with 10% binder addition had a water weight gain of only 1.2%.

The study results showed that the tested binder produces a densified and water repellent bond for torrefied wood that is helpful in producing high initial strength with this strength substantially maintained following severe water exposure is more helpful in pelletizing torrefied wood. Both initial pellet compressive strength and strength following long-term water soaking the pellet strength is a relatively linear function of proportion of binder applied. However, even pellets with only 2% binder loading produced an initial strength of 764 kg/m^2 and 68% of this initial strength was maintained following the 14 days of water soaking followed by high-temperature drying to zero moisture content. While binder loadings of 30, 20 and 15% were tested and showed superior performance it is unlikely that these loadings will be economical due to the high amount of binder required. In addition, for production of a waterproof pellet at 10% loading appears to be more than adequate. Over the 14 days of soaking only 1.2% of water absorption was observed for this loading. In addition, nearly 80% of the initial compressive strength was maintained after water exposure and oven drying.

Table 1. Pellet weight (g) and compression strength of test pellets immediately after production compared to pellets water soaked for 14 days and oven dried to oven dry water content.

Spec. no.	Weight (g)	Binder composition (%)	Compression strength (kg/m^2)	Compression strength (kg/m^2) after 14 days in water
1	16.80	30	1095	960
2	16.92	20	987	846
3	16.76	15	934	800
4	16.90	10	905	718
5	16.39	5	841	630
6	16.42	2	764	519

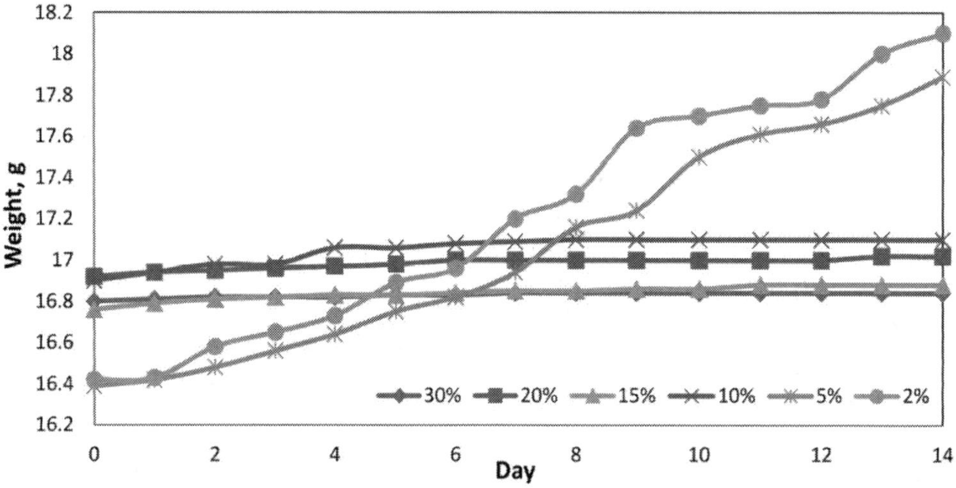

Figure 1: Daily water weight absorption values by pellet binder proportion for 14 days of soaking

For most applications it appears that a 2% loading will produce adequate initial compression strength at 764 kg/m^2. In addition, following water soaking 68% of the initial strength (519 kg/m^2) were maintained. It is not clear that the extra expense of producing a 10%, or higher, binder loading pellet would be cost effective versus the added cost of the additional binder.

References:

1. Fonseca Felfli, F., Luengo, C.A., Bezzon G. and Beaton Soler, P., 1998. *Bench unit for biomass residues torrefaction*, Biomass for Energy and Industry, Proceeding of Int. Conf., Würzburg, Germany, 8-11June, Ed. by Kopetz, Weber, Palz, Chartier and Ferrero, C.A.R.M.E.N., Rimpar Germany, 1593-1595.
2. Arcate, J.R., 2000. *New process for torrefied wood manufacturing,* Bioenergy update, vol 2, No. 4. (www.techtp.com/Bioenergy%20Update%20.pdf).
3. Wescott, J.M. and C.R. Frihart. 2004. *In* Proceedings of the 38th International Wood Composite Materials Symposium, Washington State University, Pullman, Washington. pp 199-206.

(464a) Awards Ceremony

John G. Ekerdt, Chemical Engineering, University of Texas at Austin, Austin, TX and Yueh-Lin Loo, Chemical Engineering, Princeton University, Princeton, NJ

time reserved for awards ceremony.

See more of this Session: Materials Engineering & Sciences Division Plenary Session
See more of this Group/Topical: Materials Engineering and Sciences Division

Unavailable Paper

(464b) Directed Assembly At the Nanoscale, and the Balance Between Equilibrium and Non-Equilibrium States

Juan J. de Pablo, Department of Chemical and Biological Engineering, University of Wisconsin-Madison, Madison, WI

There is considerable interest in devising nanofabrication strategies that rely on the molecular self-assembly of complex fluids and materials. Our efforts over the past several years have been focused on devising strategies to drive and direct that self assembly, largely by developing multiscale modeling models and methods capable of predicting the structure and properties of complex fluids and materials under external fields, including confinement, electric fields, or flow fields. These models and methods can vary considerably in nature and level of resolution, depending on the system and issues of interest. In this presentation I will provide an overview of various modeling approaches, along with their usefulness and limitations, in the context of two distinct nanofabrication platforms. The first is concerned with the formation of ordered, defect-free block copolymer structures on nanopatterned substrates. A new mesoscopic formalism has been developed to describe the structure and dynamics of block copolymer blends and composites, and we use it to explain the effects of surfaces and different types of confining walls on the free energy (and the concomitant stability) of a variety of morphologies of interest for lithographic fabrication. Many of these morphologies represent non-equilibrium states that are accessed by specific processing routes, and simulations can be used to discern the boundaries between such states and stable, equilibrium morphologies. The second application is concerned with the development of liquid-crystal based biosensors. A multiscale model has been used to design liquid-crystal based devices for detection of biological molecules or toxins. In one implementation, nanoscale particles self assemble into highly regular structures, including chains, upon exposure to specific chemicals. In a different implementation, liquid crystal nanodroplets are shown to adopt distinct configurations upon exposure to specific analytes. As discussed in this presentation, the models can be used to explain the defects and transmission images that arise in laboratory experiments. Theoretical calculations and simulations can then be used to identify metastable states that are particularly responsive to external perturbations, thereby offering a route for preparation of systems that offer an enhanced response to external stimuli.

See more of this Session: Materials Engineering & Sciences Division Plenary Session
See more of this Group/Topical: Materials Engineering and Sciences Division

(464c) Superhydrophobic and Superoleophobic Coatings

Di Gao, Chemical and Petroleum Engineering, University of Pittsburgh, Pittsburgh, PA

Superhydrophobic surfaces are often found on plant leaves and insect wings in nature. Water on these surfaces forms small beads and drips off rapidly. This interesting phenomenon has stimulated extensive research and commercialization effort to make artificial superhydrophobic coatings and to use them for self-cleaning, water-proof, anti-icing, anti-corrosion, and anti-fouling purposes. A key challenge in making superhydrophobic coatings is to tailor the morphology of the coatings in nanometer scales, which typically involves synthesis and fabrication of materials at nanometer dimensions. Recently, the concept of superhydrophobicity has been extended to make surfaces that repel oil in the same way as superhydrophobic surfaces repel water. Such surfaces have been referred to as superoleophobic surfaces. Apparently, due to the much lower surface tension of oil compared to that of water, making superoleophobic surfaces has been extremely challenging. In this talk, I will discuss the fundamental mechanisms underlining the superhydrophobic and superoleophobic surface phenomena as well as the commercialization progress for making such coatings for industrial applications, such as "anti-icing superhydrophobic coatings" and "superoleophobic and superhydrophilic coatings for chemical-free cleaning and for water-oil separation".

See more of this Session: Materials Engineering & Sciences Division Plenary Session
See more of this Group/Topical: Materials Engineering and Sciences Division

Paper # 464d

Synthesis, Functionalization, and Clinical Diagnostic Applications of ZnSe Nanocrystals

T.J. Mountziaris

Department of Chemical Engineering, University of Massachusetts, Amherst, MA

Abstract

Nanocrystals of II-VI compound semiconductors have unique properties, including size-tunable photoluminescence, narrow emission and broad absorption spectra, high quantum yields, high extinction coefficients, and high photochemical stability, that make them attractive for clinical diagnostic applications that combine detection sensitivity and multiplexing capabilities. Nanocrystals that are based on ZnSe are particularly appealing, because they do not contain heavy metal ions and can be easily doped with transition metals, such as Mn, to provide additional control over their electronic properties. The synthesis of ZnSe and doped ZnSe:Mn nanocrystals using a template-assisted method that employs microemulsions and liquid crystals to provide control over the size, shape, and size distribution of the nanocrystals will be discussed. Extraction of the nanocrystals from the templates, surface modification, stabilization in aqueous solution, and conjugation with probe biomolecules, such as single-stranded DNA or antibodies, has enabled the development of a new class of nanocrystal-based biomolecular sensors that can be used in the development of separation-free assays for rapid quantitative detection of biomolecular targets.

(464e) Block Copolymer Thin Films As Nanostructure Templates

Richard A. Register, Chemical and Biological Engineering, Princeton University, Princeton, NJ

Block copolymers have been extensively studied for their ability to self-assemble into microdomain morphologies such as spheres, cylinders, and lamellae, with typical periodicities of 20-100 nm. Similar structures form when block copolymers are deposited as thin films on substrates; for example, very asymmetric block copolymers form spheres of the minority component in a matrix of the majority block, while less-asymmetric materials typically form cylinders in plane. These films can serve as excellent templates for nanofabrication, where the block copolymer's nanodomain structure is faithfully reproduced in an inorganic material—but the final array of inorganic objects is, at best, only as good as the structure of the film from which it was derived. Consequently, we have worked intensively to develop methods to and manipulate the structure of the films. For example, the polygrain structure which these nanodomains normally form can be transformed to a single-crystal texture, over macroscopic areas, by a simple shearing process. Shear can also realign the domain orientation locally in films with an otherwise macroscopic orientation; to create complex orientation patterns on the millimeter scale; and even to transform spheres into cylinders. For spheres, we can exploit the selective wetting behavior of the two blocks to adjust the areal density of spheres (micelles) by at least an order of magnitude, all while keeping the micelle size fixed, simply through control of the film thickness.

We have employed these thin, substrate-supported block copolymer films to fabricate dense arrays of 20-40 nm metal or semiconductor particles: dots (from sphere-forming block copolymers) or lines (wires, from cylinder-formers), all with a size and spacing set through block copolymer molecular weight. As a particular example, we have used this approach on cylinder-forming films to fabricate centimeter-scale arrays of parallel metallic nanowires, with 33 nm pitch; due to their fine pitch, such wire grids can polarize an exceptionally broad range of wavelengths extending down into the deep ultraviolet (for 193 nm photolithography). We have also begun studying the shear-alignment of perpendicular ("standing") lamellae, which offer the possibility of higher aspect ratios.

See more of this Session: Materials Engineering & Sciences Division Plenary Session
See more of this Group/Topical: Materials Engineering and Sciences Division

(464f) Colloidal Materials for Modulating the Immune Response

Jeffrey A. Hubbell[1], Melody A. Swartz[2], Eleonora Simeoni[2], Armando Stano[2], Chiara Nembrini[2], Marie Ballester[2], Susan N. Thomas[2], Sachiko Hirosue[2], Andre J van der Vlies[2], Conlin P O'Neil[2], Diana Velluto[2], Jackson Eby[2] and Evan A Scott[2], (1)Institute of Bioengineering and Institute of Chemical Sciences and Engineering, cole Polytechnique Fdrale de Lausanne (EPFL), Lausanne, Switzerland, (2) Institute for Bioengineering, cole Polytechnique Fdrale de Lausanne (EPFL), Lausanne, Switzerland

Polymeric nanoparticles hold tremendous potential in medicine, especially if modes of biofunctionality can be incorporated into them. Here we consider different kinds of biofunctionality, focused on applications in antigen presenting cell-targeted immunotherapeutics. We consider functionality in pathways for degradation, in physiological routes of penetration of biological barriers for delivery, and in activation of dendritic cells for induction of adaptive immune responses.

Our laboratory has recently described a novel family of AB and ABA block copolymeric amphiphiles that are capable of forming micelles and vesicles [1, 2], as well as inverse emulsion-polymerized nanoparticles based on the B block using Pluronic emulsifiers [3, 4]. As a hydrophilic block A, we employ polyethylene glycol (PEG), because of its well known toxicological profile and its well-defined and low polydispersity. As a hydrophobic block, we have selected polypropylene sulfide (PPS), a low Tg polymer that can be synthesized by a ring opening living polymerization also with low polydispersity [5]. We have demonstrated that these polymers form mesoscopic aggregates that are sensitive to oxidative environments by conversion of the hydrophobic PPS to the hydrophilic polypropylene sulfone [2]; this provides a route of degradation of the micelle-forming amphiphile into a fully soluble low molecular weight polymer, sufficiently small for clearance by renal filtration. We have also sought to render these same structures sensitive to reduction, to allow destabilization of vesicles within the early endosome after endocytosis by linking the two blocks with a reduction-sensitive disulfide [6], for use in intracellular delivery, for example of adjuvant molecules targeting intracellular receptors (such as CpG DNA), antigens for targeted MHC 1 presentation, or antigen-encoding DNA. Thus, the redox sensitivity of these materials can be used to enable release of incorporated agents and ultimate elimination of the micelle-forming polymer.

With the micelle-forming AB and ABA block copolymers as well as the analogous materials formed by inverse emulsion polymerization of propylene sulfide using Pluronics as emulsifier, we are seeking nanoparticle forms that are sufficiently small as to pass barriers to efficiently access the targeted antigen-presenting cells. Of particular interest to us, is targeting dendritic cells resident in the lymph nodes after intradermal or subcutaneous injection [7]. We have shown that ultrasmall particles, ca. 25 nm, can target lymph node-resident dendritic cells very efficiently, being swept through the tissue interstitium by the slow flows that exist between the blood and lymphatic capillaries in the skin. Also with such small nanoparticles, we see very effective penetration of mucosal barriers in the nasal sinus[8] and very efficient uptake by antigen presenting cells in the lung after instillation in the mouse. We show that when antigen is reversibly coupled to these nanoparticle surfaces by a disulfide link, very potent cellular immune responses can be induced[9, 10].

We seek to be able to incorporate biofunctionality into these nanoparticles for activation of antigen presenting cells. In a first approach, we designed the nanoparticle surface for activation of complement by the alternative pathway, and we demonstrated that complement activation could serve as a very potent, particle-mediated danger signal for dendritic cell activation in situ [11, 12]. Current work also addresses incorporation of additional danger signal molecules into the nanoparticle formulation, including engineered flagellin truncation variants[8], and CpG DNA sequences incorporated into the nanoparticle surfaces and cores.

References:

1. Cerritelli, S. et al. Thermodynamic and kinetic effects in the aggregation behavior of a poly(ethylene glycol-b-propylene sulfide-b-ethylene glycol) ABA triblock copolymer. *Macromolecules* **38**, 7845-7851 (2005).

2. Napoli, A., Valentini, M., Tirelli, N., Muller, M. & Hubbell, J.A. Oxidation-responsive polymeric vesicles. *Nature Materials* **3**, 183-189 (2004).

3. Rehor, A., Botterhuis, N.E., Hubbell, J.A., Sommerdijk, N.A.J.M. & Tirelli, N. Glucose sensitivity through oxidation responsiveness. An example of cascade-responsive nano-sensors. *J. Mater. Chem.* **15**, 4006-4009 (2005).

4. Rehor, A., Tirelli, N. & Hubbell, J.A. A new living emulsion polymerization mechanism: Episulfide anionic polymerization. *Macromolecules* **35**, 8688-8693 (2002).

5. Napoli, A., Tirelli, N., Kilcher, G. & Hubbell, J.A. New synthetic methodologies for amphiphilic multiblock copolymers of ethylene glycol and propylene sulfide. *Macromolecules* **34**, 8913-8917 (2001).

6. Cerritelli, S., Velluto, D. & Hubbell, J.A. PEG-SS-PPS: Reduction-sensitive disulfide block copolymer vesicles for intracellular drug delivery. *Biomacromolecules* **8**, 1966-1972 (2007).

7. Reddy, S.T., Rehor, A., Schmoekel, H.G., Hubbell, J.A. & Swartz, M.A. In vivo targeting of dendritic cells in lymph nodes with poly(propylene sulfide) nanoparticles. *Journal of Controlled Release* **112**, 26-34 (2006).

8. Stano, A. et al. PPS nanoparticles as versatile delivery system to induce systemic and broad mucosal immunity after intranasal administration. *Vaccine* **29**, 804-812 (2011).

9. van der Vlies, A.J., O'Neil, C.P., Hasegawa, U., Hammond, N. & Hubbell, J.A. Synthesis of pyridyl disulfide-functionalized nanoparticles for conjugating thiol-containing small molecules, peptides, and proteins. *Bioconjugate Chemistry* **21**, 653-662 (2010).

10. Hirosue, S., Kourtis, I.C., van der Vlies, A.J., Hubbell, J.A. & Swartz, M.A. Antigen delivery to dendritic cells by poly(propylene sulfide) nanoparticles with disulfide conjugated peptides: Cross-presentation and T cell activation. *Vaccine* **28**, 7897-7906 (2010).

11. Reddy, S.T. et al. Exploiting lymphatic transport and complement activation in nanoparticle vaccines. *Nature Biotechnology* **25**, 1159-1164 (2007).

12. Thomas, S.N. et al. Engineering complement activation on polypropylene sulfide vaccine nanoparticles. *Biomaterials* **32**, 2194-2203 (2011).

See more of this Session: Materials Engineering & Sciences Division Plenary Session
See more of this Group/Topical: Materials Engineering and Sciences Division

(489a) A General Polymerization Algorithm for Amorphous Microporous Polymeric Materials

Lauren J. Abbott[1], Kyle E. Hart[1], Ping Lin[2] and Coray M. Colina[1], (1)Department of Materials Science and Engineering, Pennsylvania State University, University Park, PA, (2)The Materials Simulation Center, Pennsylvania State University, University Park, PA

Organic microporous materials offer a new approach to environmentally conscious and energy efficient gas storage and separation technology, such as O_2/N_2 separation, methane purification, and CO_2 capture. This work utilizes molecular simulations to study novel nanoporous compounds, such as polymers of intrinsic microporosity (PIMs) and hypercrosslinked polymers (HCPs), in order to better understand their porosity and structure.

Generating simulated structures of amorphous polymeric materials is a non-trivial task, since unlike crystalline material, no reference structure is available. In this work, we present a general procedure for developing simulated structures of amorphous polymeric materials. The proposed structure generation procedure consists of three steps: (i) ab initio calculation of the repeat unit electrostatic charge distribution, (ii) construction of a periodic system by a simulated polymerization algorithm, and (iii) relaxation of the structure by a 21-step compression protocol.

The simulated polymerization is accomplished by connecting repeat units close together in the periodic box to construct either linear polymer chains or polymer networks. This procedure enables the system to be built at realistic densities while avoiding steric overlap issues typical of large, rigid repeat units, which is shown to be crucial for obtaining accurate simulated structures. The generality of the procedure is exemplified by application to a number of linear and network PIMs, including PIM-1 and HATN-PIM, as well as poly(dichloroxylene) and poly(arylene ethynylene) networks. All simulated samples are characterized by their structure and porosity with surface areas, pore size distributions, and structure factors. The results show good agreement with available experimental data.

See more of this Session: Adsorption In Porous Polymers
See more of this Group/Topical: Materials Engineering and Sciences Division

THE JOURNAL OF
PHYSICAL CHEMISTRY C

ARTICLE

pubs.acs.org/JPCC

Effects of Residual Solvent on Membrane Structure and Gas Permeation in a Polymer of Intrinsic Microporosity: Insight from Atomistic Simulation

Liling Zhang, Weijie Fang, and Jianwen Jiang*

Department of Chemical and Biomolecular Engineering, National University of Singapore, 117576, Singapore

ABSTRACT: Microstructure and permeation in a polymer membrane are closely related to residual solvent during membrane fabrication. In this study, we conduct atomistic simulation to investigate the effects of residual solvent on free volume distribution and H_2 permeation in a polymer of intrinsic microporosity (PIM-1). The interaction energies of three solvents ($CHCl_3$, CH_3OH, and H_2O) with PIM-1 are predicted to be −16.3, −9.6, and −7.0 kcal/mol, respectively, which are in good agreement with available experimental data. On this basis, a structure−property relationship is proposed between interaction energy and the critical volume of solvent. The cyano and dioxane groups in PIM-1 interact preferentially with CH_3OH and H_2O; however, the carbon atoms interact more strongly with $CHCl_3$. The mobility of residual solvent is found to decrease in the order of $H_2O > CH_3OH > CHCl_3$. Upon comparison, the mobility of PIM-1 chains is smaller but facilitated by solvent due to the cooperative interactions between polymer and solvent. The fractional free volumes and large-size voids in PIM-1/solvent membranes are observed to decrease as $CH_3OH > CHCl_3 > H_2O$, consistent with positron annihilation lifetime spectroscopy measurements. The solubility and diffusion coefficients of H_2 decrease in the same hierarchy, and the predicted and experimental coefficients are in fairly good agreement. This simulation study provides atomistic insight into the microscopic properties of residual solvent in a polymer membrane and reveals the crucial role of residual solvent in tailoring membrane structure and gas permeation.

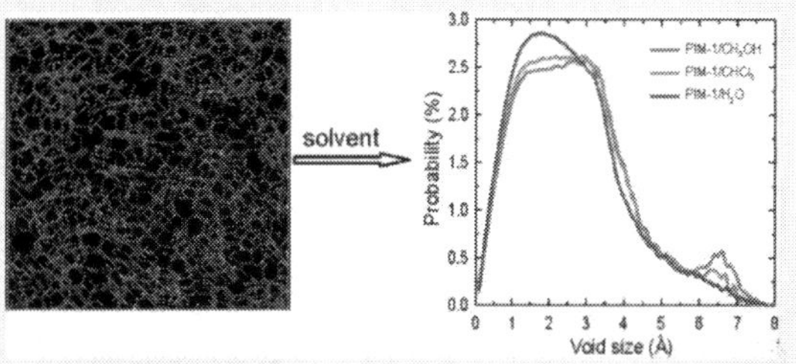

1. INTRODUCTION

Gas separation is ubiquitously involved in numerous engineering processes and applications. This is particularly important for a current topical issue, CO_2 capture to reduce the global warming effect. A number of techniques have been proposed for gas separation such as pressure swing adsorption, cryogenic distillation, and membrane permeation. In pressure swing adsorption, a gas mixture is separated due to different adsorption affinities, while separation in cryogenic distillation is based on different boiling points. Compared with these two energy-intensive techniques, separation using polymer membranes has advantages in high energy efficiency, low capital cost, large separation capability, and ease for scaling-up.[1]

In the past, extensive experimental studies have been reported for gas permeation and separation in polymer membranes.[2,3] For example, Sridhar et al. prepared polyphenylene oxide (PPO) for CO_2/CH_4 separation. A higher selectivity was observed in sulfonated PPO, while unmodified PPO showed a larger permeability.[4] Chung and co-workers reported the synthesis, cross-linking, and carbonization of copolyimides containing internal acetylene units for gas separation.[5] They also proposed a novel synthetic strategy to fine-tune the cavity sizes and free volumes of polyimide membranes with homogeneous pseudointerpenetrating networks.[6] Ribeiro and Freeman measured the sorption and dilation of CO_2 and C_2H_6 in cross-linked poly(ethylene oxide) (PEO). The dilation and sorption data were combined to calculate the partial molar volumes of gases in PEO, which were then compared against available literature data.[7] Sen and Banerjee introduced $-C(CF_3)_2-$ groups into polyetherimide (PEI) and found the modified PEI membrane has an exceptionally large permeability along with a higher CO_2/CH_4 selectivity than unmodified counterpart.[8] With ever-growing computational power, increasing molecular simulation studies have been carried out for gas permeation in polymer membranes. Theodorou and co-workers simulated the structural, conformational, and dynamic properties of amorphous poly(ethylene terephthalate) and poly(ethylene isophthalate), and predicted CO_2 sorption in glassy atactic polystyrene.[9,10] Heuchel et al. constructed atomistic packing models for glassy polyimides, polysulfone and poly(ether sulfone), and calculated gas solubility and diffusion coefficients.[11,12] Mark and co-workers used simulation results to elucidate high permeabilities in poly(dimethylsiloxane), which were attributed to the pronounced alternation in skeletal bond angles and the relatively large value of Si−O bond length.[13] These experimental and simulation studies demonstrate that gas permeation and separation in

Received: March 31, 2011
Revised: May 5, 2011
Published: May 16, 2011

dx.doi.org/10.1021/jp2029888 | J. Phys. Chem. C 2011, 115, 11233–11239

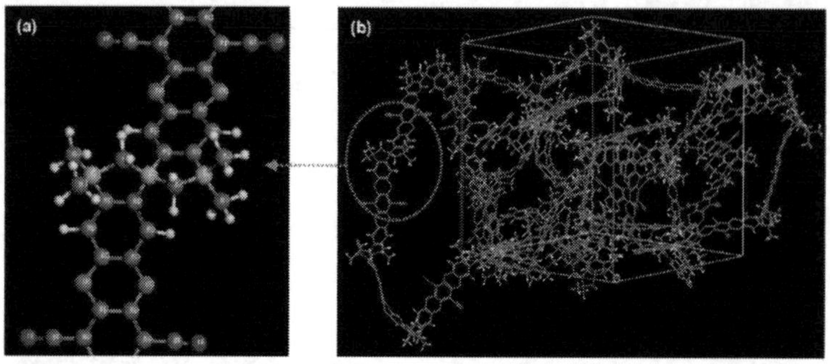

Figure 1. (a) Backbone of PIM-1 (the spiro carbons are denoted by "C" and the hydrogen atoms are in white). (b) Three dimensional simulation box of PIM-1 membrane (the box length is approximately 31.8 Å).

polymer membranes are governed by several crucial factors such as polymer structure, backbone functionalization, and chain rigidity.

A handful of experiments have revealed that the fabrication protocol of polymer membranes also affects performance. Joly et al. prepared 6FDA-*m*PDS polyimide films by spreading solutions from different solvents (CH_2Cl_2, dioxane, tetrahydrofurane, *N*-methyl-2-pyrrolidone and *N*-*N*-dimethylacetamide) and found the solubilities, diffusivities, and permeabilities of N_2 and CO_2 in the films were dependent on residual solvents.[14] Kostina et al. showed $CHCl_3$ resident in PEI could preferentially bind with oxygen atoms in membrane and lead to the conformational change of polymer chains.[15] Budd et al. observed a very strong sensitivity of gas permeability to casting method for a novel polymer of intrinsic microporosity (PIM-1). In particular, PIM-1 membrane with residual H_2O exhibits a substantially low permeability.[16]

Currently, it remains elusive how residual solvent specifically interacts with a membrane and affects membrane microstructure and performance. In this regard, fundamental understanding at a molecular level is desirable. In this work, an atomistic simulation study is reported for the effects of residual solvent on PIM-1 membrane structure and H_2 permeation. Three residual solvents (CH_3OH, $CHCl_3$, and H_2O) are considered and the simulation results are compared with the experimentally data by Budd et al.[16] The microscopic insight from simulation can facilitate the design of new membranes for gas purification. In a previous study, we have constructed an atomistic model for PIM-1 membrane and calculated gas sorption/diffusion by Monte Carlo (MC)/molecular dynamics (MD) simulations.[17] It was found that the voids in PIM-1 have diameter up to 9 Å and are largely interconnected. The simulated solubility and diffusivity of H_2, O_2, CO_2, and CH_4 are consistent with experimentally measured results. As attributed to its intrinsic microporous structure, PIM-1 appears to outperform most glassy polymer membranes in sorption and diffusion. Following this section, the models and methods are described in Section 2 including how to construct PIM-1 membranes with residual solvents and to calculate H_2 sorption and diffusion. In Section 3, the membranes are characterized in terms of fractional free volume and void distribution. The interaction and mobility of solvent and polymer are discussed. The simulated solubility and diffusion coefficients of H_2 are presented and compared with experimental data. Finally, the concluding remarks are summarized in Section 4.

Table 1. Physical Properties of Residue Solvents[22]

solvent	molecular weight	van der Waals volume ($Å^3$)	critical volume (cm^3/mol)
CH_3OH	32.04	36.82	118
$CHCl_3$	119.38	70.55	239
H_2O	18.01	19.24	55.9

2. MODELS AND METHODS

2.1. Membrane Construction. PIMs are amorphous and thermally stable glassy polymers. Remarkably, they have ultrahigh interconnected voids and are potentially useful for gas storage and separation.[18] The prototype PIM-1 was experimentally synthesized from 5,5′,6,6′-tetrahydroxy-3,3,3′,3′-tetramethyl-1,1′-spirobisindane by polycondensation reaction.[19] As illustrated in Figure 1a, the backbone of PIM-1 contains aromatic rings connected by spiro carbon atoms that are shared by two five-membered rings with a torsional angle of 90°. The rigid ladderlike polymer chains cannot pack efficiently, thus leading to a three-dimensional microporous structure.

The atomistic model of PIM-1 membrane was constructed by *Amorphous Cell* module in Materials Studio 4.3 (Accelrys Inc., San Diego, CA, U.S.A.) using Theodorou and Suter's scheme.[20,21] The model membrane was composed of three polymer chains in a periodic cubic simulation box (with a length of approximately 31.8 Å as shown in Figure 1b) and each polymer chain consisted of 15 repeat units. To prevent ring catenation during model construction, 200 CH_4 molecules were inserted into the simulation box and removed after the model was built. Ten configurations were generated and three of them were selected for equilibration by the following subsequent procedures: (1) energy minimization; (2) 500 ps NVT-MD simulation at 600 K; (3) 500 ps NPT-MD simulation at 600 K at 1 bar; (4) thermal annealing at 1 bar from 600 to 300 K with a temperature interval of 50 K; (5) 100 ps NPT-MD simulation at 10 bar with a time step of 0.1 fs; and (6) 2000 ps NPT-MD simulation with a time step of 1 fs at 300 K and 1 bar.

As in the experimental study,[16] three residual solvents are considered here including CH_3OH, $CHCl_3$, and H_2O. Table 1 lists the physical properties of the three solvents.[22] Experimental thermogravimetric analysis revealed that there were 0.5, 2.2, and 2.3% of CH_3OH, $CHCl_3$, and H_2O, respectively, in PIM-1 membrane.[16] Approximately, 4 CH_3OH, 4 $CHCl_3$, and 27 H_2O molecules existed in the simulation box shown in Figure 1b. The

Table 2. Predicted Densities and Fractional Free Volumes of PIM-1/Solvent Membranes

membrane	density (g/cm^3)	FFV (%)
PIM-1/CH_3OH	1.07	44.88
PIM-1/$CHCl_3$	1.09	44.27
PIM-1/H_2O	1.09	43.67

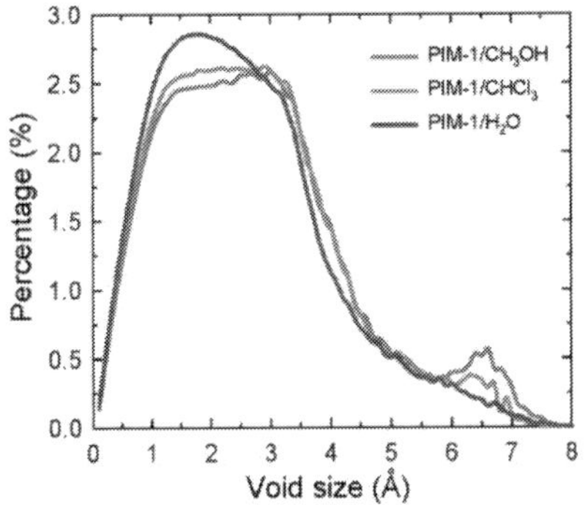

Figure 2. Void distributions in PIM-1/solvent membranes.

residual solvent molecules were randomly inserted into the equilibrated PIM-1 model membrane, then equilibrated by 8000 ps NPT-MD simulation at 300 K and 1 bar, and finally by 2000 ps NVT-MD simulation.

The energy minimization and MD simulations were conducted in DL_POLY.[23] To do this, the files created by Materials Studio were converted to DL_POLY using an in-house code. In such a way, computational time was reduced by 1−2 orders of magnitude. The polymers were mimicked by polymer consistent force field (PCFF), which has been found to be fairly accurate for PIM membranes.[17] The Lennard-Jones interactions were calculated with a cutoff of 13 Å and the Coulombic interactions were treated by the Ewald summation with a precision of 10^{-6}. The velocity Verlet algorithm was used to integrate the equations of motion. Temperature and pressure were controlled by the Berendsen method with a decay constant of 0.6 ps.

The equilibrated model membranes were characterized by fractional free volume (FFV) and void distribution. These quantities play a key role in governing gas permeation in a polymer membrane, and they were estimated here on the basis of geometric consideration using an in-house MC code. Specifically, a probe was randomly inserted into the simulation box and the insertion was considered to be successful if the probe did not overlap with any polymer atom. The ratio of successful insertions over the total number of insertions gave the FFV.[24] The probe size used in this study was set to be zero, that is, the maximum FFV was estimated. A total number of 200 000 insertions were found to be sufficient for the accurate estimation of FFV. The void size distribution was calculated by a method previously used for microporous materials.[25,26] In brief, the simulation box was divided into three-dimensional fine grids with a width of 0.12 Å. At a given grid, the void size was determined as the diameter of maximum cavity that enclosed the grid and had no overlap with any polymer atom. The percentage of a void size was normalized by the total number of grids each with a nonzero void.

2.2. Sorption and Diffusion of H_2. On the basis of solution-diffusion mechanism, gas permeability P in a polymeric membrane is equal to SD in which S is solubility coefficient and D is diffusion coefficient. In this work, S and D of H_2 in the PIM-1/solvent membranes were calculated at room temperature and infinite dilution.

The sorption was simulated by NVT MC method with a single H_2 molecule added into the membrane. The number of trial moves in the MC simulation was 2×10^7 with the first 10^7 moves for equilibration and the subsequent 10^7 moves for ensemble averages. Two types of trial moves were randomly attempted, namely, translation and rotation. In translation, H_2 molecule was translated with a random displacement in x, y, or z dimension; and the maximum displacement was adjusted to an overall acceptance ratio of 50%. In rotation, H_2 molecule was rotated around x, y, or z dimension with a random angle, and the maximum angle was adjusted to an overall acceptance ratio of 50%. The solubility coefficient was estimated from

$$S = \beta \int \exp[-\beta u_a(\mathbf{r}, \varpi)] \mathrm{d}\mathbf{r}\, \mathrm{d}\varpi \qquad (1)$$

where $\beta = 1/k_BT$ is the reciprocal temperature, $u_a(\mathbf{r},\varpi)$ is the interaction energy between the membrane and H_2 molecule at position $\mathbf{r}$ and orientation ϖ.

The diffusion was simulated by NVT MD method. To improve statistical accuracy, three H_2 molecules were inserted into the membrane. At least half of the box length existed between the H_2 molecules to minimize their interactions. The MD simulation was performed for 13 ns, in which the first 2 ns was used for equilibration and the subsequent 11 ns for production. Similar to the MD simulation used for model construction, the LJ and Coulombic interactions were also calculated with a cutoff of 13 Å and the Ewald summation with a precision of 10^{-6}. Temperature was controlled by the Berendsen method with a decay constant of 0.6 ps. The mobility was estimated by mean-squared displacement (MSD)

$$\mathrm{MSD}(t) = \frac{1}{N} \sum_{k=1}^{N} \langle |r_k(t) - r_k(0)|^2 \rangle \qquad (2)$$

where N is the number of gas molecules, $r_k(t)$ is position of the molecule k at time t. MSD was calculated from the ensemble average $\langle ... \rangle$ of the trajectory. In this study, multiple-origin method was used to calculate the MSD.

3. RESULTS AND DISCUSSION

3.1. Membrane Characterization. Table 2 lists the densities and FFVs of PIM-1/solvent membranes predicted from simulation. Compared with dry PIM-1 in the absence of solvent,[17] the density increases from 1.03 to 1.07−1.09 g/cm^3 and the FFV decreases from 47.7 to 44−45%. This is because the volume of membrane remains almost a constant in the absence/presence of solvent molecules and solvent molecules tend to occupy the voids in membrane; consequently, the density increases slightly. The increasing effect on density by residual solvent was also observed in PEI membranes.[27] It is worthwhile to note the FFVs in PIM-1 membranes are substantially larger than in common polyimide membranes (30−38%),[11] thus leading to a faster diffusion in PIM-1 membranes as discussed below.

Figure 2 shows the void distributions in PIM-1/solvent membranes. Because of the presence of solvent molecules, the

dx.doi.org/10.1021/jp2029888 | J. Phys. Chem. C 2011, 115, 11233–11239

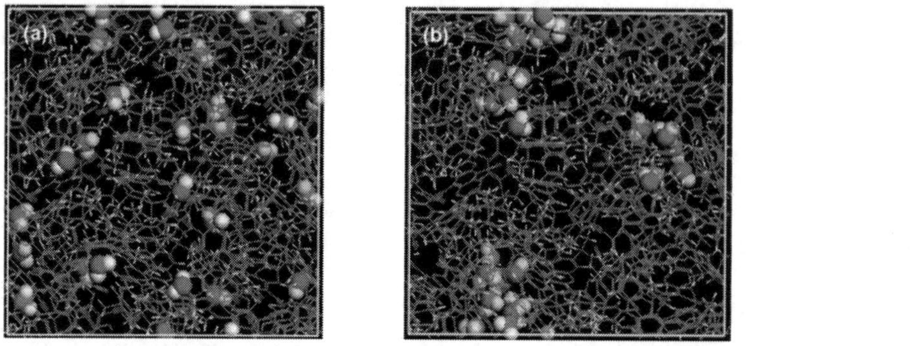

Figure 3. Simulation snapshots of PIM-1 with 27 residual water molecules. (a) Initial structure with water molecules randomly inserted in the simulation box. (b) Equilibrated structure after 5 ns simulation.

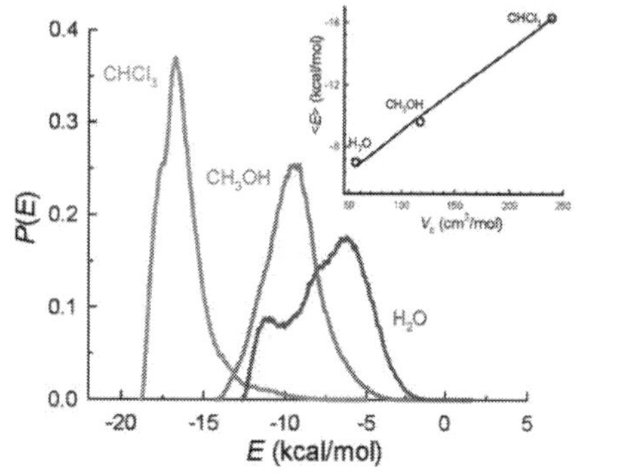

Figure 4. Distributions of interaction energy E between a single solvent molecule and PIM-1. The inset is the ensemble averaged energy $\langle E \rangle$ versus the critical volume of solvent.

largest voids (8 Å) in the three membranes are smaller than that (9 Å) in dry PIM-1.[17] The percentage of large-size voids (>6 Å) decreases following the order of PIM-1/CH_3OH > PIM-1/$CHCl_3$ > PIM-1/H_2O, which is in accord with positron annihilation lifetime spectroscopy measurements.[16] As listed in Table 1, $CHCl_3$ has a van der Waals (vdW) volume larger than that of CH_3OH; therefore, $CHCl_3$ occupies a greater void space and PIM-1/$CHCl_3$ has a smaller percentage of large-size voids. Although H_2O has the smallest vdW volume among the three solvents, 27 H_2O molecules (vs 4 CH_3OH or $CHCl_3$ molecules) exist in PIM-1/H_2O membrane and occupy a greater number of voids. Consequently, PIM-1/H_2O contains fewer large-size voids compared with PIM-1/CH_3OH and PIM-1/$CHCl_3$. Gas diffusion in a polymer membrane is largely governed by large-size voids. As we shall see, the diffusion coefficient of H_2 in PIM-1/H_2O is lower than in the other two membranes.

It is observed from simulation that 4 CH_3OH molecules in PIM-1/CH_3OH or 4 $CHCl_3$ molecules in PIM-1/$CHCl_3$ remain essentially separated during the entire simulation. Nevertheless, H_2O molecules in PIM-1/H_2O behave differently. As shown in Figure 3, initially 27 H_2O molecules are randomly distributed in the membrane, but they form into clusters after 5 ns. It was indeed observed experimentally by IR spectroscopy that alcohol in PIM-1 exists as monomers but H_2O as associates.[27] We therefore infer PIM-1 is a hydrophobic membrane in which highly polar H_2O interacts weakly with the polymer and tends to aggregate. As a result of the formation of large clusters occupying large-size voids, PIM-1/H_2O membrane possesses the lowest percentage of large-size voids among the three PIM-1/solvent membranes.

3.2. Polymer–Solvent Interaction and Mobility. To quantify polymer–solvent interaction, Figure 4 shows the distributions of interaction energy between a single solvent molecule and PIM-1. A sharp peak is seen in the distribution for $CHCl_3$ and CH_3OH but not for H_2O. The probability and energy value decrease in the order of $CHCl_3$ > CH_3OH > H_2O, which is consistent with the decreasing order in solvent hydrophobicity. Among the three solvents, $CHCl_3$ is the most hydrophobic, then CH_3OH, and finally H_2O. Therefore, the interaction between $CHCl_3$ and hydrophobic PIM-1 is the strongest and the weakest interaction is seen for H_2O. This implies that the primary interaction between solvent and PIM-1 is hydrophobic.

On the basis of the energy distribution, ensemble averaged energy $\langle E \rangle$ was evaluated from

$$\langle E \rangle = \frac{\int E(\mathbf{r}, \varpi)\exp[-\beta E(\mathbf{r}, \varpi)]\mathrm{d}\mathbf{r}\ \mathrm{d}\varpi}{\int \exp[-\beta E(r, \varpi)]\mathrm{d}\mathbf{r}\ \mathrm{d}\varpi} \quad (3)$$

For $CHCl_3$, CH_3OH, and H_2O, $\langle E \rangle$ are approximately −16.3, −9.6, and −7.0 kcal/mol, respectively. The experimentally measured values are −14.4 and −10.2 kcal/mol for $CHCl_3$ and CH_3OH (not available for H_2O).[16] Fairly good agreement is found between the predicted and experimental $\langle E \rangle$. It is recognized that hydrophobic interaction is largely proportional to molecular size.[28] Therefore, we plot $\langle E \rangle$ as a function of the critical volume of solvent V_c. As shown in the inset, $\langle E \rangle$ increases when V_c becomes larger and a good correlation is found between $\langle E \rangle$ and V_c

$$\langle E \rangle = -3.89 - 0.051V_c \quad (4)$$

This molecular-based structure–property relationship can be used to predict the interaction energies for other solvents in PIM-1.

A polymer chain contains various groups that experience different interactions with residual solvent. To elucidate this,

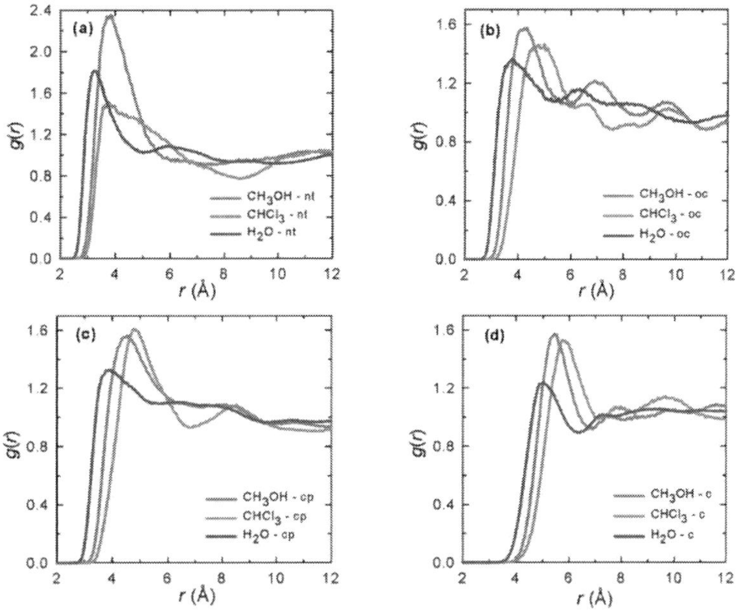

Figure 5. Radial distribution functions $g(r)$ between solvent molecules and (a) nitrogen atoms in cyano groups, (b) oxygen atoms in dioxanes, (c) carbon atoms in phenyl rings, and (d) spiro carbon atoms.

radial distribution functions $g(r)$ were calculated by

$$g_{ij}(r) = \frac{\Delta N_{ij}(r, r + \Delta r)V}{4\pi r^2 \Delta r N_i N_j} \tag{5}$$

where r is the distance between species i and j, $\Delta N_{ij}(r,r + \Delta r)$ is the number of species j around i within a shell from r to $r + \Delta r$, V is the system volume, N_i and N_j are the numbers of species i and j. Figure 5 illustrates $g(r)$ between solvent molecules and (a) nitrogen atoms in cyano groups, (b) oxygen atoms in dioxanes, (c) carbon atoms in phenyl rings, and (d) spiro carbon atoms. A common feature in the $g(r)$ is that H_2O exhibits a peak at the shortest distance because it has the smallest size among the three solvents, which is in contrast to CH_3Cl with a peak at the longest distance. More importantly, different profiles are observed in the $g(r)$ for the four different types of atoms in PIM-1. As shown in Figure 5a, the peak around nitrogen atoms in cyano groups drops in the order of $CH_3OH > H_2O > CHCl_3$, implying CH_3OH has the strongest interaction with cyano groups. A similar trend is observed in Figure 5b around oxygen atoms in dioxanes, except that $CHCl_3$ exhibits a higher peak than H_2O because of the favorable interaction between $CHCl_3$ and carbon atoms in dioxanes. It is also seen in Figure 5c,d that both $CHCl_3$ and CH_3OH have a higher peak than H_2O around carbon atoms. Overall, the structural analysis in Figure 5 reveals that polar cyano and dioxane groups interact more preferentially with hydrophilic CH_3OH and H_2O, whereas less polar carbon atoms favor the interaction with hydrophobic $CHCl_3$. This is a consistent with the interaction energies discussed above.

As evidenced from the mean-squared displacements (MSDs), residual solvents and polymer chains exhibit distinct mobility. Figure 6 shows that the MSDs of $CHCl_3$ and CH_3OH are considerably smaller than H_2O and the mobility decreases in the order of $H_2O > CH_3OH > CHCl_3$. The decreasing order could

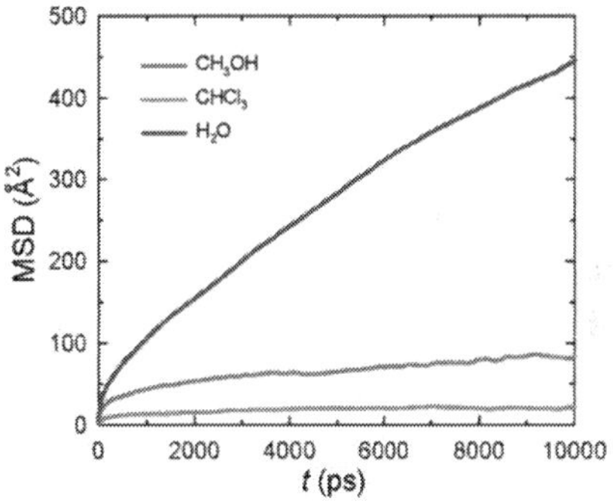

Figure 6. Mean-squared displacements of solvent molecules in PIM-1/solvent membranes.

be attributed to two factors. First, the molecular weight increases as $H_2O < CH_3OH < CHCl_3$. However, H_2O tends to form clusters with weight heavier than CH_3OH, thus the molecular weight may not be the major factor. Second, the interaction strength with polymer increases with increasing degree of hydrophobicity $H_2O < CH_3OH < CHCl_3$. Consequently, we speculate the second factor may play a more dominant role in governing the mobility of solvents.

Figure 7 shows the MSDs of polymer chains in PIM-1/solvent membranes. Compared with residual solvents, the mobility of polymer chains is negligible. Nevertheless, the mobility increases in the same order as observed in Figure 6. The reason is that solvent molecules in membrane can be considered as a part of membrane skeleton. Therefore, the mobility of solvent and polymer is cooperative and a larger mobility of the former

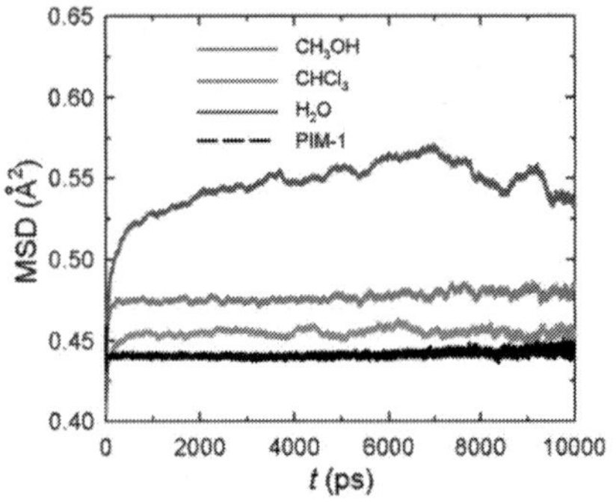

Figure 7. Mean-squared displacements of polymer chains in PIM-1/solvent membranes.

Table 3. Solubility Coefficients S [10^{-3} cm^3 (STP)/cm^3 cmHg] and Diffusion Coefficients D (10^{-8} cm^2/s) of H_2 in PIM-1/Solvent Membranes

solvent	S_{sim}	S_{exp}[16]	D_{sim}	D_{exp}[16]
PIM-1/CH_3OH	5.04 ± 0.05	6.6	6471 ± 481	5000
PIM-1/$CHCl_3$	4.75 ± 0.10	6.9	3513 ± 423	2900
PIM-1/H_2O	4.21 ± 0.03	2.7	2326 ± 604	1000

facilitates the mobility of the latter. Moreover, Figure 7 indicates the mobility of polymer chains is enhanced in the presence of residual solvent. This is because solvent–polymer interaction leads to a relatively weaker polymer–polymer interaction and a looser polymer network. Such a phenomenon is analogous to the plasticization of polymer membranes, e.g., in our recent study on CO_2-included plasticization of 6FDA-ODA polyimide.[29]

3.3. Sorption and Diffusion of H_2. Table 3 lists the solubility coefficients of H_2 in PIM-1/solvent membranes. Good agreement is found between the predicted and experimental data. The solubility decreases following PIM-1/CH_3OH > PIM-1/$CHCl_3$ > PIM-1/H_2O, which is in accord with the decreasing order of FFV in the three membranes and implies the solubility is proportional to the free volume. As observed experimentally,[16] the solubility in the presence of solvent is smaller compared against that in dry PIM-1 membrane. The reason is that solvent molecules can bind/block sorption sites and thus lead to a reduction in H_2 sorption. This effect induced by residual solvent was also observed for the solubility of N_2 in 6FDA-*m*PDA.[14] Furthermore, it is noteworthy that PIM-1 exhibits the largest solubility than other polymers reported to date, due to the existence of microporous structure and polar sorption sites. This is unique for glassy PIM-1 because a large solubility is usually observed in rubbery polymers.

In a polymer membrane, a gas molecule is trapped in a void and may jump into a neighboring void if the energy barrier is sufficiently low. The jumping occurs rapidly with a time frame shorter than the residence time in a void. The frequency of jumping depends on the energy barrier, which in turn is governed by the intrinsic properties of membrane and gas molecule. This process proceeds repeatedly as a result of the opening and close of voids in polymer matrix and leads to gas diffusion. In general, diffusion can be described by mean-squared displacement (MSD) versus time t as MSD(t) $\propto t^{\gamma}$, where γ is the signature

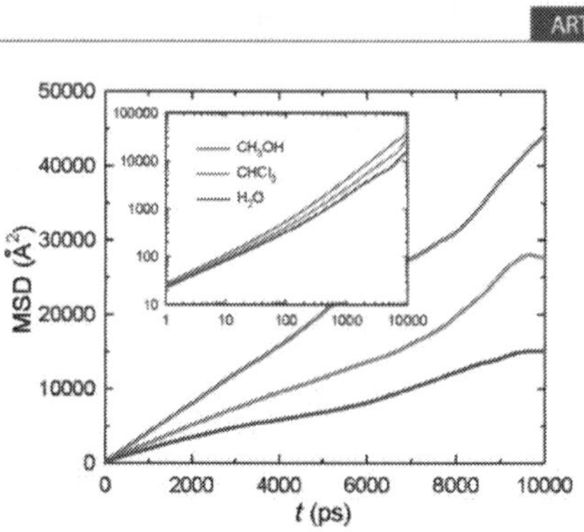

Figure 8. Mean-squared displacements of H_2 in PIM-1/solvent membranes. The inset is in log–log scale.

of different types of diffusion. Specifically, $\gamma < 1$, > 1, and $= 1$ correspond to sub-, super-, and normal-diffusion, respectively. Subdiffusion is usually observed within a short time and attributed to the structural correlation of immediate environment that retards diffusion. Superdiffusion occurs under convective or hydrodynamic transport. Normal (also called Einstein) diffusion takes place if molecules move randomly.

Figure 8 shows the MSDs of H_2 in PIM-1/solvent membranes. As indicated by the inset, the MSDs scale with t after approximately 1000 ps. This implies that H_2 reaches normal diffusion in the three PIM-1/solvent membranes after 1000 ps. Consequently, diffusion coefficient can be estimated using the Einstein equation

$$D = \frac{1}{6N}\frac{d}{dt}\lim_{t\to\infty}\sum_{k=1}^{N}\langle|r_k(t) - r_k(0)|^2\rangle \qquad (6)$$

It is commonly recognized that diffusion is faster in a membrane with a larger FFV, which is observed in Table 3. The diffusion coefficients in the PIM-1/solvent membranes decrease in the hierarchy of PIM-1/CH_3OH > PIM-1/$CHCl_3$ > PIM-1/H_2O. The trend is consistent with the decreasing order of FFV. The presence of solvent molecules reduces free volume and diffusion coefficient. Particularly, H_2O clusters occupy the large-size voids in PIM-1/H_2O and cause the largest reduction in free volume and diffusion. For the same reason, the diffusion in wet PIM-1 membrane is slower than that in dry PIM-1 membrane as experimentally observed.[16] It is interesting to compare gas diffusion in PIM-1 with that in other polymer membranes and porous structures. As a result of its microporous feature, PIM-1 has diffusion coefficient larger than most glassy polymers, though smaller than extremely permeable poly[1-(trimethylsilyl)-1-propyne], poly(4-methyl-2-pentyne) and Teflon AF.[19] However, the diffusion coefficients in PIM-1 are 2–3 orders of magnitude lower than in zeolites and MOFs.[30,31] This is because the pores/voids are irregular in amorphous PIM-1, but well-defined and highly ordered in zeolites and MOFs.

Table 3 also compares the simulated and experimental diffusion coefficients of H_2. Fairly good agreement is found despite one-fold overestimation by simulation in PIM-1/H_2O membrane. There are a number of factors that would cause the discrepancy between simulation and experiment. (1) Simulation predicts self-diffusion coefficient, whereas experiment typically reports transport diffusion coefficient by time-lag measurement. The self- and transport diffusion coefficients are identical only at infinite dilution, but experimental condition is usually at finite pressure/concentration. (2) Accurate prediction of diffusion coefficient requires not only an accurate force field, but also a well-equilibrated model membrane. It was found that the equilibration procedure to construct a model membrane may affect gas diffusion.[32] (3) With self-avoiding random-walk method to build a membrane, anisotropy may exist and leads to nonuniform distribution of polymer chains. Even if simulated density is close to experimental value, the void distribution in a model membrane may differ from that in a real membrane. (4) Polymer chain length also plays a role in gas diffusion.[33] A model membrane built with short polymer chains tends to have a larger fractional free volume than that with long polymer chains. Consequently, diffusion coefficient predicted in the former is greater. The average molecular weight of PIM-1 samples was measured to be 370 000 g/mol,[16] which implies the degree of polymerization is approximately 797. In our simulation, however, the PIM-1 model has 15 monomers and this would lead to faster diffusion than experimentally observed. In addition, short polymer chains have a greater mobility, which in turn facilitates gas diffusion.

4. CONCLUSIONS

We have examined how residual solvents (CH_3OH, $CHCl_3$, and H_2O) in PIM-1 affect the microscopic structure of membrane and gas permeation. It is observed that the fractional free volumes and large-size voids in PIM-1/solvent membranes decrease following $CH_3OH > CHCl_3 > H_2O$, which is in accordance with the measurements by positron annihilation lifetime spectroscopy. Among the three solvents, $CHCl_3$ interacts most strongly with hydrophobic PIM-1. However, H_2O does not preferentially interact with PIM-1 and thus aggregates into clusters. The average interaction energies predicted for $CHCl_3$, CH_3OH, and H_2O are −16.3, −9.6, and −7.0 kcal/mol, respectively, which agree well with available experimental data. Consequently, a structure−property relationship is proposed between the critical volume of solvent and interaction energy, and might be useful for the prediction of interaction energies for other solvents. The analysis of radial distribution functions reveals that cyano and dioxane groups in PIM-1 interact preferentially with hydrophilic CH_3OH and H_2O, whereas carbon atoms favor the interaction with hydrophobic $CHCl_3$. It is found that the mobility of residual solvent decreases as $H_2O > CH_3OH > CHCl_3$, opposite to the increasing hierarchy of molecular weight and interaction strength with PIM-1. The mobility of polymer is vanishingly small but facilitated by solvent due to the cooperative interactions between polymer and solvent.

Both solubility and diffusion coefficients of H_2 in PIM-1/solvent membranes decrease in the order of $CH_3OH > CHCl_3 > H_2O$, which follows the decreasing order of free volume. The predicted coefficients are in fairly good agreement with experimental results. The solubility and diffusion coefficients are smaller compared with those in dry PIM-1 membrane in the absence of solvent. This is because solvent molecules bind/block sorption sites and decrease free volume, thus leading to a reduction in both sorption and diffusion. Finally, it is interesting to note that gas solubility/diffusion in PIM-1 appear to be larger/faster than in many other polymer membranes, as a consequence of the intrinsic microporous structure.

■ AUTHOR INFORMATION

Corresponding Author

*E-mail: chejj@nus.edu.sg. Tel: (65) 65165083. Fax: (65) 67791936.

■ ACKNOWLEDGMENT

The authors gratefully acknowledge the National Research Foundation of Singapore (R-279-000-261-281) for support.

■ REFERENCES

(1) Bernardo, P.; Drioli, E.; Golemme, G. *Ind. Eng. Chem. Res.* **2009**, *48*, 4638.
(2) Ockwig, N. W.; Nenoff, T. M. *Chem. Rev.* **2007**, *107*, 4078.
(3) Shao, L.; Low, B. T.; Chung, T.-S.; Greenberg, A. R. *J. Membr. Sci.* **2009**, *327*, 18.
(4) Sridhar, S.; Smitha, B.; Ramakrishna, M.; Aminabhavi, T. M. *J. Membr. Sci.* **2006**, *280*, 202.
(5) Xiao, Y. C.; Chung, T. S.; Guan, H. M.; Guiver, M. D. *J. Membr. Sci.* **2007**, *302*, 254.
(6) Low, B. T.; Chung, T. S.; Chen, H.; Jean, Y.-C.; Pramoda, K. P. *Macromolecules* **2009**, *42*, 7042.
(7) Ribeiro, C. P.; Freeman, B. D. *Macromolecules* **2008**, *41*, 9458.
(8) Sen, S. K.; Banerjee, S. *J. Membr. Sci.* **2010**, *350*, 53.
(9) Karayiannis, N. C.; Mavrantzas, V. G.; Theodorou, D. N. *Macromolecules* **2004**, *37*, 2978.
(10) Spyriouni, T.; Boulougouris, G. C.; Theodorou, D. N. *Macromolecules* **2009**, *42*, 1759.
(11) Heuchel, M.; Hofmann, D.; Pullumbi, P. *Macromolecules* **2004**, *37*, 201.
(12) Heuchel, M.; Böhning, M.; Hölck, O.; Siegert, M. R.; Hofmann, D. *J. Polym. Sci., Part B* **2006**, *44*, 1874.
(13) Abou-Hussein, R.; Wu, S.; Zhang, L.; Mark, J. E. *J. Inorg. Organomet. Polym. Mater.* **2008**, *18*, 100.
(14) Joly, C.; Le Cerf, D.; Chappey, C.; Langevin, D.; Muller, G. *Sep. Purif. Tech.* **1999**, *16*, 47.
(15) Kostina, Y.; Bondarenko, G.; Alentiev, A.; Yampolskii, Y. *Vysokomol. Soedin.* **2006**, *48*, 41.
(16) Budd, P. M.; McKeown, N. B.; Ghanem, B. S.; Msayib, K. J.; Fritsch, D.; Starannikova, L.; Belov, N.; Sanfirova, O.; Yampolskii, Y.; Shantarovich, V. *J. Membr. Sci.* **2008**, *325*, 851.
(17) Fang, W. J.; Zhang, L. L.; Jiang, J. W. *Mol. Simul.* **2010**, *36*, 992.
(18) McKeown, N. B.; Budd, P. M. *Chem. Soc. Rev.* **2006**, *35*, 675.
(19) Budd, P. M.; Msayib, K. J.; Tattershall, C. E.; Ghanem, B. S.; Reynolds, K. J.; McKeown, N. B.; Fritsch, D. *J. Membr. Sci.* **2005**, *251*, 263.
(20) Theodorou, D. N.; Suter, U. W. *Macromolecules* **1985**, *18*, 1467.
(21) Theodorou, D. N.; Suter, U. W. *Macromolecules* **1986**, *19*, 139.
(22) *NIST Chemistry WebBook*; Linstrom, P. J., Mallard, W. G., Eds.; National Institute of Standards and Technology: Gaithersburg, MD, 2011.
(23) Smith, W.; Yong, C. W.; Rodger, P. M. *Mol. Simul.* **2002**, *28*, 385.
(24) Gelb, L. D.; Gubbins, K. E. *Langmuir* **1998**, *14*, 2097.
(25) Bhattacharya, S.; Gubbins, K. E. *Langmuir* **2006**, *22*, 7726.
(26) Ban, S.; Vlugt, T. J. H. *Mol. Simul.* **2008**, *35*, 1105.
(27) Yampolskii, Y.; Alentiev, A.; Bondarenko, G.; Kostina, Y.; Heuchel, M. *Ind. Eng. Chem. Res.* **2010**, *49*, 12031.
(28) Dill, K. A.; Bromberg, S. *Molecular Driving Forces*; Garland Science: New York, 2010.
(29) Zhang, L. L.; Xiao, Y. C.; Chung, T. S.; Jiang, J. W. *Polymer* **2010**, *51*, 4439.
(30) Smit, B.; Maesen, T. L. M. *Chem. Rev.* **2008**, *108*, 4125.
(31) Keskin, S.; Liu, J.; Rankin, R. B.; Johnson, J. K.; Sholl, D. S. *Ind. Eng. Chem. Res.* **2009**, *48*, 2355.
(32) Meunier, M. *J. Chem. Phys.* **2005**, *123*, 134906.
(33) Gestoso, P.; Meunier, M. *Mol. Simul.* **2008**, *34*, 1135.

(489c) A Predictive Model for Solubility of Gases and Vapors In Swelling Glassy Polymers

Matteo Minelli, University of Bologna, Bologna, Italy and Ferruccio Doghieri, Dicma, University of Bologna, Bologna, Italy

The solubility of gases and vapors is of interest for many different applications, ranging from the fabrication of gas separation membrane to the development of barrier materials for packaging or sensors.

For rubbery polymers, different equations of state have been developed following different approaches and provide reliable means for the calculation of the solubility of gases and vapors. On the other hand, glassy polymers are nonequilibrium systems and the same thermodynamic tool is not suitable. The general approach called Non Equilibrium Thermodynamics of Glassy Polymers (NET-GP) coupled to the Lattice Fluid (LF) was showed to be able to predict the solubility of gases and vapors in glassy polymers at various pressures based on pure component parameters. The NELF model relies on the hypothesis that the Gibbs free energy of a glassy polymer can be described by considering the density of the polymer as an internal state variable for the penetrant-polymer system. Therefore, in order to describe the complete sorption isotherm of a species in a glassy polymer, the model requires some information on its volumetric behavior during sorption, which might be a serious limitation to its predictive ability.

In the present work, a new rheological assumption has been used to evaluate the swelling induced by the penetrating species, considering that the polymer relaxation occurs involving two characteristic times on fairly different scales. If temperature is appreciably below the glass transition of the polymer, only the short-time relaxation phenomena are involved during sorption within the usual experimental times, in which local rearrangements of the macromolecules take place. Much longer time would be instead required to have a noticeable relaxation for the second mode of rearrangement that involves the whole polymer chain, and therefore, in most of the cases this is not observed.

In this concern, the polymer phase can be conveniently assumed as formed by two elements corresponding to the two characteristic times of relaxation; a soft part for which the equilibrium condition holds and a glassy hard part that does not contribute to the swelling.

This assumption allows the estimation of the polymer swelling at increasing penetrant activity without specific experimental data and the calculation of the complete sorption isotherm by the NELF model in a purely predictive way.

This tool was satisfactorily applied for many sorption isotherms of swelling gases (such as CO_2 or C_2H_4) and organic vapors (xylene, toluene, or hydrocarbons) in glassy polymers such as polystyrene, polycarbonate or amorphous Teflon. The model was also able to capture complex behaviors such as those given by non linear trends of dilation or when a change in concavity of the sorption isotherm is observed due to a penetrant-induced glass transitions of the polymer. The experimental dilation isotherms, for those systems for which they were available, confirmed the reliability of the present model.

See more of this Session: Adsorption In Porous Polymers
See more of this Group/Topical: Materials Engineering and Sciences Division

(489d) Modeling Transient Diffusion of Gases In Glassy Polymers by the Dual Mode Model: Is the Local Equilibrium Hypothesis Necessary?

Lei Wang[1], Jean-Pierre Corriou[2], Christophe Castel[2] and Eric Favre[2], (1)Ensic, LRGP CNRS, Nancy, France, (2)LRGP CNRS, Nancy, France

The description of the mass transfer of gaseous species in glassy polymers is of major interest for numerous industrial applications (packaging, adhesives, controlled release, membrane processes..). A quantitative description of penetrant solution diffusion in glassy polymers has been developed during 60s and 70s, known as the Dual Mode Model. This theory decomposes the sorbed molecules into two species. The molecules obeying Henry law are considered as mobile and the molecules following Langmuir law are considered as immobile in a first approach. Later, this theory has been extended to consider the immobilization as partial for molecules following Langmuir law order to obtain a closer approach to the experimental results.

The partial immobilization model achieves an acceptable time lag prediction (i.e. transient diffusion data) by assuming a local equilibrium everywhere in the membrane. Thus the partial immobilization model is considered nowadays as the most appreciate and relevant transport model for the gas diffusion in glassy polymers.

A common feature of the total and partial immobilization models is to make use of the so-called local equilibrium hypothesis. This assumption was initially applied in the total Immobilization Model as a necessary and sufficient condition for its mathematical description. Although the Immobilization Model was later shown not be able to quantitatively represent the transport data, the hypothesis has been maintained in The Partial Immobilization Model. This assumption is of interest because it offers the possibility to largely decrease the complexity of the mathematical framework: it enables indeed the ratio of the Langmuir and Henry population to be calculated throughout the polymer matrix thickness, leading to one partial differential equation to be solved in place of two. Surprisingly, to our knowledge, no research has addressed the analysis of the incidence of the local equilibrium hypothesis on the prediction performances of the Dual Mode Model. This question has been addressed through a systematic simulation study and the main results will be reported in this communication.

The experimental time-lag data previously reported for a series of ten different systems (one glassy polymer/ one gaseous penetrant) by different authors have been compared to the predictions of three different models: the total immobilization model (TIM), the partial immobilization model (PIM) and the dual mode model without the local equilibrium hypothesis (named here after the Dual Diffusion Model, DDM).Remarkably, based on exactly the same set of dual mode parameters(sorption and diffusion data), the latter one (DDM) offers the best global fit when compared to the two classical dual mode approaches.

Two examples of comparative curve fits are shown in figure 1, for the N2/Polycarbonate and CO2 /PET systems. It has to be stressed that the DDM model leads to the same permeability expression as the partial immobilization model under steady state conditions. Additionally, the general framework of the dual mode model is maintained for the DDM, except the need to make use of a local equilibrium relationship within the polymer matrix. The predictions obtained by the three different approaches on the different set of experimental data will be shown and discussed. The implications offered by the DDM, especially with regard to the no more required local equilibrium assumption will be more specifically analyzed.

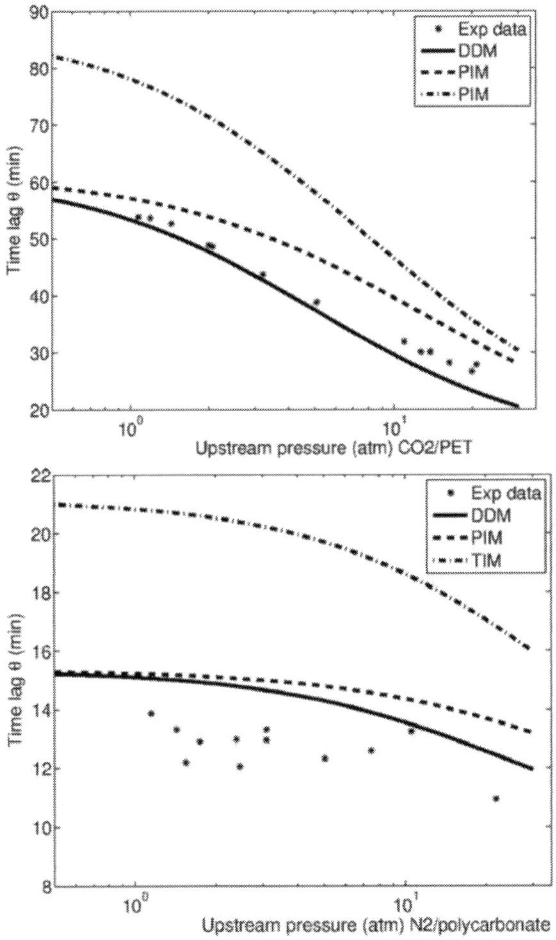

See more of this Session: Adsorption In Porous Polymers

See more of this Group/Topical: Materials Engineering and Sciences Division

(489e) Network Polymers of Intrinsic Microporosity

Flor R. Siperstein, School of Chemical Engineering and Analytical Science, University of Manchester, Manchester, United Kingdom

Tryp-PIMs are a new and interesting type of porous material. They are a member of the so called polymers of intrinsic microporosity, that have a fairly rigid backbone and contorted sites that give rise to the microporosity. In particular, Tryp-PIM is a network forming polymer that is expected to have a sheet-like structure. This material has some properties that resemble activated carbons and others that resemble typical porous polymers. In this work we present an initial characterisation and properties of Tryp-PIM.

See more of this Session: Adsorption In Porous Polymers
See more of this Group/Topical: Materials Engineering and Sciences Division

(489f) Fabrication of Functional Polymeric Nanofoams with An Bottom-up Strategy

Yingwu Luo, The State Key Laboratory of Chemical Engineering, Department of Chemical Engineering and Bioengineering, Zhejiang university, Hangzhou, China and Changhuai Ye, Chemical Engineering and Bioengineering, Zhejiang university, Hangzhou, China

The introduction of nanovoids to the polymeric material can finely tune the material properties like dielectric constant, refractive index, density, thermal conductivity, leading to many emerging functional materials like anti-reflective coatings, opaque polymers and super thermal insulation materials. Though polymeric foams have been widely used in industry, the preparation of polymeric nanofoam of high porosity presents a great challenge.

We will present a novel bottom-up strategy to construct polymeric nanofoams. Firstly, the non-collapsed hollow polymeric nanoparticles with shell thickness on the order of 10 nm were prepared by the interfacially confined reversible addition fragmentation transfer (RAFT) miniemulsion polymerization. The void fraction and average diameter of the hollow polymeric nanoparticles could be largely tuned up to 0.58 and from 68 nm to 180 nm, respectively. The non-collapsed hollow polymeric nanoparticles were then used to construct polymeric nanofoams, which hold promise in many application fields such as anti-reflection coatings, ultra-thermal insulation materials.

See more of this Session: Adsorption In Porous Polymers
See more of this Group/Topical: Materials Engineering and Sciences Division

AUTHOR INDEX

AUTHOR INDEX

AUTHOR INDEX

AUTHOR INDEX

AUTHOR INDEX

AUTHOR INDEX

AUTHOR INDEX

AUTHOR INDEX

AUTHOR INDEX

AUTHOR INDEX

AUTHOR INDEX

AUTHOR INDEX

AUTHOR INDEX

AUTHOR INDEX

AUTHOR INDEX

AUTHOR INDEX

AUTHOR INDEX

Materials Engineering and Sciences Division

Core Programming Topic at the 2011 AIChE Annual Meeting

Minneapolis, Minnesota, USA
16-21 October 2011

Volume 2 of 2

ISBN: 978-1-61839-729-4

Printed from e-media with permission by:

Curran Associates, Inc.
57 Morehouse Lane
Red Hook, NY 12571

Some format issues inherent in the e-media version may also appear in this print version.

Printed by Curran Associates, Inc. (2012)

AIChE
3 Park Avenue
New York, NY 10016-5991

Phone: (203) 702-7660
Fax: (203) 775-5177

www.aiche.org

Additional copies of this publication are available from:

Curran Associates, Inc.
57 Morehouse Lane
Red Hook, NY 12571 USA
Phone: 845-758-0400
Fax: 845-758-2634
Email: curran@proceedings.com
Web: www.proceedings.com

TABLE OF CONTENTS

VOLUME 1

BIO-REFINERY BASED ON INDIAN PAPER INDUSTRY WASTES

Sanjay Tyagi*, Narayan Chandra Mishra, and A.K.Ray
Indian Institute of Technology, Saharanpur Campus, Saharanpur, India
*Central Pulp and paper Research Institute, Saharanpur, India

ABSTRACT

Paper Industry is one of the major industries which discharge some wastes that can produce many valuable by-products. These are black liquor, steam and electrical energy, wood/ bark wastes, sawdust/ bamboo dust, solid wastes like lime mud, grits, dregs, ETP sludges and silica. These wastes can be converted to many value added chemicals, such as activated carbon from black liquor, sawdust, bamboo dust, broken chips and partly pulverized bark.

Black liquor /sulphite liquor produces ligno sulphones from rice straw, wheat straw, bagasse, bamboo, sarkanda and hardwood. It has diversified uses e.g. as a base for industrial detergents(washing powder and liquid soaps), oil-well drilling mud, cement and concrete additives, dispersants for a variety of chemicals, emulsifiers and stabilizers, brightening agents for electrolytic refining during electroplating, binders and adhesives for pellets and briquettes chemicals such as clay, calcium carbonate and titanium di-oxide resin ingredients, lignin-phenol-formaldehyde mixtures as adhesives for ply wood and other wood based panels, rubber additives, adsorbent/desorbent, micronutrient for agricultural applications, asphalt emulsions. Besides black liquor can also be converted to lignin-poly isocyanate foam, lignin reinforced polymers, lignin adducts, chlorobrominated lignins as effective fire retarding agents, composites of alkali lignin and biologically active agents such as fungicides, insecticides, nematocides, herbicides etc. ,activated carbon by coking or gasification of black liquor, hydrolytic degradation of ligno-cellulosic waste to sugars such as xylose and xylitol used as food stuff as sweeteners, synthetic fuel using solar furnace, silica from straw black liquor, hydrogen gas by electrolysis, vanillin, DMSO, methanol, acetate, formate, actate, saccharinic acids, lactones, turpentine and tall oil, alkali lignin, guaiacol, pyrocatechol, acetoguaiacone, biogas(methane) etc.

Cellulosic wastes can also produce cellulosic bio-ethanol using thermophilic bacterial species like Clostridium thermocellum. In the similar manner, the sludge can produce glucose by treatment with cellulase enzymes. Syngas from gasification of pith of some nonwood based plants can also be produced.

From straw soda black liquor, silica can be precipitated for industrial applications such as amorphous silica (for high quality synthetic pigments as paper fillers which can replace soap stone and TiO_2), filler, retention aid and as flocculating agent, for production of ceramics. The later can be employed for household china, fine earthen ware, concrete block, electric porcelain, silica light weight brick, clay brick, calcium silicate thermal insulating materials, reinforcement in rubber, in paint manufacture, adsorbent, grease thickener, insecticide and cosmetic industry.

An attempt has been made in this paper to review the existing status of Indian industry to examine up to what extent they are able to exploit the potential of aforesaid cellulose and lignin by-products. Further exploratory investigations were made to look into the feasibility of their manufacture in plant practice in India for its sustainability.

INTRODUCTION:

Pulp and paper Industry is the World's largest non-food biomass collection system. India is the 15th largest paper producer in the World. The installed capacity of Indian paper industry is 12 million tonnes and 11 million tonnes is in operation, however annual production is around 10 million tonnes with per capita consumption of 10.5 million tonnes. Now the Indian paper industry is growing at 9.5%CAGR per annum along with booming economy with GDP growth rate 8.6 % .It is projected a demand of paper and paper board in the country will be 15 million tonnes by the year 2015 and 22 million tonnes by the year 2020[1]. On the utilization of recovered paper side, India imports around 4.5-5 million tonnes of waste paper as the recovery rate at India rests at meager value of 20-22% [1]. In the coming decade Asia will be major hub for paper production and trade, particularly in a time when drastic paper production cut is taking place in Europe (minus 0.3%),North America (minus 0.5%) and other parts of the World**[2]**.As a major economy of Asia, compounded with the booming economy of India and increasing demand of paper and paper board, India will provide consistent growth for decades to come. Together India and China are estimated to make up 59% of global incremental paper and paper board demand during 2009-2020. At the same time , North America and Western Europe is forecast to decline from 46% to 38% of global production by 2020.However, Indian manufacturers have to be more conscious on the emerging opportunities and growing market size as stringent competitions are ahead for them which have to be met by adopting clean technologies and quality products [2].

Pulp and paper industry is capital-intensive industry, intensive in terms of energy, chemical, water, labor and raw materials concomitant with high pollution load generation. Besides it demands various large expensive pieces of equipments for its manufacture to produce quality products. Because of its capital intensiveness, investment in the paper industry is always a long-term proposition. The technology of operations is also very much intricate. Today's plant personnel need the concept of production integrated with energy-environmental-productivity-quality issues. The discharges from paper industry include solid waste, liquid effluent and gaseous emissions. There are regulatory limits enforced on the discharge of these pollutants by the regulatory authorities and the industry is expected to treat its effluents to meet the standards. In some of the states zero effluent concepts are being implemented in waste paper based mills. The new process must induct lower energy, water consumption along with little or no impact on environment. In order to follow this strategy the technology has been continuously passing through many revolutionary changes over the last decades to achieve low manufacturing costs, to earn profits derived from the ability of the producer and to face international competition for quality standards in the global market.

PAPER MILL BASED BIOREFINERY:

The pulp and paper industry is at the forefront of the concept of bio-refinery since it is already a biorefinery -1st generation of Biorefinery –enabling sustainable production of pulp/ paper enhanced by the coproduction of bioenergy and biofuels and bioproducts, through matured technological processes that maximizes the value chain of pulpwood **[3]**. However, the decreasing profitability of standard pulp and paper mills and the world's demand for a non-fossil Carbon Economy resulted a new biorefinery platform based upon complex technology concepts with higher revenue potential [3]. Biorefinery is an overall concept of an integrated and diversified processing plant where biomass feedstocks are converted into a wide range of valuable products, much likewise to petroleum refineries. Integrated birefinery is a processing facility that extracts carbohydrates, oils, lignin, and other materials from biomass, converts them into fuels, high value chemicals and other materials, with a near zero waste draw [4]. Paper Industry is one of the major industries which discharges some wastes as given below that can produce many valuable by-products [5-26].These are black liquor, steam and electrical energy, tall oil, wood/ bark wastes, sawdust/ bamboo dust, solid wastes like lime mud, grits, dregs, ETP sludge and silica. These wastes can be converted to many value added chemicals, such as activated carbon from black liquor, sawdust, bamboo dust, broken chips and partly pulverized bark, turpentine and tall oil. Important chemicals like methanol, acetate, formate, lactate, saccharinic acids, lactones, alkali lignin, guaiacol, pyrocatechol, acetoguaiacone can also be prepared. These are discussed as under:

Solid waste from big paper mills[12]

(Avg. data based on data collected from 8 paper mills)

Organic Waste: (T/T of Paper)

1.	Raw Material Preparation (Wood /bamboo dust)	0.1
2.	Pulp mill rejects	0.01
3.	Paper machine cleaner rejects	0.1
4.	ETP Sludge	0.2
	Sub-Total	0.41

Inorganic Waste: (T/T of Paper)

1.	Ash (Cinder + Fly ash) from boilers	0.80
2.	Lime sludge and grits (Recovery + Bleach liquor preparation)	0.70
	Sub-Total	1.50
	Total	1.91

Utilization of wastes from a paper industry can have enough potential for a perfect biorefinery in India. Production of many value added products can be explored. There are many Indian industries engaged in establishing biorefineries for producing multiple products. Some are mentioned below:

UTILIZATION OF WASTES FOR BIOREFINERY IN INDIA

1. Utilization of wood/nonwood wastes:

Large amounts of ligno cellulosic waste are generated through forestry and agricultural practices , pulp and paper industries, timber and agro-industries.

1.1 Production of bioethanol:

Cellulosic wastes, solid wastes like wood dust, bagasse pith, waste paper and even sludge can produce cellulosic bioethanol [8]. Alcohol can be produced from cellulosic wastes using thermophilic anaerobic bacteria species like Clostridium thermocellum at 60°C and pH of 7.5 under control anaerobic conditions [14]. Bioethanol from bagasse pith has been produced in the laboratory using thermophilic ethanologen, yeast strain at 50°C through sachharification both chemically and enzymetically (using hydrolytic enzymes like celluloses, laccases and xylanase) followed by fermentation of the hydrolysate. It is reported that almost 80% glucose (30 g/l)of enzyme hydrolysate was fermented to ethanol(9.5gm/l) within 40 h and thereafter some of the oligosachharides also converted to ethanol bringing the final ethanol concentration 11.5 gm/l in the broth [6,7]. Naithani et al. [9] describes utilization of Lantana Camara and pine needle as lignocellulosic wastes for ethanol production with a yeast culture of Saccharomyces Cerevisiae NCIM-3570 with final pH to 4.5 at a temperature of 30°C. The acid hydrolysate of hemicelluloses extracted bagasse with the yeast S.Cerevisiae for fermentation can produce ethanol after concentration and supplementation can produce 8.3% alcohol with fermentation efficiency of 83.3% compared to 7.8% for molasses medium of 28°Bx with FE of the order of 83.3%[25].

2. Utilization of fiber rich Sludge:

In the similar manner, the sludge can produce glucose by treatment with cellulase enzymes [10]. Treatment of cellulosic sludge with cellulase enzymes rapidly shortens the fiber length and then dissolves the fibers in to glucose. The glucose can then be sent to an ethanol plant for fermentation to ethanol. The economics are based on sludge containing 50 tonnes per day of cellulose. The enzyme cost is projected at $6,000 per day. Water treatment cost is $1,000 per day (including capital and O &M) assuming reverse osmosis will be needed. Thus the total costs are $7,100. The value of the glucose produced (assuming 70% conversion from fiber) is $13,000 per day. The value of sludge disposal avoidance is taken at $3,000 per day for a total benefit of $16,000 per day. Addition of c-PAM accelerates the rate of fiber hydrolysis. The other costs considered for investment are due to some equipment to handle the residual solids,acid and power. The cost-benefits have been found favorable [10]

3. Biomass Gassification:

Syngas from gasification of pith of some nonwood based plants such as bagasse can be produced [11]. Biomass thermo-conversion for energy production from bagasse has been attempted by Umrao et.al. [11] for pyrolysis in a free fall reactor followed by gasification in a quartz reactor to obtain a more efficient and complete conversion of the bagasse as well as production of a gas with high heating value or with a high H_2 + CO (Syngas) content to be used as a fuel for chemical industry. Pyrolysis of bagasse is done on a fluidized bed at high temperature due to which charcoal is formed and the volatile components are released. The charcoal is sent to the two stage gasifier. As a result solid ashes, and a gas mixture which is composed of CO_2,CO,H_2, water vapour, pyrolysis product , tar and hydrocarbons. This process saves one liter of liquid fuel by using 4-5 kg of bagasse.

4. **Bio-gas production [21, 24]**

Primary and secondary (biological) sludges generated in pulp and paper industry can be digested anaerobically to produce bio-gas through acid fermentation followed by conversion to methane and carbon di oxide by methane forming bacteria. Benefits of anaerobic digestion are:

- Production of Bio-gas which can be utilized as a fuel
- Reduction of organic solids (and stabilization)
- Sludge can be used as a soil conditioner or fertilizer.
- Sludge can be treated without pre-dewatering.

Typical values of bio-gas production and organic solid waste reduction using anaerobic digestion of pulp and paper mill sludges are:

Reduction of organic solids : 45-75%)

Bio-gas produced : 0.5-0.7 m^3/kg organic solids

The process details for a typical 30TPD bagasse based soda pulp mill in India producing 0.45 - 0.50 m^3/kg COD gas (75% CH_4 + 25% CO_2) with calorific value of 5400 kcals/m^3 is available. The process removes 90% BOD, and 70% COD at temperature range 35-37°C with saving of 4000kWh /day power and 5 tons per month nutrients [24].

5. **Particle board**: Pith + Lignin can make particle board.
6. **Other important value added products:**

Many valuable chemicals can be manufactured from paper industry wastes. Majority of the processes for chemical manufacture from paper industry wastes are patented [22].These are excellently reviewed by a number of investigators [19-23]. Some of the patented processes are highlighted below:

6.1 Activated carbon:

Ligno-cellulosic material such as saw dust, bamboo dust, broken chips, and partly pulverized bark is reported to have been converted into activated carbon.

Method I: The material was first treated with inorganic acid to remove sugars at low temperature range. Acid extracted material was then heated at **450°C for nine hours and** for one hour at 800°C. While steam was being injected, the tar was removed as condensate in this high temperature range. About 22 % yield of activated carbon was reported.

Method II:

Continuous production of activated carbon can be possible by carbonization of cellulosic material impregnated with activating agent such as zinc chloride. The process involved continuous supply of activating agent-impregnated material into carbonizing zone containing previously carbonized cellulosic material at such constant flow rate as to allow the impregnated material to reside for fifteen minutes in the carbonizing zone. The impregnated material was carbonized with the carbonized cellulosic material heated to 350°C by counter currently supplied hot gases and which formed an agitated fluidized bed by being fluidized with hot gas stream and being stirred by a mechanical means. The total weight of the product was two third of the original weight of the wood material supplied.

6.2 Chemicals from Bark:

Bark along with other refuse creates a disposal problem for paper mills using hard wood as raw material. Utilization of bark can be developed on following lines.

I. As a source of fuel for steam generation
II. As a source of chemicals, particularly tannins and tannin like compounds, which would be recovered by water/acid/alkali extraction of bark, prior to burning.
III. Marginal utilization as soil conditioner, extender for thermosetting resins' oil well mud additive, reinforcing and stabilizing agent for polypropylene moulding compounds.

6.3 Hydrolytic Degradation of Lignocellulosic waste to sugars, Xylose and Xylitol:

During acid hydrolysis of lignocellulosic biomass a number of chemicals are produced. These are sugars (Xylose), aliphatic acids (acetic, formic and levulinic), furan derivatives, furfural and 5-hexamethyl furfural(HMF) and phenolic compounds. Hydrolytic degradation of lignocellulosic waste to sugars such as xylose and xylitol used as food stuff as sweeteners can be potential chemicals. Hardwood and agricultural residues are rich in pentosans which mainly contain xylan. Waste and refuse generated from hard woods/straw/bagasse etc. while preparing these materials for pulping, could be degraded hydrolytically to D (+) Xylose and Xylitol, which are used in food stuff as sweeteners.

Hydrolysis of waste was carried out in first stage with sodium hydroxide of 4% concentration which removes acetic acid chemically bonded in the wood as sodium acetate. The liquid extract of this stage can be re-used after it has been brought to the alkali concentration to be used.

In the second stage of processing the residue from the first stage containing pentosan and cellolignin was further hydrolyzed with mineral acid e.g. sulfuric acid. When working without pressure, 1.5 to 6 % strength of H_2SO_4, and a ratio of liquid to solid of 3 to 6 parts by volume was preferred. Temperature was preferably 50 to 125°C and time was 2-4 hours. After completion of the second stage, the batch was filtered; the filtrate could be worked to Xylose or directly to Xylitol.The insoluble residue from the second stage could next be used to obtain organosolve lignin which was thermoplastic and very reactive; the material, this lignin preparation could be used as a base material for dyestuffs and pesticides. The residue remained after removal of lignin, on treatment with sodium chlorite followed by sodium hydroxide could be converted to practically white cellulose which could be used for industrial applications. Thus all wood waste could be converted to useful chemicals.

6.4 Synthetic Fuel using Solar Furnace:

A process of producing gases such as CO_2, CO, CH_4 and H_2 from pyrolysis and gasification of solid organic waste contained within a reactor has been patented [22]. The process comprises solar heating of working fluid (steam, CO_2 or a mixture of gases) to 600°C to 700°C by means of solar top furnace. The working fluid was contained in the lower section of the reactor situated at the focus of the solar top furnace. The heated working fluid was injected under pressure into porous reactant bed on which the catalyst ($NaHCO_3$) - mixed organic waste was situated. The catalyst-mixed organic waste was pyrolyzed causing the formation of CO_2, CO, CH_4 and H_2 gases and char residue.The hydrogen produced could be utilized to replace natural gas in a hydrogen economy, to make fertilizer or as a chemical feed stock. It is claimed that process did not require a source of pure oxygen, and provided an attractive method for using solar energy.

7.0 Basic chemicals from wastes

Wood waste has a complex carbohydrate lignin structure requiring severe conditions to reduce it to such basic chemicals as Methanol, Ethanol, Furfural and Phenols. Chemical production costs are directly related to scale and complexity of operation **[22]**. Utilization of pine needles and bagasse as bed material in solid state fermentation for production of lactic acid by Lactobacillus strains has been an active research in IITRoorkee[26].

8.0 Cellulose nano crystals and nanocomposites:

Cellulose nanocrystals have been largely applied as reinforcing fillers in the preparation of nano composites materials with improved mechanical and barrier properties.

9.0 Utilization of black liquor and sulphite waste liquor

Pulp and Paper industry has a very huge lignin potential. According to rough estimate for every ton of bleached pulp produced, about 0.5 ton of lignin is either wasted to sewer or used as a low value fuel. If the means could be found to isolate it from spent pulping liquor and to convert it to usable chemicals, the financial returns might be sufficient to support separation and processing lignin preparations from spent liquor, thus helping abatement of stream-pollution. In many ways Kraft lignin is a versatile raw material. It can be easily isolated by acidification, preferably carried out with carbon dioxide (stack gas), whereby phenolic group are liberated and lignin precipitates. The precipitated lignin can be removed by filtration at elevated temperature and purified further by re-dissolution in water and re-precipitation by sulphuric acid.

9.1 Lignin Isolation and separation:

Lignin compounds from sulphate black liquor could be precipitated by bubbling CO_2 into it at about 60°C. The precipitated lignin was separated by filtration and purified by dissolving in dry hexane and re-precipitating in ether. Alkali lignin could be isolated, from soda process black liquor by neutralizing it with CO_2 from stack gases to pH 8.3 to 8.5; about 40% of the lignin was reported[9] to have been precipitated. The mix was heated to 85°C – 90°C to coagulate the lignin, was filtered. Residual lignin could be further precipitated by addition of sulfuric/hydrochloric acid. Isolated lignin was washed first with cold water and then with dilute sulfuric acid and dried on double drum drier.

The slimy, gelatinous and hard to filter form of the precipitated lignin from black liquor could be changed to a crystalline, easily filterable and quick settling precipitate by mixing small amount of water-immiscible solvent such as chloroform, methelyene bromide, l-nitropropane or the like, with lignin containing liquid such as Kraft/soda black liquor. The chloroform, or other agent, may then be recovered from the filtered liquid by vacuum- or steam distillation.

Chemical modification of Kraft lignin drastically changes its properties and product obtained are of commercial value. Common modifications can be summarized as follows:

1. Formation of ethers by reaction with alkyl/aryl halides
2. Formation of esters by, reaction with carboxylic acid
3. Formation of polymeric condensation product with phenols, aldehydes and amines
4. Conversion to sparingly soluble phenol derivatives in reaction with various salts of heavy metals
5. Sulphonation to Kraft lignosulphonates
6. Nitration carried out with nitric acid
7. Chlorination with aqueous chlorine
8. Oxidation, with air or oxygen produces derivatives with changed solubility and melting point
9. Demethylation of lignin can be accomplished by acidic and alkaline treatment and results in the formation of catechol group increasing the reactivity in condensate e.g. with formaldehyde.

9.2 Lignin derivatives:

Lignin thus isolated could be converted to large surface area. Lignin is useful in drug formulations for adsorbing acids in the digestive system, due to their extremely high surface area and their acid-alkali resistance. Other suggested uses include carriers for adhesives, catalyst substrate, reinforcement for plastics and carriers for insecticides and herbicides.

A method to produce biodegradable uncolored lignin solutions was described wherein aqueous lignin solutions such as pulp mill effluent were treated with polyvalent metal cations. The amount of cations should be sufficient to react with lignin, and the aromatics to produce a water insoluble precipitate which was separated from remaining liquor. The precipitate was subsequently acidified with acid solution and was allowed to stand for sufficient time to permit digestion. The precipitate thus formed was separated by centrifugation; and was treated with alkali to produce aqueous basic solution. This solution was found to be effective in controlling Scale formation in boiler water systems, when added to these systems in amount of 5-50 ppm. Thus method not only recovers valuable product from effluent but also reduces the amount of non-biodegradable and toxic materials prior to discharge. In addition color of the final discharge is also reduced.

Sulpho methylated alkali lignins mixed with aqueous ferrous salt and dichromate and spray dried to a powdered product was reported to be useful as drilling mud thinner.

9.3 Lignin Reinforced Polymers:

A method for forming lignin reinforced polymers was developed by co-precipitating lignin/latex slurry and gas drying the slurry to leave a lignin reinforced polymer free of inorganic salts. In a process, alkali lignin was dissolved in ammonium hydroxides; and the ammonium lignate thus formed was then modified by adding hexamethylenetetramine to the lignate and heating to 90°-$110^{o}C$ for two: hours. Thus treated lignin was mixed with latex to give emulsion, which was precipitated by CO_2. When the precipitate was dried at $300^{o}F$ or below, inorganic salts formed in precipitation, sublimed leaving lignin/latex polymer.

9.4 Lignin- polyisocyanate/polyurethane foam:

These foams are useful as plastics and adhesives. Lignin-polyisocyanate foam can be prepared from polyethylene glycol solution of lignin when the later is treated with toluene di-isocyanate at $115^{o}C$ which afforded foam having good compressive strength and 74% recovery. Polyurethane type foam from lignin preparations and organic isocyanates are also reported.

9.5 Lignin as dispersants:

Method: preparation of ozone treated lignin from Kraft black liquor by treating lignin with oxygen containing 3% ozone at $50^{o}C$ for two hours; and spray drying the resulting solution has been reported. Oxidized lignin was found to be useful as dispersant for a variety of inorganic materials such as Clay, Cement, Calcium Carbonate, and Titanium Dioxide.

9.6 Lignin Adducts as surfactants:

Lignin adducts with 3-chloro-2-hydroxypropane-l-sulphonate, chloromethane sulphonate, acrolein and the like compounds are also reported to be useful as surfactants in dyestuff compositions.

9.7 Adhesive from lignin: Lignin-phenol-formaldehyde mixtures as adhesives are used for ply wood and other wood based panels:

Method: An adhesive from Kraft or Soda black liquor was prepared by reacting the black liquor (dry matter more than 25%) with formaldehyde and combining the product with separately prepared phenol-formaldehyde adhesive.

9.8 Lignin as coagulants and flocculants:

Lignin preparations from pulping liquor chemically modified by reaction with compounds such as formaldehyde, triazines, alkylamines were found [to] be useful as coagulants in treatment of sewage and other wastes. Cationic water soluble lignin amines prepared by reacting Kraft/Soda lignins with the reaction product of epichlorohydrin and a tertiary amine were suitable for a variety of uses such as flocculating agents, coagulants, retention aids, dispersing agents, asphalt emulsifiers and emulsion stabilizers.

9.9 Fire retarding agents:

Chlorobrominated lignins were highly effective as fire retarding agents. In one method, the inflammable material was coated or soaked with an aqueous slurry or alkaline solution of chlorobrominated lignin followed by drying. In addition chlorobrominated lignins could be combined with bonding compositions used for the production of bonded articles from particulate material, such as fibers, chips, granules, saw dust etc. eg. in the formation of chip board, particle board and the like. Chlorobromination was affected by adding chlorine and bromine concurrently or in two step operation. After completion the reaction, the chlorobrominated lignin was filtered, washed and dried.

9.10 Composites

Composites of alkali lignin and biologically active agents such as fungicides, insecticides, nematocides, herbicides and the like are useful as controlled release pesticides. These composites were prepared by intermixing ammonium salt of lignin with active agent under conditions of high shear mixing to form stable emulsion; acidifying the emulsion with mineral acid at pH 4-5 to precipitate the active agent/lignin emulsion with the active agent being physically absorbed within the lignin and drying the precipitate if desired.

9.11 Coke and activated carbon from black liquor

Process for activated carbon by coking or gasification of black liquor: In a process black liquor was heated in absence of free oxygen to an elevated temperature (450°-625°F) and pressure to prevent vaporization of the water, for a period sufficient to produce granular coke. The coke was separated from the aqueous phase to yield an aqueous effluent of low COD. The coke produced might be burned to supply heat, or used as a soil conditioner or carbonized to produce high grade coke for metallurgical purposes or used to produce activated carbon.

Organic values of black liquor have been converted into heat energy and activated carbon by a method which eliminates the recovery boiler from pulping liquor regeneration systems. Black liquor containing 16% solids or even lower or at any solids content up to 65% was hydropyrolyzed under very high pressure at 550°-650°F. The organic matter separates as a water-insoluble precipitate along with inorganic constituents such as Na_2CO_3, and minor amounts of Na_2SO_4, Na_2S. The organic matter could be utilized to produce heat energy or activated carbon which itself could be used for purifying effluents from various section of paper mill.

10.0 By-products from lignin

Generally by-products from lignin can be categorized as follows:

1. Low molecular weight compounds
2. Polymeric products

Table 1 gives a detailed picture of by-products (both commercial and non-commercial) obtained from Kraft cooking of wood using different methods of isolation. It can be seen from the table that the main by-products of Kraft cook are the degradation products of carbohydrates and lignin. The most consistent efforts on dry distillation of the alkaline black liquor powdered together with excess lime or alkali were made to obtain methanol, acetone, methylethylketone, higher ketones and other components of light and heavy oils in appreciable quantity (60-70 kg/ton of pulp each of the low molecular and high molecular compounds).

Low molecular weight compounds

The most practical approach to synthesis low molecular weight compounds from lignin is to heat the black liquor for 10-20 minutes at 250-290°C together with small amount of alkali and sodium sulphide. Table 2 covers the chemicals obtained during this process with their corresponding yield in kg/PTP. Under optimum conditions (291°C, 20.4 % of organic substance, 3.2 % Na_2S and 1.6 % of the NaOH based on total liquor) a maximum yield of 33 % of the organic products can be obtained as ether soluble phenols.

Of all chemicals listed in Table 2 the following chemicals are most important.

Use of lignin and lignin derivatives: Lignin Sulphonates and Kraft Lignin, Vanillin, Dimethyl sulphonates, Dimethyl Sulphoxide, Furfural, Methyl Mercaptan, Phenolics,thermosetting resins and graft polymers including polyurethanes, polyesters, polyamines and epoxies[13].

1. Vanillin
2. Furfural
3. Organosulphur compounds
   a) Dimethyl sulphide (DMS)
   b) Methyl mercaptans (MM)
   c) Dimethyl sulphoxide (DMSO)
   d) Dimethyl sulphone ($DMSO_2$)
4. Low molecular weight Phenolic compounds

10.1 Vanillin

Vanillin has a wide range of industrial application. Vanillin is the most widely flavoring agent,, commonly known as vanilla, due to its pleasant aroma and its property of enhancing other flavors.. It is universally used as flavoring agent with a property to mask other odors. Vanillin also finds use as a reductant and as material absorbing ultraviolet light. It can be incorporated into various plastics such as regenerated cellulose, polyvinyl ethers and polyamides to protect them against ultraviolet radiation. Ointments for protection against sunburn on the skin can be prepared with vanillin as active agent. Polymeric materials prepared from vanillin and vanillic acid have potential as fibres and films.

Generally vanillin is synthesized from Guaiacol or Eugenol.

It can be produced from lignin through oxidation of spent liquor with suitable oxidising agent or by extraction by a suitable solvent. There are a few methods for manufacturing Vanillin.

Method I: Catalytic oxidation under pressure in pressure vessel by prior raising pH of black liquor from 9.5 to 13 by sodium hydroxide and then heating for about 2 hours followed by extraction with chloroform. The aqueous layer is acidified with sulphuric acid to bring down the pH to 3.0 and the filtrate is again extracted with chloroform.

Commercial process based on alkaline oxidation of lignin with oxidising agent as cupric oxide (2% wt/wt.), various nitrobenzene compounds, mercuric oxide etc. have reported a better yield of vanillin from lignosulphonates. For example cupric hydroxide oxidation of lignin gives 22 % of vanillin at 160°C and 12.5 % at 102°C. In comparison a maximum yield of 23 % is obtained from nitrobenzene oxidation under optimum conditions of 1 hr at 180°C. Table No.2 gives the product obtained by alkaline oxidation of lignin. While vanillin is the most commonly reported simple compound obtained by alkaline oxidation of lignin, there are a number of other products that are formed which may eventually prove to be of commercial interest.

Method II: The black liquor is dissolved in chloroform and then extracted with saturated sodium bisulphate solution.

Purification of vanillin is an essential step when synthesized from lignin. Most practical way to purify it is to distillate under reduced pressure or co-distillation with saturated hydrocarbon.

10.2 Organo sulphur compounds from lignin

The most important products are DMS and DMSO.

DMSO is used as a catalyst and organic solvent for dyes and other chemical reactions. Since DMSO is an excellent solvent for many carbohydrate materials, it has found application in preparation of sugar derivatives. DMSO also has found utility as a material to protect cellulose tissue from freeze damage. In this case, it has been shown to penetrate-the cell walls better than glycerin and thus prevents excessive concentration of electrolytes in the cell during freezing. It has also been used successfully to protect red blood carpusals against freezing. Widespread interest in DMCO as a solvent for carrying drugs across cell membrane and as a drug itself have been developed.

Depending upon the source of lignin whether softwood or hardwood, the methoxyl contents of Kraft lignin varies from 14-20 %. Since methyl group represents about half of the weight of the methoxy group, it is theoretically possible to recover 7-10 % of lignin calculated as methyl radicals.

Synthesis of DMS

Method of production of DMS and MM is by heating Kraft spent liquor with sodium sulphide at 200°C and to purify the product obtained. The yields of DMS are about 3 % and MM 0.2-3 per cent on lignin solids.

Commercial process consist of heating concentrated spent liquor with elemental sulphur to a temperature of 200-250°C. The DMS formed is flashed from the reactor along with some water, MM and H_2S and collected by means of a condenser. The crude DMS is scrubbed by passing it through Sodium hydroxide and distilled.

DMS itself does not find any application but is a precursor for its halogenated and oxidized derivatives.

Synthesis of DMSO

This is a compound obtained from lignin with a variety of applications in chemical and pharmaceutical industries. This can be commercially obtained by oxidizing DMS by nitrogen-dioxide under optimum conditions. The reaction can be summarized as

$$3Me_2S + 2HNO_3 \rightarrow 3Me_2SO + 2NO + H_2O$$

The application of DMSO has extended in many fields. It has been used as a outstanding medium for promoting organic reactions and as a reactant solvent itself. The following points outline some of the most potential application of DMSO.

10.3 Polymeric compounds from lignin:

Lignosulphonates:

Bulk of the materials are imported from Russia and China **[19]**

This class of lignin polymers has attracted a great deal of interest because of their diversified uses. They can also be utilized as the base for-industrial detergents such as washing powders and liquid soaps.

10.3.1 Synthesis of lignosulphonates

Lignosulphonates can easily be prepared from purified Kraft lignin by sulphonation with solutions of alkali sulphites at elevated temperature and pressure. Sulphite spent liquor contains lignosulphonates which are formed during pulping reactions. Practical approach to isolate and purify this compound is to remove most sulphite and sulphate by lime addition upto pH of 10.5, more lime is added to the filtrate to give basic calcium lignosulphonate which precipitates in the pH interval of 10.5 - 12.2. More than half the lignosulphonate is thereby removed and separated by filtration. By the addition of sulphuric acid and magnesium and sodium lignosulphonates can be obtained.

10.3.2 Application of Lignosulphonates:

Black liquor /sulphite liquor produces lignosulphones from rice straw, wheat straw, bagasse, bamboo, sarkanda etc. and also from hardwood. Fundamental studies on the colloidal properties of purified lignosulphonates demonstrated their specific value for emulsification, defloccuation, chelation, adsorption and adhesion uses. It has diversified industrial applications; some major ones are as under:

i. Oil well drilling muds: In the drilling of oil wells, water and dispersants are continuously fed into the hole, to form, together with suspended fine clay, called as 'drilling mud'. Oil in water emulsions in a drilling fluid has been in use since the 1940's. Lignin sulphonates are valuable for stabilizing the oil in water emulsions which give the mud improved properties in control of water loss, reduction of torque on the drill stem, increase of bit life and general improvement of bore hole condition.

ii. Cement and concrete additives: Lignosulphonates are used as grinding aids in Portland cement and concrete. Their function is to reduce the agglomeration of the-ground particles, to reduce surface-tension at the solid-air interference of the particles-and to keep the surface of the grinding media clean & free. It reduces settling time; water reduction and these can also be used as air entraining agents in concretes. The entrained air gives increased resistance to spalling and cracking of the finished concrete. The dispersive action of ligno-sulphonates is useful in mixing, and placing concrete. Incorporation of 0.07 - 0.15 kg of lignosulphonates-concrete admixture per sack of cement will allow water reduction up to 20%. This results in better compressive strength and durable concrete. In oil well cementing it thus controls settling time and strength.

iii. Dispersants for variety of Chemicals: Lignin sulphonate dispersants usually have a molecular weight in the range of 10,000 - 40,000. The colloidal lignin coats the surface imparting a negative charge to the particle. The similarly charged particles now repel one another and a good dispersion is obtained. This is used as dispersants in refractories, tiles, absorbants, cosmetics, crayons, medicines, paint, paper coatings, tooth powder and dyes. It can also be base for industrial detergents (washing powder and liquid soaps

iv. Ore floatation: Grinding aids, refining of ores as floatation agent, improves filtering properties in Pb-Zn system.

v. Emulsifiers and stabilizers: Lignin emulsifiers are quite competitive with other commercial emulsifiers stabilizing oil-in-water emulsions. Emulsions formed are very stable even under frozen or boiling stage. The stabilization of fire fighting foams that results from the interaction of sodium bicarbonate solution and alum is achieved by an addition of sodium salt of Kraft lignin. Better results can be obtained by using lignosulphonates.

vi. Electrolytic Refining: Lignosulphonates have been successfully used as 'Brightening' agent for the plated surface preventing pitting and deposition of rough surface in electroplating.The electrolytic refining of copper , zinc etc. are some examples.

vii. Binders and adhesives: Lignosulphonate enhances the properties of the specific products e.g., pellets, briquettes. Large tonnage of spent liquor solids are used as binders for pellets and briquettes. 2% calcium lignin sulphonate in iron ore gives a pellet ten times as hard as that made with bentonite binder. The most familiar adhesive product is linoleum cement containing principally lignin sulphonates and clay. These cements are also used as adhesive for ceiling tiles, road surfacing, hard durable surface and laying of dust, chemicals such as clay, calcium carbonate and titanium di-oxide resin ingredients**,**

viii. Resin ingredients: Combination of lignin sulphonates, with low molecular weight urea and phenol formaldehyde resins have received attention as adhesive or binding agent for fibrous materials such as plywood, particle and chip board and paper laminates. Many good quality glues can be produced by incorporating lignin sulphonates with polyvinyl acetate or alcohol.

ix. Rubber additives: As extender, modified reinforcing pigment. Many types of lattices such as neutral nitrile and neoprene are compatible with lignin either in the acid or sodium salt form. Mixture of rubber with Kraft lignin is also suitable for use as rubber to glass adhesives.

x. Protein precipitants: Forms insoluble complex with proteins.

xi. Tanning agents: Forms more soluble tanning solution which facilitates recovery procedures.

xii. Sequestering agents: Ca, Ni, Sn, Zn & Al can be sequestered by lignosulphonates above pH 7.

xiii. Miscellaneous uses of lignosulphonates:

Other uses of lignosulphonates are lime plastering medium, storage battery plates-as expander and component of separator., lime plaster-use of less water with improved settling properties, crystal growth inhibitor, Ingot mold wash, and many more.

11.0 Other uses of black liquor: **Hydrogen gas by electrolysis,** Gasification of black liquor, Biogas from black liquor, Preparation of rubber additives, adsorbent/desorbent, micronutrient for agricultural applications, asphalt emulsions,

12. Utilization of lime sludge:

As lime sludge contain high silica content, so its burning is not considered economical. However, the following approach can be considered.

a. Burning to get low grade lime
b. In the cement industry for partial substitution of lime stones
c. As a conditioner for soil especially acidic one.
d. Burning after desilication
e. Lime mud comprising calcium carbonate was found to be useful in removing colour from pulping liquid waste effluents. In a process[32] effluents were treated with lime mud followed by clarification, separation of effluents from classifiers to remove colour bodies.
f. Puzzulanna Cement

13.0 Silica recovery from from straw black liquor:

Amorphous silica (for high quality synthetic pigments as paper fillers which can replace soap stone and TiO_2), filler, retention aid and as flocculating agent, for production of ceramics. The later can be employed for household china, fine earthen ware, concrete block, electric porcelain, silica light weight brick, clay brick, calcium silicate thermal insulating materials, reinforcement in rubber, in paint manufacture, adsorbent, grease thickener, insecticide and cosmetic industry.

Silica from boiler Fly ash, and Sludge are also used for manufacturing light weight bricks.

14. Industry Status in India:

- **Mission New Energy** : Uses agri-residues using enzymatic hydrolysis process produces 0.07Ml/year of cellulosic ethanol[3]
- **Utilization of bamboo dust:** Gasfication, Syn Gas by Hindustan Paper Corporation(HPC), **Nogaon,** Assam
- **Pith + Lignin to particle board**: Kailash Paper Mill , Awagwanpur, Moradabad, India
- **Biomethanation Process**: Satia Paper Mills, Pudumjee Paper Mills
- **Biomass cogeneration:** ABC paper mill

- **Sludge and Boiler Fly Ash to light weight Bricks**: ITC, Bhadrachalam, Rainbow Papers, Ahmedabad.
- **Corrugated roof sheets and Diesel from waste plastics in recycled paper:** Rainbow Papers, Damanganga
- **Producer gas plant using wood chips :** Andhra Pradesh Paper Mill (APPM), now International Paper, USA
- **Biogas, cogeneration and biomass gasification:**TNPL
- **PAFC :** BHEL, India[15].
- **Installation of biomethanation plant in pulp and paper industry with UASB reactor:** United Envirotechier
- **Biomass gassifier :** W2E projects and Systems
- **Lignosulphonates:** ABC paper mill, SailaKhhurd, Punjab

15. Status in Research Institutions

ETP sludge, pith and paper to Bio ethanol: Indian Institute of petroleum (IIP, Dehradun), Central pulp and paper research institute (CPPRI), Cellulose and paper branch,Forest research institute(FRI), National Sugar Institute , Kanpur,Basant Dada Patil Sugar Institute, Pune.

Cellulosic waste to alcohol: College of technology,Osmania University, Hyderabad, National Sugar Institute, Govt. of India, Indian Institute of Petroleum, Dehradun, Vasant Dada Sugar Institute, Pune, IITRoorkee.

Utilization of lime sludge,Fly ash :Central building research Institute, National thermal power corporation(NTPC)

Lignin grafted copolymers:Regional Research Laboratory(RRL at Bhubaneswar,Jammu, Jorhat)

Biogas –fuel cell based co-generation: Tata Energy Research Institute (TERI)

Production of lactic acid from waste sugarcane bagasse derived cellulose,cellulase production with Tricoderma citriviridae on solid bed ,use of acid hydrolysates of pine needles,and bagasse for lactic acid production using various strains such as Lactobacillus delbrueckii or lactobacillus pentosus have been explored in Department of paper technology, IITRoorkee( ), National sugar Institute , Kanpur, H.B.T.I., Kanpur

Ministry of New and Renewable Energy(MNRF) is keen to provide technical and financial assistance for setting up of energy generating plant based on biogas, cogeneration and biomass gasification

The technical parameters of a typical HRAD-PAFC (high rate anaerobic digester with phosphoric acid fuel cell) cogeneration system for a 50 TPD paper mill have been estimated by

TERI and development of a prototype PAFC have been explored by Bharat Heavy Elecricals Ltd.[15].

Table 1: By-products (both commercial & non-commercial) of the kraft cook

Compound	Origin	Quantity kg per ton of pulp
Methanol	Methoxyl groups of lignin & possibly glucuronoxylan	5
Acetate	Acetyl groups of the xylan as well as degraded carbohydrates	100-200
Formiate	Degradation of carbohydrates	40
Lactate	Degradation of carbohydrates	100
Saccharinic acids & lactones	Degradation of carbohydrates	250
Aliphatic sulfur compounds (methyl mercaptan, sulfide, disulfide)	Methoxyl groups of lignin and possibly glucuronoxylan	1
Turpentine	Terpenes	8-10
Tall oil (fatty & resin acid salts, phytosterols, etc.)	Wood resin components	20-100
Alkali lignin	Lignin	400-600

Table 2: By-products of the Kraft cook and from subsequent pressure heating of the black liquor with alkali and sulfide at 250-285°C (48 atm)

Before Pressure Heating	kg ptp	After Pressure Heating	kg ptp
Tall Oil	50	Acetic acid	60
Turpentine	17	Formic acid	60
Acetic acid	40	Methyl sulfide	40
Formic acid	40	Methyl mercaptan	8

Methanol	5	Pyrocatechol	30
Vanillin	2	Methyl catechol	11
Acetoguaiacone	2	Ethyl catechol	9
Guaiacol	2	Other ether-soluble phenols	150
Pyrocatechol	1	Homoprotocatechuic, protocatechuic and other phenol carboxylic acids, perhaps	150
Other phenols	42	Ether soluble acids of carbohydrate origin	150
Phenol carbonic acids	45	Demethylated lignin of novalac properties	100
Non-volatiles and lactic acids	145	Butanol-soluble lactones	35
TOTAL	**391**	**TOTAL**	**803**

CONCLUSIONS

Paper industry wastes can be properly utilized to produce value added products in order to act as second generation biorefinery. Lignin in Black Liquors has excellent potential for reuse in Fine Chemical Manufacture. The Indian Paper Industry is not yet ready for such a drastic transformation, since it will jeopardize the Chemical Recovery Processes. The Paper Industry is totally dependent upon the lignin black liquors as a source of energy fuel in Recovery Boilers. Serious attempts have to be made for medium and small scale manufacturing process. Some initiatives have been taken by Indian pulp and paper industry in recent years. A few research institutions have started research activities for establishing biorefinery based on paper/ sugar mill wastes.

It is necessary to determine the quantum of the wood waste/ agri-residue facility for the maximum size consistent with availability of raw material. A single product facility appears not to be competitive. In order to be economic utilization of the wood waste, multiple product facility is essential.

REFERENCES

1. Mukundan, P.G., Editorial, Inpaper International, Vol.12, No.2, Jan.-March 2010, Vol.13, No.1, Oct.-No.2010, www.inpaper.com.
2. Inpaper India, Vol.13(3), 14(2), March-April,2011, pp.24, 30.

3. Gabriel D.A., Sousa, Biorefinery development pathways: A survey for the pulp and paper industry, Proceedings of XXI Encontro Nacional da TECNICELPA/VI CIADICYP 2010, 12-15 Outubro 2010-Lisboa , Portugal.
4. Florbela Carvalheiro, Luis C.Duarte and Francisco M Girio, Hemicellulose biorefineries: a review on biomass pretreatments, JSIR, Vol.67, November, 2008, pp.849-864.
5. 56th Chemical weekly, July 05, 2011.
6. Thakur Vasanta V, Diwakar Pandey, R.K.Jain, R.M.Mathur, Second generation biofuels: Bioethanol from lignocellulosic materials, Inpaper India, Vol. 14(2), March-April, 2011, pp.4-12.
7. Jain, R.K., Thakur Vasanta V., Pandey Diwakar, Adhikari, Dixit A.K. and Mathur R.M., Bio ethanol from Bagasse Pith-A lignocellulosic waste biomass from paper / sugar Industry, IPPTA J., Vol.23, No.1, Jan-March, 2011, 169-173.
8. Mark E. Reno, Monique STREFF, Andrew Hird and Anjan Ray, Conversion of biomass to fuel as an environmental impact reduction opportunity for the pulp and paper industry, Inpaper International, Vol.13, No.3, April-June, 2011, pp.18-20.
9. Naithani Sanjay, Ganesh Joshi and Alok Kumar Dubey, Detoxification of Acid Catalyzed Lignocellulosic Hydrolyzate to enhance Fermentation Efficiency for Ethanol Production, INPAPER INDIA, Vol.14(2), March –April 2011, pp.14-22.
10. Banerjee Sujit, Energy from Pulp Mill Sludge,INPAPER INTERNATIONAL, Vol.13, No.1, Oct.-Nov. 2010, pp.10-12.
11. Umrao S., Pankaj Kumar and Sadhana Sachan, Method of producing Syngas from gasification of bagasse, Proceedings on "Advances in Chemical Engineering"-MacMillan, Advanced Research series February 27-28, 2009, Patiala, pp. 445-450.
12. Dhingra H.K. and Mohendru V.K., Status of Solid Waste Pollution From Indian Paper Industry and Future Prospects of its Management, IPPTA Vol.8,No.1, March 1996, pp.49-55.
13. Sahoo N, Rohella R.S.,Choudhury S.,Paul S.C.,and Chakravortty, V., Preparation of Value-Added Products From Black Liquor,IPPTA, Vol.8(3), September, 1996,pp.45-50.
14. Padmaja A, Venkateshwar S,Seenayya G and Bhagavanth Rao M, Evaluation of cellulosic substrates in the conversion of biomass to ethanol,Indian Chemical Engr., Section A . Vol.37 No.3, Jul-Sep(1995),pp.140-144.
15. Kurup ,G. Rajendra Chawla S.K.,Mathur A., Biogas –fuel cell based cogeneration in the pulp and paper Industry, IPPTA convention issue ,1992,pp.25-30.
16. Singh S.P., Recent developments in characterization and utilization of black liquor lignins from small pulp and paper mills, IPPTA, Vol.2 (3), 1990, pp.11-15.
17. Adsul, M.G., Verma A.J. and Ghokle, D.V., Lactic acid production from waste sugarcane bagasse derived cellulose, Green chemistry, 9, 58-62, 2007.
18. Guerra, G.,Casado, M.R.L.G., Arguelles,M.I., Sanchez M.I.,Manzano, A.M., and Guzman,T., Cellulase production with sugarcane straw by Trichoderma citrinoviridae on solid bed, sugar Tech.,8(1),30-35.

19. Managaonkar , N.D. and Kulkarni, Y.S.,Preparation and utilization of lignin by-products and lignosulphonates obtained from wood black liquor,IPPTA convention Issue,1987 pp.40-52
20. Ramaswamy, V. B iotechnology –applications in waste utilization and pollution abatement, ibid, pp.1-13.
21. Jain Sailendra, Environmental protection in Paper Industry, ibid, pp.53-61.
22. Dhake, J.D., Utilization of pulp and paper industry waste for chemicals,ibid, pp.19-29.
23. Trivedi, M.K., Lignin recovery from alkaline pulping of bagasse, ibid, pp.30-35.
24. NIEM (UNEP) Project Under Phase II- Activity III, October, 1990, Institute of Paper Technology (University of Roorkee), Saharanpur-247001 (India), pp.64-65, 66-67.
25. Lehri Madhu, Vartika Saxena and P.K. Agarwal, Preliminary studies on ethanolic fermentation of acid hydrolysates of hemicelluloses-extracted bagasse,pp.G587-G593, , Proceedings of the 57th Annual Convention of STAI, New Delhi, 13th -15th October,1995
26. Manoj Kumar Ghosh, Uttam K. Ghosh, Utilization of pine needles as bed material in solid state fermentation for production of lactic acid by Lactobacillus strains, Bio Resources, Vol.6, No.2,2011, pp.1556-1575.

Lignin – A Renewable Precursor for Carbon Nanofibers

John F. Kadla,[1] Ian Dallmeyer,[1] and Frank Ko[2]
[1]Biomaterials Chemistry Lab, Dept. of Wood Science, Faculty of Forestry
[2]Advanced Fibrous Materials Lab, AMPEL, Dept. of Materials Engineering
[1,2]University of British Columbia, Vancouver, BC

Abstract

Here we report on our research efforts aimed at developing carbon nanofibers through electrospinning of Kraft lignin. The research described herein focuses on understanding the effect of polymeric additives in influencing viscoelastic fluid properties which control the fiber morphology and diameter of fibers electrospun from blend solutions of softwood Kraft lignin (SKL) and poly(ethylene oxide) (PEO). To characterize the relevant fluid properties we report on the solution shear properties as well as the extensional properties of spinning solutions. We demonstrate that viscoelasticity can be varied in order to control the spinning process in our system by changing the concentrations of SKL and PEO as well as the PEO molecular weight. We also report on the mechanical properties of CNF precursor fibers as a function of thermal processing parameters. The results of these studies indicate that electrospinning is a promising platform for the development of micro- and nanofibrous carbon materials from technical lignins.

Introduction

Lignin is a naturally abundant biopolymer which reinforces the carbohydrate matrix, facilitates water and nutrient transport, and protects against invasion by microorganisms in the stems of trees and other terrestrial plants.[1,2] In terms of its chemical structure, lignin is an aromatic heteropolymer consisting of an irregular sequence of phenylpropane units connected by a variety of interunit linkages.[1,2] The renewable nature, abundance, and availability of lignin from the pulp and paper industry[3] are driving forces for the development of value-added products from lignin. Carbon fibres (CFs) are one example of value-added products that can be derived from lignin.[4] CFs have numerous applications including reinforcement in advanced light weight composite materials,[5] catalyst substrates,[6] adsorbents,[7] electrodes,[8] and tissue engineering scaffolds.[9] The most common precursors for commercial CFs are petroleum pitch, synthetic copolymers based on polyacrylonitrile (PAN), and rayon.[5] While PAN-based CFs have superior tensile strength, the high cost of PAN and non-renewable nature of petroleum-derived precursors are disadvantages.

Lignin is a potential renewable precursor for CFs, but the complex, heterogeneous chemical structure, the variability in properties among different types of isolated lignins, and the comparably poorer mechanical properties of lignin-based CF are barriers to their application.[4,10,11] Nevertheless, significant research efforts are being devoted to the production of CFs based on lignin, for example, at Oak Ridge National Laboratory[12]. Researchers in our lab have reported on thermal extrusion of lignin and lignin-polymer blends and conversion of these precursors to CFs by thermal treatment during the last decade.[4,13,14] Polymer blending was employed as a strategy to improve the spinnability of lignin[4], increase the tensile strength[13], or modify the surface porosity[14], depending on the blending polymers and their extent of interactions with lignin.[15-18] While studies conducted to date have reported improvements in fibre properties, a deeper understanding of the relationship between the complex macromolecular structure of lignin, interactions with additives, processibility, and material properties is needed to expand into a wider range of applications for lignin-based CFs. In addition, novel processing strategies must be explored to improve properties for specific applications and overcome some of the difficulties associated with lignin mentioned above.

One strategy to improve the properties of lignin-based CFs is to reduce the fiber diameter, since the tensile strength of CFs is known to increase with decreasing fibre diameter,[4,19,20] and applications such as filtration and catalysis depend on high specific surface area, which increases with decreasing diameter. Electrospinning is a robust strategy to produce fibers with diameters in the range of a few hundred nanometers or less.[21-23] Electrospinning of CF precursors such as PAN,[24-26] phenolic resin,[27] and pitch[28] has been reported. Electrospinning of organosolv lignin[29,30] and blends of sulfonated alkali lignin with PAN[31] have also been reported in the literature. However, lignin in general has been observed to display a high tendency to electrospray into droplets instead of forming fibers. To overcome this limitation, we have reported on the electrospinning of blends of several technical lignins with small amounts of poly(ethylene oxide) (PEO).[32] The advantages of our system compared to other reported systems are applicability to different technical lignins, the ability to spin fibers with greater than 99% lignin content with a single spinneret, and the ability to control the fiber diameter by varying the viscoelasticity of the spinning solution. Our current research in the area of lignin electrospinning is aimed at two main objectives: 1) To understand the role of lignin structure and interactions with additives in determining the viscoelastic fluid properties which govern the spinning process, and 2) To thermally process the electrospun fabrics and characterize the material properties as a function of processing parameters. In the following sections we provide an overview of our research efforts in the areas of rheology and its effect on the electrospinning process and processing and characterization of nonwoven fabrics produced from fractionated softwood Kraft lignin.

Materials and Methods

Lignin and solution preparation:

Softwood Kraft Lignin (SKL, Indulin-AT, Meadwestvaco, Glen Allen, VA, USA) was washed with dilute HCl (pH 2) to remove Na^+ ions and fractionated by sequential extraction with organic solvents according to a procedure based on that of Morck et al.[33] The F_4 fraction (according to the nomenclature of Morck et al[33]) was selected for rheological and electrospinning experiments. Polyethylene oxide (PEO) with M_v of 1000 and 5000 kDa (denoted PEO1M and PEO5M) were obtained from Sigma-Aldrich. N,N'-dimethylformamide (DMF, ACS Reagent grade) was obtained from Fisher Scientific (Ottawa, ON) and used as received.

Solution preparation:

SKL and SKL:PEO solutions in DMF were prepared as follows: SKL solutions were first prepared by adding dry lignin powder to the solvent. In the case of SKL:PEO solutions, PEO was first dissolved by briefly heating the solution at 80°C for 10-15 minutes, then SKL was added to the PEO solution. The mixture was heated at 80°C for 2 hours with intermittent vortexing every 30 minutes for 1-2 minute intervals. Solutions without PEO were prepared at concentrations of 25, 30, 35, 40, 45, and 50% by mass. Solutions with PEO were prepared with 25-45% SKL and 0.1-0.4% PEO (M_v = 1 or 5 x 10^6 g/mol).

Rheology:

The shear viscosity η was obtained from small amplitude oscillatory shear as the limit of $G''(\omega)/\omega$ at low ω or from steady shear measurements increasing shear rate from 1-1000 s^{-1} using an AR2000 torsional rheometer (TA Instruments, Grimsby, ON) with 40 mm diameter cone geometry and cone angle of 2°. A capillary breakup extensional rheometer (Haake CaBER 1, Thermo Fisher Scientific, Waltham, MA) was used to characterize the extensional fluid properties. In CaBER experiments, plate diameters of 4 mm, initial plate separation of 2 mm, final plate separation of 12 mm, and plate separation time (strike time) of 50 ms were used. Where applicable (for elastic solutions), the characteristic time scale of viscoelastic stress growth in uniaxial extensional flow λ (herein referred to as the relaxation time) was obtained by fitting the portion of the thinning profile where exponential thinning was observed as described by Rodd et al[34] using the equation:

(1) $D_{mid}(t) = D_1((G^*D_1)/4\sigma)^{1/3}\exp(-t/3\lambda)$

Where $D_{mid}(t)$ is the midpoint filament diameter, D_1 is the filament midpoint diameter just after the upper plate stopped moving, G is the elastic modulus, and σ is surface tension.

Electrospinning:

Electrospinning experiments were conducted using two different apparatuses. One was a self-built apparatus previously described[32] with a flat-tip needle as spinneret (gauge 18, 21, 22, or 25), operating voltage of 15 kV, flow rate of 0.03 mL/min, and spinneret-to-collector distance of 15-20 cm. The second was a Kato Tech Nanofiber Electrospinning Unit (NEU) (Kato Tech Co., LTD, Kyoto, Japan) using a horizontal configuration, rotating drum collector, and the same operating parameters listed above.

Thermal processing of electrospun fabrics:

Electrospun fabrics were heated under different conditions in a Hewlett Packard 5890 Series II gas chromatography oven under air. The samples were mounted on a Teflon-wrapped glass sample holder and heated from 25°C at a specified heating rate up to a specified final temperature and holding at the maximum temperature for 1 hour. Heating rates of 0.5, 5, and ~40 °C/min and final temperatures of 150, 200, and 250°C were used. Each condition was repeated three times and tensile testing was carried out for each of the three runs using 20 samples per run. The results were reported as the average of the three runs +/- the standard deviation. Following thermostabilization, carbonization was carried out in a GSL 1100X2 tube furnace (MTI Corporation, Richmond, CA) under N_2 heating at a rate of 10°C/min to 1000°C and holding at 1000°C for 1 hour.

Characterization of fibers and fabrics:

Optical and scanning electron microscopy was used to characterize fiber appearance using and Olympus BX41 Microscope (Richmond Hill, ON) and Hitachi S3000N SEM using gold-coated samples and accelerating voltage of 5 kV. Fiber diameters were measured using ImageJ analysis software counting a minimum of N = 100 fiber diameters per sample. The tensile properties of electrospun nonwovens fabrics were measured on a KES-G1 Kawabata microtensile tester as described elsewhere.[35] A density of 1.3 g/mL was obtained from an MSDS for alkali lignin and used to convert tensile data from units of N/tex to units of MPa.

Results and Discussion

Shear rheometry:

Shear rheometry has been used to characterize the viscosity of electrospinning solutions in the literature and correlations between shear viscosity, electrospinnability, and fiber diameter have been reported for a variety of different systems.[36-38] We first reported on the shear properties of lignin solutions with and without PEO in DMF in a published study.[32] In general, the shear properties of lignin solutions was Newtonian, with a viscosity independent of shear rate. However, it was clear in previous work that different lignins have different viscosities.[32] The observed differences underscore the inherent variability in lignins obtained by different isolation procedures or from different botanical origins. To focus our study, we selected to characterize SKL in greater detail.[32] The viscosity of SKL solutions varied over roughly 3 orders of magnitude when increasing from 25-50%. The rather strong dependence of viscosity on SKL concentration $\eta \sim c^7$ indicated that intermolecular interactions between SKL molecules increased the viscosity. Interestingly, solutions with PEO displayed a similar dependency of viscosity with concentration, with slightly higher values of viscosity but overall showing the same trend. However, SKL:PEO solutions were electrospinnable at shear viscosities far lower than those of SKL solutions. This result indicated that shear viscosity was not the key fluid property influencing the electrospinning behavior, since typically higher viscosity promotes fiber formation.

CaBER:

While viscosity affects the electrospinning process, it is also known that the elastic properties of polymer solutions strongly affect the ability to form fibers and the fiber diameter.[39-43] Based on comparison of our results with reports from the literature, we hypothesized that elasticity resulting from PEO addition may be responsible for the improved electrospinnability of lignin:PEO solutions. To test our hypothesis we varied the elasticity by varying the PEO molecular weight and the concentrations of SKL and PEO and investigated the ability to form fibres and the fibre diameter. Furthermore, we characterized the extensional properties of our solutions using a capillary breakup extensional rheometer (CaBER).

Figure 1 shows the thinning profile of the SKL solutions without PEO for concentrations of 40, 45, and 50%. Time is on the abscissa and D_{mid} as recorded by a laser micrometer is shown on the ordinate normalized by the initial filament diameter D_1. In general SKL solutions at or below 35% concentration could not be measured using the CaBER due breakup occurring before the end of the initial step. In Figure 1 it can be observed that the filament thinning profiles were linear in time, consistent with the behavior of a Newtonian fluid, in agreement with shear measurements.[44]

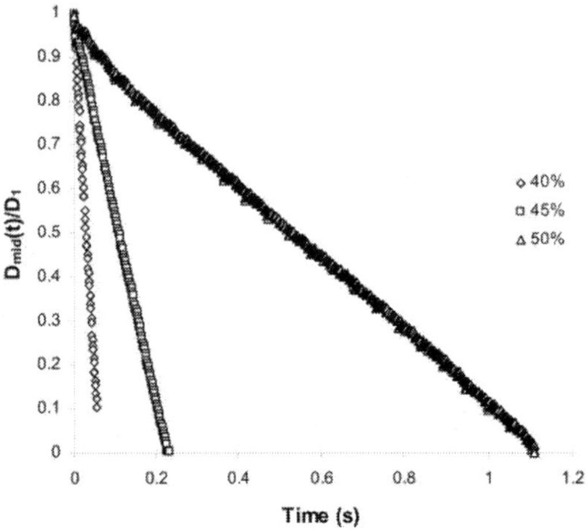

Figure 1: Thinning profiles of SKL solutions without PEO from CaBER experiments

In contrast, solutions with PEO displayed very different thinning behavior. Figure 2 shows the thinning profiles of several solutions with different SKL concentration, PEO concentration, and PEO molecular weight. Solutions with PEO display exponential thinning behavior, which indicates elasticity.[44] Elasticity implies that the extensional viscosity η_{ext} is not constant during thinning but increases with decreasing filament diameter (strain hardening). The relaxation time λ was determined from the CaBER data. Table 1 summarizes the compositions of the fluids and the values of λ obtained from the data. It can be seen in the table that λ increases with SKL concentration, PEO concentration, and PEO molecular weight, in agreement with published results.[45] Taken together these results show that the viscoelasticity of the SKL:PEO spinning solution can be varied by changing the relative amounts of SKL and PEO as well as the PEO molecular weight. The results also suggest that strain hardening occurring in extensional flow (which is absent in solutions without PEO) may provide the stability previously observed for lignin:PEO solutions in electrospinning.[32] We now discuss the effect of viscoelasticity on electrospinning.

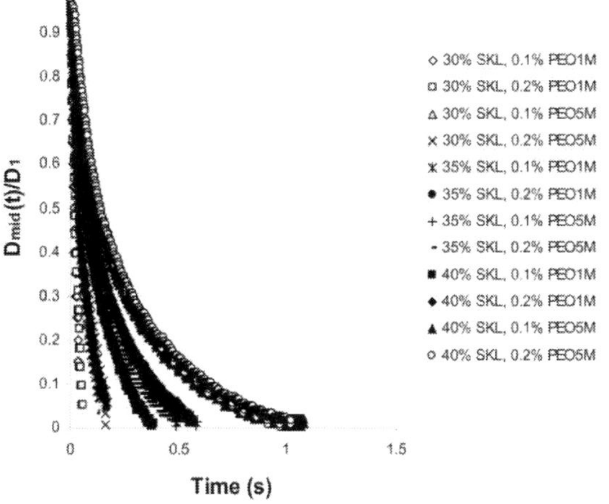

Figure 2: Thinning profiles of SKL:PEO solutions from CaBER experiments

Table 1: Fluid compositions and relaxation times

SKL concentration (%)	PEO concentration (%)	PEO mol. weight (g/mol)	λ (ms)
30	0.1	1 x 10^6	7.3
30	0.2	1 x 10^6	11.4
30	0.1	5 x 10^6	21.7
30	0.2	5 x 10^6	24.7
35	0.1	1 x 10^6	19.1
35	0.2	1 x 10^6	21.1
35	0.1	5 x 10^6	50.6
35	0.2	5 x 10^6	49.5
40	0.1	1 x 10^6	39.9
40	0.2	1 x 10^6	63.1
40	0.1	5 x 10^6	92.3
40	0.2	5 x 10^6	99.3
45	0.1	1 x 10^6	134.8
45	0.2	1 x 10^6	150.6
45	0.1	5 x 10^6	240.8
45	0.2	5 x 10^6	408.8

Electrospinning:

Example SEM images are shown in Figure 3 to illustrate the results of spinning experiments. Similar to our previous observations, only electrospray was formed without PEO below a concentration of 45-50%, with beaded fibers forming at 45%. In contrast, nearly all of the solutions with PEO were capable of forming fibers even though the amount of PEO in solution was less than 1% relative to SKL, producing fibers with SKL contents greater than 99%. The ability to achieve processibility with low PEO content is an advantage of the SKL:PEO system. Comparing rheology and electrospinning experiments, it can clearly be seen that fiber diameter increases with increasing λ. Figure 4 shows a plot of diameter vs. λ. The mean fiber diameter correlates well with λ, indicating that elasticity is correlated with the fiber diameter. Furthermore, the error bars show the standard deviations of the measured fiber diameters. It is also clear from the plot that the width of the diameter distribution tends to increase with increasing λ. This is partially due to the flattened, twisted morphology which tended to occur in solutions with higher viscoelasticity. The flattened morphology may be due to unevaporated solvent trapped in the core of the fiber followed by collapse of the fiber during the bending instability. The results presented here demonstrate that by controlling the viscoelasticity of the spinning solution, the mean diameter and width of the diameter distribution can be controlled.

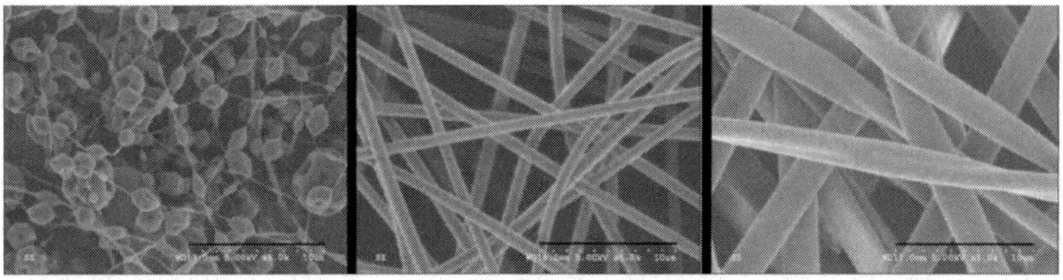

Figure 3: SKL:PEO fibers produced from solutions with different viscoelasticities (Scale bar = 10 μm)

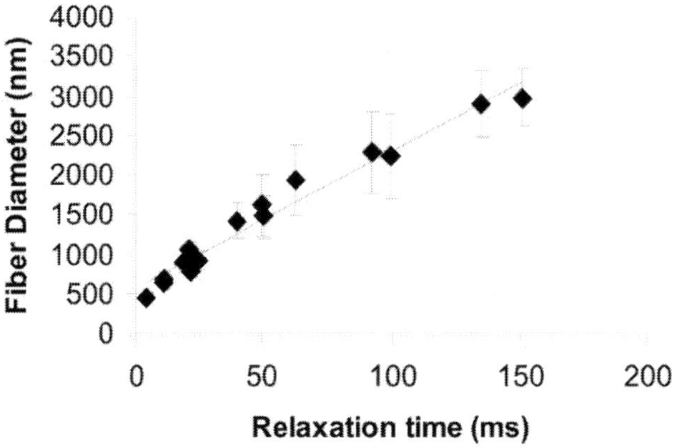

Figure 4: Fiber diameter vs. relaxation time for SKL:PEO solutions

Thermal Processing:

To prepare carbon fibers from lignin, an oxidative thermostabilization stage is necessary to prevent fibers from fusing during carbonization.[4] In previous work a slow heating rate was required to prevent fusion of melt-spun hardwood Kraft lignin (HKL) fibers.[4] The necessity of using a slow heating rate can be considered a disadvantage since it increases the time and energy required to generate carbon fibers. Due to its different chemical structure, SKL is expected to be more thermally stable.[4] However, previous attempts at melt spinning SKL were unsuccessful due to the tendency of SKL to char, so the thermostabilization of SKL fibers has not been reported in the literature. An SEM of a sample carbonized with and without thermostabilization is shown in Figure 5, showing fusion of fibers on the left and non-fused fibers on the right. To investigate thermostabilization of SKL fibers, we subjected our electrospun fabrics to different heating rates and final heating temperatures. It was observed that in contrast to melt-spun fibers or non-thermostabilized samples, none of the samples fused, indicating good stability. These samples were therefore suitable for conversion to carbon nanofibers. An SEM of carbonized 99% SKL fibers is also shown in Figure 5.

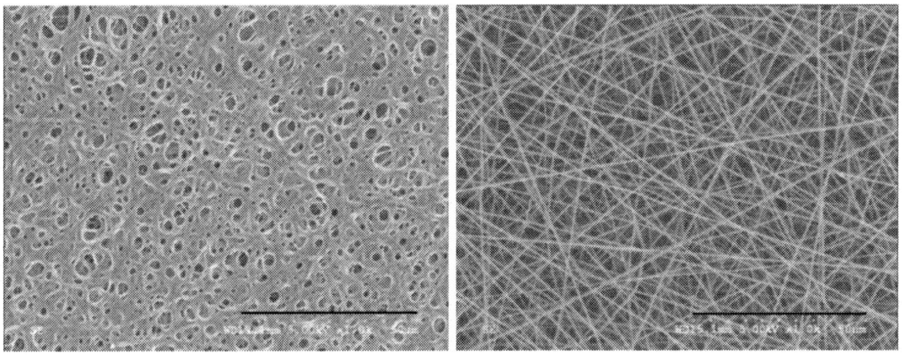

Figure 5: Carbonized 99% SKL fibers without thermostabilization (left), and after thermostabilization at 5°C/min to 250°C (right). Scale bar = 50 μm

Tensile testing:

To investigate the mechanical properties of SKL nonwoven mats, samples thermostabilized under different heating rates and temperatures but with similar diameters on the order of 1 μm were subjected to microtensile testing. The modulus and tensile strength values for different heating rates and thermostabilization temperatures are shown in Table 2. A steady increase in tensile strength was observed as the maximum treatment temperature increased from roughly 2-4 MPa for fibers heated to 150°C to values of more than 20 MPa for samples heated to 250°C. A similar trend was observed for elastic modulus, which increased from roughly 300 MPa to values of 1 GPa for fibers treated at 150 and 250°C, respectively. Interestingly, fibers heated at the lowest heating rate had a lower tensile strength, while comparable tensile strength was observed for samples heated at 5°C/min or faster. This could indicate that longer heating times lead to degradation of the mechanical properties of SKL fibers. The results presented here demonstrate the feasibility of generating precursors for carbon nanofibers from SKL by electrospinning.

Table 2: Mechanical properties of thermostabilized SKL:PEO nonwoven fabrics

Heating rate (°C/min)	T_{max} (°C)	Tensile strength (MPa)	Standard Deviation (MPa)	Elastic Modulus (MPa)	Standard Deviation (MPa)
As-spun	n/a	4.0	1.1	327.6	67.0
0.5	150	2.9	1.0	324.8	147.3
0.5	200	12.4	3.4	938.9	199.0
0.5	250	14.5	2.9	818.2	160.1
5.0	150	1.7	0.6	231.7	117.1
5.0	200	13.3	3.0	966.1	147.0
5.0	250	24.0	5.9	1055.9	269.5
47.0	150	3.7	1.0	322.6	121.6
41.0	200	10.7	1.9	849.4	101.6
39.0	250	22.1	3.7	1131.4	149.9

Summary:

The results from rheological experiments have demonstrated for the first time that while SKL solutions without PEO behave like Newtonian fluids while undergoing extensional flow, SKL:PEO solutions display non-Newtonian elastic behavior. We have demonstrated that SKL:PEO solution viscoelasticity strongly influences the fiber diameter and the width of the diameter distribution using shear rheometry and CaBER coupled with electrospinning experiments. By varying the SKL concentration, PEO concentration, and PEO molecular weight, we have demonstrated the ability to vary the viscoelastic properties over a wide range with less than 1% PEO added to the spinning solution (relative to SKL mass). With a sufficient understanding of SKL:PEO solution viscoelasticity, we then carried out thermostabilization experiments and measured the tensile properties of nonwoven SKL fabrics as a function of temperature and heating rate. In contrast with some previous reports on melt-spun fibers produced from HKL, electrospun SKL fibers did not fuse under any of the thermostabilization conditions tested, whereas carbonization performed without thermostabilization resulted in fiber fusion. In future work we hope to further optimize the process of lignin electrospinning to move toward toward carbon nanofiber precursors with properties suitable for real-world applications.

Acknowledgements

We would like to acknowledge North Carolina State University, the U.S. Department of Agriculture, and the Natural Sciences and Engineering Research Council of Canada (NSERC) for supporting this research.

References

1) Boerjan, W., Ralph, J., Baucher, M. (2003) Lignin Biosynthesis. *Annu. Rev. Plant Biol.* 54; 519-546
2) Ralph, J., Lundquist, K., Brunow, G., Lu, F., Kim, H., Schatz, P.F. Marita, J.M., Hatfield, R.D., Ralph, S.A., Christensen, J.H., Boerjan, W. (2004) Lignins: Natural polymers from oxidative coupling of 4-hydroxyphenylpropanoids. *Phytochemistry Reviews* 3; 29-60

3) Brodin, I., Sjoholm, E., Gellerstedt, G. (2009) Kraft lignin as feedstock for chemical products: The effects of membrane filtration. *Holzforschung*: 63, 290-297.
4) Kadla, J.F., Kubo, S., Venditti, R.A., Gilbert, R.D., Compere, A.L. (2002) Lignin-based carbon fibers for composite fiber applications. *Carbon* 40: 2913–2920
5) Morgan, P (2005). Carbon fibers and their composites Taylor and Francis, Boca Raton
6) Im, J.S., Park, S.J., Kim, T., Lee, Y-S. (2009) Hydrogen storage evaluation based on investigations of the catalytic properties of metal/metal oxides in electrospun carbon fibers International Journal of Hydrogen Energy; 34: 3382-3388
7) Osmond, N.M. (2000) Activated carbon fibre adsorbent materials. *Adsorption science and technology* 18, 6, 529-539
8) Sharma, C.S., Katepalli, H., Sharma, A., Madou, M. (2011) Fabrication and electrical conductivity of suspended carbon nanofiber arrays. *Carbon*; 49: 1727-1732
9) Aoki, K., Usui, Y., Narita, N., Ogiwara, N., Iashigaki, N., Nakamura, K., Kato, H., Sano, K., Ogiwara, N., Kametani, K., Kim, C., Taruta, S., Kim, Y.A., Endo, M., Saito, N. (2009) A thin carbon-fiber web as a scaffold for bone-tissue regeneration. *Small* 5: 13, 1540-1546.
10) Kubo S, Ishikawa N, Uraki Y, Sano Y. (1997) Preparation of lignin fibers from softwood acetic acid lignin—Relationship between fusibility and the chemical structure of lignin. *Mokuzai Gakkaishi* 43, 8, 655–662.
11) Dave,V., Prasad, A., Marand, H., Glasser, W. (1993) Molecular organization of lignin during carbonization. *Polymer* 34, 15, 3144-3154.
12) http://www.ornl.gov/ornlhome/news_items/news_100920.shtml
13) Kubo, S., Kadla, J.F. (2005a) Lignin-based Carbon Fibers: Effect of Synthetic Polymer Blending on Fiber Properties. *Journal of Polymers and the Environment*, Vol. 13, No. 2. 97-105.
14) Kubo, S., Yoshida, T., Kadla, J.F. (2007) Surface Porosity of Lignin/PP Blend Carbon Fibers. *J. Wood Chemistry and Technology* 27; 257-271.
15) Kubo, S., Kadla, J.F. (2005b) Kraft Lignin/Poly(ethylene oxide) Blends: Effect of Lignin Structure on Miscibility and Hydrogen Bonding. *Journal of Applied Polymer Science* 98; 1437–1444
16) Kadla, J.F., Kubo, S. (2003) Miscibility and Hydrogen Bonding in Blends of Poly(ethylene oxide) and Kraft Lignin. *Macromolecules 36:* 7803-7811
17) Kadla, J.F., Kubo, S. (2004) Lignin-based polymer blends: analysis of intermolecular interactions in lignin-synthetic polymer polymer blends. *Composites, Part A: Applied Science and Manufacturing* 35; 395-400
18) Kubo, S., Kadla, J.F. (2006) Effect of poly(ethylene oxide) molecular mass on miscibility and hydrogen bonding with lignin. *Holzforschung*. 60; 245–252.
19) Tagawa, T., Miyata, T. (1997). Size effect on tensile strength of carbon fibers. *Materials Science and Engineering* A238, 336–342
20) Tanaka, T., Nakayama, H., Sakaida, A., Horikawa, N., (1999) Estimation of tensile strength distribution for carbon fiber with diameter variation along fiber. *Materials Research Science International* 5, 2, 90-97.
21) Rutledge, G.C., Fridrikh, S.V. (2007) Formation of fibers by electrospinning. *Advanced Drug Delivery Reviews* 59, 1384-1391.
22) Reneker, D.H., Yarin, A.L. (2008) Electrospinning jets and polymer nanofibers. *Polymer* 49, 2387-2425

23) Greiner, A., Wendorff, J.H. (2007) Electrospinning: A fascinating method for the preparation of ultrathin fibers. *Angew. Chem. Int. Ed.* 46, 5670-5703.
24) Joshi, P., Zhang, L., Chen, Q., Galipeau, D., Fong, H., Qiao, Q. (2010) Electrospun Carbon Nanofibers as Low-Cost Counter Electrode for Dye-Sensitized Solar Cells. *ACS Applied Materials and Interfaces*; 2: 12, 3572-3577.
25) Arshad, S.N., Naraghi, M., Chasiostis, I. (2011) Strong carbon nanofibers from electrospun polyacrylonitrile. *Carbon*; 49: 1710-1719
26) Moon, S., Farris, R. (2009) Strong electrospun nanometer-diameter polyacrylonitrile carbon fiber yarns. *Carbon* 47: 2829-2839
27) Imaizumi, S., Matsumoto, H., Suzuki, K., Minagawa, M., Kimura, M., Tanioka, A. J. (2009) Phenolic Resin-Based Carbon Thin Fibers Prepared by Electrospinning: Additive Effects of Poly(vinyl butyral) and Electrolytes. *Polymer Journal* 41: 12, 1124-1128.
28) Park, S.H., Kim, C., Yang, K.S. (2004) Preparation of carbonized fiber web from electrospinning of isotropic pitch. *Synthetic Metals* 143: 175-179.
29) Lallave, M., Bedia, J., Ruiz-Rosas, R., Rodríguez-Mirasol, J., Cordero, T., Otero, J.C., Marquez, M., Barrero, A., and Loscertales, I.G. (2007) Filled and Hollow Carbon Nanofibers by Coaxial Electrospinning of Alcell Lignin without Binder Polymers. *Advanced Materials* 19; 4292-4296
30) Ruiz-Rosas, R., Bedia, J., Lallave, M., Loscertales, I.G., Barrero, A., Rodrı´guez-Mirasol, J., Cordero, T. The production of submicron diameter carbon fibers by the electrospinning of lignin. *Carbon* 48 (2010) 696–705
31) Seo, D.K., Jeun, J.P., Kim, H.B., Kang, P.H. (2011) Preparation and characterization of the carbon nanofiber mat produced from electrospun PAN/lignin precursors by electron beam irradiation. *Rev. Mater. Sci.*; 28: 31-34.
32) Dallmeyer, I., Ko, F., Kadla, J.F. (2010) Electrospinning of technical lignins for the production of fibrous networks. *Journal of Wood Chemistry and Technology* 30: 315-329.
33) Morck, R., Yoshida, H., Kringstad, K.P., Hatakeyama, H. (1986) Fractionation of Kraft lignin by successive extraction with organic solvents. *Holzforschung* 40(suppl.): 51-60.
34) Rodd, L.E., Scott, T.P. Cooper-White, J.J. McKinley, G.H. (2005) Capillary breakup rheometry of low viscosity elastic liquids. *Applied Rheology* 15: 12-27.
35) Gandhi,M., Yang,H., Shor, L., Ko, F. (2009) Post-spinning modification of electrospun nanofiber nanocomposite from Bombyx mori silk and carbon nanotubes. *Polymer* 50: 1918-1924.
36) Gupta, P., Elkins, C., Long, T.E., Wilkes, G.L. Electrospinning of linear homopolymers of poly(methyl methacrylate): exploring relationships between fiber formation, viscosity, molecular weight and concentration in a good solvent. *Polymer* 46 (2005) 4799–4810
37) McKee, M.G., Wilkes, G.L., Colby, R.H. Long, T.E. (2004) Correlations of solution rheology with electrospun fiber formation of linear and branched polyesters. *Macromolecules* 37, 1760-1767
38) Shenoy, S.L., Bates, W.D., Frisch, H.L., Wnek, G.E. (2005). Role of chain entanglements on fiber formation during electrospinning of polymer solutions: good solvent, non-specific polymer–polymer interaction limit. *Polymer* 46, 3372–3384
39) Feng, J.J. (2002) The stretching of an electrified non-Newtonian jet: A model for electrospinning. *Physics of Fluids* 14, 11, 3912-3926.
40) Feng, J.J. (2003) Stretching of a straight electrically charged viscoelastic jet. *J. Non-Newtonian Fluid Mechanics* 116, 55-70.
41) Carroll, C.P. Joo, Y.L. (2006) Electrospinning of viscoelastic Boger fluids: modeling and experiments. *Physics of Fluids* 18, 053102-1-14.

42) Yu, J.H., Fridrikh, S.V., Rutledge, G.C. (2006). The role of elasticity in the formation of electrospun fibers. *Polymer* 47, 4789–4797

43) Helgeson, M.E., Grammatikos, K.N., Deitzel, J.M. Wagner, N.J. (2008) Theory and kinematic measurements of the mechanics of stable electrospun polymer jets. *Polymer* 49, 2924-2936

44) Entov, V.M., Hinch, E.J. (1997) Effect of a spectrum of relaxation times on the capillary thinning of a filament of elastic liquid. *J. Non-Newtonian Fluid Mechanics* 72, 31-53.

45) Clasen, C., Plog, J.P., Kulicke, W.-M., Owens, M., Macosko, C., Scriven, L.E., Verani, M., McKinley, G.H. (2006) How dilute are dilute solutions in extensional flows? *Rheology* 50: 6, 849-881.

(499c) Effect of Lignin Obtained From Ionic Liquid Pretreated Poplar Wood On the Thermomechanical Properties of Polypropylene

Srikanth Pilla, Wisconsin Institute for Discovery, University of Wisconsin-Madison, Milwaukee, WI, Anantharam P. Dadi, SuGanit Systems, Toledo, OH, Balakrishna Maddi, Chemical Engineering, University of Toledo, Toledo, OH, Craig Clemons, USDA-Forest Products Laboratory and Joseph Lawrence, Bioengineering, University of Toledo, Toledo, OH

PP-Lignin Composites

See more of this Session: Biobased Materials II: Lignin-Based Materials
See more of this Group/Topical: Forest and Plant Bioproducts Division

Unavailable Paper

(499d) Development of Nylon Biocomposites Through the Torrefaction of Waste Stream Agricultural by-Products

Jessica L. Lattimer and Chad A. Ulven, Mechanical Engineering, North Dakota State University, Fargo, ND

In the last several years the use of biobased materials for fillers in thermoplastics has seen a remarkable increase. The desire to produce lighter, stronger, and more ecologically friendly materials being key driving forces behind the movement. While the focus has thus far been on producing biobased composites out of commodity polyolefins or bio-derived resins, little work has been done in the realm of engineering thermoplastics. For engineering thermoplastics, the increased processing temperatures are unfavorable for biocomposite production, causing significant damage to the biobased fillers. As the automotive industry is one of the largest consumers of nylon and the ever increasing push to make cars green, there is a natural fit for research into producing biobased nylon composites. Finding a way to drive off the constituents within the biomass that degrade above the traditional processing temperatures of polyolefins would provide a new biomass that could withstand the processing of engineering thermoplastics.

From previous studies into the use of biobased materials for energy production it is known that torrefaction, conducted between 220-300 °C, will drive off the hemicellulose, fats, waxes, and other constituents yielding a biomass consisting of mostly cellulose, degrading between 300-375 °C, and lignin, degrading slowly over 250-500 °C. As cellulose is the constituent in biomass that acts as a reinforcing agent in the biocomposite, this is a potential method to prepare biobased fillers for introduction in to engineering thermoplastics. This study describes the development of nylon based biocomposites using torrefied sunflower hulls (TSFH), oat hulls (TOH), and flax shive (TFS).

To prepare the biocomposites, nylon was blended with each of the torrefied biomasses analyzed based on weight percentage. This study looks at three grades of the biocomposite; neat nylon-6, 16 wt% filler in nylon-6, and 20 wt% filler in nylon-6. The biocomposite samples prepared were tested for static and dynamic properties in addition to characterizing their resistance to moisture uptake as a function of torrefied biomass loading. Correlations between biomass input and loading are discussed.

See more of this Session: Biobased Materials II: Lignin-Based Materials
See more of this Group/Topical: Forest and Plant Bioproducts Division

(499e) Lignin Based BioFoam Composites From Functionalized Soy Oil Based Biopolyurethane

Manju Misra, School of Engineering and the Department of Plant Agriculture, University of Guelph, Guelph, ON, Canada

uploading in progress

See more of this Session: Biobased Materials II: Lignin-Based Materials
See more of this Group/Topical: Forest and Plant Bioproducts Division

Unavailable Paper

(499f) Lignin-Based Carbonaceous Nano-Fibrous Felts

Chuilin Lai[1], Lifeng Zhang[1], Hao Fong[1] and **Lew P. Christopher**[2], (1)Chemistry, South Dakota School of Mines & Technology, Rapid City, SD, (2)Center for Bioprocessing Research and Development and Department of Chemical and Biological Engineering, South Dakota School of Mines & Technology, Rapid City, SD

Lignin is a naturally existing biopolymer widely found in lignocellulosic plant biomass at an average of 20-30%. It is also a high-volume byproduct of wood pulping which exceeds 15 million tons per year in the US. Nonetheless, lignin has been utilized mainly as a low-grade fuel for production of heat and power to date. There is however a growing demand to identify applications with high economic value for lignin. Although production of carbon fibers from lignin has been reported in literature, the low mechanical properties obtained so far have prevented their commercial use in reinforcement and/or composites applications. Nano-fibrous felts consisting of porous carbon nanofibers derived from lignin, however, have potential applications that do not rely on mechanical properties but on large specific surface area. The use of low-cost lignin as the precursor to develop carbonaceous nano-fibrous felts is anticipated to have a great impact on these applications. The nanomaterials processing technique of electrospinning provides a straightforward approach for convenient preparation of fibers with diameters ranging from tens to hundreds of nanometers. In this research, innovative carbonaceous nano-fibrous felts with high specific surface area were developed by electrospinning aqueous mixtures of alkali lignin and poly(vinyl alcohol) with varied mass ratios followed by heat treatments for stabilization and carbonization. The developed carbonaceous nano-fibrous felts had a large pore volume and specific surface area and are therefore expected to find important applications as catalyst support, supercapacitor electrode, and hydrogen storage materials.

See more of this Session: Biobased Materials II: Lignin-Based Materials
See more of this Group/Topical: Forest and Plant Bioproducts Division

(501a) Peptoid Based Slide Coatings for Disease Detection Via ELISA Microarray Analysis

Melissa L. Hebert, Jeremiah Born and Shannon L. Servoss, Chemical Engineering, University of Arkansas, Fayetteville, AR

Pursuit of an efficient, non-invasive, and economical tool for the early detection of diseases has been the passion of many researchers. A large factor in the ability to create one of these tools is the platform on which to build a reliable test. The use of peptoid-based slide coatings for sandwich enzyme-linked immunosorbent assay (ELISA) microarrays is a promising start to detecting small concentrations of protein biomarkers in biological solutions. Peptoids are believed to have the structure necessary to achieve the critical properties of specific binding sites, high surface area, retention of antibody structure, long shelf-life, and low non-specific protein binding. Preliminary results of recently designed peptoids have shown to form microspheres as surface structures which gives high probability for peptoids achieving these necessary surface properties. The peptoid coated slides will be compared to the currently used lysine-based slides which have shown good performance in past ELISA microarray studies. The comparisons between the slides will be based on spot background, slide noise, spot size and morphology, consistency between slides, and lower limit of detection. We believe peptoid-based slide coatings will create the platform needed to build a reliable test for the early detection of cancer, leading to an increased survival rate.

See more of this Session: Biomaterials for Sensing and Responding
See more of this Group/Topical: Materials Engineering and Sciences Division

(501b) Ultra-Low Fouling and Antibody Functionalization Properties of Zwitterionic Carboxybetaine Polymer Films for Biosensing Applications

Norman D. Brault, Jordan E. Krause, Qiuming Yu and Shaoyi Jiang, Chemical Engineering, University of Washington, Seattle, WA

Recent advancements in biosensing technologies have enabled detection limits down to unprecedented sensitivities. However, a frequent disconnect arises between theoretical biosensor capabilities and the actual assay performance in real-life applications. The primary reason for this difference is due to the necessity of biosensing devices to both prevent the background noise associated with non-specific protein adsorption but also to enable the efficient immobilization of molecular recognition elements (i.e. typically antibodies) which is necessary for high sensitivity and selectivity. Zwitterionic carboxybetaine polymers (pCB) have been shown to be ultra-low fouling to undiluted human plasma and serum as well as the ability to maintain these properties following antibody functionalization. In order to improve the control over pCB film "quality", several surface-initiated polymerization methods (i.e. atom transfer radical polymerization (ATRP) and photo-polymerization) with different antibody immobilization environments (i.e. for activation, protein attachment, and deactivation) were compared using a surface plasmon resonance (SPR) biosensor. The adjustment of pCB film parameters (thickness and packing density) enabled the combined relationship between non-fouling and functionalization to be determined. Finally, the subsequent affect on the bioactivity of an antibody functionalized sensor surface was also characterized. The result of this work provides a fundamental understanding between polymer surface morphology and its subsequent affect on the performance characteristics of biomaterials for biosensing devices.

See more of this Session: Biomaterials for Sensing and Responding
See more of this Group/Topical: Materials Engineering and Sciences Division

(501c) Surface-Mediated Release of Small-Molecule Modulators of Bacterial Quorum Sensing: Toward a New Approach to the Design of Surfaces That Prevent Formation of Bacterial Biofilms

Adam H. Broderick[1], Tony S. Breitbach[2], Helen E. Blackwell[2] and David M. Lynn[1], (1)Department of Chemical and Biological Engineering, University of Wisconsin-Madison, Madison, WI, (2)Department of Chemistry, University of Wisconsin-Madison

The ability of bacteria to communicate and function as a group is central to the development of many bacterial infections. Over the past 40 years, it has become clear that bacteria communicate by assessing their local population densities through a phenomenon known as 'quorum sensing' (QS). Many of the most notorious human pathogens use this sensing mechanism to organize into structured communities called 'biofilms' and activate virulence pathways that are the basis for acute and chronic infections. As a result, QS is a relevant target for the development of new anti-infective strategies. In disease-oriented contexts, the development of small molecule-based inhibitors of QS (QSIs) is also particularly attractive because it provides a potential means to avoid evolved-resistance mechanisms that plague currently used bactericidal agents.

One challenge to the application of QSIs as antivirulence agents lies in the development of methods for the administration of these compounds in ways that can be tailored in a range of different clinical and health-oriented contexts. Bacterial colonization and the formation of biofilms on the surfaces of indwelling medical devices, for example, represent two primary points of entry for bacteria into the body. Inhibiting or attenuating QS in bacteria locally (i.e., at or near the surfaces of these objects) presents a significant challenge. In this presentation, we describe an approach that addresses this broad challenge through the design of thin polymer films that provide temporal control over the surface-mediated release of novel small molecules that inhibit bacterial QS. We demonstrate that this approach can be used to control and activate a QS phenotype in a model bacterial system, and that it has several potential advantages relative to methods for the direct (i.e., solution-based) administration of QSIs used in past studies. Although many different approaches have been reported for the design of materials that control the release of antibiotics and other bacteriocidal agents, the controlled release of agents that disrupt bacterial communication represents, to our knowledge, a fundamentally new approach to the localized and surface-mediated prevention of bacterial virulence.

We will present the results of recent proof-of concept studies based on the encapsulation and release of a novel, synthetic *N*-acylated L-homoserine lactone (AHL) QS modulator from thin films of hydrolytically degradable polymers. We selected *N*-(3-nitrophenylacetanoyl)-L-homoserine lactone (**1**) for these studies because past work in our group has shown that this non-native AHL is a potent modulator of QS in bacteria, most notably as an inhibitor in the pathogen *Pseudomonas aeruginosa* and as a "super-activator" in the bioluminescent symbiont *Vibrio fischeri*. We selected poly(lactide-*co*-glycolide) (PLG) as a matrix for encapsulation and surface-mediated release because this polymer is both biocompatible and biodegradable, and it has a well-documented history of use in drug delivery and other biomedical applications. In addition, the environments within matrices of PLG have been demonstrated to be acidic (and could thus serve to stabilize AHL-type inhibitors against hydrolysis upon prolonged exposure to aqueous media).

Our results demonstrate that AHL **1** can be loaded into solvent-cast thin films of PLGA, and that it is released (over periods ranging from several days to several months; depending on the composition of the copolymer) in a form that is active and able to modulate (turn on) QS in the marine symbiont *V. fischeri*, a bacterial model used widely for fundamental studies of QS. Our results also demonstrate that this polymer-based approach can be used to prolong the activities of the AHL relative to one-time (bolus) treatments with AHL in solution. This result is consistent with the view that the acidic microenvironment inside these films can prevent (or minimize) the hydrolysis, and therefore deactivation, of the AHL prior to release from the film.

The results of this study, when combined, suggest a basis for the design of coatings that intercept bacterial communication in ways that are important in both fundamental and applied contexts (e.g., surfaces that prevent the formation of biofilms). Results of additional physicochemical characterization and bacterial assays will be presented, and opportunities for the application of this approach to prevent bacterial colonization and the formation of biofilms on the surfaces of indwelling devices (e.g., catheters) will be discussed.

See more of this Session: Biomaterials for Sensing and Responding
See more of this Group/Topical: Materials Engineering and Sciences Division

(501d) Cellular Analogues: Mimicking Cell Adhesion Via Temperature Triggered Antibody Organization

Dariela Almeda, Jin-Oh You and Debra. T Auguste, School of Engineering and Applied Sciences, Harvard University, Cambridge, MA

Atherogenesis is initiated by the adhesion of leukocytes to the endothelial surface of arteries followed by migration beneath the intima. Current therapies to combat atherosclerotic plaque, such as statins, work by inhibiting cholesterol production; however, they do not specifically target inflamed vasculature or improve the vascular condition. Few studies have focused on antibody organization and mobility as an approach to improve drug delivery vehicle binding and uptake. Here, we present a biomimetic liposomal design targeting vascular adhesion molecule (VCAM) and endothelial leukocyte adhesion molecule (ELAM), which are upregulated on inflamed endothelial cells and localize within lipid rafts. The unique feature of this drug delivery strategy is that it allows for rearrangement of antibodies, which closely mimics in vivo antibody-receptor interactions. Temperature-induced antibody mobility is achieved by the liposome formulation; liposomes are comprised of the lipids 18:1 (Δ9-trans) 1,2-dielaidoyl-*sn*-glycero-3-phosphoethanolamine (DEPE) (T_m = 38°C) and 1,2-dioleoyl-sn-glycero-3-phosphoethanolamine-N-dodecanoyl (N-dod-DOPE). At temperatures below 38°C, DEPE:N-dod-DOPE liposomes are in the liquid-ordered state and the conjugated antibodies have limited mobility; above 38°C, DEPE:N-dod-DOPE liposomes enter the liquid-disordered state, and the conjugated antibodies exhibit higher lateral mobility. Thus, liposomes initially have a low binding affinity (liquid-ordered) that may be triggered by a temperature increase to have a strong binding affinity (liquid-disordered). We hypothesize that an increase in antibody mobility will allow antibodies to re-organize to complement cell surface receptors and will increase cellular uptake in inflamed areas at temperatures exceeding 38°C. Our results show that activated HUVECs incubated with 1:1 anti-VCAM:anti-ELAM immunoliposomes at 41°C exhibit a higher uptake than at 37°C. This trend was observed for both 6 hr and 24 hr activation periods. We have developed a temperature-responsive drug delivery system that enhances endothelial cellular uptake due to antibody synergy and antibody mobility. Drug delivery vehicles with triggered molecular organization may be useful for increasing circulation time and increasing uptake in sites of inflammation.

See more of this Session: Biomaterials for Sensing and Responding
See more of this Group/Topical: Materials Engineering and Sciences Division

(501e) Zwitterionic Nanogel Based Smart Controlled Release of Insulin

Bin Cao, Qiong Tang and Gang Cheng, Department of Chemical and Biomolecular Engineering, University of Akron, Akron, OH

Currently, 285 million people are suffering from diabetes around globe, and this number is expected to increase to more than 450 million by 2030. It is reported that over 25.8 million of children and adults in the United States, 8.3% of the population, have diabetes, and that the total US economic impact was nearly $174 billion in 2007. Since insulin became available, subcutaneous injection has been for the standard treatment of diabetes for over 80 years. However, this method is painful to patients, especially for those requiring multiple injections per day. Although several new concepts were explored to deliver insulin using oral, pulmonary, nasal, ocular and rectal routes, several limitations, including slow deliver rate, no controlled release, systemic toxicity and others have prevented significant advances.

Zwitterionic carboxybetaine(CB)-based materials have been extensively studied on the biointerface and demonstrated to be highly resistant to nonspecific protein adsorption even from undiluted blood plasma and serum. Our objective is to generate a zwitterionic polymeric nanogel based smart system, which will load, deliver insulin with the nanocapsules and release insulin responding to the glucose concentration in human blood. To obtain a prolonged blood circulation time, nonfouling zwitterionic materials will be utilized to reduce nonspecific protein adsorption from blood. The blood stability and release rate will be first tested.

See more of this Session: Biomaterials for Sensing and Responding
See more of this Group/Topical: Materials Engineering and Sciences Division

(501f) Development of Biocathode Using Laccase and Redox-Polymer-Grafted Carbon

Tomoharu Sugiyama, Takanori Tamaki, Hidenori Ohashi and Takeo Yamaguchi, Chemical Resources Laboratory, Tokyo Institute of Technology, Yokohama, Japan

Enzymatic electrodes have been used for biosensors. Technologies of biosensors are the basis of biofuel cells which use enzymes as catalysts to convert chemical energy to electrical energy. Enzymes are able to oxidize safety fuels, such as glucose and ethanol. However, the biofuel cells have serious issues to be solved. One of the issues is low power density caused by low current density. In the biofuel cell electrode, one of the important factors is the electron transfer between electrodes and enzymes. For effective electron transfer, mediators are generally used to shuttle electrons between electrodes and enzymes. When the electrodes are constructed only from the redox-polymer in which mediators are immobilized, the current density is restricted, because a redox-polymer has low electric conductivity and the conventional redox-polymer layer is thick. To solve this problem, our group has developed a new biofuel cell electrode which consists of redox-polymer-grafted carbon black for glucose-oxidizing anode. This electrode divides conduction of electrons into the redox-polymer and carbon, and thus, the electron conduction distance in the redox-polymer was decreased. Additionally carbon black has large surface area. Effectiveness of the electrode was verified experimentally and by simulation. The model calculation shows the possibility of the further increase in current density by two orders of magnitude. The main reason of a low current density obtained in the experiment was caused by deactivation of enzymes during adsorption on carbon black. On the other hand, oxygen-reducing cathode still has the issue about rate-controlling electron transfer in the redox-polymer.

In this research, a novel biofuel cell cathode was developed using a hyperthermophilic laccase and redox-polymer-grafted carbon with the mediator for a cathode. The hyperthermophilic enzyme with the optimal temperature above 90°C has a potential to reduce deactivation of the enzyme during adsorption on carbon black because the enzyme has rigid structure. The redox-polymer immobilized 2,2'-azinobis (3-ethylbenzotiazoline-6-sulfonate)(ABTS) as the mediator to polyacrylic acid grafted on carbon black. Enzyme activity on carbon black showed that laccase from *Trametes versicolor* with the optimal temperature around 60°C has relative activity of 22% in comparison with the laccase in solution, and the hyperthermophilic laccase has relative activity of 94%. This result suggests that the hyperthermophilic laccase keeps activity during adsorption on carbon black. Immobilization of ABTS to polyacrylic acid was confirmed by NMR and IR using a linear-polymer without grafting. TG-MS of the redox-polymer-grafted carbon black showed SO and SO_2 signals from ABTS, while these peaks were not observed before immobilization of ABTS. Cyclic voltammetry of the redox-polymer-grafted carbon black showed redox peaks. Additionally, the redox potential calculated from the peaks shifted from dissolved ABTS, and the shift corresponds to the immobilization of ABTS. These results suggest that synthesis of the redox-polymer-grafted carbon black for biocathode was succeeded for the first time. The electrode using the hyperthermophilic laccase and the redox-polymer-grafted carbon with ABTS is expected to achieve improved performance on biocathode.

See more of this Session: Biomaterials for Sensing and Responding
See more of this Group/Topical: Materials Engineering and Sciences Division

(502a) Use of Catechol-Modified Polymers In Layer-by-Layer Assembly to Enhance Its Stability and Sustained Release of Biomolecules: A Bio-Inspired Approach

Younjin Min and Paula Hammond, Chemical Engineering, Massachusetts Institute of Technology, Cambridge, MA

Although Layer-by-Layer (LbL) assembly technique has been successfully used in various areas of nanotechnology and biotechnology, some LbL-assembled nanostructures have suffered from lack of stability when they are exposed to any changes in aqueous environments (e.g., pH and ionic strength). In addition, this technique has been limited because of the diffusion of polyelectrolytes throughout the film during the deposition process, resulting in poorly organized architectures, thus limiting the control of release characteristics. To improve the stability of LbL-assembled nanostructures, and to control interlayer diffusion, we have developed a strategy to conjugate catechol groups, which are largely present in mussel adhesive proteins, to conventional polyelectrolytes such as branched poly (ethyleneimine) (BPEI) and poly(acrylic acid) (PAA). Only a fraction of amine or acid groups were modified with catechol using this synthetic approach, thereby preserving the original cationic or anionic nature of the polymers for use in LbL assembly while integrating beneficial adhesive features of catechol groups into LbL films. The structure, physico-chemical properties, and stability of LbL films composing BPEI and PAA without and with catechol modifications were compared. The incorporation of catechol groups led to a significant change of the LbL growth curve from exponential to a clearly linear increase of thickness with number of layers, reflecting the nondiffusing characteristic of the catechol-modified polyelectrolytes. When LbL films were exposed to PBS pH 7.4, we observed that the catechol containing LbL films underwent far fewer changes in the degree of ionization and film thickness and exhibited stronger mechanical properties, indicative of their enhanced film stability. When LbL films with catechol modifications were used as physical barrier layers to prevent the undesired initial burst release of radiolabeled ^{14}C-dextran sulfate sodium salt (^{14}C-DS), a clear sustained release was achieved over a period of 40 days. When the bottom layer containing ^{14}C-DS is separated from radiolabeled ^{3}H-heparin sodium salt (^{3}H-HS) by catechol modified polyelectrolytes, we observed two different release rates composing an abrupt release from the surface ^{3}H-HS, together with a sustained release from the underlying ^{14}C-DS. Overall, we conclude that when catechol groups are present in LbL films, linear film growth, enhanced film stability, and extended release times are attributed to the introduction of non-ionic types of interactions such as hydrogen bonding and *Π-Π* stacking, as well as the formation of covalent diphenyl ether bonds, thus enable these films to provide a general platform for the systematic incorporation and assembly of biological therapeutics into controlled release films at mild conditions for drug delivery.

(502b) Dynamic Regulation of Bioactive Ligands Presented On Gold Particles

Wei Shen, Department of Biomedical Engineering, University of Minnesota, Minneapolis, MN, Xintong Wang, Biomedical Engineering, University of Minnesota, Minneapolis, MN and Jeremiah Riesberg, Department of Chemical Engineering and Materials Science, University of Minnesota, Minneapolis, MN

We have developed a platform to dynamically regulate the presentation of bioactive ligands immobilized on gold particles to cell surface receptors. The cell-adhesive peptide Arg-Gly-Asp (RGD) was chosen as a model ligand. A polypeptide cysA-RGD, in which A was a leucine zipper domain and cys was a cysteine residue, was synthesized. Gold particles were functionalized with cysA-RGD through the thiol-gold reaction, and the functionalized gold particles were immobilized on *N*-hydroxysuccinimide(NHS)-functionalized substrates through the amine-NHS reaction. Presentation of RGD to cell surface receptors, as indicated by cell adhesion, could be reversibly regulated by co-immobilization and removal of B-PEG conjugate, in which PEG was polyethylene glycol and B was another leucine zipper that could heterodimerize with A. Cells adhered on these substrates were detached upon addition of B-PEG, which was co-immobilized through A/B heterodimerization and shielded RGD from access to cell surface receptors. The shielded RGD could be converted back to cell-accessible state by the addition of soluble A, which competed with the immobilized A for binding to B and removed B-PEG from the substrates. The platform can be readily adapted to dynamically regulate the presentation of other bioactive ligands of interest.

See more of this Session: Biomimetic Materials I
See more of this Group/Topical: Materials Engineering and Sciences Division

(502c) Characterization of Modular Resilin-Based Artificial Protein Matrices for Cartilage Tissue Engineering

Julie N. Renner, Yeji Kim and Julie C. Liu, School of Chemical Engineering, Purdue University, West Lafayette, IN

Recombinant proteins have been explored for numerous tissue engineering applications. In particular, artificial proteins based on structural repeats of elastin and silk have been investigated for use in soft tissue and orthopedic applications. Compared to synthetic or natural materials, recombinant proteins have a number of advantages that include: exquisite control over sequence and composition; precise molecular weight; and a modular nature, which allows for the functional domains to be easily altered.

We are developing an artificial protein matrix based on a structural repeat found in resilin, an elastic protein from insect joints and tendons that contributes to flight and jumping. Resilin has a high strain and can efficiently store energy. In addition, it efficiently recovers after deformation and has a high fatigue lifetime, which makes it ideal for use in repetitive environments. Based on the natural properties of resilin, we believe that these resilin-based proteins are promising candidates for cartilage engineering. Our modular protein design includes bioactive domains and sites for either chemical or photochemical crosslinking. We adapted a recursive cloning technique and show that the number of resilin repeats in our protein can be precisely and easily tuned. The Studier auto-induction method was used to express the proteins, but had to be modified due to protein degradation. Protein expression and purity are assessed by Western blot, amino acid analysis, and mass spectrometry.

See more of this Session: Biomimetic Materials I
See more of this Group/Topical: Materials Engineering and Sciences Division

(502d) Melanic Hydrogels As Novel Biomaterials

Omar Z. Fisher[1], Robert Langer[2] and Daniel G. Anderson[2], (1)Koch Institute for Integrative Cancer Research, Massachusetts Institute of Technology, Cambridge, MA, (2)Chemical Engineering, Massachusetts Institute of Technology, Cambridge, MA

Melanins are a broad range of macromolecules derived from the oxidation of phenolic compunds. Their unique properties have proven useful in both medical and industral applications. Natural melanin formation is driven by the activity of oxidoreductases on phenolic substrates. For example, melanins in human skin are formed from tyrosine and catecholamine subtrates via tyrosinase activity. Similar reaction schemes result in the brown blemishs observed on the surfaces of bruised fruits and vegetables, and some shellfish. Despite their ubiquity in nature melanins remain poorly understood. Enzymatic melanogenesis is poorly controlled and the resulting melanins are difficult to characterize. They are amorphous insoluble polymer networks with a high density of zero length crosslinks. We theorized that by introducing hydrophilic spacers between crosslinks a novel hydrogel could be obtained with melanic properties.

We designed synthetic poly(ethylene glycol derived macromers with terminal gallate groups. Melanogenesis was achieved by oxidizing these macromers to obtain hydrogels. Polyphenols are oxidized to highly reactive quinones during melanin formation. These intermediates can form conjugate adducts with nucleophilic functional groups on proteins, such as thiols and amines. The reaction scheme allows the introduction of additional heteroatoms into a melanin network. We exploited this quinoid reactivity as a simple way to introduce cell adhesive ligands into the hydrogel networks. These substrates provide a versatile tool for controlling cell adhesion, adding biofunctionality to surfaces, and potentially for regenerating tissue.

See more of this Session: Biomimetic Materials I
See more of this Group/Topical: Materials Engineering and Sciences Division

(502e) Biomimetic Self-Assembling Copolymer-Hydroxyapatite Nanocomposites with the Nanocrystal Size Controlled by Citrate

Xunpei Liu[1], Yan-Yan Hu[2], Xing Ma[3], Adu Rawal[2], Tanya Prozorov[4], Mufit Akinc[3], Surya Mallapragada[1] and Klaus Schmidt-Rohr[2], (1)Department of Chemical and Biological Engineering and Ames Laboratory, Iowa State University, Ames, IA, (2)Department of Chemistry and Ames Laboratory, Iowa State University, Ames, IA, (3)Department of Materials Scinence and Engineering and Ames Laboratory, Iowa State University, Ames, IA, (4)Ames Laboratory, Ames, IA

Bone is a nanocomposite that mainly contains nano-sized hydroxyapatite (HAp) crystals and the fibrous protein collagen. Citrate binds strongly to the surface of HAp nanocrystals in bone and is thought to prevent crystal thickening. In this work, biomimetic self-assembled copolymer-HAp nanocomposites were synthesized using a bottom-up approach from aqueous solutions. Amphiphilic block copolymer Pluronic F127 was used as a template for the growth of HAp nanocrystals. Citrate was added as a biomimetic regulatory element to control of the size and stability of HAp nanocrystals in synthetic nanocomposites. The decrease of the HAp crystal size within the polymer matrix with increasing citrate concentration was confirmed by solid-state nuclear magnetic resonance (NMR) techniques and wide-angle X-ray diffraction (XRD). The shapes of HAp nanocrystals were determined by transmission electron microscopy (TEM). Small angle neutron scattering (SANS) results indicated that the nanocrystals are incorporated into the polymer matrix in a manner that does not disrupt the self-assembled structure of the pluronic micelles. The surface-to-volume ratios determined by advanced NMR spectroscopy show that 0, 2, 10, and 40 mM citrate changed the thicknesses of the HAp crystals from 4 nm without citrate to 2.9, 2.8, and 2.3 nm, respectively. With citrate concentrations comparable to that in body fluids, HAp nanocrystals of similar sizes and morphologies as in avian and bovine bones have been fabricated *in vitro*.

See more of this Session: Biomimetic Materials I
See more of this Group/Topical: Materials Engineering and Sciences Division

(502f) SIBLING Protein Induced Biomineralization

Kevin Zurick and Matt Bernards, Chemical Engineering, University of Missouri - Columbia, Columbia, MO

SIBLING Protein Induced Biomineralization

Native bone tissue is composed of a matrix of collagen,non-collagenous proteins, and mineral, mostly hydroxyapatite (HAP). Bonesialoprotein (BSP), osteopontin (OPN), and dentin phosphoprotein (DPP) are allmembers of the SIBLING (small integrin-binding ligand, N-linked glycoprotein)family of proteins, which are primarily found in mineralized tissues. In thiswork, the mineralization induction capabilities of BSP, OPN, and DPP areexplored. Radiolabeled adsorption isotherms were constructed for BSP, OPN, andDPP on a biomimetic collagen substrate.The results showed that BSP exhibited the highest binding capacity forcollagen, followed by DPP, and then OPN. The adsorption isotherms were thenused to study the effects of equal amounts of adsorbed proteins on a collagensubstrate with respect to HAP formation and morphology. Mineralization studieswere performed, and the adsorption isotherms were used to determine solutionconcentrations that would result in equal amounts of adsorbed protein for eachof the three proteins. Substrates were placed in simulated body fluid for 5h,10h, and 24h in an incubation chamber at 37 ^{o}C.Photochemical assays were performed on demineralized substrates to determine aCa/P ratio for minerals formed with each of the proteins used at all timepoints considered. Additionally, SEM images were taken of the substrates toprovide a qualitative comparison of mineral surface coverage at each timepoint.

See more of this Session: Biomimetic Materials I
See more of this Group/Topical: Materials Engineering and Sciences Division

(502h) Peptoids: A Potential Therapeutic Agent Against Alzheimer's Disease

James Phillip Turner Jr., Jennifer Herrera, Amanda Compean and Shannon Servoss, Ralph E. Martin Department of Chemical Engineering, University of Arkansas, Fayetteville, AR

Alzheimer's disease (AD) is a brain disease that slowly develops, leading to dementia and eventually death. AD is characterized by the build-up of the amyloid-beta protein. As the amyloid-beta proteins begin to aggregate, plaques form in the spaces between nerve cells. Currently, there is no medication to dissolve the build-up of amyloid aggregates. However, research has found that certain molecules can bind to the amyloid-beta protein and prevent aggregation. The purpose of this research project is to develop a therapeutic treatment for AD with the use of poly-N-substituted molecules (peptoids). Peptoids form strong helices, have high bioactivity, and are easy and cost-effective to synthesize. Peptoids are ideal for this research project due to the spacing of the beta sheets and the peptoid helices matching, as well as being protease resistant.

For this project, three different peptoids have been designed and synthesized with differing chemistries at the sixth side chain only. The purpose for this is to provide a strong basis for developing novel molecules that can withstand the harsh nature of the neural environment as well as pass through the blood brain barrier (BBB). Upon completion of the design stage, the peptoids will undergo rigorous tests to help provide valuable information to finding a therapeutic agent against AD.

See more of this Session: Biomimetic Materials I
See more of this Group/Topical: Materials Engineering and Sciences Division

(511a) Rational Biomaterials Design and Synthesis for Gene Delivery

Surya Mallapragada, Department of Chemical and Biological Engineering and Ames Laboratory, Iowa State University, Ames, IA

Polymeric vectors have shown enormous promise for gene delivery, as a safer alternative to viral vectors. However, the transfection efficiencies of polymeric vectors are comparatively lower. The polymeric vectors have largely been cationic because of their ability to complex with DNA. However, cationic polymers can suffer from cytotoxicity issues, and/or aggregation with serum proteins that can decrease the transfection efficiency. There is a need for the development of new biomaterials that serve as efficient gene delivery vectors with minimal cytotoxicity. We have developed pentablock copolymers with cationic and neutral blocks, and the ratio of the various blocks can be adjusted by controlled radical polymerization methods. The polymeric blocks can be selected to allow the vector to promote endosomal escape to improve transfection efficiencies. In addition, the rational design of gene delivery vectors also allows for polymer chemistry to be used to target specific cell types. For instance, small intracellular differences in pH between normal and cancer cells are being exploited in our work to design vectors that can selectively transfect cancer cells as opposed to normal cells. This provides a paradigm shift from regular targeting methods that rely on ligand attachment, and relies instead on the polymer chemistry instead to achieve targeting.

See more of this Session: Challenges In Biomaterial Synthesis
See more of this Group/Topical: Materials Engineering and Sciences Division

(511b) Synthesis and Self-Assembly of Bio-Responsive Block Copolymers

Millicent O. Sullivan, Elizabeth G. Kelley and Thomas H. Epps III, Chemical Engineering, University of Delaware, Newark, DE

We have designed responsive polymeric micelles for targeted drug delivery applications by taking advantage of the self-assembly of amphiphilic block copolymers containing protease-sensitive linkages embedded within their hydrophilic backbones. These micelles are capable of responding to biological stimuli, as cleavage of the peptide linkages should destabilize the assembled structures and result in the efficient delivery of therapeutic molecules to specific cellular locations. Our strategy is based on the solution assembly of poly(butadiene-*b*-ethylene oxide) (PB-PEO). A non-peptide containing PB-PEO model polymer with M_n = 12,600 g/mol and hydrophilic weight fraction w_{philic} = 0.66 self-assembles into spherical micelle structures in aqueous solutions. Because the chain characteristics of this PB-PEO polymer induced micelles formation, we used similar guiding principles in the synthesis of our micelle-forming PB-PEO-peptide conjugates. Polymer-peptide conjugates were created through a combination of anionic polymerization and solid phase peptide synthesis, and the resulting copolymers were characterized with nuclear magnetic resonance spectroscopy (^{1}H NMR), gel permeation chromatography (GPC), and mass spectrometry (MALDI-TOF MS). The aqueous solution-assembly of the PB-PEO-peptide amphiphiles was probed with dynamic light scattering (DLS) and cryogenic transmission electron microscopy (cryo-TEM) experiments.

See more of this Session: Challenges In Biomaterial Synthesis
See more of this Group/Topical: Materials Engineering and Sciences Division

(511c) Clickable Synthetic Polypeptides As a Tranformable Backbones for Biomaterials

Paula T. Hammond, Chemical Engineering, Massachusetts Institute of Technology, Cambridge, MA

One of the key challenges in the field of polymer chemistry is the ability to generate macromolecules with the degree of control over composition and diversity of function that nature is able to exhibit with biomacromolecular species. Although it is possible to generate the very specific sequences of naturally occurring polypeptides, these approaches are limited in practice to relatively small quantities and short peptide sequences that can generally only be provided at high cost due to low yields, time consuming process steps and costly purifications. Synthetic polypeptides introduce a powerful capability to generate macromolecular species using the amino acid backbone found in nature, thus providing a route to biocompatible polymeric systems with programmable function. A number of exciting developments have occurred with the advent of controlled polymerization of N-carboxyanhydrides (NCA), which has been extended considerably with the introduction of new catalyst systems that enhance control and block distribution. Although these methods can enable design with a number of naturally occurring amino acids, the NCA process can be restrictive with regard to the functionality of the monomer side group allowed, thus preventing the direct incorporation of various amine, acid, thiol, and other functional sidegroups that dictate polymer function, structure, and responsive behavior to temperature or pH among many other properties. Steric effects also limit the incorporation of amino acids with large macromolecular side chains. There is a need to create a more readily adaptable platform to attach a broader range of groups at the amino acid backbone with ease and control. Highly quantitative functionalization chemistries have been developed and evolved for use with polymers over the past several years, such as the Sharpless 'click' chemistry involving the near-quantitative reactions of propargyl and azide groups.

To broaden the range of capabilities possible with the NCA chemistry platform, our research group recently introduced the first NCA polymerized synthetic polypeptides with 'clickable' side groups through the introduction of a new NCA polymer, poly(γ-propargyl L-glutamate) (PPLG), which contains a pendant alkyne group that can be reacted with an azide. We were able to demonstrate key advantages to the use of a clickable backbone system with a nearly quantitative post-polymerization functionalization step, including: a) The ability to directly attach functional groups that are ordinarily difficult due to cross-reaction or the need for exhaustive protection-deprotection strategies that greatly lower yield and impact function, including amines, imines, acids, alchohols, thiols and combinations of polar and nonpolar groups that can exhibit mixed charge or function and b) Unusually high grafting-on density with oligomeric grafts in a range of molecular weights, including up to 99% functionalization with polyethylene glycol (PEG) macromolecules up to 5000 daltons. This capability provides an opportunity to create new types of graft copolymers that would not be accessible with other methods to generate biomimetic and bioinspired polymers such as artificial proteoglycans and densely grafted polymers with unique swelling, anti-fouling or pH responsive properties.

Ease of synthesis with both small molecular and macromolecular side groups that exhibit a broad range of polarity and charge provides a key platform for the generation of families of synthetic polypeptides that more directly mimic the adaptive function and responsive behavior of naturally occurring polypeptides while introducing the opportunity to incorporate new function. Examples of the use of this platform from the generation of responsive polymeric micelles that are triggered at key biological pH conditions, and new hydrogel materials, to polymers that can act as native antimicrobial peptides with much lower toxicity will be discussed.

See more of this Session: Challenges In Biomaterial Synthesis
See more of this Group/Topical: Materials Engineering and Sciences Division

(511d) Synthesis of Renewable and Degradable Block Polymers

Marc A. Hillmyer, Chemistry, University of Minnesota, Minneapolis, MN

Block copolymers with the ABA triblock architecture are useful as thermoplastic elastomers and as toughened plastics depending on the relative content of the glassy and rubbery components. These materials can also be blended with other materials and find commercial utility as additives, toughening agents, compatibilizers, and pressure sensitive adhesive formulations. Nearly all of the block copolymers currently on the market are derived from petroleum. Renewable alternatives are attractive considering the finite supply of fossil resources harbored by Earth, the overall economic and environmental cost associated with the recovery and use of oil. Furthermore, the design and implementation of materials with programmed end-of-life is an attractive approach to tomorrow's sustainable materials. With these motivations on board, we have focused our attention on the preparation and evaluation of new classes of renewable ABA triblock copolymers and other, architecturally more complex hybrid materials. We have emphasized the use of carbohydrate-derived polylactide as the high glass transition temperature and/or high melting temperature hard phase given its widespread availability and acceptance. For the soft components, we have focused on other renewable, low glass transition, amorphous aliphatic polyesters and hydrocarbon polymers. In this talk I will highlight our synthetic strategies to linear and graft copolymers containing polylactide as both the major and minor component. In several cases, the combination of multiple controlled polymerization mechanisms has proven to be a powerful enabling approach. Precision synthesis of these new hybrid macromolecules combined with detailed molecular and morphological characterization has lead to the development of new elastomers and tough plastics that marry renewability, biodegradation and performance.

See more of this Session: Challenges In Biomaterial Synthesis
See more of this Group/Topical: Materials Engineering and Sciences Division

(511e) Synthetic Challenges for Zwitterionic-Based Biomaterials

Shaoyi Jiang, Chemical Engineering, University of Washington, Seattle, WA

An important challenge in many applications, ranging from biomedical devices to drug delivery carriers, is the prevention of nonspecific biomolecular and microorganism attachment on surfaces. Over the last few years, we have demonstrated that zwitterionic and mixed charge materials and surfaces are highly resistant to nonspecific protein adsorption, even from complex media such as undiluted blood plasma and serum. In addition to their excellent nonfouling properties, zwitterionic carboxybetaine-based materials have functional groups for direct protein immobilization while the cationic precursors of zwitterionic materials have self-sterilizing capabilities. Due to the uniqueness of zwitterionic materials such as superhydrophilic and anti-polyeletrolyte properties, their synthesis poses unique challenges. In this work, challenges for the synthesis of zwitterionic materials will be discussed for a number of applications in the forms of surface coatings, hydrogels and nanoparticles. At present, zwitterionic materials have been applied to many applications, including implantable medical devices, early cancer diagnostics, drug/gene delivery, antimicrobial coatings, and marine coatings.

See more of this Session: Challenges In Biomaterial Synthesis
See more of this Group/Topical: Materials Engineering and Sciences Division

(511f) Development of Scaffolds with Deterministic Microstructure for the Guidance of Tissue Growth In Vitro and In Vivo

Abraham D. Stroock[1], Ying Zheng[1], Peter W. Henderson[2], Michael Craven[1], John P. Morgan[1], Alyssa Reiffel[2], Jason A. Spector[2] and Lawrence J. Bonassar[3], (1)Chemical and Biomolecular Engineering, Cornell University, Ithaca, NY, (2)Surgery, Weill Cornell College of Medicine, New York, NY, (3) Department of Biomedical Engineering, Cornell University, Ithaca, NY

An outstanding challenge in the field of regenerative medicine is the growth of microphysiological structure within macroscopic, target tissues. In our labs, we are investigating the use of microstructure with 3-D scaffolds to stimulate and direct the formation of appropriate physiology. In this talk, I will discuss the challenges and opportunities presented by the fusion of microfabrication with tissue engineering approaches both in vitro and in vivo. I will illustrate these points with the following two studies: 1) *Model microvessels in vitro.* The recapitulation of microvascular structure *in vitro* would provide a basis for dissecting angiogenic processes in development and in clinical contexts such as wound healing and tumor vascularization. I will present a summary of our effort to grow physiologically appropriate microvessels in the lab. In particular, I will show that we can use microfluidic structure formed in cell-remodelable matrices to initiate the growth of vessels with appropriate physiology, permeability, and response to biochemical and mechanical stimuli. 2) *Invasion of porous matrices in vivo.* Biomaterials with random porous structure are used successfully in the clinic to encourage the regrowth of the dermis in deep wounds. Yet roles of the pores have not been elucidated and these scaffolds fail to provide fast enough invasion of the host vasculature for many applications. I will present the development of implantable scaffolds with pores and networks of conduits of well-defined size and shape formed by microfabrication. Based on multi-week implantation of these scaffolds in a murine (mouse) wound model, I will show that the geometry and dimension of the pores have a dramatic impact on the invasion of tissue and blood vessels. I will discuss our insights into the cellular, chemical, and mechanical processes that dictate this process. Finally, I will show that appropriate deterministic paths guide rapid invasion over macroscopic distances, leading to significantly higher densities of tissue and vascular structure than in random matrices.

See more of this Session: Challenges In Biomaterial Synthesis
See more of this Group/Topical: Materials Engineering and Sciences Division

(511g) Synthetic Approaches to Designing Responsive Hydrogels for Dynamic Cell Culture

April M. Kloxin, Department of Chemical Engineering and Department of Materials Science and Engineering, University of Delaware, Newark, DE

Physical and chemical cell microenvironment properties have become increasingly recognized as regulators of cellular functions[1-4] such as migration, cytoskeletal organization, and differentiation. While advances have been made in the fabrication of materials for cell culture, the properties of these material systems typically are fixed upon formation. In situ control of material properties is needed to study their effect on dynamic cellular processes such as differentiation and migration. Here, we will discuss synthetic approaches for stimuli triggered control of material properties, specifically hydrogel modulus and epitope presentation, via the incorporation of photolytically[5, 6] or enzymatically[7] cleavable sequences within the gel-forming macromers and resulting hydrogels. Utilizing these materials, we will examine the influence of microenvironment structure on cell function, specifically epithelial cell phenotype and assembly. Incorporation of photolytically or enzymatically degradable sequences within monomers and hydrogels enables the design of unique, dynamic cell culture platforms for exploring how changes in properties influence cell fate.

References

1. Chen, C.S. Mechanotransduction - a field pulling together? *J. Cell Sci.* **121**, 3285-3292 (2008).

2. Tibbitt, M.W. & Anseth, K.S. Hydrogels as Extracellular Matrix Mimics for 3D Cell Culture. *Biotechnol. Bioeng.* **103**, 655-663 (2009).

3. Keung, A.J., Kumar, S. & Schaffer, D.V. in Annual Review of Cell and Developmental Biology, Vol 26, Vol. 26 533-556 (Annual Reviews, Palo Alto; 2010).

4. Lin, C.C. & Anseth, K.S. PEG Hydrogels for the Controlled Release of Biomolecules in Regenerative Medicine. *Pharm. Res.* **26**, 631-643 (2009).

5. Kloxin, A.M., Kasko, A.M., Salinas, C.N. & Anseth, K.S. Photodegradable Hydrogels for Dynamic Tuning of Physical and Chemical Properties. *Science* **324**, 59-63 (2009).

6. Kloxin, A.M., Tibbitt, M.W. & Anseth, K.S. Synthesis of photodegradable hydrogels as dynamically tunable cell culture platforms. *Nat. Protoc.* **5**, 1867-1887 (2010).

7. Fairbanks, B.D. et al. A versatile synthetic extracellular matrix mimic via thiol-norbornene

photopolymerization. *Adv. Mater.* **21**, 5005-5010 (2009).

See more of this Session: Challenges In Biomaterial Synthesis
See more of this Group/Topical: Materials Engineering and Sciences Division

(530a) Physical Vapor Transport of Aluminum Nitride On Silicon Carbide Substrates: Parameters Affecting Nucleation

J. H. Edgar and Li Du, Chemical Engineering, Kansas State University, Manhattan, KS

Employing native substrates in group III nitride semiconductor devices will reduce or eliminate strain in epitaxial layers caused by the mismatch of lattice constants and coefficients of thermal expansion. Lowering strain reduces threading dislocation densities in epitaxial films. These defects severely degrade the semiconductor's optical and electrical properties, and thus the performance of devices employing these layers. Thus, single crystal aluminum nitride is an attractive substrate for improving the performance of deep ultraviolet light emitting diodes and laser diodes with AlN and $Al_xGa_{1-x}N$ epitaxial layers. The quality of these semiconductors layers are particularly sensitive to defects.

In this study, the parameters affecting the nucleation of AlN crystals grown by physical vapor transport on silicon carbide single crystals wafers is reported. Employing 4H-SiC and 6H-SiC seed crystals offers a means of rapidly scaling bulk AlN crystals to large diameters. In addition, the choice of the SiC crystal plane determines the crystal plane of the AlN formed. Nucleation is critically important, as this step defines the ultimate defect densities in the AlN crystals.

Aluminum nitride crystals were grown from solid AlN sources at temperatures between 1400 °C to 1650 °C in a nitrogen ambient. Silicon carbide substrates with different planes and tilts were employed as seed crystals. A comparison was made of the properties and growth rates of aluminum nitride crystals grown from oxygen-rich (AlN powder) and oxygen-depleted (dense, sintered polycrystalline) sources.

Regardless of the SiC substrate plane (*c*- and *m*-planes), misorientation tilt direction (toward [1-100] and [1-210] directions), or tilt magnitude (0, 7°, and 8° off-axis), the fastest growth direction was perpendicular to the *c*-axis. This produced AlN platelet crystals with high *a*-axis to *c*-axis aspect ratios. The AlN crystals nucleated as separate, individual islands on the SiC substrates, and grew parallel to the substrate's *c*-plane (ie the (0001) plane). This caused AlN to laterally overgrow the SiC substrate (separate from the substrate), resulting in the formation of voids at the SiC/AlN interface.

At high (1 wt%) oxygen concentrations in the source, severe anisotropic etching of the SiC substrate occurred simultaneous with AlN deposition. Deep etch pits were formed in the SiC substrate, presumably at dislocations. In addition, AlN-SiC alloy hillocks formed, on which AlN platelets would subsequently grow. Consequently, the AlN platelets were attached to the substrate at discrete points corresponding to the top of the AlN-SiC hillocks. Furthermore, a high oxygen concentration enhanced the growth rate of the AlN crystals. For growth from a low oxygen concentration source, no etching of the SiC substrate was observed.

Thus, by careful selection of the process conditions, silicon carbide can be used as a seed crystal for bulk AlN crystal growth. The AlN can be grown in a manner to reduce the degree it is strained by the SiC, and by long term continuous growth, free-standing AlN wafers can be produced.

See more of this Session: Gas Phase Deposition
See more of this Group/Topical: Materials Engineering and Sciences Division

(530b) Hot Wire Assisted Chemical Vapor Deposition (HWCVD) of Niobium Nitride Thin Films

Alejandro Martnez, Daniel F. Jaramillo and Pablo Ortiz, Chemical Engineering, Universidad de los Andes, Bogot, Colombia

Niobium nitride films are very interesting materials due to their extraordinary properties and their potential applications in corrosion inhibition, wear prevention, superconduction and catalysis, among others.While there have been several efforts in the past to obtain thin films of this material by different techniques (e.g. PVD, thermal atmospheric CVD, PECVD, and ALD methods), yet the prior art clearly fails to disclose a method to deposit niobium nitrides via hot-wire assisted chemical vapor deposition. Furthermore, the mechanisms associated with HWCVD are, to certain extent, still not very well understood.

This study evaluates two HWCVD experimental approaches differing mainly in the method of supplying niobium species in the reaction zone. The first of which comprises feeding $NbCl_5$ precursor (99%) along with different N_2/H_2 ratios into a hot-filament assisted reactor operating under mid-vacuum. Alternatively, the reaction vessel is fed with N_2/H_2 atmospheres and a pure niobium filament (99.99%) is used as metal source.

Specifically, the effect of substrate temperature is evaluated in the range 700-1100°C at total pressures between 10 and 100 mbar, for deposition on Si [100] treated with ethanol/acetone solutions. Deposit samples have been characterized in their composition and morphology by SEM-EDS, XRD and AFM finding that niobium compounds grow both on filaments and on silicon substrates.

It is apparent from our preliminary results, however, that homogeneous, well-adhered films are obtained at low growth rates. Scanning electron micrographs reveal the formation of nano-structured intergrown multifaceted rodlike crystal arrays. The inhibition of nitride formation induced by oxide growth is a problem that remains to be overcome.

In view of the foregoing, it is the objective of the work at hand to gain insight into the mechanisms of film growth and elaborating on the phenomena governing heterogeneous nucleation by comparing different chemical routes and process conditions, to obtain conformal niobium nitride thin films.

See more of this Session: Gas Phase Deposition
See more of this Group/Topical: Materials Engineering and Sciences Division

(530c) Effect of Process Variables On Microestructure and Composition of HWCVD Tungsten Oxides Thin Films

Daniel F. Jaramillo, Alejandro Martnez, Ana M. Araujo and Pablo Ortiz, Chemical Engineering, Universidad de los Andes, Bogot, Colombia

This research work is focused on obtaining process maps to elaborate tungsten and tungsten oxides thin films, which allows relating structural and composition characteristics with the operating conditions of the HWCVD reactor. In the experimental set-up a 0.5 mm diameter filament was placed at a distance of 10 mm from substrate surface and was employed both as precursor source and as species decomposition mechanism. The filament temperature was regulated by applying different voltage to current ratios, and determined by optical pyrometry. An independent heating system was used to set substrate temperature. Films were deposited over silicon (100) and glass substrates at reduced pressure. The substrate temperature was modified in the range of 400°C to 600°C and the oxygen partial pressure was varied from $1x10^{-1}$ to $1X10^{-2}$ mbar, by using a rotary vane pump and controlled air input. Homogeneous and well adhered deposits with high growth rates were obtained and investigated by scanning electron microscopy (SEM), X-ray diffraction (XRD) and atomic force microscopy (AFM). Significant structure and composition differences were found as a function of PO_2 and substrate and filament temperature. Nanostructured films comprising needles, rods and/or agglomerated sphere-like geometries were found. The resulting composition ranges from highly metallic deposits to W_xO_y oxides with diverse stoichiometries. While rich-tungsten films were obtained at deposition rates of 0.05 µ m/min, oxides growth rate were about 1µ m/min. Finally, this study proposes possible formation mechanisms and discusses the influence of process parameters and substrate nature on growth and nucleation phenomena.

See more of this Session: Gas Phase Deposition
See more of this Group/Topical: Materials Engineering and Sciences Division

(530d) High-Pressure Chemical Deposition of Silicon In Extreme Aspect-Ratio Micro-Capillaries

Banafsheh Keshavarzi[1], Neil Baril[2], John V. Badding[2] and Ali Borhan[1], (1)Department of Chemical Engineering, The Pennsylvania State University, University Park, PA, (2)Department of Chemistry, The Pennsylvania State University, University Park, PA

Semiconductor waveguides of nanoscale to microscale dimensions have recently been made by chemical vapor deposition within extreme aspect-ratio microcapillaries under high pressure flow of a silane/helium mixture. The microcapillary is heated in a furnace, causing silane to decompose at the capillary wall to produce silicon and hydrogen. Experimental observations show that the thickness of the deposited silicon layer varies greatly with axial position within the microcapillary. As the annular deposited film grows thicker, the central hole formed at the cross-section with the highest film growth rate becomes smaller until it is completely plugged. Complete filling of the plugged central channel over lengths of centimeters or more is challenging in view of its extreme aspect ratio. Assuming silane decomposition to be a first-order reaction, we use a compressible flow model to examine the observed non-uniformity in the thickness of the deposited semiconductor film. The process to completely fill the pore consists of three phases. In phase I, there is little or no constriction of the pore, and semiconductor deposition has little impact on fluid flow. In phase II, the narrowing of the pore cross-section produces supersonic flow beyond the constriction at the high operating pressures of about 35 MPa. The coupling between flow and temperature-dependence of the rate of deposition gives rise to an instability that eventually leads to plugging of the pore. In phase III, the pore is largely or completely closed, and a large axial concentration gradient arises within the pore as silica selectively filters silane from helium and hydrogen, due to its permeability to the latter components at high temperature. The precursor filtering effect favors deposition several centimeters upstream of the closed end of the pore, leading to complete filling of the large aspect-ratio pore.

See more of this Session: Gas Phase Deposition
See more of this Group/Topical: Materials Engineering and Sciences Division

(530e) Initiated Chemical Vapor Deposition of Polymer Thin Films for Photolithography Applications

Christy D. Petruczok and Karen K. Gleason, Chemical Engineering, Massachusetts Insitute of Technology, Cambridge, MA

The use of initiated chemical vapor deposition (iCVD) to synthesize thin films of poly(4-vinylpyridine) (P4VP) is reported. The films are functionalized post-deposition with carboxylic acid-containing diacetylene monomers. Exposure of the functionalized films to UV light (λ=254 nm) polymerizes the diacetylene, making the exposed regions insoluble in common organic solvents.

Functionalization can be achieved by dissolving the diacetylene monomer in toluene and using a dip-coating process. Fourier Transform Infrared spectroscopy confirms the presence of hydrogen bonding between the carboxylic acid moieties and the nitrogen atoms of the pyridine rings. For P4VP films with thicknesses less than 35 nm, sublimation can be used; use of a solvent-less process results in improved pattern resolution and uniformity and reduction of the polydiacetylene crystalline domains observed as a result of the dip-coating process. Polymerization of the diacetylene is verified using UV-visible spectroscopy. Patterns with features smaller than 5 μ m have been defined on planar substrates using a contact photolithography mask.

The use of iCVD affords some unique attributes to the photosensitive film. In this process, a thermally-labile gas phase initiator is decomposed into radicals as it travels through an array of heated filaments. Free radical polymerization occurs as the initiator radicals and monomer adsorb onto a cooled substrate located beneath the filament array. iCVD is a low-energy technique that results in excellent control over film thickness and conformality. The library of iCVD polymers is substantial and includes a wide range of chemical functionalities, as well as several insoluble or infusible polymers that cannot be easily synthesized using other methods.

The conformality attainable using iCVD makes it possible to pattern micro-scale features on a variety of non-planar surfaces. Glass rods (D=2-6 mm) were coated with iCVD P4VP and functionalized using the methods described above. TEM grids (Cu, 45 μ m parallel bars, spacing 80 μ m) were used as contact masks; the films were exposed for three minutes, then developed in ethanol. The resulting patterns were well-defined—the 2 mm rod exhibited the entirety of the grid pattern (wrapped halfway around its circumference), with excellent agreement observed between the mask and the patterned polymer.

This process can also be used to develop multifunctional surfaces. 200 nm of iCVD P4VP was deposited on top of 35 nm of iCVD poly(1H,1H,2H,2H-perfluorodecyl acrylate) (PPFDA). Functionalization of the P4VP, followed by application of a mask, UV exposure, and developing, yielded a patterned surface with regions of contrasting hydrophobicity. This method is facilitated by the fact that the unexposed P4VP is removed during the developing process, which utilizes a non-solvent for the underlying PPFDA layer.

See more of this Session: Gas Phase Deposition
See more of this Group/Topical: Materials Engineering and Sciences Division

(531a) Solution-Based Assembly of Large-Scale, Unidirectionally-Aligned Carbon Nanotubes for High-Performance Transistors

Guihua Yu[1], Melburne C. LeMieux[1], Benjamin C-K. Tee[2], Eric S.G. Shaqfeh[3] and Zhenan Bao[1], (1) Chemical Engineering, Stanford University, Stanford, CA, (2)Electrical Engineering, Stanford University, Stanford, CA, (3)Chemical Engineering and Mechanical Engineering, Stanford University, Stanford, CA

Central to many proposed electronic device-based applications of carbon nanotubes (CNTs) is the development of assembly methods that enable organization of these nanoscale building blocks over large areas with controlled orientation and density. In this talk, we will present a novel solution-based technique that combines spin assembly and molecular patterning for preparing large-scale, ultrathin CNT films with unidirectional alignment and controlled density. Guided by our simulation studies of particle adsorption to investigate both effects of the hydrodynamic force (spin rate) and surface interaction force (CNT-molecules), we have assembled well-ordered, unidirectionally-aligned CNTs over the wafer scale and directly integrated them as thin-film transistors (TFTs) that show promising device characteristics with large on/off ratio and high mobility. The scalable solution-based process for large-scale assembly of well-aligned CNT thin films can lead to many exciting applications ranging from large-scale, highly sensitive biological/chemical nanosensors to large-area flexible displays and solar cells.

See more of this Session: Graphene and Carbon Nanotube Based Devices
See more of this Group/Topical: Materials Engineering and Sciences Division

(531b) Inkjet Patterned Carbon Nanotube Multilayer Devices

Christine M. Andres and Nicholas A. Kotov, Chemical Engineering, University of Michigan, Ann Arbor, MI

Layer-by-layer assembly (LBL) of carbon nanotubes serves as a method for the creation of unique combinations of electrical, mechanical and optical material properties allowing for flexible electronic applications with integrated strength and conductivities. Inkjet printing serves as a continuous direct-write approach to patterning LBL films for the simple, low-cost integration into advanced functional devices. The strictly additive approach allows for the incorporation of a variety of different materials and substrates for a versatile, multicomponent and noncontact patterning technique. Here we will demonstrate the use of this technique for the development of a wide range of application from electrical actuators as building blocks for complex micromechanical devices to low cost paper sensors for rapid detection and diagnostics.

See more of this Session: Graphene and Carbon Nanotube Based Devices
See more of this Group/Topical: Materials Engineering and Sciences Division

(531c) Structural Stability of Transparent Conducting Films Assembled From Single-Wall Carbon Nanotubes Purified by Electronic Type

John M. Harris[1], Jeffrey A. Fagan[2], Steven D. Hudson[3], Christopher M. Stafford[2] and **Erik K. Hobbie**[1], (1)NDSU, Fargo, ND, (2)Polymers Division, National Institute of Standards and Technology, Gaithersburg, MD, (3)NIST, Gaithersburg, MD

Strain-induced structural changes in thin polymer-supported membranes of purified single-wall carbon nanotubes (SWCNTs) are evaluated through the wrinkling instabilities that develop under uniaxial and isotropic compression. Nanotubes that have been separated by electronic type using density-gradient ultracentrifugation are assembled as surfactant-free films on prestrained polydimethylsiloxane (PDMS) substrates, and the strain response is measured using a broad range of techniques. The small-strain behavior is inferred from kinetic changes in the wrinkling topography during the slow drying of pre-swelled polymer supports. Our measurements reveal a remarkable degree of strain softening that couples strongly to the sheet resistance of the membranes through microscale anisotropy. By depositing comparable films on quartz, we use optical (UV-Vis-NIR) absorption spectroscopy to compute the London dispersion spectra of the purified materials, and from these we compute the attractive part of the van der Waals potential between nanotubes of identical electronic type. We find significant differences in the strength of the contact potential between metallic vs. semiconducting nanotubes, which in turn are evident in the modulus and yield strain of the films. Supported by the NSF through CMMI-0969155.

See more of this Session: Graphene and Carbon Nanotube Based Devices
See more of this Group/Topical: Materials Engineering and Sciences Division

RESONANT QUANTUM TUNNELING IN CARBON-NANOTUBE-BASED DEVICES

Meng-Mu Shih, University of Florida, Gainesville, FL

Abstract

This work discusses CNT-based electronic devices by using the resonant quantum tunneling in double-barrier quantum-well structures. This quantum model and numerical results demonstrate how the CNT diameter, quantum well width, barrier height, and barrier thickness affect the transmission probability through the double-barrier resonant tunneling structure. The energy of tunneling electrons needs to match with one of the resonant energies existed in the resonant tunneling quantum well in order to achieve sharp resonance and high transmission probability, which ideally approaches to unity. The number and the energy levels of resonant energy states can be affected and tuned by the CNT-based quantum structure. These results can be used to further explore applications such as in resonant tunneling diodes.

Model Description

Figure 1 shows the schematic structure for the resonant quantum tunneling [1]. This figure shows the cross-section view of a double-barrier quantum-well formed by semiconducting carbon nanotubes (CNTs). Each barrier can be formed by the junction of two crossed CNTs [2]. Key geometrical parameters in this model are CNT diameter (d), quantum well width (w), and barrier thickness (a). Physical parameters are energy (E), barrier height (Vo), and CNT effective mass (m_e) [3,4]. If the tunneling electron energy approaches tone of the resonant energies existed in the quantum well, the transmission probability or transmission probability (T_2) can approach to unity. With proper design, this T2 can have a very sharp response to the tunneling electron energy.

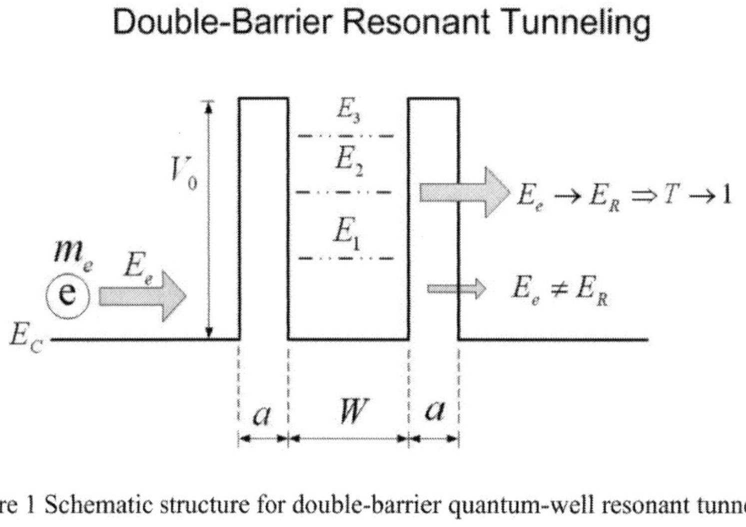

Figure 1 Schematic structure for double-barrier quantum-well resonant tunneling

Numerical Results and Discussion

The calculation results are based on the resonant model [1-5] described in Figure 1. Figure 2 shows the transmission coefficients versus the energy of tunneling electrons for CNT diameter d = 1 nm and 10 nm, when barrier height Vo = 0.3 eV, quantum well width w = 12 nm, barrier thickness a = 2.5 nm. Since the 10 nm CNT has smaller effective mass, the tunneling is easier and the overall tunneling probability is larger than the 1nm CNT. However, the 1 nm CNT has shaper response to the energy of tunneling electrons and will be used in the Figures 3-5 in order to demonstrate how the parameters affect the transmission probability. Figure 3 shows the transmission coefficients versus the energy of tunneling electrons for two quantum well width w = 12 nm and 6 nm, when CNT diameter d = 1 nm, barrier height Vo = 0.3 eV, and barrier thickness a = 2.5 nm. Smaller well width can have higher resonant energy in the well. Figure 4 shows the transmission coefficients versus the energy of tunneling electrons for two barrier height Vo = 0.3 eV and 0.15 eV, when CNT diameter d =1 nm, quantum well width w = 12 nm, and barrier thickness a = 2.5 nm. Smaller barrier height can reduce the number of resonant energy states. Figure 5 shows the transmission coefficients versus the energy of tunneling electrons for two barrier width a = 0.25 nm and 0.65 nm, when CNT diameter d = 1nm, quantum well width w = 12nm, and barrier height Vo = 0.3 eV. The larger barrier thickness will make the transmission probability curve has shaper peaks at the resonant energy levels. Since the thicker barrier will make the tunneling difficult, the electron energy needs to be more specific and closer to resonant energy in order to tunnel trough this structure.

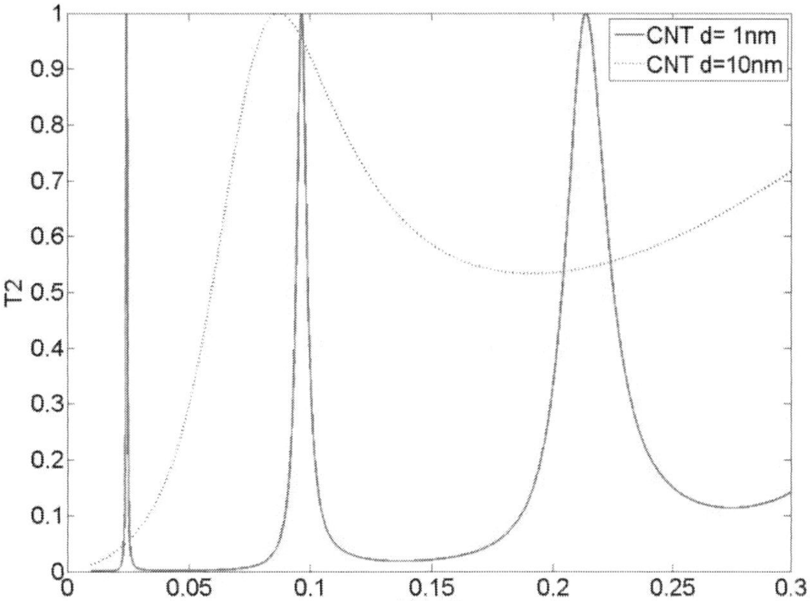

Figure 2 Transmission coefficients versus energy of tunneling electrons

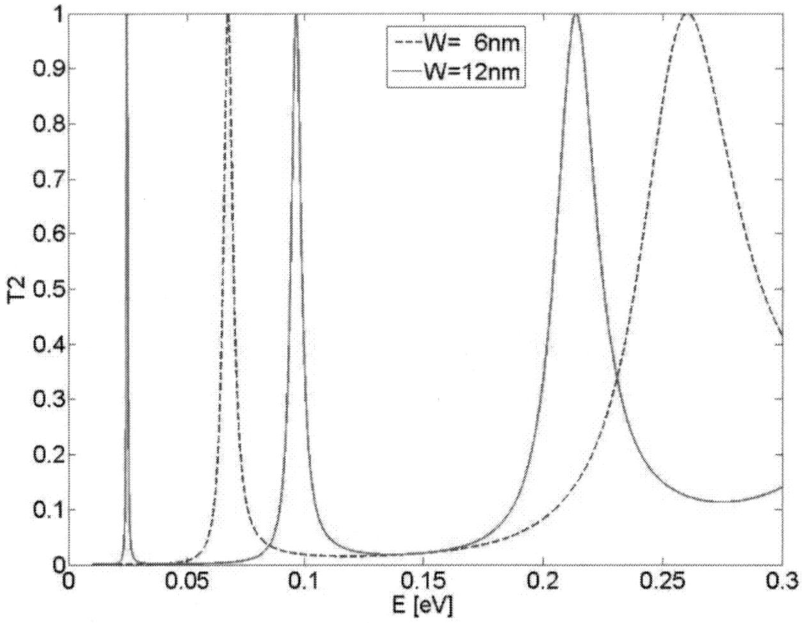

Figure 3 Transmission coefficients versus energy of tunneling electrons

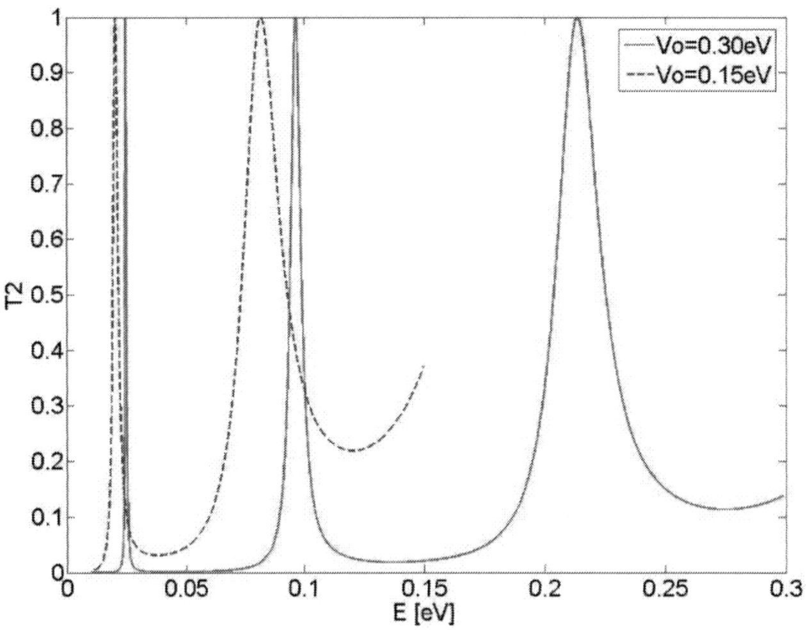

Figure 4 Transmission coefficients versus energy of tunneling electrons

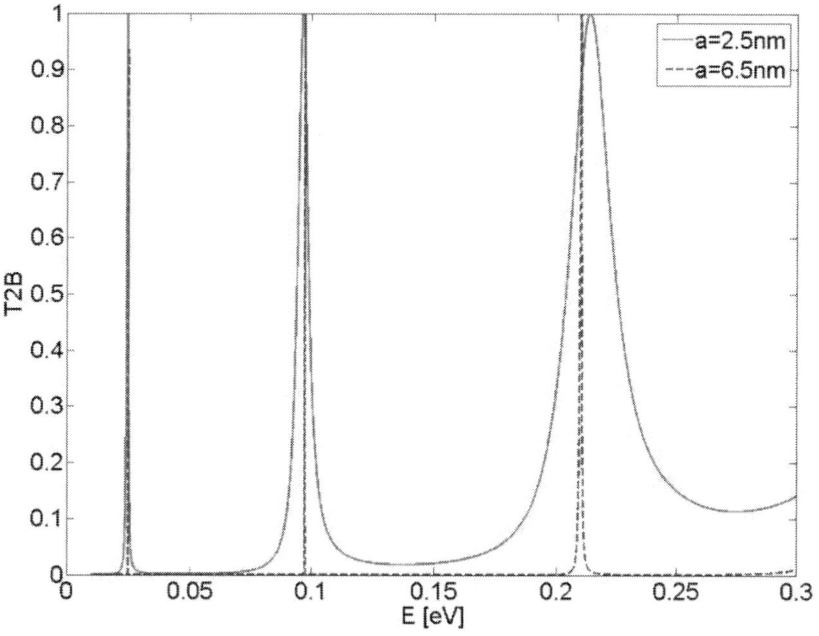

Figure 5 Transmission coefficients versus energy of tunneling electrons

References

1. Esaki, Leo (1973), "Long Journey to Tunneling," http://www.nobelprize.org/nobel_prizes/physics/laureates/1973/esaki-lecture.pdf
2. Biswas, Sujit et al. (2005), "Room-temperature Resonant Tunneling of Electrons in Carbon Nanotube Junction Quantum Wells," *Applied Physics Letters*, 86, 183101, pp. 01-03.
3. Pennington, Gary and Goldsman, Neil (2003), "Theory and Design of Field-Effect Carbon Nanotube Transistor," *International Conference on Simulation of Semiconductor Processes and Devices*, pp.167-170.
4. Marulanda, Jose M. and Srivastava, Ashok (2008), "Carrier Density and Effective Mass Calculations in Carbon Nanotubes," *Physica Status Solidi*, (b) Vol. 245, No. 11, pp. 2558–2562.
5. Singh, Jasprit (2004), "Modern Physics for Engineers," Weinheim, Germany, WILEY-VCH Verlag GmbH & KGaA.

QUANTUM TUNNELING IN CARBON-NANOTUBE TRANSISTORS

Meng-Mu Shih, University of Florida, Gainesville, FL

Abstract

A modified model is used to discuss the quantum tunneling effects in carbon nanotube (CNT) transistors. In this work, CNTs with nano-meter diameters are studied. This modified prototype model can consider several CNT-related parameters such as CNT diameter, applied electrical field, effective mass, tunneling barrier height affected by metal contacts or strain, and tunneling width tuned by dopants. Numerical results will be discussed for design and performance issues of transistors. Some of the results show the possibilities of obtaining high currents more easily. However, some currents are not desirable or should be kept as low as possible. The on/off tunneling ratio can evaluate the transistor performance in circuit design. Some further results can be used to explain the working principles of CNT-transistor sensors.

Model Description

CNT–based transistors are proposed and modeled [1]. Figure 1 shows the schematic structure for the quantum tunneling at the metal-semiconductor interface [2, 3]. The modified CNT-based quantum tunneling model has parameters: CNT diameter (d), effective mass (m_e), barrier height (ϕ_b), tunneling width (w), and electrical field (E). The CNT effective mass can be calculated [4, 5]. The barrier height is assumed by mid-gap type of CNT material.

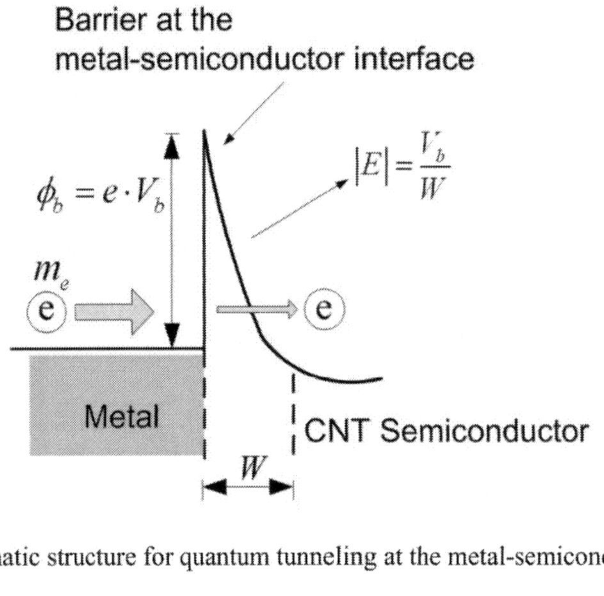

Figure 1 Schematic structure for quantum tunneling at the metal-semiconductor interface

Numerical Results and Discussion

The calculation results are based on the quantum tunneling model described in Figure 1. Figure 2 shows the tunneling probability (T) versus CNT diameter (d) for electrical field E = 5×10^9 V/m and 10^9 V/m. Since the CNT with larger diameter has smaller tunneling barrier height and smaller effective mass, the tunneling is easier and the tunneling probability is higher than the CNT with smaller diameter. The larger electrical field can assist the tunneling and cause higher tunneling probability. However, the CNT with larger diameter can have larger leakage tunneling. Figure 3 shows the tunneling probability (T) versus electrical field (E) for CNT with diameter d = 10 nm and 2 nm. The CNT with larger diameter can reach tunneling probability of unity faster with smaller electrical field. Figure 4 shows the tunneling probability (T) versus tunneling width (w) for CNT with diameter d = 1, 2, 4, 8 nm. This tunneling width can be affected by doping. The smaller tunneling width can make electrons easier to tunnel through and cause the tunneling probability to be higher. The CNT with larger diameter can make quantum tunneling much easier.

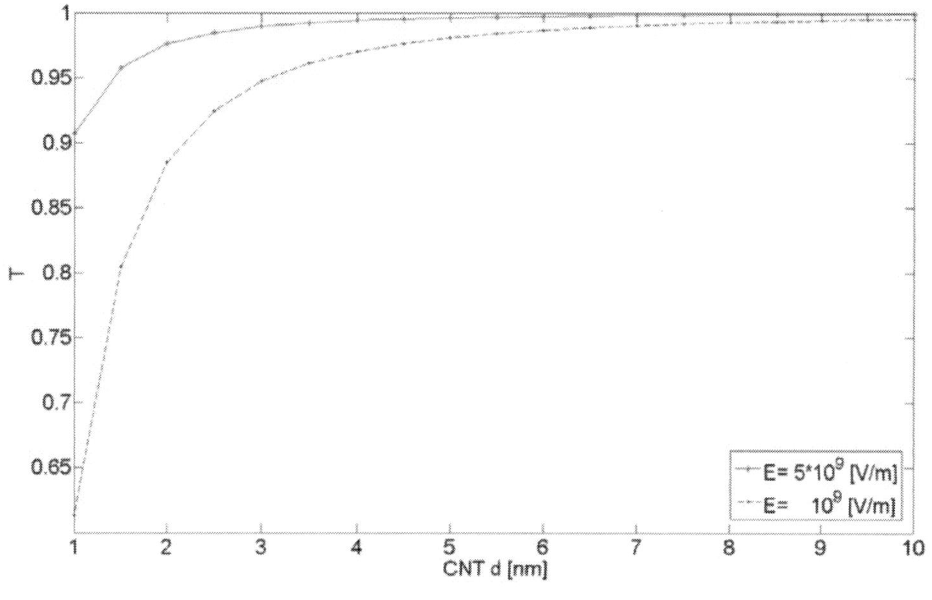

Figure 2 Tunneling probability (T) versus CNT diameter (d) for electrical field E = 5×10^9 V/m and 10^9 V/m

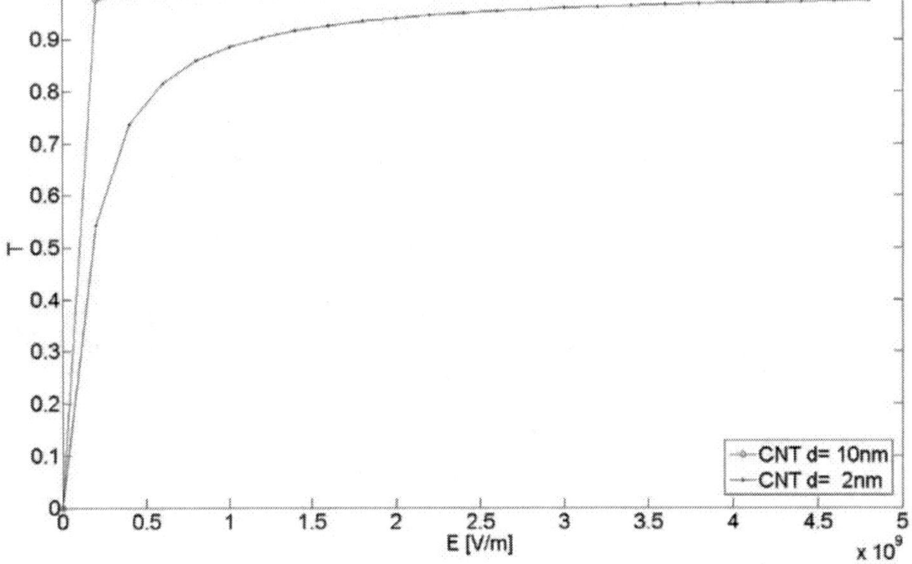

Figure 3 Tunneling probability (T) versus electrical field (E) for CNT with diameter d = 10 nm and 2 nm

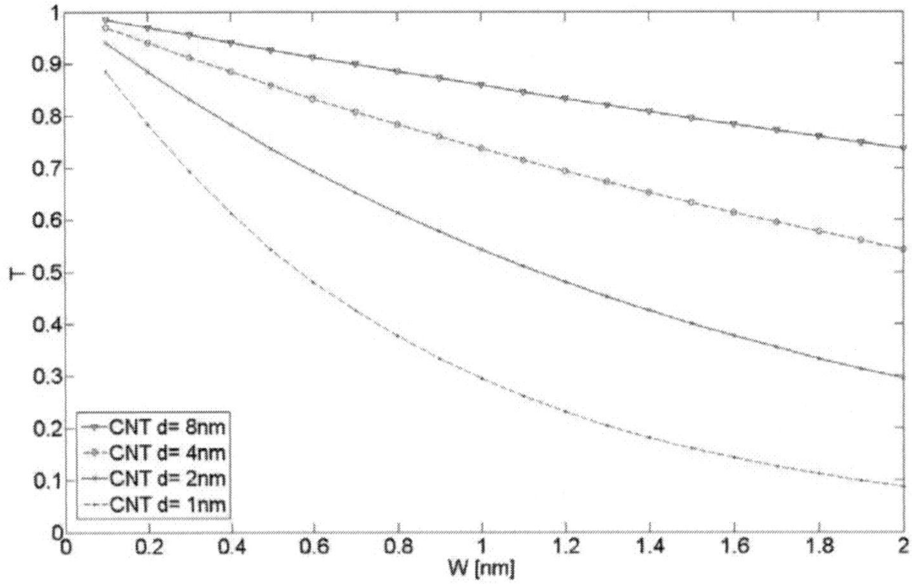

Figure 4 Tunneling probability (T) versus tunneling width (w) for CNT with diameter d = 1, 2, 4, 8 nm.

References

1. Guo, Jing et al. (2004), "A Numerical Study of Scaling Issues for Schottky-Barrier Carbon Nanotube Transistors," *IEEE Transaction on Electron Devices*, Vol. 51, No. 2, February, pp. 172-177.
2. Esaki, Leo (1973), "Long Journey to Tunneling," http://www.nobelprize.org/nobel_prizes/physics/laureates/1973/esaki-lecture.pdf
3. Singh, Jasprit (2004), "Modern Physics for Engineers," Weinheim, Germany, WILEY-VCH Verlag GmbH & KGaA.
4. Pennington, Gary and Goldsman, Neil (2003), "Theory and Design of Field-Effect Carbon Nanotube Transistor," in *International Conference on Simulation of Semiconductor Processes and Devices*, pp.167-170.
5. Marulanda, Jose M. and Srivastava, Ashok (2008), "Carrier Density and Effective Mass Calculations in Carbon Nanotubes," *Physica Status Solidi*, (b) Vol. 245, No. 11, pp. 2558–2562.

Graphene-Based Molecular-Machine: Reversible and Robust Carrier Doping in Graphene *via* Mechanical Actuation of Tethered Azobenzene

*Phong Nguyen, Kabeer Jasuja, Nihar Mohanty, and Vikas Berry**

Department of Chemical Engineering, Kansas State University

1005 Durland Hall, Kansas State University, Manhattan, KS 66506

*vberry@ksu.edu

ABSTRACT

Molecular machines – molecules capable of mechanically responding (isomerizing) to external stimuli – are important components of molecular-actuation-based nanodevices. Now we demonstrate that ultrathin-graphene's sensitivity to interfaced-molecules' orientation, surface-vicinity, electronegativity, and density-of-states, can be employed to reversibly dope carriers in it *via* the mechanics of molecular machines tethered on its surface. Here, graphenic sheets are functionalized with electronegative and isomerizable azobenzene-molecules, which are optically transformed from their *trans* conformation (π–cloud-head ~ 1.1 nm away from graphene-surface) to *cis* conformation (π–cloud-head ~ 0.7 nm from graphene). This 0.42 nm displacement of the azo-head dopes 7.5 X 10^3 holes/μm^2 in the underlying graphene. We determine that ~20 azobenzene molecules produce 1 hole (hole-mobility of 301 $\mu m^2/V/s$). Due to their sub-nanoscale actuation, the molecular-machines are typically interfaced within nanoscale openings (nanopores or tunneling-junctions), which are challenging to fabricate. Here, we show that macroscale, sensitive and easy-to-fabricate graphene platform to achieve molecularly-actuated electrical-response. To compare azobenzene with directly doped graphene, we π–functionalized graphene with electronegative perylene tetracarboxylic acid (PCA), which led to ~3 fold increase in the hole density.

(531g) Impermeable Graphenic Wrapping of Bacteria

Nihar Mohanty[1], Ashvin Nagaraja[2], Monica Fahrenholtz[1], Daniel L. Boyle[3] and **Vikas Berry**[1], (1) Chemical Engineering, Kansas State University, Manhattan, KS, (2)Electrical Engineering, Kansas State University, Manhattan, KS, (3)Division of Biology, Kansas State University, Manhattan, KS

Transmission electron microscopy (TEM) of hygroscopic, permeable, and electron-absorbing biological cells has been an important challenge due to the volumetric-shrinkage, electrostatic charging, and structural degradation of cells under high vacuum and fixed electron beam. In this talk we would demonstrate that the bacterial cells can be encased within a graphenic chamber to preserve their dimensional and topological characteristics under high vacuum (10-5 Torr) and beam current (150 A/cm2). The strongly-repelling p-clouds in the interstitial sites of graphene's lattice reduces the graphene-encased-cell's permeability from 7.6 - 20 nm/s to 0 nm/s. The C-C bond flexibility enables conformal, hermetic encasement of cells. Additionally, graphene's high Young's modulus retains cell's structural integrity under TEM conditions, while its high electrical and thermal conductivity significantly abates electrostatic-charging. This graphenic encasement approach will facilitate real-time TEM imaging of fluidic samples and potentially biochemical activity.

See more of this Session: Graphene and Carbon Nanotube Based Devices
See more of this Group/Topical: Materials Engineering and Sciences Division

(536a) Molecular-Ink Route to Thin Film CZTSSe Solar Cells

Hugh W. Hillhouse, School of Chemical Engineering, Purdue University, West Lafayette, IN and Wooseok Ki, Chemical Engineering, University of Washington, Seattle, WA

Given the terawatt scale of future energy needs, the most promising future photovoltaic materials should be Earth abundant with their primary mineral resources distributed across several geographic regions and their supply chains robust to reduce concerns of price volatility. In addition, the process of forming the solar cell should be scalable, low-cost, and not utilize dangerous or toxic materials. The strongest initial candidate appears to be Cu_2ZnSnS_4 (CZTS), but there are other potential material systems such as FeS_2, CuO, and PbS that are just beginning to be re-examined with solution phase processing. Most solution-phase CZTS thin film solar cells synthesized to date were obtained from deposition of metals (Cu, Zn, & Sn) or metal precursors first, followed by sulfurization with H_2S. More recently, two approaches that form the chacogenide directly (CZTS) have yield substantially higher efficiency devices from solution processing (even higher than the best vacuum deposited absorbers). Both new approaches incorporate Se to form $Cu_2ZnSn(S_x,Se_{1-x})_4$ (CZTSSe). Power conversion efficiencies have reached 7.2% for a nanocrystal-ink route and 9.6% for a route that uses molecular precursors dissolved in hydrazine, which is toxic and carcinogenic. Here we report a new, facile, and scalable chemical route to Earth abundant element thin film solar cells by coating a solution of highly soluble, inexpensive, and commercially available precursors in an environmentally friendly non-toxic solvent to form device quality films. Air-stable CZTS photovoltaic devices with 4.1% total area power conversion efficiency are obtained. We have generalized the chemical route and have used it to also fabricate PbS devices that are 1.5% efficient. The importance of this new route is that it: (1) avoids toxic solvents, (2) has a metal yield of 100% (as compared to 70-90% from nanocrystal-ink routes), (3) has no liquid wastes, and (4) yields good electronic quality absorber material.

See more of this Session: Materials and Architectures for Thin-Film Photovoltaics
See more of this Group/Topical: Materials Engineering and Sciences Division

(536b) Copper Zinc Tin Sulfide Solar Cells

Ankur Khare[1], Yulong Li[2], B. Selin Tosun[1], David J. Norris[1] and Eray S. Aydil[1], (1)Chemical engineering and materials science, University of Minnesota, Minneapolis, MN, (2)Physics, University of Minnesota, Minneapolis, MN

Although thin film solar cells based on copper indium gallium diselenide (CIGS) and CdTe have already achieved impressive power conversion efficiencies ranging between 15 and 20%, the semiconductor materials commonly used for their production are either toxic (e.g., cadmium) or rare in the earth's crust (e.g., indium, tellurium). Copper zinc tin sulfide (Cu_2ZnSnS_4; commonly known as CZTS) is emerging as a photovoltaic material composed of elements that are nontoxic and abundant. CZTS has a band gap of ~1.4 eV, the ideal value for converting the maximum amount of energy from the solar spectrum into electricity, and a high absorption coefficient (> 10^4 cm^{-1} in the visible region of the electromagnetic spectrum). We are exploring various methods for depositing thin CZTS films from liquid precursors (e. g., "inks"). These CZTS "inks" can be used in conjunction with high throughput roll-to-roll deposition techniques such as slot coating under atmospheric conditions and may offer alternative low-cost manufacturing routes to making thin film CZTS solar cells.

Specifically, we have developed a novel and facile synthesis method for making CZTS nanocrystals and stable colloidal nanocrystal dispersions. Our nanocrystals were synthesized from copper, zinc, and tin diethyl dithiocarbamate complexes dissolved in octadecene. Presence of oleylamine in octadecene reduces the decomposition temperature of these metal dithiocarbamate complexes to a narrow temperature range (170-220 °C). Thus, nucleation and growth of phase-pure CZTS nanocrystals can be initiated by rapidly injecting oleylamine into a hot solution of tin, zinc and copper metal complexes in octadecene. The nanocrystal size can be controlled between 2 to 7 nm by changing the synthesis temperature and time. We also used colloidal dispersions of these nanocrystals to cast thin (0.5-2 micron) CZTS films. Heating these films in argon atmosphere to temperatures ranging from 500-700 °C forms continuous CZTS layers with larger grains than the starting nanocrystals. We found that the CZTS nanocrystals melt significantly below the bulk melting temperature. Larger grains grow upon recrystallization. This process of forming CZTS films is fast compared to other techniques, which require several hours of annealing for grain growth. Solar cells based on these CZTS films will be reported.

See more of this Session: Materials and Architectures for Thin-Film Photovoltaics
See more of this Group/Topical: Materials Engineering and Sciences Division

(536c) Combinatorial CdZnS Thin Films Deposited with a Continuous Flow Microreactor

Kevin M. McPeak, ETH- Zurich, Zurich, Switzerland and **Jason B. Baxter**, Department of Chemical and Biological Engineering, Drexel University, Philadelphia, PA

Chemical bath deposited CdS thin films are commonly used as buffer layers in CdTe and CIGS photovoltaics because they form a high quality *p-n* junction with the absorber. However, light absorption by the CdS reduces solar cell efficiency. $Cd_{1-x}Zn_xS$ has a wider band gap than CdS, offering the potential to reduce deleterious absorption of light in the 300-550 nm range and improve current densities by over 2 mA/cm^2. To maximize energy conversion efficiency, the $Cd_{1-x}Zn_xS$ should be a single-phase ternary alloy and the stoichiometry should be optimized to ideally position the conduction band edge for increased transmissivity while retaining good junction properties.

We report on chemical bath deposition of combinatorial $Cd_{1-x}Zn_xS$ thin films using a continuous flow microreactor. The microreactor uses a sub-millimeter reaction channel, with the substrate acting as one reactor wall. The microreactor behaves like a plug flow reactor whereby the bath composition changes as material is deposited on the substrate. While the bath composition changes with respect to position down the reaction channel, the composition at any position is time-invariant. The graded bath composition results in deposition of graded thin films whose stoichiometry and optoelecronic properties change significantly over length scales of millimeters to centimeters. Spatially resolved characterization of the substrate enables rapid and direct correlation of material properties to growth conditions, which is not possible with a batch reactor where bath composition changes with time.

Graded $Cd_{1-x}Zn_xS$ films grown with 1:200 Cd:Zn bath composition at the inlet had composition that varied from x=0.0 to x=0.17 over a distance of 12 mm on a single substrate. Film stoichiometry was determined by x-ray photoelectron spectroscopy (XPS) mapping. Stoichiometry changes because of differences in speciation and reaction kinetics of the cation species as they flow down the channel. XPS and XRD confirm that CdZnS is a single phase material. Deep level emission in photoluminescence and XPS indicate that O and OH is also incorporated into the film and is bound to Zn, with amount increasing further down the reaction channel. The graded stoichiometry causes the absorption edge to blue-shift by over 80 nm, as determined by UV-vis transmission and reflectance measurements. Blue-shifting band edge and changes in defect density are also seen by photoluminescence. The continuous flow microreactor offers new potential for deposition of graded thin films that act as combinatorial libraries for high throughput screening and accelerated materials discovery.

See more of this Session: Materials and Architectures for Thin-Film Photovoltaics
See more of this Group/Topical: Materials Engineering and Sciences Division

(536d) Sputter Deposition of Semi-Crystalline Tin Dioxide Films for CIGS Solar Cells

B. Selin Tosun[1], Rebekah K. Feist[2], Stephen A. Campbell[3], Eray S. Aydil[4], Aloysius Gunawan[5] and K. Andre Mkhoyan[5], (1)Chemical Engineering and Materials Science, University of Minnesota, Minneapolis, MN, (2)Dow Solar Solutions, The Dow Chemical Company, Midland, MI, (3)Department of Electrical and Computer Engineering, University of Minnesota, Minneapolis, MN, (4)Department of Chemical Engineering and Materials Science, University of Minnesota, Minneapolis, MN, (5)Materials Science and Engineering, University of Minnesota, Minneapolis, MN

Tin dioxide (SnO_2) is emerging as an important material for use in copper indium gallium diselenide (CIGS) based solar cells. Amorphous SnO_2 may be used as a glass overlayer for the entire device and protecting it against water permeation. SnO_2 is also a viable semiconductor candidate to replace the wide band gap zinc oxide (ZnO) window layer to improve long-term device reliability. The film properties required by these two applications are different. Amorphous films have superior water permeation resistance while polycrystalline films generally have better charge carrier transport properties. Thus, it is important to understand how to tune the structure of SnO_2 films between amorphous and polycrystalline. Using X-ray diffraction and Hall-effect measurements, we have studied the structure and electrical properties of SnO_2 films deposited by radio frequency (RF) magnetron sputtering as a function of deposition temperature, sputtering power, feed gas composition and film thickness. Films deposited at room temperature and thinner than 200 nm tend to be amorphous. Film crystallinity increases with film thickness and deposition temperature but is not affected significantly by sputtering power. Films with resistivities ranging between 20 mΩ cm to 800 mΩ cm are deposited. The films are n-type with carrier concentrations in the 3´10^{18} cm^{-3} to 3´10^{20} cm^{-3} range. Carrier concentration decreases with oxygen concentration in the feed gas. Electron mobilities range from 1 to 10 cm^2/V s and increase with increasing film thickness, oxygen addition to the feed gas and film crystallinity. Electron mobilities in the 1-3 cm^2/V s range are obtained even in amorphous films. When we apply these differently deposited SnO_2 layers for the CIGS solar cell damp-heat performance studies. Approximately 0.2 micron and thicker amorphous tin dioxide layers deposited on top of the completed CIGS solar cells can significantly increase the device lifetime by forming a barrier against water diffusion. The SnO_2 overlayer protect the ~ 93 % of the initial solar cell efficiency even after 150 hours of damp-heat treatment (85 °C/85 % RH) when the solar cells without the SnO_2 layer lost nearly 80 % of their initial efficiency within 24 hours of commencing the test. The replacement of water sensitive and water permeable ZnO window layer of CIGS solar cells is provided with an amorphous or polymorphous (mixture of amorphous and polycrystalline material) tin dioxide layer. Using identical CIGS layer and fabrication, solar cells made with ZnO or SnO_2 window layers give similar overall power conversion efficiencies. We demonstrate an 8.3 % efficient CIGS solar cell with a SnO_2 window layer. Same solar cell fabrication process and CIGS film with ZnO window layer resulted in 8.7 % overall power conversion efficiency. The open circuit voltages of the two cells were the same indicating that the band alignment with the SnO_2 film is suitable for CIGS.

See more of this Session: Materials and Architectures for Thin-Film Photovoltaics
See more of this Group/Topical: Materials Engineering and Sciences Division

(536e) A Novel Polymer-Assisted Hydrothermal Approach to Metal-Oxide Thin Films

Qianglu Lin, Yun Xu, Ling Fei, Joshua Hill, Shuguang Deng and Hongmei Luo, Chemical Engineering Department, New Mexico State University, Las cruces, NM

A Novel Polymer-Assisted Hydrothermal Approach to Metal-Oxide Thin Films

Qianglu Lin, Yun Xu, Ling Fei, Joshua Hill, Shuguang Deng, Hongmei Luo

Department of Chemical Engineering, New Mexico State University, Las Cruces, NM88003

Metal-oxide thin films exhibit a wide range of electronic, magnetic, optical, and thermal properties that conventional metallic elements and semiconductors do not possess. Numerous methods like molecular beam epitaxy, sputtering, pulsed-laser deposition, sol-gel, and polymer-assisted deposition have been used to deposit epitaxial oxide films. Crystalline anatase or rutile titanium dioxide (TiO_2) have low band gap (3.2 eV), high photocatalytic activity, and good bioactivity. We will report epitaxial anatase TiO_2 and rutile TiO_2 films on (001) $LaAlO_3$ and R-cut sapphire prepared by a novel chemical solution approach, polymer-assisted hydrothermal method. It provides a facile way to prepare epitaxial films at low temperature of 200 °C with varied film thickness control (from 10 nm to 2 μm). The structure and morphologies of thin films are studied by X-ray diffraction (XRD), scanning electron microscopy (SEM) and transmission electron microscopy (TEM) analyses.

See more of this Session: Materials and Architectures for Thin-Film Photovoltaics
See more of this Group/Topical: Materials Engineering and Sciences Division

(536f) Enhancement of Light Harvesting In Flexible Photovoltaics

Ying Liu[1], Michelle Casper[2], Omer Gozen[2], Sharvil Desai[3], Ethan Klem[4], Jay Lewis[4], Jan Genzer[1], Jon-Paul Maria[2] and **Michael D. Dickey**[1], (1)Chemical and Biomolecular Engineering, North Carolina State University, Raleigh, NC, (2)Materials Science and Engineering, North Carolina State University, Raleigh, NC, (3)North Carolina State University, Raleigh, NC, (4)Research Triangle Institute, Durham, NC

This talk describes a simple, low-cost technique to improve light harvesting in thin-film photovoltaics. Spontaneous surface buckling is a low-cost strategy for producing hierarchically corrugated surfaces while avoiding lithographic patterning and is attractive for use in mechanically flexible, thin-film technologies. Our studies use organic photovoltaics (OPVs) as model systems and our preliminary results suggest that thin film photovoltaic efficiencies experience improvements when prepared on corrugated topographies, which is supported by recent theories of light trapping in solar cells. Thus, understanding the origins of buckling instabilities and controlling such topographies is of importance for the engineering of functional textured surfaces. We investigate the origins of buckling instabilities in indium tin oxide (ITO) – polymer multilayer systems and explore strategies for controlling the morphology with the potential for reversibility and tunability. We identify the key processing parameters and also identify several origins of stress generation. By varying the thickness of the ITO and polymer layers we can prepare buckled topographies with characteristic wavelengths of 2-12 μm. We have deposited known absorber materials onto both buckled and planar surfaces and investigated the change in absorbance. Preliminary results of organic solar cells fabricated on buckled ITO/PS indicate a 20% greater power conversion efficiency compared to a comparable planar device.

See more of this Session: Materials and Architectures for Thin-Film Photovoltaics
See more of this Group/Topical: Materials Engineering and Sciences Division

(536g) Room Temperature, Aqueous Deposition of Si Thin Films

Ian Ivar Suni, Department of Chemical and Biomolecular Engineering, Clarkson University, Potsdam, NY and Aarti Krishnamurthy, Materials Science and Engineering, Clarkson University, Potsdam, NY

We report combined galvanic and electroless deposition of Si onto Al alloys from 10 mM HF and 20 mM Na_2SiF_6 in 80 wt% formic acid. Exposed Al acts as a reducing agent for Si deposition, but Al alone produces only nanoporous Si deposits. Several low pH reducing agents were tested, but only formic acid allows deposition of compact Si thin films, which grow to 7-10 μm thickness after 30 hr. These Si films contain significant Cu and Al contamination, both arising from the Al alloy substrate. Compact Ge films can also be deposited from similar solutions. Possible methods for obtaining photovoltaic grade thin film Si and Ge will be discussed. Studies by voltammetry, electrochemical impedance spectroscopy (EIS), and spectroscopic ellipsometry will also be discussed.

See more of this Session: Materials and Architectures for Thin-Film Photovoltaics
See more of this Group/Topical: Materials Engineering and Sciences Division

(536h) Interplay Among Shape and Magnetic Properties of Core-Shell FePt-MgO Nanomagnets for Spin-Torque Transfer Memory Devices

Domingo Ferrer, Samaresh Guchhait, Andres Guevara, Helen Xu, Urmimala Roy and Sanjay Banerjee, Microelectronics Research Center, University of Texas at Austin, Austin, TX

Colloidal assemblies of FePt nanostructures surrounded by crystalline MgO provide a solution-phase, highly-scalable method to create individual atom-sized magnetic bits for spintronics devices such as Magnetic Random Access Memories (MRAMs) or Spin Transfer Torque Random Access Memory (STTRAM). FePt alloys are particular promising candidates for ultrahigh density recording media owing to their good chemical stability and high magnetocrystalline anisotropy, leading to a superparamagnetic limit as small as 3 nm, such that they are thermally stable over typical data storage periods of 10 years.[1] Conformal deposition of crystalline MgO tunnel dielectrics on FePt nanostructures has been achieved. Low resistance-area values obtained with MgO-based magnetic tunnel junctions (MTJs) allow switching of the magnetization of the electrodes not only by external magnetic fields but also by spin-polarized tunnel currents. Here, FePt nanocrystals were synthesized by simultaneous reduction of platinum acetylacetonate and thermal decomposition of iron pentacarbonyl in properly chosen conditions of solvent/surfactant proportions and temperature for rational design of their shape and magnetic properties.[2] Distinctive FePt morphologies (particles, obloids and rods) were sequentially coated with MgO by vigorously refluxing a benzyl ether solution containing portions of the fabricated nanostructures in the presence of magnesium acetylacetonate, which acted as precursor for MgO. High Resolution Transmission Electron Microscopy (HRTEM) analysis revealed 2 nm coating of crystalline MgO without the need of annealing. Spatially-resolved energy dispersive X-ray (EDX) spectroscopy confirmed the stoichiometry of the oxide shell. A heat treatment was needed to transform the crystal structure of the core from the disordered fcc FePt phase to the L10 phase with high uniaxial magnetocrystalline anisotropy, which is desirable for STTRAM applications. SQUID magnetometry revealed higher coercive fields for obloids and rods, when compared with spherical particles in the core-shell FePt-MgO system. The magnetization and switching behavior were found to be dominated by shape anisotropy according to our experimental results and corresponding theoretical verification.[3,4]

References

[1] Plummer van Ek Weller (ed.), The Physics of Ultra-High-Density Magnetic Recording, Springer, Berlin, 2001

[2] D. A. Ferrer, S. Guchhait, C. Corbet and S.K. Banerjee, Origin of Shape Anisotropy on Wet-chemistry Synthesized Nanostructures, under review

[3] http://math.nist.gov/oommf

[4] S. K. Banerjee, S. Ganguly, D. A. Ferrer, "Templated self-assembly for nanomagnetic memory", submitted to Office of Technology Commercialization (OTC) of The University of Texas at Austin

See more of this Session: Materials and Architectures for Thin-Film Photovoltaics
See more of this Group/Topical: Materials Engineering and Sciences Division

(549a) Image Formation In Thin Films of Chemically-Amplified Photoresists

Gila Stein and Ginusha Perera, Chemical and Biomolecular Engineering, University of Houston, Houston, TX

Next-generationnanofabrication requires economical lithographic processes that achieve sub-22nm resolution. Current manufacturing practices are based on projectionlithography with chemically-amplified (CA) photoresists.CA resists are comprised of a polymer resin (reactant) blended with a photoacid generator (catalyst); a coupledreaction-diffusion mechanism drives image formation, where resolution iscontrolled by slow diffusion of the acid catalyst. Extending the lifetime of CAresists requires a comprehensive understanding of the physical and chemicalvariables that control image formation in ultrathin films. Challengesassociated with thin films include surface segregation of acid catalyst,adsorption of airborne contaminants, and changes in polymer mobility due tothin film confinement. Currently, there are no methods that offer spatiallyresolved feedback for the reaction-diffusion mechanism. In this talk, wedemonstrate that image formation in CA resists can be accurately measuredthrough the film thickness with small-angle X-ray diffraction. Proof-of-conceptexperiments are based on poly(4-hydroxystyrene-co-*tert*butylacrylate) resin loaded with 6.5wt% triphenylsulfonium sulfonatephotoacid generator. The CA resist is cast in 80 nmfilms on silicon/silicon nitride substrates, and the spatial distribution ofacid catalyst is generated with electron-beam patterning. Image formation is completed at temperaturesin the range of 130-140 C, which are near the glasstransition temperature of the reactant. Preliminary results demonstrate thatspatial extent-of-reaction can vary through the film thickness, andmeasurements of the thin film glass transition suggest that enhanced dynamicsat the free surface are responsible for this behavior.

See more of this Session: Polymer Thin Films and Interfaces IV
See more of this Group/Topical: Materials Engineering and Sciences Division

(549b) Mechanical Properties of Thin Polymer Films At the Nanometer Scale

Jung-Hyun Lee, Jun Young Chung and Christopher M. Stafford, Polymers Division, National Institute of Standards and Technology, Gaithersburg, MD

Nanometer-thin polymer films are critical to a variety of technologies, such as microelectronics, optics, energy production, and biomedical devices. These thin films must retain a high degree of mechanical durability for the best performance under their operating conditions. In particular, the dependence of mechanical properties on film thickness is of fundamental importance and practical interest with rapidly growing nanotechnologies. The mechanical behavior of materials can be described by a full set of key mechanical properties, such as stiffness, strength, and ductility. However, there is a dearth of knowledge about these key mechanical properties of thin polymer films due to the lack of appropriate measurement tools at the nanometer scale. In this presentation, we describe a measurement technique of a combined wrinkling and cracking that enables the determination of the modulus, fracture strength, and onset fracture strain of thin films. We first demonstrate the general applicability of our methodology to a diverse range of thin films, and then apply it to investigate the effect of film thickness on the collective mechanical properties of polystyrene thin films with thickness ranging from 10 nm to 300 nm.

See more of this Session: Polymer Thin Films and Interfaces IV
See more of this Group/Topical: Materials Engineering and Sciences Division

(549c) Non-Contact Measurements of "Stiffness" In Confined Polymer Films by Fluorescence and X-Ray Photon Correlation Spectroscopy

Christopher M. Evans[1], Suresh Narayanan[2], Zhang Jiang[2] and John M. Torkelson[1], (1)Chemical and Biological Engineering, Northwestern University, Evanston, IL, (2)X-ray Science Division, Argonne National Laboratory, Argonne, IL

Over the past 15 years, deviations in the glass transition temperature (Tg) of thin polymer films from bulk Tg have been characterized extensively. Considerably less work has been published concerning the elastic modulus of thin films, and there have been mixed results concerning the direction of deviations from bulk behavior. Film wrinkling and AFM based approaches have indicated a reduction in modulus with decreasing thickness while bubble inflation, nanoindentation, and tensile testing of nanofibers suggest an increase in modulus upon confinement. Brillouin light scattering (BLS) showed no change in modulus in one study, while a second BLS study reported a large increase in modulus for thin films. We have applied two new non-contact measurements to investigate the stiffness of thin and ultrathin polymer films. X-ray photon correlation spectroscopy (XPCS) studies have been performed on polystyrene (PS) and poly(2-vinyl pyridine) (P2VP) single layer films supported on silica. Polystyrene has also been studied in a bilayer geometry with three polymer underlayers, poly(4-vinyl pyridine) (P4VP), poly(isobutyl methacrylate) (PiBMA) and crosslinked poly(dimethyl siloxane) (PDMS). The XPCS technique probes the relaxation times of thermally induced capillary waves at the surface of a film, and the electric field of the x-rays decays evanescently in the top ~10 nm of the film due to the experimental geometry. Studies of PS single layer films on silica at 110 °C (Tg,bulk + 10 °C) show a stiffening behavior upon confinement in terms of the capillary wave relaxation times. For sufficiently large values of the in-plane scattering wavevector (q), the relaxation times of films ranging from 29-120 nm can be rescaled by monitoring τ/h as a function of qh, where τ is the measured capillary wave relaxation time and h is the film thickness. For qh > 1, all of the film relaxation times can be superposed, in accord with predictions from capillary wave theory. For qh < 1, stiffening behavior is observed where relaxation times are disproportionately longer in the thinner films, beyond what can be accounted for by capillary wave theory. Stiffening is not observed at 120 or 140 °C, where all of the film relaxation times superpose over the entire accessible range of qh. The same behavior is observed in P2VP supported on silica, a system which shows an increase in Tg upon confinement. Thus, this stiffening behavior appears uncorrelated with Tg-confinement effects.

Bilayers of thin and ultrathin PS on different bulk substrates were also studied, and revealed a connection between substrate modulus and PS surface relaxation times. The relaxation times of a 120 nm thick PS layer on the hardest substrate (silica) were roughly two orders of magnitude longer than for PS on the softest substrate (PiBMA). The order of relaxation times followed the order of substrate modulus: silica > P4VP > PDMS > PiBMA in order of decreasing modulus at 110 °C. The difference in relaxation times for PS on silica and PiBMA grew to roughly three orders of magnitude for 30 nm thick PS layers indicating a growth of this effect upon confinement. Interestingly, the effect of substrate modulus is not observed at 150 °C. The stiffening behavior observed for PS single layer films is also observed in the bilayer films, where a 30 nm PS layer on P4VP is stiffer than a 120 nm PS layer on P4VP. Thus, at temperatures near Tg we see that films stiffen upon confinement, and soften with decreasing substrate modulus. Confinement effects and substrate modulus effects are not observed at higher temperature. The XPCS measurements represent the first reported confinement effects by this technique and are the first experiments to indicate a pronounced dependence of thin film modulus on supporting material at thicknesses greater than 100 nm.

Fluorescence is also used to probe stiffness in confined polystyrene (PS) films through the intensity ratio (I3/I1) of the dye molecule pyrene. Pyrene fluorescence is sensitive to the polarity of its local environment because in a polar environment, excited state pyrene undergoes dipole-dipole coupling with the polar solvent molecules. These molecules then orient around the dye molecule which leads to a less mobile environment for pyrene and a lower I3/I1 value is measured. Recently, our research group has used I3/I1 to measure Tg in PS films, and I3/I1 decreases with decreasing temperature as the system becomes more dense. Thus, we hypothesize that lower I3/I1 values correspond to a stiffer environment. Films of PS were investigated on two substrates, glass and crosslinked PDMS, as well as in the free standing geometry. For all films exceeding ~400 nm, I3/I1 measured at 60 °C is independent of film thickness indicating a bulk response. With decreasing film thickness, free-standing PS shows softening (an increase in I3/I1) below ~400 nm. The I3/I1 value reaches a maximum at ~200 nm, at which point stiffening behavior (a decrease in I3/I1 with film thickness) is observed in films less than 200 nm thick. Silica- and PDMS-supported PS films show no softening, but report stiffening for films less than ~200 nm thick. For sufficiently thin films (less than 30 nm), I3/I1 is the same within experimental error for both supported and free standing PS. The stiffening behavior observed is not in accord with the Tg reductions seen for PS on silica; however, a stiffening in terms of elastic modulus and rubbery compliance has been reported by other research groups. Our XPCS results also suggest stiffening upon confinement in PS films or bilayers. This work represents the first use of I3/I1 to specifically monitor stiffness, and this technique can be extended to probe the distribution of stiffness within polymer films. By strategically placing labeled layers at different points (e.g. free surface or substrate) within an unlabeled film, it may be possible to quantify a gradient of stiffness within a polymer film.

See more of this Session: Polymer Thin Films and Interfaces IV
See more of this Group/Topical: Materials Engineering and Sciences Division

(549d) Molecular Simulations of Confined Glass-Forming Polymers

Amit Shavit and **Robert Riggleman**, Chemical and Biomolecular Engineering, University of Pennsylvania, Philadelphia, PA

For many technological applications ranging from semiconductor manufacturing to flexible displays, the properties of glass-forming materials when they are confined to nanoscopic dimensions are of critical importance. For example, common semiconductor manufacturing techniques rely on the mechanical integrity of amorphous polymer nanostructures. Experiments over the past several years have demonstrated that many of the properties of polymer glasses (e.g., the glass transition temperature, Tg, and the elastic constants) can change significantly and in unintuitive ways when confined to dimensions below approximately 100 nm. These confinement effects depend strongly on the detailed polymer chemistry. In this talk, I will describe our recent work using molecular simulations to study the effects of nanoscale confinement on both the dynamic and mechanical properties of a series of coarse-grained polymer models. Our coarse-grained models systematically explore the effects of polymer chemistry by gradually changing the polymer backbone stiffness and analyzing the strengths of the confinement effects. We characterize confinement effects through a detailed analysis of the local dynamics and local mechanical properties for each polymer confined to free-standing thin films, and the results are compared to bulk simulations. All of our polymer chain lengths are above the bulk entanglement chain length, and we explore how the chain dynamics and the entanglement network are affected by nanoscale confinement. The results of these simulations provide significant new insights on the relationship between polymer chemistry and the strength of the effects of nanoscale confinement.

See more of this Session: Polymer Thin Films and Interfaces IV
See more of this Group/Topical: Materials Engineering and Sciences Division

(549e) Surface Segregation In Blends of Comb and Linear Polymers

Mark D. Foster[1], Boxi Liu[1], Roderic Quirk[1] and David T. Wu[2], (1)Department of Polymer Science, The University of Akron, Akron, OH, (2)Chemical Engineering and Chemistry Departments, Colorado School Of Mines, Golden, CO

Blending polymers with different chain architectures may prove useful in controlling interfacial properties by controlling interfacial segregation. A linear response theory by Wu *et al.* predicts that a long-chain branched polymer blended with its linear analog will be preferentially segregated to the surface and interface of the blend film. The comb architecture is promising for achieving substantial surface segregation. Its high degree of branching, which makes it more prone to separate from linear analogs in the bulk, also provides a substantial driving force for surface segregation when chain ends prefer the surface. In this study comb polystyrenes with well-defined architectural details were prepared by living anionic polymerization. DSC indicates that phase separation occurs for some blends of the linear and comb analogs, and this is also seen by SANS study of these blends. Surface segregation of a comb polystyrene when blended with its linear analog has been quantified using the complementary techniques of neutron reflectivity and SIMS.

Acknowledgements: NSF support (CBET 0730692 & 0731319) is gratefully acknowledged.

See more of this Session: Polymer Thin Films and Interfaces IV
See more of this Group/Topical: Materials Engineering and Sciences Division

(549f) Relaxation of Polymer Melts In Nanofilms At the Polymer-Solid Interface

Yongjin Wang[1], Jianing Sun[2] and **Lei Li**[1], (1)University of Pittsburgh, Pittsburgh, PA, (2)J.A. Woollam Co., Inc., Lincoln, NE

It is well known that the relaxation of polymer chains in bulk melt is fast, e.g., on the order of milliseconds to seconds. However, the relaxation could be orders of magnitude slower for some polymer melts in nanometer-thick films on solid substrates. To date, the governing mechanism behind this phenomenon remains unclear. Such a slow relaxation at molecular level could cause certain properties of the polymer nanofilms to change with time after the polymer is applied on a solid surface. This is a serious concern for the long-term performance and the lifetime of the polymer thin films in various applications. Without understanding the thermodynamics and the kinetics governing the relaxation, it is impossible to rationally design more robust thin-film materials.

In this presentation, we report the experimental results indicating that the polymer-solid interfacial interaction, which does not have a bulk counterpart, plays a key role in both thermodynamic driving force and kinetics of the relaxation. Two perfluoropolyethers (PFPEs) with the same backbone and different endgroups, one polar and the other non-polar, have been studied. The relaxation of the nanometer-thick PFPEs on a hydrophilic solid substrate was characterized by the surface energy analysis via the contact angle measurement. For the PFPE with polar endgroups, the surface energy "relaxes" with time and the relaxation time constant, obtained from Kohlrausch-Williams-Watts (KWW) model, is ten orders of magnitude higher than that of bulk polymer. However, for the PFPE with non-polar endgroups, the surface energy relaxation was not observed at all. Based on the experimental results, a molecular-level model has been proposed. According to this model, the observed relaxation is thermodynamically driven by the attractive interaction between the polar endgroups of the polymers and the polar sites on the solid substrate. The very slow kinetics of the relaxation has been attributed to the heterogeneity of the polymer-solid interaction at the interface and the cooperative nature of the molecular motions in the relaxation.

See more of this Session: Polymer Thin Films and Interfaces IV
See more of this Group/Topical: Materials Engineering and Sciences Division

(550a) Design for Six Sigma (DFSS) Approach to Copolymer Process Development

Mohan Khadilkar, SABIC Innovative Plastics, Mt. Vernon, IN and William Hollar, Sabic Innovative Plastics, Mount Vernon, IN

The SABIC Innovative Plastics LEXAN* polycarbonate business has introduced several new copolymer platforms into existing resin manufacturing assets in the past decade. These new copolymers expand the application space of traditional polycarbonates and allow business growth into new applications. Commercialization of new copolymers involves developing processes that address challenges due to copolymer chemistry, new product requirements, and fit with existing assets and investment constraints. Having a systematic and proactive methodology that provides "designed in" process capability and product quality is critical to successful process and product scale-up. The Design for Six Sigma (DFSS) methodology provides such a framework and set of tools to ensure success by design. This paper will survey critical tools for managing scale-up and product performance requirements. An example of development and scale-up of one of these copolymers will be reviewed.

* LEXAN is a registered trademark of SABIC Innovative Plastics LLC

See more of this Session: Process and Properties In Polymers
See more of this Group/Topical: Process Development Division

(550b) Thermal Degradation of Nylon 66: Modeling and Parameter Estimation

Kim B. McAuley, Hadis Karimi and Mark A. Schaffer, Chemical Engineering, Queen's University, Kingston, ON, Canada

A fundamental dynamic model is developed to predict the influence of reactor operating conditions on the rates of undesirable reactions that can occur in the low-moisture, high-temperature finishing stage of commercial nylon 66 production. Kinetic parameters and initial values for unknown concentrations are estimated using experimental data. The proposed model is an advance over previous models because it accounts for the effect of melt-phase water concentration on thermal degradation. Estimability analysis and a mean-square error (MSE) criterion are used to determine which parameters can and should be estimated using the available data, and which parameters should remain at their nominal values. Parameters are ranked from most estimable to least estimable based on their influence on the model predictions, correlation with other parameters and uncertainty in initial guesses. Subsequently, the MSE criterion is used to determine that 48 of the 56 parameters should be estimated. The remaining 8 parameters are fixed at their initial guesses. The fitted model matches the data well, with typical errors of 1.6% for amine ends, 1.5% for carboxyl ends, 7.6% for gas evolutions rate of CO_2, 15.3% gas evolutions rate of NH_3 and 18.1% for gas evolutions rate of cyclopenatone. This model will help engineers to select temperature and moisture level setpoints that lead to high polymerization rates while ensuring good product quality.

See more of this Session: Process and Properties In Polymers
See more of this Group/Topical: Process Development Division

FLOW ACCELERATES THE REACTION KINETICS OF MULTIPHASE POLYMER SYSTEMS

Jie Song[1], *Randy H. Ewoldt*[2], *Christopher W. Macosko*[1*]

[1]*Department of Chemical Engineering and Materials Science, University of Minnesota, Minneapolis, Minnesota, 55455*

[2]*The Institute for Mathematics and its Applications, University of Minnesota, Minneapolis, Minnesota, 55455*

Coupling reactions between functional group grafted macromolecules have been proven to be a powerful way to facilitate compatibilization and adhesion. Yet their reaction kinetics requires more fundamental understanding. The common wisdom for reactions between *small molecules* is that external mass transfer increases the reaction rate by increasing concentration of reactive species without affecting reaction mechanism and transition states at a certain temperature. However, for reactions between *macromolecules*, the kinetics is more complicated because of their relatively slow diffusion and complex configurations. It is generally difficult to study the effects of flow on macromolecules reactions during mixing because interfacial area generation is always convoluted and hard to isolate.

In this paper, a reactive bilayer system was created to explore this issue. In an effort to study the reaction between polyethylene (PE) and thermoplastic polyurethane (TPU), maleic anhydride (MA), hydroxyl (OH) and secondary amine (NHR) functionalized PEs were synthesized via reactive extrusion and blended into non-modified PE. We bonded these PEs to TPU via lamination and coextrusion. In order to compare reaction rates of the two processes we determined the interfacial copolymer density Σ considering both advection and interfacial area generation. We found that the reaction rate in coextrusion was much faster in comparison with lamination at the same temperature. This difference was attributed to the extensional and compressive flow in coextrusion overcoming the diffusion barrier at the interface and forcing reactive species to penetrate into interface. The effects of functional group reactivity and processing variables on adhesion were correlated with interfacial copolymer coverage. Amine functionalized PE showed dramatic adhesion improvement even at 1 wt%. We also determined that flow accelerates the reaction between polyamide 6 and maleic anhydride high density polyethylene. New die design is investigated to isolate the effect of extensional and compressive flow.

(550d) Continuum Lumping Modelling for Polymerisation Kinetics

Mustafa Adam[1], **Raffaella Ocone**[1] and Valeria Arrighi[2], (1)Chemical Engineering, Heriot-Watt University, Edinburgh, United Kingdom, (2)Chemistry, Heriot-Watt University, Edinburgh, United Kingdom

Polymerisation occurs when two “units”, called monomers, join by either chain addition or step-growth process leading to a product with a higher molecular weight characterised by repeating chemical units. A model based on the continuum lumping methodology is proposed and implemented to predict the molecular weight distribution during polymerisation. The work is aimed at assessing the benefit of the continuum lumping approach to obtain information about some specific features of the reaction. The model is tested for two selected case studies furnishing the product distribution at various times. Since many properties of a polymer depend on the molecular weight distribution (MWD) and on the average of such distribution, the results will pave the way for future work assessing the predictive capability of the lumping methodology as applied to polymerisation. The aim is twofold: to develop a model which could predict the polymer MWD at various times, and to assess the capability of the continuum lumping methodology as applied to polymerisation. The work focuses on step growth polymerisation only, since we think that the underlying kinetic mechanism is amenable to be described through a continuum lumping procedure. We show that the polymerisation yields distribution can be obtained through the methodology. The model is based on the assumption that the MWD evolves in time as described by an integro-differential equation. A Skewed Gaussian distribution function for determining product yield distribution of the polymerisation reaction is introduced. The model is based on the assumption that the kinetics is liner and first order, and the reaction is irreversible. The model captures successfully the effect of the reaction time and reaction rate constant on the MWD.

See more of this Session: Process and Properties In Polymers
See more of this Group/Topical: Process Development Division

(550e) Stress Cracking of Poly(ethylene terephthalate) Beverage Bottles: Mechanism, Structure/Property Relationships, and Methods of Prevention

Eric Morrison[1], Megan W. Malvey[1], Richard D. Johnson[1], Jeffrey S. Hutchison[1], Jessica L. Anacker[1] and Keith A. Brown[2], (1)Ecolab, Eagan, MN, (2)Plastic Technologies, Inc, Holland, OH

Starting in 1978 with the introduction of one piece poly(ethylene terephthalate) (PET) soft drink bottles with molded-in feet, there has been the problem of environmental stress cracking in bottles used for packaging carbonated beverages. Environmental stress cracking is delayed failure in which a large crack forms in the amorphous base portion of the bottle through which all of the liquid contents are lost. Using a full package test method first described in 2008, the mechanism of stress cracking was determined to be alkaline hydrolytic cleavage of ester bonds in the PET backbone and not a Volynskii crazing mechanism as was previously thought. Hydrolysis leading to failure does not happen in the case that naturally occurring water alkalinity is precipitated as carbonate minerals during evaporation. For this reason, eliminating contact of filled PET bottles with softened water has proven to be a highly effective means of stress crack prevention. Physical aging greatly increases amorphous PET susceptibility to hydrolytic stress crack failure and minimizing exposure of empty bottles to environmental conditions which promote physical aging provides another important strategy for the elimination of stress crack failure.

See more of this Session: Process and Properties In Polymers
See more of this Group/Topical: Process Development Division

(550f) A Polymer Isolation Process for Powders with Designed Properties

William Hollar, Sabic Innovative Plastics, Mount Vernon, IN and Viswanathan Kalyanaraman, SABIC Innovative Plastics, Mount Vernon, IN

Most polymer process development programs for new products focus on molecular structure to achieve desired product properties. This relationship tends to lead to a reaction focus for polymer process development. There are cases where the final properties of the polymer powder also determine its functionality in a final product. This is the case for polymer additives used for advanced composites. A new process for producing fine, porous engineering thermoplastic powders has been developed for a range of these applications. The powder formation process determines the functionality and desirability of the powder and leads to powders with unique properties.

This paper will focus on some of the key chemical engineering aspects of the process conditions that can be used to tailor properties such as porosity and particle size distribution. Such changes can enable or enhance product performance. Process alternatives for powder production will be presented and contrasted. A two-fluid atomization process and its applicability to tailored polymer powders will be described. A review of applications and desired product characteristics will be discussed.

See more of this Session: Process and Properties In Polymers
See more of this Group/Topical: Process Development Division

(558a) Photofixation of DielsAlder Networks

Chistopher N. Bowman, Department of Chemical and Biological Engineering, University of Colorado, Boulder, CO

Photoresists generally utilize a photochemical reaction that results in a solubility change in the resist. In the case of a negative resist where a material is polymerized by irradiation the fabrication of complex three dimensional shapes is hindered by the liquid nature of the resist that allows unsupported features to float out of position. This problem can be avoided by the use of a reversible polymer scaffold to support the desired shape. Reversibly crosslinked networks can be formed by the Diels–Alder reaction of furan and maleimide. Spatio-selective fixation of the crosslinks by radical reactions renders them irreversible within a well-defined region generated using 2-photon photolithography. Subsequent reversion of the crossli---nks in the unexposed regions to monomer yields the patterned image. This allows complex shapes, such as sets of interlocked rings to be readily fabricated. This approach provides several key advantages over current techniques including solventless preparation, the ability to pattern arbitrary three dimensional shapes, and reduced complexity.

See more of this Session: Adaptive and Self-Healing Polymeric Materials
See more of this Group/Topical: Materials Engineering and Sciences Division

(558b) Self-Healing Through Mechanochemistry

Paul V. Braun, Department of Materials Science and Engineering, University of Illinois at Urbana-Champaign, Urbana, IL

Spiropyran mechanophores (mechanochemically reactive units) can impart the unique functionality of visual stress detection to polymers and have potential for use in smart materials with self-sensing capabilities. Other mechanophores may enable mechanical induced self-healing. These mechanophores were incorporated into polyurethane and related polymers via step growth polymerization. Polyurethane, which is inherently a versatile engineering polymer, possesses an optimized balance of mechanical toughness and elasticity to allow for investigation of the kinetics of the mechanochemical response of the mechanophores in the bulk polymer. For example, the stress-induced 6-π electrocyclic ring-opening to the colored merocyanine form of the spiropyran mechanophore was quantified by measuring the change in absorbance of the polymer, while it was held at constant strain. The closing kinetics of the mechanophore was also studied by fluorescence imaging. Finally, the effects of mechanical strain on the equilibrium between the two forms are reported and discussed.

See more of this Session: Adaptive and Self-Healing Polymeric Materials
See more of this Group/Topical: Materials Engineering and Sciences Division

(558c) Mechanophotopatterning of a Covalent Adaptable Network

Timothy F. Scott[1], Christopher J. Kloxin[2], Hee Young Park[3] and Christopher N. Bowman[3], (1) Chemical Engineering, University of Michigan, Ann Arbor, MI, (2)Chemical Engineering and Materials Science and Engineering, University of Delaware, Newark, DE, (3)Chemical and Biological Engineering, University of Colorado at Boulder, Boulder, CO

The fabrication of well-defined topographical features or textures is a dynamic area of research owing to their extraordinary potential in applications ranging from complex optics to tunable and responsive surfaces. Here, mechanophotopatterning (MPP) on a covalent adaptable elastomeric network is introduced and enables for the first time the ability to precisely and simultaneously manipulate both material shape and surface topography by exposure to light without the need for solvents, molding, or physical contact. MPP is a unique patterning approach whereby deformation is applied to an elastomeric material capable of photoinduced structural modification to alter its equilibrium geometry, elegantly complementing previous mechanically assisted patterning techniques. This material responds to MPP by continuously and locally deforming via polymer network connectivity rearrangement, which enables 3D control of its geometry. Unlike conventional photolithographic approaches, MPP forgoes any wet chemistry or surface deposition/modification processing and simplifies multi-tiered feature fabrication. Furthermore, in contrast with mechanically assisted patterning techniques that utilize buckling phenomena, MPP on a photoresponsive elastomer allows an arbitrary-sized and designed feature to be written into the material multiple times, while also being able to change the overall shape of the material.

See more of this Session: Adaptive and Self-Healing Polymeric Materials
See more of this Group/Topical: Materials Engineering and Sciences Division

From Nano-Responsive Shapes to Self-Repairing Polymer Networks

Biswajit Ghosh, Dhanya Ramachandran, and Marek W. Urban*
The University of Southern Mississippi
School of Polymers and High Performance Materials, Hattiesburg, MS 39406

Continuous needs for engineering of new materials and devices, with distinct properties and functions that will revolutionize future technologies, require developments of unique and controllable shapes at nano- or smaller length scales. To achieve these attributes, orchestrated heterogeneous regions within engineered polymer networks are necessary for localized structural and morphological variations to occur, thus enabling favorable spatial and energetic conditions for spontaneous responses induced by minute external or internal stimuli. These generalized needs and concepts formulated the principles for the developments of new generations of stimuli-responsive nano-materials/devices and self-repairing networks. This presentation will discuss recent advances in and developments of controllable shape asymmetric nanoparticles, nanotubes, and nanorods, enabling applications in diversified stimuli-responsive devices, ranging from unique optical properties to cilia-like behavior, or nano-tubes assembled to function as oxygen generating reactors. Orchestrated network heterogeneities below nano-scale lengths provided an opportunity for the development of self-healing polyurethane networks that upon mechanical damage and subsequent exposure to UV light, exhibit self-repairing attributes via free radical crosslinking. Using a combination of spectroscopic and thermo-mechanical analytical tools the correlation between molecular changes and macroscopic network remodeling will be discussed.

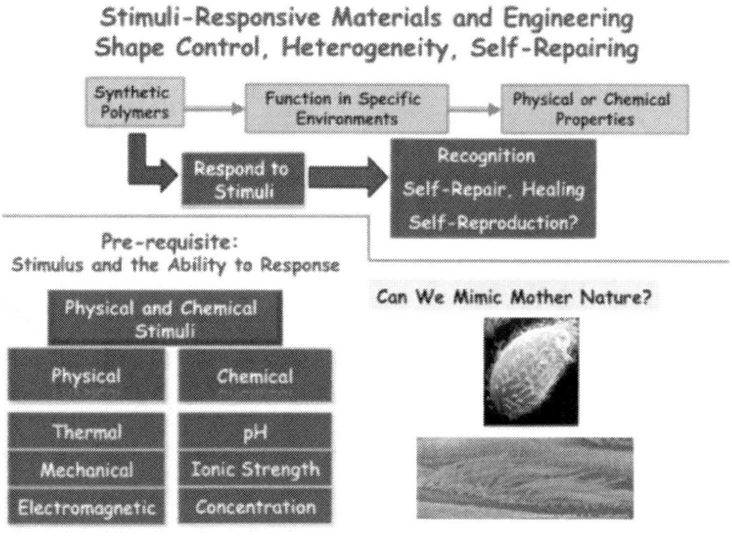

Cohen, Genzer, Luzinov, Mueller, Ober, Stamm, Szleifer, Zauscher, Urban, Winnik, Minko, **Nature Materials**, 2010, 9, 101-113

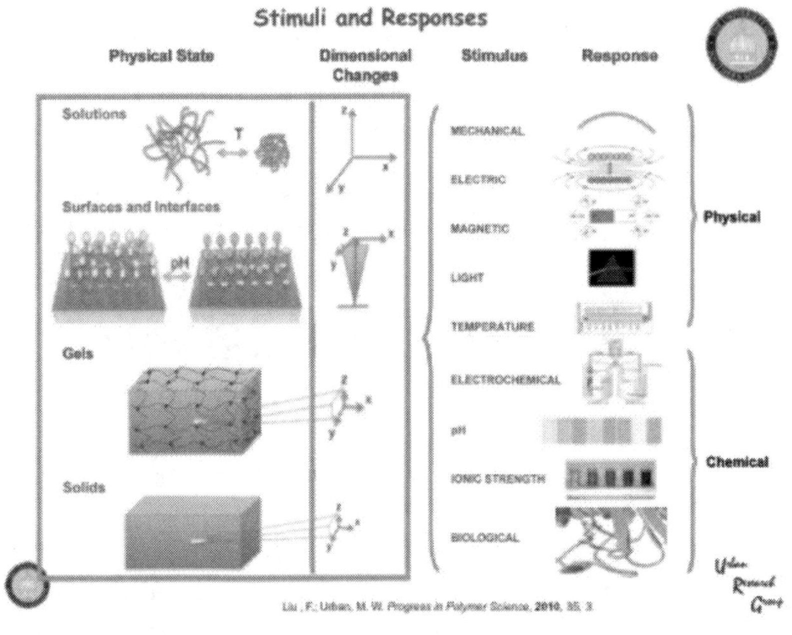
Stimuli and Responses
Physical State
Dimensional Changes
Stimulus
Response
Solutions
Surfaces and Interfaces
pH
Gels
Solids
MECHANICAL
ELECTRIC
MAGNETIC
LIGHT
TEMPERATURE
ELECTROCHEMICAL
pH
IONIC STRENGTH
BIOLOGICAL
Physical
Chemical
Liu, F.; Urban, M. W. Progress in Polymer Science, 2010, 35, 3.

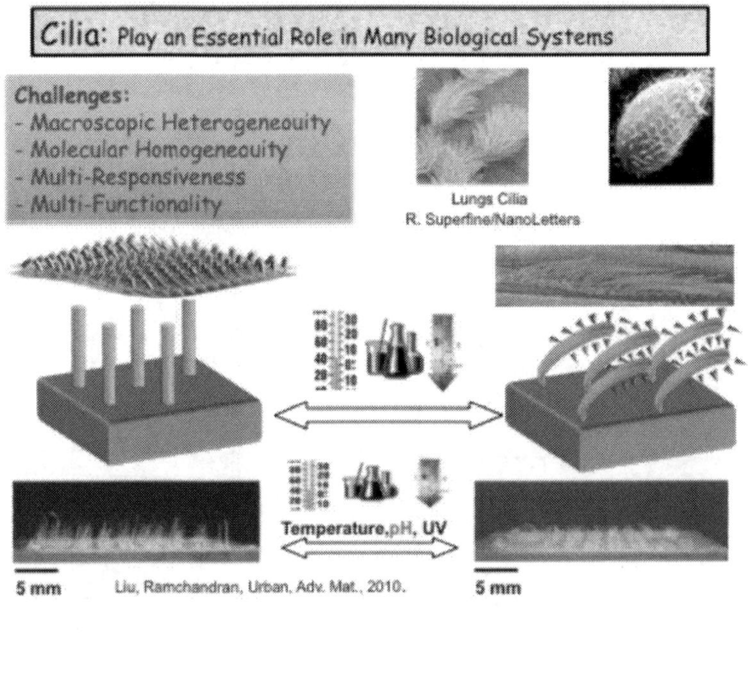
Cilia: Play an Essential Role in Many Biological Systems
Challenges:
- Macroscopic Heterogeneouity
- Molecular Homogeneouity
- Multi-Responsiveness
- Multi-Functionality
Lungs Cilia
R. Superfine/NanoLetters
Temperature,pH, UV
5 mm
Liu, Ramchandran, Urban, Adv. Mat., 2010.
5 mm

Self-Repairable Polymer Networks
Oxetane-Chitosan-Polyurethane (OXE-CHI-PUR) Networks

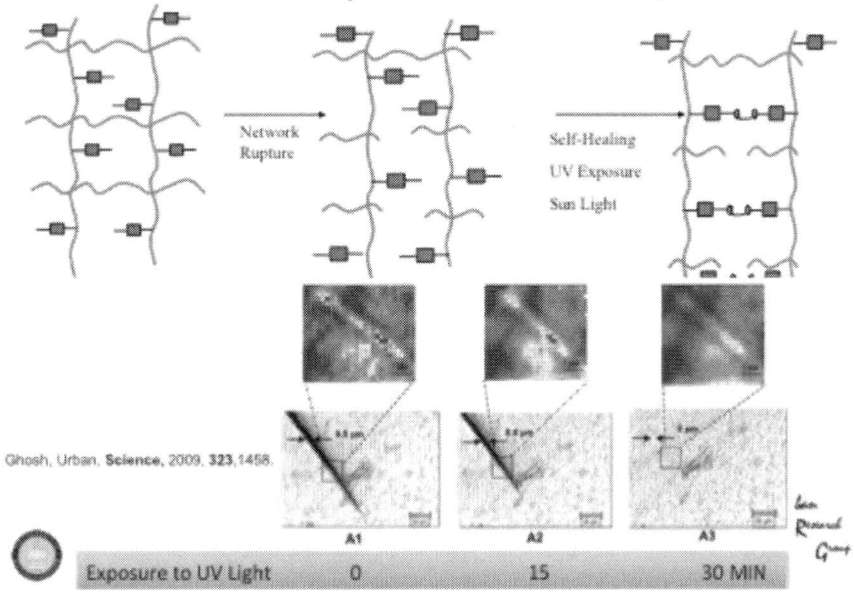

Unchartered Avenues
Self-Repairing Thermoplastic Polymer Networks:
Repair of Mechanical Repair by Active Nano-Components

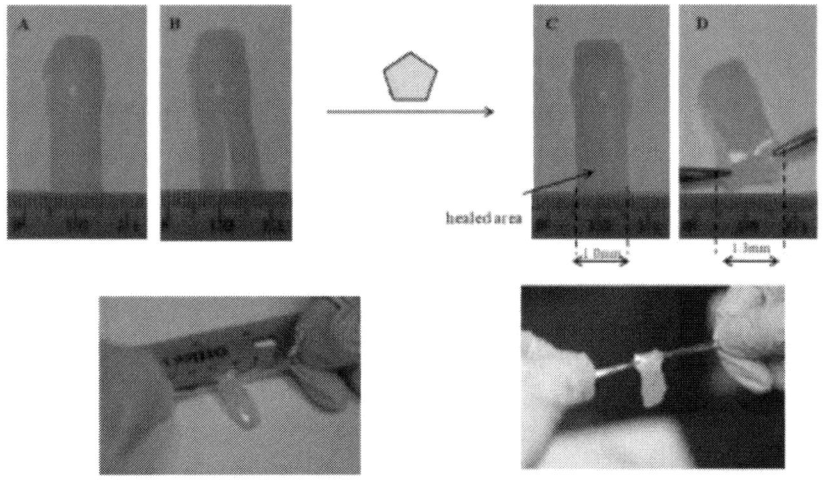

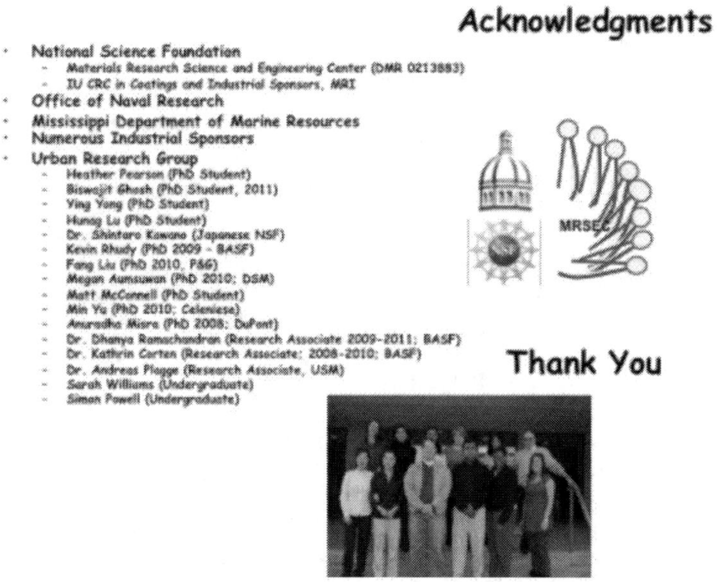

M.W. Urban, Handbook of Stimuli-Responsive Materials, 2011, Wiley-VCH (ISBN: 978-3-527-32700-3).

(558e) Solvent Response of Mixed Polymer Brushes

Kai Gong and Walter G. Chapman, Chemical and Biomolecular Engineering, Rice University, Houston, TX

"Smart" materials, also known as self-responsive materials, has attracted increasing attention in recent years due to its broad application such as controlled drug-delivery, micro- and nanofluidics, biocompatible material, chemical sensors, antifouling coatings. These materials can respond to the external signal or environment such as temperature, light, pH, solvent selectivity, electric field, etc.[1-3]

Polymer brushes,[4] especially copolymer brushes or mixed polymer brushes, is considered as a novel and robust way of creating smart surfaces.[1-3] Recent advances in the synthesis of polymer brushes has made the control of the chain length, the monomer fraction, and the grafted density become possible.[5-8] The self-assembly of mixed polymer brushes towards the solvents has special interest because of vast parameter space including molecular weight, grafted densities, solvents selectivity can be easily employed to create rich phase diagram in that system.[7,8]

We have performed density functional theory (DFT) calculation to study solvent response of mixed polymer brushes in 1-D (Dimension). To be simple, we assume no lateral segregation happens and only consider the implicit solvent here. We set the different interaction energy for each polymer brush to simulate their different response towards the solvents. A transition was found by increasing the chain length of one component while fixing the other's. We also study the solvent selectivity, the composition effect on the transition.

References

1 I. Luzinov, S. Minko, and V. V. Tsukruk, Soft Matter **4**, 714 (2008).

2 M. A. C. Stuart, W. T. S. Huck, J. Genzer, M. Müller, C. Ober, M. Stamm, G. B. Sukhorukov, I. Szleifer, V. V. Tsukruk, M. Urban, F. Winnik, S. Zauscher, I. Luzinov, and S. Minko, Nature Materials **9**, 101 (2010).

3 T. P. Russell, Science **297**, 964 (2002).

4 S. T. Milner, Science **251**, 905 (1991).

5 P. Mansky, Y. Liu, E. Huang, T. P. Russell, and C. Hawker, Science **275**, 1458 (1997).

[6] C. Xu, T. Wu, C. M. Drain, J. D. Batteas, M. J. Fasolka, and K. L. Beers, Macromolecules **39**, 3359 (2006).

[7] B. Zhao, Langmuir **20**, 11748 (2004).

[8] B. Zhao, R. T. Haasch, and S. Maclaren, J. Am. Chem. Soc **126**, 6124 (2004).

See more of this Session: Adaptive and Self-Healing Polymeric Materials
See more of this Group/Topical: Materials Engineering and Sciences Division

(558f) Can Supramolecular Interactions Be Used to Create Stable Polymer Nanoparticles?

Colin C. Neikirk, Jae Woo Chung and Rodney D. Priestley, Chemical and Biological Engineering, Princeton University, Princeton, NJ

Polymeric nanoparticles have attracted significant attention due to their promising applications in the fields of drug transportation and medical imaging. Generally, nanoparticles consist of a stabilizing hydrophilic outer shell and a hydrophobic inner core. In this talk, we report on the ability of supramolecular interactions to stabilize hydrophobic polymers in an aqueous solution. Furthermore, we illustrate that the responsive nature of supramolecular interactions may be used to impart responsiveness to the polymeric nanoparticle.

See more of this Session: Adaptive and Self-Healing Polymeric Materials
See more of this Group/Topical: Materials Engineering and Sciences Division

(567a) Biomimetic Stem Cell Culture: Toward Mimicking Natural Microenvironments

William L. Murphy, Biomedical Engineering and Orthopedics, University of Wisconsin, Madison, WI

Abstract to follow.

See more of this Session: Biomaterials for Stem Cell Expansion and Differentiation
See more of this Group/Topical: Materials Engineering and Sciences Division

Unavailable Paper

(567b) Hydrolytically Degradable Affinity PEG Hydrogel Scaffolds for Neural Stem Cell Delivery

Yunqian Wei and Jennie B. Leach, University of Maryland, Baltimore County, Baltimore, MD

Cell fate is determined by cell-cell interactions as well as response to soluble and matrix molecules in the cellular microenvironment. The decision of neural/glial fate is made during the last cell cycle of neural stem cells (NSC) in the proliferative zones of the brain1. Multiple cell-secreted molecules, e.g., epidermal growth factor, fibroblast growth factor and hepatocyte growth factor (HGF), control NSC proliferation2-4. HGF exhibits multiple activities in both neural5 and nonneural tissues6 in a precise concentration and cell cycle dependent manner; thus because of their powerful effects, we hypothesize that the precise presentation, release rate and concentration of these factors are required in order to control cell response. Three-dimensional hydrolytically-degradable polyethylene glycol (PEG) hydrogels were designed and were based on the selective cross-linking between 4-arm PEG-vinyl sulfone backbone and PEG-dithiol cross-linkers7. Our previous work demonstrated that the PEG hydrogels had tunable mechanical properties (i.e., shear modulus, degradation rate) and showed great potential for use as cell culture scaffolds as well as implantable scaffolds for stem cell therapies. In this work, we explored several approaches for presenting HGF in this 3D gel with the ultimate goal of determining which approach best promotes NSC proliferation and differentiation to the glial fate. New cross-linkers, thiolated heparin and peptide sequences with affinity to HGF were synthesized, used in conjunction with the PEG-based hydrogels and applied to NSC culture. This work provides a novel basis to present and control the release of key soluble factors through a highly tunable 3D hydrogel matrix to control NPC viability and differentiation.

References:

1. P Levitt, et al. Dev Neurosci 1997 19: 6. 2. A Chenn, et al., Science 2002 297:365. 3. R Dono, et al. EMBO J1998 17:4213. 4 B Reynolds, et al. J Neurosci 1992 12:4565.

5. F Maina, et al. Neuron 1998 20:835. 6. Y Liu, et al. Kidney Int 1999 55:442 7. Zustiak S, et al. Biomacromol 2010 11:134

See more of this Session: Biomaterials for Stem Cell Expansion and Differentiation
See more of this Group/Topical: Materials Engineering and Sciences Division

(567c) Controlling Stem Cell Differentiation with Stiffening Hydrogels

Murat Guvendiren and Jason A. Burdick, Bioengineering, University of Pennsylvania, Philadelphia, PA

Stem cells respond to many microenvironmental cues towards their decisions to spread, migrate, and differentiate and these cues can be incorporated into materials to control stem cells for regenerative medicine. As an example, matrix mechanics alone has been shown to regulate stem cell morphology, spreading, migration, proliferation and differentiation. Although many hydrogel systems have been explored with tunable mechanics by controlling the degree of crosslinking (e.g., polyacrylamide gels), these systems primarily exhibit static mechanical properties. However, the native tissue environment is often dynamic and these dynamic changes in matrix mechanics play critical roles in many biological events such as during tissue development and regeneration, wound healing, fibrosis and tumor progression. In this work, we investigated the effect of *in situ* matrix stiffening on human mesenchymal stem cell (hMSC) behavior *in vitro* on methacrylated hyaluronic acid (MEHA) substrates. MEHA (modified with RGD ligands) was designed to allow sequential crosslinking via Michael-type addition using a dithiol crosslinker (initial gel formation) and then radical photopolymerization (temporal stiffening) via UV exposure, where the secondary step could be performed with adhered cells present. Hydrogel mechanics (from ~3kPa to ~100kPa) were tuned precisely by controlling the initial methacrylate consumption during addition crosslinking, and following UV exposure time. In particular, we investigated the hMSC response to substrate stiffening from soft (~3kPa) to stiff (~30kPa) in terms of spreading, traction forces (via dynamic traction force microscopy), migration, proliferation and differentiation (using a mixed adipogenic/osteogenic bipotential media). We found that hMSCs respond to stiffening by increases in cell area, traction forces, migration and proliferation, particularly on a time scale of hours. We also observed that hMSCs differentiate based on how long they are on a substrate of a specific mechanics – e.g., more adipogenic the longer the cell is on the softer substrate and more osteogenic the longer the cell is on the stiffer substrate. After 14 days of culture ~98% of the cells stained positive for adipogenic markers on soft substrates and ~98% stained positive for osteogenic markers on stiff substrates. When the soft substrate was stiffened at day 7, ~45% of the cells were adipogenic and the remaining ~55% were osteogenic when investigated at day 14.

See more of this Session: Biomaterials for Stem Cell Expansion and Differentiation
See more of this Group/Topical: Materials Engineering and Sciences Division

(567d) Gradient Hydrogel System to Decode Paracrine Signaling Between Hematopoietic Stem Cells and Their Bone Marrow Niche

Bhushan Mahadik[1], Tobias Wheeler[1], Paul J.A. Kenis[2] and Brendan Harley[1], (1)Chemical and Biomolecular Engineering, University of Illinois at Urbana-Champaign, Urbana, IL, (2)Chemical and Biomolecular Engineering, University of Illinois, Urbana Champaign, Urbana, IL

Hematopoietic stem cells (HSCs)[1] are responsible for the formation and development of all blood and immune cells in the body. They are primarily located in the bone marrow (BM) within specific microenvironments (niches). It is composed of other cell types, an extracellular matrix (ECM) and soluble biomolecules that influence HSC fate decisions[2]:*quiescence, self renewal, differentiation and apoptosis*. However, little is known about the mechanisms underlying niche regulation, particularly the paracrine-mediated signaling between HSCs and surrounding bone marrow niche cells. We hypothesized that the cues provided by a variable yet systematic gradient of niche cells, the microenvironment and paracrine signaling between HSCs and niche cells will significantly impact HSC fate. To probe this question, we have created a microfluidic tool that enables creation of a 3D hydrogel with opposing gradients of cell and matrix content within a microfluidic chamber (~180 μL). A herringbone microfluidic mixer[3] is used to combine two hydrogel solutions to generate a variable density ratio of the constituents at distinct regions within the final hydrogel. This tool allows high throughput analyses to investigate the role cell-cell interactions and diffusion mediated paracrine signaling pathways play on HSC fate decisions.

We have previously shown that opposing gradients of multiple cell types (i.e. HSCs, osteoblast niche cells) can be created within individual multi-gradient hydrogels ('microgels'). The local ratio of HSCs to niches cells was determined via populational (fluorescent slide scanner) and individual (FACS, confocal imaging of local cell microenvironment) cell measurements. These microgels support cell culture with no observed increase in cell death due to the microfluidic mixing. Ongoing work is determining local small molecule biotransport properties within the microgels using fluorescently labeled Dextran molecules for estimating the efficiency of paracrine signaling processes. We have also created preliminary HSC:niche cell cultures to assess changes in the frequency of key HSC fates in response to defined microenvironments. By varying the type and density of niche cells as well as the density of the hydrogel, we can control the kind of paracrine signals the HSCs are exposed to and their diffusion characteristics. Here, HSCs are isolated as $Lin^{-}Sca\text{-}1^{+}c\text{-}Kit^{+}$ (LSK) from murine bone marrow; niche cell populations chosen for this assay are un-enriched Lin+ bone marrow cells, CXCL12+ bone marrow stromal cells, and murine osteoblasts. We are determining the relative frequency of HSC viability/ proliferation (MTT assay) vs. differentiation (surface antigen expression, MethoCult functional assays) within these defined heterotypic cell microenvironments. Improved understanding of HSC fate decision is critical for optimizing biomaterial systems for *ex vivo* expansion of clinically relevant hematopoietic cells and for studying the etiology, niche regulation and treatment of hematopoietic pathologies.

References

1. Wilson, A. and A. Trumpp, *Bone-marrow haematopoietic-stem-cell niches.* Nat Rev Immunol, 2006. **6** (2): p. 93-106.

2. Tokoyoda, K., et al., *Cellular niches controlling B lymphocyte behavior within bone marrow during development.* Immunity, 2004. **20**(6): p. 707-18.

3. Stroock, A.D. and G.J. McGraw, *Investigation of the staggered herringbone mixer with a simple analytical model.* Philosophical Transactions of the Royal Society of London Series a-Mathematical Physical and Engineering Sciences, 2004. **362**(1818): p. 971-986.

See more of this Session: Biomaterials for Stem Cell Expansion and Differentiation
See more of this Group/Topical: Materials Engineering and Sciences Division

(567e) Pluripotent Stem Cell Colony Formation and Cardiogenesis within PEG-Fibrinogen Microspheres

Samuel S. Chang, Auburn University, Auburn, AL, Bianca A. Williams, Chemical Engineering, Auburn University, Auburn, AL and **Elizabeth A. Lipke**, Department of Chemical Engineering, Auburn University, Auburn University, AL

Heart disease is the greatest single cause of death worldwide, and treatments currently rely on drug therapies that are limited in their capacity to restore or improve function of damaged heart tissue. Cardiomyocytes (CM) are highly specialized electrically active cells which contribute to the heart's ability to contract, but are irreversibly damaged during ischemic events and do not regenerate. Because adult CMs are difficult to expand *in vitro* while maintaining their specialized phenotype, it is desired to engineer cardiac tissue derived from pluripotent stem cells. The development of functional engineered heart tissue can greatly benefit heart disease patients by doubling as a source for cell therapy and also serving as a robust drug screening platform. However, because of the current nascent understanding of biochemical and biophysical cues that influence stem cell differentiation fate, homogeneous differentiation of CMs has yet to be accomplished. Encapsulation of stem cell colonies using hydrogel biomaterials allows for uniform temporal and spatial presentation of biochemical cues, as well as providing the cells with structural support to facilitate their differentiation. A common *in vitro* method of mesodermal differentiation begins with the hanging drop method of embryoid body (EB) formation, which produces multicellular EBs of relatively uniform size, but provides limited control over the microenvironment because of diffusion gradients. This study examines the utility of a unique mouse embryonic stem cell (mESC) encapsulation system using poly(ethylene glycol) (PEG) hydrogel-based microsphere scaffolds as differentiation vehicles. PEG hydrogels are a promising synthetic biomaterial due to their high biocompatibility, high propensity for functional modification, and *in situ* crosslinking capability. This study has utilized a PEG diacrylate-fibrinogen conjugate as the microsphere material. By using a material containing both natural and synthetic components, the cells are provided with a natural substrate for adhesion and biological recognition while also being exposed to a material with tunable mechanical characteristics. Briefly, mouse embryonic stem cells (mESCs) were suspended within the copolymer solution. Encapsulation and sphere formation was achieved via single-emulsion method, where the cell-copolymer solution was mixed with mineral oil and vortexed. Spheres were simultaneously photo-crosslinked by exposure to a metal halide lamp source using the photoinitiators Irgacure 2959 and 2-2-dimethoxy-2-phenyl acetophenone. The resulting cell-laden microspheres were cultured in suspension, allowing for colony formation and differentiation as well as matrix remodeling. Cell viability was assessed via fluorescent microscopy and Live/Dead viability staining. Differentiation was functionally assessed by monitoring cell beating, and quantitatively assessed via immunostaining and gene expression analysis for cardiac markers and germ layer markers. The effects of microsphere size, mechanical characteristics, material, and processing methods on stem cell differentiation were assessed. This study focused on understanding EB formation within PEG-fibrinogen microspheres and developing a correlation between microsphere size and differentiation capacity.

See more of this Session: Biomaterials for Stem Cell Expansion and Differentiation
See more of this Group/Topical: Materials Engineering and Sciences Division

(567f) Protein and Mineral Composition of An Extracellular Matrix Scaffold for Osteogenic Differentiation of Bone Marrow Stromal Cells

Richard A. Thibault, F. Kurtis Kasper and Antonios G. Mikos, Bioengineering, Rice University, Houston, TX

Previous studies have demonstrated that electrospun poly(ε-caprolactone) scaffolds can be coated with an osteogenic extracellular matrix generated by bone marrow stromal cells (PCL/ECM scaffold) under engineered conditions. The PCL/ECM scaffolds have been shown to retain the osteogenic differentiation of bone marrow stromal cells cultured without an osteogenic differentiation factor. Although the extracellular matrix is known to contain collagen and a calcium-based mineral, the presence of other proteins found in minor quantities in mature bone has not been elucidated. This research investigated the composition and temporal deposition of the proteins and minerals of osteogenic extracellular matrix scaffolds. PCL/ECM scaffolds with different extracellular matrix maturities were generated utilizing a flow perfusion bioreactor and analyzed for their protein and mineral composition by liquid chromatography-tandem mass spectroscopy (LC-MS/MS), energy dispersive x-ray diffraction (EDX), x-ray diffraction (XRD), and calcium and phosphate assays. The analysis revealed that at short culture durations cellular adhesion proteins were deposited, while at longer durations the scaffolds contained collagen 1, hydroxyapatite, matrix remodeling proteins, and regulatory proteins.

See more of this Session: Biomaterials for Stem Cell Expansion and Differentiation
See more of this Group/Topical: Materials Engineering and Sciences Division

(567g) Factorial Screening of Self-Assembling Peptide Matrix for Stem Cell Proliferation and Differentiation

Ying Chau and Qianqian Li, Chemical and Biomolecular Engineering, The Hong Kong University of Science and Technology, Hong Kong, Hong Kong

Stem cell proliferation and differentiation is directed by a complex orchestration of different signals [1, 2]. To create a suitable artificial environment, self-assembling peptides provide the versatile medium since oligopeptide signals can be readily presented in a well-defined and well-controlled manner. Exploiting this material property and using factorial analysis, we developed a quantitative and high-throughput screening platform to investigate the interaction between stem cells and multiple extracellular matrix signals under different culture dimensionality. We found that 3D culture system provided a more suitable environment for mouse embryonic stem (ES) cells proliferation. Interestingly, epitopes derived from two different proteins (fibronectin and collagen) showed a synergistic effect on ES cells proliferation. Experiments are ongoing to determine the specific lineage of cellular differentiation under this condition. The results from this study will provide details about a cocktail of oligopeptide epitopes useful for controlling the stem cell fate, and the information will shed light on the molecular mechanism involved in stem cell growth and differentiation.

[1] Cukierman E, Pankov R, Stevens DR, Yamada KM. Taking cell-matrix adhesions to the third dimension. Science 2001;294:1708-12.

[2] Geiger B. Cell biology - Encounters in space. Science 2001;294:1661-+.

See more of this Session: Biomaterials for Stem Cell Expansion and Differentiation
See more of this Group/Topical: Materials Engineering and Sciences Division

(567h) Endothelial Progenitor Cell Adhesion and Growth On Peptide-Linked Scaffolds with and without Flow

Xin Wang, Rustin Shenkman and Daniel Heath, Chemical and Biomolecular Engineering, The Ohio State University, Columbus, OH

Endothelial Progenitor Cell Adhesion and Growth on peptide-linked Scaffolds with and without flow

Xin Wang[1],Rustin Shenkman[1], DanHeath[1], and Stuart Cooper[1]

[1]The Ohio State University, Dept. Chemical andBioMolecular Engineering.

Statement ofPurpose: Recent literature(Hirschi K. Arterioscler ThrombVasc Biol. 2008; 28:1584-1595) hassuggested that a key mediator of the endothelium repair mechanism is the bonemarrow-derived mononuclear cell, endothelial progenitor cell (EPC). But EPC are especially rare, difficult todefine by surface antigens, and expand to colonies very slowly *in vitro*.

A possible clinicalapplication might be to adhere EPC and expand them quickly on 3Dscaffolds embedded with growth factors and/or affinity tags specific for EPC.Affinity tags now exist in the form of peptides, isolated via phage display,that bind specifically to EPC but not adult endothelial cells (Veleva A.Biomaterials. 2008; 29:3656-3661). Tentative names are based on peptidesequences, e.g. "TPS", and "GHM". Candidate peptides have been incorporatedinto a novel non-fouling polymer. This polymer may be cast into films orelectrospun into 3D nonwoven meshes. Such templates are being studied for theirspecific affinity for EPC with and without shear stress.

Methods: Nonfouling terpolymer is produced by free radicalpolymerization of methacrylate monomers, methacrylated PEG, and a succinimide-or maleimide-methacrylate. EPCs are either procured from adult humanperipheral blood (HBOECs) or directly purchased from Lonza (ECFCs).

EPCs and Human Umbilical Vein Endothelialcells (HUVECs) are cultured on the polymer to assess adhesion and growthresponses to the peptides. Adhesion is measured by DNA assay and actin stainingfor static culture experiments. In dynamic experiments, cell adhesion wascarried out in a radial flow chamber allowing real time quantification of celladhesion as a function of shear rate. The radial flow chamber was mounted on aNikon inverted microscope, and ImagePro software was used to automaticallyselect observation fields and count cell number (DickinsonRB, et al, Bioengineering, Food and Natural Products. 1995; 41:2160-2174).

Results: Phage display peptides have beensynthesized and incorporated into the methacrylate terpolymer in addition to a positivecontrol, RGD, and a negative control , RGE. Preliminary data demonstrateselective EPC adhesion to scaffold linked with the promising peptides,especially GHM (Figure 1C). HUVEC shows very low adhesion to surfaces exceptRGD and TCPS.

A comparison of EDTA-lifted HUVEC and EPCattachment at 10 minutes on TCPS is shown in Figure 2. The number of adherentHUVECs is less than the number of EPCs for the range of shear rates examined (5to 30s-1). At higher shear rates, the adhesion ability of EPCs is higher by afactor of 3 to 4 compared withthat of HUVECs.

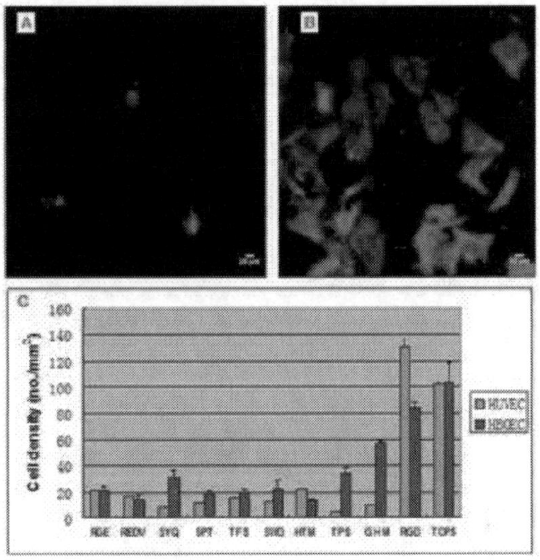

Figure 1. (A)&(B)Fluorescence images of HUVECsadherent to RGE and RGD after 2 hours of culture (C) HUVEC & EPC static adhesionon peptide-linked scaffolds

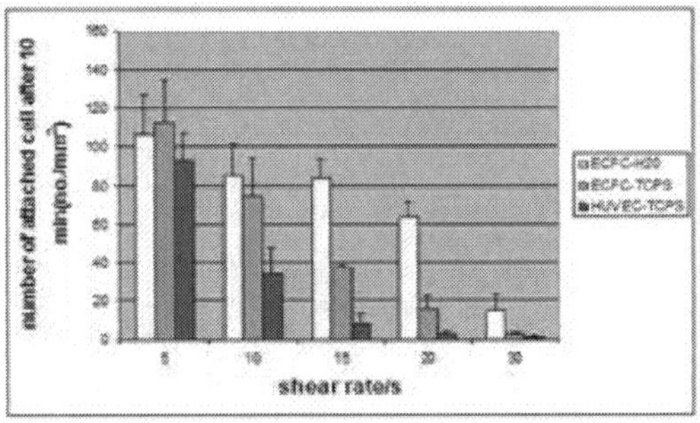

Figure 2. Dynamic adhesion of HUVECs and EPCsto TCPS and base polymer (H20) after 10 minutes

Conclusions: Methacrylate terpolymers are amenable toprocessing, and can incorporate virtually any short peptide or combination ofpeptides. Phage display peptides with specific affinity for human peripheralEPC have been synthesized and incorporated into substrates for cell adhesion:cast films, and electrospun scaffolds. Increased adhesion of cells to the RGDbinding sequence proves that the peptides remain active. Low HUVEC adhesion andrelatively high attachment of EPC on GHM illustrates specific binding abilityof the substrate.

The highlyproliferative EPCs (Lin Y, etal, J of Clin Invest. 2000; 105: 71-77) also have higher adhesion propensity than EC underflow and thus may have the potential to contribute to endothelialization of adevice Ongoing work aims to demonstrate selective EPC affinity for electrospunpeptide scaffolds. This would permit rapid EPC expansion directly from bloodsamples and support vascular graft endothelialization.

See more of this Session: Biomaterials for Stem Cell Expansion and Differentiation
See more of this Group/Topical: Materials Engineering and Sciences Division

(568a) Developing a Bio Inspired Gecko Adhesive System

Jing Yu[1], Sathya Chary[2], Saurabh Das[1], John Tamelier[2], Kimberly Turner[2] and Jacob N. Israelachvili[1], (1)Chemical Engineering, University of California, Santa Barbara, Santa Barbara, CA, (2)Mechanical Engineering, University of California, Santa Barbara, Santa Barbara, CA

The gecko can rapidly attach and detach from almost any kind of surface. This ability is attributed to the hierarchical structures of their toe pads, which generate both strong adhesion and friction forces. In this study, large arrays of micron-scale rectangular flaps composed of polydimethylsiloxane (PDMS) have been fabricated using massively parallel microelectromechanical systems (MEMS) fabrication techniques with the intention of creating a responsive, high friction, high adhesion, anisotropic material similar to that found in geckos. Friction and adhesion tests demonstrate that the tilted angle and specialized structure allow us to optimize both friction and adhesion forces along different directions during movement. These properties, when coupled with suitable articulation mechanisms, can have important implications for designing reversible adhesion systems for climbing robotic applications.

See more of this Session: Biomimetic Materials II
See more of this Group/Topical: Materials Engineering and Sciences Division

(568b) The Fabrication of Dry Adhesives Mimicking the Gecko Adhesive System

Kejia Jin, Chemical and Biomolecular Engineering, Tulane University, NEW Orleans, LA and Noshir Pesika, Chemical and Biomolecular Engineering, Tulane University, New Orleans, LA

In this study, we fabricate dry adhesives mimicking the gecko fibrillar adhesive system using conventional as well as unconventional techniques. Synthetic dry adhesives with varying tilt angles in the fibrillar structures are fabricated and the resulting adhesion and friction forces are measured and analyzed using a tribometer. We find that the tilt in the fibrillar structures provides anisotropy in the response of the surface as well as determines the magnitude of the adhesion and friction forces upon shearing. Our results show good agreement with the peel zone model for gecko adhesion. Potential applications of such a synthetic dry adhesive include reusable adhesive tapes and sticky pads on robotic appendages.

See more of this Session: Biomimetic Materials II
See more of this Group/Topical: Materials Engineering and Sciences Division

(568d) Wettability Enhancement of Bio-Inspired Micro-Textured Polymeric and Carbon Surfaces

Chandra S. Sharma[1], Kumar Abhishek[2], Hari Katepalli[2] and Ashutosh Sharma[2], (1)Department of Chemical Engineering, Indian Institute of Technology, Hyderabad, Yeddumailaram, India, (2) Department of Chemical Engineering and DST Unit on Nanosciences, Indian Institute of Technology, Kanpur, Kanpur, India

Polymeric surfaces with enhanced wettability have been a subject of growing interest and research because of a wide variety of applications in fabricating anti reflective, low friction, water repellent and anti-biofoulant surfaces. Further, carbon surfaces with ultra hydrophobic response add some more applications in the list such as in Bio-MEMS, and microfluidics because of some intrinsic properties associated with carbon as a material itself. We present a simple and robust method to enhance the wettability of polymer surfaces by micro-patterning. The patterns found on two different leaves, Elephant creeper (Argyreia Nervosa) and Nasturtium (Tropaeolum Majus) which show superhydrophobic behaviour because of hairy and micro textured surfaces respectively were bio-mimicked by micro-molding technique in different polymers, polydimethylsiloxane, polystyrene, and resorcinol-formaldehyde (RF) gel. Bio-mimicked polymer surfaces because of roughness induced my micro-texturing shows significant the enhancement in the wettability characteristics. Moreover, since RF gel is a polymer precursor to carbon, micro-patterned RF gel surfaces yielded ultra-hydrophobic carbon surfaces upon pyrolysis. The methodology presented here in this work can be easily applied for fabricating large area stable ultra-hydrophobic polymeric as well as carbon surfaces directly by biomimicking the patterns found on natural plant leaves. Furthermore, the nano-imprint lithography was also used to diversify this novel approach of fabricating the ultra-hydrophobic surfaces by transferring the leaf patterns on various other polymers like poly ethylene terephthalate and poly methyl methacrylate surfaces.

See more of this Session: Biomimetic Materials II
See more of this Group/Topical: Materials Engineering and Sciences Division

(568e) Data Mining Nature to Design New Peptide Based Biomaterials

Andrew D. White and Shaoyi Jiang, Chemical Engineering, University of Washington, Seattle, WA

One strategy for developing new biomaterials is creating a self-assembling peptide surface. Peptides have the flexibility of many monomer units to choose from and are easy to synthesize. However, the design space of peptides has a combinatorial explosion at any significant size. One approach to solving this problem is to use structural bioinformatics to utilize sequence patterns seen in Nature. Starting from a simple structural hypothesis, we've designed ultra-low fouling inert biomimetic surfaces from patterns observed in structural protein data. We first used protein ontologies to find relevant proteins and then used a combination of structural and sequence conservation metrics to gain an understanding of common features. This led to rationally designed peptides that can obtain ultra-low fouling when immobilized on self-assembled monolayers. One particular combination of peptides was found to be frequently occurring in Nature and repeats of that sequence created inert surfaces. Additionally, we fit Markov models to the protein ontologies and used these models to design combinatorial peptide libraries that "look like", for example, blood or the cytoplasma.

See more of this Session: Biomimetic Materials II
See more of this Group/Topical: Materials Engineering and Sciences Division

(568f) Templating Mechanism for Self-Assembly of Large Peptide Structures

Devin Ridgley and Justin R. Barone, Biological Systems Engineering, Virginia Tech, Blacksburg, VA

Large peptide fibers can be spontaneously assembled from combinations of short, hydrophobic peptides (the "template") and longer, alpha-helix containing peptides (the "adder"). Nanometer-sized cross-beta "elementary units" form early and can continue to grow into micron-sized fibers at the correct ratio of template to adder. The self-assembled fibers resemble silk and keratin in size, morphology, and properties but do not require extrusion or spinning to form. A thermodynamic model is presented to highlight the peptide properties necessary for assembly over 3 orders of magnitude of scale. Finally, analogies are drawn between *in vitro* systems and fibrous structures found in bacteria and mussels that show nature may utilize the templating mechanism to form high performance materials for improved modulus, strength, or catalytic rate.

See more of this Session: Biomimetic Materials II
See more of this Group/Topical: Materials Engineering and Sciences Division

Modeling Reactive, Electrodiffusion-Driven, Reversible Crosslink Gradients in Polysaccharides

Matthew A. Reilly, University of Texas at San Antonio, San Antonio, TX

Abstract

Previously, we presented a novel experimental method for generating reversible crosslink density gradients in a polysaccharide system. This system is projected as a vehicle for the study of cellular biomechanics in mechanically graded systems, and for functionally graded prostheses. These experiments were modeled using a single component, 1-D, transient Fickian diffusion model, neglecting the reactions which result in crosslink formation.

The present study updates ongoing modeling efforts to explain experimental findings. An electrodiffusion driving force is considered in a multicomponent, reactive system of simple electrolytes and polyelectrolytes. Reversible reactions govern the equilibrium between simple ionic species and polyions, thereby dictating the degree of local crosslinking. Inclusion of these modeling advances yields good agreement with experimental results.

Introduction

Hydrogels with mechanical property gradients are used for studying cell behavior and may have application in functional prosthetics. Cells may move, differentiate, or function based on substrate stiffness. The function of some tissues may require the presence of an appropriately-designed mechanical property gradient.

We previously demonstrated an experimental model of a polysaccharide which forms a mechanical property gradient in the presence of a cation gradient.[1] This gradient is caused by a crosslink density gradient: as cations diffuse into the polysaccharide solution, the polysaccharide changes from a random coil conformation to a rigid rod conformation. This conformational shift is associated with increasing intermolecular crosslinking.

Unfortunately, a simple Fickian model which attempted to predict the elasticity profile by accounting for only cation mass transfer failed to predict the complex behavior observed experimentally. Here, we expand upon this study by increasing the model complexity to account for electrodiffuion-based mass transport of cations, anions, and the polysaccharide, as well as the crosslinking and acid-base equilibrium reactions. Inclusion of these parameters considerably improved the predictive ability of the model.

Methods

Seven coupled, transient mass conservation equations were solved simultaneously to determine the crosslink density profile within the 1-D sample. A Poisson-Nernst-Planck constitutive relation was adopted to model the driving force of mobile ions. Reversible reaction rate expressions were taken from the literature for the crosslinking reaction as well as the acid-base equilibrium of the polysaccharide in an ionic solution. The anticipated elastic modulus profile of the gel was then calculated using a previously-determined constitutive power law relationship. The model's predicted elastic modulus profile was compared to experimental measurements in a least-squares sense. The model was solved iteratively to estimate diffusivity parameters for each ion in the solution.

Results and Discussion

While the model was able to predict many experimentally-observed trends, it still failed to yield quantitative agreement in all cases. Future work will attempt to develop both more robust models and improve the quality of experimental data to determine the source of the discrepancy. Development of a technique to measure ion diffusivity in the presence of polysaccharide solutions would also improve confidence in these findings.

References

1. Reilly, Matthew A., Zhang, Jingjie, and Ravi, Nathan, Transient Elasticity Gradients for Studying Cell Mobility, AIChE Annual Conference, Nashville, TN, 2009.

Techniques for Hierarchical Bio-Inspired Vascular Networks: Electrohydrodynamic Viscous Fingering and Electrical Treeing

Kristopher D. Behler, Zachary R. Melrose, Andrew Schott and Eric D. Wetzel
U.S. Army Research Laboratory, Aberdeen Proving Grounds, MD

Vascular networks provide a method to distribute fluid throughout a system. Artificial vascular materials with enhanced properties are being developed that could ultimately be integrated into systems reliant upon fluid transport while retaining their structural properties. An uninterrupted and controllable supply of liquid is optimal for many applications such as continual self-healing materials, *in-situ* delivery of optically index matched fluids, thermal management (sweating) and drug delivery systems could benefit from a bio-inspired vascular approach that combines complex network geometries with minimal processing parameters. Two such approaches to induce vascular networks whilst mimicking nature's design are electrohydrodynamic viscous fingering (EHVF) and electrical treeing (ET).

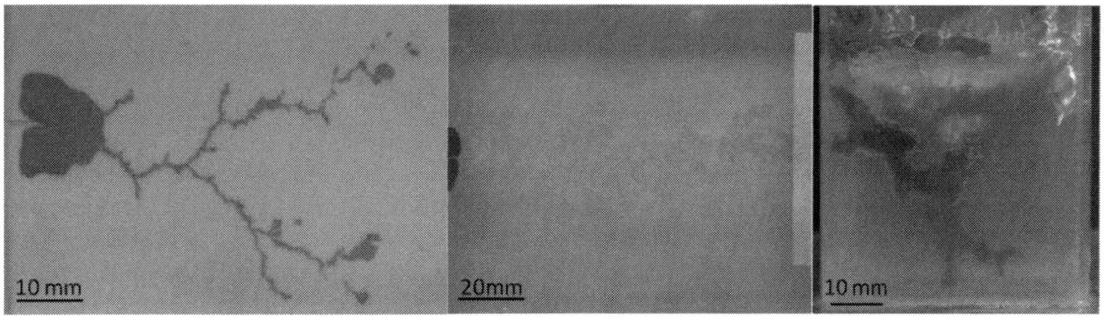

(a) (b) (c)

Figure 1. Optical images of EHVF (a) in a 1,000 cSt silicone oil system containing ~ 60 v/v% glass beads, (b) interfacial polymerization of of hexamethylene diamine and sodium chloride in water and sebocyl chloride in 10,000 cSt silicon oil, and in three dimensions (3DEHVF) using fumed silica in an index matched fluid.

Viscous fingering (VF) is a phenomenon that occurs when a low viscosity liquid is forced through a high viscosity fluid or matrix. The flowing liquid will branch, or form fingers due to capillary and viscous forces in the high viscosity material. EHVF is a modification on viscous fingering in which a DC voltage is applied to the low viscosity conductive fluid (**Fig. 1a**) and forced through a dielectric matrix material. The application of a large electrical potential, 10-60 kV, induces fingers with a reduction in size and an increased branching behavior. The ensuing patterns mimic those found in biology and geology (lung tissue and plants as well as river beds). Observation of VF and EHVF requires Hele-Shaw conditions in which a 2D system must possess a thin gap (0.8 mm used in these experiments) or in a porous system. In the 2D instance a silicone oil system is used as the matrix material, the surfactant concentration was optimized, through a reduction in the interfacial tension, thereby producing a branched pattern of small diameter fingers while still maintaining continuity when dyed water is pushed through the system. Various loadings of glass beads where subsequently used to represent a more 3D system. Typically, in a two fluid system, the fingers relax as soon as the applied voltage is removed. Addition of glass beads, up to 60 v/v%, aids in a retardation of finger relaxation while producing fine channels throughout the porous system. Delayed relaxation allows for greater control of the curing process in UV-curable systems, such as polydimethylsiloxane (PDMS). Fabrics, woven and random glass fiber mats were also investigated as a matrix to provide a different porosity to the system. Matrix filling was used as a method to reduce finger relaxation and allow for curing to occur after the voltage was turned off. Interfacial polymerization EHVF (IPEHVF), a technique in which the polymerization occurs at the interface of two materials, was also studied in producing fingers in the silicone oil phase. For EHVF to be studied in a similar system, hexamethylene diamine (C6H16N2) and sodium chloride were dissolved in the water phase, while sebacoyl chloride (C10H16Cl2O2) was disperse in the silicone oil phase. Robust, polymerized, fingers were formed (**Fig. 1b**) and remained after the voltage was turned off. These fingers were subsequently filled with water to show fluid transport. In moving toward a true 3D system, materials such as fumed silica (**Fig. 1c**) and crushed glass were investigated under 3D (porous) Hele-Shaw conditions.

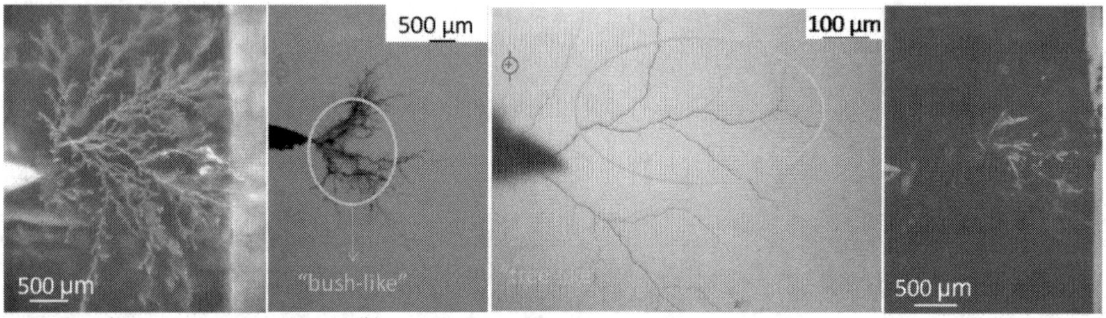

Figure 2. Optical image of ET (a) in a EPON 828/PACM system, (b) under AC driven electrical current showing "bush-like" features, (c) under DC driven electrical current showing "tree-like" features and (d) filling of an ET grown vascular network with a UV visualization dye.

Electrical treeing (ET) is the result of partial discharges in a dielectric material. In the vicinity of a small diameter electrode, the local electric field is greater than the global dielectric strength, causing a localized, step-wise, breakdown to occur forming a highly branched interconnected structure (**Fig. 2a**). The growth of these structures is influenced by the configuration of the electrodes, with geometries of a point lead electrode to a point or plane ground electrode being of most interest. ET is a viable method to produce networks in 2D systems and in more robust 3D systems on a smaller, micron, scale than the products of the EHVF method. AC driven electrical current (**Fig. 2b**), harnessing a sine wave at 100 Hz, grows a "bush-like" structure with many branches and therefore a larger volume within the epoxy samples. DC driven electrical current (**Fig. 2c**) produces a more "tree-like" structure with fewer branches and bifurcations. The surface of the electrodes were modified with dispersed multi-walled carbon nanotubes (MWCNTs) to aid in increasing the local electric field, and thus enable a higher rate of tree initiation and growth. Inclusion of particles was investigated to determine if the growth direction can be manipulated. The use of self-clearing electrodes (as a grounding material) was investigated with the infiltration of a UV dye through the hollow channels produced by ET resulting in a vascularized network capable of repeated fillings and evacuations. Fluid delivery (**Fig. 2d**) can be tailored through the applications of different electrode and ground manufacturing techniques for optimized flow rates for a given application.

(577a) Responsiveness of Polyelectrolyte Brushes to Multi-Valent Counterions

Robert Farina, Nicolas Laugel and Matthew Tirrell, Bioengineering, University of California, Berkeley, Berkeley, CA

While polyelectrolyte brushes are more classically known for their colloidal stabilization or high lubrication properties, their uses have grown as a result of their unique ability to respond to changes in their environment. Potential stimuli, which range from electrical and mechanical to environmental changes in pH and salt concentrations, have propelled applications of polyelectrolyte brushes into fields of multilayer formation and behavior, chemical gating, and drug delivery among others. The responsiveness of a polyelectrolyte brush to a given environmental change is dependent upon its structure. This work will discuss polyelectrolyte brushes comprised of the strong polyelectrolyte, poly (sodium styrenesulfonate) which has shown a strong dependence upon the quantity and valency of salt in a surrounding solution.

When these charged polymer chains are attached to a surface at one end and have a high enough tethering density, the chains of the polymers extend into solution creating a "brush-shaped" structure. The chain elasticity of the polyelectrolytes is resisted by both electrostatic interactions and the osmotic pressure of counterions associated with the charged monomer segments. When these counterions inside the brush, which begin as mono-valent sodium ions, are replaced by multi-valent counterions, polyelectrolyte brush properties are dramatically changed. Brushes which were originally extended, exhibiting purely repulsive forces when forced into contact, become collapsed and adhesive when multi-valent counterions replace their mono-valent counterparts. These results combine intermolecular force data measured when two brushes were brought together using the Surface Forces Apparatus (SFA), and the quantity of multi-valent ions inside a polylelectrolyte brush at a given time and salt concentration determined via cyclic voltametry.

See more of this Session: Charged and Ion-Containing Polymers
See more of this Group/Topical: Materials Engineering and Sciences Division

(577b) Phase Behavior and Coacervation of Aqueous Polyelectrolyte Solutions

Dimitrios Priftis[1], Katie Megley[2] and Matthew Tirrell[1], (1)Bioengineering, UC Berkeley, Berkeley, CA, (2)Bioenginnering, University of California, Berkeley, Berkeley, CA

Phase behavior and coacervation of aqueous polyelectrolyte solutions

Dimitrios Priftis, Katie Megley, Matthew Tirrell*.

Department of Bioengineering, University of California, Berkeley

E-mail: mvtirrell@berkeley.edu

Mixing of oppositely charged polyelectrolytes in aqueous solutions may result in the formation of polyelectrolyte complexes (PECs). Phase separation and complex coacervation of different polyelectrolytes was investigated in such solutions. The formation of PECs is strongly affected by the polyelectrolyte mixing ratio (stoichiometry), the ionic strength (salt concentration), the total polymer concentration, the pH and the temperature. Sample turbidity, which is an indicator of complex formation, was utilized for the study of the above factors. Discrimination between coacervate and precipitate was achieved by optical microscopy. The critical salt concentrations, required for the transition from coacervate to solution, were used for establishing phase diagrams as a function of polybase content and salt concentration. The viscosity of coacervates, formed from various polyelectrolytes mixtures, was measured with the use of a rheometer. Coacervates of PEI and PGANa or PAAsNa showed the typical viscoelastic behavior of polymer solutions, while coacervate viscosities depent on the salt concentration of the mixtures. The polymer rich coacervate phase was tested as a potential substrate for cell proliferation of fiber blast cells (NIH-3T3). Confocal microscopy was utilized to visualize cell interaction and proliferation on the coacervate surface. To optimize NIH-3T3 cell attachment specific bioactive peptide sequences such as $(Lys)_5$(Arg-Gly-Asp) were introduced in the coacervate substrates.

See more of this Session: Charged and Ion-Containing Polymers
See more of this Group/Topical: Materials Engineering and Sciences Division

(577c) Field-Theoretic Simulations of Triblock Polyelectrolyte Gels

Debra J. Audus and Glenn H. Fredrickson, Chemical Engineering, University of California, Santa Barbara, Santa Barbara, CA

Rather than forming gels using the traditional methods of covalent crosslinks or hydrophobic/ hydrophillic interactions, we study polyelectrolyte gels that are formed using electrostatic interactions. In particular, we use a system of oppositely charged, triblock polyelectrolytes with charged endblocks and neutral, hydrophilic midblocks in solution. Under the right conditions, the charged endblocks can aggregate together to form dense, endblock rich micelles called coacervates. The neutral midblocks are excluded from the micelles, but due to chain connectivity, they can form bridges between the coacervate micelles, thus creating a microphase separated gel. Since the coacervate domains are formed by electrostatic interactions, the strength of the coacervate, and therefore the properties of the gel, can be tuned by varying salt concentration, pH and temperature. We investigate the microstructures and solution sensitivity of the triblock polyelectrolytes in an implicit solvent using field-theoretic simulations (FTS). Since the formation of the coacervate domains is driven by charge correlations, the mean-field assumption fails, and it is necessary for us to study the system beyond the level of self-consistent field theory (SCFT).

See more of this Session: Charged and Ion-Containing Polymers
See more of this Group/Topical: Materials Engineering and Sciences Division

(577d) Manipulating Polyelectrolyte Conformations In Solution and At Surface by AC-Electric Fields

Shengqin Wang[1], Richard C. Gurtowski[2] and **Yingxi Elaine Zhu**[2], (1)University of Notre Dame, Notre Dame, IN, (2)Chemical and Biomolecular Engineering, University of Notre Dame, Notre Dame, IN

Recently, we have studied the conformational dynamics of polyelectrolytes under AC-electric fields of varied AC-field frequency and strength. By using florescence correlation spectroscopy (FCS), we examine the structural dynamics of poly(2-vinyl pyridine) (P2VP) at a single molecular level in response to uniform AC-fields between two extended conducting plates, where no net force is imposed on P2VP chains. In stark contrast to an abrupt first-order coil-to-globule transition (CGT) by tuning solution pH, we surprisingly observe a gradual and hysteretic CGT of P2VP conformations within an optimum AC-frequency window and beyond a critical AC-field strength of molecular-weight dependence. The hysteresis is attributed to an asymmetric bistable energy landscape of a single hydrophobic polyelectrolyte chain, whose electrostatic barrier between coil and globule conformations can be reduced below thermal fluctuation energy by AC-field induced polarization due to counterion migration and condensation. Currently, we further explore the AC-field induced conformational dynamics of surface-tethered P2VP brush chains to thereby control surface hydrophobicity and morphology.

See more of this Session: Charged and Ion-Containing Polymers
See more of this Group/Topical: Materials Engineering and Sciences Division

(577e) Reversibly Coagulatable and Re-Dispersable Polystyrene Latex Prepared by Emulsion Polymerization of Styrene Containing Switchable Amidine

Qi Zhang[1], **Wen-Jun Wang**[1], Yangyang Lu[2], Bo-Geng Li[1] and **Shiping Zhu**[3], (1)Department of Chemical & Biological Engineering, Zhejiang University, Hangzhou, China, (2)Department of Polymer Science & Engineering, Zhejiang University, Hangzhou, China, (3)Department of Chemical Engineering, Mcmaster University, Hamilton, ON, Canada

Millions of tons of polymer latexes are produced annually by emulsion polymerization, offering a variety of applications in industry. Both destabilizing latex particles (coagulation) and re-dispersing the coagulated particles are challenging. The advent of long-chain alkyl amidine compounds as switchable surfactant offers an alternative control on emulsion stability. Surface activities of the surfactants were turned on and off by bubbling CO_2 and N_2 (or Ar) gases. CO_2 in water reacts with amidine to form bicarbonate salt that stabilizes emulsion; bicarbonate salt decomposes to amidine to break the emulsion in the presence of N_2 or Ar. We hypothesize that a reversibly coagulatable and re-dispersible latex product is achievable if amidine moieties can be covalently bounded to the particle surfaces.

In this work, we have successfully developed a coagulatable and re-dispersible polystyrene latex system through a soap-free emulsion copolymerization of styrene with a styrene derivative bearing amidine group. The amidine-containing comonomer was reversibly switchable between ionic and neutral states by alternatively bubbling CO_2 and N_2. Such a comonomer incorporated into the polystyrene chains acted as an effective stabilizer during the emulsion polymerization. The latex particles thus formed could be easily coagulated by adding small amount of caustic soda. The coagulated particles could be re-dispersed into water to prepare a stable latex without adding extra stabilizer. The coagulation and re-dispersion processes were repeatable. It was particularly plausible that the aggregated particles after washing and drying were still re-dispersible. Such coagulatable/re-dispersible latex prepared directly from an emulsion polymerization is believed to have high potential in cost savings in the areas of separation, storage, and transportation and thus to benefit our economy and environment.

See more of this Session: Charged and Ion-Containing Polymers
See more of this Group/Topical: Materials Engineering and Sciences Division

(577f) Thermal Transitions In Dry and Hydrated Layer-by-Layer Assemblies

Jodie Lutkenhaus, Ajay Vidyasagar and Choonghyun Sung, Department of Chemical Engineering, Texas A&M University, College Station, TX

Polyelectrolyte multilayer thin films have received significant attention for assembling various nanostructured coatings, but their thermochemical properties are often challenging to measure. Here, we present results regarding the thermochemical properties of a "model" layer-by-layer (LbL) assembly. The LbL process involves alternate deposition of positively and negatively charged polymers resulting in an interpenetrating network of polyelectrolytes layers with fine structural control. Films may grow linearly or exponentially, and each type of growth is expected to give varied internal structure. Poly (allylamine hydrochloride)/poly(styrene sulfonate) (PAH/PSS) LbL assemblies created from solutions of varying ionic strength are examined in their dry and hydrated states using both modulated differential scanning calorimetry and quartz crystal microbalance with dissipation monitoring. We show that both techniques can accurately measure glass transitions in PAH/PSS LbL assemblies. These techniques are then applied to evaluate how Tg-value changes with assembly condition, film thickness, and hydration. Results are compared to analogous polyelectrolyte complexes. Understanding the thermal behavior of both LbL films and complexes is important for designing thermally responsive materials, which will be a future endeavor.

.

See more of this Session: Charged and Ion-Containing Polymers
See more of this Group/Topical: Materials Engineering and Sciences Division

(591a) Ag-Cu Bimetallic Nanoparticles for Metal Enhanced Luminescence (MEL)

Debosruti Dutta[1], Chi Ta Yang[1], Babu Joseph[2] and Venkat R. Bhethanabotla[3], (1)Department of Chemical and Biomedical Engineering, University of South Florida, Tampa, FL, (2)Chemical and Biomedical Engineering, Clean Energy Research Center, University of South Florida, Tampa, FL, (3) Department of Chemical Engineering, University of South Florida, Tampa, FL

Metal enhanced luminescence (MEL) of certain luminophores in the presence of metallic nanoparticles can help enhance the fluoroscence detection of a wide variety of biomolecules. Using bimetallic nanoparticles is an attractive option because important optical properties, like the wavelength corresponding to the surface plasmon resonance (SPR) peak and the scattering efficiency that dictate the enhancement/quenching of luminophore molecules, can easily be tuned by adjusting the composition, morphology, size and shape of the nanoparticles. In this study, we explore the luminescence enhancement of Cy3 dye in the vicinity of Ag-Cu bimetallic nanoparticles. We used Finite Difference Time Domain (FDTD) calculations to study the influence of composition on the Ag_xCu_{1-x} bimetallic nanoparticles. Nanoparticles of various structures like alloys, core-shell and mixed metal substrates with spherical and cubical shapes, and in the size range of 25-200 nm are studied with a varying fluorophore-particle separation distance. Optimizing these parameters provides a guiding principle for the synthesizing the bimetallic nanoparticles that can yield high fluoroscence enhancements. The calculations were validated using bimetallic particles synthesized using the polyol synthesis process. The fluoroscence enhancements were measured using a confocal microscope.

See more of this Session: Integration of Biomolecules and Biological Systems with Electronics and Photonics
See more of this Group/Topical: Materials Engineering and Sciences Division

(591b) Synthesis and Functionalization of Gold Coated Magnetic Nanoparticles for Surface Enhanced Raman Scattering Biosensors

Patrick A. Johnson, Hao Zhang, Jing Neng and Ashley J Driscoll, Chemical & Petroleum Engineering, University of Wyoming, Laramie, WY

The development and characterization of colloidal-based surface enhanced Raman scattering (SERS) bioassays for portable diagnostics will be presented. DNA, antigen and antibody detection systems have been developed using paramagnetic capture assay schemes. Recent development of gold-coated magnetic nanoparticles provides an alternative, single nanoparticle assay scheme. The nanoparticles function as SERS substrates while simultaneously providing a separation / concentration mechanism. Considerations of the advantages of using colloidal systems over surface-based systems for biological assays will be discussed. Using the DNA derived from the Rift Valley Fever Virus (RVFV) and West Nile Virus (WNV) RNA genome as the model targets, multiplex SERS detection were designed and demonstrated. Additionally, multiplexed assays for antibodies and antigens to both target viruses were also demonstrated.

See more of this Session: Integration of Biomolecules and Biological Systems with Electronics and Photonics
See more of this Group/Topical: Materials Engineering and Sciences Division

(591c) Degenerate Si As a Biosensor Substrate Material

Ian Ivar Suni, Department of Chemical and Biomolecular Engineering, Clarkson University, Potsdam, NY and Yin Huang, Dept. of Chemical and Biomolecular Engineering, Clarkson University, Potsdam, NY

We report the use of degenerate (heavily doped) Si as an electrode material for impedance detection of peanut protein Ara h 1, an allergenic food protein. Following immobilization of the mouse monocloncal antibody onto degenerate Si by amide bond formation, detection limits in the pM range can be obtained. Although p- and n-type Si have previously been reported as biosensor substrate materials, they typically form space charge layers that make electrochemical detection methods difficult. One advantage of degenerate Si is much easier formation of ohmic contacts at both sides of the substrate material, which allows for "reversible" electrochemistry, as well as simpler equivalent circuits during impedance detection. Other advantages of degenerate Si include easier incorporation of electrical biosensors into ULSI devices, and improved interfacial stability due to the strength (520 kJ/mole) of the Si-C bond.

See more of this Session: Integration of Biomolecules and Biological Systems with Electronics and Photonics
See more of this Group/Topical: Materials Engineering and Sciences Division

(591d) Biomimetic Soft-Matter Memristors Composed of Hydrogels

Ju-Hee So[1], Hyung-Jun Koo[2], Orlin D. Velev[2] and **Michael D. Dickey**[2], (1)North Carolina State University, Raleigh, NC, (2)Department of Chemical and Biomolecular Engineering, North Carolina State University, Raleigh, NC

This talk will describe the use of bio-compatible hydrogels as the basis for a new type of memory architecture composed entirely of soft materials that mimic synapse operation. New types of electronic devices and circuits based on soft materials have potential applications in bio-electronic circuits, artificial neural networks, and brain-machine interfaces. We will present a new class of diodes and memristors (i.e., memory resistors) composed entirely of soft, liquid-based materials formed by combining a moldable liquid metal and hydrogel doped with polyelectrolytes. Memristors are next generation memory devices that mimic the function of synapses. The electronic functionality of these soft devices originates from the ability to control the electronic and ionic transport at the interface between the metal and the hydrogel. The metal is a eutectic alloy of gallium and indium (75 % Ga, 25 % In). At room temperature it is a low viscosity liquid with a high conductivity ($\sigma = 3.4 \times 10^4$ S/cm). Its surface is coated with a thin, native skin of gallium oxide, which is a wide band-gap semiconductor. The oxide skin passivates the liquid metal; that is, the oxide does not grow thicker with time without any additional driving force. The electrical resistance through the oxide skin depends on its thickness, which can be controlled using pH and/or electrical bias to oxidize or reduce the skin. The agarose gel, which interfaces with the liquid electrodes, is composed of more than 90% water and offers a soft aqueous media into which polyelectrolytes may be doped to control the local pH. We fabricated soft and quasi-liquid electronic devices that mimic solid-state, semiconductor devices, such as diodes and memristors, by sandwiching two hydrogel films: one doped with polyacrylic acid (PAA) and one doped with polyethyleneimine (PEI, which is basic). We contacted both sides of the film stack with liquid metal electrodes. The PEI-doped hydrogel suppresses the oxidation reaction at the metal electrode interface, whereas the thickness of the metal oxide layer at the interface with the PAA-doped hydrogel can be controlled by applying an electric bias. The direction and the magnitude of the electric bias pre-applied to the electrodes controls the anisotropy of current conductance, resulting in tunable rectification ratios. The pre-programmed anisotropy is preserved for more than an hour, which enables the fabrication of memory storage devices (i.e., memristors). We will discuss the rectification characteristics of diodes and the on/off switching behavior of memristors with different geometries composed entirely of soft materials.

See more of this Session: Integration of Biomolecules and Biological Systems with Electronics and Photonics
See more of this Group/Topical: Materials Engineering and Sciences Division

(591e) Engineering Biological Conjugates of Polyethylene Oxide for Lithium Cation Transference

Sean C. O'Neill, Dan Steingart and Raymond Tu, Chemical Engineering, The City College of New York, New York, NY

Engineering Biological Conjugates of Polyethylene Oxide Design for Lithium Cation Transference

Sean O'Neill[1], Raymond Tu Ph.D[1] and Dan Steingart Ph.D[1]

[1]Department of Chemical Engineering – City College of New York

Solid organic electrolytes offer a viable option for electrode separation due to their low cost and toxicity, as well as high stability. A limitation to such electrolytic systems however are, various transport principles, conductivity and selectivity. The fundamental goal of this research project is to engineer an ion transport system that facilitates electrode separation while maintaining the prevention of crystallization.

In order to achieve the primary goal of interest, we first intend to construct polyethylene oxide polymers with well-defined chemical composition to observe the influence of ion:polymer complexation on molecular order. In doing so, we will engineer PEO constructs consisting of ion-coordinating groups and low molecular weight polydispersities. Second we will quantify the influence of polymer charge and structure on ion conductivity and transference number. Lastly we will examine dynamics of polymer nanostructure using ellipsometry, x-ray scattering and microrheology. Doing so will allow us to connect fundamental phase behavior and polymer structure to organic electrolyte performance. We hope that we can achieve our goal as far as electrolyte design by structural optimization of mean free path and Lithium cation transference.

See more of this Session: Integration of Biomolecules and Biological Systems with Electronics and Photonics
See more of this Group/Topical: Materials Engineering and Sciences Division

(591f) Electrostatic Assemblies of Virus-Templated Titania Nanowires for Dye-Sensitized Solar Cells

Rebecca Lynn Ladewski[1], Rebekah A. Miller[2], Forrest W. Liau[2], Angela M. Belcher[3] and Paula T. Hammond[1], (1)Department of Chemical Engineering, MIT, Cambridge, MA, (2)Department of Biological Engineering, MIT, Cambridge, MA, (3)Departments of Biological Engineering and Materials Science, MIT, Cambridge, MA

Layer-by-layer assembly allows for the incorporation of a wide range of functional materials into structured thin films based on the alternate adsorption of cationic and anionic species. Biomolecules, and in particular viruses, show great potential as components of functional materials due to their capacity for molecular recognition and self assembly. Here we report that by substituting a negatively charged variant of M13 bacteriophage for the negatively charged polymer during the assembly process, M13 phage can be incorporated into the film, resulting in a hybrid material with characteristics of both its biological and polymeric components. The resulting mesoporous polymer films can be used as a template for the construction of dye sensitized solar cells (DSSCs) with a novel nanowire architecture to enhance electron transport within the photoanode. The effect of the biotemplated nanowires on device performance has been analyzed using impedance spectroscopy techniques. Design, water-based assembly methods, and materials characterization of these systems will be discussed, as well as device characterization methods.

See more of this Session: Integration of Biomolecules and Biological Systems with Electronics and Photonics
See more of this Group/Topical: Materials Engineering and Sciences Division

(600a) Coating Process Regimes In Particulate Film Production by Forced Convection-Assisted Drag-Out

Damien D. Brewer, **Satish Kumar** and Michael Tsapatsis, Department of Chemical Engineering and Materials Science, University of Minnesota, Minneapolis, MN

Operating conditions for deposition of monolayer and bilayer particulate coatings from aqueous silica dispersions are identified in the context of a drag-out operation assisted by forced convection. Dry film thickness, uniformity, and morphology are assessed within an operating window parameterized by the capillary number and silica dispersion weight fraction. Three film deposition regimes with respect to the capillary number are observed: convective film deposition at low process rates, film entrainment at moderate process rates, and a thin-film transition regime at intermediate process rates. Locally ordered particulate films of variable layering thickness, including discontinuous submonolayer or mixed submonolayer and monolayer, mixed monolayer and bilayer, and multilayer are dominant under convective deposition conditions. A map of morphologies is presented within the capillary number – weight fraction operating window, where monolayer and mixed monolayer/bilayer films are demonstrated in the thin-film transition regime at intermediate dispersion weight fraction. These operating maps are constructed using a model silica dispersion, and are therefore relevant to particulate coatings of other inorganic materials.

See more of this Session: Nanostructured Thin Films
See more of this Group/Topical: Materials Engineering and Sciences Division

(600b) High-Performance Randomly Oriented Zeolite Membranes Using Brittle Seeds and Rapid Thermal Processing

Won Cheol Yoo[1], Jared A. Stoeger[1], Pyung-Soo Lee[1], Andreas Stein[2] and Michael Tsapatsis[1], (1) Department of Chemical Engineering and Materials Science, University of Minnesota, Minneapolis, MN, (2)Department of Chemistry, University of Minnesota, Minneapolis, MN

To exhibit high flux and high selectivity for energy efficient membrane separation processes, thin and defect-free zeolite membranes on a porous support must be fabricated. Experimental efforts have been directed to the confined synthesis of zeolite particles to be used as a seed layer for the secondary hydrothermal growth of membranes, primarily due to controllable size and good monodispersity of the seed particles. Here, we present a facile method to fabricate MFI zeolite membranes with high flux (ca. 2.6×10^{-7} $molm^{-2}s^{-1}Pa^{-1}$) and high separation factors (SF, 123–139 and in one case as high as 335) for xylene isomer separation by optimizing seed deposition and calcination processes. Zeolite nanoparticles prepared by aforementioned confined syntheses were deposited as randomly oriented seed layers on porous supports by rubbing and leveling methods to obtain thin zeolite films (400–500 nm) after secondary hydrothermal growth. Rapid thermal processing (RTP) was used in a single calcination step that substantially reduced processing times and energy usage and improved the separation efficiency of the membranes.

See more of this Session: Nanostructured Thin Films
See more of this Group/Topical: Materials Engineering and Sciences Division

(600c) Stability and Separation Performance of Highly c-Oriented AFI-Type Aluminophosphate Membranes

Jared A. Stoeger[1], Miguel Palomino[2], Charitomeni M. Veziri[3], Avelino Corma[2], Nick K. Kanellopoulos[3], Georgios N. Karanikolos[3] and Michael Tsapatsis[1], (1)Chemical Engineering and Materials Science, University of Minnesota, Minneapolis, MN, (2)Instituto de Tecnologia Quimica (CSIC-UPV), Universidad Politecnica de Valencia, Valencia, Spain, (3)Institute of Physical Chemistry, Demokritos National Research Center, Athens, Greece

Continuous,highly *c*-oriented aluminophosphate films(structure code AFI) were grown on porous α-Al_2O_3supports in order to investigate their potential for molecular separationapplications. Variations in metalsubstitution ($AlPO_4$-5, CoAPO-5, CoSAPO-5), growth method (forced airconvection oven, microwave), and calcination technique (conventional, rapidthermal processing) were chosen to examine the conditions by which membranequality was optimized. Followinghydrothermal growth in a forced air convection oven, $AlPO_4$-5 andCoAPO-5 membranes underwent densification to an AlPO-tridymitephase following heat treatment, resulting in reduced membrane functionalitywhich may be partially avoided through the use of rapid thermal processing. Onthe other hand, membranes prepared by an optimized method were found to bethermally stable with high membrane quality as inspected through pervaporationmeasurements consisting of a liquid hydrocarbon feed of n-heptaneand 1,3,5-triisopropylbenzene. We will discuss the new aspects thatallowed fabrication of oriented, continuous, high-quality aluminophosphatemembranes and we will provide useful insight into their potential forapplications such as molecular separations, catalysis, or host-guestassemblies.

See more of this Session: Nanostructured Thin Films
See more of this Group/Topical: Materials Engineering and Sciences Division

(600d) Synthesis of Zeolite Imidazolate Framework Films and Membranes On Metal Modified Supports

Miral N. Shah, Victor Varela-Guerrero and Hae-Kwon Jeong, Chemical Engineering, Texas A & M University, College Station, TX

Zeolitic imidazolate frameworks (ZIFs) are an exciting new class of hybrid organic-inorganic materials, known as metal-organic frameworks (MOFs), due to their exceptional thermal and chemical stability[1]. Their regular microporous structure makes them promising for a variety of gas separations[1-2].

To demonstrate this gas permeation performance in case of membranes, a general technique to fabricate a variety of ZIFs or MOFs membranes is required, thereby minimizing the time and resources required to develop a different method for each MOF[3]. We have previously reported a generally applicable thermal ligand modification technique which gives rise to continuous, well intergrown films and membranes in relatively short time (few hours as compared to days)[4]. Here we report a novel and general method for synthesizing MOF membranes. In this technique zinc nitrate solution is dropped on hot bare porous alumina support giving rise to metal modified supports (thermal metal modification). Zinc modified support promotes heterogeneous nucleation and growth and leads to continuous well intergrown films of ZIF-8. This technique offers certain advantages over the previously reported thermal ligand modification technique, wherein the necessity of ligand stability at high temperatures for thermal ligand modification is eliminated. Also it provides an environmentally greener route to membrane fabrication, due to reduced emissions from ligand decomposition at high temperature. Further, metal modification of support can be extended to the fabrication of any other ZIF membrane, or for any other MOF, in general. Here we investigate the gas permeation performance of a membrane grown on a zinc modified support as compared to a ligand modified support.

References:

1. Park, K. S.; Ni, Z.; Cote, A. P.; Choi, J. Y.; Huang, R. D.; Uribe-Romo, F. J.; Chae, H. K.; O'Keeffe, M.; Yaghi, O. M., Exceptional chemical and thermal stability of zeolitic imidazolate frameworks. *Proceedings of the National Academy of Sciences of the United States of America* **2006,** *103* (27), 10186-10191.

2. Liu, Y. Y.; Hu, E. P.; Khan, E. A.; Lai, Z. P., Synthesis and characterization of ZIF-69 membranes and separation for CO2/CO mixture. *Journal of Membrane Science* **2010,** *353* (1-2), 36-40.

3. Haldoupis, E.; Nair, S.; Sholl, D. S., Efficient Calculation of Diffusion Limitations in Metal Organic

Framework Materials: A Tool for Identifying Materials for Kinetic Separations. *Journal of the American Chemical Society* **2010,** *132* (21), 7528-7539.

4. McCarthy, M. C.; Varela-Guerrero, V.; Barnett, G. V.; Jeong, H. K., Synthesis of Zeolitic Imidazolate Framework Films and Membranes with Controlled Microstructures. *Langmuir* **2010,** *26* (18), 14636-14641.

See more of this Session: Nanostructured Thin Films
See more of this Group/Topical: Materials Engineering and Sciences Division

(600e) Control of Nano-Porosity In Plasma Deposited Low-k Diffusion Barrier and Inter-Layer Dielectrics for Nano-Electronic Applications

Sean King[1], Ebony Mays[1], Jeff Bielefeld[2], Ming Liu[3] and David Gidley[3], (1)Logic Technology Development, Intel Corporation, Hillsboro, OR, (2)Components Research, Intel Corporation, Hillsboro, OR, (3)Department of Physics, University of Michigan, Ann Arbor, MI

As the semiconductor industry strives to keep pace with Moore's Law, new materials with extreme properties are increasingly being introduced and tighter control of these material properties is being demanded. Low dielectric constant (i.e. low-k) materials are one specific example. Lower k materials are desired to replace SiO_2 (k=4.0) as the interlayer dielectric (ILD) and SiN_x:H (k = 7.0) as the Cu capping diffusion barrier layer in order to reduce resistance-capacitance (RC) delays in nano-electronic Cu interconnect structures. Typical methods for producing low-k materials consist of introducing controlled levels of nano-porosity via carbon doping during plasma enhanced chemical vapor deposition (PECVD) of SiO_2 and SiN_x:H matrix materials. While lowering k, the introduction of nano-porosity can seriously compromise the performance of these layers in their respective applications. In this presentation, we will demonstrate that critical thresholds in nano-porosity exist for the diffusion of water and solvents through low-k materials. Specifically, we utilize Fourier Transform Infra-Red (FTIR) spectroscopy, to show that the concentration and size of nano-pores formed in low-k a-SiC(N):H dielectric materials is controlled by the concentration of terminal Si-CH_3 bonding versus Si-C/N network bonding. We further combine moisture / solvent diffusivity measurement with x-ray reflectivity (XRR) and positron annihilation lifetime spectroscopy (PALS) to demonstrate that low-k a-SiC(N):H dielectrics become poor moisture diffusion barriers at mass densities < 2.0 g/cm^3 and when the pore size approaches that for the molecular diameter of water. Similarly, we show that low-k a-SiOC:H ILDs become easily penetrable by solvents and susceptible to downstream processing damage when the pore diameter approaches the size of the solvent and pores become interconnected. The implications of these critical nano-porosity thresholds on continued scaling of low-k diffusion barrier and ILD materials will be discussed as well as methods for overcoming these limitations.

See more of this Session: Nanostructured Thin Films
See more of this Group/Topical: Materials Engineering and Sciences Division

(600f) Zeolite Thin Films Prepared From Exfoliated MFI and MWW Nanosheets On Non-Porous Substrates

Christopher M. Lew, Kumar Varoon, Xueyi Zhang, Bahman Elyassi and Michael Tsapatsis, Chemical Engineering & Materials Science, University of Minnesota, Minneapolis, MN

We report on the synthesis of MFI- and MWW-type zeolite thin films prepared on non-porous substrates. Plate-like nanosheets were used as hierarchical building blocks for these films and were isolated from lamellar precursor materials by recently developed exfoliation methods. Film morphology was characterized by scanning electron microscopy (SEM), and X-ray diffraction (XRD) and transmission electron microscopy (TEM) were used to evaluate the zeolite structure. The use of lamellar nanosheets may create unique film properties and open new opportunities for thin film fabrication. These films are potentially useful in low-dielectric constant (low-*k*) applications for computer microprocessors.

See more of this Session: Nanostructured Thin Films
See more of this Group/Topical: Materials Engineering and Sciences Division

(600g) Assembly and Scaffolding Strategies for Realizing Ultra-Thin Inorganic Films with Tunable Porosity

Shih-Chieh Kung, Zheng Tian and **Mark A. Snyder**, Department of Chemical Engineering, Lehigh University, Bethlehem, PA

Porous thin films scalable to tens of nanometers in thickness with tunable hierarchy in pore size and topology hold exciting technological implications spanning, among others, high-selectivity and high-flux membrane separations to rapid-response sensors selective for large and small target molecules. Owing to their inherent microporosity, inorganic materials such as zeolites, carbons, and some nanoparticle assemblies, hold specific promise in such applications where molecular selectivity is critical. Yet, control over porous film thickness has been a persistent challenge for films composed of such materials and supported on porous or non-porous substrates, leading in some cases to stress-induced crack formation and sluggish molecular flux. We address the challenge of controlling film thickness of porous inorganic films, and in some cases controlling underling porosity of these films, through a hierarchical templating strategy rooted in convective deposition of tunable nanoparticles and subsequent replication and materials scaffolding techniques. Specifically, we will address three classes of porous thin films scalable to just tens of nanometers in thickness: convectively deposited nanoparticulate colloidal crystalline films, carbon replica films characterized by either ordered mesoporosity or asymmetric mesopore-supported ultra-thin microporous layers, and (sacrificial) carbon-scaffolded ultra-thin zeolite films (e.g., silicalite-1, SAPO-34) composed of extensive high-aspect-ratio single-crystal domains. We will describe stages in film formation, sensitivity of film properties to synthesis conditions, transition from non-porous to highly-porous substrates as film supports, and the role of pore topology in thin film scaffolding. We will describe critical structure-property relations governing porous film performance with regards to achievable molecular selectivity and flux towards establishing proof-of-principle application of this synthesis approach to a broader class of porous inorganic materials.

See more of this Session: Nanostructured Thin Films
See more of this Group/Topical: Materials Engineering and Sciences Division

(600i) A Finite-Element Based Global Model for Multiphase Flows In a Convective Assembly System

Gaurab Samanta[1], Andrew Yeckel[1], Satish Kumar[2] and Jeffrey J. Derby[1], (1)Chemical Engineering and Materials Science, University of Minnesota, Minneapolis, MN, (2)Department of Chemical Engineering and Materials Science, University of Minnesota, Minneapolis, MN

Convective assembly is a promising method for manufacturing self-assembled, colloidal crystals. In this process, a moving meniscus and fluid flow is employed to achieve self-assembly of particles from a suspension --- mono-disperse spheres crystallize near the moving contact line on a stationary substrate immersed in an evaporating medium. In this process, a thin fcc crystal is deposited on the substrate in a matter of few hours. In order to understand the mechanism behind such fast deposition of ordered arrays of particles, Norris et al. put forth the convective steering hypothesis, which posits that solvent flow through pore spaces of close-packed spheres could preferentially direct advancing spheres into "clear or fcc niches" rather than into the "obstructed niches". Although calculations based on porous-media network theory suggested that flow is up to 33% higher into the clear niches than into obstructed niches, subsequent calculations based on elaborate three-dimensional finite element analysis of flows into an fcc structure showed that inertial effects of the global flow may overwhelm the local steering effects induced by the particle arrangement.

In this presentation, we focus on the character on the global flows in convective assembly system by employing a two-dimensional, finite element analysis of the temporal flows in the system. We account for time-dependent flows in the solvent as well as the gas phase above it, including a rigorous accounting for evaporation, capillary, and Marangoni forces at the liquid-gas interface. We study the effect of varying gas inflow over the liquid surface, substrate and growth front inclinations, and solute concentration in the suspension on the flow behavior of the system. We validate our model against prior experimental results.

See more of this Session: Nanostructured Thin Films
See more of this Group/Topical: Materials Engineering and Sciences Division

(603a) Fundamental Aspects of Organic Heterostructure Formation Examined Using Supersonic Molecular Techniques and In Situ Real Time X-Ray Synchrotron Radiation

James R. Engstrom[1], Edward R. Kish[2], Tushar V. Desai[2] and Arthur R. Woll[3], (1)School of Chemical and Biomolecular Engineering, Cornell University, Ithaca, NY, (2)Chemical and Biomolecular Engineering, Cornell University, Ithaca, NY, (3)Cornell High Energy Synchrotron Source, Cornell University, Ithaca, NY

Over the past several years significant advances have been made concerning our understanding of the growth of crystalline small molecule organic thin films consisting of a single component. An important challenge in organic electronics, photonics and photovoltaics is to develop and improve methods to integrate both *p*-type and *n*-type small molecule organic semiconductors into the same device microstructure. Thus, developing an understanding of the molecular scale events that lead to heterojunction formation is essential in these systems consisting of multiple components. Here we report on our examinations of the nucleation, growth, and dynamics of adsorption of a *n*-type organic semiconductor, *N,N*•-ditridecylperlyene-3,4,9,10-tetracarboxylic diimide (PTCDI-C_{13}), on SiO_2 surfaces modified by self-assembled monolayers (SAMs) and on a pre-deposited monolayer of pentacene using supersonic molecular beam techniques, *in situ* synchrotron x-ray scattering and *ex situ* atomic force microscopy. From real-time x-ray scattering we find that PTCDI-C_{13} exhibits prolonged layer-by-layer growth for approximately the first 10 monolayers (MLs) of deposition on all three SAMs examined. Concerning the kinetics of growth we find that the adsorption probability of PTCDI-C_{13} on itself is similar to that observed on two SAMs that possess aromatic endgroups, but it differs significantly to that observed on a relatively short, methyl-terminated SAM. These differences could reflect mechanisms such as direct molecular insertion of PTCDI-C_{13} into either the existing PTCDI-C_{13} film, or the longer chain SAMs with aromatic endgroups. Concerning growth in the submonolayer regime, we find that nucleation is homogeneous, and that the absolute density of islands depends on the nature of the surface, while the relative change of the island density with increasing growth rate is essentially independent of the underlying SAM. From the latter we find that a critical island size of a single molecule of PTCDI-C_{13} can describe all the data. Finally, we will discuss our most recent results concerning the growth of heterostructures composed of a few to several monolayer stacks of PTCDI-C_{13} and pentacene. In this work we find that PTCDI-C_{13} grows in a smooth layer-by-layer fashion on pentacene, but the opposite is not true—pentacene grows in a purely 3D mode when deposited on PTCDI-C_{13}. We will discuss the implications of this observation concerning the growth of organic heterostructures for applications in electronics, photonics and photovoltaics.

See more of this Session: Organic Materials for Electronics, Photonics, and Photovoltaics
See more of this Group/Topical: Materials Engineering and Sciences Division

(603b) Engineering Nanoscale Film Morphology In Organic Photovoltaic Cells Using Graded Donor-Acceptor Heterojunctions

Richa Pandey and Russell J. Holmes, Chemical Engineering and Materials Science, University of Minnesota, Minneapolis, MN

Organic photovoltaic cells (OPVs) have received considerable attention due to their potential low-cost, and compatibility with roll-to-roll fabrication and flexible substrates. The excitonic nature of OPVs necessitates the use of a donor-acceptor (D-A) heterojunction in order to dissociate the exciton into its constituent charge carriers. While the optical absorption length in most organic semiconductors is ~100 nm, excitons may typically diffuse only over a distance of ~10 nm. This trade-off between absorption and exciton diffusion is often overcome by increasing the area of the dissociating D-A interface using engineered film morphologies. This work presents an approach to realize optimized film morphology in OPVs based on small molecule organic semiconductors using a highly tunable, continuously graded D-A heterojunction (GHJ). The GHJ leads to an increase in the D-A interface area allowing for an enhanced exciton diffusion efficiency, while also preserving the charge collection efficiency. This simultaneous optimization leads to a significant improvement in device performance relative to that of conventional planar and uniformly mixed OPVs. Here, the performance of GHJs containing boron subphthalocyanine chloride (SubPc) and C_{60} as the electron donating and accepting materials, respectively, is optimized based on supporting electrical and morphological characterization of uniform mixtures. Electrical characterization suggests optimum charge transport in mixed films containing 80 wt.% C_{60}. The origin of this C_{60}-rich optimum composition is explained in terms of morphological changes in the active layer upon diluting SubPc with C_{60}. While neat SubPc is found to be amorphous, mixed films containing 80 wt.% C_{60} show clear nanocrystalline domains of SubPc. Using this optimum composition in GHJ OPVs, we demonstrate a power conversion efficiency of η_P=(4.5±0.1)% under 100 mW/cm^2 AM1.5G simulated solar illumination.

See more of this Session: Organic Materials for Electronics, Photonics, and Photovoltaics
See more of this Group/Topical: Materials Engineering and Sciences Division

(603c) Understanding Charge Transport Behavior In Ionic Liquid Gated Semiconducting Polymers

Bryan D. Paulsen and C. Daniel Frisbie, Chemical Engineering & Materials Science, University of Minnesota, Minneapolis, MN

For the past decades, the prospect of cheap, solution processable, and flexible electronic devices and circuits has driven organic electronics research. Recently, low voltage flexible printed polymer transistor based circuits have been realized by utilizing ionic liquids in place of traditional gate dielectric materials. Ionic liquid gating of polymer semiconductors functions via bulk electrochemical doping as opposed to the strictly interfacial field effect, which allows for low voltage high channel current transistors. While promising, in order for this technology to reach its full potential, a fundamental study of the charge transport in ionic liquid gated polymer semiconductors is necessary. Working with the bench mark polymer poly(3-hexylthiophene) (P3HT) gated with the ionic liquid, 1-ethyl-3-methylimidazolium tris(pentafluoroethyl)trifluorophosphate (EMIFAP), an accurate electrochemical density of states was determined by cyclic voltammetry, which when coupled with *in situ* conductivity measurements allowed an accurate determination of charge carrier mobility. Charge carrier mobility was found to be heavily dependent on charge carrier density, increasing to ~0.8 $cm^2 V^{-1} s^{-1}$ with increasing charge density before falling back to zero, resulting in a finite potential window of electrical conductivity. Variable temperature measurements were carried out over the whole window of finite conductivity, revealing a thermally activated charge transport mechanism, which became very weakly activated in the region of high carrier mobility. The ionic liquid anion and cation were systematically varied to determine their individual effects on charge transport and chemical stability. Additionally, poly (2-methoxy-5-(2′-ethylhexyloxy)-1,4-phenylenevinylene) (MEH-PPV) and poly(9,9-dioctylfluorene-co-benzothiadiazole) (F8BT) were investigated in a similar manner revealing the generality of this charge transport behavior amongst ionic liquid gated polymer semiconductors.

See more of this Session: Organic Materials for Electronics, Photonics, and Photovoltaics
See more of this Group/Topical: Materials Engineering and Sciences Division

Reduced Dimensionality from Nanoengineering of Soft Matter Organic Electrooptic Materials

Stephanie J. Benight, Lewis E. Johnson, Daniel B. Knorr, Jr., Rene M. Overney, Bruce H. Robinson, and Larry R. Dalton, University of Washington, Seattle, WA, USA

Electro-Optic (EO) materials are vitally important to the progress of photonic, telecommunications, computing, sensing, and imaging applications.[1] Examples in which EO materials offer tremendous advantages to these applications include electrical to optical signal transduction and signal routing, input/output (I/O) interconnects for all types of computing and various classes of sensing including electromagnetic, biochemical, and physical sensing through sensitivity to refractive index changes, pressure or temperature.[2] Traditionally, inorganic materials grown from crystals are utilized to exploit Pockels effect for these applications, however such materials (e.g. lithium niobate) are limited in their performance ($r_{33} \approx 32$ pm/V) (based on ionic displacement mechanisms) and produce large amounts of optical loss. Because of these limitations, organic materials have been a primary focus of materials research in the improvement of EO material performance in devices. Organic EO (OEO) materials offer several advantages compared to their conventional inorganic counterparts, namely: (1) ultrafast response times on the order of tens of femtoseconds which permit higher device operational bandwidths (2) lower dielectric constants leading to lower power consumption (3) the potential for light weight and flexible devices due to inherent material properties (4) adaptability to several processing techniques (e.g. lithography and liquid or vapor phase processing) and (5) significantly higher electro-optic coefficients (r_{33}).[2]

In order to improve the r_{33} of an OEO material, the bulk material must be noncentrosymmetric (molecules unidirectionally aligned) to obtain a large net EO effect. Toward this effort, self-assembly (Langmuir-Blodgett, self-assembly, vapor phase), crystal growth, and electric field poling are techniques that can be used to induce a bulk material acentricity. The latter of these techniques, electric field poling, has solely been utilized in inducing acentric order in the materials to be discussed here. Conventional poled polymer systems consist of a guest chromophore with high hyperpolarizability doped into host amorphous polymer systems such as APC or PMMA and are poled with an applied poling field electric strength (E_p) of 100 V/μm in the z-axis direction to achieve r_{33} values on the order of 50-100 pm/V.[3] However, such systems are limited to number densities (amount of chromophore loading) of 20%, limiting the amount of r_{33} being able to be achieved. Recently, we have developed an approach referred to as matrix-assisted poling (MAP) which consists of using the host of the chromophore system to aid in the poling induced acentric order of the chromophore to induce EO activity. Toward this effort, we have synthesized a high hyperpolarizability chromophore with two coumarin-based pendant groups covalently attached to the chromophore core. This molecule is denoted as C1 and shown in Figure 1. The coumarin pendant groups, when attached to a long (8 carbons or great) carbon chain, exhibit a highly centrosymmetrically-ordered liquid crystalline smectic-A phase.[4]

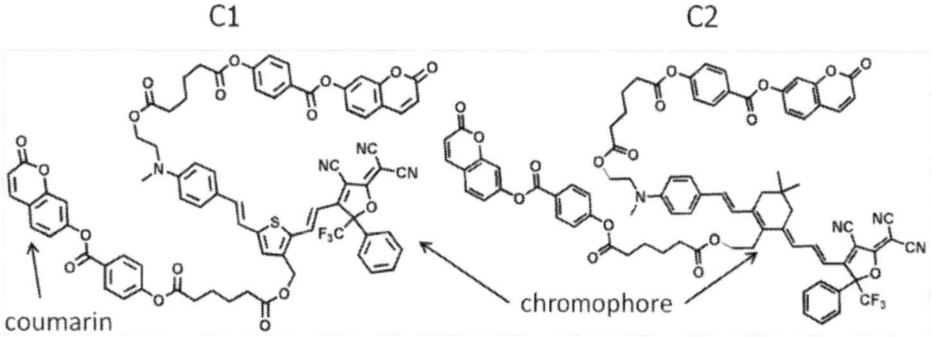

Figure 1. C1 and C2 chromophores are shown with chromophore and coumarin moieties labeled.

Upon conducting poling and r_{33} measurements, C1 exhibited superior EO activity, r_{33}/E_p = 1.9 (>140 pm/V), a 2-3 fold increase in macroscopic EO activity compared to poled polymer conventional systems. In investigation of the poling-induced order through Variable Angle Polarization Referenced Absorption Spectroscopy (VAPRAS)[5] and Variable Angle Spectroscopic Ellipsometry (VASE), C1 exhibited significantly improved centrosymmetric order of the chromophore cores ($\langle P_2 \rangle$ = 0.24) and of the coumarin groups ($\langle P_2 \rangle$ = -0.19) with the coumarin pendant groups specifically oriented in an plane orthogonal to the poling plane (xz,yz). Moreover, nanoviscoelastic experiments (methods capable of detecting intermolecular association and interaction and quantitative energetics of these interactions) further demonstrate intermolecular association of the coumarin units. The specific methods that show this phenomenon are shear-modulation force microscopy (SM-FM)[6], intrinsic friction analysis (IFA)[7], and dielectric relaxation spectroscopy (DRS); results from these methods will be discussed. These experimental results are corroborated by theoretical modeling involving Monte Carlo and Molecular Dynamics methods.[8]

These comprehensive experimental and theoretical methods illustrate that the coumarin pendant groups inflict a lattice dimensionality restriction on the EO chromophore.[9] A quantitative assessment of the dimensionality can be computed from the equation below, where *M* is the dimensionality of the system

$$\langle \cos^3 \theta \rangle_{MD} \approx \sqrt{\left(\frac{9-2M}{2+M}\right)\left(\langle P_2 \rangle_{MD} - \frac{3-M}{2M}\right)}$$

The dimensionality of the C1 material is $M = 2.2$.[9]

With the MAP concept yielding promise in nanoengineering of soft matter intermolecular interactions to assist in the acentric ordering of chromophores and therefore improvement in EO activity, we synthesized and characterized a material incorporating a chromophore with higher hyperpolarizability and the analogous coumarin pendant groups used in the C1 compound. This molecule is referred to as C2. Upon performing the same experiments on the C2 system, C2 was found to have a centrosymmetric order of $\langle P_2 \rangle$= 0.12 for the chromophore component and $\langle P_2 \rangle$ = -0.08 for the coumarin units for representative samples poled at the same applied poling field strength. The dimensionality, using the equation above was calculated to be M = 2.5 for the C2 system.[10] Because of the doubling in hyperpolarizability and therefore increase in dipole moment of the chromophore, it is likely that the coumarins are not as effective in inflicting an improved order and EO activity upon the chromophores.

These results reiterate the delicacy of intermolecular interactions and lattice dimensionality restriction in soft matter materals. Furthermore, the comprehensive characterization and theoretical modeling of the C1 and C2 systems altogether comprise a combined method for nanoengineering of soft matter interactions to create and exploit molecular order in EO materials. This concept of matrix assisted poling can easily be applied to other classes of materials with unique applications such as organic electronics, organic photovoltaic devices, and organic field effect transistors, etc.

References

1. Dalton, L. R.; Benight, S. J.; Johnson, L. E.; Knorr, D. B., Jr.; Kosilkin, I.; Eichinger, B. E.; Robinson, B. H.; Jen, A. K.-Y.; Overney, R. M., Systematic Nanoengineering of Soft Matter Organic Electro-optic Materials. *Chem. Mater.* **2011,** 23, (3), 430-445.

2. Dalton, L. R.; Sullivan, P. A.; Bale, D. H., Electric Field Poled Organic Electro-Optic Materials: State of the Art and Future Prospects *Chem. Rev.* **2010,** 110, 25-55.

3. Benight, S. J.; Bale, D. H.; Olbricht, B. C.; Dalton, L. R., Organic electro-optics: Understanding material structure/function relationships and device fabrication issues. *J. Mater. Chem.* **2009,** 19, 7466-7475.

4. Tian, Y.; Kong, X.; Nagase, Y.; Iyoda, T., Photocrosslinkable liquid-crystalline block copolymers with coumarin units synthesized with atom transfer radical polymerization. *J. Polym. Sci., Part A: Polym. Chem.* **2003,** 41, 2197-2206.

5. Olbricht, B. C.; Sullivan, P. A.; Dennis, P. C.; Hurst, J. T.; Johnson, L. E.; Benight, S. J.; Davies, J. A.; Chen, A.; Eichinger, B. E.; Reid, P. J.; Dalton, L. R.; Robinson, B. H., Measuring Order in Contact-Poled Organic Electrooptic Materials with Variable-Angle Polarization-Referenced Absorption Spectroscopy (VAPRAS). *J. Phys. Chem. B* **2011,** 115, (2), 231-241.

6. Ge, S.; Pu, Y.; Zhang, W.; Rafailovich, M.; Sokolov, J.; Buenviaje, C.; Buckmaster, R.; Overney, R. M., Shear modulation force microscopy study of near surface glass transition temperature. *Phys. Rev. Lett.* **2000,** 85, (11), 2340-2343.

7. Knorr, D. B., Jr.; Gray, T.; Overney, R. M., Intrinsic Friction Analysis - Novel Nanoscopic Access to Molecular Mobility in Constrained Organic Systems. *Ultramicroscopy* **2009,** 109, (8), 991.

8. Benight, S. J.; Knorr, D. B., Jr.; Sullivan, P. A.; Sun, J.; Kocherlakota, L. S.; Robinson, B. H.; Overney, R. M.; Dalton, L. R., Nano-Engineering of Soft Matter Lattice Dimensionality: Measurement of Nanoscopic Order and Viscoelasticity. *Submitted* **2011**.

9. Benight, S. J.; Johnson, L. E.; Barnes, R.; Olbricht, B. C.; Bale, D. H.; Reid, P. J.; Eichinger, B. E.; Dalton, L. R.; Sullivan, P. A.; Robinson, B. H., Reduced Dimensionality in

Organic Electro-Optic Materials: Theory and Defined Order. *J. Phys Chem. B.* **2010,** 114, (37), 11949-11956.

10. Benight, S. J. Nanoengineering of Soft Matter Interactions in Organic Electro-Optic Materials. University of Washington, Seattle, 2011.

(603e) Tuning Morphology and Packing of TIPS-Pentacene with Enhanced Charge Transport Using Solution Shearing

Gaurav Giri[1], Eric Verploegen[1], Stefan Mannsfeld[2], Michael Toney[2] and Zhenan Bao[1], (1)Chemical Engineering, Stanford University, Stanford, CA, (2)SLAC National Accelerator Laboratory, Menlo Park, CA

Solution deposition of organic semiconductors (OSC) is a leading contender for producing large-area, inexpensive, and flexible organic electronics. We recently developed a solution deposition method for OSCs resulted in better organic field effect transistor (OFET) performance compared to those fabricated from simple drop casting or spin casting methods. We found that this method has achieved a wide range of reproducible morphologies, which results in differing FET performance. Certain conditions have yielded remarkable FET performance, yielding large area films (> 4 cm^2) on which FETs show an average charge carrier mobility around 2 $cm^2 V^{-1}s^{-1}$ with some transistors showing mobilities as high as 4 $cm^2 V^{-1}s^{-1}$, a current On/Off ratio of > 10^6, low hysteresis and a low threshold voltage. We discuss the characteristics of this method and its applicability to a wide range of OSCs. We quantitatively measure the molecular ordering resulting from different deposition conditions and show effective tuning of molecular packing that yields higher charge carrier performance without changing the molecular structure. We investigate the stability of this molecular packing structure and also measure the relationship between the morphological features and its impact on charge carrier transport. We also study the general applicability of this method onto various OSCs.

See more of this Session: Organic Materials for Electronics, Photonics, and Photovoltaics
See more of this Group/Topical: Materials Engineering and Sciences Division

(603f) Enhancement of Molecular Fluorescence by Excitonic Coupling to a J-Aggregate Critically Coupled Resonator

Gleb M. Akselrod[1], Brian J. Walker[2], **William A. Tisdale**[3], Moungi G. Bawendi[2] and Vladimir Bulovic[1], (1)Organic and Nanostructured Electronics Lab, MIT, Cambridge, MA, (2)Department of Chemistry, MIT, Cambridge, MA, (3)Organic and Nanostructured Electronics Lab, and Department of Chemical Engineering, MIT, Cambridge, MA

The efficient absorption and reemission of light is of fundamental importance to the operation of optoelectronic devices in applications such as high efficiency lighting, lasers, chemical sensing, solar concentrators, and optical imaging. However, the properties that determine the brightness of molecules and other luminescent species – the absorption cross section and emission quantum yield – are not easily altered without affecting spectral properties. In cases where the excitation intensity is limited and/or the luminophore is expensive, a strategy for enhancing effective absorption cross section would be highly valuable. With this motivation in mind, we report a device for the scalable, tunable, and homogeneous enhancement of molecular fluorescence. The operation of this device, called a J-aggregate critically coupled resonator (JCCR), is based on excitonic energy transfer from a highly absorptive thin film of J-aggregating dye molecules positioned at the anti-node of an optical half-cavity. We demonstrate the feasibility of this scheme through enhancement of fluorescence from the laser dye DCM. A sub-monolayer equivalent of DCM molecules is shown to absorb and reemit 2.2% of incident photons when coupled to the JCCR enhancement structure, compared to 0.1% for the bare film. We discuss quantitatively the role of optical interference, Förster resonance energy transfer, and exciton diffusion in optimizing the device geometry.

See more of this Session: Organic Materials for Electronics, Photonics, and Photovoltaics
See more of this Group/Topical: Materials Engineering and Sciences Division

(604a) Engineering AlN Thin Films by Atomic Layer Deposition On Wide Bandgap Semiconductors As Gate Dielectric

Ya-Chuan Perng and Jane P. Chang, Department of Chemical and Biomolecular Engineering, University of California, Los Angeles, Los Angeles, CA

Wide bandgap semiconductors, such as SiC and GaN, are known as base materials in electronic devices operating under high temperatures and high electric field. Aluminum nitride (AlN) is a promising interfacial layer or dielectric material for SiC and GaN due to its material properties, such as wide bandgaps, similar atomic arrangement and small lattice mismatch (1.3% and 2.6%). In order to grow high quality AlN thin films for the applications, atomic scale controllability during the growth dictate the electrical performance of the synthesized materials. To achieve atomic controllability, atomic layer deposition (ALD) is used to synthesize AlN thin layer on SiC, GaN and Si to assess the possibility to extend this deposition method to nitride growth and how the substrate properties affect the interface quality and corresponding electrical properties of the synthesized film

The AlN thin films were synthesized by using trimethylaluminum (TMA) and blue grade ammonia (NH_3) at 400~600°C, with the surface structure during the deposition monitored via *in-situ* RHEED. The growth rate was 0.5~2Å/cycle as a function of the growth temperature. The surface composition, including the impurities, was related to the deposition temperature, verified by *in-situ* x-ray photoelectron spectroscopy (XPS). The surface of the films was studied by transmission electron microscopy (TEM) and atomic force microscope (AFM) and found to be conformal because the rms value was found to be about the same as that of the substrate surface roughness. The capacitance/conductance-voltage characteristics were measured to determine the dielectric constant of the ALD film and interpret the interface states density, which were 8 and on the order of 10^{12} $cm^{-2}eV^{-1}$, respectively, comparable to that reported for AlN grown by MBE.

See more of this Session: Oxide Materials: Synthesis, Characterization, and Application
See more of this Group/Topical: Materials Engineering and Sciences Division

(604b) Atomic Layer Deposition and Characterization of $Er_xTi_{1-x}O_y$ Films

Runshen Xu, Chemical Engineering, University of Illinois at Chicago, Chicago, IL and Christos Takoudis, Departments of Chemical Engineering and Bioengineering, University of Illinois at Chicago, Chicago, IL

High dielectric constant (k) materials with superb scalability and negligible leakage current are attractive for use in future nanoelectronic devices, especially in sub-45 nm metal oxide semiconductor field effect transistor (MOSFET) and dynamic random access memories (DRAM). So far, a variety of high-k candidate dielectrics, including La_2O_3, Y_2O_3, ZrO_2, HfO_2 and TiO_2, have been investigated for the replacement of traditional SiO_2 (k=3.9). From all these materials, TiO_2 is of great importance in terms of its high dielectric constant which is 80 or higher; however, issues such as small conduction band offset to Si (1.0 eV) and instability on Si substrate lead to more leakage current and limit the use of TiO_2 for MOS application.

In past years, much attention has been devoted to heterogeneous materials in order to modify film properties and overcome the drawbacks of TiO_2 dielectrics. Among these composite films, lanthanide doped TiO_2 films exhibit excellent electrical and chemical properties such as high dielectric constant, low leakage current and good stability on Si. A challenge in the implementation of Ln dopped TiO_2 dielectric thin films is the development of suitable route for thin film growth; few techniques, such as e-beam evaporation, reactive rf puttering and magnetron sputtering, have been used to deposit these materials. Compared to these techniques, atomic layer deposition (ALD) has been identified as a preferred means of fabricating nanoscale features in semiconductor industry, because it offers distinct advantages and meets the challenge requirements, such as precise thickness control at the required dimensions within transistors and conformal coverage in the deep trenches of DRAM.

In this work, we report on the co-ALD of TiO_2 and Er_2O_3 to grow $Er_xTi_{1-x}O_y$ thin films on silicon substrates with tunable amounts of Er and Ti; different ligand precursors are used in the deposition of these films, i.e., $(CpMe)_3Er$ and TDEAT. The composite films are analyzed, characterized and discussed in terms of structure and surface morphology as a function of composition in order to evaluate the influence of Er composition on the properties of the resulting $Er_xTi_{1-x}O_y$ films. Also, post-deposition high temperature annealing of the resulting films is studied and its effects on the properties and performance of the $Er_xTi_{1-x}O_y$ thin films are presented.

(604c) Engineering LixAlySizO Ionic Conductive Thin Films by Atomic Layer Deposition for Lithium-Ion Battery Applications

Ya-Chuan Perng[1], Jea Cho[1], Daniel Membreno[2], Nick Cirigliano[2], Bruce Dunn[2] and Jane P. Chang[1], (1)Department of Chemical and Biomolecular Engineering, University of California, Los Angeles, Los Angeles, CA, (2)Department of Materials Science and Engineering, University of California, Los Angeles, Los Angeles, CA

Lithium (Li)-ion batteries have drawn much attention for their outstanding performance in portable electronics applications. However, formation of the solid electrolyte interphase (SEI) layer on the surface of electrodes during the charge-discharge cycling can reduce battery capacity and the long-term reliability of current battery technology. The use of solid electrolyte layers can effectively suppress formation of the SEI. Another application for thin solid electrolytes is in microbatteries, especially those based on the engineering of electrodes into 3D architectures involving high aspect ratio pillars. To realize this potential, an ultra-thin and highly conformal electrolyte layer is needed to coat the 3D electrode array. The ionic conductor lithium aluminosilicate ($LiAlSiO_4$) synthesized by atomic layer deposition (ALD), is a promising candidate for these battery applications. The material exhibits high ionic conductivity along its c-axis because of channels formed by the alternating tetrahedra of aluminum-oxygen (Al-O) and silicon-oxygen (Si-O).

The self-limiting characteristic of ALD allows for precise control of thickness and composition of complex oxides and results in a highly conformal and pinhole-free coating suitable in 3D micro-battery applications or electrolyte surface coatings. The metal precursors used in this work are tetraethyl orthosilicate (TEOS), trimethylaluminum (TMA) and lithium t-butoxide (LTB). These precursors, along with water vapor as the oxidant, were used to deposit SiO_2, Al_2O_3 and Li_2O, with deposition rates in the range of 0.8~2Å/cycle, respectively. The deposition rate of stoichiometric $LiAlSiO_4$ was ~5Å/cycle at a temperature of 290°C. The concentration of each metal element in $Li_xAl_ySi_zO$ (LASO) thin films is found to correlate closely to ALD cycles and the associated incubation times. Complex impedance measurements show that the ionic conductivities of as-deposited LASO films are in the range of 10^{-7} to 10^{-8} S/cm and directly related to the lithium content in the film. The LASO ALD coating on 3D features, such as NWs and nanopaticles (NPs), were confirmed to be conformal and uniform by transmission electron microscopy (TEM) imaging. Further electrochemical tests to check pinhole-free coatings on 2D and 3D features were performed as well. The crystallinity of the films after post-deposition rapid thermal annealing (RTA) was a function of cation concentration. The epitaxial relation of LASO to silicon, a potential anode material, was found to be beta-LiAlSiO4 (-1 2 -1 0) || Si (400) and beta-LiAlSiO4 (1 0 -1 0) || Si (004).

(604d) Engineered Multiferroic PZT-CFO and BFO-CFO Thin Films for Large Magnetoelectric Coefficient by Atomic Layer Deposition

Ju H. Choi, Calvin D. Pham and Jane P. Chang, Department of Chemical and Biomolecular Engineering, University of California Los Angeles, Los Angeles, CA

Multiferroic materials simultaneously possess two or more of ferroic order parameters such as ferroelectricity, ferroelasticity, ferromagnetism, and antiferromagnetism. They hold considerable promise because of their potential applications in information storage and spintronics, such as magneto-electric sensors, magneto-capacitive devices, and electrically driven magnetic data storage. The revival of modern multiferroics was possible not only due to recent theoretical breakthroughs in understanding the coexistence of magnetic and electrical ordering but also advances in thin film growth techniques. The ultimate goal for practical device application of multiferroic materials is dependent on how to create strong coupling between different types of ferroic order. In order to explore new degrees of freedom for achieving stronger magnetoelectric coupling, the research has been directed towards designing engineered multiferroic composite materials in the form of horizontal multilayer (2-2), vertical superstructures (3-1) or other nanoparticle composite structures (3-0) in a precise controlled manner.

In this work, the intrinsic multiferroic materials such as $BiFeO_3$ (BFO) and ferroelectric materials Pb $(ZrTi)O_3$ (PZT) thin film were synthesized by RE-ALD. RE-ALD is a gas-phase technique in which precursor vapors are pulsed alternately into the reaction chamber and the thin film growth proceeds through surface reactions in a self limiting manner. The advantages of ALD include excellent conformality, simple and accurate thickness control and good uniformity on large areas. More effort on nanostructure composites have been made to produce large magnetoelectric (ME) coefficients through a strictive interaction between the piezoelectricity (PE) of ferroelectric (FE) phase and the magnetostriction of ferromagnetic (FM) or antiferromagnetic (AFM) phase. In order to demonstrate conformal deposition of engineered multiferroic materials in the form of 3-0 configuration, PZT (FE, PE) and BFO (FE, AFM) was deposited onto a mesoporous $CoFe_2O_4$ (CFO, AFM) substrate by radical enhanced atomic layer deposition (RE-ALD).

The mesoporous CFO films were found to be fully filled by ALD PZT and BFO. The composition and crystal structure of PZT-CFO and BFO-CFO were confirmed by X-ray Photon Spectroscopy and X-ray Diffraction (XRD), respectively. More detail crystal structure were investigated synchrotron XRD and extended x-ray absorption fine structure spectroscopy (EXAFS). The magnetic and ferroelectric properties for the PZT-CFO or BFO-CFO thin film were characterized by a superconducting quantum interference device (SQUID) magnetometer and piezoresponse force microscopy (PFM). Composite systems were found to have higher coercive magnetic field and lower saturation moment than that of the pure CFO substrate.

See more of this Session: Oxide Materials: Synthesis, Characterization, and Application
See more of this Group/Topical: Materials Engineering and Sciences Division

(604e) Control of Oxygen Self-Diffusion In Metal Oxides for Nanoelectronics

Prashun Gorai, Kristine Pangan-Okimoto, Alice Hollister and Edmund G. Seebauer, Department of Chemical and Biomolecular Engineering, University of Illinois, Urbana, IL

Oxygen defects play a central role in various physical phenomena in semiconductor metal oxides that are technologically relevant. Metal oxide properties like gas sensing, photoluminescence, bipolar switching and photocatalysis are strongly influenced by the concentration and diffusion of native oxygen defects. Previous work done in our group in silicon (Si) and titanium dioxide (TiO_2) identified the semiconductor surface as an efficient pathway for the generation and annihilation of point defects in the underlying bulk. Such surface pathways should play an especially important role in nanoelectronic and nanophotonic devices where the surface to volume ratio is high. It is expected that such surface effects should apply to other semiconductors as well. In an attempt to generalize the surface effects, the present work extends these findings for Si and TiO_2 to zinc oxide (ZnO), wherein we have identified unique surface pathways for generation of oxygen point defects. Oxygen diffusion rates were measured by exposing natural-abundance single-crystal ZnO <0001> to isotopically labeled oxygen gas. The resulting profiles were measured by secondary ion mass spectrometry (SIMS) and modeled with mass transport equations for the reaction and diffusion of mobile and stagnant oxygen point defect species. The effective oxygen diffusivity is determined by fitting this model to experimental SIMS profiles. In addition, the detailed reaction-diffusion network of point defects in ZnO is modeled with continuum equations to simulate the self-diffusion of oxygen, explicitly incorporating the oxygen adsorption on to the <0001> surface, defect generation near the surface and the subsequent diffusion of the defect through the bulk. Such a model will help us understand mechanistically the effect of each of these elementary steps on the oxygen self-diffusion in ZnO besides extracting important kinetic parameters.

See more of this Session: Oxide Materials: Synthesis, Characterization, and Application
See more of this Group/Topical: Materials Engineering and Sciences Division

(604f) MBE Growth of Fe3O4 and Fe2O3 for Both Spin Injection Layers and Templates for Spinel Complex Oxides

Katherine Ziemer, Department of Chemical Engineering, Northeastern University, Boston, MA and Bing Sun, Chemical Engineering, Northeastern University, Boston, MA

Effective integration of functional oxides (magnetic, ferroelectric, piezoelectric and other multi-functional materials) with semiconductors will lead to next-generation devices such as: new architectures that enable multiple and simultaneous interactions with the environment for multifunctional active sensors and controllers; energy harvesting and conversion devices as part of everyday items from clothing to sidewalks; integrated nonvolatile memories for harsh environments; and paradigm-shifting spintronicsbased circuits. Many of the functional properties of complex oxides are directional, thus requiring a specific plane alignment in a device, and most oxides have multiple stable structures for a given stoichiometry. In addition, small changes in stoichiometry with the same unit cell structure can produce easily measurable differences in performance properties. While this sensitivity enables the possibility of tuning complex oxides for different and novel applications, precise control of stoichiometry and structure (practically at the atomic-level for nanoscale thin film heterostrucutres) is required. This is particularly true for the interfaces and near-interface layers between the semiconductor and oxide as well as between subsequent functional oxide layers, because surface quality impacts chemistry, structure, and morphology of next-layer film deposition that, in turn, impacts the functionality of the film layer and the coupling effects across layers through critical interfaces. Through molecular beam epitaxy (MBE), the use of a magnesium oxide (MgO) template layer and the interface formation mechanisms of an oxygen bridge have been investigated for effective heteroepitaxy of high-quality ferroelectric barium titanate (BTO) and ferrimagnetic barium hexaferrite (BaM) on 6H-SiC. The results suggest strategies applicable to many functional oxide integrations and many processing techniques. Particular emphasis will be on the magnetic BaM and various forms of iron oxide.

See more of this Session: Oxide Materials: Synthesis, Characterization, and Application
See more of this Group/Topical: Materials Engineering and Sciences Division

(618a) Use of Crosslinked Hydrogels for the Control of Macromolecular Crystallization

Madeline Torres-Lugo[1], Andry Cera[1], Katia Ortiz-Lugo[1], Melanie Santos-Roman[1], Janet Mendez[1] and Jose A. Gavira[2], (1)Chemical Engineering, University of Puerto Rico, Mayaguez Campus, Mayaguez, PR, (2)Laboratorio de Estudios Cristalogrficos, Instituto Andaluz de Ciencias de la Tierra, Armilla, Spain

The potential of crosslinked hydrogels in the crystallization of proteins is an area that has not been exploited. The technique of counter-diffusion is a novel method for the crystallization of proteins which entails the minimization of convection effects resulting in high quality crystals. The implementation of this technique in capillary tubes with different diameters has been accepted for the crystallization of proteins, by using a precipitating solution which diffuses through an agarose polymer that controls the diffusion process. The diffusional process using such systems can be controlled but cannot be widely manipulated. The use of crosslinked hydrogels provides better diffusional and environmental control.

This research focuses on the role of crosslinked hydrogels on the nucleation process involved in protein crystallization. The crystallization process was performed using crosslinked Poly(ethyleneglycol) hydrogels and lysozyme as a model protein. The effects of polymer concentration and hydrogel pore size on crystallization were studied. To accomplish this, different dilutions of pre-polymer solutions were prepared using the protein as solvent. These solutions were poured in capillaries of 0.8 millimeters in diameter. Also a membrane or plug of different polymer percentages using water as solvent to control diffusion was employed using in this case, capillaries of 4 millimeters in diameter. Both of the solutions were exposed to a UV light source to induce polymerization. A fluorescent monomer was also employed to determine if the crystal structure is formed around or if it pushes polymer chains apart. Confocal Laser Scanning Microscopy was employed to assess the fluorescence throughout the crystal structure.

Results indicated that the diffusion coefficient calculated from the measurement of the crystallization displacement as a function of time was significantly different for each of the plug conditions including the 80%, 65%, and 50% dilutions. Mesh sizes were also measured and they varied from 10-7nm. Moreover, when the polymer concentration was varied qualitative differences in crystal structures were observed. Studies using a fluorescent crosslinker indicated that the polymer was incorporated into the crystal structure. These results indicated that nucleation and crystal growth can be controlled by the manipulation of polymer composition which cannot be easily obtained with current physically crosslinked hydrogels.

See more of this Session: Mesd Poster Session
See more of this Group/Topical: Materials Engineering and Sciences Division

(618aa) Enzyme Stabilized Systems Using Organic/Inorganic Hybrid Materials

Ee Taek Hwang, Rameshwar Tatavarty, Hyun Lee, Ji Hoon Kim, Jinyang Chung, Haemin Gang and Man Bock Gu, College of Life Sciences and Biotechnology, Korea University, Seoul, South Korea

In this study, we report highly stable enzyme immobilized systems - using nano/micro-sized hybrid materials. Hydrolase enzyme, esterase from *Rhyzopus oryzae,* has been studied as a model enzyme in aqueous condition for increasing storage and recycling stability. This esterase enzyme has been subjected to be immobilized in a form of entrapment and/or coating on the surfaces by using calcium phosphate nanoparticles or organic-inorganic hybrid microspheres prepared as porous support materials for good accessibility of substrate. These nano/micro-sized hybrid materials were successfully characterized by using Fourier transform infrared (FT-IR) spectroscopy, scanning electron microscopy (SEM), Transmission electron microscopy (TEM), and energy-dispersive X-ray spectroscopy (EDX). As results, the nano/micro-sized hybrid materials were found to be efficient enzyme immobilizing carriers. The immobilized esterase composite systems have shown the good storage stability retained in terms of its enzyme activity preserved at the repeated usages for more than ten times. Therefore, it is strongly anticipated that this enhanced strong stability obtained from these enzyme hybrid systems will boost the employment of these systems in various fields of enzyme related applications, such as bioprocessing, bioremediation, biosensing, or enzyme reactors.

See more of this Session: Mesd Poster Session
See more of this Group/Topical: Materials Engineering and Sciences Division

(618ab) Multifunctional Magnetic Graphene Nanocomposites for Environmental Remediation

Jiahua Zhu[1], Suying Wei[2] and Zhanhu Guo[1], (1)Integrated Composites Laboratory (ICL), Dan F Smith Department of Chemical Engineering, Lamar University, Beaumont, TX, (2)Chemistry and Biochemistry, Lamar University, Beaumont, TX

A one-pot synthesis method to obtain multifunctional graphene nanocomposites decorated with monodispersed core@double-shell nanoparticles is developed in this work. High-resolution transmission electron microscopy and energy filtered elemental mapping revealed a core@double-shell structure of the nanoparticles with the core and inner shell being crystalline iron and iron oxide, and the outer shell of amorphous Si-S-O compound. These magnetic graphene nanocomposites show an extremely fast chromium (VI) removal efficiency (completely removed in 5 min) from the waste water and a significantly high adsorption capacity (20.9 mg/g from 10 wt% nanoparticle loading). This material shows great advantages as compared to traditional adsorbents, such as active carbon, biomass, iron oxide that often require hours even days to remove the heavy metal ions.

See more of this Session: Mesd Poster Session
See more of this Group/Topical: Materials Engineering and Sciences Division

(618ac) Deposition Time Effect On Nanosilicate Layer-by-Layer Film Growth and Gas Barrier

You-Hao Yang[1], Frank A. Malek[1] and Jaime C. Grunlan[2], (1)Chemical Engineering, Texas A&M University, College Station, TX, (2)Mechanical Engineering, Chemical Engineering, Texas A&M University, College Station, TX

Deposition Time Effect on Nanosilicate Layer-by-Layer Film Growth and Gas Barrier

The influence of deposition time on layer-by-layer (LbL) film growth of montmorillonite (MMT) or laponite (LAP) nanosilicate paired with branched polyethylenimine (BPEI) was studied. Deposition times, from 5 seconds to 300 seconds in each ingredient, were evaluated with respect to thickness, mass deposited and film structure. The BPEI/LAP system is more dip-time dependent than the BPEI/MMT system because LAP platelets have a much smaller aspect ratio than MMT (40 vs. 200~1000). XRD suggests the gallery spacing between clay layers is insensitive to dip time, which is a key structural parameter linked to both mechanical and transport properties of these types of films. Oxygen transmission rates (OTR) are below the detection limit of commercial instrumentation (<0.005 cc/m^2 · day)) in both 10 sec/300 sec and 300 sec/300 sec BPEI/MMT assemblies. This study suggests LbL assemblies could be produced very quickly, which should make this thin film coating technology more commercially appealing.

See more of this Session: Mesd Poster Session
See more of this Group/Topical: Materials Engineering and Sciences Division

(618ad) The Incorporation of Electrically Conducting Polymer within Natural-Polymer Hydrogels

Richelle C. Thomas[1], Shan P. Modi[2] and Christine E Schmidt[2], (1)Chemical Engineering, University of Texas at Austin, Austin, TX, (2)Biomedical Engineering, University of Texas at Austin, Austin, TX

Hydrogels that incorporate water content, internal architecture and porosities similar to that of native tissue have shown promise in the field of tissue engineering and regenerative therapies. The ability to leverage these properties to the fullest extent can only happen if the scaffolds embody many of the known attributes beneficial to wound healing: electrical conductivity, internal architecture, and embodying chemical makeup mimic of the native extracellular matrix. Thiol "click" chemistry has been incorporated into hyaluronic acid hydrogel films to exploit the hydrolysable bonds at physiological pH. The thiol groups are used to crosslink the hydrogel and allow additional functionalites to be incorporated within the polymer matrix.

This work attempts to evaluate the ability to incorporate an electrically conducting polymer within an anionic hydrogel. Electrical conductivity is expected to increase the efficacy of wound healing obtained and by permitting local, modulated drug release in combination with the regenerative properties inherent to the bulk polymer. Pyrrole, a biocompatible polymer, has been extensively used in neural applications such as neural electrodes and nerve regeneration. High electrical conductivity and ease of synthesis make this polymer a viable candidate for this work. We hypothesize that the successful inclusion of this polymer within the bulk hydrogel will give rise to electrical conductivity—a necessity in maximizing the hydrogels' wound healing capabilities.

Various methods to incorporate the conducting polymer will be investigated. A suspension of pyrrole monomer within a bulk polymer solution functionalized with methacrylate and pyrrole groups to create a liquid bi-phase system will be evaluated. The pyrrole monomers are expected to coalesce to form domains within the bulk hydrogel that behave as conductivity channels within the hydrogel by connecting to the conductive surface. A second proposed method uses a four-armed thiol to react a methacrylated pyrrole- hyaluronic acid hydrogel solution with the conductive substrate. That is, the hydrogel will be chemically crosslinked and covalently bound to the conductive substrate simultaneously. A final method to incorporate electro conductivity within the hydrogel is to leverage a surfactant to catalyze conductive domain formation within the gel. Each of these methods leverages biocompatibility of natural, biodegradable polymers with the advantages of conductivity and cleavable bonds.

See more of this Session: Mesd Poster Session
See more of this Group/Topical: Materials Engineering and Sciences Division

(618ae) Synthesis of Carbonaceous Nano-Materials and Their Solid-Phase Adsorptivity

Walter Den, Gang Chen and Chia-Hung Yeh, Environmental Science and Engineering, Tunghai University, Taichung City, Taiwan

This study was intended to develop a solid-phase adsorbent based on carbonaceous materials embedded on zirconia (ZrO_2) spherical carriers. This study used thermal chemical vapor deposition (TCVD) process at atmospheric pressure for growth of carbon nano-material, including carbon nanosphere (CNS), carbon nanocapsule (CNC), carbon onion, and carbon web-like structures. In the preliminary study, wet impregnation process was conducted to coat nickel (Ni) as catalyst precursor on the surface of substrates, followed by using TCVD process with methane as carbon source, argon as carrier gas, and hydrogen as initiator. The structural and physical characteristics of the carbon products were determined using scanning electron microscopy, transmission electron microscopy, energy dispersive spectrometry, X-ray diffractometry, and Raman scattering spectroscopy. The results of these micro-analyses were used to evaluate the effects of individual process parameter on the carbonaceous material characteristics. The ensuing solid-phase adsorption studies were then performed to characterize the sorption properties of the materials associated with their physical characteristics. In this work, phthalic acid esters (PAEs) were used as the compounds of interest for the solid-phase adsorption studies. PAEs are commonly used in plastic industry, and are most widely used as plasticizer in products such as plastic toys, medical devices, food wrap, perfume and nail polisher. They are categorized as endocrine disrupting compounds (include sex hormone and the thyroxine) and some of them are suspected carcinogens. In Taiwan, PAEs reportedly exist in seawater and freshwater in very low concentration range (0.1 ~ 300 μ g/L). The work therefore uses a representative PAE, namely diethyl phthalate (DEP) as the testing compound. Preliminary adsorption test results indicate that CNC and carbon onion materials both achieved DEP adsorptivity in the range of 80 μ g/g at DEP solution concentration of 5 mg/L, and 110 μ g/g at DEP solution concentration of 10 mg/L. The web-like carbon structure attained approximately 50 μ g/g at DEP solution concentration of 5 mg/L. Additionally, DEP adsorption capacities of CNC and carbon onions were both higher (12.0 μ g) than that of commercial adsorbent Tenax TA (4.1 μ g).

See more of this Session: Mesd Poster Session
See more of this Group/Topical: Materials Engineering and Sciences Division

(618af) Influence of Crosslinking On Gas Barrier Behavior of Polymer Multilayers

You-Hao Yang, Chemical Engineering, Texas A&M University, College Station, TX, Merid Haile, Chemistry, Texas A&M University, College Station, TX and Jaime C. Grunlan, Mechanical Engineering, Chemical Engineering, Texas A&M University, College Station, TX

Influence of crosslinking on gas barrier behavior of polymer multilayers

Layer-by-layer assemblies of cationic polyethylenimine (PEI) and anionic poly (acrylic acid) (PAA) were crosslinked in an effort to improve their gas barrier behavior and reduce moisture sensitivity. Glutaraldehyde (GA), 1-[3-(dimethylamino)propyl]-3-ethylcarbodiimide methiodide (EDC) and thermal crosslinking were compared with regard to film thickness, microstructure and oxygen permeability. Crosslinking, in general, was found to increase multilayer density and improve barrier at 100% RH. This relatively simple recipe may be of use for a variety of packaging applications, including flexible electronics that require high flexibility and transparency.

See more of this Session: Mesd Poster Session
See more of this Group/Topical: Materials Engineering and Sciences Division

COPOLYMERS OF STYRENE, METHYL METHACRYLATE, AND ETHYLENE WITH A SERIES OF VINYL ESTERS: IMPACT OF VINYL ESTERS MOLECULAR WEIGHT ON THERMAL MECHANICAL AND RHEOLOGICAL PROPERTIES

Hai Wang, Dr Edward Kolodka,*
Chemical Engineering Department,
University of North Dakota, Grand Forks, ND

A number of new processes are being developed which produce bio-sourced fatty acids of various molecular weights. These fatty acids can easily be converted into vinyl esters using common processes. However, vinyl esters larger than vinyl propionate are not suitable for homopolymerization due to the low glass transition temperatures (T_g) of the homopolymers. Therefore, this work will focus on the copolymerization of vinyl esters of various molecular weights with both styrene and ethylene. Particular emphasis will be placed on the influence of vinyl ester molecular weight on the polymerization kinetics and thermomechanical characteristics of the resultant copolymers.

In this study copolymers of styrene (St) and methyl methacrylate (MMA) with a series of vinyl esters (VEs) including vinyl acetate (VAc), vinyl propionate (VPr), vinyl hexanoate (VHe), and vinyl decanoate (VDe) were prepared by bulk copolymerization initiated by azobisisobutyronitrile (AIBN) at 60 °C. Copolymers of ethylene with the VEs were synthesized by solution copolymerization initiated by AIBN also at 60 °C. The compositions and molecular weight distributions (MWD) of all the copolymers were determined by GPC. The impact of compositions and branch lengths on the thermal mechanical properties of copolymers including glass transition temperature (T_g), melting point (T_m), storage modulus (G') and loss modulus (G") was determined by dynamic mechanical analysis. In addition, the rheological responses (complex viscosity versus shear rate) of the copolymers of MMA-VEs and ethylene-VEs were studied.

Current results showed that the molecular weights (MWs) of the copolymers decreased as the mole fraction of VEs in the feed increased. Meanwhile, copolymers with similar VE incorporations had higher MWs for poly (St-co-VHe) and poly (St-co-VDe) than those of poly (St-co-VAc) and poly (St-co-VPr). It could be explained by the significantly higher reactivity ratios of both VHe and VDe than those of VAc and VPr. Incorporating VEs with different chain lengths from 1 mol% to 19 mol% reduced the T_g of PS by up to 32 °C. The T_g of poly (St-co-VEs) decreased as the mole fraction of VEs increased. Poly(St-co-VEs) with longer branches, such as poly(St-co-VHe) and poly(St-co-VDe), exhibited a larger drop in T_g than VAc and VPr copolymers at similar compositions. However, the branch effects on T_g were not significant for copolymers containing VEs less than 5%.

Dynamics mechanical analysis of poly(St-co-VHe) and poly(St-co-VDe) showed that storage modulus decreased with increasing of the mole fractions of linear VEs or the decreasing of MWs. The loss modulus increased with the growing of mole fractions of linear VEs firstly, and then dropped sharply after the mole fractions of linear VEs were over 17%, around which the MWs of both poly(St-co-VHe) and poly(St-co-VDe) were very low. Mole fractions of linear VEs, MWs, and the side chain lengths of resultant copolymers all had effect on the thermomechanical properties of the resultant St-VE copolymers.

BIOPOLYMER/ METAL NANOPARTICLES COMPOSITE; CONDUCTIVITY AND ANTIBACTERIAL ACTIVITY RELATIONSHIP

J. Betzabe Gonzalez-Campos[1], Montserrat Hernández-Iturriaga[2], Zaira Y. García-Carvajal[3], Evgen Prokhorov[4], Gabriel Luna-Barcenas[4] and Rosa E. del Río[1]

[1]Institute of Chemical and Biological Researches, Universidad Michoacana de San Nicolás de Hidalgo, Morelia, Mich. Mex.
[2]Chemistry Department, Universidad Autónoma de Querétaro, Qro., Mex.
[3]Instituto Nacional de rehabilitación, Mexico D.F., Mex.
[4]Centro de Investigación y de Estudios avanzados del IPN, Querétaro, Qro., Mex.

Abstract

Nowadays metal–polymer nanocomposites are the subject of increased interest due to their potential to combine the features of polymers with those of inorganic materials. Specifically, the combination of a natural polymer (biopolymer) such as chitosan (CS) and silver nanoparticles (AgnP) is highly appealing because of the individual antibacterial activity of both components, and the possibility to generate a biodegradable and biocompatible composite. The incorporation of AgnP to CS can improve its mechanical properties and enhance its antibacterial activity, which can extend its applications, specifically in those where hygiene is a key factor. In this work, the antibacterial activity of CS/AgnP films against two gram-positive and two gram-negative bacteria is evaluated. The films were obtained by the solvent cast method. The AgnP were incorporated into CS/acetic acid aqueous solutions by ultrasonic dispersion at concentrations varying from 0.5% wt to 20% wt. The contact times between the films and strains were 0. 5. 24 and 48 h with initial inocula of 9.12 UFC/cm^2. A potential increase in the antibacterial activity of chitosan with the addition of silver nanoparticles was observed; the inhibiting growth of the tested microorganisms clearly increases with increasing AgnP content and it is more effective against gram-positive bacteria, however, at 20% wt, in both, gram-positive and gram-negative this effect decreases in comparison to films with lower AgnP content, this decline is ascribe to the formation of agglomerates generating larger metal particles.

On the other hand, a percolation threshold in the electrical conductivity was observed at 2% wt of AgnP; the same percentage by which the anti-bacterial effect is the highest. This effect is related to the presence of protonated amino groups and the size of metal nanoparticles which strongly influences both the conductivity and the antibacterial effect. Thus, these compounds can be used as antibacterial sensors from conductivity measurements.

Introduction

Different investigations have been carried out to show the antibacterial activity of silver nanoparticles (AgnP) against Gram-positive and Gram-negative bacteria [1-4] and even their antiviral action against HIV [5, 6]. In biomedicine, they have been applied to treat burns, cancer and a variety of infections and to prevent bacteria colonization in different devices and human skin [7]. Several AgnP composites have been synthesized in order to expand their applications [8-11]. Among them, chitosan (CS)/silver nanoparticles composites have attracted special attention because its biocompatibility and antibacterial tests against Gramnegative and Grampositive bacteria [12-14]. In all the cases, the methodologies used for CS/AgnP composite films preparation produce chitosan with the amino group protonated (NH_3^+), and consequently chitosan in a non-neutralized state. The positively charged backbone of chitosan plays a crucial role on its interaction with other molecules, its conductivity and backbone motion [15, 16].

On the other hand, studies carried out on molecular relaxations of pristine chitosan and chitosan/AgnP composites by dielectric spectroscopy show evidence about the influence of moisture content on pure biopolymers and composites molecular relaxations [17, 18]. This effect was chiefly observed on the "*low frequency-temperature*" relaxation (the α-relaxation), since for certain conditions of moisture, this relaxation cannot be detected. In this regard, dielectric, conductivity and chemical interaction studies on CS/AgnP composites are scarce [19], and the influence of AgnP inclusion on chitosan water absorption capacity had not been previously reported. Silver nanoparticles interaction with chitosan could affect inter and intramolecular hydrogen bonds in the biopolymer and consequently its water absorption capacity, causing major changes on its physical properties and even changing its antimicrobial properties. The main objective of this work is to investigate the influence of AgnP inclusion on chitosan molecular dynamics, accounting for differences on chitosan neutralized and non-neutralized states and to establish the relationship between dielectric properties and molecular dynamics changes with the antibacterial activity. Dielectric Spectroscopy (DS), Thermogravimetric Analysis (TGA), Fourier Transform Infrared (FTIR) spectroscopy and are the experimental techniques used for this purpose.

Materials and methods

CS/AgnP composite film preparation

Chitosan medium molecular weight (Mw= 150,000 g/gmol) and 82 % of degree of deacetylation (DD) reported by the supplier, was purchased from Sigma-Aldrich. Carbon coated silver nanoparticles powder 25 nm average particle size was purchased from nanotechnologies, Inc., acetic acid from J.T. Baker were used as received. CS/AgnP films were obtained by dissolving 1 wt % of chitosan in a 1 wt % aqueous acetic acid solution with subsequent stirring to promote dissolution. This preparation technique, give rise to chitosan films with the amino side group protonated (NH_3^+groups) and in this state, chitosan is called

chitosan acetate. This solution is poured into a Petri dish and allowed to evaporate at 60°C to obtain a pristine chitosan film.The proper quantity of silver nanoparticles powder (1, 3, 5, 10, 15 and 20% w/w respect to chitosan dry based) was poured into a 1 wt % aqueous acetic acid solution and further sonicated during 30 min to obtained the nanoparticle solution. Afterward, the pristine chitosan film prepared before was dissolved into the silver nanoparticle solution and further sonicated for 10 min to eliminate bubbles formation. CS/AgnP films were also obtained by the solvent cast method by pouring the final solution into a plastic Petri dish and allowing the solvent to evaporate for 24 h at 60°C. These Cs/AgnP films are non-neutralized films, and their neutralized counterpart were prepared by soaking non-neutralized films into a 0.2 M NaOH aqueous solution during 30 min followed by several washing cycles with distilled water until pH 7.
A thin layer of gold was vacuum-deposited onto both film sides to serve as electrodes. Rectangular small pieces (5 mm × 4 mm) of these films were prepared for measurements.

Infrared measurements and Morphology analysis

Chemical analysis of CS/AgnP composites was performed by Fourier-Transform Infrared spectroscopy (FTIR) on a Perkin Elmer spectrophotometer using an ATR accessory in the range 4000-650 cm^{-1}, resolution was set to 4 cm^{-1} and the spectra shown are an average of 32 scans. CS/AgnP films morphology was analyzed by JEOM JSM-7401F Field emission scanning electron microscope.

Dielectric Measurements

Dielectric measurements in the frequency range from 100Hz-110MHz were carried out using an Agilent Precision Impedance Analyzer 4294A. The amplitude of the measuring signal was 100 mV. A peltier and a resistance heating element were used to perform measurements from 2 to 250°C. Each sample was left at each measurement temperature for 3 min to ensure thermal equilibrium.

Thermal Measurement

Free water content was determined by thermogravimetric analysis (TGA). The amount of free water may be evaluated by the decrease of sample weight during the heating scans with heating rate 5°C/min.

CS/ AgnP composite Bacterial Activity

Strains examined. Strains of *Escherichia coli* (ATCC 25922), *Listeria monocytogenes* (serotype 4b), *Salmonella* thyphimurium (ATCC 14028), and *Sthapylococcus aureus* (ATCC 8923) resistant to rifampicin were examined. Stock cultures were maintained at 4°C on tryptic soy agar (Difco, Becton Dickinson, Sparks, Md.).

Preparation of inocula. Cultures were loop transferred three times at 24-h intervals before cells from 20-h cultures were collected by centrifugation (2,000 x *g*, 15 min, 22°C) and re-suspended in saline solution (0.85% NaCl).

Inoculation of CS/AgnP films. Squares of each film (1 x 1 cm) were aseptically obtained with a sanitized scalpel. Each square was inoculated with 10 µl (8 log CFU/square), placed inside sterile glass tubes, and incubated at 22°C for up to 48 h. At 0, 5, 24 and 48 h films were removed from storage and placed in tubes containing 3 ml of Dey-Engley (DE) neutralizing broth (pH 7.6; BBL/Difco) and homogenate using vortex during 1 min.

Microbiological analysis. Inocula and inoculated silver films were analyzed for populations of *E. coli*, *L. Monocytogenes*, *S.* Thyphimurium, and *S. Aureus*. Cell suspensions were serially diluted in sterile 0.1% peptone and surface plated (0.1 ml) on tryptic soy agar supplemented with rifampicin (TSAR, 100 mg/ml). Inoculated plates were incubated at 35°C for 24 h before colonies were counted. Colonies from each sample were chosen randomly and subjected to biochemical confirmation.

Results and discussion

Chemical interaction/FTIR analysis

Infrared spectra for pristine neutralized (NCS) and no neutralized chitosan (NNCS) as well as for CS/AgnP (1,3 5, 10 AND 20 wt %) show that the stretching and bending vibrations bands corresponding to amine and hydroxyl groups clearly change indicating strong interactions with AgnP. The broad vibration band at 3000-3500 cm^{-1} corresponds to OH stretching and indicates intermolecular hydrogen bonding; and it also overlaps in the same region of NH stretching. The 1654 cm^{-1} band for amide I group, 1594 cm^{-1} for bending vibrations of NH_2, 1383 cm^{-1} for amide III as well as the 1326 and 1261 cm^{-1} bands for OH bending vibrations, 1157 cm^{-1} for anti-symmetrical stretching of C-O-C bridge, and finally, 1084 and 1028 cm^{-1} for skeletal vibration of C-O stretching can also be identified. All these bands have significant changes in position, broad and intensity, evidencing that the NH_2 and OH groups of chitosan interact with silver nanoparticles. Additionally, the 1654 and 1383 cm^{-1} bands of $CONH_2$ groups also shift to higher frequencies because of their interaction with silver nanoparticles. AgnP-NH_2, OH and $CONH_2$ interactions involve evident redistributions of the vibration bands for these groups. These interactions could strongly modify chitosan water absorption ability and for instance the T_g of neat chitosan as previously reported by our group [17].

Dielectric analysis

DC electrical conductivity measurements at room temperature as a function of AgnP wt% show an abrupt change in conductivity as silver nanoparticles wt% increases, and a subsequent saturation is observed about 3 wt% as it is shown in Figure 1. This is a clear characteristic of the percolation phenomena [20]. The percolation threshold for this system is 2% volume fraction, which was previously calculated by our group from the power-law dependence, implying that the CS/AgnP system is two-dimensional [20]. The higher *dc* conductivity values of CS/AgnP films compared to pristine chitosan are ascribed to the motion of Ag mobile charge carriers. However the low conductivity after the percolation threshold is ascribe to the presence of a thin polymer layer that coexists between silver nanoparticles [20].

On the other hand, the *DC* conductivity plot (σ_{dc}) as a function of temperature, can give important information about the nature of molecular motion. In our case, *dc* conductivity (σ_{dc}) was calculated by dielectric measurements using the methodology previously described by our group [17, 18]. From this dependence it was observed that non-neutralized chitosan sample reveals higher conductivity values (not shown) due to both higher mobility and number of conducting species (H^+ and acetate anions).

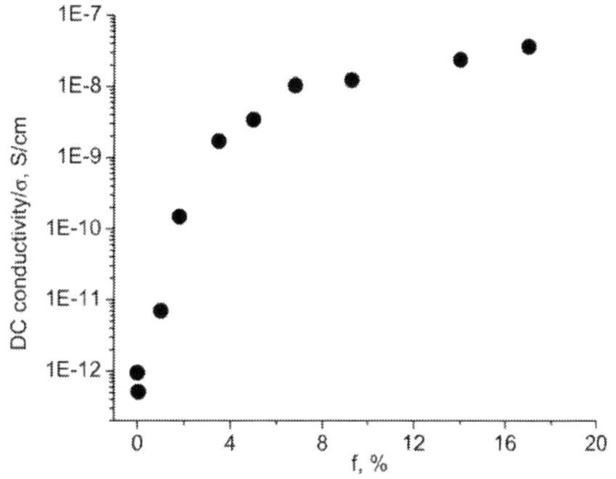

Figure 2. DC conductivity (σ_{dc}) *versus* AgnP wt%, dry films (moisture content ca. 0.3 wt. %) at room temperature.

For NCS, NNCS and CS/AgnP composites dry films (*annealed at 120°C*, and moisture content near zero), above 80°C, the conductivity plot disclose a linear dependence as a function of temperature (see window insert of Figure 2), from this linear dependence the activation energies can be calculated for each film from the Arrhenius equation, $\sigma = \sigma_0 \exp(-E_{a\sigma}/RT)$ where $E_{a\sigma}$ is the activation energy. The activation energies are between 79 and 86 kJ/mol, and this linear behavior is related to the well known σ-relaxation process. Figure 2 show the activation energy trend as a function of AgnP content and it also includes the corresponding values for NCS and NNCS. Note that non-neutralized films exhibit lower activation energy than the neutralized ones, which indicates an excess number of protons in the non-neutralized films (NH_3^+ groups are present). This linear relaxation is related to ion motion, and lower values of the activation energy implies higher number of these species present in the system, this facilitates this relaxation process and gives rise to these lower activation energy values. In CS/AgnP composites this trend continues up to 3% wt, and the lower activation energy could be a synergistic effect from the combination of the presence of more NH_3^+ groups and the mobility of Ag ions which also leads to higher conductivity values.

On the other hand, for AgnP contents above 5 wt%, the activation energy for this process increases, this could be ascribe to the formation of agglomerates above 5 wt% as it was obserbed by SEM images (not shown), consequently these agglomerates difficult the ion motion along composites directly affecting the σ-relaxation process. Therefore the particle size of AgnP affects the conductivity and ion motion in composites, which in turns may affect their antibacterial activity to be discussed below.

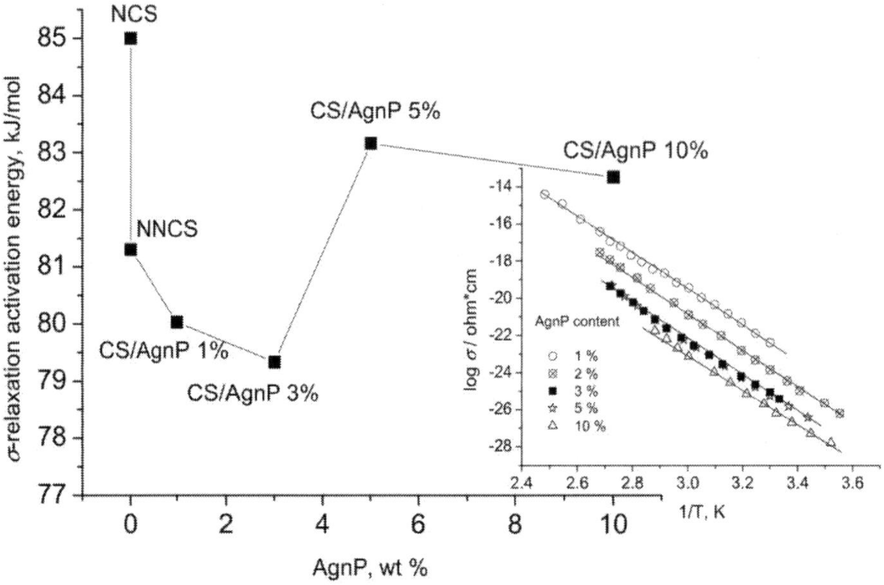

Figure 2. Activation energy for the σ-relaxation process (ion motion) as a function of AgnP content. Note: above 3 wt% of AgnP the activation energy increases as a consequence of the formation of silver nanoparticles agglomerates.

Antibacterial analysis

The antibacterial analysis against *Escherichia coli*, *Listeria monocytogenes*, *Salmonella* thyphimurium and *Sthapylococcus aureus* for neutralized and non-neutralized chitosan films without silver nanoparticles (i.e. 0 wt% of AgnP) is shown in Figure 3. This analysis was carried out at four different times; 0, 5, 24 and 48 h. The time of exposure to chitosan has an important effect, showing that regardless the type of microorganism the number of inactivated cells increases at higher exposure time. After 48 hrs of contact the population of the four microorganisms was reduced. The data dispersion increased as the contact time elapsed.

There is a clear difference in the antibacterial activity effect between NNCS and NCS on the tested bacteria, since NNCS shows higher antibacterial activity on both Grampositive and Gramnegative bacteria compared to NCS. And in turns, NNCS show higher activity against Grampositive bacteria, especially at 48h.

In this case, neutralized chitosan certainly exhibit considerable antibacterial effect after 24 h, however, it can be seen that it is substantially lower than its peers non-neutralized chitosan. The NCS is more effective against *S. aureus* after 48 h of exposure.

Our dielectric results showed that NNCS has an excess number of NH_3^+ groups compared to NCS, offering stronger antibacterial activity. This is in agreement with the hypothesis of an electrostatic interaction between chitosan and the cell wall; as cell attachment is attributed to electrostatic interactions between these cationic sites of chitosan and the negatively charge carboxylate and sulphate groups in cell-surface proteoglycans, in this case it is shown that Grampositive bacteria are more susceptible to establish electrostatic interactions with antimicrobial compounds because it has phosphate groups with negative charge giving rise to a more electronegative wall [21, 22].

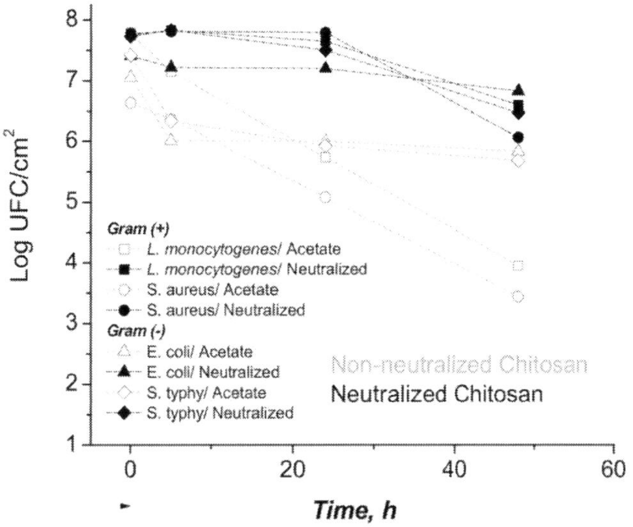

Figure 3. *logUFC/cm^2 versus Time* for neutralized and non-neutralized chitosan at different exposure times. Note: non-neutralized chitosan antibacterial activity against all the bacteria tested is higher than its counterpart.

Also, chitosan antibacterial activity seems to be highly dependent on pH, since it has higher activity at lower pH as it is shown for NNCS. The physiology of the cell is more prone to suffer damage at lower pH, therefore, carboxyl and phosphate groups of the bacterial surface are anionic and they provide potential sites for electrostatic biding of chitosan at lower pH [23]. In this case, acetic acid was used to dissolve or disperse chitosan, and some reports state that it is one of the most suitable organic antimicrobial agents [23]. However Fernandez et al [24] found no inhibitory effect in acetic acid solutions with the same pH of the chitosan solution with microbial lethality. Another important factor influencing the antibacterial activity of the agents irrespective of the form or quantity is that they require water for activity, and totally dry samples are virtually incapable of releasing the energy stored in chemical bonds to initiate reaction [25]. In this regard, our TGA analysis shown that NNCS has higher water biding ability than NCS, therefore, higher water content is present in the NNCS form and it contributes to enhance its antibacterial activity.

In this study, AgnP of 25 nm particle size were used. The antibacterial test for neutralized and non-neutralized CS/AgnP films at different AgnP concentrations after 48 h is shown in Figures 4 a and b. It can be observed that with the inclusion of silver nanoparticles, the antibacterial effect of pure chitosan is highly enhanced, especially in the case of pure neutralized chitosan films. In this case, both neutralized and non-neutralized films show the higher inhibitory effect against *Salmonella thyphimurium,* and the most resistant organism seems to be *Sthapylococcus aureus.*

Our bacteriological results are in agreement with the results presented by Morones et al. [3]: 1) higher particles size related with agglomerates formation attach to the surface cell membrane and disturb its proper function, like permeability and respiration (3-20% CS/AgnPs films); 2) smaller particles size will have a direct interaction with cell membrane and could penetrate them and cause damage (0.5 and 1% CS/AgnP films).

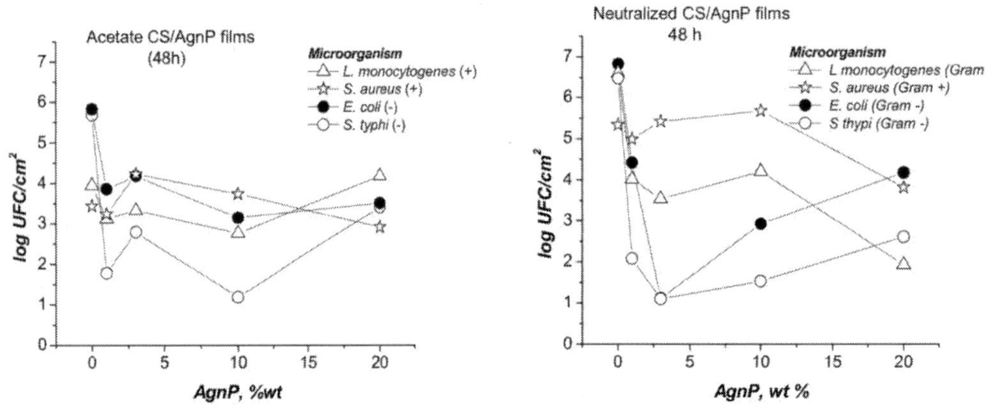

Figure 4. *logUFC/cm² versus AgnP concentration* for neutralized and non-neutralized CS/AgnP composites at 48h.

As well as for the dielectric results showed above, CS/AgnP neutralized composites against Gramnegative bacteria shows a particular trend at 48 h of contact;, 3 wt% of AgnP seems to be a turning point, since below this concentration the antibacterial effect is more effective as AngP increases, while above this weight percentage the effectiveness decreases. The same trend is observed in Grampositive bacteria up to 5 wt%, however it does not applied to 10 wt%. The lower antibacterial activity of the films as AgnP concentration increases could be ascribe to the presence of larger agglomerates above 3 wt% (this was observed by SEM, not shown), since the inhibitory effect is dependent on the size of AgnP; the smaller the AgnP particle size, the higher the antibacterial activity [3]. Also, moisture content could be affecting this inhibitory effect, given that TGA measurements showed that moisture content decreases as AgnP content increases (see Figure 3b). 3 wt% of AgnP is also a turning point for moisture content since no drastic change is observed above this concentration. Bacteria need water to grow, and in the case of CS/AgnP composites against Gramnegative bacteria, composites with higher moisture content has higher inhibition of growth. A synergic effect of silver nanoparticles size as well as moisture content is clearly observed.

In the case of CS /AgnP acetate composites, there is not a clear trend dependent on the nature of bacteria. However, as well as in CS/AgnP neutralized composites, the higher inhibitory effect is on *Salmonella thyphimurium*. In this case, for all bacteria, the higher effect is shown at 5 wt% of AgnP. In contrast to the results obtained in pristine neutralized and non-neutralized chitosan films, the antibacterial effect of CS/AgnP acetate composites is less effective compared to their neutralized peers (except in the case of the analysis against *Listeria monocytogenes*). From these results, it seems that the neutralization of CS/AgnP composites produces higher number of NH_2 groups leading to higher interaction between them and silver nanoparticles, this could decrease the number and/or size of agglomerates which in turns directly affects the antibacterial activity.

These results lead us to the conclusion that the neutralization of CS/AgnP composites favors the inhibitory effect of bacteria growth in composites, and that at this point, silver nanoparticles particle size is the most important variable affecting the antibacterial effect of composites. The inactivation of cells increases as silver nanoparticles concentration increases, however the number of cells inactivated with concentrations above 10 wt% of AgnP is similar to that at 3 wt%. This could be explained by the presence of higher number of agglomerates in the film. The optimum concentration of AgnP for the reduction microorganisms population seems to be 3 wt%.

Several studies have been carried out to demonstrate that AgnP can be successfully used in medical devices for microorganisms' growth control. In this regard, CS/AgnP films seem to be a good option to be used for the control microorganisms' growth in medical devices as well as for food safety concerns. In this work it was observed that the dielectric properties of CS and CS/AgnP films can be a measure of their antibacterial activity.

Conclusions

Chitosan-silver nanoparticles interactions were studied by dielectric spectroscopy, infrared analysis and scanning electronic microscopy. AgnP-NH_2, OH and $CONH_2$ interactions involve evident redistributions of IR vibration bands for these groups. These interactions could strongly modify chitosan water absorption ability which in turns affects the antibacterial ability of CS/AgnP composite films. Above 3 wt% of AgnP, the number of agglomerates is significant, and this directly changes the antibacterial effect of composite films.

Our dielectric results show that pristine-chitosan acetate films have an excess number of NH_3^+ groups compared to its neutralized peer, offering stronger antibacterial activity, since cell attachment is attributed to electrostatic interactions between cationic sites of chitosan. Grampositive bacteria are more susceptible to establish electrostatic interactions with antimicrobial compounds because it has phosphate groups with negative charge giving rise to a more electronegative wall. In non-neutralized films a synergic effect of NH_3^+ groups and lower pH produce a stronger bactericidal effect.

In case CS/ AgnPs composites, the bactericidal effect of chitosan is less pronounced and a strong influence of AgnPs particle size and agglomerates is observed. Silver nanoparticles act against Gram-negative and Gram-positive bacteria attaching to the surface of the cell membrane and disturbing its proper function; they are able to penetrate inside the bacteria causing DNA damage. The optimum concentration of AgnP for the inhibition of microorganisms´ growth seems to be 3 wt%. This AgnP concentration is the turning point for conductivity, agglomerates formation and the activation energy for ion motion. This fact can be significant to relate the antibacterial effect with the conductivity of films.

Acknowledgements

This work was partially supported by CONACYT and PROMEP of Mexico. The authors are grateful to J.A. Muñoz-Salas for technical assistance in electrical measurements and R. A. Mauricio Sánchez for assistance in FTIR measurements.

REFERENCES

[1] Singh, M.; Singh, S.; Prasad, S.; Gambhir, I. S. Journal of Nanomaterials and Biostructures 2008:3:115-122.

[2] Sharma, V. K.; Yngard, R. A.; Lin, Y. Advance in colloid and Interface Science 2009:145:83-96.

[3] Morones, J. R.; Elechiguerra, J. L.; Camacho, A.; Holt, K.; Kouri, J. B.; Tapia R. J.; Yacaman, M. J. Nanotechnology 2005:16:2346-2353.

[4] Fernández, E. J.; García-Barrasa J.; Laguna, A.; López-de-Luzuriaga, J. M.; Monge, M.; Torres C. Nanotechnology 2008:19:185602

[5] Lara H.H., Ayala-Nuñez N.V., Ixtepan-Turrent L., Rodríguez-Padilla C. J Nanobiotechnology 2010:80:1.

[6] Elechiguerra J.L., Burt J.L., Morones J.R., Camacho-Bragado A., Gao X., Lara H.H., Yacaman M.J. J Nanobiotechnology 2005:3:6.

[7] Panáček, A.; Kvítek, L.; Prucek, R.; Kolář, M.; Večeřová, R.; Pizúrová, N.; Sharma, V.K.; Nevěčná, T.; Zbořil, R. Journal of Phisical Chemeistry 2006:110:33:16248.

[8] Hong,K.H.; Park, J. L.; Youk, J. H.; Kang, T. J. J Polym Sci:Part B: Polym Phys 2006:44:2468.

[9] Burridge, K.; Johnston, J.; Borrmann, T. Journal of Materials Chemistry 2011:21:734.

[10] Kelly, F.M.; Johnston, J. H. ACS Appl Mater Interfaces 2011:3:1083.

[11] Zhang, Z.; Zhang, L.; Wang, S.; Chen W.; Lei Y. Polymer 2001:42:8315.

[12] Wei, D., Sun, W.; Qian, W.; Ye, Y.; Ma, X. Carbohydrate Polymers 2009:344:2375.

[13] Sanpui, P.; Murugadoss, A.; Prasad, P.V. D.; Ghosh, S.S.; Chattopadhyay, A. International Journal of Food Microbiology 2008:124:142-146.

[14] Thomas, V.; Yallapu, M.M.; Sreedhar, B.; Bajpai, S.K. Journal of Biomaterials Science 2009:20:2129-2144.

[15] Quemeneur, F.; Rinaudo, M.; Pépin-Donat, B. *Biomacromolecules* 2008:9:396-402.

[16] Viciosa, M. T.; Dionisio, M.; Silva, R. M.; Mano, J. F. Biomacromolecules 2004, 5, 2073–2078.

[17] González-Campos, J. B.; Prokhorov, E.; Luna-Bárcenas G.; Fonseca-García A.; Sanchez, I.C. J Polym Sci: PartB: Polym Phys 2009:47:2259.

[18] González-Campos, J. B.; Prokhorov, E.; Luna-Bárcenas G.; Sanchez, I.C.; Lara-Romero, J.; Mendoza-Duarte M. E.; Villaseñor F.; Guevara-Olvera, L. J Polym Sci:Part B: Polym Phys 2010:48:739.

[19] Psarras, G. C. Comp:Part A: Appl Sci Manuf 2006:37:1545.

[20] Prokhorov, E.; González-Campos, J. B.; Luna-Bárcenas, G. Mol Cryst Liq Cryst 2011:536:24/256.

[21] Chatelet, C.; Damour, O; Domard A. Biomaterials 2001:22:261.

[22] Amaral, I. F.; Granja, P. L.; Barbosa, M. A. J. Biomater. Sci. Polymer 2005: 16:1575.

[23] Fernandez-Saiz, P.; Lagaron J. M.; Ocio, M. J. Food Hydrocolloids 2009:23:913.
[24] Fernandez-Saiz. J.; Lagaron, J. M. Biopolymers 2006:83:577
[25] Gartner,C.; López, B. L.; Sierra, L.; Graf, R.; Spiess, H. W.; Gaborieau, M.; Biomacromolecules. 2011 April 11; 12(4): 1380–1386.

(618ai) Fluorescent Electrospun Quantum Dots/ Poly (Vinyl Alcohol) Nanocomposite Fibers

Emre Atabey[1], Jiahua Zhu[1], Suying Wei[2] and Zhanhu Guo[1], (1)Integrated Composites Laboratory (ICL), Dan F Smith Department of Chemical Engineering, Lamar University, Beaumont, TX, (2) Chemistry and Biochemistry, Lamar University, Beaumont, TX

Polyvinyl alcohol (PVA) and its nanocomposites filled with different loading level of the quantum dots (QDs) were prepared by the electrospinning process. With the photoluminescence from semiconductor quantum dots embedded in the fibers, the electrospun fibers could have many applications ranging from filtration to cell and tissue engineering. The electrospinning conditions including the polymer loading, quantum dots loading, working distance, electric voltage are optimized to obtain the uniform and bead-free fibers. The morphology of the nanocomposite fibers were characterized by scanning electron microscopy (SEM), thermal stability was measured by thermogravimetric analysis (TGA), the rheological behavior of the solutions was tested by the Rheometer.

See more of this Session: Mesd Poster Session
See more of this Group/Topical: Materials Engineering and Sciences Division

(618aj) Nano-Scale Cobalt Ferrite and Nickel Ferrite On High Surface Area ZrO2 Particles for Chemical Looping Hydrogen Production

Victoria J. Aston[1], Anthony H. McDaniel[2], Brian W. Evanko[1], Jonathan Scheffe[3], Mark D. Allendorf[2] and Alan W. Weimer[4], (1)Chemical and Biological Engineering, University of Colorado at Boulder, Boulder, CO, (2)Sandia National Laboratories, Livermore, CA, (3)Chemical & Biological Engineering, University of Colorado, Boulder, CO, (4)Department of Chemical and Biological Engineering, University of Colorado at Boulder, Boulder, CO

The chemical looping hydrogen production process uses synthesis gas or methane to reduce a metal oxide material, forming a waste stream of CO_2 and H_2O from which CO_2 can be easily sequestered. The reduced metal oxide is then oxidized with steam to produce pure H_2 and regenerate the original metal oxide. This system requires a metal oxide material that can achieve high H_2 and CO conversions to H_2O and CO_2 during the reduction step while maintaining fast oxidation and reduction kinetics that do not degrade over time. In this study, the use of nano-scale cobalt- and nickel-ferrite ($CoFe_2O_4$ and $NiFe_2O_4$) based materials was investigated for chemical looping hydrogen production. Nano-thin films of ferrite were deposited on a high surface area ZrO_2 support material using the atomic layer deposition (ALD) process to minimize sintering and diffusion limitations on reaction kinetics. The kinetics and cyclability performance of these thin film materials were analyzed and compared with the performance of conventionally made impregnation ferrite samples.

See more of this Session: Mesd Poster Session
See more of this Group/Topical: Materials Engineering and Sciences Division

(618ak) Preparation of Polycaprolactone Foms by Supercritical Carbon Dioxide Foaming

Ruey Chi Hsu, Chemical and materials engineering, Chang Gung University, Kweo-Shan,Tao-Yuan, Taiwan and LU-Wei Li, Chemical and Materials Engineering, Chang Gung University, Kweo-Shan,Tao-Yuan, Taiwan

ABSTRACT

The study is about the preparation of biocompatible and biodegradable polymer, Polycaprolactone (PCL), foams by supercritical carbon dioxide and finding the optimun foaming conditions.

PCL, a semi-crystalline polyester is subjected to biodegradation because of the susceptibility of its aliphatic ester linkage to hydrolysis. Matrices with porous surfaces are often desirable in tissue engineering applications. PCL are also used frequently for this purpose. In many studies, in order to enhance the interconnectivity of the pores and the average-size cell of the foamed material, the Supercritical Carbon Dioxide Foaming technique was applied in methods of preparing porous structures.

There is great potential for utilizing supercritical carbon dioxide as the medium in all these processes, because the material may be thermally labile, therefore should be free from any organic solvent residues and artifacts. Supercritical carbon dioxide is an attractive solvent or antisolvent in these processes, owing to the tunability of its solvent power with small variations in pressure and temperature and to its excellent transport properties(e.g., viscosity and diffusivity); in addition, it is nontoxic, nonflammable and environmentally friendly. The major advantages of these methods include single-step operation, a mild operating temperature, and a very narrow size distribution of matrices with controlled morphology, unlike the conventional techniques. All considered, supercritical carbon dioxide is one of the best choice in many applications.

The effects of saturation time, foaming temperature, saturation pressure and depressurization rate on the foam structure and volume expansion ratio were investigated, with the intention of finding the most influencial terms.

In this study, PCL (M_n= 80,000) was used to point out the effect of the operation conditions on the PCL porous structures, such as the temperature, the pressure, the depressure time and the soaking time. According to the experimental results, at the temperature of 40°C and pressure of 2000psi., with the soaking time of 120 minutes, and the depressization time of 2 seconds, we can obtain ideal PCL porous structures (i.e., with the average cell size of the foamed material 76 μ m).

The supercritical carbon dioxide foaming usually accompany with the morphology and crystallization behavior change. DSC was used to examine the morphology and crystallization situation. In the experiment, PCL get more crystallization region after foaming.

See more of this Session: Mesd Poster Session
See more of this Group/Topical: Materials Engineering and Sciences Division

(618al) Characterization of Chemical Information and Morphology In Decarburized Electrical Steel with Glow Discharge Sputtering

Min Serk Kwon, Technical Research Laboratory, POSCO, Pohang, South Korea

Electrical steel is extensively used for transformation and generators, and the need for more efficient electrical steel has been increased due to the global movement of energy saving and environmental protection. The electrical steel is made up 3% silicon and a majority of iron matrix for increasing specific resistance. The surface of electrical steel is covered with a film composed of mainly Mg_2SiO_4, which is a reaction product of internal oxidation precipitates formed during decarburization annealing and MgO affects not only the electrical insulation properties but also the magnetic properties of electrical steel. Therefore, it is very important to know the morphology and chemical properties of the internal oxidation layer for controlling and understanding high temperature oxidation in the manufacture of electrical steel. Although many investigations on the internal oxidation of 3% silicon steel have been conducted, there have seldom been reports on the growth and behavior of internal oxidation depend on the evaluating temperature in electrical steel sheets. In this study, the internal oxidation layers have been investigated as a function of heating temperature, annealing time and oxygen potential in order to survey the growth and formation of internal oxides during decarburization annealing in the electrical steel. The morphology and growth of oxides were observed by SEM, TEM, FT-IR, and GDS.

See more of this Session: Mesd Poster Session
See more of this Group/Topical: Materials Engineering and Sciences Division

(618am) Combination of Reversible Addition Fragmentation Chain Transfer (RAFT) and Click Chemistry for Surface Polymerization of Fluorinated Methacrylate On Iron Particles

Joko Sutrisno[1], Robert Fyda[2] and **Alan Fuchs**[1], (1)Chemical Engineering, University of Nevada, Reno, Reno, NV, (2)Chemical and Materials Engineering, University of Nevada, Reno, Reno

The combination of reversible addition fragmentation chain transfer (RAFT) and click chemistry has been successfully used for surface grafting poly(tetrafluoropropyl methacrylate) on the iron particles. 3-benzylsulfanylthiocarbonylsufanyl propionic acid has been synthesized and used as a chain transfer agent. The chain transfer agent was functionalized using alkyne and iron particles were functionalized azide groups. The alkyne functionalized chain transfer agent was used for polymerization of tetrafluoropropyl methacrylate monomer. The synthesized poly(tetrafluoropropyl methacrylate) end-capped with alkyne group was then "clicked" with azide terminated iron particles. The CTA molecular weight, surface morphology and thermal properties of grafted polymer were characterized. From the electron microscope, the grafted poly(tetrafluoropropyl methacrylate) resulted in uniform thin coating. The grafted polymer–iron particles showed a higher thermal transition temperature compared to non-grafted. The covalent bond between the polymer and the surface of the iron particles restricted the molecular mobility.

See more of this Session: Mesd Poster Session
See more of this Group/Topical: Materials Engineering and Sciences Division

(618an) Low-Temperature Synthesis of Titanium Dioxide and Its Characterization

Yuya Takubo, Junichi Ida, Tatsushi Matsuyama and Hideo Yamamoto, Department of Environmental Engineering for Symbiosis, Soka University, Hachioji, Tokyo, Japan

Crystalline titanium dioxide (TiO_2) powders have been prepared by the modified alkoxide method at low temperatures (45-180°C). In the experiment, the influences of reaction temperature, pH value and type of acid catalyst on the formation of TiO_2 were systematically investigated. The resulting TiO_2 powders were characterized with the X-ray diffraction (XRD), specific surface area measurement, and Field Emission Scanning Electron Microscopy (FE-SEM). The photocatalytic activities of the resulting samples were also evaluated by photocatalytic decomposition of 2, 4-dinitrophenol (DNP). The result indicated that the obtained TiO_2 powders possessed high specific surface area which was about three times greater than commercial photocatalyst, Degussa P-25. The phase composition, morphology, and specific surface area of TiO_2 changed significantly, depending on the pH value, reaction temperature, type of acid catalyst in the preparation process. DNP decomposition experiments showed that photocatalytic activity was affected by the balance between specific surface area and crystallinity. As to the effect of acid catalyst on the DNP decomposition activity, the activity was in the order of: HCl > HNO_3 > no acid catalyst > H_2SO_4.

See more of this Session: Mesd Poster Session
See more of this Group/Topical: Materials Engineering and Sciences Division

(618ao) Preparation and Characterization TiO_2 Thin Film On Glass Substrate for Photocatalytic Hydrogen Production

Njideka H. Okoye[1], Pedro E. Arce[1] and Dennis George[2], (1)Chemical Engineering, Tennessee Technological University, Cookeville, TN, (2)Civil Engineering/Center for the Management, Utilization and Protection of Water Resources, Tennessee Tech University, Cookeville, TN

Transparent titanium dioxide (TiO_2), anatase form, thin films were prepared and deposited on microscopic glass substrate via the sol-gel method and calcined at 500°C. The number of coating cycles varied from 6, 8 and 10 to assess the effect on hydrogen gas production. TiO_2 deposit layer produced was characterized with X-ray diffraction technique (XRD), scanning electron microscopy (SEM), transmission electron microscopy (TEM) and UV-Visible optical fiber spectrophotometer. Hydrogen was produced using the 8th coating substrate that gave consistent low absorption at wavelengths of 320nm to 350nm.This absorption wavelength range is needed for effective TiO_2 excitation. The effect of hydrogen production rate was analyzed using different ethanol ratio and also different active surface area for reaction.

See more of this Session: Mesd Poster Session
See more of this Group/Topical: Materials Engineering and Sciences Division

(618ap) Applications of Anodized Aluminum Oxide At Mainstream Engineering Corporation

Ted Amundsen[1], **Justin J. Hill**[1], Kevin Coffey[2] and Nicholas Schwartz[1], (1)Mainstream Engineering Corporation, Rockledge, FL, (2)Department of Mechanical, Materials, and Aerospace Engineering, University of Central Florida, Orlando, FL

We are using anodized aluminumoxide (AAO) for a variety of energy-related applications. Among these is as asupport for amorphous metal alloy hydrogen separation membranes. The ordered nanopore array found in AAO allows for the fabrication of amembrane support with nanometer-scale surface roughness and a tunable, yetuniform, porosity. This architecture not only enhances adhesion of the membraneto the support but also facilitates the facile deposition of *ultra*-thin membranes without loss offilm cohesion. Other benefits are the scalability and tunability of theelectrochemical self-assembly mechanism used to fabricate AAO. A square foot ofthe support is as easily fabricated as a square inch and the uniform diametersof the pores within the array are easily tunable simply by adjusting theapplied voltage that drives the electrochemical oxidation reaction. The poresare non-tortuous, channeling directly through the support which creates adirect path for hydrogen diffusion. Additionally, the pore diameter can beadjusted from 10 to 100s nm, thus avoiding the Knudsen diffusion limit. Thoughmild anodization can be quite slow (2 10 µm/hr),we have developed a process to form high□density pores using a hardanodization technique which was recently discovered to form low□densitypores under strict temperature/acid/voltage control at rates as high as 300 µm/h. Not only were the pores fabricated for thisstudy of high□density, but the AAO template supports were also up to 1.6mm thick (Figure 1). To the best of our knowledge, this is the thickest AAOtemplate ever fabricated (which only took 23 h of anodization), exceeding theclosest example in the literature by almost 20□fold.

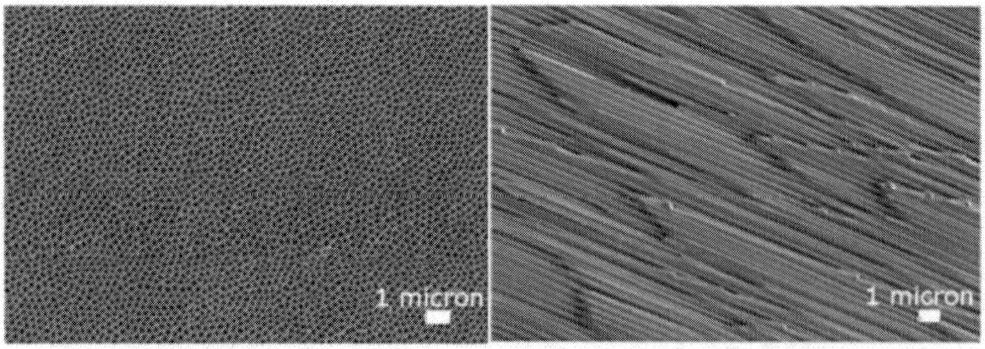

Figure 1. SEM (left: top-down, right: cross-section) of the hard anodized AAO support

The membranes were fabricated ascontinuous thin films of amorphous metal alloys (Figure 2) and were thin enoughto be surface reaction limited, exhibiting a linear increase in hydrogen fluxwith $\Delta P_{hydrogen}$. Sub-micron Pd-alloycatalytic layers are currently being investigated to improve hydrogendissociation and reassociation rates.

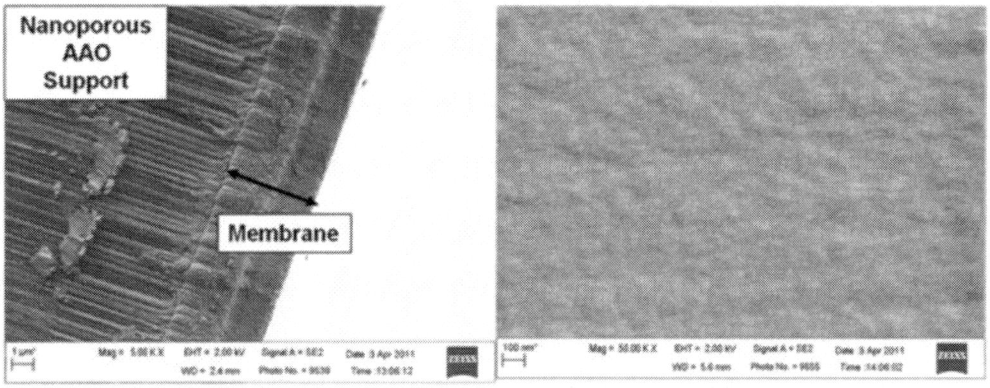

Figure 2. SEM (left: cross-section, right: membrane face) of 3 μm $Ni_{30}Ta_{70}$ membrane deposited on an AAO support

See more of this Session: Mesd Poster Session
See more of this Group/Topical: Materials Engineering and Sciences Division

(618aq) Controlled Self-Assembly of Pillars Modified with Vapor Deposited Polymeric Nanocoatings

Benny Chen, Scott Seidel and Malancha Gupta, Mork Family Department of Chemical Engineering and Materials Science, University of Southern California, Los Angeles, CA

In this talk, we will present a capillary force induced self-assembly technique to create large area arrays of complex three-dimensional microstructures. In our method, initiated chemical vapor deposition (iCVD) is used to modify the surface properties of micropillared arrays without changing the bulk mechanical properties of the pillars. The iCVD process is an all-dry vapor-phase free radical polymerization technique that is an environmentally benign alternative to traditional liquid-phase polymerization processes because no organic solvents are necessary. We used the iCVD process to sequentially deposit functional and responsive organic polymers onto elastomeric pillars in one reaction chamber without removing our substrate between steps. Capillary forces were used to self-assemble the pillars into ordered nanostructures by changing the surface properties of the pillars from hydrophobic to hydrophilic. The height of the pillars was varied in order to study the relative effect of elastic forces on the self-assembly process.

See more of this Session: Mesd Poster Session
See more of this Group/Topical: Materials Engineering and Sciences Division

(618ar) Gecko-Inspired Dry Adhesive for Robotic Applications

Saurabh Das[1], Jing Yu[1], Sathya Chary[2], John Tamelier[2], Noshir Pesika[3], Kimberly Turner[2] and Jacob N. Israelachvili[4], (1)Chemical Engineering, University of California, Santa Barbara, Santa Barbara, CA, (2)Mechanical Engineering, University of California, Santa Barbara, Santa Barbara, CA, (3)Chemical and Biomolecular Engineering, Tulane University, New Orleans, LA, (4)Chemical Engineering, University of California at Santa Barbara, Santa Barbara, CA

Geckos have the supreme ability to rapidly attach and detach from almost any kind of surface. This ability is attributed to the hierarchical structure of their feet and its articulation to generate strong adhesion and friction forces on gripping cycle and comes off the surface within milli seconds during the releasing cycle. This phenomena has inspired the fabrication of structured surfaces for robotic applications. In this study, micron-sized rectangular flaps composed of polydimethylsiloxane (PDMS) were fabricated using massively parallel micro-electromechanical systems (MEMS) techniques with the intention of creating directionally responsive, high-to-low frictional-adhesion toe pads exhibiting properties similar to those found in geckos. Using the Surface Forces Apparatus (SFA), we investigated the friction and adhesion forces of both vertical (symmetric) and angled/tilted (x-y-z asymmetric) micro-flaps under various loading, unloading and shearing conditions. We find that the anisotropic structure of tilted micro-flaps gives very different adhesion and tribological forces when articulated along different x-y-z directions: high friction and adhesion forces when articulated in the y-z plane along the tilt (+y) direction, which is also the direction of motion, and weak friction and adhesion forces when articulated against the tilt (-y) direction. Our results demonstrate that asymmetric angled structures, as occur in geckos, are required to enable the gecko to optimize the requirements of high friction and adhesion on gripping, and low frictional-adhesion on releasing. These properties, when coupled with suitable articulation mechanisms, produce a reversible adhesive system.

See more of this Session: Mesd Poster Session
See more of this Group/Topical: Materials Engineering and Sciences Division

(618as) Aligned Deposition of Polyurethane Fiber Arrays In Single and Multiple Layers

Ji Wang, Engineering Science and Mechanics, Virginia Tech, Blacksburg, VA and **Amrinder S. Nain**, Mechanical Engineering, Virginia Tech, Blacksburg, VA

In this study, polyurethane (PU)fibers were manufactured using our recently reported Spinneret based Tunable Engineering Parameter (STEP) technique. Mixture of N-Methyl-2-pyrrolidone(NMP) and Tetrahydrofuran(THF) was chosen as solvent for PU fiber deposition. Initially the volume ratio of NMP/THF was set at 6:4 and beaded fibers were obtained at 24% wt of PU. With further increase in PU concentration, beads were observed to gradually stretch along fiber axis and finally smooth fibers (diameter ranges from 400nm to 800nm) were obtained at 39% wt. Solvent effect on PU fiber morphology was investigated by varying the NMP/THF. Eight different ratios (volume ratio of NMP/THF= 1/9,2/8,3/7,4/6,5/5,6/4,7/3,8/2 respectively) with PU dissolved at 30% wt., were investigated to determine PU spinnability. PU fibers exhibited ultimate elongation up to 453.5%, which makes it possible to make braids of few PU fibers: PU fiber braids were made with two and three fibers respectively. The braiding angle was controlled by the number of twists added to the braids. Bio-scaffolds of polyurethane fibers were fabricated through continuous deposition of PU fiber arrays and the fiber spacing can be well controlled through STEP technique. Stress-strain analysis demonstrate typical sigmoidal shape characteristic of elastomeric materials. Current ongoing efforts are devoted towards embedding multi wall carbon nanotubes in polyurethane fibers and understanding the reinforcement mechanisms and polymer molecular arrangements under high elongation.

See more of this Session: Mesd Poster Session
See more of this Group/Topical: Materials Engineering and Sciences Division

(618at) Electrospun Chitosan Nanofibers for the Controlled Released of Bioactive FGF-2 and Mesenchymal Stem Cell Osteogenic Differentiation

Jorge Almodovar[1], Fabio Zomer Volpato[2], Samantha Bacon[1], John D. Kisiday[3], Claudio Migliaresi[2] and Matt J. Kipper[4], (1)Chemical and Biological Engineering, Colorado State University, Fort Collins, CO, (2)University of Trento, Trento, Italy, (3)Clinical Sciences and School of Biomedical Engineering, Colorado State University, Fort Collins, CO, (4)Chemical and Biological Engineering and School of Biomedical Engineering, Colorado State University, Fort Collins, CO

Scaffolds prepared from the polysaccharide chitosan have gained considerable attention in the past several years for tissue engineering applications because of the many attractive properties of chitosan. Electrospinning of polymeric solutions has becoming increasingly popular as a simple and reproducible technique for creating nanofibrous scaffolds with a wide range of geometry, porosity, and alignment. In this work, chitosan nanofibers were created via electrospinning and they were investigated to be used as a platform for growth factor delivery and mesenchymal stem cell (MSC) activation. Basic fibroblast growth factor (FGF-2) was adsorbed on polyelectrolyte complex nanoparticles (PCN) and these were successfully adsorbed and released from the nanofibers. The release of the PCN was controlled by using layer-by-layer assembly of poyelectrolyte multilayers (PEM). Lastly, the nanofiber scaffolds were tested in vitro using bone marrow-derived ovine and equine MSCs for applications in bone regeneration. Controlled release of FGF-2 is achieved over a period of 27 days and its biological activity is preserved. MSCs were successfully seeded on chitosan nanofibers and both equine and ovine MSCs demonstrated a higher alkaline phosphatase production when seeded on fibers as compared to flat polystyrene surfaces. Chitosan nanofibers promote osteogenic differentiation even in the absence of chemical cues. Future work will be performed by delivering growth factors that promote MSC differentiation using chitosan nanofibers and polysaccharide-based nanomaterials.

See more of this Session: Mesd Poster Session
See more of this Group/Topical: Materials Engineering and Sciences Division

(618au) Morphological and Quantitive Characterization of Wear Particles Generated From Nanocomposites

Aydar Akchurin[1], **Songbo Xu**[1], **Annie X.W. Tangpong**[1], Iskander Akhatov[1], Tian Liu[2], Weston Wood[2] and Wei-Hong Zhong[2], (1)Department of Mechanical Engineering, North Dakota State University, Fargo, ND, (2)School of Mechanical and Materials Engineering, Washington State University, Pullman, WA

Wear debris of materials used in joint replacements is known to be a major cause of the joint loosening and osteolysis that eventually lead to failure of the joint and the revision procedure. Research has shown that the degree of biological reaction of the immune system to the wear debris generated in the joint strongly depends on the size and shape of the wear particles. In this research, debris particles generated from a series of wear testing of carbon nano fiber (CNF) reinforced high density polyethylene (HDPE) samples were characterized both quantitatively and morphologically. The wear tests were conducted using a pin-on-disk wear tester (CETR UMT-2). The nanocomposite sample was submerged in a saline bath during the wear test. A variety of nanocomposite samples fabricated with different CNFs (pristine and silane-coated) were tested, and the weight concentrations of CNFs in the samples varied from . The primary debris particles were isolated following a sophisticated particle isolation protocol. The isolated particles were then characterized using a dynamic light scattering (DLS) particle sizer, as well as a scanning electron microscope (SEM). The primary sizes of the wear particles measured by the DLS technique were consistent with the SEM observations.

See more of this Session: Mesd Poster Session
See more of this Group/Topical: Materials Engineering and Sciences Division

(618av) Parallel Synthesis and Screening of Polymers with Endotoxin Binding Activities

Gabriela Montanez[1], Divya Nair[1], **Thrimoorthy Potta**[1], Lucas Vu[1], Amrita Mallik[2] and Kaushal Rege[1], (1)Chemical Engineering, Arizona State University, Tempe, AZ, (2)Biomedical Engineering, Arizona State University, Tempe, AZ

A component of the outer membrane of Gram-negative bacteria, lipopolysaccharides are toxic to the health of humans and animals. Lipopolysaccharides are endotoxins and cause strong immunological responses in normal individuals. Responses include inflammation and fever, leading to septic shock, which can result in death. Sequestration of lipopolysaccharides (endotoxin) is therefore of critical importance in therapeutic interventions and decontamination applications. A library of polymers was screened in order to rapidly identify lead candidates that bind endotoxin. A fluorescence-based parallel screening assay was employed to screen the polymer library. In this assay, the displacement of a fluorescent dye that binds specifically to the toxic lipid area of a lipopolysaccharide was used as an indicator of polymer-endotoxin binding. In addition to polymer type and dose, the role of different physicochemical factors including salt concentration and pH on polymer-endotoxin binding was investigated. Several polymers that demonstrated high efficacies for endotoxin binding were identified. Lead polymers were further characterized using FT-IR and NMR spectroscopy, and their ability to bind endotoxin was demonstrated additional experimental methods. Our results indicate that parallel synthesis and screening approaches can lead to the rapid identification of polymers for endotoxin sequestration.

See more of this Session: Mesd Poster Session
See more of this Group/Topical: Materials Engineering and Sciences Division

(618aw) Mechanical Testing of Nanocomposites Using An AFM-Based Micro/Nano Tensile Tester

Wyatt Leininger, Marshall McNea, **Xinnan Wang** and **Annie X.W. Tangpong**, Department of Mechanical Engineering, North Dakota State University, Fargo, ND

Nanoscale materials have applications across many engineering fields including structural composites, surface coatings, electronics, bio materials, and chemical catalysts, due largely to the significantly different properties these structures exhibit versus than their large scale counterparts. Understanding the mechanical properties of nanoscale structures is thus vital to utilize these materials effectively. A micro/nano tensile tester was designed for use with an atomic force microscope (AFM) to capture mechanical behavior mechanisms throughout load application (such as nano-crack propagation, and composite reinforcement re-orientation and deformation). This tester is able to offer great flexibilities in application with options of added functionalities to the control system as the requirement arises. The range of applications is extended through interchangeable load cells and sample mounting fixtures. Carbon nanotube-reinforced polymer composites were fabricated and tensile/compressive tests were performed using the newly designed tester in conjunction with an AFM.

See more of this Session: Mesd Poster Session
See more of this Group/Topical: Materials Engineering and Sciences Division

(618ax) Characterization of Bacterial Polymer Formation Using Vibrational Spectroscopy

Santanu Kundu[1], Michael S. Waters[2], Irene Calizo[3], Angela R. Hight Walker[3] and Kathryn L. Beers[4], (1)Polymers Division, National Institute of Standards and Technology (NIST), Gaithersburg, MD, (2) Polymers Division, Biomaterials Group, National Institute of Standards and Technology, Gaithersburg, MD, (3)Optical Technology Division, National Institute of Standards and Technology (NIST), Gaithersburg, MD, (4)Polymers Division, National Institute of Standards and Technology, Gaithersburg, MD

Polyhydroxyalkanoates (PHAs) are increasingly used in various applications ranging from degradable packaging to biomedical applications because of their biodegradability and biocompatibility. In addition, these materials have properties similar to conventional plastics and are easily processable in common processing equipment. PHAs are produced by different strains of naturally occurring and genetically modified bacteria in unfavorable growth conditions such as limited nitrogen and oxygen supply and the presence of excess carbon. Gas chromatography (GC) is the most common method for quantitative composition determination of PHAs formed in bacterial cells; however, the technique is time consuming and cannot be used online. We developed a sampling method which enables use of Raman spectroscopy to characterize PHA formation within a single bacterium of *Cupriavidus necator*. We are currently developing a technique to quantify PHA formation using this new method.

See more of this Session: Mesd Poster Session
See more of this Group/Topical: Materials Engineering and Sciences Division

(618ay) In Situ Measurement of Chemical and Mechanical Properties of Acidogenic Biofouling Biofilms

Michael S. Waters[1], Santanu Kundu[2], Irene Calizo[3], Angela R. Hight-Walker[4], Nancy J. Lin[5] and Sheng Lin-Gibson[5], (1)Polymers Division, Biomaterials Group, National Institute of Standards and Technology, Gaithersburg, MD, (2)Department of Polymer Science and Engineering, University of Massachusetts-Amherst, Amherst, MA, (3)Optical Technology Division, National Institute of Standards and Technology (NIST), Gaithersburg, MD, (4)National Institute of Standards and Technology, Gaithersburg, MD, (5)NIST, Gaithersburg, MD

Diagnostic measurements of biofilm-material interactions currently struggles to identify changes in biofilms beyond species presence and live/dead. However, most bacteria responsible for disease are intermittently pathogenic, which are controlled by distinct signals. In fact, many bacteria associated with pathogenesis exist in a non-pathogenic form in the natural human microbial flora. Current methods in evaluating bacterial biofilm pathogenicity are focused on evaluating overall biofilm presence and/or viability. This approach implements bulk evaluation that limits the ability discern if a live biofilm is exhibiting pathogenic properties, or is surviving in a non-pathogenic capacity. In an attempt to evolve analytical techniques to distinguish living biofilms in pathogenic and non-pathogenic modes, acidogenic dental pathogen *Streptococcus mutans* was used as a model. In this study, sucrose was used to control the exhibition of *S. mutans* pathogenic properties, which were then evaluated by surface enhanced Raman spectroscopy (SERS) to identify unique chemical signatures and rheology to evaluate the mechanical properties of biofilms grown directly on the instrument in a specially designed bioreactor. These techniques require minimal sample preparation and offer a rapid, low-cost, high throughput system for clinical evaluations, or evaluating of the capacity of various materials to affect bacterial-mediated biofilm pathogenesis.

See more of this Session: Mesd Poster Session
See more of this Group/Topical: Materials Engineering and Sciences Division

(618az) New Approaches for Directing the Route of Cell- and Tissue-Level Drug Delivery Using Synthetic Materials

Christopher M. Jewell, Departments of Bioengineering, Materials Science and Engineering, and The Ragon Institute of MGH, MIT, and Harvard, Massachusetts Institute of Technology, Cambridge, MA and Darrell J. Irvine, Departments of Bioengineering, Materials Science and Engineering, Koch Institute for Integrative Cancer Research, Ragon Institute of MGH, MIT, and Harvard, and the Howard Hughes Medical Institute, Massachusetts Institute of Technology, Cambridge, MA

A major challenge facing drug delivery is efficient targeting of bioactive cargo to specific cells or tissues. Delivery of macromolecular cargo (e.g., nucleic acids, proteins) to cells for example, is hindered by the inability of these therapeutics to escape endosomes following cell internalization, ultimately leading to cargo degradation. Additionally, directing cargo to the correct tissues (e.g., a tumor or infection site) is necessary for maximum drug efficacy and to minimize side effects. Thus, developing new clinically-relevant treatments requires drug delivery systems that address both cellular- and tissue-level barriers. This poster will describe several biomaterial-based strategies we are developing to overcome these challenges, emphasizing materials design and characterization. One strategy seeks to bypass endocytosis to achieve efficient cellular delivery of large, membrane-impermeable macromolecular cargo (e.g., DNA or RNA oligonucleotides) using cell-penetrating gold nanoparticle carriers that display amphiphilic mixed ligand shells composed of 11-mercapto-1-undecanesulphonate (MUS) and 1-octanethiol (OT). On a tissue-delivery level, *in vivo* strategies for increasing the localization and persistence of nucleic acids and small molecule vaccine components in secondary lymph organs – the key site of immune response generation – will be described. Polymeric particles displaying amphiphilic lipids as stabilizers at the polymer/aqueous interface allow controlled release of these vaccine components, and the ability to tune materials properties (e.g., size, aggregation control, biodistribution/kinetics) plays a critical role in vaccine effectiveness during *in vivo* studies with the polymeric vaccine particles.

See more of this Session: Mesd Poster Session
See more of this Group/Topical: Materials Engineering and Sciences Division

(618b) Analysis of the In Vitro Swelling Behavior of Poly(vinyl alcohol) Hydrogels In Osmotic Pressure Solution for Soft Tissue Replacement

Julianne L. Holloway, Anthony M. Lowman and Giuseppe R. Palmese, Chemical and Biological Engineering, Drexel University, Philadelphia, PA

A number of osmotic solutions ranging from 0 atm to 4 atm were used to evaluate poly(vinyl alcohol) (PVA) hydrogels as potential nondegradable soft tissue replacements *in vitro*. Osmotic solutions were necessary in order to mimic the swelling pressure observed *in vivo* for soft tissues present in load bearing joints. *In vitro* studies indicated that PVA hydrogels experience minimal changes in swelling with a polymer concentration of 20 wt% PVA in phosphate buffered saline solution (0 atm) and between 30 and 35 wt% PVA in osmotic solution with a pressure of 0.95 atm. Swelling in osmotic pressure solutions caused decreases in the equilibrium hydrogel hydration. A strong correlation between PVA polymer concentration and swelling ratio was observed; however, swelling ratio was found to be independent of freeze-thaw cycles. This allows for swelling to be tailored through polymer concentration while modulus and porosity is modified using freeze-thaw cycles creating a suite of materials to match a wide range of properties depending on the application desired. An investigation of hydrogel compressive modulus indicated that PVA hydrogels are within the range of articular cartilage, meniscal tissue, and the TMJ disc. The molecular weight between cross-links and microstructure of the PVA hydrogels were also evaluated as a function of freeze-thaw cycles and polymer concentration to lend insight into the processes occurring during synthesis and swelling in osmotic solutions.

See more of this Session: Mesd Poster Session
See more of this Group/Topical: Materials Engineering and Sciences Division

(618ba) Biodegradable Graft Polymer-Drug Conjugate

Yun Yu[1], Jiong Zou[1], Lu Yu[1], Wei Ji[2], Yukun Li[1], Wing-Cheung Law[2] and **Chong Cheng**[1], (1) Department of Chemical and Biological Engineering, SUNY-Buffalo, Buffalo, NY, (2)Department of Chemistry, SUNY-Buffalo, Buffalo, NY

Functional polylactide-*g*-paclitaxel-poly(ethylene glycol), a novel graft polymer-drug conjugate (GPDC) with paclitaxel (PTXL) as the divalent agent to bridge between the degradable polylactide (PLA)-based backbone and hydrophilic poly(ethylene glycol) (PEG) side chains, were prepared by the copper-catalyzed azide-alkyne cycloaddition reaction of acetylene-functionalized polylactide (PLA) with azide-functionalized PTXL-PEG conjugate. The acetylene-functionalized PLA was prepared by ring-opening copolymerization (ROCP) of acetylene-functionalized LA monomer with L-lactide (LA). The azide-functionalized PTXL-PEG conjugate was prepared by multi-step organic synthesis. The well-controlled chemical structures of the GPDC and its precursors were verified by ^{1}H NMR and GPC characterizations. DLS analysis indicated that GPDC molecules assembled in water to form nanoparticles with sizes of 8-40 nm. GPC analysis of buffer solutions (pH = 5.5 and 7.4) of the GPDC indicated the occurrence of multiple hydrolysis reactions under the experimental conditions, which resulted in the release of PTXL moieties and the cleavage of PLA-based backbone.

See more of this Session: Mesd Poster Session
See more of this Group/Topical: Materials Engineering and Sciences Division

EXPERIMENTAL AND THEORETICAL STUDY OF THE BULK POLYMERIZATION OF STYRENE USING THE SYMMETRICAL CYCLIC TRIFUNCTIONAL INITIATIOR DIETHYL KETONE TRIPEROXIDE (DEKTP)

E. Berkenwald[1], C. Spies[2], J. R. Cerna Cortez[3], G. Morales[3], D. Estenoz[2]

[1]Department of Chemical Engineering, Instituto Tecnológico de Buenos Aires (ITBA) Buenos Aires, Argentina.
[2]Instituto de Desarrollo Tecnológico para la Industria Química, INTEC (Universidad Nacional del Litoral - CONICET), Santa Fe, Argentina.
[3]Centro de Investigaciones en Química Aplicada (CIQA), Saltillo, México.

Abstract

This work experimentally and theoretically investigates the use of the symmetrical cyclic trifunctional initiator diethyl ketone triperoxide (DEKTP) in the bulk polymerization of styrene (St). The experimental work consisted on a series of isothermal batch polymerizations at different temperatures, 120 and 130°C, with different initiator concentrations: 0.005, 0.01 and 0.02 mol/L. A mathematical model was developed in order to predict the evolution of the reacting chemical species and the produced molecular weights distributions (MWDs). The kinetic model includes chemical and thermal initiation, propagation, transfer to the monomer, termination by combination and reinitiation reactions. Simulation results predict the concentration of di- and monoradicals as well as polymeric chains, characterized by the number of undecomposed peroxide sites. Experimental results show that, at reaction temperatures of 120-130°C, DEKTP initiation produces an increase in polymerization rates and average molecular weights, depending on the initiator concentration, due to sequential decomposition of the initiator molecule. The optimal DEKTP concentration for St bulk polymerization is found to be 0.01 mol/L. The mathematical model was adjusted and validated using the experimental data. Theoretical predictions are in excellent agreement with the experimental results.

Keywords: DEKTP, Mathematical model, Polystyrene, Trifunctional initiator

Introduction

The industrial use of multifunctional initiators in the bulk polymerization of styrene (St) has been increased due to the possibility of obtaining simultaneously high reaction rates and molecular weights, as well as enhanced mechanical properties compared to polystyrene (PS) obtained by using traditional initiators.[1-3]

The use of peroxide type initiators has been preferred over the use of azo or C-C initiators in the polymerization of vinyl monomers, due to their relative ease of synthesis and lower rate of self-induced decomposition reactions.[4,5] However, the use of monofunctional initiators is limited due to the difficulty of obtaining an adequate balance between reaction rate, molecular weight, polidispersity and monomer conversion.[6,7]

In the last few decades, there has been a growing interest in the industrial use of multifunctional initiators in order to avoid these limitations. With the use of bifunctional initiators (with two reacting peroxide groups), it was possible to achieve optimization of polymerization processes, while enhancing the final properties of the product (i.e. molecular weight, monomer conversion and polydispersity).[1,2,7-10] Both symmetrical and asymmetrical initiators were used, showing that it was possible, under the right temperature and initiator concentration, to obtain high reaction rates and high molecular weights simultaneously. Reaction temperature affects the thermal stability of peroxide groups, while initiator concentration affects the rate of polymerization and reduces polymerization times without lowering the molecular weight of the final product.[9]

A few works have theoretically studied the synthesis of PS using bifunctional initiators. Mathematical models were developed, allowing prediction of reacting species' concentrations, as well as molecular structure of the obtained polymer during the course of the polymerization reaction. Choi and Lei[1] and Kim and Choi[3] developed detailed kinetic models for the bulk St homopolymerization with symmetrical and asymmetrical diperoxyester initiators. They showed that operating at high temperatures, it is possible to obtain high reaction rates and molecular weights, with relatively narrow molecular weight distributions (MWDs). Villalobos[9] theoretically and experimentally investigated the polymerization of St with the following bifunctional initiators: 2,5-dimethyl-2,5-bis(2-ethylhexanol peroxy) hexane (Lupersol-256, L-256); 1,1-di(t-butyl-peroxy) cyclohexane, (Lupersol-331-80B); and 1,4-bis(t-butyl peroxycarbo) cyclohexane, (D-162). Compared to the standard monofunctional case, bifunctional initiators reduced polymerization times as much as 75%, without substantial changes in the final product properties. In Estenoz et al.[2] a detailed mathematical model was developed to simulate the synthesis of high-impact polystyrene (HIPS) using symmetrical bifunctional initiators such as L-256 and L-118 [2,5-dimethyl-2,5-bis(benzoyl peroxyl) hexane]. High reaction rates and high molecular weights were observed, and the phase inversion periods were produced at lower conversions.

There are a limited number of works dealing with the use of multifunctional initiators.[11-16] Cerna et al.[11] studied the bulk polymerization of St using initiators with different functionalities (mono-, di- and trifunctional initiators). In the particular case of the diethylketone triperoxide (DEKTP), a cyclic trifunctional peroxide initiator, at 120-130° C, it yielded polymers with high molecular weight (250,000-450,000 g/mol, depending on the DEKTP concentration) at relatively short polymerization times (4-6 h). The radical decomposition mechanism of DEKTP was found to be as complex as the one of a lineal bi-functional initiators, but DEKTP showed greater activity.[13] Two mechanisms have been proposed for the decomposition of DEKTP: a total decomposition of the 3 O-O groups at temperatures higher than 130°C, and a sequential decomposition of the O-O groups at temperatures between 110-130°C. In the last case, Cerna et al. were able to synthesize PS with undecomposed O-O in the polymer backbone.[11]

In Scorah et al.[14,15], the bulk polymerization of St and methyl methacrylate in the presence of the tetrafunctional initiator polyether tetrakis(tert-butylperoxy carbonate) (JWEB50) was experimentally and theoretically studied. The developed model allowed the calculation of monomer conversion, average molecular weights and polymer structure and was validated with experimental results. Sheng et al.[16] studied the use of the cyclic trifunctional intiator 3,6,9-triethyl-3,6,9-trimethyl-1,4,7-triperoxonane in the bulk polymerization of St. Experimental results showed that it was possible to produce polymers with higher molecular weights and lower polydispersites at a higher rate. The obtained PS had O-O bonds in the molecular chains.

This works deals with the experimental and theoretical study of the DEKTP used in the bulk polymerization of St. A mathematical model is developed that allows the prediction of the evolution of the reacting species' concentration, monomer conversion and detailed polymer molecular structure. The effect of reaction conditions on process productivity and final product molecular characteristics is investigated.

Experimental Work

DEKTP initiator was synthesized according to the following procedure[17]: in a 250 mL Erlenmeyer flask 4.6 mL of hydrogen peroxide and 7.3 mL of sulfuric acid (70% v/v) were added and the temperature was kept at -10°C. Then, 5.6 mL of diethyl ketone was added into the reaction medium and the reaction remained under agitation for 3 h. Thereafter, the organic phase was extracted with petroleum ether and dried with sodium sulfate for 12 hours. The petroleum ether was removed under distillation and the product was crystallized from methanol yielding a white power.

The molecular structure of DEKTP is shown in Figure 1.

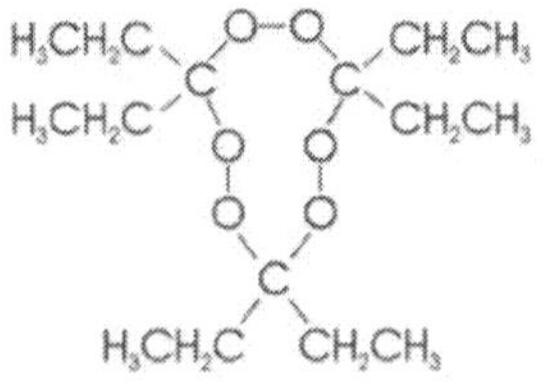

Figure 1. Molecular structure of DEKTP

Several bulk polymerizations of St were carried out in the presence of DEKTP at different temperatures and different initiator concentrations. In each experiment, a certain amount of DEKTP was dissolved in 2 mL of previously distilled St. The solution was placed in clean and dry test tubes, degassed by successive cycles of freezing-cooling from -180 ° C (using liquid nitrogen) to room temperature, at reduced pressure. The tubes were sealed under vacuum and immersed in an oil bath. The selected polymerization temperatures were 120 °C, and 130 °C to assure sequential decomposition of the initiator molecule. The decomposition reaction for DEKTP is presented in Figure 2.

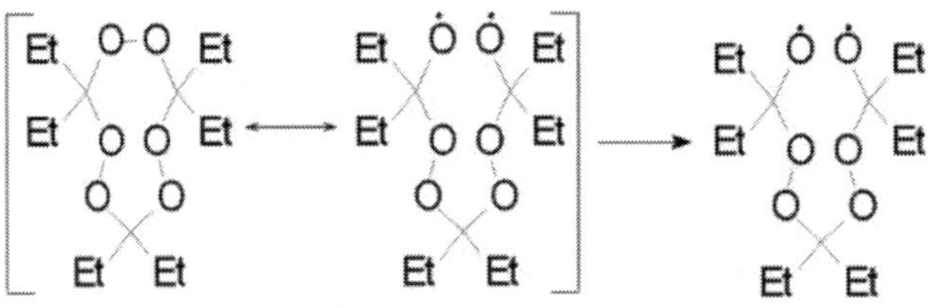

Figure 2. Sequential decomposition of DEKTP

Initiator concentrations were varied in the range of 0.005 and 0.02 mol/L. Monomer conversion was determined gravimetrically and molecular weights by gel permeation chromatography (GPC) at different reaction times. The equipment used was a Hewlett-Packard HPLC 1050 with software from Polymer Laboratories for Chemstation. Refraction Index and UV detectors were used, with ultrastyragel columns (10^5, 10^4, 10^3, A). Polystyrene standards (580 – 3,900,000 g/mol) were used, at ambient temperature. To measure conversion, the obtained polymer was directly precipitated in methanol (volume ratio 1:10), filtered and dried under vacuum until constant weight (approximately 24 hours). The final experimental results are presented in Table 1.

Experimental results show that initiation by DEKTP allows for high reaction rates and high average molecular weights simultaneously. As expected, increasing the initiator concentration increases reaction rates but produces lower molecular weights, due to a higher rate of termination reactions. An initiator concentration of 0.01 mol/L is optimal for obtaining high reaction rates and molecular weights at a reaction temperature of 120 °C.

Figures 3 and 4 show the evolution of conversion and average molecular weights for the reactions using a 0.01 mol/L concentration of DEKTP. In Figure 3, an increase in the polymerization rate is observed as temperature increases. For the period of time evaluated, the molecular weights increase along the reaction (Figure 4). The calculated polydispersities are around 2 in agreement to the reported results for multifunctional initiators.[1,9] As it was informed in Cerna et al.[11], this behavior in the temperature range of 120-130 ° C indicates that the initiator is decomposed by a sequential mechanism of rupture of the peroxide sites, by which the diradicals formed by initiator decomposition have 2 undecomposed peroxide groups.

Table 1. Experimental results for the bulk polymerization of St in the presence of DEKTP.

T (°C)	[I][a] (mol/L)	Time (min)	Conversion (%)	$\bar{M}_n \times 10^5$ (g/mol)	$\bar{M}_w \times 10^5$ (g/mol)
120	0.005	15	2.78	2,61	5.33
		60	13.05	2.17	4.27
		180	38.66	2.19	4.30
		240	42.58	2.39	4.57
		360	46.27	2.28	4.46
120	0.01	15	4.35	2.32	3.60
		60	16.23	1.75	3.38
		180	42.14	2.00	3.81
		240	62.12	2.18	4.21
		360	92.05	2.50	5.32
120	0.02	15	3.83	2.23	4.30
		60	17.88	2.21	4.84
		180	55.16	1.94	3.65
		240	71.22	2.00	3.63
		360	98.22	1.74	3.16
130	0.01	15	4.42	1.86	3.52
		30	11.70	1.47	2.81
		45	20.04	1.41	2.61
		60	26.64	1.50	2.67
		90	40.48	1.47	2.67
		180	86.51	1.50	2.77
		240	98.10	1.57	3.30

[a][I]: Molar concentration of DEKTP

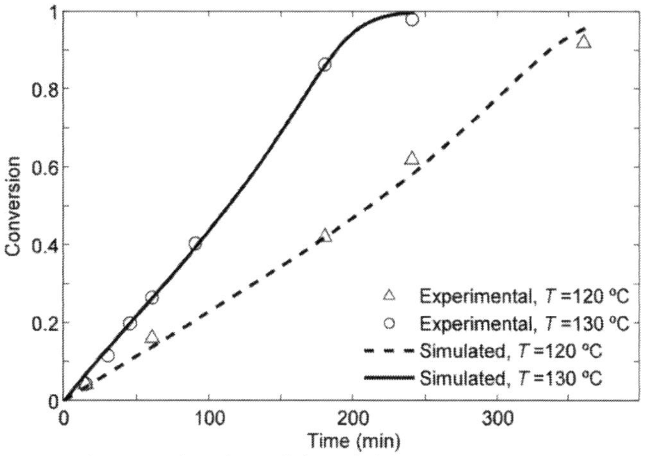

Figure 3. Conversion as a function of time, initiator concentration 0.01 mol/L

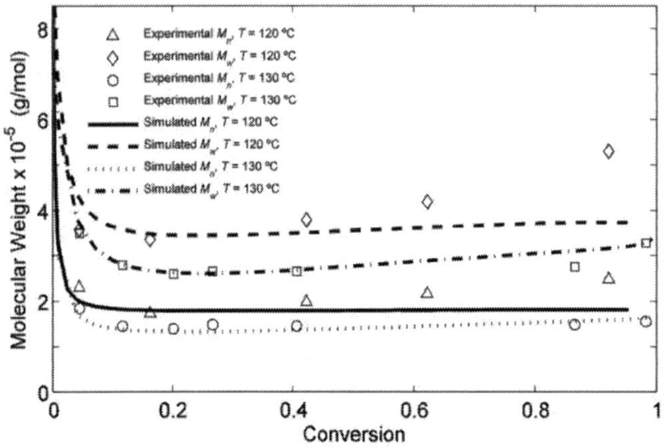

Figure 4. Average molecular weights as a function of conversion, initiator concentration 0.01 mol/L

Mathematical Model

Kinetic Mechanism

The proposed kinetics involves initiation via a symmetrical cyclic trifunctional initiator, thermal initiation, propagation, transfer to the monomer, combination termination and re-initiation. Consider the global kinetic mechanism presented in Table 2. The following nomenclature is adopted:

St	Styrene.
I	Trifuncional initiator (DEKTP).
$\bullet I \bullet^{(2)}$	Initiator diradical with 2 undecomposed peroxide groups.
$\bullet S_1 \bullet^{(2)}$	Monomer diradical with 2 undecomposed peroxide groups.
$S_1 \bullet^{(0)}$	Monomer monoradical without undecomposed peroxide groups.
$S_n \bullet^{(i)}$	PS monoradical of chain length n containing i undecomposed peroxide groups.
$\bullet S_n \bullet^{(i)}$	PS diradical of chain length n containing i undecomposed peroxide groups.
$S_n^{(i)}$	Polymer with n repetitive units of St and i undecomposed peroxide groups.

Table 2. The Adopted Kinetic Mechanism

Initiation

$$3\text{St} \xrightarrow{k_i} 2\text{S}_1\bullet^{(0)}$$

$$\text{I} \xrightarrow{k_{d1}} \bullet\text{I}\bullet^{(2)}$$

$$\bullet\text{I}\bullet^{(2)} + \text{St} \xrightarrow{k_{p1}} \bullet\text{S}_1\bullet^{(2)}$$

Propagation ($n = 1,2,3,...;\ i = 0,1,2...$)

$$\text{S}_n\bullet^{(i)} + \text{St} \xrightarrow{k_p} \text{S}_{n+1}\bullet^{(i)}$$

$$\bullet\text{S}_n\bullet^{(i)} + \text{St} \xrightarrow{2k_p} \bullet\text{S}_{n+1}\bullet^{(i)}$$

Transfer to Monomer ($n = 1,2,3,...;\ i = 0,1,2...$)

$$\text{S}_n\bullet^{(i)} + \text{St} \xrightarrow{k_{fM}} \text{S}_n^{(i)} + \text{S}_1\bullet^{(0)}$$

$$\bullet\text{S}_n\bullet^{(i)} + \text{St} \xrightarrow{2k_{fM}} \text{S}_n\bullet^{(i)} + \text{S}_1\bullet^{(0)}$$

Re-initiation ($n = 2,3,...;\ m = 1,2,...,n-1;\ i = 2,3,...;\ j = 1,2...i-1$)

$$\text{S}_n^{(i)} \xrightarrow{k_{d2}} \text{S}_{n-m}\bullet^{(i-j)} + \text{S}_m^{(j-1)}\bullet$$

Combination Termination ($n, m = 1,2,3,...;\ i, j = 0,1,2...$)

$$\bullet\text{S}_n\bullet^{(i)} + \bullet\text{S}_m\bullet^{(j)} \xrightarrow{4k_{tc}} \bullet\text{S}_{n+m}\bullet^{(i+j)}$$

$$\bullet\text{S}_n\bullet^{(i)} + \text{S}_m\bullet^{(j)} \xrightarrow{2k_{tc}} \text{S}_{n+m}\bullet^{(i+j)}$$

$$\text{S}_n\bullet^{(i)} + \bullet\,\text{S}_m^{(j)} \xrightarrow{k_{tc}} \text{S}_{n+m}^{(i+j)}$$

The following has been assumed: (a) at the temperature employed, the initiator decomposition is due exclusively to sequential rupture; (b) intramolecular termination is neglected; (c) disproportion termination is negligible; (d) all peroxide groups present in the trifunctional intiator and in the accumulated polymer exhibit the same thermal stability; (e) propagation and transfer reactions are unaffected by chain length or conversion; (g) degradation reactions are negligible.

Note the following: (i) when two monoradicals with i and j undecomposed peroxide groups terminate, the formed polymer will contain $i + j$ undecomposed peroxide groups; (ii) diradicals only have an even number of peroxide groups, as they are generated only by propagation of the initiator diradical (with only 2 peroxide groups), and by combination termination of other diradicals, all of which have an even number of peroxide groups.

Homogeneous Model

From the kinetics of Table 2 and assuming homogeneous bulk polymerization, the mathematical model of Appendices A and B was developed. This model is based on the mass balances for the chemical species present in the reaction system. In order to model the reinitiation reactions, polymeric chains were assumed to have uniformly distributed peroxide groups. A random chain scission can then be performed by using a uniformly distributed random variable.

The Basic Module of Appendix A allows the prediction of the evolution of the global chemical species along the reaction (total mono- and diradicals, total polymer). Appendix B simulates the evolution of all chemical species characterized by their chain length and their number of undecomposed peroxide groups, and allows the estimation of the evolution of the molecular weight distribution –MWD- of each species and the number of undecomposed peroxide groups.

The proposed model considers constant temperature, assuming that the reactor cooling/heating system is ideal in the sense that it is capable of removing/providing the exact amount of heat in order to keep the temperature constant. However, it is possible to simulate reactions at different temperatures through the use of Arrhenius expressions for the kinetic constants.

The Basic Module is self-sufficient from the calculation point of view, and for its resolution equations (A.1), (A.2), (A.5), (A.11), (A.12), (A.16), (A.23), and (A.25), must be simultaneously solved. The equations of the Distributions Module must be integrated using the Basic Module results.

The Basic Module is solved by a standard stiff differential equation numerical method based on a modified Rosenbrock formula of order 2. In the Distributions Module, a large number of equations (more than 250,000) must be integrated. For this reason, a finite difference based method was used, using the time intervals obtained from resolution of the Basic Module. A typical simulation requires less than 1 second for the Basic Module and more than 8 hours for the Distribution Module using an AMD Turion 64 based processor with 1.80 GHz.

Simulation Results

The adopted kinetic parameters used for the simulation are presented in Table 3.

Table 3. Employed Kinetic Constants

k_{d1}, k_{d2}	$\left[s^{-1} \right]$	$9.47 \cdot 10^{12} e^{-31819/RT}$	Adjusted in this work
k_i	$\left[\frac{L^2}{\text{mol s}} \right]$	$1.1 \cdot 10^5 e^{-27340/RT}$	Peng (1990)[19]; Yoon and Choi (1992)[8]
k_{p1}, k_p	$\left[\frac{L}{\text{mol s}} \right]$	$1.74 \cdot 10^6 e^{-6347/RT}$	Adjusted in this work
k_{fM}	$\left[\frac{L}{\text{mol s}} \right]$	$2.93 \cdot 10^{10} e^{-19939/RT}$	Adjusted in this work
k_{tc}	$\left[\frac{L}{\text{mol s}} \right]$	$1.7 \cdot 10^9 e^{-(1667.3/RT)-2(C_1x+C_2x^2+C_3x^3)}$ *	Friis and Hamielec (1976)[20]
f		0.98	Cerna (2000)[21]

$C_1 = 2.75 - 0.00505T$; $C_2 = 9.56 - 0.0176T$; $C_3 = -3.03 + 0.00785T$, with x monomer conversion

R = 1,986 cal/molK

The expressions for k_{d1}, k_p and k_{fM} were adjusted in this work. Since peroxide groups in the polymer chains was assumed to have the same thermal stability as peroxide groups in the initiator, $k_{d2} = k_{d1}$. Due to the short simulation times required, a fast interactive adjustment was possible. The adjusted values for k_{dI} are in agreement with the reported results for DEKTP decomposition in toluene.[11] The values for k_p and k_{fM} are within the ranges reported for St polymerization.[18] The Basic Module allowed the adjustment for k_{dI} and k_p from the estimation of the monomer conversion. The parameter k_{fM} was adjusted to fit the experimental data of average molecular weights using the Distributions Modules.

The expression for k_{tc} and k_i were directly adopted from the literature[7,19,20] and the value for f was obtained from experimental results by Cerna.[21] Simulations were carried out at the optimal conditions for DEKTP initiation, that is an initiator concentration of 0.01 mol/L, and temperatures of 120 and 130 °C, where initiator decomposition is mostly sequential.

Figures 3 and 4 show the simulation results corresponding to the evolution of conversion and average molecular weights. An excellent agreement between predictions and measured values is observed. The change of the slope in the conversion curve in Figure 3 at approximately and 200 min at 120 °C and 150 min at 130 °C, indicates a second initiation due to decomposition of the peroxide sites in the polymer chains, producing an increase in the rate of polymerization. This second initiation is also responsible for the high molecular weights observed experimentally and predicted by the model, as the scission of polymer chains can generate long chain monoradicals, capable of further growing by propagation reactions.

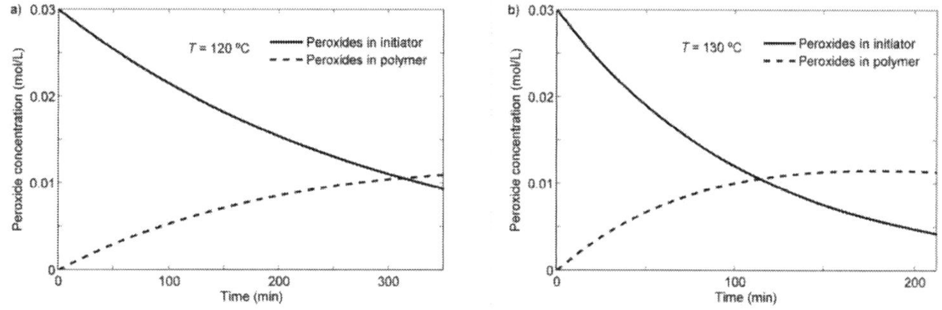

Figure 5. Evolution of peroxide concentration in DEKTP and in polymeric chains: a) T = 120°C and b) T = 130°C

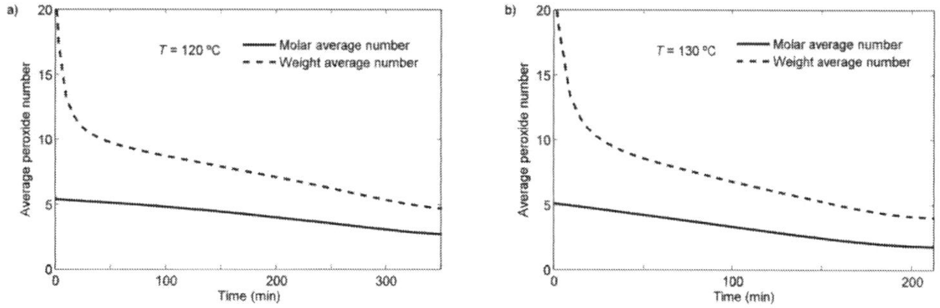

Figure 6. Evolution of average number of peroxide groups and in polymeric chains: a) T = 120°C and b) T = 130°C

Aside from the prediction of experimental results presented in Table 1, the model was used to simulate other variables. In Figure 5, the evolution of the concentrations of undecomposed peroxide groups in the initiator and in the polymer chains are presented. It can be observed that the undecomposed peroxide groups are not totally consumed during the reactions in the period of time evaluated. While peroxides in the initiator are consumed by decomposition reactions only, peroxide concentration in the polymer tends to increase at the beginning of the reaction due to the formation of chains. As the undecomposed peroxides decompose in the formed chains, peroxide concentration tends to attain a maximum value and could eventually decrease. This is reflected in Figure 6, which shows the evolution of the average number of undecomposed peroxide sites in polymer chains. It is possible to observe that, while the initiator has only 3 peroxide sites, polymer chains may have a greater number. This is due to propagation, transfer and termination of diradicals and monoradicals containing undecomposed peroxide groups. The weight averages are higher than the corresponding molar averages, because, as expected, polymers with a greater number of undecomposed peroxide groups are of greater chain length.

Figure 7 shows the evolution of the different polymer species, characterized by the number of undecomposed peroxide groups within the polymeric chains. Only polymer chains with 0, 1, 2, 3, 4, 10, 11, 12, 15 and 20 undecomposed peroxide groups are considered in the figure for clarity reasons. The model allows for prediction of the evolution of all polymeric species with up to 30 undecomposed peroxide sites. Dead polymer (i.e. no undecomposed peroxide groups) concentration monotonically increases along the reaction time, exhibiting a slope that increases at around 200 min at 120 °C and 150 min at 130 °C as seen in Figure 7 (a) and (b), coinciding with the time at which polymerization rate increases (as indicated for Figure 3). As the number of peroxide groups in the polymer increases, decomposition reactions for the polymer become more important, and the concentration of polymer with a high peroxide group number may decrease with time (Figure 7 (c) and (d)).

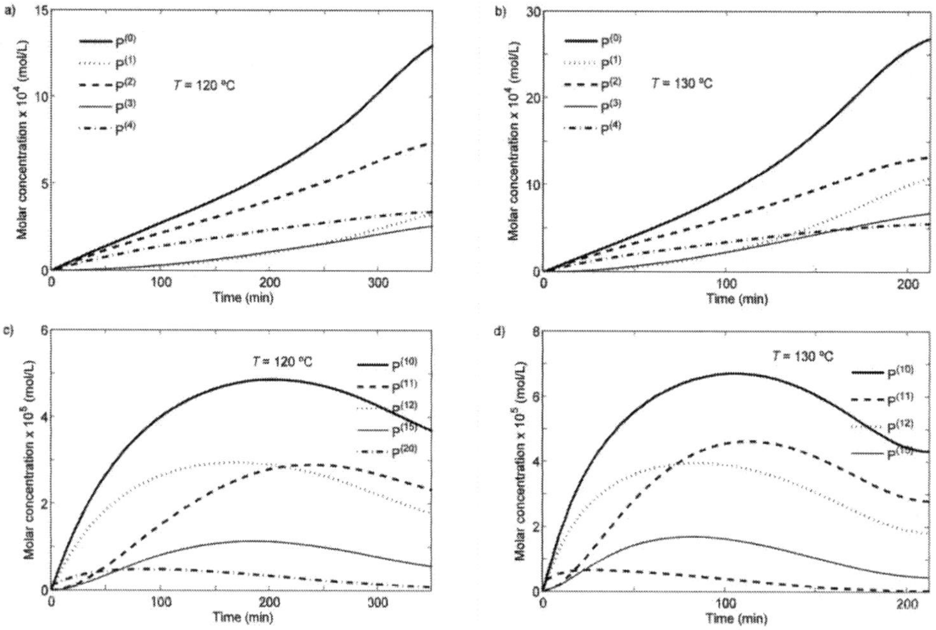

Figure 7. Evolution of polymeric species characterized by the number of undecomposed peroxide groups

It is possible to observe that the molar concentration of the polymer decreases with increasing peroxide group number per chain when considering polymers with even number of peroxide groups. The same can be stated for polymers with an odd number of peroxide groups. This is because a high number of peroxides groups per chain leads to a higher rate of decomposition reactions. The evolution of the concentration of the different polymers species is highly dependent on the undecomposed peroxide numbers. This is due to the fact that peroxide groups in the polymer are a result of propagation, transfer and termination of diradicals generated by initiator decomposition, but also to the decomposition of polymeric chains with a higher peroxide number. A great number of monoradicals without peroxide groups is also produced by thermal initiation of the monomer. For polymers with an even number of peroxide groups, the concentrations show a similar evolution, indicating that most of the peroxide groups in polymer chains are generated by initiator decomposition. This is not the case for polymers with an odd number of peroxide groups, because radicals with an even number of peroxide groups are only generated by decomposition reactions. These types of behaviors are a clear result of the sequential decomposition of the initiator molecule.

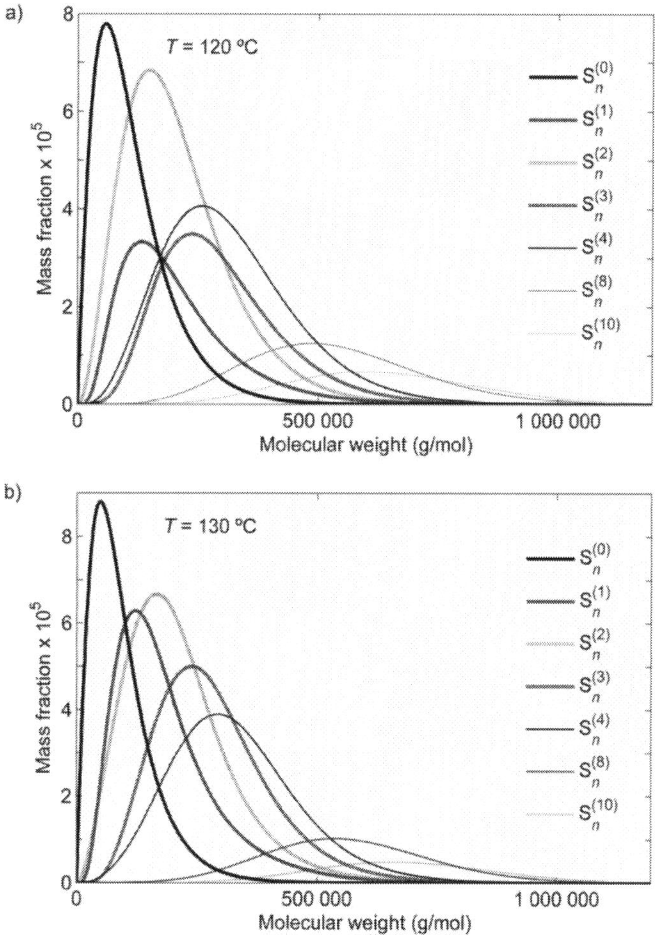

Figure 8. MWD of the polymeric species at the end of polymerization: a) $T = 120°C$ and b) $T = 130°C$

Figure 8 shows the MWD of the formed polymer species once the reaction is completed. Only MWDs of a few polymeric species are represented in order to keep the picture clear but still representative. The obtained MWD shows that polymer with a high number of undecomposed peroxide groups have a wider distribution, over higher molecular weights. This is consistent with the fact that high peroxide group numbers arise from termination combination or transfer of monoradicals that have been generated by diradicals of initiator decomposition. These diradicals can propagate before terminating, so it is expected that termination of two diradicals will generate a long chain monoradical, which can itself terminate with another long chain monoradical. As expected, polymer chains with 1 undecomposed peroxide group are in average shorter that the ones from the polymer with 2 undecomposed peroxide groups, because the former is generated mostly by decomposition of the latter, after transfer to the monomer or termination with a radical without peroxide groups. This is also true for the polymer with 3 undecomposed peroxide groups, but in this case the polymer can be formed either by decomposition of the polymer with 4 undecomposed peroxide groups or combination of radicals with 1 and 2 peroxide groups. Similar arguments can be developed for polymers with a greater number of undecomposed peroxide groups. Polymers with a high number of peroxide groups (12 or more) have a relatively low mass fraction, but enough to significantly increase the average molecular weight. By comparing the MWDs obtained at different temperatures, it is possible to see that lower temperatures tend to increase the mass fraction of polymer with undecomposed peroxide groups compared to total polymer (this was also evidenced in Figure 6, as the average peroxide group number is higher). Comparison of the MWDs for the polymers with 1, 2, 3 and 4 undecomposed peroxide groups shows that at 130 °C, decomposition reactions for polymers with 2 and 4 undecomposed peroxide groups are responsible for the high mass fractions of polymers with 1 and 3 undecomposed peroxide groups, relative to the rest of the species' mass fractions. At 120 °C, the rate of these decompositions reactions is decreased in such a way that these relative mass fractions of polymers with 1 and 3 undecomposed groups are lower, which increases the average molecular weights. This, together with the lower rate of transfer reactions, accounts for the high molecular weights obtained at 120 °C. The values obtained for average molecular weights are higher than those expected using a purely thermal initiation of the St polymerization reaction.

Conclusions

This work experimentally and theoretically investigates the use of the symmetrical trifunctional initiator DEKTP in the bulk polymerization of St. The experimental work consisted on a series of isothermal batch polymerizations at different temperatures, in which conversion and average molecular weights were determined. For the theoretical analysis, a new mathematical model was developed, that requires following the evolution of the chemical species characterized by the number of undecomposed peroxide groups within the chains, including di- and monoradicals as well as polymeric species. Reinitiation reactions are accounted for by using a uniformly distributed random variables in order to simulate random chain scission.

The main result is that the use of DEKTP allows obtaining high polymerization rates and high molecular weight simultaneously, because of its sequential decomposition. This leads to the formation of di- and monoradicals with undecomposed peroxide groups. These peroxide groups are then found in the polymeric chains, which are susceptible of reinitiation. Simulation of the proposed kinetic mechanism provides values which are in excellent agreement with experimental results.

Reaction temperature affects the MWD as it has an influence both on transfer and reinitiation reactions. However, average molecular weights are much higher than those expected in an exclusively thermally initiated polymerization.

In future works, the kinetic mechanism could be extended to include the total decomposition of the initiator molecule at higher reaction temperatures.

Appendix A: Basic Module

Balances for the Non-polymeric Reagents and Products

Initiator (DEKTP)

$$\frac{d}{dt}\left([\mathrm{I}]V\right) = -3k_d[\mathrm{I}]V \tag{A.1}$$

Monomer

Assuming the "long chain approximation" (by which propagation is the only monomer-consuming reaction), one can write:

$$\frac{d}{dt}\left([\mathrm{St}]V\right) = -R_p V = -k_p[\mathrm{St}]\left([\mathrm{R}\bullet]+2[\bullet\mathrm{R}\bullet]\right)V \tag{A.2}$$

where R_p is the global St polymerization rate, and

$$[\mathrm{R}\bullet] = \sum_{i=0}^{\infty}\sum_{n=1}^{\infty}\left[\mathrm{S}_n\bullet^{(i)}\right] \tag{A.3}$$

$$[\bullet\mathrm{R}\bullet] = \sum_{i=0}^{\infty}\sum_{n=1}^{\infty}\left[\bullet\mathrm{S}_n\bullet^{(i)}\right] \tag{A.4}$$

represent the total concentrations of mono- and diradicals respectively.

Radical Species

Consider the mass balances of all free radical appearing in the global kinetics. Such balances provide:

$$\frac{d}{dt}\left(\left[\bullet\mathrm{I}\bullet^{(2)}\right]V\right) = \left(3fk_d[\mathrm{I}] - 2k_{p1}\left[\bullet\mathrm{I}\bullet^{(2)}\right][\mathrm{St}]\right)V \tag{A.5}$$

$$\begin{aligned}\frac{d}{dt}\left(\left[\mathrm{S}_1\bullet^{(0)}\right]V\right) = \Big\{&2k_i[\mathrm{St}]^3 - \Big(k_p\left[\mathrm{S}_1\bullet^{(0)}\right] + k_{fM}\left([\mathrm{R}\bullet]+2[\bullet\mathrm{R}\bullet]\right)\\ &-k_{fM}\left[\mathrm{S}_1\bullet^{(0)}\right]\Big)[\mathrm{St}] - k_{tc}\left[\mathrm{S}_1\bullet^{(0)}\right]\left([\mathrm{R}\bullet]+2[\bullet\mathrm{R}\bullet]\right)\Big\}V\end{aligned} \tag{A.6}$$

$$\begin{aligned}\frac{d}{dt}\left(\left[\bullet\mathrm{S}_1\bullet^{(2)}\right]V\right) = \Big\{&\Big(2k_{p1}\left[\bullet\mathrm{I}\bullet^{(2)}\right] - 2k_p\left[\bullet\mathrm{S}_1\bullet^{(2)}\right] - 2k_{fM}\left[\bullet\mathrm{S}_1\bullet^{(2)}\right]\Big)[\mathrm{St}]\\ &-2k_{tc}\left[\bullet\mathrm{S}_1\bullet^{(2)}\right]\left([\mathrm{R}\bullet]+2[\bullet\mathrm{R}\bullet]\right)\Big\}V\end{aligned} \tag{A.7}$$

$$\begin{aligned}\frac{d}{dt}\left(\left[\bullet\mathrm{S}_n\bullet^{(i)}\right]V\right) = \Big\{&2k_p[\mathrm{St}]\left(\left[\bullet\mathrm{S}_{n-1}\bullet^{(i)}\right] - \left[\bullet\mathrm{S}_n\bullet^{(i)}\right]\right) - 2k_{fM}[\mathrm{St}]\left[\bullet\mathrm{S}_n\bullet^{(i)}\right]\\ &-2k_{tc}\left([\mathrm{R}\bullet]+2[\bullet\mathrm{R}\bullet]\right)\left[\bullet\mathrm{S}_n\bullet^{(i)}\right] + 2k_{tc}\sum_{j=1}^{i}\sum_{m=1}^{n-1}\left[\bullet\mathrm{S}_{n-m}\bullet^{(i-j)}\right]\left[\bullet\mathrm{S}_m\bullet^{(j)}\right]\Big\}V\end{aligned} \tag{A.8}$$

$(n \geq 2,\ i = 2, 4, 6\ldots)$

$$\frac{d}{dt}\left(\left[S_n\bullet^{(i)}\right]V\right)=\Big\{\Big(k_p\left(\left[S_{n-1}\bullet^{(i)}\right]-\left[S_n\bullet^{(i)}\right]\right)+k_{fM}\left(2\left[\bullet S_n\bullet^{(i)}\right]-\left[S_n\bullet^{(i)}\right]\right)\Big)[St]$$

$$-k_{tc}\left([R\bullet]+2[\bullet R\bullet]\right)\left[S_n\bullet^{(i)}\right]+2k_{tc}\sum_{j=1}^{i}\sum_{m=1}^{n-1}\left[\bullet S_{n-m}\bullet^{(i-j)}\right]\left[S_m\bullet^{(j)}\right] \qquad (A.9)$$

$$+k_{d2}\sum_{j=i+1}^{\infty}\sum_{m=n+1}^{\infty}p_{mj}(n,i)\,j\left[S_m^{(j)}\right]\Big\}V \qquad (n\geq 2,\ i=1,2,3...)$$

Where $p_{mj}(n,i)$ is the probability that a scission of a chain of dead polymer of length m and j peroxide groups yields a monoradical of chain length n with i peroxide groups. Adding this probability over all i s and n s, the following can be proven:

$$\sum_{i=1}^{\infty}\sum_{n=1}^{\infty}\sum_{j=i+1}^{\infty}\sum_{m=n+1}^{\infty}p_{mj}(n,i)\,j\left[S_m^{(j)}\right]=2\sum_{i=1}^{\infty}\sum_{n=1}^{\infty}i\left[S_n^{(i)}\right]=2\left[Pe_P\right] \qquad (A.10)$$

where $\left[Pe_P\right]$ is the concentration of peroxide groups in the polymer chains. The 2 in equation (A.10) arises from the fact that the scission of any polymer chain with peroxide groups produces 2 monoradicals.

From equations (A.6) through (A.10) the total concentration of mono- and diradicals may be obtained:

$$\frac{d}{dt}\left([R\bullet]V\right)=\left\{2k_i[St]^3+4k_{fM}[St][\bullet R\bullet]-k_{tc}[R\bullet]^2+2k_{d2}[Pe_P]\right\}V \qquad (A.11)$$

$$\frac{d}{dt}\left([\bullet R\bullet]V\right)=\left\{\left(2k_{p1}\left[\bullet I\bullet^{(2)}\right]-2k_{fM}[\bullet R\bullet]\right)[St]-2k_{tc}[\bullet R\bullet]\left([R\bullet]+2[\bullet R\bullet]\right)+2k_{tc}[\bullet R\bullet]^2\right\}V \qquad (A.12)$$

Balances for the Polymeric Species

Polymer without peroxide groups ($i=0$)

$$\frac{d}{dt}\left(\left[S_n^{(0)}\right]V\right)=\left\{k_{fM}[St]\left[S_n\bullet^{(0)}\right]+\frac{k_{tc}}{2}\sum_{m=1}^{n-1}\left[S_{n-m}\bullet^{(0)}\right]\left[S_m\bullet^{(0)}\right]\right\}V \qquad (n\geq 2) \qquad (A.13)$$

Polymer with peroxide groups ($i\neq 0$)

$$\frac{d}{dt}\left(\left[S_n^{(i)}\right]V\right)=\left\{k_{fM}[St]\left[S_n\bullet^{(i)}\right]+\frac{k_{tc}}{2}\sum_{j=1}^{i}\sum_{m=1}^{n-1}\left[S_{n-m}\bullet^{(i-j)}\right]\left[S_m\bullet^{(j)}\right]-k_{d2}i\left[S_n^{(i)}\right]\right\}V$$

$$(n\geq 2,\ i=1,2,3...) \qquad (A.14)$$

The concentration of the polymer with i undecomposed peroxide sites can be defined as

$$\left[P^{(i)}\right]=\sum_{n=1}^{\infty}\left[S_n^{(i)}\right] \qquad (A.15)$$

Let us define the total polymer concentration $[\mathrm{P}]$

$$[\mathrm{P}]=\sum_{i=0}^{\infty}\left[\mathrm{P}^{(i)}\right]=\sum_{i=0}^{\infty}\sum_{n=1}^{\infty}\left[\mathrm{S}_n^{(i)}\right] \tag{A.16}$$

By adding up equation (A.13) over all n's and (A.14) over all n's and all i's, the balance for the molar concentration of polymer can be written:

$$\frac{d}{dt}([\mathrm{P}]V)=\left\{k_{fM}[\mathrm{St}][\mathrm{R}\bullet]+\frac{k_{tc}}{2}[\mathrm{R}\bullet]^2-k_{d2}[\mathrm{Pe_P}]\right\}V \tag{A.17}$$

Peroxide Groups

The total concentration of peroxide groups is

$$[\mathrm{Pe}]=3[\mathrm{I}]+2\left[\bullet\mathrm{I}\bullet^{(2)}\right]+[\mathrm{Pe_{R\bullet}}]+[\mathrm{Pe_{\bullet R\bullet}}]+[\mathrm{Pe_P}] \tag{A.18}$$

with

$$[\mathrm{Pe_{R\bullet}}]=\sum_{i=1}^{\infty}\sum_{n=1}^{\infty}i\left[\mathrm{S}_n\bullet^{(i)}\right] \tag{A.19}$$

$$[\mathrm{Pe_{\bullet R\bullet}}]=\sum_{i=1}^{\infty}\sum_{n=1}^{\infty}i\left[\bullet\mathrm{S}_n\bullet^{(i)}\right] \tag{A.20}$$

$$[\mathrm{Pe_P}]=\sum_{i=1}^{\infty}\sum_{n=1}^{\infty}i\left[\mathrm{S}_n^{(i)}\right] \tag{A.21}$$

where $[\mathrm{Pe_{R\bullet}}]$, $[\mathrm{Pe_{\bullet R\bullet}}]$ and $[\mathrm{Pe_P}]$ respectively represent the molar concentration of peroxide groups accumulated in monoradicals, diradicals and polymer species. From equation (A.18) and assuming pseudosteady-state for radical species (by which all time derivatives may be set to zero), it can be written:

$$\frac{d[\mathrm{Pe}]}{dt}=3\frac{d[\mathrm{I}]}{dt}+\frac{d[\mathrm{Pe_P}]}{dt} \tag{A.22}$$

Peroxide groups are consumed only by decomposition reactions in the initiator and the polymer chains. Therefore:

$$\frac{d[\mathrm{Pe}]}{dt}=-k_d[\mathrm{I}]-k_{d2}[\mathrm{Pe_P}] \tag{A.23}$$

By considering equations (A.1) and (A.23) and from equation (A.22), the total peroxide groups contained in the polymer chains can be calculated from the following equation:

$$\frac{d[\mathrm{Pe_P}]}{dt}=2[\mathrm{I}]-k_{d2}[\mathrm{Pe_P}] \tag{A.24}$$

Molar Average Number of Peroxide Groups in Polymer Chains

The average number of peroxide groups in polymer chains in a molar basis may be calculated from

$$\bar{p}_n = \frac{\sum_{n=1}^{\infty}\sum_{i=0}^{\infty} i\left[S_n^{(i)}\right]}{\sum_{n=1}^{\infty}\sum_{i=0}^{\infty}\left[S_n^{(i)}\right]} = \frac{\left[Pe_P\right]}{\left[P\right]} \tag{A.25}$$

Conversion

The monomer conversion can be calculated from

$$x = \frac{\left[St\right]^0 V^0 - \left[St\right] V}{\left[St\right]^0 V^0} \tag{A.26}$$

where the superscript "0"indicates initial conditions. In this model we neglect the effect of volume contraction and therefore

$$x = \frac{\left[St\right]^0 - \left[St\right]}{\left[St\right]^0} \tag{A27}$$

Equations (A.1), (A.2), (A.5), (A.11), (A.12), (A.17), (A.24), (A.26) can be simultaneously solved to find the evolutions of $\left[I\right]$, $\left[\bullet I\bullet^{(2)}\right]$, $\left[St\right]$, $\left[R\bullet\right]$, $\left[R\bullet\right]$,$\left[P\right]$, $\left[Pe_P\right]$,and x.

Appendix B: Distributions Module

Consider equations (A.8) and (A.9). Assuming pseudosteady-state, all time derivatives may be set to zero and the following recurrence formulas can be obtained:

$$\left[\bullet S_n\bullet^{(i)}\right] = \frac{k_p\left[St\right]\left[\bullet S_{n-1}\bullet^{(i)}\right] + k_{tc}\sum_{j=1}^{i}\sum_{m=1}^{n-1}\left[\bullet S_{n-m}\bullet^{(i-j)}\right]\left[\bullet S_m\bullet^{(j)}\right]}{k_p\left[St\right] + k_{fM}\left[St\right] + k_{tc}\left(\left[R\bullet\right] + 2\left[\bullet R\bullet\right]\right)} \quad (n \geq 2,\ i = 2,4,6\ldots) \tag{B.1}$$

$$\left[S_n^{(i)}\bullet\right] = \frac{\left(k_p\left[S_{n-1}\bullet^{(i)}\right] + 2k_{fM}\left[\bullet S_n\bullet^{(i)}\right]\right)\left[St\right] + 2k_{tc}\sum_{j=1}^{i}\sum_{m=1}^{n-1}\left[\bullet S_{n-m}\bullet^{(i-j)}\right]\left[S_m\bullet^{(j)}\right]}{k_p\left[St\right] + k_{fM}\left[St\right] + k_{tc}\left(\left[R\bullet\right] + 2\left[\bullet R\bullet\right]\right)}$$
$$+ \frac{k_{d2}\sum_{j=i+1}^{\infty}\sum_{m=n+1}^{\infty} p_{mj}(n,i)\, j\left[S_m^{(j)}\right]}{k_p\left[St\right] + k_{fM}\left[St\right] + k_{tc}\left(\left[R\bullet\right] + 2\left[\bullet R\bullet\right]\right)} \quad (n \geq 2,\ i = 1,2,3\ldots) \tag{B.2}$$

Define the following molar ratio:

$$\sigma_2 = \frac{\left[\bullet S_1^{(2)}\bullet\right]}{[R\bullet]+2[\bullet R\bullet]} \tag{B.3}$$

and the following dimensionless kinetic parameters:

$$\tau = \frac{k_{fM}}{k_p} \tag{B.4}$$

$$\beta = \frac{k_{tc}R_p}{\left(k_p[St]\right)^2} = \frac{k_{tc}\left([R\bullet]+2[\bullet R\bullet]\right)}{k_p[St]} \tag{B.5}$$

$$\alpha = \tau + \beta \tag{B.6}$$

$$\gamma = \frac{k_{d2}}{k_p} \tag{B.7}$$

By replacing the definitions in (B.3), (B.4), (B.5), (B.6) in equation (B.1) and solving the recurrence formula, the following explicit expression for diradicals MWD can be obtained:

$$\left[\bullet S_n^{(2k)}\bullet\right] = \frac{1}{(1+\alpha)^{n-1}}\left(\beta\sigma_2\right)^{k-1}\frac{(n+k-2)!}{(n-k)!k!(k-1)!}\left[\bullet S_1^{(2)}\bullet\right] \qquad (n \ge 2,\ k = 1,2,3...) \tag{B.8}$$

Equation (B.2) can be written

$$\left[S_n^{(i)}\bullet\right] = \frac{1}{1+\alpha}\left(\left[S_{n-1}\bullet^{(i)}\right]+2\tau\left[\bullet S_n\bullet^{(i)}\right]+\frac{2\beta}{[R\bullet]+2[\bullet R\bullet]}\sum_{j=1}^{i}\sum_{m=1}^{n-1}\left[\bullet S_{n-m}\bullet^{(i-j)}\right]\left[S_m\bullet^{(j)}\right]+\frac{\gamma}{[St]}\Psi(n,i)\right) \tag{B.9}$$

$$(n \ge 2,\ i = 1,2,3...)$$

where

$$\Psi(n,i) = \sum_{j=i+1}^{\infty}\sum_{m=n+1}^{\infty} p_{mj}(n,i)\,j\left[S_m^{(j)}\right] \tag{B.10}$$

Equations (B.8) and (B.9) must be solved by using equations (A.6) and (A.7).
The Number Chain Length Distribution (NCLD) for the PS species can then be found by integrating equations (A.13) and (A.14) with the expressions found for $\left[S_n^{(i)}\bullet\right]$.

In order to account for the formation of monoradicals from random scission of the polymer chains containing peroxide groups, consider a polymer chain with length n and i peroxide groups, all of which have the same thermal stability. Let m be a uniformly distributed random variable whose value ranges from 1 to $n-1$. The polymer chain may form 2 monoradicals, one with length m, and the other one with length $n-m$. These chains will have $i-j$ and $j-1$ undecomposed peroxide groups respectively.If the peroxide groups are uniformly distributed in the formed monoradicals, the following relation must hold:

$$\frac{j-1}{n-m} = \frac{i-j}{m} \tag{B.11}$$

Therefore,

$$j = \left[\frac{i(n-m)+m}{n} \right] \tag{B.12}$$

where the brackets indicate the integer part of the expression.

The scission has then generated two monoradicals, one with length m and $i-j$ peroxide groups, the other one with length $n-m$ and $j-1$ peroxide groups.
This scission must be performed for all polymer chains whose peroxide group number is greater than 0.

Average Molecular Weights

The number average molecular weight may be calculated with:

$$\bar{M}_n = M_{\text{St}} \frac{\sum_{i=0}^{\infty}\sum_{n=2}^{\infty} n\left[\text{S}_n^{(i)}\right]}{\sum_{i=0}^{\infty}\sum_{n=2}^{\infty}\left[\text{S}_n^{(i)}\right]} \tag{B.13}$$

and the weight average molecular weight with:

$$\bar{M}_w = M_{\text{St}} \frac{\sum_{i=0}^{\infty}\sum_{n=2}^{\infty} n^2\left[\text{S}_n^{(i)}\right]}{\sum_{i=0}^{\infty}\sum_{n=2}^{\infty} n\left[\text{S}_n^{(i)}\right]} \tag{B.14}$$

Weight Average Number of Peroxide Groups in Polymer Chains

The average number of peroxide groups in polymer chains in a mass basis may be calculated from

$$\bar{p}_w = \frac{\sum_{n=1}^{\infty}\sum_{i=0}^{\infty} in\left[\text{S}_n^{(i)}\right]}{\sum_{n=1}^{\infty}\sum_{i=0}^{\infty} n\left[\text{S}_n^{(i)}\right]} \tag{B.15}$$

References

1. Choi K.Y., Lei G.D., AIChE. J.33(12), 2067 (1987)

2. Estenoz D. A., Leal G. P., López Y. R., Oliva H. M., Meira G. R., J. Appl. Polym. Sci. 62, 917(1996)

3. Kim J., Liang W., Choi K., Ind. Eng. Chem. Res.28, 131 (1989)

4. Priddy D.B., Recent Advances in Styrene Polymerization (1994)

5. Drumright R.E., Kastl P.E., Priddy D.B., Macromolecules 26, 2246 (1993)

6. González I. M., Meira G. R., Oliva H., J. Appl. Polym. Sci, 59, 6 (1996)

7. Yoon W.J., Choi K.Y., J.Appl. Polym. Sci. 46, 1353 (1992).

8. Yoon W.J., Choi K.Y., Polymer 33, 4582 (1992).

9. Villalobos M.A., Hamielec A.E. , Wood P.E.; J. Appl. Polym. Sci. 42, 629 (1991).

10. Choi K.Y., Liang W.R., Lei G.D., J. Appl. Polym. Sci. 35, 1547 (1988)

11. Cerna, J., Morales, G., Eyler, G. N., Canizo, A. I., J. Appl. Polym. Sci, 83, 1 (2002)

12. Cavin, L., Rouge, A., Meyer, T., Renken, Polymer, 41, 3925 (2000)

13. Mateo, C. M.; Eyler, G. N., Alvarez, E. E.; Cañizo, A. I. Información Tecnológica, 9 (2), 19 (1998)

14. Scorah, M.J. Doctoral Thesis, University of Waterloo, Canada (2005)

15. Scorah, M.J., Dhib R., Penlidis A., Chem. Eng. Sci. 61, 4827(2006)

16. Sheng, W.C., Wu, J.Y., Shan, G.R., Huang, Z.M., Weng, Z.X., J.Appl. Polym. Sci. 94 (3), 1035–1042 (2004)

17. Eyler, N.; Cañizo, A. I., Alvarez, E. E.; Cafferata, L. F. R. Tetrahedr Letts 1003, 34, 1745.

18. Meyer, Th., Keurentjes J. Handbook of Polymer Reaction Engineering, Wiley (2005)

19. Peng, F. M. J. Appl. Polym. Sci. 40, 1289 (1990)

20. Friis, N, Hamielec, A. E. ACS Symposium Series 24, 82 (1976)

21. Cerna, J., Doctoral Thesis, CIQA, Mexico (2002)

(618bc) Lipase-Catalyzed Synthesis of Poly (butylene succinate)

Jingying Zhu, Siyuan An, Diannan Lu and Zheng Liu, Chemical Engineering, Tsinghua University, Beijing, China

Catalytic synthesis of Poly(butylene succinate) (PBS) by lipase was carried out through condensation polymerization of diethyl succinate and 1,4-butanediol using Diphenyl ether as solvent. Using Novozyme 435 as catalyst, the condensation polymerization started by 2 h pre-polymerization under atmospheric pressure, followed by 20 h polymerization under high vacuum (<10 mmHg) at 80~95 °C. The product was characterized by FT-IR, ^{1}H-NMR, ^{13}C-NMR and gel permeation chromatography (GPC). The effects of monomer, the molar ratio of reactants and solvents on the polymerization were examined, respectively. The final product PBS with Mw of 44000 g/mol and Mn of 27,000 g/mol while a melting point of 114 °C, is similar to that obtained by chemical synthesis, as well as thermal and mechanical properties determined by DSC and direct tension test. The biodegradability of this new PBS by lipase was also confirmed by scanning electron microscope (SEM). This novel property would ignite a great interest in the invention and application of new materials involving this new PBS synthesized with lipase as catalyst.

See more of this Session: Mesd Poster Session
See more of this Group/Topical: Materials Engineering and Sciences Division

UO_2 MICROSPHERE PREPARATION for a HTGR NUCLEAR FUEL in KOREA

Kyung-Chai Jeong* and Moon-Seoung Cho
Division of HTGR Fuel Technology Development, KAERI,
Daejeon 305-353, Korea, kcjeong@kaeri.re.kr

Abstract

This paper describes aspects of an R&D project of an HTGR nuclear fuel particle preparation in Korea. In this study, to obtain a spherical UO_2 kernel which is the core material of the TRISO coated particle, the most important factors for the spherical UO_2 kernel preparation were the changes of the composition ratio, the viscosity of the broth solution, the frequency/amplitude of the vibrator in the wet process part, and the heating modes and gases atmosphere in the dry process part. The uranium concentration and the viscosity are 0.5~0.8 mol/L and 50~80 cP of the broth solution respectively, and the vibration frequency of the dropping system is about 100~120Hz for the most thoroughly prepared condition. Also the optimum heating rate in the calcination process of the dried-ADU gel particles must be kept below 3°C/min. in an air atmosphere, and the optimum conditions for the sintering process of reduced UO_3 were in a H_2 gas atmosphere and a 3 steps heating mode up to 1700°C.

Key words : Sol-gel, UO_2 microsphere, Vibration

Introduction

The weak points of fossil fuel utilization include its limited supply, air pollution, and carbon dioxide emissions. Carbon dioxide emissions, are thought to be responsible for global warming, and are now international issues[1,2]. Also, energy demand is continually increasing because of the increased energy usage in emerging nations and the Third World. With the international increase of energy requirements and the clean production of energy, nuclear energy has, nowadays, been in the spotlight as a countermeasure for the massive supply of combined electricity and clean hydrogen gas production in the next decades[3].

With these changing circumstances, concentrated studies and discussions have already progressed on the R&D of an HTGR(High Temperature Gas Reactor) due to its inherent safety and its capability to generate the high temperatures needed for producing hydrogen or heat.

Generally, the nuclear fuel of an HTGR uses TRISO(TRistructural ISOtropic) coated fuel particles which are enclosed by three layers of coating materials, such as pyrolytic carbons(PyC), as a buffer layer, inner dense PyC(IPyC) layer, silicon carbide(SiC) layer, and an outer PyC layer(OPyC) on the surface of the spherical UO_2 kernels as shown in fig.1[4,5].

The technologies for an HTGR nuclear fuel have already been developed in the early 1980s in advanced nations such as Germany, Japan, France, USA, etc.[6], but taking into consideration that the research on an HTGR nuclear fuel is at a conceptual stage in Korea, the establishment of this country of the nuclear fuel preparation process with Korea's own abilities is essential to relieve the technical subordination of advanced nations. The sol-gel processes in the field of nuclear fuel preparation have many advantages in a high purity atmosphere and a low processing temperature. However, there are only a few reports or technical reviews on the preparation of a spherical UO_2 kernel by the sol-gel method.

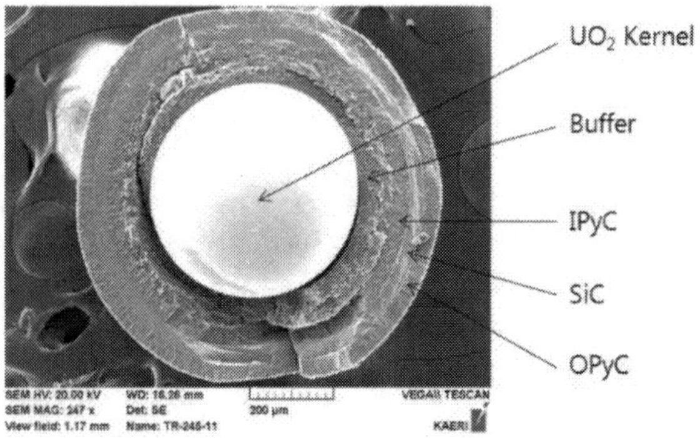

Fig.1. General concept of TRISO coated particle.

In this study the preparation methods for spherical ADU(Ammonium DiUranate) particles, an intermediate compound, and the first material of an HTGR nuclear fuel were investigated and the thermal characteristics for ADU→UO_3 and UO_3→UO_2 conversions were studied through heat treatments in calcination and sintering furnaces, respectively.

Technical Background

The important characteristics of the HTGR are its ability to produce energy by nuclear fission reaction with coolant outlet temperatures of up to 900~950°C. Therefore, the fuel element surface temperature of an HTGR necessitates temperatures higher than 1000°C. These conditions almost inevitably lead to the use of He gas as a coolant and to a fuel element made exclusively of ceramic materials. The TRISO type uranium oxide nuclear fuel, as a fission material for the HTGR, is among the nuclear ceramic materials such as uranium carbide, uranium nitride, and uranium oxy-carbide, that is often selected in many developed countries. This is primarily because of its leak-tightness to fission materials by three layers of coating materials. Its other advantages include the inherent safety of its high melting point, and adequate experiences and accumulated know-how in its fuel manufacturing technologies[7].

Preparation of the HTGR nuclear fuel is usually divided into three parts: UO_2 kernel preparation, the coating process of the prepared UO_2 kernel, and the spherical fuel element fabrication using graphite[8,9]. This study mainly discusses the UO_2 kernel preparation in Korea. The preparation of the spherical UO_2 kernel, mentioned above, for an HTGR nuclear fuel can be carried out by a modified sol-gel method, among other wet chemical processes, based on a solidification of spherical liquid droplets which contain nuclear fission material, prepared by a vibrating dropping system.

Fig.2 depicts a simple flow diagram for the overall process of UO_2 kernel sphere production. The detail descriptions and mutual relations on each unit process are very complicated and also there are many connected parameters between each. A typical example is the sphericity of the UO_2 kernel, because the shape of the liquid droplets formed in a vibrating dropping system of an initial stage are kept until they are converted to UO_2 in a final sintering stage. Of course, other factors, such as the pH adjustment in the pre-neutralization step, the vibration frequency and amplitude in the liquid droplets preparation, and the heating rates in the calcining, reducing and sintering processes are co-related in the whole process, too.

In this research, the UO_2 kernel manufacturing method is based on the German HTGR technology[10]. The UO_2 kernel production process is based on the gel-supported precipitation process (a kinds of an external gelation process), and the feed material is the UN (Uranyl Nitrate, $UO_2(NO_3)_2$) solution. Some points of each unit step are the followings:

▶ UN solution preparation and its pre-neutralization step. The preparation of the feed solution is the simple chemical reaction on the uranium oxide powder dissolution with concentrated nitric acid,

$$3\ U_3O_{8(s)} + 20\ HNO_{3(aq)} \rightarrow 9\ UO_2(NO_3)_{2(aq)} + 10\ H_2O + 2\ NO\uparrow_{(g)}$$

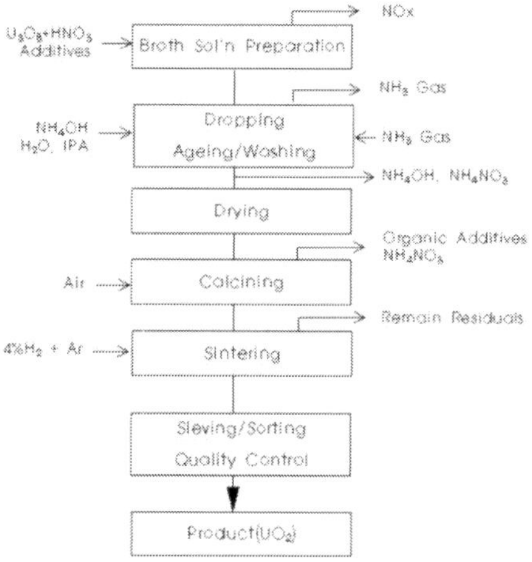

Fig.2. Overall folw chart of UO_2 kernel.

and the pre-neutralized by an ammonium hydroxide solution with an UN solution

$$2\ UO_2(NO_3)_{2(aq)} + NH_4OH_{(aq)} \rightarrow 2\ UO_2(NO_3)_{1.5}(OH)_{0.5(aq)} + NH_4NO_{3(aq)}$$

▶ Broth solution is prepared after a through mixing by adding suitable amounts of organic additives, and H_2O to the pre-neutralized UN solution

▶ Spherical liquid droplets are prepared by the vibrating dropping system. Individual and continuous liquid droplets from supply of the broth solution are prepared by the break-up of the natural laminar jet stream of the broth solution in air space, and the NH_3 gas layers with a mechanical vibrating system[11]. The droplets first go through an air layer where they form a spherical shape because of the surface tension of organic additives. Then the spherical liquid droplets pass through an ammonia gas layer, where the surface of the spherical liquid droplets are changed to a solid state by a pre-hardening of a liquid droplets' surface by a chemical reaction between an uranyl ion in the droplet and an ammonium ion from the ammonia gas.

The pre-hardened uranium complex compound is called an ADU (this word is commonly used for various uranium oxide compounds such as $UO_3{\cdot}2H_2O$, $3UO_3{\cdot}NH_3{\cdot}5H_2O$, $2UO_3{\cdot}NH_3{\cdot}3H_2O$, and $3UO_3{\cdot}2NH_3{\cdot}4H_2O$)[12], and the microsphere enables them to retain their spherical form by this precipitated outer layer on impacting the ammonia solution without a deformation.

▶ The ADU gel microspheres are transferred to an ageing/washing/drying apparatus. During ageing, the ADU gel microspheres are completely maturated to solid ADU particles in ammonia solution. After ageing, the solid ADU microspheres are separated from the ageing solution, washed with demi-water and IPA

(Iso-PropylAlcohol) to remove the residual ammonium nitrate and water, respectively. The final step is to dry the solid ADU gel microspheres in a low vacuum atmosphere.

▶ Dried ADU gel microspheres are calcined in air heated up to 500°C in a calcination furnace. Firstly, the remaining organic additives (organic additives, residual water, impurities) and reaction product in the ADU gel microspheres are decomposed off during the temperature increase. Then the calcining reaction is as follows:

$$UO_3 \cdot xNH_3 \cdot yH_2O_{(s)} + O_{2(g)} \rightarrow 2\ UO_{3(s)} + 2H_2O_{(g)} + NO_{(g)}$$

▶ After calcining, converted UO_3 particles are reduced and sintered in a high temperature sintering furnace to remove any final remaining trace impurities and for densification of the microspheres. The sintering process is carried out in a 100% H_2 gas atmosphere.

$$UO_{3(s)} + H_{2(g)} \rightarrow UO_{2(s)} + H_2O_{(g)}$$

3. Experimental

The production of a spherical UO_2 kernel can be carried out by a wet chemical process, such as a sol-gel method, based on a solidification of spherical liquid droplets prepared by a vibrating dropping system. Generally, in the case of a HTGR nuclear fuel, the production process of a UO_2 kernel uses the gel-supported precipitation method, among the many techniques. After making a broth solution containing an uranyl nitrate as a uranium source, the spherical liquid-ADU droplets were prepared by using a vibrating dropping system.

Fig. 3 shows the experimental apparatus for the preparation of the spherical liquid-ADU droplets and the ADU gel particles[13]. The uranium concentration of the broth solution was 0.5~0.8 mol-U/ℓ, and a 1mm Φ diameter dropping nozzle, a 100~120 Hz frequency vibration, a 7~10 mol-NH_4OH/ ℓ solution in the gelation system were used for the spherical ADU gel particles preparation. The ADU gel particles obtained from the above experiments were transported to the fluidized bed column (or AWD apparatus) for the ageing and washing of the liquid ADU particles by using an ammonia solution, distilled water, and isopropyl alcohol, respectively. After the solid-liquid separation, finally the dried ADU particles were finished in a slightly vacuumed condition air atmosphere.

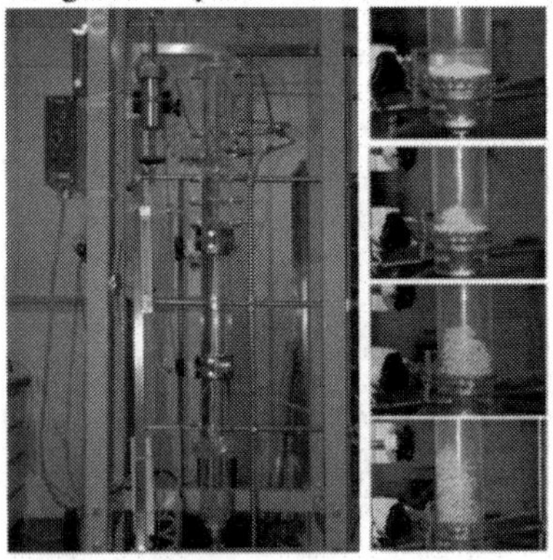

Fig.3. Experimental apparatus.

Also thermal treatments of the dried ADU particles converted to a UO_3 particle under an air atmosphere were carried out using a calcination furnace at 500℃ with a heating rate of 0.5~5°C/min. Then, the UO_3 particles were converted again to a UO_2 particle in a sintering furnace at 1700℃ with a heating rate of 3~5 steps, controlled in a 100%H_2 gas atmosphere.

In the analysis, the shapes of the liquid ADU droplets, dried ADU gel particles, UO_3, and UO_2 particles were observed using a Streoscope microscope. The characteristics of the kernel particles were also analyzed by FT-IR spectroscopy and TG/DTA.

Results and Discussion

Spherical Liquid ADU Droplets Preparation

To obtain the spherical shape UO_2 particles, it is very important that the preparation of an initial liquid ADU droplets be a spherical shape, because the liquid ADU droplets of an initial spherical shape should continuously maintain the final spherical form of a UO_2 kernel particles after the thermal treatment steps in which it is converted from an intermediate state, dried ADU→ UO_3 → UO_2 particles.

Generally, a spherical UO_2 particles preparation by wet chemical processes uses a gel-supported precipitation method[14]. This method consists of very complex chemical processes such as the dissolution of uranium oxide powder, broth solution preparation, gelation, washing and drying, calcination, and sintering of the final product. Here, spherical liquid-ADU droplets were prepared by using a vibrational dropping system with the made-broth solution among the above steps. The vibrational dropping system consisted of a vibrator of a horizontal or vertical frequency type, a flow-meter supplying the broth solution, and a nozzle for making the fine spherical droplets. Spherical liquid droplets were made by an elaborate controlling of the feed rate, an appropriate viscosity of broth solution, and a suitable frequency and amplitude of the vibrator. To improve the surface tension and viscosity properties of the liquid droplets, a organic additives were added to the UN raw material solution to achieve a controlled shrinkage of the ADU liquid-droplets during the gelation, ageing, and washing processes[15].

The intermediate ADU compound preparation step is a chemical reaction step, whereby the liquid-droplets of the broth solution are transformed into $UO_3 \cdot 1.5H_2O \cdot 0.5NH_3$ by a reaction with NH_4OH in a gelation column. In our case, the reaction is represented by the following stoichiometry:

$$2\ UO_2(NO_3)_{1.5}(OH)_{0.5} + 2\ NH_4OH \rightarrow UO_3 \cdot xNH_3 \cdot yH_2O_{(s)} + NH_4NO_3 + H_2O$$

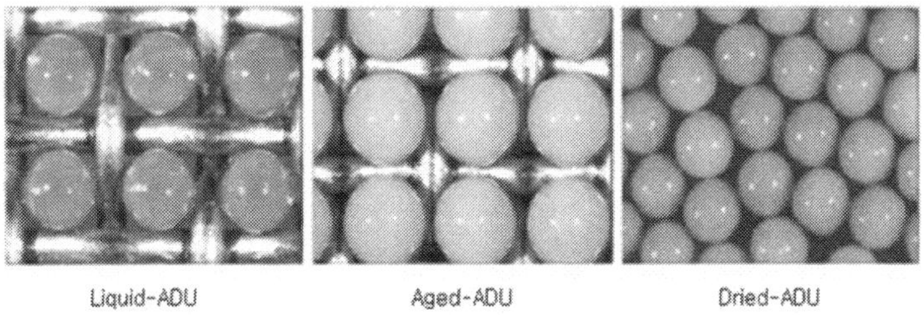

Fig.4. Spherical ADU droplets, aged gel, and dried particles.

Fig.4 shows the spherical shape liquid-droplets and the dried ADU gel particles made from our experimental system. Nozzle size was a 1Φ mm diameter, the viscosity range of the broth solution was about 50~80 cP, and the vibrating frequency was changed from 100 to 120 Hz. Sphericity of UO_2 particles obtained by our experiments was observed at about +1.06 that was calculated by a ratio of the vertical and horizontal lengths by a Streoscope observation.

To obtain the spherical intermediate ADU particles, the uranium concentration and the viscosity of the broth solution, the organic additive's ratio, and the vibrating frequency and the flow-rate of the feeding solution must be controlled precisely. If they are not controlled, the obtained ADU gel particles were as shown in fig. 5.

Fig.5. Crushed and cracked particles.

Particle Characteristics

FT-IR has proved to be the most useful tool for a structural determination and identification of organic and inorganic compounds[16]. In obtaining infrared spectrum of inorganic solids, an experimental complication arises from a possible chemical reaction between inorganic compounds and the infrared window material or support medium. Fig. 6 shows the FT-IR spectra of the raw material solution (UN solution) and the broth solution (mixed solution of an UN + organic additives) obtained by our experimental system. This spectrum showed that new absorption bands were formed at a nearly 3000cm^{-1} wavelength. It is supposed that the complex compounds are perhaps formed because of the very complicated reactions of an uranyl ion(UO_2^{2+}) from the UN solution, hydroxyl ion(OH^-) from additives, and NH_4^+ ion added at the pre-neutralization processing stage using a low concentration ammonium hydroxide. The absorption bands in the other wavelengths are similar with that of the UN solution when compared with the spectra of UN solution and a mixed solution.

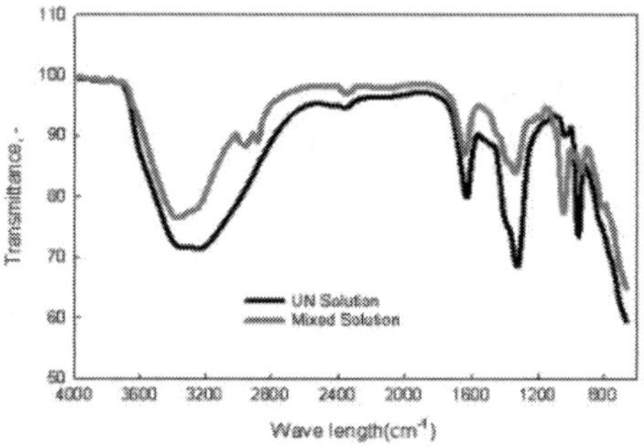

Fig.6. FT-IR spectra of UN and broth solutions.

Otherwise, fig. 7 shows the FT-IR spectra of the intermediate compound and ADU particles, prepared by a gel-supported precipitation method in our experiment, in the KBr pellet. Generally, the ADU compound particle exhibited absorption bands associated with a stretching and the bending vibrations of H-O-H in the coordinated water and hydrated water at nearly 3470 and 3225 cm^{-1}, also, stretching bands of the UO_2^{2+} ion, peaks by NH_3 rock, stretching bands of U←O or U←N appeared at nearly 930, 828, 450 cm^{-1}, respectively[17].

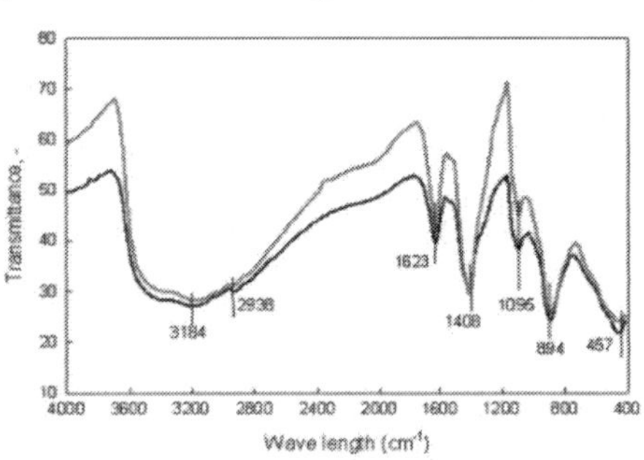

Fig.7. FT-IR spectra of ADU gel particles.

According to the bands of a general ADU, an ADU compound obtained from our experiment was different to that of the above mentioned. In our case, the obtained ADU compound particle exhibited absorption bands associated with a stretching and the bending vibrations of H-O-H in the coordinated water and hydrated water at nearly 3370 and 3150 cm^{-1}, and the stretching and bending vibrations of N-H in NH_4^+(sharp) and U-O(middle), one sharp peak and another middle peak at nearly 1400 and 1090 cm^{-1}, respectively. Also, a stretching vibration of U-O(strong) and a bending vibration of O-H(middle) containing a organic-precipitate appeared at 450 cm^{-1} and 1627 cm^{-1}, respectively.

The reason for the slightly different peaks seems to be that the sample preparation methods and processes are different in the two samples. In the general sample case, the ADU compound was prepared by the chemical reaction of UNH and ammonium hydroxide, however, our ADU compound was prepared by a reaction of the broth solution (mixed solution of UNH + organic additives) and the ammonium hydroxide solution. So the constitutes of an ADU compound prepared in our sample are predicted to be a complex form compared to a general ADU particle.

Thermal Treatments

Calcination and sintering furnaces are commonly used in the dried ADU→UO_3 and UO_3→UO_2 conversion steps, respectively. Our heat treatment experiments for a calcination of the dried ADU particles were carried out by raising the temperature up to 500℃ in an air atmosphere. Heating rate is a very important factor for obtaining no cracked spherical particles[18]. If the heating rate is faster than 5℃/min., the dried ADU gel particles were cracked or broken in our experiments. This is caused by the exothermic heat which occurs from a rapid thermal decomposition of the organic material among the ADU constituents. So the heating rate in the calcination process must be kept below nearly 3℃/min.

The heat treatment for the sintering process was carried out up to 1700℃ in a 100%-H_2 atmosphere with the sintering furnace. The sintering furnace reduces UO_3 to UO_2 using a hydrogen reduction atmosphere. Furthermore, the furnace sinters the UO_2 in order to achieve dense UO_2 kernels.

The following chemical reduction reaction takes place:

$$UO_3(s) + H_2(g) \rightarrow UO_2(s) + H_2O(g)$$

To obtain the optimum condition of the sintering process, pre-sintering experiments were carried out by TG/DTA instruments. Test conditions were in a 100%H_2 gas atmosphere, and 3~5 step heating modes which progressed up to 1700℃. Fig. 8 and 9 show the heating modes for a calcining and a sintering of our heat treatment experiments and the shapes of the UO_3 and UO_2 particles converted according to changes of the heat treating modes(here, 3 step mode), respectively.

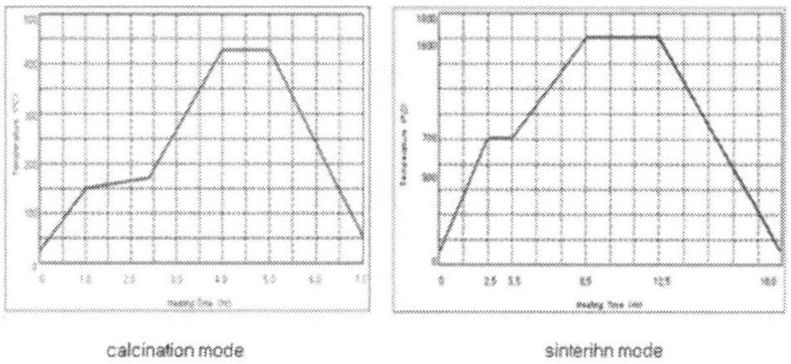

Fig.8. Heating profiles for calcinations and sintering processes.

- 1st step : reduction : 25→700℃(5℃/min. heating) and 1 hr isotherm
- 2nd step : heating and sintering : 700→1700℃, (3℃/min. heating) and ~4 hrs isotherm
- 3rd step : cooling : 1700℃→ room temperature, (5℃/min. cooling rate)

In the thermal treatments of the dried ADU or UO_3 particles, the elucidation of a suitable heating mode, heating rate and atmosphere, were primary factors in our experiments. Fortunately, we could find the heating mode where the heating rate and atmosphere in the calcination and sintering processes were kept below a 3℃/min. heating rate and in an air atmosphere up to 500℃, and a 3 step heating rate control and a 100%-H_2 gas atmosphere up to 1700℃.

UO_3 particles UO_2 Particles Ceramograph of UO_2

Fig.9. Well-prepared UO_3, UO_2 particles and its ceramograph.

Conclusions

In this study, to obtain a spherical UO_2 kernel, the most important factors for the droplets preparation are the composition ratio and viscosity of the broth solution, and the frequency/amplitude of the vibrator. Uranium concentration is 0.5~0.8mol/L, viscosity is 50~80 cP, vibration frequency is about 100~120Hz. Also, the heating rate in the calcination process must be kept below 3°C/min., and the optimum conditions of the sintering were in a 100%H_2 gas atmosphere and a 3 step heating mode up to 1700 °C.

References

[1] SR/OIAF/99-02, Analysis of the Impacts of an Early Start for Compliance with KYOTO PROTOCOL : The Energy Information Administration of the Department of Energy (1999), www.eia.doe.gov.

[2] SR/OIAF/98-03, Impacts of the KYOTO PROTOCOL on US Energy Markets and Economic Activity : The Energy Information Administration of the Department of Energy (1998), www.eia.doe.gov.

[3] INEEL/EXT-05-02581, Next Generation Nuclear Plant Research and Development Program Plan, Idaho National Engineering and Environmental Laboratory, p.1, (2005).

[4] W.Heit, H.Huschka, W.Rind, and G.G.Kaiser, "Status of Qualification of High-Temperature Reactor Fuel Element Spheres", *Nucl. Tech.*, **69**, pp. 44(1985).

[5] K.Minato, H.Kikuchi, T.Tobita, K.Fukuda, and M.Kaneko, "Improvements in Quality of As-Manufactured Fuels for High-Temperature Gas-Cooled Reactor", *J, Nucl. Sci. & Tech.*, **34**, pp.325 (1997).

[6] D.A.Petti, J.Buongiorno, J.T.Maki, and G.K.Miller, Key Differences in the Fabrication of US and German TRISO-Coated Particle Fuel, and Their Implications on Fuel Performance, INEEL/EXT-02-00300, INL and Environmental Lab., Bechtel BWXT Idaho, LLC (2002).

[7] H.Nickel, H.Nabielek, G.Pott, and A.W.Mehner, "Long Time Experience with the Development of HTR Fuel Elements in Germany", *Nucl. Eng. & Design,* **217**, pp. 141 (2002).

[8] G. Brambilla, P. Gerontopulos, and D. Neri, "The SNAM Process for the Preparation of Ceramic Nuclear Fuel Microspheres : Laboratory Studies", *Nucleare*, **17**(4), pp.217 (1970).

[9] P.A Hass, J.M.Begovich, A.D.Ryon, and J.S.Vavruska, Consolidated Fuel Recycle Program-Refabrication, Chemical Flowsheet Conditions for Preparing Urania Spheres by Internal Gelation, ORNL/TM-6850, (1979).

[10] M.Kadner and J.Baier, "Production of Fuel Kernels for High-Temperature Fuel Elements", *Kerntechnik*, **18**, pp.413 (1976).

[11] P. A. Hass, "Formation of Uniform Liquid Drops by Application of Vibration to Laminar Jets", *Ind. Eng. Chem. Res.*, **31**, pp.959 (1992).

[12] E.H.P.Cordfunke, "On the Uranates of Ammonium-I, The Ternary System NH_3-UO_3-H_2O", *J. Inorg. Nucl. Chem.*, **24**, pp.303 (1962).

[13] K.C.Jeong, S.C.Oh, Y.K.Kim, and Y.W.Lee, "ADU Compound Particle Preparation for HTGR Nuclear Fuel in Korea", *J. Ind .Eng. Chem.*, **13**(5), pp.744 (2007).

[14] E.Zimmer, P.Naefe, and H.Ringel, "Aqueous Chemical Processes for the Preparation of High Temperature Reactor Fuel Kernels", *Radiochimmica Acta*, **25**, pp.161 (1978).

[15] Y.Tomita, M.Morohira, Y.Kihara, and Y.Tamaki, "Fuel Microsphere Fabrication Tests for Sphere-Pac Fuel by the External Gelation Process", Proceedings of GLOBAL 2005, Tsukuba, Japan, Oct 9-13, Paper No.198, (2005).

[16] R.A.Nyquist and R.O.Kagel, Infrared Spectra of Inorganic Compounds, Academic press Inc., NewYork, (1971).

[17] A.M.Deane, "The Infrared Spectra and Structures of Some Hydrated Uranium Trioxides and Ammonium

Diuranates", *J. Ind. Eng. Chem.*, **21**, pp.238 (1961).

[18] B.Weimin and W.Xuejun, "Preparation of Dense UO_2 Fuels Kernels by External Gelation Process(II)", *Atomic Energy Sci. and Tech.*, **249,** pp.10 (1990).

(618be) Mixing Distribution Pattern In a Glass Packed Bed As a Function of Hydrophobic/ Hydrophilic Surface Treatment

Daniel F. Lopez[1], Jhon A Triana[2], Alejandro Martnez[1], Marcela Hernandez[2] and **Pablo Ortiz**[1], (1) Chemical Engineering, Universidad de los Andes, Bogot, Colombia, (2)Systems and Computing Engineering, Universidad de los Andes, Bogot, Colombia

In many petrochemical and biological processes packed beds play an important role increasing the efficiency of the separation or reaction mechanisms. Besides the geometry, distribution and pellets size, interfacial characteristics of the packing are critical. This study evaluates the effect that hydrophobic/ hydrophilic properties of the packing have in the flow pattern distribution when a dyed drop impacts a water saturated glass bed. To produce hydrophilic surfaces, glass spheres of 0.5 mm diameter were cleaned with a HCl (30%)-Methanol (1:1 v/v) by ultrasonication to activate silanol functional groups followed by an ethanol washing in order to eliminate organic contaminants. On the other hand, hydrophobic surfaces were obtained carrying out the process mentioned hereinbefore, followed by a silanization reaction with thrimethylclorosilane as the active compound. Subsequently, surface properties were characterized by angle contact measurements. A Petri dish of 11.0 cm diameter was randomly packed with a monolayer of treated spheres and flow evolution was monitored after the release of the dyed drop from 1.0 cm above the surface. Image data was collected with a 30 fps digital video camera and analyzed with a custom designed image processing software. Dyed flow intensity, flow front velocity and flow distribution were measured at several angle orientations in order to determine pattern symmetry and to obtain diffusion parameters. The results indicated that dyed fluid diffuses at very high rates with a star-shape pattern for the hydrophilic packing, whereas for the hydrophobic case the distribution is completely different, obtaining a highly circular front which evolves 600 times slower. The study analyzes the obtained data in terms of convection/diffusion models and discusses the surface interaction mechanisms.

See more of this Session: Mesd Poster Session
See more of this Group/Topical: Materials Engineering and Sciences Division

(618bf) On Multiplicity In Conversion of Monomers In Transient CSTR

Kal Renganathan Sharma, Department of Chemical Engineering, Prairie View A & M University, Prairie View, TX

Transient analysis of continuous polymerization by free radical methods in a CSTR was completed. The system was characterized by Damkohler number for initiator, Da_D and Damkohler number for monomer, Da_M .The initiator concentration can be described using a simple ordinary differential equation. The governing equation to describe the monomer concentration was non-linear. This equation was solved for by 4th order Runge Kutta method using MS Excel in a desktop computer. Multiplicity was found at transient coversion of monomer X =0.84. There are two dimensionless times associated with this conversion. These can be seen from Figure 2.0 to be t = 1.26 and t = 2.6. The maximum transient conversion of momomer X for the parameters used in the simulation study was found to be 0.863.

See more of this Session: Mesd Poster Session
See more of this Group/Topical: Materials Engineering and Sciences Division

(618bg) A Solid Polymer Electrolyte with Comprehensively Enhanced Performance for Lithium Ion Battery

Jianying Ji, Bin Li and Katie Zhong, School of Mechanical and Materials Engineering, Washington State University, Pullman, WA

It is a compelling challenge to both improves ionic conductivity and mechanical properties in polymeric electrolytes, and to accomplish it through a simple processing method. This study reveals that poly (ethylene oxide) -*block*- polyethylene (PEO-*b*-PE) possesses a capability for enhancing the multiple performances of PEO-based polymer electrolyte, i.e. it functions as a multi-functional miscible filler for obtaining poly(ethylene oxide)(PEO)/lithium perchlorate ($LiClO_4$) having such enhanced performances. Analyses revealed that the ionic conductivity of a ternary polymer electrolyte, PEO/PEO-*b*-PE/$LiClO_4$, was enhanced by an order of magnitude with the addition of this copolymer. Tensile tests showed the ultimate stress of the ternary polymer electrolyte increased to 7.29 MPa, modulus 114.21 MPa, and breaking strain 1142.53%, which are 2252%, 688% and 274% higher than that of PEO/$LiClO_4$, respectively. Scanning electron microscope (SEM), optical micrograph and X-ray diffraction (XRD) results indicate the addition of the copolymer, PEO-*b*-PE, decreased the defects of the PEO electrolyte films. Good dimensional stability was observed by dynamic rheological techniques up to 100 °C (higher than the melting point of PEO, 65 °C). A bi-continuous phase structure, i.e. crystalline PE domains within a matrix of PEO/salt, was proposed as the mechanism for such comprehensive enhancements in the ionic conductivity, mechanical properties and dimensional stability obtained simultaneously in this study through a facile approach based on incorporation of a copolymer.

See more of this Session: Mesd Poster Session
See more of this Group/Topical: Materials Engineering and Sciences Division

The Effects of the Nitrogen Gas and Pressure Cycling for the Hydrogen Storage Capacity on Li-Based Hydrides

Wen-Ming Chien*, Joshua H. Lamb and Dhanesh Chandra

Chemical and Metallurgical Engineering / MS388, University of Nevada, Reno, Reno, NV 89557

Abstract

Lithium imide/amide hydriding/de-hydriding pressure cycling has been studied to determine the hydrogen storage capacity by using different kinds of the hydrogen gases. Pressure equilibrium isotherms were obtained at 255°C after cycling between $Li_2NH \leftrightarrow LiNH_2$. The results showed that the nitrogen addition both allowed more of the theoretical hydrogen capacity to be storage as well as improved the stability of the hydrides after cycling. Other results by using different kinds of gas additions to hydrogen showed the loss of the hydrogen storage capacity after cycling. After 560 pressure cycles, 100 ppm O_2 in hydrogen were most detrimental (0.4 wt%H_2 capacity remaining out of ~5.6 wt.%) as compared to 100 ppm H_2O in hydrogen gas (2 wt%H_2). It was found that 1.65 wt.% hydrogen remained after 500 cycles by using industrial hydrogen. X-ray diffraction results of the products after pressure cycling showed mainly Li_2NH and LiH phases, and the impurity Li_2O phase. The phase analysis results showed the Li_2NH phase reduced from 77% to 13%, and LiH phase increases from 18% to 57% after 1100 cycles. The results of the effect of the impurity gases in hydrogen showed that the N_2 addition will improve the hydrogen storage capacity, and the O_2 has the most detrimental effect.

In this study, the losses of hydrogen storage capacities after pressure cycling were determined. The pressure-composition isotherm results shown that after 560 pressure cycles, 100 ppm O_2 in hydrogen were most detrimental (0.4 wt%H_2 capacity remaining out of ~5.6 wt.%) as compared to 100 ppm H_2O in hydrogen gas (2 wt%H_2). It was found that 1.65 wt.% hydrogen remained after 500 cycles by using industrial hydrogen. Pressure isotherms after ~100 pressure cycles showed 3.6 and 3.3 wt%H_2 capacity remained by using 100 ppm CH_4 and industrial hydrogen, respectively. X-ray diffraction results of the products after pressure cycling showed mainly Li_2NH and LiH phases, and the impurity Li_2O phase. The phase analysis results showed the Li_2NH phase reduced from 77% to 13%, and LiH phase increases from 18% to 57% after 1100 cycles. A summary plot pressure-composition isotherms (at 255°C) before and after Li_2NH-$LiNH_2$ cycling (no catalyst) with industrial hydrogen and 100 ppm levels of O_2, H_2O, CH_4 in UHP H_2 shows in Figure 1. The results showed that the nitrogen addition both allowed more of the theoretical hydrogen capacity to be storage as well as improved the stability of the hydrides after up to 1359 cycles. X-ray diffraction data shows that more Li_2NH phase (less LiH phase) is present in the desorbed hydride after cycling with the nitrogen addition than is present without. The results of the effect of the impurity gases in hydrogen showed that the N_2 addition will improve the hydrogen storage capacity, and the O_2 has the most detrimental effect.

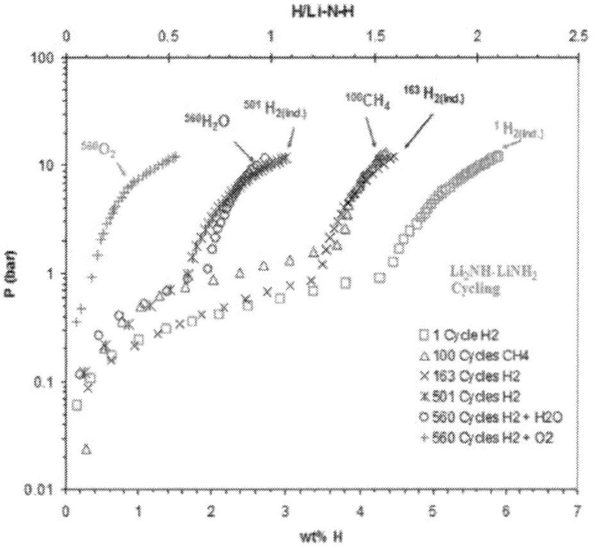

Figure 1. Summary plot of pressure-composition isotherms (at 255°C) before and after Li_2NH-$LiNH_2$ cycling (no catalyst) with industrial hydrogen and 100 ppm levels of O_2, H_2O, CH_4 in UHP H_2.

References:

[1] Y. Nakamori and S. Orimo, *Materials Science and Engineering*, B108, 2004, p 48-50.

[2] Y. Nakamori and S. Orimo, *J. Alloys Compd.*, 370, 2004, p 271-275.

[3] T. Ichikawa, S. Isobe, N. Hanada and H. Fujii, *J. Alloys Compd.*, 365, 2004, p 271-276.

[4] J.W. Wiench, V.P. Balema, V.K. Pecharsky and M. Pruski, *J. Solid State Chemistry*, 117, 2004, p 648-653.

[5] P. Chen, Z. Xiong, J. Luo, J. Lin and K.-L. Tan, *Nature*, 420, 2002, p 302-304.

[6] P. Chen, Z. Xiong, J. Luo, J. Lin and K.-L. Tan, *J. Phys. Chem.*, B 107, 2003, p 10967-10970.

[7] T. Ichikawa, N. Hanada, S. Isobe, H. Leng and H. Fujii, *J. Phys. Chem.*, B108, 2004, p 7887-7892

[8] Y. Hang and E. Ruckenstein, *Ind. Eng. Chem. Res.*, 44, 2005, p 1510-1513.

[9] A. Huq, J.W. Richardson Jr., E. Maxey, D. Chandra, and W. Chien, *J. Alloys Compd.*, 436, 2007, p 256-260.

[10] A. Huq, J.W. Richardson Jr., E. Maxey, D. Chandra, and W. Chien, *J. Phys. Chem. C*, 111 (28), 2007, p 10712-10717.

[11] W. Chien, J. Lamb, D. Chandra, A. Huq, J. Richardson Jr., and E. Maxey, *J. Alloys Compd.*, 446-447, 2007, p 363-367.

(618bi) Micromachined High Surface Area Structures for Energy Storage Applications

Sue Ann Bidstrup-Allen[1], Andac Armutlulu[1], Yichen Fang[1] and Mark G. Allen[2], (1)Chemical & Biomolecular Engineering, Georgia Institute of Technology, Atlanta, GA, (2)Electrical and Computer Engineering, Georgia Institute of Technology, Atlanta, GA

A micron-scale, high surface area metallic structure was built based on a sequential multilayer robotic electroplating process. Alternating Ni and Cu layers were electrodeposited from their respective plating baths, followed by selective and partial chemical etching of the Cu layers to obtain a final high surface area Ni structure which was intended to serve as a backbone for the anode electrode in Zn-air microbatteries. The last step of anode fabrication process involves electrodeposition of Zn layer onto the Ni backbone structure. This fabrication resulted in a high surface area electrode with a reduction in internal resistance, which are crucial for high performance batteries, particularly at high discharge rates. Using this approach, an electrode composed of 10 pairs of Ni and Cu layers was built and tested. The surface area of the electrode was found to be more than 6 times higher than that of its footprint. Based on the discharge tests through various electrical loads, the areal capacity and energy density were measured as ~2.5 mAh/cm^2 and 2.5-3mWh/cm^2, respectively.

See more of this Session: Mesd Poster Session
See more of this Group/Topical: Materials Engineering and Sciences Division

(618bj) Versatile Low-Cost Air-Gap Structures for MEMS Packaging

Sue Ann Bidstrup-Allen, School of Chemical and Biomolecular Engineering, Georgia Institute of Technology, Atlanta, GA, Nathan Fritz, Department of Biomolecular and Chemical Engineering, Georgia Institute of Technology, Atlanta, GA, Rajarshi Saha, Department of Chemical and Biomolecular Engineering, Georgia Institute of Technology, Atlanta, GA and Paul Kohl, Chemical and Biomolecular Engineering, Georgia Institute of Technology, Atlanta, GA

Recent advances in microelectromechanical systems (MEMS) technology have expanded their possible applications and potential market use. However, MEMS packaging presents a pivotal challenge to the evolving MEMS industry because it is costly and overly specialized to each MEMS device or family of devices. For today's typical MEMS-based products, packaging expenses can be up to 20% to 40% of the product's total material and assembly cost. [1] The aim of this project is to create a wafer-level packaging technology for families of MEMS devices that would transform the MEMS device into a form compatible with IC packaging processes. In this work, a polymer-based Air-gap structure process is being investigated to provide a cost efficient method for lead frame packaging of MEMS devices.

The air gap process uses an overcoat material to cover a patterned sacrificial polymer, polycarbonate (PC), which protects the MEMS device during packaging. Once the overcoat is in place, the sacrificial polymer is thermally decomposed freeing the MEMS structure while the overcoat dielectric provides mechanical and chemical protection from the environment. Improvements in the sacrificial materials and the introduction of an inorganic/organic epoxy-POSS overcoat has improved cavity design for clean removal of the PC material without damaging the cavity formed during the process. These developments have demonstrated that air-gaps with rigid overcoats can be used to encapsulate MEMS devices in a wafer-level packaging process.

The packaging structures can be designed for a range of MEMS device sizes and operating environments including fluidic, hermetic and vacuum conditions. However, the air-cavity structures need additional rigidity to withstand chip level packaging conditions, which are dependent on cavity size and molding pressure. Current work is focused on implementing a wafer level air-cavity package into a lead frame packaging scheme for MEMS devices. Metalized air-gap packages have been tested and optimized to withstand injection molding conditions necessary for chip-level packaging. A second approach will look at a size independent, semi-hermetic package (no metal). This process will look to combine the thermal steps of the sacrificial release and the molding process in to one step. This design will provide improved mechanical stability by prevention of cavity deformation during decomposition and molding simultaneously.

References:

1. Grace, R.; Maher, M. A. "MEMS: Think outside the chip... at the package level," Hearst Electronic Products 2010, http://www2.electronicproducts.com/PageSearch.aspx?FName=farc_softmems_nov2010.html.

See more of this Session: Mesd Poster Session
See more of this Group/Topical: Materials Engineering and Sciences Division

(618bk) Synthesis and Characterization of Surface Coated Silicotungstic Acid/Nafion Composite Membrane for Direct Methanol Fuel Cells

Joko Sutrisno, Irawan Pramudya and **Alan Fuchs**, Chemical Engineering, University of Nevada, Reno, Reno, NV

A composite proton exchange membrane (PEM) from surface coated silicotungstic acid (SiWA) and Nafion^{TM} for direct methanol fuel cells have been synthesized and characterized. Surface polymerization of sodium 4-vinyl benzene sulfonate (S4VBS) on SiWA has been accomplished by using atom transfer radical polymerization (ATRP). The ion exchange of sodium with hydrogen has been done by immersing the composite membrane in 2M sulfuric acid solution and sonicated for 10 minutes. The incorporation of surface coated SiWA in composite membrane resulted in homogenous membranes. It is expected to suppress the methanol crossover within PEM. The DMFCs have been fed with 5, 10, and 15 vol.% methanol solution and tested at room temperature. From polarization curve, DMFCs is fed with methanol solution 15 vol.% generated highest power output.

See more of this Session: Mesd Poster Session
See more of this Group/Topical: Materials Engineering and Sciences Division

(618bl) Significant Enhancement of Lithium Intercalation Into Mesoporous Titania with Trace TiO2-B On Surface

Xin Feng, Wei Zhuang, Linghong Lu, Yudan Zhu, Xinbing Wu, Meng Meng and Xiaohua Lu, State Key Laboratory of Materials-oriented Chemical Engineering, Nanjing University of Technology, Nanjing, China

TiO_2 is expected to play an important role in addressing the problems of clean energy supply and global warming on account of its promising applications in photovoltaics, photocatalytic oxidation and splitting of water. It is well recognized that in order to implement these applications, TiO_2 should own a large surface area, a high crystallinity and high thermal stability simultaneously. Our group has developed a template-free method to fabricate mesoporous Titania (m-TiO_2) nanofibers with all these properties. Trace TiO_2-B was proved by high resolution transmission electron microscopy and Raman spectra. However, its contribution in terms of performance was not perfectly specified.

In this work, we report on lithium-ion intercalation/deintercalation reaction in two differrent m-TiO_2 with or without trace TiO_2-B and find significant intercalation increasement of m-TiO_2 with trace TiO_2-B on surface. TiO_2-B is a metastable monoclinic polymorph of TiO_2 different from anatase. Its open structure, characterized as large and continuous channels, makes it well suited as a host for Li^+ ion intercalation. It can be seen that the cyclic voltammetry of m-TiO_2 exhibits a pair of cathodic/anodic peaks at 1.70 and 2.05V, which are attributed to lithium-ion intercalation/deintercalation reaction in anatase phase. In addition to that, the voltammogram of m-TiO_2 with trace TiO_2-B shows another two pairs of peaks with formal potentials in the range of 1.5-1.6 V vs. Li^+/Li, respectively. This notation can be assigned to the reaction of lithium-ion intercalation/ deintercalation reaction in TiO_2-B phase lattice. The first discharge capacity of m-TiO_2 with trace TiO_2-B can reach up to 247 $mA{\cdot}h{\cdot}g^{-1}$ at the current rate of 130 $mA{\cdot}g^{-1}$ which is about 50 $mA{\cdot}h{\cdot}g^{-1}$ higher than that of m-TiO_2 without TiO_2-B. Stable cycling discharge capacity attains 170 $mA{\cdot}h{\cdot}g^{-1}$ at the discharge rate of 155 $mA{\cdot}g^{-1}$ and 110 $mA{\cdot}h{\cdot}g^{-1}$ even up to 4000 $mA{\cdot}g^{-1}$. These findings indicate that m-TiO_2 with trace TiO_2-B on surface represents a promising material in lithium ion delivery and storage.

Hierarchical Bio-inspired Vascular Networks: Electrical Treeing

Kristopher D. Behler, Zachary R. Melrose and Eric D. Wetzel
U.S. Army Research Laboratory, Aberdeen Proving Ground, MD

Vascular networks provide a method to distribute fluid throughout a system. Artificial vascular materials with enhanced properties are being developed that could ultimately be integrated into systems reliant upon fluid transport while retaining their structural properties. An uninterrupted and controllable supply of liquid is optimal for many applications such as continual self-healing materials, *in-situ* delivery of optically index matched fluids, thermal management (sweating) and drug delivery systems could benefit from a bio-inspired approach that combines complex network geometries with minimal processing parameters. One such approach to induce vascular networks whilst mimicking nature's design is electrical treeing (ET).

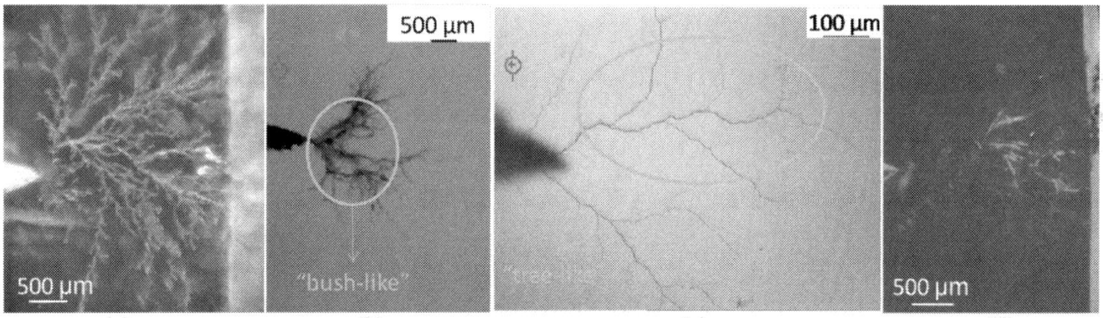

(a) (b) (c) (d)

Figure 1. Optical image of ET (a) in a EPON 828/PACM system, (b) under AC driven electrical current showing "bush-like" features, (c) under DC driven electrical current showing "tree-like" features and filling of an ET grown vascular network with a UV visualization dye.

Electrical treeing (ET) is the result of partial discharges in a dielectric material. In the vicinity of a small diameter electrode, the local electric field is greater than the global dielectric strength, causing a localized, step-wise, breakdown to occur forming a highly branched interconnected structure (**Fig. 1a**). The growth of these structures is influenced by the configuration of the electrodes, with geometries of a point lead electrode to a point or plane ground electrode being of most interest. ET is a viable method to produce networks in 2D systems and in more robust 3D systems on a smaller, micron, scale than the products of the EHVF method. AC driven electrical current (**Fig. 1b**), harnessing a sine wave at 100 Hz, grows a "bush-like" structure with many branches and therefore a larger volume within the epoxy samples. DC driven electrical current (**Fig. 1c**) produces a more "tree-like" structure with fewer branches and bifurcations. The surface of the electrodes were modified with dispersed multi-walled carbon nanotubes (MWCNTs) to aid in increasing the local electric field, and thus enable a higher rate of tree initiation and growth. Inclusion of particles was investigated to determine if the growth direction can be manipulated. The use of self-clearing electrodes (as a grounding material) was investigated with the infiltration of a UV dye through the hollow channels produced by ET resulting in a vascularized network capable of repeated fillings and evacuations. Fluid delivery (**Fig. 1d**) can be tailored through the applications of different electrode and ground manufacturing techniques for optimized flow rates for a given application.

Bio-inspired Hierarchical Vascular Networks: Electrohydrodynamic Viscous Fingering

Kristopher D. Behler, Andrew Schott and Eric D. Wetzel
U.S. Army Research Laboratory, Aberdeen Proving Grounds, MD

Vascular networks provide a method to distribute fluid throughout a system. Artificial vascular materials with enhanced properties are being developed that could ultimately be integrated into systems reliant upon fluid transport while retaining their structural properties. An uninterrupted and controllable supply of liquid is optimal for many applications such as continual self-healing materials, *in-situ* delivery of optically index matched fluids, thermal management (sweating) and drug delivery systems could benefit from a bio-inspired vascular approach that combines complex network geometries with minimal processing parameters. One such approach to induce vascular networks whilst mimicking nature's design is electrohydrodynamic viscous fingering (EHVF).

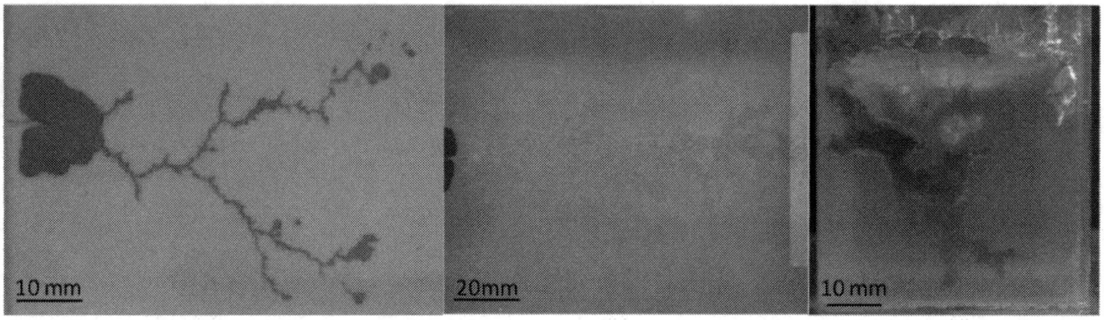

(a) (b) (c)

Figure 1. Optical images of EHVF (a) in a 1,000 cSt silicone oil system containing ~ 60 v/v% glass beads, (b) interfacial polymerization of of hexamethylene diamine and sodium chloride in water and sebocyl chloride in 10,000 cSt silicon oil, and in three dimensions (3DEHVF) using fumed silica in an index matched fluid.

Viscous fingering (VF) is a phenomenon that occurs when a low viscosity liquid is forced through a high viscosity fluid or matrix. The flowing liquid will branch, or form fingers due to capillary and viscous forces in the high viscosity material. EHVF is a modification on viscous fingering in which a DC voltage is applied to the low viscosity conductive fluid (**Fig. 1a**) and forced through a dielectric matrix material. The application of a large electrical potential, 10-60 kV, induces fingers with a reduction in size and an increased branching behavior. The ensuing patterns mimic those found in biology and geology (lung tissue and plants as well as river beds). Observation of VF and EHVF requires Hele-Shaw conditions in which a 2D system must possess a thin gap (0.8 mm used in these experiments) or in a porous system. In the 2D instance a silicone oil system is used as the matrix material, the surfactant concentration was optimized, through a reduction in the interfacial tension, thereby producing a branched pattern of small diameter fingers while still maintaining continuity when dyed water is pushed through the system. Various loadings of glass beads where subsequently used to represent a more 3D system. Typically, in a two fluid system, the fingers relax as soon as the applied voltage is removed. Addition of glass beads, up to 60 v/v%, aids in a retardation of finger relaxation while producing fine channels throughout the porous system. Delayed relaxation allows for greater control of the curing process in UV-curable systems, such as polydimethylsiloxane (PDMS). Fabrics, woven and random glass fiber mats were also investigated as a matrix to provide a different porosity to the system. Matrix filling was used as a method to reduce finger relaxation and allow for curing to occur after the voltage was turned off. Interfacial polymerization EHVF (IPEHVF), a technique in which the polymerization occurs at the interface of two materials, was also studied in producing fingers in the silicone oil phase. For EHVF to be studied in a similar system, hexamethylene diamine (C6H16N2) and sodium chloride were dissolved in the water phase, while sebacoyl chloride (C10H16Cl2O2) was disperse in the silicone oil phase. Robust, polymerized, fingers were formed (**Fig. 1b**) and remained after the voltage was turned off. These fingers were subsequently filled with water to show fluid transport. In moving toward a true 3D system, materials such as fumed silica (**Fig 1c**) and crushed glass were investigated under 3D (porous) Hele-Shaw conditions.

(618bo) Silver Nanowire Embedded In P3HT: PCBM for High Efficiency Hybrid Photovoltaic Device Applications

Sang-Ho Cha, Chemical Engineering, University of Michigan, Ann Arbor, Ann Arbor, MI, Sung-Ho Jin, Department of Chemistry Education, Interdisciplinary Program of Advanced Information and Display Materials, Pusan National University, Pusan, South Korea and Nicholas A. Kotov, Chemical Engineering, University of Michigan, Ann Arbor, MI

A systematic approach has been followed in the development of a high efficiency hybrid photovoltaic device which has the blend of poly(3-hexylthiophene) (P3HT), [6,6]-phenyl C_{61}-butyricacid methyl ester (PCBM) and silver nanowires (Ag NWs) in the active layer using the bulk heterojunction concept. The active layer is modified by utilizing a binary solvent system for blending. In addition, the solvent evaporation process in after spin-coating is changed and an Ag NWs is incorporated to improve the performance of the hybrid photovoltaic device. Hybrid photovoltaic devices were fabricated by using 1:0.7 weight ratio of P3HT to PCBM in a 1:1 weight ratio of *o*-dichlorobenzene and chloroform solvent mixture, in the presence and absence of 20 wt% of Ag NWs. We also compared the photovoltaic performance of Ag NWs embedded in P3HT:PCBM to that of silver nanoparticles (Ag NPs). Atomic force microscopy (AFM), scanning electron microscope (SEM), transmittance electron microscope (TEM), UV-visible absorption, incident photon-to-current conversion efficiency (IPCE), and time-of-flight (TOF) measurements are performed in order to characterize the hybrid photovoltaic devices. The optimal hybrid photovoltaic device composed of Ag NWs generated in this effort exhibits a power conversion efficiency (PCE) of 3.91 %, measured by using an AM 1.5G solar simulator (Oriel 300 W) at 100 mW/cm^2 light illumination intensity.

See more of this Session: Mesd Poster Session
See more of this Group/Topical: Materials Engineering and Sciences Division

(618bp) Theoretical Studies of Ion Transport and Surface Reactions In Solar Thermal Water Splitting by Mixed Metal Ferrites

Christopher Lawrence Muhich[1], Charles Bruce Musgrave[2] and Alan W. Weimer[2], (1)Department of Chemical and Biological Engineering, University of Colorado, Boulder, Bouler, CO, (2)Chemical and Biological Engineering, University of Colorado, Boulder, Bouler, CO

We have used quantum simulations based on density functional theory within the periodic boundary condition approach to study the solar thermal water splitting (STWS) by mixed metal ferrites. STWS occurs by cycling metal oxides through a high temperature reduction step and a lower temperature oxidation step where H_2O oxidizes the oxide and produces hydrogen gas. We investigate the reduction surface reactions that form O_2, and the transport of ions from the reduced/oxidized metal oxide interface to the surface in the STWS cycle. We investigate the effects of doping on the ion transport activation barrier. We have found that, in addition to lowering the reduction temperature, dopant ions has the added benefit of accelerating ion movement through the reduced phase. We compare the kinetics of the surface reactions to those of ion transport to determine the STWS bottleneck.

See more of this Session: Mesd Poster Session
See more of this Group/Topical: Materials Engineering and Sciences Division

(618bq) Atomic Layer Deposition of Bimetallic Nickel/Nickel Oxide and Platinum Catalysts on Alumina for Electrochemical Devices

Alia M. Lubers, Troy D. Gould, Kelly M. Anderson and Alan W. Weimer, Chemical and Biological Engineering, University of Colorado at Boulder, Boulder, CO

Platinum-nickel bimetallic, oftenin the form of nanoparticles, is one of the most effective catalysts for boththe hydrogen oxidation reaction (HOR) and oxygen reduction reaction (ORR) inhydrogen fuel cells. Issues arise with use of nanoparticles due to theiraggregation at higher temperatures, such as when used in phosphoric acid fuelcells. With larger particles, expensive bulk atoms go unutilized whichincreases the cost of the device and decreases its lifetime. Atomic layer deposition (ALD) is a scalable depositionmethod that can create materials with tunable sizes. Using either oxygen or hydrogen as aprecursor, nickel particles and nickel oxide films were grown on alumina usingALD. Then platinum was deposited usingALD on both materials using either the oxygen and hydrogen chemistry. These catalysts were investigated for HORactivity and were inspected using TransmissionElectron Microscopy (TEM), and Inductively Coupled Plasmon Mass Spectroscopy(ICP-MS).

See more of this Session: Mesd Poster Session
See more of this Group/Topical: Materials Engineering and Sciences Division

(618br) Electrically Conductive Polymers As Electrode Coatings for Entrapped Enzyme Biosensors

Linda J. Steele, Department of Chemistry, University of Kansas, Lawrence, KS and Stevin H. Gehrke, Chemical & Petroleum Engineering and Pharmaceutical Chemistry, University of Kansas, Lawrence, KS

A biosensor has several key components such as sample delivery, sensing material, transducer and data processor. We are focused on the sensing material and transducer, as it is imperative to the operation of the device that they communicate efficiently. In a biosensor incorporating an enzyme which catalyzes a redox reaction of a substrate, a sensing material made of a electrically conductive polymer is adhered to a physical transducer to convert change in oxidation into a physical signal (a current). The electrically conductive polymer transports the electrons from the redox reaction of the embedded enzyme to the electrode.

For such a biosensor to work reliably, a uniform and reproducible coating of conductive polymer must be laid down upon a base electrode, such as a carbon fiber. In this work, cyclic voltammetry is used to carry out the electropolymerization of monomers such as aniline and 8-anilino-1-napthalene sulphonic acid (PANSA). Uniformity of coating is assessed as a function of reaction conditions such as pH, catalyst, and monomer concentration. Such a coating process will ultimately be used to entrap substrate-sensing enzymes, thus providing the base platform for a enzyme-mediated single-use disposable biosensor.

See more of this Session: Mesd Poster Session
See more of this Group/Topical: Materials Engineering and Sciences Division

(618bs) Novel Fluorescent Nanofibrous Membrane As Sensory Materials - towards Landmine Detection

Ying Wang[1], Yu Ding[1], Anthony La[1], Christian Brckner[2] and Yu Lei[1], (1)Chemical, Materials & Biomolecular Engineering, University of Connecticut, Storrs, CT, (2)Department of Chemistry, University of Connecticut, Storrs, CT

The worldwide landmine problem is one of driving forces for the development of explosive detection. The fluorescence sensor described here is developed *via* simple and cost-effective routes for landmine signature vapors detection in vapor phase at ultratrace concentration levels. Electrospinning is adapted to generate highly porous nanofibrous morphology, providing large surface-to-volume ratio and free channels to expedient analytes adsorption and diffusion. The binding between the sensing film with electron-deficient explosive molecules results in a dramatic decrease in fluorescence emission (e.g. 90% quenching by 2,4-dinitrotoluene (DNT) vapor within 6 min). In addition, this sensor shows a remarkable selectivity towards nitrated explosives vapor, such as DNT, 2,4,6-trinitrotoluene (TNT), 1,3,5-trinitro-1,3,5-triazacyclohexane (RDX), 2,4,6-trinitrophenylmethylnitramine (Tetryl), pentaerythritol tetranitrate (PETN) and nitroglycerin (NG). A portable sensor prototype with a handheld UV lamp has been demonstrated for the detection of DNT buried in soil, indicating its potential application for visualizing landmines by naked eyes.

See more of this Session: Mesd Poster Session
See more of this Group/Topical: Materials Engineering and Sciences Division

SYNTHESIS OF GADOLINIUM OXIDE NANOCOMPOSITE PARTICLES FOR MRI AND NCT APPLICATIONS[†]

Inwoo Cheong and Heuikyoung Cho

Kyungpook National University, Daegu, South Korea (inwoo@knu.ac.kr)

[†]This work was recently published in J. Mater. Chem. (2011) DOI: 10.1039/c1jm11608h

Summary

MRI-active Gd_2O_3 core/SiO_2 shell/poly(2-methacryloyloxyethyl phosphorylcholine) corona composite nanoparticles (Gd_2O_3@SiO_2@PMPC NPs) have a number of attractive features in cancer diagnosis and neutron capture therapy (NCT).[1] Monodisperse and highly crystalline Gd_2O_3 NPs were prepared using a polyol protocol to control the average particle size and surface properties. The Gd_2O_3 NPs were then functionalized with SiO_2 and a biomimetic layer of PMPC, to achieve reduced toxicity and enhanced nucleus affinity, for use as an MRI-active Gd-NCT agent.

Introduction

Gadolinium (Gd) has been developed as an MRI contrast agent for positive intensity images and has also been investigated as a therapeutic agent for neutron capture therapy (NCT).[1, 2] NCT is a promising cancer therapeutic approach because it can effectively eliminate cancer cells using a low-energy neutron beam. To eliminate tumor cells, Gd-NCT uses emitted photons and electrons, especially Auger electrons, resulting from thermal neutron capture reactions of gadolinium-157 (^{157}Gd). There are considerable similarities between the design criteria for MRI and Gd-NCT, such as biocompatibility, shelf-life stability, effective delivery, and sufficient accumulation of Gd in tumor sites. In addition, one of the key factors for Gd-NCT is effective biodistribution of the Gd agent in close proximity to the DNA in a cancer cell because the Auger electrons emitted display high ionization over a short range of several tens of nanometers.[3] To achieve these goals, 2-methacryloyloxyethylphosphorylcholine (MPC) was selected to endow Gd-NCT agents with nucleus affinity as well as biocompatibility. It has been proved that MPC reduces biofouling, such as protein adsorption and adhesion of bacteria, in biomembrane structures because of its biomimetic features.[4, 5]

Herein, it was focused on the synthesis and characteristics of functionalized gadolinium oxide NPs for both MRI diagnosis and potential Gd-NCT therapy. In order to achieve these goals, monodisperse and highly crystalline Gd_2O_3 NPs were synthesized and functionalized with SiO_2 shells and PMPC corona layers; these can effectively reduce the cytotoxicity of Gd(III) and increase the affinity for cell nuclei. Cell viability, *in vitro* MRI performance, and *in vivo* MRI performance were then investigated to demonstrate the performance of the functionalized Gd_2O_3 NPs.

Experimental

Nanocomposite particles consisting of a Gd_2O_3 core/SiO_2 shell/poly(MPC) corona (Gd_2O_3@SiO_2@PMPC NPs) were prepared by a two-step reaction. First, monodisperse Gd_2O_3 nanoparticles of high crystallinity were synthesized by the polyol method, and then SiO_2 shells were introduced by a sol–gel reaction of TEOS and APTES-Br. Secondly, the corona layer of PMPC was formed by atom-transfer radical polymerization (ATRP). The thicknesses of the SiO_2 shell and PMPC layers were controlled by varying the amounts of TEOS and MPC, respectively.

Results and Discussion

To prepare stable gadolinium oxide nanoparticles, we introduced PMPC, which has been shown to have phosphorus groups with the highest affinity for nucleic acids.[6] An affinity for nucleic acids is regarded as a critical function in Gd-NCT since the short emission distance of Auger electrons from Gd can be compensated for by intra-nuclear localization of Gd_2O_3 NPs. In the preparation, Gd_2O_3@SiO_2-Br was synthesized as an ATRP initiator and the PMPC corona layer was introduced by ATRP directly from the surface of the NPs, i.e., the "grafting from" method.[7] Our approach facilitates a robust process without purification by either dialysis or solvent exchange. The synthetic conditions and characteristics of the nanoparticles are summarized in Table 1.

Table 1. Synthetic conditions and properties of gadolinium nanoparticles

Samples	TEOS [mL]	APTES-Br [mL]	MPC monomer [mL]	Gd(III) conc.[a] [mM]	T_1[b] [ms]
NP1	0.4	0.4	4	119	33
NP2	0.4	0.4	2	142	35
NP3	0.4	0.1	4	185	NIL
NP4	0.4	0.1	2	215	NIL
NP5	0.8	0.4	4	139	52

[a] Gd concentrations were measured by ICP-MS analysis.
[b] Magnetic properties were measured at an external field of 4.5 T.

In vitro cytotoxicity tests were performed in mouse fibroblast cells (NIH3T3) and two cancer cells (U-87 MG and B16F10) with Gd_2O_3 NPs coated with SiO_2 and PMPC. As we expected, the three kinds of gadolinium NPs coated with SiO_2 and PMPC exhibit no serious cytotoxicity in both incubation periods. In addition, no significant difference is found in cell viability over the entire concentration range (~1 mM).

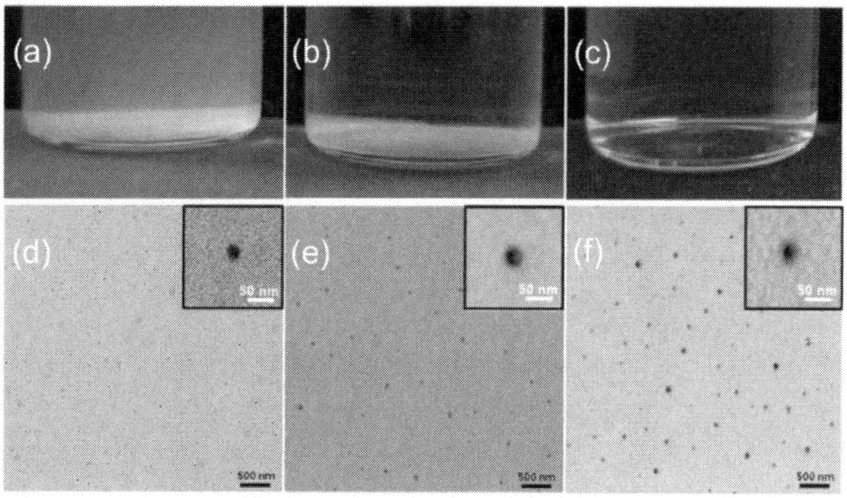

Figure 1. Photographic images (a-c) and TEM images (d-f) of Gd_2O_3, Gd_2O_3@SiO_2-Br, and Gd_2O_3@SiO_2-Br@PMPC (NP1 in Table 1).

Gd_2O_3@SiO_2@PMPC were dispersed in DI water at a concentration of 2.5×10^{-3} g/mL using a magnetic stirrer, and then kept for more than one week to observe dispersion stability of the prepared nanoparticles. As shown in Figure 1(a) and (b), one can judge that Gd_2O_3 and Gd_2O_3@SiO_2 NPs have poor stability from the precipitates at the bottom. In contrast, Gd_2O_3@SiO_2@PMPC NP solution is transparent without precipitation as shown in Figure 1(c). These results imply that the incorporation of PMPC as a corona enhances the colloidal stability of Gd_2O_3@SiO_2 NPs. We investigated the particle sizes and morphologies of Gd_2O_3, Gd_2O_3@SiO_2, and Gd_2O_3@SiO_2@PMPC using TEM because once recovered in the form of powder, Gd_2O_3 and Gd_2O_3@SiO_2 are not re-dispersed in any solvent. To present overall particle size, the low magnification TEM image of corresponding NPs are shown in Figure 1 (d)-(f). The average particle sizes of Gd_2O_3, Gd_2O_3@SiO_2, and Gd_2O_3@SiO_2@PMPC were 28, 66, and 75 nm, respectively, and the thickness of the SiO_2 shells and the PMPC layers were ca. 42 nm and 9 nm, respectively. The shape of the Gd_2O_3 NPs is spherical and the core@shell@corona morphology of Gd_2O_3@SiO_2@PMPC can be confirmed by inset images in Figure 1.

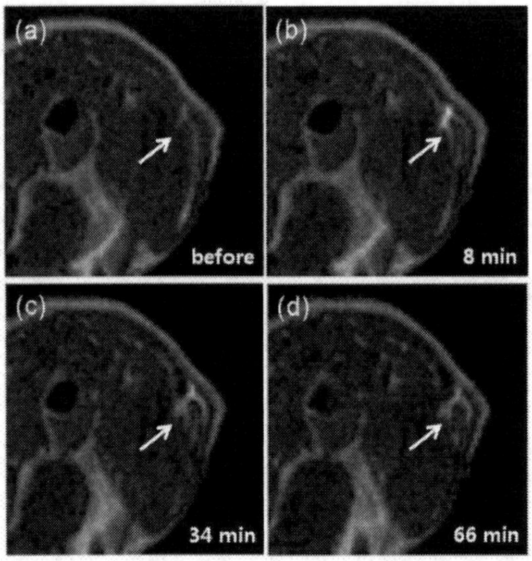

Figure 2. Series of *in vivo* T_1 MR images of melanoma cancer of a nude mouse (marked by a white arrow) before and after injection of NP1 in Table 1.

To evaluate the contrast ability and retention time in a live animal model, *in vivo* MRI experiments were performed using a B16F10-tumor-xenografted mouse model. In the experiment, 0.02 g of NPs were dissolved in PBS (pH 7.2) and this NP solution (1 mM with respect to Gd(III) ions) was intratumorally injected. In general, MRI contrast agents are injected intravenously and then allowed to accumulate in the tumor. However, in several reports,[3, 8] the agents were directly injected into tumor sites in order to deliver a sufficient amount of Gd. The nanoparticles started to accumulate in the tumor site 8 min after injection; the tumor site became brighter, as shown Figure 2(b). The total *in vivo* scan can be completed within 1 h. The margin of the tumor tissue is clearly seen in the bright T_1-weighted image at 34 min after injection. These findings indicate that Gd_2O_3@SiO_2@PMPC NPs are able to stay long enough to be detectable by MRI and to be used to treat tumor tissue by Gd-NCT; the half-lives of commercially available contrast agents are approximately 10 min.[9, 10] We cannot be sure that the enhanced retention of the NPs originated from interactions between the phosphorus atoms of the MPC groups on the NPs and nucleic acids in the cell nuclei;[6] however, the results suggest that the prepared NPs accumulate in the tumor site for a longer time than that achieved with other agents, and that they can be available during neutron irradiation in Gd-NCT.

This work was supported by Basic Science Research Program through the National Research Foundation of Korea (NRF) funded by the Ministry of Education, Science and Technology (2010-0012479). This work was also supported by the Pioneer Research Center Program through the National Research Foundation of Korea funded by the Ministry of

Education, Science and Technology (2010-0019308) and the Fundamental R&D Program for Core Technology of Materials funded by the Ministry of Knowledge Economy, Republic of Korea (K0006005).

References

1. Miyamoto, M.; Hirano, K.; Ichikawa, H.; Fukumori, Y.; Akine, Y.; Tokuuye, K. (1999) Biol. Pharm. Bull. , 22, 1331.
2. Barth, R. F.; Coderre, J. A.; Vicente, M. G. H.; Blue, T. E. (2005) Clin. Cancer Res. , 11, 3987.
3. Oyewumi, M. O.; Yokel, R. A.; Jay, M.; Coakley, T.; Mumper, R. J. (2004) J. Controlled Release 95, 613.
4. Iwasaki, Y.; Uchiyama, S.; Kurita, K.; Morimoto, N.; Nakabayashi, N. (2002) Biomaterials 23, 3421.
5. Patel, J. D.; Iwasaki, Y.; Ishihara, K.; Anderson, J. M. (2005) J. Biomed. Mater. Res., Part A 73A, 359.
6. De Stasio, G.; Rajesh, D.; Casalbore, P.; Daniels, M. J.; Erhardt, R. J.; Frazer, B. H.; Wiese, L. M.; Richter, K. L.; Sonderegger, B. R.; Gilbert, B.; Schaub, S.; Cannara, R. J.; Crawford, J. F.; Gilles, M. K.; Tyliszczak, T.; Fowler, J. F.; Larocca, L. M.; Howard, S. P.; Mercanti, D.; Mehta, M. P.; Pallini, R. (2005) Neurol. Res. , 27, 387.
7. Cho, H. K.; Cheong, I. W.; Lee, J. M.; Kim, J. H. (2010) Korean J. Chem. Eng. , 27, 731.
8. Oyewumi, M. O.; Mumper, R. J. (2002) Bioconjugate Chem. , 13, 1328.
9. Kuennemeyer, J.; Terborg, L.; Nowak, S.; Brauckmann, C.; Telgmann, L.; Albert, A.; Tokmak, F.; Kraemer, B. K.; Guensel, A.; Wiesmueller, G. A.; Karst, U. (2009) Electrophoresis, 30, 1766.
10. Hong, K. S. (2009) Kobunja Kwahak Kwa Kisul 20, 239.

(618bu) Syntheses, Acidichromic and Hg2+-Sensing Properties of Novel Naphthalimide Derivatives

Pei-Yu Kuo, Ya-Chi Jang and **Walter Den**, Environmental Science and Engineering, Tunghai University, Taichung City, Taiwan

Two novel fluorescent naphthalimide-based sensors as ion sensor and acidichromic colorant are synthesized. This naphthalimide fluorophore is functionalized via a condensation reaction in which dithiodioxomonoaza-based crown-containing phenyl units are conjugated to the naphthalimide core as a trapper for detecting acids in the gas phase and heavy metal ions Hg^{2+}. It shows a selective chromogenic behavior toward Hg^{2+} ion by changing the color of the solution. It also can be found that both on a solid PC (polycarbonate) film and in solution show acidichromism, i.e. a reversible color change upon alternatively exposing the films or solution to HCl and NH_3 gases, respectively. In this proposal, the designed sensors have been developed for the colorimetric and fluorescent determination by UV-Vis and fluorescence spectroscopy, and visual detection by the naked-eye. The visual detection provides a simple, rapid and sensitive method and was used for the detection of gaseous acids and Hg^{2+} ion in environmental applications.

See more of this Session: Mesd Poster Session
See more of this Group/Topical: Materials Engineering and Sciences Division

(618bv) Photo-Responsive Trithiocarbonate

Diana Leung, Chemical and Biological Engineering, University of Colorado, Boulder, CO and Chistopher N. Bowman, Department of Chemical and Biological Engineering, University of Colorado, Boulder, CO

Abstract

Title: Photo-responsive Trithiocarbonate

Author: Diana Leung; Christopher N. Bowman

Presenting Author: Diana Leung

The ability to createwell-defined topography is useful in responsive surfaces and optics. To allowfor photoinduced structural adaptation in elastomeric networks, a functionalgroup that can rearrange in order to relieve stress is needed. Here, we used **1**, which contains acrylate functionalgroups that undergo a Michael addition with PETMP and a trithiocarbonate corethat subsequently is used for addition-fragmentation chain transfer (AFCT) withthe acrylate in the presence of a photointiator. By incorporating thetrithiocarbonate backbone, stress relaxation by rearrangement can occur, sothat upon stretching followed by irradiation of the photoinitiator, the AFCTcan occur, to produce a well-defined and controlled topography in theirradiated region.

1

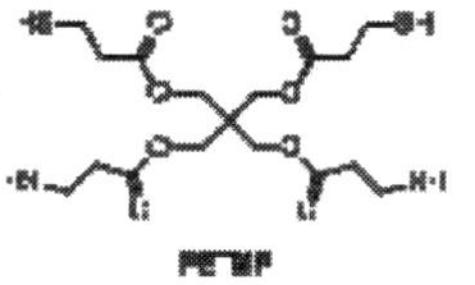

PETMP

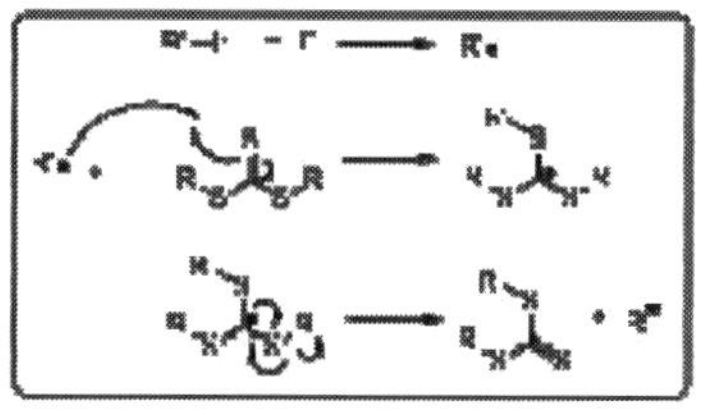

See more of this Session: Mesd Poster Session
See more of this Group/Topical: Materials Engineering and Sciences Division

(618bw) Metal Oxide Nanofibers for Oxygen Detection At High Temperature

Yixin Liu[1], Yu Ding[1], Haiyong Gao[2], Puxian Gao[2] and Yu Lei[3], (1)Department of Chemical, Materials and Biomolecular Engineering, University of Connecticut, Storrs, CT, (2)Institute of Materials Science, University of Connecticut, Storrs, CT, (3)Chemical Engineering, University of Connecticut, Storrs, CT

A high temperature oxygen sensor was developed based on electrospun metal oxide nanofibers. The morphology of the nanofibers was investigated by scanning electron microscopy and transmission electron microscopy. The as-prepared nanofibers were employed to detect oxygen at 800 - ^{o}C. The structure and composition of nanofibers before and after oxygen detection was characterized by X-ray diffraction. The conductivity of the metal oxide nanofibers was hugely enhanced at high temperature. The sensor showed good sensitivity, fast response and concentration dependence towards oxygen detection. In addition, the device has good thermal stability and reproducibility after operating at 800 ^{o}C for several days. These good results indicate the great potential of as-prepared metal oxide nanofibers in the application of high temperature gas sensors.

See more of this Session: Mesd Poster Session
See more of this Group/Topical: Materials Engineering and Sciences Division

Tungstate-Based Electrochromic LbL Thin Films

Yong Tae Park and Jaime C. Grunlan*

Department of Mechanical Engineering, Texas A&M University, College Station, TX 77843-3123

Layer-by-layer assembly was used to generate tungstate-based electrochromic thin films that transition from transparent to dark blue in their oxidized and reduced states, respectively. Tungstate is a good electrochromic material because it is completely colorless in its deposited state, however, it is plagued by long switching time due to lack of electrical conductivity. In the absence of ITO, these films take 30–60s to completely switch and exhibit reduced contrast with each switch. ITO-containing films, with ITO in every other bilayer, fully switch in 3–10s and do not exhibit the same drift in transmittance with repeated switching.

Layer-by-layer Thin Film Fabrication

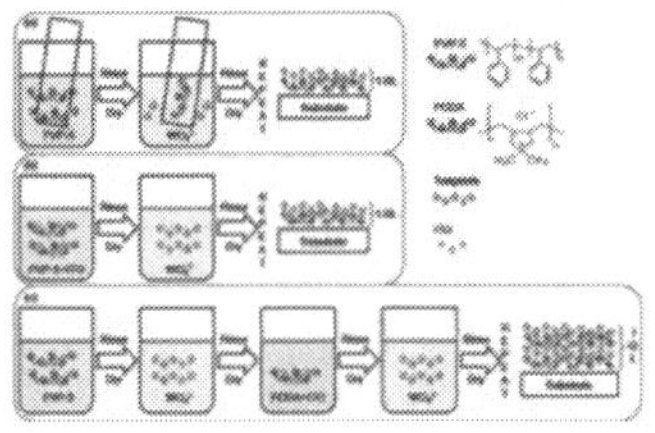

Schematic of the Layer-by-Layer Process

- ITO-coated glass or polystyrene slides were alternately exposed to aqueous solutions containing charged ingredients. Each deposition of a positive and negative pair of ingredients is referred to as a *bilayer* (BL).
- Substrates were first immersed into a positively charged aqueous solution (PVP-S, PVP-S+ITO or PDDA+ITO), followed by rinsing with deionized water and blow drying. Immersion into a (tungstate, WO_4^{2-}) aqueous solution came next and the process was repeated following rinsing and drying.

Growth of Thin Film Assemblies

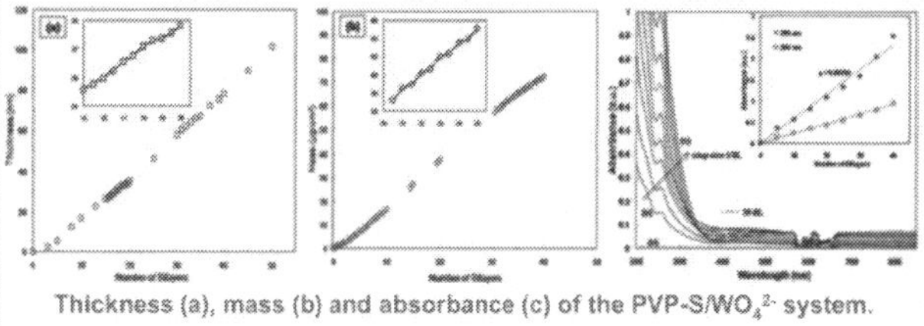

Thickness (a), mass (b) and absorbance (c) of the PVP-S/WO_4^{2-} system.

- PVP-S/WO_4^{2-} films increase linearly in thickness by approximately 2.0 nm per BL.
- Mass growth data yields a PVP-S/WO_4^{2-} film density of 8.25 g/cm³, which is close to their sum. It is likely that WO_4^{2-} penetrates into, and adsorbs onto, an underlying PVP-S layer to form a PVP-S/WO_4^{2-} complex.
- The increase in absorbance of the film with increasing number of BLs is due to increasing film thickness.

Thin Film Cross-Sections

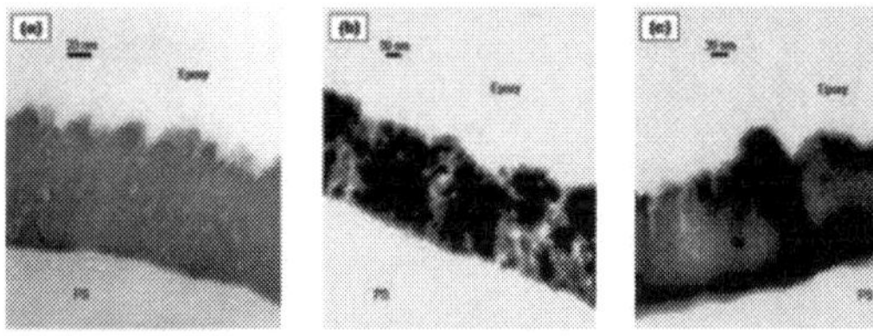

TEM cross-sections of 60 BL PVP-S/WO_4^{2-} (a), 60 BL PVP-S+ITO/WO_4^{2-} (b), and 30 QL PVP-S/WO_4^{2-}/PDDA+ITO/WO_4^{2-} (c).

- The darkest black spots are ITO particles and darker and lighter grey areas are WO_4^{2-} and PVP-S (or PDDA)-rich regions, respectively.
- TEM images show a highly inter-diffused nanostructure and support the suggestion that WO_4^{2-} penetrates into and adsorbs onto each underlying PVP-S layer.

Thin Film Surface Images

TEM surface

AFM height

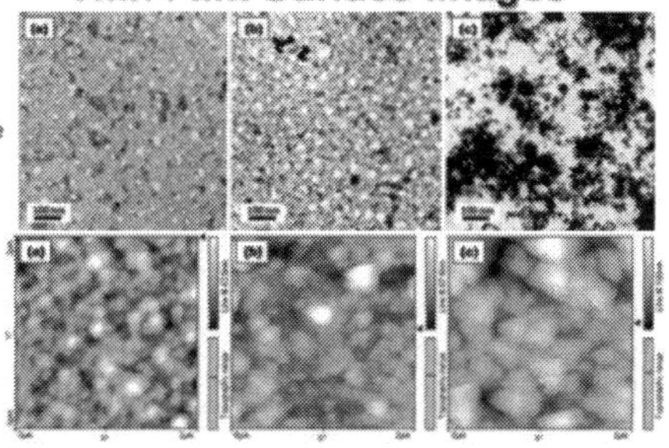

- These images show greater ITO aggregation with higher concentration, which produces greater surface roughness. ITO aggregation forms a network within the PVP-S/WO_4^{2-}/PDDA+ITO/WO_4^{2-} system, whereas ITO particles are isolated from each other (i.e., no network) in the PVP-S+ITO/WO_4^{2-} system.

Cyclic Voltammetry

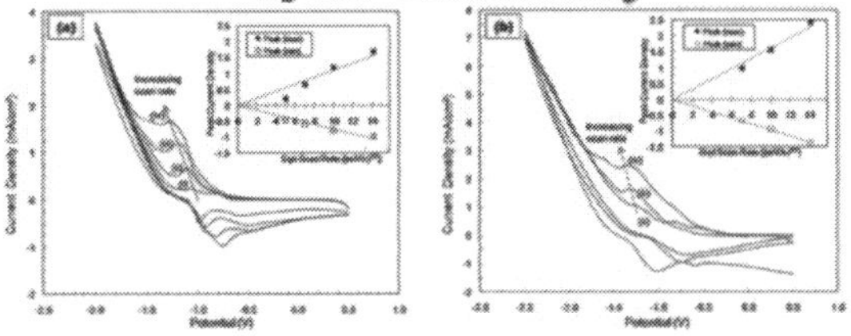

Cyclic voltammetry of 60 BL PVP-S/WO_4^{2-} (a) and 30 QL PVP-S/WO_4^{2-}/PDDA+ ITO/WO_4^{2-} film (b) at 25 (i), 50 (ii), 100 (iii), and 200 mV/s (iv) rates.

- During the reduction scan, the CV profile of PVP-S/WO_4^{2-} exhibits a peak near −1.0 V, which is the first reduction ($WO_4^{2-} \rightarrow WO_4^{3-}$). The thin film keeps a deep blue color up to −1.3 V, but changes to pale yellow beyond −1.3 V.
- The PVP-S+ITO/WO_4^{2-} system shows the same CV trend as the PVP-S/WO_4^{2-}.
- In the PVP-S/WO_4^{2-}/PDDA+ITO/WO_4^{2-} system, however, the peaks during both reduction and oxidation scans are not as pronounced.

Electrochromic Performance

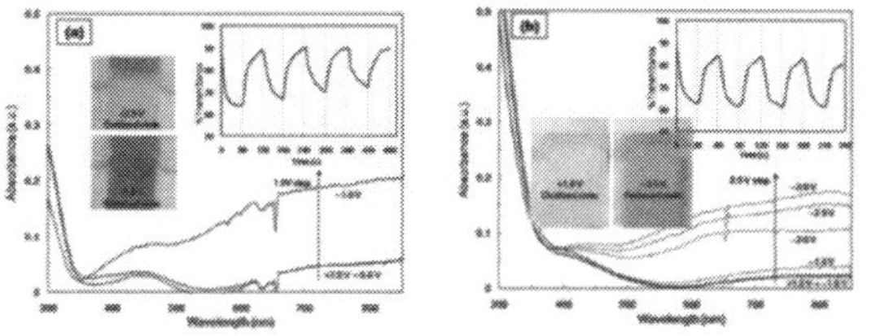

Absorption spectra at 750 nm during potential steps for 60 BL PVP-S/WO_4^{2-} (a) and 30 QL PVP-S/WO_4^{2-}/PDDA+ITO/WO_4^{2-} (b).

- The contrast in (a) is from 25.2 % to 17.7 % because of decreased tungstate coloration in the reduced state. PVP-S/WO_4^{2-} switching time increased to 40 seconds at the third cycle.
- The 20% contrast in (b) is nearly constant. This system fully oxidizes and reduces, unlike the PVP-S+ITO/WO_4^{2-} system, with a switching time of 14 s.

TEXAS A&M UNIVERSITY

Polymer Nanocomposites Lab

This study demonstrates the ability of the layer-by-layer technique to produce fast-switching and stable contrast tungstate-based electrochromic thin films. The use of an electrically conductive particles to speed switching and improve stability could potentially be used in a variety of electrochromic systems.

* Author to whom correspondence should be addressed (jgrunlan@tamu.edu) – PNC Lab (http://nanocomposites.tamu.edu)

(618by) Development of DDGS Based Renewable Composites

Long Jiang, Mechanical Engineering, North Dakota State University, Fargo, ND

Development of DDGS based renewablecomposites

Long Jiang (long.jiang@ndsu.edu)

Mechanical Engineering Department,North Dakota State University, Fargo, North Dakota, 58108

Distiller'sDried Grains with Solubles (DDGS) is a byproduct ofcorn ethanol production. It is primarilyused as a livestock feed. The main constituents of DDGS are zeinprotein and carbohydrate fibers. Zein is a strongadhesive and is able to bind the DDGS fibers to form a fiber composite. Thisposter presents the development of a DDGS based zeincomposite. The effects of processing and formulation on the microstructure andmechanical properties of the composites were discussed. Its potentialapplications were identified.

See more of this Session: Mesd Poster Session
See more of this Group/Topical: Materials Engineering and Sciences Division

(618c) Enhanced Loading of Proteins Into Hydrogel Delivery Vehicles

Erik Van Kampen, Department of Chemical and Petroleum Engineering, University of Kansas, Lawrence, KS and Stevin H. Gehrke, Chemical & Petroleum Engineering and Pharmaceutical Chemistry, University of Kansas, Lawrence, KS

Hydrogels are widely used as vehicles for the controlled release of therapeutic proteins because of their ability to control the location and rate of release of a protein to the body. In most cases, cross-linked gels must be formed prior to loading with the protein to avoid side reactions with the protein, or inability to remove unreacted byproducts without also leaching out the protein. Loading of macromolecules into pre-formed gels, however, can be limited by the phenomenon of size exclusion. However, we have shown that entropic size exclusion of proteins from polymer networks can be overcome using the thermodynamic principles of aqueous two phase extraction. This concept, typically used for protein purification, uses the fact that solutions of two water soluble polymers typically split into two immiscible phases, and that proteins partition unevenly between these phases. Using dextran and poly (ethylene glycol) diacrylate (PEGDA) gels, we have shown that by adding a thermodynamically incompatible polymer to the loading solution in which a gel is immersed, the protein partition coefficient into the gel can be increased over two orders of magnitude, and loading over 20 wt% (dry basis) protein can be achieved by this simple method.

See more of this Session: Mesd Poster Session
See more of this Group/Topical: Materials Engineering and Sciences Division

(618d) Biaxial Mechanics of Biocomposite Thin Films Supported On Polydimetylsiloxane (PDMS)

Hallie Zeller, Department of Mechanical Engineering, Lafayette College, Easton, PA and James K. Ferri, Department of Chemical and Biomolecular Engineering, Lafayette College, Easton, PA

We discuss biaxial mechanical measurements of force-displacement and digital image correlation for lamellar composites of biocomposite thin films supported by PDMS. Biofilms of the E. Coli mutant d21g were growth on 5x5 mm PDMS substrates under different environmental conditions and age. We show that by modulating the cross-link density of PDMS, the biolfilm elasticity can be estimated from the force-displacement data. Additionally, by measured the strain field of the composite system, the relative important of interlayer adhesion can also be determined. In the systems we studied, adhesive interactions between PDMS and d21g play a larger role in the displacement of individual bacteria in the biofilm under tension.

See more of this Session: Mesd Poster Session
See more of this Group/Topical: Materials Engineering and Sciences Division

(618e) Microfluidic Preparation of Multicompartment Microcapsules for Co-Encapsulation and Controlled Release of Multiple Components

Wei Wang, Rui Xie, Xiao-Jie Ju, Tao Luo, Li Liu and Liang-Yin Chu, School of Chemical Engineering, Sichuan University, Chengdu, China

Microfluidic Preparation of Multicompartment Microcapsulesfor Co-encapsulation and Controlled Release of Multiple Components

Wei Wang, Rui Xie, Xiao-Jie Ju,Tao Luo, Li Liu, Liang-Yin Chu[*]

School of Chemical Engineering, Sichuan University, Chengdu, China, 610065

Microcapsules are widely usedas encapsulation systems for protection of active species[1], controlledrelease of various substances[2] and confined microreaction of chemicals[3],etc. However, most of these microcapsules can only encapsulate one contentwithin the same structure. How to co-encapsulate multiple incompatible or functional substances in one single microcapsule,especially with precise control over the encapsulation level of each substance,is still a major challenge[4,5]. Our previously reported monodisperse multicomponent multipleemulsions[6], with co-encapsulated different dropletsin the same inner structure, offer noveland diverse templates to createmicrocapsules with differentcompartments forco-encapsulation of multiple incompatible actives or functional components. Theprecise and individuall control over the number, ratio and size ofdifferent inner droplets enablethe optimization of the encapsulation of each active or functional component.

Herewe present the microfluidicpreparation of thermo-responsive multicompartment microcapsules for co-encapsulationand controlled release of differentlipophilic components with multicomponent-double-emulsions as templates. Each of the microcapsule comprises different oil cores and a single thermo-responsive shell composed of poly(*N*-isopropylacrylamide) (PNIPAM). The different oil cores can be used as separate compartments for encapsulation of distinctlipophilic components. The precise manipulation of different oil cores afforded by our microfluidicdevice enables the fabrication of microcapsules with select number of each core, for optimizingthe encapsulation of different lipophiliccomponents. Because of the shell shrinking during heating process due to the thermo-responsive PNIPAM network,different lipophilic components co-encapsulated in the microcapsule can be simultaneously released via an external thermal trigger, as shown in Fig.1. The thermo-responsive multicompartment microcapsule provides an efficient system forco-encapsulation and controlledrelease of multiple lipophilic components. The thermo-responsivemulticompartment microcapsulescan be converted into other stimuli-responsive ones such as pH-responsive,molecular-recognizable, and glucose- responsive ones, by simply changing the shell materials. The microfluidic preparationapproach presented here shows exciting potential in the design and fabricationof functional multicompartment microcapsules for co-encapsulation techniques.

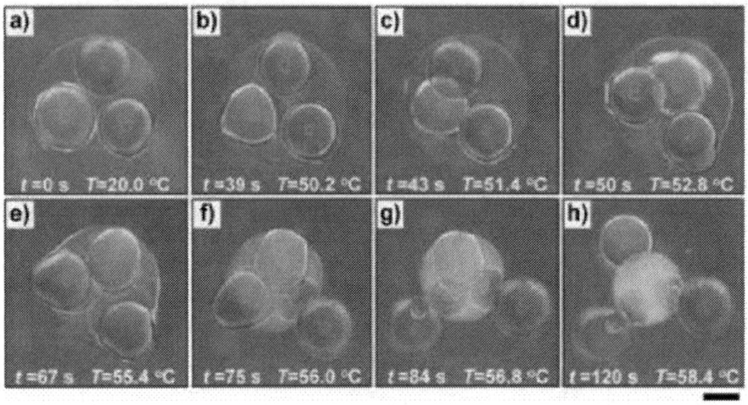

Fig 1. Temperature-controlled release of theco-encapsulated lipophilic substances from the thermo-responsivemulticompartment microcapsule. (a) Dark-field microscopeimage of multicompartment microcapsule co-encapsulated with onetransparent oil core and two red oil cores. (b)-(h) The temperature-controlled release of different oil cores from the microcapsule whentemperature is increased from 20 °C to 60 °C. Withincrease in temperature, the thermo-responsive PNIPAM shell of the microcapsuleshrinks dramatically. Since the oil cores are incompressible but the internalpressure in oil cores keep increasing due to the shell shrinkage, the PNIPAMshell finally ruptures because of the limited mechanical strength, whichresults in burst releasing of the oil cores. Scale bar is 100 μm.

References

1 R. De Rose, A. N. Zelikin,A. P. R. Johnston, A. Sexton, S. F. Chong, C. Cortez, W. Mulholland, F. Caruso, S. J. Kent, *Adv.Mater.*, **2008**, 20, 4698.

2 L. Y. Chu, T. Yamaguchi, S. Nakao,*Adv. Mater.*, **2002**, 14, 386.

3 O. Kreft, M. Prevot, H. Möhwald, G. B. Sukhorukov, *Angew. Chem.Int. Ed.*, **2007**, 46, 5605.

4 H. Chen, Y. Zhao, Y. Song, L. Jiang, *J. Am. Chem. Soc.*, **2008**, 130, 7800.

5 B. J. Sun, H. C. Shum, C. Holtze, D. A. Weitz, *ACS Appl. Mater. Interfaces*, **2010**, 2, 3411.

6 W.Wang, R. Xie, X. J. Ju, T. Luo, L. Liu, D. A. Weitz, L. Y. Chu, *LabChip*, **2011**, 11: 1587.

See more of this Session: Mesd Poster Session
See more of this Group/Topical: Materials Engineering and Sciences Division

(618f) Self-Assembling Biomimetic Hydrogels with Bioadhesive Properties for Tissue Engineering Applications

Bryan Johnson[1], Jennifer Kadlowec[2], Cristina Iftode[3], Thomas Tulenko[1] and **Jennifer Vernengo**[4], (1) Department of Surgery, Cooper University Hospital, Camden, NJ, (2)Department of Mechanical Engineering, Rowan University, Glassboro, NJ, (3)Department of Biological Sciences, Rowan University, Glassboro, NJ, (4)Chemical Engineering, Rowan University, Glassboro, NJ

Tissue engineering is a multidisciplinary field that aims to repair or regenerate lost or damaged tissues and organs in the body. A key aspect of tissue engineering is the use of a scaffold, or three-dimensional biomaterial that houses stem cells and provides biochemical cues to promote the formation of new tissue. However, for certain applications, success of a tissue engineering strategy depends on having a scaffold that adheres with adjacent tissue, anchoring it in place, and thereby potentially improving outcomes. Such integration may be achieved with the use of a bioadhesive polymer that covalently bonds with tissue. However, current bioadhesive polymers suffer from poor biocompatibility. In this collaborative project among faculty at Rowan University and Cooper University Hospital, we aim to design and characterize a bioadhesive polymer so that, in addition to bonding with tissue, it can support cell survival and differentiation post-adhesion with surrounding tissue. In the current work, we present the characterization of the adhesive strength of a multi-component "smart" hydrogel system. We also demonstrate the thermally triggered release of ECM-derived components, which become integrated into the gel, allowing for the assembly of a biomimetic matrix. In future work, we will characterize the long-term viability of adipose-derived stem cells encapsulated within the scaffold. This work will have therapeutic potential for several tissue engineering applications, such as fistula and intervertebral disc repair.

See more of this Session: Mesd Poster Session
See more of this Group/Topical: Materials Engineering and Sciences Division

(618g) A Microfluidic Method to Evaluate Small Molecule Uptake and Elution In Hydrogels

Erin L. Jablonski, Chemical Engineering, Bucknell University, Lewisburg, PA

Small molecule elution from a hydrogel is characterized using a microfluidic technique that allows for both image and effluent analyses. For elution experiments, the microfluidic device has channels defined by dye-loaded hydrogel. Photolithography is used to cure a solution of poly(ethylene glycol) diacrylate (PEG-DA), photoinitiator, and dye inside a microfluidic device with a channel through the center of the gel. As an aqueous solution is pumped through the channel, dye diffuses out of the hydrogel and into the water. Channel sizes within the devices range from 300-1000 μ m to simulate vessels within the body. The properties of the PEG hydrogel were characterized by the extent of PEG-DA crosslinking, the swelling ratio, and the mesh size of the gel. The structure of the hydrogel was related to the UV exposure dosage and the initial water and dye content in the PEG-DA solution.
Three analysis techniques for measuring diffusion coefficients were used to quantify the diffusion of a solute in a hydrogel, including optical microscopy, characterization of device effluent, and NMR analyses. The optical microscopy technique analyzes images of the dye diffusion captured by a stereomicroscope to generate dye concentration v. position profiles. The data was fit to a diffusion model to determine diffusion coefficients and the dye release profile. Analysis of the device effluent was performed using ultraviolet-visible (UV/Vis) spectroscopy and high performance liquid chromatography (HPLC) to determine a short-time diffusion coefficient. Nuclear magnetic resonance (NMR) was used to determine a free diffusion coefficient of molecules in hydrogel without the effect of a concentration gradient.

Diffusion coefficients for methylene blue dye in PEG hydrogel calculated using the three analysis methods all agree well. It was determined that utilizing a combination of the three techniques offers greater insight into molecular diffusion in hydrogels than employing each technique individually. The optical and effluent methods were determined to be very effective for visualizing the diffusion and screening solutes to confirm that the solute is not just able to move within the hydrogel, but actually elute as well.

Separately, to study uptake, the neat hydrogel is contacted by dye-laden fluid. Implications for using hydrogels for controlled drug delivery are presented.

See more of this Session: Mesd Poster Session
See more of this Group/Topical: Materials Engineering and Sciences Division

(618h) Glucose Sensitivity of Hydrogels Containing Phenylboronic Acid and a Prototype of Biosensor for the Diabete

Arum Kim, Biomedical Engineering, University of Minnesota, Minneapolis, MN and Ronald A. Siegel, Pharmaceutics, University of Minnesota, Minneapolis, MN

Hydrogels are three dimensional polymeric networks which swell in aqueous solution without dissolution. With manipulation of chemical structure hydrogels can be designed to undergo volume changes in response to environmental stimuli such as temperature, pH, and electric field. In particular, glucose-sensitive hydrogels that swell or shrink according to blood glucose concentration are very useful for developing a glucose sensor. We synthesized glucose-sensitive hydrogels by using methacrylamidophenylboronic acid (MPBA), acrylamide (AAm), with/without N,N'-dimethylaminopropyl methacrylamide (DMP). MPBA-co-AAm hydrogels bind sugar molecules such as fructose and glucose, and change their volume with change in sugar concentrations at higher pH values than pKa of MPBA, pH 8.8. DMP lowers pKa by interaction with MPBA moieties, allowing binds of sugar molecules to MPBA at physiological pH value. Swelling behaviors were studied by measuring changes in diameter of capillary-shaped hydrogels at different pH values and at different sugar concentrations. In addition, we conducted compression test to see relaxation modulus related crosslink density. This experiment confirmed that only glucose molecules could make reversible crosslinks with MPBA moieties. Furthermore, we developed a prototype of a glucose sensor based on a micro-sized MPBA-co-AAm hydrogel. PRINT (Particle Replication In Non-wetting Templates) technique was used for mass production of the micro-sized hydrogels that allow quick responses in diffusion based system. The prototype of a glucose sensor consisted of nanoporous aluminum oxide membrane, a piece of the hydrogel, and a pressure sensor. Swelling pressure changes with changes of sugar concentration were monitored by a voltage meter over time. This glucose-sensitive hydrogel based prototype would be applied to development of an implantable glucose sensor along with insulin delivery system.

See more of this Session: Mesd Poster Session
See more of this Group/Topical: Materials Engineering and Sciences Division

(618i) Relationship Between Dielectric Properties and Wear Processes In Nanocomposites: Detecting Wear-Related Damage In Joint Replacement Polymer Materials

Tian Liu[1], Weston Wood[1], Wei-Hong Zhong[1], Md Mehedi Hasan[2] and Annie X.W. Tangpong[2], (1) School of Mechanical and Materials Engineering, Washington State University, Pullman, WA, (2) Department of Mechanical Engineering, North Dakota State University, Fargo, ND

ABSTRACT

Weexamined the correlation of wear effects with dielectric properties of carbonnanofibers (CNFs; untreated and organosilane-treated)-reinforced high-densitypolyethylene (HDPE) composites. Wear testing for the nanocomposites over up to120 hr was carried out, and then, dielectric permittivity and dielectric lossfactor of the polymer composites with the increased wear time were studied.Scanning electron microscope and optical microscope observations were made toanalyze the microstructure features of the nanocomposites. The results revealthat there exist approximate linear relationships of permittivity with wearcoefficient for the nanocomposites. Composites containing silanized CNFs withthe sufficiently thick coating exhibited high wear resistance. The change inpermittivity was more sensitive to the increased wear coefficient for thenanocomposites with lower wear resistance. This work provides potential forfurther research on the application of dielectric signals to detect the effectsof wear process on lifetime of polymeric materials.

See more of this Session: Mesd Poster Session
See more of this Group/Topical: Materials Engineering and Sciences Division

(618j) Synthesis and Characterization of Phase Transition of Aqueous Poly-(N-isopropylacrylamide-ran-acrylic acid) Solutions As a Function of pH

Isha Koonar and Ronald A. Siegel, Pharmaceutics, University of Minnesota, Minneapolis, MN

The synthesis and characterization of a random copolymer of N-isopropylacrylamide and acrylic acid (pNIPAAm-ran-AA) has been presented. The polymer is synthesized by RAFT (radical addition fragmentation chain transfer) polymerization. Liquid state 1H-NMR is used to determine extent of monomer conversion. Molecular weight and weight distribution has been estimated by MALDI-MS. The polymer molecular weights agree with those predicted and have narrow weight distributions with PDIs less than 1.1. Copolymer composition has been determined via estimation of acrylic acid content by potentiometric titration. The estimated copolymer composition matches fairly well with the feed composition. RAFT polymerization thus provides good control over polymer molecular weight and weight distribution. Also, the monomer incorporation is governed by the relative abundance of monomers in the feed.
NIPAAm undergoes an inverse phase solubility transition with a rise in temperature at 32 degrees C. This transition for NIPAAm by itself, is pH insensitive but polymerization with an ionizable comonomer causes the transition to become pH-dependent. The effect of acrylic acid, on polymer critical solution temperature has been studied by transmittance measurements on samples varying in copolymer composition and degree of ionization (a). The lower critical solution temperature (LCST) of pNIPAAm-ran-AA increases with increasing acrylic acid content and ionization. Also, the rate of change of LCST with acrylic acid ionization is higher for samples containing higher amounts of acrylic acid. This suggests that the phase transition of pNIPAAm-ran-AA can be set to a desired temperature by changing solution pH and the degree of pH dependence can be modified by changing the acrylic acid incorporation in the copolymer.

See more of this Session: Mesd Poster Session
See more of this Group/Topical: Materials Engineering and Sciences Division

(618k) Drug Delivery Nanoparticles Modified by Zwitterionic Polymers

Lei Zhang[1], Zhiqiang Cao[2] and Shaoyi Jiang[1], (1)Chemical Engineering, University of Washington, Seattle, WA, (2)Department of Chemical Engineering, University of Washington, Seattle, WA

Zwitterionic polymers such as poly(carboxybetaine) (PCB) have shown excellent stealth and functioanlization properties and are potential alternatives to poly(ethylene glycol) (PEG) for a broad range of applications due to the instability of PEG. However, it is very difficult to combine PCB to any hydrophobic domains due to its super-hydrophilic property. This work introduces a generic method to covalently attach super-hydrophilic PCB to hydrophobic moieties by adopting the hydrolysable hydrophobic precursor of zwitterionic PCB. Base on this method, poly(lactic-co-glycolic acid) (PLGA)-PCB block copolymers were synthesized. We further demonstrate that drug delivery nanoparticles (NPs) assembled from such "sharp contrast" macromolecules exhibit unusual properties that are not found in any of current NP systems. Other types of NPs modified by PCB will also be discussed.

See more of this Session: Mesd Poster Session
See more of this Group/Topical: Materials Engineering and Sciences Division

CONTROLLED SYNTHESIS OF GOLD BRANCHED AND POROUS PLATES IN POLYETHYLENEGLYCOL SOLUTIONS

*Chunrong Wang**[1], *Jinrong Liang*[1] *and Yun Fang*[2]

[1]*School of Food Science and Engineering, Jiangxi Agricultural University, Nanchang, Jiangxi, P. R. China*

[2] *School of Chemical and Material Engineering, Jiangnan University, Wuxi, Jiangsu, P. R. China*

Abstract. Gold branched and porous plates were prepared by the reduction of $HAuCl_4$ in the solution of polyethyleneglycol (PEG). PEG acted as both reducer and protective agent in reacted solutions. The crystal structures and the influence of PEG on the growth of gold branched and porous plates have been investigated by X-ray diffraction (XRD), transmittance electron microscopy (TEM) and UV-vis spectra. The gold branched plates had polygonal head with tortuous tail (even above 3 μm). The gold porous plates prepared had equivalent diameter around 2.5 μm.

Keywords: porous plate; gold; PEG

Introduction

In the past few years, metal nanostuctures have been the focus of intensive research due to their novel electronic, optical, magnetic, thermal, catalytic, and other properties that are distinctly different from their bulk counterparts, and considerable attention from both fundamental and applied research has been paid to synthesizing and characterizing metal nanostructures are closely related to their size and shape.1-3In these metal nanostructures, special anisotropic noble metal nanoparticles has attracted lots of science researchers because of their dramatic effect on the electric, optical, and catalytic application3-4.

To prepare 1-D nanostructures is a challenging task, especially for structurally isotropic metals such as gold with face-centered cubic (fcc) symmetry. There have been several recent reports describing differently shaped gold nanostructures including nanowire, nanocubes, nanoplates, nanobelts and nanorings made through various chemical approaches. For preparing these nanostructures, macromolecules method has been proved to be effective5-6.

PEG is a nontoxic macromolecule can be used to synthesizing nanostructures7-8. In this paper, we attempted to use the PEG to synthesize gold nanostructures.

Experimental section

Chemicals

Hydrogen tetracholoaurate ($HAuCl_4 \cdot 4H_2O$) and PEG (Mw=8000) were purchased from China Medicine (Group) Shanghai Chemical Reagent Corporation. All reagents were of analytical grade and used as received without further purification. The water used was deionized water.

Preparation of gold nanoparticles

The water solution containing the reagents ($HAuCl_4$ and PEG) was placed in a 100 mL round-bottomed flask in a water bath maintained at a constant temperature at 50±1 °C for 20 h. The products were separated by centrifugation (TGL-16G, Anting Scientific Instrument Factory), washed with deionized water several times. Optical spectra were recorded using a SPECORD200 spectrophotometer (Analytik Jena AG, Germany). TEM images and electron diffraction patterns (ED) were recorded on JEM-2010 microscopy operated at 200 kV and H-600 microscopy operated at 100 kV. X-ray powder diffraction (XRD) patterns were recorded using a Rigaku D / max-B X-ray diffractometer with high-intensity Cu Kα radiation (λ=1.54178Å) and a graphite monochromator.

Results and Discuss

In this work, Au nanostructures were synthesized by reducing $HAuCl_4$ with PEG in water at 50 °C. Compared to nonaqueous systems, water-based synthesis can offer an environmentally benign route to the production of metal nanoparticles because it does not involve toxic organic solvents.

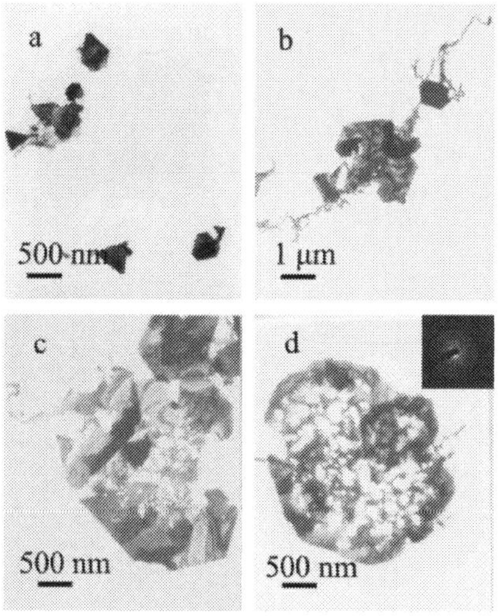

Figure 1 TEM images of gold nanostrurctures

Figure 1 shows TEM images of a sample that was prepared in a solution containing 0.2 mM

$HAuCl_4$ and 1.25 mM PEG at 50 °C. Under this reaction condition, the resulting nanostructures changed from ~500 nanoplates (Figure 1a), branched plates with a remarkably enlarged polygonal heads and tortuous tail (even above 3 μm) (Figure 1b), to porous microplates equivalent diameter around 2.5 μm (Figure 1c and d) as the reaction proceeding. The reaction time is 5 h, 10 h, 20 h respectively. The electron diffraction of gold poruous plate (inset of the Figure 1d) shows the diffraction rings.

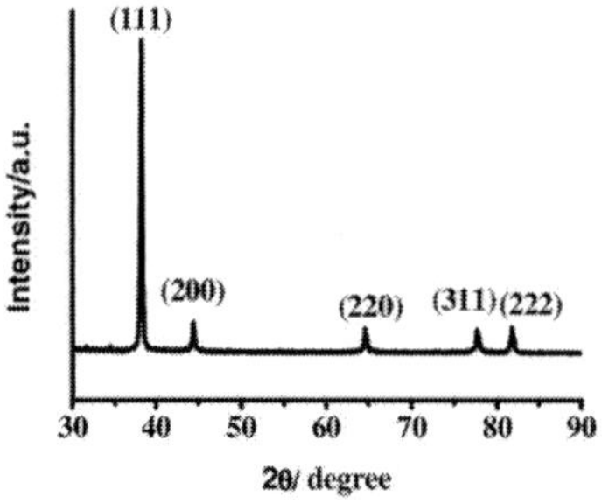

Figure 2 XRD pattern of gold porous plates

Figure 2 shows the XRD pattern of gold porous plates. Judging from the XRD patterns, the products have very high crystallinity. It is worth noting that the ratio between the intensity of (111) and (200) peaks exhibits a relatively higher value than the standard file (the standard file value is 1.92), indicating the enrichment of {111} crystalline planes in the porous structure.

Gold nanoparticles dispersed in a liquid medium usually display a very UV-vis absorption spectrum. Figure3 shows the UV-vis absorption spectrum of gold porous plates. It is clearly that the spectrum above 700 nm has stronger absorption indicating there are large size gold structures formed.

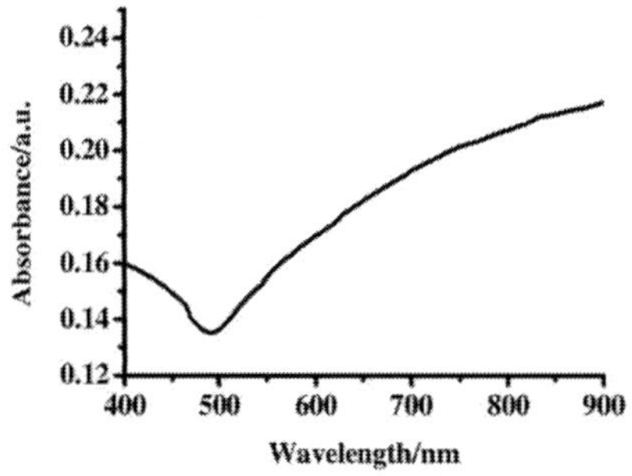

Figure 3 UV-vis spectra of gold hydrosols

Conclusions

Gold branched plates with polygonal head and tortuous tail (even above 3 μm) were prepared in PEG solution after 5h at 50 °C. As the reaction proceeding, gold porous plates with equivalent diameter around 2.5 μm were prepared in the PEG solution after 20 h. In the solution,

PEG acted as both reducer and protective agent. XRD pattern shows the enrichment of {111} crystalline planes in the porous structure.

Acknowledgement

We thank the National Science Foundation (31000820), the Jiangxi Provincial Office of Education Science Foundation (GJJ10109) and Jiangxi Agricultural University Doctor Start Foundation (2313) for funding this research.

References

1 Tan, Y. N.; Su, X.; Liu, E. T.; Thomsen, J. S.(2010), "Gold-Nanoparticle-Based Assay for Instantaneous Detection of Nuclear Hormone Receptor-Response Elements Interactions", *Analytical Chemistry*, 82, (7), pp2759-2765.

2. Qu, S.; Li, H.; Peng, T.; Gao, Y.; Qiu, J.; Zhu, C.(2009), "Optical nonlinearities from transverse plasmon resonance in gold nano-rods", *Materials Letters*, 58, (9), pp1427-1430.

3. Ko, K. D.; Kumar, A.; Fung, K. H.; Ambekar, R.; Liu, G. L.; Fang, N. X.; Toussaint, K. C.(2011), "Nonlinear Optical Response from Arrays of Au Bowtie Nanoantennas", *Nano Letters*, 11, (1), pp61-65.

4. Wu, B.; Boland, J. J.(2006), "Synthesis and dispersion of isolated high aspect ratio gold nanowires", *Journal of Colloid and Interface Science*, 303, (2), pp611-616.

5. Pardinas-Blanco, I.; Hoppe, C. E.; Pineiro-Redondo, Y.; Lopez-Quintela, M. A.; Rivas, J.(2008), "Formation of Gold Branched Plates in Diluted Solutions of Poly(vinylpyrrolidone) and Their Use for the Fabrication of Near-Infrared-Absorbing Films and Coatings", *Langmuir*, 24, (3), pp983-990.

6. Park, J.; Atobe, M.; Fuchigami, T.(2006), "Synthesis of multiple shapes of gold nanoparticles with controlled sizes in aqueous solution using ultrasound", *Ultrason. Sonochem.*, 13,, pp237-241.

7. Longenberger, L.; Mills, G.(1995), "Formation of Metal Particles in Aqueous Solutions by Reactions of Metal Complexes with Polymers", *Journal of Physical Chemisty*, 99, (2), pp475-478.

8. Leontidis, E.; Kyprianidou-Leodidou, T.; Caseri, W.; Kyriacou, K. C.(1999), "From Beads-on-a-String to Colloidal Aggregation: Novel Crystallization Phenomena in the PEO-SDS System", *Langmuir*, 15, (10), pp3381-3385.

(618m) Multi-Component Continuum-Based Single Particle Model for Tri-Calcium Silicate Hydration

Manohar Gottapu and Joseph J. Biernacki, Chemical Engineering, Tennessee Technological University, Cookeville, TN

Recent modeling efforts illustrate the utility of single particle continuum-based models for the elucidation of complex hydration processes such as that of tri-calcium silicate, e.g. alite, the major constituent of ordinary portland cement [1]. The current work demonstrates a single particle modeling with rigorous multi-ionic transport and particle proximity and volume filling effects based on ensemble averaged methods. An Avramian based representative volume concept is thus used to introduce the influence of neighboring particles. The model also includes multiple mechanisms; nucleation and growth, surface area reduction, and a two-step densification process [1, 2]. Such is able to explain observed hydration behaviors including dissolution, induction, acceleration, and deceleration.

References:

1. J. J. Biernacki and T. Xie "An Advanced Single Particle Model for C_3S and Alite Hydration", *J. Am. Ceram. Soc.*, (2011).
2. S. Bishnoi and K. L. Scrivener "Studying nucleation and growth kinetics of alite hydration using ," *Cem. Concr. Res.*, **39** 849-860 (2009).

See more of this Session: Mesd Poster Session
See more of this Group/Topical: Materials Engineering and Sciences Division

(618n) Versatile Preparation of Monodisperse Carbon Microspheres In a Simple Microfluidic Device

Lixiong Zhang, Sate Key Laboratory of Materials-oriented Chemical Engineering, Nanjing University of Technology, Nanjing, China

Versatile Preparation of Monodisperse CarbonMicrospheres in a Simple MicrofluidicDevice

Yichang Pan,Minhua Ju, Ya Liu, ChongqingWang, Lixiong Zhang,* and Nanping Xu

State Key Laboratoryof Materials-Oriented Chemical Engineering and Collegeof Chemistry and Chemical Engineering,Nanjing Universityof Technology, Nanjing210009, P. R. China. Fax: 86 25 8317 2263; Tel: 86 25 8317 2265; E-mail: lixiongzhang@yahoo.com

Microspherescomposed of polymeric or inorganic materials are of great interests because oftheir potential application as catalyst supports, carriers for cosmetics,drugs, protein and DNA delivery, electrodes, sensors and so on. They areprepared by various methods such as emulsion/micelles/vesicle polymerization,spray drying, template methods, and self°Ctemplate method. Recently, a newtechnique was developed to synthesize uniform-sized hollow microspheres such asnylon, biopolymer, organosilicon and TiO_2 using microfluidicemulsion technique via interfacial polymerization at the surface of themicrodroplet.

In this report, wefabricated hollow and solid poly(furfuryl alcohol) (PFA) microspheres withdiameters in the range from tens to several hundreds of micrometers by thephase transfer of FA and polymerization in the aqueous droplets usingmicrofluidic devices. Wefed an oil solution containing FA, fatty acid methyl ester, and a surfactant(span 80) and an sulfuric acid aqueous solution separately into the microfluidicdevices to form an aqueous solution microdroplets dispersed in the oil phase.FA in the oil phase would diffuse into the droplets, and consequently start topolymerize at the catalysis of sulfuric acid. Thus, a layer of PFA would beformed at the surface of aqueous droplet, resulting in formation of PFAmicrocapsules containing aqueous solution. The PFA microcapsules in-turn couldbe transformed to microporous carbon hollow microspheres by carbonization.Solid PFA microspheres could also be prepared by prolonging the reaction time.

We first preparedPFA microspheres with diameters of several hundreds of micrometers in amicrofluidic device assembled by inserting a syringe needle perpendicularlyinto a polyvinyl chloride (PVC) tube. By changing the flow rates of oil andaqueous solutions, residence time of aqueous microdroplets in the fluidicdevice, FA concentration in oil solution and curing temperature, the particlesize and wall thickness of PFA hollow microspheres can be easily adjusted. Inaddition, a magnetic precursor could be simply incorporated in PFAmicrospheres, leading to formation of magnetic carbon hollow microspheres. Byintroducing silica sols in the aqueous phase, PFA-SiO_2 compositemicrospheres were obtained, which were then converted into carbon-silicacomposite microspheres after carbonization and finally the carbon microspheresafter HF etching. The resultant carbon microspheres were mesoporous becausein-situ polymerized silica gel networks served as hard template, and their mesoporositydepended on the flow rate of the oil phase and the residence time of themicrodroplets in the device during the preparation of PFA-SiO_2microspheres. The highest mesoporous volume and mesoporosity of carbonmicrospheres reached 1.06 $cm^3.g^{-1}$and 82 %, respectively, and the mesopores were in the range of 2~20 nm. Mesoporous silica microspheres embeded with submicron-sized cavities could alsobe prepared by burning-off the carbon in the PFA- SiO_2 microspheres.

See more of this Session: Mesd Poster Session
See more of this Group/Topical: Materials Engineering and Sciences Division

(618o) 3D Multi-Scale Arrangement of Nanomaterials for Advanced Materials and Devices

Christine M. Andres and Nicholas A. Kotov, Chemical Engineering, University of Michigan, Ann Arbor, MI

Actual application of the highly desired properties of nanomaterials into functional devices requires radically new approaches to fabrication which achieve multiple combinations of diverse materials and complex multi-scale 3D architectures in a simple, low cost and versatile approach. Layer-by-layer assembly (LBL) is an excellent technique for the fabrication of an extensive variety of nanocomposites with exceptional properties unavailable by other means. However, while being low cost and simple, the technique significantly limits integration of such materials into functional devices. We have developed a toolbox of versatile methods to provide micro and macroscale control of LBL materials for the simple transition into advanced technologies. As the ability to precisely tune material properties is critical for successful development of novel and improved functionalities, the developed LBL techniques based on inkjet printing, elastomeric molds and colloidal crystal templates, provide systematic control of chemical and structural features from the molecular scale through the macroscale. In addition, the morphological control of the techniques allows for uniquely high contents of nanophase material as well as uniform composites or gradient structures to be integrated with traditional fabrication techniques or fabricated into freestanding, 3D functional structures.

See more of this Session: Mesd Poster Session
See more of this Group/Topical: Materials Engineering and Sciences Division

(618p) A Simple Mechanical Evaluation of Polymer Encapsulant Toughness

Preejith Ambuken[1], Holly A. Stretz[2], Hardik Patel[3] and Dustin Cannon[3], (1)Chemical Engineering, Tennessee Technological University, Cookeville, TN, (2)Department of Chemical Engineering, Tennessee Technological University, Cookeville, TN, (3)Dept. of Chemical Engineering, Tennessee Technological University, Cookeville, TN

A number of commercial substances from foods to drugs to pesticides are available as micro-encapsulated formulations in today's markets to provide, for example, for slow/sensitive-release or protection against post-processing degradation (UV protection being one example). The polymer encapsulant cannot perform its function to release or protect the host substance, however, if during a spraying process or similar downstream operation the polymeric encapsulant's mechanical integrity is breached. The authors report a statistical method for determining the toughness of the encapsulated material using a combination of constant shear rate rheometery and light microscopy. Breakage rate was found to correlate linearly with shear rate for water-based formulations.

See more of this Session: Mesd Poster Session
See more of this Group/Topical: Materials Engineering and Sciences Division

(618q) New Self-Healing Coatings

Hossein Birjandi Nejad, Biomedical and Chemical Engineering, Syracuse University, Syracuse, NY, Xiaofan Luo, Flow Polymers, LLC, Celeveland, OH and Patrick Mather, Biomedical and Chemical Engineering Department, Syracuse University, Syracuse, NY

Most self-healing materials to date show limited healability and high material/fabrication process costs. In our group, we have developed thermally triggered self-healing coating materials in which crack-closing occurs through a shape memory effect of a thermoset matrix and crack re-bonding occurs by diffusion of an interpenetrating thermoplastic component. In order to apply the coatings on metal substrates, two methods were used. In the first method, poly(e-caprolactone) (PCL) was electrospun onto the substrate, yielding a fiber web, and then an epoxy copolymer formulation (DGEBA/NGDE/ Jeffamine D-230) was spin-coated onto the fiber web. Following epoxy cure, the interpenetrating composite coating exhibits the described self-healing characteristics. Our second method utilized polymerization-induced phase separation (PIPS) of epoxy and PCL, yielding a distinct morphology also amenable to self-healing. Phase separation kinetics and blend morphology was investigated using optical microscopy, turbidity measurement, and differential scanning calorimetry. For both approaches, structural self-healing (crack closure) was observed using microscopy, while functional recovery (corrosion resistance restoration) was investigated using linear sweep voltammetry on samples exposed to NaCl/ solution. Both approaches led to excellent structural and functional self-healing. From potentiostat results, corrosion current density and consequently, corrosion rate was calculated. Results indicate enhanced corrosion density for healed samples. Moreover, self-healing during immersion in water was proved to be effective, indicating applicability to naval and biomedical applications.

Acknowledgment: *This project is sponsored by DOE through collaboration with NEI Corporation.*

See more of this Session: Mesd Poster Session
See more of this Group/Topical: Materials Engineering and Sciences Division

(618r) Mechanical Characterization of Friction and Wear Properties of Nanocomposites

Songbo Xu[1], **Aydar Akchurin**[1], **Annie X.W. Tangpong**[1], Iskander Akhatov[1], Tian Liu[2], Weston Wood[2] and Wei-Hong Zhong[2], (1)Department of Mechanical Engineering, North Dakota State University, Fargo, ND, (2)School of Mechanical and Materials Engineering, Washington State University, Pullman, WA

Mechanical characterization of friction and wear properties of nanocomposites

Songbo Xu[1], Aydar Akchurin[1], Annie X.W. Tangpong[1], Iskander Akhatov[1], Tian Liu[2], Weston Wood[2] and Wei-Hong Zhong[2]

[1]Department of Mechanical Engineering

North Dakota State University, Fargo, ND 58108

[2]School of Mechanical and Materials Engineering

Washington State University, Pullman, WA 99164

With the average lifespan of artificial joint replacements being about 20 years, there is a critical need for inventing new materials that will increase the service life of those joints. Carbon nanofiber (CNF) reinforced composites offer the promise as materials with less friction and lower wear rate compared to pure polymers. High density polyethylene (HDPE) nanocomposite samples were fabricated with various weight concentrations of carbon nanofibers (0.5%, 1% and 3%) that were treated with different silane coating thicknesses through melt-mixing and compressive processing. Wear and friction tests were performed on a pin-on-disc tribometer under phosphate buffered saline lubricated condition, as well as the dry condition. The coefficients of friction of the nanocomposites were found to decrease significantly compared to the pure HDPE. The method of calculating the wear rate was discussed and the wear rates of the samples were compared. The temperature change of the sample surfaces was monitored during the dry wear testing process and the results contribute to understanding of the thermal properties of CNF-reinforced nanocomposites.

See more of this Session: Mesd Poster Session
See more of this Group/Topical: Materials Engineering and Sciences Division

(618s) Electrochromical Window and Spectroelectrochemical Study of Diphenylamine End-Group Polymer/WO3 Nanoparticles Composite Films

Yunfeng Li[1], Vivek Iyer[1], Rahul Patil[2], Jiahua Zhu[1], Xi Zhang[1], Suying Wei[3] and Zhanhu Guo[1], (1) Integrated Composites Laboratory (ICL), Dan F Smith Department of Chemical Engineering, Lamar University, Beaumont, TX, (2)Integrated Composites Laboratory (ICL), Dan F. Smith Department of Chemical Engineering, Lamar University, Beaumont, TX, (3)Chemistry and Biochemistry, Lamar University, Beaumont, TX

Polymer nanocomposite films incorporated with WO3 nanoparticles were electropolymerized through monomer and WO_3 suspended solution on Indium Tin Oxide (ITO) coated glass. The monomer has a structural motif of R-X-R, with R= Diphenylamine end-group and X=Naphthalene diimide group. The UV-Visible incorporated electrochemical spectroscopies coupled with amperometry were also studied on the composite films as they display different colors under different applied potentials in the range of 0-2 V vs. Ag/AgCl. These films were also characterized by FTIR, and UV-Vis spectroscopies, the scanning electron microscopy (SEM). It was shown that many of these WO_3 nanoparticless were uniformly distributed in the polymeric matrix by microscopy and SEM. An air-stable Electrochromical window was also assembled and obtained by a home-made electrochemical cell and it shows stable electrochromis even after 100 times scan.

See more of this Session: Mesd Poster Session
See more of this Group/Topical: Materials Engineering and Sciences Division

(618t) Synthesis and Properties of Polyurethane-Vermiculite Nanocomposites

Yuqiang Qian[1], Andreas Stein[1] and Christopher W. Macosko[2], (1)Chemistry, University of Minnesota, Minneapolis, MN, (2)Chemical Engineering and Materials Science, University of Minnesota, Minneapolis, MN

Natural vermiculite was modified by cation exchange with long-chain quaternary alkylammonium salts and then dispersed in a polyether-based polyol suitable for elastomeric polyurethane synthesis. The dispersions were evaluated by X-ray scattering and rheology. Polyol was intercalated into the vermiculite interlayer space and a large interlayer spacing of ~90 Å was achieved. Significant shear thinning behavior was also observed. In polyurethane composites, clay platelets were extensively intercalated or exfoliated. The composites showed a >270% increase in tensile modulus, >60% increase in tensile strength, and a 30% reduction in N2 permeability with a loading of 5.3 wt% clay in polyurethane.

See more of this Session: Mesd Poster Session
See more of this Group/Topical: Materials Engineering and Sciences Division

(618u) Synthetic Process Engineered Conductive Polyaniline Nanostructures

Xi Zhang[1], Jiahua Zhu[1], Suying Wei[2] and Zhanhu Guo[1], (1)Integrated Composites Laboratory (ICL), Dan F Smith Department of Chemical Engineering, Lamar University, Beaumont, TX, (2)Chemistry and Biochemistry, Lamar University, Beaumont, TX

Conductive polymer nanocomposites have a broad range of applications. There are two different categories. One is the conductive fillers percolated insulating polymer system. The other is the conductive polymer based nanocomposites. In this presentation, we demonstrated that the conductive polymers can be tuned by varying the synthetic parameters for various morphologies. The related nanocomposites filled with WO_3 will be included. The unique negative permittivity is presented. The application of the conducive polymer filler in fabricating epoxy nanocomposite system is discussed as well.

See more of this Session: Mesd Poster Session
See more of this Group/Topical: Materials Engineering and Sciences Division

(618v) POSS-Sorbitol Interactions and Their Effects On the Reinforcement of Polypropylene Spun Fiber

Sayantan Roy, Byoung J. Lee and Sadhan C. Jana, Department of Polymer Engineering, University of Akron, Akron, OH

Templated, self assembly of molecules polyhedral oligomeric silsesquioxane (POSS) was used to obtain polypropylene (PP) nanocomposite fibers. A nucleating agent 1,3:2,4-dibenzylidenesorbitol (DBS) compatible with phenyl silanol-POSS promoted dispersion of POSS molecules and provided templates for self-assembly into cylindrical nanoparticles having close to 100 nm in diameter and 200 nm in length in the spun fibers. It was found that at the optimum combination of phenyl-POSS and DBS, PP can be spun into fibers with close to 40% reduction in diameters compared to unfilled PP and 60% improvement in tensile strength and modulus, and 100% improvement in yield stress. Optimum formulation was established with 1% of DBS and 5% of tri-silanol phenyl-POSS (Tri-POSS), and 1% of DBS and 2% of tetra-silanol phenyl-POSS (Tetra-POSS). Rheological studies revealed phenyl silanol-POSS prevented fibrillation of DBS. The interactions between silanol POSS and DBS was investigated using DSC, FT-IR, XRD, NMR and mass spectrometry. The mixtures of tri-POSS and tetra-POSS with DBS when heated at the processing temperature of 200°C, formed a transparent liquid whereas tri-silanol cyclopentyl POSS (cyclo-POSS) failed to do so. The transparent liquid was amorphous as confirmed by XRD with a glass transition temperature above the room temperature as identified by DSC. The amorphous material was believed to be consisted of supramolecular complexes which failed to form crystalline structures. The cylindrical shaped nanoparticles as found in spun fibers were believed to have resulted from the stretching of this liquid complex. The compounds formed between cyclo-POSS and DBS were crystalline in nature as confirmed from XRD. No chemical reaction occurred between DBS and POSS as identified from the solid state ^{29}Si and ^{13}C NMR and complexes of different molecular ratios were identified from mass spectrometry (Electrospray Ionization) study. Only complexes with 1:1 molecular ratio were identified between cyclo-POSS and DBS, whereas higher molecular ratio complexes were observed for tri-POSS and tetra-POSS with DBS. From the above observations it can be concluded that although hydrogen bonding interactions was the prime force for the complex formations, intermolecular π- π stacking between the phenyl rings play a major role towards the formation of supramolecular complexes.

See more of this Session: Mesd Poster Session
See more of this Group/Topical: Materials Engineering and Sciences Division

(618w) Effect of Dual Templates On Texture Properties of Mesoporous Alumina Prepared by Spray Pyrolysis

Kyeong Youl Jung and Hye Ran Jang, Chemical Engineering, Kongju National University, Cheonan, South Korea

ABSTRACT

Porous alumina is one ofimportant engineering materials due to its catalytic, electronic, andbiomedical properties. Mostly, mesoporous alumina has been synthesized by aliquid-phase reaction method using surfactant micelles or organic additives asstructure-directing agents. In most applications, the mesoporous alumina shouldhave a high surface area, well-defined pore structure, and a narrow pore sizedistribution. So, many researchers have been devoted to synthesize high porousalumina with regularity in pore size. Recently, an aerosolroute using spray-drying and self-assembly was reported as a potential methodto prepare mesoporous particles with high surface area. Most of researchesusing the aerosol route are focusing on silica or metal/silica.

Inthis work, a conventional spray pyrolysis process was applied to preparemesoporous alumina particles with spherical shape. In special, various dualtemplates such as CTAB, CTAC, P123, CTAB/P123, and CTAC/P123 were used toinvestigate the changes in the texture properties of alumina. Fig.1 showsnitrogen-adsorption–desorption isotherms of alumina particlesprepared by the spray pyrolysis with the CTAB and CTAB/P123 mixture as atemplate. The prepared alumina shows a typical type IV isotherm. The BET surfacearea of the aluminaprepared by using only CTAB was 233 m^2/g. Also, the sample prepared fromthe dual template, CTAB/P123, had the BET surface area of 282m^2/g. In terms of pore size and pore volume, however, thealumina particles prepared from the dual template (CTAB/P123) were larger thanthe alumina prepared from using the only CTAB template. On the base of BJHdesorption isotherms, the pore size distribution was calculated and shown inFig. 1. When the only CTAB was used, a broad pore size distribution wasobserved compared with that for the alumina sample prepared from the CTAB/P123dual template. From this result, it was clear that using the dual template ishelpful to improve the pore properties of alumina particles when they are preparedby the spray pyrolysis. The TEM images of the mesoporous alumina prepared usingthe CTAB/P1223 template were shown in Fig. 2. Also,the SAXS result was inserted in Fig. 2. The prepared particles have sphericalmorphology and a clear SAXS peak indicating the presence of mesopores with auniform size which can be confirmed by the TEM image. As the same vein, weinvestigated the effect of different dual templates such as CTAC/P123 on thetexture properties. More details in analysis and discussion will be presented.

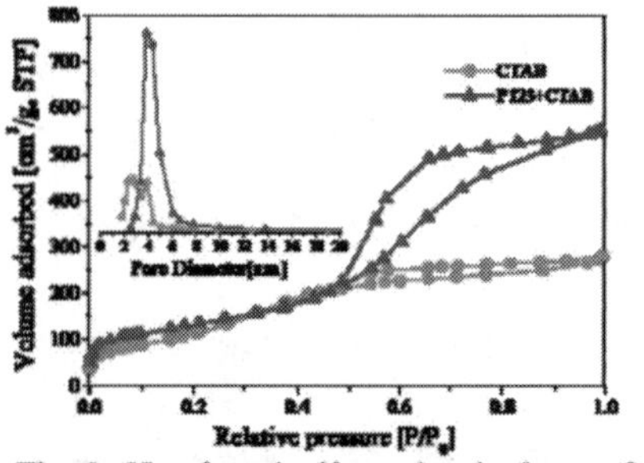

Fig. 1. N_2 adsorption/desorption isotherm of prepared alumina with different templates.

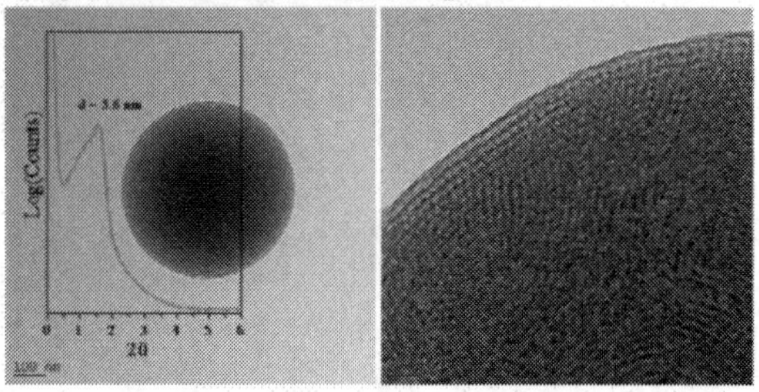

Fig. 2. TEM images of alumina particles prepared by spray pyrolysis different templates.

See more of this Session: Mesd Poster Session
See more of this Group/Topical: Materials Engineering and Sciences Division

(618x) Synthesis of Anisotropic Porous Titanium Dioxide Thin Films by Using Scaffolds of Magnetically Aligned Microparticles

Sergio Mendez, Hamad Nasoordeen and Daniel Hernandez, Chemical Engineering, California State University, Long Beach, Long Beach, CA

The motivation of this research project is to develop a simple, low-cost method to make titanium dioxide (TiO2) thin films with improved mechanical and electrical properties for use in dye-sensitized solar cells (DSSC). The aim is to make films with mechanical integrity, high porosity, and the ability to conduct electrons in a direction perpendicular to the substrate. The innovation is that we utilized magnetic particles to make anisotropic films of TiO2. First super-paramagnetic nanoparticles were embedded in ~0.5 micrometer polymer microspheres. In the presence of an external magnetic field, the particles form chain-like structures and, when on a surface, they form miniature pillars. Sol-gel chemistry was performed to deposit TiO2 around the polymer microspheres. Upon heating at sufficiently high temperature, the polymer was burned off, and the microstructure of the remaining TiO2 was almost uni-directional. The chemical process at the various steps was measured with reflectance infrared spectroscopy. To ascertain the uniformity and amount of cracks, these thin films were imaged with scanning electron microscopy as well as atomic force microscopy. The porosity of the TiO2 was determined with nitrogen absorption techniques. We expect that this method will result in DSSC's that can produce larger amounts of electrical power than those made with nanopowders.

See more of this Session: Mesd Poster Session
See more of this Group/Topical: Materials Engineering and Sciences Division

(618y) Preparation and Characterization of Conductive Polyaniline-Magnetite Nanocomposites

Hongbo Gu[1], Jiahua Zhu[1], Yudong Huang[2], Suying Wei[3] and Zhanhu Guo[1], (1)Integrated Composites Laboratory (ICL), Dan F Smith Department of Chemical Engineering, Lamar University, Beaumont, TX, (2)School of Chemical Engineering and Technology, State Key Laboratory of Urban Water Resource and Environment, Harbin Institute of Technology, Harbin, China, (3)Chemistry and Biochemistry, Lamar University, Beaumont, TX

In order to obtain conductive and magnetic polymer nanocomposites (PNCs) (1-4), polyaniline (PANI) nanocomposites reinforced with magnetite (Fe_3O_4) nanoparticles (NPs) have been fabricated using surface-initiated-polymerization (SIP) method. The PANI/Fe_3O_4 nanocomposites are characterized by Fourier transform infrared spectroscopy (FTIR). The thermal properties of PANI/ Fe_3O_4 nanocomposites are characterized by thermogravimetric analysis (TGA) and differential scanning calorimetry (DSC). Both Transmission electron microscopy (TEM) and Scanning electron microscope (SEM) are used to characterize the dispersion of Fe_3O_4 NPs and the morphology of PANI/ Fe_3O_4 nanocomposites. The effect of the Fe_3O_4 NPs on the crystallization behavior of the PANI during polymerization is studied. The electrical conductivity, dielectrical and magnetic properties of these nanocomposites are investigated and will be reported.

See more of this Session: Mesd Poster Session
See more of this Group/Topical: Materials Engineering and Sciences Division

(618z) Chemical and Mechanical Analysis of Green Composites Materials Made From Natural Fibers of Hemp, Flax and Hay

Kevin Cheng, Daniel Bahr, Roger C. Lo and Sergio Mendez, Chemical Engineering, California State University, Long Beach, Long Beach, CA

The goal of this project was to develop green composites that could potentially serve as an alternative to materials derived from forest and non-renewable sources. The two systems that were investigated are woven fabrics and chopped fibers imbedded in a resin. Flax and hemp woven fabrics were chosen because of their lightweight and exceptional mechanical properties. To make these woven composites withstand moist environments, a commercially available marine resin was utilized as the matrix. A hand lay-up process was utilized to make multi-layers of fabric imbedded in the resin. The chemical properties were measured with FTIR spectroscopy, and the morphology was imaged with scanning electron microscopy. The tensile, three-point bending, and edgewise compression strengths of these green textile/woven composites were measured using ASTM protocols. We also studied composites made from hay straw fibers. Various chemical treatments of the raw hay were investigated to enhance the moisture resistance and mechanical strength. Our vision for this work is to minimize deforestation by using sustainable agricultural resources to fabricate new green composites for the construction building industry.

See more of this Session: Mesd Poster Session
See more of this Group/Topical: Materials Engineering and Sciences Division

(643a) Comparison of Polymer-Microcrystalline Cellulose Composites with Polymer Nanocomposites Made with Carbon Nanotubes or Graphite: Processing by Solid-State Shear Pulverization

Philip J. Brunner[1], Amanda M. Flores[2], Katsuyuki Wakabayashi[2], Jun'ichi Masuda[3] and **John M. Torkelson**[4], (1)Materials Science and Engineering, Northwestern University, Evanston, IL, (2)Chemical and Biological Engineering, Northwestern University, Evanston, IL, (3)Department of Chemical and Biological Engineering, Northwestern University, Evanston, IL, (4)Departments of Chemical and Biological Engineering, Materials Science and Engineering, Northwestern University, Evanston, IL

Solid-state shear pulverization (SSSP) is used to make two types of composites. In the first case, composites of relatively well-dispersed microcrystalline cellulose (MCC) are produced using polypropylene (PP) or poly(lactic acid) (PLA) as the matrix. In the case of PP, 20 wt% addition of MCC yields a material with a 100% increase in modulus relative to neat PP. This composite can be easily processed via injection molding into products using approximately the same molding conditions as for neat PP. Such materials retain the "infinite hinge" characteristics of neat PP and may allow for the production of thinner parts than with neat PP. Composites of relatively well-dispersed MCC (at 1 wt % content) are made in PLA as a means of dramatically increasing the crystallization rate of PLA. The MCC/PLA interfaces serve as nucleation sites for crystallization, which can overcome the grudgingly slow crystallization rate of PLA during injection molding. Such semi-crystalline materials have high-temperature mechanical robust due to their semi-crystalline nature. These systems will be contrasted with well-dispersed PP-carbon nanotube and PP-graphite nanocomposites made by SSSP. Such materials yield major increases in mechanical properties, e.g., 50-100% increases in modulus at 1.0-2.5 wt% nanofiller content, and also provide for the possibility of multifunctionality in terms of both mechanical properties and electrical conductivity

See more of this Session: Composites III
See more of this Group/Topical: Materials Engineering and Sciences Division

(643b) Polymer Nanocomposites Comprised of Diphenylamine End-Group Polymer and WO3 Nanoparticles

Yunfeng Li[1], Jiahua Zhu[1], Xi Zhang[1], Rahul Patil[2], Suying Wei[3] and Zhanhu Guo[1], (1)Integrated Composites Laboratory (ICL), Dan F Smith Department of Chemical Engineering, Lamar University, Beaumont, TX, (2)Integrated Composites Laboratory (ICL), Dan F. Smith Department of Chemical Engineering, Lamar University, Beaumont, TX, (3)Chemistry and Biochemistry, Lamar University, Beaumont, TX

Polymer nanocomposites comprising of diphenylamine end-group polymer and WO_3 nanoparticles were fabricated electrochemically through its monomer and WO_3 suspended solution on an indium tin oxide (ITO) coated glass. The film growth was investigated by electrochemical methods such as cyclic voltammetry (CV). These films were also characterized by FTIR, and UV-Vis spectroscopies, and scanning electron microscopy (SEM). It was shown that many of these WO_3 nanoparticless were uniformly distributed in the polymeric matrix by microscopy and SEM. The electrochromical window was also assembled by a home-made electrochemical cell and found to be air-stable even after 100 times scan. The UV-Visible incorporated electrochemical spectroscopies were also studied on the composite films as they display different colors under different applied potentials in the range of 0-2 V vs. Ag/AgCl.

See more of this Session: Composites III
See more of this Group/Topical: Materials Engineering and Sciences Division

(643c) Significant Enhancements In Bio-Compatibility of a Soy-Polymer Blend Via a Facile Approach

Jianying Ji and Katie Zhong, School of Mechanical and Materials Engineering, Washington Stte University, Pullman, WA

Compatibility among the components is required for an effective polymer blending system. Though soy protein products have great potential to be applied in polymer blends, the poor interaction with polymers such as poly (ethylene glycol) (PEG) limits the broad applications. This study reveals a new facile approach to effectively enhancing the compatibility of SP with PEG by introducing a PE block, i.e. applying a co-polymer, polyethylene-*block*-poly (ethylene glycol) (PE-*b*-PEG), mixed with soy protein isolate (SPI) a SP product with 90%, SPI. This was shown to possess a capability for enhancing the compatibility of PEG/ SPI. In this paper, the polarized optical microscope (POM), scanning electron microscope (SEM), differential Scanning Calorimetry (DSC), and dielectric spectrum analyses are deployed to analyze the interaction between PEG and SPI. We found that the interaction between PEG and SPI enhanced significantly due to the introduction of the PE block. The strong interaction resulted in a reduction in the crystallinity of the PEG and decrease in the dielectric constant in copolymer/SPI blending. All these results indicate that good compatibility has been achieved by introducing a non-polar hydrophobic block.

See more of this Session: Composites III
See more of this Group/Topical: Materials Engineering and Sciences Division

(643d) Matrix and Glass-Resin Interface Optimization In Composite Materials for Structural Applications

Giacomo Perfetti, DSM Composite Resins, Zwolle, Netherlands

The research aims to understand the effect of some constituents (resin matrix, fiberglass reinforcement, resin-fibre interface) on the performances of vinyl ester glass reinforced composites used for structural applications. The effects of chemical modifications in the resin backbone, of the air release additives, of different curing systems on the fibre wet-out are investigated. Moreover the effect of different glass/sizing on the efficiency of resin-fibre interface is investigated by mechanical characterization and microscopy observation (a series of different techniques are proposed) of the laminates produced by vacuum infusion.

See more of this Session: Composites III
See more of this Group/Topical: Materials Engineering and Sciences Division

(643e) Nanocomposites of Multi-Walled Carbon Nanotube (MWCNT)/Polyimide

Lu Shao, Xiquan Cheng and Hongguang Sun, Harbin Institute of Technology, Harbin, China

Multi-walled carbon nanotubes (MWCNTs)/Polyimide(PI) nanocomposite films have been prepared by a simple solution casting method with the pre-modified MWCNTs. The dispersion of MWCNTs in PI matrix was observed by Atomic Force Microscopy (AFM) which illustrated the improved dispersion for acid treated MWCNTs in PI. The adding of MWCNTs in PI decreased the dispersive component of surface energy and increased the polar component of surface energy, which resulted in the decrement of film surface energy. Differential scanning calorimetry showed that the glass transition temperature of PI increased by about 4°C after the introduction of MWCNTs. This improvement was related to the better affinity between MWCNTs and PI matrix, which also resulted in the improvement of mechanical strength in MWCNTs/PI nanocomposites.

See more of this Session: Composites III
See more of this Group/Topical: Materials Engineering and Sciences Division

(643f) Microfluidic Synthesis of Non-Spherical Polymeric and Composite Particles

David Baah, Materials Science and Engineering Program, Tuskegee University, Tuskegee, AL, Julaunica Tigner, Mechanica Engineering, Tuskegee University, Tuskegee, AL and Tamara Floyd-Smith, Chemical Engineering, Tuskegee University, Tuskegee, AL

ABSTRACT

Colloidal particles in the nano- or micrometer size range are some of the most commonly encountered forms of materials in nature. Such monodisperse spherical and non-spherical colloidal suspensions and their assemblies have diverse applications in optic-electronics, photonics, catalysis, drug delivery, and field responsive rheological fluids. However, the synthesis of such highly monodispersed particles with tunable functionalities has been a great challenge. Several reports have documented the synthesis protocols of various morphologies of inorganic nano- and micro-sized particles. Heterophase polymerization methods including microemulsion and suspension polymerization have been developed and used to synthesize both polymeric particles and their composites. This polymerization approach does not offer control over the size and morphology and has been primarily used to synthesize spherical particles.

However, non-spherical particles have applications in a number of areas including liquid body armor applications due to potential shape-induced enhancements in shear thickening behavior. Microfluidics technology presents an alternative approach to synthesizing non-spherical particles with tunable functionalities. Using this approach, it is possible to synthesize colloidal based nano/microparticles and their composites with predetermined shapes usable in sensors, optical devices, microelectromechanical systems (MEMS) and field-responsive rheological fluids for liquid body armor applications.

The microfluidics method is dependent on the use of UV-curable prepolymer, with an appropriate photo-initiator. In this study, a photo mask patterned with pores defining the shape of the particles blocks the UV light while the pores allow the passage of light thereby cross linking the liquid prepolymer in the light path. The prepolymer solution flows through a microfluidic device fabricated by bonding poly (dimethylsiloxane), PDMS, molds to a glass slide. An Olympus BX51WI Microscope focuses the UV light to synthesize micron sized organic particles. Micron-sized organic particles in cubic, pentagonal, triangular, hexagonal, and tetragonal shapes were synthesized with this method using poly(ethylene glycol diacrylate) as the precursor. These particles were characterized, and it can be shown that the extruded depth of the particle is a function of the depth of the microfluidic channel. Also, the cross sectional area of the particle is dependent on the mask pore size and the magnification of the microscope objective. In addition to filled polymeric particles, composite particles and polymeric particles with vias have also been synthesized.

(643g) Synthetic Process Engineered Polyaniline Nanostructures

Xi Zhang[1], Jiahua Zhu[1], Yunfeng Li[1], Suying Wei[2] and Zhanhu Guo[1], (1)Integrated Composites Laboratory (ICL), Dan F Smith Department of Chemical Engineering, Lamar University, Beaumont, TX, (2)Chemistry and Biochemistry, Lamar University, Beaumont, TX

Polyaniline as a unique conjugated polymer has attracted much interest due to its tunable doping level and its wide applications such as gas sensor, anticorrosion, and toxic reduction in water. However, the application is largely restricted by its limited solubility in common solvents. Even worse, the lower thermal degradation temperature than the melting temperature prevents the melting process as used for thermoplastics. Researches on the nanostructure polyaniline significantly enhance the applications of polyaniline, however, the traditional synthetic ways to obtain polyaniline nanostructures are limited by the used template and the low yield cannot meet the demand for industrial-scale applications. Interfacial polymerization is a more convenient way to produce polyaniline nanofibers, without using any template or additional reagent. Here, three methods are used to synthesize polyaniline at different synthetic conditions, which affect the quality of polyaniline nanostructures. Polyaniline nanofibers obtained by interfacial polymerization is studied and compared with polyaniline prepared by ultrasonic stirring and polymer acid methods. For polyaniline nanofibers, the effects of reaction time, interfacial area size, scale and concentration of the reactant on the crystalline structure, thermal stability, morphology, conductivity and dielectrical property are systematically studied and will be reported.

See more of this Session: Composites III
See more of this Group/Topical: Materials Engineering and Sciences Division

(645a) Single Molecule Diffusion On Stimuli-Responsive Polymer Brush Surfaces

Yingxi Elaine Zhu, Chemical and Biomolecular Engineering, University of Notre Dame, Notre Dame, IN and Shengqin Wang, University of Notre Dame, Notre Dame, IN

How molecules and macromolecules diffuse at polymer surface remain inadequately understood. In this talk, we present a recent single-molecule spectroscopy study of fluorescent probes molecules at surface-tethered polymer brush surfaces. We focus on the dynamics of fluorescent molecules, Rhodamine110 (RG110) and protein macromolecules on thermo-responsive poly(N-isopropylacrylamide) (PNIPAM) brush thin films as a model system of tunable molecule-polymer surface interactions by varying temperature (T) across the LCST of PNIPAM brushes. The diffusion coefficient, *D* of RG110, measured by fluorescence correlation spectroscopy (FCS) at a single molecule resolution, decreases as increasing the solution T across the LCST (~32 ^{o}C) of PNIPAM brush thin films, due to the enhanced hydrophobic interaction at the molecule-polymer interfaces. However, it is surprising to observe the faster diffusion of RG110 on the self-assembled monolayer of octadecyltriethoxylsilane (OTE) than that on the PNIPAM brush surface, despite the stronger RG110 interaction with OTE than PNIPAM. Additionally, we observe that reduced PNIPAM brush thickness lead to further slowing down the diffusion dynamics of RG110 on PNIPAM brush surfaces at constant T. We thus speculate that the retarded diffusion process of small molecules on soft polymer brush surfaces, in comparison to the faster diffusion on hard surfaces, is a result of the coupling between molecule surface diffusion and the relaxation of wiggling polymer brush chains at the interface. Currently we further examine the surface diffusion of protein macromolecules on responsive polymer brushes.

See more of this Session: Diffusion In Polymers
See more of this Group/Topical: Materials Engineering and Sciences Division

(645b) Orders of Magnitude Decrease In Dye Diffusion In Nanoconfined Polymer Films: Fluorescence Nonradiative Energy Transfer Technique

Hui Deng[1], Manish K. Mundra[2] and John M. Torkelson[1], (1)Chemical and Biological Engineering, Northwestern University, Evanston, IL, (2)Materials Science and Engineering, Northwestern University, Evanston, IL

Upon nanoscale confinement, properties of supported polymer films, such as the glass transition temperature (Tg), often deviate from those of the bulk material. In confined, supported polymer films, a reduction in Tg has been associated with enhanced mobility near the air-polymer surface while an increase in Tg has been connected to attractive polymer-substrate interactions that are stronger than free-surface effects. Polymer properties other than Tg can also be impacted by confinement, such as mobility associated with diffusion. Compared to the many previous Tg-nanoconfinement studies, relatively few studies on diffusion in confined polymer systems have been reported.

Here we demonstrate the dramatic impact of confinement on the out-of-plane diffusion of dyes in supported polymer films. By employing a fluorescence nonradiative energy transfer/break-through time method, we have determined the diffusion coefficients of the dyes Disperse Red 1 (DR1), 9,10-bis (phenylethynyl)anthracene (BPEA) and decacyclene in polystyrene (PS) as a function of supported film thickness. When film thickness is reduced from bulk to 60 nm, up to a three-order-of-magnitude decrease in the diffusion coefficient of the dyes is observed in PS films at Tg,bulk + 3 K. There is a greater reduction in the diffusion coefficient of the smaller dyes upon confinement, with a factor of 10 decrease in the diffusion coefficient of decacyclene to a factor of 1000 decrease in the diffusion coefficient of DR1. The onset thicknesses for deviation from bulk diffusion coefficients are also greater for the smaller dyes, ranging from 150 nm for decacyclene to 450 nm for DR1. These film thicknesses are all above the onset thickness for Tg reduction in PS. Furthermore, while PS exhibits a reduction in Tg as film thickness decreases, associated with enhanced mobility, dye diffusion coefficients decrease upon confinement. Therefore, we conclude that there is no direct correlation between Tg-confinement effects and dye diffusion-confinement effects.

These results can be explained by the fact that Tg reflects the average, i.e., longer time side, of the relaxation time distribution of cooperative segmental mobility while probe diffusion reflects the shorter time side of the relaxation time distribution. We hypothesize that confinement suppresses the shorter time side of the relaxation distribution, resulting in the decrease in dye diffusion coefficients, and also the longer time side of the relaxation distribution, leading to the decrease in Tg. This agrees with our findings that the smallest dye exhibits the largest onset thickness for diffusion coefficient reduction and also the largest diffusion coefficient reduction upon confinement.

See more of this Session: Diffusion In Polymers
See more of this Group/Topical: Materials Engineering and Sciences Division

(645c) A Molecular Insight Into the Translocation Mechanism of Solvent Molecules In Amorphous Polymer Matrices

Li Xi and Bernhardt L. Trout, Department of Chemical Engineering, Massachusetts Institute of Technology, Cambridge, MA

Diffusion of small molecules in glassy amorphous polymer is known to be dominated by the so-called "hopping" motion. Penetrant molecules are usually constrained in small cavities within the polymer matrix for extended periods of time, before sudden translocation from one cavity to another takes place. The time scale between consecutive hopping events can be a few nanoseconds or even longer for certain penetrant-polymer systems, limiting classical atomistic molecular dynamics simulations to few realizations of such pathways. We apply transition path sampling techniques to harvest a large number of these events, and systematically investigate their mechanism. Effects of penetrant-polymer interactions, as well as polymer chain configuration and its fluctuations, on the translocation process will be quantitatively examined. The goal is to obtain a molecular level insight into the mechanism of penetrant movements in a polymer matrix. This knowledge could provide theoretical guidelines in various applications, such as solvent selection in the coating of polymer films and design of polymer membranes for liquid or gas separations.

See more of this Session: Diffusion In Polymers
See more of this Group/Topical: Materials Engineering and Sciences Division

(645d) Non-Equilibrium Sorption and Diffusion of Water In Polylactide: Experiments and Model Predictions

Eric M. Davis[1], Matteo Minelli[2], Marco Giacinti Baschetti[3], Giulio Sarti[4] and Yossef A. Elabd[1], (1) Chemical and Biological Engineering, Drexel University, Philadelphia, PA, (2)Department of Chemical Engineering, Mining and Environment Technology, Alma Mater Studiorum - University of Bologna, Bologna, Italy, (3)Department of Chemical Engineering, Mining and Environmental technology, University of Bologna, Bologna, Italy, (4)Chemical Engineering, Mining and Environmental Technology, University of Bologna, Bologna, Italy

Polylactide (PLA) is actively being pursued as a sustainable replacement to widely used commodity plastics because it is produced from renewable sources and readily degrades into benign natural small molecule products. For many applications (e.g., medical implants, packaging), an accurate assessment and fundamental understanding of water sorption and diffusion in PLA is of great interest. Previous investigations report low water solubilities in PLA (e.g, < 1 wt%). However, there are only a few reports on water sorption and diffusion in PLA and the values reported vary among these studies.

In this study, sorption and diffusion of pure water in amorphous PLA films were measured over a wide range of vapor activities and temperatures using several experimental techniques, including quartz spring microbalance (QSM), quartz crystal microbalance (QCM), and time-resolved Fourier transform infrared attenuated total reflectance (FTIR-ATR) spectroscopy. Non-Fickian behavior (diffusion-relaxation phenomena) was observed by all three techniques, while FTIR-ATR spectroscopy also provides information about the distribution of the sates of water and water transport mechanisms on a molecular level. This non-Fickian behavior highlights the non-equilibrium nature of the water-glassy polymer system and provides insight into the variability in the sorption isotherms reported in the literature.

In addition to experimental measurements, a purely predictive non-equilibrium lattice fluid (NELF) model was applied to predict the sorption isotherms of water vapor in amorphous PLA. The NELF model accurately predicts the experimental results over a wide range of temperatures and water vapor activities as high as 0.7. Interestingly, at different water vapor activities, different states of water were observed with FTIR-ATR spectroscopy. Water dimers were exclusively observed at activities < 0.7, while larger hydrogen-bound water clusters were noticed at activities > 0.7. The deviation between the NELF model prediction and the experimental results at high water vapor activities can be ascribed to the states of water, where the NELF model does not accurately account for the self-association of water.

See more of this Session: Diffusion In Polymers
See more of this Group/Topical: Materials Engineering and Sciences Division

(645e) Super Gas Barrier of All-Polymer Layer-by-Layer Assemblies

Jaime C. Grunlan, Department of Mechanical Engineering and Department of Chemical Engineering, Texas A&M University, College Station, TX and **You-Hao Yang**, Chemical Engineering, Texas A&M University, College Station, TX

Layer-by-layer assembly of cationic polyethylenimine (PEI) and anionic poly (acrylic acid) (PAA) was investigated with regard to pH of each polymer during deposition. Thickness and surface roughness can be altered dramatically due to the pH-responsive morphology of PEI and PAA. The oxygen permeability of an 8-bilayer, glutaraldehyde-crosslinked PEI/PAA film (305 nm thick) is below 3.2×10^{-21} $cm^3 \cdot cm/cm^2 \cdot s \cdot Pa$, which is believed to be the lowest permeability of an all-polymer system ever reported. It is believed that this transparent thin film has a structure akin to a highly interpenetrating network that contributes to its impressive oxygen barrier behavior. Even at 100% RH this film is able to maintain high gas barrier. This relatively simple recipe may be of use for a variety of packaging applications, including flexible electronics that require high flexibility and transparency.

See more of this Session: Diffusion In Polymers
See more of this Group/Topical: Materials Engineering and Sciences Division

(645f) Fundamental Studies of Gas Diffusivity and Solubility On Molecular Transport In Thermally Rearranged Polyimides

Zachary P. Smith[1], David Sanders[1], Claudio Ribeiro[1], Benny D. Freeman[1], Donald Paul[1], Ruilan Guo[2] and James E. McGrath[2], (1)Department of Chemical Engineering, The University of Texas at Austin, Austin, TX, (2)Department of Chemistry, Virginia Polytechnic Institute and State University, Blacksburg, VA

Polyimides containing *ortho*-functional hydroxyl groups undergo thermal rearrangement to form polybenzoxazoles between 350°C and 450°C. For certain separations, such as CO_2/CH_4 and O_2/N_2, these thermally-rearranged (TR) polymers have attractive combinations of permeability and selectivity. This study seeks to develop a more detailed understanding of gas diffusivity and solubility in TR polymers, using a model *ortho*-functional polyimide synthesized from 3,3'-dihydroxy-4,4'-diamino-biphenyl diamine and 2,2'-bis-(3,4-dicarboxyphenyl) hexafluoropropane dianhydride (HAB-6FDA).

By treating the polyimide at different thermal conversion temperatures between 350°C and 450°C for varying lengths of time, the sample was partially converted to its corresponding polybenzoxazole structure. The diffusivity of CO_2 in a series of partially converted TR polymers based on HAB-6FDA increased by over an order of magnitude as the conversion of HAB-6FDA increased. At the same extent of conversion, the solubility of CO_2 increased by a factor of approximately two. Thus, a substantial increase in CO_2 permeability was observed, driven primarily by an increase in gas diffusion coefficients.

Gas permeability and sorption were determined as a function of temperature and used to obtain activation energies of diffusion and enthalpies of sorption for CO_2 and CH_4. Further polymer characterization tests including thermogravimetric analysis coupled with mass spectrometry and differential scanning calorimetry were used to determine the regions of thermal rearrangement and thermal degradation, and to monitor the change in glass transition temperature at the onset of thermal rearrangement, respectively.

See more of this Session: Diffusion In Polymers
See more of this Group/Topical: Materials Engineering and Sciences Division

(645g) High Gas Barrier Polymer-Clay Nano Brick Wall Thin Films

Jaime C. Grunlan, Department of Mechanical Engineering and Department of Chemical Engineering, Texas A&M University, College Station, TX and Morgan Priolo, Materials Science and Engineering, Texas A&M University, College Station, TX

Thin films of sodium montmorillonite clay and weak polyelectrolytes were prepared by alternately dipping a PET substrate into four different dilute aqueous mixtures (polyethylenimine, poly(acrylic acid), polyethylenimine, and montmorrilonite clay). After depositing four of these quadlayers (QL), the resulting transparent film exhibits an oxygen transmission rate below the detection limit of commercial instrumentation (< 0.005 cm^3/m^2 day). This level of oxygen barrier, which is unprecedented for a clay-filled polymer composite, is believed to be due to a nano-brick wall microstructure comprised of completely exfoliated clay bricks in polymeric mortar and clay layer spacing on the order of tens of nanometers. This 4 QL film has a thickness of only 51 nm and an optical transparency of 95%. When multiplying OTR by film thickness to achieve oxygen permeability, we find that these films are a better barrier than SiO_x and also rival metalized plastic film. Some all-polymer assemblies with nearly equivalent gas barrier will also be mentioned. These thin film composites could be a microwaveable foil replacement for a variety of food packaging applications.

See more of this Session: Diffusion In Polymers
See more of this Group/Topical: Materials Engineering and Sciences Division

(651a) Invited: Hydrogel Systems to Evoke Physiological Cellular Programs

Jennie Leach, Chemical and Biochemical Engineering, University of Maryland Baltimore County, Baltimore, MD

Much of our current understanding of neurobiology relies on disrupted tissues, laboratory studies in artificial environments, and clinical observations. We hypothesize that the next generation of nerve repair therapies relies on the design of materials that better replicate the three-dimensional (3D) structure and physiology of native tissues. Our ongoing work aims to advance the understanding of neuronal response to 3D environments and to provide new improved materials and tools to study and repair neurons. This presentation will describe our efforts to delineate how ramifications of the 3D microenvironment (e.g., altered integrin signaling, diffusion of soluble molecules and matrix remodeling) impact neuronal survival and outgrowth.

Our findings indicate that 3D culture imposes changes in β1 integrin signaling that result in patterns of neurite outgrowth that better mimic *in vivo* morphology compared to culture on two-dimensional surfaces. To study this effect in more detail, we have established two novel tools to provide quantitative data in a physiologically relevant 3D system. First, we have developed a 3D culture system with controllable physical and biochemical material properties. Second, we have developed novel fluorescent oxygen-sensing microparticles to detect spatial and temporal changes in dissolved oxygen content. The microparticles demonstrate sensing performance comparable to traditional electrochemical probes, but are biocompatible and allow rapid, automated and non-invasive measurements local to cells and without consuming oxygen. Such enabling technologies pose vast potential for deciphering both the cellular morphogenic and molecular response to engineered scaffolds, and ultimately, the *in vivo* extracellular matrix.

See more of this Session: Hydrogel Biomaterials I
See more of this Group/Topical: Materials Engineering and Sciences Division

(651b) Multilayered, Self-Assembled Polycaprolactone and Gelatin-Chitosan Scaffold

Seokwon Pok and Jeffrey G. Jacot, Bioengineering, Rice university, Houston, TX

Multilayered, Self-Assembled Polycaprolactone and Gelatin-Chitosan Scaffold for Congenital Heart Defect Repair

Seokwon Pok[1], Jeffrey G. Jacot[1,2]

[1]Rice University, department of Bioengineering; [2]Texas Children's Hospital Division, of Congenital Heart Surgery

Introduction: Surgical repair of a ventricular septal defect (VSD) involves placement of a patch or baffle across the defect. Currently, surgeons use various forms of knitted polyester (most commonly Dacron), expanded polytetrafluoroethylene (such as Gore Tex) and autologous or bovine pericardium for VSD repair. However, these materials have significant drawbacks, including the inability to grow with the patient's heart, loss of mechanical strength over time and little ability to remodel, or regenerate as well as an increased risk of infection and aneurysm (Mirensky and Breuer 2008).

We have recently reported a novel process of generating flat PCL matrices in aqueous medium, which decreased hydrophobic surface properties and maintained high mechanical strength (Pok, Wallace et al. 2010), though hydrophobicity remained too high for effective cell attachment. While gelatin has excellent biocompatibility and bioresorbability and induced fibrous tissue in-growth in the right ventricular outflow tract of rats (Gelfoam, Pharmacia & Upjohn Co, Kalamazoo, Mich), it is too fragile for repair of VSD (DeLeon, LoCicero et al. 1986). One study demonstrated the feasibility of using a chitosan scaffold for the creation of contractile cardiac tissue constructs for use as a cardiac patch. However, the chitosan scaffolds did not promote the binding of cardiomyocytes (CMs) (Blan and Birla 2008). In this study, we developed a method of synthesizing flat, self-assembled PCL with immobilized natural polymers gelatin and chitosan, in order to promote functional interaction with primary cardiac cells, while maintaining sufficient mechanical properties to ensure feasible surgical handing, and function.

Methods and Preliminary Results: 10% (wt/v) PCL (80 kDa) solution was prepared in glacial acetic acid, and matrices formed by self-assembled method in aqueous media. Ratios of chitosan containing gelatin were also prepared. 1 M of glacial acetic acid was added to the gelatin-chitosan solution to enable immobilization with acetic acid based PCL matrices. A multilayered hydrogel was formed successfully using a previously described procedure with modifications (Madihally and Matthew 1999).

Briefly, formed PCL matrices were submerged into a gelatin-chitosan solution for pre-conditioning during one hour. Gelatin-chitosan solution poured into custom made Teflon well plates, and then pre-conditioned PCL matrices were sandwiched with the gelatin-chitosan solution then frozen immediately using dry ice. Samples were lyophilized overnight at – 30°C. Formed hydrogels showed no layer separation during re-hydration and neutralization process. Different MW of PCL were used to find optimal porous and degradation characteristics for neonatal rat cardiomyocytes (CMs). Blending with low MW of PCL (50:50 of 80 and 10 kDa) showed an increase in pore size and degradation rate compared to pure 80 kDa PCL. In addition, tensile stress and elastic modulus were decreased by a presence of 10 kDa PCL, but still showed strong mechanical properties (~1.5 MPa of tensile stress and ~2.0 MPa of elastic modulus). Cellular activity analysis of neonatal rat CMs on novel synthetic patches include colonization, survivability and in-growth are ongoing. In summary, mechanical properties of hydrogels were improved by a presence of self assembled PLC matrices. We highly expected that multilayered PCL matrices with gelatin-chitosan will have strong mechanical properties due to presence of PCL layer, and also they will have improved cellular activity due to gelatin-chitosan layer. Hence, this multilayered self assembled PCL with gelatin-chitosan hydrogels have significant potential in design of synthetic cardiac patch for VSD repair.

See more of this Session: Hydrogel Biomaterials I
See more of this Group/Topical: Materials Engineering and Sciences Division

(651c) PEG Based Hydrogels with Permeability Gradients for Enhanced Cell Function

Tugba Bal[1], Burcu Kepsutlu[1], Riza Kizilel[2] and **Seda Kizilel**[1], (1)Chemical and Biological Engineering, Ko University, Istanbul, Turkey, (2)Chemical and Biological Engineering, Koc University, Istanbul, Turkey

The development and synthesis of a bioartificial ECM-like network is currently challenged by incorporation of biomolecules with a gradient. To this end, we have investigated strategies to control incorporation of biomolecules within poly(ethylene glycol) (PEG) hydrogels. The biomimetic PEG-based hydrogels enable high level of control for the design of a bioartificial ECM-like network. In this study, we used biofunctional PEGDA hydrogels formed via bulk and surface initiated photopolymerization as environments to enhance cell survival and function. Swelling experiments were performed to quantify number average molecular weight between crosslinks (Mc) for hydrogels formed by both bulk and surface initiated polymerization. The effects of PEGDA molecular weight and concentration on hydrogel crosslink density and on insulin secreting MIN6 cell viability/functionality have also been investigated in these functional PEG hydrogel systems. These platforms allow for incorporation of ligands and growth factors with a gradient, and include covalently bonded cell adhesion sequences and specific growth factors to promote cell survival and function. Results demonstrate the significance of tailoring the permeability structure to enhance the survival and function of encapsulated insulin secreting cells.

See more of this Session: Hydrogel Biomaterials I
See more of this Group/Topical: Materials Engineering and Sciences Division

(651d) Peptide-Amphiphile Hydrogels Formed by Mechanical Shear for Use As Injectable Tissue Engineering Matrices

Katie Megley[1], Won Hyuk Suh[2], Seema Desai[3] and Matthew Tirrell[2], (1)The UC Berkeley - UCSF Graduate Program in Bioengineering, University of California, Berkeley, Berkeley, CA, (2) Bioengineering, University of California, Berkeley, Berkeley, CA, (3)Molecular and Cell Biology, University of California, Berkeley, Berkeley, CA

Shear induced, self assembly has been extensively studied using model surfactant systems which form long extended micelles at a critical shear rate. These polymer based extended micelles are able to entangle, giving the system viscoelastic properties. Here we present a peptide amphiphile, a short peptide sequence attached to a fatty acid tail, which transforms into extended worm-like micelles under shear force. Peptide amphiphiles are easily synthesized and modular in design, allowing a range of bioactive peptides to be introduced into the system. The particular peptide sequence used in this study is alanine (A) rich with interspersed lysine (K) residues, which gives it a strongly alpha helical secondary structure. The peptide amphiphile initially forms spherical micelles in aqueous solution and then undergoes a switch to form long cylindrical micelles when the correct shear force is applied. The cylindrical micelles entangle and create a gel with viscoelastic mechanical properties. The physical transition from spherical to worm-like micelle coincides with a secondary structure transition from alpha helix to beta sheet which is monitored using circular dichroism. As in the surfactant system, our peptide amphiphile solution initially behaves as a Newtonian fluid and then transitions to a shear thinning gel once a critical shear rate is surpassed. The resulting gel is self healing, and stable at 4°C on the order of weeks to months. The material properties can also be tuned to achieve the modulus of various biological tissues based on the environmental conditions such as salt concentration and content. Additionally we have shown that bioactive peptides, such as RGDS, can be added to the end of the base peptide group without affecting the modulus of the material. This material was tested for biocompatibility using a fibroblast cell line and was shown to support cell growth in 3D over a 3 day time period. The versatile nature of this system makes it an attractive material for use as an injectable tissue engineering matrix.

See more of this Session: Hydrogel Biomaterials I
See more of this Group/Topical: Materials Engineering and Sciences Division

(651e) Biomolecular Templated Assembly of Droplet-Derived Hydrogel Microtissues

Cheri Y. Li[1], David K. Wood[2], Caroline M. Hsu[1] and Sangeeta N. Bhatia[2], (1)Chemical Engineering, Massachusetts Institute of Technology, Cambridge, MA, (2)Health Sciences & Technology, Massachusetts Institute of Technology, Cambridge, MA

Paracrine and autocrine cell signaling are critical factors guiding tissue development and maintenance, and dysregulation of these cues contributes to the pathogenesis of diseased states such as cancer. Patterning multiple cell types is thus a critical step for engineering functional tissue [1], but current "top-down" approaches such as dielectrophoresis and photopatterning are challenging to scale-up for the construction of mesoscale tissues. On the other hand, "bottom-up" methods wherein small tissue building blocks are assembled into larger structures have potential for constructing multicellular tissues in a facile, scalable fashion. Synthetic microtissues comprised of cell-laden hydrogels in this size range represent appropriate fundamental building blocks of such bottom-up methods [2].

To specify the placement of many different microtissues relative to one another, we have developed a "bottom-up" approach for fabricating multicellular tissue constructs that utilizes specific biomolecular interactions to template the assembly of 3D cell-laden hydrogel microtissues. A flow focusing-generated emulsion of photopolymerizable prepolymer is used to produce 100 μm monodisperse microtissues at rates of 30 Hz (10^5/hr). Multiple cell types, including suspension and adherently cultured cells, can be encapsulated into the microtissues with high viability (~97%). We then use a surface encoding scheme to self-assemble microtissues "bottom-up" from a template that is defined using "top-down" techniques. The microtissues are derivatized with interactive functional groups using a biotin-streptavidin linkage to the polymer network, and are then assembled by specific binding onto patterned arrays. Using orthogonal interactions, we have achieved multiplexed patterning of multiple microtissue types with high binding efficiency and >90% patterning specificity. We have also demonstrated the ability to organize multicomponent constructs composed of epithelial and mesenchymal microtissues while preserving each cell type in a 3D microenvironment. The combination of high throughput microtissue generation with scalable surface-templated assembly offers the potential to dissect mechanisms of cell-cell interaction in three dimensions in healthy and diseased states as well as provide a framework for templated assembly of larger structures for implantation.

[1] S. March, E. E. Hui, G. H. Underhill, S. Khetani, and S. N. Bhatia, "Microenvironmental regulation of the sinusoidal endothelial cell phenotype in vitro," *Hepatology*, vol. 50, pp. 920-8, 2009.

[2] A. A. Chen, G. H. Underhill, and S. N. Bhatia, "Multiplexed, high-throughput analysis of 3D microtissue suspensions," *Integr Biol (Camb)*, vol. 2, pp. 517-27, 2010.

(651f) Inherent Mechanical Gradients In 3D Matrigel Cultures Influence Tumor Cell Behaviors

Shreyas S. Rao[1], Sarah Bentil[2], Jessica DeJesus[3], John Larison[1], Alex Hissong[4], Rebecca Dupaix[2], Atom Sarkar[5] and Jessica O. Winter[6], (1)Chemical and Biomolecular Engineering, The Ohio State University, Columbus, OH, (2)Department of Mechanical and Aerospace Engineering, The Ohio State University, Columbus, OH, (3)Department of Neurological Surgery, The Ohio State University, Columbus, OH, (4)Department of Biomedical Engineering, The Ohio State University, Columbus, OH, (5)Department of Neurosurgery and Laboratory for Nanomedicine, Geisinger Health System, Danville, PA, (6)Chemical and Biomolecular Engineering and Biomedical Engineering, the Ohio State University, Columbus, OH

It is widely appreciated that cells can sense stiffness of their environment and alter their response accordingly [1]. To examine the influence of stiffness on cell behaviors, hydrogels with independent and varying mechanical properties are required. Investigators have recently begun to incorporate stiffness gradients in hydrogel systems using photo patterning and microfluidic platforms [2]. Whereas most of these methods provide well defined gradients and control over material properties, they are usually time intensive often requiring several steps to generate gradients. Additionally, most are designed to examine cell behaviors in 2D cultures. Here, we exploit edge effects in existing 3D culture systems (gels supported on rigid substrates) to investigate the influence of stiffness gradients on tumor cells using Matrigel as a model hydrogel material. As a model tumor system, we investigated glioblastoma multiforme (GBM), an astrocyte-derived brain cancer and one of the most deadly forms of human cancer affecting ~22500 individuals in the United States annually. GBMs are highly infiltrative and inevitably recur both locally and distantly within the brain; median survival times are extremely low (~12-15 months) [3].

Whereas the bulk hydrogel modulus does not vary, in regions close to the rigid support (i.e., glass), edge effects create an inherent gradient in mechanical stiffness that varies between that of the support and the hydrogel. A finite element model was employed to confirm the presence of varying mechanical environments in 3D cultures. To investigate GBM response to these environments, patient derived GBMs (OSU-2) were encapsulated in different concentrations of Matrigel (40, 55, 70, 85 % v/v). In particular, we examined cell morphology and spreading, intracellular morphology (actin organization) and migration capacity of OSU-2 cells at the lowest gel positions and highest gel positions. OSU-2 cells close to the bottom of the substrate showed highly elongated, bipolar morphologies and had statistically different cell areas compared with those more distant from the rigid support, which showed mostly rounded morphologies (with short processes in some cases). This was also reflected in actin organization, with actin filaments being highly organized and forming stress fibers in cells close to the substrate whereas those nearer to the gel surface displayed poorly defined actin architecture. OSU-2 cells also migrated at significantly higher speeds close to the bottom of substrate in a mesenchymal fashion displaying long processes compared to those nearer to the gel surface that migrated more slowly while displaying rounded or ellipsoid cell bodies with short processes. This system is, by far, one of the simplest systems to study the influence of stiffness gradients on the behavior of tumor cells in a single 3D hydrogel system and also provides insights as to how far cells can "sense" their microenvironments in 3D.

References:

[1] D. E. Discher et al., Science. 2005. 310; 1139-43.

[2] S. Sant et al., The Canadian Journal of Chemical Engineering. 2010.88; 899-911.

[3] P. Y. Wen et al., N Engl J Med. 2008. 359 (5); 492-507.

See more of this Session: Hydrogel Biomaterials I
See more of this Group/Topical: Materials Engineering and Sciences Division

TUNING THE DEPENDENCY BETWEEN STIFFNESS AND PERMEABILITY OF CELL-ENCAPSULATING HYDROGEL WITH HYDROPHILIC PENDANT CHAINS

Chaenyung Cha, Jae Hyun Jeong, Hyunjoon Kong
University of Illinois, Urbana, IL

Introduction

Hydrogels have been increasingly studied as in vitro cell culture platforms and cell transplantation devices because they are structurally similar to natural extracellular matrices (ECM) and provide insoluble signals to regulate diverse cell phenotypes.[1,2] The mechanical stiffness of hydrogel acts as an important insoluble signal in regulating the phenotypes of cells that adhere to its surface.[3] The role of mechanical stiffness of hydrogel matrices on cell phenotypes has been largely studied using cells adherent to two dimensional (2D) hydrogel surfaces, even though cells in vivo are surrounded by three dimensional (3D) ECM.

Therefore, extensive efforts are now being made to understand the role of 3D cell microenvironments on cellular activities, usually by encapsulating cells in hydrogel matrices with varying chemical and mechanical properties. However, the effect of hydrogel stiffness on the cells cultured within its matrix is not well understood because of the intrinsic inverse dependency between the permeability and stiffness of hydrogels.[4] For example, most efforts to adjust the mechanical stiffness of a hydrogel are accompanied by changes in the permeability of the hydrogel, both of which potentially influence cellular activities. So it is difficult to discriminate between the individual and combined effects of matrix stiffness and permeability on cells cultured in a 3D matrix.

This study therefore presents an advanced biomaterial design strategy to decrease the inverse dependency between permeability and stiffness of a cell-encapsulating hydrogel. We hypothesize that the incorporation of pendant hydrophilic polymers in cross-linked networks will allow us to control the stiffness via concentration of pendant polymers while limiting changes in permeability. This hydrogel system would enable us to address the individual and combined effects of matrix stiffness and permeability on viability and gene expression levels of cells within a 3D matrix. This hypothesis was examined by cross-linking poly(ethylene glycol) diacrylate (PEGDA) and poly(ethylene glycol) monoacrylate (PEGMA) to form hydrogels, with PEGMA acting as a pendant polymer chain. The mass fraction, molecular weight, and end group of PEGMA were controlled to tune the inverse dependency between permeability and stiffness of the hydrogels and also to address the underlying mechanism. Pure PEGDA hydrogels with varying molecular weights, which show the same range of stiffness but a much greater range of permeability, were used as a control. Mechanical stiffness of the hydrogels was evaluated by measuring the elastic modulus, and permeability of the hydrogels was evaluated by measuring the swelling ratio and diffusion coefficient of biomacromolecules within each hydrogel. Finally, both PEGDA-PEGMA hydrogels and pure PEGDA hydrogels were used to uncover the effects of matrix stiffness and permeability on the viability and endogenous vascular endothelial growth factor (VEGF) expression levels of encapsulated cells.

Results

Mechanical Properties of Poly(ethylene glycol) (PEG) Hydrogels with Varying Numbers of Pendant PEG chains

Hydrogels with varying numbers of cross-links and pendant PEG chains were prepared by chemically cross-linking a mixture of PEG diacrylate (PEGDA) and PEG monoacrylate (PEGMA).

The mass percentage of PEGMA (Φ_{PEGMA}) in the PEGDA-PEGMA hydrogel was varied while keeping the total concentration constant at 20 wt % (**Figure 1a**). Separately, pure PEGDA hydrogels with varying numbers of cross-links were prepared by cross-linking PEGDA with different molecular weights ($M_{W,PEGDA}$) while keeping the PEGDA concentration constant at 20 wt % (**Figure 1b**). Increasing both Φ_{PEGMA} of PEGDA-PEGMA hydrogels and $M_{W,PEGDA}$ of pure PEGDA hydrogels led to the exponential decrease of the elastic modulus (E) from 41 to 3 kPa. The ranges of E attained with the PEGDA-PEGMA hydrogels and pure PEGDA hydrogels were equivalent.

(a)

THE GRAPHIC ON THIS PAGE WAS NOT AVAILABLE ON THE E-MEDIA VERSION OF THIS PAPER.

(b)

THE GRAPHIC ON THIS PAGE WAS NOT AVAILABLE ON THE E-MEDIA VERSION OF THIS PAPER.

Figure. 1. Schematic description of (a) PEGDA-PEGMA hydrogels with varying mass percentage of PEGMA (Φ_{PEGMA}), and (b) PEGDA hydrogels with varying M_W of PEGDA ($M_{W,PEGDA}$). In (a), increasing Φ_{PEGMA} of PEGDA-PEGMA hydrogels (from left to right) was expected to increase the number of pendant PEGMA chains, while reducing the number of cross-links. In (b), increasing $M_{W,PEGDA}$ of PEGDA hydrogels (from left to right) was expected to decrease the number of cross-links and increase the pore diameter.

In contrast, Φ_{PEGMA} of PEGDA-PEGMA hydrogels and $M_{W,PEGDA}$ of PEGDA hydrogels influenced the swelling ratio (Q) of PEG hydrogels differently. Increasing Φ_{PEGMA} of the PEGDA-PEGMA hydrogels from 0 to 50 led to a 1.8-fold increase in Q. However, increasing $M_{W,PEGDA}$ of pure PEGDA hydrogels from 3400 to 14000 g mol^{-1} resulted in a 2.8-fold increase in Q. These changes in Q related to the changes in E were fitted to a power law, described by $Q \propto E^{-m}$. The inverse dependency between the swelling ratio and the elastic modulus, represented by m, was smaller with the PEGDA-PEGMA hydrogel than with the pure PEGDA hydrogel. Permeability of hydrogels was further characterized by measuring diffusion coefficient of fluorescently labeled dextran. As with swelling ratio, increasing Φ_{PEGMA} of the PEGDA-PEGMA hydrogels from 0 to 50 only resulted in 1.15-fold increase in diffusion coefficient, whereas increasing $M_{W,PEGDA}$ of pure PEGDA hydrogels resulted in 1.3-fold increase in diffusion coefficient. This result demonstrates that by using PEGDA-PEGMA hydrogel system, the elastic modulus could be modulated while minimizing the change in permeability.

Effects of Length and Hydrophobicity of Pendant PEG Chains on Hydrogel Properties

The role of pendant PEG chains on the stiffness and permeability of PEGDA-PEGMA hydrogels was further explored by modulating the length and hydrophobicity of the pendant chains. First, the length of pendant PEG chains was modulated by varying the M_W of PEGMA from 750 to 5000 g mol^{-1}. The elastic moduli of the PEGDA-PEGMA hydrogels decreased with Φ_{PEGMA} regardless of M_W of PEGMA. However, the extent of decrease in elastic modulus became greater with increasing M_W of pendant PEG chains. In contrast, the swelling ratio of the hydrogels was only minimally influenced by the M_W of pendant PEG chains at any given Φ_{PEGMA}.

Next, the end groups of pendant PEG chains were modified with hydrophobic phenyl groups in order to examine the role of the hydrophobicity of pendant chains on the dependency between the permeability and stiffness of the hydrogel (**Figure 2**). Interestingly, increasing the fraction of pendant PEG chains substituted with phenyl groups ($\varphi_{PEG\text{-}Ph}$) while keeping the total number of pendant PEG chains constant (Φ_{PEGMA} at 50 %) resulted in an almost ten-fold increase in the elastic modulus and a decrease in the swelling ratio of the hydrogel by a factor of two.

THE GRAPHIC ON THIS PAGE WAS NOT AVAILABLE ON THE E-MEDIA VERSION OF THIS PAPER.

Figure 2. Schematic description of PEGDA-PEGMA hydrogels with increasing hydrophobicity of pendant PEG chains (left to right), by modifying the end groups of pendant PEG chains to present hydrophobic phenyl groups. The total number of pendant PEG chains were fixed (Φ_{PEGMA} at 50), while changing the fraction of pendant PEG chains with phenyl end groups ($\varphi_{PEG\text{-}Ph}$).

The hydrophobicity of hydrogels modulated with the amount of pendant PEG chains with hydrophobic functional groups incorporated into PEGDA-PEGMA hydrogels was analyzed by encapsulating pyrene within the hydrogels. Pyrene monomers are well known to associate with hydrophobic domains in the surrounding environment and generate a fluorescent emission spectrum which correlates with the hydrophobicity of the surrounding environment.[5] Increasing φ_{PEG-Ph} of PEGDA-PEGMA hydrogels resulted in the fluorescent emission spectrum of encapsulated pyrene with increasing hydrophobic characteristics. These results demonstrate that the chain length and hydrophobicity of the pendant PEG chain are important factors that can be used to control the dependency between elastic modulus and permeability of hydrogels.

Effects of Hydrogel Properties on Viability and Vascular Endothelial Growth Factor (VEGF) Expression of Encapsulated Cells

NIH3T3 fibroblasts were encapsulated in PEGDA-PEGMA hydrogels with varying Φ_{PEGMA}, and in PEGDA hydrogels with varying $M_{W,PEGDA}$ to study the effects of hydrogel stiffness on cell viability and VEGF expression level. Since PEGDA-PEGMA hydrogels and pure PEGDA hydrogels have the same range of elastic moduli but differ only in swelling ratios, the different cellular phenotypes between these two systems would reveal the individual and combined effects of stiffness and permeability on cellular activities.

First, the viability of cells was determined by staining the cells with MTT 24 hours after their encapsulation into hydrogels (**Figure 3**). In PEGDA-PEGMA hydrogels, the cell viability profile showed a biphasic dependence on the elastic modulus, displaying maximal viability at an elastic modulus of 26 kPa (Φ_{PEGMA} = 12.5). Above 10 kPa, the viability profile of pure PEGDA hydrogels was similar to that of PEGDA-PEGMA hydrogels, with the maximal viability value at 20 kPa ($M_{W,PEGDA}$ = 6000 g mol^{-1}). However, below 10 kPa, cell viability increased significantly with decreasing elastic modulus.

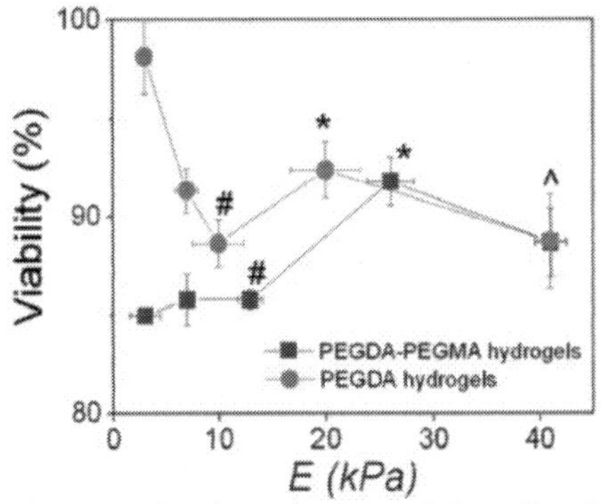

Figure 3. The viability of cells was related to elastic modulus (*E*) of PEGDA-PEGMA hydrogels with varying Φ_{PEGMA} (■) and pure PEGDA hydrogels with varying $M_{W,PEGDA}$ (●). ($p<0.05$ between * and ^, and also between * and #)

Next, cellular expression of vascular endothelial growth factor (VEGF) within the hydrogel was examined by measuring the concentration of VEGF in the cell culture media (**Figure 4**). Cells were stimulated to up-regulate VEGF by exposure to 12-O-tetradecanoylphorbol-13-acetate (TPA) for 24 hours. TPA is a potent activator of protein kinase C, known to stimulate VEGF expression in NIH3T3 fibroblasts.[6] The fibroblasts not exposed to TPA made minimal expression of VEGF. The measured VEGF expression level was normalized to the viability of cells to evaluate the VEGF expression by individual viable cell. In PEGDA-PEGMA hydrogels with varying Φ_{PEGMA}, the VEGF expression level displayed a biphasic curve, similar to the viability profile. The VEGF expression level was maximal at the elastic modulus of 13 kPa (Φ_{PEGMA} = 25 %). The VEGF expression level of cells encapsulated in pure PEGDA hydrogels with varying $M_{W,PEGDA}$ was also maximal when elastic modulus was 10 kPa ($M_{W,PEGDA}$ = 8000 g mol^{-1}). However, the biphasic dependency was much smaller than that of PEGDA-PEGMA hydrogels.

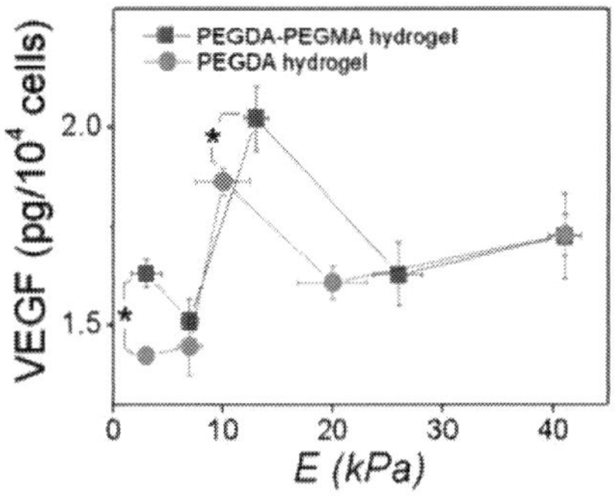

Figure 4. VEGF expression level from encapsulated fibroblasts was related to the elastic modulus (*E*) of PEGDA-PEGMA hydrogels with varying (■) and pure PEGDA hydrogel with varying $M_{W,PEGDA}$ (●).(*P<0.05)

Conclusion

The results of this study demonstrate an advanced biomaterial design strategy which allows us to move one step closer to the independent control of stiffness and permeability in a cell-encapsulating hydrogel. Hydrophilic pendant polymer chains incorporated in the hydrogel networks enable us to decrease the inverse dependency between stiffness and permeability of the hydrogels conventionally used for 3D cell culture. It is suggested that the pendant polymer chains solely influence the pore diameter and molecular transport in the hydrogel, while only minimally affecting its stiffness. The hydrogels formulated to present varying stiffness with only limited changes in permeability illuminate the role of matrix stiffness on the viability and VEGF expression of encapsulated fibroblasts. Overall, this biomaterial design strategy will be highly useful to better understand the effect of the extracellular microenvironment on a wide array of cells and to advance the quality of 3D cell culture platforms and implantable cell-encapsulating hydrogels.

References

1. Jen, A. C.; Wake, M. C.; Mikos, A. G., Review: Hydrogels for cell immobilization. *Biotechnol Bioeng* **1996**, 50, 357-364.
2. Drury, J. L.; Mooney, D. J., Hydrogels for tissue engineering: scaffold design variables and applications. *Biomaterials* **2003**, 24, 4337-4351.
3. Brandl, F.; Sommer, F.; Goepferich, A., Rational design of hydrogels for tissue engineering: Impact of physical factors on cell behavior. *Biomaterials* **2007**, 28, 134-146.
4. Karande, T. S.; Ong, J. L.; Agrawal, C. M., Diffusion in musculoskeletal tissue engineering scaffolds: Design issues related to porosity, permeability, architecture, and nutrient mixing. *Ann Biomed Eng* **2004**, 32, 1728-1743.
5. Glushko, V.; Thaler, M. S. R.; Karp, C. D., Pyrene fluorescence fine structure as a polarity probe of hydrophobic regions: Behavior in model solvents. *Arch Biochem Biophys* **1981**, 210, 33-42.

6. Wang, D.; Huang, H-J. S.; Kazlauskas, A.; Cavenee, W. K., Induction of vascular endothelial growth factor expression in endothelial cells by platelet-derived growth factor through the activation of phosphatidylinositol 3-kinase. *Cancer Res* **1999**, 59, 1464-1472.

(662a) Engineering of Nanocrystals for Clean Energy Applications

Michael Z. Hu, Energy & Transportation Sciences Division, Oak Ridge National Laboratory, Oak Ridge, TN

Nanocrystals such as quantum dots are important for electronic/optoelectronic energy technologies. In the course of developing energy devices or energy process technologies, chemical synthesis and process engineering play significant roles in controllably achieving designed tailoring of materials nanostructures, as well as drastically improving the process production yield and nanomaterials properties. Distinct from other nanotechnology applications, energy processes or devices (such as large-dimension panels for solar cells or for solid state lighting) could consume large quantities of high-quality, but low-cost nanomaterials (such as nanocrystals, films, or membranes). Thus, new engineering principles must be developed to enable high performance/cost ratio and the nanoscience-to-nanomanufacturing transition in order to meet the energy demands. Energy efficiency performance is determined by the high quality (e.g., well controlled fine nanostructures) of the produced materials. The speaker will focus on chemical/molecular process engineering perspectives by presenting a few examples of energy nanomaterials being developed at the Oak Ridge National Laboratory, including quantum-dot nanocrystals for solid-state lighting and biofermentation-produced CIGS nanoparticles for thin film solar cells.

See more of this Session: Nanocrystal Science and Technologies
See more of this Group/Topical: Materials Engineering and Sciences Division

(662b) Connected but Confined: Surface Functionalization of Semiconducting Nanocrystals

Haitao Zhang[1], Bo Hu[1], Tobias Hanrath[2] and **Richard D. Robinson**[1], (1)Materials Science and Engineering, Cornell University, Ithaca, NY, (2)Chemical and Biomolecular Engineering, Cornell University, Ithaca, NY

Connected but Confined: Surface Functionalization of Semiconducting Nanocrystals

Semiconductor colloidal nanocrystals (NCs) have attracted much attention in the past decade due to their unique size and shape dependent properties. In most of the successful synthetic routes to colloidal NCs, the use of bulky hydrocarbon (C_8-C_{18}) molecules with coordinating functional groups (such as -COOH, -NH_2, etc) as surfactant ligands is crucial for stabilization, as well as for size and shape control of NCs. The presence of these large organic molecules, however, creates highly insulating barriers which block the electronic communication between NCs, limiting the applications of assemblies of colloidal NCs. We have developed a novel chemical method to efficiently remove these surfactant ligands. This unique surface modification process produces bare nanocrystals without any surfactant ligands, which has been confirmed by FTIR and elemental analysis. GISAX study also reveals a significant decrease on inter-particle spacing. The bare nanocrystals are attached but maintain quantum confinement. Such "connected but confined" nanocrystal assemblies exhibit much stronger inter-particle coupling, and are promising new materials for fabricating electronic and optoelectronic devices.

See more of this Session: Nanocrystal Science and Technologies
See more of this Group/Topical: Materials Engineering and Sciences Division

(662c) Seeded Growth of Shape-Controlled Wurtzite CdSe Nanocrystals: Cubes, Hexagonal Platelets, and Bullets

Katherine P. Rice, Mark P. Stoykovich and Aaron E. Saunders, Chemical and Biological Engineering, University of Colorado, Boulder, CO

Previous investigations into the synthesis of wurtzite CdSe nanocrystals have given rise to well-developed methods for producing particles with anisotropic shapes such as rods, tetrapods and wires; however, the synthesis of other anisotropic shapes has proved challenging. Here, we demonstrate the use of a seeded-growth approach to produce colloidal, shape-controlled wurtzite CdSe nanoparticles with previously unobserved morphologies. The synthesis, which makes use of small CdSe nanocrystals as nucleation sites for subsequent growth, can be tuned to selectively yield colloidal CdSe cube- and hexagonal platelet-shaped nanocrystals, in addition to the previously observed rod- and bullet-shaped particles. We characterize the structure and discuss possible growth mechanisms for these new shapes, and demonstrate a quantitative analysis technique for shape classification based on Fourier descriptors obtained from transmission electron micrographs.

See more of this Session: Nanocrystal Science and Technologies
See more of this Group/Topical: Materials Engineering and Sciences Division

(662d) A Method to Form Molecular Nanorods Using Inorganic Nanoparticles As Nucleation Seeds

Li Li, Sunxi Wang and Guangzhao Mao, Chemical Engineering and Materials Science, Wayne State University, Detroit, MI

This paper describes a method to nucleate molecular crystalline nanorods from inorganic nanoparticles. The method is demonstrated using *n*-carboxylic acids that crystallize from organic solvents, such as ethanol and isopropanol, by solvent evaporation. Thiolate-protected inorganic nanoparticles including CdSe, CdS, and Au are used as nucleation seeds. *n*-Carboxylic acids form highly ordered 2-D crystalline layers on highly oriented pyrolytic graphite (HOPG). However, when crystallization occurs in the vicinity of the nanoparticles, the persistent HOPG-dominated nano-stripe pattern is disrupted, and nanorods attached to the nanoparticles become the dominant feature. In order to understand the underlying mechanism, the effects of film-forming conditions, carboxylic acid chain length, nanoparticle size, and chemical composition of the nanoparticle are examined. It is determined that high colloidal stability of the nanoparticles is a prerequisite for seed-mediated nucleation. A carboxylic chain length range amenable to the nanorod formation is identified, as is the relationship between the nanoparticle size and the number of nanorods per nanoparticle. This study contributes to solution-based bottom-up processes of nanocrystals that may be applicable to a variety of molecular crystalline compounds.

See more of this Session: Nanocrystal Science and Technologies
See more of this Group/Topical: Materials Engineering and Sciences Division

(662e) Electroluminescence From Colloidal Nanocrystals (Quantum Dots) Via Field-Driven Ionization

Matthew J. Panzer[1], Vanessa Wood[2], Deniz Bozyigit[2], Yasuhiro Shirasaki[3], Ian Rousseau[3], Scott M. Geyer[4], Moungi G. Bawendi[4] and Vladimir Bulovic[3], (1)Chemical & Biological Engineering, Tufts University, Medford, MA, (2)Information Technology and Electrical Engineering, ETH Zurich, Zurich, Switzerland, (3)Electrical Engineering, Massachusetts Institute of Technology, Cambridge, MA, (4) Chemistry, Massachusetts Institute of Technology, Cambridge, MA

Colloidal semiconductor nanocrystals, or quantum dots (QDs), are a promising class of materials for the development of next-generation solid-state lighting technologies due to their facile solution-based syntheses, widely tunable and narrowband emission characteristics, and high photoluminescence efficiencies. However, obtaining efficient and sustained electroluminescence from QD thin films remains a significant challenge. A high degree of morphological and energetic disorder inherent to many nanosized lumophores, including QDs, places severe limitations on the charge injection into and transport rates through thin films of these materials. Here, we demonstrate that sustained electroluminescence from QD thin films can be achieved by the local generation and recombination of charges induced by the application of an alternating electric field, thereby eliminating the need for injection of charge carriers from the device electrodes and long-range carrier transport. We show electroluminescence from novel light-emitting structures containing thin films of QDs and other nanoscale emissive materials that do not support direct current excitation, and suggest a mechanism for the charge generation and electroluminescence that is consistent with our time-averaged and time-resolved experimental observations. The proposed field-driven ionization phenomenon may provide a new route toward the realization of efficient lighting and display platforms incorporating colloidal QDs.

See more of this Session: Nanocrystal Science and Technologies
See more of this Group/Topical: Materials Engineering and Sciences Division

(662f) Manipulating Crystal Growth and Polymorphism by Confinement In Nanoscale Crystallization Chambers

Benjamin Hamilton, Artie McFerrin Department of Chemical Engineering, Texas A&M University, College Station, TX, Jeong-Myeong Ha, Department of Chemical Engineering, University of California, Berkeley, CA, Marc A. Hillmyer, Chemistry, University of Minnesota, Minneapolis, MN and Michael D. Ward, Molecular Design Institute, Department of Chemistry, New York University, New York, NY

The phase behaviors of crystalline solids embedded in nanoporous matrices have been studied for decades. Classic nucleation theory conjectures that phase stability is determined by the balance between a unfavorable surface free energy and stabilizing volume free energy. The size constraint imposed by nanometer-scale pores during crystallization results in large surface-to-volume ratios, which are reflected in crystal properties – for example, melting points and enthalpies of fusion – that differ drastically from their bulk-scale counterparts. Moreover, confinement within nanoscale pores can dramatically influence crystallization pathways and crystal polymorphism, particularly when the pore dimensions are comparable to the critical size of an emerging nucleus, at which the surface and volume free energies are in delicate balance and polymorph stability rankings may differ from bulk. Recent investigations have demonstrated that confined crystallization can be used to screen for and control polymorphism, which can impact food, pharmaceutical, explosive, and dye technological sectors, for which polymorph identity is critical for regulatory compliance and function.

We review our recent studies of the polymorphic and thermotropic properties of crystalline materials embedded in the nanometer-scale pores of porous glass powders and polymer monoliths, the latter generated from selective chemical etching of one component from shear-aligned diblock or triblock samples. The embedded nanocrystals exhibit an array of phase behaviors, including the selective formation of metastable amorphous and crystalline phases, thermodynamic stabilization of normally metastable phases, size-dependent polymorphism, formation of new polymorphs, and shifts of thermotropic relationships between polymorphs. Size confinement also permits the measurement of thermotropic properties that cannot be measured in bulk form using conventional methods. The well-aligned cylindrical pores of the polymer monoliths also allow determination and *manipulation* of nanocrystal orientation, wherein the constraints imposed by the pore walls result in a Darwinistic competition between nuclei that favors alignment of the fastest growth direction with the pore axis. Collectively, the examples described provide substantial insight into crystallization at a size scale that is difficult to realize by other means. Moreover, the behaviors resulting from nanoscopic confinement are remarkably consistent for a wide range of compounds, suggesting a reliable approach to studying the phase behaviors of compounds at the nanoscale. Newly emerging classes of porous materials promise expanded explorations of crystal growth under confinement and new routes to controlling crystallization outcomes.

Authors B.D. Hamilton and J.-M. Ha participated in this work while at the University of Minnesota Department of Chemical Engineering and Materials Science.

See more of this Session: Nanocrystal Science and Technologies
See more of this Group/Topical: Materials Engineering and Sciences Division

(664a) Peptide-Based Supramolecular Materials for Cancer Therapeutics

Lye Lin Lock, Yi-an Lin, Andrew G. Cheetham and **Honggang Cui**, Chemical and Biomolecular Engineering, Johns Hopkins University, Baltimore, MD

Peptides offer great molecular attributes as building units to construct biomaterials that incorporate explicit bioactivities into well-defined structural features at the nanoscale. Here we present our recent efforts in the design, synthesis and self-assembly of peptide-based molecules for applications in drug delivery, cancer diagnosis and imaging, and tumor microenvironment mimicking. Specifically, we take advantage of the design versatility of peptide primary and secondary structures to produce 0-D (spheres), 1-D (fibrils), 2-D (sheets and vesicles) nanostructures with the aim of controlling their interactions with cells of interest. Our preliminary results have shown that we can not only control size and shape of these nanoarchitectures, but also their surface chemistries as well as the mechanical properties of the resulting materials. Interfacing of these nanostructures with cancer cells will also be reported.

See more of this Session: Nanostructured Biomaterials
See more of this Group/Topical: Materials Engineering and Sciences Division

(664b) Self-Assembly of Peptide-Like Tubes

Katherine J. Harvey and **Justin R. Barone**, Biological Systems Engineering, Virginia Tech, Blacksburg, VA

Ethyl vinyl sulfone (EVS)-substituted lysine can spontaneously form tubes at optimal substitution levels and solution conditions. The tubes can have diameters from tens of nanometers to several microns and lengths up to millimeters. EVS-substituted lysine aggregates into beta-sheet mimics through an amine-sulfonyl hydrogen bond that is of similar length to an amine-carbonyl hydrogen bond in natural peptide beta-sheets. The nanometer-sized beta-sheet mimics form larger sheets that then roll into tubes to minimize surface area. The sheet size can be controlled by concentration and rate of addition so tubes of varying size can be built. The tubes could find use in nano/microdevices or as cellular substrates where defined fluid transport would be needed.

See more of this Session: Nanostructured Biomaterials
See more of this Group/Topical: Materials Engineering and Sciences Division

(664c) Intracellular Delivery of Transcription Factors Using Polymeric Nanocapsules

Anuradha Biswas, Chemical and Biomolecular Engineering, UCLA, Los Angeles, CA, Guoping Fan, Human Genetics, University of California, Los Angeles, Los Angeles, CA and Yi Tang, Chemical and Biomolecular Engineering, University of California, Los Angeles, Los Angeles, CA

Transcription factors have the unique ability to regulate gene expression and thus determine cellular fate. Intracellular delivery of transcription factors has important therapeutic implications and can be a mode to treat loss-of-function diseases or initiate cellular reprogramming. DNA-based methods raise safety concerns because of the potential for unexpected genetic modifications by exogenous sequences in target cells. From a therapeutic perspective, protein-based approaches are safer than gene therapy because no random or permanent genetic changes are involved, and only transient actions of proteins are needed for the desired results. We have developed degradable polymeric nanocapsules (NCs) which can release encapsulated cargo intracellularly in active form. Nuclear proteins were encapsulated in NCs synthesized with monomers and either endoprotease-degradable or redox-responsive crosslinkers. Release of cargo was demonstrated successfully *in vitro* and elucidated the importance of the degradability of the nanocapsules for observed nuclear localization. To demonstrate delivery of biologically relevant nuclear cargos, we have prepared NCs encapsulating transcription factors relevant to cellular reprogramming and differentiation. We show that these transcription factors delivered *via* NCs show enhanced nuclear localization compared to those delivered *via* protein transduction domains. We also demonstrate that NC-delivered proteins are able to deliver active proteins through functional biological assays based on the specific activity of the transcription factors. This degradable polymeric nanocapsule intracellular delivery system may be generalized to effectively deliver many biological components.

See more of this Session: Nanostructured Biomaterials
See more of this Group/Topical: Materials Engineering and Sciences Division

(664d) Biosynthesis Strategies for Controlling the Nanostructure of Polyhydroxyalkanoate Block-Copolymers

John S. F. Barrett, Chemical Engineering and Material Science, University of Minnesota, Minneapolis, MN and Friedrich Srienc, Biotechnology Institute / Chemical Engineering, University of Minnesota, St Paul, MN

Polyhydroxyalkanoates (PHAs) are microbially-synthesized polymers that are both biodegradable and produced from renewable feedstocks. Despite these favorable characteristics commercial demand for PHAs remains limited. This is partly due to a greater cost advantage for petroleum-derived polymers but also because of the small number of unique material properties which are currently available from PHAs. In response to this need for further technological development of PHAs, our laboratory is exploring novel materials with controlled polymer nanostructure. Already we have shown how mass spectroscopy-controlled biosyntheses can be used to create PHA, core-shell, nanoparticles as well as PHA block-copolymers which show superior toughness compared to random copolymers. In this paper we describe the continuing work of our laboratory to create novel PHAs with controlled nanostructure. We have engineered new polymer synthase enzymes with expanded substrate specificity to allow the synthesis of unique block-copolymers from structurally dissimilar monomers. Additionally we have engineered new *E. coli* host strains with altered metabolic functionality to increase the flux of monomer supply pathways and reduce the time required to switch between different monomer types. With these new capabilities along with mass-spectroscopy controlled biosynthesis, we aim to explore how the physical interaction between dissimilar monomers as well as the size of separate polymer domains affects the material properties of this important class of biopolymers.

See more of this Session: Nanostructured Biomaterials
See more of this Group/Topical: Materials Engineering and Sciences Division

(664e) Antimicrobial Biomaterials Based On Single Walled Carbon Nanotubes and Charged Polymers

Seyma Aslan[1], Nan Li[1], Mathilde Hindie[2], Sara Hashmi[1], Lisa Pfefferle[1], Menachem Elimelech[1], Emmanuel Pauthe[2] and Paul Van Tassel[1], (1)Chemical & Environmental Engineering, Yale University, New Haven, CT, (2)University of Cergy-Pontoise, Cergy-Pontoise, France

Biomedical implant infection represents a significant health care problem. We investigate here the possibility of single walled carbon nanotubes (SWNT), together with aqueous polymers, as antimicrobial thin film biomaterials. To address the SWNT solubility problem, we investigate two methods: i) assembly with the non-toxic surfactant Tween-20 and ii) encapsulation within polysiloxane shells. SWNT in aqueous solution is then combined with linear polyelectrolytes to form thin films via layer-by-layer (LbL) assembly. SWNT solubility and SWNT-polymer film properties are characterized via transmission electron microscopy (TEM), scanning electron microscopy (SEM), zeta potential analysis, optical waveguide lightmode spectroscopy (OWLS), and dissipative quartz crystal microgravimetry (QCMD). We observe both surfactant and polysiloxane coatings to result in highly soluble single and small bundle SWNT, that readily assemble via LbL with charged polymers to form films of controlled nanoscale thickness and nanotube content. We find films containing SWNT to be highly antimicrobial and cytocompatible: > 90% of *E. coli* are inactivated on SWNT containing films (vs. ca. 10% for control films), and MC3T3-E1 osteoblast-like cells exhibit increased attachment and metabolic activity on SWNT containing versus control films.

See more of this Session: Nanostructured Biomaterials
See more of this Group/Topical: Materials Engineering and Sciences Division

(664f) Synthesis of Chitosan Based Hyper-Branched Polymers for Biomedical Applications

Laura W. Place, School of Biomedical Engineering, Colorado State University, Fort Collins, CO and Matt J. Kipper, Chemical and Biological Engineering and School of Biomedical Engineering, Colorado State University, Fort Collins, CO

Aggrecan is an important molecule in the extracellular matrix of articular cartilage; it contributes to the high compressive modulus of cartilage and to joint lubrication. Aggrecan has a bottlebrush structure made up of a core protein and highly charged glycosaminoglycan (GAG) side chains. One such GAG is chondroitin sulfate (CS), which is a strong polyanion and is very hydrophilic. Chitosan has many unique properties that make it an ideal material for tissue engineering applications. It is a natural polysaccharide, structurally similar to GAGs, it is antimicrobial, and promotes wound healing. CS side chains are grafted onto chitosan to form a nanostructured polymeric bottlebrush that mimics aggrecan. These nano-assemblies are synthesized for use as viscosupplements and as carriers to stabilize and deliver therapeutic proteins. Pendant amine groups along the chitosan chain are thiolated and reacted with a maleimide-activated CS. This strategy enables one to tune the degree of substitution of chitosan with CS side chains. The effects of these parameters on solution dynamics and emergent biochemical properties of these new aggrecan-mimetic polymers can then be studied.

See more of this Session: Nanostructured Biomaterials
See more of this Group/Topical: Materials Engineering and Sciences Division

(664g) Breath Figure PLGA Films for Controlled Drug Delivery Applications

Thiruselvam Ponnusamy, Chemical & Biomolecular Engineering, Tulane university, New orleans, LA, Louise B. Lawson, Department of microbiology and immunology, Tulane university, New orleans, LA and Vijay T. John, Chemical & biomolecular engineering, Tulane university, New orleans, LA

Breath figure PLGA films for controlled drug delivery applications

Thiruselvam Ponnusamy[1], Louise B. Lawson[2], Vijay T. John[1]

[1]Dept of Chemical & Biomolecular Engineering, [2]Dept of Microbiology and Immunology, Tulane University, New Orleans

Biodegradable polymers are of significant interest in the design of controlled drug delivery application. We describe the breath figure method[1,2] of synthesizing highly porous poly lactic-co-glycolic acid (PLGA) films for the controlled release of drug compounds. The breath figure method is a versatile and simple way of generating porous structures by evaporating solutions of a polymer in a volatile solvent in a humid environment. Condensation of water droplets on the polymer due to the cooling produced by evaporating solvent leads to patterned pore formation upon eventual droplet removal. The facile removal of the pore template is a specifically appealing aspect of breath film technology.

Because of the highly porous nature of breath figures, it is possible that release characteristics of encapsulated pharmaceutical compounds can be significantly modified in comparison to nonporous films of the same material. It is therefore important to characterize such release especially in connection with the fact that the polymer film degrades with time. This work describes the synthesis of drug-loaded porous PLGA films prepared using the breath figure method and attempts to correlate the release kinetics to the degradation characteristics. A single stage process combining spin coating and the breath figure technique was used to obtain drug incorporated porous thin films. In addition to PLGA, poly (ethylene glycol) (PEG) was used as a plasticizer to modify pore structures and release characteristics through incorporation of the hydrophilic polymer into the film. Salicylic acid and Ibuprofen were used as model drug compounds. A five-week film degradation study showed a gradual deterioration of the pore structure. *In vitro* drug release experiments for the PLGA film in pH 7.4 phosphate buffered saline solution (PBS) showed a sustained release of drug for a month with initial burst release. The release profile showed that the drug compounds released markedly higher for the porous films compared to the non-porous films. The addition of hydrophilic PEG accelerated the drug release from porous films.

References:

1) Mohan Srinivasarao, *et al*, *Science*, **292**, 79(2001)

2) U.H.F.Bunz, *Advanced Materials*, **18,** 973-989(2006)

See more of this Session: Nanostructured Biomaterials
See more of this Group/Topical: Materials Engineering and Sciences Division

(664h) Real-Time SE/QCM-D Characterization of Biomolecule Adsorption within Sculptured Thin Films

Tadas Kasputis[1], Daniel Schmidt[2], Keith Brian Rodenhausen[3], Mathias Schubert[2] and Angela K. Pannier[1], (1)Biological Systems Engineering, University of Nebraska - Lincoln, Lincoln, NE, (2) Electrical Engineering, University of Nebraska - Lincoln, Lincoln, NE, (3)Chemical and Biomolecular Engineering, University of Nebraska - Lincoln, Lincoln, NE

Recent improvements in nanofabrication strategies have led to the development of precisely crafted nanostructures with intricate features. Incorporating biomolecules such as proteins, DNA, drugs, and even whole cells could allow for functionalization of nanostructured surfaces for biological applications including biosensing, tissue engineering scaffolds, as well as drug and gene delivery. Along with the nanofabrication of biological devices, there is a need to develop instruments that are able to probe and characterize the dynamic evolution of these bio-functionalized interfaces. Spectroscopic ellipsometry combined with quartz crystal microbalance with dissipation (SE/QCM-D) is a non-destructive optical/ mechanical characterization technique that when run in tandem, reveal dynamic characteristics of functional nanostructure development including film thickness, adsorbed mass, and porosity with sub-angstrom resolutions. Nanostructures in the form of sculptured thin films (STF) were fabricated by glancing angle deposition via electron beam evaporation of titanium onto gold-coated QCM-D crystals. The crystals were then mounted within the QCM-D liquid chamber and proteins of varying sizes were deposited and characterized *in situ*. Protein adsorption was detected shortly after introducing the protein solutions with both QCM-D and SE as an increase of mass and change in the optical response, respectively. The QCM-D reports greater adsorbed mass for larger proteins (fibronectin) than smaller proteins (bovine serum albumin). The total adsorbed mass of proteins within the nanostructured scaffold exceeded the total absorbed mass of their respective depositions on flat surfaces, confirming that these nanostructures are capable of trapping proteins. Analysis of the anisotropic optical response from the nanostructures, which is very sensitive to environmental changes, adds complementary information on protein adsorption and the optical quantification is in agreement with QCM-D results. In addition, deposition of other biomolecules such as cells and DNA complexes have also been probed on these surfaces. The use of combinatorial SE/QCM-D to characterize and monitor the deposition of biomolecules on complex nanotopographies will improve the design and fabrication strategies for a wide array of biotechnological devices.

See more of this Session: Nanostructured Biomaterials
See more of this Group/Topical: Materials Engineering and Sciences Division

COPOLYMERS OF STYRENE, METHYL METHACRYLATE, AND ETHYLENE WITH A SERIES OF VINYL ESTERS: A STUDY OF THERMAL MECHANICAL AND RHEOLOGICAL PROPERTIES

Hai Wang, Dr Edward Kolodka,*
Chemical Engineering Department,
University of North Dakota, Grand Forks, ND

A number of new processes are being developed which produce bio-sourced fatty acids of various molecular weights. These fatty acids can easily be converted into vinyl esters using common processes. However, vinyl esters larger than vinyl propionate are not suitable for homopolymerization due to the low glass transition temperatures (T_g) of the homopolymers. Therefore, this work will focus on the copolymerization of vinyl esters of various molecular weights with both styrene and ethylene. Particular emphasis will be placed on the influence of vinyl ester molecular weight on the polymerization kinetics and thermomechanical characteristics of the resultant copolymers.

In this study copolymers of styrene (St) and methyl methacrylate (MMA) with a series of vinyl esters (VEs) including vinyl acetate (VAc), vinyl propionate (VPr), vinyl hexanoate (VHe), and vinyl decanoate (VDe) were prepared by bulk copolymerization initiated by azobisisobutyronitrile (AIBN) at 60 °C. Copolymers of ethylene with the VEs were synthesized by solution copolymerization initiated by AIBN also at 60 °C. The compositions and molecular weight distributions (MWD) of all the copolymers were determined by GPC. The impact of compositions and branch lengths on the thermal mechanical properties of copolymers including glass transition temperature (T_g), melting point (T_m), storage modulus (G') and loss modulus (G'') was determined by dynamic mechanical analysis. In addition, the rheological responses (complex viscosity versus shear rate) of the copolymers of MMA-VEs and ethylene-VEs were studied.

Current results showed that the molecular weights (MWs) of the copolymers decreased as the mole fraction of VEs in the feed increased. Meanwhile, copolymers with similar VE incorporations had higher MWs for poly (St-co-VHe) and poly (St-co-VDe) than those of poly (St-co-VAc) and poly (St-co-VPr). It could be explained by the significantly higher reactivity ratios of both VHe and VDe than those of VAc and VPr. Incorporating VEs with different chain lengths from 1 mol% to 19 mol% reduced the T_g of PS by up to 32 °C. The T_g of poly (St-co-VEs) decreased as the mole fraction of VEs increased. Poly(St-co-VEs) with longer branches, such as poly(St-co-VHe) and poly(St-co-VDe), exhibited a larger drop in T_g than VAc and VPr copolymers at similar compositions. However, the branch effects on T_g were not significant for copolymers containing VEs less than 5%.

Dynamics mechanical analysis of poly(St-co-VHe) and poly(St-co-VDe) showed that storage modulus decreased with increasing of the mole fractions of linear VEs or the decreasing of MWs. The loss modulus increased with the growing of mole fractions of linear VEs firstly, and then dropped sharply after the mole fractions of linear VEs were over 17%, around which the MWs of both poly(St-co-VHe) and poly(St-co-VDe) were very low. Mole fractions of linear VEs, MWs, and the side chain lengths of resultant copolymers all had effect on the thermomechanical properties of the resultant St-VE copolymers.

(668b) Poly (ethylene oxide) (PEO) Nanofiber Design Space Using Spinneret-Based Tunable Engineered Parameters (STEP) Technique

Mohammad Khan, Department of Mechanical Engineering, Virginia Tech, Blacksburg, VA, Ji Wang, Engineering Science and Mechanics, Virginia Tech, Blacksburg, VA and **Amrinder S. Nain**, Mechanical Engineering, Virginia Tech, Blacksburg, VA

Poly(ethylene oxide) (PEO) nano fibers as 1D building blocks in polymeric composites have potential applications in tissue engineering, pharmaceuticals and structural components. Furthermore, addition of PEO improves the spinnability of otherwise difficult to spin bio-polymeric systems like chitosan and silk. However, a comprehensive design strategy to increase the spinnability of polymer systems using PEO as an additive is currently missing due to incomplete knowledge on spinnability of stand-alone PEO system. Currently state-of-the-art fiber manufacturing platforms of electrospinning, template synthesis, and phase separation are capable of depositing fibers of different polymers. However, depositing smooth, uniform diameter fibers of sizeable length in aligned configurations has been challenging using aforementioned techniques In this respect, we have recently reported the pseudo-dry spinning based STEP technique, which deposits aligned polymeric fibers with control on fiber dimensions (diameter: sub 100nm-micron, length: mm-cm) in single and multiple layers at tunable porosity. Using the STEP technique, this study explores the design space of PEO fibers for achieving control on fiber diameter by varying the molecular weight, solution concentration and solvent (i.e. ethanol) concentration in water. We also demonstrate fabrication and tensile strength characterization of PEO fibers' bundles composed of approximately fifty thousand of PEO individual nanofibers, which can then be used as 1D structural component in a hierarchical composite design. PEO of M_w 100K (g. mol^{-1}), 400K (g. mol^{-1}), and 900K (g. mol^{-1}) (Scientific Polymer Products, USA), were dissolved without further purification in deionized water/ethyl alcohol (Decon labs, Inc, USA) in varying concentrations. PEO was dissolved in ethanol solution (wt.-% of 20-80%) at wt.-% concentrations ranging from 3-12% and used to explore the diameter design space. Bundles were made by using PEO (M_w 400 000) dissolved in 70% ethanol solvent with 6 %(w/w) PEO concentration. Bare fibers' bundles were braided at different twist turns (3, 6, 9, and 12) using an in-house custom braiding device. The effect of strain rate on tensile behavior and fracture mechanisms is observed at varying rates of 1mm/s, 2.5 mm/s, and 5 mm/s. The tensile strength increases gradually with increasing twist angle and peaks at 6 twists compared with parallel fibers bundles, and decreases with further increase in twist angle. Strain rate variation effects (1-5 mm/s) on tensile properties for 6 twists of 6% PEO (M_w 400K (g. mol^{-1})) fibers' bundles resulted in a 33.8% increase of tensile loads and 30.8% decrease of ultimate strain. It is envisioned that the fiber design space of PEO presented in this study will help in integration of PEO in polymeric systems with defined properties, thus, leading to improved spinnability and increased applications.

(668c) Producing Next-Generation, High-Porosity, Nanocellular Polymer Insulations by the Gas Foaming Process

Allan E. David, Department of Pharmaceutical Sciences, The University of Michigan, Ann Arbor, MI and Arthur J. Yang, Industrial Science & Technology Network, Inc, York, PA

We have utilized both modeling as well as experimental prototypes to demonstrate that the thermal conductivity of pore-entrapped air can be substantially reduced if pore size is reduced close to, or smaller than, the mean free path of air (~ 100 nm). Since good insulations normally contain more than 95% air, this lowering of air thermal conductivity led to superior nanopore thermal insulations. The production of super insulation by creating nanoporous structures, however, has been only marginally successful due to the cost-prohibitive, supercritical drying process.

Here we present a new method for achieving sub-micron pore structures in high-porosity foams using a gas blowing process. By substantially increasing the nuclei population through homogeneous nucleation in a gas foaming process, we expect to reduce the average pore size, within the foam, by two orders of magnitude compared to current state-of-the-art. A rigorous kinetic and thermodynamic nucleation theory is derived to verify the feasibility of such an approach. Results have shown that by reducing the dynamic interfacial tension during the bubble formation process, we are able to lower the critical radius of nucleation to much below 5 nanometers, and thus promote the creation of nanobubbles by the gas nucleation process. The concept of utilizing polymer side-chain groups as "doped" nucleation centers will be discussed along with experimental analysis of foaming experiments done in a high-pressure foaming vessel.

See more of this Session: Polymer Processing and Rheology I
See more of this Group/Topical: Materials Engineering and Sciences Division

(668d) PE/PEO Cocontinuous Polymer Blends with Application In Gas Separation Membranes

Milana Trifkovic[1], Aaron Hedegaard[1] and Christopher W. Macosko[2], (1)Chemical Engineering and Materials Science, University of Minnesota, Minneapolis, MN, (2)Department of Chemical Engineering and Materials Science, University of Minnesota, Minneapolis, MN

The microporous support layer for gas separation membranes provides mechanical strength and prevents the rupture of the selective layer under the incoming gas pressure. Therefore, average pore size, pore size distribution, pore wall functionality and thickness of the support layer play a vital role in the overall performance of the membrane. This study will focus on the design of a microporous membrane support layer derived from cocontinuous polymer blends. The effect of blend composition, shear rate, residence time, and annealing time on the cocontinuous morphology of polyethylene (PE)/ polyethylene oxide (PEO) blends was investigated. Porous PE sheets were generated by solvent extraction of PEO and used as a support layer for gas separation membranes.

The PE/PEO blends using non-functional and maleic-anhydride functional PE (PE-g-MA) were mixed in batch microcompounder and pilot plant scale co-rotating twin-screw extruder. The effect of residence time on the cocontinuous morphology in the twin-screw extruder was studied by varying screw geometry, mixing speed, and the zone of PEO introduction.

The morphology and phase size of the blends was examined using SEM (scanning electron microscopy) and LSCM (laser scanning confocal microscopy). In order to satisfy the LSCM requirements of specimen transparency and fluorescence of one phase, hydroxyethyl(methacrylate) (HEMA) stained with rhodamine B dye was polymerized inside of the pores of the porous PE sheet. Approximately 100 2D images per sample were acquired, deconvolved and binarized prior to the 3D image reconstruction. A triangular mesh representation of the interface was generated using marching cubes algorithm, and the specific interfacial area and characteristic length were calculated.[1]

Rheological characterization of the blends with droplet and cocontinuous morphologies was performed. The rheology of droplet blends was fit to the Palierne viscoelastic droplet model to quantify the equilibrium interfacial tension of PE/PEO blends using both PE and PE-g-MA. Cocontinuous blend rheology was investigated over annealing time to predict coarsening behavior.

The gas separation membranes were prepared by solvent casting a THF solution of PS-PIL-PS ionic gel into porous PE sheets.[2] The effect of type of processing, pore size, film thickness, and pore wall functionality on the performance of the membranes is discussed and compared.

1. Lopez-Barron, C.; Macosko, C. W. *Langmuir* **2009**, *16*, 9392-9404.

2. Gu, Y.; Lodge, T.P. *Macromolecules* **2011**, *DOI/ma2001838.*

See more of this Session: Polymer Processing and Rheology I
See more of this Group/Topical: Materials Engineering and Sciences Division

(668e) Fabrication of Thin Films by Drying Aqueous Polymer Solutions

Ramana Susarla[1], Yueyang Shen[1], Phong T. Huynh[2], Paul Takhistov[2] and **Boris Khusid**[1], (1)Chemical, Biological & Pharmaceutical Engineering, New Jersey Institute of Technology, Newark, NJ, (2)Food Science, Cook College, Rutgers, The State University of New Jersey, New Brunswick, NJ

We developed a lab-scale setup for fabrication of thin films (~0.1mm) by drying multicomponent polymer solutions over a wide range of drying regimes. The drying chamber is equipped with a heater, an air blower with adjustable flow rate and temperature, a humidifier to vary the partial pressures of water and organic solvent, a balance to automatically monitor the sample weight, and sensors to control the temperature, airflow, and the atmosphere composition in the process of drying. We will report on the drying kinetics and mechanical properties of films prepared by drying aqueous HPMC solutions without and with additions of sodium dodecyl sulfate and polyethylene glycol over a broad range of operating regime.

The work was supported by NSF Engineering Research Center on Structured Organic Particulate Systems.

See more of this Session: Polymer Processing and Rheology I
See more of this Group/Topical: Materials Engineering and Sciences Division

(668f) Nanofibers by a Gas Jet Method

Rafael E. Benavides[1], Sadhan C. Jana[1] and D. H. Reneker[2], (1)Polymer Engineering, The University of Akron, Akron, OH, (2)Department of Polymer Science, The University of Akron, Akron, OH

High voltage electric field and low production rate limit electro-spinning for industrial production of nanofibers. In this work, a new process -Gas Jet Fibers (GJF)- that uses a high speed jet of gas to produce nanofibers from a nozzle is presented. In operation, a continuous layer of fiber precursor is fed on a flat surface, whereby the material is stretched and fibers are launched by the action of the high speed air flow field of the jet. Fiber morphology, diameter, and length can be easily controlled by varying the jet conditions, nozzle geometry, and fiber precursor physical properties. The GJF process can be easily adjusted to produce mono and multi-component fibers with morphologies varying from side-by-side to coaxial. We demonstrated the utility of this process by producing a wide range of fibers from different polymer solutions including polyethylene oxide (PEO), polyvinyl acetate (PVAc), and gelatine. Fibers with mean diameter varying between 400 nm and 3 μm and lengths around a few millimeters were found. These fibers can find range of applications including in the production of scaffolds, reinforced fibers, encapsulated materials, and controlled drug delivery systems.

See more of this Session: Polymer Processing and Rheology I
See more of this Group/Topical: Materials Engineering and Sciences Division

(681a) Colloidal Array of Silica Nanospheres As Template for Fabricating Three-Dimensionally Ordered Metal Oxides

Toshiyuki Yokoi, Ryota Watanabe and Takashi Tatsumi, Chemical Resources Laboratory, Tokyo Institute of Technology, Yokohama, Japan

We have developed a simple and novel liquid-phase method for preparing uniform-sized silica nanospheres (SNSs) *ca.* 12 nm in size. The SNSs can be synthesized through hydrolysis and condensation reactions of TEOS in the presence of basic amino acids such as lysine and arginine under weekly basic conditions. Recently, the particle size control of SNSs ranging from 14 to 550 nm was achieved through a process of regrowing the SNSs. These silica spheres easily self-assemble into a cubic closed packed (ccp) colloidal array by solvent evaporation or centrifugation. Very recently, we have found out that the colloidal array of the SNSs can be re-dispersed into polar solvents and the re-dispersed SNSs can be re-assembled without adding any chemicals by simple solvent evaporation or centrifugation. The colloidal array of the SNSs with a morphological reversibility is expected to be suitable for template for ordered nanomaterials.

Here we report the synthesis of the colloidal array of hollow-structured zirconia nanospheres (ZNSs) with a size of less than 100 nm by using the SNSs with a morphological reversibility as a template. The synthetic procedures consist of the five major steps: (1) preparation of colloidal array of the SNSs, (2) re-dispersion of the SNSs into the ethanol - water solution, (3) coating the SNSs with zirconia, (4) self-assembly of the zirconia-coated SNSs spheres, (5) removal of the silica template to obtain the colloidal array of hollow-structured ZNSs. Note that the shell of the hollow structure is composed of nanoparticles ca. 4 nm in size. Therefore, the hollow-structured ZNSs possess high surface area and bimodal pore system arising from the hollow structure and interparticle voids of zirconia nanoparticles. The size of hollow void can be tuned from 50 to 500 nm by changing the particle size of the SNSs. Finally, the crystallization of hollow-structured ZNSs was also demonstrated.

See more of this Session: Advances In the Applications of Porous Materials
See more of this Group/Topical: Materials Engineering and Sciences Division

(681b) Realizing Three-Dimensionally Ordered, Hydrothermally Stable Mesoporous Catalysts Via Assembly and Nanotemplating Strategies

Qianying Guo and **Mark A. Snyder**, Department of Chemical Engineering, Lehigh University, Bethlehem, PA

Realization of high-selectivity catalytic conversions under the hydrothermal conditions of the biorefinery demands the development of next-generation, hydrothermally stable heterogeneous catalysts bearing three-dimensionally ordered mesoporous (3DOm) and hierarchical (i.e., macro-meso-microporous) pore structure that can accommodate bulky sugar molecules and their derivatives. The enhanced hydrothermal stability and reducible structure of titania establishes its promise as a robust substrate for liquid phase biofuel catalysis. Through a hierarchical nanotemplating strategy involving the infiltration of assembled, size-tunable silica nanoparticles with titanate solutions, we have successfully fabricated three-dimensionally ordered mesoporous titania structures with tunable pore size ranging from nanometers to tens of nanometers. The mesopore structure during high-temperature calcination, treatment intended for titania polymorph interconversion and enhanced crystallinity, is bolstered by the presence of the hard silica template. The resulting nanoparticulate walls and ordered mesostructure lead to BET surface areas as high as 289 m^2/g. The confinement imparted by the sacrificial template enables fundamental insight into titania polymorphism as it relates to its sensitivity to the degree of confinement and surface chemistry imparted by the template. We will employ glucose-fructose isomerization and dehydration as a probe reaction-diffusion system to elucidate structure-function relations linking pore topology, pore size, and systematic tuning of both organic and inorganic pore space functionalization (e. g., Pt, SO_3H) to catalytic activity and selectivity.

See more of this Session: Advances In the Applications of Porous Materials
See more of this Group/Topical: Materials Engineering and Sciences Division

(681c) Synthesis of Ordered Porous Graphitic-C3N4 and Regularly Arranged Ta3N5 Nanoparticles by Using Self-Assembled Silica Nanospheres As a Primary Template

Tatsuya Okubo[1], Yuki Fukasawa[1], Kazuhiro Takanabe[1], Atsushi Shimojima[1], Kazunari Domen[1] and Markus Antonietti Antonietti[2], (1)Department of Chemical System Engineering, The University of Tokyo, Tokyo 113-8656, Japan, (2)Department of Colloid Chemistry, Max Planck Institute of Colloids and Interfaces in Potsdam-Golm, 14424, Potsdam, Germany

Uniform-sized silica nano- spheres (SNSs) assembled into close-packed structures were used as a primary template for ordered porous graphitic carbon nitride (g-C_3N_4), which was subsequently used as a hard template to generate regularly arranged Ta_3N_5 nanoparticles of well-controlled size. Inverse opal g-C_3N_4structures with the uniform pore size of 20–80 nm were synthesized by polymerization of cyanamide and subsequent dissolution of the SNSs with an aqueous HF solution. Back-filling of the C_3N_4 pores with tantalum precursors, followed by nitridation in an NH_3 flow gave regularly arranged, crystalline Ta_3N_5 nanoparticles that are connected with each other. The surface areas of the Ta_3N_5 samples were as high as 60 m^2 g^{-1},and their particle size was tunable from 20 to 80 nm, which reflects the pore size of g-C_3N_4. Polycrystalline hollow nanoparticles of Ta_3N_5 were also obtained by infiltration of a reduced amount of the tantalum source into the g-C_3N_4 template. An improved photocatalytic activity for H_2 evolution on the assembly of the Ta_3N_5 nanoparticles under visible-light irradiation was attained as compared with that on a conventional Ta_3N_5 bulk material with low surface area.

See more of this Session: Advances In the Applications of Porous Materials
See more of this Group/Topical: Materials Engineering and Sciences Division

(681d) Dispersible Exfoliated Zeolite Nanosheets and Their Application As Membrane

Kumar Varoon, Xueyi Zhang, Lorraine F Francis and Michael Tsapatsis, Chemical Engineering & Material Science, University of Minnesota, Minneapolis, MN

Highaspect ratio zeolite particles with thickness in the 2-3 nanometer range(zeolite nanosheets) can be derived from exfoliation of their lamellarprecursors (layered zeolites) and can be used as catalysts and building blocksfor fabrication of thin films. Thin zeolite films made using exfoliated zeolitenanosheets could show improved performance and processing advantages asmembranes for separations, low dielectric coatings, corrosion protection,chemical sensors, membrane reactors, etc. Here we report exfoliation of amultilamellar zeolite (silicalite-1) and subsequent processing to obtain adispersible coating suspension of exfoliated silicalite-1 sheets followed bydemonstration of its utility as a gas separation membrane. Exfoliation and structurepreservation of multilamellar silicalite-1 was confirmed by Transmissionelectron microscopy (TEM). Scanning probe microscopy (SPM) was used tocharacterize the thickness of the exfoliated silicalite-1 nanosheets. A thinb-oriented film of exfoliated nanosheets was coated on top of a porousα-alumina support. The coating morphology and cross section wascharacterized by scanning electron microscopy (SEM) and TEM respectively. Thedefects in the coating were removed by hydrothermal treatment of the silicalite-1in presence of a growth solution. The silicalite-1 membrane was characterizedby xylene isomer vapor permeance in Wicke-Kallenbach mode at 150°C, where aseparation factor of 67 was obtained between p-xylene and o-xylene. Thepermeance of p-xylene through the membrane (3.1E-07 mol/m^2/s/Pa) wasclose to that of the alumina support (4.2E-07 mol/m^2/s/Pa).

See more of this Session: Advances In the Applications of Porous Materials
See more of this Group/Topical: Materials Engineering and Sciences Division

(681e) Pure-Silica Zeolite MFI Crystals with Organic-Modified Interiors

Mohamad H. Kassaee, David Sholl and Sankar Nair, School of Chemical and Biomolecular Engineering, Georgia Institute of Technology, Atlanta, GA

Zeolites with organic-modified interiors can significantly widen the range of the applications of molecular sieving materials in separations, catalysis and other areas. Internal surface functionalization of zeolites can be accomplished by covalently binding organic groups to silanols or other reactive sites in the internal pore structure. The functionalized zeolite would then become an organic-inorganic hybrid with a potential for a diverse range of new applications. In this presentation, we describe the functionalization of pure-silica MFI zeolite crystals with selected functional groups that confer different types of chemical functionality to the zeolitic material. Detailed characterization by a number of techniques including NMR, TGA/DSC, and N_2 physisorption suggests that these functional groups are chemically bound to the internal pore sites and that the functional group loading is strongly dependent on the concentration of internal silanol defects. We then discuss the gas adsorption behavior of the functionalized materials, and show that their adsorption properties are significantly different from those of bare MFI zeolite materials. These results are explained in terms of the specific interactions between the organic-modified zeolite material and the gas molecules of interest.

See more of this Session: Advances In the Applications of Porous Materials
See more of this Group/Topical: Materials Engineering and Sciences Division

(681f) Ultrastable and Redispersible Mesoporous Silica Nanoparticles for Targeted Anticancer Drug Delivery

Yu-Shen Lin, Nardine Abadeer and Christy L. Haynes, Chemistry, University of Minnesota, Minneapolis, MN

Mesoporous silica nanoparticles (MS NPs) are attracting much attention of late due to their unique characteristics, including a large internal surface area and volume, making them promising candidates for biomedical applications. However, the large size and low stability of these porous silica NPs in biological media result in rapid unintentional uptake by the reticuloendothelial system and limit their potential for *in vivo* biological use.

Recently, we have developed the synthesis of size-tunable and multifunctional MS NPs and shown that MS NPs have lower hemolytic activity and cytotoxic effect compared to their nonporous counterparts.[1,2,3] Additionally, we found that small MS NPs (<50 nm) with polyethylene glycol (PEG) modification and hydrothermal treatment are shown to exhibit high long-term stability in biological media at 37 °C. These hydrothermally treated PEGylated particles are highly biocompatiable, resistant to degradation and nonspecific protein adsorption, and less efficiently taken up by macrophages.[4] In this work, PEGylated MS NPs with a secondary hydrophobic silane (chlorotrimethyl silane or 3,3,3-trifluoropropyltrimethoxysilane) modification show further increase in stability in biologically relevant media like phosphate buffered saline and simulated body fluid, compared to bare and PEGylated MS NPs. Two different anticancer drugs (doxorubicin and 5-fluorouracil) are used to study the loading capacity and delivery kinetics of these co-modified MS NPs. In addition, the drug-loaded MS NPs are conjugated to a cancer-specific ligand (transferrin) to assess targeting and drug delivery to cancerous cells.

References:

1. Lin, Y.-S.; Haynes, C. L. *Chem. Mater.* **2009,** *21,* 3979-3986.

2. Lin, Y.-S.; Haynes, C. L. *J. Am. Chem. Soc.* **2010**, *132*, 4834-4842.

3. Mauer-Jones, M. A.; Lin, Y.-S.; Haynes, C. L. *ACS Nano* **2010**, *4*, 3363-3373.

4. Lin, Y.-S.; Abadeer, N.; Haynes, C. L. *Chem.Commun.* **2010**, *47,* 532-534.

Acknowledgments:

This work was funded by a grant from National Science Foundation (NSF, CHE-0645041). Y.-S.L. acknowledges financial support from Taiwan Merit Scholarship (NSC-095-SAF-I-564-052-TMS).

See more of this Session: Advances In the Applications of Porous Materials
See more of this Group/Topical: Materials Engineering and Sciences Division

(681g) SBA-15 - Polymer Hybrid Materials for Selective Capture From Model Cellulose Hydrolysis Mixtures

Yong-Hong Zhao and Daniel F. Shantz, Artie McFerrin Department of Chemical Engineering, Texas A&M University, College Station, TX

The synthesis of SBA-15 / hybrid materials and their use for capturing target compounds from aqueous mixtures will be summarized. The organic groups attached to SBA-15 are made via ATRP chemistry and subsequent side chain functionalization of the polymer brush formed. These materials have been characterized using methods including spectroscopy, porosimetry, and microscopy. These well-defined hybrids possess a high density of functional groups that can be used to selectively capture compounds from solution. The majority of the talk will focus on our efforts investigating materials with pendant phenyl boronic acid groups which can be used to selectively capture sugars from aqueous media. Uptake data and equilibrium adsorption properties of both single component sugar/water mixtures as well as multicomponent mixtures will be reported. The current work demonstrates that it is possible to utilize ATRP chemistry on high surface area supports to make hybrid materials that can be tailored to selectively capture compounds from solution. These materials represent one route to achieve separations of complex mixtures produced by, for instance, biomass hydrolysis in aqueous solutions.

See more of this Session: Advances In the Applications of Porous Materials
See more of this Group/Topical: Materials Engineering and Sciences Division

(685a) INVITED: Using Antigenic Disguise to Inhibit Protein and Cellular Interactions with Surfaces

Kimberly Anderson, Chemical and Materials Engineering, University of Kentucky, Lexington, KY, **Matthew T. Dickerson**, Department of Chemical & Materials Engineering, University of Kentucky, Lexington, KY and Marc Knecht, Chemistry, University of Kentucky, Lexington, KY

When a foreign body enters the bloodstream it elicits a swift immune response predicated on the adsorption of plasma proteins followed by a cascade of events that results in thrombosis. Though it is a critical function of the body's immune system, when blood contacting medical devices are involved it can lead to many undesirable outcomes including heart attack or stroke. To resolve this issue new ways to enhance material hemocompatibility are necessary. One such technique is the use of antigenic disguise proteins found in various microorganisms. Antigenic disguise is a term sometimes used to describe the ability of a microorganism to circumvent the host immune system. The exact method varies widely in each organism. One example is *Treponema pallidum*, which is believed to bind host plasma fibronectin (FN) and use it to shield potential antigen sites on its surface from attack. This ability is thought to originate in a number of surface proteins, one of which is labeled Tp0483. This presentation will focus on the use of antigenic disguise to inhibit protein and cellular interactions with surfaces. Specifically, the interactions of Tp0483/FN with self-assembled monolayers will be discussed and the effects of these adsorbed proteins on plasma protein binding and cell interactions will be discussed.

See more of this Session: Cell-Biomaterial Interactions
See more of this Group/Topical: Materials Engineering and Sciences Division

Assembly of 3D Tumor Spheroids by Tuning Matrix Mechanics

Youyun Liang[1,2], Jae Hyun Jeong[1], Ross J. DeVolder[1], Chaenyung Cha[1], Fei Wang[1], Yen Wah Tong[2] and Hyunjoon Kong[1]
(1)Department of Chemical and Biomolecular Engineering, University of Illinois at Urbana-Champaign, Urbana, IL, USA
(2) National University of Singapore, Singapore, Singapore

Three dimensional (3D) tumor spheroid models are becoming important biomedical tools for both fundamental and applied cancer studies, but current models do not account for different levels of cancer malignancy. Several studies have reported that the mechanical rigidity of a hydrogel plays a significant role in regulating the phenotypes of cancer cells adhered to the gel surface. This finding suggests that matrix rigidity should also modulate the malignancy of 3D tumor spheroids. However, the role of matrix stiffness is often confounded by concurrent changes in 3D matrix permeability. This study reports an advanced strategy to assemble 3D liver tumor spheroids with controlled intercellular organization, phenotypes, and angiogenic activities using hydrogels with controlled stiffness and minimal differences in molecular diffusivity. The elastic moduli of cell-encapsulated collagen gels were increased by stiffening interconnected collagen fibers with varied amounts of poly(ethylene glycol) di-(succinic acid N-hydroxysuccinimidyl ester) (PEG-diNHS) (Fig 1). Interestingly, hepatocellular carcinoma cells encapsulated in a fat-like, softer hydrogel formed malignant cancer spheroids, while cells cultured in a liver-like, stiffer gel formed compact hepatoids with suppressed malignancy (Fig. 2). Overall, both the hydrogel and the 3D tumor spheroids developed in this study will be greatly useful to better understand and regulate the emergent behaviors of various cancer cells.

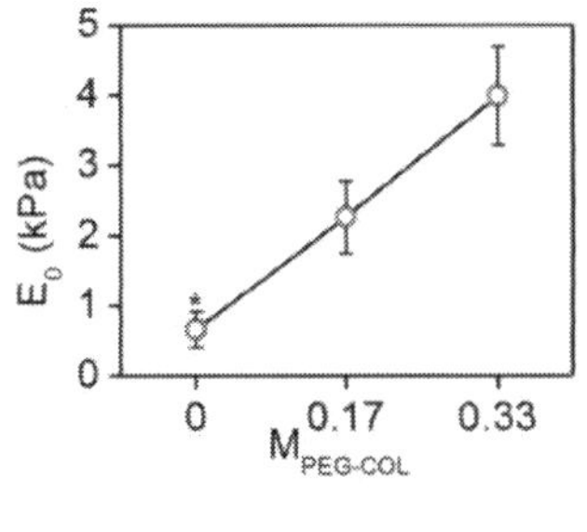

Fig 1. *Increasing the mass ratio of PEG-diNHS to collagen ($M_{PEG-COL}$) resulted in significant increase in the compressive elastic moduli (E_0). In the absence of PEG-diNHS, the E_0 is 0.7 kPa, similar to the stiffness of fat. When $M_{PEG-COL}$ is 0.33, the E_0 is 4.0 kPa, similar to the stiffness of healthy liver tissue.*

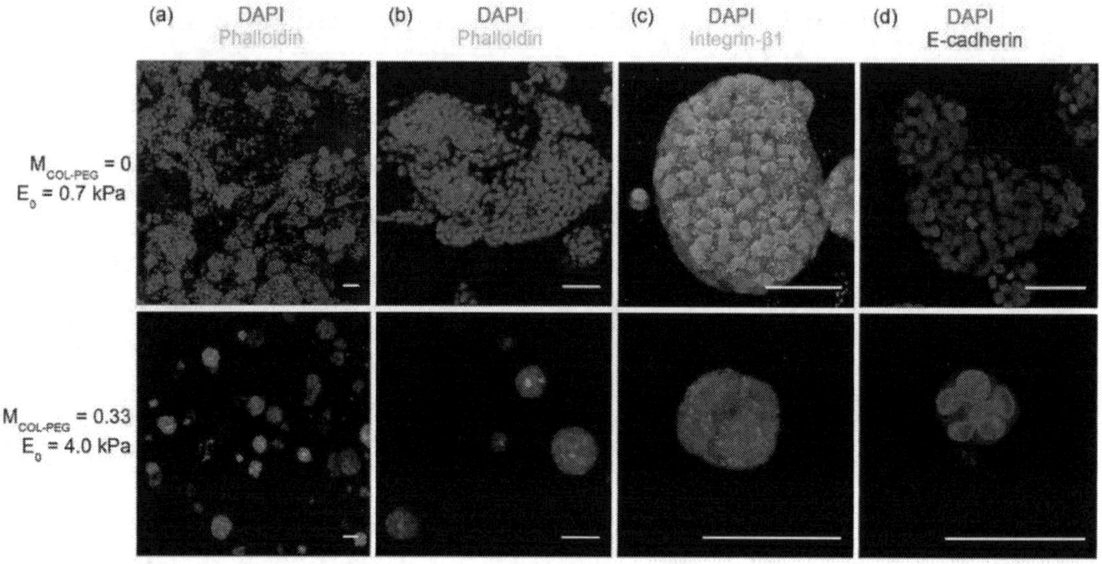

Fig 2. ***(a)*** *HepG2 cells encapsulated in 0.7 kPa gels for 9 days proliferated into large, loose cell clusters while HepG2 cells encapsulated in 4.0 kPa gels formed small, compact spheroids. The large, loose clusters are characteristic of higher malignancy while the small, compact spheroids are indicative of suppressed malignancy.* ***(b)*** *represents magnified views of (a).* ***(c)*** *Integrin-β1 expression of HepG2 cells were higher in the softer gel while* ***(d)****E-cadherin expression was higher in the stiffer gel.*

(685c) INVITED: Design and Function of Programmable Biomaterials for the Study and Application of Mechanobiology

James H. Henderson, Syracuse Biomaterials Institute and Department of Biomedical and Chemical Engineering, Syracuse University, Syracuse, NY

In vitro studies have begun to elucidate the principles through which extracellular matrix (ECM) behavior supports and regulates morphogenesis. In recent years, engineered two- and three-dimensional substrates and scaffolds have provided increasingly powerful tools with which to investigate the relationships between cell mechanical behavior and ECM composition and organization. As engineered *in vitro* environments become more accurate biochemical and biophysical tools for investigating and modeling *in vivo* environments, the critical next step for many areas of cell biomechanics and mechanobiology will be incorporation of increased programmable physical functionality into the environments.

Motivated by the potential to improve human health through tissue engineering and regenerative medicine, we have been investigating programmable biomaterials for the study and application of mechanobiology [1]. This presentation will present ongoing work on programmable cell culture substrates, which we refer to as active cell culture substrates. These substrates, which exploit the functionality of shape memory polymers, can be programmed to change surface topography during cell culture.

The results to be presented demonstrate control of cell behavior through shape-memory-activated topographic changes and introduce the use of active cell culture substrates for investigation of mechanotransduction, cell biomechanical function, and cell soft-matter physics. The talk will discuss the potential for these new approaches to be extended to tissue engineering and regenerative medicine. Particular emphasis will be placed on the role of cell-biomaterial interactions in the design and function of these new materials.

[1] Davis, et al., Biomaterials 32(9): 2285-2293, 2011.

See more of this Session: Cell-Biomaterial Interactions
See more of this Group/Topical: Materials Engineering and Sciences Division

(685d) Matrix-Induced Alignment and Shear Flow: Effects On Endothelial Cells

Edwina Lai[1], Ngan F. Huang[2], Claire M. Anderson[1], John P. Cooke[2] and Gerald G. Fuller[1], (1) Chemical Engineering, Stanford University, Stanford, CA, (2)Department of Cardiovascular Medicine, Stanford University, Stanford, CA

In straight segments of the blood vessel, the endothelium is aligned and the endothelial cells are resistant to inflammation and thrombosis. At bends and branches of the vessel where the endothelium is not aligned, however, the endothelial cells express adhesion molecules and chemokines that promote the adherence of platelets and leukocytes, leading to atherosclerosis. In this study, we aim to investigate the effects on the morphology and biology of endothelial cells due to the presence of an aligned matrix, with and without the presence of shear flow. Although it is known how shear flow can influence endothelial cells, the combination of matrix-controlled alignment and shear flow on endothelial biology (and morphology) is relatively unknown. While the phalloidin alignment and assembly of focal adhesions will help characterize the morphological effects on the cells, their biological function will also be evaluated. Proliferation, monocyte adhesion and integrin signaling are a few of the biological indicators of the endothelial cells that can indicate the effects of matrix-induced alignment and shear flow.

Endothelial cells are cultured onto aligned and unaligned collagen substrates and then exposed to a continuous period of shear flow, perpendicular to the direction of collagen alignment. Afterwards, the morphology and biology of the endothelial cells are evaluated. Cells cultured on aligned collagen demonstrate a resistance to re-aligning themselves along the shear flow direction, while cells on unaligned collagen tend to elongate along the direction of shear flow. It is also expected that the endothelial cells on aligned collagen will express less adhesion molecules that prevent the adherence of monocytes. These observations may be especially significant in the development of vascular conduits, since endothelial cells are generally healthier if they can maintain their aligned orientation despite the turbulence of surrounding blood flow.

See more of this Session: Cell-Biomaterial Interactions
See more of this Group/Topical: Materials Engineering and Sciences Division

(685e) Patterning Biomaterials to Engineer Cell Fate

Brendan A. Harley, Chemical & Biomolecular Engineering, University of Illinois at Urbana-Champaign, Urbana, IL

The extracellular matrix (ECM) is a complex organization of structural proteins such as collagens and proteoglycans. Heterogeneous tissues with spatially and temporally modulated properties and their biomaterial mimics play an important role in organism physiology and regenerative medicine. With the understanding that the microstructure, mechanics, and composition of the ECM is dynamic and often spatially patterned or heterogeneous over the length-scale of traditional biomaterials, there has recently been significant effort aimed at moving away from static, monolithic biomaterials towards instructive biomaterials that provide specialized cell behavioral cues in spatially and temporally defined manners.

We have been developing patterned, tunable biomaterial systems to explore the practical significance of how cell/matrix cues can be optimized to improve biomaterial regenerative potential and the mechanistic details of how individual (stem) cells sense, integrate, and respond to multiple microenvironmental signals. Here I will present the development of biomaterials for traditional regenerative medicine applications as well as stem cell fate engineering. We are integrating anisotropic and multi-compartment collagen scaffolds with photolithography-based biomolecule patterning tools for the regenerative repair of orthopedic defects. We are also creating multi-gradient and combinatorial biomaterials for rigorous investigation of fundamental questions regarding niche-mediated regulation of hematopoietic stem cell (HSC) behavior. Here microfluidic tools aid our investigation of the role played by matrix elasticity, ligand presentation, and paracrine-mediated signaling on HSC fate.

See more of this Session: Cell-Biomaterial Interactions
See more of this Group/Topical: Materials Engineering and Sciences Division

(685f) A Cell-Biomaterial Feedback Loop for Neural Tissue Engineering

Kyle J. Lampe[1], Karen S. Straley[2] and Sarah C. Heilshorn[1], (1)Materials Science and Engineering, Stanford University, Stanford, CA, (2)Chemical Engineering, Stanford University, Stanford, CA

Neural regeneration within the central nervous system (CNS) is a critical unmet challenge as brain and CNS disorders continue to be the leading cause of disability nationwide. Common tissue engineering goals seek to customize cell-biomaterial interactions and guide cell behavior. Here we have developed a material that can be both cell instructive, as well as cell responsive, thus creating a dynamic interplay between cells and their engineered extracellular matrix. Using recombinant protein technology, we have engineered a family of elastin-like, protein-based hydrogels with multiple independently tunable properties. This allows molecular-level design of our protein polymers and the incorporation of bioactive peptide sequences directly into the polymer backbone. With these materials, we can investigate individual and synergistic effects of elastic modulus, degradation rate, and cell adhesivity on cell behavior in a tunable microenvironment. The protein-polymers are covalently crosslinked into hydrogel scaffolds with elastic moduli from 1-100 kPa. Unlike natural matrices, the concentration of adhesive ligands such as the fibronectin-derived sequence RGD, laminin-derived YIGSR, or neural cell adhesion molecule (NCAM) domain, can be precisely controlled without altering the mechanical properties of the hydrogel. Increased RGD density was found to increase the adhesion of neuronal-like PC12 cells more than 10-fold and increase neurite outgrowth nearly 4-fold.

We have designed the proteins with a second set of bioactive sequences that specifically respond to changes in cell phenotype. By incorporating cell-mediated degradable subunits and adhesive sequences into the elastin-like proteins, we are able to mimic the natural remodeling of the extracellular matrix. We have found that neural stem cells (NSCs) undergoing differentiation increase their production of the protease urokinase plasminogen activator (uPA), which has previously been found at the growth cones of extending neurites. We engineered multiple uPA degradable bioactive sites with different degradation kinetics into the elastin protein to allow neural cell-mediated control of the scaffold degradation dynamics. This strategy was also used to enhance the functionality of the polymer by controlling delivery of multiple molecules with distinct release kinetics. One molecule was tethered to the matrix via a fast-degrading uPA-responsive sequence and fully released in 48 hours, while another molecule tethered by a more slowly degrading uPA-responsive sequence was continually released for greater than 240 hours.

These crosslinked scaffolds are useful for directing the growth and differentiation of multiple cell types including clinically relevant NSCs. Adult murine NSCs were capable of proliferation and differentiation into neurons and glia when seeded on top of RGD-containing scaffolds. These tunable scaffolds are responsive to neural cells which may be able to specifically self-modulate the release of multiple bioactive factors while undergoing differentiation. This work demonstrates the versatility and responsiveness of our modularly-designed protein hydrogels for neural cell culture and encourages continued development as a biomaterial tissue construct for treating spinal cord injury.

See more of this Session: Cell-Biomaterial Interactions
See more of this Group/Topical: Materials Engineering and Sciences Division

(685g) Motor Neuron Neurite Outgrowth and ECM Production In 3D Microenvironments

Naga Shilpa Vadala, Swarnalatha Balasubramanian and Jennie B. Leach, Department of Chemical & Biochemical Engineering, University of Maryland Baltimore County, Baltimore, MD

Despite great strides in the field of tissue engineering, we have yet to design responsive guidance channels for neuronal repair that are superior to autograft tissue. Three-dimensional (3D) culture models with improved physiological relevance are expected to bridge the gap between two-dimensional (2D) studies on petri-plates and the complex environment inside the body. We have recently demonstrated that the morphology and signaling pathways of sensory neurons are dramatically altered when the cells are placed in a 3D matrix, with the cellular response in vitro 3D environments being a better representation of that which occurs in vivo compared to 2D models. One of the fundamental differences between 2D and 3D culture is the distribution of cell-cell and cell-matrix interactions, which can alter signaling mechanisms regulating cell response. Moreover, we and others have demonstrated that neuronal response in 3D culture is dependent on cues from their local environment, including substrate physical properties as well as ligand concentration and geometry, indicating that neuronal growth and survival are improved by specific cell-matrix interactions that are different in 2D when compared to 3D. Cells alter these local cues as they secrete soluble factors and remodel their surrounding extracellular matrix (ECM). To date, the majority of neural tissue engineering studies have been carried out with model systems of neuronal differentiation (e.g., PC12 cells) or primary sensory neurons derived from dorsal root ganglia. To advance our understanding of how to design improved therapeutics for nerve regeneration, the specific goal of this work is to delineate mechanisms by which motor neurons respond to 3D environments. We are using a NCS-34 cell line, which is derived from a fusion of motor neuron enriched embryonic day 12-14 mouse spinal cord cells with mouse neuroblastoma. Herein we report the baseline characterization of NCS-34 morphology and ECM synthesis evoked by 2D and 3D collagen-based culture environments. In combination with investigations of cell-matrix signaling mechanisms evoked by motor neurons in 3D environments, this work will lead towards design rules for engineering new biomaterials with improved efficacy for treating nerve repair in vivo.

See more of this Session: Cell-Biomaterial Interactions
See more of this Group/Topical: Materials Engineering and Sciences Division

(690a) Development of New Redox Flow Battery Systems

Liyu Li[1], Wei Wang[2], Soowhan Kim[2], Vijayakumar Murugesan[2], Zimin Nie[2], Baowei Chen[2], Qingtao Luo[2], Feng Chen[2], Jun Liu[2], Jianzhi Hu[2] and Gary Yang[2], (1)Hydrocarbon Processing Group, Pacific Northwest National Laboratory, Richland, WA, (2)Pacific Northwest National Laboratory, Richland, WA

Concerns over the environmental consequences of burning fossil fuels and their remaining reserves have led to an increasing use of renewable energy generated from sources such as solar and wind. However, the intermittent, varied nature of the renewable resources makes it difficult to integrate these valuable energies into the electrical grids. A solution to this problem is to employ large-scale electrical energy storage (EES), which is widely considered as an effective approach to improve the reliability, power quality, and economy of these renewable energies. Among the most promising large-scale EES technologies are the redox flow batteries (RFBs). RFBs are special electrochemical systems that can repeatedly convert and store multi-megawatt-hours (MWhs) of electricity energy to chemical energy and convert the chemical energy back to electricity energy when needed. The advantages of RFBs include short response time, low self discharge, long life-time, and independent tunable power and storage capacity, which make it possible to independently scale up the electricity storage capacity and power generation capacity. With the support of US Department of Energy's Office of Electricity, PNNL has carried out extensive efforts in the development of advanced redox flow batteries. Substantial progress has been achieved in the past two years, including two new redox chemistries (V-Fe system and all vanadium mixed acid system), which will be reported in this presentation

See more of this Session: Energy Storage and Electrochemical Processes I
See more of this Group/Topical: International Congress on Energy 2011

Study on preparation and characterization of the composite bipolar plate for Vanadium Redox Flow Battery

XIE Xiao-Feng[1], WANG Wen-Pin[2] , WANG Shu-Bo[1], WANG Jin-Hai[1], LV Ya-Fei[2], MATHUR V. K[3]

1 Institute of Nuclear and New Energy Technology, Tsinghua University, Beijing 100084, China

2 College of Material Science And Engineering, Beijing University of Chemical Technology, Beijing 100029,China

3.Department of Chemical Engineering, University of New Hampshire, NH 03824, USA

Bipolar plate, as one of the main components of all-vanadium redox flow battery (VRB), can collect the current generated by electrochemical reaction, and can separate positive and negative electrolyte. The electrolyte is a sulfuric acid solution containing vanadium ions (V(IV/V,II/III)/ V(V)). Among these ions, the V(II) has the most strong oxidation ability. Therefore, the bipolar plate not only should have good conductivity, but also have high resistance to chemical and electrochemical corrosion.

Bipolar plate is a carbon composite material, which is a mixture of carbon materials and a polymer. The polymers used are polyethylene, nylon, polypropylene (PP), rubber modified polypropylene, polytetrafluoroethylene and polyvinylidene fluoride, etc. Polymer is helpful to form the conductive network, and improve the corrosion resistance of bipolar plate.

Maleic anhydride grafted polypropylene(g-PP) is a non-polar molecules in the main chain with a strong polar side groups maleic anhydride. It can increase the bonding and the compatibility between polar materials and non-polar materials for enhancing the dispersion between the conductive filler and the polymer matrix, and to improve conductive property and corrosion resistance of the bipolar plate.

In this work, composite bipolar plates, by polypropylene (PP), polypropylene-maleic anhydride graft copolymer (g-PP) as polymer matrix, and flake graphite(GC) as conductive filler, were prepared using a solution intercalation method. Four probes method, potentiodynamic experiments, potentiostatic and cyclic voltammetry(CV) were used to investigate the conductivity, corrosion resistance and electrochemical performance of bipolar plates. The results show that g-PP made the graphite uniformly dispersed in the polymer, which significantly improved compatibility between the conductive filler and the polymer matrix. The conductivity of the composite bipolar plate was about 77.8 $S\cdot cm^{-1}$. When the voltage was 1.2 V, the corrosion current was about $5.74\times10^{-2} A\cdot cm^{-2}$. The prepared bipolar plate could meet the requirements of VRB.

EXPERIMENTALS

The conductivity of samples (5 cm×1.5 cm) was measured by a four probe resistance tester. Other tests carried out were: Cyclic voltammetry: -1.5~2.0V, scanning speed

10mv·s-1; potentiodynamic polarization curves: -1.5~2.0V, scanning speed 5v·s-1; potentiostatic test (1.2 V, scan time 3600 s). Electrochemical measurements used three-electrode cell with the sample as the working electrode. A Pt electrode as the counter electrode and a saturated calomel electrode as reference electrode (SCE) were used. A homemade VRB electrolyte was used for the experiments.

A certain amount of GC, PP, g-PP and xylene were placed in three-necked bottle, stirred, heated to reflow temperature and kept under reflux for a period of time. The product was precipitated by a solvent, leached, dried, and hot-pressed to form the composite conductive plate.

RESULTS AND DISCUSSION

GC is the most familiar of the natural graphite materials. It has good conductivity. For a GC content of 70wt%, bipolar plates were prepared by using the g-PP content 10wt%, 15wt%, 20wt% or 25wt%. The effect of g-PP contents on conductivity of bipolar plates are shown in Fig. 1. It can be seen that with an increase in g-PP content, the conductivity of bipolar plate increases steeply. It means that g-PP addition contributes to the uniform distribution of the GC in the matrix, resulting in the conductive network structure formation and thus conductive property improvements. Polyacrylonitrile graphite felt (PAN-based) as electrode materials is widely used in VRB. Its conductivity commonly is about 50 $S\cdot cm^{-1}$. It is preferred to have the g-PP content greater than or equal to 20 wt% for collecting current.

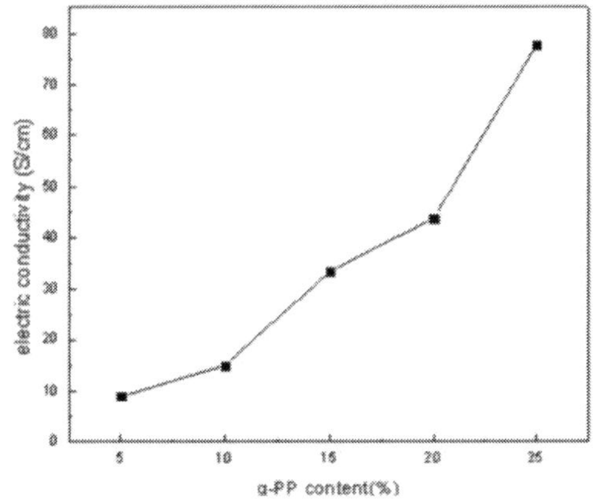

Fig.1 Effect of g-PP contents on conductivity

Fig.2 shows the potentiodynamic polarization curves of bipolar plates. It can be seen that as the voltage increases, anode passivation of bipolar plate occurs. That means the corrosion of bipolar plate is mainly due to oxygen corrosion. As the g-PP content increases, the

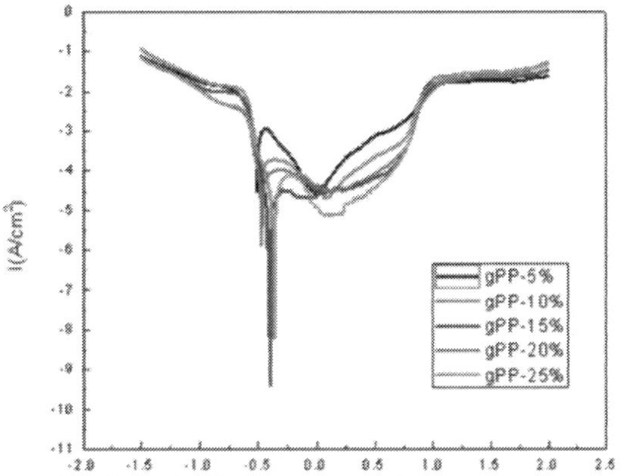

Fig.2 Potentiodynamic polarization curves

corrosion current of bipolar plate decreases first and then increases. When the g-PP content of bipolar plate is 20 wt%, the corrosion current is the lowest.

The effect of g-PP contents on corrosion current are showed in Fig.3. It can be seen that with the increase in g-PP content, the electrochemical corrosion current of bipolar plate decreases rapidly, and then runs through the transition state into the steady-state. When g-PP contents is about 5%, the corrosion current density is 1.8×10^{-2} A/cm^2. With the g-PP content increasing, the corrosion current density reduces to 9.1×10^{-3} A/cm^2. Moreover, once corrosion occurs, it will cause intergranular corrosion or pitting corrosion, and then the corrosion rate further increases. When the g-PP content is 20wt%, the corrosion current of bipolar plate is close to the minimum, which is approximately 8.9×10^{-3} A·cm^{-2}.

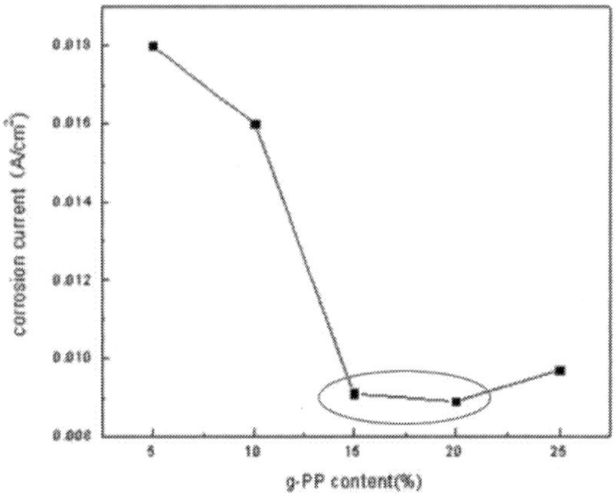

Fig.3 Effect on g-PP contents on corrosion current

Fig.4 shows cyclic voltammetry of bipolar plates under various g-PP contents. With the g-PP content increasing, more V(II) and V(III), V(IV) and V(V) transformation peaks were observed with reversibility. When g-PP is about 10%, it has the high activity. If it was used as bipolar plate in battery, the bipolar plate was easily corroded. For oxygen evolution peak of bipolar plate at anode, the higher the oxygen evolution potential, the more stable the battery is. When the g-PP content is 20wt%, there is low activity and high oxygen evolution potential of the bipolar plate providing the best the corrosion resistance of the bipolar plate.

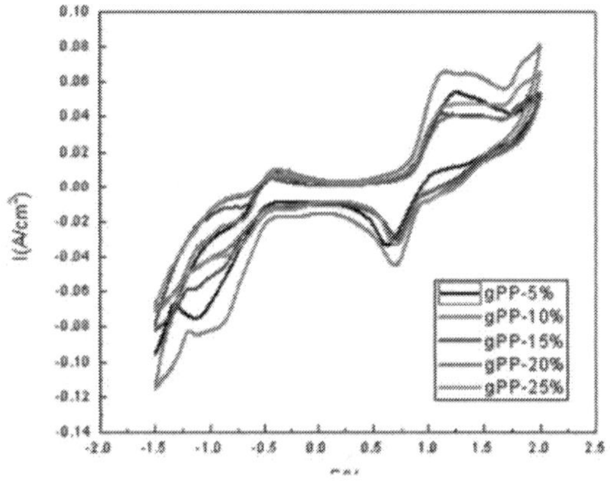

Fig.4 Cyclic voltammetry of bipolar plates

With the increase of g-PP content, the dispersion of GC in the polymer matrix is more uniform, the holes in the bipolar plate surface are reduced, and vanadium ions have difficulty in attaching to the surface of bipolar plate. As the g-PP content increases to 25wt%, the functional group COO^- content of the bipolar plate surface increases. COO^- can contribute to vanadium ion adsorption from solution to the surface. With the reaction of vanadium ion adsorption and desorption, the charge transfer should obviously change. It makes the electrochemical reaction of the vanadium ions on the surface of the bipolar plate to easily happen. These results are in agreement with the previous investigation.

CONCLUSIONS

Analysis of potentiodynamic, potentiostatic experiments and cyclic voltammetry shows when the GC content is constant, within a certain range, conductivity of bipolar plate is proportional to the g-PP content. The g-PP is added to make the dispersion between GC and the polymer matrix and improve the compatibility between the conductive filler and the polymer matrix significantly. Addition of g-PP leads to improving the corrosion resistance of bipolar plate. Conductivity and corrosion resistance of composite bipolar plate can meet the requirements of VRB.

ACKNOWLEDGEMENTS

The work was funded by Tsinghua University Initiative Scientific Research Program (201010800623), State Key Laboratory for Automotive Safety and Energy(KF10132), State Key Lab of Power Systems(SKLD10KZ03).

REFERENCES

[1] WANG Wenpin , WANG Jinhai, WANG Shubo, XIE Xiaofeng , LV Yafei. Properties of Conductive Bipolar Plate for All-Vanadium Redox Flow Battery [J]. The Chemical Industry and Engineering Society of China, 2011, 62(SI): 203-207

[2] **http://www.asbury.com/Natural-Flake-Graphite.html**

[3] Xia Liping, Lin Changwu, Luo Dongmei, Zhang Aimin. Study on properties of fluoro rubber composite current collector for vanadium redox battery [J]. Power Sources, 2010, 134(7): 672-675

[4] Chen maobin, Zhang Shengtao, Li Xiaobing, Liu Lian, Liu Xiaojiang, Meng Fanming. Preparation and performance of polytetrafluoroethylene conducting plastic on vanadium redox flow battery collector [J]. Polymer Materials Science And Engineering, 2009, 25(6): 121-124

[5] Chen Maobin, Lan Wei, Zhang Shengtao, Cui Yanhua, Meng Fanming, Liu Xiaojiang. Performance of Conductive Polyethylene Composite [J]. Applied Chemistry, 2009, 26(5): 619-621

Investigation on electrode reaction mechanism for vanadium redox flow battery

XIE Xiaofeng[1], LI Xiaoshan[1], YANG Chun[1], WANG Jinhai[1], WANG Shubo[1], Shang Yuming[1], MATHUR V. K[2]

1. Institute ofNuclear andNew EnergyTechnology, Tsinghua University, Beijing100084, China

2.Department of Chemical Engineering, University of New Hampshire, NH 03824, USA

Abstract:

The effect of glycerine addition to the anodic electrolyte, on the anodic reaction, was investigated. The results show that 1% glycerine added to the electrolyte had positive effect on the performance of electrolyte reaction on the surface electrode. The solution resistance, capacitance value and polarization resistance also improved. The reaction of VO_2^+ to VO^{2+} was helped as a result of lower polarization resistance.

Key words: Vanadium Redox Flow Battery ; Glycerine ; Electrochemical impedance spectroscopy; Electrolyte, Cyclic voltammetry; Potentiodynamic

1. Intruduction

Vanadium redox flow battery (VRB) employs vanadium ions of different valences as liquid electrolytes. It is independent of storage capacity, output power, normal atmospheric temperature operation, fast response, and long operating life. The VRB system can widely be used for load leveling, distributing power, standby power supply, and municipal transport. Cyclic voltammetry and Electrochemical impedance spectroscopy (EIS) methods are useful in characterizing the polarization resistance corrosion rate and electrochemical mechanism.

When the positive electrolyte vanadium concentration is up to 2M or the temperature is elevated above 310K, the fully charged electrolyte has poor stability. The precipitate of V(V) would block the electrode and prevent the electrolyte from flowing through the pipe. In order to keep the electrolyte from precipitating, and improve the energy storage density of electrolyte used in VRB, one method is to increase the concentration of vanadium ions. Many workers have chosen inorganic and organic additives to be added to the electrolyte for solving this problem. Some additives added to electrolyte can broaden the electrochemical window and increase electrochemical activity. The dissolved V(V) in concentrated sulfuric acid solution becomes more stable and effective at high concentration of vanadium. Glycerine has been used to enhance the conductivity and stability of anodic electrolyte for improving the performance of VRB. In this work, the effect of glycerine addition, to the anodic electrolyte, on the mechanism has been studied.

2 EXPERIMENTS

2.1 Preparation of anodic electrolyte

$VOSO_4 \cdot 3H_2O$ was dissolved in the deionized water, and H_2SO_4 was added to form $VOSO_4$+3 M solution. The solution was added to anodic electrolytic cell, and H2SO4 of the same concentration was added to cathodic electrolytic cell. The electrolytic cell was separated by an ion exchange membrane (Nafion 117, DuPont), and the lead electrodes were used for anode and cathode electrolysis. The prepared electrolyte was 1.8 mol/L, with the active area of $16cm^2$ (4cm × 4cm) and electrolytic current density of 20 ~ $200mA/cm^2$.

2.2 Electrochemical tests

A three-electrode electrochemical cell was used for electrochemical tests. The graphite electrode (with 5mm diameter) was a working electrode. The reference electrode was SCE, and a platinum mesh was used as auxiliary electrode. Different concentration (v/v) of glycerine was added in the anodic electrolyte.

The electrode characterization included the determination of the polarization resistance and electrochemical mechanism. A Zanher IM6ex instrument was used to conduct the potentiodynamic, cyclic voltammetry and EIS measurements. Potentiodynamic tests were used to measure the corrosion resistance. The initial potentials were -0.15 V, and the final potentials were 2.0 V, the scan rate was 1 mVs-1. Cyclic voltammetry was used to measure the electrochemical activities. The scan range was -0.8 ~ 1.6 V. The EIS measurements were performed at different potential and the perturbation amplitude was 5 mV. The applied frequencies ranged from 100K to 10^{-1} Hz.

3. RESULTS AND DISCUSSION

A potentiodynamic test is commonly employed to measure a material's resistance to corrosion. Potentiodynamic tests were conducted to estimate the corrosion resistances. The "corrosion resistances" here was considered as the electrolyte's ionization after the glycerine was added. Fig. 1 presents the potentiodynamic polarization scans at different glycerine contents. From the Tafel curves in Fig. 1, the related polarization kinetic parameters were analyzed by potentiodynamic polarization model in Zanher IM6ex electrochemical workstation. From the potentiodynamic polarization results and using Eq. (1), polarization resistance, Rp was estimated. Table 1 presents the polarization parameters of anodic electrolyte. The polarization resistances for 0.1% and 10% glycerine were 44744 Ω·cm2 and 13027 Ω·cm2, respectively.

$$R_P = \frac{\beta_a \beta_c}{2.3 i_{corr}(\beta_a + \beta_c)} \quad (1)$$

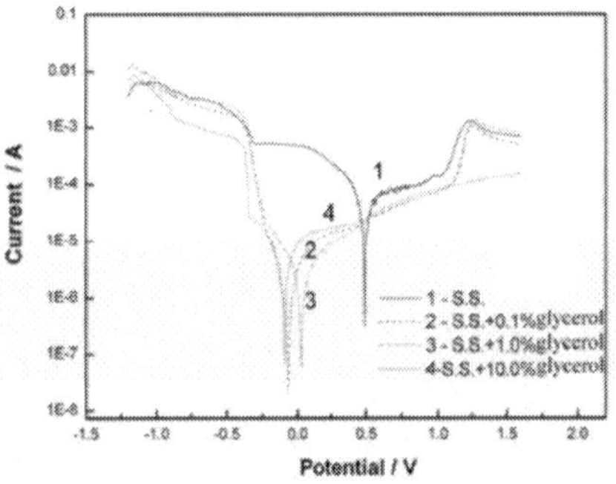

Fig. 1 Potentiodynamic polarization scans of anodic electrolyte

Table 1 Polarization parameters of anodic electrolyte

	βa/mV	-βc/mV	Rp/Ω	Icorr/μA
Non glycerin	86.2	79.9	1654	10.9
0.1% glycerin	64.1	62.3	44744	0.307
1.0% glycerin	54.0	52.9	28617	0.406
10% glycerin	57.4	60.7	[illegible]	[illegible]

Cyclic voltammetry or CV is a type of potentiodynamic electrochemical measurement. Cyclic voltammetry was used to study the electrochemical properties of the glycerine in solution. The results are shown in Fig. 2. When glycerine was added to the anodic electrolyte, the quantity and status of vanadium is changing between the electrolyte and electrode. With the glycerine content increased from 0.1% to 10%, the oxidation peak current decreased. The hydroxy groups on glycerine are bound with the vanadium ions, resulting in the decrease of surface electrode charge, transferring more vanadium ions to the bulk solution.

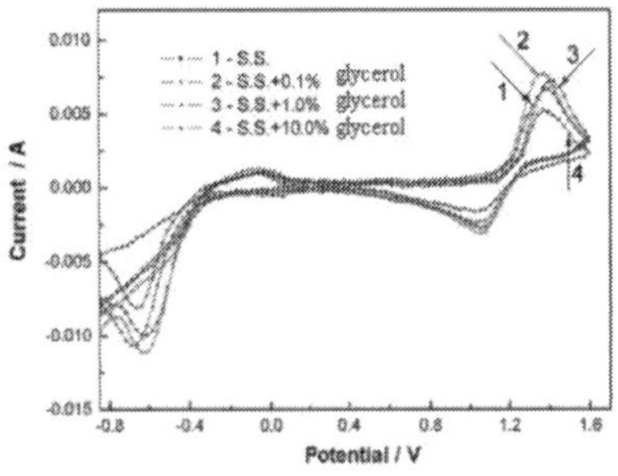

Fig. 2 Cyclic voltammetry scans of anodic electrolyte

EIS reveals information about the reaction mechanism of an electrochemical process: different reaction steps will dominate at certain frequencies, and the frequency response shown by EIS can help identify the rate limiting step. The EIS was employed to investigate the electrode reaction mechanism in anodic electrolyte containing glycerine as additive. The spectrums were simulated and analyzed through the equivalent circuit of $R_\Omega(C_d(R_pW))$. The reaction mechanism of glycerine added to electrolyte is shown in Fig 3.

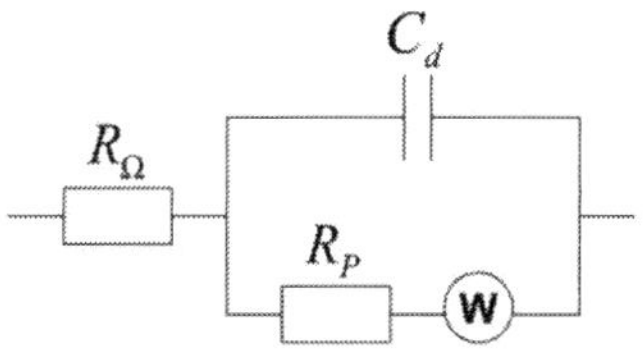

Fig 3. Equivalent circuit using RΩ(Cd(RpW))

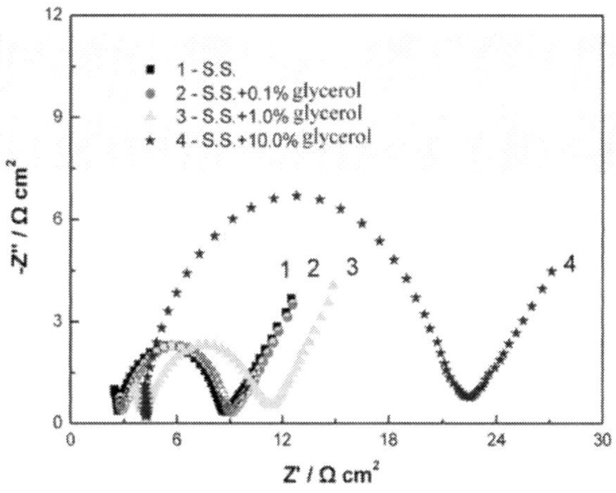

Fig.4 Impedance spectra for anodic electrolytes

The results of EIS are shown in Fig 4. From the impedance curve, with the glycerine concentration increasing, the impedance seems to increase. It means there is an optimal amount for glycerine that can be added to the electrolyte. Therefore, we should conduct the simulation of equivalent circuit to explore the mechanism of reaction with different concentration of glycerine added in. The results show the solution resistance increases by the addition of glycerine. It is difficult for glycerine to ionize in water. When the non-conducting glycerine is added to the electrolyte, the solution resistance should increase. With the increase in glycerine concentration, the resistance of electrolyte increases from 2.795Ω·cm^2 to 4.16 Ω·cm^2. In fact, the molecular structure of glycerine has three hydroxyls,

Table 2 Parameters of the equivalent circuit

	RΩ/(Ω·cm^2)	Cd/(S·cm^2)	Rp/(Ω·cm^2)	W/(S·cm^2)
Non glycerin	2.572	7.735	6.106	0.2478
0.1% glycerin	2.795	8.809	6.25	0.2676
1.0% glycerin	3.855	4.04	7.413	0.245

which are prone to bind with vanadium ions. With the glycerine content increasing from 0% to 10%, the polarization resistance increased. It means the glycerine added to electrolyte is unfavorable for reaction VO_2^+ to VO^{2+}. When the concentration exceeds 1%, the unnecessary glycerine might occupy the reaction site which can catalyze the redox reaction of vanadium ions. Therefore, when more glycerine is added to electrolyte, the reaction impedance increases. It means less glycerine is helpful for VO_2^+ to react to VO^{2+}. The results are in agreement with potentiodynamic test and Cyclic voltammetry. It indicates that glycerine contents when increased beyond certain limits will decrease the reaction rate.

4 . CONCLUSIONS

The results show that glycerine added to anodic electrolyte greatly improved oxidation peak current, reaching a maximum when glycerine was about 1%(v/v). Meanwhile, with the increase in glycerine content, the solution resistance (R_Ω), the double electric layer capacitor (C_d) and the polarization resistance (R_p) reached their minimum. However the concentration polarization reached its maximum when glycerine added was about 1%(v/v).

The addition of glycerine increased solution resistance (R_Ω) which is disadvantageous to the vanadium ion movement in electrolyte solution. Moreover, the glycerine could greatly decrease the corrosion of electrolyte materials. With the increase in additive, the anti-corrosion tends to decrease. When the addivties was up to 0.1%(v/v), the anti-"corrosive ability" was the best.

ACKNOWLEDGEMENTS

The work was funded by Tsinghua University Initiative Scientific Research Program (201010800623), State Key Laboratory for Automotive Safety and Energy(KF10132).

REFERENCES

1) Bard, Allen J.; Larry R. Faulkner (2000-12-18). Electrochemical Methods: Fundamentals and Applications (2 ed.). Wiley. ISBN 0471043729.

2) http://en.wikipedia.org/wiki/Electrochemical_impedance_spectroscopy

(690d) Active Species Concentration Effects On Performance for a Non-Aqueous All-Vanadium Redox Flow Battery

Aaron A. Shinkle, Alice E. S. Sleightholme, Levi T. Thompson and Charles W. Monroe, Department of Chemical Engineering, University of Michigan, Ann Arbor, MI

Redox flow batteries(RFBs) are being developed for large-scale energy storage, uninterruptablegrid-independent power supplies and load-leveling systems for solar or windpower. Aqueous chemistries are used for all current commercial RFBs, therefore, cell potentials are limited to thestability range of water (01.23 V vs. SHE). Since a battery's cellpotential determines its energy and power density, aqueous chemistries limitthe possibilities for the performance of RFBs.

Several non-aqueous systems have beenreported recently, including systems based on ruthenium, uranium, chromium,manganese, and vanadium coordination complexes [1-5]. In the case ofvanadium acetylacetonate, the use of a non-aqueous electrolyte results in a stablecell potential window of 2.2 V [5], more than 50% wider than the aqueoussingle-metal vanadium chemistry (1.26 V [6]). However, very few resultsfrom battery charge/discharge experiments have been shown for this non-aqueousvanadium system.

This talk will examine thecharge/discharge profile for the vanadium acetylacetonate ($V(acac)_3$)RFB system in acetonitrile, with tetraethylammonium tetrafluoroborate ($TEABF_4$)support. The effects of active-species concentration on batteryperformance were examined. Linear sweep voltammetry will be used todetermine the maximum charge/discharge rate of the battery cell as a functionof electrolyte composition. Figure 1 shows the first five cycles of acharge/discharge result for 0.4 M $V(acac)_3$,0.5 M $TEABF_4$ in acetonitrile. Coulombic, energy, and powerefficiency for charge/discharge experiments were examined as a function ofconcentration to illustrate the effects of active species crossover. Efficiency and energy density were balanced to determine an optimal activespecies concentration range for operation of the battery. Lastly, thenon-aqueous system was compared to the optimized aqueous all-vanadium system toidentify areas of future work.

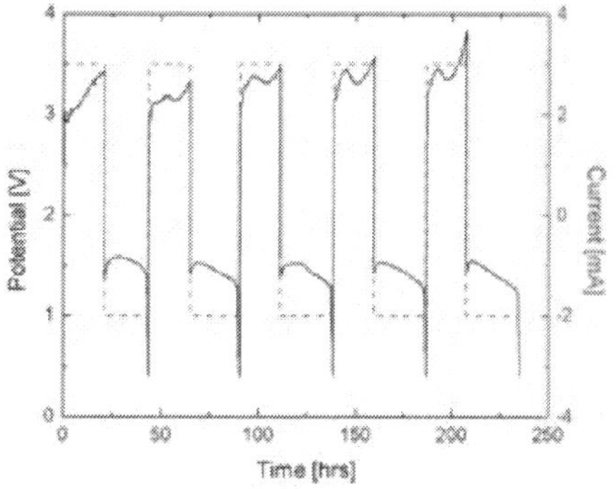

Figure1: First five cycles of a charge/discharge experiment for an all-vanadiumnon-aqueous redox flow battery chemistry. The system was charged at 3 mA until 50% theoretical stateof charge and discharged at 2 mA until 0.4 V. The system shown is 0.4 M vanadium acetylacetonate supported by 0.5 Mtetraethylammonium tetrafluoroborate in acetonitrile solvent. An H-cellconfiguration was used with graphite plate electrodes and a NeoseptaAHA anion exchange membrane.

REFERENCES

1. Matsuda Y, Takasu Y, Morita M, Tanaka K, Okada M, Matsumura-Inoue T(1985) Denki Kagaku 53:632

2. YamamuraT, Shiokawa Y, Yamana H,Moriyama H (2002) Electrochim Acta48:43

3. LiuQ, Shinkle AA, Li Y, Monroe CW, Thompson LT, Sleightholme AES (2010) Electrochem Comm 12:1634

4. SleightholmeAES, Shinkle AA, Liu Q, Li Y, Monroe CW, Thompson LT (2011) J Power Sources196:5742-5745

5. LiuQ, Sleightholme AES, Shinkle AA, Li Y, Thompson LT (2009) ElectrochemComm 11:2312

6. Poncede Leon C, Frias-Ferrer A, Gonzalez-Garcia J, Szanto DA, Walsh FC (2006) J Power Sources 160:716-732

See more of this Session: Energy Storage and Electrochemical Processes I
See more of this Group/Topical: International Congress on Energy 2011

(690e) Energy Storage and Generation From Thermopower Waves: Covalently Functionalized Thermal and Electrical Conduits

Joel T. Abrahamson[1], Changsik Song[2], Jennifer H. Hu[1], Jared M. Forman[1], Sayalee G. Mahajan[1], Nitish Nair[1], Wonjoon Choi[3], Eun-Ji Lee[2] and Michael S. Strano[1], (1)Department of Chemical Engineering, Massachusetts Institute of Technology, Cambridge, MA, (2)Department of Chemistry, Sungkyunkwan University, Suwon, South Korea, (3)Mechanical Engineering, Massachusetts Institute of Technology, Cambridge, MA

We have demonstrated through simulation and experiment that the nonlinear coupling between an exothermic chemical reaction and a nanowire or nanotube with large axial heat conduction accelerates the thermal reaction wave along the nano-conduit [1-3]. The thermal conduit rapidly transports energy to unreacted fuel regions, and the reaction wave induces a concomitant thermopower wave of high power density, producing electrical current in the same direction. At up to 7 W/g, this can be substantially larger than the power density offered by current micro-scale power sources (e.g. fuel cells, batteries) and even about seven times greater than that of commercial Li-ion batteries. MEMS devices and wireless sensor networks would benefit from such high power density sources to enable functions such as communications and acceleration hampered by present power sources.

Recently, we have covalently attached energetic molecules of mono-, di-, and trinitrobenzenes to single-walled carbon nanotubes via diazonium chemistry. Although covalent functionalization introduces defects that can scatter electrons and phonons, thermopower waves can still rapidly propagate and produce electricity on these decorated nanotubes. Differential scanning calorimetry confirms that the energetic molecules release heat, but thermopower waves require additional fuel adsorbed on the nanotubes to propagate.

[1] W. Choi, S. Hong, J. T. Abrahamson, *et al. Nature Materials*, 9, (2010), 423.

[2] W. Choi, J. T. Abrahamson, J. M. Strano, M. S. Strano. *Materials Today,* 13, (2010), 22.

[3] J. T. Abrahamson, *et al. ACS Nano* 5, (2011), 367.

See more of this Session: Energy Storage and Electrochemical Processes I
See more of this Group/Topical: International Congress on Energy 2011

(695b) Invited: Thiol-Ene Click Hydrogels for Cytocompatible Generation and Rapid Recovery of Insulin-Secreting Cell Spheroids

Chien-Chi Lin, Biomedical Engineering, Indiana University-Purdue University at Indianapolis, Indianapolis, IN

Poly(ethylene glycol) or PEG-based hydrogels have been widely used for cell delivery and as a platform for 3D cell culture. Recently, PEG hydrogels formed by step-growth thiol-ene photopolymerizations have emerged as attractive and versatile matrices for tissue regeneration. Compared to chain-growth PEG diacrylate (PEGDA) hydrogels, gels prepared from thiol-ene click reactions have homogeneous network structures, higher monomer conversion, and enhanced mechanical properties. Thiol-ene click reactions also preserve all of the preferential properties offered by photopolymerizations, specifically rapid and spatial-temporal controlled over gelation kinetics. In addition, thiol-ene PEG hydrogels can be rendered biodegradable by using enzyme sensitive peptide sequences as crosslinkers. Here, we compare properties of chain-growth and step-growth photopolymerizations in the context of pancreatic beta cell encapsulation. In addition to the above-mentioned benefits, we found that step-growth thiol-ene click reactions are highly cytocompatible for pancreatic beta cells, even when cells were encapsulated at very low density. Beta cells (MIN6) not only had high viability in thiol-ene click hydrogels, they also proliferated to form narrowly dispersed cell spheroids. Furthermore, when a chymotrypsin sensitive peptide sequence was used as gel crosslinker, insulin-secreting cell spheroids can be rapidly and safely recovered (within 5min) from the thiol-ene hydrogels via a surface erosion mechanism. These studies demonstrate the advantages and versatility of thiol-ene photopolymeirzations in tissue engineering and regenerative medicine applications.

See more of this Session: Hydrogel Biomaterials II
See more of this Group/Topical: Materials Engineering and Sciences Division

(695c) Invited: Bioinspired Hydrogels That Regulate Growth Factor Signaling

William L. Murphy[1], Gregory Hudalla[2], Michael Toepke[2], Nicholas Impellitteri[2], Sheeny Lan Levengood[2] and Justin Koepsel[2], (1)Biomedical Engineering and Orthopedics, University of Wisconsin, Madison, WI, (2)Biomedical Engineering, University of Wisconsin, Madison, WI

Control over the soluble signals that cells encounter in their local environment is a common theme in natural tissue formation, and also an emerging theme in functional tissue engineering strategies. This concept is particularly important in stem cell-based applications, in which local soluble signals can dictate cell fate decisions. Nature often achieves intricate control over local soluble signaling via specific, non-covalent interactions. Inspired by these natural interactions, we are interested in creating biomaterials that actively regulate soluble signaling. For example, our recent studies show that biomaterials can be designed to bind specific growth factor signaling components, including growth factors, growth factor receptors, and proteoglycans. We have used this approach to up- or down-regulate growth factor signaling in vitro and in vivo. One example uses soluble proteoglycan sequestering as a broad mechanism to regulate human mesenchymal stem cell proliferation and osteogenic differentiation in standard cell culture conditions. Another example explores the utility of injectable, growth factor-sequestering hydrogels in tissue engineering applications. In summary, our recent studies show that tailored non-covalent interactions between growth factors and biomaterials may provide an adaptable mechanism to control growth factor signaling and, in turn, stem cell behavior and new tissue formation.

See more of this Session: Hydrogel Biomaterials II
See more of this Group/Topical: Materials Engineering and Sciences Division

(695f) Modeling and Experimental Design of Poly (ethylene glycol) Hydrogel Extracellular Matrix Mimics Formed by Free-Radical Photopolymerization

Chu-Yi Lee[1], Michael Turturro[2], Josha James[2], Fouad Teymour[1] and Georgia Papavasiliou[2], (1) Chemical and Biological Engineering, Illinois Institute of Technology, Chicago, IL, (2)Biomedical Engineering, Illinois Institute of Technology, Chicago, IL

Synthetic hydrogels of poly(ethylene glycol) diacrylate (PEGDA) formed by free-radical photopolymerization have been extensively investigated as extracellular matrix (ECM) mimics for tissue engineering applications. However, the ability to predict and control the spatial and temporal presentation of biochemical and mechanical properties for guiding cell behavior still remains a significant challenge. To this end, we are coupling experimental data with computational modeling in order to determine unknown kinetic parameters and to optimize hydrogel crosslink density and the final biochemical composition of ECM molecules immobilized in the crosslinked network.

The hydrogels were synthesized by visible light free-radical photopolymerization using an Argon Ion laser (λ= 514 nm) in the presence of the crosslinking agent PEGDA, N-vinyl pyrrolidone (NVP), Acrylate-PEG-YRGDS, using the photoinitiator eosinY and the coinitiator triethanolamine (TEA). The photoinitiation mechanism of this photopolymerization system was investigated experimentally through photopolymerization of NVP and precipitation of poly(NVP). The rate of polymerization was measured gravimetrically and the kinetic parameters for this initiation system are obtained from a series of experiments, and applied in the computational models.

Two variants of the mathematical model are presented; the first predicts the molecular weight distribution of the the primary polymer chains that result in hydrogel formation, and the second is a full crosslinking model that predicts hydrogel crosslink density and the incorporation of the cell adhesion ligand YRGDS. Predictions from the first model provide the kinetic parameters that can be used in the crosslinking model. Both models implement the kinetic approach based on the method of moments. The crosslinking model utilizes the Numerical Fractionation technique to circumvent the issue of numerical divergence at the gel point in order to predict the hydrogel crosslink density and the final concentration of YRGDS in the hydrogel. Model predictions agree with experimental data of hydrogel YRGDS incorporation (measured by radiolabeling experiments) and crosslink density (obtained via swelling experiments). The predictive capability of the computational and experimental studies presented will provide invaluable insight for designing scaffolds that dictate cell behavior.

See more of this Session: Hydrogel Biomaterials II

(695d) Photopolymerized Carboxybetaine Hydrogels with a Carboxybetaine Dimethacrylate Crosslinker for Functionalization and High Mechanical Strength

Louisa R. Carr, Lei Zhang, Yibo Zhou, Jordan B. Krause and Shaoyi Jiang, Chemical Engineering, University of Washington, Seattle, WA

Hydrophilic crosslinked polymer networks, or hydrogels, are commonly studied systems for many biological applications ranging from corneal implants to drug delivery and wound dressings. The key attributes of hydrogels that make them particularly well-suited to biological applications are their high water content, biological stability, optical transparency, permeability to metabolites, and sufficient crosslinking to prevent dissolution. In order to expand the scope of hydrogels' impact in the biomedical arena, several challenges must be addressed, primarily improving mechanical properties, without reducing its high water content and permeability.

Hydrogels made from zwitterionic materials have many interesting properties. Zwitterionic hydrogels have been studied as superabsorbants[1], with particular attention being paid to their swelling properties [2]. We have previously reported low protein adhesion on sulfobetaine methacrylate and mixed charge hydrogels[3] and low cell adhesion on carboxybetaine methacrylate gels[4]. The zwitterionic gels studied so far have shown low mechanical strength[5],[3], which limits their potential biological uses, but we have achieved high mechanical properties by designing and synthesizing a carboxybetaine dimethacrylate crosslinker. Thermally polymerized hydrogels with a compressive modulus of 8 MPa and 60% water content were made[6].

Photopolymerizing these hydrogels, on the other hand, improves the uniformity of the polymer network and results in an order-of-magnitude improvement in compressive modulus. With 50% water, hydrogels were made with a compressive modulus of 90 MPa. Furthermore, a redesign of the crosslinker to enable it to be more functionalizable allows for uniformly functionalizable hydrogels regardless of crosslinker content or physical properties, with a completely nonfouling background. These hydrogels were implanted subcutaneously into mice for 4 weeks to monitor their ability to reduce the foreign body response and minimize capsule formation. pHEMA hydrogels were used as a control.

[1]Slaughter, B.V., Khurshid, S.S., Fisher, O.Z., Khademhosseini, A. & Peppas, N.A. Hydrogels in Regenerative Medicine. *Advanced Materials* **21**, 1-23 (2009).

[2]Didukh, A.G., Koizhaiganova, R.B., Khamitzhanova, G., Bimendia, L.A. & Kudaibergenov, S.E. Stimuli-sensitive behaviour of novel betaine-type polyampholytes. *Polymer International* **52**, 883-891

(2003).

[3] Chen, S. & Jiang, S. A New Avenue to Nonfouling Materials. *Advanced Materials* **20**, 335-338 (2008).

[4] Zhang, Z., Chen, S. & Jiang, S. Dual-Functional Biomimetic Materials: Nonfouling Poly (carboxybetaine) with Active Functional Groups for Protein Immobilization. *Biomacromolecules* **7**, 3311-3315 (2006).

[5] Kabiri, K., Faraji-Dana, S. & Zohuriaan-Mehr, M.J. Novel sulfobetaine-sulfonic acid-contained superswelling hydrogels. *Polymers for Advanced Technologies* **16**, 659-666 (2005).

[6] Carr, L., Xue, H. & Jiang, S. Functionalizable and nonfouling zwitterionic carboxybetaine hydrogels with a carboxybetaine dimethacrylate crosslinker. *Biomaterials* **32**, 961-968 (2011).

See more of this Session: Hydrogel Biomaterials II
See more of this Group/Topical: Materials Engineering and Sciences Division

(695e) Molecular Engineering of Stem Cell Surfaces Via Biomembrane Fusion Transfer From Proteolipobead/Matrix Hybrid Materials

M. Lane Gilchrist, Chemical Engineering, City College and the Graduate Center of the City University of New York, New York, NY and Bin He, Department of Chemical Engineering, City College and Graduate Center of the City University of New York (of CUNY), New York, NY

The presentation of ligands to stem cells in order to recreate the complex 3-D microenvironment of the stem cell niche is a major challenge in biomaterials. The overall aim of our effort is to test the hypothesis that the introduction of biomembrane- microsphere assemblies into 3D scaffolds is a viable biomimetic means to present ligands/bound factors to stem cells and mimic cellular communication in the stem cell niche. We have developed a new platform to present molecules to stem cells within a biomimetic architecture: as laterally-mobile molecules embedded in the context of a tailored biomembrane structure (protetolipobead: PLB) . Confocal microscopy studies were conducted to visualize proteolipobead-displayed N-cadherin engaged in interactions with human mesenchymal stem cells (MSCs) *in situ*, from 3D CLSM reconstruction of hMSCs in N-cadherin proteolipobead/Collagen-I 3D constructs. In a fraction of the MSCs interacting with the PLBs, images consistent with PLB-to-MSC biomembrane fusion were evidenced. In essence, molecular engineering of the MSC surface was evidenced where N-Cadherin from the engineered PLBs diffused onto the surface of live MSCs. This work constitutes a new method for introducing a wide range of complex membrane proteins and signaling molecules to live cells without genetic manipulation. We present further investigations of this fusion phenomenon under a range of conditions and surface characteristics of the PLB substrates.

See more of this Session: Hydrogel Biomaterials II
See more of this Group/Topical: Materials Engineering and Sciences Division

(695g) Improved Interfacial Shear Strength In UHWMPE-PVA Hydrogel Composites Following Glutaraldehyde Grafting for Soft Tissue Applications

Julianne L. Holloway, Anthony M. Lowman and Giuseppe R. Palmese, Chemical and Biological Engineering, Drexel University, Philadelphia, PA

A fiber-reinforced hydrogel-based synthetic biomaterial allows tailoring of the mechanical properties to match the anisotropic property distribution of many soft fibrous tissues and molding of the implant to the size and shape of the tissue being replaced. The mechanical and swelling properties of this composite have been evaluated previously showing the ability of this material to be tailored across a wide range of properties matching that of many soft tissues. Poor adhesion between UHMWPE fibers and the hydrogel, however, has been an issue. Oxygen plasma and glutaraldehyde treatments were performed to increase adhesion at the fiber-matrix interface. The degree of adhesion was quantified by calculating interfacial shear strength at the fiber-hydrogel interface using single fiber pull-out tests. Scanning electron microscopy was used to show fiber morphology and grafted thickness for various treatment times. Titrimetric methods were used to determine grafting yield. Fiber surface treatments successfully increased fiber-matrix interfacial shear strength from 11.5 kPa without any treatment to above 200 kPa following chemical grafting. Fiber-Hydrogel interfacial adhesion was dependent upon both oxygen plasma treatment time and exposure to glutaraldehyde, indicative of control over both the number of sites available for grafting and grafting efficiency respectively. In some cases, calculated fiber stresses exceeded 2 GPa during debonding and sample failure occurred within the fiber phase showing successful utilization of fiber strength after fiber surface modification.

See more of this Session: Hydrogel Biomaterials II
See more of this Group/Topical: Materials Engineering and Sciences Division

(695h) Nanostructured Photo Cross-Linked Biopolymers in Wound-Healing and Drug Delivery Applications

Kristan L. Sorenson[1], C. Allan Guymon[1] and Aliasger K. Salem[2], (1)Chemical and Biochemical Engineering, University of Iowa, Iowa City, IA, (2)Department of Chemical Engineering, Division of Pharmaceutics, Holden Comprehensive Cancer Center, University of Iowa, Iowa City, IA

Chitosan materials have attracted significant attention in the past decade due to their inherent antimicrobial and haemostatic properties, biocompatibility, and renewable nature. Wound dressings containing these cationic polysaccharide fibers have been shown to promote wound healing and prevent infection while also slowing blood loss[1, 2]. In addition to the already-promising potential of chitosan, further improvement can be accessed by more closely mimicking the structure of the extra cellular matrix. Increased cell survival and attachment have been demonstrated on nanostructured synthetic polymers relative to their isotropic counterparts[3]. The overall goal of this work is to determine if sub-micron order, as caused by surfactant templating of the solution prior to cross-linking, can improve the transport properties and wound-healing capabilities of cross-linked chitosan materials. Low-molecular weight chitosan was functionalized with photopolymerizable groups via Michael-addition of 3-(acryloyloxy)-2-hydroxypropyl methacrylate (AOHPMA)[4]. Various surfactants were added to induce nanostructure in the solution and subsequent hydrogel polymers. Nanostructure was characterized using small-angle x-ray scattering and polarized light microscopy. The transport properties of the hydrogels, including water uptake and drug release rates, were measured gravimetrically and spectroscopically, respectively. Although only small indications of nanostructure presence were observed, very significant differences in transport properties resulted from the use of varying types and concentrations of surfactant. The results thus indicate that the sub-micron order of these chitosan materials can be manipulated and tailored to specific applications using surfactant templating. The cytotoxicity of the templated hydrogels and non-cross-linked solutions was evaluated using a colorimetric cell viability assay with human fibroblast cells. The cytotoxicity results further enforced the potential of the material as biocompatible. Future studies will investigate the mechanical properties of chitosan hydrogels as well as their effect on the wound-healing process in vivo. Better control of polymer nanostructure and further understanding of its role in cell-material interactions are crucial to the advancement of chitosan biomaterials for use in many applications, including wound-healing, drug delivery, and tissue scaffolding applications.

References:

(1) Muzzarelli, R. A. A. Chitins and chitosans for the repair of wounded skin, nerve, cartilage and bone. *Carbohydr. Polym.* **2009**, *76*, 167-182.

(2) Jayakumar, R.; Prabaharan, M.; Sudheesh Kumar, P. T.; Nair, S. V.; Tamura, H. Biomaterials based on chitin and chitosan in wound dressing applications. *Biotechnol. Adv.* **2011**, *29*, 322-337.

(3) Clapper, J. D.; Pearce, M. E.; Guymon, C. A.; Salem, A. K. Biotinylated biodegradable nanotemplated hydrogel networks for cell interactive applications. *Biomacromolecules* **2008**, *9*, 1188-1194.

(4) Jiang, M.; Wang, K.; Kennedy, J. F.; Nie, J.; Yu, Q.; Ma, G. Preparation and characterization of water-soluble chitosan derivative by Michael addition reaction. *Int. J. Biol. Macromol.* **2010**, *47*, 696-699.

See more of this Session: Hydrogel Biomaterials II
See more of this Group/Topical: Materials Engineering and Sciences Division

(703a) Scalable Nanomanufacturing of Millimeter Length 2D Nanosheets of Thermoelectric $Na_{0.7}CoO_2$

Mahmut Aksit, David P. Toledo and Richard D. Robinson, Material Science and Engineering, Cornell University, Ithaca, NY

A scalable nanomanufacturing technique is reported for batch fabrication of electrically-conducting 2D metal-oxide nanosheets of $Na_{0.7}CoO_2$. We report a sol-gel based, high temperature bottom-up synthesis which is a cost-effective route capable of producing tens of thousands of nanosheet layers packed into a macro-scale pellet. The synthetic procedure consists of sol-gel coordination of metal ions, auto-combustion, pressurized pellet formation, kinetic demixing, and calcination. The nanosheets are uniform in length and shape with highly anisotropic dimensions of nanometer sheet thickness and millimeter lateral lengths (10^{-5}:1:1), and are readily delaminated into free-standing nanosheets. According to GIXRD, the nanosheet stacking has turbostratic arrangement with nanosheets misaligned rotationally with respect to stacking axis. $Na_{0.7}CoO_2$, in bulk form, is already a surprisingly good thermoelectric material with a thermoelectric (TE) power factor as high as the industry standard Bi_2Te_3 (50 vs. 40 $\mu W/K^2cm$ respectively). Also, it has been reported in the literature that 2D nanosheets are observed to reduce thermal conductivity of materials with a factor $\sim 10^2$ due to phonon confinement, which provides a significant increase in the thermoelectric efficiency. Therefore, in conjunction with low fabrication cost, the reported millimeter length $Na_{0.7}CoO_2$ nanosheets are expected to provide high thermoelectric efficiency with practical importance. Preliminary thermoelectric measurements have been obtained for turbostratically aranged millimeter-length nanosheet stackings.

See more of this Session: Nanomanufacturing
See more of this Group/Topical: Materials Engineering and Sciences Division

(703b) Self-Assembled, Nanostructured Carbon for Energy Storage and Water Treatment

Richard T. Mayes[1], James O. Kiggans[2], Costas Tsouris[3], Sheng Dai[1] and **David W. DePaoli**[3], (1) Chemical Sciences Division, Oak Ridge National Laboratory, Oak Ridge, TN, (2)Materials Science and Technology Division, Oak Ridge National Laboratory, Oak Ridge, TN, (3)Energy and Transportation Science Division, Oak Ridge National Laboratory, Oak Ridge, TN

This talk will present progress in efforts aimed at translating a unique approach for synthesis of self-assembled nanostructured carbon [1] into industrially viable technologies for two important applications: electrochemical double-layer capacitors for energy storage, and capacitive deionization systems for water treatment. Self-assembly through templating is one way to tailor carbon-based materials, enabling the formation of a uniform porous network in the mesopore size regime via manipulation and control of macroscopic process variables. These tailorable, nanostructured materials are synthesized by self assembly in solution, followed by heat treatment. Further activation results in a bimodal pore structure where micropores are present with the mesopores. The larger mesopores allow for faster electrolyte transport. Characteristics of templated carbons having relatively high surface areas and hierarchical pore structures will be presented, as well as measurements of their specific capacitance under conditions relevant to applications.

A focus of the presentation is on recent efforts to convert the mesoporous carbon into electrode materials that may be used practically in capacitive deionization devices. Previous work had indicated that ORNL's mesoporous carbon (OMC) had attractive electrical performance; bench scale testing of the material showed significantly better performance in terms of capacity than commercial aerogel material [2]. However, a previous approach to generate electrodes suffered from variability and mechanical problems. More recent advances have generated materials that utilize the mesoporous carbon in a form more suitable for CDI electrodes in industrial application. Results of prototype testing will be presented.

1. C. Liang and S. Dai, *J. Am. Chem. Soc.*, **2006**, *128* (16), pp 5316–5317. **DOI:** 10.1021/ja060242k
2. Mayes, R.T., C. Tsouris, J.O. Kiggans, S.M. Mahurin, D.W. DePaoli, and S. Dai, *J. Mater. Chem.*, 2010, **20**(39), 8674-8678; **DOI:** 10.1039/C0JM01911A.

ACKNOWLEDGMENT: This research was supported by the Industrial Technologies Program of the U. S. DOE Office of Energy Efficiency and Renewable Energy (EERE), under Contract DE-AC05-00960R22725 with Oak Ridge National Laboratory, managed by UT-Battelle, LLC.

See more of this Session: Nanomanufacturing
See more of this Group/Topical: Materials Engineering and Sciences Division

(703c) Towards a Mechanistic Growth Model for Ionic Crystals

Preshit Dandekar, Department of Chemical Engineering, University of California Santa Barbara, Santa Barbara, CA and Michael F. Doherty, Department of Chemical Engineering, University of California, Santa Barbara, Santa Barbara, CA

The relative surface areas and shapes of the various faces of ionic crystals significantly affect their functionality in areas like catalysis, photovoltaics, zeolites, etc. The layered growth rates of the crystal faces determine the steady-state shape of the crystal. For crystals grown from solution, the growth rate depends on the rates of attachment and detachment of solute units on to and from the crystal surfaces. These rates are determined by the solid-solid as well as the solid-solvent interactions. A mechanistic understanding of these interactions is instrumental in engineering the shapes of ionic crystals to achieve better functionality. The Periodic Bond Chain (*PBC*) theory has been used here to study the solid-state interactions in ionic crystals by identifying directions of strong bonding. The *PBC*s are parallel to the edges of the growth spirals and their identification provides insight into the spiral growth mechanism of crystals. Bond chain networks that are stoichiometric and do not have any out-of-plane dipole moment have been identified for anatase (TiO_2) and calcite ($CaCO_3$) crystals using the crystal structure as input.

The electrostatic interactions in the *PBC*s were calculated by a simple columbic potential and the Madelung approach for a 1D ionic chain. The bond chain energies for all the *PBC*s in anatase have been calculated and the lattice energy has been accurately predicted. The kink and edge energies will be calculated for the spiral growth on these crystal faces to estimate the relative growth rates and eventually predict steady-state shapes of ionic crystals.

See more of this Session: Nanomanufacturing
See more of this Group/Topical: Materials Engineering and Sciences Division

(703d) The Structural Evolution and Diffusion During the Chemical Transformation From Cobalt to Cobalt Phosphide Nanocrystals

Don-Hyung Ha, Liane M. Moreau, Clive R. Bealing, Haitao Zhang, Richard G. Hennig and Richard D. Robinson, Materials Science and Engineering, Cornell University, Ithaca, NY

Nanoscalesystems can display interesting and unpredictable transformation kinetics that increase the structural complexity of the original material. Characterization of thetransformation routes to form the complex final structure is one of the majorchallenges in nanoscience. A more completeunderstanding of the transformation pathways would provides a means to improve synthesis techniques aswell as an insight into the control of chemical and physical properties, in orderto optimize the nanocrystals (NCs) for use in applications. We report the structural evolution and the diffusionprocesses which occur during the phase transformation of nanocrystal ε-Co to $Co_2P$to CoP, from a reaction with tri-*n*-octylphosphine (TOP). Extended X-ray absorption fine structure (EXAFS) investigations wereused to elucidate the changes in the local structure of cobalt atoms whichoccur as the chemical transformation progresses. Results from EXAFS show boththe Co_2P and CoP phases contain excess Co. Results from EXAFS, transmission electron microscopy, X-ray diffraction, and density functionaltheory calculations reveal that the inward diffusion of phosphorus is morefavorable at the beginning of the transformation from ε-Co to $Co_2P$by forming of an amorphous Co-P shell, while retaining a crystalline cobaltcore. When the major phase of the sample turns to Co_2P, thediffusion processes reverse and cobalt atom out-diffusion is favored, leaving ahollow void, characteristic of the nanoscale Kirkendalleffect. For the transformation from Co_2P to CoP theory predicts anoutward diffusion of cobalt while the anion lattice remains intact. In realsamples, however, the Co-rich nanocrystals continue Kirkendallhollowing.

See more of this Session: Nanomanufacturing
See more of this Group/Topical: Materials Engineering and Sciences Division

(703f) High Efficiency Rare Earth Doped Core-Shell Nanophosphors for Energy Applications

James Dorman[1], Gregory Kuzmanich[2], Abhijeet Joshi[3], Ju H. Choi[1] and Jane P. Chang[4], (1) Department of Chemical and Biomolecular Engineering, University of California at Los Angeles, Los Angeles, CA, (2)Chemisty and Biochemistry, University of California at Los Angeles, Los Angeles, CA, (3)Electrical Engineering, University of California at Los Angeles, Los Angeles, CA, (4)Department of Chemical and Biomolecular Engineering, Chemical Engineering Department, Cambridge, MA

The development of rare-earth (RE) doped phosphors allows for the conversion of photons to energies that are more usable for the desired application. Additionally, these RE phosphors have long lifetimes, on the order of ms, which offer potential in many energy conversion and energy transfer devices. Currently, RE phosphors are used in fiber optics amplifiers, modulated hybrid laser, optical interconnects and switches, optical displays, broad absorption solar cells and various other lighting applications. Energy transfer mechanisms of the excited RE states, such as defect quenching and sensitizer/emitter interactions, must be understood in order to achieve high efficiency energy conversion and propagation for future applications.

In order to synthesize high efficiency phosphors, trivalent RE ions are being doped into a core-shell metal oxide host lattice. The role of the core-shell structure reduces the effect of the surface quenching sites by increasing the distance between active ions and the surface hydroxyl groups. Secondly, the luminescent fingerprint can be further controlled with proper doping of the shell structure by either increasing the absorption spectrum or adding additional emission peaks. Primarily, this work focuses on the emission of visible photons through upconversion in Y_2O_3:Er^{3+}, Yb^{3+} nanophosphors, making them ideal components in broad absorption solar cells. By spatially controlling the position and concentration of the RE ions within the nanostructure, increased luminescence is observed due to energy transfer between the dopant ions within a critical interatomic distance. Passitvation of surface sites with increasing shell thickness was shown to increase luminescent lifetimes up to 53%, with a critical shell thickness of 8 nm, while lowering the theoretical lifetimes extracted from Judd-Ofelt parameters. The effect of the spatially controlled Yb ions was probed through the extraction of the upconversion photon requirement, showing a statistical decrease in photons from 2.16 to 1.43, or ~30 %. Finally, the effective energy transfer distance and energy transfer coefficients were studied by selectively exciting the Yb $^4F_{5/2}$-$^4F_{7/2}$ transition with a 904 nm laser diode while probing the Er^{3+} emission at 1540 nm as a Y_2O_3 spacer layer is added to the system. Measured results show the occurrence of energy transfer between the ions with a ~3 nm spacer layer, confirming the prediction of the Föester-Dexter theory. Additionally, these results are applied to a RE doped $LaPO_4$ system, producing white light from a single, high efficiency core-shell phosphor.

(703g) A Comparative Investigation of CO2 Removal In Two Different Hydrogen Redistribution Strategies for a Two-Stage Hydrogen-Permselective Membrane Methanol Reactor

M.S. Baktash[1], S. Mazinani[1], M. Zare[1], A. Najafi[1] and Mohammad Reza Rahimpour[2], (1)Department of Chemical engineering, Chemical and Petroleum Engineering School, Shiraz University, Shiraz, Iran, (2)School of Chemical and Petroleum Engineering, Shiraz University, Shiraz, Iran

One of the major problems facing mankind in 21st century is the global warming which is induced by the increasing concentration of carbon dioxide (CO_2) and other greenhouse gases in the atmosphere. One of the most promising processes for controlling the atmospheric CO_2 level is conversion of CO_2 to methanol by catalytic hydrogenation. In this paper, two different hydrogen redistribution strategies on CO_2 removal rate along a two-stage hydrogen-permselective membrane reactor have been compared. In the first strategy (co-current mode), fresh synthesis gas flows in the tube side of the second stage in same direction with reacting gas stream in the shell side, so that more hydrogen is provided in the first sections of the reactor. In the second strategy (counter-current mode), the gas streams are in the opposite direction, so that more hydrogen is provided in the last sections of the reactor. For this system, a dynamic plug flow one-dimensional model in the presence of catalyst deactivation was employed to compare both strategies in terms of activity, temperature, CO_2 removal and water production. It was recognized that the reactor in counter-current mode has great potential to higher conversion of CO_2 whereas in co-current mode, profile of temperature is more favorable and water production is lower. Enhancement of the CO_2 removal in counter-current mode causes a lower environmental impact. The lower water production rate in co-current mode reduces catalyst recrystallization.

See more of this Session: Nanomanufacturing
See more of this Group/Topical: Materials Engineering and Sciences Division

(705a) Fundamental Studies On the Radical Initiated Grafting of Vinyltrimethoxysilane Onto Polymeric Models

Swetha Sivaswamy[1], Emily Nixon[2], Farhana Momim[1], Rani Jha[2], Manjusha Verma[2], Leslie Gelbaum[2], Pamela Pollet[2], Charles Eckert[1], Charles Liotta[2], Bharat Chaudhary[3] and Jeffrey Cogen[3], (1)School of Chemical and Biomolecular Engineering, Georgia Institute of Technology, Atlanta, GA, (2)School of Chemistry and Biochemistry, Georgia Institute of Technology, Atlanta, GA, (3)The Dow Chemical Company, Piscataway, NJ

Cross-linked polyethylenes have widespread applications such as insulators for current-carrying wires/ cables and for hot water pipes, owing in part to their high upper use temperatures compared to non-crosslinked polyethylenes. Industrially, cross-linked polyethylene is often prepared via radical initiated grafting of polyethylene with vinyltrimethoxysilane (VTMS) followed by moisture curing. The efficiency of the grafting reaction is not optimal, as a large portion of the VTMS remains un-reacted and controlling the distribution and extent of grafting is difficult. We present our studies of VTMS grafting reactions on polymeric models to gain improved fundamental understanding of the radical initiated grafting mechanism. In particular, we will present experimental results using dodecane and heptane as model compounds.

See more of this Session: New Trends In Polymerization and Chain Modification
See more of this Group/Topical: Materials Engineering and Sciences Division

(705b) The Development of a Paradigm for the Facile Incorporation of Diels-Alder Moieties

Kenneth Christopher Koehler[1], Christopher N. Bowman[1] and Christopher J. Kloxin[2], (1)Chemical and Biological Engineering, University of Colorado at Boulder, Boulder, CO, (2)Chemical Engineering and Materials Science and Engineering, University of Delaware, Newark, DE

The Development of a Paradigm for the Facile Incorporation of Diels-Alder Moieties

Kenneth C. Koehler, Christopher J. Kloxin and Christopher N. Bowman

Department of Chemical and Biological Engineering

University of Colorado at Boulder

424 UCB Room ECCH 111

Boulder, CO 80309

Abstract

One of the reactions associated with the "Click" paradigm, the Diels-Alder reaction has come to use in a variety of applications pertaining to polymer chemistry including protecting groups, mendable materials and covalent adaptable networks. Unfortunately, the creation of material making use of the Diels-Alder reaction typically relies upon a trial and error process to select the proper diene/dienophile pair for a given application. To address the downfall of the trial and error method, this work aimed to develop a library containing the thermodynamic and kinetic parameters of select Diels-Alder reactions, allowing for the *a priori* selection of a diene/dienophile pair for a given application. Using HNMR spectroscopy, this study focused on the determination of thermodynamic and kinetic properties of the Diels-Alder reaction between maleimide and furan carboxylic acids. The impact of changing the substitutional position and proximity of the carboxylic acid group to the furan and maleimide on the thermodynamics and kinetics of the Diels-Alder reaction was investigated. It was found that direct conjugation of the carboxylic acid moiety into the furan ring at the two position allowed for a fast Diels-Alder reaction with a low equilibrium conversion. Substituting the carboxylic acid group at the three position resulted in a slower Diels-Alder reaction with a much higher equilibrium conversion. Placement of the carboxylic acid in the two position while removing conjugation to the furan ring with a aliphatic spacer granted both, a fast reaction and high conversion. The impact of changing the length of the carbon chain separating the maleimide ring from the carboxylic acid group was also studied. It was found that as the chain between the maleimide ring and carboxylic acid group increased, the Diels-Alder reaction rate remained relatively unchanged, however the equilibrium conversion increased. With the over-arching goal of using maleimide and furan moieties to create materials, the impact of incorporating the furan and maleimide carboxylic acids onto another species, t-butyl glycine, had on the thermodynamic and kinetic properties was studied. In general, it was found that maleimide and furan carboxylic acids had similar reaction rates and thermodynamic properties as compared to their t-butyl glycine coupled counterparts.

See more of this Session: New Trends In Polymerization and Chain Modification
See more of this Group/Topical: Materials Engineering and Sciences Division

(705c) Functionalization of Syndiotactic Polystyrene Aerogels by Incorporation of Functionalized Atactic Polystyrene

Xiao Wang and Sadhan C. Jana, Polymer Engineering, The University of Akron, Akron, OH

Functionalization of syndiotactic polystyrene (sPS) aerogel was achieved by incorporation of functionalized atactic polystyrene (aPS) into the gel system. Firstly, functionalization of aPS was performed via chloromethylation reaction, which provides access to further functionzalization. Thanks to the high reactivity of chloromethyl groups, various functional groups are able to be grafted onto aPS, e.g., chelating group, 2-mercaptobenzothiazole (MBZ), which is capable of adsorbing heavy metal ions. Secondly, a functionalized gel was formed by thermo-reversible gelation, whereby a solution of sPS and functionalized aPS was cooled to room temperature. Finally, the functionalized aerogel was obtained by drying the gel under supercritical condition. The morphology of functionalized aerogel was investigated by scanning electron microscopy. It was seen that the functionalized aerogel presented uniform porous morphology as those observed in native sPS aerogel, indicating uniform dispersion of functionalized aPS onto sPS network. The heavy metal adsorption behavior was studied using elemental analysis technique to further demonstrate the efficiency of this approach.

See more of this Session: New Trends In Polymerization and Chain Modification
See more of this Group/Topical: Materials Engineering and Sciences Division

One pot synthesis of ethoxylated bisphenol A acrylate macro monomer from bisphenol A acrylate and ethylene oxide

Xueyi HU, School of chemical and material engineering, Jiangnan University, Wuxi,China
Yun FANG, School of chemical and material engineering, Jiangnan University, Wuxi,China
Xin JIN, School of chemical and material engineering, Jiangnan University, Wuxi,China
Yongji GU , School of chemical and material engineering, Jiangnan University, Wuxi,China

Abstract

Ethoxylated bisphenol A acrylate, containing symmetric rigid aromatic rings and flexibility ethoxy groups, is a polyfunctional macro monomer. The homopolymer or copolymer well used as biomedical materials such as medical membrance, self-healing dental restorative formulation. To our knowledge, the monomer could be synthesized by either esterification of ethoxylated bisphenol A and acrylic acid or etherification of ethoxylated hydroxyethyl acrylate and bisphenol A. We developed a more atom economic route to produce the functional macro monomer. The inherent atom economy of the developed route is 100% which is much higher than that of the former multi-steps process. Catalysted by a heterogeneous acid-base bifunctional catalyst, ethylene oxide reacts with an active hydrogen free raw material, bisphenol A acrylate, to produce the target molecule. The inhibitors and some other reaction conditions were well discussed here. The conversion of bisphenol A acrylate was 90% at 140 °C for 8 h with 1% wt catalyst and 0.5% wt composite inhibitors in a steal autoclave.

Keywords: Ethoxylated bisphenol A acrylate, Macro monomer, Heterogeneous acid-base bifunctional catalyst, Ethylene oxide

Introduction

Ethoxylated bisphenol A acrylate (EBAA) was first prepared by Schmitt[1] , and acted as tooth filling compounds. After that, the molecular or the homologues were used as an effective monomer of the dental ingredients [2-4]. Very recently, self-repairing dental compositions containing the EBAA were patented by Perichaud [5], Ragain [6] and Gross [7].

The homopolymer or copolymer of the multifunctional molecular was also used as liquid-crystal material [8-11] and photo sensitive resin [12-14]. The beta-amino ester derivatives showed excellent biocompatibility, and could be applied in the field of gene delivery [15-18].

The multifunctional monomer was used to be synthesized by esterification of ethoxylated bisphenol A and acrylic acid [1, 8, 19]or etherificationof ethoxylated hydroxyethyl acrylate and bisphenol A [8] with low conversion and target product yield. Furthermore, some water or other small molecular was produced as byproduct during the reaction. Here we describe a more atom economic route to prepare the functional macro monomer.

Experimental

Chemicals Bisphenol A acrylate (BAA) , was laboratory made and recrystalized before use. Ethylene oxide (EO) was purchased from Sinopec Yangzi Petrochemical comp Ltd. Nanjing, China. All the inhibitors were recrystalized before use. All the other chemicals were regent grade and used as received. Heterogeneous acid-base bifunctional catalyst were prepared elsewhere[20] with Al/Mg molar ratio 0.7.

Synthesis procedure 0.01 mol BAA, 0.03 g bifunctional catalyst and 0.01 g inhibitor or composite inhibitors were added into the 20 mL autoclave under a nitrogen atmosphere. The desired amount of EO was feed into the vessel at lower temperature. The mixture was reacted at 140 °C for 10 h. Cool down the reactants, and a grey solid product was obtained.

Charactrization The product was identified by preparativeTLC and FT-IR.

Result and discussion

Inhibitor screening

To avoid undesirable polymerizing during the ethoxylation process, some active hydrogen-free inhibitors were suggested. 0.01 mol BBA and 0.01 g inhibitor were ground and mix well, and packaged in a closed container under nitrogen atmosphere. Then the solid mixtures were heated to 140 °C for 24 h with magnetic stirring, cool down and monitored by TLC. The result showed that none new compound was discovered with NNSP-1, TEMPO, BQ or those complexes as inhibitors, while some unwanted polymer was found with no inhibitor.

Synthesis of EBAA

EBAA could be synthesized by multi-step route, see Scheme1. A novel route was developed, see Scheme 2. Compare with those two process, the novel route produce none apparent byproduct with only one steps.

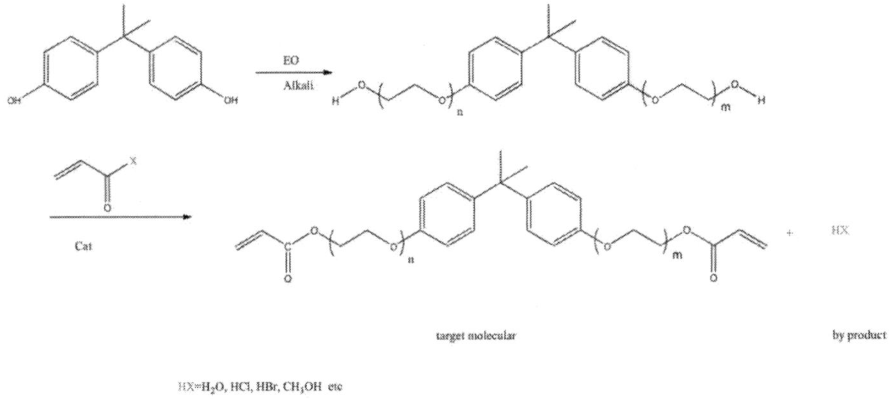

Scheme 1. Traditional route to synthesis of EBAA

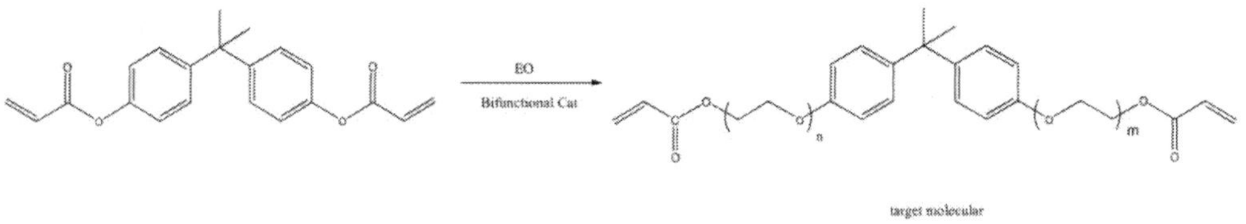

Scheme 2. Novel route to synthesis of EBAA

The inhibitors and some other reaction conditions were optimized. The optimal conditions were achieved at 140 °C for 8 h with 1% wt catalyst and 0.5% wt composite inhibitors in a steal autoclave. At the condition, monoethylene derivate (m=1, n=0), symmetric diethylene derivate (m=n=1) and polyethylene derivate (m+n>2) were found in the reaction mixture with 28.8 wt %, 30.7 wt % and 35.0 wt % respectively.

Conclusion

Ethoxylated bisphenol A acrylate could be synthesized on a bifunctional catalyst with bisphenol A acrylate as raw material.

References

[1] W. Schmitt, R. Purrmann, P. Jochum, et al. Diacrylic Acid Ester of Bifunctional Alcohols[P]. DE, **1969**: 1921870.

[2] M. J. M. Abadie and B. K. Appelt. Photocalorimetry in the Study of Photopolymerizable Dental Resins. Preliminary Study[J]. *J. Biomater. Dent.* **1987,** 3(1), 17-27.

[4] A. Ogliari Fabricio, C. Ely, H. Zanchi Cesar, et al. Influence of Chain Extender Length of Aromatic Dimethacrylates on Polymer Network Development[J]. *Dent Mater.* **2008**, 24(2), 165-171.

[5] A. Perichaud and M. Devassine. Self-Repairing Compositions, Self-Repairing Materials, Self-Repairing Methods and Applications[P]. WO, **2009**: 2009115671.

[6] J. C. Ragain, D. Charlton and A. Tiba. Multifunctional Acrylates Used as Cross-Linkers in Dental and Biomedical Self-Etching Bonding Adhesives[P]. WO, **2009**: 2009097152.

[7] S. M. Gross and M. A. Latta. Self-Healing Dental Restorative Formulations and Related Methods[P].US, **2011**: 20110082232.

[8] D. J. Broer, J. Boven, G. N. Mol, et al. In-Situ Photopolymerization of Oriented Liquid-Crystalline Acrylates. 3. Oriented Polymer Networks from a Mesogenic Diacrylate[J]. *Makromol. Chem.* **1989**, 190(9), 2255-2268.

[9] I. Heynderickx and D. J. Broer. Thermo-Optical Properties of a Polymer Dispersed Liquid Crystalline Polymer[J]. *Mol. Cryst. Liq. Cryst. Sci. Technol., Sect. A.* **1995**, 263

[10] F. Ciuchi, M. Giocondo and R. Barberi. Electrically Controlled Defects at a Liquid Crystal-Polyimide Interface[J]. *Liq. Cryst.* **2008**, 35(1), 99-102.

[11] A. Hinojosa and S. C. Sharma. Effects of Gold Nanoparticles on Electro-Optical Properties of a Polymer-Dispersed Liquid Crystal[J]. *Appl. Phys. Lett.* **2010**, 97(8), 081111-081113.

[12] M. Miyasaka and Y. Muramatsu. Photosensitive Resin Composition, Photosensitive Element, Method of Forming Resist Pattern and Method of Producing Printed Wiring Board[P]. WO, **2009**: 2009145120.

[13] M. Miyasaka, M. Kaji, J. Yin, et al. Photosensitive Resin Composition, Photosensitive Element Utilizing the Composition, Method for Formation of Resist Pattern, and Process for Production of Printed Circuit Board[P].WO, **2010**: 2010126006.

[14] M. Miyasaka and Y. Muramatsu. Photosensitive Resin Composition, Element, Resist Pattern Formation and Manufacture of Printed Circuit Board[P]. JP, **2011**: 2011034005.

[15] D. G. Anderson, A. Akinc, N. Hossain, et al. Structure/Property Studies of Polymeric Gene Delivery Using a Library of Poly(Beta -Amino Esters)[J]. *Mol. Ther.* **2005**, 11(3), 426-434.

[16] J. R. Greenland, H. Liu, D. Berry, et al. Beta -Amino Ester Polymers Facilitate in Vivo DNA Transfection and Adjuvant Plasmid DNA Immunization[J]. *Mol. Ther.* **2005**, 12(1), 164-170.

[17] J.-S. Lee, J. J. Green, K. T. Love, et al. Gold, Poly(Beta -Amino Ester) Nanoparticles for Small Interfering Rna Delivery[J]. *Nano Lett.* **2009**, 9(6), 2402-2406.

[18] D. G. Anderson, D. M. Lynn and R. Langer. Semi-Automated Synthesis and Screening of a Large

Library of Degradable Cationic Polymers for Gene Delivery[J]. *Angew. Chem., Int. Ed.* **2003**, 42(27), 3153-3158.

[19] W. Schmitt, R. Purrmann, P. Jochum, et al. Diacrylic and Dimethacrylic Acid Esters[P]. US, **1974**: 3810938.

[20] L. Hu, X. Hu and Y. Fang. Synthesis of Polyoxyethylene Methacrylate Macromonomer by Ethoxylation[J]. *AIChE Spring Meet. 6th Global Congr. Process Saf., Conf. Proc.* **2010**, hu3/1-hu3/5.

Presenting Author
Xueyi Hu
PhD candidate
School of chemical and material engineering
Wuxi,
China
Phone Number: 86-13382888458
Email: xueyihu@163.com

Yun Fang
Professor
School of Chemical and material engineering
Jiangnan University
Wuxi,
China
Phone Number: 86-18906180600
Email: yunfang@126.com

Xin Jin
Undergraduate
School of Chemical and material engineering
Jiangnan University
Wuxi,
China
Phone Number: 86-15861445028
Email: jx881117@126.com

Yongji Gu
Undergraduate
School of Chemical and material engineering
Jiangnan University
Wuxi,
China
Phone Number: 86-15861416638
Email: yongjigu1989@126.com

(705e) Micromolding Surface-Initiated Polymerization: A Versatile Route for Microscale Replication Onto a Solid Support

Carlos A. Escobar, Juan C. Tuberquia, Nabijan Nizamidin and G. Kane Jennings, Chemical and Biomolecular Engineering, Vanderbilt University, Nashville, TN

This presentation will introduce the use of confined surface-initiated ring-opening metathesis polymerization (SI-ROMP) of perfluoroalkyl or alkyl norbornene monomers from solid substrates to synthesize surface-bound polymer structures with tunable physical and chemical properties that accurately replicate those exhibited by Nature's engineered, microscopically rough, and highly functional surfaces. This approach not only allows mimicking of highly evolved and functional surface architectures but also provides versatility in that it introduces a wide variety of chemical compositions available in materials chemistry, including partially fluorinated polymers with ultralow critical surface tensions. Optical microscopy and scanning electron microscopy confirm growth of the polymer structures and the precise replication of the microscale and nanoscale features exhibited by the target natural surface with the added freedom to expand beyond Nature's chemical building blocks. Contact angle measurements show that the surface architectures exhibit both hydrophobic and oleophobic behavior, and in some cases, superhydrophobic properties. This approach is not limited to natural surfaces and could be applied in a straightforward manner to a variety of synthetic surfaces that have microscale features.

See more of this Session: New Trends In Polymerization and Chain Modification
See more of this Group/Topical: Materials Engineering and Sciences Division

(705f) Photo-Enforced Stratification

Clinton J. Cook and C. Allan Guymon, Chemical and Biochemical Engineering, University of Iowa, Iowa City, IA

Polymerizationtechniques typically result in a homogenous final polymer. If surface and/orsubstrate interface chemistries that differ from the homogenous composition aredesired for applications such as coatings and adhesives, a multiple stepprocesses is necessary to achieve the desired chemistries. It could bepossible to overcome these obstacles with photo-enforced stratification; amethod to allow for the production of a polymeric thin film with differentsurface and/or substrate interface chemistries in a single reaction step. Photo-enforcedstratification is a method of photopolymerization designed to induce acompositional gradient in a polymer thin film upon co-photopolymerization inthe presence of a light gradient by taking advantage of the temporal andspatial control that photopolymerization affords. The light gradient isestablished in the film by having a sufficiently high concentration of achromophore in the pre-polymer mixture such that illumination of the mixtureresults in attenuation of the light as a function of depth in the mixture. Thelight gradient will establish a gradient in the reaction rate ofpolymerization. A co-photopolymerization in the presence of a reactiongradient results in the monomer with the higher reactivity ratio to reactfaster than the other monomer in the high light intensity regions at the top ofthe film where the rate of reaction is the fastest. As the monomer with thehigher reactivity ratio is depleted at the surface of the film, more of thatmonomer will diffuse from the bulk to the surface of the film. At the sametime, a counter diffusion of the lower reactivity ratio monomer from thesurface into the bulk will occur. This results in the surface of the thin filmbeing enriched in the monomer that has the highest reactivity ratio.

See more of this Session: New Trends In Polymerization and Chain Modification
See more of this Group/Topical: Materials Engineering and Sciences Division

(706a) Free Surface Electrospinning From a Wire Electrode

Keith M. Forward and Gregory C. Rutledge, Department of Chemical Engineering, Massachusetts Institute of Technology, Cambridge, MA

The needle or nozzle-based electrospinning process has long been explored for its capability to produce unique nanofiber materials with desirable properties such as high surface area and high porosity. These materials are of interest in a wide range of fields such as textiles, filtration, tissue engineering, drug delivery systems, nanocomposites, and alternative-energy generation systems such as solar cells, fuel cells, and energy storage devices. However, one of the perceived drawbacks of the method for industrial purposes is its low production rate. A typical production rate from a single spinneret is 0.1 -1 g of fiber per hour, depending on the solution properties and operating parameters. To overcome the low productivity of nozzle-based electrospinning, we consider "free surface electrospinning" (also referred to as "needleless electrospinning") where electrohydrodynamic jets self-organize spontaneously on a free liquid surface. It has been estimated that the concentration of the jets can be increased by an order of magnitude or more compared to conventional electrospinning by employing free surface electrospinning.

In this report, we analyze free surface electrospinning from a thin wire electrode. In this process, metal wire electrodes mounted on a spindle are drawn through an electrified fluid bath in a direction perpendicular to the wire axes. As a wire moves through the fluid/air interface, fluid is entrained on the wire, resulting in a thin film of fluid coating the wire. Due to a Plateau-Rayleigh instability, the coating breaks up into individual droplets of charged fluid on the metal wire. At sufficiently high local electric field, the individual drops deform and jets are produced from the droplets, giving rise to a form of free surface electrospinning. As the spindle rotates, electrospinning continues to occur until the supply of fluid is exhausted or the required electric field conditions are no longer met. By mounting several wires on a rotating spindle, the process of immersion, entrainment, dewetting and jetting can be performed repeatedly in a simple manner.

The processes of charging, entrainment, droplet breakup, and jetting are all coupled in this process. Here we examine how the fluid properties (i.e. surface tension, viscosity, density, concentration) and wire electrode rotation rate affect fluid entrainment and drop breakup. Applied potential and rotation rate of the spindle are varied to study the effects of these operating parameters. Current measures are preformed to examine the effects of fluid properties and operating parameters on the fluid charge density. The productivity is determined for a range of applied voltages and electrode rotation rates and compared to the theoretical limit of fluid entrainment. In addition, scanning electron micrographs are used to investigate the fiber diameter and distribution of the electrospun mats.

See more of this Session: Polymer Processing and Rheology II

(706b) Dispersion and Orientation of Nanoparticles within Nylon6 Polymer Matrix by Novel Fabrication Method

Ilchgerel Dash, Chemical and Biological Engineering, SDSMT, Rapid city, SD and Robb M. Winter, Department of Chemical and Biological Engineering, S.D. School of Mines and Technology, Rapid City, SD

In this research, Organically Modified Montmorillonite (OMMT) and Multiwall Carbon Nanotube (MWCNT) were dispersed and oriented within Nylon6 utilizing a new processing technique to create thin film nano-materials. Nylon6 nanocomposites containing 0.1wt% to 10wt% of different fillers (OMMT or -MWCNT) were directly prepared in high shear thin film device (HSTFD) which is mounted to HAAKE Polylab single screw. The HSTFD produced thin film nanocomposite sheets with uniform dispersion and good orientation. Mechanical properties of the Nylon6 nanocomposites were studied through tensile and flexural tests. The rheological behavior of the nanocomposites was determined by capillary and rotational rheological measurements. X-ray Diffraction (XRD) and Transmission Electron Microscopy (TEM) were used characterize the dispersion, exfoliation and orientation of the OMMT or MWCNT in the Nylon6 matrix. The thermal properties of Nylon6 nanocomposites were characterized by Dynamic Mechanical Analysis (DMA). The results obtained from XRD and TEM analyses of the OMMT and MWCNT fillers in Nylon6 matrix will be presented. In addition, the mechanical properties and thermal properties of the both Nylon6 nanocomposites compared with other experimental data will be discussed in detail. A COMSOL Multiphysics model of this new technique to form thin films of dispersed and oriented nano-materials will be presented.

See more of this Session: Polymer Processing and Rheology II
See more of this Group/Topical: Materials Engineering and Sciences Division

Modulus Enhancement and Orientation in Die-Drawn PP-Talc Composites

Rahul H. Rane, K. Jayaraman and T. R. Bieler, Michigan State University, MI, USA
K. L. Nichols, The Dow Chemical Company, Midland, MI, USA
(ranerahu@msu.edu)

Introduction

Solid phase die drawing of heated polymer composites filled with micron sized particles is carried out to produce expanded or voided composites with a high level of molecular orientation. These composites with enhanced moduli and reduced density are used for structural applications in the construction industry [1].

It is well known that die drawing of unfilled semi-crystalline polymers at elevated temperature can produce up to 10-20 fold increase in the modulus [2-5]. The spherulitic structure with radially oriented crystalline lamellae in the undrawn specimen changes upon drawing to a fibrous morphology and block lamellar structure oriented along the draw direction is formed [2-5]. X-Ray diffractometry has been used frequently to investigate development of crystal structure or texture in processed isotactic polypropylene (i-PP) [4-6]. Die drawing of neat i-PP at high temperature has been reported [3-5] to result in orientation of the crystalline c-axis along the machine direction. The objective of this work was to study the extent of molecular orientation and modulus achieved at various draw rates in die drawing of polypropylene (PP)-talc composite.

An extruded billet of PP-talc composite filled with 20 wt% (7 vol %) talc was die-drawn at 145^{0}C with the objective to study the fundamental aspects of microstructure development during the process of solid phase die-drawing of PP-composite. A low level of filler loading was selected in order to track the crystal orientation and texture using X-ray diffraction (XRD) technique. A billet of neat PP was also drawn under the same processing conditions for comparison.

For the range of draw rates studied actual draw ratios (ratios of final haul-off velocity to intake velocity) above 6 are obtained for both neat PP and its composite away from the die exit. Both the materials showed a high degree of crystalline orientation on die drawing. The crystal orientation saturated with draw ratio. Lower values of draw ratios and crystal orientation were attained for the drawn composite. The void fraction in the drawn composites also leveled at the same draw ratio where the crystal orientation is saturated. Despite the leveling of the crystal orientation and void fraction the tensile modulus of the drawn material increased with increase in draw ratio.

Results and Discussion:

Draw ratios of drawn samples:

The actual draw ratio is given as:

$$D.R. = (\rho A)_{initial} / (\rho A)_{final} \quad (1)$$

Where ρ=density and A= area.

The draw ratio achieved for die drawn composite at each draw rate was lower than that of drawn neat PP as shown in **Fig.1**. The nominal draw ratio $(DR)_N$ defined as the ratio of area of the material at the die entrance to that of die exit is also shown in the figure.

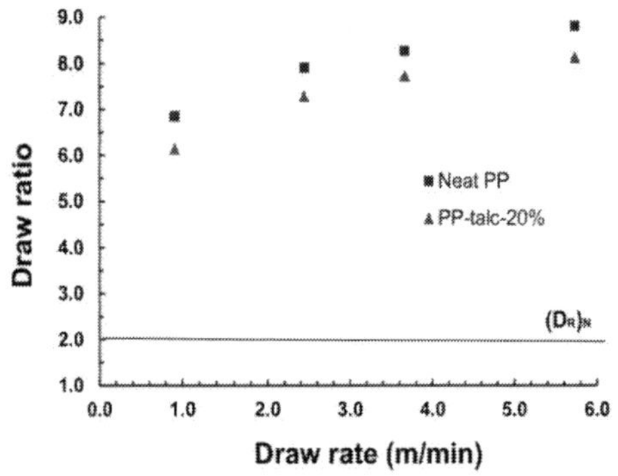

Fig.1: Draw ratio of drawn materials with draw rate

Crystal orientation from XRD:

Fig.2 (b) shows normal XRD scans for undrawn neat PP and PP-talc-20 wt% composite obtained using Ni-filtered monochromatic beam of Cu Kα X-rays. The crystalline reflection peaks for the monoclinic α-crystals (**Fig.2 (a)**) of i-PP are indicated in the figure. The broad peak observed away from the sharper peaks is due to the diffraction of the amorphous phase of the polymer. Normal scan of the composite shows the crystalline peaks for talc particles as well. The (040) and (110) crystal planes of PP are not interfered by any of the talc peaks; therefore pole figures of these planes were measured to study the orientation of the PP-matrix.

The normal scans here show that the relative intensity of (040) crystal planes w.r.t. (110) crystal planes has increased with the addition of talc suggesting enhanced allignment of the b-axes of PP perpendicular to the surface.

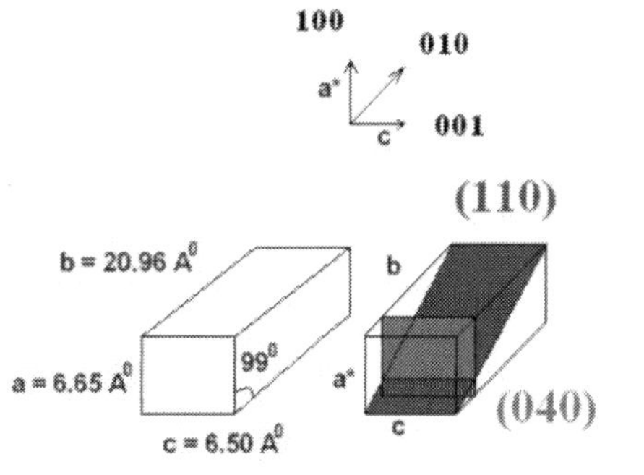

(a)

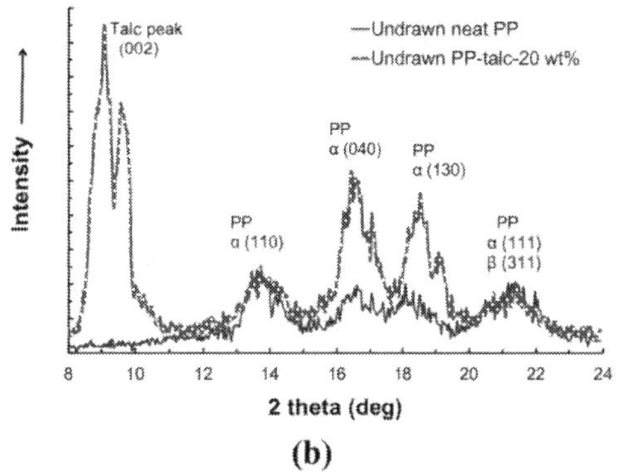

(b)

Fig.2: **(a)** Schematic of α-monoclinic crystal of PP and **(b)** XRD patterns for undrawn materials

Using the pole figure data of crystal planes: α(040) and α(110) and the modified unit cell parameters of i-PP, inverse pole figures for crystalline a*, b and c-axis of PP w.r.t. process coordinates were obtained for the two materials before and after the die drawing process.

From **Fig.3** it can be seen that the presence of talc particles in the composite material has resulted in preferential orientation of the b-axis of PP crystals in the normal direction (ND), this is consistent with the literature [6]. For both the extruded or undrawn materials the c-axis is oriented in the MD-TD plane perpendicular to the ND, whereas the a* axis is oriented both in the MD-TD and TD-ND planes at about 35-60^0 away from the ND. In case of composite the maxima for a*-axis are also oriented in the TD-MD plane perpendicular to the ND.

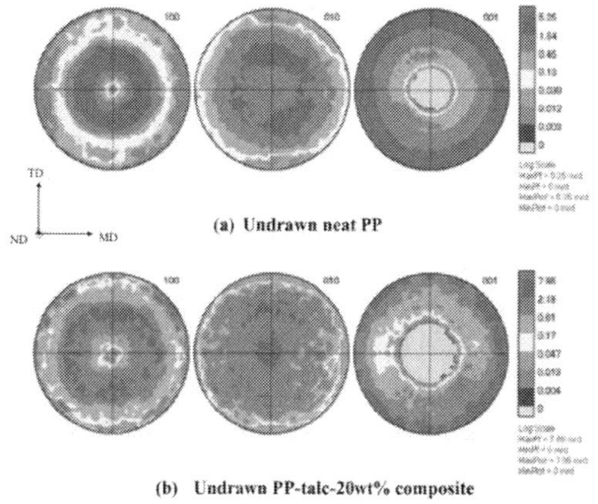

Fig.3: Pole figures for a*,b and c- axes of PP crystals in undrawn materials.

This suggests that in the undrawn composite although the talc particles serve as nucleating sites for the PP crystals, it has majority of lamellar morphology similar to that of undrawn neat PP. Naiki et.al [7] have shown that during crystallization of PP-talc composites, crystalline lamellae grow orthogonal to the faces of the plate-shaped talc particles until they impinge on spherulites nucleated in the bulk.

Fig.4 shows representative pole figures for die drawn neat PP and PP-talc-20wt% composite each drawn a draw rate of 3.66 m/min. From this figure it can be seen qualitatively that die drawing has led to enhanced orientation of the crystalline c-axis of i-PP in the MD for both the materials. The b-axis has got oriented in ND and in the TD-ND plane at about 25-75^0 from the ND. The a*-axis is oriented in TD as well as in the plane of TD-ND at ~12-60^0 away from the ND. Both neat PP and PP-talc-20wt% composite show similar type of crystal orientation however for the same draw rate the relative intensity of the maxima obtained for neat PP are higher than that of the composite.

Herman orientation parameters for the crystalline c-axis were also computed from the pole figures of α (040) and α (110) crystal planes of i-PP using well know procedures [4].

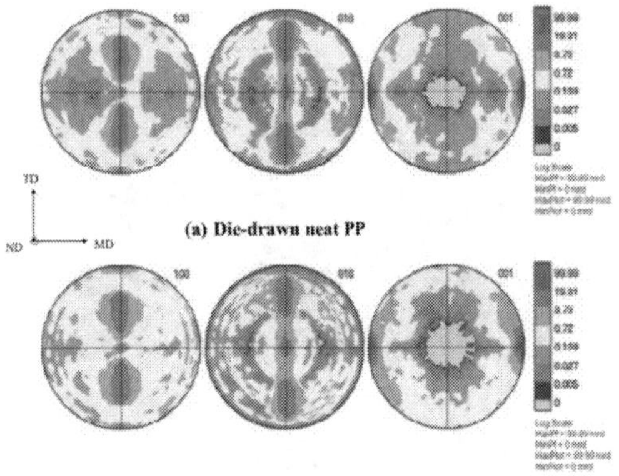

Fig.4: Pole figures for a*,b and c- axes of PP crystals in drawn materials-each drawn at 3.66 m/min.

The range of values for f_{cM} is between -0.5 and 1. When it is -0.5, the chain axes are all aligned perpendicular to the draw direction. When f_{cM} is 0, the orientation is random; when f_{cM}=1, the chain axes are all aligned with the draw direction. The c-axis orientation parameters obtained for die drawn neat PP and its composite are shown in **Fig.5**. Within experimental error the maximum crystalline orientation in drawn composite was reached at the draw ratio of 7.3 whereas for neat PP it was reached at a draw ratio of 6.8.

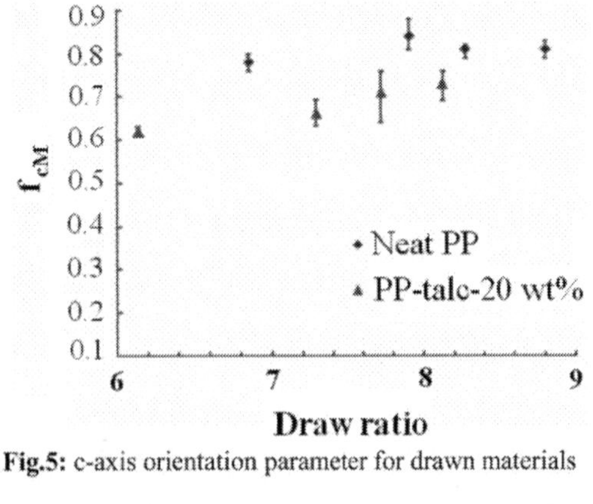

Fig.5: c-axis orientation parameter for drawn materials

The tensile modulus of the drawn materials continued to increase with the draw ratio (**Fig.6**). This can be attributed to the progressively increasing orientation of the amorphous phase and also to the progressively increasing aspect ratio of crystalline lamellae.

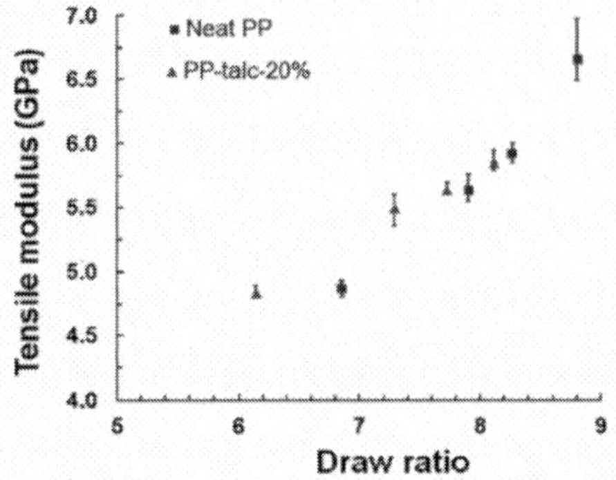

Fig.6: Tensile modulus of drawn materials as a function of draw ratio.

Conclusions:

Die drawing with a wedge die of nominal draw ratio 2 led to a lower extent of crystal orientation in PP-talc composite than in neat PP. The crystalline orientation of the matrix phase saturated at a draw ratio of 7.3. Despite leveling of the crystal orientation and of the void fraction in the drawn composite, the tensile modulus continued to increase beyond this draw ratio to a maximum value that was 3 times that of undrawn neat PP at the highest draw rate studied.

References:

1. Nichols, K.L. and B.M. Birchmeier, U.S. Patent Application US20080111278A1, October 15, 2007.
2. A. Cifferri and I.M.Ward, ed., Ultra High Modulus Polymers, Appl. Science Publishers Ltd, 1, (1979)
3. Ward, I.M., Coates, P.D. and Dumoulin, M.M., Solid Phase Processing of Polymers, Hanser Publishers, Munich (2000)
4. I. Karacan, A.K. Taraiya, D.I. Bower and I.M. Ward, Polymer, **34** (13), 2691 (1993)
5. C. Chaffey, A. Taraiya and I. M. Ward, Polym. Eng. And Sci. **37**, 1774, (1997)
6. M.C. Barnciforti, C. A. Oliveira and J.A. De Sousa, Polym. Adv. Technol., **21**(5), 322 (2010)
7. M. Naiki, Y. Fukui, T. Matsumura, T. Nomura and M. Matsuda, J. Appl. Polym. Sci. **79**,1693, (2001)

(706d) Surface Infusion Micropatterning of Elastomeric Substrates: A Novel Processing Method for Microfluidic Device Fabrication

Ronald C. Hedden[1], Huipeng Chen[1], Ziniu Yu[1], Daniel M. Lentz[2] and Ryan K. Nunley[1], (1)Chemical Engineering, Texas Tech University, Lubbock, TX, (2)3M Corporate R&D, Saint Paul, MN

Thelast decade has witnessed revolutionary advances in the fields ofmicrofabrication and microfluidics because of a number of attractive attributesof miniaturization, such as conservation of material in biological analyses. Microfluidic devices have found applicationsin a variety of fields including cell and molecular biology, medicine,synthesis of novel materials, biosensors and food safety. In recent years, microfabrication procedureshave shifted from materials such as silicon or glass toward inexpensivepolymers, which may be more amenable to low-cost mass production of disposabledevices. The development of accessible processing methods for microscalepatterning of soft elastomeric materials such as poly(dimethylsiloxane) (PDMS)has led to an explosive growth in research in microfluidic devices andmicrocontact printing devices.

Currentmicrofabrication processes for polymers include photolithography, casting,injection molding, microthermoforming, and hot embossing. Soft lithography'has revolutionized research in academic laboratories, where fabrication of PDMSdevices by simple casting protocols proves cost effective. On the other hand, injection molding,microthermoforming, and hot embossing are more adaptable to high-throughputmanufacturing. However, these processesare arguably less attractive to labs engaged in rapid prototyping of newdevices because of the prohibitive costs associated with repeated moldtooling. As a result, scale-up of newdevices from the prototype stage to the mass production stage can require achange in both materials and equipment, a hindrance to commercialization. A single process that allows rapid, cost-efficient production of micropatterned polymer devices at both the labscale and the industrial scale could prove to be a critical advance inmicrofabrication science and industry.

SurfaceInfusion Micropatterning (SIM) is a new microfabrication process for surfacerelief patterning of elastomeric substrates, including both chemicallycrosslinked or vulcanized materials (e.g. polydimethylsiloxane, PDMS) andthermoplastic elastomers (e.g., thermoplastic polyurethanes, TPUs). The starting material (substrate) is anoptically transparent elastomeric material, which can be formed into flatsheets by standard processing techniques prior to SIM. The substrate may be either free-standing ormounted as a surface layer on glass or another rigid material. Patterning of the substrate by SIM is athree-step process: 1) monomer infusion, 2) photopatterning, and 3) drying. For the first time, SIM has been demonstratedwith two elastomeric materials, a crosslinked polydimethylsiloxane (PDMS) and athermoplastic polyurethane (TPU), to create micropatterned surface relieffeatures including a "checkerboard" well pattern and a microfluidicchannel system. Surface micropatterns werecharacterized by scanning electron microscopy, optical microscopy, and opticalprofilometry to examine channel depth and shape. High-resolution surface features ofdimensions as small as 10 micrometers can be obtained using an uncollimated,broadband ultraviolet light source, making the technique amenable to low-costlab production. SIM is most suitable forproducing shallow (low aspect-ratio) microfluidic channels in soft elastomericsubstrates. After considering processadvantages and limitations, the potential for SIM to impact emerging microfluidicmanufacturing technologies will be discussed.

See more of this Session: Polymer Processing and Rheology II
See more of this Group/Topical: Materials Engineering and Sciences Division

(706e) Designed Experimental Study of Vapor-Grown Carbon Nanofiber/Vinyl Ester Nanocomposites with Focus On the Effects of Formulation and Processing Factors On the Nanocomposite Dynamic Mechanical Properties Over a Wide Temperature Range

Sasan Nouranian, Center for Advanced Vehicular Systems (CAVS), Mississippi State University, Mississippi State, MS, Hossein Toghiani, Dave C. Swalm School of Chemical Engineering, Mississippi State University, Mississippi State, MS, Thomas E. Lacy, Department of Aerospace Engineering, Mississippi State University, Mississippi State, MS, Janice L. DuBien, Department of Mathematics and Statistics, Mississippi State University, Mississippi State, MS and Charles U. Pittman Jr., Department of Chemistry, Mississippi State University, Mississippi State, MS

Polymer nanocomposites have been considered for automotive applications, but commercialization has not yet been fully realized. This is primarily due to processing and fabrication issues such as nanoreinforcement dispersion in the polymer matrix and poor nanoreinforcement-matrix interfacial adhesion. A designed experimental study was conducted on vapor-grown carbon nanofiber (VGCNF)/vinyl ester (VE) nanocomposites to better understand the effects of formulation and processing factors on the nanocomposite dynamic mechanical (viscoelastic) properties. VGCNF/VE nanocomposites represent a class of thermoset nanocomposites that are suitable for use in many automotive structural parts. The objective of the study was to identify a statistically reliable and reproducible method to fabricate these materials with a focus on prediction and optimization of their mechanical performance through response surface modeling. For engineering purposes, the thermal behavior of polymers must be characterized as part of component design and during assessments of the material's in-service performance. The inclusion of nanoreinforcements may significantly affect the thermo-mechanical behavior of the nano-phased matrix, including its glass transition temperature (T_g) and thermal expansion behavior. By including temperature in the design, the influence of its variation on the nanocomposites' viscoelastic properties (stiffness and energy dissipation characteristics) during service were analyzed and incorporated into the response surface models. The design factors considered were 1) the use of oxidized versus pristine VGCNFs, 2) the use of dispersing agent versus none, 3) the mixing method (ultrasonication, high-shear mixing, and a combination of both), 4) the nanofiber weight fraction (0, 0.25, 0.50, 0.75, and 1.00 parts per hundred parts resin (phr)), and 5) temperature (30, 60, 90, and 120 °C). Viscoelastic properties (storage and loss moduli) were selected as the mechanical responses since these are measures of stiffness and energy dissipation, respectively. These measures are of interest for automotive applications. Statistical analysis was used to construct response surface predictive models and to determine the nanocomposite formulation and mixing method leading to optimal viscoelastic properties. The predicted storage modulus was a function of three variable combinations, i.e., high-shear mixing (or high-shear mixing/ultrasonication), ultrasonication with oxidized VGCNFs in the presence of a dispersing agent, and ultrasonication with other VGCNF type/dispersing agent combinations over the entire temperature range (30-120 °C). A similar but simpler dependence was observed for the loss modulus, where the ultrasonication combinations all had the same effect on the predicted loss modulus. The response surface models were used to identify combinations of formulation and processing factors that would lead to optimal predicted nanocomposite viscoelastic properties over the entire temperature range. The use of high-shear mixing, oxidized VGCNFs without a dispersing agent or pristine VGCNFs with a dispersing agent, and a VGCNF weight fraction of ~0.50 phr are recommended for maximizing the storage modulus at low, moderate, and high temperatures. For a low loss modulus over the same temperature range, the combination of oxidized VGCNFs with no dispersing agent is recommended. However, the combination of pristine VGCNFs with a dispersing agent would yield moderate energy dissipation. To maximize the loss modulus over the entire temperature range, the use of ultrasonication with ~0.25 phr VGCNF is recommended. Nevertheless, this will result in lower predicted storage moduli compared to the high-shear mixing. These predictive models will enable composite manufacturers to tailor nanocomposites for different applications through the manipulation of design factors.

See more of this Session: Polymer Processing and Rheology II
See more of this Group/Topical: Materials Engineering and Sciences Division

(706f) Nonlinear Viscoelasticity and Recovery of Polypropylene-Clay Nanocomposite Melts

Krishnamurthy Jayaraman and Tanmay J. Pathak, Chemical Engineering & Materials Science, Michigan State University, East Lansing, MI
See more of this Session: Polymer Processing and Rheology II
See more of this Group/Topical: Materials Engineering and Sciences Division

Unavailable Paper

Aqueous fiber spinning from a renewable protein waste material

Jason W. Soares, Steven Arcidiacono, Charlene M. Mello
U.S. Army Natick Soldier Research, Development & Engineering Center, Natick, Massachusetts 01760

Madhusudan Vasudevamurthy, Juliet Gerrard
School of Biological Sciences, The University of Canterbury, Christchurch NZ

Extended Abstract

Crystallin proteins from animal eye lenses represent a renewable, environmentally responsible biopolymer source for bio-derived fibers. The crystallin protein family is a diverse and highly conserved set of proteins that form the macromolecular structure of the lenses of many vertebrates. Despite variability in lens anatomy across organisms, these proteins are consistently organized in a supramolecular structure that is optically clear, preserving the ability of the lens to refract light. Due to their natural functions, these proteins possess a range of interesting features which make them attractive for the production of new biomaterials such as self-assembly and the formation of stable, well ordered structures. These structures include soluble protein, nanofibrils, and amorphous aggregates, each derived from native conformations. Here, we describe the aqueous processing and spinning of crystallin proteins generated from a waste material, namely fish eye lens, into durable fibers. Crystallin protein was isolated from North Atlantic Haddock (*Melanogrammus aeglefinus*) eye lenses through a simple homogenization technique. Intact lenses were extracted and gently stirred at room temperature in an aqueous homogenization buffer designed to minimize protein aggregation and stabilize intermediate assembly states. After 24 hrs, the fish lens homogenate (FLH) is clarified by centrifugation and collected as an opaque suspension in which crystallin protein is present in both amorphous and structured forms. The FLH is comprised of a series of related structural proteins, which are used for subsequent investigations without further purification.

Aqueous spinning of fibers from the FLH entails a delicate balance of processing, spinning and post-spinning variables. The desired aqueous processing approach creates a uniform, soluble protein solution removing insoluble aggregates. The processing approach must stabilize solution dynamics to facilitate proper ordering and organization of the protein, at a molecular level. Minimization or suppression of aggregation during solution preparation and aging at relatively high concentrations is required to generate a reproducible spin solution for fiber spinning. Toward designing suitable processing conditions, FLH was dialyzed into biological buffers and concentrated through ultrafiltration. Several variables throughout the process were considered including buffer composition, protein concentration and spin solution aging. Dialysis buffer composition was varied by supplementation with a number of amino acid stabilizers (i.e. aggregation suppressors) at varying concentrations including glycine, lysine and arginine. Buffer exchange was also conducted under varying temperature and pH conditions. While glycine and lysine enhanced protein solubility as amino acid concentration increased, the influence on protein molecular assembly was not sufficient since gelation occurred relatively rapidly during spin solution aging. However, supplementation with arginine improved protein solubilization and stabilization, enhancing protein yields and allowing for the preparation of concentrated spin solutions (~300-400 mg/ml). Arginine resulted in reproducible, stable solutions which, after sufficient aging, facilitated suitable protein assembly states for subsequent formation of intact, durable fibers. Interestingly, the arginine supplemented buffer was basic in nature (pH = 10.5) whereas glycine and lysine buffers were more acidic (pH 5 and 5.2 respectively). The influence of pH relative to the presence of arginine on protein solubilization and stabilization is still under investigation. Optimal processing temperature was determined as 4°C.

Spin solutions of various concentrations and stages of supramolecular assembly were spun into fibers through a patented aqueous spinning approach. The spinning technique was inspired by nature and, more specifically, the process by which a spider spins native dragline silk. The design of the spinneret is based on the gradually tapered gland within the spider, which causes a shear-induced transition of self-assembled silk protein into a fiber. Our lab-scale technique has previously been demonstrated in spinning recombinant spider silk proteins into fibers with mechanical integrity equivalent to natural dragline silk. Spin solutions are loaded into the micro-spinneret, which has a volume capacity of 500 ul although typical spin trials are 50-75 ul. A syringe pump applies a constant force at a constant rate (2-10 ul/min) to force the spin dope through PEEK tubing with I.D. ranging from 0.15-0.25 mm into a coagulation bath. The fibers are allowed to coagulate within the bath and are collected after removal from the coagulation bath and submersion into a water bath. Numerous spinning and post-spinning variables must be considered to optimize fiber formation, durability and mechanical integrity from concentrated spin dopes. Here, this spinning technique was used to generate fibers from FLH solutions containing arginine at various protein concentrations and stages of spin solution aging, which correlate to various intermediate assembly states. Fiber spinning variables considered include coagulation bath composition and spinneret I.D. while post-spinning variables include coagulation time and draw ratio. Coagulation tests were conducted to choose optimal coagulation conditions for the various spin solutions. Fibers up to 12 inches were generated although typically fibers ranged from 2-3 inches in length. Fibers were of sufficient durability to withstand not only natural draw due to surface tension while removing from the coagulation bath but also extension to twice the original length prior to collection. Drawn fibers were, on average, ~17 um in diameter while undrawn fibers were ~28 um. Light microscopy of the as-spun fibers under white light was conducted to investigate surface and core defects along the length of the fiber. Molecular alignment along the fiber axis was analyzed by investigating the birefringent nature of the single fibers under polarizing light with a first order red plate at 530 nm. In general, fibers exhibited birefringence, which was indicative of some level of molecular order. However, the fibers did contain a variety of defects that limited subsequent mechanical testing.

Analysis of a compilation of spin trials enabled the identification of integral processing, spinning and post-spinning variables essential for formation of fibers from crystallin proteins. Spin solution aging is dependent on protein concentration - higher concentration requires less aging to reach the appropriate conditions to spin fibers. Increasing draw decreased fiber diameter, but molecular order was not enhanced at the current extensibility (~2-fold). Optimal coagulation time was determined as 15-20 min for formation of the most durable fibers. The information derived is being considered for optimization of processing and spinning processes to produce fibers with enhanced mechanical integrity. Also under investigation is the use of nano-reinforcements, specifically nanofibrils, to enhance fiber properties. Crystallin proteins readily form nanofibrils in solution, with branching and length being controlled by processing conditions and temperature. Nanofibrils occur by subjecting the proteins to mildly destabilizing conditions that allow the proteins to partially unfold and refold. Recent advances in our lab enable the control of nanofibril structure and morphology (e.g. length, bundling and branching). Nanofibrils of up to 10 um in length with varying degrees of bundling and branching have been created and may serve as an additive that could further facilitate formation of fibers with increasing mechanical integrity from concentrated soluble protein spin solutions. Alternatively, fibers can be spun directly from solutions containing nanofibrils.

Fibers from crystallin proteins are a naturally-derived biomaterial that could be used in composite or textile applications including new fiber reinforcing material for (bio)composites or as a substitute for Nylon and polypropylene-based textiles. Furthermore, the fibers represent a chemically diverse (defined by the amino acid side chains) platform for the future design of new multifunctional materials for use in military and civilian applications. This work represents a significant advancement in the state-of-the-art regarding spinning of bio-derived fibers from a renewable waste protein material and offers a basis for spinning a range of biopolymers into useful materials.

(708a) Rapid Exfoliation and Physical Gelation of a Clay-Polymer Nanocomposite

H. Henning Winter, Division of Chemical, Bioengineering, Environmental and Transport Systems, National Science Foundation, Arlington, VA and Brian Momani, Chemical Engineering, University of Massachusetts, Amherst, MA

A new exfoliation mechanism is proposed for organically modified clay (layered silicate) in a matrix of end-functionalized ("sticky") macromolecules. The evolving clay structure gives rise to physical gelation which can be monitored rheologically. When mixing the clay particles into the polymer, sticky macromolecules anchor on the outer clay surfaces and generate an entopic pulling force that peels clay sheets away from the stack. The molecular force is due to the restricted free energy in the presence of the solid surface. The assembly of anchored macromolecules (brush) on the outside of clay particles can overcome the internal cohesion since the organically modified clay sheets are only weakly connected with each other. The entropic force increases with temperature due to increased thermal motion, but then decreases again when macromolecules begin to detach at high temperature. The experimental protocol includes time-resolved rheometry (Rheol Acta 33:385-397, 1994) and rescaling of data (Rheol Acta 45:331-338, 2006). Surprisingly, the product of equilibrium modulus and longest relaxation time, $G_e(t_r)$ $l_{max}(t_r)$, was found to remain constant throughout the late stages of physical gelation. Parameter is the duration of the isothermal ripening process ("ripening time", t_r). Acknowledgment: NSF support through CBET-0651888.

See more of this Session: Processing of Nanocomposites
See more of this Group/Topical: Materials Engineering and Sciences Division

(708b) Polypropylene Nanocomposites Reinforced with Different Carbon NanoStructures

Yunfeng Li[1], Jiahua Zhu[1], Suying Wei[2], Luyi Sun[3] and **Zhanhu Guo**[1], (1)Integrated Composites Laboratory (ICL), Dan F Smith Department of Chemical Engineering, Lamar University, Beaumont, TX, (2)Chemistry and Biochemistry, Lamar University, Beaumont, TX, (3)Department of Chemistry and Biochemistry, Texas State University-San Marcos, San Marcos, TX

Polypropylene (PP) nanocomposites reinforced with different carbon nanosize fillers (carbon nanotubes, carbon nanofibers, graphite nanoplatelets, active carbon nanoparticles) were fabricated via an ex-situ method.[1,2] High quality filler dispersion in the PP matrix was obtained via PP/xylene solution strategy and verified by the scanning electron microscopy (SEM). The thermal properties of PP/CNFs nanocomposites were characterized by thermal gravimetric analysis (TGA) and differential scanning calorimetric (DSC) and shown to exhibit improved thermal stability and higher crystallinity. The prepared PP nanocomposites were further processed into pellets by compression molding. The electrical conductivity and dielectric property of the PP nanocomposite were studied. Both the electric conductivity and real permittivity increase with the increase of the nanofiller loadings. The electrical conductivity percolation is observed at the filler loading of 15.0, 12.0, 5.0 and 3.0 wt% for graphite nanoplatelets, carbon nanofibers and carbon nanotubes, respectively. The rheological behaviors of the nanocomposite melts are also investigated and the results are compared with different carbon nanocomposite systems. These results verify the percolation percentage is related to the material aspect ratio. It is found that a small concentration of filler could affect the modulus and viscosity of PP melt significantly.

1) Poly(propylene)/Carbon Nanofiber Nanocomposites: Ex Situ Solvent-Assisted Preparation and Analysis of Electrical and Electronic Properties, *Macromolecular Materials and Engineering*, DOI:10.1002/mame.201000341, (2011)

2) Ex-Situ Solvent-Assisted Prepared Magnetic Polypropylene Nanocomposites Filled with Fe@FeO Nanoparticles, *Macromolecular Materials and Engineering*, DOI: 10.1002/mame.201100010, (2011)

See more of this Session: Processing of Nanocomposites
See more of this Group/Topical: Materials Engineering and Sciences Division

(708c) Magnetic Polyacrylonitrile-Fe@FeO Nanocomposite Fibers - Electrospinning, Stabilization and Carbonization

Jiahua Zhu[1], Yutong Li[2], Suying Wei[3] and Zhanhu Guo[1], (1)Integrated Composites Laboratory (ICL), Dan F Smith Department of Chemical Engineering, Lamar University, Beaumont, TX, (2)Western Digital Corporation, Fremont, (3)Chemistry and Biochemistry, Lamar University, Beaumont, TX

Uniform and bead-free pure polyacrylonitrile (PAN) and its magnetic polymer nanocomposites (PNC) fibers with different core-shell Fe@FeO nanoparticles (NPs) loadings are prepared using electrospinning method. The morphology of the resulting products is correlated to the corresponding rheological behaviors of the pure PAN and PAN/Fe@FeO solutions. The diameter of the PAN fibers is linearly related to the polymer solution concentration. However, with a fixed PAN concentration of 10 wt%, the Fe@FeO NP loading shows negligible effect on the morphology of the PNC fibers. Thermogravimetric analysis (TGA) results indicate an enhanced thermal stability of the PNC fibers than that of the pure PAN fibers. Magnetic carbon nanocomposite (MNC) fibers are prepared through the stabilization and carbonization of the electrospun PNC fibers. The effects of the heating procedures, including the stabilization and carbonization temperature and time, on the fiber morphology are systematically investigated. Both short and long MNC fibers could be easily produced by changing the heat procedures. Room temperature magnetic properties of the CNC fibers based on different heating procedures are also studied in this work.

Magnetic Polyacrylonitrile-Fe@FeO Nanocomposite Fibers - Electrospinning, Stabilization and Carbonization, Jiahua Zhu, Suying Wei, Dan Rutman, Neel Haldolaarachchige, David P. Young, Zhanhu Guo; Polymer, doi:10.1016/j.polymer.2011.04.034 2011

See more of this Session: Processing of Nanocomposites
See more of this Group/Topical: Materials Engineering and Sciences Division

(708d) Influence of Polymerizable Organoclays On Thiol-Acrylate Photopoltmerization System

Soon Ki Kim and Allan Guymon, Chemical and Biochemical Engineering, The Universsity of Iowa, Iowa city, IA

Photopolymerizable organoclays in producing clay nanocomposites can react with monomers and incorporate into polymer matrices during the UV curing process. This research investigates the effects of polymerizable organoclays on the reaction behavior as well as ultimate composite properties based on *in-situ* thiol-ene photopolymerization. To this end, different types of polymerizable organoclays prepared by changing the functional group structure were examined in various thiol-acrylate systems. Higher chemical compatibility of components induced enhanced clay exfoliation resulting in significant increase in both photopolymerization rate and overall toughness of ultimate nanocomposites compared to those with non-reactive organoclays. In addition, by affecting the stoichiometric ratio of functional groups in the clay gallery, thiolated organoclays facilitate great degree of step-growth thiol-ene reaction with higher thiol conversion while acrylated organoclays encourage acrylate homopolymerization. The enhanced thiol-ene reaction by adding thiolated organoclays also induces less oxygen inhibition with increased flexibility of nanocomposites. Finally, based on increased interaction with polymer networks during the photopolymerization, all types of polymerizable organoclays significantly reduce the polymerization shrinkage and shrinkage stress compared to non-reactive organoclays.

See more of this Session: Processing of Nanocomposites
See more of this Group/Topical: Materials Engineering and Sciences Division

ENHANCED ELECTRICAL CONDUCTIVITY OF TEXTILE FABRICS VIA POLYMERIC NANOCOMPOSITES

Cem Güneşoğlu [a,b], Sinem Güneşoğlu[a], Jiahua Zhu[b], Emre Atabey[b], Rahul Patil[b], Suying Wei[c], Zhanhu Guo[b]

[a] Textile Engineering Department, Gaziantep University, Gaziantep, 27310, Turkey
[b] Integrated Composites Laboratory, Dan F. Smith Department of Chemical Engineering, Lamar University, Beaumont, TX, 77710, USA
[c] Department of Chemistry and Physics, Lamar University, Beaumont, TX, 77710, USA

General approaches to enhance selected properties of textile fabrics include the embedment of relevant non-textile particles within structure or cover the surface with those particles via various techniques. Electrical conductivity of textile fabrics among those properties has growing interest for researchers; traditional fibers that are used in the textile fabrics are electrically insulating materials and to add some conductivity will bring new usage areas of textile fabrics, especially for health-related topics like antistatic and electromagnetic shielding. The widely accepted technical approach for increased conductivity of textile fabrics is incorporation of stainless steel and cooper wire as conductive fillers; however the existence of metal wires will increase fabric thickness which is accepted as variation source on surface resistance. Therefore coating conductive polymers (CP) onto textiles would allow power to flow through without incorporation of any wiring throughout the fabric. There are some studies on coating textile fibrous surfaces with conductive polymers like polyaniline (PANI) and polypyrrole (PPy) via electrochemical and chemical polymerization (block or graft) upon insulating fiber or preparation of composite conductive fibers by chemical or electrochemical routes. Metallic based CP nanocomposites are explored in view of great variety of applications and improved electron transfer and eventually achieved synergistic effect due to the combination of the electrocatalytic performances of the composite constituents are expected. Therefore, one particular research objective has been the dispersion of metal nanoparticles within the polymer matrix, since the incorporation of metal nanoparticles is known to enhance the conductivity of the polymer. Additionally, metal nanoparticles imbedded in CP matrix is claimed to serve as efficient electrocatalysts during oxidation and reduction of monomer and arrange CP into well-defined structure, avoid aggregation and give controllable particle size of nanocomposites. As a result, the textile surfaces covered with CP / metal nanoparticle would give a good level of electrical conductivity which is a key factor for EMR shielding property. This study has been conducted to investigate the enhancement at electrical conductivity of 100% cotton knitted textile fabric after covering it with conductive polymer / metallic particle nanocomposites via completing the polymerization of CP at the presence of fabric. The study was run for PANI as CP. The oxidant selected was ammonium peroxydisulfate (APS) and the protonic doping agent was p-toluene sulphonic acid (PTSA). Metal nanoparticles were selected as Fe@C and Co@C as widely investigated magnetic encapsulates with average particle size of 20 nanometers. The oxidant: aniline: doping agent ratios were maintained as 1 : 1,25 : 1 where molar volumes of aniline were varied for several applications. The metal nanoparticles had weight ratios to textile fabric to be covered as 1% and they were subjected to silanization with (3-aminopropyl)-triethoxysilane (APTES) before polymerization. The polymerization cycle of aniline were completed in accordance with the procedures in literature.

The volume resistivities of covered samples were measured with a resistivity cell (HP Agilent 16008B) satisfying insulation measurement standard ASTM D-257. The morphology of nanocomposite over cotton fibre were evaluated via SEM (Hitachi 3400 N), size distribution and cluster formation were also assessed. FT-IR characterizations were carried out using a Bruker Inc. Tensor 27 FT-IR spectrometer with Hyperion 1000 attenuated total reflection (ATR) microscopy accessory. The magnetic properties of covered samples were recorded at room temperature in a 9T physical properties measurement system (PPMS) by Quantum Design and the conditions which give magnetically harder covering were obtained.

The results showed that different volumes of CP / metal particle nanocomposites gave statistically significant volume resistivity values of covered cotton fabrics; where 0,5 M PANI + Co@C (or Fe@C) covering decreased volume resistivity of fabrics at most (from $9{,}5 \times 10^9$ Ω.cm to $1{,}5 \times 10^4$ Ω.cm). FT-IR figures pointed the chemical bonding between fibre and nanocomposite and magnetic properties of covered samples were observed only after using metal nanoparticle at polymerization.

Thermogravimetric Analysis of Linear Low-Density Polyethylene (LLDPE) and Maleated Linear Low-Density Polyethylene (LLDPE-g-MA) Infused with Nanoclay in Supercritical Carbon Dioxide

Matthew Factor[1] and Sunggyu Lee[2]

[1]Department of Chemical & Biological Engineering
Missouri University of Science & Technology
Rolla, MO 65409-1230
[2]Department of Chemical & Biomolecular Engineering
Ohio University
Athens, OH 45701

Abstract

The addition of clay filler to polymers can have a desirable effect on the mechanical, thermal and gas-barrier properties of the polymer. To maximize the effect of clay on the polymer, a high level of clay surface area should be in contact with the polymer matrix. To accomplish this, the clay is intercalated or delaminated into individual platelets and uniformly dispersed throughout the polymer matrix. The processing of clays and polymers in supercritical carbon dioxide can increase intercalation, delamination and dispersion of clays into a polymer matrix. Linear low density polyethylene (LLDPE) and maleated linear low-density polyethylene (LLDPE-g-MA) were subjected to supercritical carbon dioxide in the presence of nanoclay and the resulting nanocomposites were analyzed to determine the extent of infusion and delamination. The X-ray diffraction (XRD) and Fourier Transform Infrared (FTIR) analyses confirmed infusion of clay into the polymer matrix. In this work, thermogravimetric analysis (TGA) was employed to determine the thermal stability of the samples and the amount of clay infused into the polymers.

Introduction

Polymers augmented with clay filler are of great interest to researchers since they can exhibit superior enhancement in gas-barrier, mechanical and thermo-mechanical properties as compared to their unfilled counterparts[1-5]. Clays also have the advantage of being inexpensive, abundant and environmentally benign. Typically, polymer property enhancement increases as the level of mixing between the polymer and clay increases[1,2,6]. The lowest level of mixing occurs when the clay is immiscible in the polymer and the clay agglomerates. Miscible mixtures of polymer and clay occur when the polymer penetrates between stacked clay platelets, creating an intercalated structure, or when clay platelets completely separate to individually mix with the polymer matrix, exhibiting an exfoliated structure. However, intercalation and, to a greater extent, exfoliation are difficult to achieve due to the inherent immiscibility between a hydrophilic clay and an organophilic (hydrophobic) polymer[1,7].

One method to increase mixing between polymer and clay is to increase their compatibility with each other through modification[1,3,7]. Sodium cations present within the clay interlayer (region between clay platelets) are typically exchanged with organic cations (e.g., alkylammonium salts) to modify clay, converting it to an organoclay. The organic cations increase the organophilicity of the clay and expand the interlayer spacing to make it easier for polymer chains to diffuse between platelets. Modifications of organophilic polymers, such as polyolefins, to make them more hydrophilic can be accomplished by grafting them with hydrophilic moieties, like maleic anhydride. Modifying polymer and clay can greatly enhance mixing, but typical processing conditions for the melt mixing of polymer and clay can thermally degrade the clay modification and once again lead to immiscible mixtures.

Processing polymer and clay in supercritical carbon dioxide ($scCO_2$) can circumvent the issue of thermal degradation of the clay organic modifier. Thermal degradation is avoided due to the low critical temperature of CO_2 existing at 31.1°C (critical pressure is 7.4 MPa)[1,8,9]. Carbon dioxide in its supercritical state possesses unique properties that give it a gas-like diffusivity and liquid-like density[1,8]. The liquid-like density allows $scCO_2$ to behave as an organic solvent to dissolve and swell the polymer, while the gas-like diffusivity permits quick mass transfer of the clay particle into the polymer matrix[1,10]. In addition, CO_2 is inexpensive, abundant, non-combustible and environmentally benign. The applications of $scCO_2$ include extraction, polymer grafting and depolymerization. Furthermore, $scCO_2$ has shown promising results in exfoliating clay platelets[1]. The exfoliation occurs when CO_2 molecules between platelets, reduced in size from being in a pressurized environment, expand upon rapid depressurization to break up stacks of clay platelets.

The following experiments were conducted to determine if clay could be infused into linear low-density polyethylene with $scCO_2$ processing. In order to increase the probability of infusion and clay dispersion, modified clays and polymers were utilized. After processing in $scCO_2$, the samples were analyzed by thermogravimetric analysis (TGA) to determine if their thermal stability was altered and how much clay was infused.

Experimental

Materials

Experiments employed either linear low-density polyethylene (LLDPE) or maleated linear low-density polyethylene (LLDPE-g-MA) as the polymer to undergo $scCO_2$ processing with clay. Exhibiting its melting point at 124°C, the LLDPE has a specific gravity of 0.917 and sells by the trade name Dowlex 2517. Similarly, the LLDPE-g-MA, a.k.a. Polybond 3109, has a melting point of 123°C and a density of 0.926 g/cm^3. Maleated LLDPE has a higher compatibility with the clay[3,5-7] and $scCO_2$ as compared to pure LLDPE. The LLDPE-g-MA has a maleic anhydride graft level of 1 wt%. Both polymers were processed in $scCO_2$ in pellet form.

The clay Cloisite 93A (C93A) employed in this experiment is an organoclay created by modifying natural montmorillonite with methyl, dihydrogenated tallow ammonium (quaternary ammonium salt). The modifier concentration is 90 meq/100 g of clay and density of the clay is 1.88 g/cm^3 at room temperature. The organoclay C93A was provided by Southern Clay Products, Inc. (located in Gonzales, Texas). Cloisite 93A has been reported to have a higher compatibility with carbon dioxide than other organoclays[1]. An oven set to 80°C was used to dry the clay for a period of 24 hours before experiments were performed.

Processing Procedure

The production of polymer/clay composites via $scCO_2$ processing occurred in a stainless steel autoclave reactor (300-cc volume) over a period of 3 hours in a batch mode. In order promote a uniform temperature throughout the reactor, a magnetically driven impeller was rotated within the reactor at about 300 rpm and cooling water was passed through the reactor head. Upon reaching the operating pressure and temperature, the experiment was started. At the end of the 3-hour run time, a fast depressurization of the reactor was performed. Each experiment was conducted with 10 g of polymer well coated with 2.5 g of C93A by manually mixing the polymer and clay. Four experiments were conducted per polymer, two experiments at 60°C (one experiment at a pressure of 10.3 MPa and another at 17.2 MPa) and two experiments at 98.9°C (one experiment at a pressure of 10.3 MPa and another at 17.2 MPa). Labels for runs are listed in Table 1.

Table 1. $ScCO_2$ processing conditions for polymer and clay

	LLDPE		**LLDPE-g-MA**	
Run	**T (°C)**	**P (MPa)**	**T (°C)**	**P (MPa)**
1	98.9	17.2	60	10.3
2	98.9	10.3	98.9	10.3
3	60	10.3	60	17.2
4	60	17.2	98.9	17.2

Characterization

Thermogravimetric analysis (TGA) was conducted with samples weighing from 40 to 65 mg. The device used for TGA was a Netzsch STA 409 C/CD. Samples were heated from 30°C to 600°C at a rate of 10°C/min. Analyses were conducted under nitrogen atmosphere with a flow rate of 20ml/min.

Results and Discussion

Pure LLDPE and processed LLDPE/C93A samples underwent thermogravimetric analysis (TGA) to determine their thermal stability and the resultant curves are displayed in Figure 1. The onset temperature of decomposition for pure and processed samples varies between 455°C and 460°C. Decomposition ends in the range of 488°C to 493°C, while weight loss is between 98.8 and 99.4%. In comparing the amount of residue remaining after polymer decomposition, processed samples left a residue from 0.61 to 0.82 wt% while pure LLDPE left a higher residue of 1.19 wt%. Residual weights for individual processing conditions are displayed in Figure 2. As all LLDPE/C93A sample residuals are between 30 to 50% lower than the residual for pure LLDPE, processing the samples in $scCO_2$ appears to alter the polymer, increasing the degree of decomposition. Shieh et al.[11] also observed this phenomenon to a larger extent after processing PET in a compressed CO_2 environment. Since pure LLDPE did not completely decompose and had a greater residual weight than the $scCO_2$-processed LLDPE/clay samples, a precise amount of clay infused into LLDPE could not be confidently reported without first comparing the TGA results to the results obtained by another method. In previous work with Fourier Transform Infrared (FTIR) analysis[12], LLDPE/C93A samples were determined to have a near zero clay infusion. Using the FTIR results and the residual weight of pure LLDPE as a reference, it can reasonably be proposed that since the processed samples possess a lower residual weight than the pure LLDPE, no clay was infused. Polymer weight that was not accounted for as degraded polymer or residue was attributed to water evaporation. Pure LLDPE contained no water whereas the processed samples contained between 0.01 and 0.08 wt% water.

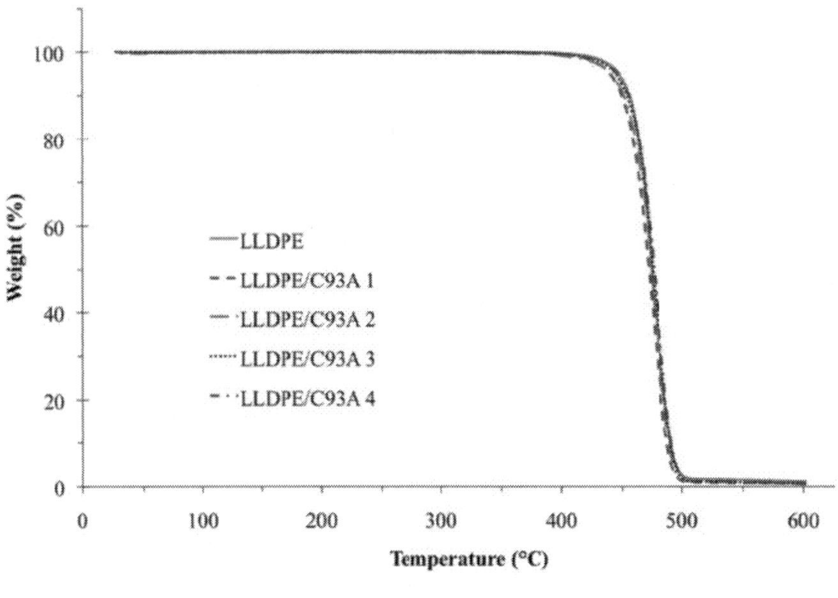

Figure 1. TGA of pure LLDPE and LLDPE processed with C93A in $scCO_2$

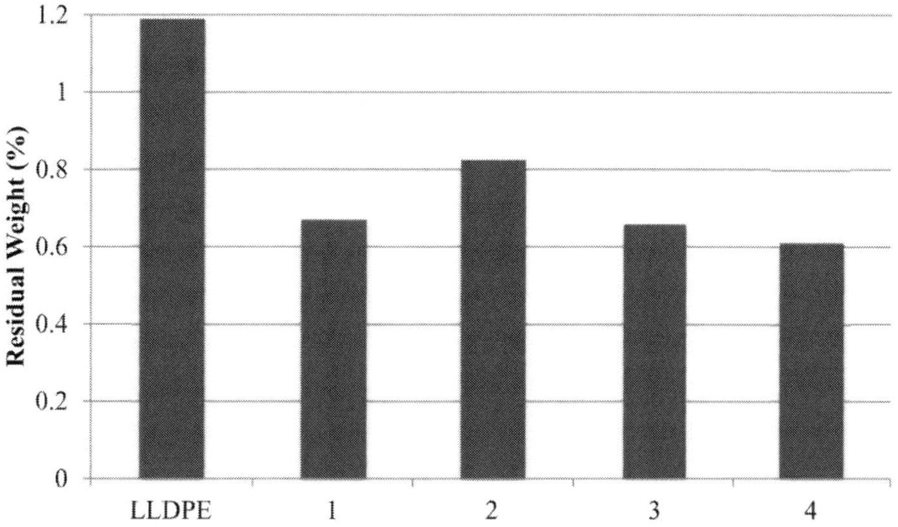

Figure 2. Residual weight of pure LLDPE and LLDPE processed with C93A in $scCO_2$

Pure LLDPE-g-MA and samples containing LLDPE-g-MA and C93A began to lose weight at about 147°C which is associated with a loss of bound water[13]. Similar to the above TGA curves for LLDPE samples, the main decomposition for LLDPE-g-MA/C93A starts at about 459°C and finishes at 492°C. (TGA curves for LLDPE-g-MA samples were not included for brevity.) The residual weight left by pure and processed maleated samples (Figure 3) is between 0.82 and 2.19 wt%. The higher residual weight of the LLDPE-g-MA samples compared to the LLDPE samples is attributed to a larger amount of clay infusion into the polymer since both pure samples of LLDPE and LLDPE-g-MA left the same residual weight of 1.19 wt%. Applying the same logic as above to the LLDPE-g-MA/C93A samples, the residual weight of sample 3 is below that of pure LLDPE-g-MA, suggesting that no clay infusion occurred. In order of increasing residual weight, samples 1, 2 and 4 have residual weights above that of pure LLDPE-g-MA, indicating that C93A was infused into the polymer. The previously conducted FTIR analysis[13] also exhibits the same order of increasing clay amounts for samples 1, 2 and 4 with sample 3 exhibiting a near zero clay infusion. Using pure LLDPE-g-MA as a basis for the amount of residual polymer that remains after degradation, samples 1, 2 and 4 respectively contain 0.32, 0.49 and 1 wt% clay. Sample 4 is within the range of the clay quantity attained from FTIR analysis, whereas samples 1 and 2 are respectively 0.12 and 0.04 wt% below the FTIR range. The disparity between results could be due to $scCO_2$ altering the samples to allow more polymer degradation during TGA or due to a difference in sample preparation between the two methods. Namely, sampling for TGA involves only one polymer pellet to contribute to clay infusion whereas FTIR analysis involves many melt-mixed polymer pellets to contribute to an average clay infusion within a film. The higher amount of C93A in LLDPE-g-MA as compared to LLDPE is associated with the higher compatibility of the maleated polymer with both the modified clay and the $scCO_2$ medium. Weight loss from water and, to a much lesser extent, from degradation of the organic modifier of C93A was determined to vary from 0.14 to 0.26

wt%. The degradation temperature of the organic modifier was undetectable as a result of the low weight of clay infused, corresponding to an even lower (i.e., undetectable) organic modifier weight. As compared to LLDPE samples, LLDPE-g-MA samples contain higher water content due to the polar nature of the maleic anhydride interacting favorably with the polar water molecule.

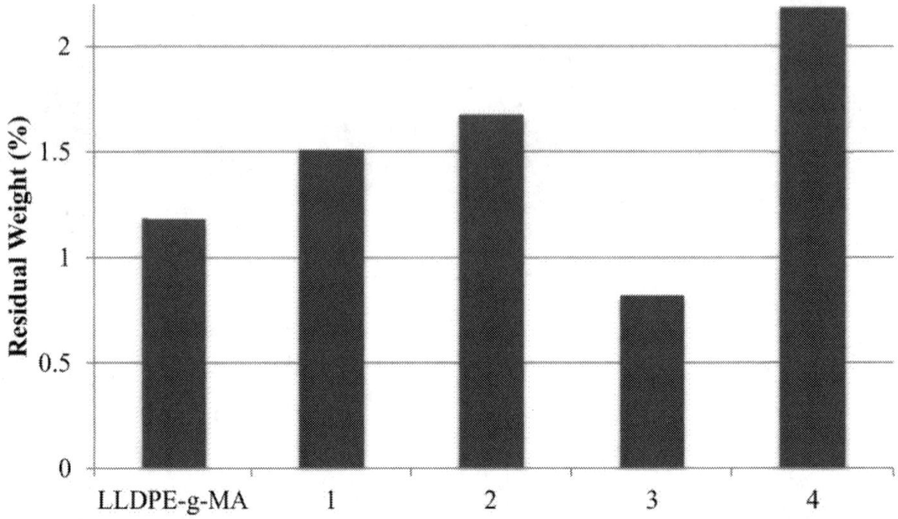

Figure 3. Residual weight of pure LLDPE-g-MA and LLDPE-g-MA processed with C93A in $scCO_2$

Conclusion

The LLDPE/C93A and LLDPE-g-MA/C93A samples underwent $scCO_2$ processing and were determined by TGA to have similar thermal stabilities. Maleated LLDPE/C93A samples possessed a higher residual weight that was attributed to a higher clay infusion amount as compared to LLDPE/C93A samples. The amount of clay infusion into samples was determined by TGA and was confirmed by previously conducted FTIR analysis[12]. Future work with transmission electron microscopy (TEM) will help to determine the morphology of clay within the polymer matrix.

References

1. Horsch, S.; Serhatkulu, G.; Gulari, E.; Kannan, R. M. Supercritical CO_2 Dispersion of Nano-Clays and Clay/Polymer Nanocomposites. *Polymer* **2006**, 47, 7485-7496.
2. Morgan, A. B.; Gilman, J. W. Characterization of Polymer-Layered Silicate (Clay) Nanocomposites by Transmission Electron Microscopy and X-Ray Diffraction: A Comparative Study. *Journal of Applied Polymer Science* **2003**, 87, 1329-1338.
3. Wang, K. H.; Choi, M. H.; Koo, C. M.; Choi, Y. S.; Chung, I. J. Synthesis and Characterization of Maleated Polyethylene/Clay Nanocomposites. *Polymer* **2001**, 42, 9819-9826.
4. Ma, X.; Yu, J.; Wang, N. Production of Thermoplastic Starch/MMT-Sorbitol Nanocomposites by Dual-Melt Extrusion Processing. *Macromolecular Materials and Engineering* **2007**, 292, 723-728.
5. Luyt, A. S.; Geethamma, V. G. Effect of Oxidized Paraffin Wax on the Thermal and Mechanical Properties of Linear Low-Density Polyethylene-Layered Silicate Nanocomposites. *Polymer Testing* **2007**, 26, 461-470.
6. Chen, B. Polymer-Clay Nanocomposites: An Overview with Emphasis on Interaction Mechanisms. *British Ceramic Transactions* **2004**, 103, 6, 241-249.
7. Ryu, S. H.; Chang, Y. W. Factors Affecting the Dispersion of Montmorillonite in LLDPE Nanocomposite. *Polymer Bulletin* **2005**, 55, 385-392.
8. Xu, Q.; Pang, M.; Li, J.; Jiang, Y. Application of Supercritical Carbon Dioxide in the Preparation of Biodegradable Polylactide Membranes. *Journal of Applied Polymer Science* **2004**, 94, 2158-2163.
9. Shieh, Y. T.; Su, J. H.; Manivannan, G.; Lee, P. H. C.; Sawan, S. P.; Spall, W. D. Interaction of Supercritical Carbon Dioxide with Polymers. I. Crystalline Polymers. *Journal of Applied Polymer Science* **1996**, 59,695-705.
10. Leavitt, L. The Development of VCI-Infused Biodegradable Polymeric Resin Using Supercritical Carbon Dioxide. Personal Communication, University of Missouri – Columbia, Columbia, MO, 2006.
11. Shieh, Y. T.; Su, J. H.; Manivannan, G.; Lee, P. H. C.; Sawan, S. P.; Spall, W. D. Interaction of Supercritical Carbon Dioxide with Polymers. I. Crystalline Polymers. *Journal of Applied Polymer Science* **1996**, 59,695-705.
12. Factor, M. J.; Lee, S. Quantitative Analysis of Cloisite 93A Infused into Linear Low-Density Polyethylene (LLDPE) and Maleated Linear Low-Density Polyethylene (LLDPE-g-MA) in a Supercritical Carbon Dioxide Medium. *Journal of Applied Polymer Science* (forthcoming).
13. Cervantes-Uc, J. M.; Cauich-Rodriguez, J. V.; Vazquez-Torres, H.; Garfias-Mesias, L. F.; Paul, D. R. Thermal Degradation of Commercially Available Organoclays Studied by TGA-FTIR. *Thermochimica Acta* **2007**, 457, 92-102.

(708g) Initial Stage of Foaming of Poly(methyl methacrylate)/ Carbon Dioxide / Tetraethoxysilane Ternary System

Masaaki Tanaka[1], Daisuke Kobayashi[2], Tomoki Takahashi[2], Atsushi Shono[2], Katsuto Otake[2], Takeshi Furuya[3] and Satoshi Yoda[3], (1)Graduate School of Chemical Science and Technology, Department of Chemical Sciences and Technology, Tokyo University of Science, Tokyo, Japan, (2)Faculty of Engineering, Department of Chemical Engineering, Tokyo University of Science, Tokyo, Japan, (3) Nanosystem Research Institute, National Institute of Advanced Industrial Science and Engineering, Ibaraki, Japan

Polymer foams are well known to have high shock absorbing ability, good handling ability and many other good characteristics. Further, with carbon dioxide (CO_2) as a foaming agent, microcelluler foams that have high foam density and mechanical strength could be obtained. However, CO_2 have also the weak point that the adiabaticity is comparatively low. In this study, to overcome the weak point of the CO_2 as the foaming agent, addition of third component which will fill inside of the foam was examined. In this study, tetraethoxysilane (TEOS), was chosen as an additive for the Poly(methyl methacrylate) (PMMA)/CO_2 system due to the possibility of the formation of silica aerogel in the foam bubble to form composite material. In this study, foaming behavior, or the leading time of foaming after pressure release and bubble diameters growth rate as well as number density increase of bubbles, was examined with PMMA/CO_2 and PMMA/TEOS/CO_2 systems. With increasing the quantity of TEOS, the bubble diameters of foamed PMMA became smaller and the bubble density became larger. On the other hand, the leading time for foaming after the pressure release on PMMA/TEOS/CO_2 mixtures became longer. It is well known that the foaming of the polymers could be explained by the uniform nucleation theory. Based on the theory, from results suggests that the addition of TEOS lowers the interfacial tension however, latter fact implies that the increase in surface tension. From an experimental evidence of the PMMA/TEOS/CO_2 phase behavior, it became clear that the foaming behavior of the ternary systems could be explained from the phase separation pressure of TEOS/CO_2 binary system.

See more of this Session: Processing of Nanocomposites
See more of this Group/Topical: Materials Engineering and Sciences Division

(718a) Lifetime and Efficiency Studies of Zinc/ Bromine Electrochemical Energy Storage In Large Scale Renewable Energy Applications

Alex Bistrika[1], Eunice Naswali[2], Chianna Alexander[2], Hai-Yue Han[2], David Naviaux[2], Annette von Jouanne[2], Ted Brekken[2] and Alexandre Yokochi[1], (1)School of Chemical, Biological and Environmental Engineering, Oregon State University, Corvallis, OR, (2)School of Electrical Engineering and Computer Science, Oregon State University, Corvallis, OR

As the world's power needs grow, the demand for power from renewable resources, such as wind or solar, is increasing. One major drawback associated with these renewable resources is that the power output is dependent on environmental factors, such as cloud cover and wind speeds. This allows the possibilities of either power output exceeding or falling short of forecast levels which may lead to grid instabilities. Therefore, Large Scale Energy Storage (LSES) systems are critical to store excess power when the output exceeds demand in order to supplement output power when it falls short of demand.[1] Some promising LSES contenders are: 1) Zinc bromide flow cells (FC), 2) Vanadium redox FC, 3) Sodium sulfur batteries, and 4) Lead acid batteries.[2] Of these available technologies the zinc bromide technology was selected for further investigation because of previously reported long cycle-life (CL) and high efficiencies.[3]

In this work, the cell efficiency dependence on temperature of a bench scale Zinc Bromide battery manufactured in our lab was investigated to confirm the theoretical efficiency. Efficiency cycling was voltage limited on the discharge cycle at a 1.5 / V cell cut-off potential. Standard method of measuring columbic efficiency was implemented. At room temperature, ~80% efficiency was achieved when cycling the system between 20 - 100% State of Charge (SOC), which is in good agreement with published values.[4] The cell efficiency was demonstrated at over 90% at 50°C and cycling between 20 - 100% SOC.

In addition, given that deep discharge is perceived as highly damaging to electrochemical energy storage, the battery was cycled through the Zero-Point (ZP) 10,000 times. The cell was continuously cycled between 0 - 5% (SOC). Battery performance was evaluated from transfer rate and exchange current density, determined using Tafel scans. No degradation of the electrodes was found, and in fact a slight increase in the exchange current density was observed. We interpret this exchange current increase as a result expected for electrode conditioning of the polished carbon electrode substrates used in our test cells.[5]

These along with further more recent results will be presented.

1. Dell, R. M.; Rand, D. A. J. *Energy storage: a key technology for global energy sustainability. J. Power Sources* **2001,** *100*, 2-17.

2. (a) Schoenung, S. M. *Characteristics and Technologies for Long- vs. Short-Term Energy Storage: A Study by the DOE Energy Storage Systems Program.* Sandia National Laboratories SAND2001-0765 **2001** p 59; (b) Schoenung, S. M.; Eyer, J. *Benefit/Cost Framework for Evaluating Modular Energy Storage.* Sandia National Labratories SAND2008-0978 **2008** p 40.

3. Manla, E.; Nasiri, A.; Hughes, M. *Modeling of zinc energy storage system for integration with renewable energy. Industrial Electronics* **2009**, IECON '09 35th Annual Conference of IEEE, 3-5, p 3987-3992.

4. Lex, P.; Jonshagen, B. *The zinc/bromine battery system for utility and remote area applications. Power Engineering Journal* **1999**, *13*, 142-148.

5. Kautek, W.; Conradi, A.; Fabjan, C.; Bauer, G. *In situ FTIR spectroscopy of the Zn-Br battery bromine storage complex at glassy carbon electrodes. Electrchimica Acta* **2001**, *47*, 815-823.

See more of this Session: Sustainable Electricity: Generation, Transmission, and Storage
See more of this Group/Topical: Sustainable Engineering Forum

(718b) Parametric Sensitivity of Reverse Electrodialysis for Electricity Production Using Salinity Gradient

Rahul Patil and Glenn Lipscomb, Department of Chemical and Environmental Engineering, University of Toledo, Toledo, OH

Global population growth has led to increased energy demand and stress on energy supplies. Concerns over the stability, sustainability, and environmental impact of global energy consumption have generated renewed interest in alternative energy sources. The ideal source would be locally distributed and renewable with minimal economic and environmental impact on our current lifestyle.

The sun is the source of virtually all energy. Solar energy can be utilized in the form of fossil fuels, wind energy, photovoltaic or thermal solar, and hydroelectric. Growing interest exists in using the salinity difference between fresh water and sea water to produce power from mixing the two streams. Membrane processes such as Reverse Electrodialysis (RED) and Pressure Retarded Osmosis (PRO) can be used to produce electricity [1, 2]. Literature reports the effects of membrane type, membrane and gasket thickness, electrode systems and flow rates on power production [3].

In the current study, we demonstrate the optimization of power output using RED. The effects of operating parameters such as current density, concentration of process streams and the concentration and composition of electrode rinse are investigated. We also examined the relationship between electrons generated (current) and ion transfer between the solutions. The efficiency of RED can be determined from the ratio of current to ion transfer. We examined possible sources of energy loss and inefficiency. Shunt currents and membrane area occupied by spacer mesh are believed to be significant contributors. The origin and reduction of these factors to increase the efficiency is discussed as well as the effect of other cell design variables.

References:

1. Post, J. et. al., *Salinity-gradient power: Evaluation of pressure-retarded osmosis and reverse electrodialysis*, Journal of Membrane Science 288 (2007) 218–230.
2. Dlugolecki, P. et. al., *Practical Potential of Reverse Electrodialysis As Process for Sustainable Energy Generation*, Environmental Science and Technology 43 (2009) 6888–6894.
3. Veerman, J. et. al., *Reverse electrodialysis: evaluation of suitable electrode systems,* Journal of Applied Electrochemistry 40 (2010) 1461–1474.

See more of this Session: Sustainable Electricity: Generation, Transmission, and Storage
See more of this Group/Topical: Sustainable Engineering Forum

(718c) Energy Storage with Molten Metal Oxide

Sounak Roy[1], Ashay D. Javadekar[1], Abhimanyu Jayakumar[2], Douglas Buttrey[1], John Vohs[2] and Raymond J. Gorte[2], (1)Chemical Engineering, University of Delaware, Newark, DE, (2)Chemical and Biomolecular Engineering, University of Pennsylvania, Philadelphia, PA

Energy storage will play a critical role for electrical power generation as we move towards using renewable resources, since many renewable energy sources are intermittent in nature. Reversible fuel cells have been proposed as a possible way to store energy; however, energy losses associated with the conversion between H_2O and H_2 and the volumes required for storing H_2 are serious impediments to this technology. To solve these problems, we propose to use the electrochemical oxidation of Sb and reduction of Sb_2O_3 in a solid oxide fuel cell (SOFC) as a means of energy storage. Both Sb (m.p. 903K) and Sb_2O_3 (m.p. 929K) are molten at typical SOFC operating temperatures (e.g. 973K). Furthermore, initial data for Sb electrode overpotentials indicates that these losses are very low. Although the Nernst potential for Sb-Sb_2O_3 equilibrium is relatively low, this is inconsequential for the energy-storage application. Finally, the volume associated with 1MW-h of energy is only 45 liters, based on cell performance that has already been achieved. The main goals of work in this area are to understand how cell performance is affected by Sb-Sb_2O_3 conversion. Mixing the metal and oxide phases is found to be a critical factor in determining cell performance.

See more of this Session: Sustainable Electricity: Generation, Transmission, and Storage
See more of this Group/Topical: Sustainable Engineering Forum

(718d) Evaluation of Organic Rankine Cycle Geothermal Power Plant and Considerations

Kirtipal Barse[1], Michael D. Mann[1], William D. Gosnold[2] and Hossein Salehfar[3], (1)Chemical Engineering, University of North Dakota, Grand Forks, ND, (2)Geology, University of North Dakota, Grand Forks, ND, (3)Electrical Engineering, University of North Dakota, Grand Forks, ND

The objective of this work is to evaluate the feasibility of an Organic Rankine Cycle (ORC) power plant to generate electricity from low flow rates of a geothermal resource (~875 gpm) at low temperatures (T~100°C). A model of ORC was developed using HYSYS process simulator. DOE's Cost of Renewable Energy Spreadsheet Tool (CREST) model was used for economic modeling of the power plant. First law efficiency of the power plant was studied using energy analysis. Exergy analysis was used to study the second law efficiency of the power plant. A comparison was made between shell and tube heat exchangers and plate and frame heat exchangers for performance and cost evaluation. Plate and frame heat exchangers have advantages such as higher overall heat transfer co-efficient and smaller footprint compared to shell and tube heat exchangers. The quality of water used with respect to geothermal development was investigated, including dissolved solids, toxic minerals, additives, scale formation, corrosion and non-condensable gas. This paper will summarize the results of this study.

See more of this Session: Sustainable Electricity: Generation, Transmission, and Storage
See more of this Group/Topical: Sustainable Engineering Forum

(725a) Synthesis of Zeolites from an Energy Conserving Route Using Seed Directed Crystallization

Joshua Hill[1], Qianglu Lin[2], Hongmei Luo[3], Shuguang Deng[1], Baraka Lwoya[4] and Marco Dunwell[1], (1) Chemical Engineering, New Mexico State University, Las Cruces, NM, (2)Chemical Engineering Department, New Mexico State University, Las cruces, NM, (3)Chemical Engineering, New Mexico State University, Las cruces, NM, (4)Chemical Engineering, New Mexico State University, Las Cruces

Synthesis of Zeolite Y From an Energy Conserving Route Using Seed Directed Crystallization.

The need for renewable energy is of great importance as the supply of petroleum based fuels dwindle. Due to their porous structure and Brönsted acidity, zeolite Y is an excellent candidate for the conversion of fatty acid methyl esters produced from algal biomass to more energy dense hydrocarbons. We will report on a seed directed synthesis technique for the crystallization of zeolite-Y without the use of organic templates. This synthesis is controlled by the composition of the solution, followed by growth of the zeolite seeds to larger sizes by incorporation of precursors from the solution. The exclusion of organic templates allows for the elimination of the calcination step to remove organics from the framework of the material. Compared to the more traditional synthesis route involving organotemplates, seed directed synthesis via hydrothermal treatment at 200°C is a less energy intensive process, which makes it more appealing for an industrial application. The structure of the synthesized zeolite was characterized using x-ray diffraction. Topology and morphology was investigated using both scanning and transmission electron microscopy.

See more of this Session: Advances In the Synthesis of Porous Materials
See more of this Group/Topical: Materials Engineering and Sciences Division

(725b) Synthesis of Uniform Silica Nanospheres In Liquid-Liquid Biphasic Systems Using Amines or Ammonia Catalyst and Their Controlled Self-Assembly

Junzheng Wang[1], Ayae Sugawara-Narutaki[1], Masashi Fukao[1], Toshiyuki Yokoi[2], Atsushi Shimojima[1] and Tatsuya Okubo[1], (1)Department of Chemical System Engineering, The University of Tokyo, Tokyo, Japan, (2)Chemical Resources Laboratory, Tokyo Institute of Technology, Yokohama, Japan

A significant progress has recently been made in the synthesis of uniform silica nanospheres (SNSs) less than 30 nm in diameter by using basic amino acids (e.g., lysine) as a base catalyst for hydrolysis of silicon alkoxide.[1,2] Alternatively, this talk describes a more versatile and economical amino acid-free method to synthesize uniform SNSs with low polydispersity (<12%) in organic–aqueous biphasic systems containing tetraethoxysilane (TEOS), water, and primary amine (or ammonia) under precisely controlled pH conditions (pH 10.8–11.4).[3] The diameter of the SNSs determined from scanning electron microscopy (SEM) can be tuned from ~12 to ~36 nm by simply changing the initial pH of the aqueous phase in the reaction mixtures. Moreover, the as-synthesized silica sol was taken as the starting material for studying the influences of the type of base catalysts on the solvent evaporation-induced three-dimensional (3D) self-assembly of SNSs. X-ray diffraction (XRD) and nitrogen adsorption–desorption were used to characterize the degree of packing of the resulting 3D SNSs arrays. The SNSs arrays that possess large interparticle mesopores with the diameter of ca. 8.1 nm and low packing fraction of ca. 66.1% are observed upon solvent evaporation of as-synthesized sol in the presence of primary amine. This indicates that SNSs are loosely packed, compared with the packing fraction of 74% for a face-centered cubic array of ideal hard spheres. In contrast, with the aid of an organic buffer or lysine as additives, the SNSs arrays having smaller mesopores (ca. 3.9 nm) and higher packing fraction of 70.5 –71.5% are achieved. It is suggested that the chemical additives with the ability to maintain relatively strong repulsive interaction until the final stage of evaporation play a vital role in the fabrication of well-ordered SNSs arrays.

References

(1) Yokoi, T.; Sakamoto, Y.; Terasaki O.; Kubota, Y.; Okubo, T.; Tatsumi, T. *J. Am. Chem. Soc.* **2006,** *128,* 13664–13665.

(2) Yokoi, T.; Wakabayashi, J.; Otsuka, Y.; Fan, W.; Iwama, M.; Watanabe, R.; Aramaki, K.; Shimojima, A.; Tatsumi, T.; Okubo, T. *Chem. Mater.* **2009,** *21,* 3719–3729.

(3) Wang, J. Z.; Sugawara-Narutaki, A.; Fukao, M.; Yokoi, T.; Shimojima, A.; Okubo, T. *ACS Appl.*

Mater. Interfaces, **2011,** doi: 10.1021/am200104m.

See more of this Session: Advances In the Synthesis of Porous Materials
See more of this Group/Topical: Materials Engineering and Sciences Division

(725c) Porous Materials From Time Dependent Magnetically Assisted Self-Assembly of Nanoparticles

Marco Furlan, Institute for Chemical and Bioengineering, Department of Chemistry and Applied Biosciences, ETH Zurich, Zurich, Switzerland and **Marco Lattuada**, Department of Chemistry and Applied Biosciences , Institute for Chemical and Bioengineering, ETH Zurich, Zurich, Switzerland

There is a big interest in the production of porous materials with a well organized pore structure for applications such as chromatography, membrane separation, catalyst supports, bio-inspired materials and scaffolds for cell growth. In many cases, porous materials can be prepared from colloidal suspensions and have usually a random porous structure.

We present a method to create porous ceramic materials with a complex structure. In its simplest realization, the method makes use of magnetic polymer composite nanoparticles, synthesized via free radical miniemulsion, as templates in the formation of silica monolith via the sol gel process. When the nanoparticles are mixed with a silica precursor, a catalyst and a porogen, silica nucleates and condenses on top of the nanoparticles, which are aligned in the direction of the applied magnetic field. After heat treatment the particles are removed and porous anisotropic silica monoliths are obtained. Monoliths with a more complex morphology can be prepared using time-dependent configuration of the magnetic field applied during the sol-gel process. The sol-gel process kinetics has first been investigated by means of rheology, in order to precisely identify the gel time. Then, simple time dependent configurations of magnetic field have been applied. In one case, the sol-gel process was started in the presence of a magnetic field for a time shorter than the gel point, and then and the reaction completed without any magnetic field. In another case, the reaction was started in the absence of a field, which was then turned on at a later time point. An additional variation of the procedure has been explored, in which the reaction is started in the presence of a magnetic field, and the sample is then subject to strong mixing before letting gelation reach completion in the absence of a field. Finally, materials obtained from sol-gel processes carried out in the presence of pulsating magnetic fields with different frequencies have been also prepared. All these procedures aim at controlling the structure through a competition between magnetic dipolar interactions, which align the nanoparticles into strings, and diffusion or shear forces, which disrupt the alignment. All samples have been characterized using SEM microscopy, torque magnetometry, mechanical compression and mercury porosimetry.

See more of this Session: Advances In the Synthesis of Porous Materials
See more of this Group/Topical: Materials Engineering and Sciences Division

(725d) Template-Directed Synthesis of Porous Alumina Particles Via Atomic Layer Deposition

Xinhua Liang and Alan W. Weimer, Chemical and Biological Engineering, University of Colorado at Boulder, Boulder, CO

Highly porous alumina particles with precise wall thickness control were synthesized by atomic layer deposition (ALD) of alumina on highly porous poly(styrene-divinylbenzene) (PS-DVB) particle templates. Alumina ALD was carried out using alternating reactions of trimethylaluminum and water at 33 °C. The growth rate of alumina was ~0.3 nm per coating cycle. After deposition, the polymer was removed by air oxidation leaving a highly porous alumina skeletal structure. The wall thickness can be precisely controlled by adjusting the number of ALD coating cycles. Thermogravimetric analysis, X-ray diffraction, nitrogen adsorption, scanning electron microscopy, and transmission electron microscopy were used to characterize the fabricated porous alumina particles. The effect of number of ALD coating cycles and calcination temperature on the mesoporous structure of the alumina particles was investigated. γ-alumina was formed at temperature above 600 °C. Porous crystalline alumina particles with a surface area of 80-100 m^2/g were obtained and thermally stable at 800 °C. The pore volume of the porous particles can be as high as 1 cm^3/g after calcination at 800 °C. Control of the wall thickness to within ~0.3 nm provides for ultimate control of surface area. Such porous alumina particles may find wide application in nanostructured materials fabrication, including novel catalyst supports.

See more of this Session: Advances In the Synthesis of Porous Materials
See more of this Group/Topical: Materials Engineering and Sciences Division

(725e) Synthesis of Hierarchically Porous Materials From Bicontinuous Interfacially Jammed Emulsion Gels (Bijels)

Matthew N. Lee and **Ali Mohraz**, Department of Chemical Engineering & Materials Science, University of California, Irvine, Irvine, CA

Materials possessing hierarchical porosity across the nanometer and micrometer scales are of value in several emerging technologies such as catalysis, separations, tissue engineering, and electrochemical systems, to name a few. In an effort to further both the breadth of applications and the performance of these types of materials, robust methods for their synthesis with tunable chemistry and pore sizes are needed. The ability to prescribe the size separation between large and small pore networks may provide an additional avenue for optimization. In this work, a novel family of macroporous and hierarchically porous materials are synthesized from bicontinuous interfacially jammed emulsion gels (bijels), which are a recent class of soft materials formed by arrested spinodal decomposition of partially miscible fluids by neutrally wetting colloids. Through selective polymerization of one fluid phase, a fluid-fluid bijel is transformed into a crosslinked polymer template, which provides a versatile platform for synthesis of several types of porous materials, including ceramics and metals. All the materials share a 3D bicontinuous, spinodal-like morphology, inherent to their bijel parent. Processing of materials through nanocasting is of particular utility here, and allows for tuning of hierarchical pores over several orders of magnitude, from tens of nanometers to over a hundred microns. Application of these novel microstructures in several emerging fields, including tissue engineering and electrochemical systems, will be discussed.

See more of this Session: Advances In the Synthesis of Porous Materials
See more of this Group/Topical: Materials Engineering and Sciences Division

(725f) Microporous Aluminophosphate Synthesis In Ionic Liquids

Xin Sun[1], Sean Tomlinson[1], John R. Schlup[2] and **Jennifer L. Anthony**[1], (1)Chemical Engineering, Kansas State University, Manhattan, KS, (2)Department of Chemical Engineering, Kansas State University, Manhattan, KS

Zeolitic molecular sieves have extensive applications in catalysis and separations. These stable, crystalline materials with nanometer-sized pores can have a wide range of chemical compositions, such as silicates, phosphate metal oxides, or all-carbon molecular sieves. Subsequently, there has been extensive research on the synthesis of new frameworks as well as trying understanding their crystallization mechanism.[1] Crystalline molecular sieves traditionally are prepared using a hydrothermal synthesis method. A typical zeolite synthesis would consist of a mixture of water, a source of the framework atoms, a mineralizing agent (e.g. OH^- or F^-), and a structure-directing agent (SDA). In the recent literature, it has been shown that aluminophosphate molecular sieves can be prepared ionothermally.[2] In ionothermal synthesis, an ionic liquid serves as the solvent and often the structure directing agent as well. Our group has focused on the ionothermal synthesis of porous, crystalline aluminophosphates and silicates with the emphasis on determining how the ionic liquid properties influence the final materials. In particular, we are investigating the interactions between the ionic liquid solvent and the molecular sieve precursors prior to the synthetic reaction through solubility and spectroscopic measurements. In this presentation, we will discuss the effects of reaction composition and conditions on structures of the porous aluminophosphate molecular sieves (AlPOs) and compare to those using traditional hydrothermal techniques. We also will discuss the influence of the ionic liquid structure on the resulting materials and will present our results from investigating the influence of the ionic liquid structure on the solubility of the precursors and the ionic liquid-precursor complexes seen by NMR and FTIR spectroscopy.

1. J. Yu, *Stud. Sur. Sci. Catal.*, 2007, 168, 39-104.

2. Cooper, E. R.et. al., *Nature* 2004, **430**, 1012-1016.

See more of this Session: Advances In the Synthesis of Porous Materials
See more of this Group/Topical: Materials Engineering and Sciences Division

CREATION AND CHARACTERIZATION OF MAGNESIUM OXIDE MACROPOROUS CERAMICS

Lisa A. Mondy[1], Christopher Diantonio[2], Tom Chavez[2], Lindsey Gloe[1], Anne Grillet[1], Christine Roberts[1], David Ingersoll[3]

[1]Thermal & Fluid Experimental Sciences
[2]Organic Materials Department
[3]Advanced Power Sources R&D
Sandia National Laboratories
Albuquerque, New Mexico

Summary

We examine several methods to create a sheet of magnesium oxide (MgO) macroporous ceramic material via tape casting. These methods include the approach pioneered by Akartuna et al.[1] in which an oil/water emulsion is stabilized by surface-modified metal oxide particles at the droplet interfaces. Upon drying, a scaffold of the self-assembled particles is strong enough to be removed from the substrate material and sintered. We find that this method can be used with MgO particles surface modified by short amphiphilic molecules. This approach is compared with two more traditional methods to induce structure into a green ceramic: 1) creation of an MgO ceramic slip with added pore formers, and 2) sponge impregnation of a reticulated foam with the MgO slip. Green and sintered samples made using each method are hardness tested and results compared for several densities of the final ceramics. Optical and SEM images of the materials are shown.

Introduction

Highly porous, mechanically robust ceramic foams are useful for a wide variety of applications ranging from catalyst supports, filters for liquid metals, tissue engineering, high temperature thermal insulation, and lightweight structural supports.[2,3] Drying particle stabilized oil-water emulsions, termed Pickering emulsions,[4] is a promising method for fabricating such a structure in a way that is conducive to scale-up. Pickering emulsions can be created using a variety of particle shapes and chemistries[1,5-7] and, once formed, are remarkably stable against droplet coarsening and drainage[8-10]. Droplet diameters are often on the order of microns and can be controlled by varying process parameters, allowing a direct design of the final porosity of the product. Wet emulsions have a measurable yield stress, allowing them to be shaped or extruded into forms or coated into sheets.[1,5,11,12] These forms retain their macroscopic shape upon drying and sintering.[7] Their microstructures have been shown to be open cell foams that can have compressive strengths up to 13 MPa and porosities above 70 vol%.[1]

The effectiveness of particles in stabilizing an oil-water emulsion depends on the solids shape[13], size[14], level of flocculation[15,16], concentration[17,18], surface charge[19], and surface wettability[20]. Finkle[21] is credited with recognizing that particles stabilize emulsions when they are equally wet by both the oil and the aqueous phase. Short chain carboxylic acid amphiphiles have been employed to modify the wetting behavior of various ceramic oxide particles in order to make them viable for use as emulsion stabilizers.[1,11] In this way, particles with any ceramic chemistry can potentially be used to create a Pickering emulsion given that a suitable amphiphile can be attached to its surface.

This research aims to create a highly porous magnesium oxide (MgO) ceramic foam that can be cast onto a substrate using a blade coating technique and dried and sintered to a desired thickness. The final product must be permeable, crack free, and also have a high compressive modulus. To create such a foam, MgO particles, surface modified by short chain carboxylic acid amphiphiles, will be used to stabilize a water/octane emulsion. A statistical design of experiments allows connections to be made between the emulsion processing parameters, emulsion droplet size distribution, green ceramic properties, and the final integrity and performance of the piece. Samples created using the Pickering emulsion templating method will be compared to samples fabricated by two other common templating techniques. These include the addition of a polymeric pore-former to a conventional MgO dispersion with subsequent burnout of the pore-former, and the impregnation of a reticulated foam with an MgO dispersion with subsequent burnout of the foam.

Experimental Materials and Methods

Preparation of Pickering Emulsions

Aqueous dispersions of 2.5 – 10 vol% MgO particles (Fisher Scientific) are ball milled with 10 mm zirconium milling media for more than 20 hours in baffled jars, creating a well dispersed particle mixture with an average particle size of 2 μm. Hydrochloric acid (1N, Fluka Analytical) is used to maintain a pH between 9-10 while milling. After milling, an amphiphile is added at concentrations ranging from 150-500 mM, and the aqueous dispersion is stirred for 3 minutes. Amphiphiles explored include propionic acid (99%, Sigma Aldrich), butyric acid (99%, Sigma Aldrich), valeric acid (99%, Acros Organics), and hexanoic acid (99%, Sigma Aldrich).

Up to 80 vol% n-octane (>99%, Sigma Aldrich) is added slowly to the aqueous MgO dispersion as the liquids are mixed using a Cowles mixer (Dispermat, VMA-Getzmann GMBH). Emulsification is accomplished by vigorously mixing at speeds ranging from 1,000 – 7,000 RPM for three minutes after the addition of the oil. If desired, 0.1-0.5 wt% polyvinyl alcohol (PVA) (100,000 g/mol, 87% hydrolyzed, Fisher Scientific) is added and mixed into to the emulsion. Polyvinyl alcohol concentrations are reported with respect to the entire volume of the emulsion, typically 100 mL.

Emulsions are dried in ambient conditions on a silicone coated mylar substrate. The emulsion can also be cast into Teflon molds (3 cm diameter, 0.5 cm thick). Once dry, the samples are sintered in a furnace (Thermolyne 46200). Polyvinyl alcohol burnout is accomplished with a 2 hr hold at 650 °C, and then the samples are ramped to final sintering temperatures ranging from 1250 °C to 1675 °C over 12 hrs, with a 2 hr hold at temperature.

Preparation of Pore-Former Slips – Sacrificial Template Method

The pore-former technique consists of preparing a biphasic composite comprising a continuous matrix of ceramic particles and a dispersed sacrificial phase that is initially homogeneously distributed throughout the matrix and then extracted to generate pores within the microstructure. This method leads to porous materials displaying a negative replica of the original sacrificial template. The biphasic composite is prepared by forming a two-phase suspension that is then processed using a wet colloidal route and casting tape. Tape casting suspensions containing variations of nanocrystalline size MgO

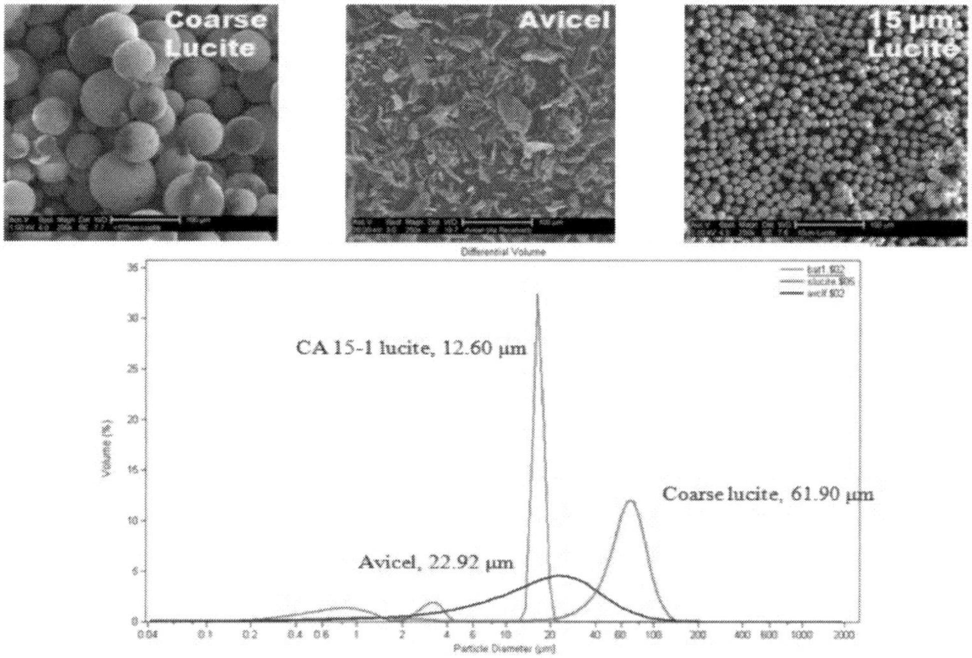

Figure 1. Particle size distribution analysis and SEM images of organic poreformers, 'Coarse Lucite', '15um Lucite' and 'Avicel'.

particles (Inframat), sub-micron crystalline size MgO particles (Inframat) and/or organic poreformers (Lucite and Avicel) are prepared using a B73305 Ferro tape cast vehicle, hypermer KD1 dispersant and toluene. The suspensions varied between 10-20 wt% MgO and 4-8 wt% organic poreformer and were ball milled 24-48 hours in baffled nalgene bottles using YSZ milling media. Figure 1 shows the particle size distribution analysis and SEM images of organic poreformers, 'Coarse Lucite', '15um Lucite' and 'Avicel'. Lucite particles tend to be spherical, whereas Avicel particles are more elongated in shape.

Impregnation of Reticulated Foams – Replica Technique with Synthetic Templates

In order to produce a macroporous ceramic that exhibits the same morphology as the original porous material a replica method is employed where a cellular structure is impregnated with a ceramic suspension. A highly porous polymeric sponge, a polyurethane sponge in this case, is soaked in a ceramic suspension so that the internal pores are filled in with the ceramic material. The excess ceramic suspension is then removed from the sponge, enabling the formation of a thin ceramic coating over the struts of the original cellular structure. Reticulated bulk ceramics have been prepared from ceramic suspensions that are 25-50 wt% ceramic powder, 2-10 wt% polyvinyl alcohol (Fisher) and 0.5-1 wt% dolapix dispersing agent. Polyurethane foams are infiltrated with the ceramic suspension by desiccation, dried, pyrolyzed and sintered.

Sample Characterization

Droplet size distributions of wet Pickering emulsions are measured by viewing the suspensions using an inverted optical microscope (Nikon Eclipse TE2000-5) in transmission mode. Images are captured using a camera (Phantom Miro-4M, Vision Research) and analyzed using Image Pro Plus 7.0 software. The yield stress of wet Pickering emulsions are also measured using a TA ARG2 rheometer with a vane geometry at room temperature. Emulsion stability can be checked by allowing the suspension to settle over days in a glass vial.

Dried green samples are measured using a micrometer to estimate the drying shrinkage. Green strength is evaluated using a TA.XT*Plus* texture analyzer (Stable Micro Systems) with either a 1/8" diameter stainless steel ball indenter or a 2 mm diameter stainless steel punch.

The final density and shrinkage of the sintered sample is measured by estimation of the sample dimensions and weight (Mettler Toledo, XS1003S). The fraction of open pore volume can be estimated by an Archimedes test by measuring the fraction of pores that can be filled with water. A more quantitative measure of the permeability is obtained by measuring the electrical impedance across the thickness of a sample filled with an aqueous electrolyte solution. The final strengths of sintered samples are tested with the texture tester using both indenter geometries. Finally, the microstructures are imaged using a Zeiss Ultra Plus FESEM.

Results and Discussion

Magnesium oxide successfully stabilizes octane/water emulsions for all amphiphiles that are investigated. Figure 2 shows a sample optical image of a typical emulsion. Emulsion droplet sizes are polydisperse, ranging from 1 – 100 μm in diameter depending on preparation conditions. The continuous phase of all emulsions was water, and free MgO particles were sometimes found in this phase.

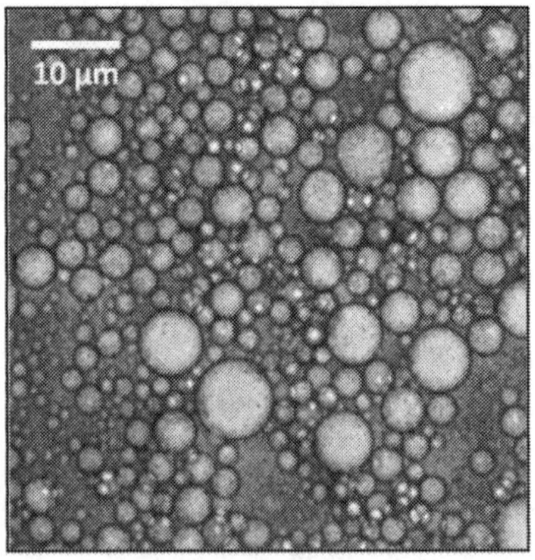

Figure 2. Typical MgO, water, octane Pickering emulsion containing 0.3% PVA and 350 mM propionic acid.

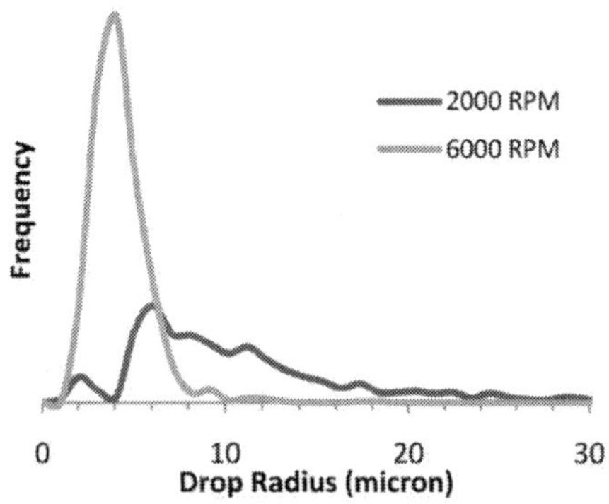

Figure 3. Effect of mixing speed on Pickering emulsion droplet size distribution. Emulsion contains 0.3% PVA and 350 mM propionic acid.

Mixing speed had one of the greatest effects on droplet size as compared to the other experimental factors. Figure 3 shows the difference in size distribution between two dispersions of equal composition but varying mixing speed. The average droplet size is halved by adding more mixing energy into the dispersion preparation. This control of droplet size without altering the emulsion composition will be advantageous for the eventual design of a final product; however, mixing speeds cannot be increased indefinitely or the emulsion will break and the octane and water will separate. The upper bound on the mixing speed varies with the composition of the emulsion.

MgO Pickering emulsions shrink to about 50 - 80% of their initial volume during drying. This extreme drying shrinkage promotes cracking, which was shown to be reduced with the addition of only small amounts of polyvinyl alcohol binder. Figure 4 shows two dry coatings containing different amounts of PVA. In this case, 0.1 vol% PVA was insufficient to prevent cracking, whereas 0.3 vol% PVA produces an acceptable coating. Figure 4 also demonstrates that the wet Pickering emulsions can be coated onto a substrate at a substantial wet thickness (in this case 5 mm) and will dry without running due to their significant yield stress. Yield stresses for typical emulsions are between 50-200 Pa.

Upon sintering, the final pore size is a function of the initial emulsion droplet distribution. Figure 5 is an SEM image of the final sintered microstructure, demonstrating that the final product is an open cell foam with large pores between chambers that will allow transport. Furthermore, at some sintering temperatures a nanoscale porosity also exists between partially sintered MgO particles (see inset). At higher sintering temperatures the nanoscale porosity is eliminated. The optimal sintering temperature that will provide sufficient strength and permeability has yet to be determined.

Sintered porosities were measured at 60 – 95% with 50 – 75% of those pores being networked. The honeycombed structure of the ceramic permits this remarkable porosity in a structure that is still able to support a compressive stress.

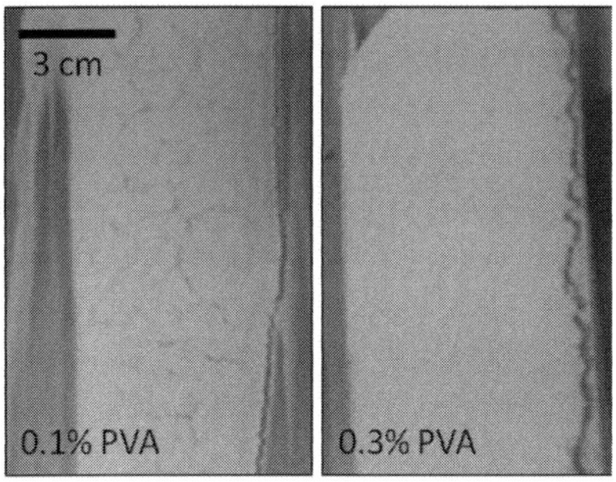

Figure 4. Effect of polyvinyl alcohol concentration on the extent of cracking. Coatings have a wet thickness of 5 mm.

Figure 5. Cross section of a Pickering emulsion sample sintered at 1400 °C. Inset: high magnification image of strut showing nanoporosity.

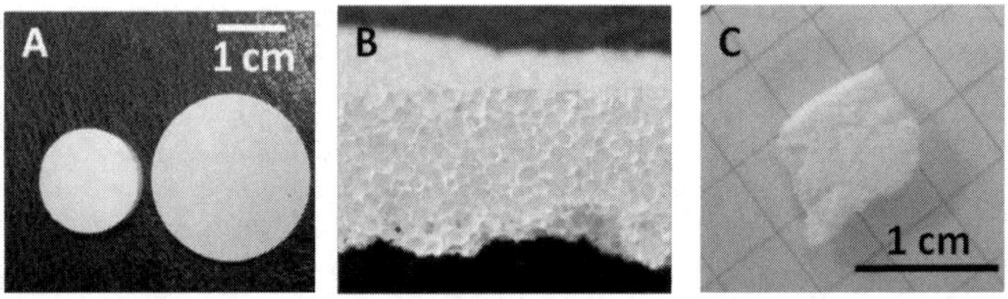

Figure 6. Macroscopic appearance of sintered porous MgO structures formed using A) pore-former, B) reticulated foam, C) Pickering emulsion synthesis techniques

Other templated synthesis techniques produced porous MgO structures compared in Figure 6. Slips including pore-former particles are coated, dried and stacked to form thick sheets that are able to be cut into pellets. The porosity of these samples can be adapted by the concentration of the pore former that is added. Porosities ranging from 30-60 vol% have been demonstrated. Reticulated sponge structures have also been used as a template; the connectedness of the original sponge structure guarantees the open-foam nature of the final product.

Conclusions

Magnesium oxide, water, octane Pickering emulsions were successfully created with the use of short chain carboxylic acid amphiphiles to control the wettability of the ceramic. A study of emulsion processing parameters was undertaken in order to determine how best to control emulsion droplet size distributions. Mixing speed was found to be one parameter that greatly affects the average droplet size. Once formed, the emulsions had a significant yield stress, allowing them to be coated into thick sheets or shaped. Although considerable shrinkage was observed during drying and sintering, the final product was an open foam structure with observed porosities as high as 95%. Other templating techniques also produced porous MgO structures of lower porosity, but experienced less shrinkage and cracking during sintering and expressed greater compressive strength.

Acknowledgements

Sandia National Laboratories is a multi-program laboratory managed and operated by Sandia Corporation, a wholly owned subsidiary of Lockheed Martin Corporation, for the U.S. Department of Energy's National Nuclear Security Administration under contract DE-AC04-94AL85000. This research is supported by the Laboratory Directed Research and Development program at Sandia National Laboratories.

References

1. Akartuna I, Studart AR, Tervoort E, Ludwig J Gauckler. Macroporous ceramics from particle stabilized emulsions. *Advanced Materials*. 2008;20(24):4714-4718.
2. Nettleship I. Applications of Porous Ceramics. *Key Engineering Materials*. 1996;122-124:305-324.
3. Studart AR, Gonzenbach UT, Tervoort E, Gauckler LJ. Processing Routes to Macroporous Ceramics: A Review. *Journal of the American Ceramic Society*. 2006;89(6):1771-1789.
4. Pickering SU. Emulsions. *J. Chem. Soc., Trans.* 1907;91:2001.
5. Akartuna I, Tervoort E, Wong JCH, Studart AR, Gauckler LJ. Macroporous polymers from particle-stabilized emulsions. *Polymer*. 2009;50(15):3645-3651.
6. Bon SAF, Chen T. Pickering Stabilization as a Tool in the Fabrication of Complex Nanopatterned Silica Microcapsules. *Langmuir*. 2007;23(19):9527-9530.
7. Wong JCH, Tervoort E, Busato S, et al. Macroporous polymers from particle-stabilized foams. *J. Mater. Chem.* 2009;19(29):5129.
8. Stocco A, Garcia-Moreno F, Manke I, Banhart J, Langevin D. Particle-stabilised foams: structure and aging. *Soft Matter*. 2011.
9. Schelero N, Lichtenfeld H, Zastrow H, et al. Single particle light scattering method for studying aging properties of Pickering emulsions stabilized by catanionic crystals. *Colloids and Surfaces A: Physicochemical and Engineering Aspects*. 2009;337(1-3):146-153.
10. Sacanna S, Kegel WK, Philipse AP. Thermodynamically Stable Pickering Emulsions. *Phys. Rev. Lett.* 2007;98(15):158301.
11. Akartuna I, Studart AR, Tervoort E, Gonzenbach UT, Gauckler LJ. Stabilization of Oil-in-Water Emulsions by Colloidal Particles Modified with Short Amphiphiles. *Langmuir*. 2008;24(14):7161-7168.
12. Blaker JJ, Lee K-Y, Li X, Menner A, Bismarck A. Renewable nanocomposite polymer foams synthesized from Pickering emulsion templates. *Green Chem.* 2009;11(9):1321.
13. Madivala B, Vandebril S, Fransaer J, Vermant J. Exploiting particle shape in solid stabilized emulsions. *Soft Matter*. 2009;5(8):1717.
14. Binks BP, Lumsdon SO. Pickering Emulsions Stabilized by Monodisperse Latex Particles: Effects of Particle Size. *Langmuir*. 2001;17(15):4540-4547.
15. Binks BP, Lumsdon SO. Stability of oil-in-water emulsions stabilised by silica particles. *Phys. Chem. Chem. Phys.* 1999;1(12):3007-3016.
16. Thieme J, Abend S, Lagaly G. Aggregation in Pickering emulsions. *Colloid Polym Sci.* 1999;277(2-3):257-260.
17. Moore WC. Emulsification of water and of ammonium chloride solutions by means of lamp black. *Journal of the American Chemical Society*. 1919;41(6):940-946.
18. Frelichowska J, Bolzinger M-A, Chevalier Y. Effects of solid particle content on properties of o/w Pickering emulsions. *Journal of Colloid and Interface Science*. 2010;351(2):348-356.
19. Leunissen ME, van Blaaderen A, Hollingsworth AD, Sullivan MT, Chaikin PM. Electrostatics at the oil–water interface, stability, and order in emulsions and colloids. *Proceedings of the National Academy of Sciences*. 2007;104(8):2585 -2590.
20. Binks BP, Clint JH. Solid Wettability from Surface Energy Components: Relevance to Pickering Emulsions. *Langmuir*. 2002;18(4):1270-1273.
21. Finkle P, Draper HD, Hildebrand JH. The theory of emulsification. *Journal of the American Chemical Society*. 1923;45(12):2780-2788.

(730a) Enhancing Nonviral Gene Delivery by Manipulating Cell-Material Interactions

Tadas Kasputis and Angela K. Pannier, Biological Systems Engineering, University of Nebraska - Lincoln, Lincoln, NE

Gene delivery approaches serve as a platform to modify gene expression of a cell population with applications including functional genomics, tissue engineering, and gene therapy. The delivery of exogenous genetic material via nonviral vectors has proven to be less toxic and to cause less of an immune response in comparison to viral vectors, but with decreased efficiency of gene transfer. Attempts have been made to improve nonviral gene transfer efficiency by modifying physicochemical properties of gene delivery vectors as well as developing new delivery techniques. In order to further improve and understand nonviral gene delivery mechanisms, our approach focuses on the cell-material interface, since materials are known to modulate cell behavior, potentially rendering cells more responsive to nonviral gene transfer. In this study, self-assembled monolayers of alkanethiols on gold were employed as model biomaterial interfaces with varying surface chemistries. NIH/3T3 mouse fibroblasts were seeded on the modified surfaces, transfected after 18 hours, and assayed for transfection efficiency 24 hours thereafter. In comparison to polystyrene and gold surface controls, transfection efficiency was increased on charged surfaces presenting $-COO^-$ and $-NH_3^+$ terminal functional groups, while hydrophilic and hydrophobic uncharged surfaces (-OH and $-CH_3$) demonstrated dramatically reduced transfection efficiencies. Cellular cytoskeleton, focal adhesions, nuclei, and cytoplasm were labeled to further characterize cell-surface interactions. We demonstrate that cellular morphological characteristics, such as focal adhesions and cytoskeletal stress fibers, that have been shown to be influenced by surface characteristics, could be related to transfection efficiency. We demonstrate that simple modifications of surfaces lead to increased efficiency of nonviral gene delivery. Understanding the cell-biomaterial interface as it pertains to enhancing transfection efficiency heralds the development of precisely tunable substrates for biotechnological and tissue engineering applications.

See more of this Session: Building Drug and Gene Delivery Into Tissue Engineering
See more of this Group/Topical: Materials Engineering and Sciences Division

(730b) Investigating the Interaction Between Virus Vectors and Adhesive Proteins for Substrate-Mediated Gene Delivery

Kellie I. McConnell[1], Sibani L. Biswal[2], Jonathan J. Silberg[3] and Junghae Suh[1], (1)Bioengineering, Rice University, Houston, TX, (2)Chemical & Biomolecular Engineering, Rice University, Houston, TX, (3)Biochemistry & Cell Biology, Rice University, Houston, TX

Complex tissue development requires precise control over the spatial organization of gene expression. We have developed a platform to allow for both viral gene delivery vectors and adhesive proteins to be co-patterned to control both cellular attachment and gene expression. Our approach, based on adeno-associated virus (AAV), should enable tissue engineers to obtain specific, spatially organized expression of genes in a population of cells. Polydimethylsiloxane (PDMS) stamps were first used to micropattern protein-adhesive alkanethiols onto gold-coated substrates. Incubation of the self-assembled monolayers (SAMs) with extracellular matrix (ECM) proteins and AAV resulted in co-patterning of cell-adhesive proteins and gene delivery vectors. We have previously observed a synergistic effect between the patterning of human fibronectin (HFN) and AAV for the delivery of genes to HeLa cells. Here, we further explore this technology by examining additional adhesive proteins and demonstrate that collagen I, elastin, laminin, and poly-l-lysine are also compatible with our platform. Cell adhesion and gene expression were monitored over three days and while all proteins and surfaces tested were successful, we observe that laminin pre-adsorption yields the highest gene delivery efficiency. BrU incorporation tests indicate this increase in gene expression is not due to increased HeLa cell transcriptional activity when seeded on laminin surfaces. Quantitative analysis of AAV binding/dissociation kinetics from various ECM surfaces may shed light on why laminin is able to mediate the most effective gene delivery. Therefore, quartz crystal microbalance (QCM) and surface plasmon resonance (SPR) were used to investigate the interaction between AAV and these adhesive proteins. By combining microcontact printing with viral gene delivery vectors, we have successfully patterned both cell adhesion and gene expression.

See more of this Session: Building Drug and Gene Delivery Into Tissue Engineering
See more of this Group/Topical: Materials Engineering and Sciences Division

(730c) Effects of Scaffold Mechanical Properties On the Delivery of Stromal Cell-Derived Factor-1 From Polyurethane Scaffolds In Rat Cutaneous Wounds

Elizabeth J. Adolph[1], Fang Yu[2], Lillian B. Nanney[3], Jeffrey M. Davidson[2] and Scott A. Guelcher[1], (1) Chemical and Biomolecular Engineering, Vanderbilt University, Nashville, TN, (2)Pathology, Vanderbilt University, Nashville, TN, (3)Pathology, Cell & Developmental Biology, Vanderbilt University, Nashville, TN

In cutaneous wounds, the biochemical and mechanical environment of the defect affects the wound healing response. Stromal cell-derived factor-1 (SDF-1) is a chemokine that has been shown to recruit mesenchymal stem cells, promote angiogenesis, and decrease inflammation *in vivo*. Furthermore, mechanical forces generated in the wound site can affect myofibroblast differentiation and matrix production and organization. In this study, we investigated the effects of SDF-1 delivered from scaffolds with varying mechanical properties in subcutaneous and excisional wounds in rats. Polyurethane (PUR) scaffolds were synthesized by reacting lysine triisocyanate with a hardener component comprising a polyester triol with a backbone composed of 60% caprolactone, 30% glycolide, and 10% lactide; triethylenediamine catalyst; water; and calcium stearate. The mechanical properties of the scaffolds were altered by varying the molecular weight of the polyester triol (450, 900, or 3000 Da), and SDF-1 was incorporated in the scaffolds as a lyophilized powder. The scaffold pore structure was analyzed using scanning electron microscopy, and the thermal transitions of the materials were evaluated by differential scanning calorimetry. Mechanical properties of the scaffolds were measured using a dynamic mechanical analyzer. The release kinetics of SDF-1 from the scaffolds was measured *in vitro* by incubating samples of scaffolds containing SDF-1 in buffer, collecting the releasates and replacing release media every day, and measuring SDF-1 concentration in the releasates using an ELISA kit. The effects of the scaffold mechanical properties and delivery of SDF-1 on wound healing were investigated using subcutaneous and excisional wound models in rats. The study design comprises 6 treatment groups: 450 Da PUR scaffold, 900 Da PUR scaffold, 3000 Da PUR scaffold, 450 Da + SDF-1, 900 Da + SDF-1, and 3000 Da + SDF-1. Time points include 7, 14, 21, 28, and 35 days. The Young's modulus is 9000 +/- 3000 kPa for the 450 Da scaffold, 270 +/- 40 kPa for the 900 Da scaffold, and 110 +/- 40 kPa for the 3000 Da scaffold. The pore size of the scaffolds ranges from 200 – 500 μ m and is not significantly affected by polyester triol molecular weight. The *in vitro* release experiments show a burst release of 40% of the SDF-1 in the first 4 days followed by a sustained release that results in 50% release by day 14. There are no substantial differences in the release profiles from the three scaffolds. Preliminary results from the rat subcutaneous wound study show that delivery of SDF-1 enhances cellular infiltration into the scaffolds and that mechanical properties of the scaffolds affect the cell and extracellular matrix organization in the wound site.

(730d) Elastin-Based Scaffolds for Local Delivery of Antibiotics and Proteins

Shruti Amruthwar and **Amol V. Janorkar**, Biomedical Materials Science, University of Mississippi Medical Center, Jackson, MS

Tissue engineering applications utilizing polymeric scaffolds benefit from the sustained and controlled release of signaling molecules, growth factors for cellular proliferation, and antibiotics essential for bactericidal activity to obtain optimal healing of tissues. Elastin-like polypeptides (ELPs) are biopolymers with pentapeptide repeat unit, VPGXG where V = valine, P = proline, G = glycine, and X, the guest residue, is any amino acid other than proline. ELPs can be genetically engineered, providing precise control over their molecular weight and structure. They possess good biocompatibility, are biodegradable, exhibit mechanical properties similar to natural elastin, and can be further cross-linked for additional manipulation of properties. This research investigated the release profiles of a model protein (albumin) and a commonly used antibiotic in periodontal therapy (doxycycline) from ELP hydrogel scaffolds. The amount of ELP, doxycycline, or albumin released was analyzed. Both albumin and doxycycline showed a gradual time dependent release over the 500h and 200h period, respectively. Albumin had a slower rate of release from ELP as compared to doxycycline and no burst release, suggesting that molecular size and weight are key factors in determining release rate. Both doxycycline and albumin showed a trend of higher release fractions at higher loading doses. These experiments demonstrated that ELP scaffolds, the synthetic form of elastin, can provide for sustained and controlled release of proteins and antibiotics critical to tissue engineering.

See more of this Session: Building Drug and Gene Delivery Into Tissue Engineering
See more of this Group/Topical: Materials Engineering and Sciences Division

Silica-hydrogel microcapsules for controlled release applications

Petra Haufová, František Štěpánek

Chemical Robotics Laboratory, Department of Chemical Engineering,
Institute of Chemical Technology Prague, Technická 5, 166 28 Prague 6, Czech Republic;
Telephone: (+420) 220 443 048; e-mail: petra.haufova@vscht.cz

Abstract: We present a methodology for the fabrication of silica-coated calcium alginate microcapsules (hydrogel core with a permeable protective membrane). The microcapsules were synthesised in two steps. In the first step, alginate gel cores were prepared by the drop-on-demand InkJet technology and in the second step the organo-silica shell has been formed by a sol-gel process using alkoxysilane precursors. Incorporation of nanoparticles or biomolecules inside the alginate core mimics the further encapsulation of active substances (drug) or immobilization of proteins. The resulting microcapsules have a mean diameter in the range of 40-80 μm and the silica surface layer thickness is in the range of a few μm's.

Key words: Calcium Alginate, Core-shell microcapsule, Hydrogel, Silica.

1 Introduction

Micro-encapsulation is a process by which tiny solid or liquid particles are surrounded by a continuous film of material (polymers, metals, ceramics) to give small core-shell microcapsules of useful properties. These microcapsules have a number of benefits such as converting liquids to solids, separating reactive compounds, providing environmental protection, improved material handling properties etc. [1]. Almost every area in the process industries could benefit from micro-encapsulation technologies. Especially in biotechnology, it can be found in e.g. encapsulation of living cells [2], immobilisation of enzymes [3], or controlled release applications [4].

In the first part of this contribution we present a methodology for the fabrication of silica-coated calcium alginate microcapsules for the encapsulation of active substances. Sodium alginate is a natural polymer extracted from marine algae. It is widely used for the encapsulation due to its biocompatibility and simple gelation with divalent cations such as Ca^{2+}, Ba^{2+}, etc [5, 6]. As it can be easily degraded by a change of the environment, silica coating (SiO_2) can provide a protection against chemical and mechanical stresses. Apart from increasing the stability of the alginate capsules, the surface layer can also provide additional functionality, namely the possibility of surface modification or the attachment of biomolecules (ligands) for controlling adhesion to specific substrates. Ability to adhere to specific substrate (e.g. cells) is a key parameter for successful application of such microcapsules. Microcapsules were characterized in detail by image analysis (Optical Microscopy, Scanning Electron Microscopy, Laser Scanning Confocal Microscopy) and their basic parameters: shape, size, thickness of the layer, were described.

In the second part of the work we studied the adhesion of microparticles to cell substrates, particularly the antibody-antigen specific interactions. Dynabeads®, magnetic microparticles with protein G on the surface, were used. First the magnetic microparticles were modified with specific receptors (monoclonal antibody) and then the interactions of microparticles with prostatic cancer cells were monitored. Potential binding of particles to cells was evaluated by image analysis.

2 Experimental methods

2.1 Synthesis of silica-hydrogel microcapsules

The core-shell microcapsules were synthesized in two steps. In the first step, alginate gel cores were prepared by the drop-on-demand InkJet technology and in the second step the silica shell has been formed by a sol-gel process using alkoxysilane precursors – namely tetramethoxysilane (TMOS) and aminopropyltrimethoxysilane (APTrMOS).

Using the InkJet Drop-on-Demand technique we can produce small microdrops with diameters between 20 and 100 μm depending on the setting of the generating conditions (voltage, etc.) [7]. A piezoelectric drop-on-demand print-head type M5-ABP-01-80-6MX supplied by Microfab, Inc. (Plano, Texas, USA) was used, coupled with a control unit type JetDrive III and a pressure controller type CT-PT-01 also supplied by Microfab, Inc. A 1.0 wt.% sodium alginate solution (with or without suspended nanoparticles – fluorescent silica or iron oxide) was fed into the print-head and dropped into a 1.0 wt.% aqueous solution of $CaCl_2$. The ejected droplets were collected in a beaker under constant stirring to avoid their coalescence. Then the capsules were separated from the $CaCl_2$ solution by filtration through a 25μm seed and washed by deionised water for further use. The resulting microcapsule size was in the range of 40-80 μm.

The organo-silica layer was formed according to the procedure proposed by Sakai et al. [8]: the alginate capsules were mixed with *n*-hexane and kept at 4°C in an ice-bath. APTrMOS and TMOS were then added to *n*-hexane containing the Ca-alginate capsules. APTrMOS was added first and stirred for 1 min, followed by TMOS and stirring for another 1 min. A constant volume ratio of 10:14:0.8:0.6 for alginate:*n*-hexane:APTrMOS:TMOS was used. After the sol-gel process, the capsules were rinsed with a $CaCl_2$ solution (1.0 wt %), filtered through a 25μm seed and then suspended in deionised water. The microcapsules with doubled reaction time (2 min) were also prepared.

The microcapsule morphology and the thickness of the silica surface layer were characterized by Optical Microscopy (OM, Olympus BX41), Scanning Electron Microscopy (SEM, Jeol JCM-5700) and Laser Scanning Confocal Microscopy (LSCM, Olympus Fluoview FV1000 with Olympus IX71 microscope).

2.2 Adhesion studies

Magnetic microparticles Dynabeads® Protein G (size 2.8 μm in diameter) were modified with a monoclonal antibody (Mouse monoclonal GCP 05) according to the protocol proposed by Invitrogen [9]. Then the modified as well as unmodified microparticles were added to prostatic cancer cells (LNCaP cell line with prostatic memgrane specific antigen) and incubated for 1 hour. As a blank test a different type of cells was used (HEK 293T). The interactions of microparticles with cells were monitored by optical microscope and their potential binding was evaluated by image analysis (software ImageJ 1.44).

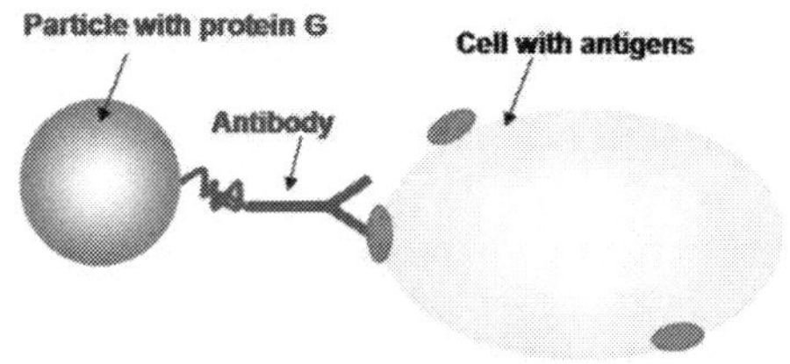

Figure 1: Illustration of antibody-antigen interaction.

3 Results and discussion

3.1 Morphology characterization of silica-hydrogel microcapsules

Silica-coated as well as uncoated capsules were air-dried and their microstructure was observed by OM and SEM, as shown in Fig. 2 and Fig. 3. The difference in the morphology of the particles after drying clearly indicates the presence of the silica shell. While the uncoated particle retains its spherical shape in course of drying, only decreasing its volume, the coated one shrinks by a completely different mechanism. This is in agreement with our previous studies, where silica-coated alginate macrocapsules (size: 1-2 mm in diameter) were tested. A remarkable ability of the silica layer to recover from extreme local deformations (up to 180 degrees folding) without sustaining defects or cracks was shown by performing repeated drying and rehydration experiments [10]. The thickness of the silica coating was estimated to be in the range of 1-4 μm.

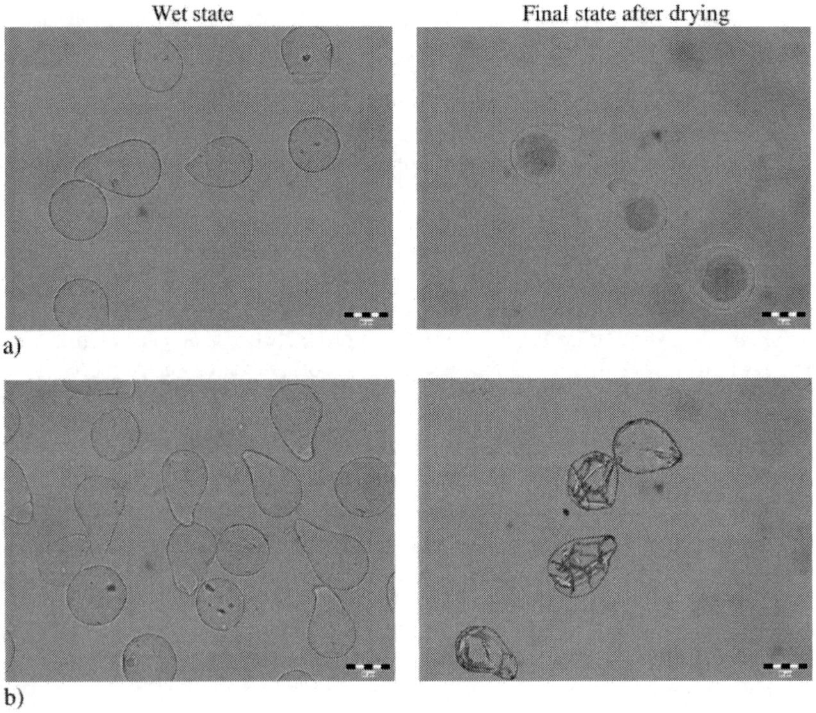

Figure 2: Optical Microscopy images of the morphology of (a) uncoated and (b) silica-coated alginate capsules (image b is for 2 min reaction time) during air drying at room temperature; the scale bar represents 50 μm.

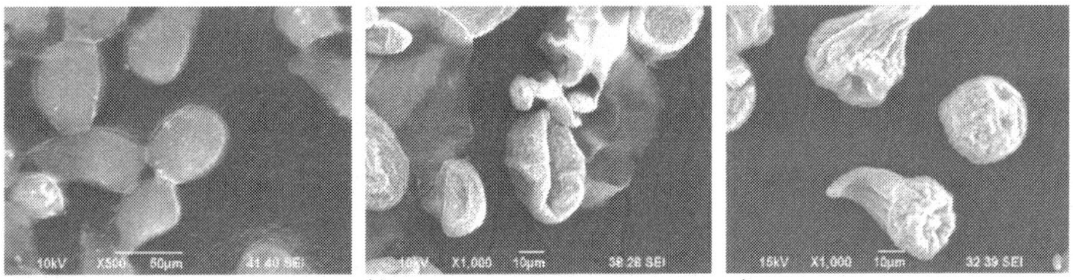

a) b) c)

Figure 3: SEM images of uncoated (a) and silica-coated alginate capsules with 1 min (b) and 2 min (c) reaction time after air-drying.

For detailed visualisation of the organo-silica layer in the wet state, we have used Laser Scanning Confocal Microscopy. The organo-silica layer was fluorescently stained by adding SiO_2 nanoparticles containing FITC (fluorescein isothiocyanate) to the alkoxysilane precursors prior to their hydrolysis. The silica layer is more visible in the dry state where the individual SiO_2 particles immobilised in the shell are closer together (Figure 4b).

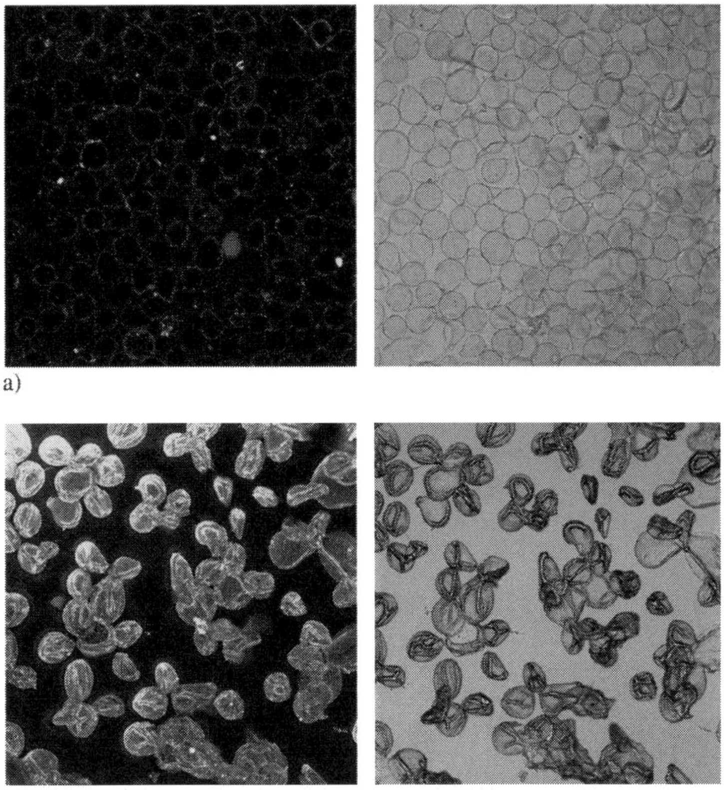

a)

b)

Figure 4: Laser Scanning Confocal Microscope images of silica-coated capsules with SiO_2 stained by a fluorescent dye in the wet state (a) and in the dry state (b).

Encapsulation of fluorescent silica nanoparticles and iron oxide nanoparticles can be seen in Fig. 5 and Fig. 6. The nanoparticles are uniformly distributed inside the hydrogel core. However, sometimes nanoparticle agglomeration occurred. The benefit of iron oxide nanoparticles is their magnetic behaviour, which makes them easy to handle and suitable for further diffusion studies as well as radiofrequency heating.

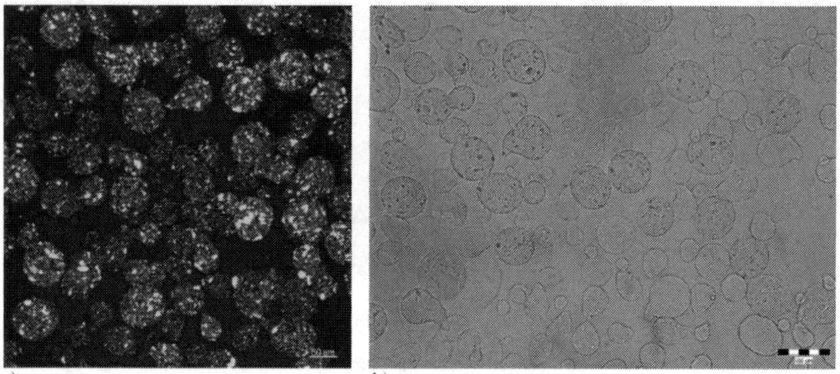

a) b)

Figure 5: Laser Scanning Confocal Microscope (a) and Optical Microscope (b) images of alginate microcapsules with fluorescent SiO_2 nanoparticles inside.

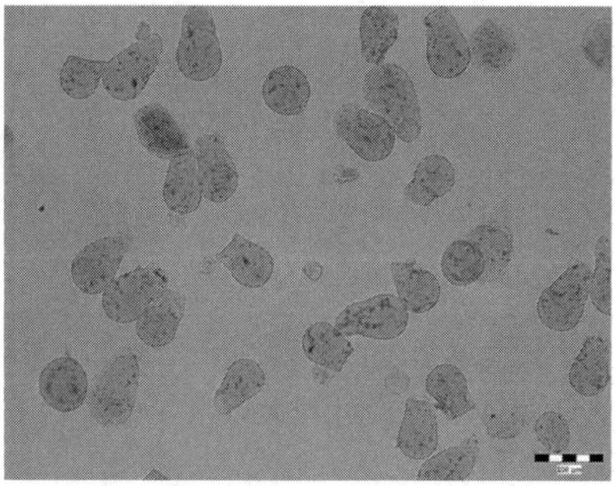

Figure 6: Optical Microscope image of alginate microcapsules with iron oxide nanoparticles inside.

3.2 Adhesion studies

Interactions between microparticles and cells were studied (Fig. 7). It was found that modified as well as unmodified microparticles interact with both types of cells. There are clear interactions between the microparticles and LNCap cells. There are only few interactions with HEK 293T cells. The interactions are specific as well as nonspecific. Further studies have to be done for a better understanding of the antibody-antigen binding process.

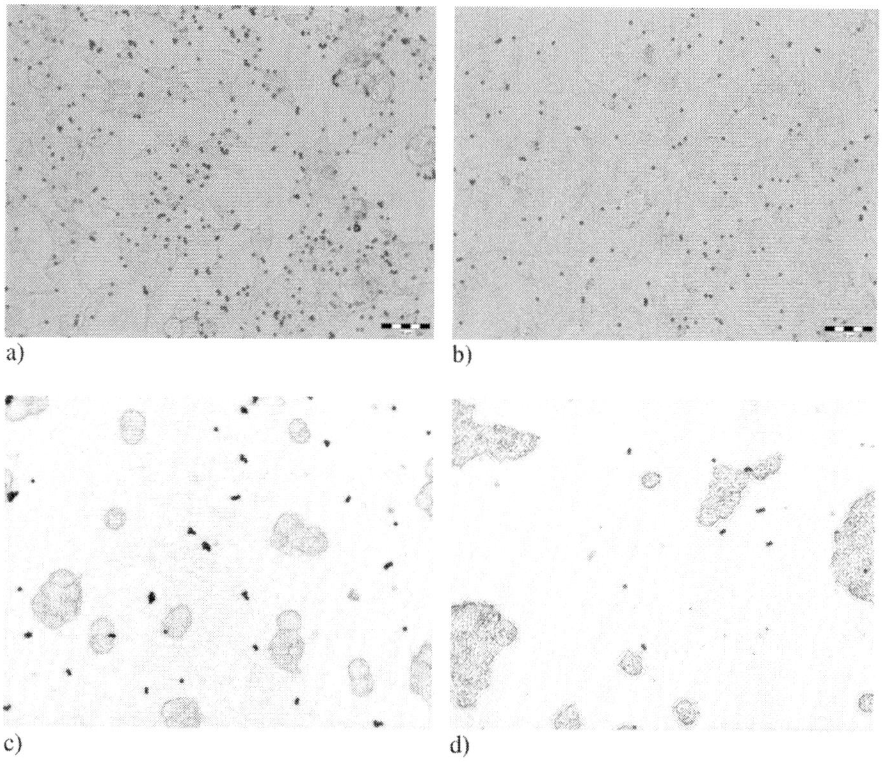

Figure 7: Optical Microscope image of cells – microparticles interactions. a) LNCap cells with unmodified microparticles, b) LNCap cells with microparticles modified by antibody, c) HEK 293T cells with unmodified microparticles, d) HEK 293T cells with microparticles modified by antibody.

4 Conclusions

Composite microcapsules (silica-coated alginate microcapsules) were prepared by the drop-on-demand InkJet method followed by sol-gel deposition. We have shown that the silica membrane is flexible and porous. Thickness of the membrane is in the range of 1-4 μm. Nanoparticles (silica / iron oxide) can be incorporated in the silica membrane as well as encapsulated inside the alginate core.

First adhesion studies were done. Specific as well nonspecific interaction between cells and microparticles were observed.

Further research will be focused on microcapsules surface modifications and additional adhesion studies to cells and biologic substrates.

Acknowledgement

This project is supported by the European Research Council (200580 CHOBOTIX), by the Czech Science Foundation (104/08/H055) and by the grant of the Specific University Research (MSMT no. 21/2011).

References

[1] Lim, F.; Moss, R. D. *J. of Pharmaceutical Science* **1981**, *70*, 4, 351–354.

[2] Carturan, G.; Toso, R. D.; Boninsegna, S.; Monte, R.D. *J. Mater. Chem.* **2004,** *14*, 2087–2098.

[3] Perullini, M.; Jobbágy, M.; Mouso, N.; Forchiassin, F.; Bilme, S. A. *J. Mater. Chem.* **2010**, *20*, 31, 6479–6483.

[4] Polk, A.; Amsden, B.; Yao, K. D.; Peng, T.; Goosen, M. F. A. *J. Pharm. Sci.* **1994,** *83*, 2, 178–185.

[5] Georgie, M.; Abraham, T. E. *Journal of Controlled Release* **2006**, 114, 1-14.

[6] Drury, J. L.; Money, D. J. *Biomaterials* **2003**, 24, 4337–4351.

[7] Dohnal, J.; Štěpánek, F. *Powder Technol.* **2010**, 200, 254-259.

[8] Sakai, S.; Ono, T.; Ijima, H.; Kawakami, K. *Biomaterials* **2001**, *22*, 2827-2834.

[9] http://tools.invitrogen.com/content/sfs/manuals/100%2003D_04D(rev006).pdf

[10] Haufová, P.; Knejzlík, Z.; Hanuš, J.; Zadražil, A.; Štěpánek, F. *J. Coll. Interf. Sci.* **2011**, 357, 109-115.

(730f) Angiogenic Microfiber Patches for Directed Blood Vessel Growth

Ross J. DeVolder[1], Harim Bae[2], Jonghwi Lee[2] and Hyunjoon Kong[1], (1)Chemical and Biomolecular Engineering, University of Illinois, Urbana, IL, (2)Chemical Engineering and Materials Science, Chung-Ang University, Seoul, South Korea

Introduction: Therapeutic angiogenesis has emerged as a promising strategy to treat various acute and chronic vascular diseases, and to enhance tissue repair and regeneration. Common revascularization therapies include the administration of angiogenic factors, such as vascular endothelial growth factor (VEGF). These therapies greatly rely on the ability to engineer mature and functional neovessels, uniformly distributed within target tissues. The objective of this study was to develop an angiogenic patch that releases angiogenic growth factors and ultimately regulates the directional growth of mature and functional blood vessels.

Materials and Methods: Angiogenic microfiber patches were prepared by integrating electrospinning and electrospraying processes. PLA microfibers with negative surface chargers were extruded through an electrohydrodynamic nozzle, and VEGF encapsulating PLGA microparticles with positive surface charges were simultaneously sprayed over the fiber surfaces. The microparticle covered fibers were collected on Teflon frames in a directionally aligned manner. The proangiogenic patches were individually implanted on the chorioallantoic membrane (CAM) of chicken embryos and examined over a week for blood vessel formation. The CAMs were excised and histological CAM cross-sections were analyzed for blood vessel densities, distributions, and alignment.

Results and Discussion: The electrospinning/electrospraying technique produced an angiogenic patch that sustainably released VEGF from degrading PLGA microparticles, while the PLA microfibers act as a stable structure preventing the dislocation of microparticles at the delivery site. The temporal control of electrostatic interactions using this technique has the advantage of improving bond strength between fibers and particles compared to previously reported techniques. Cells adhered to the patch anisotropically, directed by the alignment of the microfiber patches. The implantation of angiogenic patches on CAMs of chicken embryos resulted in a significant increase in the size and density of mature blood vessels compared to control samples. The formation of blood vessels was evenly distributed and aligned parallel to the microfiber alignment. We suggest that the VEGF released from the microparticles locally distributes around and associates with the PLA microfibers, subsequently presenting concentration gradients that direct blood vessel formation along the microfibers.

Conclusion: We have demonstrated that the angiogenic patch fabricated by integrating an electrospinnning and elecrospraying process allows the control of both the density and spatial organization of blood vessels at an implant site. Overall, the angiogenic patch developed in this study will be broadly useful for improving the quality of treatments for various vascular diseases and tissue defects.

See more of this Session: Building Drug and Gene Delivery Into Tissue Engineering
See more of this Group/Topical: Materials Engineering and Sciences Division

(730g) DNA Nanoparticles - Structure, Assembly and Osmotic Properties

Preethi L. Chandran, NICHD/Nibib, National Institute of Health, Bethesda, MD, Emilios K. Dimitriadis, Laboratory of Bioengineering and Physical Science, National Institute of Biomedical Imaging and Bioengineering, National Institute of Health, Bethesda, MD and Ferenc Horkay, Section on Tissue Biophysics and Biomimetics, Program in Pediatric Imaging and Tissue Sciences, NICHD, National Institute of Health, Bethesda, MD

DNA is an anionic polyelectrolyte, which occupies a large volume in salt free solution due to the coulomb repulsion between the charged groups. In the presence of multivalent cations, DNA condenses into compact nanostructures. There has been considerable interest in the DNA condensation process for understanding its packaging in viruses and for creating nano-scale self-assembled structures. Importantly, the Human Genome project and the prospect of combating cancer and hereditary diseases by delivery of therapeutic genes has fuelled interest in nanoparticles formed by condensation of DNA with polycations. Delivering genes by nanoparticles has the advantages of reduced antigenicity and improved target specificity compared to viral modes of gene delivery. In this talk we focus on three aspects of DNA nanoparticle formation and stability. (1) Counterion valence as a tool of modulating DNA assembly into nanoparticles. We discuss how ion valence influences the pathway of DNA condensation into nanoparticles. (2) Organization of DNA strands within the nanoparticles. We determine the morphology of the DNA within the nanoparticles by the Atomic Force Microscope. (3) Stability of DNA in different ionic environments. We estimate the effect of counterions on the osmotic pressure of DNA. This information is important because the interaction between DNA and cations control the efficacy of gene release within the cell.

See more of this Session: Building Drug and Gene Delivery Into Tissue Engineering
See more of this Group/Topical: Materials Engineering and Sciences Division

(733a) Renewable Fuel From Activated Sludge Using Fluidized-Bed Catalytic Cracking (FCC) Process

Emmanuel D. Revellame, William Holmes, Rafael Hernandez and William T. French, Dave C. Swalm School of Chemical Engineering, Mississippi State University, Mississippi State, MS

Annual production of activated sludges in the United States ranges from 7 – 8 million tons. Currently, disposal technologies of these sludges include incineration and landfilling, both of which are heavily regulated and could create adverse environmental impacts. A fraction of the activated sludges can be used as fertilizer once it complies with regulations for Class A biosolids. Activated sludges contain microbial consortia known to produce lipidic materials and many other organic compounds, which could be used to produce biofuels and specialty chemicals. Several studies showed that reasonable yields (2 – 6% weight) of biodiesel could be produced from activated sludge. At these yields however, activated sludge biodiesel is less economical than petroleum diesel. This study describes the production of renewable fuel from activated sludge using Fluidized Catalytic Cracking (FCC). This process could convert most of the compounds present in sludge into fuel, and thus, increase fuel yields and reduce cost.

Several catalysts, sludge to catalyst ratios, sludge moisture content, fluid flow rates and temperatures were tested. Initial results using ZSM-5 at 500°C with air showed considerable production of aromatic compounds, from benzene to methylnaphthalene.

See more of this Session: Conversion of Solid Wastes to Energy and/or Product
See more of this Group/Topical: International Congress on Energy 2011

(733b) Simultaneous Extracellular Polymeric Substance (EPS) and Lipid Production by Activated Sludge Via Fermentation of Glucose

Patrisha J. Pham, Rafael Hernandez, Hien Nguyen, Andro H. Mondala and W. Todd French, Dave C. Swalm School of Chemical Engineering, Mississippi State University, Mississippi State, MS

The use of activated sludge microflora to convert sugars, derived from the hydrolysis of lignocellulose biomass, into lipids for biofuel production is of current research interest in our group. Lipid accumulation in this system has been enhanced by manipulating the carbon-nitrogen (C:N) ratio and initial sugar loading on activated sludge bioreactors with glucose as the main carbon source. Activated sludge is a flocculated suspension composed of microorganisms that can also produce extracellular polymeric substances (EPS) *in situ* during fermentation. Microbial EPS which are secreted largely by bacteria in harsh environments are a matrix rich in polymers including polysaccharides, glycoproteins, proteins, nucleic acids and phospholipids. Thus, studying EPS production along with lipid formation is essential for understanding the technical and economic value of the whole process. In this study, samples were obtained from aerobic batch fermentation inoculated with 20 % (v/v) activated sludge. The total incubation time for the batch fermentation experiment was seven days. EPS from the collected samples were extracted using a Dowex cation exchange resin in phosphate buffered saline solution and precipitated with 95% v/v cold ethanol and dialyzed. Fourier transform-infrared spectroscopy of freeze-dried EPS revealed characteristic IR peaks for polysaccharides and proteins. Total polysaccharide and protein concentrations were estimated in each EPS sample using phenol-sulfuric and modified Lowry methods, respectively. An increase in lipid % (g/g cell dry weight) and a decrease in specific EPS % (g/g cell dry weight) was observed from day 0 to 7 in all treatments. Increasing the carbon loading while maintaining the same C:N ratio increased lipid % and decreased specific EPS %. However for a low C:N ratio, lipid % remained constant while specific EPS % decreased. These results will be useful for manipulation of EPS and lipid production and consequently enable development of valuable strategies for improving the process.

See more of this Session: Conversion of Solid Wastes to Energy and/or Product
See more of this Group/Topical: International Congress on Energy 2011

(733c) The Hydrolysis of Nylon6 In Ionic Liquid to Recycle Nylon Waste

Xiao-Li Xia and Hai-Jun Wang, Jiangnan University, Wuxi, China

To date, several processes for Nylon6 depolymerization have been put forward. Herein, we first examined several ionic liquids for the hydrolysis of Nylon6. These ionic liquids are Brønsted acidic and they have been used both as a catalyst and environmental solvent for the hydrolytic reaction of Nylon6. The final result showed that using 1-butyl-3-methylimidazolium chloride ([Bmim]Cl) as solvent and catalyst for hydrolysis of Nylon6 ,the high conversion of Nylon6 and yield of Aminocaproic acid were obtained. In order to confirm the product, FT-IR and ^{1}H NMR were performed. After the experimental, theoretically, we also used high level ab initio methods with the 6-31G(d) basis sets to calculate the interaction between [Bmim]Cl and Nylon6. Based on our calculated interaction, we conclude that hydrogen bonds exist between [Bmim]Cl and Nylon6 which explain that Nylon6 dissolve in [Bmim]Cl.

See more of this Session: Conversion of Solid Wastes to Energy and/or Product
See more of this Group/Topical: International Congress on Energy 2011

(733d) Process Development for Efficient Utilization of Waste Plastic to Liquid Product for Rural Development In Nigeria: Gasification Option

Bamikole Amigun, Environmental Biotechnology and Conservation Department, National Biotechnology Development Agency (NABDA), Abuja, Nigeria, Christie Onyia, Environmental Biotechnology and Conservation Department, National Biotechnology Development Agency, Abuja, Nigeria and Bamidele Ogbe Solomon, Director General's Office and consultancy, National Biotechnology Development Agency (NABDA), Abuja, Nigeria

Waste production is a very important problem in all countries especially in developed countries and Nigeria is no exception. Waste plastics create a very serious environmental challenge because of their huge quantities and their disposal problems, due to their non-biodegradable attributes (or the biodegradation process is very slow). Plastic waste poses a serious environmental problem that is not addressed by recourse to landfill. Open dumping has been the predominant solid waste disposal option in Nigeria. Their destruction by incineration poses serious air pollution problems due to the release of airborne particles and carbon dioxide into the atmosphere. Only a limited amount of plastics can be recycled, while most plastics could be used, for example in energy production. The management of plastic waste using gasification in fluidized bed technology is a technically feasible option. Apart from eliminating the waste, a chemical recycling is produced during this process, obtaining a gas (syngas) which can be used in different applications such as ethanol through bacteria conversion of the syngas intermediate. Ethanol can be further converted to ethanol gel fuel to substitute the use of paraffin and traditional firewood in rural communities. This is doubly environmentally friendly since it will reduce the volume of plastic waste being disposed of in landfill whilst producing green fuel without generating green house gases. This application could give some prospect of self-reliant energy supplies at local levels, with potential economic, ecological, social, and security benefits; thereby improving the socio-economic and political wellbeing of the rural people.

See more of this Session: Conversion of Solid Wastes to Energy and/or Product
See more of this Group/Topical: International Congress on Energy 2011

(733e) An Anaerobic Bacterial Treatment of Corn Stover to Prepare a Biorefining Feedstock

James MacLellan, Biosystems and Agricultural Engineering, Michigan State University, East Lansing, MI

Anaerobic Bacterial Treatment of Corn Stover to Prepare a Biorefining Feedstock

James MacLellan, Zhengbo Yue, Robert Kraemer, Wei Liao*

Department of Biosystems and Agricultural Engineering, Michigan State University

East Lansing, Michigan 48823

Abstract

The objective of this study is to apply an anaerobic bacterial consortium to uptake the hemicellulose in corn stover to prepare a relatively cellulose-enriched feedstock for biofuel production. Previous research demonstrated that submerged anaerobic bacterial culture is capable of converting hemicellulose in raw lignocelluloses to methane gas. The in-depth study is needed to optimize the culture, enhance the hemicellulose utilization efficiency, and improve the quality of treated cellulose as a biorefining feedstock. An anaerobic bacterial consortium from animal manure will be screened and enriched as the seed for this study. Two feedstocks of raw corn stover and steam-treated corn stover will be used to compare the effects of structure on the anaerobic culture of hemicellulose uptake. A completely Randomized Design (CRD) on seed-to-feedstock ratio, culture temperature, and culture time will be carried out. Fiber composition, methane production, and microbial consortia will also be analyzed. An optimal condition can then be established based on the quality and quantity of cellulose in the treated corn stover. The ethanol production on the treated corn stover will be fulfilled in order to elucidate the effects of anaerobic bacterial treatment on biofuels production.

*: Corresponding author

Dr. Wei Liao

Department of Biosystems & Agricultural Engineering

Michigan State University

202 A.W. Farrall Hall

East Lansing, MI 48824-1323

Tel: 517-432-7205

Fax: 517-432-2892

E-mail: liaow@msu.edu

See more of this Session: Conversion of Solid Wastes to Energy and/or Product
See more of this Group/Topical: International Congress on Energy 2011

(733f) Towards Developing Anaerobic Digestion Based Biorefinery - Research On Food Wastes As Feedstock

Shulin Chen[1], Liang Yu[2], Jingwei Ma[2] and Craig Frear[2], (1)Department of Biological Systems Engineering, Washington State University, Pullman, WA, (2)Washington State University, Pullman, WA

Anaerobic digestion (AD) is a proven technology for waste treatment. However, as the organic material becomes more valuable and being treated as resources, traditional AD process is no longer adequate. The Bioprocessing and Bioproduct Engineering Laboratory at Washington State University has been devoting its effort in developing the second generation of AD technology with the concept of biorefinery, featured by producing various products from improved AD processes. Key to an AD biorefinary is innovative design of the digestors. This presentation reports the recent results from a study aimed at evaluating a two stage-digester designed for treating food wastes operated within a solid concentration range of 15-20%. The digester had a unique two-stage configuration that resulted in high biogas production rate than the existing technologies based on the laboratory data. The experimental observations also led to a simple mixing design that has low capital and operational costs. The information provides some new insights in AD research and development related to municipal solid waste applications.

See more of this Session: Conversion of Solid Wastes to Energy and/or Product
See more of this Group/Topical: International Congress on Energy 2011

(741a) Dynamic and Rheological Properties of Hydrogels/Bioactive Ceramics Based On Degradable Poly(2-hydroxyethyl methacrylate)

Jijun Huang[1], Antoni P. Tomsia[1] and Eduardo Saiz[2], (1)Materials Sciences Division, Lawrence Berkeley National Laboratory, Berkeley, CA, (2)Center for Advanced Structural Ceramics, Department of Materials, Imperial College London, United Kingdom, London, United Kingdom

Dynamic and rheological properties of hydrogels/bioactive ceramics based on degradable poly(2-hydroxyethyl methacrylate)

Jijun Huang[1], Eduardo Saiz[2], Antoni P. Tomsia[1]

[1]Materials Sciences Division, Lawrence Berkeley National Laboratory, Berkeley, California 94720

[2]Center for Advanced Structural Ceramics, Department of Materials, Imperial College London, United Kingdom

Hydrogel-based organic-inorganic bioactive composites have attracted increasingly attention in bone tissue engineering. The hydrogel/bioactive ceramics materials are expected to exhibit cell adhesion, calcium-binding, and mechanical properties as tissue scaffolds. Hydrogels composites based on poly(2-hydroxyethyl methacrylate), pHEMA, have been recently investigated to create elastomeric bonelike materials that may be machined, cut, and bent for low- to medium-load bearing applications. However, the non-degradability of the pHEMA hydrogel limits its applications.

We recently initiate our efforts to create biocompatible, biodegradable pHEMA hydrogels and their biocomposites with hydroxyapatite and bioglass. Our goal was to create a series of biodegradable pHEMA/hydroxyapatite and pHEMA/bioglass materials with various concentrations (30%, 50%, and 70 wt%) of hydroxyapatite and bioglass with tuning mechanical properties. These novel materials have a variety of applications in bone repair, reshaping, and regeneration. We purposely synthesized a hydrolytically cross-linker, N, O-dimethacryloyl hydroxylamine through which degradable pHEMA hydrogels were synthesized. Degradable pHEMA/hydroxyapatite and pHEMA/bioglass biocomposites were formed by pre-dispersing the biocermics into the structural components of the hydrogel prior to cross-linking. In this work, we will show the effect of bioceramics centration on dynamic mechanical properties, rheological properties, stress relaxation, creep behavior, and Payne effect. We will compare hydroxyapatite to bioglass in these properties. All these properties are important to bone regeneration. This work was supported by the National Institutes of Health/National Institute of Dental and Craniofacial Research (NIH/NIDCR) Grant No. 1 R01 DE015633.

(741b) Interfacial Templating of Metallic Nanostructures Using a Rationally Designed Peptide

Lorraine F. Leon Gibbons, Chemical Engineering, The Graduate Center of the City University of New York, New York, NY and Raymond Tu, Chemical Engineering, City College of New York (of CUNY), New York, NY

In nature, biological molecules form interfaces that assemble patterns of chemical functionality with exceptional precision. These interfaces serve a key role in the templating of inorganic material in biological systems. Our work applies model sheet-forming peptides at interfaces to explore the dynamics of assembly in order to template inorganic growth of metallic structures. The peptide molecules are rationally designed to have amphiphilic properties and a propensity for sheet like secondary structure. These designed peptides are deposited at the air water interface to explore the dynamics of their self-assembly and investigate their 2D order. To characterize the phase behavior, techniques such as Langmuir Blodgett and Brewster Angle Microscopy are used. In addition we verify our proposed sheet forming design using both Circular Dichroism and Attenuated Total Reflection Fourier Transform Infrared Spectroscopy. Thermodynamic analysis of structure formation with increasing pressure allows us to understand the nature of self-assembly with iterative changes in the peptide sequence. Additionally, we look at the dynamics of the self-assembled state, where the organic phase switches between short- and long-range order as a function of surface pressure. This model system allows us to explore our underlying hypothesis that the time scale of the phase-transitions of the peptide at the interface defines the length-scale of the crystalline phase. This is in contrast to a system that starts with a well-ordered preformed template that defines the epitaxial growth of the mineral phase. Versions of our model peptides are modified to include histidine in order to nucleate Au nanocrystals in both the short and long range ordered organic matrix. We have found the shape and crystallinity of the resulting nanocrystals are surface pressure dependant. The characterization of the metallic nanocrystals is done using Transmission Electron Microscopy, Electron Diffraction, and Atomic Force Microscopy.

See more of this Session: Hybrid Biomaterials
See more of this Group/Topical: Materials Engineering and Sciences Division

Bioactive Electroconductive Hydrogels

Christian Kotanen[1,2], *Chaker Tlili*[1,3], *and Anthony Guiseppi-Elie*[1,2,3,4,5,*]

[1]Center for Bioelectronics, Biosensors and Biochips (C3B), Clemson University Advanced Materials Center, 100 Technology Drive, Anderson, South Carolina 29625, USA

[2]Department of Bioengineering, [3]Department of Chemical and Biomolecular Engineering, [4]Department of Electrical and Computer Engineering, Clemson University, Clemson, South Carolina 29634, USA

[5]ABTECH Scientific, Inc., Biotechnology Research Park, 800 East Leigh Street, Richmond, Virginia 23219, USA.

*Correspondence: Anthony Guiseppi-Elie, guiseppi@clemson.edu

Abstract

A class of bioactive, stimuli-responsive co-joined interpenetrating networks of inherently conductive polymers and highly hydrated hydrogels are being developed for implantable biodevice interfaces and for field induced release of elutable drugs. Polymers were formed from tetraethyleneglycol diacrylate (TEGDA) cross-linked (3–12 mol%) hydroxyethylmethacrylate (HEMA) and *N*-[tris(hydroxymethyl)methyl]-acrylamide (HMMA) that served as the principal hydrophilic monomers. Polyethylene glycol methacyrlate ($PEG_{200}MA$; 0.3 – 0.5 mol%) and 2-methacryloyloxyethyl phosphorylcholine (MPC; 5 – 10 mol%) were introduced to provide *in vivo* bioactive biocompatibility. Also included was 3-sulfopropyl methacrylate potassium salt (SPMA) that served as the dopant counter anion for oxidatively or electrochemically incorporated polypyrrole (PPy) or polyaniline (PAn). PPy provides interference screening of endogenous interferents in biosensor applications while PAn served as an electro-actuatable polymer for electro-stimulated release of bioactive agents stored within the hydrogel. These bio-device interfaces demonstrate reduced electrical impedance, high RMS 13 and PC12 cell viability (> 80%) and low proliferation (< 40%). Deposition of these electroconductive hydrogel *co-networks* onto microdisc electrode arrays resulted in enhanced sensitivity and dynamic range for the detection of electroactive species. Bioactive electroconductive hydrogel co-networks were made responsive to glucose and lactate by the inclusion of glucose oxidase and lactate oxidase. Enzyme biosensors showed similar sensitivity but increased current and minimum interference for endogenous interferents. *Keywords:* Biological Engineering, Fuel Cells, Nanotechnology, Polymers, Transport Phenomenon

Introduction

Simple Clark style amperometric biosensors[1] continue be highly investigated because of their many potential applications in the field of clinical chemistry, bioprocess monitoring, biotechnology, and in the food industry[2]. Such devices integrate the molecular recognition of enzymes such as glucose oxidase and lactate oxidase with solid state electrodes that may be coinage metals, carbonaceous materials or semiconductors. Recently, another new application of glucose and lactate biosensors was developed for the management of hemorrhagic shock in critically injured patients[3]. Hemorrhage brought on by trauma can lead to multiple organ dysfunction syndrome (MODS) and/or eventual death[4]. To increase the survival rate of the injured who may be at risk for hemorrhagic shock, it is imperative that we develop and deploy simple, rapid, sensitive, continuous and selective on-site platform that can quantify glucose and lactate levels in a suitable tissue bed[5]. Both glucose and lactate have been identified as trauma-associated biomarkers. The continuous monitoring of glucose and lactate levels in injured patients becomes a significant parameter in evaluating the severity of injury and understanding the progress toward hemorrhagic shock. Both are intended to aid the clinician in optimizing resuscitation and reanimation.

The immobilization technique to localize the enzyme at the surface of an electrode is an engineering challenge in the fabrication of glucose and lactate biosensors. A major challenge exist in the fabrication of fully implantable multi-analyte biosensors when the electrode sites are micron dimensions and are adjacent to each other such as in a CMOS device[6]. Various immobilization procedures have been developed for macro-electrodes or single microelectrodes. Among these are enzyme entrapment and encapsulation within polymer membranes that may be spun applied, dipped-coated, sprayed, covalently immobilized using homo- or hetero-bifunctional conjugation agents, or adsorbed via cross-linking with bifunctional cross-linkers such as glutaraldehyde. None of these additive techniques are however appropriate for closely spaced microelectrodes on a CMOS device. The additive placement and entrapment of enzymes on such devices may be achieved by electropolymerization. Electropolymerization produces controllable thin films of inherently conductive electroactive polymers (CEP) deposited at electrode sites[7]. During formation such polymers entrap counter anions that serve as "dopants". In the presence of enzymes, which are often negatively charged, electropolymerization leads to formation of an enzyme enriched membrane layer deposited directly on the electrode[8]. Two challenges exist with this approach. Firstly, the rapid denaturation of the enzyme in the dense hostile environment of the CEP. Secondly, the high background currents associated with the CEP at the electrode. We have introduced two major innovations to address these limitations. Firstly, the electropolymerization is performed at hydrogel-coated electrodes thereby placing the enzyme within a water rich milieu that sustains its bioactivity[9]. Secondly, following electropolymerization, the electroactive polymer is over-oxidized[10], eliminating its contribution to a high background amperometric current during interrogation. These innovations offer the advantage of high efficiency of enzyme loading, ease of preparation, stability of the enzyme/CEP/hydrogel membrane, interference suppression[11], and enables control of the hybrid polymer membrane properties.

Experimental

Chemicals and reagents

Electroconductive hydrogels of poly(pyrrole-*co*-4-(3-pyrrolyl) butyric acid) formed within poly(2-hydroxyethyl methacrylate)-based hydrogels were prepared as previously described (Ref). Briefly, pyrrole and its co-monomer, 4-(3-pyrrolyl) butyric acid, were formulated as components of a hydrogel dope and cast onto platinum microelectrodes (ϕ = 100µm) (BAS, Wes Lafayette, IN) or onto an electrochemical cell-on-a-chip (ECC) that uses a co-planar arrangement of the microdisc electrode array (MDEA, n = 37, ϕ = 50µm)) working electrode, a large area counter electrode and a reference electrode (MDEA 5037-Pt, ABTECH Scientific, Richmond, VA).

Surface modification and application of hydrogel to platinum microelectrode surface

Following mechanical polishing with 0.05 µm alumina for 10 min (Pt microelectrode), solvent cleaning (boiling TCE, Acetone, IPA), treatment with RCA clean (Ref) and anodic cleaning (0.5M H2SO4, repeated cycling between -0.20 to 1.40 V *vs.* Ag/AgCl, 3 M KCl at 100 mV/s) the bare platinum microelectrodes were silanized (0.1 wt% 3-aminopropyl-trimethoxysilane in anhydrous toluene, 1 hr), functionalized (1 mM acryloyl(polyethyleneglycol)-N-hydroxysuccinamide (ACRL-PEG-NHS, MW 400) in 0.1 M HEPES buffer, pH = 8.5 for 2 hr), ultrasonicated in DI-water for 5 min and then blown dried with ultrahigh purity nitrogen. The candidate microelectrode was dipped in a 3 mol% TEGDA cross-linked p(HEMA)-based hydrogel cocktail that was prepared by mixing HEMA, TEGDA, PEG(400)MA, MPC, HMMA, p(NVP), DMAEMA, and DMPA in typical ratio 78:3:5:1:5:2:5:1 mol% and 1:1 (v/v) solution of ethylene glycol/water corresponding to 20 % of the volume of the monomer cocktail (Ref). The thin film, ca. 5μm, was then crosslinked with UV light for 5min (CX-2000 Crosslinker, UVP, Upland, CA) under an inert nitrogen atmosphere.

Conferring biospecificity to hydrogel-coated microelectrodes

Before electropolymerization of pyrrole, the hydrogel coated electrode was immersed and incubated for 1 h in 5.0 ml of an aqueous Py (0.4 M) / GOx or Lox (1 mg/ml) (GOx = Glucose oxidase; E.C. 1.1.3.4 from Aspergillus niger; LOx = Lactate oxidase;) prepared in nanopure water. Finally, potentiostatic electropolymerization was commenced by the application of 0.75V *vs.* Ag/AgCl, 3 M KCl to the Pt W.E leading to polypyrrole formation within the hydrogel membrane [12] and the concomitant immobilization of GOx within the hydrogel. This produced an enzyme-loaded, electroconductive PPy-hydrogel composite. The enzyme electrode, or biotransducer, was subsequently washed with nanopure water and by PBS to remove any unbound GOx from hydrogel film. Several electropolymerization charge densities were explored. Before use, the modified enzyme electrode was over-oxidized by repeatedly cycling the electrode in PBKCl (100 mM, pH 7.0) between 0 to 1.2 V *vs.* Ag/AgCl, 3M KCl for 40 cycles.

Characterization of the enzyme electrodes

For immobilized enzyme kinetics, biosensor sensitivity, response time and biosensor stability studies, amperometric measurements were performed at 0.4 V *vs.* Ag/AgCl, 3 M KCl and by sequential injection of appropriate aliquots of glucose solution into 5.0 ml of 100 mM PBKCl (pH 7.0) under continuous stirring to provide for convective transport. Enzyme electrodes were stored in PBKCl solution at 4 °C over a period up to 18 days. The kinetic parameters were determined using Lineweaver-Burk analysis of biosensor amperometric response every few days during the storage period. Electrochemical experiments were performed using a BAS-100B/W (West Lafayette, USA) Electrochemical Analyzer with a BAS PA-1 preamplifier module used to amplify the current and to filter out noise. All experiments were carried out in three-electrode with the platinum microelectrode (West Lafayette, USA) as the working electrode (W.E.), a Ag/AgCl 3M KCl reference electrode (R.E.) (ABTECH Scientific, Inc., Richmond, Virginia) and a platinum mesh counter electrode (C.E.).

Results and Discussion

Microdisc electrode arrays (MDEAs) have been shown to have higher sensitivities than planar metal electrodes [13]. Sensitivity gains while maintaining constant surface area are due to the increasing contribution of radial diffusion of analytes to electrode surfaces as electrode dimensions decrease resulting in higher current densities (**Table 1**). Electroconductive hydrogels (**Figure 1**) utilizing polypyrrole networks have been shown to decrease limits of detection and increase sensitivity to analytes even further but at the cost of narrowing the linear dynamic range of biosensor systems. The electrical impedance between the electrode surface and the solution is lowered by polypyrrole. Upper limits of analyte detection are lost due to exceedingly high background current from oxidation of polypyrrole networks [14]. Overoxidation of the electroconductive gels using cyclic voltammetry greatly reduces the background current.

Table 1. Effects of microdisc electrode feature size (left) and effects of different electrode and hydrogel membrane types containing polypyrrole (right) on biosensor performance [12].

MDEA (Diameter)	Linear Range (mM)	Sensitivity (uA/mM)	Detection Limit (mM)
MDEA 50 Pt-Pt [r=25μm, 5184 discs]	0.025 - 4.0	1.99	0.012
MDEA 100 Pt-Pt [r=50μm, 1296 discs]	0.025 - 3.0	2.30	0.015
MDEA 250 Pt-Pt [r=125μm, 207 discs]	0.05 - 3.0	1.21	0.045
SDE Pt-Pt [r=1800μm, 1 disc]	0.10 - 100.0	0.10	0.08

Electrode Type	Membrane Type	Linear range (mM)	Sensitivity ($uA\ mM^{-1}$)	Detection Limit (mM)
PME-Au	P(HEMA)-PEGLOx	0.1 - 3.0	0.0662	0.0900
MDEA50-Au	P(HEMA)-PEGLOx	1.0 - 90.0	0.0033	0.0000
MDEA50-Au	P(HEMA)-PPy-PEGLOx	0.002 - 0.5	1.4122	0.0018

Fig 1

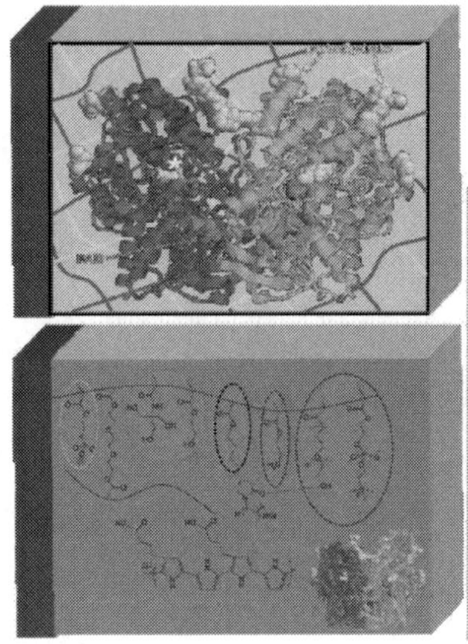

Fig 2

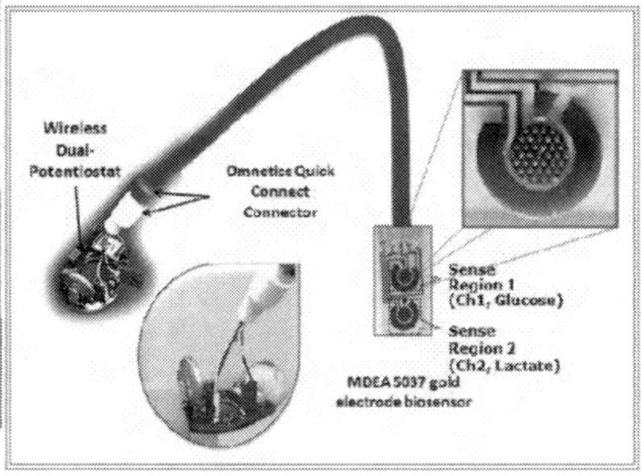

Figure 1. Schematic illustration of on electroconductive hydrogel membrane (*left*).
Figure 2. The dual responsive biotransducer the ECC MDEA 5037 has two independently operating channels for sensing two different analytes (*right*) [12].

The Electrochemical Cell-on-a-Chip Microdisc Electrode Array (ECC MDEA 5037) has been designed for use as a minimally invasive, dual analyte, wireless biotransducer for the monitoring of glucose and lactate (**Figure 2**)[15]. Qualification of unmodified MDEA 5037 biotransducers in ferrocyanide has shown monotonically increasing current as a function of analyte concentration, **Figure 3**. Both channels of the MDEA 5037 operate independently and may be used simultaneously to address different dynamic ranges of the same analyte or different analytes. The Pinnacle 8151 dual potentiostat (Pinnacle Technologies, Lawrence, Kansas) has an inherent 6.5% discrepancy between current readings of its two channels when measuring real electrochemical systems (resistor-capacitor networks) compared to pure resistors. This discrepancy can be taken into account during use *in vitro* and *in vivo*.

Amperometric enzyme-based biosensors used here are Type 1 devices that involve the measurement of the rate of hydrogen peroxide formation. The first two reactions [Equations (1) and (2)] represent the enzyme (glucose oxidase, GOx) redox reactions, wherein FAD is the oxidized form of the prosthetic group, flavin adenine dinucleotide:

$$\beta\text{-D-glucose} + \text{GOx/FAD} \rightarrow \text{D-glucono-}\delta\text{-lactone} + \text{GOx/FADH}_2 \quad (1)$$
$$\text{GOx/FADH}_2 + \text{O2} \rightarrow \text{GOx/FAD} + H_2O_2 \quad (2)$$
$$H_2O_2 \rightarrow O_2 + 2H^+ + 2e^- \quad (3)$$

Fig 3

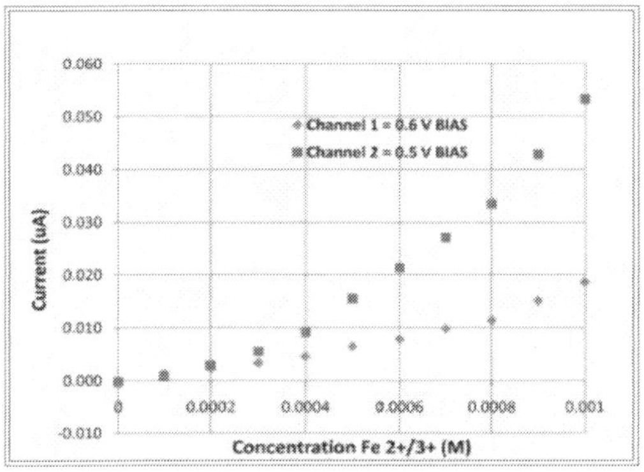

Fig 4

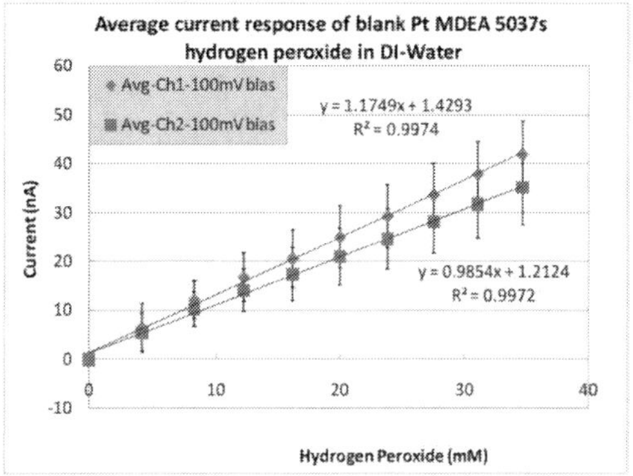

Figure 3. Current calibration curves of wireless MDEA 5037 biotransducers as a function of ferro/ferricyanide concentration at two different voltages 0.6V and 0.5V [12].

Figure 4. The current response of wireless MDEA 5037 biotransducers (n=4) to hydrogen peroxide in DI-water at 100 mV bias of onboard working vs. counter electrode. Ch1: WE-A vs. CE-A and Ch2: WE-B vs. CE-B.

The MDEA 5037 transducers were cleaned and interrogated in hydrogen peroxide (H_2O_2), the electrochemically dischargeable analyte of Type 1 oxidoreductase biosensors, and with ferrocene monocarboxylic acid (FcCOOH), a well-known reversible redox probe molecule. Using the dual responsive wireless potentiostat system (Pinnacle) and the on-chip working and counter electrodes dose response curves to hydrogen peroxide were obtained and showed good reproducibility amongst four cleaned platinum MDEA 5037s (Ch1 $R^2 = 0.9974$; Ch2 $R^2 = 0.9972$), **Figure 4**. A high sensitivity to hydrogen peroxide was noted. To avoid exceeding the resolution maximum of 250 nanoamperes of the 8151 potentiostat, low bias potentials of 100 mV were utilized. While there is excellent reproducibility, the performance of Chanel 2 shows a different slope (sensitivity = 1.18 nA/mM) than Channel 1 (sensitivity = 0.99 nA/mM). This is a concern that goes to the performance of the dual potentiostat itself. The dose response curve of cleaned platinum MDEA 5037 transducers to FcCOOH likewise showed an expected linear response, (Ch1 $R^2 = 0.9909$; Ch2 $R^2 = 0.9859$), **Figure 5**. In this case, the performance of Channel 2 (sensitivity = 298.2 nA/mM) and Chanel 1 (sensitivity = 305.1 nA/mM) are quite similar. We have concluded that for very facile electrochemical reactions (large heterogeneous rate constants) where the device under test behaves essentially as a resistor, the potentiostat appears quite capable of delivering desired channel reproducibility. However, for real electrode reactions that may be quasi-reversible and have low heterogeneous rate constants, the potentiostat is unable to respond uniformly between the channels. We propose to take up these issues with our collaborating manufacturer, Pinnacle Technologies, Inc.

Polypyrrole was electrodeposited to the working electrodes of MDEA 5037-Pt transducers and overoxidized using cyclic voltammetry in PBS (0.0-750 mV, 20 cycles at 100 mV/s). A dose-response interrogation in aqueous 0.5 mM FcCOOH was then performed. The dose-response of the transducer modified with overoxidized polypyrrole (OO-PPy) had a significantly decreased current sensitivity (Ch1 sensitivity = -0.299 nA/mM, R^2 = 0.881; Ch2 Sensitivity = 0.781 nA/mM, R^2 = 0.995) when compared to bare electrode (Ch 1 sensitivity = 305.1 nA/mM, R^2 = 0.991 and Ch 2 sensitivity = 298.2 nA/mM, R^2 = 0.986). The drastic loss of current response is likely due to anion exclusion of the negatively charged $FCCOO^-$ as the anionic form of FcCOOH (pKa = 4.2) [16] is electrostatically repulsed by dopant anions within the positively charged OO-PPy film. This represents effective interference shielding (a rejection ratio of 390:1) of anionic interferents, such as ascorbic acid, that plague amperometric biosensors. These results can be confirmed through amperometric response of OO-PPy transducers to FcCOOH as a function of buffer pH that result in shifts of anionic to neutral forms of FcCOOH.

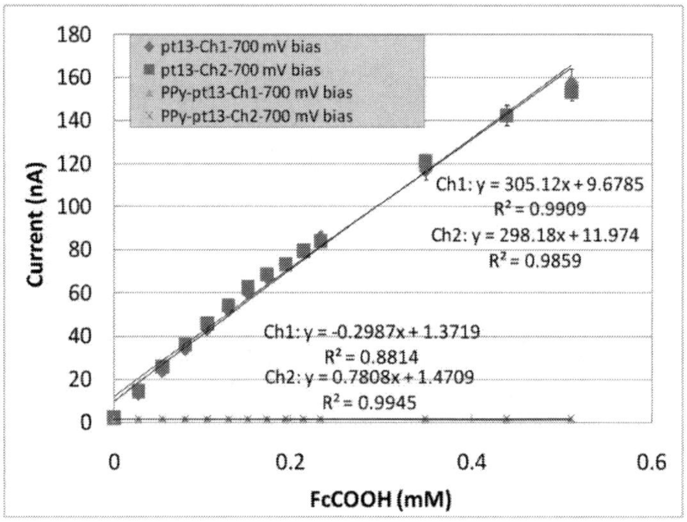

Figure 5. The effects of overoxidized polypyrrole applied to working electrodes of a MDEA 5037 transducer on the current response to FcCOOH in PBS buffer (pH = 7.4), Ch1: WE-A vs CE-A, Ch2: WE-B vs CE-B.

In vitro testing of the MDEA 5037 biosensor system have shown a linear response to glucose and lactate concentrations of 0.1-13.0 mM and 1.0-7.0 mM respectively, **Figure 6 and Table 2.** Stress induced hyperglycemia in trauma victims can lead to arterial blood glucose levels of 6-10 mmol/L [17]. Our current biosensor design is adequate for monitoring the clinically relevant range of glucose during hemorrhage. However, blood lactate levels in trauma victims have been observed to increase to 13-14 mM [18]. It is possible to accurately monitor the initial rate of change interstitial lactate, but there will be inaccuracies in measuring extreme levels of lactate that trauma can induce. Moderate and high risk trauma patients can be defined as having blood lactate levels of 2.3-5.0 mM and >5 mM respectively [18]. Therefore the current linear dynamic range for lactate is adequate for potential use in management and risk categorization of trauma patients.

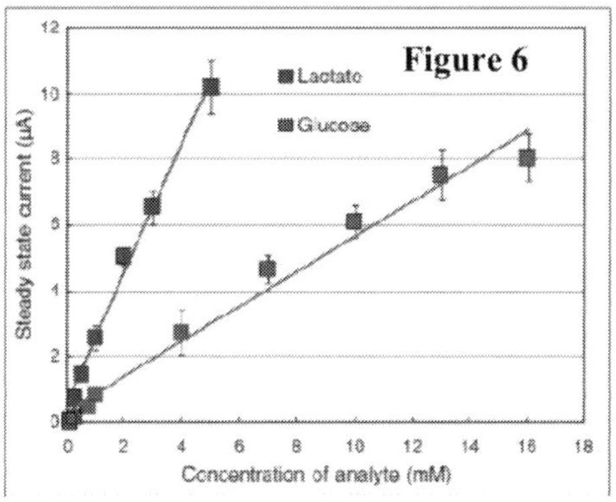

Table 2

	Glucose	Lactate
Linear dynamic range	0.1 - 13.0 mM	1.0 - 7.0 mM
Sensitivity	0.59 µA/mM	2.11 µA/mM
Detection limit	0.2 mM	0.4 mM
Response time (t95)	50 s	35-40 s

Figure 6. (A) *In vitro* characterization of MDEA biosensors for both glucose and lactate [19].

Table 2. The bioanalytical performance of the bioactive electroconductive hydrogel for detection of glucose and lactate [12].

Conclusions

A dual responsive amperometric electrochemical transducer that uses the microdisc working electrode array format has been designed, fabricated, packaged and now tested *in vitro*. This device is a prototype scheduled for use in an AUP approved (AUP 2011-013) *in vivo* study of hemorrhage in Sprague Dawley rats. The electrochemical transducer was shown to be highly reproducible when analyzing hydrogen peroxide, the product of immobilized oxidoreductase (glucose oxidase and lactate oxidase) enzyme activity. The biotransducer was also shown to be highly reproducible with analyzing ferrocene monocarboxylic acid, (FcCOOH), a facile redox robe. When modified with over-oxidized polypyrrole (OO-PPy), the transducer was shown to reject access of the negatively charged FcCOOH, analogous to interference shielding of ascorbic acid.

Acknowledgements

The authors acknowledge support from the US Department of Defense (DoDPRMRP) grant PR023081/**DAMD17-03-1-0172** and the Consortium of the Clemson University Center for Bioelectronics, Biosensors and Biochips (C3B).

References

1. John W. Severinghaus and P.B. Astrup, *History of blood gas analysis. IV. Leland Clark's oxygen electrode.* Journal of Clinical Monitoring, 1986. **2**(2): p. 125–139.
2. Wang, J., *Electrochemical Glucose Biosensors.* Chemical Reviews, 2008. **108**(2): p. 814-825.
3. Guiseppi-Elie, A., *An implantable biochip to influence patient outcomes following trauma-induced hemorrhage.* J Anal and Bioanal Chem, 2010. **339**(1): p. 403-419.
4. Ward, K.R., R.R. Ivatury, and R.W. Barbee, *Endpoints of resuscitation for the victim of trauma.* J. Intensive Care Med., 2001. **16**: p. 55–75.

5. Barbee, R.W., P.S. Reynolds, and K. Ward, *Assessing shock resuscitation strategies by oxygen debt repayment.* Shock, 2010. **33**(2): p. 113-122.
6. Rahman, A.R., G. Justin, and A. Guiseppi-Elie, *Towards an implantable biochip for glucose and lactate monitoring using micro-disc electrode arrays (MDEAs).* Biomed Microdevices, 2009. **11**(1): p. 75-85.
7. Guiseppi-Elie, A., S. Brahim, and A. Wilson, *Biosensors based on electrically conducting polymers*, in *Handbook of Conducting Polymers: Conjugated Polymer Processing and Applications, 3rd edition*, T. Skotheim and J.R. Reynolds, Editors. 2006, Marcel Dekker: New York. p. 435-479.
8. Sharon A. Rothwell, Sarah J. Killoran, and R.D. O'Neill, *Enzyme Immobilization Strategies and Electropolymerization Conditions to Control Sensitivity and Selectivity Parameters of a Polymer-Enzyme Composite Glucose Biosensor.* Sensors, 2010. **10**: p. 6439-6462.
9. Brahim, S., D. Narinesingh, and A. Guiseppi-Elie, *Bio-smart materials: Kinetics of immobilized enzymes in p(HEMA)/p(pyrrole) hydrogels in amperometric biosensors.* Macromolecular Symposia, 2002. **186**: p. 63-73.
10. Duffitt, G.L. and P.G. Pickup, *Permselectivity of polypyrrole in acetonitrile.* J. Phys. Chem., 1991. **95**(24): p. 9634-9635.
11. Brahim, S., D. Narinesingh, and Anthony Guiseppi-Elie, *Interferent suppression using a novel polypyrrole-containing hydrogel in amperometric enzyme biosensors.* Electroanalysis, 2002. **14**(9): p. 627-633.
12. Kotanen, C., O. Karunwi, and A. Guisepp-Elie, *Physiological Status monitoring for glucose and lactate during the onset of hemorrhagic shock.* American Society of Gravitational and Space Biology Bulletin 2010. **23**(2): p. 55-63.
13. Justin, G., A.R. Abdur Rahman, and A. Guisepp-Elie, *Bioactive hydrogel layers on microdisc electrode arrays: Cyclic voltammetry experiments and simulations.* Electroanalysis, 2009. **21**(10): p. 1125-1134.
14. Justin, G. and A. Guiseppi-Elie, *Characterization of Electroconductive Blends of p(HEMA-co-PEGMA-co-HMMA-co-SPMA) Hydrogels and p(Py-co-PyBA).* Biomacromolecules, 2009. **10**(9): p. 2539-2549.
15. Rahman, A.R.A., et al., *Fabrication and Packaging of a Dual Sensing Electrochemical Biotransducer for Glucose and Lactate Useful in Intramuscular Physiologic Status Monitoring.* IEEE Sensors Journal 2009. **9**(12): p. 1856-1863
16. Skeika, T., et al., *Preparation and Electrochemical Characterization of a Carbon Ceramic Electrode Modified with Ferrocenecarboxylic Acid.* Sensors, 2011. **11**(2): p. 1361-1374.
17. McNAMARA, J.J., et al., *Hyperglycemic Response To Trauma in Combat Casualties.* The Journal of Trauma, 1971. **11**(4): p. 337-339.
18. Karon, B.S., et al., *Comparison of Lactate Values Between Point-of-Care and Central Laboratory Analyzers.* American Journal of Clinical Pathology, 2007. **128**(1): p. 168-171.
19. Guiseppi-Elie, A., *An implantable biochip to influence patient outcomes following trauma-induced hemorrhage.* Analytical and Bioanalytical Chemistry, 2011. **399**(1): p. 403-419.

(741d) On-Demand, Targeted Drug Delivery Using Magnetic Thermosensitive Nanocomposites

Scott B. Campbell, Elysia Jellema and Todd R. Hoare, Chemical Engineering, McMaster University, Hamilton, ON, Canada

On-demand,targeted drug delivery using magnetic thermosensitive nanocomposites

Scott Campbell, Elysia Jellema, and Todd Hoare

Department of Chemical Engineering, McMasterUniversity, Hamilton, ON, Canada

INTRODUCTION: Controlof both the site and rate of drug release is a persistent challenge inmedicine. Several polymer-based delivery systems have been able to achieveconstant or sustained drug release for up to several months, but devicesdelivering drugs via pulsatile release have proven difficult to achieve.[1]Specifically, in cases where the disease progression is uncertain (e.g.chemotherapy or chronic pain management) or pulsatile, on-off release of drugis clinically required (e.g. insulin delivery), current treatments are largelylimited to external pumps that can be up or down-regulated or multiple,sequential injections to provide bolus drug amounts. The development of adevice that can be externally and non-invasively triggered to deliver high/lowor on/off doses of a drug locally inside the body would address this challenge.Such a delivery system would have the potential to improve drug safety, reducethe risk of systemic side effects, reduce the cost and prolong the effectiveduration of action of a given drug delivery vehicle. In order to effectivelyprovide for "on-demand" or pulsatile delivery, a drug delivery vehicle mustcontain materials with two key properties: a switching material that canmodulate drug diffusion *in vivo* bychanging its bulk size, pore size, or affinity for a target drug upon astimulus and a triggering material that can modulate an external trigger (e.g.light or an electromagnetic field) into a stimulus recognized by the triggeringmaterial. Recently, a composite membrane consisting of ethyl cellulose (thematrix), thermoresponsive microgels based onpoly(N-isopropylacrylamide) (PNIPAM, the switching material), and magnetitenanoparticles (the triggering material) was reported that demonstrated effective, on-off pulsatile drug delivery upon the application of an oscillating magneticfield.[2] While this membrane was highly effective for pulsatilerelease, the macroscopic size of the membrane-based devices used requiressurgical implantation for their effective use. The development of injectablematerials that can provide similar release profiles would be highly beneficialto expand the potential applications and patient convenience of such a device. This research focuses on the development ofdrug-loaded, injectable, "smart" nanocomposites that can both repeatedlydeliver a drug "on-demand" and have site-specific functionality.

EXPERIMENTAL: Weare have entrapped microgels and magnetite nanoparticles inside an *in situ*-injectable hydrogel prepared bymixing hydrazide-functionalized PNIPAM and aldehyde-functionalized dextran. Themicrogels consist of copolymers of poly(N-isopropylacrylamide) (PNIPAM) andpoly(N-isopropylmethacrylamide) (PNIPMAM); suchmicrogels are thermosensitive such that they reversibly decrease in size whenthe temperature of their environment exceeds their volume phase transitiontemperature (VPTT), which is designed to be just above physiologicaltemperature (~40°C).[3] The magnetite nanoparticles can generate heatwhen placed in an oscillating magnetic field (OMF) via hysteresis losses andprovide targeting functionality using an external permanent magnet.[2]When microgels and magnetite nanoparticles are combined in the samenanocomposite, heating of the nanoparticles induces a phase transition in themicrogels, creating free volume in the microgel-templated pores and thusincreased drug release.

Site-specificityis achieved by injecting the composite at the desired site, where it quicklygels *in vivo*. The thermosensitive nature of the both themicrogels and the surrounding hydrogel can be adjusted, along with numerousother parameters (ferrofluid and microgel concentration, functionalization andconcentration of hydrogel precursor polymers, etc.) to attempt to control therate of drug release and the on-demand control of drug release under thepresence of an OMF.

RESULTS: Awide range microgel-magnetite-hydrogel nanocomposites, with variations in theratio and concentration of the two hydrogel polymeric components and microgel content,have been characterized for their swelling, degradation and drug releasecharacteristics at 37°C and 43°C, as well as their drug release under thepresence of an OMF. Using a non-thermosensitive and non-swelling bulk phase toencapsulate the microgels and the magnetite nanoparticles, a threefold increasein drug release rate (assayed using sodium fluorescein as a model drug) can beachieved in the presence of an oscillating magnetic field, illustrating theutility of this approach as an injectable and externally controllable drugdelivery vehicle. The composites havehigh mechanical strength and can be programmed to degrade either at controlledrates via hydrolysis of the bulk phase or catastrophically via magneticablation. MTT assays and live/deadassays will also be discussed that assessed cell biocompatibility.

CONCLUSIONS: Thecombination of thermosensitive polymers with magnetite nanoparticles is apowerful tool for in the fabrication of composites for drug delivery. Theinjectable and externally-triggerable drug deliverynanocomposite offers significant advantages over current drug deliverytechniques in terms of facilitating triggered changes in drug release using apatient-friendly and non-invasive triggering technology.

REFERENCES: (1)Saltzman, W. M. (2001). *Drug Delivery:Engineering Principles for Drug Therapy*. New York: Oxford University Press;(2) Hoare, T.; Santamaria, J.; Goya, G.F.; Irusta, S.; Lin, D.; Lau, S.; Padera,R.; Langer, R.; Kohane D.S. (2009). A Magnetically TriggeredComposite Membrane for On-Demand Drug Delivery. *Nano Letters, 9(10)*, 3651-3657; (3) Pelton, R. (2000). Temperature-sensitive aqueous microgels. *Advances in Colloid and InterfaceScience*, *85*, 1-33.

ACKNOWLEDGEMENTS: Thisresearch is funded by the Natural Sciences and Engineering Research Council ofCanada (NSERC) and the J.P. Bickell Foundation(Medical Research Grant).

COMPOSITE IRON OXIDE/SILICA MICROPARTICLES FOR DRUG DELIVERY SYSTEMS

Pavel Kovacik, ICT Prague, Czech Republic

Zuzana Kremlackova, ICT Prague, Czech Republic

Frantisek Stepanek, ICT Prague, Czech Republic

Abstract

Our aim is to fabricate hollow composite iron oxide/silica microparticles in the size range of micrometers. Microparticles have porous silica (SiO_2) shells in which nanoparticles of superparamagnetic iron oxide (Fe_2O_3/Fe_3O_4) are dispersed. Such prepared microparticles can be applied for targeted delivery systems for field of applications as cleaning substances. We prepared the composite microparticles via soft shell templating route. Tetraethyl orthosilicate and octylamine droplets are used as silica source and liquid template, respectively. This method has good reproducibility it is fast and very simple. The resulting particles were studied and characterized in size, shape and ability to heat up in the presence of an alternating magnetic field radiofrequency magnetic field (400 kHz). We have shown that particle size can be systematically varied in the range of 20-100 μm. The hollow core of the microparticles was loaded with model substance (vitamin B12) and release kinetic was measured by time-dependent UV/VIS spectrophotometery. Consequently, the sorption capacity and efficiency of encapsulation of prepared microparticles was calculated.

1. Introduction

The synthesis of hollow silica spheres has received considerable attention in recent years because of their proper attributes such as high surface area, low density, surface permeability, biocompatibility and good mechanical stability [1]. The potential applications include areas as pharmaceuticals [2], biomedical [3], chemical catalysis [4], cosmetics [5] and others [6]. Lots of efforts have been devoted to prepare hollow silica nano-/micro- spheres by a variety of chemical methods [7]. Structures with hollow cores are commonly prepared by spray drying techniques (nanospheres) or templating procedures (nano-/micro- spheres). In templating methods two different templates is possible to use, a soft template (liquid droplets) or a hard one (monodisperzed particles). But the procedure is similar to both of them and includes following steps: i) coating the template's surfaces by layers of the desired materials or its precursors (e.g. by precipitation), ii) selective removal of the templates (e.g. by calcinations).

As silica source is often used tetraethyl orthosilicate (TEOS), tetramethyl orthosilicate (TMOS) or sodium silicate.

In contrast with preparation of hollow silica spheres only a few research groups have reported the preparation of iron oxide/silica composite hollow spheres up to now. Two main approaches have been used. The first approach is the deposition of silica on the magnetic hollow spheres. According this approach different methods were reported including layer-by-layer techniques [8], coprecipitation [9] or surface reaction [10]. Second approach is the opposite process, the deposition of iron oxide materials on silica hollow spheres. Abe et al. [11] used ferrite plating to coat the surface of silica microshpheres with magnetite. Yuan et al. [12] prepared superparamagnetic hollow silica/magnetic composite spheres using sulfonic acid-functionalized hollow silica spheres as templates.

Herein, we report a fast and simple synthesis of composite iron oxide/silica microspheres by using liquid templates suitable for targeting delivery systems. Resulting particles are able to heat up in the alternating magnetic field. This property can be used as a tool for controlling a release kinetic of trapped chemical material inside hollow cores.

2. Experimental section

The steps used herein combine two previously published methods for preparation magnetite nanoparticles [13] and hollow silica spheres [14].

2.1 Materials

Tetraethyl orthosilicate (TEOS, 98%), n-octylamine (99%), oleic acid ($C_{18}H_{34}O_2$, 65-88%), acetone (99%), ammonium hydroxide solution (NH_4OH, 25%), iron(III) chloride hexahydrate and Iron(II) chloride tetrahydrate were purchased from Aldrich. Nitric acid (p. a., 65%) and ethanol were purchased from Penta. All chemicals were used as received. Deionized water was used for all reactions and treatment processes (Aqual25, 0,07µS/cm).

2.2 Synthesis of Fe_3O_4 nanoparticles

The iron oxide nanoparticles were synthesized by the co-precipitation method and stabilized by oleic acid to make them hydrophobic. The method involves co-precipitation from Fe^{2+} and Fe^{3+} aqueous salt solutions by addition of a base. The chemical reaction may be written as follows:

$$(1)\ Fe^{2+} + 2\ Fe^{3+} + 8\ OH^- \rightarrow Fe_3O_4 + 4\ H_2O$$

In a typical synthesis 0.25 g of $FeCl_3.6H_2O$ and 0.125 g of $FeCl_2.4H_2O$ was dissolved in 40 ml of water in a three-neck flask equipped with a reflux condenser and heated to 80°C for 30 min. Then 0.2 ml of oleic acid and 5 ml of acetone was added. After 20 min 5 ml of NH_4OH was added and the mixture was stirred at 500 rpm for the next 10 min. Further 0.2 ml of oleic acid was added (five times in five minutes interval). Finally the mixture was stirred at 500 rpm for 30 min and cooled down to ambient temperature. Dry iron oxide nanoparticles were obtained by freeze-drying.

2.3 Synthesis of iron oxide/silica microparticles

The hollow composite iron oxide/silica microparticles with a porous shell are prepared via a soft templating route. We used tetraethyl orthosilicate (TEOS) as a silica source and octylamine (OA) as a liquid template. Proposed scheme of the method is shown in the Figure 1.

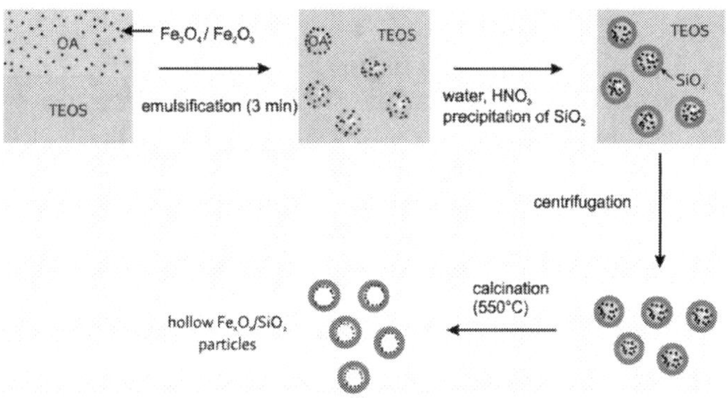

Figure 1: Scheme of preparation of hollow composite iron oxide/silica microparticles with porous shell via soft templating route.

In a typical synthesis various amount of prepared iron oxide nanoparticles (stabilized by oleic acid) were dispersed in octylamine (10 ml) folowing by sonification for 5 min. In the next step TEOS (10 ml) was added into this solution and the mixture was stirred (800 rpm) for 3 min at room temperature followed by a quick addition of water (100 ml), in which nitric acid was added (0.2 ml) to promote the precipitation of silica. The mass ratio of iron oxide nanoparticles to silica was 0%, 0.5%, 1%, 2%, 3% and 5% (in weight 0.000, 0.013, 0.027, 0.054, 0.081 and 0.135 mg of iron oxide nanoparticles). The reaction was kept for 5 min and after that the resultant microparticles were collected by centrifugation (5 min, 6000 rpm), washed with acetone, again centrifuged and then air-dried. The molar ratio of $TEOS:OA:H_2O:HNO_3$ was 1:1.36:113:0.05. The dried sample was calcined from room temperature to 600 °C with a heating rate of 1 °C /min and kept at 600 °C for 9 h, followed by cooling down in the oven to ambient temperature.

2.4 Characterisation

Jeol JCM-5700 device was used for scanning electron microscopy (SEM) imaging. The samples were covered with a 5 nm thin film of gold. To observe the hollow structure of silica microparticles, mortar with pestle was used to break them apart.

The particle size and the particle size distribution were evaluated by the laser diffraction particle size analyzer (Horiba Partica LA 950/V2). The sample size was measured in flow cell and Fraunhofer diffraction was used for the calculation.

Release kinetic was measured by using time-dependent UV/VIS spectrophotometry (Specord 205 BU). Fabricated composite microparticles with different mass fraction of iron oxide nanoparticles were used as drug carrier to load vitamin B12. The loading process was i) calcined microparticles (0.1 g) were steamed by hot water steam, ii) they were quickly inserted into cold water solution of vitamin B12 (1 mg/1 ml, 4°C, 3 ml), iii) and kept in the solution for one week by room temperature. After that the particles were centrifuged, dried by nitrogen gas and inserted into special cuvette. The experiment set up is shown in Figure 2. In typical experiment 3 mg of composite nanoparticles were attached on the tape which was inserted into the cuvette filled with 2.5 ml of deionized water. The cuvette was placed in temperature control holder and stirred by magnetic stirrer. The release kinetics was observed at 361 nm and the release process took around 50 hours.

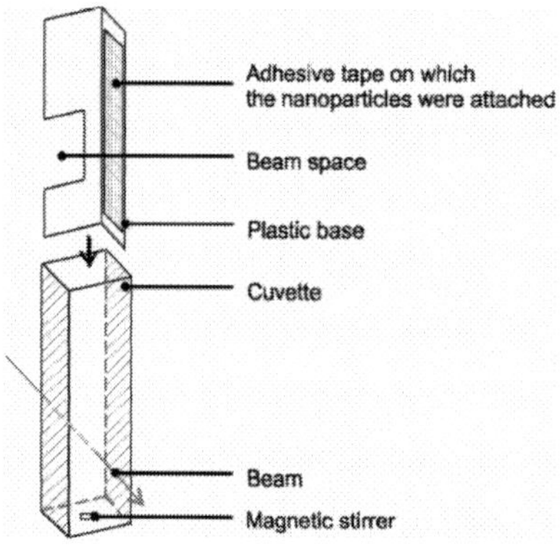

Figure 2: Experiment set up for measurement of release kinetics of model substance from prepared composite microparticles.

Prepared composite microparticles were inserted into an external radiofrequency magnetic field and the temperature rise with time was recorded. The field was generated by CEIA PowerCube 32/400 working at 400 kHz with variable output at 0.25-2.5 kW. A two-loop water cooled coil with diameter of 20 mm was used. Apart from the maximum temperature which can be reached, the specific absorption rate number (SAR number) was evaluated. This number quantifies the power absorbed per mass of nanoparticles [W/g]. In typical experiment 0.4 g of the sample was placed inside the coil and power output of the CEIA PowerCube and it was set to 50% power output. The SAR of the prepared composite microparticles was calculated according to the Eq.2

$$(2)\ SAR = \frac{\bar{c}_p}{w_{Fe_3O_4}} \cdot \frac{\Delta T}{\Delta t}$$

Where c_p [J/g/K] is average heat capacity of measured sample, w_{Fe3O4} [-] is mass fraction of iron oxide nanoparticles dispersed in silica microparticles and $\Delta T/\Delta t$ is the slope of temperature during the linear growth in first three minutes [K/s].

3. Results and discussion

3.1 Particles size and structure

The fabricated microparticles possess a good spherical shape with narrow size distribution (Fig 3). SEM images of broken particles proved the hollow core (Fig. 4-A, 4-B). In the figure 2-A is silica particle without iron oxide nanoparticles and in the figure 2-B is silica particle with iron oxide nanoparticles (1% mass fraction). On the inner side one can see clusters of iron oxide nanoparticles.

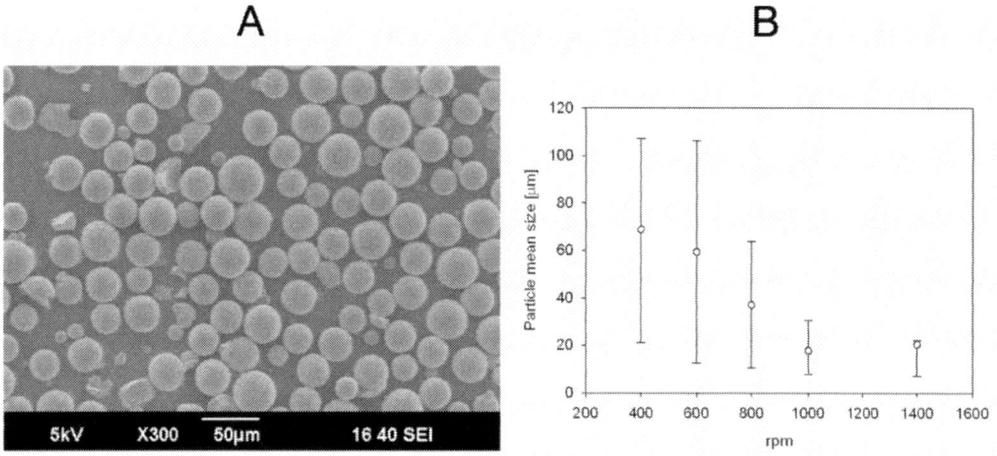

Figure 3: A) Silica microparticles (800 rpm). B) The mean size of prepared microparticles depends on velocity of magnetite stirrer. The bars represent 90% and 10% quantiles.

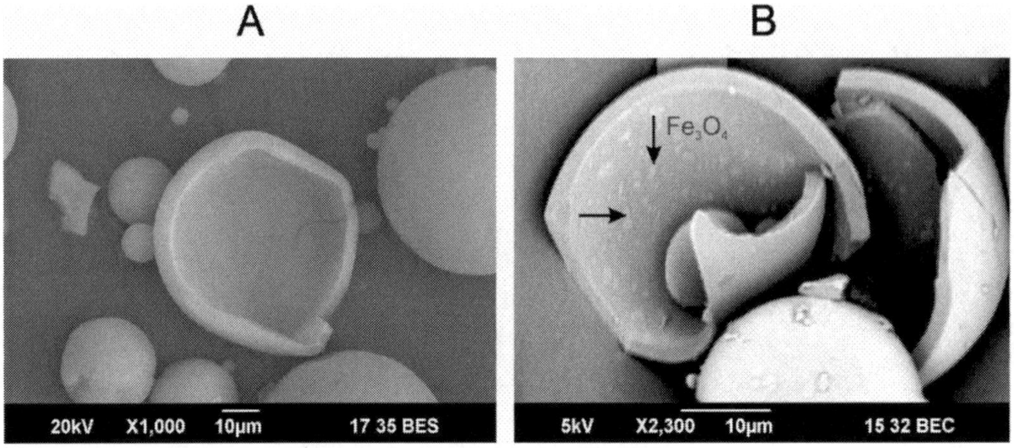

Figure 4: A) SEM image of broken particles without iron oxide nanoparticles. B) SEM image of broken particles with iron oxide nanoparticles (1% mass fraction).

The shell thickness was found by image analysis of broken particles using the NIS elements software (Tab. 1).

Table 1: The shell thickness of prepared composite iron oxide/silica microparticles.

w (iron oxide) [%]	Shell Thickness [µm]	Standard Deviance [µm]
0	**5.2**	**2.3**
0.5	**6.3**	**2.3**
1	**6.4**	**2.9**
2	**4.9**	**2.0**
3	**3.0**	**1.4**

3.2 Heating in magnetic field

The temperature rise with time of various prepared microparticles is shown in the Figure 6. Microparticles with higher content of iron oxide nanoparticles are heated up faster and higher maximum temperature is reached. This proves the proposed mechanisms of method: the iron oxide nanoparticles prefer to stay in the octylamine phase during the whole preparation and after calcinations process they are present inside the hollow core. The linear growth in first three minutes is highlighted in the Figure 6 – B. From the Figure 6-B was calculated SAR number of prepared composite microparticles (Table 2).

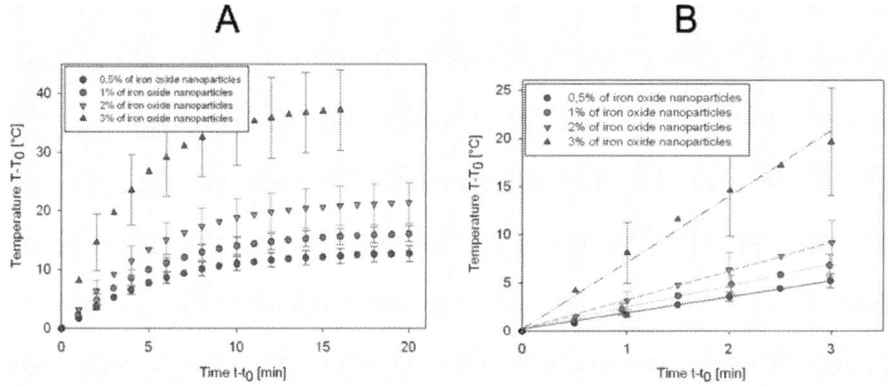

Figure 5: A) Heating up rate of prepared hollow composite iron oxide/silica microparticles with 0.5%, 1%, 2% and 3 % of iron oxide nanoparticles. B) Linear growth during the first minute of heating.

Table 2: SAR number of prepared composite microparticles.

	Calcined particles		Non calcined particles	
w (Fe_3O_4)	Slope $\Delta T/\Delta t$ [K/s]	SAR [W/g]	Slope $\Delta T/\Delta t$ [K/s]	SAR [W/g]
0.005	0,0337	2.49	0.004	3.54
0.01	0,0707	1.62	0.08	3.30
0.02	0,0604	1.40	0.011	2.48
0.03	0,0621	2.23	0.011	1.71

3.3 Release kinetics

The release profiles of vitamin B12 from prepared microparticles are shown in Figure 7 (A - for temperature 25°C, B – for temperature 40°C). The release was much slower if the temperature was 25°C than 40°C and also the presence of iron oxide nanoparticles slowed down the release kinetic (Table 3). However it was not found any trend which is dependent on mass fraction of iron oxide nanoparticles. We assume that the iron oxide nanoparticles are not present inside the silica shell but they are only inside the hollow core. And after calcination process they are sticked on the inner side of silica shell. Due this they block a part of pores in silica shell and the release kinetic is slowed down.

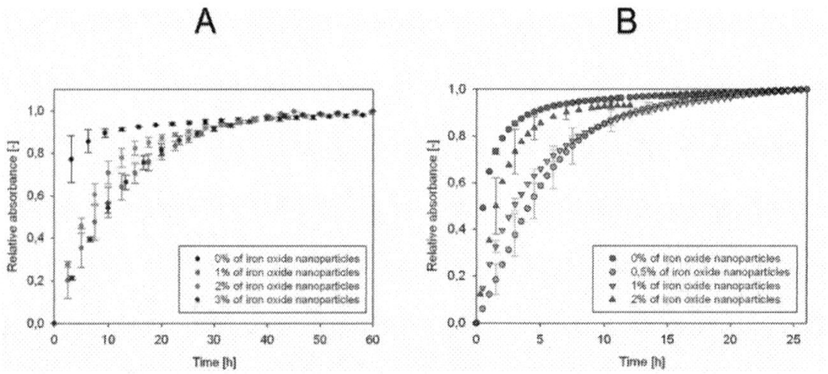

Figure 6: Vitamin B12 profiles for loaded microparticles: A) Released by temperature 25°C, B) Released by temperature 40°C.

Table 3: Times necessary to release 50% and 90% of trapped vitamine B12.

w (Fe_3O_4)	t50 [h] 25°C	t50 [h] 40°C	t90 [h] 25°C	t90 [h] 40°C
0.000	1.3	0.5	9.8	4.5
0.005	-	4.1	-	12.2
0.01	8.1	3.0	28.0	12.8
0.02	5.7	1.5	24.8	7.0
0.03	8.9	-	27.6	-

4. Conclusion

Hollow composite iron oxide/silica microparticles with a porous shell have been synthesized by a soft templating route. The method is fast, simple and reproducible. Properties of prepared composite microparticles were characterized via different methods. Hollow core was confirmed by SEM and confocal microscopy. Particle size distribution and mean size distribution were measured by laser scattering method. The mean size of resulting microparticles can be controlled through the process of mixing and it is possible to fabricate microparticles in the size range 20-70 μm with narrow size distribution. The composite particles are able to heat up in presence of an alternating magnetite field with temperature increase from 3 to 9 °C/min. It was shown it can be utilized for affecting the release kinetics of encapsulated active substance (vitamin B12) from the particles.

Acknowledgments

This work has been supported by the European Research Council (grant no. 200580-Chobotix), Czech Science Foundation (104/08/H055), and Grant of the Specific University Research (MSMT no. 21/2011).

References

1. Liu, J., F. Liu, K. Gao, J. Wu, and D. Xue, *Recent developments in the chemical synthesis of inorganic porous capsules.* Journal of Materials Chemistry, 2009. **19**(34): p. 6073-6084.
2. Mohamed, F. and C.F. van der Walle, *Engineering biodegradable polyester particles with specific drug targeting and drug release properties.* Journal of Pharmaceutical Sciences, 2008. **97**(1): p. 71-87.
3. Réthoré, G. and A. Pandit, *Use of Templates to Fabricate Nanoscale Spherical Structures for Defined Architectural Control.* Small. **6**(4): p. 488-498.
4. Lee, J., J.C. Park, and H. Song, *A Nanoreactor Framework of a Au@SiO2 Yolk/Shell Structure for Catalytic Reduction of p-Nitrophenol.* Advanced Materials, 2008. **20**(8): p. 1523-1528.
5. Jones, S.R., B.D. Grey, K.K. Mistry, and P.G. Wildgust, *The development of colour-encapsulated microspheres for novel colour cosmetics.* 2009. p. 325-333.
6. Ma, H., et al., *Nest-like Silicon Nanospheres for High-Capacity Lithium Storage.* Advanced Materials, 2007. **19**(22): p. 4067-4070.
7. Guerrero-Martínez, A., J. Pérez-Juste, and L.M. Liz-Marzán, *Recent Progress on Silica Coating of Nanoparticles and Related Nanomaterials.* Advanced Materials, 2010. **22**(11): p. 1182-1195.
8. Sadasivan, S. and G.B. Sukhorukov, *Fabrication of hollow multifunctional spheres containing MCM-41 nanoparticles and magnetite nanoparticles using layer-by-layer method.* Journal of Colloid and Interface Science, 2006. **304**(2): p. 437-441.
9. Wu, W., et al., *Synthesis of magnetic hollow silica using polystyrene bead as a template.* Journal of Magnetism and Magnetic Materials, 2007. **311**(2): p. 578-582.
10. Shen, S.-L., W. Wu, K. Guo, H. Meng, and J.-F. Chen, *A novel process to synthesize magnetic hollow silica microspheres.* Colloids and Surfaces A: Physicochemical and Engineering Aspects, 2007. **311**(1-3): p. 99-105.
11. Abe, M., *Ferrite plating: a chemical method preparing oxide magnetic films at 24-100°C, and its applications.* Electrochimica Acta, 2000. **45**(20): p. 3337-3343.
12. Yuan, J., X. Zhang, and H. Qian, *A novel approach to fabrication of superparamagnetite hollow silica/magnetic composite spheres.* Journal of Magnetism and Magnetic Materials, 2010. **322**(15): p. 2172-2176.
13. Maity, D. and D.C. Agrawal, *Synthesis of iron oxide nanoparticles under oxidizing environment and their stabilization in aqueous and non-aqueous media.* Journal of Magnetism and Magnetic Materials, 2007. **308**(1): p. 46-55.
14. Cheng, X., et al., *Fast fabrication of hollow silica spheres with thermally stable nanoporous shells.* Microporous and Mesoporous Materials, 2007. **98**(1-3): p. 41-46.

(741f) Gold Nanoparticle Layer by Layer Assembled Nanostructures for Neural Interface

Huanan Zhang and Nicholas Kotov, Chemical Engineering, University of Michigan, Ann Arbor, MI

Electrical stimulation is a technique for diagnosis and therapy of many neurological diseases. We often utilize a neural prosthetic device (NPD) to deliver the electrical stimulation.[1, 2] The common clinical NPDs have several problems, such as low charge storage capacity, large size , chronic inflammatory response, and mismatch of mechanical properties with brain tissue.[3] In order to improve the current design of NPDs, we need to design nanocomposite materials that exhibit superior properties and functionalities than current materials. Layer by layer (LBL) assembled gold nanoparticle (Au NP) multilayers are composite materials made from gold nanoparticles and polyelectrolyte polymers. Nanocomposite materials made from combining nanoparticles and LBL assembly technique have superior electrochemical properties, long term chemical stability,and tunable mechanical properties. These remarkable properties are very attractive for chronically implanted NPDs.[4] The versatility of the LBL assembly enables us to easily modify current clinical NPDs to create multifunctional devices. As a coating material, LBL assembled Au NP nanocomposite can potentially improve the performance of implantable NPD by increasing its charge injection capacity and decrease electrical impedance. In addition, we could also create freestanding LBL assembled Au NP film to directly fabricate flexible implantable electrodes. Reference: 1. Merrill, D.R., M. Bikson, and J.G. Jefferys, Electrical stimulation of excitable tissue: design of efficacious and safe protocols. J Neurosci Methods, 2005. 141(2): p. 171-98. 2. Benabid, A.L., Deep brain stimulation for Parkinson's disease. Curr Opin Neurobiol, 2003. 13(6): p. 696-706. 3. Kotov, N.A., et al., Nanomaterials for Neural Interfaces. Advanced Materials, 2009. 21 (40): p. 3970-4004. 4. Cogan, S.F., Neural stimulation and recording electrodes. Annu Rev Biomed Eng, 2008. 10: p. 275-309.

See more of this Session: Hybrid Biomaterials
See more of this Group/Topical: Materials Engineering and Sciences Division

(741g) Nanoparticle Netpoints for Shape Memory Polymers

Praveen Agarwal, Chemical and Biomolecular Engineering, Cornell University, Ithaca, NY and Lynden A. Archer, Chemical & Biomolecular Engineering, Cornell University, Ithaca, NY

Shape memory polymers (SMP's) are a class of stimuli responsive materials which have the capability to remember a permanent shape; can be deformed to impart a desired temporary shape; and recover their permanent shape when acted upon by a stimulus, e.g. heat, light or magnetic field. SMPs have received a lot of attention recently because of their potential use in minimally invasive surgery and as biomedical implants. We have created a new family of hybrid, inorganic-organic SMPs employing inorganic nanoparticles as netpoints where each netpoint is a junction for hundreds of polymer chains. We find that this network design leads to dramatic increase in the elastic modulus and yet sharp transition temperature and excellent shape memory properties of SMPs. Moreover this platform is capable of synthesizing SMPs with highly tunable mechanical properties and transition temperature.

See more of this Session: Hybrid Biomaterials
See more of this Group/Topical: Materials Engineering and Sciences Division

(745a) Vapor-Phase Transport As a Novel Route to Prepare Hyperbranched Aminosilica Hybrids

Watcharop Chaikittisilp, Wen Li, Stephanie A. Didas, Hyung-Ju Kim and Christopher W. Jones,
School of Chemical & Biomolecular Engineering, Georgia Institute of Technology, Atlanta, GA

Silica-based inorganic–organic mesoporous hybrids are of growing interest due to their wide range of potential applications. Incorporation of organic functionalities into the silica framework has been achieved mostly via three synthetic routes: i) post-synthesis grafting of functional silanes, ii) direct co-condensation during synthesis, and iii) use of bridged organosilanes as precursors. In particular, functionalization of mesoporous silica with amino groups has been intensively studied because these materials can be applied for base-catalyzed reactions, adsorption of heavy metal cations, immobilization of bio-molecules, and CO_2 capture. Among others, our group has developed a simple way to prepare hyperbranched aminosilica (HAS) materials by ring-opening polymerization of aziridine in the presence of porous silicates in organic solvent media [1,2]. The HAS materials showed excellent performance in CO_2 capture from humid gas streams. In the previous report, the results suggested that the mesopores can become blocked or closed by aminopolymers, especially at the high organic loading [2]. In addition, functionalization of structural objects such as supported membranes and monoliths via liquid-phase polymerization is considered to be difficult to carry out on a large scale. In this presentation, we report a novel synthetic route to prepare HAS materials through vapor-phase transport. In this method, the aziridine monomers are transported into the solid in the vapor phase and subsequently polymerization is initiated at the silica surface. The obtained HAS materials had organic contents comparable to materials prepared via the conventional liquid-phase reactions. The amount of aminopolymers formed inside the pores was found to be affected by several synthesis parameters including temperature and reaction time. This novel method provides opportunities for efficient introduction of aminopolymers into other forms of supports including silica membranes and monolith.

References

[1] J. C. Hicks, J. H. Drese, D. J. Fauth, M. L. Gray, G. G. Qi, C. W. Jones, *J. Am. Chem. Soc.* **2008**, *130*, 2902-2903.

[2] J. H. Drese, S. Choi, R. P. Lively, W. J. Koros, D. J. Fauth, M. L. Gray, C. W. Jones, *Adv. Funct. Mater.* **2009**, *19*, 3821-3832.

See more of this Session: Nanostructured Organic / Inorganic Hybrid Materials
See more of this Group/Topical: Materials Engineering and Sciences Division

(745b) Synthesis of Solvent Dispersed Ultrathin Sheets of Boron Nitride

Kabeer Jasuja, Kayum Ayinde, Christina Davis, Myles Ikenberry, Keith L. Hohn and **Vikas Berry**, Chemical Engineering, Kansas State University, Manhattan, KS

In this talk, we demonstrate a chemical method that enables high-throughput synthesis of solvent dispersible ultrathin hexagonal Boron Nitride Nanosheets (UTBNSs) with highest yields reported so far. UTBNSs are the lesser explored structural isomorphs of graphene, and offer immense promise to complement graphene by virtue of their analogous electrical nature and profound chemical stability. We report a synthetic route which utilizes the strongly protonating ability of superacid to chemically functionalize the intermittent *h*-BN layers (by introducing –NH and $–NH_2$ functionalities), which not only enables their layer-by-layer exfoliation (1-5 layer thick sheets at a yield of ~23%), but is also shown to result in an efficient aqueous phase stabilization without the use of polymeric surfactants. The availability of surfactant-less high yield UTBNS dispersions will pave the path for using conventional solution processing techniques to facilely create new BN based materials and hence broaden the scope of their applications. We also show that the amine-functionalities on the UTBNSs can interface effectively with electron-rich gold nanoparticles and negatively charged dye molecules to form inorganic composites. The results presented here will evolve research on atomic-thick BN; and will open avenues for next-generation applications of insulating 2D planar nanocrystals including piezoelectric devices, UV-emitters, transistor gate barriers, polymeric composites with reinforced strength, luminescent nanomaterials, single-atom-tunneling-barrier devices, impermeable membranes, and graphene-UTBNS interfaced devices.

See more of this Session: Nanostructured Organic / Inorganic Hybrid Materials
See more of this Group/Topical: Materials Engineering and Sciences Division

(745c) Hypercrosslinked Siloxane-Organic Hybrid with Hierarchical Pores and Ultrahigh Surface Area Via Self-Condensation of Functionalized Cubic Siloxane Cages

Watcharop Chaikittisilp[1], Ayae Sugawara[2], Atsushi Shimojima[2] and Tatsuya Okubo[2], (1)School of Chemical & Biomolecular Engineering, Georgia Institute of Technology, Atlanta, GA, (2)Department of Chemical System Engineering, The University of Tokyo, Tokyo 113-8656, Japan

Designed synthesis of *a priori* nanoporous materials with tunable structures and properties is a scientific and technological challenge in chemistry and engineering. A bottom-up self-assembly of well-defined molecular building blocks has settled at the center of synthetic chemistry as a strategic route to construct novel nanoporous architectures. The double four-ring (D4R) siloxane cage, also known as cubic polyhedral silsesquioxane (POSS), and its organic derivatives have been utilized as building units to produce porous materials; however, control over molecular homogeneity of the obtained porous materials is still a great challenge. Recently, we succeeded in syntheses of novel microporous hybrids possessing narrow pore size distribution and high surface area via covalent-linking of the D4R cages and rigid aromatic linkers [1,2]. In this presentation, we have focused on the increase in surface area of the siloxane–organic hybrid through the increase in the degree of cross-linking. Specifically, a novel hierarchically-porous siloxane–organic hybrid has been synthesized by the Friedel–Crafts self-condensation of benzyl chloride-terminated D4R siloxane cages as a singular molecular precursor. The resulting material exhibits ultrahigh surface area (BET surface area of 2330 m^2 g^{-1}), large pore volume, and high H_2 uptake capacity.

References

[1] W. Chaikittisilp, A. Sugawara, A. Shimojima, T. Okubo, Chem. Eur. J. 2010, 16, 6006-6014.

[2] W. Chaikittisilp, A. Sugawara, A. Shimojima, T. Okubo, Chem. Mater. 2010, 22, 4841-4843.

See more of this Session: Nanostructured Organic / Inorganic Hybrid Materials
See more of this Group/Topical: Materials Engineering and Sciences Division

(745d) 3D Graphene Inverted Colloidal Crystal Through Vacuum Assisted Assembly

Jian Zhu, Department of Chemical Engineering, University of Michigan, Ann Arbor, MI and Nicholas A. Kotov, Chemical Engineering, University of Michigan, Ann Arbor, MI

Graphene 3D porous structure is a newly emerging morphology of graphene based materials. This type of material can be advantageous in fields such as energy storage, flexible electronics, and new cushion techniques. Most of the reported graphene 3D structures were made by hydrogel-aerogel transition or template methods, which usually result in unconnected pores with a large size distribution or only nanosize monolayer honeycomb structures. Here, we demonstrate that a novel graphene 3D structure can be fabricated via polystyrene colloidal crystal templates through vacuum assisted assembly. The organized porous structures afford novel mechanical and thermal properties. In addition, the connected pores in the material plus conductive nature of graphene can be potentially applied in neural cells culture and stimulation. Finally, this material can be good platform to grow various nanowires or nanoparticles to form active sites for catalysis.

See more of this Session: Nanostructured Organic / Inorganic Hybrid Materials
See more of this Group/Topical: Materials Engineering and Sciences Division

(745e) Nanoreinforced Cellular Foam Materials for Improved Energy Absorption Capacity

Wei Wang[1], Yunfei Qu[2], Jung-Wuk Hong[2], Rigoberto Burgueo[2] and Ilsoon Lee[1], (1)Department of Chemical Engineering and Materials Science, Michigan State University, East Lansing, MI, (2) Department of Civil and Environmental Engineering, Michigan State University, East Lansing, MI

Cellular foam materials have been widely used in aerospace and automotive fields for a long time. One of these applications is energy absorption, primarily due to their unique stress-strain response. Because of this feature, cellular foams can be efficient structural enclosures for vehicles and buildings. A longer and higher plateau stress can further enhance the energy absorption capacity of cellular foams. Efforts have been made to provide reinforced cellular foams with improvement energy absorption capacity, but success has been limited. Meanwhile, nano-scale reinforcement strategies exhibit the potentials to provide superior mechanical properties over conventional through molecular level interactions both at interfaces and material continuum. In our study, open cell foams were reinforced at the nano-scale by stiffer metals to enable an enhanced energy absorption capability. Featured electrodeposition systems for direct current and pulsed current were built up and with which the porous 3D cellular structure was successfully coated. Parameters such as relative density, topology of foam (struts thickness, pore size), processing conditions and their effects on the strain-stress responses will be presented.

See more of this Session: Nanostructured Organic / Inorganic Hybrid Materials
See more of this Group/Topical: Materials Engineering and Sciences Division

(745f) Synthesis, Functionalization and Application of Novel Porous Microclusters for Protein Separation

Alexandros Lamprou, Betrand de Neuville, **Miroslav Soos**, Giuseppe Storti and Massimo Morbidelli, Institute of Chemical and Bioengineering, ETH Zurich, Zurich, Switzerland

Separation of various biomolecules from fermentation broth can be very efficiently done by high performance liquid chromatography (HPLC).

The separation speed and efficiency strongly depends on the active surface and functionalization of the stationary phases used. Although spherical particles packed into a chromatographic column is the simplest solution, they are not always the optimal one. On one hand, chromatographic columns packed with small particles provide sufficient surface for good separation, however due to small size and therefore high pressure drop, their application is restricted to very low flow rates. On the other hand, large particles could ensure low pressure drop and therefore rather high flow rates, up to tens of ml per minute, however their active surface is limited.

Objective of this work is to synthesize highly porous particles to maintain low pressure drop, however with a sufficient number of active sites available for separation. The presented approach is based on an aggregation and breakage process to form porous microclusters combined with "reactive gelation" to increase mechanical rigidity of the prepared microclusters. Atom transfer radical polymerization is used to growth functional brushes on the internal surface of formed microclusters.

The effects of parameters such as size and crosslinking degree of primary particles, as well as shear rate applied during microcluster preparation are investigated. Various analytical techniques, namely light scattering, N_2 sorption, mercury porosimetry, micro- and nano-tomography, size-exclusion chromatography are used to characterize the size and internal structure of the obtained microclusters, which are subsequently applied for protein separation, with their performance evaluated against other commercial stationary phases.

See more of this Session: Nanostructured Organic / Inorganic Hybrid Materials
See more of this Group/Topical: Materials Engineering and Sciences Division

(745g) Preparation and Characterization of Sodium Sulfate/Silicon Dioxide Composite As a Shape-Stabilized Solid-Liquid Phase Change Material

Qiang Guo Sr., State Key Lab of Chemical Engineering, Department of Chemical Engineering, Tsinghua University, Beijing, China and **Tao Wang Sr.**, Chemical Engineering, Tsinghua University, Beijing, China

The composite of sodium sulfate (Na_2SO_4)/silicon dioxide (SiO_2) as a shape-stabilized solid–liquid phase change material (PCM) was prepared by sol-gel procedure using rice husk ash as silicon source. Inside the composite, the in-situ generated Na_2SO_4 acts as the solid-liquid phase change latent heat storage substance. As the supporting material, SiO_2 provides structural strength and prevents the leakage of the melted Na_2SO_4. The structure and composition of the prepared composite PCM were characterized using a scanning electronic microscope (SEM), Fourier transformation infrared spectroscope (FTIR) and x-ray diffraction (XRD). The results showed that the prepared Na_2SO_4/SiO_2 PCM was the hybrid of Na_2SO_4 and SiO_2, in which no new substance was produced during the phase change of Na_2SO_4. The in-situ generated Na_2SO_4 was fixed in the network of porous SiO_2. Shown by differential scanning calorimeter (DSC) analysis, the prepared Na_2SO_4/SiO_2 shape-stabilized PCM has a phase change enthalpy of 75-85 J/g with phase change temperature between 878-883°C. The Na_2SO_4/SiO_2 composite was thermal stable and shape-stabilized during the melting-condensing cycles.

See more of this Session: Nanostructured Organic / Inorganic Hybrid Materials
See more of this Group/Topical: Materials Engineering and Sciences Division

(749a) Polylactic Oligomers Degradation Kinetics

Fabio Codari[1], Stefano Lazzari[1], Miroslav Soos[1], Davide Moscatelli[2] and Massimo Morbidelli[1], (1) Institute of Chemical and Bioengineering, ETH Zurich, Zurich, Switzerland, (2)Chimica, Materiali ed Ingegneria Chimica G. Natta, Politecnico di Milano, Milano, Italy

Polylactic acid (PLA) is a biodegradable and biocompatible, thermoplastic polymer produced from lactic acid (LA) as derived from renewable sources. Degradation studies have been reported in literature for both large scale applications, such as composting processes, and niche applications, such as in drug delivery systems. In the latter case, the degradation kinetics has a direct impact on the rate of drug release and system morphology and, therefore, on the final performance. The aim of the present work is to investigate experimentally the degradation kinetics of low molecular weight PLA as a function of selected key parameters, such as polymer chain length, pH, temperature and chain chiral composition.

PLA oligomers with different chiral composition were produced by polycondensation of DL-LA and L-LA and collected by semipreparative HPLC method. Hydrolysis experiments at different conditions, ranging from pH 1 to pH 7.4 and from 30 up to 120°C, were carried out in batch and characterized by HPLC analysis.[1] Following the time evolution of the concentration of both initial oligomer species and all degradation products it was possible to identify different reaction mechanisms among the different explored conditions. Finally, a comprehensive kinetic scheme has been developed and the corresponding rate coefficients were estimated by minimizing the error between model predictions and experimental data. In particular, it was found that hydrolysis occurs preferentially through chain end-scission and backbiting mechanism at acid and basic pH, respectively, and that the overall degradation constant presents a minimum at the pKa of the carboxylic acid. Moreover, LL-oligomers were found to have larger reactivity compared to DL ones.

See more of this Session: Polymer Reaction Engineering Kinetics & Catalysis
See more of this Group/Topical: Materials Engineering and Sciences Division

(749b) Pseudo-Solid State Polymerization In Amorphous Polymer Micro-Layers: A Novel Route to Produce Ultra-High Molecular Weight Polycarbonate

In Hak Baick[1], Carla Luciani[1], Woojic Yang[1] and Kyu Yong Choi[2], (1)Chemical and Biomolecular Engineering, University of Maryland, College Park, MD, (2)Department of Chemical and Biomolecular Engineering, University of Maryland, College Park, College Park, MD

A novel polymerization technique is presented to produce ultra-high molecular weight bisphenol A polycarbonate via step-growth polymerization in micro-layers using low molecular weight polymer precursors produced by melt polycondensation of bisphenol A and diphenyl carbonate. The polymerization is carried out at low pressures and at temperatures close but below the polymer melting point to ensure the reaction proceeds in amorphous state. It was found that the polymerization rate is much faster than in a conventional solid-state polymerization using pre-crystallized prepolymer particles. Transparent polycarbonates of molecular weights as high as 700,000 Da were obtained. Since the reacting polymer micro-layers are neither in the solid state nor in the melt phase, the proposed polymerization is called *pseudo-solid state polymerization (p-SSP)*. The relatively high mobility of polymer chains in the amorphous state, the efficient removal of the condensation by product (*i.e.*, phenol) from the confined geometry, the free radical-induced reactions via thermal decomposition of the residual casting solvent or Fries rearrangement are hypothesized to be the main responsible for the departure of the proposed polymerization technique from the conventional solid state polymerization. Experimental and theoretical investigations to elucidate the key mechanisms governing the p-SSP in amorphous polymer micro-layers will be presented at the meeting.

See more of this Session: Polymer Reaction Engineering Kinetics & Catalysis
See more of this Group/Topical: Materials Engineering and Sciences Division

(749c) Computational Study of Chain Transfer to Monomer Reactions In Thermal Polymerization of Methyl Acrylate

Nazanin Moghadam, Chemical and Biological Engineering, Drexel University, Philadelphia, PA, Masoud Soroush, Department of Chemical and Biological Engineering, Drexel University, Philadelphia, PA, Sriraj Srinivasan, Arkema, Philadelphia, PA, Andrew M. Rappe, Makineni Theoretical Laboratories, Department of Chemistry, University of Pennsylvania, Philadelphia, PA and Michael C. Grady, DuPont Marshall Lab, Philadelphia, PA

North American regulations [1, 2] have required volatile organic content (VOC) of coatings to be less than 300 ppm by 2010. These and similar regulations [3] have been responsible for the changes in the paint and coatings industries in the last sixty years. Low temperature (50-80°C) polymerization processes producing high molecular weight, low solids polymer resins have been replaced with high temperature (>100°C) polymerization processes producing low molecular weight, high solids polymer resins [4]. It has also been reported that secondary reactions, such as backbiting, β-scission and chain transfer reactions, have stronger impacts on the dynamics of the processes at higher temperatures.

Previous studies [5-9] have shown the successful use of computational quantum chemistry to better understand self-initiation and propagation reactions in polymerization systems. To the best of our knowledge, no computational study of chain transfer to monomer reactions in polymerization of alkyl acrylates has been reported as of yet.

This paper presents a computational study of chain transfer to monomer reactions in self-initiated high-temperature polymerization of methyl acrylate. Density functional theory calculations are used to predict energy barriers and rate constants of the chain transfer to monomer reactions. The effect of polymer chain-length and radical centers (e.g., tertiary vs. secondary radical) on the energy barriers and rate constants of the reactions are studied. Intrinsic reaction coordinate calculations are carried out to determine the pathways for the mechanisms of interest.

References:

[1] Superintendent of Documents, Title 1, US Government Printing Office, Washington, DC, 1990, p.1.

[2] Superintendent of Documents, Clean Air Act Amendments of 1990, Title 111, US Government Printing Office, Washington, DC, 1990, p.236.

[3] VOC's Directive, *EU Committee of the American Chamber of Commerce in Belguim*, ASBL/VZw, Brussels, July 8, 1996.

[4] Adamsons, K.; Blackman, G.; Gregorovich, B.; Lin, L., Matheson, R.; Oligomers in the Evolution of Automotive Clearcoats : Mechanical Performance Testing as a Function of Exposure, *Progress in Organic Coatings* 1998, 34, 64-74.

[5] Heuts, J.P.A.; Gilbert, R.G.; Radom, L.; A Prior Prediction of Propagation Rate Coefficients in Free Radical Polymerizations: Propagation of Ethylene, Macromolecules, 1995, 28, p. 8771-81.

[6] Khuong, K.S.; H. Jones, W.H.; Pryor, W.A.; Houk, K.N.; The Mechanism of Self-Initiated Thermal Solution of a Classic Problem, J. Am. Chem. Soc. 2005. 127, 1265-77.

[7] Yu, X.; Pfaendtner, J.; Broadbelt, L.J.; Ab Initio Study of Acrylate Polymerization Reactions: Methyl Methacrylate and Methyl Acrylate Propagation, J. Phys. Chem. A. 2008, 112, 6772-82.

[8] Srinivasan, S.; Lee, M.W.; Grady, M.C.; Soroush, M.; Rappe, A.M.; Computational Study of the Self-Initiation Mechanism in Thermal Polymerization of Methyl Acrylate, J. Phys. Chem. A 2009, 113, 10787-94.

[9] Srinivasan, S.; Lee, M.W.; Grady, M.C.; Soroush, M.; Rappe, A.M.; Self-Initiation Mechanism in Spontaneous Thermal Polymerization of Ethyl and n-Butyl Acrylate: A Theoretical Study, J. Phys. Chem. A 2010, 114, 7975-83.

See more of this Session: Polymer Reaction Engineering Kinetics & Catalysis
See more of this Group/Topical: Materials Engineering and Sciences Division

(749d) Synthesis and Characterization of Hyperbranched Polyacrylamide Using Semi-Batch RAFT Copolymerization of Acrylamide and N,N'-Methylenebisacrylamide

Dunming Wang[1], Wen-Jun Wang[1], Bo-Geng Li[1] and Shiping Zhu[2], (1)Department of Chemical & Biological Engineering, Zhejiang University, Hangzhou, China, (2)Department of Chemical Engineering, McMaster University, Hamilton, ON, Canada

Abstract Highly branched acrylamido polymers have been widely used in drug delivery systems, catalyst supports, viscosity modifiers, biomedicines, etc. In this work, hyperbranched polyacrylamides (b-PAMs) were prepared via a semi-batch reversible addition-fragmentation chain transfer (RAFT) polymerization of acrylamide (AM) using N,N'-methylenebisacrylamide (BisAM) as branching agent (BA) and 3-benzyltrithiocarbonyl propionic acid (BCPA) as chain transfer agent (CTA). Water soluble b-PAMs were successfully produced at low [BCPA]/[BisAM] ratios of 1/20 ~ 1/50 in comparison with gel formation in RAFT batch copolymerization of AM and BisAM at the same condition, indicating the crosslinking reaction of BisAM was effectively suppressed at low BCPA concentration using the semi-batch technique.

A systematic kinetic study on the semi-batch RAFT copolymerization of AM with BisAM was conducted. Effect of [BCPA]/[BisAM] ratios and feeding rates on the evolution of molecular weight (MW) and distribution (PDI), branching density (BD), and cyclization density (CD) in the polymerization were investigated. It was observed that the b-PAM samples collected at different time during the RAFT copolymerization had similar BD, suggesting that the b-PAMs prepared via the semi-batch RAFT copolymerization technique having homogeneous branching structure. High BisAM feeding rate or BisAM usage benefitted to branching formation. The cyclization reaction was suppressed at low instantaneous BisAM concentration in the copolymerization system.

See more of this Session: Polymer Reaction Engineering Kinetics & Catalysis
See more of this Group/Topical: Materials Engineering and Sciences Division

(749e) Photoinitiators As Copper(II) Reductants: A New Approach to the Copper-Catalyzed AzideAlkyne Cycloaddition

Brian Adzima, Dept. of Chemical and Biological Engineering, University of Colorado, Boulder, CO, Christopher J. Kloxin, Chemical Engineering and Materials Science and Engineering, University of Delaware, Newark, DE and Chistopher N. Bowman, Department of Chemical and Biological Engineering, University of Colorado, Boulder, CO

The copper catalyzed azide–alkyne cycloaddition (CuAAC reaction) click reaction is utilized in a wide range of synthetic strategies, from dendrimer and block copolymer synthesis to bioconjugation. Traditional methods for catalyzing the CuAAC reaction include the direct addition of Cu(I) salts, the reduction of copper sulfate via sodium ascorbate, and the comproportionation reaction of copper metal and Cu(II). Recently, we have demonstrated that the CuAAC reaction can be catalyzed using conventional photoinitiators to reduce Cu(II). This approach allows for spatial control of the CuAAC reaction, as demonstrated by patterning of hydrogel formation and labelling of hydrogels with features as small as 25 microns. This photochemical approach not only simplifies patterning surfaces, compared to traditional approaches such as dip pen lithography, but also allows the CuAAC reaction to be used for spatially selective coupling reactions within a material. Herein, we discuss recent kinetic experiments suggesting that undesirable catalyst consuming reactions, such as disproportionation, further reduction, and oxidation are suppressed, and that the ligands present play a key role in the reaction. Accordingly, like the conventional approach of reducing Cu(II) using sodium ascorbate this process is robust and does not require purging to eliminate oxygen. However, unlike conventional approaches this methodology readily allows for spatial control of the reaction using conventional photolithographic techniques.

See more of this Session: Polymer Reaction Engineering Kinetics & Catalysis
See more of this Group/Topical: Materials Engineering and Sciences Division

(749f) Conversion Control In Acrylate-Epoxide Hybrid Photopolymerizations with Hydroxyl-Containing Acrylates

Gbenga I. Ajiboye and Julie L.P. Jessop, Chemical & Biochemical Engineering, University of Iowa, Iowa City, IA

Photopolymerization uses light, instead of heat, to initiate reactions that convert monomer to polymer. The technique is increasingly used in industry due to its high levels of spatial and temporal control, low reaction temperatures, fast curing speeds, and virtually no emission of volatile organic compounds. Free-radical and cationic photopolymerizations have been implemented in many applications, such as coatings, electronics, biomedicine, and adhesives. Although free-radical polymerizations of acrylates are most common, inhibition by oxygen remains a problem. Oxygen reacts with the excited-state photoinitiators and the free-radical active centers to prevent chain propagation. Cationic ring-opening reactions of epoxides, while not inhibited by oxygen, generally exhibit low reaction rates and are sensitive to moisture and alcohol. Epoxide-acrylate hybrid systems promise to reduce the sensitivity of the reaction to oxygen and moisture by combining the advantages inherent to each reaction. However, the much faster acrylate reaction interferes with epoxide conversion because the epoxide chain growth takes place via the relatively slow active chain end (ACE) mechanism. In order to increase epoxide conversion, the faster activated monomer (AM) mechanism, a chain transfer reaction, is promoted in this research by pairing hydroxyl-containing acrylates with a diepoxide monomer.

Hybrid formulations with varying concentrations of 2-hydroxyethyl acrylate (HEA, acrylate) and 3,4-epoxycyclohexane carboxylate (EEC, epoxide) were photopolymerized using the cationic photoinitiator diaryliodonium hexafluoroantimonate (DAI). During photolysis, DAI also produces free radicals capable of initiating the acrylate polymerization, thereby removing the need for a separate free-radical photoinitiator. Conversions of acrylate and epoxide functional groups were obtained using real-time Raman spectroscopy. As the acrylate monomer concentration increased from 50 to 80%, the epoxide conversion increased from 55 to 85%, indicating that the hydroxyl group did indeed facilitate the AM mechanism. In a control experiment to verify that the increased epoxide conversion was not due to a dilution effect, HEA was replaced with a similar non-hydroxyl-containing acrylate, ethylene glycol methyl ether acrylate (EGMEA). The maximum EGMEA and EEC conversion obtained for an 80:20 mixture after 5-minutes illuminations was 86% and 35%, respectively. These results further confirm that the hydroxyl group in HEA promotes the AM mechanism. Physical and mechanical properties (e.g., glass transition temperature, modulus, and phase separation) for these systems are being investigated using dynamic mechanical analysis to determine how the AM mechanism affects the final hybrid polymer properties.

See more of this Session: Polymer Reaction Engineering Kinetics & Catalysis

IN LINE MONITORING OF RESIDUAL MONOMER BY NIR SPECTROSCOPY DURING STYRENE-DIVINYLBENZENE SOLUTION POLYMERIZATION REACTIONS

Dennis Chicoma Lara, María Verónica Carranza Oropeza,
Leandro Gonçalves de Aguiar and Reinaldo Giudici,

University of São Paulo, São Paulo, SP, Brazil.

Abstract

In this work, the sensitivity of NIR spectroscopy was studied for in-line, in-situ monitoring of residual monomer concentrations in solution copolymerization of styrene and divinylbenzene (S-DVB). Reactions were carried out in a glass reactor (stirred tank) to gather data used to develop a NIR calibration model. Experimental data were collected via off-line measurements of monomer conversion by gravimetry and they were previously validated with a kinetic model of the process. These data were correlated with the NIR spectra collected by an immersion probe (transflectance mode) during the reaction at the same sampling times. Spectral changes in the ranges 4460-4750 cm^{-1} and 6212-6187 cm^{-1} were related to the variations of monomer concentration. The results show good agreement between the experimental data and in-line monitoring predictions from the NIR spectra and the calibration model.

Keywords

Chemical Reactions, In Line Monitoring, NIR Spectroscopy, Polymers.

Introduction

Solution polymerization processes have been widely used for the production of polymer resins and other commercial products. In these processes, monomer, initiator, solvent and polymer are all soluble in the reaction medium, leading to the formation of a homogeneous reacting mixture. This technique show, as advantages, the easy-control of temperature and the solution can be directly used (according to the desired end-product). However, the solvent reduces the polymer molecular weight and reaction rate, promoting some control over these variables. It is well known that the main limitation for the efficient control of polymerization processes is the lack of robust sensors, which provide quick and reliable information on the properties of the polymer being produced [1,2]. Actually, the NIR spectroscopy is enabling a great development in both pharmaceuticals and food industries, as well as, in agriculture area and cosmetic industry. In the polymer field, this technique is used successfully in monitoring in bulk [3,4], suspension [5-7] and in emulsion [8,9] polymerization process. For the solution polymerization, there are also some published works, such as, Long et al. [10], Fontoura et al. [11], Chefri and Févotte [12], Chefri et al.[13] and Beyers et al.[14]. The success of NIR spectroscopy is possible through the acquisition of spectra assisted with chemometric techniques that relate the spectral information with the reference variable (residual concentration of comonomers, for example). The application of these techniques in polymerization process has been studied, with special attention to the analytical and quantitative results due to spectral variations between the peaks of the monomers and solvent during the reaction. As such, the final spectrum displays broad, overlapping peaks and bands with presence of shoulders, indicating that, although "rich" in information, the spectrum does not allow a direct measurement of the variables of interest, thus requiring appropriate calibration. In the present work, in-line and in-situ monitoring using a NIR probe coupled with fiber optics was used to monitor styrene and divinylbenzene (S-DVB) residual monomer concentrations in solution copolymerization reactions. The aim of this work was to produce copolymers with pre-specified properties that are able to attend the increasingly exigent demands of the polymer market.

EXPERIMENTAL

The following reactants were used in the solution copolymerization reactions: industrial grade styrene (BASF-Brazil) and divinylbenzene (Sigma Aldrich) monomers, toluene as solvent and AIBN as initiator. Methanol was used as precipitation agent at the end of the reaction. Polymerization reactions were carried out in a jacketed glass stirred-tank reactor using different ratios between styrene and toluene (see Table 1).The temperature of the reactor content was controlled within 85±1 °C and the agitation speed was kept at 200 rpm. All NIR spectra were collected with an IFS 28/N spectrometer (Bruker Optics), equipped with a quartz bean splitter, using a probe (Hellma 661.622 NIR, with a transflection system with overall light path equal to 1mm), immersed into the reaction medium. Off-line measurements of monomer conversion were obtained by gravimetry and they were previously validated with a kinetic model of the process. These measurements were used as reference for the development and validation of the multivariate NIR calibration model. This model was developed based on Partial Least Squares (PLS) using the software of the equipment (OPUS, program QUANT2).

Table 1. Polymerization recipes

Reactant	Run 1 (g)	Run 2 (g)	Run 3 (g)	Run 4 (g)
Toluene	200.36	280.58	481.83	440
Styrene	193.5	115.73	309	346.6
Divinylbenzene	7.453	4.467	11.911	13.4
AIBN	3.4	3.472	6.765	6.744

In order to obtain the global conversion by gravimetry, an empty aluminum capsule was weighted, then 3 mL of methanol (responsible for the precipitation of the polymer) was added and weighted again. This content was weighted after receiving 1.5 mL of reaction sample. This procedure was done periodically throughout the reaction and the capsules were immediately conducted to an oven at 70 °C, proceeding a drying process during 24 hours. The experimental measurements were validated with the results predicted by the mathematical model of the process based on the work of Ortiz et al. [15] and Gonçalves et al. [16]. In the model, the following assumptions are considered: pseudo-steady state for the radicals, each polymer molecule has only one radical center; pseudo-homopolymerization approach; similarity between the average size of the chain for dead and dormant polymer during the process. Figure 1 shows the comparison between the experimental and the predicted results of global monomer conversion evolution. As can be seen, there is a good agreement on all reactions. Based on these results, the model is used to estimate the concentration of residual monomer during the reaction and used as reference set for the NIR calibration model. For the elaboration this NIR model, the following known steps were performed [17]: identification and selection of the spectral region representative of the conversion of monomer into polymer, spectral pre-processing (necessary to remove unwanted variations such as noise and fluctuations in the baseline), calibration set (evaluation parameters) and validation (internal and external).

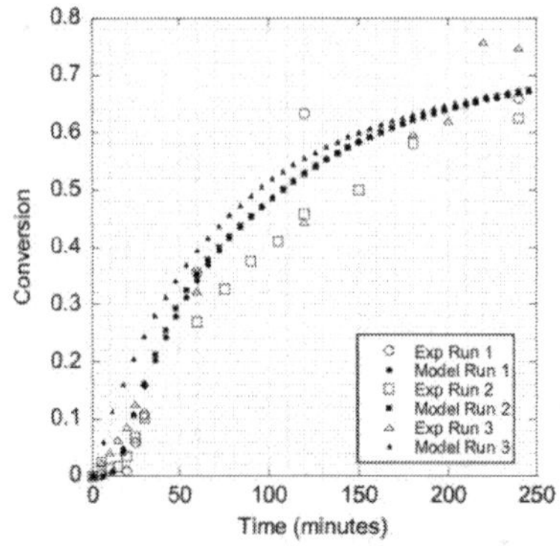

Figure 1. Comparison of the evolution of global conversion between the experimental results (Exp Run) and the ones predicted by the kinetic model (Model Run).

RESULTS

Figure 2 shows the NIR spectral evolution during the polymerization process. It can be observed that the decreasing of the peaks at 6134, 4720 and 4487 cm^{-1} are assigned to stretching vibrational motions of the vinyl group and region of combination bands C–H + C=C, respectively.

As such, the spectral regions chosen for residual monomer were the combination (4460-4750cm^{-1}) and the first overtone region (6212-6187cm^{-1}), wherein the bands are attributed to the monomer concentration. Spectral pre-processing was performed using second derivative with 25 points smoothing and the internal validation was made by using conventional "leave-one-out" cross-validation. In this procedure, parts of the calibration set are left out of the calculation used to fit the model (with a certain number of latent variables –LVs) to the data. Then the model is used to estimate the values of the left-out samples. This procedure is repeated several times for different sets and the root mean square error (RMSECV) between the estimated and real values is calculated. The number of LVs is chosen as the minimum of the plot of RMSECV versus the number of LVs. The resulting calibration model uses three principal components (latent variables) and presents a high correlation coefficient (0.986).

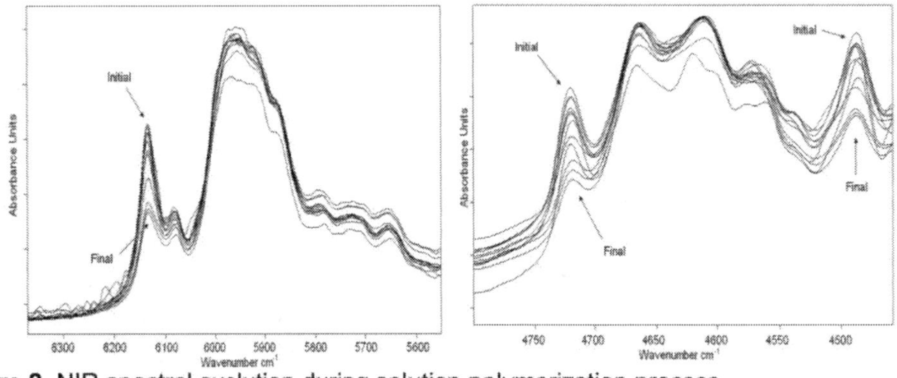

Figure 2. NIR spectral evolution during solution polymerization process.

Figures 3a, 3b and 3c show the comparison for monomer content between the off-line measurements by reference methods and the predictions from the on-line measured NIR spectra with the calibration model (internal validation). These results exhibited good agreements. Then, the NIR calibration model was tested with a new reaction data that were not used in calibration step (external validation), the results of prediction showed good agreement between mathematical model data and the NIR prediction data.

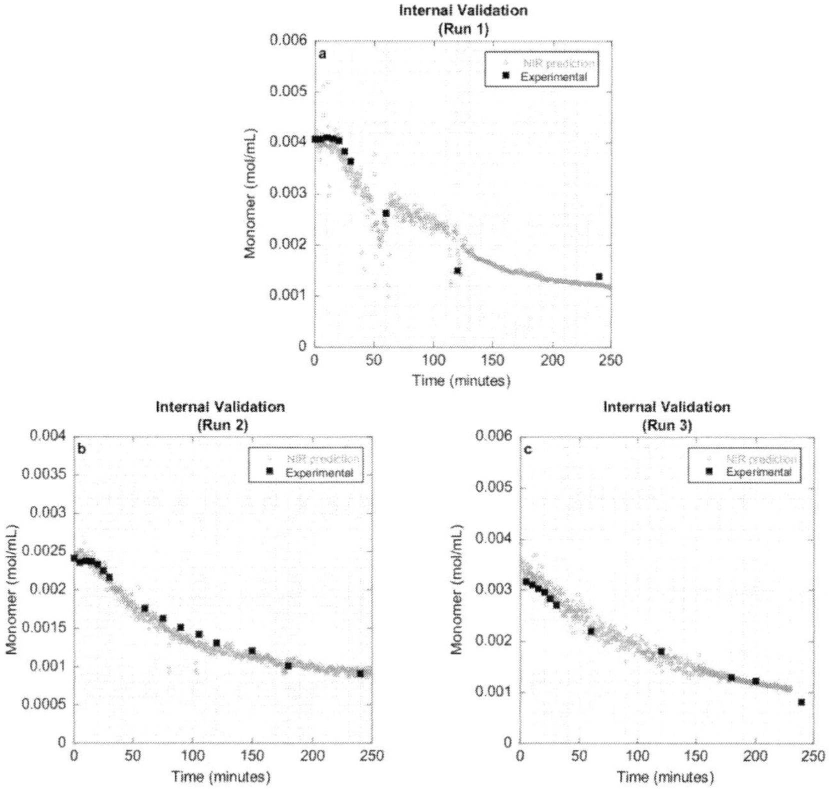

Figure 3. Comparison between the estimates from the NIR spectra and the experimental monomer concentration by internal validation. (a) Run 1, (b) Run 2 and (c) Run 3.

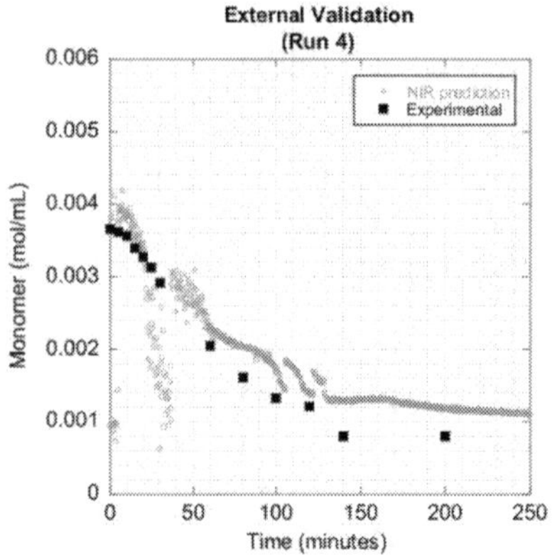

Figure 4. Comparison between the estimates from the NIR spectra and the experimental monomer concentration by external validation.

CONCLUSION

The present approach for the in-line and in-situ monitoring of monomer concentration during solution copolymerization of styrene and DVB was based on a NIR calibration model. This model was developed and validated with NIR spectra (from external validation), off-line gravimetric measurements and a kinetic model of the process. The results showed good agreement between the experimental data and the in-line monitoring predictions from the NIR spectra. The calibration model can enable, in a near future, the implementation of closed-loop control strategies and real-time decision-taking during polymerization processes.

ACKNOWLEDGEMENTS

The financial support from FAPESP (Fundação de Amparo à Pesquisa do Estado de São Paulo) and CNPq (Conselho Nacional de Desenvolvimento Científico e Tecnológico) is gratefully appreciated. The authors also thank BASF Brazil for providing monomer.

REFERENCES

1. MacGregor, J. F.; Penlidis, A.; Hamielec, A. E. (1984), "Control of Polymerization Reactors. A Review", *Polym. Proc. Eng.*, v. 2 (2-3), pp. 179-206.

2. Giudici, R. (2000), "Polymerization Reaction Engineering. A Personal Overview of the State of Art", *Latin American Applied Research*, v. 30(4), pp. 351-356.

3. Aldridge, P. K.; Kelly, J. J.; Callis, J. B.; Burns, D. H. (1993), "Non Invasive Monitoring of Bulk Polymerization Using Short-Wavelenght Near Infrared Spectroscopy", *Anal. Chem.*, v. 65, pp. 3581-3585.

4. Lousberg, H. H. A.; Boelens, H. F. M.; Le Comte, E. P.; Hoefsloot, H. C. J.; Smilde A. K. (2002), "On Line Determination of the Conversion in a Styrene Bulk Polymerization Batch Reactor using Near-Infrared Spectroscopy", *Journal of Applied Polymer Science*, v. 84 (1), pp. 90-98.

5. Santos, A. F.; Lima, E. L.; Pinto, J.C. (1998), "In-line Evaluation of Average Particle Size in Styrene Suspension Polymerization Using Near-Infrared Spectroscopy", *J. Appl. Polym. Sci.*, v. 70, pp. 1737-1745.

6. Santos, A. F.; Lima, E. L.; Pinto, J.C. (2000), "Control and Design of Average Particle Size in Styrene Suspension Polymerization Using NIRS", *J. Appl. Polym. Sci.*, v. 77, pp. 453-462.

7. Machado, F.; Lima, E. L.; Pinto, J. C. (2007), "Uma Revisão Sobre os Processos de Polimerização em Suspensão", *Polímeros*, v. 17, 2, pp. 166-179.

8. Reis, M. M.; Araujo, P. H. H.; Sayer, C.; Giudici, R. (2004[b]), "In situ near infrared spectroscopy for simultaneous monitoring of multiple process variables in emulsion copolymerization", *Ind. Eng. Chem. Res.*, v. 43, pp. 7243-7250.

9. Chicoma, D.; Sayer, C.; Giudici, R. (2011), "In-Line Monitoring of Particle Size during Emulsion Polymerization under Different Operational Conditions using NIR Spectroscopy", *Macromolecular Reaction Engineering*, v. 5, Issue 3-4, pp. 150–162, April 13.

10. Long, T. E.; Liu, H. Y.; Schell, B. A.; Teegarden, D. M.; Uerz, D. S. (1993), "Determination of Solution Polymerization Kinetics by Near-Infrared Spectroscopy 1: Living Anionic Polymerization Processes", *Macromolecules*, v. 26 (23), pp. 6237–6242.

11. Fontoura, J. M. R.; Santos, A. F.; Silva, F. M.; Lenzi, M. K.; Lima, E. L.; Pinto, J. C. (2003) "Monitoring and Control of Styrene Solution Polymerization Using NIR Spectroscopy", *Journal of Applied Polymer Science,* v. 90, pp. 1273-1289,.

12. Cherfi, A.; Févotte, G. (2002), "On-Line Conversion Monitoring of the Solution Polymerization of Methyl Methacrylate Using Near-Infrared Spectroscopy", *Macromol. Chem. Phys.*, v. 203, 9, pp. 1188-1193.

13. Cherfi, A.; Févotte, G.; Novat, C. (2002), "Robust On-Line Measurement of Conversion and Molecular Weight Using NIR Spectroscopy During Solution Polymerization", *Journal of Applied Polymer Science*, v. 85, pp. 2510-2520.

14. Beyers, C. P.; Boelens, H. F. M.; Klumperman, L.; Westerhuis, J. A. (2004), "In-line reaction monitoring of the methyl methacrylate and N,N-dimethylacrylamide copolymerization reaction employing near infrared spectroscopy", *Applied Spectroscopy*, v. 58, 7, pp. 182A-208A.

15. Ortiz, J.C.H.; Lima, E.V.; Lona, L.M.F.; Mccmanus, N.T.; Penlidis, A. (2009), "Modeling of the Nitroxide-Mediated Radical Copolymerization of Styrene and Divinylbenzene", *Macromol. React. Eng.*, v. 3, pp. 288–311.

16. Gonçalves, M.A.D.; Trigo, I.M.R.; Dias, R.C.S; Costa, M.R.P.F.N. (2010), "Kinetic Modeling of the Molecular Architecture of Cross-Linked Copolymers Synthesized by Controlled Radical Polymerization Techniques", *Macromol. Symp.*, v. 291, pp. 239–250.

17. Silva, W.; Chicoma, D.; Giudici, R. (2011), "In-situ Real-time Monitoring of Particle Size, Polymer, and Monomer Contents in Emulsion Polymerization of Methyl Methacrylate by Near Infrared Spectroscopy". *Polymer Engineering Science*. doi: 10.1002/pen.22100.

(750a) Polymers for Biomass Energy Conversion: Porous Scaffold Materials for a Continuous-Flow, Immobilized-Cell Fermentation Process

Ronald C. Hedden, Lan Ma, Jun Zhao, **Seunghyun Ryu** and M. Nazmul Karim, Chemical Engineering, Texas Tech University, Lubbock, TX

Due to limited global supplies of fossil fuel resources,rising prices, and environmental problems, future energy needs must be met inpart or entirely by fuels derived from renewable resources. Production of ethanol (EtOH) from non-foodfeedstocks, namely cellulosic biomass, has received much attention. If cellulosic EtOH is to become aneconomically sustainable source of fuel in the future, drastic improvements toprocess economics must be realized, however.Part of the cost reduction can be realized by lowering the cost of rawmaterials and increasing the efficiency of conversion to EtOH, while otherreductions must be made in the areas of capital equipment and operating costsby design of more efficient fermentation processes.

Our work therefore focuses on design of an efficientcontinuous-flow bioconversion process for production of EtOH, with emphasis oncellulose-derived feedstocks, which may contain particulate material andinhibiting compounds. Sponge-likepolymeric materials having micrometer-scale porosity are a key element of anew immobilized-cell, packed-bed fermentation process that already exhibitsvolumetric productivity *more than anorder of magnitude higher* than comparable batch fermentations. Synthetic porous polymer scaffolds (SPPS)are fabricated in the form of irregular particles of typical mesh size 1 to 2mm, which are used to immobilize fermenting microorganisms as a packed bedwithin a columnar reactor. Design ofoptimal (SPPS) for the immobilized cell reactor process is an emerging materialschallenge within the broad field of biomass energy conversion.

Optimal SPPS are inexpensive polymers or gels of relativelylow water content that contain pores and channels of typical diameter 1 to30 micrometers and have a high pore volume fraction (0.4 to 0.8). Pores should have extensive surfaceconnectivity, facilitating entry and exit of both reactants and fermenting micoorganisms. Ethanologensof typical size 1 to 10 micrometers are targeted, such as *Escherichia coli, Zymomonas mobilis, or Saccharomyces cerevisiae*,and maximizing the cell density within the SPPS material is critical. Twogeneral approaches to SPPS design are being pursued:

1) Preparation of a bicontinuous blend with one crosslinkedcomponent and one extractable component, which is removed to generate randomlystructured, interconnected pores.

2) Crosslinking of amatrix polymer around extractable polymer microfibers, which are removed toleave well-defined cylindrical pores.

In both cases, the extractable phase is removed prior tofermentations by leaching in a solvent, preferably water, leaving pores orchannels in the SPPS with a high degree of connectivity to the surface. SPPS materials are colonized naturally bycells during the exponential growth phase in batch operation, after which continuousreactor operation is initiated.

Our work has so far examined conversion of glucose at aninlet feed concentration of 1.0 to 5.0 % w/w by *E. coli* strain LY01. Acontinuous-flow column ICR packed with an SPPS material achieved volumetricproductivity 14 times higher than that of a comparable batch fermentation,while the porous structure of the SPPS bed mediated problems with CO_2holdup. SPPS materials are a greatimprovement over traditional polymer gel materials such as calcium alginate,carageenan, and polyacrylamide because the pores greatly reduce mass transferresistance and allow facile venting of CO_2 bubbles duringfermentation. In addition, packed bedsof SPPS particles allow small particulate solids to pass through the bed,lowering the risk of clogging. Ourpresentation will discuss the advantages and limitations of the SPPS columnreactors, materials design issues, and progress in the conversion of sugars toEtOH.

See more of this Session: Polymeric Materials for Energy Conversion
See more of this Group/Topical: Materials Engineering and Sciences Division

(750b) Well-Ordered Poly(3-alkylthiophene) Diblock Copolymers for Organic Photovoltaic Applications

Bryan W. Boudouris, Victor Ho, Bryan L. McCulloch and Rachel A. Segalman, Department of Chemical and Biomolecular Engineering, University of California, Berkeley, Berkeley, CA

Conjugated,semiconducting macromolecules have risen to the fore as electron-donatingmaterials in organic photovoltaics (OPVs). In fact, advances in the fabricationand post-processing of polymer-fullerene bulk heterojunction solar cells haveallowed for devices with power conversion efficiencies approaching those ofamorphous silicon devices to be realized. Better knowledge of how exciton (abound electron-hole pair) dissociation and the internalmorphology of the active layer affects device performance should facilitateprocessing optimization and, ultimately, lead to devices with higher powerconversion efficiencies. Because block copolymers self-assemble on thesame length scale as the exciton dissociation length in organic materials (~10nm), diblock copolymers containing a rigid, semiconducting moiety (rod)covalently linked to a flexible block (coil) have attracted an ever-increasingamount of attention. In contrast to coil-coil diblock copolymers, conjugatedrod-coil diblock copolymers self-assemble due to a balance of liquidcrystalline (rod-rod) and enthalpic (rod-coil) interactions. Previous work hasdemonstrated that while coil-coil diblock copolymers self-assemble into a hostof nanostructures with varying degrees of interfacial curvature, when rod-rodinteractions dominate self-assembly in rod-coil block copolymers, lamellarstructures with little curvature are preferred.

Here, it is demonstrated thatnon-lamellar, potentially more useful nanostructures can be formed when liquidcrystalline and enthalpic interactions are more closely balanced throughprecise molecular design of the conjugated moiety. Specifically,hexagonally-packed cylinders of the polylactide (PLA) coil block embedded in asemiconducting poly(3-alkylthiophene) (P3AT) matrix are shown. Note that thisparticular semiconducting moiety, P3AT, is of great interest as polythiophenesare used commonly as active layer components in plastic electronic devices.Previous efforts to generate this phase in polythiophene-based block copolymershave failed due to the high driving force for P3AT crystallization at roomtemperature. By carefully designing the molecular architecture of the P3ATmoiety, we are able to strike a balance between crystallization and microphaseseparation due to chemical dissimilarity between copolymer segments. Inaddition to hexagonally-packed cylinders, P3AT-PLA block copolymers formnanostructures (*i.e.*, lamellae) withlong-range order at almost all block copolymer compositions. Importantly, anddespite the presence of large weight fraction of PLA present in the diblockcopolymers (w_{PLA}> 0.50), the conjugated moiety of the P3AT-PLA block copolymers retains thecrystalline packing structure and characteristic high time-of-flight (TOF)charge transport of the homopolymer polythiophene (μ_h ~10^{-4} cm^2V^{-1} s^{-1}) in the confined block copolymer domains.

(750c) Building Polymer Solar Cells with Top-Down Lithography

Gila Stein, Suchanun Moungthai and Nikhila Mahadevapuram, Chemical and Biomolecular Engineering, University of Houston, Houston, TX

Solar cellsbased on π-conjugated polymers show promise as sources of clean energy:Manufacturing costs are low, and devices can be fabricated in light-weight,flexible, and durable plastic modules. Currently, the active layer in polymersolar cells is based on the bulk heterojunction (BHJ) design. BHJs are preparedby arresting the phase separation of a polymer/fullerene blend to produce a nanoscale, interpenetrating network. BHJs are cheap tofabricate and offer a large interfacial area for charge generation. However,such non-equilibrium structures are very difficult to control and reproduce, posinga significant challenge for fundamental structure-property investigations. In this work, we demonstrate a new approachto build polymer solar cells with controlled morphologies based onelectron-beam patterning of π-conjugated polymers. Heterojunction designis easily varied with a two-step fabrication process: First, a thin film of poly(3-hexylthiophene) (P3HT) is exposed to a periodicpattern of electron-beam radiation, which cross-links P3HT into stablestructures at moderate radiation doses (ca. 300 μ C/ cm^2at 30 keV).[1]Nanodots and nanolinesdown to 40 nm in size are reliably printed with a 30 keVscanning electron microscope. Second, [6,6]-phenyl-C_{61}-butyricacid methyl ester (PCBM) is spun-cast on top of the patterned polymer tocomplete the heterojunction. As a proof-of-concept demonstration, solar cellswith bilayer P3HT/PCBM heterojunctionswere fabricated with electron-beam patterning.Significantly, irradiated P3HT films retain their optoelectronicproperties, and these bilayer devices yield power-conversionefficiencies in the range of 0.5-1%. Preliminary data for nanostructureddevices will also be presented in this talk.

[1] S. Holdcroft, Adv. Mater. 2001, 13, 1753-1765.

See more of this Session: Polymeric Materials for Energy Conversion
See more of this Group/Topical: Materials Engineering and Sciences Division

(750d) Structure-Property Relationships In Organogels of Polythiophenes

Danilo Pozzo, Greg Newbloom and Kathleen Weigandt, Chemical Engineering, University of Washington, Seattle, WA

Conjugated polymers, such as polythiophenes (PTs), are known to form fibrillar structures when solvent quality is reduced. At high concentrations, PTs will self-assemble and form semiconducting elastic networks of interconnected fibers. The interconnectivity of these networks lead to a continuous, 3-dimensional path for charge transport which is ideal for the bulk heterojunction structure of organic photovoltaic devices. This talk will describe our recent systematic evaluation of these networks over multiple length scales (0.1 – 10,000 nm) using small angle neutron scattering (SANS). SANS is used in combination with *in situ* rheology (Rheo-SANS) and *in situ* electrochemical analysis (Dielectric-SANS) to characterize the development of structural, mechanical and electrical properties. This allows us to formulate additional structure-property relationships for this important class of materials. This talk will also address the implications of our findings on the development of novel photovoltaic technology.

See more of this Session: Polymeric Materials for Energy Conversion
See more of this Group/Topical: Materials Engineering and Sciences Division

(750e) Systematic Molecular Weight Study of Poly (thienylene vinylene)/Fullerene Bulk Heterojunction Photovoltaic Cells

Bryan D. Paulsen[1], Josh C. Speros[2], Marc A. Hillmyer[2] and C. Daniel Frisbie[1], (1)Chemical Engineering & Materials Science, University of Minnesota, Minneapolis, MN, (2)Chemistry, University of Minnesota, Minneapolis, MN

Molecular weight is among the most important properties governing polymer behavior. In the rapidly developing field of polymer solar cells, understanding molecular weight effects is necessary to maximizing device efficiency and realizing commercialization. Semicrystalline poly(3-hexadecyl-2,5-thienylene vinylene) (P3HDTV), with a low band gap of 1.65 eV, was synthesized by acyclic diene metathesis polymerization with tuned chain length. Neat polymer samples of various molecular weight were characterized by size-exclusion chromatography, cyclic voltammetry, differential scanning calorimetry, UV-Vis spectroscopy, and wide-angle X-ray scattering. P3HDTV was blended with the electron acceptor methanofullerene [6,6]-phenyl C61-butyric acid methyl ester (PCBM) to form the bulk heterojunction (BHJ) photovoltaic cell active layer. Completed cells were characterized with diode measurements in the dark and under simulated solar illumination, and with monochromatic external quantum efficiency (EQE) measurements. BHJ cells were optimized for active layer thickness and P3HDTV/PCBM composition. In a series of 100 nm thick, constant composition devices, increasing molecular weight produced a modest drop in cell open circuit voltage (V_{oc}), saturating at 0.40 V. This was overcome by large increases in cell short circuit current (J_{sc}) with increasing molecular weight, resulting in increased power conversion efficiency (PCE) with increased molecular weight. Cells incorporating high molecular weight P3HDTV showed PCE of ~1%.

See more of this Session: Polymeric Materials for Energy Conversion
See more of this Group/Topical: Materials Engineering and Sciences Division

(750f) Understanding Self-Assembly of All-Conjugated Block Copolymers

Rafael Verduzco, Kendall Smith and Yen-Hao Lin, Chemical and Biomolecular Engineering, Rice University, Houston, TX

Blends of conjugated polymers have significant potential for solution-processed optoelectronic applications such as organic photovoltaics (OPVs) and organic light-emitting diodes (OLEDs). Such blends can have broad absorbance spectra and significantly larger open circuit voltages compared with polymer-fullerene OPVs. However, a significant drawback of conjugated polymer blends is their tendency to phase separate. Here, we explore two methods for creating self-assembled mixtures of two or more conjugated polymers. In a first method, copper-catalyzed azide-alkyne "click" chemistry is used to synthesize all-conjugated block copolymers with a flexible linker. This versatile synthetic method gives access to all-conjugated block copolymers which self-assemble into nanoscale donor/acceptor structures. In a second approach, conjugated polymers are end-capped with a small molecule capable of forming hydrogen bonds. We show that conjugated polymers with hydrogen bonding end groups can form uniformly mixed polymer films. We will focus on the synthesis and structure of polymer films created using these two approaches and present data on OPVs based on these materials.

See more of this Session: Polymeric Materials for Energy Conversion
See more of this Group/Topical: Materials Engineering and Sciences Division

(751a) Reconfigurable Bacterial Polyester Nanocomposites Exhibiting Soft Shape Memory

Kazuki Ishida, Rebecca A. Lyons, Xiaofan Luo and Patrick T. Mather, Department of Biomedical and Chemical Engineering, Syracuse University, Syracuse, NY

Shape memory polymers (SMPs) with low stiffness in the "fixed" temporary shape (below the transition temperature, T_{tr}) as well as in the "relaxed" permanent shape (above T_{tr}) have tremendous potential in biomedical applications owing to better modulus match to soft tissues. However, for most existing SMPs, the conventional use of main-chain vitrification or crystallization to fix temporary shapes limits the opportunity to obtain soft "fixed" SMPs. In this paper, we report the preparation of novel soft SMP nanocomposites featuring reconfigurable nanostructures. The materials are derived from a bacterial medium-chain-length polyhydroxyalkanoate (mcl-PHA), poly(3-hydroxyoctanoate-*co*-3-hydroxyundecenoate) (PHOU), functionalized with polyhedral oligomeric silsesquioxane (POSS), a crystallizable nano-filler that is composed of an inorganic core and organic vertex groups. The PHOU–POSS nanocomposites exhibit excellent shape memory performance, and behave elastomerically both above and below T_{tr}. We attribute these properties to the ability of segregated POSS nano-domains to structurally reconfigure the amorphous PHOU network under different thermomechanical conditions. Owing to their biocompatibility, PHOU–POSS nanocomposites are promising materials for a variety of biomedical applications.

See more of this Session: Properties and Characterization of Nanocomposites
See more of this Group/Topical: Materials Engineering and Sciences Division

(751b) Thermal, Mechanical Properties, and Fracture Toughness of Surface Modified Graphene Epoxy Nanocomposites

Mitra Yoonessi, Materials and Structures Division, NASA Glenn Research Center/OAI, Cleveland, OH, Eileen Boyd, NASA GRC, Cleveland, OH and Derek J. Quade, Materials and Structures Division, NASA Glenn Research Center, Cleveland, OH

Addition of nanoparticles to the host polymer matrix has resulted in improvements of stiffness and moduli, increased glass transition temperature, enhanced thermal stability, improved electrical and thermal conductivity, enhanced barrier properties, and thermal barrier coating. Graphene is SP^2 hybridized 6-membered carbon ring monoatomic thick with micron-sized lateral dimension exhibited superior electrical, thermal, and mechanical properties when incorporated in polymer matrices, such as polyvinyl alcohol, polystyrene, and polycarbonate.

Series of low concentration graphene bisphenol-A epoxy nanocomposites, (0.01-1 wt%), with reduced graphene, oxygenated graphene, and aminopropyl terminated polydimethylsiloxane modified graphene were prepared using solvent mixing method. The attachment of aminopropyl terminated polydimethylsiloxane to the graphene was verified with IR and TGA.

Thermal stability of all nanocomposites was higher with increased onset of thermal degradation compared to the neat epoxy. Glass transition temperature of all graphene nanocomposites exhibited 5-10 ^{o}C increase compared to the neat epoxy resin. The storage tensile modulus of the all nanocomposite were higher than the neat epoxy where the moduli of the graphene with oxygenated functional groups increased with increasing graphene concentration. The moduli of the reduced graphene epoxy nanocomposites had an initial increase and then remained constant, where the moduli of the PDMS-graphene epoxy nanocomposites exhibited an initial increase followed by a decrease with increasing graphene concentration. Fracture toughness of graphene epoxy nanocomposites were 30-40% higher than the neat epoxy resin with only addition of 0.01-0.05 wt% graphene. Fractured surface of the nanocomposites were examined by scanning electron microscopy.

See more of this Session: Properties and Characterization of Nanocomposites
See more of this Group/Topical: Materials Engineering and Sciences Division

(751c) Controlling Carbon Nanotubes Dispersion Using Stimuli-Responsive Polymers

Krishna Etika, Texas A&M University (Currently Process TD Engineer at Intel Corporation), Hillsboro, OR and Jaime C. Grunlan, Mechanical Engineering, Chemical Engineering, Texas A&M University, College Station, TX

Despite their immense potential, the ability to control the dispersion and microstructure of carbon nanotubes remains a hurdle for their widespread use in nanotube-filled polymer composites. Stimuli responsive polymers exhibit conformational changes as a function of applied stimulus (pH, temperature, light etc.) which can be used to tailor SWNT microstructure. In this work, several pH-responsive polymers(i.e., weak polyelectrolytes) such as, poly(acrylic acid), poly(methacrylic acid), poly (allylamine),polyethleneimine and temperature responsive poly(N-cyclopropylacrylamide), poly(N-isopropylacrylamide) functionalized with varying concentrations of pyrene were used to stabilize aqueous SWNT suspensions. Studies performed on aqueous SWNT suspensions stabilized with these pH and temperature responsive polymers suggest exfoliated and highly networked nanotube microstructure a function of pH and temperature, respectively. Cryo-TEM, viscosity and UV-Vis spectroscopy were used to characterize the nanotube dispersion state in these suspensions. Furthermore, studies performed on the nanocomposite formed by drying these suspensions suggests that microstructure of nanotubes in these suspensions is preserved in the solid composite, as evidenced by SEM imaging and electrical conductivity measurements. This method of controlling the microstructure of SWNTs in liquid medium with external stimuli to tailor the properties of solid composites could have a significant impact on solution process polymer nanocomposites preparation. Many of the relationships described here could also be applied with other responsive polymers and for the tailored dispersion of other types of nanoparticles (e.g., inorganic nanotubes and nanowires).

See more of this Session: Properties and Characterization of Nanocomposites
See more of this Group/Topical: Materials Engineering and Sciences Division

(751d) Crystallization and Biodegradation of Poly (lactic acid)/Clay Nanocomposites Prepared by Solid-State Shear Pulverization

Katsuyuki Wakabayashi and Alexander S. Fielding, Department of Chemical Engineering, Bucknell University, Lewisburg, PA

Bioplastics such as polylactic acid (PLA) offer a sustainable alternative to traditional polymers from petroleum derivatives, from both the source (bio-based) and disposal (biodegradable) standpoints [1]. The ability for PLA to degrade [2] under certain natural conditions is ideal for short lifetime applications such as packaging and consumer products. Neat PLA, however, exhibits some undesirable physical characteristics, mainly due to its hydrophilicity and low glass transition and melting temperatures. Incorporating small amounts of nanoscale fillers, such as layered silicates (clay), into the PLA matrix can improve mechanical, thermal, and other physical properties to make them more viable in commercial applications. Previous studies on PLA/montmorillonite clay nanocomposites have shown significant property enhancement when high levels of exfoliation and dispersion of the clay sheets are achieved [3-5]. Common processing methods, such as *in situ* polymerization and twin-screw extrusion (TSE) processing, have been employed to used to prepare PLA-clay nanocomposites.

We have recently introduced solid-state shear pulverization (SSSP) as an alternative processing technique to exfoliate and disperse clay nanofillers in PLA; SSSP processes polymer nanocomposites without the need for chemical modification, additives, and solvents/monomers, and is simple and industrially applicable. SSSP has proven to successfully compatibilize immiscible polymer blends [6] and exfoliate sheet-like fillers in more traditional polymer-clay nanocomposites [7,8].

In this paper, commercial grade PLA and pristine and organically-modified clay were processed under different TSE and SSSP conditions, and characterized in terms of PLA crystallization and biodegradation kinetics. Both of these rates are expected to be closely tied to the molecular weight of PLA and the level of exfoliation and dispersion of the clay fillers, both of which are controlled by the processing parameters. The structural characterization of the materials was conducted by way of gel-permeation chromatography and X-ray diffraction. Differential scanning calorimetry and dynamic mechanical analysis, as well as polarized optical microscopy, were employed for crystal development monitoring. Biodegradation study was conducted using an in-house instrument based on the ISO 14855 standard method. The results demonstrate the versatility and tunability of SSSP processing parameters in PLA nanocomposites.

[1] R. E. Drumright, P. R. Gruber, and D. E. Henton, Advanced Materials, 12, 1841 (2000).
[2] M.-A. Paul, C. Delcourt, M. Alexandre, P. Degee, F. Monteverde, and P. Dubois, Polymer Degradation and Stability, 87, 535 (2005).

[3] S. S. Ray, P. Maiti, M. Okamoto, K. Yamada, and K. Ueda, Macromolecules, 35, 3104 (2002).
[4] J. A. Ratto, D. M. Steeves, E. A. Welsh, and B. E. Powell, ANTEC: Society of Plastics Engineers, 57, 1628 (1999).
[5] K. Fukushima, D. Tabuani, and G. Camino, Materials Science and Engineering, 29, 1433 (2009).
[6] A. H. Lebovitz, K. Khait, and J. M. Torkelson, Macromolecules, 35, 8672 (2002).
[7] K. G. Kasimatis, and J. M. Torkelson, Polymeric Materials: Science and Engineering, 91, 173 (2004).
[8] K. Wakabayashi, C. Pierre, D. A. Dikin, R. S. Ruoff, T. Ramanathan, L. C. Brinson, and J. M. Torkelson, Macromolecules, 41, 1905 (2008).

See more of this Session: Properties and Characterization of Nanocomposites
See more of this Group/Topical: Materials Engineering and Sciences Division

(751e) Clay and Carbon Black Synergy for Enhancing Mechanical and Electrical Properties of Epoxy Composites

Krishna Etika, Texas A&M University, Hillsboro, OR and Jaime C. Grunlan, Mechanical Engineering, Chemical Engineering, Texas A&M University, College Station, TX

Studies of acetone-based suspensions suggest a synergistic stabilization of clay by carbon black (CB) that involves a haloing effect (i.e., CB surrounding clay). This synergy results in unique microstructure development that ultimately influences the electrical and mechanical properties of epoxy composites containing both particles. With the addition of 0.5 wt% clay, electrical conductivity increases by an order of magnitude for CB-filled epoxy (relative to composites containing no clay), but no significant improvement is observed in storage modulus. Composites containing equal concentrations of CB and clay show reduced electrical conductivity, but significant improvement in storage modulus (relative to composites containing equal amount (wt%) of either CB or clay alone). Both electrical conductivity and storage modulus improve in composites containing a 1:2 clay:CB (wt/wt) ratio. This synergy between CB and clay is a useful tool for simultaneously improving the electrical and mechanical properties of solution processed composites. Similar synergy has also been observed with carbon nanotubes and clay, which resulted in a dramatic reduction in percolation threshold.

See more of this Session: Properties and Characterization of Nanocomposites
See more of this Group/Topical: Materials Engineering and Sciences Division

(751f) Characterization of Epoxy Spin-Coated Tailored Multifunctional Nanocomposite Structures

Timothy Shenk[1], Robb M. Winter[2] and Kenneth M. Benjamin[1], (1)Chemical and Biological Engineering, South Dakota School of Mines and Technology, Rapid City, SD, (2)Department of Chemical and Biological Engineering, S.D. School of Mines and Technology, Rapid City, SD

AICHE 2011 Abstract

Title: Characterizationof Epoxy Spin-Coated Tailored Multifunctional Nanocomposite Structures

Abstract

Polymer nanocompositesprovide unique solutions to industrial and scientific applications where weightmust be minimized and /or functionality maximized. Researchers are interestedin improving the ability to tailor a product to meet specific weight, thermal,optical, mechanical and electrical requirements. Historically functional composite structureshave been realized through a top-down approach.With the advent of atomic level measurement tools and experimentaltechniques a bottom-up approach to the creation of multifunctional structuresis receiving intense study. We aredeveloping unique multifunctional structures using such a bottom-up approachwith the intent of developing molecular simulations to guide such a process. Properties of polymeric nanocompositestructures are tailored and optimized through a fundamental understanding ofintermolecular forces.

While macroscopic models of bulkproperties of polymer nanocomposites have beencharacterized, much less is known on the dynamics of their interfacialcharacteristics, which must be fully developed in order for the tailorfabrication of multifunctional nanocomposites using abottom-up approach. Targeted functionalization and spin coating are used to provideconsistent means of creating multilayer multifunctional thin film composite strictires allowing for the investigation of properties ofthese multifunctional composites. Wewill use a surface response design of experiment in the characterization of oursystem. We investigate the effects of uniquecombinations of sonication, high shear mixing, surface modification and other dispersion mechanismson the ability to control dispersion to create multifunctional layers of epoxyspin coated nanocomposites in order to obtain desiredmechanical properties such as loss and storage moduli,coefficient of thermal expansion, glass transition, among other properties.

It is known that other variables,such as surface morphology effect adhesion between layers. We explore the various parameters impactingmorphology and its control, including substrate type, the impact of centripetalforces, environmental conditions and nanoparticle dispersion. Of particular interest is to betterunderstand the effect of humidity on morphology and adhesion in multifunctionalmultilayer nanocomposites. Using a controlled environment, weinvestigate the impact of humidity and spin coating on the interfacial adhesionof nanoparticles with the polymer matrix and interlayer adhesion. In addition, we investigate the impact ofother parameters, such as curing, on our systems.

Our ability to characterize andcontrol dispersion using a spin coating allows us the ability to tailor make thinfilm nanocomposites to meet certain specifiedcriteria in the production of multifunctional, multilayer nanocomposites.

See more of this Session: Properties and Characterization of Nanocomposites
See more of this Group/Topical: Materials Engineering and Sciences Division

(768a) Introducing Biomaterial Concepts Through Pharma- and Cosmeceuticals

Jennifer Fiegel, Department of Chemical and Biochemical Engineering / Department of Pharmaceutical Sciences and Experimental Therapeutics, University of Iowa, Iowa City, IA

The early experiences of children are extremely important for their later development. Research indicates that childhood experiences may influence future personality traits, including traits that affect educational and occupational choices. The age at which students determine their interest in science is in the adolescence years (~10-14 or earlier). Therefore, students who lack early exposure to positive science experiences may be at risk of developing a negative interest toward learning science. This is particularly critical for girls in technology-driven fields, who tend to express more negative views of physical sciences than boys and therefore have a lower likelihood of choosing careers in STEM fields.

To help address this issue, a week-long day camp, PharmCamp, was developed to enable middle school girls to make informed decisions about careers in science and engineering. The camp aims to help middle school girls see that science and engineering professions are exciting and rewarding and that scientists and engineers have a positive impact on society. The focus of the camp is on hands-on experience with engineering design and product development, specifically with pharmaceutical and cosmetic products. Laboratory exercises, brainstorming activities, and group discussion help the students learn about the design and use of common dosage forms (solid, semi-solid and liquid). For example, hands-on exercises in making lip balm and handcream illustrate the requirements for making semi-solids. Brainstorming sessions focus on identifying important design properties for product development (appealing smell, spreadability, etc.) and how one's choice of materials affects the final product.

See more of this Session: Outreach In Biomaterials Program
See more of this Group/Topical: Materials Engineering and Sciences Division

(768b) To Be Determined Shortly

Elizabeth Dirk, Department of Chemical and Nuclear Engineering, University of New Mexico, Albuquerque, NM

tba

See more of this Session: Outreach In Biomaterials Program
See more of this Group/Topical: Materials Engineering and Sciences Division

Unavailable Paper

(768c) Integrating Outreach Into a Joint REU IGERT Biomaterials Program

Kimberly Anderson, Chemical and Materials Engineering, University of Kentucky, Lexington, KY

To Be Updated

See more of this Session: Outreach In Biomaterials Program
See more of this Group/Topical: Materials Engineering and Sciences Division

Unavailable Paper

(768d) Introducing Girls to Chemical Engineering Through Biomaterials for Drug Delivery

Julie Champion, Chemical and Biomolecular Engineering, Georgia Institute of Technology, Atlanta, GA

The Georgia Tech Women in Engineering Program hosts TEC camp for a week every summer. Designed to offer middle school girls an early introduction to the world of **t**echnology, **e**ngineering, and **c**omputing, TEC Camp inspires middle school age girls to consider college majors and careers in these important fields. The format is highly interactive and designed to address the specific interests of middle school students. The campers apply to the program and are selected based on an interest in science, but often have little knowledge of what engineers do. The girls rotate through hands-on modules such as web design, aerospace engineering, robotics, presentation skills, and chemical engineering. The goal is to get them excited about engineering so that they maintain their curiosity and effort in science classes at a time when many girls lose interest.

The chemical engineering module is designed to introduce the girls to biomaterials and drug delivery, brand new concepts to most of them. The campers spend one hour a day in the lab. The session starts with an interactive discussion on the scientific concepts and safety issues of the day's activity. Then the girls work in pairs to complete their experiments, with help from the professor and graduate students. For example, the campers made their own polymeric drug delivery particles and investigated the ratio of polymer to crosslinker that generated particles with ideal properties. Then, they explored the concepts of saturation and crystallization by crystallizing their own "drugs". Finally, the concepts were combined and the campers encapsulated the drugs in their particles and observed drug release and particle swelling phenomena. The campers kept notes of their experiments and observations and at the final banquet presented a skit about their activities. Exit surveys indicated that the campers learned new things from the module and most importantly, greatly improved their perception of chemical engineers and what chemical engineers do.

See more of this Session: Outreach In Biomaterials Program
See more of this Group/Topical: Materials Engineering and Sciences Division

(768e) Biomaterials Reaching High Schools Through Students and Educators

Edna Margarita Prieto, Department of Chemical and Biomolecular Engineering, Vanderbilt University, Nashville, TN and Scott A. Guelcher, Chemical and Biomolecular Engineering, Vanderbilt University, Nashville, TN

The effort towards promoting dialogue between the scientific and non-scientific communities has significantly increased in the last decade. A continuous dialogue provides benefits to both groups: the non-scientific community is more informed for decision making-purposes, and scientists gain different perspectives about their research topics. High school, for example, can greatly benefit from interactions with the scientific community. Here, students are preparing to decide on a professional path, and teachers are looking for ways of presenting the different options. Our lab has been working for 2 years with high school students and teachers addressing these points. High school students, who are part of the School for Science and Math at Vanderbilt University (SSMV), have been attending our lab and completing research activities with biomaterials from the spring of their junior year through the fall of their senior year. Their time in the lab includes an introductory period focused on defining their research topic, followed by a summer internship, and a final semester of concluding work for their project. This program aims to provide students with a research-centered learning experience in which they develop their oral and writing skills as well as their research abilities. As a result, students have presented their work in poster sessions on campus, published their work in the Young Scientist journal, and have expressed high motivation to pursue a career related to the topics they have been working on. The work with high school teachers has mainly been conducted in collaboration with the Research Experiences for Teachers (RET) program at Vanderbilt. The teachers involved are assigned to labs and participate in research projects during the summer. At the end, the teachers develop a learning module for their classes in which they incorporate their research experience into the classroom. Outcomes from this collaboration include learning modules for high school chemistry, and biology classes related to mechanical properties, bone, and the profession of Chemical Engineering in the biomaterials field. Finally, working with high school students and teachers has provided our lab with new ways of viewing the biomaterials research we conduct, as well as teaching opportunities for graduate students interested in pursuing academic careers.

See more of this Session: Outreach In Biomaterials Program
See more of this Group/Topical: Materials Engineering and Sciences Division

(768f) Bridging the Gap Between Biomaterials Research and High School Students Using New Advances In Communication Technology

Cody A. Schoener, Chemical Engineering, University of Texas at Austin, Austin, TX and Molly M. Schoener, New Tech High, Akins High School, Austin, TX

High school students lack the general knowledge and understandingof what biomaterials are, what they are used for, and the potential they haveto greatly improve patient care and comfort. Furthermore, these high schoolstudents are unaware of what graduate school is, the typesof research performed, and how these graduate institutions are the pipelines ofideas and concepts that advance science and technology. This gap between highschool and post-bachelor degrees is creating educational voids which may leadto the stunted growth or deceleration of biomaterial innovation. To improve thefuture outlook and enrollment of graduate students in the field ofbiomaterials, we have implemented a program connecting high school studentsfrom Akins New Tech High School from the Austin Independent School District inTexas and chemical engineering graduate students from the University of Texasat Austin, using new advancements in communication technology.

We designed the project to be educational for both theparticipating graduate students and high school students. For the high schoolstudents, the primary focus was to promote the understanding of and spark theinterest in biomaterials. The secondary focus was to educate high schoolstudents about academic life and how to effectively prepare for their nexteducational step. Graduate students also benefited by improving their abilityto communicate research to a non-collegiate audiences and enhance managementskills.

The project was designed as follows. 1) Chemical engineeringgraduate students developed 2-3 page written summaries of their thesis research,including any embedded visual explanations, appropriate for high schoolaudiences. Subsets of these graduate students were filmed explaining andphysically illustrating their research project using a FLip^{TM}video camera. The short vignettes were compiled to create an introductory videoexplaining to the high school student what would be their responsibilities andfocus on immersion into biomaterials and graduate research. 2) A singleresearch summary was assigned to groups of 3 or 4 high school students forreading and review. The groups collectively generated a series of 10-15questions focused on the research they read as well as general overviewquestions on what is and what it is like to attend graduate school. 3) The highschool students then used these questions to interview graduate student using skype™, an online video chat application. 4)For a final evaluation of comprehension and understanding, the high schoolstudents were required to write an essay about this experience and presenttheir understanding of the research project using PREZI™, a zoomingpresenting editor.

At the completion of this program we had 78 high school studentparticipants and 9 graduate student participants. 55 of the high school studentswere underrepresented minorities, 49 were economically disadvantaged, 16 wereEnglish language deficient, and 7 needed special education services. 7 of thegraduate students were actively pursuing doctorals inthe field of biomaterials. Being the first time this was attempted, it waswidely regarded as a success by the Akins High School faculty. It successfullyenlightened students about biomaterials and graduate research. As for thegraduate students, the online interviews help them self-evaluate their researchand the ability to clearly explain research topics to a non-academic audience.This project is geographically independent as it functions on online tools,does not require significant financial support or significant time resources tocomplete; therefore, implementing this program into current biomaterialsresearch programs can be simple and be capable of outreaching many levels ofeducation including primary and secondary schools aswas as undergraduates at collegiate institutions.

See more of this Session: Outreach In Biomaterials Program
See more of this Group/Topical: Materials Engineering and Sciences Division

(770a) Distribution of Stresses within Scaffolds Used In Perfusion Bioreactors

Ngoc Hong Pham, Chemical, Biological and Materials Engineering, The University of Oklahoma, Norman, OK, Roman S. Voronov, School of Chemical, Biological, and Materials Engineering, University of Oklahoma, Norman, OK, Vassilios I. Sikavitsas, Chemical, Biological and Materials Engineering, University of Oklahoma, Norman, OK and Dimitrios V. Papavassiliou, School of Chemical Biological and Materials Engineering, The University of Oklahoma, Norman, OK

Tissue engineering is an interdisciplinary field involving the replacement of damaged tissues or organs with artificial tissue substitutes. In bone tissue engineering, stem cells are typically seeded within 3-D scaffolds, made of synthetic biodegradable polymers, and followed by dynamic culture in perfusion bioreactors to reproduce tissue. Stresses that are generated by the flow of the culture media play an extremely important role in cell proliferation and cell differentiation. A recent study (Voronov R.S., Van Gordon S.B., Sikavitsas, V.I., Papavassiliou, D.V., Appl. Phys. Let., 2010, 97:024101) discovered that normalized flow-induced stresses inside high porosity and randomly structured scaffolds follow a common gamma probability density function. This goal of the present study is to focus on the distribution of flow-induced stresses in scaffolds that can be obtained with lithography or other techniques and to further investigate whether a common distribution can describe the distribution of stresses in these kinds of structure. Different constructed scaffolds were created numerically and stresses induced by flow were obtained by solving the discrete Navier-Stokes equation. It was found that the stress distributions follow a common distribution within statistically acceptable accuracy, and that the direction of flow relative to the internal architecture of the porous scaffolds is an important factor that affects the stress distribution.

See more of this Session: Porous Scaffold Fabrication
See more of this Group/Topical: Materials Engineering and Sciences Division

(770b) Investigation of Poly(β-amino ester) Biodegradable Hydrogel Systems As Tunable Porogen Particles In Tissue Engineering Scaffolds

Ashley M. Hawkins[1], Todd A. Milbrandt[2], David A. Puleo[3] and J. Zach Hilt[1], (1)Department of Chemical and Materials Engineering, University of Kentucky, Lexington, KY, (2)Department of Orthopaedic Surgery, University of Kentucky, Lexington, KY, (3)Center for Biomedical Engineering, University of Kentucky, Lexington, KY

The goal of tissue engineering is to successfully regenerate diseased or damaged tissue by providing a scaffold on which cells and tissue can grow. Successful scaffolds must exhibit the mechanical integrity to support the new growth, degrade at a rate similar to new tissue formation, and produce degradation products that are easily metabolized by the body and non-toxic. In some cases, scaffold materials must have a 3D interconnected pore structure to allow cellular ingrowth and the movement of nutrients and waste materials. To facilitate the formation of a 3D porous scaffold, we propose using fast degrading hydrogel particles entrapped in a slower degrading outer matrix hydrogel. The introduction of the scaffold to an aqueous environment will begin the porogen degradation process, thereby opening the porous network and potentially releasing drug molecules. In this work, poly(β-amino ester) biodegradable hydrogel systems were prepared, and the bulk material properties were analyzed. Specifically, the degradation rate, compressive moduli, and cytotoxicity of the individual hydrogel systems were studied. A procedure for creating hydrogel particles of controlled particle size from the bulk hydrogel system was developed and particles of several hydrogel systems were created. Porogen particles were then entrapped in an outer slower degrading hydrogel matrix during polymerization. The resulting composite systems were degraded, and the porosity development was analyzed. The first goal was to determine the properties of successful porogen particles to better screen potential bulk hydrogel systems prior to particle creation and entrapment. Composite system properties were also studied and compared to the properties of the bulk hydrogels to determine if the resulting scaffolds would be applicable in tissue engineering applications.

See more of this Session: Porous Scaffold Fabrication
See more of this Group/Topical: Materials Engineering and Sciences Division

Preparation and characterization of superlowfouling electrospun scaffolds of zwitterionic polysulfobetaine methacrylate for biomedical applications

Zwitterionic surfaces and hydrogels have been previously studied for their superlow biofouling characteristics, a quality essential to biomaterials and tissue engineering. However in order to be useful as tissue-engineered constructs, scaffolds are needed. Electrospinning (ES) has been shown to produce continuous, porous, and fibrous matrix mimicking the structures of native extracellular matrix. Here we have prepared fibrous scaffolds of polysulfobetaine methacrylate (pSBMA) with different fiber diameters for their potential biomedical applications by ES. Zwitterionic monomer, [2-(methacryloxy)ethyl]-dimethyl-(3-sulfopropyl)-ammonium hydroxide, was polymerized to afford polysulfobetaine methacrylate (pSBMA) via free radical polymerization and crosslinked with tetraethylene glycol dimethacrylate. The biocompatibility of the scaffolds with different diameters was examined by bacterial adhesion, protein adsorption, and cell attachment to show that ES scaffolds of pSBMA possess exceptional superlow biofouling properties.

(770d) Electrospinning of Sulfobetaine Methacrylate Nanofibers

Erin E. Emerick[1], Sheila Grant[2] and **Matthew Bernards**[1], (1)Chemical Engineering, University of Missouri-Columbia, Columbia, MO, (2)Biologial Engineering, University of Missouri-Columbia, Columbia, MO

Sulfobetaine methacrylate (SBMA) is a zwitterionic polymer that has been shown to have nonfouling properties. This resistance to nonspecific protein adsorption makes the polymer a potential candidate for tissue engineering scaffolds. In this work, the preparation of polySBMA nanofibers by electrospinning was investigated. The SBMA was polymerized to different molecular weights by controlling the concentration of KCl in the polymerization solution. It was shown that by increasing the concentration of KCl, the molecular weight of the polySBMA was decreased for a given reaction time. In order to electrospin polySBMA nanofibers, solutions were prepared using aqueous NaCl with varying concentrations. As expected, it was shown that as the molecular weight of the polySBMA increases, there is a related increase in fiber diameter. The effect of the NaCl salt concentration in the electrospinning solution presented less of a trend. For larger concentrations of NaCl, the nanofibers were larger, but not as uniform. However, when the NaCl was at or below a threshold value of 0.25M, smaller and more consistent nanofibers were formed. Protein adsorption testing confirmed that the polySBMA nanofibers retained their nonfouling properties after electrospinning. Overall, polySBMA nanofibers with a range of diameters were able to be formed by controlling the electrospinning conditions, and they represent a promising nonfouling material for use as a tissue engineering scaffold.

See more of this Session: Porous Scaffold Fabrication
See more of this Group/Topical: Materials Engineering and Sciences Division

(770e) Independent Control Stiffness and Permeability of a Cell-Encapsulating Hydrogel; Integration of Bio-Inspired Material Chemistry and Microfabrication

Jae Hyun Jeong[1], Vincent Chan[2], Chaenyung Cha[3], Pinar Zorlutuna[4], Rashid Bashir[4] and Hyunjoon Kong[5], (1)Chemical and Biomolecular Engineering, University of Illinois, Urbana, IL, (2)Department of Bioengineering, University of Illinois at Urbana-Champaign, (3)Chemistry, University of Illinois Urbana-Champaign, Urbana, IL, (4)Micro and Nanotechnology Laboratory, Electrical & Computer Engineering and Bioengineering, University of Illinois, Urbana-Champaign, Urbana, IL, (5)Department of Chemical and Biomolecular Engineering, University of Illinois Urbana-Champaign, Urbana, IL

Hydrogels are being extensively used as cell encapsulation or mobilization devices for both fundamental biology studies and cell transplantation therapies because of their structural similarity to natural extracellular matrices. Successful use of a hydrogel in these applications greatly relies on an ability to control stiffness and permeability of a cell-encapsulating hydrogel, both of which significantly influence cellular phenotypes. However, conventional hydrogel design is often plagued by the inverse dependency between stiffness and permeability, thus limiting a capability to control cellular function to recreate new tissues. This study presents a novel strategy to control stiffness and permeability of a cell-encapsulating hydrogel in an independent manner by integrating bio-inspired material chemistry and microfabrication technique. The resulting hydrogel is ultimately used to promote the recreation of vascular network and also regeneration of bone tissues at tissue defects.

First, we incorporated methacrylic alginate (MA) into a poly(ethylene glycol) diacrylate (PEGDA) hydrogel to mimic the critical function of glycosaminoglycan in a natural ECM. We have found that incorporation of MA into the PEGDA hydrogel system increased both elastic modulus and swelling ratio. Cells encapsulated into the stiffer composite PEGDA-MA hydrogel remained more viable and bioactive to express angiogenic factors as compared with cells encapsulated in the pure PEGDA hydrogel. Second, we introduced microchannels with diameter of 500 μ m into a cell-encapsulating hydrogel using stereolithographic assembly (SLA) unit via *in situ* photo-cross-linking. The incorporation of microchannels into the PEGDA-MA hydrogel enhanced the water diffusivity without changing gel rigidity, as examined with MRI. These microchannels further improved the cellular secretion of VEGF. Remarkably, we have found that the cell-encapsulating PEGDA-MA hydrogel with microchannels significantly increased the density of blood vessels at implantation sites. In addition, the gel implanted at a rat calvarial defect enhanced the bone tissue formation with vascularization. Overall, the design principles established in this study will be broadly useful to regulating the emergent behavior of a wide array of cells for tissue regeneration.

(770f) Novel pH-Responsive Scaffolds with Self-Actuating Oxygen Transport

Jin-Oh You, **Dariela Almeda**, Marjan Rafat and Debra T. Auguste, School of Engineering and Applied Sciences, Harvard University, Cambridge, MA

The lack of vascularization in implanted tissue engineered constructs has been one of the principal challenges of tissue engineering. Cells residing at a critical distance (>100 microns) from the blood supply on an implanted scaffold are limited by diffusion, nutrient delivery and waste removal. In this study, we present scaffolds composed of a pH-sensitive polymer, dimethylaminoethyl methacrylate (DMAEMA) and biocompatible polymer, 2-hydroxyethyl methacrylate (HEMA). Our pH-sensitive scaffolds increase their pore size when exposed to low pH, thereby allowing an increase in flux. In vitro pO_2 measurements and cell viability studies confirm higher oxygen levels and cell proliferation in DMAEMA/HEMA scaffolds compared to nonresponsive HEMA scaffolds. We conducted in vivo studies which corroborated increased cell viability in pH-responsive scaffolds. Immunofluorescence staining of human umbilical vein endothelial cells (HUVECs) seeded in scaffolds after implantation in mice showed that the pH-sensitive DMAEMA/HEMA (30/70, mol/mol) scaffolds provided an environment more favorable to cell survival, as a significant number of cells were present after implantation compared to HEMA scaffolds. Additionally, DMAEMA/HEMA (30/70, mol/mol) scaffolds exhibited vessel formation while cells seeded onto HEMA scaffolds did not. We present a scaffold design that significantly increases cell survival by increasing oxygen availability through a self-actuating, pH-responsive mechanism.

See more of this Session: Porous Scaffold Fabrication
See more of this Group/Topical: Materials Engineering and Sciences Division

(770g) Multi-Compartment Collagen-Glycosaminoglycan Scaffolds for Tendon-Bone Interfacial Tissue Engineering

Steven R. Caliari[1], Daniel W. Weisgerber[2], Doug O. Kelkhoff[2] and Brendan A. Harley[1], (1)Chemical & Biomolecular Engineering, University of Illinois at Urbana-Champaign, Urbana, IL, (2)Materials Science and Engineering, University of Illinois at Urbana-Champaign, Urbana, IL

Tendon and ligament injuries are exceedingly common with over 32 million occurrences annually in the United States and associated costs of $30 billion [1]. Current surgical and tissue engineering approaches have had limited success, particularly when dealing with injuries at the tendon-bone interface (TBI). Failure rates for TBI repairs are as high as 94% [2]. This is largely due to the inability of available repair strategies to recapitulate the spatially-graded microstructural, mechanical, and chemical properties of the native interfacial tissue. Collagen-glycosaminolgycan (CG) scaffolds have previously been used as regenerative templates for soft tissues such as dermis [3] and more recently have been applied to bone and osteochondral defects [4]. This work details the development of a potential strategy to improve TBI injury outcomes through the application of CG scaffolds containing distinct tendinous and osseous compartments joined by a continuous interface mimicking native TBI physiology.

Multi-compartment CG scaffolds were fabricated using a previously described liquid-phase co-synthesis method [4]. Briefly, a suspension of type I collagen, chondroitin sulfate, and calcium salts (osseous compartment) was layered on top of a suspension of type I collagen and chondroitin sulfate (tendinous compartment) and allowed to interdiffuse. The suspensions were then freeze-dried in a thermally mismatched mold to promote directional solidification and the formation of aligned pore tracks to mimic the native microstructure of tendon [5]. SEM analysis demonstrated that pores were aligned in the longitudinal plane of the tendinous compartment but more isotropic in the osseous compartment. Additionally, SEM showed a continuous network of CG content at the osseous-tendinous interface. MicroCT and mechanical testing are being used to further characterize the microstructural and mechanical properties of the scaffolds. Multi-compartment CG scaffolds have been shown to support tenocyte, osteoblast, and human mesenchymal stem cell (hMSC) viability for up to three weeks of *in vitro* culture. Histology (H&E and Alizarin Red) will be used to assess the distribution of hMSCs and mineral content respectively. The alignment of hMSCs within the tendinous compartment will be measured from H&E stained histology sections using the OrientationJ plugin within ImageJ. Ongoing work is assessing the differential effects of scaffold pore size (controlled by final freezing temperature), osseous compartment mineral content, and soluble factor presentation on hMSC number, metabolic activity, soluble collagen synthesis, and expression of tendon (collagen I and III, COMP, tenomodulin, scleraxis) and bone (collagen I, ALP, osteocalcin, bone sialoprotein) markers.

References

1. Butler D.L. et al. J Orthop Res, 2008. **26**(1): p. 1-9.
2. Galatz L.M. et al. J Bone Joint Surg Am, 2004. **86-A**(2): p. 219-24.
3. Harley B.A.C. and L.J. Gibson. Chemical Engineering Journal, 2008. **137**(1): p. 102-121.
4. Harley B.A. et al. J Biomed Mater Res A, 2010. **92**(3): p. 1078-93.
5. Caliari S.R. and B.A.C. Harley. Biomaterials, in press.

See more of this Session: Porous Scaffold Fabrication
See more of this Group/Topical: Materials Engineering and Sciences Division

AUTHOR INDEX

AUTHOR INDEX

AUTHOR INDEX

AUTHOR INDEX

AUTHOR INDEX

AUTHOR INDEX

AUTHOR INDEX

AUTHOR INDEX

AUTHOR INDEX

AUTHOR INDEX

AUTHOR INDEX

AUTHOR INDEX

AUTHOR INDEX

AUTHOR INDEX

AUTHOR INDEX

AUTHOR INDEX

AUTHOR INDEX